T0226447

EUROPEAN SYMPOSIUM ON COMPUTER AIDED PROCESS ENGINEERING - 12

COMPUTER-AIDED CHEMICAL ENGINEERING

Advisory Editor: R. Gani

COMPUTER-AIDED CHEMICAL ENGINEERING, 10

EUROPEAN SYMPOSIUM ON COMPUTER AIDED PROCESS ENGINEERING - 12

35th European Symposium of the Working Party on Computer Aided Process Engineering

ESCAPE-12, 26 - 29 May, 2002, The Hague, The Netherlands

Edited by

Johan Grievink
Delft University of Technology
Faculty of Applied Sciences
Department of Chemical Technology
Julianalaan 136
2628 BL Delft, The Netherlands

Jan van Schijndel
Shell International Chemicals
Badhuisweg 3
1031 CM Amsterdam, The Netherlands

2002
ELSEVIER
Amsterdam - Boston - London - New York - Oxford - Paris -
San Diego - San Francisco - Singapore - Sydney - Tokyo

ELSEVIER SCIENCE B.V.
Sara Burgerhartstraat 25
P.O. Box 211, 1000 AE Amsterdam, The Netherlands

First edition 2002

Library of Congress Cataloging in Publication Data
A catalog record from the Library of Congress has been applied for.

ISBN: 0-444-51109-1

Printed and bound by Antony Rowe Ltd, Eastbourne
Transferred to digital printing 2005

Preface

This book contains papers presented at the 12th European Symposium on Computer Aided Process Engineering (ESCAPE-12) held in The Hague, The Netherlands, from May 26-29, 2002. ESCAPE-12 is the 35th event of the Working Party on Computer Aided Process Engineering (CAPE) of the European Federation of Chemical Engineering (EFCE). The first event on computer applications in the chemical industry organised by the CAPE Working Party was in Tutzing, Germany in 1968. The common symposium name of ESCAPE was adopted in 1992 and started at Elsinore, Denmark, that year. The most recent symposia were organised in Kolding, Denmark, 2001 (ESCAPE-11), in Florence, Italy, 2000 (ESCAPE-10) and in Budapest, Hungary, 1999 (ESCAPE-9).

The ESCAPE series serves as a forum for bringing together scientists, students, engineers and managers from academia and industry, which are interested in CAPE. The scientific aim of this series is to present and review the latest developments in CAPE or Systems Process / Product Engineering. This research area bridges fundamental sciences (physical, chemical and biological) with the various aspects of process and product engineering.

The objective of ESCAPE-12 is to highlight advances made in developments and innovative use of computing methods and information technology tools on five specific technical themes:
Integrated Product and Process Design, Process Synthesis and Plant Design, Process Dynamics and Control, Manufacturing and Plant Operations, Computational Technologies.
The advances in CAPE technology have implications for the chemical engineering profession in general; hence the sixth theme addresses: *Sustainable CAPE education and careers for chemical engineers.*

The papers at ESCAPE-12 are arranged in order of the above six themes. A total of 167 papers, consisting of 7 keynote and 160 contributed papers are included in this book. All papers have been reviewed and we are very grateful to the members of the international scientific committee for their evaluations, comments and recommendations. It was an intensive task since we started with 369 submitted abstracts. The selection process involved review of abstracts, review of manuscripts and final acceptance of the revised manuscripts.

This book is the third ESCAPE Symposium Proceedings included in the series on Computer Aided Chemical Engineering. Again, we hope it will serve as a valuable reference document to the scientific and industrial community and that it will contribute to the progress in computer aided process and product engineering.

Johan Grievink
Jan van Schijndel

International Scientific Committee

Local Organising Committee

Conference Secretariat

Contents

Keynote Papers

Contributed Papers

Integrated product and process design

Process Synthesis / Plant Design

Process Dynamics and Control

Manufacturing and Process Operations

Computational Technologies

Sustainable CAPE Education and careers for Chemical Engineers

European Symposium on Computer Aided Process Engineering – 12
J. Grievink and J. van Schijndel (Editors)
© 2002 Published by Elsevier Science B.V.

Sustainable CAPE-education and careers for chemical engineers

Bart Drinkenburg

Technische Universiteit Eindhoven, DSM Research, Geleen, The Netherlands

Abstract

Chemical engineering education is in need to include new elements. Business of the Process Industry in the developed world is shifting from specification to performance products. More attention is then required for product technology and product development.

Students are given tools for technology build-up, but not for technology handling. Therefore a good basis in technology management is also essential in the chemical engineering study.

CAPE has organically grown to be very important part in design and operation and should also be given a fair share in the programme. This all means that choices have to be made and that at least a specialized MSc in chemical product/process engineering is necessary. Such a clear cut choice might well attract more students.

In industrial practice CAPE steadily gains more importance, not only as a tool but also because plants are increasingly dependant upon CAPE and control. A CAPE background, as well as such in Process Control will in the future be very valid, perhaps even necessary, for eg. plant management jobs.

1. Introduction

It has always been point of discussion whether education, or schooling, has to follow societal demand or that society can pick its choice from whatever is offered and has to take on further schooling within the individual companies and institutes. The question arises most often within universities where the concept of academic proliferation is a hot issue in words, but seldom in contents. In our chemical world such has led to a basic structure where in the first part of the study emphasis is laid on acquiring tools and in the later part (M.Sc. and Ph.D) research is a prominent part.

In many universities, especially in Germany and the Netherlands, it has even gone so far that the M.Sc. part is merely seen as the introductory phase to do a Ph.D.

Now all over Europe the position has to be reconsidered due to the Bologna agreement and because of major changes that develop in the process industry.

I will limit myself to the chemical engineering education and mirror that to the work chemical engineers have to do within their jobs once they have left university. The latter will already tell that I see university as professional schooling and I immediately follow up this statement that this is not in contradiction with the academic stature. Or, to make it simple: an academic study not only has to follow society, industry being part of it, it has also to anticipate on changes that are likely to occur in that society.

To be more explicit: academic level has to do with analyzing complex problems and, if possible, finding solutions. Especially the first part, analyzing complex problems, is important. What is the context of the problem, how is fact-finding to be organized. Only then possible directions to solutions can be found and ranked, only then good work can be done.

Especially engineers, educated in terms of tools, experience great troubles with the analysis part. They are schooled in problems with two or three parameters that deliver answers in three decimals accuracy. But then, in practice, they are confronted with many context parameters that only can be weighted and no arithmetics are available.

It will much improve the level of creativity of our students if we introduce such problems early in the engineering studies, at the time the students are most sensitive to it!

2. Nowadays chemical engineering education

Chemical engineering has become a discipline in the twentieth century, not the least because it was largely stimulated in university by financial contributions from industrial companies. It has developed along a few lines, called by James Wei (1) paradigms. The first was the paradigm, I prefer concept, of unit operations. Unit operations like distillation, extraction, sorption, not only provided a generalized method for separation processes. Also, and more important, unit operations established firmly countercurrent processes as an effective and efficient method. Moreover, since it showed that many different types of industrial activity indeed had a common methodological background, it laid the foundation of the concept Process Industry.

In the fifties the work of Bird, Stewart and Lightfoot (2) digged deeper and the concept of physical transport phenomena was introduced, not only important for what was called during that time "physical technology" but even more for reactor calculations and reactor development. Process Integration then followed together with the extremely rapid introduction and development of Process Control, once we got rid of pneumatic control through the powerful introduction of process computers and electronic control from the seventies on. Table 1 provides the layered build-up of the chemical engineering discipline as it stands now on a strong basis of first principles.

Building on knowledge in basic sciences like chemistry, physics, etc. a first layer of thermodynamics is present, often considered to be part of layer zero, but more often than not to be redone in chemical engineering education because of two reasons: firstly because students are unacquainted with how to apply thermodynamics in practical situations like phase equilibria and flow-through systems, secondly because they have gathered the

Table I. Build up of the discipline

5. Manufacturing and technology management
4. Process integration
3. Equipment modeling
2. Kinetics

----

1. Thermodynamics

0. Basic chemistry, physics, biology, mathematics, economy

impression that thermodynamics forbid to break through equilibria while the essential work of chemical engineers is just doing that to complete conversions and separations, Nevertheless, thermodynamics supply essential data, eg. physical and chemical equilibria.

Layer 2, kinetics be it in reactions kinetics or in transport phenomena describes the basic mechanisms and through this alone already provides an rough impression of the size of the equipment needed. Thinking in time constants is a difficult but essential part to teach here, since it provides a background for process dynamics.

Layer 3, modeling, precizes the equipment in terms of design variables, both for reactors, separation equipment and some auxiliary equipment.

In the layer process integration the process layout, HES, flow sheeting, instrumentation, process dynamics, process control, logistics and many details are worked out, including start up and shut down. CAPE is extremely useful in this part.

The layer manufacturing and technology management is almost non-existent in chemical engineering as a discipline and a serious lack in the education.

Now, of course, this layered model is not representative for the way chemical engineering is applied in practice, which is problem-oriented. Then the starting point will be, especially in Research and Development, process synthesis, providing rough alternatives for process lay-out, followed by onion-skin procedures as reactor configuration, separation processes, logistics, heat/energy integration, process control. There is an inherent danger in this approach, upon I will come back later.

In general terms it can be said that the methodology and the tools are very connected to practice in the petrochemical industry, not surprising given the gigantic growth of the industry in the time period that chemical engineering developed into a discipline.

3. What is going on in the process industry?

The process industry, as said, showed a very rapid expansion after world war II, both in production volumes and in scale of the plants. Environmental problems, including safety, extended from local impact to regional and later even to global issues. Consequently great efforts and investments were done in waste abatements, energy efficiency and process mastering and control. Modeling and subsequently CAPE have played and play a very important, not to say a dominant, role in making the process industry an example of good manufacturing practice, especially in the base chemical industry.

But in the mean time there are also considerable shifts in the industry itself.

Reorientation cause major mergers in the base chemical and material industry, often called the commodity industry, leading to potential cost reductions in the business environment as marketing and sales, business development and research and development.

It means less work for professionals but at a high level.

In contrast with this is the situation in the areas of life sciences/fine chemicals and performance materials. Here there is much to do. Large companies are shedding off their divisions, others reorientate on acquiring just these parts. Splitting occurs here as often as mergers. In a technological sense these companies are on the average working at a relatively low level of technology and this is very challenging.

It is questionable whether this work will be done in house or via outsourcing even to the point that research and development and possibly manufacturing will be outsourced. Behind this thinking are a number of reasons:

- quality: in a modern company it is extremely difficult to keep up quality of the work in technological issues. Employees are expected, certainly at academic level, to broaden their view by job rotation. Youngsters enter the company, see the opportunities and after four years say goodbye to in particular Rand D, just at the moment that their knowledge becomes balanced by experience.

  On the contrary, specialized companies can offer opportunities in management while keeping the employee in the technology field. Engineering firms have been the forerunners of this trend, but others will follow.

- quantity: knowledge, also in engineering, doubles every eight years. It is not possible to keep the expertise over the whole range up to the level without becoming top-heavy. Specialized companies grow more easy to that level since their experience in different fields is greater and new developments are more often than not on the borderline between different disciplines.

- networking: due to their short residence time company workers have generally a poor network outside the company they work in. Specialized companies, and in the extreme university professors, have a much better built up network. They know how many beans make five.

The shift can already clearly be seen. Companies reduce their workforce in R and D and Engineering. They make nowadays much better use of external intelligence, the "virtual laboratory" is already existent. Of course, another reason is the much increased productivity per person. Where formerly each researcher and engineer was backed by two assistants, nowadays the backing is provided by analytical instruments and hardware/software and that at a very much increased rate of productivity. It appears to me very strange that the productivity factor is not mentioned in the lamenting about the decrease of freshmen entering university.

4. Omissions in chemical engineering education

Chemical engineering students can be provided with a lot of tools fit for integration. Luckily many universities do the integrative part in their curriculum. Nevertheless the programme is oriented upon the processing of specification products, i.e. products that are sold on basis of their chemical composition. As already told, performance products, i.e. products that are sold on basis of their mechanical, physical, chemical and not seldom esthetical characteristics, are increasingly more the basis of the chemical industry in the developed countries. The characteristics are manifold, for one type of product going up to fifty or more as for example mechanical properties, melting point, glass point, heat conductivity, optical properties as color, transparency, debility, electric and magnetic properties, resistances to fluids/gases, surface conditions as glossiness, roughness, abrasiveness and many, many more. Not to speak about all characteristics of food products during storage and consumption.

First of all we need the line of thinking about product marketing issues like performance analysis back to product measurable properties and product development and from there to process development. A good basis in physical chemistry and in equipment choice will help. I am afraid CAPE is still far away from practice in this field.

What is more intriguing: the concept of unit operations flounders. In base chemical we are used to take a unit operation for each step in the process, providing one more characteristic for the end product. E.g. taking one component out per distillation column. Would we use the same procedure in product technology, then we would have an impossible task, not only because the number of equipment items would grow out

through the roof, but also because fixing one characteristic will often degenerate another. Therefore the equipment is necessarily multifunctional and "multi" here means many more than the two, reaction and separation in a multifunctional reactor. Nevertheless, even in the last case it is clear that onion-peeling process integration comes to a natural end.
CAPERS, much to do!

The second omission in the programme is the lack of understanding for what technology means in a business environment. Universities specialize on teaching tools and doing research. But we do not tell our students how to handle technology, how to assess the technology used in the company in regard to competitors and to new developments that are in the pacing stage, how to set up a programme in technology innovation that fits into time, order and budget with a business strategy. Let alone that we teach them how to set up a business strategy considering the challenges in technology. In other words technology management is necessary for our students if we really want them to play a role again in company management as we did before in the days of yore. It will also teach our students that the world consists out of more parameters than five and that making decisions on basis of ranking might be as valuable as on numerics.

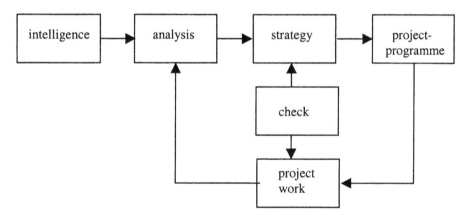

Figure. 1 The technology management cycle.

Figure 1 illustrates the basic concept. Innovation is essential in the business strategy and innovation programmes should not conflict with the strategy. It means that careful analysis is necessary on all business production factors, technology being an important one besides marketing, finances and human resources. Together they will be discussed to determine the future of the business.
Naturally, the whole cycle is non-linear. There will be outside developments, actions of competitors and customers, politics will play a role too. There is much more about it, from technology assessment to organizational details.
How well can our alumni play a role in this?
It is terra incognita, nothing about it appears in the curriculum. Nevertheless, if they don't catch up quickly, they are lost for management positions.

5. BSc – MSc - ?-Ph.D.

From the foregoing we see that we are set up with a dilemma: from one point of view we want to add to the programme product technology, and technology management and at the same time there has been such an expansion in modeling tools, CAPE being the prime example , which we want to introduce also much more firmly in the teaching programme. How to climb out of this hole? One way out is reduction in the courses of competing elements, what we call in the Netherlands the cheese-slicer method. It may help a bit, putting down courses to the essential concepts, but it will also threaten the power of application for our students. In some cases courses could disappear from the programme because the contents were taken up by others in modernization of the teaching material. Such an example is physical technology, now integrated in eg. chemical reaction engineering science.

It is my opinion that we have come to an end in trying to make from our students both full-fledged chemists and engineers, how pitiful it sounds. The tremendous influx of knowledge in both compartments cannot be absorbed adequately within the four or five years that stand for it. Trying to pursue on this road will only lead to two crippled legs to stand on for our alumni. There is another element. Already chemistry and chemical engineering is not only the privilege of chemistry and chemical engineering faculties. It has spread together with the broadening of the field. Chemical engineering has found its applications within mechanical engineering, mixing, transport, materials , biomolecular chemistry, pharmacy, meteorology, environmental technology, technical MBA-studies, medical engineering, even in geochemistry! No wonder that we experience a shortage in freshmen for the original chemistry and chemical engineering studies, young people are attracted to the NEW and the EXTRA as they are bombarded with through the media. not the least by TV-commercials. They are less attracted to the backbone left, what is merely considered as a difficult but backyard study.

Thus, let's be clear about our profession and the nice things we do in society. The Bologna agreement gives us this opportunity. A B.Sc. study that opens up creative fields more than pile up tools and after this period a well-defined MSc to professional development in much more specialized fields, one of them being product and process technology. CAPE can find there a far better and extended place for a, perhaps, minority instead of a low level introductory course for a majority. Although I keep reminding you: make CAPE much more usable for the performance product industry. A point of consideration is the feeling of university professors about this matter and then I come back to the point of academic level. Many of them relate academic level to two points: the student should have a profound knowledge of the professor's particular subject and the student must do research. The first point everyone who works at the university knows by heart to be not true. Students forget about the tools as soon as they have done the particular examination. They, as we did, only will remember the projects they have worked on and thought about, not what they have read listened to.

The second point is a misconception: design projects with their multitude of loose ends are as good or better than deep digging but narrow research assignments.

Academic level has to do with overseeing problem areas, analyzing these, fact-finding, finding and ranking solutions. Don't think I am against research, it is fun and luckily quite a few are attracted to it. But there are also students who hate it and want to be creative in a much more practical sense. And don't forget that there remains the possibility to do a Ph.D. once a MSc. is finished.

There remains a question mark in the subtitle of this chapter. In the Netherlands we give students, after a MSc. degree the opportunity to take part in a designers course of two years, paid on the same level as a four years PhD student.

The course objective is to educate chemical engineers up to the level that they can play an essential and even leading role in industrial design projects. Dutch industry explicitly asked for this education and it must be considered in the same way as the PhD study: not only it is a fulfillment of own ambitions to reach further, it is also considered as a good entrance level for designers in industry as PhD's are for research. There is considerable effort in these courses for product and process innovation. In industry the two, PhD and designer will meet and work together in the field of development. Experience with the alumni is good after approximately 10 years of practice.

6. From the industrial point of view

Here I will pinpoint more to CAPE directly. Chemical engineers work in quite a variety of jobs from general management jobs in which the knowledge of the existence of CAPE is only needed together with the idea what it may or does mean for the business, to plant engineers that feel comfortable once they can play around with a set of models that is constructed by specialists, up to the level of chemical process engineers (mark the distinction) that do the creative part in process synthesis, engineering and control. For them CAPE is not only the tool that increases tremendously their work productivity as said earlier, it also stimulates highly the creativity by making complicated situations much more evident, transparent. People can mathematically think in terms of say five parameters, but in plant design there are many more. CAPE offers the possibility to see through the complications of the process as if it were a movie shown. That certainly helps in finding Eureka-solutions to problems.

But of course, the principle of garbage in – garbage out is valid. CAP-engineers must be very careful about the reality of their models and often have to dig in substantially into the underlying physical and chemical mechanisms to have a fair idea about the model build up and how its assumptions deviate from practice.

Moreover, they can be considered to be the gatekeeper of the plant models that run parallel to the real thing. Now that we, by steadily increasing detailed knowledge of processes, are changing from feedback to model predictive process control such a gate keeping becomes ever more essential. As the physical plant integrity of the hardware is a quality issue, the same applies for the software and the CAP-engineer must be the conscience, the guard against sloppiness. Here we encounter again a dilemma: the plant manager carries the final responsibility for the integrity and safety of the plant, certainly not an easy part of his job in the base chemical industry with its huge inherently dangerous inventories. But now we see that his plant is already running as much on software as on physical hardware. Can he keep in control if he is not heavily involved in the matter? Will he need in the future a background in CAPE and control? It means more than being convinced that in case of emergencies fail-safe procedures exist: hick ups will be always there and his plant will anyway operate much more dynamically in the future.

As a round up for this part I come to the conclusion that, given the proliferation of CAPE from engineering tool to manufacturing backbone, sincere attention must be given to CAPE in the chemical process engineering study, so much that specialist designer courses on the issue, as in Delft, are very welcome and that, because of the expanding field, continuing education on the topic is also a necessity.

7. Conclusions

At the end of this lecture I will sum up:

- chemical engineering and chemistry studies need modernization on a number of issues:
    - product technology and development
    - technology management
    - modeling, CAPE, process control
- Integration of all these elements without breaking eggs is not possible
- A clear choice for a chemical product/process engineering MSc study is necessary. Such a choice will also attract more students, especially if shown that design is a hot issue
- CAPE must be an essential part of this study
- Specialized post-MSc courses have a good future
- CAPE and Process Control will further increase in importance within industry. Job experience in this will be a good background for eg. plant management
- CAPE is lingering far behind for processes that produce performance products.
- For these processes and even for processes that produce specification products the present methodology, resting on a unit operation approach, must be revised.

References

1. Wei, J. A century of changing paradigms in chemical engineering CHEMTEC 26, 16-18, (1996)
2. Bird, Stewart and Lightfoot, Transport Phenomena, Wiley (1960)
3. Drinkenburg, B. New ways to educate chemical engineers, European Conference on Chemical engineering (ECCE, Neurenberg), 121-124 (2001).
4. Wesselingh, private communication

European Symposium on Computer Aided Process Engineering – 12
J. Grievink and J. van Schijndel (Editors)
© 2002 Published by Elsevier Science B.V.

Process Synthesis and Design in Industrial Practice

Gerd Kaibel, Hartmut Schoenmakers

BASF AG, 67056 Ludwigshafen

Germany

Abstract

This contribution will demonstrate how a large chemical company, BASF, carries out process synthesis and process design in practice. First of all, the synthesis of a chemical process has to be included in the company's process chain, and the physical and chemical properties of at least the main components and their mixtures have to be known. It is then possible to formulate possible alternative solutions for the specific process. This can be done in two different ways: using a knowledge-based method with heuristic rules or using a method based on thermodynamics, often accompanied by special mathematical procedures (MINLP). The process synthesis phase is followed by a process design phase. Suggestions must be validated by means of economic comparison. Suitable tools for process synthesis and design include CAPE tools; suitable tools for validation include miniplants.

This will be demonstrated using several non-standard processes as examples. The synthesis and design of dividing wall columns and of reactive distillations will be described. Mention will be made of the limitations of this procedure, and there will be some comments on future research needs as well as some general remarks on combined fluid-solid processes and hybrid processes.

1. Introduction

This contribution will demonstrate how a large chemical company, BASF, carries out process synthesis and process design in practice. This will be demonstrated using several non-standard processes as examples. The results do not only describe BASF procedures: most of them apply to other companies in an approximately similar manner. As a prerequisite for the synthesis, a chemical process has to be included in the company's process chain, i.e. the chemical reaction path has to be specified. In addition, capacities, purities and the manner of operation – continuous or batch-wise – have to be fixed. The same applies to the site-specific conditions.

The next step, the actual process synthesis, comprises the preparation of the educts, the design of the reaction stage, downstream processing for the products including the recycle streams and, if necessary, the purge streams, each of these stages being carried out for each process step. That is what is meant by process synthesis in this contribution. Changes are made to what is included in the company's process chain only if the economic evaluation brings shortcomings to light or if there are other limitations such as disadvantageous environmental aspects. In such a case, the process synthesis must be repeated for another chemical reaction path.

Each activity leading to the synthesis of a process starts with physical and chemical properties. There is an absolute necessity to know the vapour pressures, the other physical properties, the vapour-liquid equilibria, if necessary the liquid-solid equilibria, the reaction equilibria and kinetics, special data such as gas-phase association and information on chemical and thermal stability. The crucial factor determining the quality of a process synthesis is the quality and completeness of this set of physical and chemical properties, at least for the main components. Unfortunately, in many cases the method pursued is different: Estimated data or data from non validated sources are used, the process is selected and the type of underlying property is only replaced by a better one for the final version. This procedure does not guarantee that the preferred version is really the best one. If a different type of underlying property were chosen another process might be the best.

Armed with knowledge of the physical and chemical properties, the engineer is able to carry out the next step, the actual process synthesis. The first part of this consists in clarifying the thermodynamic viability of separation options and summarizing the possible options for selection. The second part consists in finding the correct order for the appropriate unit operations and assembling a complete process. Even today this step is done more by intuition than in a systematic way. In the past, it was commonly believed that an experienced engineer or a team of experienced engineers would find an adequate solution to any synthesis problem and this opinion is still common. Brainstorming techniques involving teams of experts were used to find such solutions. As an alternative to this, in recent years some systematic approaches to process synthesis have been developed [1][2][3].

2. Systematic methods for process synthesis

Systematic methods for process synthesis are based on suggested alternative solutions to a specific process. They differ in the way they identify the different options.
Two approaches can be differentiated:
The first method is based on documented experience with processes and heuristic rules [4][5]. This is the knowledge-based approach, or the use of expert systems. This technique has become established and there is a commercially available form (PROSYN)[6].
The second method is based on thermodynamic principles, the main element being the construction of superstructures that contain the optimum solution as an element (see figure 1 [7][8]). In the case of distillative sequences, distillation lines or residue curves [9][10] can be used to identify the structures. They show feasible ways to processes. The optimum process is found by cutting some of the tie lines and re-arranging the remaining elements.
This procedure can be carried out using thermodynamic considerations [11][12] or mathematical procedures (MINLP) [13].
A common element of both procedures is that they can only identify process alternatives that fall within those suggested. In the case of expert systems, the documented experience must contain the optimum solution and, even more importantly, this solution must be identified using a heuristic rule or other rule. In the case of superstructures, it is

up to the creative powers of the engineer to ensure that the optimal solution is in principle part of the structure.

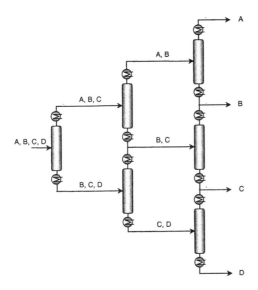

Figure 1: Superstructure for a four-component distillative separation

In more concrete terms, if, for example, an extraction step is the optimum solution, an expert system must contain rules that lead to this result or a superstructure must contain an extraction as an alternative. That is the disadvantage of both approaches. If the optimum alternative is not predefined it will not be found. There is a real problem here for processes with solvents, for example extractive or azeotropic distillations. There are very few approaches to finding such solvents. Even quantum mechanical simulations, regarded with some enthusiasm in the last decade [14][15], have not been really effective to date.

The following points must be borne in mind:

- As a prerequisite for any process synthesis, the inclusion of a process in the company's process chain and in the site conditions must be specified. The physical and chemical properties of at least the main components and their mixtures have to be known.
- After this first step, the possible alternative solutions for the specific process can be formulated. This can be done in one of two ways, using a knowledge-based method or a method based on thermodynamics, and the problematic step of finding alternatives encompassing solvents for specific process alternatives has to be included.
- It is possible to identify an optimum process consisting of some of these different options in a sequence to be specified. This can be done using heuristic rules or thermodynamic considerations. The latter considerations can be

12

replaced by mathematical procedures. The quality of the solution is heavily dependent on the quality and completeness of the process alternatives under consideration.

- The suggestions must be validated by economic comparison. This is possible only if the process synthesis phase is followed by a process design phase. Each part of the equipment has to be designed and the dimensions have to be specified. Process intensification and heat integration measures have to be included in these considerations.

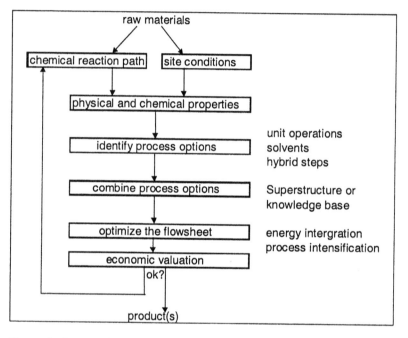

Figure 2: Steps of the process synthesis procedure

Both methods are used in the industry (see figure 2). Knowledge-based methods are mainly chosen to design new processes or search for process alternatives; methods based on residue curves or superstructures are preferred when optimizing existing processes. It must be stated, however, that neither of these methods is used regularly.

The well-known CAPE tools are suitable tools for process synthesis and design. They are available for each step: a physical properties data bank, thermodynamic programs, process synthesis aids, tools for process simulation and equipment design. Screening experiments to check the feasibility of some process options may be necessary.

In general, the process resulting from this work must be validated by means of experiments. In industrial practice integrated miniplants are used if at all possible [16][17] (see figure 3).

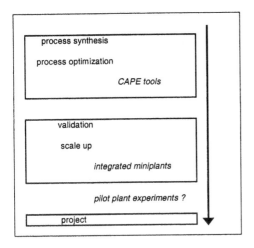

Figure 3: time scale in process synthesis

Miniplants are characterized by the following parameters:
- flow rates between 0.3 and 1.5 kg/h
- equipment on a laboratory scale, mainly made of glass
- non-explosive operation

Operation with closed recycles over a sufficient time period is required to reach a process scale-up. Expenditure on construction and operation is thus high, but it is the only way of discovering concentration increases of small impurities or complex quality problems, for example with product colour.

Equipment scale-up used to necessitate experiments on a pilot-plant scale, which is 10 times larger than miniplant scale and correspondingly more time-consuming. Today more and more of the larger chemical companies have the expertise to manage scale-up from the miniplant scale, at least for thermal separations [18][19]. This contribution does not report on this in any more detail.

Following experimental validation, a process is developed to a stage normally known as the basic engineering stage; it can then be transferred to the project department or to an engineering contractor.

Some examples that illustrate process synthesis and design for non-standard processes will now be described.

3. Examples

3.1 Partitioned distillation columns (dividing wall columns)

Many methods of developing distillation column arrangements are known and documented. The conventional method involves the application of heuristic rules. An alternative uses thermodynamic analyses. As distillation is a separation process that is determined by the parameters of entropy and exergy, an exergetic analysis can provide

14

valuable tips on developing column arrangements [20][21]. It is possible to derive options for process integration by direct steam coupling [22][23]. The results of such methods can be summarized as follows. A distillation process of good thermodynamic quality should have

-        intermediate reboilers and condensers
-        a small pressure drop
-        heat and mass transfer which are as direct as possible in order to avoid heat exchangers
-        an appropriate enthalpy of feed
-        the thermodynamically correct separation sequence.

Based on these characteristics, a general separation scheme can be developed in which there is provision for heat transfer on each tray. For practical applications, most of the intermediate condensers and reboilers must be ignored and the remainder of the design consists of coupled columns in different, but thermodynamically equivalent, configurations (see figure 4).

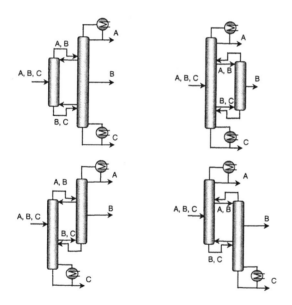

Figure 4: Column arrangements with direct heat coupling

The Brugma or Petlyuk configuration is one example, as is a dividing wall column or partitioned distillation column which combines full thermal coupling and process intensification using one column instead of two (see figure 5 [24]). Typical energy and investment savings of 30% in each case are achieved with this arrangement. BASF was the first to use this technique and operates about 30 of these columns.

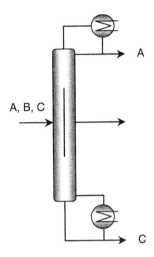

Figure 5: Dividing wall or partitioned distillation column

Other companies followed BASF's lead. Let us take as an example the biggest column of this type [25], which was built by Linde and Sasol in South Africa with BASF acting as a consultant. The task in hand was to pre-fractionate a stream from a Fischer-Tropsch plant containing 1-octene together with high- and low-boiling components. The aim was to reduce the content of high and low boilers before the high-purity distillation section of the plant was entered. The resulting design used trays instead of packing (two-pass sieve trays), the first such column and also the biggest one, with a height of 64.5 m and a diameter of between 4 and 4.5 m.

Even if this arrangement seems to be the best solution from the thermodynamic point of view, it may not in every case be the best one under practical conditions [26]. Its shortcomings include the following:

- *Wide range of boiling points among the components*: The temperature difference between condenser and reboiler may be so high that expensive heating or cooling sources may be necessary and two columns with different pressures and heat integration condenser-reboiler may be more advantageous.
- *Plant revamps*: The investment required for a dividing wall column, as opposed to combining two existing columns and integrating them, may be excessive. However, there are exceptions where it makes sense to modify an existing plant [27] (see figure 6).
- *Column height*: A dividing wall column is always higher than either of the two alternative columns, which may prove to be a drawback. However, the combined number of theoretical stages for the two alternative columns is always higher than the number of stages for the dividing wall column.

16

- Hydraulic imbalances: If the component that has to be removed in the side stream is too small, the hydraulics may be such that an equal number of streams on each side is not the optimum solution from the thermodynamic point of view. For column operation, however, an equal partition of streams is important. Additional energy could thus be necessary, diminishing the principal advantage of the dividing wall arrangement. The bigger the side-stream part is the better the dividing wall solution.

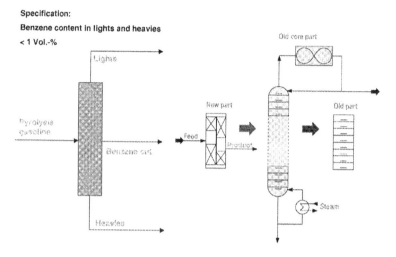

Specification:

Benzene content in lights and heavies

< 1 Vol.-%

**Implementation of the Ruhr Oel (Veba)
Münchsmünster Benzene Removal Project**

Figure 6: Krupp Uhde Revamp Project [27]

In general, reusing the heat going to and from columns via heat integration exchangers should be considered as an alternative to a dividing wall column. The cost of the total energy consumption for this alternative has to be compared with the amount of investment needed for the additional heat integration equipment.

3.2 Reactive distillation

Reactive distillation provides a key opportunity to improve the structure of a process [28][29]. However, it is naturally only possible to combine distillation and reactions if the conditions for both unit operations can be combined. That means that the reactions have to have satisfactory data for conversions at pressure and temperature levels that are compatible with distillation conditions. In reactive distillation, the reaction is superimposed on distillative separation. On the one hand, this results in synergistic effects, e.g. a shift in the chemical equilibrium as a result of products being removed and distillation limits being exceeded owing to the reaction, while, on the other hand, it is precisely these synergies which make reactive distillation so extraordinarily complex. One important objective of process synthesis is therefore to reduce the complexity in order to enable simple solutions to be recognized quickly. A comprehensive process synthesis strategy has been developed to systematically analyse processes involving reversible reactions. One element of this strategy is the analysis of reactive distillation lines [30]. Reactive distillation lines make it possible to examine the feasibility of reactive distillation processes in a simple manner. This simplicity of examination results from the fact that, according to the Gibbs phase rule, the number of degrees of freedom of a system in physical and chemical equilibrium is reduced by the number of independent equilibrium reactions.

Various authors have developed transformation methods to make it possible to handle these concentration parameters appropriately [31]. These transformation methods make it possible to describe reactive distillation using a system of equations which is known from conventional distillation (see figure 7).

Balance: Rectifying operating line of a RD column

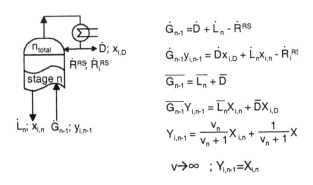

$$\dot{G}_{n-1} = \dot{D} + \dot{L}_n - \dot{R}^{RS}$$

$$\dot{G}_{n-1} y_{i,n-1} = \dot{D} x_{i,D} + \dot{L}_n x_{i,n} - \dot{R}_i^{RS}$$

$$\overline{G_{n-1}} = \overline{L_n} + \overline{D}$$

$$\overline{G_{n-1}} Y_{i,n-1} = \overline{L_n} X_{i,n} + \overline{D} X_{i,D}$$

$$Y_{i,n-1} = \frac{v_n}{v_n + 1} X_{i,n} + \frac{1}{v_n + 1} X$$

$$v \rightarrow \infty \quad ; \; Y_{i,n-1} = X_{i,n}$$

Figure 7: Transformed coordinates

The transformation converts the concentration parameters small x and y for the liquid-phase and gas-phase concentrations into the concentration parameters capital X and Y. At the same time, the transformation eliminates the reaction term in the balance equation. The operating line for the rectifying section of a reaction column is identical in form to the operating line of a non-reactive column. An infinite reflux ratio gives an expression which is identical in form to that for calculating conventional distillation lines [32][33][34][35].

These analogies become particularly clear if we take the synthesis of methyl acetate (MeAc) from methanol (MeOH) and acetic acid (HAC) as an example. The diagrams we see are essentially similar to the distillation line diagrams of non-reactive systems. As a result of the transformation, the four pure substances lie at the corners of a square and the non-reactive binary systems lie along the edges. The highest boiling point in the system is that of acetic acid, while the MeOH/MeAc azeotrope has the lowest boiling point (figure 8). The reactive distillation line diagram makes it possible to determine the product regions of a reactive distillation for infinite reflux. In a manner analogous to conventional distillation, the top and bottom products have to lie on a reactive distillation line and on the balance line. It can be seen that the desired products, that is to say MeAc and water, do not lie in the product region.

18

Reactive distillation lines: HAC + MeOH ↔ MeAc + H2O

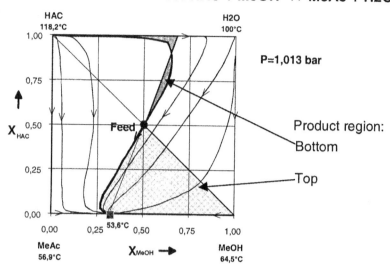

Figure 8: Reactive distillation lines

However, by analogy with extractive distillation, it can be expected that a second feed point would drastically widen the product region at a finite reflux ratio and thus also increase the conversion. Between the two feed points, the column profile is perpendicular to the distillation lines (figure 9).

Production of methyl acetate with catalytic distillation

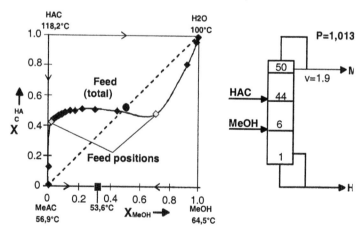

Figure 9: Concentrations along a column with finite reflux, simulation results

While process synthesis produces good qualitative reference points, for industrial implementation we need quantitative results which are as exact as possible. Process simulation programs have to be used for the detailed design of the reactive distillation process [36]. However, a prudent engineer will carry out experiments in a reaction column to verify the achievable purities and yields, and there is a scale-up problem with the results of these experiments. The reason for this is that reaction and mass transfer interact with one another in these experiments. If a development engineer has to design an industrial column purely on the basis of miniplant experiments, he has to maintain not only the separation performance but also the ratio of separation performance to reactor performance so that main and secondary reactions proceed to a comparable extent in the industrial-scale reaction column. One way of achieving this in terms of construction is by separating reaction and product separation from one another both in miniplant tests and on the industrial scale. This is possible, for example, when the reaction is carried out with a heterogeneous catalyst in the downcomer or with side reactors at the column. An alternative is to use structured packing with well-defined paths for the liquid flow. This does not provide a complete solution to the problem, the main reason for this being the lack of reference columns on an industrial scale. Nevertheless reference processes do exist; for example the well-known Eastman Kodak Process for methyl acetate very clearly shows the potential advantages of combining reaction and distillation (see figure 10).

Eastman Kodak Process for Methyl Acetate

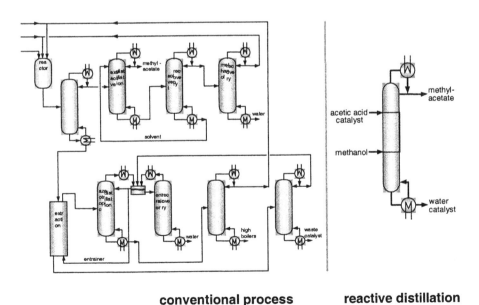

conventional process   **reactive distillation**

Figure 10: Methyl acetate, Eastman Kodak Process

Thus, design and synthesis methods have been developed to a considerable extent for reactive distillation, partly with the aid of the numerous similarities with non-reactive distillation. A major focus of research and development in future years should be the scale-up of reaction columns. This is where great deficiencies still lie [37].

3.3 Using crystallization as a separation step

Most of the published process synthesis methods focus on fluid processes. Possible options in fluid processes include column sequencing, solvent selection and heat integration. Commercial computer tools for process synthesis do not normally produce any options containing solid-phase process steps. Some university research groups are working on the synthesis of solids processes, specifically on crystallization [38] or on processes that combine fluid and solid options [39]. One supplier of knowledge-based process synthesis methods claims to have crystallization among its process options (PROSYN)[6].

The inclusion of processes containing solids in practical process synthesis is not the standard industrial procedure. Nevertheless, industrial applications show that there is great potential. BASF has some experience with processes that combine distillation with crystallization. This proves that crystallization as a separation step, not for obtaining crystals but as a purification step, may be a strong option if fluid separations are not successful or are too complicated. This may even hold true in the case of solvent crystallization where a suitable solvent has to be added, including all the steps for recycles, purge streams and fluid/solid separation units.

The reason for this is well known: The selectivity of crystallization is totally independent from vapour pressures and vapour liquid equilibria and thus provides a key function for the separation of close-boiling liquids. However, it is difficult to include crystallization in both the process synthesis and in process simulation tools and no practical solution has yet been found. The problem is that additional physical properties that contain kinetic data are necessary and that the simulation models have to include particle sizes in a time-dependent way. The mathematical structure of such tools is at least one order of magnitude more complex than the simulation of fluid processes. There is an additional, important drawback with the scale-up procedure: steps containing solids cannot yet be scaled up from miniplant dimensions. Additional pilot-plant scale experiments are thus necessary, at least for the steps containing solids, which makes the process design procedure more complex in terms of time and cost.

4. Conclusion

Process synthesis and design are generally well-established procedures in the chemical industry, whereas systematic process synthesis is not consistently used. The two methods, the knowledge-based method and the method based on thermodynamics, are used in parallel, but no real rules or reported experience exist on the question of when to choose which method. The first examples, the dividing wall column and reactive distillation, seem more suited to the thermodynamic approach. For the last example, the fluid-solid process, the knowledge-based methods could be helpful if they contain enough knowledge data. The design of such processes using CAPE tools is generally successful, but there are drawbacks caused by the lack of suitable models [40], especially in the case of solid-liquid processes.

Hybrid processes such as reactive distillation, membrane reactors, pervaporation and other combinations of well-known unit operations as well as combinations of single steps into one apparatus have great potential for saving energy and investment cost. The process synthesis methods to identify favourable hybrid processes are not yet well developed, and often the design tools do not exist or at least are not commercially available.

Future research is thus necessary to find methods of systematic process synthesis and suitable methods of process design for these non-standard processes.

5. References

[1]   Kussi, J.S., Leimkühler, H.-J., Perne, R., (2000), Chem. Ing. Techn. (72) 11, 1285

[2]   Douglas, J., (1988), Conceptual Design of Chemical Processes, McGraw-Hill, ISBN 0-07-017762-7

[3]   Blass, E., (1989), Entwicklung verfahrenstechnischer Prozesse, Salle + Sauerländer, ISBN 3-7935-5510-0

[4]   Han, C., Stephanopoulos, G., Liu, Y.A., (1996), AIChE Symp. Ser., 92, No. 312, 148

[5]   Jacobs, R., Jansweijer, W., (2000), Comp. and Chem. Eng. 24, 1781

[6]   Schembecker, G., Simmrock, K.H., (1996), AIChE Symp. Ser., 92, No. 312, 275

[7]   Kaibel, G., Blass, E., Köhler, J., (1990), Gas Separation & Purification 4, June, 109

[8]   Stichlmair, J., Herguijuela, J.-R., (1992), AIChE Journal (38) 10, 1523

[9]   Vogelpohl, A., (1993), Chem. Ing. Techn. (65) 5, 515

[10]   Jimenez, L., Wahnschafft, O., Julka, V., (2001), Comp. and Chem. Eng. (25), 635

[11]   Jaksland, C., Gani, R., Lien, K., (1995) Chem En. Science (50) 3, 511

[12]   Petlyuk, F. (1998), Theor. Found. Of Chem. Eng. (32) 3, 245

[13]   Bauer, M., Stichlmair, J., (1996), Chem. Ing. Techn. (68) 8, 911

[14]   Vrabec, J., Fischer, J., (1997), Chem. Ing. Techn. (69) 8, 1126

[15]   Clausen, I., Arlt, W., (2000), Chem. Ing. Techn. (72) 7, 727

[16]   Cozier, M., (1998), European Chem. News, 12-18 October

[17]   Steude, H., Deibele, L., Schröter, J., (1997), Chem. Ing. Techn. (69) 5, 623

[18]   Hofen, W., Körfer, M., Zetzmann, K., (1990), Chem. Ing. Techn. (62) 10, 805

[19]   Deibele, L., Goedecke, R., Schoenmakers, H., (1997), Instit. Of Chem. Eng., Symp. Ser. Nr. 142, Vol. 2, 1021

[20]   Castillo, F., Thong, D., Towler, G., (1998), Ind. Eng. Chem. Res. (37) 987

[21]   Castillo, F., Thong, D., Towler, G., (1998), Ind. Eng. Chem. Res. (37) 998

[22]   Hernandez, S., Jimenez, A., (1999), Comp. and Chem. Eng. (23), 1005

[23]   Duran, M., Grossmann, I., (1986), AIChE Journal (32) 1, 123

[24]   Becker, H., Godorr, S., Kreis, H., Vaughan, J., (2001), Chem. Eng., January, 68

[25]   Becker, H., Godorr, S., Kreis, H., Vaughan, J., (2000), LINDE Berichte aus Technik und Wissenschaft 80, 42

22

[26]   Agrawal, R., Fidkowski, Z., (1998), Ind. Eng. Chem. Res. (37), 3444
[27]   Personal Communication Krupp Uhde and BASF, Citation with allowance of
         Krupp Uhde
[28]   M. Doherty, F. Michael, G. Buzad, (1992), TransIChemE, (70), 448
[29]   J. L. DeGarmo, V. N. Parulekar, V. Pinjala, (1992), Chem. Eng. Progr., March,
         43
[30]   Ung, S., Doherty, M., (1995), Ind. and Eng. Chem. Res. (34), 3195
[31]   Barbosa, D., Doherty, M., (1987), Proc. R. Soc. Lond. A 413, 459
[32]   Stichlmair, J., (1988), Chem. Ing. Tech. 6010, 747
[33]   Stichlmair, J., Offers, H., Potthoff, R.W., (1993), Ind. Eng. Chem. Res. 32,
         2438
[34]   Beßling, B., (1998), Ph.D.Thesis (Dissertation) Universität Dortmund,
[35]   Barbosa, D., Ph.D.Thesis, University of Massachusetts, Order Number
         8727018
[36]   Tuchlenski, A., Beckmann, A., Reusch, D., Düssel, R., Weidlich, U., Janowski,
         R., (2001), Chem. Eng. Science (56), 387
[37]   Schoenmakers, H., Bessling. B., (2002), Chem. Eng. Science, in print
[38]   Cisternas, L., (1999), AIChE Journal (45) 7, 1477
[39]   Pressly, T., Ng, Ka M., (1999), AIChE Journal (45) 9, 1939
[40]   Gani, R., O'Connel, J., (2001), Comp. and Chem. Eng. (25), 3

European Symposium on Computer Aided Process Engineering – 12
J. Grievink and J. van Schijndel (Editors)
© 2002 Published by Elsevier Science B.V.

Process Software in the Chemical Industry – the Challenge of Complexity

J. Kussi, R. Perne, A. Schuppert

Bayer AG, Process Technology, Leverkusen, Germany

{juergen.kussi.jk, rainer.perne.rp, andreas.schuppert.as}@bayer-ag.de

Abstract

Process optimisation and control are widely accepted to be key technologies for the chemical industry. The technological development of process simulation and optimisation software has widely been driven by the vision of integrating the entire production process into one model and to optimise the entire process using standardised software components. Despite significant success in software development, the breakthrough of this vision cannot be observed in the chemical industry.

The product portfolio of an integrated chemical – pharmaceutical company like Bayer covers a broad range of the chemical industry products. Therefore a careful analysis of the challenges we are facing at Bayer applying process software for high end applications may be typical of a wide range of chemical companies.

We will discuss typical problems for applications in the chemical industry and present our answers to these challenges.

1. Introduction

Process optimisation and control are widely accepted to be key technologies in process industries. To use optimisation and control technologies for our processes, the implementation of an entire workflow is required to establish the basis for optimisation and control [Fig. 1].

In any case of application the analysis of the process is the **first step**. The goal of this step is to identify the key parameters which control the input - output relations of the process. Moreover a clear understanding of their mutual interactions is an additional goal of the first step.

In the **second step** quantitative models for the relevant subprocesses have to be established. For industrial applications the focus is not a thorough understanding on an ab initio basis of each subprocess, but establishing reliable, quantitative relations between the relevant input parameters and the output properties of the entire process at affordable costs.

In the **third step** the integration of these models into an entire model for the relevant part of the process is required together with the implementation of efficient numerical simulation technologies to establish a basis for all optimisation and control strategies, both off-line and on-line.

24

Since none of these preparation steps can be compensated by additional efforts elsewhere, a lack in any step is a bottleneck for the industrial application of optimisation and control.

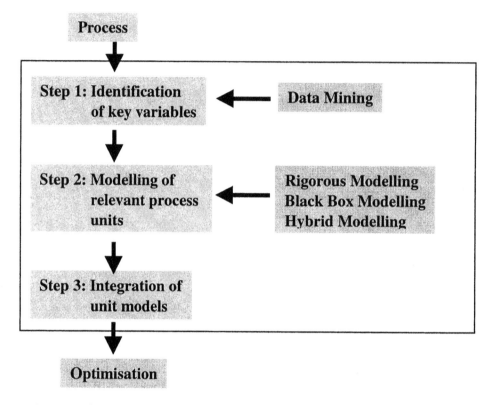

Figure. 1: Unavoidable preprocessing steps for process optimisation of existing plants

Because of the high cost saving potential in petrochemical and bulk chemical processes, separation processes have been the first type of subprocesses being optimised and controlled using simulation software. The thermodynamic models describing separation processes have been developed on a rigorous (e.g. founded on physical chemistry) model basis. The main problem in these applications is the third step: the integration of a high number of models for different separation steps into a single model as well as the efficient solution of very large DAE systems. Therefore research and development in computer based methods have been focussed on these problems with significant success in the petrochemical industry as well as in chemical commodities manufacturing.

What is the situation, however, in the chemical and pharmaceutical industry focussed on high value added specialities? Bayer as an integrated chemical – pharmaceutical company with a product portfolio ranging from pharmaceutical proteins with a production capacity of 100g/a to technical polymers with a production capacity of some 100000 tons/a seems to be an excellent example for studying the challenges on process software. Our experience is a wide acceptance of modelling and simulation of bulk processes and separation subprocesses. We find, however, more and more technical challenges with high economic potential which cannot be solved using commercial process software available today.

We observe a clear trend to business models which differ significantly from the classical bulk chemical industry with the aim to produce well defined products at minimal cost. The trend in the chemical industry tends to a more marketing driven business model with the clear focus on customers requirements:

- Production of specialities on customers demands. Critical for the business success becomes the ability to develop production processes for complex molecules fast. This business model becomes common e.g. in the pharmaceutical industry, where intermediate components for the synthesis of the drugs or the drugs itself are produced by special chemicals manufacturers.
- Development and production of specialities being designed on customers demand. In this business model the chemical companies act no longer as a provider of chemicals, but more and more as a solution provider for customers demands. A typical example for this business model is the automotive industry, where the polymer materials are designed on the request of the car manufacturers by the chemical companies.
- Just in time production on customers demands.

All these trends are accompanied by focussing on high value added products demanding more and more complicated production processes. Typical are complex catalytical processes, multiphase reactive processes or biotechnological processes. For these processes, however, we face a significant lack in quantitative process models compared to the classical processes. Therefore the bottleneck of the optimisation workflow is found in steps one and two. Step 3 is significantly less critical compared to classical applications.

Computer aided methods focusing on these business models have to support the following strategic requirements:

- Cost reduction is no longer the most dominant business factor. Even more important is the capacity to develop products and processes on customers demand within a very short timeframe as well as the ability to deliver the products just in time.
- It is obvious that the rigorous, quantitative understanding of such processes can never have the same strength as for well established, "classical" processes. Therefore efficient methods for fast detection of critical parameters in case of unforeseen problems in the process are crucial for business success.
- Efficient supply chain management for the entire process is critical for the ability of a just in time delivery.

Therefore we have focussed our efforts in R&D of modelling and simulation methods in the last years on the strategic requirements depicted above:

- Development of advanced data mining methods which are specialised on the analysis of process data
- Development of hybrid modelling and data analysis technologies allowing us to integrate available process knowledge and black box modelling and analysis methods, like neural networks into an integrated modelling framework with significant benefits
- Development of specialised modelling and data analysis methods key problems:
  - recipe – quality relations of polymers
  - modelling and analysis of biotechnological products
  - catalyst development with high speed experimentation technologies.

In this presentation we will give a short overview over these developments and present an outlook on further developments which are required to overcome the future challenges.

2. Hybrid Modelling

One of the most common bottlenecks for the application of model based process optimisation methods is the lack of reliable, quantitative models for the relevant parts of the process. Therefore advanced modelling technologies allowing the access to new areas of application show a great economic potential.
The classical technologies suffer from different technological lacks:

- Establishing rigorous models based on quantitative scientific principles requires detailed knowledge of all subprocesses involved in the process. Although highly desirable, closing the gaps in quantitative process knowledge for all subprocesses, and thereby allowing rigorous modelling techniques to be applied, is often not affordable.
- Black - box modelling techniques, like neural networks, allow the quantitative description of the input-output (i/o)-relations of the overall process, which is sufficient for industrial use. There are, however, two problems for the use of black box techniques for process modelling:

  1. The high number of statistically well distributed data sets in the input variables are required in complex processes (curse of dimensionality). Typical production data, however, are highly correlated, and the number of data sets available from comparable production conditions are comparably small.
  2. Due to the lack of extrapolability, black box models cannot be applied for optimisation.

These shortcomings in rigorous as well as in black box modelling techniques have lead to the development of modelling techniques which combine both approaches.
A significant improvement on black box methods is achieved by the additional use of a-priori-known structural information about the process. The central idea is that structural information is used to reduce the complexity of the black box submodels.

Due to a reduction in complexity compared with pure black box modells, the number of data sets required to identify the model can be reduced significantly without any loss of accuracy.

This approach is called „Structured Hybrid Modelling". A structured hybrid model (SHM) therefore consists of three components:

- rigorous submodels describing the i/o-relation of those subprocesses which are well understood
- black box submodels for those subprocesses for which no rigorous model are available
- a model flowsheet describing the i/o-structure of all the submodels, merging them by mapping the i/o-structure of the real process.

The new component is the explicit use of the structure of the process in the generation of the model flowsheet. This leads to the incorporation of additional qualitative knowledge reducing the complexity of the model significantly. The main improvement, however, is not only the simple reduction of the complextiy, but also the explicit integration of the process flowsheet structure resulting in a drastical improvement of the performance of the model.

The main improvement of SHM could be achieved by proving the extrapolability of SHM. Therefore SHM can be used for optimisation.

Because of these promising findings of our research, we launched at Bayer in 1998 a project to develop a software environment as well as numerical and analytical methods to enable us an effective application of the SHM technology at various application projects.

The SHM technology has been applied in projects focussed on the identification of the relations between

- production parameters and quality properties of materials
- material composition and quality properties

The technology as well as the results in application projects are described in detail in a special paper on this conference.

3. Data Mining in Process Analysis

For a successful development of process models the identification of the key process parameters which are relevant for the required process output is necessary. In the most cases the modelling of the full process including all parameters leads to models too complex for an efficient analysis and, even worse, to an uneconomic effort in establishing the models. Therefore efficient methods for the identification of key parameters are the key for an efficient, fast modelling of poorly understood, but economically relevant processes.

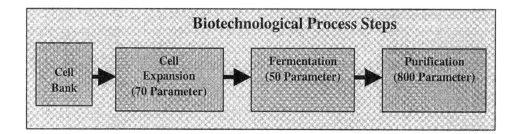

Figure. 2: Complexity of main process steps of an industrial biotechnological production plant

Good examples for this type of problems can be found in biotechnological processes. Typical biotechnological production processes start with the thawing and cell expansion [Fig. 2], until the cell densities are high enough to start the production fermentation. After fermentation the product will be purified in often very complex purification processes.

Each of these production steps are controlled by numerous process variables which are measured for regulatory purposes. Since measurements of biological parameters often are too expensive, the number of data sets may be small compared to the number of process variables. In one of our application projects in a purification process about 800 parameters are controlled, but only 150 data sets are available. In the fermentation subprocess about 50 variables are measured, similar data sizes are found in the cell expansion subprocess. Therefore the data are very sparse distributed in the variable space. To find sources for process quality variations a careful data analysis is required.

The technologies allowing to identify relevant parameter combinations from given sets of archived data are summarised in the literature as "Data Mining". Data Mining means a collection of mathematical methods allowing to extract a small set of explicit rules out of large, unstructured sets of process data. These methods have been developed on the basis of artificial intelligence research and are widely established in business applications like CRM (customer relationship management) and market forecasts as well as in biotechnology.

The available methods can be classified as follows:

- Clustering methods. These methods allow to group the data sets into clusters, such that the data sets within one cluster are more related with each other than the data sets in different clusters. The key problem in cluster methods is the appropriate choice of the neighbourhood definition. Unfortunately there are no methods to identify a priori an optimal neighbourhood definition.
- Decision trees (and other partitioning methods). These methods allow to identify a partition of the input variable space such that the process output shows maximum variation between the segments. The problem of these methods is that the identification of an optimal partition is NP hard. Therefore in any application case

appropriate heuristic methods have to be chosen a priori, which may lead to bad results. The most common methods, the tree algorithms, lead to partitions which can be described by tree structures of rules. They are very fast, but a monotonic dependency of the output variable with respect to each input variable is required. Therefore in many cases tree methods do not find efficient rules.

- Subgroup Search. These methods try to identify segments in the input variable space which can be characterised by a few input variables and show by some means extreme output values. Each segment is described by a single rules. In contrast to the partition methods, subgroup search can result in a set of rules which is not complete (some data sets are not contained in any segment) or redundant (some data sets are elements of more than one rule). These effects lead to more and more problems in the appropriate interpretation of the results, if the complexity of the rules (and the admitted segmentation) increases.

- Separation methods. Separation methods allow to split the input variable space into segments which differ in the distribution of the output values but can be separated using simple surfaces. Well established methods are separation planes or simplex methods. Neural networks can be used in the same way. Therefore complex interactions between the input variables of the process can be described which can not be mapped into sets of rules with the method described above. The problems, however, are that the admitted complexity of the separation manifolds are crucial for the quality of the result. Increasing the complexity, however, leads to the identification of a black box model with all resulting problems. Therefore the a priori characterisation of the separation manifolds by small sets of parameters is crucial for the success of these type of methods. This a priori characterisation, however, requires an understanding of the process which is often not available.

Within each class of methods there exist numerous different variations. Each of the methods has its specific benefits, but also specific problems for the applications, too. Data Mining can be compared to a house with different windows which can be used to get a view into the house. Through each window we will get different kinds of information, and only through the combination of all the different information a reliable understanding of the process can be achieved. Therefore a profound understanding of the various methods is necessary for success in applications. Compared to the classical applications in business data analysis, process data analysis leads to additional problems:

- The size of the available data sets is often very small compared to the typical data sets in classical data mining applications.
- The data are sparsely distributed and they are often correlated in the input parameter space. Moreover there are significant lacks in the data sets.
- Correlated, small, data sets lead to large sets of rules with questionable reliability. To extract reliable information from the data, an efficient postprocessing of the rules is necessary. For this purpose, however, no established methods exist.

In the example of a biotechnological process the analysis of the fermentation using decision trees (C4.5) showed no significant set of rule with respect to the yield of the process. Subgroup search, however, resulted in more than hundred rules describing process states with low and high yield. It is very likely that the most of these rules are redundant caused by the statistically bad distribution of the data.

To overcome these technical problems and to open the full potential of Data Mining to process analysis, in 2000 there has been established a three year project "Data Mining with Process Data" at Bayer. We aim to establish software solutions for engineers and scientists to be applied directly by the application engineers. Moreover we develop special tools which are dedicated to data analysis experts. Based on the problem analysis sketched above, our main focus is

- Establishment of data preprocessing methods to fill gaps in the data sets
- Development of methods to analyse large sets of redundant rules
- Integration of dynamic data sets into data analysis to analyse series of batch data

For these methods we find a broad range of applications in our chemical and pharmaceutical processes, like

- Analysis of High speed experimentation results for catalyst development
- Analysis of biotechnological processes to identify relevant process parameter combinations
- Analysis of the impact of various process parameters for trouble shooting in "classical" processes for a fast identification of causes of process disturbances.

Moreover these technologies can be applied not only in process data analysis, but also in chemical and pharmacological research. Therefore traditional benefits from such "spin-offs" could be realised which make the development and implementation of these methods more attractive even from an economic point of view.

4. Summary

The increasing demands on the chemical industry from market and environment forces us to deal with a high variety of processes with increasing complexity. This challenge we face in chemical and pharmaceutical industry requires additional methods for model based process improvement. In the paper we have described the additional demands of chemical and pharmaceutical industry as necessary steps in the workflow of process improvement. We have described shortly two technologies which we expect to be of strongly increasing importance for process. Their economic benefit is increased by significant spin-offs into applications in chemical-pharmaceutical research.

European Symposium on Computer Aided Process Engineering – 12
J. Grievink and J. van Schijndel (Editors)
© 2002 Published by Elsevier Science B.V.

Designing Industrial Processes for On-Aim Product Quality Control

Michael L. Luyben

E. I. du Pont de Nemours and Company, Inc.

Engineering Technology – Process Dynamics & Control

1007 Market St. – B7434

Wilmington, DE 19898 USA

Abstract

This paper summarizes an industrial view on incorporating controllability and developing plantwide control strategies at the stage where a new process is being designed. The use of computer-aided process engineering tools is discussed within this context and three industrial examples are given to highlight key concepts and point out areas of potential future academic research.

1. Introduction

The act of designing an industrial process demands disciplined methodology combined with inspiration. It is deeply based upon the fundamental application of scientific principles and draws upon a considerable engineering arsenal. In addition, it requires touches of imagination to create a new workable and coordinated whole out of numerous concepts, options, and possibilities. A successful new design project ultimately combines technical, economic, market, and timing factors to satisfy the necessary pieces of a commercial puzzle that is often only well-defined after the fact.

One of the key challenges in this puzzle centers on designing and operating new processes that deliver excellent dynamic performance for *on-aim product quality control* (defined as keeping product quality variables at a given target with low variability around that target). This requires understanding and minimizing variability in product quality, in order to meet customer needs. In practical terms the new process must be *capable* of operating to specific requirements that include:

- produce product consistently within required customer specifications
- reject process disturbances easily
- start up and shut down rapidly
- transition quickly between product grades to reduce off-specification production
- be robust to changes in operating conditions or process upsets

The philosophy presented in this paper is that the design of the new process (the unit operations used, the equipment parameters chosen, the equipment layout, the variables available to be measured and manipulated, etc.) plays a dominating role in determining how "controllable" and "capable" the process is. Furthermore it is economically advantageous to consider operability at the *design stage* (when changes are relatively easy to make) rather than wait until after the process is built (when changes are often quite expensive).

Over the past few years, a welcome number of academic researchers have been studying how the design of processes affects controllability and how plantwide control strategies can be designed. A partial list of more recent papers includes Groenendijk et al. (2000), Heath et al. (2000), Jorgensen and Jorgensen (2000), Kookos and Perkins (2001), Larsson et al. (2001), McAvoy (1999), Seferlis and Grievink (2001), and van Schijndel and Pistikopoulos (2000).

The purpose of this paper is to communicate an industrial view on incorporating controllability and developing plantwide control strategies (with particular emphasis on on-aim product quality control) at the stage where a new process is being designed.

The viewpoint in this work argues that controllability and control strategy design should not be simply an afterthought of the process design. Further, this viewpoint believes well-designed processes (in a way that inherently eliminates variability) can be controlled by relatively simple, straightforward, decentralized control strategies that are robust, easily understood by operators and plant engineers, and easily maintained. It seeks to avoid poorly designed processes that need "advanced" control schemes that too often turn out to be complicated and fragile, are not understood by operators and plant engineers, and require constant attention and maintenance.

This philosophy unfortunately runs counter to much of the current academic research in process control, which tends to focus almost exclusively on the development of control algorithms and ignores many of the other important parts of the control strategy (including the process design). This emphasis has the unfortunate effect of convincing people that process control algorithms (requiring no process understanding) can be developed (however complicated) to overcome any limitation or problem encountered in the operation of a plant.

Downs and Doss (1991) presented a noteworthy paper a decade ago pointing out many of these ideas and listed three main areas ripe for change: (1) move away from control algorithm toward control strategy development, (2) move toward an integration of process and control system design, and (3) move away from teaching control theory toward teaching how to control processes. These are as much true today as a decade ago, but it is reasonable to question how much significant change has actually occurred.

2. On-Aim Product Quality Control

This section discusses the conceptual steps and computer-aided process engineering tools that are often important in designing a new industrial process to achieve on-aim product quality control. These steps are not meant to be one-directional but rather are more a circular chain of thinking.

2.1 Customer Needs

The objective of a new process is to manufacture a product that will satisfy customer needs. The product must be not only suitable for the customer's purposes but its characteristics must also be consistent (constant) over time. This *consistency* is a particularly key issue for pharmaceuticals, food ingredients, specialty chemicals, monomers, polymers, fibers, etc., where it is often impossible to measure all properties and where small differences in impurities can cause a significant change in end-use characteristics. In the world of commerce, very often the product that customers receive today needs to be indistinguishable from one they received three months ago and from one they will receive three months hence.

2.2 Quality Variables

So customer needs must be related to measurable quality variables (e.g. compositions as measured by GC, pH, color, viscosity, molecular weight, bulk density, denier, etc.). This is typically done through customer feedback, scientific and engineering knowledge, and from the statistical design of experiments aimed at establishing such relationships. The allowable variation in product quality variables must also be quantified. A tool called Quality Function Deployment (QFD) is used to define and organize the relationships between product quality measurements and customer requirements (Kiemele et al., 1999). Computer-aided process engineering (CAPE) tools can be used in constructing the QFD matrices and transforming them subsequently into quantitative input-output transfer function relationships. It is important to understand, however, that some product characteristics affecting customers are unmeasurable prior to product consumption. Therefore product inspection is not always possible, and well-controlled processes are critical to ensure customer requirements are maintained.

2.3 Process Variables

This naturally shifts the focus backwards into the process itself. The design engineer can develop a process that will in principle achieve certain quality targets, but it is not self-evident how this is guaranteed to happen in practice, how consistent the product quality will be, and what economic cost will need to be paid to achieve low variability in quality performance.

The strategic step becomes establishing which small number of process variables (e.g. temperatures, pressures, flows, intermediate compositions, pH, speeds, etc.) out of potentially thousands available have the most significant effect on the key quality variables. Also required are the relationships between allowable product quality variations and allowable variations in process variables. The trick involves identifying a manageable number of important process variables (at the design stage) that need to be held "on-aim" to ensure product consistency. Of course, these process variables must be measurable or observable.

Engineering knowledge and previous operating experience are often the starting point of this activity. But these can be extrapolated only so far into a new process design that uses new technology or that produces a new product.

The rest of the knowledge, to the extent possible, must be generated from actual experiments (at both lab and pilot plant scale) and from computer modeling (where CAPE tools play a pivotal role). Ideally the modeling should be done in intimate conjunction with the experimental testing. Models are useful tools to help design experiments since a reasonable model captures not only process understanding but also process uncertainty.

Empirical models are often developed from statistically designed experiments that deliberately push the process into different operating regimes, thereby permitting quantification of the allowed variability. First-principles models (steady state and dynamic), however, often prove to be more powerful as a way to scale up and extrapolate from the pilot plant to the commercial plant. Developing first-principles models requires, first, the choice of model structure and, second, the determination of the unknown model parameters once the model structure is formulated. The model structure needs to be verified via experimental testing. The model parameters need to be regressed from experimental data. A key activity is to design experiments for the purpose of maximizing the information content aimed specifically at identifying these unknown model structures and parameters for both steady-state and dynamic models.

To use models that relate product quality to process variables, the CAPE simulation tools must permit such variables to be calculated and tracked throughout the entire process. For example, in a polymer process, the steady-state or dynamic model needs to be able to relate process conditions to quantifiable polymer properties. The tools must also provide a convenient framework for model parameter regression from experimental data. In addition, changes to the model structure and connectivity should be easy and relatively painless from a plantwide modeling framework

Models can often be used to guide the designed experiments that need to be run on the commercial process as part of the qualification effort. Out of this activity would come the aims and limits of the process variables, which either directly or indirectly are under closed-loop feedback control and some of which are then monitored by statistical process control techniques (Kiemele et al., 1999).

2.4 Control Strategies

The discussion above naturally leads into the design of the control strategies for the entire new plant. Part of the job of the base-level, regulatory, decentralized control structure will be to minimize the variability of the key process variables in the face of inevitable process disturbances. The regulatory control loops must keep the process at the specified operating conditions (stability) and must allow on-aim control of the key process variables (performance) within the allowable operating window.

Luyben et al. (1999) published a nine-step procedure to aid in the design of the regulatory control strategies for a complex integrated process that hold the system at the desired operating condition (called plantwide control). The plantwide control structure must take into account the effects of material recycle, energy integration, and the inventory of the chemical components, which are commonly not issues when focusing on the control of an individual unit operation.

From this perspective, five major objectives can be listed for any plantwide control system:

- stabilize the process
- cope with imposed constraints (safety, environmental, equipment, operational)
- balance the inventory of material and energy
- satisfy the economic objectives (including on-aim product quality) of the plant
- control the recycle structure

All else being equal, virtually all processes have fewer control degrees of freedom than objectives for the control system. Hence the only way to ensure proper satisfaction of *all* control objectives is to use the process design itself to make sure that the process can be operated effectively and that a sufficient number of manipulated variables are available. This is the theme of the paper by Tyreus and Luyben (2000).

CAPE tools are essential in the design and testing of plantwide control strategies. Rigorous nonlinear dynamic simulations form the basis of much of the analysis. Such simulators must be able to encompass large integrated plantwide flowsheets, must run with reasonable speed (20-50 times real time), and must be flexible to allow easy modifications (changing equipment sizes, adding or deleting unit operations, changing control strategies, etc.). The dynamic simulations also have to capture the essential effects of manipulated variables on the key process variables and also the effect of the process variables on product quality. Other CAPE tools are used extensively to aid the design and analysis of the control strategies, including linear transfer function, frequency domain, and data-driven empirical models. The tools need to provide transparent and convenient ways to transfer information and data among them.

If possible, control strategy tests can also be conducted in the pilot plant, with the results tied into the simulations and models. Time and cost constraints, however, typically limit how much design uncertainty can actually be effectively eliminated as a new process is being developed. Further, some unit operations cannot be reliably modeled or scaled up. In addition to the objectives listed above, Shinnar et al. (2000) point out that another objective of the control system is to allow compensation for uncertainties in the scale up of the new process design.

2.5 Process Design

As noted above, the final part of this quality puzzle rests in the process design, in the most general sense of the term. The new process design must be economically viable, technically workable, and capable of delivering on-aim product quality control. It should inherently minimize product quality variability in the face of disturbances and it should have suitable manipulated variables and suitable measurements to use for controlling the key process variables that determine product quality.

Clearly the high-level view about the process design involves the transfer of variability from where it is not desired (product quality) to parts of the process where the variability can be dissipated without serious economic or process consequences. Downs and Doss (1991) call these "process shock absorbers." Typically these locations include the plant utility system and strategically placed liquid or vapor inventories. Variability reduction can also be obtained with modifications to the flowsheet that create additional degrees of freedom, e.g. addition of bypass lines, addition of unit interconnection lines, addition of trim heat exchangers or other equipment, addition of theoretical stages in a distillation column or other overdesign, etc. Not all overdesign, however, improves controllability so this must be used judiciously.

The actual accomplishment of such design and control interactions typically results from experience and inspiration, and also from having control engineers who understand how processes are controlled and can work hand-in-hand (as well as communicate effectively) with the design engineers. Although some general understanding exists about how process design affects controllability and product variability, future academic research can certainly provide training and develop methodologies that quantify such interaction.

Improved CAPE tools are also needed to permit more seamless integration among the needs of the design and control engineers. In particular, the specifications and manipulators design engineers set in process simulators should not be limited by the solution algorithms employed, but rather should be more like the specifications and variables that would actually be used to control the process.

3. Process Examples

This section gives three industrial process examples, in order of increasing complexity, to illustrate the general ideas discussed above. Although shorn of their specifics due to space limits, each contains some key concepts to add to a general knowledge base.

3.1 Multi-effect Evaporator System

The first example is a three-stage multi-effect evaporator system (Figure 1) that operates at successively lower pressures. Three stages is an arbitrary choice for only illustrative purposes here. Additionally, the evaporators are assumed to be of the falling-film type. A liquid stream containing a higher-boiling organic species in water is fed to the first evaporator operating at the highest pressure. Steam is used to generate water vapor that leaves the first evaporator and is used as the heating medium in the second evaporator. Liquid from the first evaporator is the feed to the second evaporator, which operates at a lower pressure than the first. Liquid and vapor then flow similarly from the second to the third evaporator. Vapor flow from the third stage goes to further units, not shown, where pressure is controlled. Liquid product from the third stage also is the feed to downstream units.

The key product quality variable to be held on-aim is the water content in the third stage. In many evaporators, this would typically be measured by liquid density, but is here measured by liquid temperature. Another control objective could also be to provide as steady a liquid flow from stage 3 as possible for downstream units. The standard control strategy for such a system, shown in Figure 1, would involve the control of third stage temperature with steam flow to the first stage, and then the control of liquid level in each stage with the liquid exit flow. What makes this system so interesting is the virtually complete integration of the design with control. The sizing of the heat transfer areas, the vapor flow pressure drops, the amount of liquid holdup in each evaporator, the location of temperature measurements, and even the tuning of the level controllers all play a role in determining the controllability of the system. This is caused by dynamics resulting from vapor flow and pressure changes and by dynamics from liquid flow and composition changes.

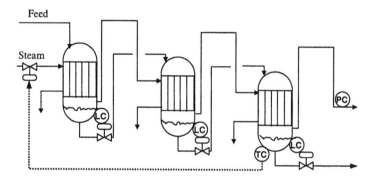

Figure 1: Multi-effect Evaporator System

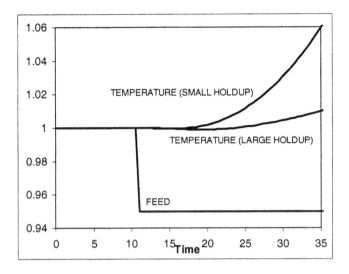

Figure 2: Stage 3 Temperature Response to Feed (Large and Small Liquid Holdups)

If not designed with care, the resulting process can exhibit significantly long dynamics between a change in steam or feed flow to stage 1 and change in stage 3 temperature, which then behaves like an integrator. Several concepts for improving the ability to control stage 3 temperature on-aim can be evaluated and tested at the design stage.

1. To compensate for load disturbances, steam flow to stage 1 can be ratioed (with proper dynamic compensation) to stage 1 feed flow (implying that a measurement of feed flow be actually in the design).
2. Another temperature measurement, in stage 2, can be included if it were found to be sensitive to changes in steam flow (this may not be feasible). This additional temperature can be used instead to control steam flow, since response will be faster, and stage 3 temperature would be used to reset the stage 2 temperature setpoint.
3. Since these are falling-film type evaporators, the amount of liquid holdup can be chosen using a basis independent of system performance (unlike other evaporator types). Normally, more liquid holdup is considered to be better for control since it allows more effective filtering of disturbances. However, in this system, the larger the liquid holdup in each stage, the slower the response of stage 3 liquid temperature is to changes in feed flow (Figure 2) or other disturbances. Minimizing the liquid holdup in each stage will speed up the composition dynamics. However, the more holdup, particularly in stage 3, the more constant the liquid exit flow can be controlled, which may be essential for flow smoothing to downstream. The solution may then be to minimize the liquid holdups in stages 1 and 2, include sufficient liquid holdup in stage 3 for disturbance rejection, and install a small liquid catch pot directly under the tubes in stage 3. The temperature thermowell, located in this catch pot rather than in the evaporator base, would allow a more direct and quicker measurement of temperature for control with steam flow.

38

4. Finally, instead of designing the vapor and liquid flows in the same direction, if the flow direction can be reversed, then steam flow can be directly manipulated in stage 3 to control stage 3 temperature. Of course this design would require the pressure in stage 1 to be lowest (which again may not be feasible).

This example illustrates a more general comparison of two approaches. One alternative is first to build a multi-effect evaporator system that is reasonably designed from a steady-state viewpoint, second to develop and apply a control algorithm to the given design, and third to hope the control algorithm will achieve the required quality control performance. Another alternative, advocated in this paper, is to perform the steady-state design and control strategy design simultaneously, where concepts such as those suggested above could influence the design of the evaporator system before it is built. This latter approach has a more likely chance to produce a new process design capable of achieving on-aim product quality control.

3.2 Distillation Column System With On-Demand Production Rate Control

The second example is a two distillation column system designed to separate desired product B from component A (Figure 3). The first is an extractive column (C1) where the A/B mixture is fed at a location below the feed point for component S. The feed rate of S to C1 is to be maintained at a fixed ratio to the feed rate of A/B. Component A goes overhead from C1 as the vapor product from a partial condenser. The bottoms product from C1 is a B/S mixture, which is then fed to the second column (C2). C2 separates component B as the vapor overhead product from a partial condenser. The bottoms from C2, component S, recycles as feed back to C1.

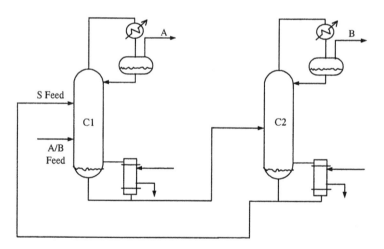

Figure 3: Two Distillation Column System Design

In the absence of other information, the control strategy design for this system would be relatively straightforward and the effect of the process design would not in itself be unusual. However, the customer needs for component B change this picture completely. Component B is a reactant fed to multiple batch reactors that operate virtually independently of each other. It is also required to be very high purity. This means that the overhead vapor flow from C2 changes instantaneously at the will of the batch reactors since no storage of B can exist. The key process variables with the most significant effect on product quality are then identified to be the reflux to feed ratio in C2, the control temperature in C2 (for composition control of S), and the control temperature in C1 (for composition control of A). The control strategy must then be designed to keep these process variables on-aim and also to satisfy the on-demand requirement for production rate.

Without enumerating all alternatives and details, the final process with control structure is shown in Figure 4. Dynamic analysis clearly shows that product B flowrate cannot be physically changed quickly enough by manipulating the feed flows to C1. The overhead flow from C2 is essentially controlled by the batch reactors, removing it as the obvious manipulator to control C2 pressure. This means that the on-demand control strategy must start with the satisfaction of product rate and work backwards (contrary to standard thinking, which usually moves forwards through a process). Since C2 pressure indicates the inventory of component B, it is controlled using the feed flow to C2. C2 reflux flow is ratioed to feed flow and C2 temperature is controlled by C2 steam flow.

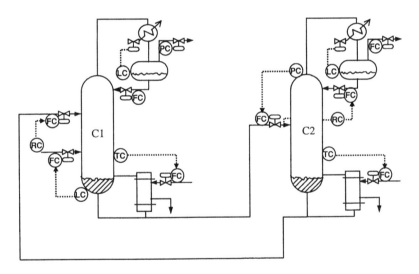

Figure 4: Design and Control Strategy

The original system design shown in Figure 3 is perfectly reasonable from a steady-state viewpoint. However, it would have been a failure in its ability to satisfy the dynamic requirements, including the manipulation of feed flow to C2 independent of C1. It is possible to envision the development and deployment of many "advanced" control algorithms on the original design to try bandaging the perceived "control" problem. Modifying the process design by adding a buffer tank to provide surge capacity between the bottom of C1 and C2, however, will solve the problem. The actual cost of a separate tank can be saved if the tank volume is simply built into the bottom of C1. Once this is done, the A/B feed flow to C1 can be used to control the base level in C1. Feed flow of S is ratioed to the A/B feed rate, which means that sufficient liquid inventory in the base of C2 must also be provided. C1 temperature is controlled by C1 steam flow. Since level in the base of C2 is not controlled, rational design criteria must be used to size the volumes of the two column bases to cover the spectrum of possible operating conditions. With this process design and control strategy in place, the key process variables can be held on-aim to satisfy the customer needs for product quality and on-demand production rate control. It is worth emphasizing here that more holdup is inherently neither good nor bad. What is good is the judicious use of the right holdup at the right place.

3.3 Bio-based 1,3-Propanediol Process

The final example is an entire process currently being developed by DuPont to produce 1,3-propanediol (PDO) via fermentation using glucose as the raw material. PDO is a key ingredient for a new polymer, polytrimethylene terephthalate or 3GT, and hence must satisfy stringent product consistency requirements. This development of new technology by the chemical industry seeks to begin producing materials with biologically-based processes that start with renewable feedstocks. The Cargill Dow process to make lactic acid and then polylactide polymer is another example.

For obvious proprietary reasons, details of a bio-based PDO process cannot be given here. However, conceptually such a process will contain sections for fermentation (to transform glucose into PDO), for separation (to remove fermentation material), and for purification (to produce quality PDO product). This example is included primarily to point out the need for academic research in the general area of designing bio-based processes to achieve on-aim product quality control. Since a fermentor uses living organisms, it will naturally have a certain amount of variability. Yet the final product from such a bio-based process needs to have the same consistency as one manufactured from a petroleum-based feedstock. This provokes questions about the need for and the location and nature of process shock absorbers (Figure 5) for a bio-based process whose final product quality must be controlled on-aim. Reasonable questions also can be asked about modeling and control needs; about assessing and quantifying variability at the design stage; and about where the focus of variability reduction should be along with additional economic costs this might require.

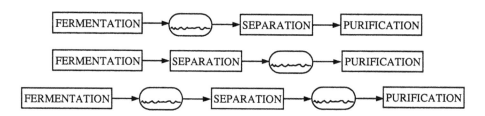

Figure 5: Process Shock Absorbers for Bio-PDO Process?

4. Conclusions

This paper has presented an industrial view of the conceptual steps taken during the act of designing a new process for ensuring capability to achieve on-aim product quality control. These steps span from the investigation of customer needs, to the definition of quality variables, to the determination of key process variables, to the design of the process control strategies, and finally to the actual design of the process itself. Because of the technical uncertainty inherent when designing a new process or deploying new process technology, two final noteworthy points are:
- avoid limiting process and control flexibility
- be creative in incorporating potential degrees of freedom

5. References

Downs, J.J. and J.E. Doss, 1991, Present status and future needs – a view from North American industry, CPC IV, 53, Eds. Y. Arkun and W.H.Ray, AIChE, New York.

Groenendijk, A.J., A.C. Dimian, and P.D. Iedema, 2000, Systems approach for evaluating dynamics and plantwide control of complex plants, AIChE J. 46, 133.

Heath, J.A., I.K. Kookos, and J.D. Perkins, 2000, Process control structure selection based on economics, AIChE J. 46, 1998.

Jorgensen, J.B. and S.B. Jorgensen, 2000, Automatic selection of decentralized control structures, Proc. of ADCHEM, Pisa, Italy, 147.

Kiemele, M.J., S.R. Schmidt, and R.J. Berdine, 1999, Basic Statistics - Tools for Continuous Improvement, 4th ed. Air Academy Press, Colorado Springs.

Kookos, I.K. and J.D. Perkins, 2001, An algorithm for simultaneous process design and control, Ind. Eng. Chem. Res. 40, 4079.

Larsson, T., K. Hestetun, E. Hovland, and S. Skogestad, 2001, Self-optimizing control of a large-scale plant: the Tennessee Eastman process, Ind. Eng. Chem. Res. 40, 4889.

Luyben, W.L., B.D. Tyreus, and M.L. Luyben, 1999, Plantwide Process Control. McGraw-Hill, New York.

McAvoy, T.J., 1999, Synthesis of plantwide control systems using optimization, Ind. Eng. Chem. Res. 38, 2984.

Seferlis, P. and J. Grievink, 2001, Process design and control structure screening based on economic and static controllability criteria, Comp. Chem. Eng. 25, 177.

Shinnar, R., B. Dainson, and I.H. Rinard, 2000, Partial control. 5. A systematic approach to the concurrent design and scale-up of complex processes: the role of control system design in compensating for significant model uncertainties, Ind. Eng. Chem. Res. 39, 103.

Tyreus, B.D. and M.L. Luyben, 2000, Industrial plantwide design for dynamic operability, FOCAPD V, 113, Eds. M.F. Malone and J.A. Trainham, AIChE Symp. Ser. No. 323, Vol. 96, New York.

Van Schijndel, J. and E.N. Pistikopoulos, 2000, Towards the integration of process design, process control, and process operability: Current status and future trends, FOCAPD V, 99, Eds. M.F. Malone and J.A. Trainham, AIChE Symp. Ser. No. 323, Vol. 96, New York.

European Symposium on Computer Aided Process Engineering – 12
J. Grievink and J. van Schijndel (Editors)
© 2002 Published by Elsevier Science B.V.

Adaptivity in Process Systems Modeling

Wolfgang Marquardt

Lehrstuhl für Prozesstechnik, RWTH Aachen

Germany

Abstract

The concept of adaptivity has gained increasing attention in numerical analysis and scientific computing in recent years, but has not yet been adequately appreciated in process systems modeling. This contribution identifies key problems in model formulation and model solving where adaptivity is supposed to play a major role in constructing attractive computational strategies. Some basic concepts of adaptivity are sketched and some applications are discussed to illustrate the potential of adaptivity.

1. Introduction

Process systems modeling has reached a high level of maturity. A variety of modeling tools are available and widely used in industrial practice to solve process systems problems by model-based approaches. However, the current scope of modeling is largely focused on intermediate length, time and chemical scales (Marquardt et al., 2000) which largely correspond to a single process unit, a plant or part of a site. Steady-state and dynamic simulation or optimization are employed routinely. An obvious challenge is to formulate and solve larger models as part of integrated design or operations problem formulations to achieve more holistic solutions of process engineering tasks. The integrated models are typically not on the same length, time or chemical scales. Instead, several scales will be covered in a model in the sense of multiscale and lifecycle process modeling (Marquardt et al., 2000, Pantelides, 2001), ranging from the molecular to the supply chain scale in extreme cases.

The solution of a multiscale model will show smooth trends along space, time or chemical coordinates, which are superimposed by local structures of significantly shorter characteristic scales. These local phenomena may for example comprise boundary layers, sharp fronts, vortices or oscillations, if high spatial and/or temporal resolution of transport and reaction phenomena is attempted in some detailed process unit model as part of a plant or even supply chain model. Further, a higher chemical resolution with a larger number of components may be required in parts of the model of interacting plants on a site to properly predict the quality of the product streams whereas only few components suffice in other parts.

If mathematical models and numerical solution algorithms would attempt to resolve the whole range of time, space or chemical scales of interest uniformly with high resolution, model complexity would quickly reach a level which cannot be handled reasonably anymore. A solution to this problem might be expected from the still breathtaking progress in computer hardware development without any effort in process systems engineering. However, given the inherent limitations of computational tractability, a complementary improvement of algorithms has to be achieved to push the limits of complexity in concert with hardware improvement in the future as in the past (Hossfeld, 2000). More specifically, sophisticated modeling and solution techniques need to be provided to predict interesting phenomena with a local resolution complying with their characteristic scales. Such an adaptation of the model formulation and the approximation of the model solution for the resolution of local phenomena has to be achieved by means of automatic adaptation algorithms to relieve the engineer from building up knowledge about the solution structure by trial and error.

Hence, adaptivity in the sense of this paper refers to all means to adjust the model or the approximation of its solution locally in one or more scales of interest to resolve some detail with a given accuracy. This paper does not attempt to provide a concise review on adaptive modeling and solution techniques under development. Rather, it aims at raising awareness on the potential of adaptivity in process systems modeling today and in the future. Therefore, it is sufficient to focus on some key concepts, which may be adopted to serve as basic building blocks for the development of adaptive techniques in process systems modeling. Consequently, we will not go into the technical details, but provide some references to the interested reader for further study.

Section 2 provides a more detailed discussion of what adaptivity is all about. Some well-known and illustrative examples are introduced. The representation of functions, which is at the core of any adaptive algorithm, is briefly reviewed in Section 3. Section 4 summarizes some key strategies to implement adaptivity. The final section 5 presents some recent developments in adaptive computational methods with applications in process engineering.

2. Perspectives of adaptivity

Two fundamentally *different perspectives* of adaptivity in process systems modeling are identified. First, adaptivity may be cast into a numerical algorithm to solve a given model. Here, the algorithm is automatically adjusted during runtime to approximate the true solution within some tolerance bound at minimal computational effort. We will refer to this class as *adaptation of the solution algorithm (AS)*. Second, the model itself, or more precisely either the number of equations or the functional expressions and parameters in the equations, may be automatically adjusted to match some reference either during the modeling process or as part of the solution algorithm. The reference may be given by experimental data or by the solution of a more detailed model. In the latter case, model order reduction or model simplification (Marquardt, 2002) can be distinguished. These classes will be called *adaptation of the model (AM)* and *adaptive model reduction (AMR)*, respectively.

Obviously, we can have a combination of *AS* and *AM* or *AMR*. A model is solved by an adaptive numerical method to a prescribed error tolerance in an inner loop *(AS)*, and the model is adjusted to match a reference in an outer loop *(AM or AMR)*. If model adaptation and solution are highly intertwined, the distinction between the classes may become blurred.

Adaptation is either carried out as part of an iteration sequence or during a transient in a dynamic problem. Typically, the adaptation is specific to a part of the model corresponding for example to different spatial locations or balance envelopes describing part of the process to be modeled.

The *implementation of adaptivity* requires three major ingredients. The first is a flexible and adjustable way to represent the solution in *AS* or the model in *AM* and *AMR*. Adjustment of the solution or the model has to be based on an estimate of quality to measure a global or a local approximation error in case of *AS* and *AMR* or the goodness of fit in case of *AM*. Such an estimate forms the second ingredient. The third ingredient is a strategy to implement the adjustment of the solution algorithm or the model itself on the basis of the estimate of quality. It is reasonable to require a largely monotonic increase of the quality measure with increasing computational effort. In the limit, we would like to reduce the approximation error to zero in case of *AS* and *AMR* and to achieve a perfect match of the reference in case of *AM*, provided the reference is free of noise (see section 5.1 for details). The computational effort should further scale moderately, if more stringent quality bounds are requested.

In the following, we present three concrete areas where adaptation has been applied in process systems modeling.

2.1 Adaptive solution of ODE and DAE models

Automatic integration algorithms for the *solution of initial value problems* comprising *ordinary differential (ODE)* or *differential-algebraic equation (DAE) systems* are classical examples for *AS*. Such adaptive methods are routinely used in industry as part of commercial process modeling environments. These algorithms march along the discretized time coordinate and determine an approximate solution at the current grid point from solutions at previous grid points. The local error is estimated from two approximations at the same grid point computed by either two different step sizes or approximation orders. The algorithm is adjusted to just match the given local error bound by reducing or enlarging the step size, the approximation order, or both simultaneously. Advanced adaptation strategies aim at a minimization of the predicted total computational effort in future time steps. Consistent algorithms recover the correct solution in the limit, if the step size approaches zero.

A detailed exposition of the various numerical integration methods and adaptation strategies can be found in many textbooks (e.g. Brenan et al., 1996). The error controlled adaptation of numerical algorithms has lead to a breakthrough for the integration of stiff ODE and DAE systems to balance numerical stability, accuracy of the approximate solution and computational effort. Such algorithms are indispensable in process systems modeling since (staged) separation as well as reaction processes often display very high degrees of stiffness.

2.2 Adaptive solution of PDE models

The most *advanced adaptation strategi*es have been developed in the context of numerical *solution algorithms for partial differential equation systems* (PDE) and are hence of *AS* type.

A popular approach almost exclusively employed in process systems modeling is the so-called method of lines or MOL for short (Liskovets, 1965, Schiesser, 1991, Vande Wouwer et al., 2001) where the spatial differential operator is discretized first typically by some finite difference or finite volume scheme on a spatial grid. The resulting DAE system is solved by an automatic integration algorithm. In classical MOL (Schiesser, 1991) the spatial grid is fixed, whereas more recent developments are aiming at an adaptation of the spatial grid (Vande Wouwer et al., 2001), in order to facilitate the resolution of fine detail in the solution or to properly track spatio-temporal phenomena such as propagating reaction or separation fronts. Adaptivity along the spatial coordinate includes static and dynamic regridding. In static regridding, the grid is typically adapted after a fixed number of time steps to reestablish the equidistribution of some functional (e.g. an average of curvature) on the grid. Grid adaptation may either rely on a movement of a fixed number of nodes or by a local refinement or coarsening of the grid by insertion or removal of nodes. Alternatively, in dynamic regridding, a fixed number of nodes is moved continuously. The criterion for grid point movement can either result from prescribed dynamics reflecting physical insight or it can stem from the minimization of some error functional.

Though, the MOL nicely integrates with the simultaneous solution strategy in equation-oriented process modeling environments and is able to successfully solve challenging problems (Vande Wouwer, 2001), there are some disadvantages of this semi-discretization approach (e.g. Wulkow, 1996, Lang, 1998). In many MOL variants, control of the spatial discretization error is not guaranteed. In case of moving grid strategies, the total number of grid points or basis functions and hence the level of the global spatial approximation error is fixed. Static regridding often does not account for spatial errors and always introduces discontinuities in the model, which reduce computational efficiency of the integration algorithm, even if they are accounted for rigorously. In dynamic regridding, the method parameters need to be properly tuned to avoid mesh tangling and to balance numerical stability and computational efficiency. Further, the MOL does not generalize easily to higher dimensional PDE models.

Consequently, alternative adaptive solution strategies for distributed process systems models have been developed. Some of them will be reviewed in section 5.4.

2.3 Model identification

Function fitting or *model identification* are classical examples for *AM*. Typically, a certain structure of a steady-state or dynamic model is assumed and the parameters in this structure are adjusted by minimizing some weighted least-squares objective of the difference between model predictions and experimental data (Bates, Watts, 1988). Adaptation is often implemented numerically by means of some mathematical programming algorithm. It may be carried out repetitively, for example in the context of a real-time application, where the model has to be adjusted continuously to the time-varying process.

Obviously, there is a fundamental difference to *AS*. An *AM* strategy can typically not drive the prediction error to zero for a given model structure regardless of the computational effort spent. In part, this is due to the typically inappropriate model structure which usually does not perfectly fit the process behavior. In addition, the reference in model identification is often corrupted with noise. Therefore, the model may interpolate the training data set with zero error, but will not properly approximate a validation data set. Last but not least, model identification is an inherently ill-posed problem (Kirsch, 1996), where small errors in the reference will be strongly amplified in the solution. A satisfactory goodness of fit not only for the training but also for the test data set can therefore only be achieved, if we move from parametric fitting to nonparametric fitting, where the model structure itself is subject to adaptation during the modeling process (Banks, Kunisch, 1989, Gershenfeld, 1999).

2.4 Discussion

The presentation in the last section indicates that an appropriate representation of functions is crucial for the construction of adaptation strategies at least for problems of class *AM*, where the *function of interest is part of the model*. Such a function may for example represent a time-varying disturbance in a dynamic model or the input-output relation of a stationary empirical model, or it may describe some reaction rate or transport coefficient in a model derived from first principles. The function to be identified has to be represented in such a way that the level of resolution can somehow be adjusted to the information content in the typically noisy data.

However, the *solution of a given model can also be viewed as a function* of at least one independent variable such as time and/or space or space-like coordinates. Examples are the concentration field in a plug flow reactor or the molecular weight distribution of a polymer in a polymerization reactor. The automatic integration algorithms for DAE models as well as the MOL for PDE models do not easily reveal the functional character of the solution because of the recursive computations on discrete grid lines.

Model fitting and model solving are considered related problems, since a typically multivariate function parameterized by a finite number of degrees of freedom has to be determined in both cases. These degrees of freedom are constrained by the data in case of model fitting and by the model equations in case of model solving. An appropriate representation is crucial to resolve the function of interest at an appropriate level of detail and to achieve a certain measure of quality. Since the measure of quality also determines the computational complexity, an automatic adjustment of the function representation is indispensable to implement an adaptation strategy.

3. Representation of functions

There are basically two different approaches to represent functions. In the first case, a mesh of grid points is introduced to cover the domain of the independent variable(s) such as time and space(-like) coordinates. The *function is sampled at these grid points* during the modeling or solution process. A uniform grid with an equidistribution of grid points is employed if no information on the structure of the solution is available. Obviously, the accuracy of the representation as well as the computational effort scale with the number of grid points. For infinitely many points, the function is recovered at arbitrary accuracy. However, no analytical representation of the function is available. As a consequence, a dense representation or derivative information can only be obtained if interpolating functions are postulated locally in the vicinity of a grid point.

Alternatively, the *function can be represented in some function space* spanned by a (potentially infinite) number of (orthogonal) basis functions (e.g. Courant, Hilbert, 1968). This way, an arbitrary function can be represented by a series expansion comprising a linear combination of the basis functions. This approach provides an analytical representation with a degree of continuity determined by the chosen basis. For appropriately chosen function spaces, the accuracy and detail of the function representation increases monotonically with the number of basis functions retained in the expansion, or, in other words, with the dimension of the finite subspace the function of interest is projected to. Hence, accuracy and computational complexity scale with this dimension. A variety of bases with local and global basis functions have been suggested.

Global basis functions are defined on the whole domain of the independent variable(s). Examples are orthogonal polynomials such as Lagrange or Chebyshev polynomials or trigonometric functions (e.g. Courant, Hilbert, 1968). By construction, the localization properties of these functions is limited, since the frequency content is identical at any point of the independent variable domain. Therefore, the resolution of a local detail cannot be accomplished without influencing the function on the whole domain. Therefore, these bases are useful for the representation of smooth functions with moderate curvature but fail to resolve local details on the back of a global smooth trend.

In contrast, local basis functions are defined on a subdomain (or finite element) of the function domain. They are restricted by certain continuity constraints at the element boundaries. The most popular functions are B-splines (de Boor, 1978), which result in piecewise polynomial representations of different order (e.g. piecewise constant, linear etc.). By construction, such bases are perfectly suited to resolve fine detail of a solution. However, a large number of basis functions is required if smooth trends have to be represented. To remedy this problem, higher order orthogonal polynomials may also be employed to approximate the function on every finite element.

In some sense, global and local bases constitute two extreme cases with respect to the representation of smooth trends and of fine detail of a function with as few degrees of freedom as possible. These extremes can be bridged by bases which incorporate basis functions with varying support ranging from the whole domain of the function to a subdomain the size of which is characteristic for the scale of the detail to be resolved. This idea is implemented by wavelet (DeVore, Lucier, 1992) and hierarchical bases (Yserentant, 1992). Both types of bases allow for a representation of a function at multiple scales with increasing resolutions. For a single independent variable x, the resulting representation is

$$f(x) = \sum_{l \in \Lambda} c_l \varphi_l(x) = \sum_{l \in \Lambda} \sum_{i \in \Gamma} d_{i,l} \psi_{i,l}(x) \tag{1}$$

The global trend is represented by few functions with large support on coarse scales (small indices l), whereas increasingly fine details can be resolved by adding functions on finer scales (larger indices l). More precisely, the expansions on every scale l represent functions in a sequences of nested spaces V_l, where detail is added from l to l+1. Hence, V_l is a subspace of V_{l+1}

Obviously, adaptation requires functions which are able to resolve local detail. Global orthogonal basis function do not serve this purpose. Local basis functions defined on finite elements such as B-splines or orthogonal polynomials are preferred. Wavelet and hierarchical bases offer a flexible way of representation, since they provide basis functions with varying support. All these functions may be used for function fitting (e.g. Gershenfeld, 1999) or for the discretization of differential equations by means of weighted residuals or Petrov-Galerkin methods (Villadsen, Michelsen, 1982, Fletcher,

1988, Dahmen, 1997). All the function representations are linear in the parameters which reduces computational complexity.

4. A general adaptation strategy

After the brief summary of various ways of function representation in the last section, we ask now the question of how to adjust the representation of a specific function of interest in an adaptive modeling or solution scheme. The objective of adaptation is to represent the function to match the given measure of quality (approximation error, goodness of fit, etc.) by a low number of degrees of freedom to achieve a minimum computational complexity. We first focus on problems with no time-like coordinate and return to models described by evolutionary equations in section 4.5.

4.1 Adaptation by successive refinement

We present first a general adaptation strategy which is useful for all three problem classes *AS, AM* or *AMR*. For discrete representations on a grid, a set of suitable grid points has to be selected whereas suitable basis functions have to be determined in case of representations in a function space to capture global trends as well as local details of the function of interest. Obviously, for a sequence of increasingly tighter measures of quality, a sequence of function representations at increasingly finer scales of resolution and accuracy can be obtained. This observation forms the basis of an adaptation strategy which is in full compliance with general problem solving principles: we start from a coarse problem and successively refine until a satisfactory solution is obtained. The resulting *multilevel adaptation strategy* comprises of the following major steps (e.g. Verfürth, 1995):

1. Set k=0 and formulate an initial problem to achieve a coarse resolution of the function of interest with "few" degrees of freedom at low computational effort.
2. Solve problem k to an accuracy corresponding to the resolution of the problem.
3. Estimate measures for assessing solution quality globally as well as locally.
4. If the global measure of quality is satisfactory, stop.
5. Set k=k+1, refine the problem formulation using local estimates of the measure of quality, and go back to step 2.

More detailed remarks on the individual steps are presented in the next subsections.

4.2 Refinement of problem formulation

We start with the refinement in step 5 which can be carried out in various ways. First of all, one may distinguish uniform from non-uniform refinement. *Uniform refinement* refers to cases where the problem formulation is updated exactly the same way on the whole domain. In particular, the number of grid points or finite elements may be doubled by interval halving, or all functions on the next scale of finer resolution may be added to the expansion (1). This way, only a rough adaptation is accomplished, which does not take into account the local details. Therefore, *nonuniform refinement* is preferred in step 5. The grid is only refined locally (e.g. by interval halving) in those domains, where local detail is supposed to be resolved in the function of interest. Analogously, if representation (1) with wavelet or hierarchical bases is employed, only those functions on a candidate scale l of higher resolution would be selected, which are expected to contribute most significantly to an improvement of the solution.

Both strategies adjust the number or degrees of freedom by increasing the number of grid points or basis functions to improve the measure of quality. In addition to this very common *h-adaptation* one may also consider to increase the number of degrees of freedom by adjusting the order of the approximation functions on a finite element or in the expansion (1). In fact, there are sophisticated finite element techniques which employ such *h-p-adaptation* combining mesh and order adaptation.

4.3 Estimation of measure of quality

Steps 3 and 4 of the multilevel strategy in section 4.1 decide on the basis of an estimated measure of quality when the refinement cycle can be stopped and where the function of interest should be refined locally. The error estimation schemes are of a very different nature in case of *AS* and *AMR* on the one and *AM* on the other hand.

In case of *AS* and *AMR*, a reasonable tolerance for the approximation error can easily be specified according to the accuracy requirements of the context the model is used in. An error estimator has to be constructed which gives tight bounds on the true error but is far less expensive than the computation of the solution. Several types of *a posteriori error estimators* have been suggested (Verfürth, 1995). All of them are based on available approximate solutions. Common approaches are built on the equation residual, on the solution of a simpler problem, on an averaged gradient or curvature, or on a comparison of two solutions of different accuracy in nested spaces. No general guidelines on the performance of a certain estimator are available. The estimators are used locally to identify the need for nonuniform refinement or globally to decide on terminating the refinement cycle.

The problem of *stopping refinement* is delicate in case of *AM* problems, because the measurements are noisy in general. The goal is to determine the model function such that only the meaningful information content in the data is captured by the model. A compromise has to be found to avoid model mismatch in case of underfitting and the modeling of the noise in case of overfitting. Furthermore, the ill-posedness of the model fitting problem (Kirsch, 1996) requires some kind of regularization to control the amplification of data errors in the solution. Since discretization also acts as a regularizing operator in addition to e.g. Tikhonov regularization, the stopping criterion in the refinement process becomes of vital importance for the quality of the solution. A better understanding of the interactions between regularization and discretization and the development of constructive criteria to decide on stopping the refinement process are still largely open research issues (e.g. Ascher, Haber, 2001a, Binder et al., 2002).

4.4 Efficient solution procedures

The multilevel strategy of section 4.2 is particularly attractive, if the effort to solve problem k+1 can be significantly reduced by using information about the solution and/or the solution process of problem k. Two successful concepts (e.g. Briggs et al., 2000) are presented next.

Often, *iterative solvers* are employed in the solution step 2 of the multilevel algorithm in section 4.1 for nonlinear as well as linear subproblems. Besides avoiding the costly LU decomposition of a direct linear method for very large problems, (linear as well as nonlinear) iterative techniques allow for prematurely terminating the iteration process to implement some regularization in *AM* problems (cf. section 4.3) or to adjust the solution accuracy to the discretization accuracy in *AS* problems. Furthermore, iterative solvers are most suitable to implement so-called *nested iteration*, where the solution of problem k with coarse resolution is used to initialize the solution process for the refined problem k+1. For example, wavelet or hierarchical bases representations of the function of interest converge very quickly with increasing scale. Hence, the values of the expansion coefficients of a solution on level k are often good initial guesses for the same coefficients on level k+1, and the unknowns newly introduced into the problem are advantageously initialized with zero due to their small contribution to the solution.

Multigrid methods are used to extend the basic multilevel scheme of section 4.1 for improving the solution efficiency. The basic multigrid algorithm employs an approximation of the solution at some resolution, say on a fine grid. After projection of the problem to a coarser grid (also called relaxation), a coarse correction of the approximate solution can be computed at moderate effort. After interpolation (or prolongation) of this coarse correction to the finer grid, an update of the previous fine grid solution can be computed before the next cycle is started. In a *V-cycle,* a sequence of relaxation and solution steps are taken to go from a fine to a coarse resolution, before a sequence of relaxation and correction steps is carried out to go back from the coarse to the fine resolution. In the so-called *full multigrid cycle,* one nested iteration is used to go from k=0 to k=1, then a first *V-cycle* with one relaxation and two subsequent nested iterations is used to go to k=2 and so on until the desired quality measure is established.

4.5 Rothe method

We are now returning to the problem of treating evolution along time-like coordinates. Obviously, it seems to be attractive to combine the extremely efficient recursive adaptation for time-stepping commonly used algorithms for the solution of ODE or DAE initial-value problems (cf. section 2.1) with the recursive refinement strategy for the space or space-like coordinates outlined in the previous sections. Conceptually, such a combined strategy can easily be motivated by an interpretation of PDE models of evolutionary type as ODE or DAE models in a state space of infinite dimension (Lang, 1998). During successive refinement, the infinite-dimensional model is projected into finite-dimensional state spaces of appropriate order. The approximation error acts as a perturbation of the time stepping algorithm.

Such a strategy is known as a so-called Rothe method (Liskovets, 1965) where time is discretized prior to space(-like) coordinates in contrast to the MOL. The same adaptive discretization schemes which have proven successful for ODE and DAE problems are applied to deal with the time coordinate. However, one-step method are preferred to multi-step methods to better cope with discontinuities and to avoid large overhead for storing and retrieving many past solutions. After time discretization, the approximated dynamic problem has basically the same structure as the stationary model. associated with it. This discretized problem has to be solved repetitively in every time step. Hence, any adaptive method for the stationary problem with only space(-like) coordinates such as the multilevel strategy of section 4.1 carries over to the transient case. The only problem to be solved is the coordinated control of the error in time and space to guarantee an overall solution quality which matches user-defined tolerances.

5. Applications

The advanced concepts of adaptivity can be employed to address a variety of process systems modeling problems. The selection of examples cover different classes of adaptation problems, but is not meant to be comprehensive. Rather, it reflects the author's experience. Sections 5.1, 5.2 and 5.3 present *AM* problems, the examples in section 5.4 are of class *AS* and in the last section 5.5 *AMR* type problems are discussed.

5.1 Function fitting

In a function fitting problem, a set of (n+1)-tuples $(y_j, x_{1,j}, x_{2,j} ... x_{n,j})$, j=1...J, of observed data is given to develop a model y = f(x). A pragmatic approach to solve the problem employs the successive refinement strategy suggested in section 4.1. The model structure is preferably chosen to facilitate the representation of f(x) on different scales of resolution (or more precisely in a sequence of nested subspaces). Wavelet or hierarchical bases are meeting this requirement. Then, a first parameter estimation problem is formulated in step 1 with a candidate model function on a coarse scale with only few degrees of freedom. Its solution in step 2 results most likely in an unsatisfactory fit. Next, basis functions are successively added on the next scale in a uniform or nonuniform manner to improve the quality of fit in steps 3 to 5 until the prediction error is smaller than a given threshold.

This procedure forms the core of Mallat's theory of multiresolution signal decomposition. A generalization to multivariate function fitting has been claimed to be straightforward (Bakshi, Stephanopoulos, 1993), but the curse of dimensionality (e.g. Gershenfeld, 1999) has not yet been successfully tackled with function representations linear in the parameters. Adaptivity seems to be the only way to address this fundamental problem. An interesting strategy for high-dimensional function fitting by means of hierarchical bases has been reported recently (Garcke, et al., 2001).

Several approaches have been suggested to deal with measurement noise including denoising as part of wavelet based regression by thresholding small expansion coefficients (e.g. Donoho, Johnstone, 1994) and wavelet versions of kernel estimators (Antoniades et al., 1994). A general framework relating the resolution of the discretization to the amount of measurement data needed is provided by statistical learning theory (Evgeniou et al., 2000). A sequence of nonparametric estimation problems is formulated for model functions of increasing resolution in nested function spaces, which are fixed a priori. The best fit is chosen adaptively by minimization of the expected risk of an inappropriate model fit.

5.2 Disturbance identification

Disturbance modeling and identification occurs in the context of model-based monitoring of processes in transient operational phases, where unmeasurable process quantities (i.e. states and disturbances) should be inferred from measurable ones. Instead of modeling the disturbance by a trend model and estimating the extended system state (e.g. Kurtz, Henson, 1998), the disturbance can be treated as an unknown time dependent function which has to be determined from an approximate inversion of the model. This problem is related to the function fitting problem in section 5.1, since an unknown function has to be inferred from observed inputs and outputs of a dynamic instead of a stationary process. Binder et al. (1998) state a nonlinear least-squares problem with a penalty on the unknown initial conditions and disturbances to be inferred from the measurements, DAE model constraints and inequality bounds. Two different adaptive numerical strategies are under development to solve this problem.

Schlegel et al. (2001) suggest to parameterize the unknown function by means of a Haar wavelet basis. The optimization problem is first solved by a sequential approach with coarse resolution of the unknown function. Then, the multilevel strategy of section 4.2 is applied to successively refine the disturbance locally to identify sharp transitions. A potential elimination of a basis function is based on thresholding the coefficients in the current solution (similar to multiresolution analysis, cf. section 5.1), whereas the addition of new basis functions builds on a sensitivity analysis of the prediction error with respect to the expansion coefficients. A sequence of dynamic optimization problems with an increasing number of degrees of freedom has to be solved.

Alternatively, Binder et al. (1998, 2001a, 2001b, 2002) suggest a full discretization approach to solve the optimization problem. Both, the states of the model as well as the unknown disturbances are discretized in time by means of a wavelet-Galerkin scheme. For stability reasons, different bases are used for states and disturbances. Tailored multilevel strategies employing nested iteration are devised to implement a specialization of the adaptive scheme presented in section 4.1. Direct and iterative methods are studied to solve the linear equation systems. Different a-posteriori error estimators are investigated for nonuniform adaptation of states and disturbances. Binder et al. (2002) study the interactions between regularization and adaptation. They suggest to use the L-criterion for deciding when to stop the refinement process.

5.3 Hybrid modelling

Hybrid modeling refers to the situation where part of a model can be formulated on the basis of first principles and part of the model has to be inferred from data because of a lack of understanding of the mechanistic details. Typically, balances for mass and energy can be formulated, but the constitutive equations to describe reaction rates, interphase and intraphase fluxes, or kinetic coefficients are often unknown and subject to nonparametric regression (e.g. Marquardt, 2002).

For lumped parameter models and assuming the unknown functions to depend on time only, the problem is very similar to the one of section 5.2 (e.g. Mhamdi, Marquardt, 1999). Such nonparametric fitting problems have also been studied for hybrid distributed models. For example, Liu (1993) and Ascher and Haber (2001a, 2001b) study the identification of the transport coefficient in stationary diffusion problems which is assumed to depend on the spatial coordinates only. Liu (1993) demonstrates the advantages of multiscale wavelet bases and advocates a multilevel identification scheme with uniform refinement. The interdependence of grid refinement and Tikhonov regularization is studied by Ascher and Haber (2001a), whereas Ascher and Haber (2001b) present a multigrid technique for solving the fitting problem.

Usually, however, we are interested in the dependence of these kinetic functions on the state variables or their spatial derivatives. A systematic procedure for their identification using multilevel strategies is the subject of current work in the author's group.

5.4 Fully adaptive solution of distributed process system models

The general adaptive scheme introduced in section 4 has been successfully applied to a few distributed process systems models. Lang (1998) developed an adaptive algorithm for a general class of models involving coupled parabolic, elliptic and differential-algebraic equations in the framework of Rothe methods. A one-step Rosenbrock method of order three for time discretization, piecewise linear hierarchical basis functions for space discretization and a multilevel strategy are the major building blocks of the algorithm. The method has been tested with various one- and two-dimensional reaction and combustion problems by the author and with computation of mass transfer through a swelling polymer membrane (Bausa, Marquardt, 2001). Most recently, Wulkow et al. (2001) have reported applications of an adaptive technique to solve dynamic population balance models of a MSMPR crystallizer. Their method uses semi-implicit Euler time discretization and Chebyshev polynomials on finite elements for discretization of the particle size coordinate. A multilevel strategy is constructed with h-p-adaptation to control the particle size distribution error.

5.5 Adaptive solution of complex multicomponent mixture models

A complex multicomponent mixture is a chemical system which comprises of very many, say a hundred or more species. Such mixtures occur in a variety of areas including polymerization, fatty alcohol chemistry, natural extracts processing or most notably in petroleum refining and petrochemical processing. Even for a small number of balance envelopes, the models of such multicomponent processes grow quickly to very large order because of the large number of material balances related to the species considered. Heuristic model reduction techniques such as pseudocomponent lumping or moment approximation are usually used to tackle the complexity of these models.

More recently, advanced adaptive numerical methods building on the principles in section 4, have been developed. They exploit the similarity of multicomponent mixture and PDE models. The spatial coordinate in PDE models is replaced by some discrete component index such as the number of monomer molecules in a polymer chain or the pure component boiling temperature of a petroleum constituent in multicomponent mixture models. The solution of these models is interpreted as a discrete distribution of concentrations of the chemical species considered.

This idea has been introduced first by Deuflhard and Wulkow (1989) to tackle the modeling of polymerization kinetics. Wulkow (1996) presents a sophisticated algorithm which significantly extends the approach of his previous work. The discrete Galerkin h-p method, which is very similar to the method reported by Wulkow et al. (2001), combines a semi-implicit Euler scheme for time discretization and a multilevel approximation of the chain length distribution by means of discrete Chebyshev polynomials on finite elements to implement a Rothe method. The size of the elements as well as the order of polynomials are adaptively adjusted based on a-posteriori error estimation. The technique is of *AMR* type because it adaptively selects those basis functions which are required to approximate the solution of the detailed, high order DAE model within a given error tolerance.

Basically the same idea of *AMR* has been picked up by von Watzdorf and Marquardt (1995) for the modeling and simulation of multicomponent distillation processes, where the component concentrations on the trays are interpreted as discontinuous distribution functions. Watzdorf (1998) presents an adaptive wavelet-Galerkin algorithm for steady-state simulation. A multilevel strategy with uniform and error controlled nonuniform adaptation of the concentration distribution is possible in every iteration resulting in different sparse bases on every tray of a distillation column. In particular, the technique can be interpreted as an adaptive pseudocomponent lumping method, since the expansion coefficients directly correspond to component concentrations if a Haar basis is employed. An extension to dynamic simulation of separation and reaction processes using a Rothe method has been reported by Briesen and Marquardt (1999, 2000).

More recent studies of Briesen (2002) show shortcomings of this *AMR* procedure, since the detail neglected in the concentration distributions due to model reduction cannot be recovered again by the algorithm. Hence, undesired error propagation between connected balance envelopes (such as trays of a distillation column) is unavoidable. These findings have lead to an improved adaptive method for simulation and optimization of multicomponent separation processes. Briesen (2002) exploits multigrid concepts to replace the multilevel *AMR* procedure of von Watzdorf by a multilevel *AS* procedure. In particular, the loss of detail is avoided by prolongating coarse grid updates to the finest grid possible, which corresponds to all the chemical species in the mixture.

6. Conclusions

The examples in section 5 show the variety of different applications of adaptivity in process systems modeling. Surprisingly, the building blocks used in the various techniques can be related to the same fundamental concepts of adaptation as presented in section 4. This observation forms the basis for potential cross-fertilization of traditionally distinct fields. Most of the current development takes place in the scientific computing community. Since the focus there is largely on well-defined and relatively simple model problems with a general physics background, there are great opportunities for transferring the techniques to tackle challenge problems in process systems engineering such as multiscale and lifecycle modeling. Conversely, the feedback of our problem formulations and solutions will foster fundamental research on new concepts to implement adaptivity in scientific computing.

Acknowledgements
The author appreciates the fruitful discussions with W. Dahmen, which have shaped many of the ideas presented in this paper over the years. The paper would not have been possible without the contributions of his students, in particular, of A. Bardow, T. Binder, H. Briesen, A. Mhamdi, and R.v.Watzdorf.

References
Antoniadis, A., G. Gregoire, and I.W. McKeague, 1994, Wavelet methods for curve estimation, J. Am. Stat. Assoc., 89, 1340

Ascher, U. M., and E. Haber, 2001a, Grid refinement and scaling for distributed parameter estimation problems, Inverse Problems, 17, 571.

Ascher, U. M., and E. Haber, 2001b, A multigrid method for distributed parameter estimation problems. Preprint, Computer Science, University of British Columbia.

Banks, H.T., and K. Kunisch, 1989, Estimation techniques for Distributed Parameter Systems, Birkhäuser, Boston.

Bates, D.M, and D.G. Watts, 1988, Nonlinear Regression Analysis and its Applications, Wiley, New York.

Bausa, J. and W. Marquardt, 2001, Detailed modeling of stationary and transient mass transfer across pervaporation membranes, AIChE Journal, 47, 1318.

Binder, T., L. Blank, W. Dahmen, and W. Marquardt, 1998, Towards multiscale dynamic data reconciliation, Nonlinear Model Based Process Control, Eds. R. Berber, C. Kravaris, NATO ASI Series, Kluwer, 623.

Binder, T., L. Blank, W. Dahmen, and W. Marquardt, 2001a, Multiscale concepts for moving horizon optimization, Online Optimization of Large Scale Systems, Eds. M. Grötschel, S.O. Krumke, J. Rambau, Springer, New York

Binder, T., L. Blank, W. Dahmen, and W. Marquardt, 2001b, Iterative algorithms for multiscale state estimation, Part I and II, J. Opt. Theo. Appl., 111, 341, 529.

Binder, T., L. Blank, W. Dahmen, and W. Marquardt, 2002, On the regularization of dynamic data reconciliation problems, J. Process Control, in press.

Binder, T., 2002, Adaptive multiscale methods for the solution of dynamic optimization problems, PhD thesis, RWTH Aachen, in preparation.

Brenan, K., S. Campbell, and L. Petzold, 1996, Numerical Solution of Initial-Value Problems in Differential-Algebraic Equations, North Holland, New York.

Briesen, H., and W. Marquardt, 1999, An adaptive multiscale Galerkin method for the simulation of continuous mixture separation processes, AIChE meeting, Dallas.

Briesen, H., and W. Marquardt, 2000, Adaptive model reduction and simulation of thermal cracking of multicomponent mixtures, Comput. Chem. Engng. 24, 1287.

Briesen, H., 2002, Adaptive composition representation for the simulation and optimization of complex multicomponent mixture processes, PhD thesis, RWTH Aachen, in press.

Briggs, W.L., Van Emden Henson, and S.F. McCormick, 2000, A Multigrid Tutorial. 2nd Ed., SIAM, Philadelphia.

Courant, Hilbert, 1968, Methoden der Mathematischen Physik I, 3rd Edition, Springer, Berlin.

Dahmen, W., 1997, Wavelet and multiscale methods for operator equations. Acta Numerica, 6, 55.

de Boor, C., 1978, A Practical Guide to Splines. Springer, New York.

Deuflhard, P., and Wulkow, M., 1989, Computational treatment of polyreaction kinetics by orthogonal polynomials of a discrete variable, IMPACT Comput. Sci. Eng. 1, 289.

DeVore, R., and B. Lucier, 1992, Wavelets, Acta Numerica, 1, 1.

Donoho, D.L., and I.M. Johnstone, 1994, Ideal spatial adaptation via wavelet shrinkage, Biometrika, 81, 425.

Evgeniou, T., M. Pontil, and T. Poggio, 2000, Statistical learing theory: a primer, Int. J. Computer Vision, 38, 9.

Fletcher, C.A.J., 1984, Computational Galerkin Methods, Springer, New York.

Gershenfeld, N., 1999, The Nature of Mathematical Modeling, Cambridge University Press, Cambridge.

56

Garcke, J., M. Griebel, and M. Thess, 2001, Data mining with sparse grids, Computing, 67, 225.

Hossfeld, F.H.W., 2000, Komplexität und Berechenbarkeit: Über die Möglichkeiten und Grenzen des Computers, Nordrhein-Westfälische Akademie der Wissenschaften, Vorträge N448, Westdeutscher Verlag, Wiesbaden.

Kirsch, A., 1996, An Introduction to the Mathematical Theory of Inverse Problems, Springer, New York.

Kurtz, M.J., and A. Henson, 1998, State and disturbance estimation for nonlinear systems affine in the unmeasured variables, Comput. Chem. Engng. 22, 1441.

Lang, J., 1998, Adaptive FEM for reaction-diffusion equations. Appl. Num. Math. 26, 105.

Liskovets, O.A., 1965, The method of lines (review), Differential Equations, 19, 1308.

Liu, J., 1993, A multiresolution method for distributed parameter estimation, SIAM J. Sci. Stat. Comput., 14, 389.

Marquardt, W., 2002, Nonlinear model reduction for optimization based control of transient chemical processes, Chemical Process Control 6, Eds. J.B. Rawlings, and J. Eaton, in press.

Marquardt, W., L. v. Wedel, and B. Bayer, 2000, Perspectives on lifecycle process modeling, Foundations of Computer-Aided Process Design, Eds. M.F. Malone, J.A.Trainham, and B. Carnahan, AIChE Symp. Ser. 323, Vol. 96, 192.

Mhamdi, A. and W. Marquardt, 1999, An inversion approach to the estimation of reaction rates in chemical reactors, Proc. ECC'99, Karlsruhe, Germany, F1004-1.

Pantelides, C.C., 2001, New challenges and opportunities for process modeling, ESCPAE-11, Eds. R. Gani, S. Jørgensen, Elsevier, 15.

Schiesser, W., 1991, The Numerical Method of Lines, Academic Press, San Diego.

Schlegel, M., T. Binder, A. Cruse, J. Oldenburg, and W. Marquardt, 2001, Component-based implementation of a dynamic optimization algorithm using adaptive parameterization, ESCPAE-11, Eds. R. Gani, S. Jørgensen, Elsevier, 1071.

Vande Wouwer, A., Ph. Saucez, and W. E. Schiesser, Eds., 2001 Adaptive Method of Lines, Chapman & Hall/CRC, Boca Raton.

Villadsen, J.V., and M. Michelsen, 1982, Solution of Differential Equations by Orthogonal Collocation, Prentice Hall, Englewood, Cliffs.

Verfürth, R., 1996, A Review of A Posteriori Error Estimation and Adaptive Mesh-Refinement Techniques. Wiley-Teubner.

von Watzdorf, R. and W. Marquardt, 1995, Reduced modeling of multicomponent mixtures by the wavelet-Galerkin method, Proc. DYCORD+, Ed. J.B. Rawlings.

von Watzdorf, R., 1998, Adaptive Multiskalenverfahren für die Modellreduktion von thermischen Vielstofftrennprozessen. Fortschritt-Berichte VDI, Reihe 3, Nr. 561, VDI-Verlag, Düsseldorf.

Wulkow, M., 1996, The simulation of molecular weight distributions in polyreaction kinetics by discrete Galerkin methods, Macromol. Theor. Simul. 5, 393.

Wulkow, M., A. Gerstlauer, and U. Nieken, 2001, Modeling and simulation of crystallization processes using Parsival, Chem. Eng. Sci. 56, 2575.

Yserentant, H., 1992, Hierarchical bases, Proc. ICIAM'91, Ed. R. E. O'Malley, SIAM, Philadelphia.

European Symposium on Computer Aided Process Engineering – 12
J. Grievink and J. van Schijndel (Editors)
© 2002 Published by Elsevier Science B.V.

Plantwide control: Towards a systematic procedure

Sigurd Skogestad

Norwegian University of Science and Technology (NTNU)
Department of Chemical Engineering, 7491 Trondheim, Norway

Abstract

Plantwide control deals with the structural decisions of the control system, including what to control and how to pair the variables to form control loops. Although these are very important issues, these decisions are in most cases made in an ad-hoc fashion, based on experience and engineering insight, without considering the details of each problem. In the paper, a systematic procedure towards plantwide control is presented. It starts with carefully defining the operational and economic objectives, and the degrees of freedom available to fulfill them. Other issues, discussed in the paper, include inventory control, decentralized versus multivariable control, loss in performance by bottom-up design, and recycle systems including the snowball effect.

1. Introduction

A chemical plant may have thousands of measurements and control loops. In practice, the control system is usually divided into several layers:

- scheduling (weeks),
- site-wide optimization (day),
- local optimization (hour),
- supervisory/predictive control (minutes)
- regulatory control (seconds)

We here consider the three lowest layers. The local optimization layer typically recomputes new setpoints only once an hour or so, whereas the feedback layer operates continuously. The layers are linked by the controlled variables, whereby the setpoints are computed by the upper layer and implemented by the lower layer. An important issue is the selection of these variables.

By the term *plantwide control* it is *not* meant the tuning and behavior of each control loop, but rather the *control philosophy* of the overall plant with emphasis on the *structural decisions* (Foss, 1973); (Morari, 1982); (Skogestad and Postlethwaite, 1996). These involve the following *tasks:*

1. *Selection of controlled variables c* (``outputs''; variables with setpoints)
2. *Selection of manipulated variables m* (``inputs'')

3. *Selection of (extra) measurements v* (for control purposes including stabilization)
4. *Selection of control **configuration*** (the structure of the overall controller *K* that interconnects the variables c_s and *v* (controller inputs) with the variables *m*)
5. *Selection of controller type* (control law specification, e.g., PID, decoupler, LQG, etc.).

A recent review of the literature on plantwide control can be found in Larsson and Skogestad (2000). In practice, the problem is usually solved without the use of existing theoretical tools. In fact, the industrial approach to plantwide control is still very much along the lines described by Page Buckley in his book from 1964. The realization that the field of control structure design is underdeveloped is not new. Foss (1973) made the observation that in many areas application was *ahead* of theory, and he stated that

> The central issue to be resolved by the new theories is the determination of the control system structure. Which variables should be measured, which inputs should be manipulated and which links should be made between the two sets.... The gap is present indeed, but contrary to the views of many, it is the theoretician who must close it.

This paper is organized as follows. First, an expanded version of the plantwide control design procedure of Larsson and Skogestad (2000) is presented. In reminder of the paper some issues related to this procedure are discussed in more detail:

- Degree of freedom analysis
- Selection of controlled variables
- Inventory control

Finally, we discuss recycle systems and the socalled snowball effect.

2. A procedure for plantwide control

The proposed design procedure is summarized in Table 1. In the table we also give the purpose and typical model requirements for each layer, along with a short discussion on when to use decentralized (single-loop) control or multivariable control (e.g. MPC) in the supervisory control layer. The procedure is divided in two main parts:

I. Top-down analysis, including definition of operational objectives and consideration of degrees of freedom available to meet these (tasks 1 and 2)

II. Bottom-up design of the control system, starting with the stabilizing control layer (tasks 3, 4 and 5 above)

Table 1: A plantwide control design procedure

STEP	Comments, analysis tools and model requirements
I. TOP-DOWN ANALYSIS:	
1. PRIMARY CONTROLLED VARIABLES: Which (primary) variables c should we control? • Control active constraints • Remaining DOFs: Control variables for which constant setpoints give small (economic) loss when disturbances occur.	*Steady-state economic analysis:* • Define cost and constraints • Degree of freedom (DOF) analysis. • Optimization w.r.t. DOFs for various disturbances (gives active constraints)
2. MANIPULATED VARIABLES Select manipulated variables m (valves and actuators) for control.	Usually given by design, but check that there are enough DOFs to meet operational objectives, both at steady state (step 1) and dynamically. If not, may need extra equipment .
3. PRODUCTION RATE: Where should the production rate be set? (Very important choice as it determines the structure of remaining inventory control system.)	Optimal location follows from steady-state optimization (step 1), but may move depending on operating conditions.
II. BOTTOM-UP DESIGN: (With given controlled and manipulated variables)	*Controllability analysis*: Compute zeros, poles, pole vectors, relative gain array, minimum singular values, etc.

4. REGULATORY CONTROL LAYER.	*4.1 Pole vector analysis* (Havre and Skogestad, 1997) Select measured variables and manipulated inputs corresponding to large elements in pole vector to minimize input usage caused by measurement noise.
4.1 Stabilization *4.2 Local disturbance rejection* *Purpose:* "Stabilize" the plant using single-loop PID controllers to enable manual operation (by the operators) *Main issue:* What more should we control? • Select secondary controlled variables (measurements) v • Pair these with manipulated variables m, avoiding m's that saturate (reach constraints)	*4.2 Partially controlled plant analysis.* Control secondary measurements (v) so that the effect of disturbances on the primary outputs (c) can be handled by the layer above (or the operators). *Model:* Tuning may be done with local linear models or on-line with no model. Analysis requires linear multivariable dynamic model (generic model sufficient).
5.. SUPERVISORY CONTROL LAYER. *Purpose:* Keep (primary) controlled outputs c at optimal setpoints c_s, using unused manipulated variables and setpoints v_s for regulatory layer as degrees of freedom (inputs). *Main structural issue:* Decentralized or multivariable control? *5a. Decentralized (single-loop) control* Possibly with addition of feed-forward and ratio control. • May use simple PI or PID controllers. • Structural issue: choose input-output pairing *5b. Multivariable control* Usually with explicit handling of constraints (MPC) • Structural issue: Chooose input-output sets for each multivariable controller	*5a. Decentralized:* Preferred for noninteracting process and cases where active constraints remain constant. *Pairing analysis:* Pair on RGA close to identity matrix at crossover frequency, provided not negative at steady state. Use CLDG for more detailed analysis *Model:* see 4 *5b Multivariable:* 1. Use for interacting processes and for easy handling of feedforward control 2. *MPC with constraints:* Use for moving smoothly between changing active constraints (avoids logic needed in decentralized scheme 5a) *Model:* Linear multivariable dynamic model (identified for each application).

6. REAL-TIME OPTIMIZATION LAYER *Purpose:* Identify active constraints and compute optimal setpoints c_s for controlled variables Main structural issue: What should c be (se step 1)	*Model:* Nonlinear steady-state model, plus costs and constraints.
7. VALIDATION	Nonlinear dynamic simulation of critical parts

The procedure is generally iterative and may require several loops through the steps, before converging at a proposed control structure.

Additional comment:
(i) *"Stabilization" (step 4).* The objective of the regulatory control layer is to *"stabilize"* the plant. We have here put stabilize in quotes because we use the word in an extended meaning, and include both modes which are mathematically unstable as well as slow modes ("drift") that need to be "stabilized" from an operator point of view. The controlled variables for stabilization are measured output variables, and their setpoints v_s may be used as degrees of freedom by the layers above.

(ii) *Model requirements:* In the *control layers* (step 4 and 5) we control variables at given setpoints, and it is usually sufficient with linear dynamic models (local for each loop) with emphasis on the time scale corresponding to the desired closed-loop response time (of each loop). The steady-state part of the model is not important, except for cases with pure feedforward control. For analysis it is usually sufficient with a generic model (which does not match exactly the specific plant), but for controller design model identification is usually required. In the *optimization layer* (steps 1 and 6) a nonlinear steady-state model is required. Dynamics are usually not needed, except for batch processes and cases with frequent grade changes.

(iii) *Decentralized versus multivariable control (step 5).* First note that there is usually some decentralization, that is, there is usually a combination of several multivariable and single-loop controllers. An important reason for using multivariable constraint control (MPC) is usually to avoid the logic of reconfiguring single loops as active constraint move. The optimization in step 1 with various disturbances may provide useful information in this respect, and may be used to set up a table of possible combinations of active constraints. MPC should be used if a structure with single-loop controllers will require excessive reconfiguration of loops.

(iv) *Why not a single big multivariable controller?* Most of the steps in Table 1 could be avoided by designing a single optimizing controller that stabilizes the process and at the same time perfectly coordinates all the manipulated variables based on dynamic on-line optimization. There are fundamental reasons why such a solution is not the best, even with tomorrows computing power. One fundamental reason is the cost of modeling and tuning this controller, which must be balanced against the fact that the hierarchical structuring proposed in this paper, without much need for models, is used effectively to control most chemical plants.

3. Degree of freedom analysis (step 1)

The first step in a systematic approach to plantwide control is to formulate the operational objectives. This is done by defining a cost function J that should be minimized with respect to the N_{opt} optimization degrees of freedom, subject to a given set of constraints. A degree of freedom analysis is a key element in this first step. We start with the number of operational or control degrees of freedom, N_m (m here denotes manipulated). N_m is usually easily obtained by process insight as the number of independent variables that can be manipulated by external means (typically, the number of number of adjustable valves plus other adjustable electrical and mechanical variables). Note that the original manipulated variables are always extensive variables.

To obtain the number of degrees of freedom for optimization, N_{opt}, we need to subtract from N_m

- the number of manipulated (input) variables with no effect on the cost J (typically, these are "extra" manipulated variables used to improve the dynamic response, e.g. an extra bypass on a heat exchanger), and
- the number of (output) variables that need to be controlled, but which have no effect on the cost J (typically, these are liquid levels in holdup tanks).

In most cases the cost depends on the steady state only, and N_{opt} equals the number of steady-state degrees of freedom. The typical number of (operational) steady-state degrees of freedom for some process units are (with given pressure)

- each feedstream: 1 (feedrate)
- non-integrated distillation column: 2 (boilup and reflux) + number of sidestreams
- absorption or adsorption column: 0
- adiabatic flash tank: 0 (assuming given pressure)
- liquid phase reactor: 1 (volume)
- gas phase reactor: 0 (assuming given pressure)
- heat exchanger: 1
- splitter: n-1 (split fractions) where n is the number of exit streams
- mixer: 0
- compressor, turbine, pump: 1 (work)

For example, for the reactor-distillation-recycle process with given pressure (Figure 4), there are four degrees of freedom at steady state (fresh feedrate, reactor holdup, boilup and reflux in distillation column

The optimization is generally subject to constraints, and at the optimum many of these are usually "active". The number of ``free" (unconstrained) degrees of freedom that are left to optimize the operation is then $N_{opt} - N_{active}$. This is an important number, since it is generally for the unconstrained degrees of freedom that the selection of controlled variables (task 1 and step 1) is a critical issue.

4. What should we control? (steps 1 and 4)

A question that puzzled me for many years was: Why do we control all these variables in a chemical plant, like internal temperatures, pressures or compositions, when there are no *a priori* specifications on many of them? The answer to this question is that we first need to control the variables directly related to ensuing *optimal economic operation* (these are the primary controlled variables, see step 1):

- Control active constraints
- Select unconstrained controlled variables so that with constant setpoints the process is kept close to its optimum in spite of disturbances. These are the less intuitive ones, for which the idea of self-optimizing control (see below) is very useful.

In addition, we need to need to control variables in order to achieve *satisfactory regulatory control* (these are the secondary controlled variables, see step 4):

Self-optimizing control (step 1).
The basic idea of self-optimizing control was formulated about twenty years ago by Morari et al.(1980) who write that "we want to find a function c of the process variables which when held constant, leads automatically to the optimal adjustments of the manipulated variables." To quantify this more precisely, we define the (economic) loss L as the difference between the actual value of the cost function and the truly optimal value, i.e. $L = J(u; d) - J_{opt}(d)$. *Self-optimizing control (Skogestad, 2000) is achieved if a constant setpoint policy results in an acceptable loss L (without the need to reoptimize when disturbances occur).* The main issue here is *not* to find the optimal setpoints, but rather to find the right variables to keep constant. To select controlled variables for self-optimizing control, one may use the stepwise procedure of Skogestad (2000):

> **Step 1.1** Degree of freedom analysis
> **Step 1.2** Definition of optimal operation (cost and constraints)
> **Step 1.3** Identification of important disturbances
> **Step 1.4** Optimization (nominally and with disturbances)
> **Step 1.5** Identification of candidate controlled variables
> **Step 1.6** Evaluation of loss with constant setpoints for alternative combinations of controlled variables (when there are disturbances or implementation errors)
> **Step 1.7** Final evaluation and selection (including controllability analysis)

This procedure has been applied to several applications, including distillation column control (Skogestad, 2000), the Tennessee-Eastman process (Larsson et al., 2001) and a reactor-recycle process (Larsson et al., 2002).

5   Production rate and inventory control (step 3)

The liquid inventory (holdup, level) in a processing unit usually has no or little on steady-state effect, but it needs to be controlled to maintain stable operation. The bottom-up design of the control system (step 4) therefore usually starts with the design of the liquid level control loops. However, one needs to be a bit careful here because the design of the level loops has a very large effect of the remaining control system design, as it consumes steady-state degrees of freedom, and determines the initial effect of feedrate disturbances. Also, because the level loops link together the transport of mass from the input to the output of the plant, the level loops are very much dependent on each other. There are many possible ways of pairing the inventory loops, and the basic issue is whether to control the inventory (level) using the inflow or outflow? A little thought reveals that the answer to this question is mainly determined by where in the plant the production rate is set, and that we should control inventory (a) using the outflow downstream of the location where the production rate is set, and (b) using the inflow upstream of this location. This justifies why there in Table 1 is a separate step called "Production rate", because the decision here provides a natural transition from step 1 (top-down economic considerations and identification of active constraints) to step 4 (bottom-up design, usually starting with the level loops).

The production rate is most commonly assumed to be set at the inlet to the plant, so outflows are used for level control. One important reason for this is probably that most of the control structure decisions are done at the design stage (before the plant is build) where we usually fix the feedrate. However, during operation the feedrate is usually a degree of freedom, and very often the economic conditions are such that it is optimal to maximize production. As we increase the feedrate we reach a point where some flow variable E internally in the plant reaches its constraint E_{max} and becomes a bottleneck for further increase in production. In addition, as we reach the constraint we loose a degree of freedom for control, and to compensate for this we have several options:

1)   Reduce the feedrate and "back off" from the constraint on E (gives economic loss).
2)   Use the feedrate as a manipulated variable to take over the lost control task (but this usually gives a very "slow'' loop dynamically because of long physical distance). To avoid this slow loop one may either:
3)   Install a surge tank upstram of the bottleneck, and reassign its outflow to take over the lost control task, and use the feedrate to reset the level of the surge tank, or:
4)   Reassign all level control loops upstream of the bottleneck from outflow to inflow (which may involve many loops).

All of these options are undesirable. A better solution is probably to permanently reassign the level loops, and we have the following rul: *Identify the main dynamic (control) bottleneck (see definition below) in the plant by optimizing the operation with the feedrate as a degree of freedom (steady state, step 1). Set the production rate at this location.* The justification for this rule is that the economic benefits of increasing the production are usually very large (when the market conditions are such), so that it is important to maximize flow at the bottleneck. On the other hand, if market conditions are such that we are operating with a given feed rate or given product rate, then the economic loss imposed by adjusting the production rate somewhere inside the plant is usually zero, as deviations from the desired feed or production rate can be averaged out over time, provided we have storage tanks for feeds or products. However, one should be careful when applying this rule, as also other considerations may be important, such as the control of the individual units (e.g. distillation column) which may be effected by whether inflow or outflow is used for level control.

We have here assumed that the bottleneck is always in the same unit. If it moves to another unit, then reassignment of level loops is probably unavoidable if we want to maintain optimal operation.

Note that we here have only considered changes in operating conditions that may lead to bottlenecks and thus to the need to reassign inventory (level) loops. Of course, other active constraints may move and the best unconstrained controlled variable (with the best self-optimizing properties) may change, but the reconfiguration of these loops are usually easier to handle locally, as they . This may also require configuration of loops, but usually may done locally.

MPC in regulatory control layer

The above discussion assumes that we use single-loop controllers in the regulatory control layer (which includes level control), and that we want to minimize the logic needed for reassigning loops. An alternative approach, which overcomes most of the above problems, is to use a multivariable model-based controller with constraints handling (MPC), which automatically tracks the moving constraints and reassigns control tasks in an optimal manner. This is many ways a more straightforward approach, but such controllers are more complex, and its sensitivity to errors and failures is quite unpredictable, so such controllers are usually avoided at the bottom of the control hierarchy.

Another alternative, which is more failure tolerant, is to implement a MPC system on top of a fixed single-loop regulatory control layer (which includes level control). As shown in Theorem 1 (below) this gives no performance loss provided we let the multivariable have access also to the setpoints of the lower-layer regulatory controllers (including the ability to dynamically manipulate the level setpoints). The regulatory layer then provides a back-up if the MPC controller fails, but under normal conditions does not effect control performance.

Definition of bottleneck

Consider a given objective function, given parameters, given equipment (including given degrees of freedom) and given constraints (including quality constraints on the products). A unit (or more precisely, an extensive variable E within this unit) is a *bottleneck* (with respect to the flow F) if

(1) With the flow F as a degree of freedom, the variable E is optimally at its maximum constraint (i.e., E= E_{max} at the optimum)

(2) The flow F is increased by increasing this constraint (i.e., $dF/dE_{max} > 0$ at the optimum).

A variable E is a *dynamic(control) bottleneck* if in addition

(3) The optimal value of E is unconstrained when F is fixed at a sufficiently low value

Otherwise E is a *steady-state (design) bottleneck*.

Remarks on definition:

1. Typically, F is the flowrate of the main feed or main product.
2. Most of the information required to identify bottlenecks follow from the optimization with various disturbances in step 1.
3. The fact that an extensive variable is at its maximum constraint does not necessarily imply that it is a bottleneck, because we may have that $dF/dE_{max} = 0$ (e.g., this may happen if the variable E is a cheap utility).
4. In many cases F is also the objective function to be maximized, and the values of dF/dE_{max} are then directly given by the corresponding Lagrange multipliers.
5. We may in some cases have several bottlenecks at the same time, and the *main bottleneck* is then the variable with the largest value of $E_{max} \cdot dF/dE_{max}$. (i.e. with the largest relative effect on the flowrate).
6. The location of the bottleneck may move with time (i.e., as a function of other parameters)
7. The concept of "bottleneck" is clearly of importance when redesigning a plant to increase capacity. It is also important in terms of operation and control, because the main bottleneck is the variable that should be operated closest to its constraint.
8. Steady-state bottlenecks may be important in terms of design, but need normally not be considered any further when it comes to deciding on a control structure (as they should always be kept at their maximum).. Examples of possible steady-state bottleneck variables are reactor volumes and heat exchangers areas.
9. A control policy based on fixing intensive variables is not steady-state optimal for systems with bottlenecks.

6. Application: Recycle systems and the snowball effect

Luyben (1993) introduced the term ``snow-ball effect'' to describe what can happen to the recycle flow in response to an increase in fresh feedrate F_0 for processes with recycle of unreacted feed (see Figure). Although this term has been useful in pointing out the importance of taking a plantwide perspective, it has lead to quite a lot of confusion.

To understand the problem, let us first consider the ``default" way of dealing with a feedrate increase, which is to keep all the intensive variables (compositions) in the process constant, by increasing all flows and other extensive variables with a steady-state effect in proportion to F_0. This is similar to how one scales the production rate when doing process simulation, and is the idea behind the ``balanced" control structures of <u>Wu and Yu (1996)</u>. Specifically, this requires that we keep the residence time M_r/F constant, that is, we need to increase the reactor holdup M_r in proportion to the reactor feedrate F.

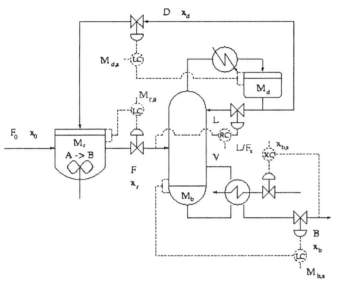

Figure 1 . *Reactor with recycle process with control of recycle ratio (L/F), M_r (maximum reactor holdup), and x_B (given product composition).*

However, changing the reactor holdup (volume) during operation is usually not possible (gas phase reactor), or at least not desirable since for most reactions it is economically optimal to use a fixed maximum reactor volume in order to maximize per pass conversion (i.e., reactor holdup is a steady-state bottleneck, see above). To increase conversion (in response too an increase in feedrate F_0) one may instead increase the concentration of reactant by recycling unreacted feed. However, the effect of this has limitations, and the snowball effect occurs because even with infinite recycle the reactor concentration cannot exceed that of pure component. In practice, because of constraints, the flow rates do not go to infinity. Most likely, the liquid or vapor rate in the column will reach its maximum value, and the result of the snowballing will be a breakthrough of component A in the bottom product, that is, we will find that we are no longer able to maintain the product purity specification (x_B).

To avoid snowballing Luyben et al. (1993, 1994) and <u>Wu and Yu (1996)</u> propose, to use the "default" approach with a varying reactor holdup, rather than a ``conventional" control structure with constant holdup. Their simulations show that a variable holdup policy works better, but these simulations are strongly misleading, because in the "conventional" structure they fix the reactor holdup at a value well below the maximum values used in the varying holdup structures. In fact, the lowest value of the recycle D for a given value of F_0 is when the reactor holdup M_r is at its maximum, so the conventional structure with maximum holdup is actually better in terms of avoiding snowballing.

A more careful analysis of the reactor with recycle process shows that there are four degrees of freedom at steady-state, including the fresh feedrate F_0. With a *fixed feedrate* F_0, there are three degrees of freedom. If the economic objective is to minimize the energy usage (i.e., minimize boilup V), then optimization with respect to the three degrees of freedom, give that M_r should be kept at its maximum (to maximize conversion), and that the product composition x_B be kept at its specification (overpurifying costs energy). These two variables should then be controlled (active constraint control). This makes the Luyben structure and the two balanced structures of Wu and Yu (1996) economically unattractive. There is one unconstrained degree of freedom left, and the issue is next to decide which variable we should select to keep constant. Alternatives are, for example, the amount of recycle D or F ("Luyben rule"), composition x_D (conventional structure), reflux L, reflux ratios L/D or L/F, etc. Larsson et al. (2002) evaluated the energy loss imposed by keeping these constant when there are disturbances in F_0 and recommended for the case with a given feedrate F_0 to use the recycle ratio structure shown in the Figure. Keeping D or F constant (Luyben rule) yields infeasible operation for small disturbances. This confirms the results of Wu and Yu (1996). This is easily explained: As the feedrate F_0 is increased, we must with constant $F=F_0+D$ reduce the recycle D to the reactor. Therefore light component A will accumulate in the distillation column and operation becomes infeasible.

For this plant the reactor holdup is a steady-state (design) bottleneck, whereas the column capacity (V_{max}) is the dynamic (control) bottleneck. Thus, if it is likely that the plant will be operated under conditions where we want to maximize production, then we should probably use a control structure where the production rate is set at the column bottleneck (V), and inventory control should use inflow upstream of this location. In Figure 4, this would involve using the feedrate F_0 to control the reactor level, using the column feed F to control bottom composition, and using the boilup V to reset the feedrate F_0 to its given value (note that F_0 is both an input and output in this case).

In summary, the ``snowball effect'' is a real operational problem if the reactor is ``too small'', such that we may encounter or get close to cases where the feedrate is larger than the reactor can handle. The ``snowball effect'' makes control more difficult and critical, but it not a control problem in the sense that it can be avoided by use of control. Rather, it is a design problem that can be easily avoided by installing a sufficiently large reactor to begin with. The Luyben rule of fixing a flow in the recycle loop seems to have little basis, as it leads to control structures that can only handle very small feedrate changes.

7. Conclusion

We have here presented a first step towards a systematic approach to plantwide control. There are many outstanding research issues related to filling in out more detailed procedures in Table 1 on what to do in each step of the procedure. For example, more work is needed in order to understand how to decompose and coordinate the layers of the control system.

References

Buckley, P.S., 1964. *Techniques of process control.* Wiley.

Havre, K. and S. Skogestad, 1998, ``Selection of variables for regulatory control using pole vectors", *Proc. IFAC symposium DYCOPS-5,* Corfu, Greece, 614-619.

Larsson, T. and S. Skogestad, 2000, "Plantwide control: A review and a new design procedure", *Modeling, Identification and Control*, **21**, 209-240.

Larsson, T., K. Hestetun, E. Hovland and S. Skogestad, 2001, "Self-optimizing control of a large-scale plant: The Tennessee Eastman process'', *Ind.Eng.Chem.Res.*, **40**, 4889-4901.

Larsson, T., M. Govatsmark, S. Skogestad and C.C. Yu, 2002, "Control of reactor, separator and recycle process'', *Submitted to Ind.Eng.Chem.Res.*

Luyben, W.L. 1993. "Dynamics and control of recycle systems. 2. Comparison of alternative process designs." *Ind. Eng. Chem. Res.* **32**, 476-486.

Luyben, W.L. 1994. "Snowball effect in reactor/separator processes with recycle". *Ind. Eng. Chem. Res.* **33**, 299-305.

Morari, M., G. Stephanopoulos and Y. Arkun 1980. "Studies in the synthesis of control structures for chemical processes..Part I'' *AIChE Journal* **26**, 220–232.

Skogestad, S. and I.Postlethwaite, 1996, *Multivariable feedback control*, Wiley.

Skogestad, S. (2000). "Plantwide control: The search for the self-optimizing control structure". *J. Proc. Control* **10**, 487-507.

Wu, K.L. and C.-C. Yu (1996). "Reactor/separator process with recycle. 1.Candidate control structure for operability". *Computers. Chem. Engng.*, **20**, 1291-1316.

European Symposium on Computer Aided Process Engineering – 12
J. Grievink and J. van Schijndel (Editors)
© 2002 Published by Elsevier Science B.V.

Decision Confidence – Handling Uncertainty Through the Plant Life Cycle using Statistics and Datamining

David Stockill

Manager, Statistics and Risk

Shell Global Solutions International

Cheshire Innovation Park,

PO Box 1, Chester, CH1 3SH, UK

email: David.J.Stockill@OPC.shell.com

Abstract

The pressures of the early 21st century economic and operating climate continue to provide challenges to the established process engineering operation. Traditional methods and approaches to Design, Operation, Management and Maintenance come under closer scrutiny, in particular their ability to handle variability and uncertainty. Simultaneously, the accessibility of integrated plant data (from process historians, maintenance systems) and records through modern desk-top facilities, together with the growing realisation that this data is an asset which can be used to provide insight into process operation, leads inexorably to the need for tools and techniques which can exploit the data in pursuit of the challenges mentioned.

No longer can plant design be considered in isolation from the variability and uncertainty that will undoubtedly impact the plant during its life-cycle. Traditional overdesign/sparing rules of thumb are challenged in a CAPEX-conscious world. Life-cycle cost and availability requirements are now specified at the Basis of Design.

Statistics (in its broadest sense) and related methodologies becomes a vehicle to address these issues. Whilst Statistics may be a mature and well established technology in areas such as product development, laboratory R&D and discrete parts manufacturing, it is a relative newcomer in the large scale process domain. From Design through Operation and Maintenance its introduction offers a new perspective into the handling of risk and uncertainty in these processes.

The Business Environment: Design, Operate & Maintain.

It is now well recognised that we need to design new plants with a much greater understanding of the ability of that plant to handle uncertainty and the realities of operation throughout its life cycle:

- The days of "Golden Rules" and standard over-design factors are gone.

- Plant availability guarantees and life-cycle cost optimisation is increasingly demanded by the end customer.

- In a capex-concious world a thorough understanding of the link between, equipment size, reliability and maintainability is needed.

We therefore look to new analytical and statistical tools to assist this process, embedding them in design philosophies such as "Availability Assurance"

Turn to the operational scenario. We've installed advanced control, our instrumentation has been replaced, we've carried out maintenance and RCM programmes. Yet still the pressures on costs and margins are ever present. Several factors start to dominate:

- We are in a data-dominated world – all parts of our business have large repositories of business and process data. Access to this data and the ability to integrate it is easier and easier. How do we turn it into Information?

- There is a clear change to a risk-based approach. For example in plant maintenance the days of rule-based activities are gone – the focus is now condition-based maintenance. We need decision support tools to help us decide when to maintain or service. This needs some statistical benchmarking and framework.

- Capital expenditure is still tight. "Sweating the Assets" is common jargon. Yet if we are really to understand the operation and behaviour of our existing assets then our prime window through which to observe them is via the process measurements and the historical records of plant performance.

We turn therefore to our process computers, historians and look for statistical techniques to help analyse the data.

Handling Uncertainty in Design

Typical of the modern tools used in availability assurance is **SPARC** (System for Production Availability and Resources Consumption) a software tool for analysing and predicting the production availability and resource requirements of production facilities. The program requires unit capacities, network configuration, component reliability characteristics and maintenance strategies as input. **SPARC** uses these data to compute the available production capacity and resources consumption, including maintenance workload, costs of materials and consumption of critical spares. These system performance predictions can be used to determine the optimal balance between initial capital expenditures, future operating expenditures and costs of deferred production due to system unavailability. What-if capabilities and zoom and compare options enable the user to interactively optimise the system.

The system supports decisions in Design and Operations. Design options, such as system configuration and required equipment reliability, are evaluated in terms of their effect on production availability. Adaptations in operational or maintenance strategies, for example on the basis of Reliability Centred Maintenance (RCM), are evaluated on their effectiveness, because the relation between component failure, associated repair behaviour and system performance is addressed explicitly.

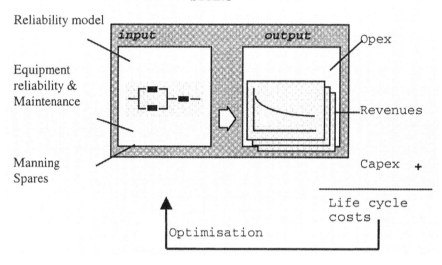

Model input

From a functional point of view, a production system consists of a network of process units such as vessels, compressors, pumps, etc. Following the RCM philosophy, each unit comprises maintainable components and failure modes. Hence, the data model of **SPARC** consists of two main elements: basic data and network configuration. The basic data part contains for each failure mode distributional information on failure and repair characteristics, maintenance policies and resource requirements associated with repairs, for example workload, material costs and usage of spares. For the network, unit capacities and by-pass capacities are required as well as the interconnections between the units (series/parallel), buffer capacity (optional), and supply and demand to the system.

Output analysis

Various performance measures relating to production availability and resource consumption are provided, both in steady state and as a function of time. These measures include distribution information on production capacity and on the expected production in relation to demand (system effectiveness), unit availability, average number of failures and repairs, and associated resource requirements in terms of manhours, costs and number of spares. The user can select the performance measure, time scale and part of the system.

Decision support

By changing input variables to the model, the user can evaluate the effect of measures to enhance production (redundancy, equipment reliability) or to reduce the costs of operation and maintenance (maintenance strategies). Alternative system configurations or operational policies can be compared on the basis of their Life Cycle Costs. Insight in the effects of uncertainties in the input data can be obtained by performing sensitivity analyses.

Operations, Maintenance and Process Data

In many ways the new generation of process computing systems are finally realising the promises of the systems we started with 30 years ago. Before the days of microprocessors and DCS the early process computers were being installed to complement pneumatic instrumentation systems. Complex electro-pneumatic convertors (eg the famous Scani-Valves) enabled early mini-computers (eg from Control Data, HP and DEC-VAX) to scan and store pneumatic measurements. Reporting was pre-VDU and limited to simple logging, shift reports, calculations and data-storage. Extravagent claims were made as to the value of "Information" and whilst these systems were clearly a quantum leap from the days of manual recording via pencil and clip-board it is doubtful whether the claims were really achieved. One of the problems was the accessibility of the data. You needed a degree in computer science to get at it.

From the mid-70's until the early 1990's process computing took a back seat – the focus was on DCS and Advanced Control. Process Computing did provide an environment for Advanced Control whilst DCS was in its infancy and many reliable systems, usually based on DEC-VAX or UNIX were installed. Operator interfaces improved with colour graphics. However in essence we still had glorified data-loggers.

The breakthrough has come in the last 5 years or so with the introduction of client-server technology and PC/Windows-based interfaces. All the major vendors now have user-friendly packages that enable easy access to the process database. Active-X components and web-connectivity allow many new options. With a few clicks of a mouse the process engineer can fill an EXCEL spreadsheet with weeks or years of tag data. Data-base connectivity allows the integration of (say) a process historian with a maintenance management system. How do we exploit this?

Statistics and Process Data

The use of specialised statistical techniques to analyse process data has not been common practice, certainly in the large-scale petrochemical and energy businesses. Its role in Product development, Laboratory and the discrete parts manufacturing environment (eg SPC) is well known. However statistical toolboxes and modelling packages are becoming available which allow the application of techniques such as Principal Component Analysis, Rank Correlation and so forth without the need to code up programs in specialised Maths packages. Increasingly the basic tools available in packages such as EXCEL offer possibilities that until a few years ago were out of the reach of the non-specialists. The issue is now the sensible and educated use of these techniques.

Unlike (say) advanced control where "standard" solutions have evolved for typical applications (eg Cat Cracker Reactor-Regenerator control) the use of statistics tends to more of a consultancy approach to problem solving. The problems tend to be unique – the approach is to bring together a skilled practitioner with his personal toolbox of skills and techniques and the business problem. This, of course, does present its own difficulties in the introduction of the technology to new users – there isn't such a simple "off the shelf" mentality.

74

Process Analysis

Typically we see a broad range of process investigations where the starting point is historical process data. 2 general classifications may be made - troubleshooting and optimisation.

In the former statistical tools are being used to aid the analysis of some problem or mal-operation. For instance on an alkylation unit excessive levels of Fluoride were being produced as an impurity in the final product. Historical analysis pinpointed certain combinations of operating conditions that were responsible for the higher levels of impurity. The consequent changes in operating conditions that were instigated resulted in improved product yield benefit of $500,000 per annum. In another case, principal component analysis (PCA) was used to pinpoint the reasons behind yield changes in Butane-Butene production following a shut-down on a Catalytic Cracker. The following graph clearly shows how PCA can segment out multivariate changes in operation, the clusters representing operation before and after shutdown. Analysis of the change in the scores plot helped restore production.

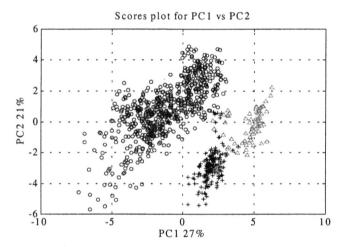

In other cases historical data can be used as a basis to optimise production. For instance on a Secondary Butyl Alcohol Unit exploratory analysis identified a key dependency between SBA production/Butene conversion and the acid settler temperature. The resultant control and optimisation of the settler temperature resulted in yield improvements worth $200,000 per annum. In a second case a higher Olefin plant was equipped with a complex triple-layer reactor bed system; each layer having 3 or 4 reactors operating in series or parallel. Analysis and modelling from historical performance data allowed the effects of short-term deactivation, long-term deactivation, catalyst effects in different beds and optimum cycle length to be investigated. The resulting optimisation of the catalyst operating and regeneration cycle produced benefits of $400,000 per annum.

In all the preceding cases the starting point was historical process data (temperatures, qualities etc) taken from a process computer.

Moving up the Supply Chain - Working Capital Analysis
In the current economic climate, reduction in working capital through minimisation of product and intermediate stock holdings is an attractive option that can realise major benefits through the release of buffer stocks and reduction in storage facilities.

In order to make the most of such opportunities, a clear picture of the efficiency of current stock handling is required. Work was carried out to examine the stock holdings of one of the finished petrochemical products at a large processing plant. The actual stock levels from historic inventory records were analysed to model the stock holding performance. Particular attention was given to modelling the important performance features: flow rates into and out of the system and the Time-to-Stock-Out distributions.

These models complement other methods of analysis of stock levels, such as simulation.

It was possible to demonstrate, through a thorough analysis of actual stock performance data from the refinery, that it would be possible to reduce contingency stock holding by around a half. The interesting feature of this work is that it takes full account of the overall reaction within the supply chain to any incidents, rather than doing an assessment of stock requirements on the basis of the refinery having to cope alone

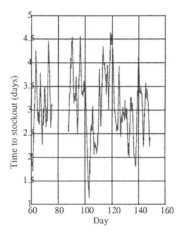 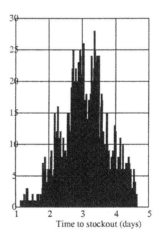

The statistical approach provides a decision-support framework for stock working capital reduction that allows:

- Current stock-handling policy to be quantified.

- Stock variability to be estimated and visualised.

- The key drivers affecting this variability to be identified.

- Different stock-handling scenarios can be explored and their impact in terms of relative risk and reliability can be quantified.

Potential reductions in working stock levels of 20%, worth $2.8M, were identified and due to the success of the initial study, the project was extended to include all finished products at the plant, with projected savings of $7M.

From this start the loop can be closed – include Availability Studies on the plant producing the product, link with analysis of working capital and a very powerful methodology for approaching risk-based stock management emerges.

Technology

As discussed already industrial statistics adopts what is predominantly a consultancy approach rather than repetitive application of a standard solution. The consultant will have at his fingertips a collection of analytical techniques ranging from simple mathematical manipulation (eg variance analysis) through filtering, correlation and regression techniques to multivariate tools such as Principal Component Analysis, Partial Least Squares and their non-linear extensions. Further, the potential of modern Bayesian methods for adaptive modelling and incorporation of expert opinion is being more fully recognised as computational horse-power disappears as a constraint.

Process data mining

Data mining is an emerging field in the retail and financial sectors, where huge quantities of data are being accumulated in the absence of good underlying physical or economic models. The vast quantities of data available call for specialist statistical tools. Some of these tools are applicable in the process arena. On occasions, we do have large amounts of data relating potential explanatory variables in the tree with the response that we're interested in; sufficient data to go further that conventional regression modelling. We can build hierarchical models similar to Bayesian Networks. In such cases, there are Graphical Modelling tools that can help us identify the variables in a tree-like model, find the important connections in the tree and estimate the parameters. Other statistical data mining tools, like logistic and multinomial regression and log-linear modelling, are often well-suited to process applications.

The Issues and Pitfalls

In many ways the issues surrounding the application and roll-out of these techniques mirror those experienced 10-15 years ago when Advanced Process Control was in its infancy. In some ways its even more difficult.

User acceptance that there are benefits to be gained is a major hurdle – just as in the early days of APC. Advanced mathematical techniques can seem rather esoteric in the day-to-day life on a manufacturing plant. Engineers who are not strong numerically may have problems in buying in to the ideas. An added complication is that there is limited possibility to "clone" applications which was a key element in the acceptance of APC – the solutions are more unique to the local process and problem. However we have strong evidence that once a user has bought into the concept and experienced the benefits then the technology does become well accepted and part of the Site problem solving portfolio.

Image is a problem area. Statistics is perceived as dull and uninteresting. Many people were not particularly happy with maths at school. There is an element of fear and lack of understanding. This can only be addressed by the continual promotion of successful and profitable solution of business problems using thses techniques – people like to be associated with success!

Finally there is the damage that can be done by DIY or black-box blind application of these techniques. Throwing a pile of data at a modelling package, picking up a regression and then using it without a full and serious verification and validation process is a recipe for disaster. The decoupling of maths techniques from the physical understanding of the process should never be attempted. In this sense the easy access to statistical functions, for instance, in the EXCEL tool bar can be dangerous in the hands of the inexperienced user. In all cases these techniques should be a quantifiable support role to a sound physical knowledge of the business or process in question.

European Symposium on Computer Aided Process Engineering – 12
J. Grievink and J. van Schijndel (Editors)
© 2002 Elsevier Science B.V. All rights reserved.

Property Integration - A New Approach for Simultaneous Solution of Process and Molecular Design Problems

Mario Richard Eden[*], Sten Bay Jørgensen[*], Rafiqul Gani[*] and Mahmoud El-Halwagi[♦]

[*] CAPEC, Department of Chemical Engineering, Technical University of Denmark

[♦] Chemical Engineering Department, Auburn University, USA

Abstract

The objective of this paper is to introduce the new concept of property integration. It is based on tracking and integrating properties throughout the process. This is made possible by exploiting the unique features at the interface of process and molecular design. Recently developed clustering concepts are employed to identify optimal properties without commitment to specific species. Subsequently, group contribution methods and molecular design techniques are employed to solve the reverse property prediction problem to design molecules possessing the optimal properties.

1. Property Integration

Aside from reaction systems, the primary task of most processing units is to tailor properties of various streams throughout the process. Furthermore, the use of streams and species throughout the process is primarily driven by the need to satisfy certain properties in the processing units. Notwithstanding the critical role of properties in designing a process, the conventional approach to process design is based on tracking mass and energy throughout the process. Properties are considered indirectly by selecting species as well as operating conditions and ensuring that the resulting properties are acceptable. This approach is quite limiting since the insights associated with properties are masked by species and operating conditions. Alternatively, properties should be tracked and integrated explicitly. In this paper, we introduce the new concept of property integration. It is a holistic approach to the tracking, manipulation, and allocation of properties throughout the process.

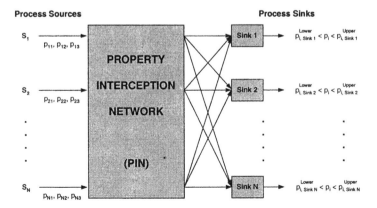

Figure 1: Schematic representation of property integration.

In order to develop a systematic framework for property integration, we first represent the process from the perspective of properties. A process can be represented through sources, sinks, and interception devices. Sources are process streams that possess certain properties. Sinks are process units that process the sources. Interception devices are additional units that can modify the properties of the sources. Property integration deals with the identification of optimal mixing, segregation, and interception of sources so as to satisfy the property constraints for the sinks. Figure 1 is a schematic representation of the property integration framework. In this work, we will exploit unique features that lie at the interface of process and product design.

2. Simultaneous Solution of Process and Molecular Design Problems

Traditionally process design and molecular design have been treated as two separate problems, with little or no feedback between the two approaches. Each problem has been conveniently isolated or decoupled from the other.

Figure 2 shows a schematic representation of the two problems, e.g. the required inputs and solution objectives of the different design algorithms.

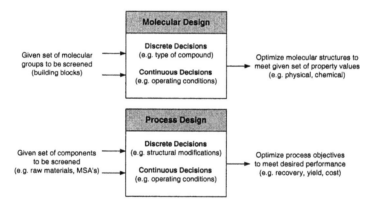

Figure 2: Conventional approach for molecular design and process design.

Both approaches have some inherent limitations due to the amount of information that is required prior to invoking the design algorithm. When considering conventional process design methodologies, the selected species are chosen from among a list of pre-defined candidate components, therefore, limiting performance to the listed components. On the other hand, with molecular design techniques, the desired target properties are required input to the solution algorithm. Once again these decisions are made ahead of design and are usually based on qualitative process knowledge and/or experience and thus possibly yield a sub-optimal design. To overcome the limitations encompassed by decoupling the process and molecular design problems, a simultaneous approach as outlined in

Figure 4 is proposed. Using this approach the necessary input to the methodology is the molecular building blocks and the desired process performance, for the molecular and process design algorithms respectively. The final outputs of the algorithm are the design variables, which facilitate the desired process performance target and the molecules that satisfy the property targets identified by solution of the process design problem.

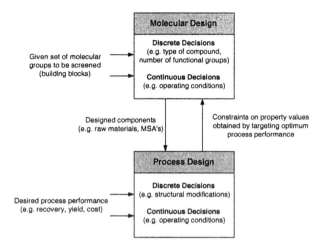

Figure 4: New approach for simultaneous molecular and process design.

The strength of this approach is to identify the property values that correspond to the optimum process performance without committing to any components at this stage. This is a critical characteristic for property integration. These property values are then used for the molecular design, which returns the corresponding components. One inherent problem with this approach is the need to solve the process design problem in terms of properties and not components. Unfortunately properties, unlike mass and energy, are not conserved; hence a framework for tracking properties among process streams and process units is needed.

3. Concepts of Property Clustering

To overcome the limitations encompassed when trying to track properties among process streams and units, the use of conserved property-based clusters has been proposed (Shelley & El-Halwagi, 2000). The objective is to replace component based material balances with cluster based material balances without any simplification or loss of information. Another advantage of the clustering concept is that, by removing the compositions as variables, even a large dimension problem solution can be visualized in a two- or three-dimensional space. The clusters are tailored to possess the two fundamental properties of inter- and intra-stream conservation, thus enabling the development of consistent additive rules along with their ternary representation. This graphical representation has been employed to obtain valuable insights with respect to optimal condensation and allocation strategies for complex hydrocarbon mixtures.

4. General Problem Statement and Solution Methodology

The overall problem addressed in this work is the recovery and allocation of Volatile Organic Compounds (VOCs). Given a process with a number S_N of gaseous streams (sources) with known properties and flowrates, these streams may be utilized in a number N of process units (sinks) if they satisfy given constraints on flowrate and property values. The Property Interception Network (PIN) can be used to change the property values of the streams e.g. mix or split streams to meet the sink constraints. Furthermore process utilities such as coolants and refrigerants are available for condensing the VOCs for possible reuse in one or more of the process sinks. The *objective is to identify the optimal strategies for the condensation of the VOCs and their subsequent utilization in the process sinks as well as the identification of target property values for external process fluids (if any) that may be required. The solution methodology for minimizing the flowrate of fresh material is presented in* Figure 6.

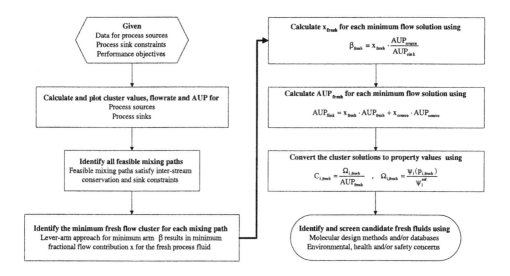

Figure 6: Method of solution for minimum flow of fresh material.

The first two steps correspond to problem initialization and visualization. The next two steps solve the process design problem (in this case with respect to minimum flow of fresh material) in terms of cluster (property) values. The next three steps convert the cluster solution to properties and the final step solves the molecular design problem. Since the process design solution is given in terms of properties only and not components, both problems are addressed at the same time, even though the minimum flow methodology given above is sequential.

5. Case Study

The metal degreasing process presented in
Figure 8 uses a fresh organic solvent in the absorption column and another one in the degreaser. The process is described by Shelley and El-Halwagi (2000). Three properties are examined to determine the suitability of a given organic process fluid for use in the absorber and/or degreaser; sulfur content (for corrosion considerations), density (for hydrodynamic aspects) and Reid vapor pressure (for volatility, makeup and regeneration). The solvents to be synthesized are pure component fluids, thus the sulfur content of these streams is zero. The constraints on the inlet conditions of the feed streams to the absorber and degreaser as well as the property operator mixing rules are given by Shelley & El-Halwagi (2000). Experimental data are available for the degreaser off-gas condensate. Samples of the off-gas were taken, and then condensed at various condensation temperatures ranging from 280K to 315K, providing

measurements of the three properties as well as the flowrate of the condensate. These data correspond to the condensation route given in

Figure 10, while the sink constraints were converted to cluster values yielding the two regions for the absorber and degreaser respectively.

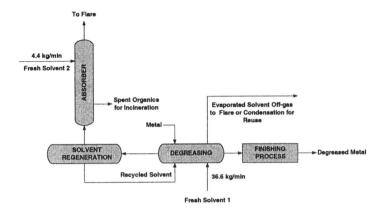

Figure 8: Original process flowsheet.

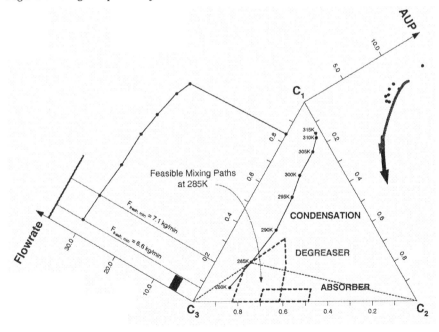

Figure 10: Source-sink mapping of diagram for metal degreasing process

Using the methodology outlined in

Figure 6 the cluster data was plotted and the feasible mixing paths identified. Since the fresh process fluids contain no sulfur, any feasible solution will be on the C_2-C_3 axis. Lever-arm analysis is employed to identify the minimum flow solutions. For a condensation temperature of 280 K (which corresponds to the highest condensate flowrate 30.0 kg/min), it turns out that the minimum feasible flowrate of the fresh is 11.8 kg/min, however the target value from the sink constraints was 6.6 kg/min, thus the same investigation was performed at 285 K. At this temperature, the condensate flowrate is slightly reduced (29.5 kg/min) however the minimum feasible flowrate corresponds to the target value of 7.1 kg/min. It should be noted that using this approach the flowrate of the fresh material has been reduced by approximately 80%. The analysis showed that the cluster solutions to the degreaser problem correspond to the degreaser points on the C_2-C_3 axis. Since all the condensate has been recycled to the degreaser, the solution for the absorber is a simple molecular design problem. Using the information obtained from the source-sink mapping analysis a computer-based tool ProCAMD (CAPEC, 2001) was invoked to synthesize candidate process fluids. Harper (2000) gives the underlying algorithm for the software tool. Not allowing phenols, amines, amides or compounds containing silicon, sulfur or halogenes, due to safety and health considerations, reduced the search space. The CAMD algorithm yielded one candidate solvent for each of the process units, iso-Pentane for the absorber and n-Butane for the degreaser.

6. Conclusions

In this work, a framework for solving process design and molecular design problems simultaneously has been introduced. By employing recently developed property clustering techniques, the process design problem is solved by targeting the property values yielding optimum process performance. The property values obtained by the process design algorithm are used as inputs to a molecular design algorithm for synthesis of the corresponding components. The usefulness of this approach has been demonstrated through a case study.

7. References

CAPEC (2001) ICAS User Manual, Department of Chemical Engineering, Technical University of Denmark.

Harper P.M. (2000) A Multi-Phase, Multi-Level Framework for Computer Aided Molecular Design, Ph.D. Thesis, Department of Chemical Engineering, Technical University of Denmark.

Shelley M.D., El-Halwagi M.M. (2000) Component-less Design of Recovery and Allocation Systems: A Functionality-based Clustering Approach, *Computers and Chemical Engineering*, 24, 2081-2091.

European Symposium on Computer Aided Process Engineering – 12
J. Grievink and J. van Schijndel (Editors)
© 2002 Published by Elsevier Science B.V.

Mass Balance and Capacity Optimisation in the Conceptual Design of Processes for Structured Products

Nilesh Y. Jadhav[1], Michel L.M. vander Stappen[2], Remko Boom[3], Giljam Bierman[1], Johan Grievink[1]

1-Department of Chemical Technology, Delft University of Technology, the Netherlands
2-Unilever Bestfoods R&D, Heilbronn, Germany
3-Food and Bioprocess Engineering, Wageningen University, the Netherlands

Abstract

To find systematic ways to justify the design decisions or evaluate alternative and novel solutions for processes for structured products, a conceptual design methodology is being developed. It gives a six-levelled top-down approach similar to the Douglas hierarchical design procedure.

The design of a food dressings plant is taken as a pilot to illustrate the application of the methodology. The synthesis, analysis and evaluation procedure at Level-2 is discussed in details. The problem is formulated as a (non-linear) optimisation. Mixed integer problem is avoided by reducing the number of flowsheet alternatives using heuristics.

The first results show that significant capacity savings (26%) can be obtained compared to the conventional design. Generally, the findings of this work illustrate that combining the more heuristics oriented hierarchical design approach with numerical optimisation on a refined subset of alternatives works very well in practice without loss of quality of the end results and maintaining insight.

1. Introduction

Structured products in the context of the paper are defined as substances, which are characterized by a specific physical microstructure that is a determining factor towards their quality. Mayonnaise for example, consists of oil droplets dispersed in a water matrix, stabilised by a protein network. Thus these products are different from conventional chemical products and cannot be characterized by composition alone. Plants producing structured products in the food industry are typically multiproduct plants processing a large number of ingredients. Currently, the design of such plants is largely based on using best-practice equipment and using design rules derived from years of experience with existing plants. There exists no systematic way to justify the design decisions or evaluate alternative solutions. This design approach inhibits the development of novel processing methods and technologies.

In view of this, development of a design methodology for structured product processes is being found necessary. With such a methodology, it will be possible to systematically generate and evaluate process flowsheets for structure product processes. It will also formalize the design procedure for structured product plants and facilitate the development of new technologies. Meeuse proposes a method for the conceptual design of processes for structured products (Meeuse *et.al.*, 2000). A hierarchical decomposition approach similar to the Douglas hierarchical design procedure is used there (Douglas, 2000). It was observed that the Douglas procedure is not applicable directly to the case of structured products. The levels in design had to be adapted so that the microstructure of the product is taken into account during design. Also the multiproduct nature of the plants must be considered.

2. Design methodology for structure product processes

The methodology used in this work is an extension of the conceptual design method proposed by Meeuse (Meeuse *et.al.*, 2000). It consists of six hierarchical design levels:
1. Input-Output Structure
2. Generation of structure blocks, based on product microstructure
3. Transformations and tasks
4. Operational units
5. Integration (of heat streams, utilities and equipment)
6. Equipment selection and design

From an input-output structure of the overall plant at Level-1, a synthesis of a block diagram is achieved at Level-2, using structure blocks which build the microstructure of the products from the ingredients. At Level-3, tasks and transformations within the structure blocks are identified. The structure blocks from Level-2 are hence further decomposed into 'task blocks' at Level-3. The required transformations that change the necessary attributes of streams occur in the task blocks. At Level-4, operational units are formed from task descriptions. An operational unit describes one or more tasks through rates of changes in an abstract geometrical subspace (volume, area), while incorporating contacting patterns and mode of operation (batch or continuous). Level-5 is reserved for integration of energy, mass and operational units. This then gives a good basis for equipment selection and finalizing the design at Level-6.

A clear sequence of design steps is used at each level of the methodology. This design structure is derived with the help of the Delft Design Matrix (Grievink, 1998). It consists of seven phases:

1. Scope of design
2. Knowledge of design building blocks (objects)
3. Synthesis
4. Analysis
5. Evaluation
6. Report of design (alternatives)
7. Go/no-go

This structure enables synthesis, analysis and evaluation of design alternatives at each hierarchical design level. Also the referencing and reporting of design details and decisions is accomplished at the same time. Thus the methodology overcomes the disadvantages of current design practices. It provides a systematic decomposition of the design problem at the conceptual design phase.

3. Scope

In this paper, the Level-2 of the design methodology will be explained in details by using the design structure as mentioned above. Only the synthesis analysis and evaluation phases are covered. The development of Level-2 is supported by means of a design case of a real multiproduct mayonnaise/dressings plant. A product typically is made from upto ten (lumped groups of) ingredients whereas the mass fractions can span a wide range (between 0.005 and 0.75). Not necessarily all ingredients are present in all products. In reality a dressings plant will have around 200 or more different ingredients. In the design case, the ingredients sharing the same function are first grouped into ten composite ingredient streams. This is only to reduce the complexity during calculations and does not affect the design or the methodology.

4. Design steps at Level-2

At Level-2 of the design methodology, the input-output structure (from Level-1) is decomposed into a block diagram. The building blocks at this level are the 'structure blocks' that are formed by explicitly considering the product microstructure. Each of these blocks builds a specific thermodynamic phase present in the microstructure. For the design case at hand, three structure blocks are identified. E.g. the V-block builds the vegetable-phase, T-block builds the thickened aqueous-phase and E-block builds the emulsion-phase for mayonnaise. In addition to these structure blocks one needs stream splitting and stream mixing blocks to manipulate the streams to and from the structure blocks.

Thickened aqueous-phase

Emulsion-phase

Vegetable-phase

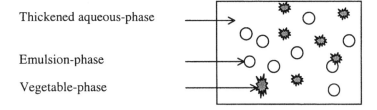

Fig. 1. Microstructure of the product (mayonnaise / dressing)

Synthesis

In the synthesis phase at this level, the various structure blocks are identified. Next, the ingredients going to each of the product phases are identified. It should be noted that some ingredients might be present in more than one phase. The major ingredients (e.g. water, oil etc.) in each of the phases are identified. All other ingredients are minor ingredients (e.g. flavours, colours etc.). A flowsheet backbone is first constructed by only considering major ingredients. Additional split blocks are considered that are necessary to split and direct the ingredients to the different structure blocks. Also a mixing block is considered downstream. Minor ingredients are later considered to complete the flowsheet. Preliminary mass balances are made. An example of a Level-2 block diagram is shown in Fig. 2 (not all ingredient streams are shown). 'S1' and 'S2' represent the split blocks and 'MIX' represents the mixing block.

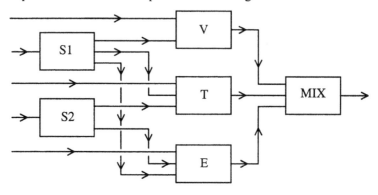

Fig. 2 Block diagram at Level-2

The block diagram in Fig. 2 shows three structure blocks in parallel. This is however not necessary and alternatives exist. (parallel, series, mixed). The process constraints/ heuristics may not permit some flowsheet arrangements. E.g. processing methods in block E is such that block V cannot be placed before it in series. All such heuristics are first checked and only feasible flowsheet alternatives are considered further.

Analysis

Preliminary mass balances during the synthesis phase will provide a design framework with structure blocks where stream routing is decided as much as possible. In the analysis phase, this framework has to be optimized to find the optimal stream routing that minimizes the investment costs. Large variations in throughputs between different production modes are to be avoided. The degrees of freedom available in terms of time allocation and flowrates are used for defining an optimization problem, which can be characterized as a complex blending problem. Minimizing the total investment costs is used as the objective for optimisation. Incorporating the relation between investment costs and the capacity of the blocks, the objective function is defined as the weighted sum of the individual block capacities:

$$\text{Minimize } \sum_{i=1}^{N} f_i C_i \tag{1}$$

C_i	= Capacity of the ith block	[kg/hr]
f_i	= Weight factor assigned to the ith block.	[-]
N	= Total number of structure blocks to be considered	[-]

The capacity used in the objective function is related to the stream flows. It equals the maximum of all throughputs (total of all incoming stream flows) of the block during different production modes. Thus, the block capacity C_i can be defined as follows.

$$C_i = \underset{j=1}{\overset{M}{Max}}(F_{i,j}) \qquad\qquad (2)$$

$F_{i,j}$ = Throughput of block i during the jth production mode   [kg/hr]

C_i = Capacity of block number i [kg/hr]

Defining the block capacity as illustrated above, introduces apparent discontinuity in the optimization objective function (Edgar *et.al.*, 1989). Introducing C_i itself as a variable and using a constraint to ensure that C_i can take values that are greater than or equal to the throughputs during different production modes tackles this.

$$C_i \geq F_{i,j} \text{ (For } j =1 \text{ to } M) \qquad\qquad (3)$$

This also facilitates the use of a linear solver in the optimization code. The flows of each of the ingredient streams, allocated time to production modes and the block capacities are thus the variables during optimization. The allocated time has to sum up to the total available time for production. This gives the constraint for time allocation. The pre-fixed stream routing (for some streams) and the mass balance models can be considered as constraints on the stream variables. Also there are constraints, which are ingredient specific (e.g. a particular ingredient must be routed to a certain block) or block specific (e.g. the water levels in the block have to be above a certain limit). These also have to be suitably defined in the problem. Further, the ingredient flows are related to their demand by equation 4.

$$F_{k,j} = \frac{1000 X_{k,j} D_j}{t_j} \qquad\qquad (4)$$

$F_{k,j}$ = mass flow of ingredient k during the jth production mode   [kg/hr]

$X_{k,j}$ = fraction of ingredient k in the product during the jth production mode   [-]

D_j = demand of product j [tpy]

t_j = the time allocated to production mode number j [hr]

This equation makes the optimization problem non-linear, as both flows and time are variables. Thus the objective function is linear but there is non-linearity in the constraints of the optimization problem.

The flows of each of the ingredient streams, allocated time to production modes and the block capacities and the throughputs of the blocks during different production modes are the variables during optimization. The allocated time has to sum up to the total available time for production. This gives the linear constraint for time allocation. The pre-fixed stream routing and the mass balance models can be considered as linear constraints on the stream variables. Also there are constraints, which are ingredient specific (e.g. a particular ingredient must be routed to a certain block) or block specific (e.g. the water levels in the block have to be above a certain limit). These also have to be suitably defined in the problem.

For programming the optimization problem, MATLAB was used. To facilitate, the efficient solution, the problem was iteratively solved using the following steps:

1. Introducing fixed time allocation and thus making the problem linear (to be solved with the 'LINPROG' subroutine in MATLAB). The flows and block capacities are defined as variables.
2. Keeping the stream routing fixed (using results from step-1) with time allocation and block capacities as variables. This still gives a non-linear problem (Solved by 'FMINCON' routine in MATLAB)
3. Using the time allocation obtained in step-2 as fixed time allocation in step-1
4. Repeat steps 1-3 until difference between solutions from two successive iterations is less than a pre-defined tolerance.

This iterative procedure enables the use of 'LINPROG' in which initial guesses for variables are not required. Moreover, it gives a structure where, the problem can be solved as a fixed time allocation problem (useful for example when the time allocation has to be fixed proportional to the demands) by stopping optimization at Step-1.

Evaluation of the resulting design case

The different alternatives generated and screened during synthesis are evaluated on basis of the total of block capacities resulting after optimization. Each option is evaluated separately instead of combining the choices into a superstructure to be solved as a larger optimization program, which can be imagined to be a complex mixed integer linear or non-linear programming (MILP/MINLP) code. For the current case, four potential block diagrams alternatives were considered.

The optimization program was run for a block diagram with parallel blocks. The results show an improvement in capacity gained by the variable time allocation. It was compared with a scenario where time allocation is kept fixed and proportional to the demands. Capacity savings of about 26 % were seen for the current case.

5. Evaluation

The first results show that significant capacity savings (26%) can be obtained compared to the conventional design. Generally, the findings of this work illustrate that combining the more heuristics oriented hierarchical design approach with numerical optimisation on a refined subset of alternatives works very well in practice without loss of quality of the end results and maintaining insight.

References

Douglas, J.M., 1988, Conceptual Design of Chemical Processes, McGraw-Hill.

Edgar, T.F., D.M. Himmelblau, 1989, Optimization of Chemical Processes, McGraw Hill International Editions.

Grievink, J., Lecture Notes with the course on Process Systems Design, Delft University of Technology, 1989.

Meeuse, F.M., J. Grievink, P.J.T. Verheijen, and M.L.M. vander Stappen, 2000, Conceptual Design of Processes for Structured Products, AIChE Symposium Series, vol 96, 323.

European Symposium on Computer Aided Process Engineering – 12
J. Grievink and J. van Schijndel (Editors)
© 2002 Elsevier Science B.V. All rights reserved.

Enzyme conformational predictions by molecular modeling

Jesus, S.S.; Franco, T.T., Maciel Filho, R.; Shimono, F.N.; Sousa, R.R.; Cappi, A.;
Stein, G.C.
Laboratory of Optimization, Design and Advanced Control – LOPCA
Department of Chemical Process; School of Chemical Engineering
State University of Campinas – UNICAMP
Cidade Universitária Zerferino Vaz, CP. 6066, Campinas, SP, Brazil
maciel@feq.unicamp.br, ssantos@lopca.feq.unicamp.br

Abstract

The main objective of this paper is to verify the influence of temperature on α-amylase conformation during the drying process by atomization and freeze drying as well as to realize predictions by molecular modeling. These allows to verify structural changes due to partial or total activity loss during either drying processes as well as to investigate, by simulation what are the better conditions of pressure and temperature. Studies at bench scale used a α-amylase from *Bacillus*. Atomization experiments used a vertical spray dryer and freeze drying of enzymes. Computational simulations were carried out using a ECEPP/3 program for IBM-AIX. Tertiary structure of the α-amylase was obtained using a SWISS PROT data bank, in which was modeled the structure concerned with the 512 aminoacids that exists in this protein. The minimized structured charatetistics were quantified using calculations from RMSD (root mean squared desviation). The characteristics were compared with the original structure using a PDB (protein data bank). The whole optimization was carried out to the 483 residues that exists in the chain. Through the obtained results it was possible to verify that the conformational structure is strongly influenced by the temperature even for relatively low values. The molecular modeling showed to be very efficient to predict the temperature which allows to carry out freeze drying experiments in such conditions to preserve the desired properties of the enzyme. The modelling predictions have shown a very good agreement with the experimental data which allow to conclude that it may be used to the search of operating conditions as well as for process optimization.
Key word: Structural molecular simulation, conformational modeling, α-amylase.

1. Introduction

Dehydrated enzymes application within textile, food and pharmaceutical industries have been continously improved. Alpha-amylase is largely used in industry and shows a great potential to have new further applications. Among all microorganisms able to produce this enzyme, emphasy is given to the *Bacillus* specie. Due to the characteristic of enzymes, it is not a trivial task to find out a suitable drying method which allows to keep the desired properties of the product. Among the conventional drying processes of biological materials, the atomizaton and the freeze drying deserve attention. The freeze drying (also called lyophilization), consists of a process of dehydration of the material through the sublimation of the frozen water at low temperatures and low pressure. It is also known that up to 20 % of enzyme activity can be lost with atomization process, however fits to stand out that the process presents low energy (Çacaloz et al.,1997). In freeze drying process conveniently operated, it is expected that desnaturation and of activity loss are minimal, however the time of process and the high energy consume can be considered as restriction for real implementation of the technique (Alcaide & Lobraña, 1983).

Nowadays computational techniques are widely spread in order to simulate protein conformation, as well as to obtain information to understand inherent processes of Chemical Engineering, for instance, drying process. Computational molecular modeling is also important for genetic engineering, pure and applied sciences, biochemistry and biophisycs.

The main objective of this paper is to verify the influence of temperature on α-amylase conformation during the drying process using atomization and freeze drying and to analyze the predictions by molecular modeling.

2. Experimental

2.1 Enzyme
Alpha-amylase bacterial enzyme (EC 3.2.1.1) from *Bacillus* (NOVO NORDISK) was used.

2.2 Drying atomization
The experiments were proceeded with spray dryer LAB-PLANT, SD-04 model.

Five hundred grams broth enzymatic were used for each drying experiment with a beak atomizer (nozzle 0,5 mm diameter). The air compression was to 0,41 bar. A peristaltic bomb operating from 1 to 100 rpm, through a tube of silicon of 4 diameter mm.

The spray dryer experiments was optimized through experimental design in star with two variables in order to investigate flow-rate and the air inner temperature upon the enzyme activity (EA).

For the statistic analysis software Statistica 5.0 was used.

2.3 Drying freeze drying
The experiments were proceeded with a vertical freeze dryer TELSTAR, Cryodos –80 model.

Initially the broth enzymatic was frozen with liquid nitrogen, dry ice and acetone and in conventional freezer.

The freezing rates were obtained by the following expression:

$$W = \frac{0 - T}{\Delta t} \tag{1}$$

where: W is thefreezing rate [^{0}C/min], T is the temperature [^{0}C] and Δt is the freezing time [min].

For each drying experiment 5.0 mL of the enzymatic broth was used, and the samples were drying in different times (from four to twenty-four hours).

The freezing temperature was measured by the device OMRON model E5CN coupled with an entrance K thermocouple.

During the lyophilization process the chamber temperature was maintained at approximately - 90^{0}C and a 0,048 mBar.

The material obtained by both techniques were analyzed through the enzyme activity (EA).

Alpha-amylase activity was assayed using start 7 % solution as the substrate, and the amount of glucose released was determined by glucose oxidase.

One unit of enzyme activity was defined as 1 μmol of glucose produced per minute under the given conditions.

3. Molecular modeling

For the studies of molecular modeling ECEPP/3 (Empirical Conformational Energy Program for Peptides) for IBM-AIX was used (Némethy et al., 1992).

The total potential energy (E_T) or steric energy of a protein, given by ECEPP/3 Program is by the sum of two fields of forces: space energy and torsional energy (E_{TOR}). The space part is expressed by the sum of the energy interactions among atoms. In this part it includes the electrostatic term(E_{EL}), repulsive and attractive Lennard-Jones (E_{LJ}) term and the potential of the connections of hydrogen (E_{HB}). The space energy depends basically on the distances among atoms. In the second part it is considered the torsional energy of a molecule, being reduced to a sum of torsionals energies for each connection. Each energy of torsional connection is only a function of the angle of corresponding diedro, being described by a periodic potential with a barrier of energy for each chemical connection (Klepeis & Foudas, 2000).

The equation resulting from these two terms has the following form:

$$E_T = E_{LJ} + E_{EL} + E_{HB} + E_{TOR} \qquad (3)$$

The tertiary structure of the α-amylase was obtained using a SWISS PROT data bank, in base on 512 aminoacids presented in this protein. The minimized structured charactetistics were quantified using calculations from RMSD (root mean squared desviation). There characteristics were compared to the original structure using a PDB (protein data bank). The whole optimization was carried out to the 483 residues that there exist in the chain.

4. Results and discussion

4.1 Experimental results

4.1.1 Atomization

Spray dryer experiments were optimized through a full experimental design with two variables in order to investigate the load flow-rate as well as the air inner temperature upon either the EA, as it shows table 1.

Table 1. Superior and inferior levels to independent variables.

Variables	Level				
	$-\sqrt{2}$	-1	0	+1	$+\sqrt{2}$
Air inner temperature T (^{0}C)	130	145	180	215	230
Enzyme load flow-rate Q (mL/s)	0.20	0.23	0.31	0.38	0.41

The experiments carried in the spray dryer were analyzed by surface response methodology. The results clearly show that the enzymatic activity is influenced by the temperature and the outflow of enzyme feed.

When the enzyme was treated with lower temperatures better outflow conditions have been obtained (Figure 1).

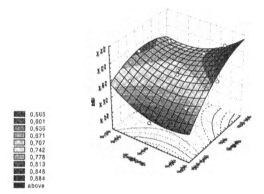

Figure 1 Surface response of the variation of the catalytic activity with the effect of the temperature and outflow.

The codified model obtained through experimental data is given by equation 2.

$$EA = 0{,}722563 - 0{,}008116Q - 0{,}073959T + 0{,}0383QT + 0{,}040161Q^2 \\ - 0{,}037853T^2 \tag{2}$$

Through the model that considers all of the effects it was obtained the surface response presented in the figure 1, allowing the analysis of temperature and load flow rate influences on AE.

The surface shows that to low flows and low temperatures there is a significant increase in the enzymatic activity, and when high flows and the temperature in the central point are considered a good enzymatic activity is also obtained. However it was observed in, practice, that the humidity tenor and the activity of water are larger than to low flows and low temperatures. To very high temperatures and low flows it is noted a drastic reduction in the enzymatic activity.

4.1.2 Freeze drying
In the table 2 is show the freezing rates obtained for the three studied techniques.

Table 2 Rate of freezing for contact with liquid nitrogen, dry ice and acetone and in freezer.

Techniques of freezing	Temperature (^{0}C)	Display time (min)	Freezing rate (^{0}C/min)
Liquid nitrogen (LIN)	-195.8	1.48	132.3
Dry ace and acetone (LIG)	-82.0	7.33	11,.8
freezer (LIF)	-35	80.00	0.38

The better performance using the technique of lyophilization were obtained when the freezing was carried out with liquid nitrogen (Figure 2). Through the results of both techniques, it was verified that lyophilization presented larger enzymatic activity.

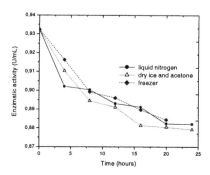

Figure 2 Influence of the time of handling with the enzimatic activity.

The figure 2 depictes that EA decreases with the time of drying. When the technique of freezing was used with freezer was observed that good results were obtained after 12 hours of process. In that period it was observed a loss of 4% of EA. In the process of freezing with dry ice and acetone better results were obtained; it was verified that after a period of eight hours of process high EA, with a loss of only 4,1%. The best results were obtained when the freezing was accomplished with liquid nitrogen, because in a period of four hours of process only 3% of EA was lost. Good results were also found with a period of eight hours, what proves that the lyophilization is favored with high freezing rates.

4.2 Molecular modeling results

In the first stage the ambient temperature was optimized (25°C), which in fact supplied an energy of -134,494.359 kcal/mol.

In the second stage, simulations with the temperatures used in the drying studies have been considered. Such temperatures are: –90°C (temperature of lyophilization), 130, 145, 180, 215 and 230°C (temperature of the spray drying). The temperature of 70°C also was evaluated, and it was found that this is the temperature, limit to do not have any conformational structure change.

It could be observed that good results have been obtained for temperatures of -90°C. Temperature higher than 70°C is already enough to cause shifts of atoms and molecules in the enzymatic strings, which reflect changes in the enzyme structure (Figure 3).

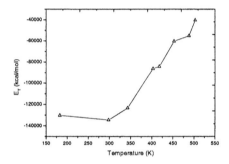

Figure 3 Total Potential Energy (E_T) as a function of temperature T.

5. Concluding Remarks

Through the obtained results it was possible to verify that the conformational structure is strongly influenced by the temperature, even for relatively low temperatures (as 70°C, for the case of α-amylase). The simulation showed to be very efficient to predict the temperature which allows to carry out freeze drying in such conditions to preserve the desired properties of the enzyme. The modelling predictions presented a very good agreement with the experimental data which allows to conclude that it may be used to the search of operating conditions as well as for process optimization.

Through molecular modeling it could also be verified that the best method of enzyme conservation is the freeze drying conclusion which was also found experimentally.

References

Alcaide, L.J, Lobraña, J.I.,1983, Influencia de las variables del proceso sobre el tiempo de secado por liofilizacion del zumo de zanahoria, Anales de Quimica de la Real Sociedad Española de Física y Química, 79, 1, 101-107.

Çakaloz, T., Akbaba, H., Yesügey, E.T., Periz, A., 1997, Drying model for α-amilase in a horizontal spray dryer, Journal of Food Engineering, 31, 499-510.

Klepeis, J.L. and Floudas C.A., 2000, Deterministic global optimization and torcion angle dynamics for molecular struture prediction, Computers chem. Eng., 24, 1761-1766.

Némethy, G., Gibson, K.D., Palmer, K.A., Yoon, C.N., Paterlini, G., Zagari, A., Rumsey, S. Sheraga, H.A., 1992, Energy parameters in polypeptides, Journal of Physical Chemistry, 10, 96, 6472.

European Symposium on Computer Aided Process Engineering – 12
J. Grievink and J. van Schijndel (Editors)
© 2002 Published by Elsevier Science B.V.

ASTRO-FOLD: Ab-initio Secondary and Tertiary Structure Prediction in Protein Folding

John L. Klepeis and Christodoulos A. Floudas
Department of Chemical Engineering
Princeton University, Princeton, NJ USA 08544

Abstract

The ability to predict protein structures from gene sequence data through computational means has significant implications in the fields of computational chemistry, structural biology and medicinal chemistry, as well as in expanding the impact of the research on the human genome. Current challenges for this task, also known as ab-initio protein structure prediction, include validity of the available molecular models and complexity of the search space. A novel four stage ab-initio approach for the structure prediction of single chain polypeptides is introduced. The methodology combines the classical and new views of protein folding, while using free energy calculations and integer linear optimization to predict helical and β-sheet structures. Detailed atomistic modeling and the deterministic global optimization method, αBB, coupled with torsion angle dynamics, form the basis for the final tertiary structure prediction. The excellent performance of the proposed approach is illustrated using the bovine pancreatic trypsin inhibitor protein.

1. Introduction

Structure prediction of polypeptides and proteins from their amino acid sequences is regarded as a holy grail in the computational chemistry, molecular and structural biology communities. According to the thermodynamic hypothesis (Anfinsen, 1973) the native structure of a protein in a given environment corresponds to the global minimum free energy of the system. Recent reviews assess the advances made in this area by researchers across several disciplines (Dill, 1999; Koehl and Levitt, 1999; Wales and Scheraga, 1999). In spite of pioneering contributions and decades of effort, the ab-initio prediction of the folded structure of a protein remains a very challenging problem. The approaches for the structure prediction of polypeptides can be classified as: (i) homology or comparative modeling methods, (ii) fold recognition or threading methods, (iii) ab-initio methods that utilize knowledge-based information from structural databases (e.g., secondary and/or tertiary structure restraints), and (iv) ab-initio methods without the aid of knowledge-based information.

In this article, we focus on the ab-initio methods. Knowledge-based ab-initio methods exploit information available from protein databases regarding secondary structure, introduce distance constraints, and extract similar fragments from multiple sequence alignments in an attempt to simplify the prediction of the folded three-dimensional protein structure. Ab-initio methods that are not guided by knowledge-based information represent the most challenging category. Orengo et al. (1999) provide a recent assessment of the current status of both types of ab-initio protein structure prediction approaches.

2. Methods

We introduce the novel ASTRO-FOLD approach for the ab-initio prediction of the three dimensional structures of proteins, a four stage approach outlined below.

2.1 Helix Prediction

The first stage involves the identification of helical segments and is accomplished by partitioning the aminoacid sequence into pentapeptides (or other oligopeptides) such that consecutive pentapeptides posses an overlap of four aminoacids; atomistic level modeling using the selected force field; generating an ensemble of low energy conformations; calculating free energies that include entropic, cavity formation, polarization and ionization contributions for each pentapeptide; and calculating helix propensities for each residue using equilibrium occupational probabilities of helical clusters. The concept of partitioning the aminoacid sequence into overlapping oligopeptides is based on the idea that helix nucleation relies on local interactions and positioning within the overall sequence. The explicit consideration of local interactions through overlapping oligopeptides allows for detection of cases in which identical aminoacid sequences adopt different conformations in different proteins (Minor and Kim, 1996). This is consistent with the observation that local interactions extending beyond the boundaries of the helical segment retain information regarding conformational preferences (Baldwin and Rose, 1999). The partitioning pattern is generalizable and can be extended to heptapeptide or nonapeptide systems (Anfinsen and Scheraga, 1975). The ab-initio prediction of helical segments encompasses the following steps (Klepeis and Floudas, 2002):

1. The overlapping pentapeptides are modeled as neutral peptides surrounded by a vacuum environment using the ECEPP/3 force field (Nemethy et al., 1992). An ensemble of low potential energy pentapeptide conformations, along with the global minimum potential energy conformation, is identified using a modification of the αBB deterministic global optimization approach (Klepeis and Floudas, 1999) and the conformational space annealing approach (Lee et al., 1998). For the set of unique conformers, Z, free energies, F_{vac}, are calculated using the harmonic approximation for vibrational entropy (Klepeis and Floudas 1999).

2. The energy for cavity formation in an aqueous environment is modeled using a solvent accessible surface area expression, $F_{cavity} = \gamma A + b$, where A is the surface area of the protein exposed to the solvent.

3. For the set of unique conformers, Z, the total free energy is calculated from

$$F_{total} = F_{vac} + F_{cavity} + F_{solv} + F_{ionize} \tag{1}$$

Here F_{solv} represents the difference in polarization energies caused by the transition from a vacuum to a solvated environment and F_{ionize} represents the ionization energy.

4.   For each pentapeptide, total free energy values (F_{total}) are used to evaluate the equilibrium occupational probability for conformers having three central residues within the helical region of the $\phi-\psi$ space. Helix propensities for each residue are determined from the average probability of those systems in which the residue constitutes a core position.

2.2 β-Strand and β-Sheet Prediction

In the second stage, β-strands, β-sheets, and disulfide bridges are identified through a novel superstructure-based mathematical framework originally established for chemical process synthesis problems (Floudas, 1995).   Two types of superstructure are introduced, both of which emanate from the principle that hydrophobic interactions drive the formation of β-structure. The first one, denoted as hydrophobic residue-based superstructure, encompasses all potential contacts between pairs of hydrophobic residues (i.e., a contact between two hydrophobic residues may or may not exist) that are not contained in helices (except cystines which are allowed to have cystine-cystine contacts even though they may be in helices). The second one, denoted as β-strand-based superstructure, includes all possible β-strand arrangements of interest (i.e., a β-strand may or may not exist) in addition to the potential contacts between hydrophobic residues.   The hydrophobic residue-based and β-strand-based superstructures are formulated as mathematical models which feature three types of binary variables: (i) representing the existence or nonexistence of contacts between pairs of hydrophobic residues; (ii) denoting the existence or nonexistence of the postulated β-strands; and (iii) representing the potential connectivity of the postulated β-strands. Several sets of constraints in the model enforce physically legitimate configurations for antiparallel or parallel β-strands and disulfide bridges, while the objective function maximizes the total hydrophobic contact energy. The resulting mathematical models are Integer Linear Programming (ILP) problems which not only can be solved to global optimality, but can also provide a rank ordered list of alternate β-sheet configurations.

We focus on the hydrophobic residue-based superstructure and its mathematical model. First, all residues not belonging to helices (except cystines) are classified as hydrophobic, bridge, turn or other residues. Each residue is also assigned a position dependent parameter, $P(i)$, which is equal to the position of the hydrophobic residue in the overall sequence and is used extensively to describe allowable contacts between hydrophobic residues. Hydrophobicity parameters, H_i, are taken from hydrophobicity scales derived either from experimental transfer free energies or from statistical data. The interaction energy for a potential hydrophobic contact is assumed to be additive, and for cases involving a cystine-cystine contact an additional energy contribution, H_{ij}^{add}, is included. The objective is to maximize the contact energy

$$\max \ \Sigma i \ \Sigma j \ (H_i + H_j + H_{ij}^{add}) \ y_{ij} \qquad \text{forall } P(i) + 2 < P(j) \qquad (2)$$

where the binary (0-1) variables, y_{ij}, represent potential hydrophobic contacts. The maximization is subject to several sets of constraints which define allowable β-sheet configurations. For instance, in the case of antiparallel β-sheets the constraints become:

$$y_{ij} + y_{kl} \leq 1 \quad \text{forall} \ \ P(i) + P(j) \neq P(k) + P(l); \ \ y_{ij} \ \text{OR} \ y_{kl} \ \text{NOT}\{Cys,Cys\} \qquad (3)$$

which require that the sum of the contact position parameters must be equal, enforce the formation of symmetric, non-intersecting loops. Note that the constraint is not enforced when the potential contact represents a disulfide bridge.

2.3 Tertiary Structure Prediction

The third stage serves as a preparative phase for atomistic-level tertiary structure prediction, and therefore focuses on the determination of pertinent information from the results of the previous two stages. This involves the introduction of lower and upper bounds on dihedral angles of residues belonging to predicted helices or β-strands, as well as restraints between the C^α atoms for residues of the selected β-sheet and disulfide bridge configuration. Furthermore, for segments which are not classified as helices or β-strands, free energy runs of overlapping heptapeptides are conducted to identify tighter bounds on their dihedral angles.

The fourth and final stage of the approach involves the prediction of the tertiary structure of the full protein sequence. The problem formulation, which relies on dihedral angle and atomic distance restraints acquired from the previous stage, is

$$
\begin{aligned}
&\text{min} && E_{ECEPP/3} && (4)\\
&\text{subject to} && E_l^{distance}(\phi) \ \leq \ E_l^{ref} && l=1,\ldots,N_{CON}\\
& && \phi_i^L \ \leq \ \phi_i \ \leq \ \phi_i^U && i=1,\ldots,N_\phi
\end{aligned}
$$

Here $i=1,\ldots,N_\phi$ refers to the set of dihedral angles, ϕ_i, with ϕ_i^L and ϕ_i^U representing lower and upper bounds on these dihedral angles. The total violations of the $l=1,\ldots,N_{CON}$ distance constraints are controlled by the parameters E_l^{ref} (Klepeis et al., 1999). To overcome the multiple minima difficulty, the search is conducted using the αBB global optimization approach, which offers theoretical guarantee of convergence to an ε global minimum for nonlinear optimization problems with twice differentiable functions (Androulakis et al., 1995; Adjiman et al., 1998; Floudas 2000). This global optimization approach effectively brackets the global minimum by developing converging sequences of lower and upper bounds, which are refined by iteratively partitioning the initial domain. The generation of low energy starting points for constrained minimization is enhanced by introducing torsion angle dynamics (Guntert et al., 1997) within the context of the αBB global optimization framework.

3. Computational Study

The proposed approach for the ab-initio secondary and tertiary structure prediction is illustrated with bovine pancreatic trypsin inhibitor (BPTI). The partitioning of the 58 residue chain into overlapping sub-sequences results in 54 pentapeptides. For each pentapeptide, an average of 14000 unique conformers are generated for free energy analysis. Initial calculations provide occupational probabilities of helical clusters for the uncharged systems, and segments of at least three pentapeptides exhibiting probabilities above about 90 % (i.e., strong helical clusters) are marked for further analysis. Probabilities are refined for the marked pentapeptides through the inclusion of solvation and ionization energies for a subset of conformers. The final assignment is made according to average probabilities. For BPTI, helical segments between residues 2-5 and 47-54 are predicted, which is in excellent agreement with experiment.

The performance of the hydrophobic residue-based formulation is also illustrated for BPTI. In the first iteration three disulfide bonding pairs and one antiparallel β-sheet are identified, while the subsequent iteration produces an additional single residue-to-residue contact. The results feature a dominant antiparallel β-sheet with varying disulfide bridge configurations. The β-structure of the globally optimal solution of the (ILP) model consists of the following contacts between hydrophobic residues: Cys5-Cys55, Cys14-Cys38, Cys30-Cys51, Ile18-Val34, Ile19-Phe33, Tyr23-Leu29, and Phe22-Phe45, and is in excellent agreement with the experimentally observed conformation.

For the BPTI tertiary structure, the a-helix and b-sheet prediction results were used to bound dihedral angles for 30 of the 58 total residues. Residues 2-5 and 47-54 were constrained to the helical $\phi-\psi$ region ([-85,-55] for ϕ, [-50,-10] for ψ), while the b-strands between residues 17-23, 29-35 and 44-46 were assigned $\phi-\psi$ bounds consistent with extended conformations ([-155,-75] for ϕ, [110,180] for ψ). Distance restraints included 32 lower and upper C^α-C^α distance restraints representing the predicted α-helix (5.5-6.5 A) and β-sheet (4.5-6.5 A) configuration, as well as six additional restraints (2.01-2.03 A for sulfur atoms) to enforce the disulfide bridge network. During the course of the global optimization search, the branch and bound tree was formed by partitioning domains belonging to selected backbone variables of the undefined (coil) residues, while the remaining variables were treated locally. A significant sample of low energy structures with C^α RMSD (root mean squared deviation) values below 6.0 A was identified along with the global minimum energy structure. The lowest energy structure, with an energy of -428.0 kcal/mol, also provided the best superposition with the crystallographic structure, with a 4.0 A RMSD (see Figure 1.

4. Conclusions

An important question regarding the prediction of the native folded state of a protein is how the formation of secondary and tertiary structure proceeds. Two common viewpoints provide competing explanations to this question. The classical opinion regards folding as hierarchic, implying that the process is initiated by rapid formation of secondary structural elements, followed by the slower arrangement of the tertiary fold. The opposing perspective is based on the idea of a hydrophobic collapse, and suggests that tertiary and secondary features form concurrently. This work bridges the gap between the two viewpoints by introducing a novel ab-initio approach for tertiary structure prediction in which helix nucleation is controlled by local interactions, while non local hydrophobic forces drive the formation of β-structure. Excellent agreement between the experimental and predicted structures through the ASTRO-FOLD method has been obtained for several proteins.

Figure 1. Comparison of predicted tertiary structure (in black) of BPTI and experimentally derived structure (in grey).

References

Adjiman, C.S., I.P. Androulakis, and C.A. Floudas, 1998, Comp. Chem. Eng., 22, 1159.

Androulakis, I.P., C.D. Maranas, and C.A. Floudas, 1995, J. Glob. Opt., 7, 337.

Anfinsen, C.B., 1973, Science, 181, 223.

Anfinsen, C.B., and H.A. Scheraga, 1975, Adv. Prot. Chem., 29, 205.

Baldwin, R.L, and G.D. Rose, 1999, TIBS, 24, 77.

Dill, K.A., 1999, Prot. Sci., 8, 1166.

Floudas, C.A., 1995, Nonlinear and mixed-integer optimization, Oxford , New York.

Floudas, C.A., 2000, Deterministic Global Optimization: Theory, Methods and Applications, Kluwer Academic Publishers.

Guntert, P., C. Mumenthaler, and K. Wuthrich, 1997, J. Mol. Biol., 273, 283.

Klepeis, J.L., and C.A. Floudas, 1999, J. Chem. Phys., 110, 7491.

Klepeis, J.L., C.A. Floudas, D. Morikis, and J.D. Lambris, 1999, J. Comp. Chem., 20, 1354.

Klepeis, J.L., and C.A. Floudas, 2002, J. Comp. Chem., 23, 1.

Koehl, P., and M. Levitt, 1999, Nature Struct. Biol., 6, 108.

Lee, J., H.A. Scheraga, and S. Rackovsky, 1998, Biopolymers, 46, 103.

Minor, D.L., and P.S. Kim, 1996, Nature, 380, 730.

Nemethy, G., K.D. Gibson, K.A. Palmer, C.N. Yoon, G. Paterlini, A. Zagari, S. Rumsey, and H.A. Scheraga, 1992, J. Phys. Chem., 96, 6472.

Orengo, C.A., J.E. Bray, T. Hubbard, L. LoConte, and I. Sillitoe, 1999, Proteins, 3, 149.

Wales, D.J., and H.A. Scheraga, 1999, Science, 285, 1368.

European Symposium on Computer Aided Process Engineering – 12
J. Grievink and J. van Schijndel (Editors)
© 2002 Elsevier Science B.V. All rights reserved.

Incorporation of Sustainability Demands into the Conceptual Process Design phase of Chemical Plants

Gijsbert Korevaar, G. Jan Harmsen, Saul. M. Lemkowitz

Hoogewerff-Chair 'Sustainable Chemical Technology', Delft University of Technology

E-mail: g.korevaar@tnw.tudelft.nl; Phone/Fax: +31 (0) 15 2784466/84452

Abstract

Incorporating sustainability issues into the chemical industry requires a society-focussed methodology for the design of processes and products. In this paper, we present our design methodology that is able to support both students and experienced process engineers. We constructed this methodology from existing chemical process design practices, insights from various engineering disciplines, and the main issues from the societal debate on sustainability. The methodology is illustrated by a conceptual design for the conversion of biomass into methanol, which is done by students trained in the methodology. We found some striking differences in comparing this case to an existing industrial case.

1. Introduction

Modern industrial civilization is *unsustainable* in terms of its social system and its interaction with ecological systems. Intensive discussion is possible about the size and effects of issues like global warming, growing rich-poor gap, ozone depletion, etc. However, we state that the development of many kinds of societal activities, including technological development, can not continue as it has done in the last decades.

Sustainable Development strives to provide an alternative. The precise definition is dealt with in many (non-) governmental institutions or business organizations, and all formulate it roughly as the integration of social issues with environmental considerations into a lasting and profitable long-term development. Incorporation of sustainability issues into chemical engineering design asks for more than engineering tools only. It requires a structured methodology that concurrently integrates societal demands, ecological restrictions, and economic performance from the earliest design stages.

In the current chemical engineering literature, we did not find design methodologies that are able to cover the whole theme of sustainability. Many methods are proposed for *waste minimisation*, *waste reduction*, *pollution prevention*, *loss prevention*, *green engineering* etc. The process designers are helped largely by those approaches, but it is still necessary to have a framework that guides the design and locates the tools. Although it seems obvious that the chemical engineering practice should have such frameworks, there is a lack of knowledge in the conceptual field. This appears in (1) the discussion on the shared responsibility of management and design teams, (2) the indistinctness in trading off various criteria during the decision making.

104

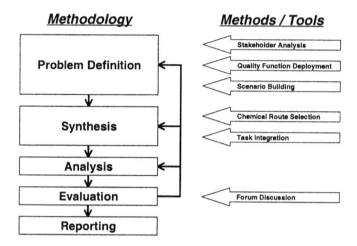

Figure 1   *General Design Procedure and the place of the tools*

Besides that, technological tools only can not solve the sustainability problem, because its main focus is a societal development. Since society-focussed tools are not available for chemical engineers, they have to be developed in such a way that process engineers remain open-minded towards societal issues during the design project and are stimulated in translating vague demands into hard engineering constraints.

In this paper, our scope is the conceptual process design phase, which is the earliest part of the development of a new chemical plant, resulting in the construction of a process flow diagram together with mass and heat balances. Since we want to present here the complete methodology, we can only briefly describe its framework, along with some practical and available tools. We trained students in the use of the methodology and their conceptual design of a biomass to methanol process is compared to that of a professional design team.

2. Design Procedure for Sustainable Chemical Processes

Just like any other design (Cross 1984), the conceptual process design of chemical plants consists in five stages, as shown in Figure 1. These steps are repeated frequently throughout this whole scheme. The complete design procedure consists of two parts:

1. The *Methodology*, (left side of Figure 1) which is the body of methods representing the common practice within an institute or company
2. The available *Methods* and *Tools* (right side of Figure 1), their use varies in every design project.

In the remaining part of this paragraph we introduce our methodology by defining every stage and the tools that can be used for the incorporation on sustainability issues. The methodology is based on intensive literature study, industrial and academic experience, and the integration of insights from various engineering disciplines.

2.1 Problem Definition

A process designer needs a clear description of qualitative and quantitative properties of the final result; only then various process options can be evaluated. This accents the crucial role of the problem definition phase in our design methodology, because here the final requirements have to be specified, together with the design and evaluation tools. The battery limits or system boundaries are defined at the start of the project in the Basis of Design. From the perspective of sustainability the process has to be closed or has to be part of a closed system. The designer is responsible for the effects and controllability of every flow entering or leaving the battery limits.

The Problem Definition stage requires an overview of domain knowledge, which determines the feasibility of a given design solution. From the perspective of sustainability, it is important to be knowledgeable of newly developing disciplines such as *Green Chemistry, Industrial Ecology, Pollution Prevention, Life Cycle Assessment,* etc. Additionally, sustainability requires an overview of societal needs at the early stage of the design.

We propose to use tools that are developed in other engineering disciplines, like *Stakeholder Analysis, Quality Function Deployment* (QFD), *Scenario Building* and *Forum Discussion.*

- Listening to stakeholders can avoid unpleasant surprises (e.g. bans, protests, changing trends etc.).
- The use of QFD 'assures that customer needs drive the design' (Sullivan 1986). The relation of QFD to sustainability is that it is a powerful tool in translating and weighing vague demands into concrete engineering constraints for the designer (Harmsen 1998, Revelle 1998).
- The tool of *Scenario Building* helps the designers in setting the framework of the design problem. Scenarios mainly are used in the evaluation phase by external parties (Schwartz 1996, Von Reibniz 1988), but also the process engineers should use them already in the process definition stage.

2.2 Process Synthesis and Analysis

Current chemical engineering methods only focus on the process synthesis stage, while they are oriented at optimizing economic profitability. Some of those methods can be used for sustainable chemical engineering under strong restrictions, others can be used for multi-objective optimization by defining sustainability as one of the objectives.

Here we plead for a tool that uses the concept of exergy from the early stage on. The exergy concept mainly is used as an evaluation tool for chemical processes (De Swaan Arons 2000), but it can also be helpful for the process designers for the selection of the chemical route and the process units. We developed a selection tool to do this based on insight from the exergy concept, which helps achieving efficient use of energy and mass. The tool consists in two steps:

1. An Exergetic Life Cycle Analysis, evaluates various chemical routes. This is done by an LCA-approach in which exergy calculations based on compound properties are done in all stages. (Cornelissen 1997, Smalberg 2001).
2. During the design, process synthesis hints indicate the choice of process units that cause a small amount of lost work. An industrial case shows that such a set of hints leads to better results (Smalberg 2000).

2.4 Evaluation and Reporting

In the evaluation and reporting phase, the final conceptual process design is presented in a process flow diagram, together with an energy and mass balance, an economic and Health, Environment, and Safety assessment. The incorporation of sustainability into the design requires a close contact with the public, so we recommend the evaluation of the design to a public forum. Such an evaluation can be done by various stakeholders from outside and inside the company.

3. Methanol Case

In this paper we use a simple conceptual design as an illustrative example. The process manufactures 250.000 ton/year of industrial grade methanol from a biological feedstock in The Netherlands. The design by a student group is made as assignment for a commercial engineering company. In the Basis of Design the following statements were explicitly mentioned:

1. Every cycle has to be closed: only CO_2 and H_2O enter and leave the process; all other material flows must be kept inside the system boundary.
2. Sustainability Criteria (including societal, ecological, and economic issues) are defined at the start of the project, and are used during the evaluation of the process.
3. Only existing knowledge is used to ensure immediate implementation.

We trained the students our methodology and they applied all steps to the Methanol Case Study. In summary the following design decisions are made:

- A direct conversion of biomass into methanol is not possible with the current knowledge (both biological and chemical), so it is decided to perform the process in two steps, with methane as intermediate.
- From the strict definition of closed cycles in the problem definition phase, only a biological process is acceptable, because in a biological process all elements present in the biomass can be put together into one stream (useful as fertilizer) without any additional step.
- Anaerobic digestion (AD) was chosen as conversion technology. AD can be used for all kinds of biomass (waste from agriculture, humans, and animals), only wood-like biomass can not be used, because of the high percentage of lignin.
- The process is carried out on a small scale near to the biomass winning, which avoids extensive transport. In the Netherlands, a very dense natural gas grid is present that can be used for the transport of the produced biogas
- The biogas is transported to a methanol plant, where it is converted into syngas and then goes into a methanol synthesis. For those steps, an existing process is used, which can be replaced and improved in a next case study, e.g. by using partial oxidation in stead of steam reforming.

The biogas (mixture of CH_4, CO_2, SO_x, and H_2S) must be purified in two gas separation steps (GS) to meet the specifications of the gas grid (PIPELINE). The CO_2 is removed from the methane by membrane separation. A biological treatment converts the sulfur compounds into elemental S_8 that is added to the fertilizer stream. Steam Reforming (SR) and subsequently Methanol Synthesis (MS) use the natural gas from the gas grid as feedstock. Half of the methane is needed for energy supply for both the digesters and the methanol plant (via a furnace, F).

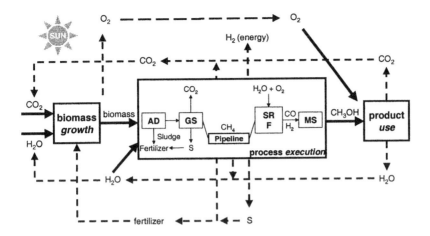

Figure 2   Methanol Case Study: *Block Diagram of Mass Flows*

The synthesis gas produced in the steam reformer ($CO + H_2$) contains more H_2 than necessary for the methanol reaction. Therefore part of the methane used for the energy supply can be saved, by using the H_2 and also by using sustainable energy conversion steps, like sunboilers or photovoltaic systems for the anaerobic digestion step. In Figure 2, the final design is presented.

All material cycles are closed including the feedstock exploitation and the use of the final product; in this way the process is driven by sunlight only on a short temporal scale. The presented process setup is economic profitable and price competitive, due to the low price of the feedstock, the low transport costs, and the avoidance of a CO_2-tax (assumed as 5$/ton C). All processing steps use well-known technology and meet HES-requirements. The design is presented to a forum consisting of NGOs, government, industry, and local public. It is possible to explain the process to a non-experienced public and clarify its sustainability.

If we compare this conceptual design to the results from a report 'Green Methanol from Groningen' (Interduct 2000), that describes a sustainable process design by a team of professional designers for the same kind of process. We would not criticize this design here, but only indicate some striking differences:

- The 'Green Methanol' Design does not mention the closing of material cycles. A furnace is used for the conversion of biomass into syngas. It is not clear how is ensured that every nutrient is recycled.
- The 'Green Methanol' Design transports all the biomass to one location
- The 'Green Methanol' Design is not confronted to a public forum and societal constraints play a marginal role.

These points make clear that the new framework with its tools leads to other concepts for existing processes, in which sustainability is incorporated to a higher degree. This is mainly reached by another kind of decision making and the use of some specific hints and heuristics.

4. Conclusions

The comprehensive methodology presented here integrates sustainable development into conceptual process design. Use is made of existing chemical engineering practice, insight from other engineering disciplines as well as industrial experience. Problem definition is crucial; it should be broad enough to deal concurrently with societal, economic, and ecological issues. Exergetical insights can aid selection of chemical route and task integration. The methodology developed is illustrated by applying it to a new process for the synthesis of methanol from biomass methodology. This results in a new concept regarding the material cycles and the choice of the process units. Comparison with an industrial case shows that the new concept meets the sustainability criteria to a higher degree. This is mainly reached by another kind of decision making and the use of some specific hints and heuristics.

References

Cornelissen, R.L (1997), *Thermodynamics and sustainable development; The use of exergy analysis and the reduction of irreversibility*. Thesis University of Twente

Cross, N. (1984), *Developments in Design Methodology*, John Wiley & Sons, Chichester

De Swaan Arons, J. (2000), Van der Kooi, H.J., *Thermodynamics: Key Discipline for the Design of a Sustainable Society*, Proc. International Symposium ECOS 2000, Enschede, p29-42

Harmsen G. J. (1998), H. van Hulst, *Reactor Selection Methodology*, CAPE symposium Process Synthesis 3, Art and Application, Amsterdam

Interduct/Delft University of Technology (2000), *Groene Methanol uit Groningen*, Novem Utrecht

Lemkowitz, S.M. (2001), G. Korevaar, G.J. Harmsen and H.J. Pasman, *Sustainability as the Ultimate Form of Loss Prevention*, Loss Prevention and Safety Promotion in the Process Industries - Proceedings of the 10th International Symposium, 19-21 June 2001, Elsevier Stockholm, page 33-52

Reibnitz, U. von (1988), *Scenario Techniques*, McGraw-Hill Hamburg

Revelle, J.B. (1998), J.W. Moran, C.A. Cox, *The QFD Handbook,* John Wiley & Sons New York

Schwartz, P. (1996), *The art of the long view; Paths to strategic insight for yourself and your company*, Doubleday New York

Smalberg, B.F. (2001), *Sustainable Process Innovation; A Thermodynamic Approach to Process Synthesis and Analysis*, Graduation Report, TU Delft

Sullivan, L.P. (1986), *Quality Function Deployment*, Quality Progress June, p39-p50

European Symposium on Computer Aided Process Engineering – 12
J. Grievink and J. van Schijndel (Editors)

Multi-Scale Modeling Strategy for Separation of Alkane Mixtures using Zeolites

R. Krishna

Department of Chemical Engineering, University of Amsterdam
Nieuwe Achtergracht 166, 1018 WV Amsterdam, The Netherlands

Abstract

Mixtures of alkanes can be separated by selective sorption and diffusion across zeolite membranes. We develop a multi-scale strategy for estimating the permeation fluxes of the components across the membrane relying almost entirely on simulation techniques. This strategy consists of the following steps:

(1) Calculating the sorption isotherms of the pure components and the mixtures using Configurational-Bias Monte Carlo (CBMC) techniques,

(2) Determination of the Maxwell-Stefan diffusivities from Molecular Dynamics or Transition State Theories,

(3) Using Kinetic Monte Carlo (KMC) simulation techniques for studying and verifying the mixture diffusion rules following the Maxwell-Stefan theory,

(4) Solving the transient equations of continuity of mass for each species with the Maxwell-Stefan equations describing intra-crystalline diffusion to obtain the transient permeation fluxes.

The applicability of our multi-scale modeling strategy is illustrated by means of a specific example for separation of methane (C1) and n-butane (nC4) by permeation across a MFI (silicalite-1) membrane.

Introduction

Zeolitic materials are used as sorbents and catalysts in a variety of processes within the chemical, petroleum, petrochemical and food industries. Zeolite crystals are incorporated into binders (such as amorphous aluminosilicate) and perhaps a diluent (typically a clay mineral), and used in the form of pellets (in fixed beds). Alternatively, zeolite crystals are coated on to a porous membrane support and used in (catalytic) membrane permeation devices. Zeolite based processes are carried out either under steady-state, unsteady-state or cyclic conditions. Fixed bed adsorbers are typically operated under transient conditions. Zeolite membrane processes typically operate under steady state conditions. Simulated moving bed adsorbers operate under cyclic conditions. The focus in this paper is on the development of a multi-scale modeling strategy for describing permeation fluxes of alkane mixtures across membranes; a similar strategy holds for estimation of breakthrough curves in fixed bed adsorbers.

Our modeling strategy is summarized in the sketch shown in Fig. 1. We shall demonstrate the utility of our approach by considering a practical example of separation of a mixture of methane (C1) and n-butane (nC4) by allowing the mixture to permeate across a MFI (silicalite-1) membrane. The major advantage of the strategy outlined in

Fig. 1 is that there is a major saving in process development times because far fewer experiments need to be performed. Also, to the author's knowledge such a multi-scale approach has never been put forward in the literature.

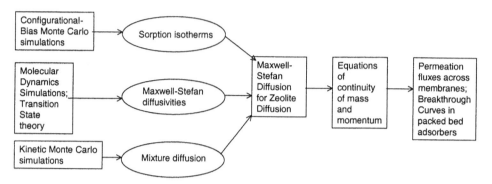

Fig 1. Multi-scale modeling strategy.

CBMC simulations of sorption isotherms

The alkanes are described with a united-atom model, in which CH3, CH2, and CH groups are considered as single interaction centres. When these pseudo-atoms belong to different molecules or to the same molecule but separated by more than three bonds, the interaction is given by a Lennard-Jones potential. The Lennard-Jones parameters are chosen to reproduce the vapour-liquid curve of the phase diagram. The bonded interactions include bond-bending and torsion potentials. We consider the zeolite lattice to be rigid and we assume that interactions of an alkane with the zeolite are dominated by the dispersive forces between alkane pseudo atoms and the oxygen atoms of the zeolite. These interactions are described by a Lennard Jones potential. Adsorption isotherms are conveniently computed using a Monte Carlo simulation in the grand-canonical ensemble. In this ensemble the temperature and chemical potentials are imposed. The average number of adsorbed molecules per unit cell of the zeolite follows from the simulations. The characteristics of these type of simulations is that during the calculations attempts are made to change the total number of particles by making attempts to insert molecule into or remove molecules from the zeolite. To make these types of moves possible for the long chain alkanes, we use the Configurational-Bias Monte Carlo (CBMC) technique. Instead of inserting a molecule at a random position, in a CBMC simulation a molecule is grown atom by atom in such a way that the "empty spots" in the zeolite are found. This growing scheme gives a bias that is removed exactly by adjusting the acceptance rules. The number of Monte Carlo cycles performed for pure component and binary mixture isotherms were, respectively, 2×10^7 and 5×10^7. Details of the CBMC simulation model are given in Vlugt et al. (1999).

The pure component sorption isotherms at 300 K for C1 and nC4 are shown in Fig. 2. A good description of the pure component isotherms can be obtained with the Dual-site Langmuir (DSL) model. In this model the loading, $\Theta_i^0(P)$, expressed in molecules per unit cell is expressed as a function of the pressure P as follows:

$$\Theta_i^0(P) = \frac{\Theta_{i,sat,A}b_{i,A}P}{1+b_{i,A}P} + \frac{\Theta_{i,sat,B}b_{i,B}P}{1+b_{i,B}P} \qquad (1)$$

The superscript 0 on $\Theta_i^0(P)$ is used to emphasize that the relation is for *pure* component loadings. In eq. (1) $b_{1,A}$ and $b_{1,B}$ represent the DSL model parameters expressed in Pa^{-1} and the subscripts A and B refer to two sorption sites within the MFI structure, with different sorption capacities and sorption strengths. The $\Theta_{i,sat,A}$ and $\Theta_{i,sat,B}$ represent the saturation capacities of sites A and B, respectively. The fitted parameters for the pure component isotherms are listed in Table 1. It is to be noted that the total saturation loading $\Theta_{i,sat} = \Theta_{i,sat,A} + \Theta_{i,sat,B}$ is not a fitted parameter but taken from the final plateau value of the sorption isotherm, estimated from CBMC simulations.

Table 1. Dual-site Langmuir parameters for pure alkanes in MFI at 300 K. The fits correspond to CBMC simulations.

Component	Dual Langmuir Parameters			
	Site A		Site B	
	$b_{i,A}$ /[Pa^{-1}]	$\Theta_{i,sat,A}$ /[molecules per unit cell]	$b_{i,B}$ /[Pa^{-1}]	$\Theta_{i,sat,B}$ /[molecules per unit cell]
C1	4.86×10^{-6}	11.0	2.38×10^{-7}	8.0
nC4	1.63×10^{-2}	9.0	1.14×10^{-5}	1.0

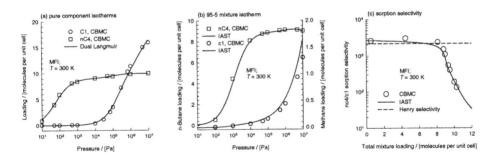

Fig 2. Pure component and 95-5 mixture isotherm for methane and n-butane in MFI at 300 K.

Consider a 95-5 mixture of C1 and nC4. The component loadings in the mixture obtained from CBMC simulations are shown in Fig. 2 (b). In Fig. 2 (c) we plot the sorption selectivity, S, defined by:

$$S_{1,2} = \frac{\Theta_1/\Theta_2}{p_1/p_2} \qquad (2)$$

where p_1 and p_2 are the partial pressures in the bulk gas phase. For mixture loadings, $\Theta_{mix} = \Theta_1 + \Theta_2$, below 8, the sorption selectivity of nC4 with respect to C1 is practically constant and equals that calculated from the corresponding Henry coefficients, i.e. 2000. However, as Θ_{mix} increases beyond 8, the sorption selectivity decreases dramatically to values just above 20. Near saturation loadings, the vacant spaces in the zeolite are more easily occupied by the smaller methane molecule. This is a size entropy effect that favours *smaller* sized molecules. It is clear that the size entropy effects counters the effect of chain length; increase in the chain length favours sorption of the *larger* sized molecule. Both component loadings and the sorption selectivity are very well predicted by the Ideal Adsorbed Solution Theory (IAST), developed by Myers and Prausnitz (1965).

Pure Component Diffusivities

The next step is to estimate the diffusivities of C1 and nC4 within the MFI matrix. The Fick diffusivities are strong functions of the molecular loading but the Maxwell-Stefan (MS), or corrected, diffusivities $Đ_i$ are practically independent of the loading (Krishna and Wesselingh, 1997). The M-S diffusivities, $Đ_i$, can be estimated with the aid of Molecular Dynamics simulations (Maginn et al., 1993) or by use of transition state theory (Vlugt et al., 2000). As shown by Vlugt et al. (2000) for diffusion of isobutane in MFI, the calculations of the $Đ_i$ are very sensitive to the choice of the Lennard-Jones potential parameters. For the calculations to be presented later in this paper we use the MS diffusivities for C1 and to be $Đ_1 = 10^{-9}$ m^2/s and for nC4 we take $Đ_2 = 10^{-11}$ m^2/s following the experimental data presented in Bakker (1999).

Mixture diffusion model; verification using KMC simulations

For *n*-component diffusion the fluxes N_i are related to the gradients of the fractional occupancies by the generalization of Fick's law:

$$(N) = -\rho[\Theta_{sat}][D]\frac{\partial(\theta)}{\partial r} \tag{3}$$

where $[D]$ is the *n*-dimensional square matrix of Fick diffusivities; $[\Theta_{sat}]$ is a diagonal matrix with elements $\Theta_{i,sat}$, representing the saturation loading of species *i*. The estimation of the $n \times n$ elements of $[D]$ is complicated by the fact that these are influenced not only by the species mobilities (i.e. diffusivities $Đ_i$) but also by the sorption thermodynamics. For estimating $[D]$ it is more convenient to adopt the Maxwell-Stefan formulation, in which the chemical potential gradients are written as linear functions of the fluxes (Krishna and Wesselingh, 1997; Kapteijn et al., 2000):

$$-\rho\frac{\theta_i}{RT}\nabla\mu_i = \sum_{\substack{j=1 \\ j\neq i}}^{n}\frac{\Theta_j N_i - \Theta_i N_j}{\Theta_{i,sat}\Theta_{j,sat}Đ_{ij}} + \frac{N_i}{\Theta_{i,sat}Đ_i}; \quad i = 1,2,\ldots n \tag{4}$$

We have to reckon in general with two types of Maxwell-Stefan diffusivities: $Đ_i$ and $Đ_{ij}$. The $Đ_i$ are the jump diffusivities that reflect interactions between species *i* and the zeolite matrix; they are also referred to as jump or "corrected" diffusivities in the literature and can be identified with the pure component transport parameters. Mixture diffusion introduces an additional complication due to sorbate-sorbate interactions. This interaction is embodied in the coefficients $Đ_{ij}$. We can consider this coefficient as representing the facility for counter-exchange, *i.e.* at a sorption site the sorbed species *j* is replaced by the species *i*. The net effect of this counter-exchange is a slowing down of a faster moving species due to interactions with a species of lower mobility. Also, a species of lower mobility is accelerated by interactions with another species of higher mobility. As shown by Paschek and Krishna (2001), $Đ_{ij}$ encapsulates the correlation effects associated with molecular jumps. The interchange coefficient $Đ_{ij}$ can be estimated by the logarithmic interpolation formula that has been suggested by Krishna and Wesselingh (1997):

$$Đ_{ij} = [Đ_i]^{\theta_i/(\theta_i+\theta_j)}[Đ_j]^{\theta_j/(\theta_i+\theta_j)}$$

(5)

It is convenient to define a n-dimensional square matrix $[B]$ with elements

$$B_{ii} = \frac{1}{Đ_i} + \sum_{\substack{j=1 \\ j \neq i}}^{n} \frac{\theta_j}{Đ_{ij}}; \quad B_{ij} = -\frac{\theta_i}{Đ_{ij}}; \quad i,j = 1,2....n$$

(6)

With this definition of $[B]$, eq. (4) can be cast into n-dimensional matrix form:

$$(N) = -\rho[\Theta_{sat}][B]^{-1}[\Gamma]\frac{\partial(\theta)}{\partial r}$$

(7)

where the matrix of thermodynamic factors $[\Gamma]$ is given by

$$\frac{\theta_i}{RT}\frac{\partial\mu_i}{\partial r} = \sum_{j=1}^{n}\Gamma_{ij}\frac{\partial\theta_j}{\partial r}; \quad \Gamma_{ij} \equiv \left(\frac{\Theta_{j,sat}}{\Theta_{i,sat}}\right)\frac{\Theta_i}{p_i}\frac{\partial p_i}{\partial\Theta_j}; \quad i,j = 1,2...n$$

(8)

The MS formulation for mixture diffusion, an in particular the validity of eq. (5), can be verified by use of Kinetic Monte Carlo simulations; this has been done by Paschek and Krishna (2001) for two-component diffusion in MFI.

Calculation of Transient and Steady-State permeation rates

The permeation flux is obtained by solving the following set of two PDEs:

$$\frac{\partial\theta_i}{\partial t} = -\frac{1}{\rho\Theta_{i,sat}}\frac{\partial N_i}{\partial z}$$

(9)

where eq. (3) describes the individual fluxes. For single component permeation of C1 and nC4 with an upstream partial pressure of 50 kPa, maintaining the downstream partial pressures at vanishing values, the transient fluxes are shown in Fig. 3 (a). At steady state, the permeation flux of methane is 19.46 mmol/m^2/s and that of nC4 is 4.65 mmol/m^2/s because of its much lower diffusivity value. The permeation selectivity, S_P, defined by

$$S_P = \frac{N_2/N_1}{p_{20}/p_{10}}$$

(10)

is 4.65/19.46 = 0.239. For binary mixture permeation, each component with 50 kPa upstream partial pressures, the transient fluxes are shown in Fig. 3 (b). The steady-state permeation selectivity is found to be 197, emphasizing that mixture permeation cannot be simply related to single-component permeation. Diffusion and sorption are closely interlinked. The more strongly adsorbed nC4 has a higher occupancy within the zeolite. The higher the occupancy the higher the permeation rate. Futhermore, due to the interchange, expressed in eq. (5), the more mobile C1 is slowed down by nC4. These two effects combine to yield a permeation selectivity of 197 in place of 0.239. In Fig. 3 (c) the calculations of the permeation selectivity are compared with the experimental results of Bakker (1999) obtained for a variety of upstream pressures; the agreement is good considering that virtually no empirical inputs were required in the model. The decrease in the selectivity with increasing butane partial pressure is a consequence of size entropy effects. Such size entropy effects are properly accounted for by the IAST which has been used to describe the mixture thermodynamics.

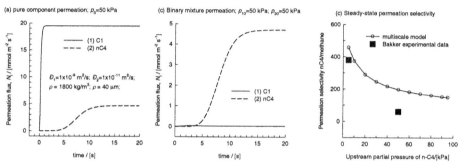

Fig 3. *Transient permeation fluxes of methane and n-butane across MFI membrane at 300 K. (c) shows the steady-state permeation selectivity.*

Conclusions

For calculation of the permeation fluxes of a binary mixture across a zeolite membrane we have developed a multi-scale modeling strategy. Molecular simulations are used to generate the sorption and diffusion parameters, the Maxwell-Stefan diffusion equations binds these together and the solution of the transient equations of continuity yields the permeation fluxes. For permeation of a mixture of methane and n-butane across an MFI membrane, good agreement is obtained when comparing with experimental data. More interestingly, the results show strong mixture diffusion and entropy effects which are properly encapsulated by the Maxwell-Stefan and IAST models. A similar strategy could be adopted for simulating the breakthrough in a packed bed adsorber.

The strategy outlined in the present paper will help to reduce process development times by reducing significantly the experimental effort required in process development.

References

Bakker, W.J.W., 1999, Structured systems in gas separation, Ph.D. dissertation, Delft University of Technology, Delft.

Kapteijn, F., Moulijn, J.A. and Krishna, R., 2000, Chem. Eng. Sci., 55, 2923 – 2930

Krishna, R. and Wesselingh, J.A., 1997, Chem. Eng. Sci., 52, 861- 911.

Paschek, D. and R. Krishna, 2001, Phys. Chem. Chem. Phys., 3, 3185 – 3191

Maginn, E.J., Bell, A.T., and Theodorou, D.N., 1993, J.Phys.Chem., 97, 4173 – 4181.

Myers, A.L. and Prausnitz, J.M., 1965, A.I.Ch.E.J., 11, 121 - 130.

Vlugt, T.J.H., Krishna, R. and Smit, B., 1999, J.Phys.Chem. B, 103, 1102 - 1118.

Vlugt, T.J.H., Dellago, C. and Smit, B., 2000, J.Chem.Phys., 113, 8791 – 8799.

European Symposium on Computer Aided Process Engineering – 12
J. Grievink and J. van Schijndel (Editors)
© 2002 Published by Elsevier Science B.V.

Simultaneous Synthesis and Design of Novel Chemicals and Chemical Process Flowsheets

Patrick Linke and Antonis Kokossis

Centre for Process and Information Systems Engineering, School of Engineering,
University of Surrey, Guildford, Surrey, GU2 7XH, U.K.

Abstract

This work introduces a framework for the simultaneous molecular design and process synthesis. The methodology integrates previous efforts in molecular design with a systematic reaction-separation synthesis scheme. The framework allows promising process design options and solvent candidates to be simultaneously screened with the use of stochastic search techniques. The simultaneous exploitation of the global systems trade-offs can offer major advantages over the conventional sequential approach of addressing the molecular design and the process design problems in isolation. A case study is presented to illustrate the use of the synthesis framework.

1. Introduction

Computer-aided molecular design (CAMD) techniques aim at screening of novel chemicals by employing a reverse engineering approach of systematically combining structural groups to yield high-performance molecules. These approaches exploit advances in the development of methods for the prediction of molecular properties as a combination of group contributions (see e.g. Fredenslund et al., 1975). A review of recent CAMD literature is given in Marcoulaki and Kokossis (2000a).

Solvents are generally designed with the aim of utilising them for a particular (reactive) separation task. These (reactive) separation systems are expected to constitute parts of an optimal overall process design. The molecules are generally designed for target ranges of thermodynamic properties, which are anticipated to have a significant effect on the process performance (Hostrup et al., 1999). This requires a range of desired properties to be specified without process design feedback so that their appropriate levels are easily misrepresented.

The isolation of synthesis stages inevitably leads to at best a set of optimal solvents with respect to a weighted objective of optimal thermodynamic properties, and a set of optimal processes for the solvents obtained. Clearly, the success of such an approach is highly sensitive to the formulation of the objective function and thermodynamic property constraints employed in CAMD. Any important thermodynamic properties that are missed as well as any misjudged property significance will inevitably lead to process designs with sub-optimal performances. Despite the potential for identification of improved designs, very few approaches have been reported that account for the consideration of molecular design and process performance interactions. Pistikopoulos and Stefanis (1998) simulate absorption processes in the course of solvent design in order to evaluate environmental impact and economic constraints. Marcoulaki and

Kokossis (2000a,b) design molecules that are simultaneously employed in short-cut absorber simulations in order to determine the best solvent for cost optimal absorber operation.

In contrast to previous work on the design of solvents and processes in isolation, this work presents an integrated process synthesis/solvent design approach. This is to be seen as a first step towards technology to simultaneously determine a set of optimal reaction-separation processes along with a corresponding set of optimal solvents.

2. Integrated process synthesis and molecular design supermodel

It is the purpose of this work to develop a representation and optimisation framework that exploits all possible molecular and process design options for solvent-processing reaction-separation systems in a unified framework. This is facilitated in a process-molecule synthesis supermodel (PMSS) generation, which combines previous developments in reaction-separation process synthesis (Linke, 2001; Linke et al., 2000) and in CAMD (Marcoulaki and Kokossis, 2000a,b). Process design options are represented by superstructures of generic units to capture all possible novel and conventional process design options. Mass separating agents are represented by molecular vectors of functional groups that capture all feasible molecular design options. Apart from the degrees of freedom associated with the process design options, the type and number of functional groups of the GC method that comprise the solvent molecules are treated as optimisation variables.

The molecular vector representation of the solvent molecules by functional groups proposed by Marcoulaki and Kokossis (2000a) is adopted in this work. The functional groups accounted for in the representation are based on UNIFAC (Fredenslund et al., 1975). For each solvent, a molecular vector is defined as the product of the functional group vector comprised of the functional groups of the GC method and the composition matrix, which represents the number of occurances of a particular group in the solvent molecule. The representation is structured through the definition of functional group subsets, termed metagroups and classified according to common properties such as the bonding potential of aromatics and nonaromatics. Structural feasibility constraints on the molecular vector include connectivity constraints to ensure zero valency. Additional constraints can limit the size of the solvent molecule through upper bounds on the number of groups, aromatic rings and aromatic bridges. This is mainly to avoid highly complex molecules that are impossible to synthesise. Supplementary complexity controls allow to further limit the occurances of only those structural groups that belong to certain classes. The CAMD methodology allows the introduction of constraints on the existence of groups so that health and safety considerations can be taken into account via the elimination of undesired groups from the search space

The process synthesis representation employed as part of the overall supermodel takes the form of the general reaction-separation superstructures developed previously (Linke, 2001; Linke et al., 2000). The superstructure generations incorporate process models to represent reaction, reactive separation and mass exchange operations in generic reaction/mass exchange units. Separation task units facilitate the conceptual representation of the tasks performed by the different separation operations that are to be included in the search. According to their purpose within the processes, mass separating agents are included into the superstructure model either as new solvent phases or as additional components into the existing liquid phases where they are anticipated to influence the phase equilibria such that the driving forces for separations employing the existing phases are increased. The solvent components are represented by molecular vectors as described above. Throughout this work, we assume that solvent molecules do not take part in any reactions, that pure solvent feeds are available in the solvent phases and that pure solvents can be recovered from loaded solvent streams. The supermodel size is user-defined through the specification of the maximum number of process synthesis units and functional groups that are to be considered.

The activity coefficients assume particular importance in solvent-based systems. For their prediction, the of the original UNIFAC (Fredenslund *et al.*, 1977) is used. The UNIFAC method has been developed with a focus on predicting vapour-liquid equilibria (VLE). Parameters for liquid-liquid (LLE) equilibria have been fitted by Magnussen (1981); however, the VLE tables are much more complete and extended compared to the LLE tables. Although the accuracy of predictions deteriorated, Pretel et al. (1994) observed that the performance rank of solvent molecules essentially remains the same when LLE are predicted using VLE tables. To demonstrate the ability of the novel synthesis tool to identify designs of good performance from the enormous search space that exists for the simultaneous process-molecule synthesis problem, the LLE are predicted using the more complete VLE tables here.

For CAMD, performance criteria and constraints are derived that are supposed to measure and guarantee the fitness of a solvent for a given task. Examples of such formulations include amongst others relative volatilities, solvent power, solvent selectivity, solute distribution coefficient, solvent losses in raffinate, and solute solubility in solvent. The above expressions give a good indication of the performance of a solvent with respect to certain aspects. Intelligent use of this knowledge is made in order to streamline the optimal search of the supermodel. Along these lines, toxicity indices (Wang & Achenie, 2001) can also be incorporated to prevent the generation of molecules that might have adverse effects on biocatalytic systems. It should be mentioned here that the synthesis methodology inherits the shortcomings of the group-contribution method used for the property predictions.

3. Stochastic optimisation scheme

In view of the model complexities and bearing in mind the computational experience gained in reactor and reaction-separation network synthesis and CAMD, the PMSS are searched using stochastic optimisation techniques in the form of Simulated Annealing in order to enable the determination of performance limits and optimal design options. The implementation of the SA algorithm comprises a state transition framework to generate states from the supermodel, a state evaluation framework to determine the quality of the states, the SA search heuristic to decide upon acceptance of states, the cooling schedule as well as termination criteria, and problem specific rules for efficient searches of the solution space. Convergence properties of the Simulated Annealing algorithm are discussed in Marcoulaki & Kokossis (1999).

4. Illustrative example

The production of ethanol by fermentation is studied as an example process, which potentially benefits from reactive separation (Fournier, 1986). The major kinetic effects are captured by an extension of the simple Monod equation and are given in Fournier (1986). The model represents the glucose (Glu) to ethanol (EtOH) reaction according to the biological scheme

$$Glu \xrightarrow{+cells} EtOH + cells$$

In view of the EtOH inhibition, Fournier (1986) advocates the use of an extractive fermentor with dodecanol as the solvent. The reactive extractor was found to have its volume requirements increased towards higher productivities as the solvent occupies part of the reaction volume. These trade-offs justify a systematic study in order to systematically address the solvent selection and to determine the optimal process configuration of fermentors, extractive fermentors and extractors. Equilibrium calculations are performed using the original UNIFAC and a feed flow rate of 400kg/hr of Glucose in 1000kg/hr of water is assumed for the study. The desired process completely converts the glucose in the aqueous phase and extracts all ethanol whilst utilising a minimum of solvent flow that dissolves as little as glucose as possible. These trends are incorporated into the dimensionless objective function according to

$$J = \frac{N_{EtOH,sol}^2}{N_{Glu,aq} N_{sol}} \tag{1}$$

where N denote molar flowrates and indices aq and sol refer to the solvent and aqueous phases, respectively. A number of cases are studied with solvent designs and process structures being considered fixed or degrees of freedom for optimisation to illustrate the benefits that can be gained from simultaneous optimisation of the entire supermodel as compared to the optimisation of only the process or the molecular sub-systems.

The classical configuration of a fermentor followed by a countercurrent liquid-liquid extractor using dodecanol as the solvent is taken as the reference design. Stochastic experiments consistently converge to a performance target of 3.66. An optimal solvent flow of around 1680kg/hr is observed in all solutions. This sequential process configuration achieves glucose conversions of 60.5%. In comparison, the performance limit of the well-mixed extractive fermentor using the same solvent is established to be 6.46, an improvement by around 75% as compared to the optimal sequential fermentor-extractor design. The performance improvements result from lower solvent flows (1396kg/hr) and higher glucose conversions (X_{Glu} = 84.2%).

The potential gains from solvent substitution for the sequential process configuration is quantified with a performance of 9.49 through the simultaneous optimisation of solvent molecules and the fixed classical process structure. In comparison to the reference case, the performance has been increased by around 160% by the new solvents. The improvements are solely down to the significant reductions in the regenerated solvent intake rates (around 760kg/hr). All solvent designs obtained from the stochastic experiments feature two aromatic rings. Hydroxyl groups are only occasionally observed although all molecules feature at least five polar elements. This contrasts to optimal solvents for extraction of ethanol from water reported in Marcoulaki and Kokossis, 2000b, who identified large aromatic alcohols as the best options.

Reaction-separation superstructures are optimised to investigate the potential performance gains through non-conventional process design options for the existing dodecanol solvent. The performance target for this system is established at around 960, i.e. a performance gain of 26,000% over the reference state is realisable through structural process changes whilst maintaining the solvent. Near complete conversions of glucose (X_{Glu} > 99%) are achieved by all stochastically optimal designs. Optimal designs range from simple structures with a plug-flow fermentor followed by an extractive fermentor as well as complex designs of three extractive fermentor and mass exchanger units. An optimal solvent flow of around 2000kg/hr is observed in all solutions.

The significant performance enhancement observed for the new process configurations utilising the dodecanol solvent can be increased by a further 325% (performance limit: 3135) through the use of new process configurations featuring simultaneously optimised solvent molecules. The designs enable a near complete conversion of glucose (X_{Glu} > 99%). The optimal process design options feature combinations of fermentor, reactive fermentor and extractors with recycles in the solvent as well as aqueous phases. All solvents observed feature one or two aromatic rings and generally feature polar groups other than the hydroxyl groups. As compared to the optimal designs utilising dodecanol, the solvent flows have been decreased by 60% on average. Figure 1 shows a typical design with a performance close to the limit. The composition of the corresponding solvent molecule is shown in Table 1. The solvent features a single aromatic ring and seven polar functional groups none of which is a hydroxyl group.

It should be noted that the illustrative example represents a large-scale *test problem* to evaluate the performance of the methodology. The molecules obtained may not be practical nor will the UNIFAC predictions be accurate. This is especially true for the solvents featuring two aromatic rings. However, studies investigating smaller, more practical molecule sizes will simplify the optimisation problem and can easily be studied using the proposed framework.

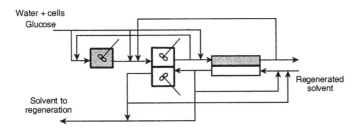

Figure 1.   Optimal fermentation-extraction process. Performance: 3136, solvent intake: 753kg/hr (optimised solvent, see Table 2)

Table 1. *Design information for process-solvent-supermodel optimisation*

Group	>CH2	CH3-O-	-CH2-O-	>CH-O-
Occurances	2	1	4	1
Group	FCH2-O-	ACH	AC- >C=CH2	
Occurances	1	5	1	

5. Conclusions

A new technology for the simultaneous design of reaction-separation processes and solvents is illustrated. The technology combines a superstructure representation of generic synthesis units to capture the process design options with a group-wise representation of the molecular solvent system into a process-molecule synthesis supermodel. Stochastic search techniques are employed to optimise the supermodel. The available solvent-process interaction models employed in CAMD are not utilised in the objective functions, thus by-passing the weighing problem inherent in sequential synthesis technology. Instead, these models along with other process constraints that can be anticipated from thermodynamic properties alone are employed to identify those molecular structures with apparently bad performance characteristics prior to any computationally demanding process simulations.

References

Wang, Y.P., & L.E.K. Achenie (2001). *Computer Aided Chemical Engineering, 9*, 585.

Buxton, A., A.G. Livingston, & E.N. Pistikopoulos (1999). *AIChE J.*, 45, 817.

Fredenslund, A., R.L. Jones, & J.M. Prausnitz (1975). *AIChE J.*, 21, 1086.

Fredenslund, A., J. Gmehling, & P. Rasmussen (1977). *Vapor liquid equilibria using UNIFAC.* Elsevier Scientific, Amsterdam.

Fournier, R.L. (1986). *Biotech. & Bioeng.*, 28, 1206.

Gmehling J., P. Rasmussen, & A. Fredenslund (1982). *Ind. Eng Chem. Proc. Des. Dev.*, 21, 118.

Hansen, H. K., P. Rasmussen, A. Fredenslund, M. Schiller, & J. Gmehling (1991). *Ind. Eng. Chem. Res.*, 30, 2352.

Hostrup M., P.M. Harper & R. Gani (1999). *Comp. Chem. Eng.*, 23, 1395.

Linke, P. (2001). PhD thesis. UMIST, UK.

Linke, P., V. Mehta & A. Kokossis (2000). *Computer Aided Chem. Engng, 8*, 1165.

Macedo, E. A., U. Weidlich, J. Gmehling, & P. Rasmussen (1983). *Ind. Eng. Chem. Proc Des. Dev.*, 22, 678.

Magnussen T., P. Rasmussen, & A. Fredenslund (1981). *Ind. Eng. Chem. Proc. Des. Dev.*, 20, 331.

Marcoulaki, E.C., & Kokossis, A.C. (2000a). *Chem. Eng. Sci.*, 55, 2529.

Marcoulaki, E.C., & Kokossis, A.C. (2000b). *Chem. Eng. Sci.*, 55, 2547.

Marcoulaki, E.C., & Kokossis, A.C. (1999). *AIChE J.*, 45, 1977.

Pistikopoulos, E.N., & S.K. Stefanis (1998). *Comp. Chem. Eng.*, 22, 717.

Pretel, E.J., P.A. Lopez, S.B. Bottini, & E.A. Brignole (1994). *AIChE J.* 40, 1349.

European Symposium on Computer Aided Process Engineering – 12
J. Grievink and J. van Schijndel (Editors)
© 2002 Elsevier Science B.V. All rights reserved.

Intensification of a Solvent Recovery Technology through the Use of Hybrid Equipment

Peter Mizsey, Agnes Szanyi, Andreas Raab, Jozsef Manczinger, Zsolt Fonyo
Chemical Engineering Department, Budapest University of Technology and Economics, Budapest, H-1521 Hungary

ABSTRACT

In printing companies often different four-component-mixtures (mixture 1: ethanol, ethyl acetate, isopropyl acetate, and water, mixture 2: ethanol, ethyl acetate, methyl-ethyl-ketone, and water) arise as waste. The recovery of the individual components is complicated by the highly nonideal feature of the mixtures, namely several binary and ternary azeotropes are formed by the components.

Based on the synthesis procedure proposed by Rev et al. (1994) and Mizsey et al. (1997), new separation processes using hybrid separation techniques are developed followed up the vapour-liquid-liquid equilibrium behavior of the mixtures. The first process (ternary-cut-system) splits mixture 1 into two ternary mixtures which are separated later in subsequent units into components of the prescribed purity (~95%). This technology needs, however, seven distillation columns and two extractors. The second separation process is based on two coupled columns and a three-phase-flash (two-column-system) which can cope with the separation of both mixtures into binary mixtures and the binaries can be easily separated further with conventional methods. Both processes are using extra water addition for the necessary separation. These processes are experimentally also verified (Mizsey et al., 1997, Raab, 2001).

A third integrated separation process is developed with evolutionary steps merging the extractor and distillation units, designed for the ternary-cut-system, into integrated extractive distillation process using water as extractive agent (integrated-system). The integrated-system can separate both mixtures into components of the prescribed purity and consists of only four distillation columns. Beside the simpler separation structure the energy consumption is also investigated and compared with the other two processes. The use of this innovative solvent recovery process makes the reduction of the number of processing units possible and operating costs can be significantly reduced compared to the other non-integrated separation technologies.

Introduction

The solvent recovery is an important task for chemical engineers to minimize burden upon the environment due to exhaustive use of solvents and the emission associated with the incineration of the used solvents. In printing companies the following typical quaternary waste mixtures of different solvents are produced, mixture 1: ethanol 30w%, ethyl acetate 25w%, isopropyl acetate 20w%, water 20w%, and mixture 2: ethanol 23w%, ethyl acetate 32w%, methyl-ethyl-ketone 28w%, and water 13w%. There are some accompanying components in less than 5w%, which are neglected in this feasibility study. Membrane separation technologies can be also included (e.g. Szitkai et al, 2001) but considering the complexity of these separation problems (four components and wide concentration ranges) in this work special attention is given to hybrid separation technologies combining the advantages of special distillation techniques, phase separation and extraction.

Rev et al. (1994) and Mizsey et al. (1997) have recommended a framework for designing feasible schemes of multicomponent azeotropic distillation. This procedure recommends to study in detail the vapour-liquid-liquid equilibrium data to explore immiscibility regions, azeotropic points of binary and ternary ones, and separatrices for ternary and quaternary regions. On the behalf of the VLLE data the set of feasible separation structures can be explored. This procedure is followed and new separation technologies are developed.

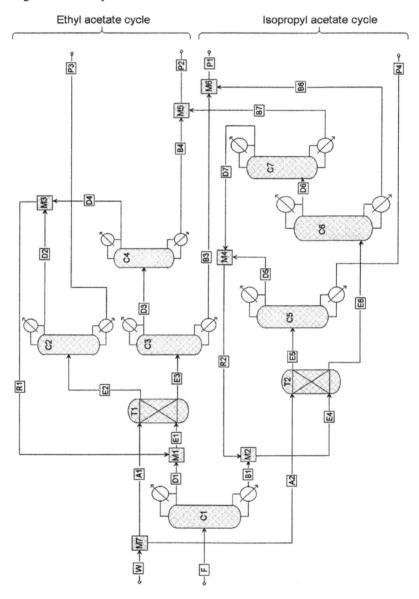

Figure 1. The separation scheme of the ternary-cut-system
(W, P2 = water, F = feed, P1 = ethanol, P3 = ethyl acetate, P4 = isopropyl acetate)

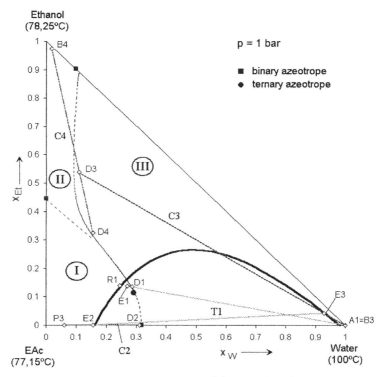

Figure 2. The VLLE data and representation of the ETAC cycle (ternary-cut-system)

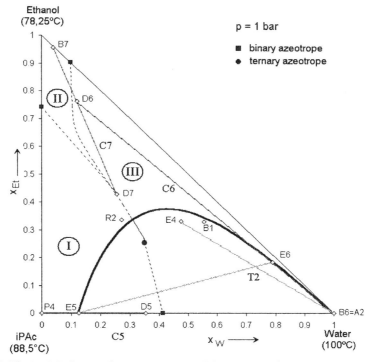

Figure 3. The VLLE data and representation of the IPAC cycle (ternary-cut-system)

Research results

Three separation technologies are developed and tested in details. The ternary-cut-system is designed specifically for the recovery of the solvents of mixture 1. Figure 1 shows the flowsheet of the technology. According to the exhaustive study of the VLLE data it is possible to split the quaternary mixture in one distillation column (C1) into two ternary ones. Afterwards the two ternary mixtures, due to their similar VLLE data, can be separated in two similar ways (ethyl acetate cycle, isopropyl acetate cycle) with the use of extractor and three distillation columns into components of the prescribed purity. Internal recycling is also needed. The VLLE data and the unit operations, extractors and columns, of these two cycles are shown in Figures 2-3.

Due to the complexity of the ternary-cut-system attempts are made to design simpler separation processes. First, an earlier separation scheme, the two-column-system, (see details in Mizsey, 1991, Rev et al., 1994, Mizsey et al., 1997), which has been designed to separate mixture 2 into two binary mixtures is considered and tested also for mixture 1 (Figure 4). The system containing two coupled distillation columns and a three phase flash proves to be successful also for mixture 1 and splits this quaternary mixture into two binary ones (ethanol-water, ethyl acetate-isopropyl acetate).

Since water is used in the previous separation technologies to exploit the advantage of the immiscibility regions allowing the step over separation boundaries, a generalization of synthesizing hybrid separation technologies for azeotropic mixtures proves to be possible. An integrated separation technology (integrated-system) is designed based on an extractive distillation unit with water as extractive agent (Figure 5). This integrates the distillation and the water addition or extraction of the previous technologies and the use of this multifunctional unit allows significant simplification. The VLLE data indicate that with subsequent distillation it is possible to obtain the individual components with the prescribed purity. Figure 6 shows the VLLE data of columns 2 and 3 and also their splitting lines.

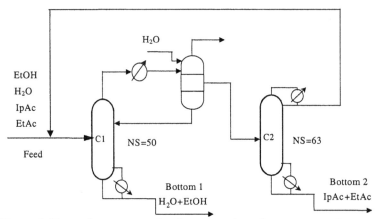

Figure 4. Two-column-system for the separation of quaternary mixtures

Experimental investigation

The core of the ternary-cut-system (column 1) and the two-columns system are verified experimentally.

First, the column 1 (C1) of the ternary-cut-system (Fig. 1), in which the split of mixture 1 takes place, is tested in a laboratory size column containing structured packing.

Steady-state continuous distillation is carried out for 6.5 hours controlling the column according to the typical top temperature. The distillate and bottom products are analyzed and compared with the simulation data (Table 1). The comparison of the data shows a good agreement (Raab, 2001) which proves the accuracy of the VLLE method and the simulated results.

Table 1. Comparison of measured and simulated data of C1 of ternary-cut-system

	Feed	Simulated data %		Measured data %	
	%	Distillate	Bottom pr.	Distillate	Bottom pr.
Water	21	8.3	26.9	8.9	29.7
ETOH	32	10.0	40.6	8.6	39.9
ETAC	26	81.3	0.6	82.1	0.2
IPAC	21	0.4	31.9	0.4	30.2

Secondly, the two-column-system is verified experimentally for mixture 2 in laboratory equipment. The results are compared with the simulated results. The experimental results are in agreement with the calculated ones.

Table 2. Comparison of measured and simulated data of two-column-system

	Feed	Simulated data %		Measured data m%	
	%	Bottom 1	Bottom 2	Bottom 1	Bottom 2
Water	36	84.91	0	83.9	0
ETOH	26	14.76	0.05	13.1	0.5
ETAC	19	0.03	49.13	1.1	48.9
MEK	19	0.29	50.81	1.9	50.6

COMPARISON OF ENERGY CONSUMPTION

The three separation technologies are investigated from the points of feasibility and energy consumption. The simplest system is the integrated-system and the total minimal energy consumption (heating and cooling) of the three systems are also determined. The new integrated solvent recovery technology using hybrid equipment proves to be not only the simplest but also the most attractive for the energy consumption (Table 3). According to estimations the savings of the solvent recovery depends on the actual prices and range between 70-95%. The energy consumption could be further minimized (Szitkai et al., 2001) if membrane separation technology is combined with distillation for the ethanol-water separation (Fig. 2, Column 4) and this option is considered in our research group as possible further process improvement.

Table 3. Comparison of separation technologies

	Integrated-system [MW]	Two-column-system [MW]	Ternary-cut-system [MW]
Mixture 1	12.04	57.86	23.26
Mixture 2	14.62	35.44	—

Discussion and conclusions

The successful application of the integrated-system obtained with evolutionary steps allows significant simplification and improves economic features of the separation of non-ideal mixtures typical for solvent recovery. With the use of this novel hybrid system a general separation technology seems to be outlined for the separation of non-

126

ideal mixtures containing heterogeneous azeotropes. Experiments also support the accuracy of the solvent recovery technologies which can realise significant saving and minimisation the burden upon the environment.

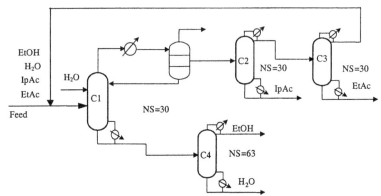

Figure 5. Integrated-system based on extractive distillation

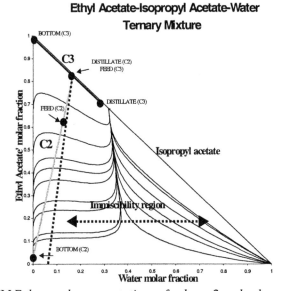

Figure 6. The VLLE data and representations of column 2 and column 3

References

Mizsey P., A global approach to the synthesis of entire chemical processes, PhD. Thesis No. 9563, Swiss Federal Institute of Technology, ETH Zürich (1991).

Mizsey, P., E. Rev and Z. Fonyo, Systematic separation system synthesis of a highly nonideal quaternary mixture, *AIChE Spring National Mtg.*, Paper 19f, Chicago (1997).

Raab, A., Separation of highly nonideal mixture for solvent recovery, Diploma work at Budapest Univ. of Technology and Economics (2001).

Rev, E. P. Mizsey and Z. Fonyo, Framework for designing feasible schemes of multicomponent azeotropic distillation, *Comp. Chem. Engng.*, 18, S43-S47 (1994).

Szitkai, Z., Z. Lelkes, E. Rev, Z. Fonyo, Optimization of an industrial scale ethanol dehydration plant: Case study, ESCAPE-11, p. 553 Elsevier (2001).

European Symposium on Computer Aided Process Engineering – 12
J. Grievink and J. van Schijndel (Editors)
© 2002 Elsevier Science B.V. All rights reserved.

A Global Optimization Technique for Solid-Liquid Equilibrium Calculations: Application to Calcium Phosphate Precipitation

L. Montastruc, C. Azzaro-Pantel, B. Biscans, A. Davin, L. Pibouleau, S. Domenech
Laboratoire de Génie Chimique- UMR CNRS/INP/UPS 5503
ENSIACET, 18 Ch de la Loge 31078 TOULOUSE Cedex 4, France
email : Ludovic.Montastruc@ensiacet.fr

Abstract

Phase equilibrium calculations constitute an important problem for designing and optimizing crystallization processes. The Gibbs free energy is generally used as an objective function to find the quantities and the composition of phases at equilibrium. In such problems, the Gibbs free energy is not only complex but can have several local minima. This paper presents a contribution to handle this kind of problems by implementation of a hybrid optimization technique based on the successive use of a Genetic Algorithm (GA) and of a classical Sequential Quadratic Programming (SQP) method: the GA is used to perform a first search in the solution space and to locate the neighborhood of the solution. Then, the SQP method is employed to refine the solution. The basic operations involved in the design of the GA developed in this study (encoding with binary representation of real values, evaluation function, adaptive plan) are presented. Calcium phosphate precipitation which is of major interest for P-recovery from wastewater has been adopted as an illustration of the implemented algorithm.

1. Introduction

Phase equilibrium calculations constitute an important class of problems in chemical engineering applications and considerable literature has been devoted to global optimization of vapor-liquid equilibrium (Mc Donald and Floudas, 1995) (Lee et al., 1999). Liquid-solid equilibrium modeling is now receiving much attention due to the advent of specialty chemicals, as well as to new constraints such as environmental considerations. An accurate knowledge of such equilibria is particularly of great opportunity for designing and optimizing crystallization processes.

This paper is devoted to calcium phosphate precipitation which has been identified as a major issue. Let us recall that phosphorus can be found under various chemical forms in urban wastewater, which represents about 30 to 50% of the total refusal of P: insoluble or dissolved organic phosphorus, orthophosphates (until 70% sometimes) and condensed inorganic phosphates. In France, the average concentration of phosphorus in domestic wastewater is within the range of 15-25 mg/L, which may strongly vary from day to day, even during day. The P-discharge in aqueous natural environment leads to an excessive development of algae and, generally to a pH increase, thus corresponding to eutrophication. Consequently, the phosphorus reduction in rivers is considered as a key factor of the fight against pollution. The principal legislative tool in Europe for fighting against eutrophication is the EC Urban Waste Water Treatment Directive (271/91/EEC). This action came into force in 1991 and enabled waterbodies to be classified as Sensitive Areas if they display symptoms of eutrophication.

Calcium phosphate precipitation involves various parameters: calcium and phosphate ion concentrations, supersaturation, ionic strength, temperature, ion types, pH and also time. In fact, the nature of the calcium phosphate precipitate depends on the supersaturation of the various species. More precisely, the co-crystallization of Amorphous Calcium Phosphate (ACP, $Ca_3(PO_4)_2$) and DiCalcium Phosphate Dihydrated (DCPD, $CaHPO_4$) which may occur in the range of pH to be considered has been taken into account in this study (Seckler, 1994).

This paper first presents the development of a simple thermochemical model, enough representative of calcium phosphate precipitation using a Debye-Huckel based approach for activity coefficient modeling. A preliminary study has shown that the initialization problem of classical optimization techniques which may be used (for instance, Sequential Quadratic Programming) is crucial for the robustness of the code. For this purpose, an optimization strategy combining a two-stage approach, i.e. a genetic algorithm for initialization purpose and identification of the search zone followed by an SQP method to refine the solution is proposed in this paper.

2. Chemical equilibrium model for calcium phosphate precipitation

To model the evolution of phosphate conversion rate as a function of pH with respect to ACP and DCPD precipitation, mass and electroneutrality conservation balances have been taken into account as well as the supersaturation relative to each species. During this precipitation, the aqueous species considered are, on the one hand, for phosphoric acid, H_3PO_4, H_2PO_4, HPO_4^{2-}, PO_4^{3-}, and on the other hand, the concentration of Ca^{2+} ion and the corresponding calcium salts.

The ACP precipitation equation can be written as follows:

$$3Ca^{2+} + 2PO_4^{3-} \rightarrow Ca_3(PO_4)_2 \tag{1}$$

The ACP surpersaturation is defined by the β parameter :

$$\beta_{ACP} = \frac{1}{5} \ln \left(\frac{\left([Ca^{2+}]\lambda_{Ca^{2+}} \right)^3 \left([PO_4^{3-}]\lambda_{PO_4^{3-}} \right)^2}{Ks_{ACP}} \right) \leq 0 \tag{2}$$

The DCPD precipitation equation can be expressed as:

$$Ca^{2+} + HPO_4^{2-} \rightarrow CaHPO_4 \tag{3}$$

The DCPD supersaturation is defined by the following constraint:

$$\beta_{DCPD} = \frac{1}{2} \ln \left(\frac{\left([Ca^{2+}] \lambda_{Ca^{2+}} \right) \left([HPO_4^{2-}] \lambda_{HPO_4^{2-}} \right)}{Ks_{DCPD}} \right) \leq 0 \qquad (4)$$

The different mass balances in the liquid phase include :
- a mass balance for calcium :

$$[Ca^{2+}] + [CaH_2PO_4^+] + [CaHPO_4] + [CaPO_4^-] + [CaOH^+]$$
$$= (Ca_{total} - \frac{3}{2} P_{total} X_{ACP} - P_{total} X_{DCPD}) \qquad (5)$$

where X_{ACP} (respectively X_{DCPD}) is the conversion ratio relative to ACP (respectively DCPD) form defined as :

$$X = \frac{P_{Total} - P_{sol}}{P_{Total}} \qquad (6)$$

- a mass balance for phosphate :

$$[H_3PO_4] + [H_2PO_4^-] + [HPO_4^{2-}] + [PO_4^{3-}] + [CaH_2PO_4^+]$$
$$+ [CaHPO_4] + [CaPO_4^-] = P_{total}(1 - X_{ACP} - X_{DCPD}) \qquad (7)$$

The electroneutrality requirement gives :

$$[H_2PO_4^-] + 2[HPO_4^{2-}] + 3[PO_4^{3-}] + [CaPO_4^-] + [Cl^-] + [OH^-]$$
$$= [CaH_2PO_4^+] + 2[Ca^{2+}] + [CaOH^+] + [H^+] + [K^+] \qquad (8)$$

The concentrations of ions and complexes are determined from chemical equilibrium relations (equilibrium constants are given for a temperature of 25°C in molar units) (see table 1):

Table 1. Equilibrium constants for the system Ca-PO_4-H_2O $K_i = \dfrac{(A_i)(B_i)}{(AB_i)}$

K_i	A_i	B_i	AB_i	K_i value
K_1	H^+	$H_2PO_4^-$	H_3PO_4	$7.1285*10^{-3}$
K_2	H^+	HPO_4^-	$H_2PO_4^-$	$6.2373*10^{-8}$
K_3	H^+	PO_4^-	HPO_4^-	$453.942*10^{-15}$
K_4	Ca^{2+}	$H_2PO_4^-$	$CaH_2PO_4^+$	$3.908*10^{-2}$
K_5	Ca^{2+}	HPO_4^-	$CaHPO_4$	$1.8239*10^{-3}$
K_6	Ca^{2+}	PO_4^-	$CaPO_4^-$	$347.536*10^{-9}$
K_7	Ca^{2+}	OH^-	$CaOH^+$	$5.8884*10^{-2}$
K_w	H^+	OH^-	H_2O	$1.004*10^{-14}$

The Debye-Huckel model giving the activity coefficient of each species is defined by :

$$Log_{10}\lambda = -A_{DH}z_i^2 \frac{\sqrt{\mu}}{1+B_{DH}\alpha\sqrt{\mu}} + C_{DH}\mu \tag{9}$$

with $B_{DH} = \sqrt{\dfrac{2e^2 N_A \rho_o}{\varepsilon k_B T}}$ and C_{DH} is a constant equal to 0,055 mol.L^{-1}.

The function to be minimized is the Gibbs free energy G of the system expressed as a linear combination of the chemical potential of each component in each phase :

$$G = \sum_{i=1}^{N}\sum_{k=1}^{\pi} n_{ik}\mu_{ik}$$

$$\mu_{iL} = \mu_{iL}(T) + RT\ln\left([x_i]\lambda_{x_i}\right) \tag{10}$$

$$\mu_{iS} = \mu_{iS}(T)$$

μ_{iL} (respectively μ_{iS}) represents the chemical potential of the species i in the liquid phase (respectively in the solid phase)

A substitution method has been applied so that the only unknowns of the system are the following concentrations [Ca^{2+}], [PO$_4^{3-}$], [H$^+$] and the phosphate conversion rate under DCPD and/or ACP forms. The system has been solved for various concentrations in KOH in order to analyze pH influence on conversion. Since calcium exists in the form of calcium chloride, this concentration has been taken equal to 2 [Ca^{2+}].

To solve the system, the two cases of calcium phosphate precipitation have been dissociated, thus leading to an optimization problem with simultaneously 4 equations and one inequality constraint. They are summarized in table 2. In the former case, DCPD (respectively in the latter case ACP) is the equilibrium component of the system.

Table 2. Different cases considered in the optimization method

	First case	Second case
ACP Supersaturation	≤ 0	$= 0$
DCPD Supersaturation	$= 0$	≤ 0

3. System resolution

Several approaches have been proposed for the computation of the solutions to the phase and chemical equilibrium problem (see the paper of Mc Donald and Floudas (1995) for a review of the different contributions). A typical feature of such problems is that the generation of starting points that are used in the search with conventional optimization methods (for example SQP) is very important to guarantee the success of the optimization procedure.

A preliminary study on the above mentioned example has shown that the only use of an SQP method (SQP package from IMSL library) (Schittkowski, 1986) is very sensitive to the choice of the initial guess and often leads to a failure. Consequently, this work is motivated by the development of a technique for the automatic generation of good starting points. For this purpose, a hybrid optimization method is proposed in this paper: a Genetic Algorithm (GA) is used to perform a first search in the space and to locate the neighborhood of the solution. Then, the derivative driven optimization tool (SQP) is used to refine the solution. Let us recall that Genetic Algorithms (Goldberg, 1989) differ from most classical optimization methods since no assumption about the problem space is required and yet produce a global search. GAs use a guided random search in which many different solutions to a problem are investigated and refined simultaneously to identify near-optimum solutions. A major interest of such methods is that they lead to reasonable solutions even with a poor initial guess. GA implementation requires the definition of parameters: population generation mode, population size, (i.e., the number of individuals forming a population which must be sufficiently large to create sufficient diversity, in order to cover the possible solution space) crossover probability, mutation rate, survival, crossover and mutation mechanisms. In this paper, the initial population has been selected randomly. Nevertheless, constraint respect at a precision of 5.10^{-6} is required. Each potential solution, i.e., a real value, has been coded as an haploid chromosome with the so-called "weight box". The weight box consists to encode each digit of real in 4 bits with respectively the following weights 1, 2, 3, 3 with a precision of 10^{-20}. By lack of place, the choice of weight is not justified here but has been achieved after a sensitivity analysis. For example the real 0.18 is encoded 1000 0111, the zero digit and the point are not encoded. Each chromosome represents the 5 unknowns. The development of the GA involves the following features: a survival rate of 0.6, a mutation rate of 0.4, a number of generations of 100 with 101 individuals in a population (the 101th place corresponds to the 'niche' of the best individual kept from a generation to the following one).

As initially proposed by Goldberg (1989) and followed by several authors (Costa and Oliviera, 2001), the penalty function method has been used to take into account the constraints in the optimization procedure. Actually, the fitness function (F) involves the objective function (Gibbs free energy) and the penalty terms (p) denoting the equality ($h_l(x)$) and inequality constraints ($g_k(x)$) violations.

$$F = -G - pC \tag{11}$$

$$C = \sum_{k=1}^{p} \left[\max\{0, g_k(x)\} \right]^2 + \sum_{l=1}^{m} \left[h_l(x) \right]^2 \tag{12}$$

The penalty coefficients have been adjusted to have a same order of magnitude for both terms G and pC.

In table 3, the Gibbs free energy and the evolution of the penalty terms obtained, on the one hand, after a GA run and on the other hand, at the end of the GA+SQP phase are presented. At step 2 of procedure, the SQP method is initialized by the best solution (i.e. best individual) found by the GA.

Table 3. Gibbs free energy and penalty term evolution values under the resolution method influence

	Gibbs free energy (J/mol)	Penalty term
Initialization for GA process	-6236.95	$5 \ 10^{-6}$
Best solution obtained by GA used as an initialization for SQP process	-4573.10	$1.2 \ 10^{-7}$
Best solution found by SQP	-5425.07	10^{-32}

4. Conclusions

In this paper, a hybrid optimization tool combining a Genetic Algorithm and an SQP method has been developed and tested on an Solid-Liquid-Equilibrium based calculations. This two-level strategy leads to very efficient search: the GA has been able to provide good starting points for the subsequent SQP method (even if the user has provided rough starting points), thus favoring the local search. Calcium phosphate precipitation which is of major interest for P-recovery from wastewater has been adopted as an illustration of the procedure. More precisely, the co-crystallization of ACP (Amorphous Calcium Phosphate) and DCPD (DiCalcium Phosphate Dihydrated) which may occur in the range of pH to be considered has been taken into account. The generic feature of the optimization strategy presented in this study finds a widespread application for solid-liquid equilibrium calculations.

References

Costa, L., Oliviera, P., Evolutionary algorithms approach to the solution of mixed integer non-linear programming problems, Computers and Chemical Engineering, Vol. 25, pp.257-266, 2001

Goldberg, D.E., Genetic algorithms in search, optimization and machine learning. Massachusetts: Addison-Wesley, 1989

Lee, Y.P. Rangaiah, G.P., Luus, R., Phase and chemical equilibrium calculations by direct search optimization, Computers and Chemical Engineering, Vol. 23, pp. 1183-1191, 1999

Mc Donald, C.M., Floudas, C.A., Global optimization for the phase and chemical equilibrium problem: Application to the NRTL equation, Computers and Chemical Engineering, Vol. 19, N° 11, pp. 1111-1139, 1995

Schittkowski, K., NLPQL: A FORTRAN subroutine solving constrained non-linear programming problems, (edited by Clyde L. Monma), Annals of Operations Research, Vol.5, pp.485-500, 1986

Seckler, M.M., Calcium phosphate precipitation in a fluidized bed, Ph.D. Thesis, Technische Universiteit Delft, 1994

European Symposium on Computer Aided Process Engineering – 12
J. Grievink and J. van Schijndel (Editors)
© 2002 Elsevier Science B.V. All rights reserved.

Multivariate analysis for product design and control

Ingela Niklasson Björn, Staffan Folestad, Mona Johansson, Lars Josefsson,
Andreas Blomberg

AstraZeneca R&D Mölndal, SWEDEN

Abstract

Fluidised bed coating of pellets, granules or particles plays a very important role in the development and production of oral controlled release formulations. Based on design of experiments for robustness testing of the manufacturing process, a 2^{4-1} fractional factorial was planned. During the experiments in-line NIR spectroscopy was combined with a theoretical film growth model in order to monitor and control the coating process.

The increase in film thickness during the process evolvement was calculated and compared with the film thickness obtained by a reference method off-line, image analysis. NIR spectra were aquired during coating by means of a fiber optic probe positioned in the fluid bed. Time series of NIR spectra were calibrated with the corresponding theoretical film thickness by a multivariate analysis method, PLS (partial least square) regression. Film thickness predicted by the PLS model was then compared with the actual median thickness, as measured by image analysis off-line.

The calibration of in-line NIR-spectra to film thickness calculated with a growth model is successful with PLS. The PLS model contains three principal components, which describe 99 % of the variance in NIR-spectra and the variation in film thickness calculated by the growth model. The root mean square error (RMSE) between the film thickness predicted by NIR-spectra and the film thickness calculated by the theoretical growth model is 1.2 μm.

1. Introduction

Design of experiments and multivariate analysis are valuable tools in the development and optimisation of a formulation and its manufacturing process. The robustness of the manufacturing process is investigated during the scale up phase to create planned disturbances of the identified critical process parameters and evaluate the impact on the product characteristics. The coating of granules are often performed in a spouted fluidised bed with a draft tube (Wurster, 1950) where the coating thickness of the individual granules increases every time they enters a draft tube and are hit by the sprayed coating solution from below.

Traditionally, the quality and properties of this type of products have been assessed after processing assuming that the process is well controlled. Particle size analysis is a common way to asses the coating thickness, typically conducted off-line. A fast, non-destructive, in-line measuring technique would be preferable. In this respect NIR (Near InfraRed) Spectroscopy can be an interesting alternative due to its ability to detect changes in physico chemical properties of the material. It is of general interest to characterize the progress of the process, i.e. to monitor the growth of the coating *in-line*, both in terms of the average and the size distribution.

Monitoring with in-line NIR spectrometry offers fast and non-destructive possibilities to acquire qualitative and quantitative information regarding the coated granules during the actual process. NIR spectroscopy has earlier been described to measure the coating thickness (Kirsch and Drennen, 1996), (Andersson et al., 1999) and (Andersson et al., 2000).

2. Particle growth model

In this study a general model for the layered particle growth during the coating process was used (Nienow,1985). The following assumptions were made in the model, particles are spherical and non-porous with an initial diameter d_{po} . Particles grow uniformly with a thickness a ,with no losses.

The film thickness growth is calculated with the following equation:

$$\frac{da}{dt} = \frac{1}{2} \frac{\dot{m}_{coat}}{\left(\dfrac{M_o \rho_{coat}}{\rho_s} \right) \left[\dfrac{3}{d_{po}} + \dfrac{12a}{d_{po}^2} + \dfrac{12a^2}{d_{po}^3} \right]} \qquad (1)$$

where ρ_s is the density of the starting material and M_0 the total mass of the starting particles . The feed rate of solid material is constant, $\dot{m}_{coat} = dM_{coat}/dt$

3. Experimental

Uncoated pellet cores containing an active drug substance with a diameter of ca 500μm were coated with an aqueous suspension of filmformer. The coating process was performed in a bottom-sprayed spouted fluidised bed with a draft tube as shown schematically in Figure 1. The NIR-instrument was connected to the bed through a fibre optic cable with an optimized interfacing (patent pending).

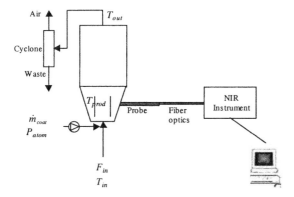

Figure 1. Fluidised bed of Wurster type monitored with in-line NIR Spectroscopy

The monitoring with NIR spectroscopy was performed with a process version of NIRSystems 6500 spectrometer (1996). The instrument has a lightsource (Tungsten-Halogen lamp), 4 lead sulphide detectors (1100-2500nm) and 2 silicon detectors (400-1100 nm). Data acquisition was made by NSAS software version 3.0. The NIR light is transferred through and from the fiber optic probe mounted in the process vessel.

Based on design of experiments nine test batches were made where the following factors were varied, the amount of starting mass M_0, the inlet temperature T_{in}, the atomising pressure P_{atom} and the fluidising flow F_{in}. The product temperature T_{prod} and the outlet temperature T_{out} were held constant throughout the design. The process was controlled by adjusting the coating flow $\dot{m}_{coat}$ to achieve a suitable product temperature T_{prod}.

To test if the growth model is valid, image analysis was performed on coated granules from each batch experiment, including the starting material (Beadcheck™ 830, 2000). Samples from every batch were withdrawn after ca 50%, 70% of consumed coating material and from the end product. All experiments have the same starting material with a median diameter measured by image analysis.

4. Calibration

The first step was to develop a theoretical model where real process data are taken into account and consequently give a description of the film growth under different operating conditions.

The second step was to test the theoretical model by using image analysis as a reference method.

The third step was to use the theoretical model as a reference method when developing a NIR multivariate batch calibration.

The forth step was to test the validity of the NIR multivariate batch calibration by comparing film thickness predicted from in-line NIR data with film thickness measured by image analysis off-line.

5. Results and discussion

The film thickness growth is calculated theoretically and is shown in Figure 2 for each batch process conditions in the experimental design. The film yield is taken into account and the coating flow $\dot{m}_{coat}$ is assumed to be constant. Batch no 7 had the slowest growth rate due to low inlet temperature in combination with a high mass and low fluidisation flow. The particles in batch no 2 grew fastest. In this case the mass was low, fluidisation flow was high and the fluidisation air had a high inlet temperature.

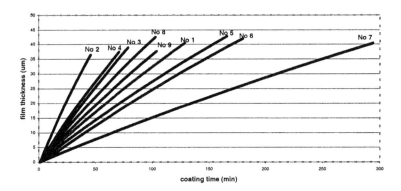

Figure 2. Theoretical film thickness vs coating time

The growth model was evaluated by comparing the calculated film thickness with image analysis, Figure 3. A correlation interval of +/- 2.0 μm was found with a target value for the coating thickness of 45 μm. The error bar shows the spread in median film thickness for the end product from batch no3.

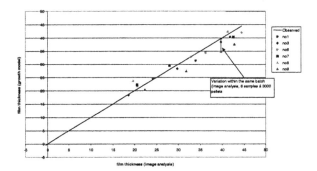

Figure 3. Observed versus theoretical film thickness. The theoretical film thickness is calculated with the growth model and the observed values are obtained by image analysis.

To test if the calibration is valid, film thickness predicted by NIR spectra is compared with observed values from the image analysis, Figure 4. As a measurement of the error between the predicted film thickness and the film thickness obtained by image analysis Root Mean Square Error (RMSE) is used. A value of 2.2 μm was found in this case, which emphasise the high precision and accuracy of the presented method. To see how the film thickness increases during the process, the predicted film thickness growth from NIR spectra and the theoretical film thickness growth are plotted versus the process time. In Figure 4, batch no 3 is shown. The observed median film thickness obtained from the image analysis is shown. The bars on the observed start film thickness and the end film thickness show max and min values for the median film thickness. The variation in median film thickness obtained from image analysis is similar to the variation in the film thickness predicted by NIR-data.

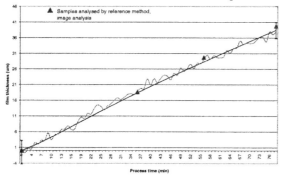

Figure 4. Predicted, theoretical and observed film thickness increase during batch no3. The curvy line is film thickness predicted by NIR-spectra and the smooth line is the theoretical film thickness.

To test the usefulness of real-time monitoring, one batch was deliberately run slightly outside normal conditions, Figure 5. It is obvious already early during processing of this batch that the film thickness does not grow as fast as expected. Monitoring the process in this way reveals that the process has to proceed further until the target film thickness (45 μm), i.e. 100% film yield, is reached.

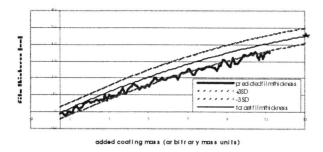

Figure 5. Control chart which shows film thickness calculated from measured NIR-spectra. Early during the process, the film thickness is under the alarm limit.

5. Conclusions

A generic method for real-time monitoring of film thickness growth in a batch spouted bed coating process is described. NIR spectroscopy were combined with a theoretical growth model and developed to control the coating process. The results emphasize that,

- Design of experiments and multivariate analysis are useful tools in the development and optimisation of a robust process and formulation.
- Quantitative determination of the coating thickness can be performed with in-line NIR.
- In-line NIR provides information regarding the underlying mechanisms of the coating process, the process dynamics .

6. References

Wurster, D.E., Air-suspension of coating drug particles, 1950, J.Am. Pharm. Assoc., 48(8), 451

Kirsch J.D., Drennen J.K., 1996, Pharmaceutical Research, 13 (2): 234-73

Andersson, M., Josefson, M., Langkilde, F.W., Wahlund, K.G., J. Pharm. Biomed. An., 1999, 20: 27-37

Andersson, M., Folestad, S., Gottfries, J., Johanson, M.O., Josefsson, M., Wahlund, K.G., 2000, Analytical Chemistry, 72(9): 2099-2108

Nienow A.W., Rowe P.N., 1985, Particle growth and coating in gas-fluidized beds, Fluidization, Academic Press, London

Patent pending.

Your NIRSystems^{TM} Instrument Performance Guide, 1996, Perstorp Analytical

BeadCheck™ 800, User's manual version 3.19, 2000, Sweden

7. Nomenclature

a	film thickness (m)
d_{po}	median start particle diameter (m)
M_0	Total mass of the start particles (kg)
M_{coat}	Total coating mass (kg)
$\dot{m}_{coat}$	coating rate (kg/s)
ρ_{coat}	density of the coating material (kg/ m^3)
ρ_s	density of the starting material (kg/ m^3)

European Symposium on Computer Aided Process Engineering – 12
J. Grievink and J. van Schijndel (Editors)
© 2002 Elsevier Science B.V. All rights reserved.

A Modelling Formalism for Multiproduct and Multiplant Batch Processes

Sebastián Oddone and Oscar A. Iribarren

Institute for Process Design and Development INGAR – Conicet

Avellaneda 3657 – 3000 Santa Fe – Argentina

Abstract

Computer aided process engineering tools devised for batch processes are still well behind the performance of those developed for steady state processes. As commercial process simulators still lack optimisation capabilities, the alternative for optimising process recipes is resorting to programming languages like Gams, Matlab or gProms. This paper proposes a modelling formalism that could be implemented in such languages and is aimed at retaining a large degree of modularity (reuse of pre programmed modules) to expedite the set up of the problems, as well as reducing the computation time by exploiting the mathematical structure of the models at hand (this is especially convenient for highly nested process structures, whose mass balances can be handled by a linear system of equations). The formalism is illustrated with a multiplant arrangement that produces sugar cane derivatives: sugar, bio ethanol and yeast.

1. Introduction

The modular approach was pioneered by Barrera and Evans (1989) and consists of individual modules for each stage, which are then connected to render a particular process. Commercial simulators like Batch Pro and Batch Design Kit work in this way but unfortunately lack optimisation capabilities to optimise the process variables (the process recipe). Academic contributions within this approach as in Salomone et al. (1997) permit optimising medium size problems (10 stages, 2 products) but demand large computation times with most of this time spent in the interactions among blocks written in different languages: C and Gams.

On the other hand, the mathematical model can be written as a unique system of equations, which are the constraints of an optimisation program. With this approach Pinto et al. (2001) solved larger problems (8 stages, 4 products) with a small computational demand. These authors only considered algebraic models, however Bhatia and Biegler (1996) solved problems with stages represented by simple dynamic models (by discretizing them).

While the modular approach permits implementing more rigorous stage models (which is in general only necessary for some selected stages) and maximizes reuse of pre programmed modules, it has a large computational demand. On the other hand single optimisation programs are computationally efficient but require much programming effort to set up the problems, which imposes a bound on the complexity of the process models that can be implemented in practice.

2. Approach Proposed in this Paper

In this paper we exploit the property that even for complex stages, the mass balances can usually be represented by a linear system of equations. This occurs if appropriate extents of reactions or extents of separations are selected as the optimisation variables. For continuous processes, Chapter 3 in the book by Biegler et al. (1997) presents the foundations for this approach.

The unit operation models are split into a linear mass balance block and a size and time factors generator for the node (which needs the plant or multiplant mass balances be solved in advance). The optimisation of process variables is an external loop: setting a value for this variables permits solving the whole plant mass balances as a linear system which includes each node balances, plus the stream splitters and mixers present in the states tasks network representation of the processes.

The whole plant mass balances information permits predicting the size and time factors of each batch or semi continuous node, which in turn permits solving the multiproduct or multiplant sizing problem (which for fixed factors becomes a geometric program). This in turn, permits computing the Objective Function value that corresponds to the given set of process variables.

The size and time factors generator may be a complex dynamic simulation module, but it is solved only once at each iteration of the process variables optimisation module (it is not involved in the convergence of mass balances).

2.1 Process example

We consider a multiplant arrangement to produce sugar cane derivatives: sugar, bioethanol and *torula utilis* yeast for cattle feed. Figure 1 schematises a simplified flow sheet for this arrangement. We will illustrate the approach by focusing on the derivative plants consisting of pre fermentors, fermentors and a distillery, and the recycle of streams that contain substrates that can be consumed in the fermentors. The set of variables selected as optimisation variables are the extents of reaction in the fermentors, the extents of recovery of the two key components alcohol and water in the distillation, and the split ratios of the substrate streams indicated with circles in Figure 1.

The streams exiting the sugar plant: filter juices (which are the solids concentrated phase of clarifications) and molasses (which are the mother liquors of crystallizations) are substrates in the plants of derivatives. Also the alcohol plant produces distillery wastes (vinasses) that can eventually be recycled to any of the plants (to the same alcohol plant or to the yeast plant) to contribute to the total substrate.

2.2 Linear mass balances models

For the batch fermentors, a reactor model with a fixed conversion of total reducing sugars renders a linear mass balance for the substrate so this conversion is selected as a

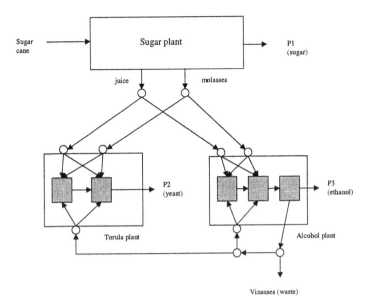

Figure 1. Simplified flow sheet for the multi plant sugar derivatives arrangement.

process optimisation variable, fixed by the outer optimisation loop:

$$m_s^{out} = (1-\eta_s)m_s^{in}$$ (1)

where m_s $[kg]$ is the mass of substrate and η_s the conversion. Other components in the fermentation broth are related to this one through stoichiometry.

For the distillation, the optimisation variables are the key components recoveries (amount in the distillate after the distillation divided by the amount present in the feed before the distillation), and the number of separation stages. So the linear mass balances for each component i are:

$$d_i = \eta_i f_i \quad \text{and} \quad w_i = (1-\eta_i)f_i$$ (2)

were f_i, d_i, w_i are the amounts in the feed, distillate, and residue respectively, and η_i are the fixed recoveries. The splitters and mixers in the states tasks network are obviously represented by also linear mass balances.

It should be pointed here, that the linear mass balances approach could not be used if more complex performance models were needed. Specifically, if maintenance coefficients in the fermentors cannot be neglected, or the relative volatilities in the distillation cannot be considered constant. In this case, the mass balances cannot be decoupled from the simulations.

2.3 Models for predicting time and size factors

The inclusion of process variables into the computation of the time and size factors has been described by Salomone and Iribarren (1992) for analytical models and Salomone et al. (1994) when dynamic simulations are required. In this paper we adopted fermentation models that require dynamic simulations (the differential equations describing the fermentation are not analytically integrable), but the simplified analytical model for distillation presented by Zamar et al. (2000). For example, in the case of the prefermentor of the ethanol plant, the size factor is:

$$S_{pe}\left[kg/m^3\right] = \left[\rho_{fb} \, x_s^{in}\right]^{-1} \tag{3}$$

where $\rho_{fb}\left[kg/m^3\right]$ is the apparent density of the aerated fermentation broth and $x_s^{in}\left[kg/kg\right]$ is the initial concentration of substrate, which is a known figure after solving the mass balances. This size factor will be used afterwards in the sizing block:

$$V_{pe}\left[m^3\right] \geq S_{pe} \, B_e\left[kg\right] \tag{4}$$

where V is the volume required by this prefermentor and B_e is the batch size at the ethanol plant. We adopted the amount of substrate entering each plant as the batch size, instead of the more traditional amount of final product: we make some comments about this in the Conclusions section. On the other hand, the fermentation time is computed (with η_s fixed) running a dynamic simulation which stops when the amount present in the prefermentor is m_s^{out}.

2.4 Implementation

The model was finally implemented in Matlab (other languages could be used, we make some comments in the Conclusions section), generating functions for the resolution of each computation task, discriminated in independent blocks.

Among them we identify the simulation block, the mass balances block, the sizing block, and the external block for the optimisation of process variables.

The simulation block includes the functions that call the dynamic simulator Simulink.

The sizing block gets as inputs, the actualised figures for the time and size factors and minimizes a "Total Annual Cost" objective function. A zero wait scheduling policy was implemented for each plant, and a UIS policy for connecting the plants among themselves. Details about this fixed factors model can be found in Chapters 6 and 22 of the book by Biegler et al. (1997).

The process variables are fixed parameters within this block, while the batch sizes and the operating time of the semicontinuous items (the distillation column, the boiler and the condenser) are the optimisation variables. Finally, the external optimisation block proposes new values for the process variables that affect the Total Annual Cost received from the sizing block.

The principal advantage of this implementation is it modular architecture which permits that changes in the model be easy to implement. This is specially important in the hierarchical design approach (Douglas, 1988) which we start with simpler models, and take into account only the stages with the highest economic impact, adding new units and more rigorous models as we advance in the design.

In every case, the inclusion of new equipment or more rigorous models is implemented by activating and deactivating subroutines in the commands line. In particular, Matlab resorts to the switch – case sentences:

```
switch method
  case 'linear', disp('Method is linear')
    .
    .
    .
  case 'non-linear', disp('Method is non linear')
    .
    .
    .
end
```

3. Conclusions

As a result of the computations, we obtain the optimal values for the set of process variables which minimize the Total Annual Cost of the multi plant arrangement. The optimal solution contains features known to be implemented in industrial practice as recycling all the vinasses, but only to the fermentors (not to pre fermentors).

The computational demand was very small as compared with the previous work that included dynamic simulations reported in Salomone et al. (1997), even when they did not include recycle streams. Both approaches permit to make changes in the model in an easy way, as well as decompose a large problem into the sequential resolution of smaller blocks, the main difference is that in the present work these blocks are part of the same environment, which saves computation time spent in interfacing the blocks.

This is the first time that we implement recycle streams, which leaded to explore decomposing the process performance models into partition function models for solving the mass balances, and time and size factor prediction models which are computed afterwards. Solving the mass balances by the sequential resolution of the stages and converging the values of tear streams, demanded in our case in the order of 10 times more computer time than solving the linear mass balances first.

This is also the first time that we implement multi plant arrangements, and we found more practical to define the size factors at each plant on the basis of the batch fed to the plant, instead of the batch of products (ethanol or yeast) exiting the plants. This is so because these exiting batches were not subject to production targets (as in the traditional model), but their amount contributed (through their selling prices) to the Total Annual Cost.

144

Another goal of this work was to explore the advantages (disadvantages) of different computational tools like gPROMS, SuperPro designer, Matlab and GAMS. Matlab was found to be adequate for the implementation of the formalism.

With respect to gPROMS, even if it has internal dynamic optimisation routines, we had troubles to implement the modular structure because the definition of variables and parameters must be specified in a global way, i.e. it does not permit to redefine the type of variables in separate blocks. Also, the scheduling is a result from the optimisation and cannot be handled from outside.

SuperPro permits the user to connect pre programmed operations and solves the heat and mass balances, but lacks an optimisation block. Future developments of this tool could probably find useful the experience reported in the present paper.

GAMS has a large resolution capability but lacks modularity and in the case of the dynamic models they must be discretized into algebraic equations (extra work for setting up the problem).

Finally, we point a disadvantage of Matlab with respect to gPROMS: while in gPROMS the definition of differential equations is done in a conventional and simple way, in Matlab it demands the connection of mathematical functions blocks which is also an extra work for setting up the problem.

Acknowledgements

The authors gratefully acknowledge Conicet and Agencia for financial support, and Universidad Nacional de Quilmes for Sebastián Oddone's fellowship.

4. References

Barrera M.D. and L.B. Evans, 1989, Optimal Design and Operation of Batch Process, Chem. Eng. Comm., 82, 45.

Bathia T. and L.T. Biegler, 1996, Dynamic Optimization in the Design and Scheduling of Multiproduct Batch Plants, Ind. Eng. Chem. Res., 85, 2234.

Biegler L.T., I.E. Grossmann and A.W. Westerberg, 1997, Systematic Methods of Chemical Process Design, Prentice Hall, New Jersey.

Douglas J.M., 1998, Conceptual Design of Processes, 1988, McGraw – Hill, New York.

Pinto J.M, J.M. Montagna, A.R. Vecchietti, O.A. Iribarren and J.A. Asenjo, 2001, Process Performance Models in the Optimization of Multiproduct Protein Production Plants, Biotechnology and Bioengineering, 74(6), 451.

Salomone H.E. and O.A. Iribarren, 1992, Posynomial Modeling of Batch Plants, Comput. Chem. Engng., 16, 173.

Salomone H.E., J.M. Montgna and O.A. Iribarren, 1994, Dynamic Simulations in the Design of Batch Process, Comput. Chem. Engng., 18, 191.

Salomone H.E., J.M. Montagna and O.A. Iribarren, 1997, A Simulation Approach to the Design and Operation of Multiproduct Batch Plants, Trans. I. Chem E., 75, A, 427.

Zamar S.D., S. Xu and O.A. Iribarren, 2000, Analytical Process Performance Model for Batch Distillations, Paper at Escape – 10 in "Computer Aided Chemical Engineering 8", Editor S. Pierucci, Elsevier, Amsterdam.

European Symposium on Computer Aided Process Engineering – 12
J. Grievink and J. van Schijndel (Editors)

Design of a Membrane Process in a Countercurrent Operation for the Treatment of Industrial Effluents

Sergio M. Corvalán[#], Inmaculada Ortiz[*] and Ana M. Eliceche[#]

[#]PLAPIQUI-CONICET, Chem. Eng. Dpt., Universidad Nacional del Sur, 8000 Bahia Blanca, ARGENTINA

[*]Dpto. de Ingeniería Química y QI. ETSIIyT, Universidad de Cantabria, 39005 Santander, SPAIN

Email: meliceche@plapiqui.edu.ar

Abstract

The design of a plant with membrane modules for effluent treatment and metal recovery in a countercurrent continuous operation is addressed for the first time in this work. The design is formulated as a nonlinear programming problem where the set of algebraic and differential equations that model the membrane separation processes are included as equality constraints. The separation objectives related to maximum contaminant concentration in the effluent and minimum contaminant composition in the product for re-use are posed as inequality constraints. The objective function to be minimised is the total membrane area required in the plant. The optimisation variables are the flowrates and membrane areas.

As a motivating example the removal and recovery of Cr(VI) is analysed, which poses a real challenge for pollution prevention and has a wide range of applications of industrial interest. The countercurrent continuous operation requires less membrane area than the cocurrent operation, indicating that this flow pattern should be further explored at the conceptual design stage.

1. Introduction

Separation processes based upon reversible chemical complexation offer possibilities of high mass transfer and selective extraction. The chemical pumping of a relatively dilute solute against its concentration gradient by a carrier, known as coupled transport (Cussler, 1971), allows the extraction of pollutants from industrial effluents and their concentration in another phase, because of the reversibility of the chemical reaction with the carrier or extractant.

Extraction processes using supported liquid membranes are of particular interest due to their versatility. While in conventional-practice solvent extraction processes rely on dispersion, in microporous membrane-based non-dispersive solvent extraction technology (NDSX) the dispersion of a phase into another is eliminated, and both are contacted through the pores of the fibre. With the NDSX technology, limitations of conventional liquid extraction such as flooding, intimate mixing and requirement of density difference are overcome. In the last years, extensive studies on dispersion-free solvent with the use of microporous membrane have been carried out by Prasad and Cussler (1988), D'Elia et al. (1986), Wickramasinghe et al.(1991), Yang and Cussler

(1986). Ho and Sirkar (1993) presented a general review of the NDSX technology, and more recently, Gabelman and Hwang (1999) provided an overview of hollow fibre contactors and de Gyves et al (1999) reviewed the applications for metal separations.

Decisions regarding the selection of the operating mode, process configuration, size of equipment and operating conditions need to be addressed in order to promote their industrial application. A large number of studies of this technology have been mentioned in the literature, but there is still little information on the analysis and optimisation of these processes.

The removal and recovery of hexavalent chromium is studied in this work as a motivating example. Cr(VI) is one of the most toxic elements to be discharged into the environment, thus strict legal restrictions apply for its release into inland waters.

Alonso and Pantelides (1996) carried out the simulation of the NDSX process to extract and re-use Cr(VI) in a batch mode. An analysis to validate the models with the performance of an NDSX pilot plant in the semicontinuous mode is reported by Alonso et al. (1999). The optimisation of the semicontinuous operation mode has been reported by Eliceche et al. (2000).

At the conceptual design stage the analysis of different configurations is important. In this work the optimum design of a NDSX plant on an industrial scale, in a continuous countercurrent operation is carried out. The optimum membrane areas of the extraction and stripping processes are evaluated to achieve the separation objectives, maximum contaminant composition in the effluent and minimum contaminant composition in the product for further re-use, at a minimum cost.

2. Effluent treatment and metal recovery process

This technology has two main processes, the extraction process where the contaminant is extracted from the effluent into an organic phase, and the stripping process where the contaminant is re-extracted from the intermediate organic phase and it is concentrated in an aqueous phase for further re-use. The membrane modules of hollow fibre that perform the extraction and stripping separations are shown in Figure 1.

The separation process involves mass transport phenomena coupled with selective chemical reactions that can lead to different operation fluxes in the extraction and back-extraction steps. In the extraction module, the effluent F_e runs by the lumen of the fibre, while the organic phase stream F_o containing the carrier flows in the shell side. The organic solution wets the hydrophobic membrane. A slight overpressure of the aqueous phase with respect to the organic phase is necessary to avoid the dispersion of the organic into the aqueous phase.

The organic phase takes the contaminant from the effluent in the extraction module and it is then fed to the stripping module, where a stripping aqueous stream F_s flows through the lumen of fibres. The aqueous solution concentrates the contaminant in the stripping process and it is then recycled to the stripping tank while the organic phase returns to the extraction process. A product stream F_p containing the concentrated product leaves the NDSX plant. This product stream can be located before or in the stripping tank, both options are analysed in this work. An aqueous make up stream F_a is necessary to balance the aqueous phase stream in the steady state operation.

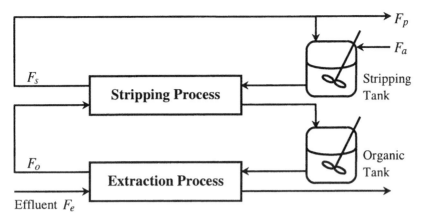

Figure 1. Schematic diagram of an effluent treatment and metal recovery plant

The organic and aqueous phases can run in a cocurrent or countercurrent mode inside the membrane modules. The continuous countercurrent operation with product extraction before the stripping tank is shown in Figure 1.

3. Model of the extraction and stripping processes

The plant is designed for the removal of hexavalent chromium and its recovery for industrial re-use. The first step to model the process involves the mathematical representation of the chemical equilibrium at the membrane interfaces, in the extraction and stripping processes. The organic and stripping reactions between the solute Cr(VI) and the organic carrier (quaternary ammonium salt Aliquat 336 dissolved in kerosene) is represented by:

$$CrO_4^{-2} + 2\overline{AlCl} \leftrightarrow \overline{Al_2CrO_4} + 2Cl^- \tag{1}$$

Aliquat 336 has complex extraction behaviour. Alonso et al. (1997) have shown that an ideal behaviour could be assumed for the aqueous phase while a non-ideal behaviour of the organic phase should be taken into account in the equilibrium expression, resulting in the following expression for the chemical equilibrium:

$$K = \frac{4C_{oi}^E(C_{in} - C_e^E)^2}{C_e^E(CT - 2C_{oi}^E)^2}(0.001CT)^{0.6} \tag{2}$$

Where C_{in} is the initial Cr(VI) in the effluent and CT is the total carrier concentration.
As the chlorine concentration in the stripping aqueous solution is maintained high (1000 mol.m^{-3}) in order to favour the stripping process, the change in Cl^- and $AlCl$ concentration can be neglected along the modules. For that reason the chemical equilibrium can be well described by a distribution coefficient defined as the ratio of the equilibrium concentrations in the aqueous and organic phases as it has been experimentally checked by Ortiz et al. (1996):

$$H = \frac{C_s}{C_{oi}^S} \tag{3}$$

The Cr(VI) mass balances in the organic and aqueous phases in the extraction and stripping process, in a countercurrent operation with product removal before the stripping tank, are modelled by the following set of differential and algebraic equations:

EXTRACTION MODULE

Aqueous Stream: $\dfrac{dC_e}{dz} = -\dfrac{A^E}{F_e L^E} K_m (C_{oi}^E - C_o^E)$, $\qquad C_e(0) = C_{in}$ $\qquad$ (4)

Organic Stream: $\dfrac{dC_o^E}{dz} = -\dfrac{A^E}{F_o L^E} K_m (C_{oi}^E - C_o^E)$, $\qquad C_o^E(0) = C_o^S(0)$ $\qquad$ (5)

STRIPPING MODULE

Aqueous Stream: $\dfrac{dC_s}{dz} = -\dfrac{A^S}{F_s L^S} K_m (C_o^S - C_{oi}^S)$, $\qquad C_s(0) = C_s^T$ $\qquad$ (6)

Organic Stream: $\dfrac{dC_o^S}{dz} = -\dfrac{A^S}{F_o L^S} K_m (C_o^S - C_{oi}^S)$, $\qquad C_o^S(L^S) = C_o^E(L^E)$ $\qquad$ (7)

STRIPPING TANK

$$(F_s - F_p).C_s(0) = F_s C_s^T \tag{8}$$

In the extraction sector, $z \in [0, L^E]$ and for the stripping sector, $z \in [0, L^S]$, being the shape factor $L^E/A^E = L^S/A^S = $ constant.

4. Formulation of the optimisation problem

The design of the plant is carried out minimising the total cost. The main contribution to the fixed cost is due to the membrane areas required in the extraction and stripping processes. Thus the objective is to minimise the total membrane area in the extraction and stripping process and the optimisation problem is formulated as:

$$\begin{aligned}
&\underset{v}{Min} \quad (A^E + A^S) \\
&s.t. \quad h(x, \dot{x}, v) \quad = \quad 0 \\
&\qquad\ I(x, v) \quad\ = \quad 0 \qquad\qquad\qquad (P1) \\
&\qquad\ g(x, v) \quad\ \leq \quad 0 \\
&\qquad\ v^L \ \leq \ v \ \leq \ v^U
\end{aligned}$$

The algebraic and differential equations h and their initial conditions I represent the plant model (Equations 1 to 8). The design and operating specifications are formulated as inequality constraints g. In particular the separation objectives, maximum contaminant concentration in the treated effluent and minimum product concentration for re-use are included in g. The organic, stripping and product flowrates and the membrane areas are the optimisation variables v. The non-linear programming problem was formulated in gPROMS (2000) and solved with the gOPT code.

5. Design of an effluent treatment and Cr(VI) recovery plant

As a motivating example this work presents the design of a plant for the treatment of wastewaters contaminated with Cr(VI) and its recovery as a product for further industrial re-use. The numerical results reported correspond to an effluent flowrate of 1 $m^3.h^{-1}$ with a Cr(VI) concentration of 13.462 $mol.m^{-3}$. The maximum Cr(VI) concentration allowed in the treated effluent is 1.923 $mol.m^{-3}$ and the minimum Cr(VI) concentration in the product for industrial reuse is 384.615 $mol.m^{-3}$. These two constraints are always active at the solution points. The lower and upper bounds for both the organic and stripping streams are 1 and 10 $m^3.h^{-1}$ respectively and F_p is positive. The system of differential equations (4) to (7) are solved using a second order orthogonal collocation over 50 finite elements in gPROMS.

Four designs obtained as solutions of problem *P1* for different structures are reported in Table 1. The second and third columns correspond to the optimal points for the countercurrent operation mode and the following to the cocurrent mode. In the second and fourth column the product stream F_p is removed before the stripping tank as shown in Fig. 1. In the third and fifth column this stream is extracted from the stripping tank. The number of iterations needed to find the solutions of problem *P1* was between 9 and 10 and the CPU time required was between 3 and 4 seconds in a Pentium III, 700 MHz workstation.

A reduction in the total required area in the order of 15% has been achieved in the countercurrent pattern compared with the cocurrent mode (as can be seen in the second and fifth columns) mainly due to a reduction in the order of 20 % in the required stripping area. About 5 % improvement can be achieved in both flow configurations by removing the product before the tank, as it decreases the Cr(VI) concentration in the stripping phase and increase the stripping driving force.

Table 1. Optimum solutions. Active lower (LB) and upper (UB) bounds.

Operating mode	COUNTERCURRENT		COCURRENT	
	Purge before tank	Purge in tank	Purge before tank	Purge in tank
TotalArea, m^2	**2862.12**	**2979.83**	**3238.51**	**3387.57**
A^E, m^2	1013.76	1033.79	1039.75	1058.87
A^S, m^2	1848.35	1946.04	2198.76	2328.71
F_o, $m^3.h^{-1}$	1.000 LB	1.000 LB	10.000 UB	10.000 UB
F_s, $m^3.h^{-1}$	1.000 LB	10.000 UB	1.000 LB	10.000 UB
F_p, $m^3.h^{-1}$	0.030	0.030	0.030	0.030

The countercurrent operation is analysed for the first time in this work and it is the best flow pattern. The area required in the extraction process is quite insensitive to the flow pattern. This option will also have the smaller operating and fixed cost of pumps because the organic and stripping flowrates are at their lower bounds.

6. Conclusions

This work presents the design of a new and cleaner technology with membrane modules which allows the extraction and recovery of pollutants, such as Cr(VI), from industrial effluents or contaminated waste waters. The optimum design of different operating modes and configurations in the continuous process is presented.

The countercurrent operation with product removal before the stripping tank is the best configuration, reducing the required membrane area in the stripping module by 20 %. The organic and stripping flowrates are at their lower bounds at the solution point, thus reducing the fixed and operating costs for pumps. Thus a significant reduction in the total cost can be achieved designing and operating optimally these new plants at the conceptual design stage, helping to improve the insight and the economic viability.

7. Notation

A	Effective surface area, m^2	**Superscripts**	
C	Solute Concentration, mol/m^3	E	Extraction module
C_{in}	Inlet Concentration in the feed, mol/m^3	S	Stripping module
CT	Total Carrier Concentration, mol/m^3	T	Tank
F	Flow rate, m^3/h.	**Subscripts**	
H	Stripping Distribution Coefficient.	a	Aqueous make-up
K	Chemical Equilibrium Constant.	e	Extraction phase
K_m	Membrane mass transfer coef., m/h.	o	Organic phase
L	Fiber Length, m.	oi	Organic interface
z	Axial distance, m.	s	Stripping phase

8. References

Alonso A.I. and C. Pantelides, 1996, J. Memb. Sci. 110, 151.

Alonso A.I., B. Galan, A. Irabien and I. Ortiz, 1997, Sep. Sci. Tech. 32, 1543.

Alonso A.I., B. Galan, A. Irabien and I. Ortiz, 1999, Ind. Eng. Chem. Res. 38, 1666.

Cussler, E.L, 1971, AIChE J. 17(6), 1300.

De Gyves J. and E.R. de San Miguel, 1999, Ind. Eng. Chem. Res. 38, 2182.

D'Elia N.A., L. Dahuron and E.L. Cussler, 1986, J. Memb. Sci. 29, 309.

Eliceche, A.M., A.I. Alonso and I. Ortiz, 2000, Comp. Chem. Engng. 24, 2115.

Gabelman A. and S. Hwang, 1999, J. Memb. Sci. 106, 61.

gPROMS Technical Document, 2000, Process System Enterprise Ltd. UK.

Ho W.S.W and K.K. Sirkar, 1992, Membrane Handbook. Van Nostrand, New York.

Ortiz, I., B. Galan, A. Irabien, 1996, Ind. Eng. Chem. Res. 35, 1369.

Prasad R. and E.L. Cussler, 1988, AIChE J. 34(2), 177.

Wickramansinghe S.R., J. Semmens and E.L. Cussler, 1992, J. Memb. Sci. 69, 235.

Yang M.C. and E.L. Cussler, 1986, AIChE J. 34(1), 1910.

European Symposium on Computer Aided Process Engineering – 12
J. Grievink and J. van Schijndel (Editors)
© 2002 Elsevier Science B.V. All rights reserved.

Feasibility of equilibrium-controlled reactive distillation processes: application of residue curve mapping

C.P. Almeida-Rivera and J. Grievink

Process Systems Engineering Group

DelftChemTech, Delft University of Technology

Julianalaan 136, 2628 BL Delft, The Netherlands

Abstract

The residue curve mapping technique (**RCM**) has been considered a powerful tool for the flow-sheet development and preliminary **design** of conventional multi-component separation processes. It does not only represent a good approximation to actual equilibrium behavior, but also it allows performing feasibility analysis of separation processes where non-ideal and azeotropic mixtures are involved (Malone and Doherty, 2000). Applications of RCM to reactive distillation processes have recently been reported, but a generalized and systematic approach is still missing for reactive feeds outside the conventional composition ranges and represents the aim of this contribution. An RCM-based feasibility analysis has been applied to the synthesis of methyl tert-butyl ether (**MTBE**) at 11 atm and in the presence of an inert component. The reaction space, defined in terms of transformed composition variables, has been divided into sub-regions characterized by separation boundaries. A feasibility analysis of the reactive distillation process has been performed based upon the location of the reacting mixture in the space and initial separation sequences have been generated according to the feed transformed-composition. In all the cases, *high* purity MTBE has been obtained as a product.

1. Introduction

The art of **process design** involves finding equipment sizes, configurations and operating conditions that will allow an economical operation, only by specifying the state of the feeds and the targets on the output streams of a system.

As reported by Fien and Liu (1994), the design task may be assisted by RCM techniques, particularly in the case of non-ideal systems. Traditionally, residue curves have been used to predict the liquid-composition trajectories in continuous distillation units in the ∞/∞ case (*i.e.* infinite number of stages and infinite reflux). Although for finite columns those profiles differ slightly compared to the residue curves under same isobaric conditions, this difference is normally considered negligible at the first stages of design. Analytically, RCM's are constructed based upon physical properties of the system (*i.e.* VL equilibrium, *LL* equilibrium and solubility data), wherein the composition of a non-reacting liquid remaining in the system may be determined by performing overall and component material balances.

When a mixture of n-components undergoes R simultaneous *equilibrium* chemical reactions, the RCM expression may be written in terms of the transformed molar compositions (Ung and Doherty 1995a; 1995b), leading to a similar expression as that obtained for the non-reactive case,

$$\frac{dX_i}{d\tau} = X_i - Y_i, \forall i = 1,...n - R - 1 \tag{1}$$

where: τ is a warped time and X_i and Y_i -the transformed molar compositions of component i in the liquid and vapour phase, respectively- are given by

$$X_i = \frac{x_i - v_i^T \cdot V_{ref}^{-1} \cdot \underline{x}_{ref}}{1 - v_{total}^T \cdot V_{ref}^{-1} \cdot \underline{x}_{ref}}, \forall i = 1,...n - R \tag{2}$$

$$Y_i = \frac{y_i - v_i^T \cdot V_{ref}^{-1} \cdot \underline{y}_{ref}}{1 - v_{total}^T \cdot V_{ref}^{-1} \cdot \underline{y}_{ref}}, \forall i = 1,...n - R \tag{3}$$

where: x_i and y_i are the molar composition of component i in the liquid and vapor phase, respectively; $\underline{x}_{ref}$ and $\underline{y}_{ref}$ are column vectors composed of the liquid and vapor molar compositions of the R reference components, respectively; v_i^T is the row vector of stoichiometric coefficients for component i in the R reactions; v_{total}^T is the row vector of the total mole change in each reaction and V_{ref} is the square matrix of stoichiometric coefficients for the R reference components in the R reactions.

The reactive problem is completed by calculating the phase and chemical equilibrium for the multicomponent VL mixture with multiple reactions:

$$y_i = f(T, P, x_i), \forall i = 1,...n \tag{4}$$

$$k_{eq,R} = \prod_{i=1}^{i=n} (\gamma_i \cdot x_i)^{v_i} = f(x_i, \gamma_i(x_i, T)) \tag{5}$$

where: $k_{eq,R}$ is the chemical equilibrium constant for reaction R and f is a non-linear function of the liquid activity coefficients (γ_i) and molar fractions of the species involved in the reaction.

2. Design problem

The number of equations involved in the reactive problem is: $(n-R-1)$ (expression 1), $(n-R)$ (expression 2), $(n-R)$ (expression 3), n (expression 4), R (expression 5) and 2 constitutive expressions $(\Sigma x = \Sigma y = 1)$, resulting in $4n-2R+1$ equations. The set of involved variables comprises: n (x_i), n (y_i), $n-R$ (X_i), $n-R$ (Y_i), P and T, resulting in $4n-2R+2$ variables. Since the problem has only one degree of freedom, all the unknown variables may be found by specifying a single variable (*e.g. P*) and solving simultaneously the set of differential and algebraic expressions (1 to 5).

2.1 Case study: synthesis of MTBE

In the early 1990's methyl tert-butyl ether (MTBE) was considered one of the most promising clean burning octane enhancers, especially promoted by the increased demand for premium-grade fuels and the elimination of lead gasoline. Although its harmlessness has been lately questioned, the synthesis of MTBE via reactive distillation still represents a major development in the field and an open-door to novel and innovative processes.

2.1.1 Description of the system

MTBE is produced from methanol and isobutene in an equilibrium-limited liquid-phase reaction (Jimenez *et al.*, 2001) catalyzed heterogeneously by a strong acid ion-exchange resin (*e.g.* Amberlyst 15) or homogeneously by sulphuric acid according to the following expression,

$$C_4H_8 + CH_3OH \rightleftarrows C_5H_{12}O \tag{6}$$

The isobutene stream is normally originated from FCC crackers or steam cracking units and therefore, it contains a complete range of butanes and butenes. For the sake of simplifying the study and provided that the reaction is highly selective for isobutene, all the organic components in the isobutene stream are lumped together as *n*-butane.

The thermodynamic model used for this study corresponds to the *Gamma-Phi* formulation, in which the fugacities of each component in the vapor phase equal those in the liquid phase ($y_i \cdot \Phi_i \cdot P = x_i \cdot \gamma_i \cdot p_i^0$). The vapor phase has been assumed to behave ideally, whereas for the liquid phase Wilson activity coefficient equation has been used. The kinetics of MTBE synthesis has been subject of intensive research during the last decade (Izquierdo *et al.*, 1992; Zhang and Datta, 1995) and without loosing generality, the chemical equilibrium constant (expression 5) can be expressed as an algebraic function of temperature.

2.1.2 Simulation data

The thermodynamic and kinetic parameters listed in Ung and Doherty (1995a) have been employed in this study after being corrected to the system pressure (11 atm).

2.1.3 Simulation problem

The simulation problem encompasses the solution of the RCM, phase and chemical equilibrium expressions for the reactive mixture isobutene-methanol-MTBE-*n*-butane at 11 atm. The criterion for choosing the pressure value has been the solubility of isobutene in the liquid phase (Zhang and Datta, 1995).

gPROMS-programming software (Process Systems Enterprise Ltd.) has been used to solve the set of simultaneous differential and non-linear expressions. The solving algorithm is composed of a DAE integrator based on variable step-size/variable order backward differentiation formulae. Initial transformed compositions have been specified to generate the residue lines, covering the complete composition space. The simulations have been performed until singular solutions are found, which correspond to pure component or (non-) reactive azeotropes (*i.e.* $X_i = Y_i$).

3. Feasibility analysis and sequence generation

Residue curve maps provide a powerful tool to represent relevant properties of the system, particularly those aiming to predict feasible design sequences. In addition, analytical material balances may be represented in a RCM, resulting in constraints to feasible product compositions and convenient operating strategies (*e.g.* direct or indirect distillation). The presence of (reactive) singular points in RCM's allows to divide the transformed composition diagram into separate reactive distillation regions by introducing reactive distillation separatrices, which connect two singular points in the composition space.

3.1 RCM-based feasibility analysis

As depicted in Figure 1(left) all the reactive curves -plotted in the transformed composition space- start at either the pure isobutene corner or the methanol-*n*-butane azeotrope (**MNAz**) and end *predominantly* at the pure methanol corner[1]. Furthermore, a saddle point (quaternary azeotrope **QAz**) and a *pseudo*-reactive azeotrope (**PRAz**) located in the hypotenuse of the transformed composition simplex have bee detected.

The pseudo-reactive azeotrope corresponds to an isobutene-methanol-MTBE mixture at high conversion and imposes the limit beyond which distillation cannot proceed (Ung and Doherty, 1995a). Furthermore, according to its stability behavior, the pseudo-reactive azeotrope allows to divide the composition simplex into zones where PRAz behaves like a saddle point (region IV in Figure 1(right)) and like a stable node (region V in Figure 1(right)).

The singular points of the MTBE system define five reactive distillation boundaries (Figure 1(right)): 1. (QAz, MNAz); 2. (QAz, isobutene); 3. (QAz, methanol); 4. (QAz, *n*-butane) and; 5. (QAz, PRAz). Furthermore, they divide the composition space into five sub-regions: I: (QAz, MNAz, *n*-butane); II: (QAz, isobutene, *n*-butane); III: (QAz, methanol, MNAz); IV: (QAz, PRAz, isobutene), and; V: (QAz, PRAz, methanol).

Based upon their contribution in the overall reaction space, regions I, II and III can be neglected at the first stages of design without incurring in measurable errors. This assumption simplifies considerably the feasibility analysis of the process, since all but one of the boundaries have been removed and, therefore, the composition simplex is divided into two well-defined zones:

---

[1] For high inert-containing mixtures the residue curves end at the *n*-butane pure corner.

- zone **A**: where residue curves start at methanol-*n*-butane azeotrope and end at pure methanol corner with a saddle point at the pseudo-reactive azeotrope, and;
- zone **B**: where residue curves start at pure isobutene and methanol-*n*-butane azeotrope and end at high purity MTBE.

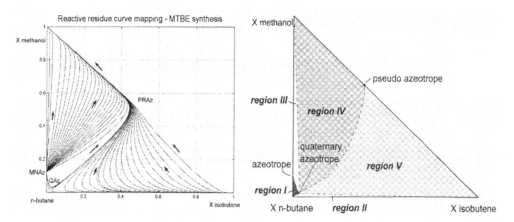

Figure 1. (left) Reactive residue curve for the MTBE synthesis at 11 atm in the transformed composition space. (right) Schematic representation of the distillation boundaries in the reaction space.

3.2 RCM-based sequence generation

The separation sequences differ depending upon the location of the feed in the composition simplex. For zone **B** the sequence is composed of a single column as depicted in Figure 2(left). In this case the fresh feed stream is mixed with a recycle stream (high purity isobutene) before entering the column, while the product stream (pseudo-reactive azeotrope) is continuously removed at the bottom of the column in an indirect separation strategy.

When the feed to the column is embedded within zone **A,** the RD sequencing is not a straightforward task and requires some additional insight originating from the residue curve map. A RD sequence has been generated, according to the approach presented by Fien and Liu (1994) for non-reactive mixtures and involves the following operations (Figure 2(center, right)):

- feed **F** (methanol:isobutene:*n*-butane) is mixed with the recycle stream **B2** to obtain the feed mixture **M1**;
- **M1** is fed to column **C1** and separated into distillate **D1** and bottoms **B1** (pure methanol);
- **D1** enters column **C2** and is divided into the stream **D2** and the recycle stream **B2** (since the distillation boundary is curved, the distillation balance line is able to cross the boundary);
- **D2** is separated following the sequence designed for zone **B** and stream **B1** may be *eventually* recycled back to the first column.

156

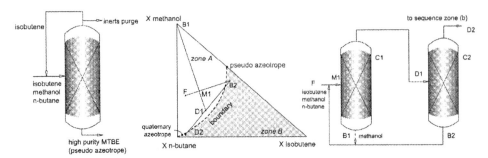

Figure 2. (left) Separation sequence for zone B. (center) RCM design of the separation sequence for zone A. (right) Separation sequence for zone A.

It can be noticed that the pseudo-reactive azeotrope is an intermediate boiling reacting mixture and behaves like a tangent pinch that prohibits designing the separation sequence towards pure MTBE.

4. Conclusions

Residue curve maps have shown to provide valuable insights and design assistance for non-ideal systems, particularly for reactive distillation. Transforming the composition variables according to Doherty's approach allows to define a *reaction space* of lower dimension, formed by attainable product compositions and where the conventional concepts for residue curves can be applied.

The location of the reactive mixture in the reaction space has been found to define the RD structure. For the MTBE synthesis two main sub-regions have been defined and two different separation sequences have been generated, covering *all* the possible reacting mixture compositions. The generated designs allow performing RD processes with reactive feeds outside the conventional composition ranges.

In all the generated sequences, *high* purity MTBE has been obtained as a product, due to the appearance of a pseudo-reactive azeotrope, which imposes limitation to the reaction-separation task.

References

Fien, G.J. and Y.A. Liu, 1994, *Ind. Eng. Chem. Res.*, 33, 2505-2522.
Izquierdo, J., F. Cunill, M. Vila, J. Tejero and M. Iborra, 1992, *J. Chem. Eng. Data*, 37, 339-343.
Malone, M. and M. Doherty, 2000, *Ind. Eng. Chem. Res.*, 39, 3953-3957.
Ung, S. and M. Doherty, 1995a, *Chem. Eng. Sci.*, 55 (1), 23-48.
Ung, S. and M. Doherty, 1995b, *Ind. Eng. Chem. Res.*, 34, 3195-3202.
Zhang, T. and R. Datta, 1995, *Ind. Eng. Chem. Res.*, 34, 730-740.

European Symposium on Computer Aided Process Engineering – 12
J. Grievink and J. van Schijndel (Editors)
© 2002 Elsevier Science B.V. All rights reserved.

Selection of Internals for Reactive Distillation Column - Case-based Reasoning Approach

Yuri Avramenko, Lars Nyström, Andrzej Kraslawski*

Department of Chemical Technology, Lappeenranta University of Technology, P.O.Box 20, FIN-53851, Lappeenranta, Finland. *E-mail: andrzej.kraslawski@lut.fi

Abstract

The design of reactive distillation (RD) systems is considerably more complex than design of the individual, conventional reactors and distillation columns. One of the main issues in the design of RD column is a definition of type and specification of geometric features of the internal devices. In this work, there is described a design supporting tool for pre-selection of packing type for RD column. Case-based reasoning has been applied as a basic method for decision support. The computer system for pre-design of reactive column internals has been built as an implementation of developed methodology. The acid catalysed reaction of 2-methylpropylacetate synthesis was used to test the system. The CBR system significantly reduces the development time for design of reactive distillation column and suggests a good start points for further design activity.

1. Introduction

The design of RD systems is considerably more complex than that of the conventional reactors and distillation columns. It includes several steps (Malone, Doherty, 2000): feasibility analysis, conceptual design, equipment selection and design, operability and control.

There is a lack of well-structured methods for the equipment design of RD column. The development of column internals for a new RD application is usually based on the complicated modelling and carrying out of expensive and time consuming sequences of laboratory and pilot plant experiments. Therefore any method and support system which could improve and speed-up RD design process are of great interest to the designers.

The objective of the presented work is the creation of a decision supporting system for the pre-selection of column internal for RD process and specifically for preliminary phase of column internal design. It provides information on the detailed features and geometric properties of column packing for a new process. Case-based reasoning (CBR) has been applied as a basic method in the system under consideration.

2. Case-based Reasoning Concept

Case-based reasoning imitates a human reasoning and tries to solve new problems by reusing solutions that were applied to past similar problems. CBR deals with very specific data from the previous situations, and reuses results and experience to fit a new problem situation.

158

The CBR process can be described as a cyclic procedure that is presented in Fig.1. The description of a new problem to be solved is introduced in the problem space. During the first step, retrieval, a new problem is matched against problems of the previous cases by computing similarity function, and the most similar problem and its stored solution are found. If the proposed solution does not meet the necessary requirements of a new problem situation, the next step, adaptation, occurs and a new solution is created. A received solution and a new problem together form a new case that is incorporated in the case base during the learning step. In this way CBR system evolves into a better reasoner as the capability of the system is improved by extending of stored experience.

CBR has been used recently in chemical engineering for equipment selection (Kraslawski et al., 1999), process synthesis (Pajula E. et al., 2001), support of preliminary design (Surma and Braunschweig, 1996; Hurme and Heikkilä, 1999), and process control (Roda et al., 1999).

3. Selection Methodology

3.1 Case representation
In this work, a case of column design is presented as a set of attribute-value pairs. Each case is described as a set of features and theirs corresponding values. Based on opinions of experts, there has been selected the set of the essential parameters for the correct identification of the design case. It has then been divided into several parts according to the group of parameters:

- reaction description: names of reactants, products, side-products, class of reaction rate, and conditions of the reaction (temperature, pressure);
- process and operating parameters: such values as flow rate, reflux ratio, description of catalyst applied (particle size range, pocket thickness, total mass etc.);
- reactive packing features: detailed data of packing geometry, its material etc.

First two groups of design parameters are used as the problem description to identify appropriate design combination from the past experience. The last group of characteristics is the solution part.

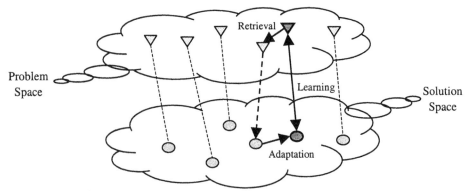

Figure 1. The illustration of CBR process.

3.2 Similarity determination
One of the main problems of CBR is calculation of the similarity between the cases. A design case contains different types of data. All of them can be divided into two classes:

numeric and symbolic. In addition, the symbolic data in this work have been divided into several subclasses: simple, structuring, and classifying.

3.2.1 Global similarity

The evaluation of the similarity between two cases is based on the computation of local similarity of each attribute. The local similarity function deals with a value of a single attribute and takes the values from the interval [0,1]. Thus, the value of the global similarity between two cases is defined as:

$$SIM(A,B) = \sum_{i=1}^{k} w_i \cdot sim(a_i,b_i) / \sum_{i=1}^{k} w_i \qquad (1)$$

where w_i -weight of importance of an attribute, taking the integer values from 0 to 10 ; $sim(a_i,b_i)_I$ -the local similarity for i-attribute of cases A and B;k - number of attributes.

3.2.2. Local similarity for numeric data

If the attribute's data is expressed in the form of numbers, then the local similarity is calculated utilizing simple distance function: the shorter a distance between two points in the problem space the bigger the similarity is.

3.2.3. Local similarity for symbolic data

If a textual attribute contains information that can be encoded in integer numbers, then it is defined as a simple symbolic. The local similarity for this type of variable takes just two values: either 1 if one attribute is completely matched to another, or 0 otherwise.

However, the design features are often expressed as multi-word terms. For such data, a set of keywords is defined to represent the most important features of the domain under consideration. It means that the most significant information is contained in the keywords, and the rest of the text is ignored during the matching. Each attribute is represented as a subset of a full list of keywords available for this type of attribute. They are so-called structuring symbolic attributes. The local similarity of them is defined as

$$sim(a,b) = \frac{|S_a \cap S_b|}{\max(|S_a|,|S_b|)} \qquad (2)$$

where a,b – structuring symbolic attributes; S_a, S_b – subset of keywords of attributes a and b.

Some of the textual data contain values that can be easily divided into classes and a hierarchical structure can be built to show the relations between classes. The local similarity between such attributes is described in a tree-like structure. This approach has been applied to evaluate a similarity between chemical compounds basing on their chemical structure (modified Pajula et al., 2001). According to this principle, so-called similarity tree, composed of the branches and nodes, was created (Fig.2). The root of the tree represents all substances. The first-level nodes in the tree correspond to a basic group of the chemical compounds (Organic/Inorganic). The daughter nodes correspond to classes/subclasses of the chemical substances (hydrocarbons, aromatics, acids, etc).

160

The value of similarity between two compounds depends on the first common level where they have met. For example, methane and propane have the nearest common level "Paraffinic", but benzene and methane have the nearest common level "Organic", which means that the similarity is bigger between methane and propane than between benzene and methane. Each node in the tree has a value that allows to determine the local similarity in a numeric form, e.g. the level "organic" has a similarity 0,1, and the last level corresponding a group of most similar individual substances has the value of local similarity equal to 0,9.

Since one attribute can contain several individual compounds, the similarity of whole attribute has to be defined. First the most similar pairs of the components in past cases and a new problem are found, basing on analysis of matrix of binary similarity of every possible pair. Next, the local similarity of classifying symbolic attribute, for example, between chemical compounds, is defined as follows:

$$sim(a,b) = \frac{1}{m} \cdot \sum_{i=1}^{m} sim_{tree}(a,b)_i \qquad (3)$$

where m - the maximum number of the components in chemical systems; $sim\_tree(n,p)$ - the value of similarity in the nearest common node of i-"best" pair of components. For a component, which has no pair due to not equal number of components, the nearest common node is the root of the tree and the similarity value equals 0.

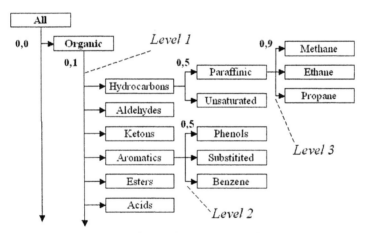

Figure 2. A fragment of the similarity tree for chemical compounds

3.3 Adaptation.

An adaptation phase is needed if the retrieved solutions do not match the expectations of the user of the system. An adaptation procedure is based on the simple algorithm, under the assumption that we can use for adaptation not only the most similar case but the set of the similar cases located near to the new problem in the problem space. The relative distances between new problem and several similar cases in the problem space are transferred into the solution space. The position of a "shadow" of new problem in the solution space identifies an adapted solution. This approach can be applied only if there are several cases in the neighbourhood of a new problem and the distances between them are small enough for linear interpolation.

4. CBR System for Selection of Internals

The described methodology has been implemented as a computer application for preliminary selection of column internals in reactive distillation. The system consists of tree components:

- decision-supporting module, using case-based reasoning for recommendation of a solution (column packing);
- similarity tree editor for assigning a similarity between chemical compounds;
- case-base editor for maintaining the historical design data.

The design data stored in the cases is composed of process data from US patents and commercial packing of the different structure (monolithic, sandwich-like bed, modular). A problem description of a case includes information about chemical reaction, process and catalyst properties. These three groups of design parameters are used to identify appropriate design combination from the past designs. As a result, the type of column internal and its detailed features are retrieved. The obtained values can be modified to fit a new problem. During the last step, learning phase, a new problem with approved solution is stored as a new case in the case base.

2-Methylpropylacetate example

The testing task was to select an appropriate type of packing for synthesis of 2-methylpropylacetate from 2-methylpropanol and acetic acid. The case library does not include any design case of this process application. The information in the system is based on design cases of the production of methyl acetate, butyl acetate, methyl tertiary butyl ether (MTBE), and TAME.

To create a problem description we used the data of pilot plant experiment (Table 1). The class of this reaction rate was defined as moderate, and general process parameters were introduced into CBR system. All chemical compounds were registered in the similarity tree and similarity values between them and others have been computed.

The local similarity values for the numeric attributes (e.g. flow rate, temperature, etc) were determined as reverse values of Eucledean distance in the problem space. Such parameters as class of reaction rate and type of catalyst were defined as simple symbolic type of attributes with a similarity value 1 (exact match) or 0 (not exact match). We used similarity tree and formula (3) to calculate the local similarity function for reactants and products The global similarity was calculated using formula (1). The values of weights are established basing on the experience of the designers. Very often it is treated as proprietary information of the company. In the presented example, we identified the first group of parameters (chemical reaction description) as the most important and set weights of importance to 9-10. The importance of the operating parameters has been lower (5-7), and importance of catalyst information was set to 1-3. As a result, the system retrieved the most similar case (with the highest value of similarity) and provided the detailed information about the suitable packing type for this process. There was selected the corrugated sheet type of packing KATAPAK® manufactured by Sulzer Chemtech Ltd. (Table 2)

Table 1. Problem description of the test case.

Parameter	Value
Reactants	2-methylpropanol; acetic acid
Products	2-methylpropylacetate
Reaction temperature, °C	118
Class of reaction rate	moderate
Feed, acetic acid, kg/h	0,797
Feed, 2-methylpropanol, kg/h	1,203
Bottom product, kg/h	1,5
Distillate (top product), kg/h	0,5
Type of catalyst	autocatalisis

Table 2. Detailed features of selected packing (corrugation sheet type)

Parameter	Value
Corrugation height, mm	14.9
Corrugation angle, mm	42.5
Corrugation length, mm	37
Mesh size, mm	0.5
Wire thickness, mm	0.25
Element diameter, mm	220
Element height, mm	290
Catalyst volume fraction, %	25
Surface area, m^2/m^3	85

5. Conclusions

CBR imitates an engineer's approach of using the past designs to a new situation. The CBR system can significantly reduce the development time for design of reactive distillation column and propose a good start point for the next design phases. The paper presents a method and CBR system for support of preliminary design of equipment. In order to cope with various design data of RD column the corresponding methods (based on keywords sets and similarity tree) have been developed.

6. References

Hurme, M., Heikkila, A-M., 1999, Proc. PRES'99, 341. Budapest

Kraslawski, A., Lyssov, I., Kudra, T., Borowiak, M., Nystrom L., 1999, Comp. Chem. Eng. 23 (Suppl.), 707. Elsevier Science.

Malone M.F., Doherty M.F., 2000, Ind. Eng. Chem. Res. 39, 3953: Reactive Distillation. American Chemical Society.

Pajula, E., Seuranen T., Hurme, M., 2001, European Symposium on Computer Aided Process Engineering, 11, 469. Elsevier Science B.V.

Roda, I. R., Poch, M., Sanchez-Marre, M., Cortes, U., Lafuente, J., 1999, Chem. Eng. Prog., June, 39.

Surma, J., Braunschweig, B., 1996, Eng. Appl. Artif. Intel., v 9, n 4, 385.

European Symposium on Computer Aided Process Engineering – 12
J. Grievink and J. van Schijndel (Editors)
© 2002 Elsevier Science B.V. All rights reserved.

An Industrial Case Study in Simultaneous Design and Control using Mixed-Integer Dynamic Optimization

V. Bansal[a,*], R. Ross[a,†], J.D. Perkins[b], E.N. Pistikopoulos[b,‡] and S. de Wolf[c]

[a] Process Systems Enterprise Ltd., 107a H'smith Bridge Rd., London W6 9DA, U.K.
[b] Dept. of Chemical Engineering, Imperial College, London SW7 2BY, U.K.
[c] Shell International Chemicals B.V., Badhuisweg 3, 1031 CM Amsterdam, The Netherlands.

Abstract

This paper describes an industrial case study concerned with the simultaneous optimization of the design of a high-purity, azeotropic distillation system and its control scheme. The resulting mathematical problem is a large-scale, mixed-integer dynamic optimization (MIDO) one, where the integer variables correspond to discrete decisions such as the determination of the optimal feed tray location. The optimal solution found has a total annualised cost that is 18% less than that of the existing design, which was obtained using a sequential design and control approach.

1. Introduction

This paper is concerned with the process design and control of a two-column distillation system that forms part of a Shell iso-propyl alchohol (IPA) separation train (see Figure 1). This system was previously studied by Ross *et al.* (1999, 2001). In that work, the design of the process equipment and control system's tuning parameters were simultaneously determined by applying a methodology developed by Pistikopoulos, Perkins and co-workers (see Mohideen *et al.*, 1996; Bansal *et al.*, 2000a). The study showed that if such a simultaneous optimization approach had been used when the distillation system was originally designed, then the operability difficulties that had been experienced during its operation would have been avoided. Furthermore, this approach led to a system with a substantially lower total annualised cost than the existing design.

The problem studied by Ross *et al.* is one of the largest dynamic optimization problems whose solution has been reported in the open literature. However, it only considered *continuous* design decisions such as determination of the optimal column diameter and PI-controller gain. The objective of this study is to demonstrate how recently developed, state-of-the-art algorithms for solving mixed-integer dynamic optimization (MIDO)

---

* Current address: Orbis Investment Advisory Limited, 5 Mansfield St., London, W1G 9NG, U.K.
† Current address: United Technologies Research Center, East Hartford, CT 06108, U.S.A.
‡ Corresponding author. Tel: +44 (0)20 7594 6620. Fax: +44 (0)20 7594 6606. E-mail: e.pistikopoulos@ic.ac.uk.

problems can be utilised to also make *discrete* process and control design decisions such as where to optimally locate the feed tray in the IPA column.

2. Problem Statement

The problem to be solved can be stated in the following terms. The objective is to design the distillation system and its required control system at minimum total annualised cost, capable of feasible operation over the whole of a given time horizon in the face of disturbances in the feed flow rate and composition. As in the study of Ross *et al.*, feasibility is defined through the satisfaction of the following constraints: (i) composition specifications on the NPA and water in the IPA/water azeotrope product from the IPA column; (ii) a composition constraint on the IPA in the bottoms stream from the NPA column; (iii) minimum column diameters to avoid flooding in each column; and (iv) fractional entrainment limits in each column. Solution of the problem thus requires the determination of:

1. The optimal process design, in terms of the locations of the feed and draw-off trays in the IPA column (discrete decisions), and the columns' diameters, reboilers' heat duties and flow rates of the draw-off stream to the NPA column and the return stream back to the IPA column (continuous decisions); and

2. The optimal control design, in terms of the tuning parameters of the control loops used. For this study, the control loops under consideration are the same as those used by Ross *et al.* and the tuning parameters to be determined correspond to the gain of the ratio controller in the feedforward loop and the gain and reset time of the PI-controller in the feedback loop. In the feedback loop, the controlled variable is the average of the temperatures on the trays two and four above the IPA feed tray.

3. Modelling Aspects

The high-fidelity, dynamic model of Ross *et al.* was also used in this study but with the introduction of binary variables yf_k and yd_k, $k=1,...,30$, in order to account for the locations of the feed and draw-off trays, respectively, within the bottom thirty trays of the IPA column. Thus, $yf_k = 1$ if the feed enters tray k and is 0 otherwise; similarly $yd_k = 1$ if liquid is drawn-off from tray k to the NPA column and is 0 otherwise. The following new equations and constraints apply:

- Only one tray receives the feed and only tray is used as a draw-off:

$$\sum_{k=1}^{30} yf_k = \sum_{k=1}^{30} yd_k = 1. \tag{1}$$

- The draw-off tray cannot be located below the feed tray:

$$yf_k - \sum_{k'=1}^{30} yd_{k'} \leq 0, \quad k = 1,\ldots,30. \tag{2}$$

- The feed and draw-off trays must be even-numbered due to the arrangement of "inner" and "outer" trays within the column:

$$yf_{2j-1} - yd_{2j-1} = 0, \quad j = 1,\ldots,15. \tag{3}$$

- The feed flow rate to tray k:

$$Feed_k = Feed \cdot yf_k, \quad k = 1,\ldots,30. \tag{4}$$

- The draw-off flow rate from tray k:

$$Draw_k = Draw \cdot yd_k, \quad k = 1,\ldots,30. \tag{5}$$

- In keeping with the original model given by Shell, the free area of the trays below the feed is different than the free area of the feed tray and above:

$$A_{free,k} = 0.079 \cdot D_{IPA}^2 \cdot \sum_{k'=k+1}^{30} yf_{k'} + 0.115 \cdot D_{IPA}^2 \cdot \left(1 - \sum_{k'=k+1}^{30} yf_{k'}\right), \quad k = 1,\ldots,29,$$
$$\tag{6}$$

$$A_{free,k} = 0.115 \cdot D_{IPA}^2, \quad k = 30,\ldots,70. \tag{7}$$

4. Problem Formulation

In this study, the simultaneous process and control design optimization problem has the following mathematical form:

$$Cost = \min_{\mathbf{d},\mathbf{y}} \left(C_{cap} + C_{op}\right),$$

$$s.t. \quad \mathbf{h_d}\left(\dot{\mathbf{x}}_\mathbf{d}(t), \mathbf{x_d}(t), \mathbf{x_a}(t), \mathbf{i}(t), \mathbf{d},\mathbf{y},t\right) = 0,$$

$$\mathbf{h_a}\left(\mathbf{x_d}(t), \mathbf{x_a}(t), \mathbf{i}(t), \mathbf{d},\mathbf{y},t\right) = 0,$$

$$\mathbf{h_0}\left(\dot{\mathbf{x}}_\mathbf{d}(0), \mathbf{x_d}(0), \mathbf{x_a}(0), \mathbf{i}(0), \mathbf{d},\mathbf{y},t = 0\right) = 0, \tag{8}$$

$$\mathbf{g_e}\left(\dot{\mathbf{x}}_\mathbf{d}(t_f), \mathbf{x_d}(t_f), \mathbf{x_a}(t_f), \mathbf{i}(t_f), \mathbf{d},\mathbf{y},t_f\right) = 0,$$

$$\mathbf{g_y}(\mathbf{y}) = 0.$$

In (8), C_{cap} and C_{op} are the total annualised capital and operating costs, respectively; $\mathbf{h_d}=0$ and $\mathbf{h_a}=0$ represent the differential and algebraic equations, respectively; $\mathbf{h_0}=0$ is the set of initial conditions; and $\mathbf{x_d}$ and $\mathbf{x_a}$ are the differential and algebraic variables.

$g_e \leq 0$ denote end-point constraints which result once the original path constraints listed in Section 2 have been converted using the method described in Section 7.2.10.2 of Bansal (2000). $g_y \leq 0$ is the set of pure integer constraints such as (1) to (3). The two disturbances in this study, $v(t)$, are a step change in the feed flow rate from 115 to 135 *t/hr.* and a sinusoidal variation in the IPA feed mass fraction (nominal 0.12, ±5% amplitude, 3 *hr.* period). d is the set of six continuous process design variables and three controllers' tuning parameters described in Section 2; and y is the set of binary variables yf_k and yd_k, $k=1,...,30$.

5. Solution and Results

(8) corresponds to a MIDO problem with 509 differential equations, 8295 algebraic equations (not including the 20,000 or so physical properties equations), 475 end-point inequalities, 32 pure binary constraints, 9 continuous search variables and 60 binary search variables. This formidably-sized problem was solved using the MIDO algorithm described in Chapter 6 of Bansal (2000). This algorithm has already been used to solve industrial-type problems (Chapter 7 of Bansal, 2000 and Bansal *et al.*, 2000b). In this study, gPROMS v1.8.4 (Process Systems Enterprise Ltd, 2000) was used to solve the primal problems and GAMS v2.5/CPLEX (Brooke *et al.*, 1992) was used to solve the master problems.

Using a termination tolerance of $\varepsilon = 0$, the MIDO algorithm converged in just 8 iterations and was able to find three structures that are cheaper than the original structure considered in the study of Ross *et al.* (1999,2001). The optimal solution was found on the fourth primal solution, and as shown in Table 1, its total annualised cost is 18% less than the existing Shell design. These savings are achieved by removing more NPA from the IPA column (increased draw-off rate) so that the IPA column capacity can be significantly reduced (smaller diameter and reboiler duty). Although the capacity and hence cost of the NPA column increase, the savings in the cost of IPA column easily outweigh this.

Figure 2 plots the mass fraction of NPA in the IPA/water azeotrope product for the optimally designed system. It can be seen that the constraint that the mass fraction is less than 800 ppm is satisfied at all times during operation. Other plots like this one are automatically given as part of the MIDO solution. Note that Figure 2 shows the dynamic operation over one week, which is more than ten times longer than the 15 *hr.* time horizon that was used for solving the primal dynamic optimization problems. The cyclical behaviour and the fact that the constraint is never violated thus suggest that the optimally designed process and control system is also stable.

6. Concluding Remarks

The previous study by Ross *et al.* (1999, 2001) illustrated the economic and operability benefits of simultaneously considering process design and control decisions within an optimization framework, compared to traditional approaches where process design and process control are considered sequentially. This study has demonstrated that further benefits are possible by also considering discrete decisions within the optimization framework with the result that a design has been found for the IPA/NPA system that is 18% cheaper than the existing design. In order to achieve this, very difficult, mixed-integer dynamic optimization problems need to be solved. This study has also shown that the state-of-the-art algorithms for solving such problems have progressed to a point where complex, industrial problems can be readily tackled.

References

Bansal, V, 2000, Analysis, Design and Control Optimization of Process Systems under Uncertainty. PhD Thesis, University of London.

Bansal, V., R. Ross, J.D. Perkins and E.N. Pistikopoulos, 2000a, The Interactions of Design and Control: Double-Effect Distillation, J. Proc. Control 10, 219.

Bansal, V., R. Ross, J.D. Perkins, E.N. Pistikopoulos and J.M.G. van Schijndel, 2000b, Simultaneous Design and Control Optimisation under Uncertainty, Comput. Chem. Eng. 24, 261.

Brooke, A., D. Kendrick and A. Meeraus, 1992, GAMS Release 2.25: A User's Guide. The Scientific Press, San Francisco.

Mohideen, M.J., J.D. Perkins and E.N. Pistikopoulos, 1996, Optimal Design of Dynamic Systems under Uncertainty, AIChE J. 42, 2251.

Process Systems Enterprise Ltd., 2000, gPROMS Advanced User Guide, http://www.psenterprise.com.

Ross, R., V. Bansal, J.D. Perkins, E.N. Pistikopoulos, G.L.M. Koot and J.M.G. van Schijndel, 1999, Optimal Design and Control of a High-Purity Industrial Distillation System, Comput. Chem. Eng. 23, S875.

Ross, R., J.D. Perkins, E.N. Pistikopoulos, G.L.M. Koot and J.M.G. van Schijndel, 2001, Optimal Design and Control of an Industrial Distillation System, Comput. Chem. Eng. 25, 141.

Table 1: Comparison of the Design obtained via MIDO with the Existing Design.

Design Variable	Optimal Design using MIDO	Existing Design
Draw-off Location	18	22
Feed Location	8	14
Q_{reb} (IPA) (MW)	14.67	19.54
D_{col} (IPA) $(m.)$	2.58	3.17
$Draw$ (t/hr)	4.20	1.69
$Return$ (t/hr)	3.43	1.20
Q_{reb} (NPA) (MW)	2.45	0.87
D_{col} (NPA) $(m.)$	1.33	0.87
Capital Cost $(\$M\ yr^{-1})$	0.56	0.63
Operating Cost $(\$M\ yr^{-1})$	3.56	4.37
Total Cost	4.12	5.00

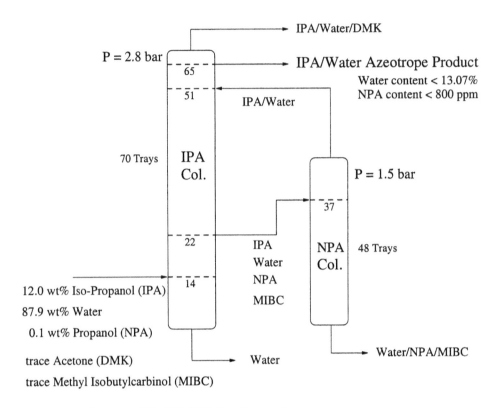

Figure 1: Schematic of the IPA/NPA Distillation System.

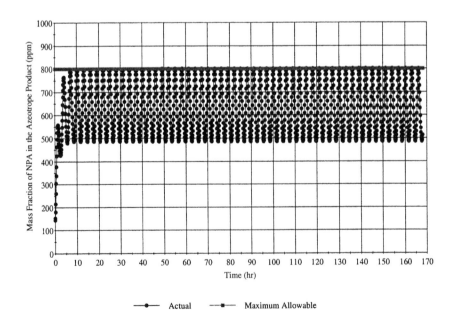

Figure 2: Mass Fraction of NPA in the Azeotrope Product.

European Symposium on Computer Aided Process Engineering – 12
J. Grievink and J. van Schijndel (Editors)
© 2002 Published by Elsevier Science B.V.

Logic-Based Methods for Generating and Optimizing Thermally Coupled Distillation Systems.

José A. Caballero[+] and Ignacio E. Grossmann[*]

+ Departamento de Ingeniería Química Universidad de Alicante. Ap. Correos 99 E-03080 SPAIN
* Chemical Engineering Department. Carnegie Mellon University. Pittsburgh PA, 15213. USA

Abstract

The separation of a non-azeotropic mixture of N components in their pure components can be performed using from conventional to fully thermally coupled distillation sequences, while for conventional columns there is a one to one match between columns and separations tasks this is not true for (partially or fully) thermally coupled distillation systems where it is possible to find a large number of thermodynamically equivalent configurations for a single sequence of separation tasks. A systematic way of generating all those thermodynamically equivalent configurations is presented based on logic propositions. An equation for calculating the number of possible thermodynamically equivalent alternatives is presented as well.

1. Introduction

In 1949 Wright patented a fractionation apparatus for ternary distillation. This is the first known fully thermally coupled distillation configuration for ternary distillation. It is now known as a divided wall column. Later Petlyuk (1965) studied a particular class of fully thermally coupled (FTC) distillation systems for separating an N component mixture. The most important aspect of this new configuration is that only one reboiler provides the heating demand of the system. For a mixture of three components the Petlyuk configuration reduces the total vapor flow by 10 to 50% compared to conventional systems using direct and indirect split arrangements (Triantafyllou and Smith, 1992). This result can in principle be extended to separations of N components.

Between the fully thermally coupled distillation (only one reboiler and one condenser) and the sequences of conventional columns (each column with condenser and reboiler) there are a large number of 'partially thermally coupled configurations' including for example side strippers and side rectifiers among other. A systematic for generating all the possible sequence of separation tasks from conventional systems to fully thermally coupled, including all the partially thermally coupled configurations can be found in Caballero and Grossmann (2001). Agrawal (1996) showed a procedure for drawing superstructures for fully thermally coupled distillation systems.

A given sequence of separation tasks can be performed by N-1 distillation columns (being N the number of components – divided wall columns are not considered here), however, only in sequences using conventional columns there is one to one match between separation task and column that perform that separation. For fully or partially thermally coupled distillation sequences there are different rearrangement of columns all of them performing the same sequence of tasks. These are thermodynamically equivalent configurations. Figure 1 illustrate this point for a mixture of 3 components.

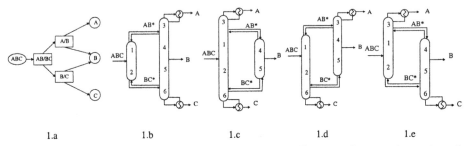

| 1.a | 1.b | 1.c | 1.d | 1.e |

Figure 1. Sequence of tasks (1.a) and thermodynamically equivalent configurations for that sequence (1b to 1e).

In this work we present a systematic way of generating all the thermodynamically equivalent structures for a given sequence of separation tasks. Agrawal (2000) presented a procedure for drawing 'at hand' the thermodynamically equivalent structures, but it is not clear how to introduce it in an optimization framework.

2. Thermodynamically equivalent structures

The problem addressed in this section can be stated as follows: given is a sequence of tasks for separating a mixture of N components in N pure streams. The objective is to obtain all the possible thermodynamically equivalent arrangements of column sections in N-1 thermally coupled distillation columns. For a mixture of three components Figure 1 illustrates the objective of this part.

The starting point in this case is a sequence of feasible tasks (and states) to perform a given separation (a solution of the problem outlined in previous section). As an illustrative example consider the feasible sequence of tasks and states presented in Figure 2. This example corresponds to the separation of a mixture of 5 components using 20 column sections – N(N-1) – (Note that one possible rearrangement is that proposed by Sargent and Gaminibandara, 1976).

The problem of finding all the possible rearrangements in N-1 columns can be reduce to solve a set of logical relationships among tasks and states.

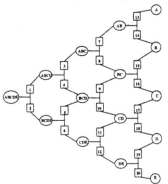

Figure 2. Sequence of separation tasks and states (circles) for separating a mixture of 5 components using 10 separation tasks. Each square represent a column section. A tasks is performed by 2 column sections.

Let us introduce the following index sets

SECTION	$\{s \mid s$ is a column section$\}$
COLUMN	$\{c \mid c$ is a column $\}$
STATE	$\{e \mid e$ is a state$\}$
PURE	$\{k \mid k$ is a pure product$\}$

REC_e { sections that give rise to state e coming from a rectifying section}
STR_e { sections that give rise to state e coming from a stripping section}
T_REC_e { rectifying section produced by state e}
T_STR_e { stripping section produced by state e}

Note that pure products can be considered also as a state, the set PURE can be considered a subset of STATE if desired.

Let now define the Boolean variable $P_{s,c}$ that takes the value of True if the section s is assigned to the column c, and False otherwise. The following set of logical relationships define completely the problem:

1.- Each column has at least one section

$$\bigvee_{s \in SECTION} P_{s,c} \quad \forall c \in COLUMN$$

2.- A given section can only be assigned to one column

$$\underline{\bigvee_{c \in COLUMN}} P_{s,c} \quad \forall s \in SECTION$$

3.- Connectivity relationships.
3.1.- There are no heat exchanger associated to state e. These relations are deduced directly from the structure of tasks and states. It can be stated as follows: Among all the sections that reach or exit from a given state only two, a rectifying section and a stripping section, must be assigned to a given column c.

$$P_{r,c} \Rightarrow P_{l,c} \vee P_{s,c} \quad r \in T\_REC_e \; ; \; l \in T\_STR_e \; ; \; s \in REC_e; \quad \forall c \in COLUMN$$

$$P_{l,c} \Rightarrow P_{r,c} \vee P_{s,c} \quad r \in T\_REC_e \; ; \; l \in T\_STR_e \; ; \; s \in STR_e; \quad \forall c \in COLUMN$$

$$P_{r,c} \Rightarrow P_{l,c} \vee P_{s,c} \quad r \in REC_e \; ; \; l \in T\_REC_e \; ; \; s \in STR_e; \quad \forall c \in COLUMN$$

$$P_{l,c} \Rightarrow P_{r,c} \vee P_{s,c} \quad r \in T\_STR_e \; ; \; l \in STR_e \; ; \; s \in REC_e; \quad \forall c \in COLUMN$$

$$\neg P_{r,c} \vee \neg P_{l,c} \vee \neg P_{s,c} \quad r \in REC_e; \; l \in T\_REC_e \; ; \; s \in T\_STR_e \quad \forall c \in COLUMN$$

$$\neg P_{r,c} \vee \neg P_{l,c} \vee \neg P_{s,c} \quad r \in STR_e; \; l \in T\_REC_e \; ; \; s \in T\_STR_e \quad \forall c \in COLUMN$$

The first four relations are feasible relations among sections for a given state. The other two equations assure that only two of those sections are assigned to column c.

3.2.- There is heat exchanger associated to state e. In this case the stripping and rectifying sections (T_REC$_e$. T_STR$_e$) of state e must be in the same column.

$$P_{r,c} \Leftrightarrow P_{s,c} \quad r \in T\_REC_e, \ s \in STR_e \quad \forall c \in COLUMN$$

4.- The two sections produced by the feed state must be assigned to the same column:

$$P_{s1,c} \Leftrightarrow P_{s2,c}$$

In some special cases this last relation can be relaxed (the feed is introduced by the top or the bottom of a column).

5.- For the products of intermediate volatility that leave the system, the rectifying and stripping section must be assigned to the same column. Of course this relationship will hold only if there is no heat exchanger associated to that pure product.

$$P_{s,c} \Leftrightarrow P_{l,c} \quad s \in REC_e \ ; \ l \in STR_e; \ e \in PURE; \ \forall c \in COLUMN$$

Previous logical relationships can be if desired transformed in a algebraic equations and solve it like an MILP (Raman and Grossmann, 1991, 1993, 1994).

Another interesting point is knowing what is the total number of thermodynamically equivalent configurations for a given sequence of tasks. Agrawal (2000) presented a procedure for calculating the number of sequences, but for systems with more that 4 components it seems that forget some configurations. The following formula allows calculate the total number of rearrangements.

$$N.C = 2^{\left(NT - NHE_I - 1\right)}$$

where NT is the number of separation tasks, NHE$_I$ is the number of heat exchangers of class I (associated to non pure products –heat exchanger associated to a stream that do not leave the system-) in the structure, and N.C is the total number of configurations. Some special cases are of interest: First, when only two heat exchangers are used and the number of column sections is N(N-1) (like in example of Figure 2) In this case the total number of configurations can be related to the number of components to be separated (N):

$$N.C. = 2^{\left(\frac{N(N-1)}{2} - 1 \right)}$$

And second, If we want only sequences of tasks using the minimum number of sections then the total number of possible rearrangements as a function of the total number of components and heat exchangers, is:

$$N.C. = 2^{\left(\frac{4N-6}{2} - NHE + 1 \right)}$$

Due to the obvious lack of space in this paper cannot be presented the 512 alternatives of the example, however as an example Figure 3 presents the 8 alternatives for the separation of a mixture of 5 components using 5 heat exchangers and Figure 4 the 16 possible rearrangements of a mixture of 4 components for a sequence of tasks using the minimum number of column sections.

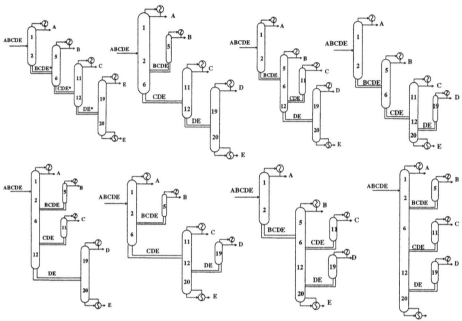

Figure 3.(cont) Thermodynamically equivalent configurations for the sequence of tasks {A/BCDE, B/CDE, C/DE, D/E} using 5 heat exchangers.

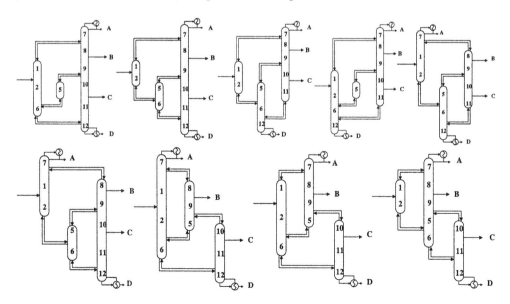

Figure 4. All the possible rearrangements in 3 columns for a mixture of 4 components for a sequence of tasks* using the minimum number of column sections and only two heat exchangers

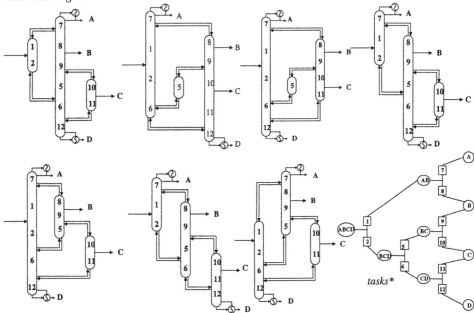

Figure 4 (cont): All the possible rearrangements in 3 columns for a mixture of 4 components for a sequence of tasks* using the minimum number of column sections and only two heat exchangers

References

Agrawal, R. 1996. Synthesis of distillation column configurations for a multicomponent separation. Ind. Eng. Chem. Res. 35. 1059-1071.

Agrawal, R. 2000. A Method to draw fully thermally coupled distillation column configurations for multicomponet distillation. Trans. Inst. Chem. Eng. 78A 454-464

Caballero J.A. And Grossmann I.E.; 2001. Generalized Disjunctive Programming Model for the Optimal Synthesis of Thermally Linked Distillation Columns. Ind. Eng. Chem. Res. 40(10) 2260-2274.

Petlyuk, F.B; Platonov, V.M.; Slavinskii, D.M; 1965. Thermodynamically optimal method for separating multicomponent mixtures. Int. Chem. Eng. 5(3) 555.

Raman R. and Grossmann I.E.; 1991. Relation Between MILP modeling and logical inference for chemical process synthesis. Comput. Chem. Eng. 15. 73;

Raman R. and Grossmann I.E.; 1993. Symbolic integration of logic in mixed integer linear programming techniques for process synthesis. Comput. Chem. Eng. 17.909.

Raman R. and Grossmann I.E.;1994. Modeling and computational techniques for logic based integer programming. Comput. Chem. Eng. 18. 563.

Sargent, R. W. H. and Gaminibandara K. Introduction: Approaches to Chemical Process Synthesis. In Optimization in action.1976. Dixon L.C. Ed. Academic Press: New York.

Trantafyllou C; Smith, R.; 1992. The design and optimization of fully thermally coupled distillation columns. Trans. Inst. Chem. Eng. 70A. 118-132.

European Symposium on Computer Aided Process Engineering – 12
J. Grievink and J. van Schijndel (Editors)
© 2002 Elsevier Science B.V. All rights reserved.

Entrainer-Assisted Reactive Distillation:
Application to Process Synthesis for Fatty Esters

Alexandre C. DIMIAN, Florin OMOTA and Alfred BLIEK

Department of Chemical Engineering, University of Amsterdam,
Nieuwe Achtergracht 166, 1018 WV Amsterdam, The Netherlands
Correspondent: alexd@science.uva.nl

Abstract

A new approach in designing reactive distillation processes is presented: the use of an entrainer. This can help to overcome limitations due to distillation boundaries, and in the same time to increase the degrees of freedom in design. As example the catalytic esterification of fatty acids with 1-propanol is studied. The entrainer enhances the water removal and increases the amount of alcohol recycled inside the reaction space. A major benefit is a significant reduction of the amount of catalyst compared with the situation without entrainer.

Introduction

Previous research (Omota, Dimian, and Bliek, 2001) demonstrated the feasibility of a continuous process for fatty acid esterification based on catalytic Reactive Distillation (RD) in a multi-product configuration. As substrates we considered the lauric acid (C12), as well as 2-ethyl-hexanol (C8) and methanol (C1), the highest and the lowest boilers in the series C1-C8 alcohols. We found that various fatty esters may be produced in the same RD column, but with different operation policies: alcohol reflux for high-boiling alcohol (C8), and acid reflux for low-boiling alcohol (C1). The last strategy could be in principle applied for intermediate alcohols, as C3-C4, but problems might arise because of high sensitivity of purity with the reflux ratio.

The use of a Mass Separation Agent (entrainer) in RD can enhance simultaneously both water removal and the internal alcohol recycle. A second unit for alcohol recovery is no longer necessary, or reduced to a simple stripper. In this way the same flowsheet configuration may be preserved for different substrates.

Literature search showed that the use of an entrainer in reactive distillation processes has been not reported so far. Only Bock et al. (1997) described a continuous process for the esterification of myristic acid with iso-propanol in a two-columns set-up. The acid used as homogeneous catalyst was lost, and the recovery column recycled in fact the azeotrope water/iso-propanol.

The conceptual design of reactive distillation systems for fatty acid esterification arises to a number of scientific issues. The Residue Curve Methods (RCM) developed for the separation of azeotropic mixtures cannot be directly applied, because of the chemical reaction, as well as the complex three-phase behaviour of the reaction mixture. Simultaneous phase and chemical equilibrium is a computational problem in itself. We should also add the absence or the uncertainty of some physical properties, which could compromise the feasibility of the proposed design. That is why a minimum of experimental research is necessary, but this should be guided by systematic conceptual methods. Rather than an algorithm, we should speak about a methodology in developing innovative ideas in this area.

Problem definition

As reactants we consider lauric (dodecanoic) acid with 1-propanol, and sulphated zirconia as solid catalyst. Homogeneous reaction conditions and high degree of water removal from the reaction medium is a key feature, necessary both to shift the equilibrium reaction to completion, and to protect the catalyst against deactivation. Some important thermodynamic properties are given in Table 1. The components form azeotropes with each other. The ester is the highest boiler, followed by the acid, and at a large distance by water and 1-propanol. We expect that the RD column should operate under vacuum, or/and the product be diluted with some alcohol, which can be recycled after product conditioning. Without further treatment, the top distillate is the azeotrope 1-propanol/water. Thus, a solution for the separation of water product and the recycle of alcohol should be found.

Table 1. Key thermodynamic properties

Normal boiling points	Dodecanoic acid	1-propyl dodecanoate	1-propanol	Water
T, K	571.75	574.95	370.35	373.15
Azeotrope	Acid/Ester	Acid/Water	Ester/Water	Alcohol/Water
	Homogeneous	Heterogeneous	Heterogeneous	Homogeneous
T, K	568.9	373.15	373.13	360.86
Composition	$y_{az}=0.5784$	$x_1=0$, $x_2=0.7195$ $y_{az}=0.0001$	$x_1=0.0746$ $x_2=1$ $y_{az}=0.9996$	$y_{az}=0.4330$

Methodology

The methodology consists of an evolutionary research of the feasible design space by a systematic combination of thermodynamic analysis, computer simulation and experimental research. The environment of constraints ensures a proper bounding of the feasible design space. The approach can be decomposed in the following steps:
1. Property generation and estimation when missing.
2. Simultaneous Chemical and Phase equilibria, and reduction of the design space.
3. Preliminary equilibrium-based design: solvent selection, feasibility by RCM analysis and finite reflux calculations, rigorous simulation, and analysis of data uncertainty by sensitivity analysis.
4. Thermodynamic experiments identified by simulation.
5. Reviewed equilibrium-based design.
6. Catalyst study and kinetic experiments.
7. Kinetic design and optimisation.
The main elements of the approach will be illustrated in the next sections.

Entrainer in Reactive Distillation

Fig. 1 illustrates conceptual alternatives of a RD process by the fatty acids' esterification with light alcohols that may form homogeneous azeotropes. The bottom product of the RD column, the fatty acid ester, might contain small amount of alcohol and fatty acid. The presence of alcohol is preferred, because it can be easily separated. Contrary, the fatty acid cannot be removed by distillation due to its high boiling point, comparable with the fatty ester. Three situations may be distinguished:

Case 1: Acid reflux (A_1), L-L separation with quantitative water removal (A_2), and acid recycle (A_R). The alcohol is completly converted in the reactive zone (Omota et al., 2001), such as no auxiliary column is needed for alcohol recovery.

Case 2: Alcohol/water top product (B_1), homogeneous azeotropic distillation for water removal (B_2) and azeotrope recycle (B_R). The process requires an additional distillation column for alcohol recovery as homogeneous azeotrope. Recycling the alcohol carries with a large amount of water in the reaction zone.

Case 3: Reactive distillation with excess of alcohol (C_1), heterogeneous azeotropic distillation (C_2), liquid decanting (C_3) and entrainer recycle (C_R). This case will be investigated in the present research.

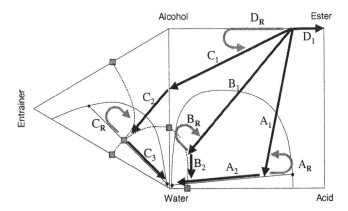

Figure 1. Three conceptual design alternatives for fatty acid esterification with light alcohols forming homogeneous azeotropes.
◼ *azeotrope;* (·······) *distillation boundary;* (⟵) *two-liquid phase splitting.*

Entrainer selection

Pham and Doherty (1990), and recently Stichlmair and co-workers (1999) proposed generic rules for entrainer selection in azeotropic distillation. The peculiar physical properties of the components implied in the fatty acid esterification allows the formulation of the following three conditions for the entrainer selection:

1. Form *heterogeneous* azeotrope (lowest boiler) with alcohol and water, or a binary heterogeneous azeotrope with water in the most preferable case.
2. Low solubility of both entrainer and alcohol in the water-phase after L-L split.
3. Give acceptable impurity in the final product.

Preliminary (non-exhaustive) investigation showed that some oxygenated components, as esters, as well as hydrocarbons might be candidates. Fig. 2 presents three representative RCM's calculated with Aspen Split^{TM} 10.2.

Case (a) makes use of 1-propyl acetate (nbp 374.6 K). This forms homogeneous azeotrope with 1-propanol (nbp 368.4 K), heterogeneous azeotrope with water (nbp 355.6 K), as well as a heterogeneous ternary azeotrope (nbp 355.2 K), which is the lowest boiler. From the ternary azeotrope the water can be separated quantitatively by decantation, while carrying with only a very limited amount of alcohol.

178

Case (b) employs a very light hydrocarbon, n-pentane. Water may be separated as binary azeotrope followed by decantation. Low miscibility ensures quantitative water removal, but low top temperature is not convenient. In the case (c) a higher boiler hydrocarbon is used, o-xylene. This time there is a ternary heterogeneous azeotrope, but after decantation the water-phase has a composition close to the homogeneous binary water-alcohol, so there is no advantage. Hence, 1-propyl acetate fulfills at best the above conditions.

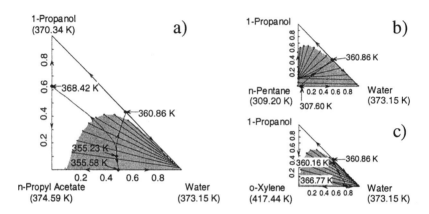

Figure 2. Solvent selection for 1-propanol/water separation, based on ternary diagrams: a) n-Propyl Acetate, b) n-Pentane, and c) o-Xylene.

Chemical and phase equilibrium

Fig. 3a presents the true thermodynamic equilibrium constant K_a, as well as K_x, and K_y determined by UNIQUAC. Thermo-chemical data and interaction parameters have been calibrated by means of experimental data, which finally verified the results predicted by simulation. Fig. 3b displays in transformed co-ordinates X_1 (acid +water), X_2 (acid+ ester) the simultaneous chemical and phase equilibrium diagram calculated by Gibbs free energy minimization. A temperature higher than 373 K is necessary to ensure that the reaction takes place only in a homogeneous system.

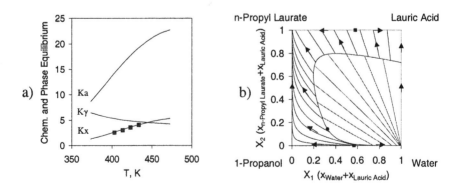

Figure 3. a) Chemical and phase equilibrium (■ experimental data), and b) RCM.

Kinetics based design

Fig. 4 presents the flowsheet. Liquid acid and vapour alcohols are introduced at the top and at the bottom of the reaction zone at the temperatures of 453 K and 380 K, respectively. The product containing 3% 1-propanol is sent to an evaporator, from which high purity fatty ester is obtained, while the alcohol is recycled. The vapour leaving the reaction zone enters a distillation section in counter-current with the entrainer-rich reflux. The top column has the composition of a binary or a ternary azeotrope. After condensation and decantation, the water-phase is removed as by-product, or sent to a simple stripping device with few stages. Note that the recovered alcohol with some water is recycled to the decanter and not to the RD column. This is a key difference with the homogeneous azeotropic set-up mentioned (Bock and al., 1997), where a substantial amount of water was recycled into the RD column.

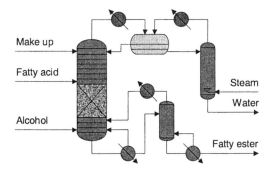

Figure 4. The esterification of lauric acid with 1-propanol using entrainer.

Fig. 5a presents profiles based on kinetic model simulated with Aspen Plus™ 10.2. We considered kinetic data by analogy with the system lauric acid/1-ethyl-hexanol (Bliek et al., 2001). It may be observed that the entrainer enhances the water extraction from the reaction zone, and increases the amount of alcohol recycled. Higher reaction rate reduces significantly the amount of catalyst, up to 40%, compared with the acid reflux and total alcohol consumption (Case 1 in Fig. 1).

The top composition is placed in the heterogeneous ternary mixture of alcohol-water-entrainer. By liquid decantation at 353 K the mixture splits into an aqueous-phase containing 98 % water and a recycled organic phase.

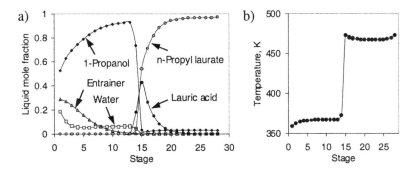

Figure 5. Profiles in RD column: a) Concentrations, and b) Temperature.

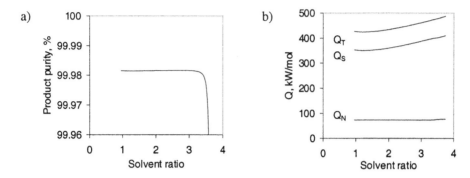

Figure 6. Effect of solvent ratio on product purity and energetic consumption.

In the reactor zone the acid concentration decreases sharply after the first 7 stages (Fig. 5a). The remaining 7 reactive stages are still required to achieve high purity. The temperatures of the reaction and distillation zones are quite different (Fig. 5b), corresponding to reaction and separation requirements. Therefore, at the junction of the two zones a heater is needed to compensate the enthalpy difference.

Product purity remains high over a large interval of the entrainer/reactant ratio. Minimum and maximum values have been identified (Fig. 6a). The first ensures a stable L-L split, the second is reached when the entrainer contaminates the bottom product. Energy increases with the solvent/reactant ratio (Fig. 6b). While the reboiler duty remains constant, the side-heater duty is proportional. Therefore, solvent/reactant ratio turns out to be an optimisation variable between purity and energy requirements. In this case the optimum entrainer/reactant ratio is about 1.3. The simulation shows that 40% less catalyst is necessary when using an entrainer compared with the alternative with total alcohol consumption (Case 1 above).

Conclusions

The use of an entrainer as Mass Separation Agent in reactive distillation can bring significant advantages, as demonstrated by the catalytic esterification of lauric acid with 1-propanol. The simultaneous removal of alcohol with water from the reaction space is prevented. A second distillation unit for alcohol recovery becomes unnecessary, or limited to a simple stripping column. The increased amount of alcohol recycled in the reaction zone has a positive effect on the reaction rate. Therefore, the catalyst loading can be reduced substantially.

References

Bock, H., G. Wozny and B. Gutsche, 1997, Design and control of a reaction distillation column including recovery system, Chem. Eng. & Proc., 36, 2, 101-109.

Omota, F., A. C Dimian and A. Bliek, 2001, Design of reactive distillation process for fatty acid esterification, Proceedings Escape-11, Elsevier, 463-469.

Bliek, A., A. C. Dimian and F. Omota, 2001, Kinetics based design of reactive distillation for esterification of lauric acid with 2-ethylhexanol, ISMR-2, Nuremberg

Warter, J., R. Duessel, J. Stichlmair and U. Weidlich, 1999, Eintrainerauswahl bei der azeotroperektifikation, Chem. Ing. Tech., 71, 385-389.

Pham, H. and M. F Doherty, 1990, Design and synthesis of heterogeneous azeotropic distillations - I, II, III, Chem. Eng. Sci., 45, 1823-1854.

European Symposium on Computer Aided Process Engineering – 12
J. Grievink and J. van Schijndel (Editors)

Catalytic Distillation Modeling

Florence Druart[1], David Rouzineau[2], Jean-Michel Reneaume[1] and Michel Meyer[2]

[1] Laboratoire de Génie des Procédés de Pau, Ecole Nationale Supérieure en Génie des Technologies Industrielles, Université de Pau et des Pays de l'Adour, (France)
Mail: jean-michel.reneaume@univ-pau.fr
[2]Laboratoire de Génie Chimique, Ecole Nationale Supérieure des Ingénieurs en Arts Chimiques et Technologiques, Institut National Polytechnique de Toulouse, (France)

A non-equilibrium (NEQ) model including heterogeneous reaction is developed. Mass transfer rates are described using Stefan-Maxwell equations. In a first section, model equations are described. Then the resolution strategy is presented in more details. In the third section, an illustrative example is considered: acetic acid esterification. The system is solved and results are presented: molar liquid fraction profiles in the column. Those results are compared to those obtained with a more classical model (equilibrium model): fraction profiles are quite different. Reaction rates are influenced by this variation.

1. Introduction

Catalytic distillation combines in situ two unit operations: distillation and reaction. This association brings numerous advantages, thanks to interaction between physical equilibrium and chemical reaction. Separation could enhance conversion reaction or chemical reaction could resolve azeotropic distillation. For heterogeneous reaction, solid catalyst could be integrated in distillation column as well. Then an optimal localization of catalyst could exist. The introduction of reaction in distillation column leads to difficulties in modeling (Taylor and Krishna, 2000). The presence of chemical reaction often involves multicomponent mixture and the model must describe numerous interactions: interactions between physical equilibrium, mass transfer and chemical kinetics (homogeneous and heterogeneous). We choose a NEQ model and Maxwell-Stefan equations are used for mass transfer (Krishna and Wesselingh, 1997) to transcribe as well as possible physical phenomena. The aim of our work is create an effective tool for simulation and optimization of catalytic distillation.

2. Models

The multicomponent separation process is described by NEQ model where three phases are considered: liquid, gas and solid which represent the catalyst. We focus on the diffusional layer near the interface gas-liquid and liquid-solid, where we describe complete multicomponent reactive mass and heat transfer (see fig. 1). The interaction solid-gas is neglected.

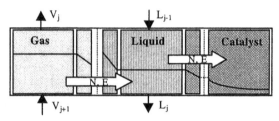

Fig. 1: Non equilibrium model : three phases considered

2.1 Maxwell-Stefan Equations for mass transfer
2.1.1 Gas-Liquid transfer

Concerning the vapor-liquid transfer, a novel model is used to compute heat and mass transfer through the diffusion layer considered in the film theory. Indeed, the fluid is considered as a Nc components reactive non-ideal mixture. The balance equations for simultaneous heat and mass transfer are written in steady state, taking account of the reactions (eq 1,2,3,4 and 5). For mass transfer, the Maxwell Stephan diffusion law is used in a novel formulation (eq 1).

$$\sum_{j=1}^{Nc}\left(\delta_{ij} + x_i \frac{\partial \ln \gamma_i}{\partial x_j}\right)_{T,P} \frac{dx_j}{dz} = -\sum_{j=1}^{Nc}\frac{(x_i N_j - x_j N_i)}{c_t D_{ij}} \qquad i \in [1,Nc] \qquad (1)$$

$$-\frac{dN_i}{dz} + \sum_{j=1}^{NRC} \upsilon_{ij} R_j + \sum_{j=1}^{NRE} \upsilon_{ij}'' \xi_j = 0 \qquad i \in [1,Nc] \qquad (2)$$

$$K_j = \prod_{i=1}^{Nc} a_i^{\alpha_{ij}} \qquad j \in [1,NRE] \qquad (3)$$

$$E = -\lambda \frac{dT}{dz} + \sum_{i=1}^{Nc} N_i H_i \qquad (4) \qquad\qquad \frac{dE}{dz} = 0 \qquad (5)$$

Neither the diffusion coefficients, nor the molar flux due to the reaction (eq.2), are considered constant. In our model, a Nc component formulation is proposed without using the summation equation. The major advantage of this formulation is that the molar fluxes in fixed referential are directly derived. For these calculations, no additional equation is used, as in a more traditional model. No assumption is made on the type or the number of reactions; thus they can be controlled by kinetics or equilibrium. Equation 3 represents the chemical equilibrium constant for the NRE equilibrium reactions. For the heat transfer, the *Dufour* and *Soret* effects are neglected and the diffusion heat rate is evaluated by *Fourier*'s law (eq. 4). Liquid / vapor film thickness and interfacial area are calculated using both AICHE and Zuiderwerg correlation.

2.1.2 Liquid-Solid transfer

A heterogeneous model without effectiveness factor describes liquid-solid transfer. Catalyst model includes both liquid-solid film diffusion and intra-particle diffusion. Thus the model is divided in two domains. The first domain is the liquid diffusion film at the catalyst surface and the second domain is the porous media of catalyst particle. Variables are continuous between the two domains.

Model of external diffusion in liquid film is similar to the model of liquid diffusion at liquid / vapor interface described in the previous paragraph. Film thickness is calculated by appropriated mass transfer correlation (Van Krevelen and Krekels for sphere).

In porous media domain, mass (eq.6) and energy (eq.7) fluxes are written according to the model of pseudo homogeneous media with averaged effective coefficients. We suppose that the pore diameter (250Å) is wide enough to consider ordinary bulk diffusion in pore. Then the Maxwell-Stefan equations are used with averaged effective coefficients and molar fractions of the liquid phase.

$$\sum_{j=1}^{Nc}(x_i\frac{\partial \ln \gamma_i}{\partial x_j}+\delta_{ij})\frac{dx_j}{dr}=\sum_{j=1}^{Nc}\left\{\frac{x_iN_j-x_jN_i}{c_t\left(Ð_{ij}^{EFF}\right)}\right\} \qquad i \in [1,Nc] \qquad (6)$$

$$E=-\lambda^{EFF}\frac{dT}{dr}+\sum_{i=1}^{Nc}N_iH_i \qquad (7)$$

Porosity ε_P and tortuosity τ characterize porous media. Thus, effective diffusion coefficients are defined (eq 8). An effective conductivity is also defined (eq 9).

$$D_{ij}^{EFF}=D_{ij}\left(\frac{\varepsilon_p}{\tau}\right) \qquad (8)$$

$$\lambda^{EFF}=(1-\varepsilon_p)\lambda+\varepsilon_p\lambda_c \qquad (9)$$

Mass and energy balances in spherical model are expressed as equations 10 and 11.

$$+r^2\frac{d\left(r^2N_i\right)}{dr}=\sum_{j=1}^{NRH}v'_{ij}R'_j+\sum_{j=1}^{NRC}v_{ij}(\varepsilon_pR_j) \qquad (10) \qquad r^2\frac{d\left(r^2E\right)}{dr}=0 \qquad (11)$$

2.2 Column modeling

On each section, balances for liquid bulk and vapor bulk are written. Mass rates from liquid-vapor interface, mass rates from solid catalyst, and source terms of homogeneous chemical reactions are included in mass balances. The interface equations (physical equilibrium and the mass and energy transfer rate continuities) link liquid and vapor phases. This set of equation is so called "Column model".

3. Resolution

3.1 Gas-liquid and liquid-solid mass transfer

The description of the mass and heat transfer near the two interfaces involved, for each one, a differential and algebraic equation (DAE) system. The same resolution method is used for the two systems: DAE integration based on Gear method (Le Lann, 1998). With this integration, we obtain the molar and energy flux and the compositions in the diffusional layer. Different key points are achieved for each of these problems.

3.1.1 Gas-Liquid interface

To efficiently use a DAE integrator there are two main problems to overcome. First of all, a robust procedure leading to a coherent initial state (ie. all algebraic equations must be satisfied) before starting the integration has to be used. And secondly, an automatic substitution procedure is used to reduce the number of mass balances in order to take into account the chemical equilibrium constraints and to reduce the differentiation index to 1. This resolution has been described in a previous paper (Rouzineau and Co., 2001). The initial values are the interface values.

3.1.2 Liquid-Solid interface

For symmetric considerations, the mass and energy fluxes are zero at the center of the catalyst (r=0). For continuity considerations, molar fractions (x) and temperature (T) at the outer surface of liquid layer (r=dp/2+δ_{LS}) are equal to those of the liquid bulk (xl$_i$,Tl). Such a problem is solved using an iterative procedure (fig. 2).

For a given values of x and T at the center of the catalyst (x0 and T0), the DAE system is solved: first in the porous media (eq. 6 – 11), next in the external liquid film. Then x, T, N and E are computed at the outer surface of liquid layer (r=dp/2+δ_{LS}). Actually x and T are implicit functions of x0 and T0 (mx$_i$(x0,T0) and mT(x0,T0)) and the following equations must be satisfied:

$$xl_i - mx_i\left(x0,T0\right) = 0 \tag{12}$$

$$Tl_i - mT\left(x0,T0\right) = 0 \tag{13}$$

This system (eq 12 and 13) is solved using a Newton-Raphson method. Variables are x0 and T0.

3.2 "Column model" resolution

The general balances in both phases and at the interface leads also to a DAE system. Due to boundary condition, this system is solved by a discretisation method. Then, the discretised system is solved by a traditionnal Newton's method. The values born of the integration in the diffusionnal layers are used in the general balances.The resolution results are the molar fraction, temperature and mass rates profiles in the column for each phases and interfaces.

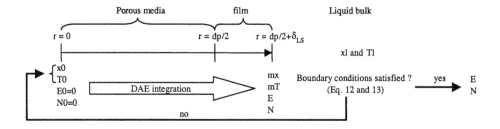

Fig. 2: Catalyst model resolution by an iterative procedure

Table 1. Catalyst and plate characteristics

Catalyst		Plate	
		Column Diameter	44d-3 m
Particle Size (diameter)	dp =1 mm	Bubbling Area	8.63d-4 m^2
Particle Porosity	εp=0.23	Weir Length	8.1d-3 m
Particle Tortuosity	τ=1.2	Exit weir height	2.13d-3 m
Bed Porosity	εl=0.52	Average liquid length	3.08d-2 m
Bed Height	h=5 cm	Total Hole Area	2.12d-4 m^2
		Sieve Tray Hole Pitch	1.5d-3 m

4.Results from Simulation

4.1 Description of one example: acetic acid esterification
The simulation of an esterification illustrates the catalytic distillation model.

Acid Acetic (AA) + Methanol (M) ⇔ Methyl Acetate (MA) + Water (W)

$$R'=3.7048x10^9 exp(-47980/RT)x_{AA} x_M - 2.5542x10^{10} exp(-58600/RT)x_{MA} x_W \qquad (14)$$

The column is composed of 9 sieve plates with 3 catalyst baskets. The acid acetic feed is set on sieve 4 (400 g.h^{-1}: 78.5 w% acid acetic and 21.5w% water) and the pure methanol feed is set on sieve 7 (200 g.h^{-1}). The reboiler heat duty is 640 W and top stream is fixed at 0.0022 mol.s^{-1}. Catalyst is spherical particles of exchange resin. Equation 14 gives intrinsic kinetic. The characteristics of catalyst and sieve plates are presented in table 1. Catalyst is in basket full of liquid and there is no liquid / vapor transfer in basket. The Thermodynamic model used is UNIQUAC. Fig. 3 shows liquid composition profiles in column for several locations of catalyst. MA appears in the reaction section. Rate of conversion (ratio of molar MA production to molar AA feed) varies between 40 % and 34 %. Those results illustrate the fact that with the same mass of catalyst, the conversion of acid could be different. For best conversion, catalyst must be placed in optimal condition of composition and temperature.

4.2 Comparison with equilibrium model
For comparison purposes, we also carry out one simulation of the equilibrium model with ProSim^{TM}. Global plate efficiency is 60 %. A catalytic bed efficiency is defined as ratio of reaction production in catalytic bed to production without diffusion limitation. Reaction terms are computed considering the previous reaction rate (eq 14), a liquid hold up equal to the basket volume and a catalyst efficiency derived from the previous simulation (43 %). As shown on fig. 4, profiles inside column vary from NEQ model results. Thus rate of conversion is different: 44 % for the equilibrium model.

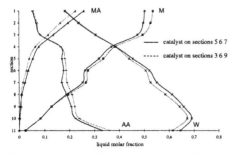 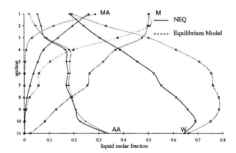

Fig. 3: non equilibrium model profile   *Fig. 4: Comparison of liquid molar profile*

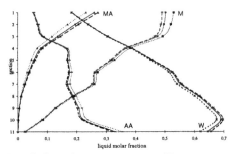

Fig. 5: liquid composition profile
... dp=0.1 mm ; ___ dp=1mm; - - - dp=2 mm

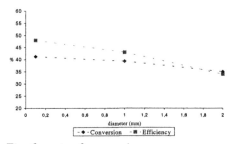

Fig. 6 : rate of conversion
and catalyst efficiency

4.3 Particle diameter variation

Fig. 5 shows results from simulation with few different catalyst diameters. When diameter increase internal diffusion limitation increase. Fig 6 illustrates the consequence on catalyst efficiency and rate of conversion. Use a smaller particle enable to increase rate of conversion.

5.Conclusion and perspectives

This paper describes a NEQ model that integrate homogeneous and heterogeneous reaction with intra-particle diffusion. This model takes non-ideality and transfer in multicomponent mixture into account by including Nc Maxwell-Stefan equations. Simulation of acetic acid esterification points out that catalyst place and characteristics are sensible parameters for conversion rate. Variation between equilibrium model and our model demonstrates that for proper simulation or optimization, NEQ model is needed. To complete this study and validate model, we develop an experimental area. Two laboratory units are placed at our disposal. The first is a single section of column. Inlet flux are controlled, we measure local accurate temperature and composition in the section. The second is a distillation pilot 7m height. Macroscopic results (top and bottom rate, temperature profile in column) are obtained.

References

[1] Taylor R., Krishna R. (2000), Chem. Eng. Science, 55,pp.5183-5229
[2] Krishna R., Wesselingh J. A . (1997),Chem.Eng.Science,52,pp.861-911
[3] Le Lann J. M.,(Février 1998), Habilitation à diriger les recherches,80 pages.
[4] Rouzineau D., Prevost M., Meyer M., (2001),Escape 11,pp.267-272

Notation

c_t : Total concentration (mol.m^{-3})
D_{ij} : Maxwell Stephan diffusion coefficient binaries i-j (m^2.s^{-1})
E : Energy flux (J.m^{-2}.s^{-1})
H_i : Enthalpy of specie i (J.mol^{-1})
K_j : Equilibrium constant for reaction j
N_i : Molar flux specie i (mol.m^{-2}.s^{-1})
Nc : Number of constituents
NRE, NRC, NRH : Number of Equilibrium, Control, Heterogeneous reaction
$R_{J,}$: Rate of control reaction j (mol.m^{-3}.s^{-1})

R_j' : Rate of heterogeneous reaction j (mol.m^{-3}.s^{-1})
T : Tempreature (K)
x_i : Molar fraction specie i
z,r : Space reference (m)
α_{ij} : Order of specie i in equilibrium j
δ_{LS} : Liquid-Solid film thickness (m)
γ_i : Activity coefficient specie i
v_i, v'_i, v''_i : Stœchiometric coefficients of specie i for control, heterogeneous, equilibrium reactions j
ξ_i : Advancement of equilibrium reaction j (mol.m^{-3}.s^{-1})
λ, λ_C : Thermal Conductivity of liquid, catalyst (J.K^{-1}.m^{-1}.s^{-1})

European Symposium on Computer Aided Process Engineering – 12
J. Grievink and J. van Schijndel (Editors)
© 2002 Elsevier Science B.V. All rights reserved.

Integration of Reaction and Separation in a Batch Extractive Distillation Column with a Middle Vessel

José Espinosa

INGAR - Avellaneda 3657 - 3000 Santa Fe - Argentina

E-Mail address: destila@ceride.gov.ar

Abstract

In this work, the integration of reaction and separation in a batch extractive distillation column with a middle vessel/reactor is analyzed for azeotrope-forming mixtures. This equipment configuration has the potential to promote complete conversion of reactants and therefore, the main process characteristics are investigated. A mixture showing several azeotropes and suffering an esterification reaction was selected as an academic example.

1. Introduction

A middle vessel batch distillation column consists of a rectifier, a stripper and a feed vessel in between. The feed mixture is loaded into the middle vessel and the products are simultaneously obtained from the top and bottom of the apparatus whilst the middle vessel contents are increasingly purified (Robinson and Gilliland, 1950). An ideal ternary mixture can be separated into a light-component rich fraction as distillate, a heavy-component rich fraction as bottom and an intermediate-component rich fraction that accumulates in the middle vessel. With this equipment configuration, however, minimum boiling azeotropic mixtures can not be separated. Batchwise extractive distillation in a column with a middle vessel has the potential to break minimum boiling azeotropic mixtures (Warter and Stichlmair, 1999). A heavy entrainer is fed close to the top of the rectifier in order to overcome the azeotrope. The entrainer absorbs one of the azeotropic components of the mixture from the rising vapor making the other azeotropic species the distillate product. The entrainer is recovered as bottom product and recycled into the entrainer vessel once it has been cooled. The main advantage of this configuration is to avoid filling the middle vessel (Hilmen et al., 1997). A middle vessel batch column can be also used to improve the performance of reactive batch distillation when the reaction products are the light- and heavy-component of the mixture (Mujtaba and Macchietto, 1992). In the case of an equilibrium reaction taking place in the middle vessel, it is possible to achieve complete conversion of the reagents. In azeotropic mixtures, azeotropes normally consist of both reagents and products and therefore, the middle vessel configuration can not be used to simultaneously enhance conversion and product separation. In this work, the integration of reaction and separation in a batch extractive distillation column with a middle vessel is analyzed for azeotrope-forming mixtures. This equipment configuration has the potential to promote complete conversion of reactants and therefore, the main process characteristics are investigated. The esterification of ethanol and acetic acid was selected as an academic example because the mixture exhibits a highly nonideal phase behavior and it has several azeotropes. Moreover, acetic acid can be used as solvent and hence, it is not necessary to add new species to the multicomponent mixture. In order to provide insights for exploring process alternatives and examining the influence of operating and process parameters on the operation performance CAPE tools are employed. The phase equilibrium is modeled with the Wilson equation and the reaction is confined to the middle vessel for convenience.

2. Phase Equilibrium Analysis and Equipment Configuration

In the reaction mixture, there are three minimum-boiling binary azeotropes (ethyl acetate and water, ethyl acetate and ethanol, ethanol and water) and one ternary azeotrope (ethyl acetate, ethanol and water). The azeotrope between ethyl acetate and water is heterogeneous. Figure 1 shows the composition tetrahedron for the quaternary system. The Figure presents the eight singular points of the system. The system presents one unstable node (the ternary azeotrope, light component) and one stable node (pure acetic acid, heavy component) and therefore, there is only one distillation region although the system presents four azeotropes. All the binary azeotropes and the remaining pure components are saddle points (intermediate components). Azeotropic compositions, node stabilities and distillation regions were calculated as explained in Espinosa et al. (2000) and Salomone and Espinosa (2001) with the aid of CAPE tools (AzeoPredictor, CBD Toolkit). The ternary azeotrope is the unstable node and hence, it will be recovered as distillate during the first cut in a normal batch rectification. In order to overcome the azeotrope, an appropriate entrainer should be selected. An appropriate entrainer, following thermodynamic rules, must be the stable node of the multicomponent mixture where the azeotrope to be broken acts as unstable node. As it was stated above, acetic acid is the stable node of the multicomponent mixture and hence, the criterion for entrainer selection in processes without distillation boundaries (one distillation region) is obeyed (Stichlmair and Fair, 1998).

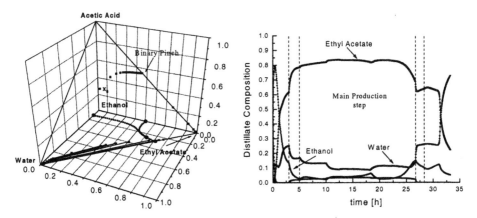

Figure 1. Composition Tetrahedron   *Figure 2. Composition vs Time*

Bearing these ideas in mind, the esterification of ethanol and acetic acid could be batchwise performed in a middle vessel column where acetic acid is used as entrainer to break the homogeneous ternary azeotrope. The entrainer is either totally or partially recovered in the stripper section and recycled to the entrainer vessel. At the end of the main production step, the composition of the middle vessel consists of pure water or a mixture of water and acetic acid and the ethanol conversion approaches unity. Although there exist several possible equipment arrangements, the configuration where the vapor stream leaving the lower column section is passed by to the upper column section was selected because the practical feasibility of this column configuration was studied for non-reactive mixtures by Barolo *et al.* (1996). Kinetic data from Bogacki et al. (1989) were adopted to explore integration possibilities.

3. Operating Sequence

Step 1 Startup) Total reflux- and reboil ratios without solvent feeding. The composition of the reflux drum changes until it achieves the composition corresponding to the ternary azeotrope.

Step 2 Eliminate off-spec. product) Total reflux- and reboil ratios with solvent feeding. Due to the absorption effect of the solvent, a product richer in the reaction products than in the ternary azeotrope accumulates in the reflux drum. The level in the middle vessel increases to accommodate the volume of the entrainer added during this period.

Step 3 Production Period) Operation under finite reflux- and reboil ratios with solvent feeding. A product rich in the reaction products is collected as distillate. The solvent is either totally or partially recovered at the column bottom. The level in the middle vessel decreases from its maximum value at the end of Step 2. High limiting reagent conversion can be achieved depending on operating and process parameters.

Step 4 Eliminate off-spec. product) Total reflux operation without solvent feeding. At the end of Step 3, the middle vessel composition is mainly formed by acetic acid and water. Ethyl acetate and ethanol are only present in small amounts. The solvent feeding is stopped. Total reflux operation is implemented to achieve the composition of the ternary azeotrope in the reflux drum to avoid off-specification product.

Step 5 Recovery of remaining ester as ternary azeotrope) Operation under finite reflux- and reboil ratios. The azeotropic cut is diverted to the main product tank. The entrainer is removed at the column bottom.

Step 6 Shutdown) The reflux drum content is diverted to the product tank. The middle vessel content, on the other hand, can be processed in a conventional batch rectifier to remove water as distillate product and leave excess solvent in the still for recycle to the next batch. This operation can be also performed in the middle vessel column. Depending on the reflux- and reboil ratio policy, either acetic acid or water is collected in the middle vessel at the end of the operation. Figure 2 shows a typical evolution of the compositions in the reflux drum.

4. Influence of operating and process parameters on operation performance

Table 1 presents the results of a systematic investigation with Hysys (Hyprotech, 1999) of the influence of operating and process parameters on the operation performance. The recovery rate of the ester, ethanol conversion and product composition are adopted as measures of the operation performance. The recovery rate of the ester is the ratio of the amount of ethyl acetate in the distillate to the amount of ethyl acetate produced by chemical reaction.

Table 1. Feed Data and Results

Feed Amount [172.8 kmol]	x_{Feed} [0.45 AcOH, 0.45 EtOH, 0.00 EtAc, 0.10 W]					
	x_{EtOH}	σ_{EtAc}	AcOH	EtOH	EtAc	W
Feed Stage [1-10] [from top to down]	Dec	Max	Dec	Inc	Max	Inc
Solvent Flow Rate [20-30] [kmol/h]	Inc	Inc	Inc	Dec	Inc	Dec
Distillate Flow Rate [2.5-6] [kmol/h]	Max	Inc	Inc	Inc	Max	Min
Entrainer Purity [0.9-1]	Inc	Inc	Inc	Inc	Inc	Dec

Dec = decreases, Inc = increases, Max = maximum, Min = minimum

Influence of the Feed Stage

The conversion of the limiting reagent decreases with the feed stage moving downwards in the column. The ester recovery rate and the composition of ethyl acetate in the product, on the other hand, present a maximum when the solvent is fed at an intermediate stage. In the studied range of feed locations, the solvent mainly absorbs the alcohol. Therefore, the distillate products are rich in ethyl acetate and water. The internal composition profile presents a ternary pinch in the absorption zone. Both ethanol and water compositions in the distillate increase their values as the absorption zone diminishes because the absorption capacity of the solvent diminishes. Moreover, an inversion in the volatility order between the ester and both the alcohol and water takes place in the rectifying zone above the feed stage. A feeding of the solvent that is too close to the top of the column prevents the necessary removal of acetic acid from the vapor stream in the rectifying section of the column. Two or three stages are required in the rectifying section to get a product with small amounts of solvent. As a consequence, a maximum in the ester concentration is expected. Since the product amount is the same for all simulations, an increase in the ethanol loss at the column top as the location of the feed stage is increased produces a smaller conversion of the alcohol. The recovery rate of the ester follows the same trend as the composition of ethyl acetate in the distillate. The optimal location of the feed stage should consider the trade-off between separation and reaction targets.

Influence of the solvent Flow Rate

The increment in the entrainer flow rate involves increments in the conversion of the limiting reagent, the recovery rate of the ester and the composition of the ester in the distillate. However, the solvent recovery rate as pure acetic acid at the stripper bottom decreases whilst the remaining mixture in the middle vessel at the operation end increases both its amount and its acid mole fraction as the entrainer flow rate increases. The improvement in the performance for increased values of the entrainer flow rate is based on two reasons: a better mean reaction rate and a more effective separation between the ester and both water and alcohol in the absorption zone. The enhancement in the reaction rate is mainly a consequence of the higher concentrations of acetic acid achieved in the middle vessel. To demonstrate the influence of high solvent flow rates in the absorption capacity of the entrainer, a simulation was performed at a high entrainer flow rate. Both ethanol and water are completely returned to the reaction zone and therefore, the ester and acetic acid mainly form the distillate product. The acid composition in the distillate can be sharply reduced by increasing the number of stages in the rectifying section above the feed stage but in this case, a loss of the other components should be expected as explained in the previous section. Note that for this example, the internal profile is controlled by a binary pinch. The optimal solvent flow rate should consider the trade-off between both reaction and separation targets in the first part of the process (until the end of the main production step) and the separation of the mixture acetic acid-water. Also, a constraint on the maximum feasible entrainer flow rate should be taken into account. If the amount of solvent required for a given operation is more than what the column can accommodate without flooding, the middle vessel should be charged to a certain fraction of the maximum capacity.

Influence of the Distillate Flow Rate

An increment in the distillate flow rate enhances the recovery rate of ethyl acetate and produces a sharply increment in the throughput. The throughput is defined here as the ratio of the amount of distillate to the operation time. Separation water/acetic acid is also facilitated because the remaining amount of moles in the middle vessel sharply decreases. However, there should be an optimum value for the distillate flow rate since both the conversion of the limiting reagent and the ester composition present a maximum. The maximum in the alcohol conversion can be explained with the aid of the mean reaction versus distillation ratio, RD defined as $rV/(Dx_{EtAc}^{D})$. This ratio relates the rate of ethyl acetate production by chemical reaction with the rate of removal of the ester by distillation. The ratio RD diminishes its value starting from a maximum value below zero as the distillate flow rate increases. This behavior indicates that the rate of removal is greater than the rate of production and therefore, both alcohol conversion and reaction rate should also increase because the ester concentration in the middle vessel is maintained as low as possible. However, for high values of the distillate flow rate both conversion and reaction rate decrease because the low reflux ratio is not enough to maintain the purity of the product and therefore, both reaction products and reagents are carried away together. This is also the cause of the sharp increase in water composition. The same study was done for different entrainer flow rates. The results were qualitatively similar. The same ester composition can be obtained either with a low entrainer flow and a low distillate flow or with a high entrainer flow and a high distillate flow. Even if the entrainer flow rate is increased it is possible to diminish the final amount of material in the middle vessel at the operation end. Moreover, operation at both high entrainer and distillate flow rate enhances alcohol conversion, ester recovery and throughput. However, the mole fraction of acid in the middle vessel at the end of the main production period will be larger for the operation at high solvent flow rate.

Influence of the Entrainer Purity

Since water is a reaction product, as the water content in the entrainer stream increases a lowering in both reaction rate and alcohol conversion takes place. Moreover, the ethyl acetate recovery rate suffers a diminution because the water concentration in the pinch region below the feed point increases, giving rise to a greater loss (lower recovery rate) of water (ethyl acetate) in the column top. Remember that an inversion in the volatility order occurs in the rectifying section above the feed stage. Distillate products with the same ester composition can be obtained either with a high distillate flow rate combined with a pure entrainer or with a low distillate flow rate combined with a high water impurity in the entrainer. However, the performance variables indicate the convenience of operating with a pure entrainer.

5. Conclusions

The strong mutual influence among operating parameters is a consequence of the reaction and separation steps being integrated in the same equipment. Both pure rectification and absorption take place in the rectifier whilst a chemical reaction occurs in the middle vessel and the solvent is recovered in the stripper section. Two different operating regimes are found with their corresponding feasible distillate products. Regimes controlled by a ternary pinch point give rise to distillate products rich in ethyl acetate and water. High purity ethyl acetate distillate products are achieved when the geometry of the internal profiles is controlled by a binary pinch. The reaction versus distillation ratio RD plays an important role during the main production step because it greatly influences both the limiting reagent conversion and the product recovery rate. The main conclusion of this work is that the integration of reaction and separation in a batch extractive distillation column with a middle vessel is a promissory technology for multi-azeotropic mixtures and therefore additional research efforts should be driven in this area.

6. References

AzeoPredictor 1.0. Copyright ©2000 Enrique Salomone & José Espinosa.

Barolo, M.; Guarise, G. B.; Rienzi, S. A.; Trotta, A. and S. Macchietto, 1996, Running Batch Distillation in a Column with a Middle Vessel, Ind. Eng. Chem. Res. 35, 4612.

Bogacki, M. B., Alejski, K. and J. Szymanowski, 1989, The Fast Method of the Solution of a Reacting Distillation Problem, Comput. Chem. Eng. 13 (9), 1081.

CBD Toolkit 3.0. Copyright © 2001 José Espinosa & Enrique Salomone.

Espinosa, J.; Salomone, E. and S. Xu, 2000, Using Conceptual Models for the Synthesis and Design of Batch Distillations, European Symposium on Computer Aided Process Engineering-10 8, S. Pierucci, Ed., Elsevier, 1033.

Hilmen, E. K.; Skogestad, S.; Doherty, M. F. and M. F. Malone, 1997, Integrated Design, Operation and Control of Batch Extractive Distillation with a Middle Vessel, Presented at AIChE Annual Meeting, Los Angeles CA (USA).

Hyprotech (1999). HYSYS user manual.

Mujtaba, I. M. and S. Macchietto, 1992, Optimal Operation of Reactive Batch Distillation, Presented at AIChE Annual Meeting, Miami Beach (USA).

Robinson, C. S. and E. R. Gilliland, 1950, Elements of Fractional Distillation, McGraw-Hill Book Company, Inc., New York, 4th edition.

Salomone, E. and J. Espinosa, 2001, Prediction of Homogeneous Azeotropes with Interval Analysis Techniques Exploiting Topological Considerations. Ind. Eng. Chem. Res. 40 (6), 1580.

Stichlmair, J. and J. Fair, 1998, Distillation: Principles and Practice. Wiley-VCH.

Warter, M. and J. Stichlmair, 1999, Batchwise Extractive Distillation in a Column with a Middle Vessel, Computers and Chemical Engineering, S915.

7. Acknowledgments

This work has been partly supported by ANPCyT under grant PICT '99 14-05263 and UNL under grant CAI+D No. 146.

European Symposium on Computer Aided Process Engineering – 12
J. Grievink and J. van Schijndel (Editors)
© 2002 Elsevier Science B.V. All rights reserved.

MINLP Heat Exchanger Network Design incorporating Pressure Drop Effects

S. Frausto-Hernández, V. Rico-Ramírez, S. Hernández-Castro[†] and A. Jiménez
Instituto Tecnológico de Celaya, Av. Tecnológico y García Cubas S/N, Celaya, Gto.
C.P. 38010, México
[†]Universidad de Guanajuato, Facultad de Química, Col. Noria Alta S/N, C.P. 36000,
México

Abstract

Successful work has been done on heat exchanger network synthesis (HEN) using both Pinch Technology and MINLP techniques. However, most of the design procedures reported to date assume constant stream heat transfer coefficients. Motivated by the fact that detailed heat exchanger design is based on pressure drop of the streams, in this paper we extend the simultaneous MINLP model for the design of heat exchanger networks (Yee and Grossmann, 1990) by removing the assumption of constant film heat transfer coefficients and incorporating instead the effect of allowable pressure drop. An illustrative example is used to show the relevance of the approach.

Keywords: Heat Exchanger Network, Pressure Drop

1. Introduction

Recent approaches for the synthesis of heat exchanger networks with both Pinch Technology and MINLP techniques have shown to be capable of synthesizing near optimal networks for real industrial problems. Some of the design procedures can be reviewed, for instance, in Linhoff and Flower (1978), Papoulias and Grossmann (1983) and Yee and Grossmann (1990).

However, although most of the current synthesis techniques are based on the assumption of constant film heat transfer coefficients, detailed heat exchanger design is based on the satisfaction of three major objectives (Polley and Panjeh Shahi, 1991): *i)* Transfer of required heat duty, *ii)* Tube side pressure drop below a maximum allowed value, and *iii)* Shell side pressure drop below a maximum allowed value. Hence, since synthesis and detailed design are not conducted on the same basis, there is no guarantee that the values assumed for the heat transfer coefficients in the synthesis stage are the same as those actually achieved in equipment design.

Polley and Panjeh Shahi (1991) proposed that one way of making consistent network synthesis and detailed exchanger design is to base network synthesis on allowable stream pressure drop rather than on constant film transfer coefficients. Since then, a number of applications have been reported which suggest the incorporation of pressure drop effects into the synthesis stage based on pinch technology (Serna, 1999). Further applications of this approach have also been considered for network retrofit based on MINLP techniques (Nie and Zhu, 1999). In this paper, the simultaneous MINLP model for heat exchanger design proposed by Yee and Grossmann (1990) has been extended to incorporate pressure drop considerations into the synthesis stage. So, the network is synthesized based on allowable pressure drop rather than on constant film heat transfer coefficients and it is, therefore, made closer to industrial reality.

2. MINLP Synthesis Incorporating Pressure Drop Effects

In this section we present the equations that have been incorporated to the simultaneous MINLP formulation of Yee and Grossmann (1990) in order to consider pressure drop effects during the synthesis stage. The main issue here is that the heat exchanger network to be synthesized must satisfy not only a specified heat recovery target but also pressure drop constraints.

2.1 Simultaneous MINLP approach to heat exchanger network design
The MINLP model proposed by Yee and Grossmann (1990) is intended to provide an appropriate trade-off between utility consumption, number of units and exchanger areas. Such a model is based on a stage-wise superstructure with the temperature driving forces as optimization variables. The superstructure is constructed so that each cold stream can be potentially matched with each hot stream in each of the stages. In this model, all the constraints but the objective function are linear. The objective function consists of the annual cost for the network and is nonlinear because of the heat exchanger area calculation (the Chen approximation is used, (Chen, 1987)). Also, the overall heat transfer coefficients required by the area calculation are obtained in terms of the constant (assumed as given) film heat transfer coefficients of the streams.

2.2 Pressure drop considerations
Polley et al. (1990) developed a general relationship between frictional pressure drop and convective film heat transfer coefficients as follows:

$$\Delta P = K A h^m \tag{1}$$

where ΔP is the exchanger pressure drop, A is the heat transfer area and h is the film heat transfer coefficient. Serna (1999) developed equations of the type of Equation (1) based on the detailed Bell-Delaware method for the design of heat exchangers. Hence, for turbulent flow in shell and tube exchangers, the equation for the tube side is:

$$\Delta P_T = K_{PT} A h_T^{3.5} \tag{2}$$

and for the shell side is:

$$\Delta P_S = K_{PS} A h_S^{5.109} \tag{3}$$

where S and T stand for shell and tube, correspondingly. The parameters K_{PT} and K_{PS} depend on the physical properties of the streams and on the geometry. The equations derived by Serna (1999) are used in this wok. Expressions such as (2) and (3) apply for those streams which do not experiment change of phase. Hence, it will be assumed that the heat transfer coefficients for heating and cooling utilities are still given constants. Furthermore, in order to be consistent with the assumptions of the simultaneous MINLP approach described above, only shell and tube heat exchangers will be considered here.

2.2.1 Pressure drop calculations for heat exchanger networks

When a process stream exchanges heat in more than one unit, individual and overall pressure drops must be calculated. We use a simplified approach presented by Shenoy (1995) for the calculation of the pressure drop in each of the stages of a network superstructure. That approach assumes that the pressure drop of a stream is linearly distributed according to the surface area of the exchange units.

2.2.2 Calculation of Cost of Power

In order to incorporate the cost of power to the objective function of the synthesis problem, the expression proposed by Serna (1999) can be applied:

$$Cost\ of\ Power = CW\ Q\ \Delta P \qquad (4)$$

where CW is a cost coefficient (unit cost of power, $/KW-year), Q is the volumetric flow rate and ΔP is the pressure drop of the stream.

2.3 The extended model equations

Seeking concreteness, this section presents only the equations that have been added to the simultaneous MINLP formulation for HEN synthesis presented by Yee and Grossmann (1990). It will be assumed that hot streams flow in the shell side and the cold streams flow in the tube side of the heat exchangers of the network. Hot streams are represented by index i ($i \in HP$) and cold streams by index j ($j \in CP$). Index k denotes the k-th stage of the superstructure ($k \in ST$). cu stands for cooling utility (cooling water) and hu for hot utility. The equations are presented next.

$$A_{ijk} = \left(\frac{q_{ijk}}{\left[\left(\Delta T_{ijk} \right)\left(\Delta T_{ijk+1} \right)\left(\frac{\Delta T_{ijk} + \Delta T_{ijk+1}}{2} \right) \right]^{1/3}} \right)\left(\frac{1}{h_i} + \frac{1}{h_j} \right) \qquad \forall i \in HP, j \in CP, k \in ST \qquad (5)$$

$$A_{icu} = \left(\frac{q_{cui}}{\left[\left(\Delta T_{icu} \right)\left(T_{OUT\,i} - T_{IN\,cu} \right)\left(\frac{\Delta T_{icu} + T_{OUT\,i} - T_{IN\,cu}}{2} \right) \right]^{1/3}} \right)\left(\frac{1}{h_i} + \frac{1}{h_{cu}} \right) \qquad \forall i \in HP \qquad (6)$$

$$A_{huj} = \left(\frac{q_{huj}}{\left[\left(\Delta T_{huj} \right)\left(T_{IN\,hu} - T_{OUT\,j} \right)\left(\frac{\Delta T_{huj} + T_{IN\,hu} - T_{OUT\,j}}{2} \right) \right]^{1/3}} \right)\left(\frac{1}{h_{hu}} + \frac{1}{h_j} \right) \qquad \forall j \in CP \qquad (7)$$

$$A_{Ti} = \sum_k \sum_j A_{ijk} + A_{icu} \qquad \forall i \in HP \tag{8}$$

$$A_{Tj} = \sum_k \sum_i A_{ijk} + A_{huj} \qquad \forall j \in CP \tag{9}$$

$$\Delta P_i = K_{Pi} A_{Ti} h_i^{5.109} \qquad \forall i \in HP \tag{10}$$

$$\Delta P_j = K_{Pj} A_{Tj} h_j^{3.5} \qquad \forall j \in CP \tag{11}$$

$$\Delta P_i \le \Delta P_{Pi} \qquad \forall i \in HP \tag{12}$$

$$\Delta P_j \le \Delta P_{Pj} \qquad \forall j \in CP \tag{13}$$

$$\Delta P_{ik} = \frac{\sum_j A_{ijk}}{A_{Ti}} \Delta P_i \qquad \forall i \in HP, k \in ST \tag{14}$$

$$\Delta P_{jk} = \frac{\sum_i A_{ijk}}{A_{Tj}} \Delta P_j \qquad \forall j \in CP, k \in ST \tag{15}$$

$$\begin{aligned}
\min \quad & \sum_i C_{CU} q_{cui} + \sum_j C_{HU} q_{huj} + \sum_i \sum_j \sum_k CF_{ij} z_{ijk} + \sum_i CF_{icu} z_{cui} + \sum_j CF_{jhu} z_{huj} \\
& + \sum_i \sum_j \sum_k C_{ij} \left(A_{ijk} \right)^\beta + \sum_i C_{icu} \left(A_{icu} \right)^\beta + \sum_j C_{huj} \left(A_{huj} \right)^\beta + \sum_i CW_i \, Q_i \Delta P_i \\
& + \sum_j CW_j \, Q_j \Delta P_j
\end{aligned} \tag{16}$$

Equations (5) through (7) correspond to the explicit calculation of heat exchanger area of each unit. Evaluation of the total contact area for hot and cold streams are provided by equations (8) and (9). Equations (10) and (11) relate the film heat transfer coefficients to the pressure drop for each hot and cold stream. Besides, these values of pressure drop are related to the maximum allowable values for the pressure drop of each stream in equations (12) and (13). Pressure drop for hot and cold streams in each stage of the superstructure is determined by using the Shenoy (1995) approximation, equations (14) and (15). Equation (16) is the objective function which consists of the annualized cost of the network. Compared to the objective provided by Yee and Grossmann (1990), there are two extra terms which are incorporated to calculate the cost of power for hot and cold streams.

3. Illustrative Example

An example has been developed in order to show the effect of pressure drop in the synthesis stage of heat exchanger networks. The example corresponds to the problem described by Polley and Panjeh Shahi (1991) for an HRAT of 20 °C. Specifications of the problem are given in Table 1. Two designs obtained with the proposed MINLP approach are compared. Figure 1 shows the network configuration, which is the same for both designs. The network consists of 7 units with minimum utility consumption (1075 KW and 400 KW for heating and cooling, correspondingly). Hence, the utility cost is $122,249.999/year. The designs, however, are different from each other with respect to the values of heat duty, exchanger areas, heat transfer coefficients and pressure drops. Such values are shown in Table 2. For the design 1, the total exchanger area is 567.36 m^2, with an investment cost of $236,474. For this case the cost of power is $23,310/year. Therefore, the total cost of the network is $382,034/year. For design 2, the total exchanger area is 618.24 m^2, with an investment cost of $249,937, and the cost of power is $9,044/year. Therefore, the total cost of the network is $381,231/year. Note that in Table 2 the summation of the values corresponding to the contact area is not equal to the total area of the network. That is because the exchange area of a given match is considered as contact area for both the hot and the cold streams of the match. Observe that, although the total heat exchanger area for design 1 is smaller than that of design 2, the total cost of design 1 is larger that that of design 2. As a matter of fact, this example has been selected to show that the cost of power might be important in the determination of an optimal design. Also, since pressure drop has been considered into the synthesis stage, it can be expected that these designs are consistent with what is finally achieved in terms of industrial hardware.

4. Discussion

In this work, the simultaneous MINLP formulation developed by Yee and Grossmann (1990) has been extended by incorporating equations that relate the overall pressure drop of a stream to its film heat transfer coefficient. Hence, the optimization of the problem not only minimizes capital but also satisfies both the transfer of required heat duty and the pressure drop constraints imposed by practical considerations. Also, the cost of power has been incorporated into the objective function of the optimization problem. The numerical disadvantages of the proposed approach can be easily identified (non-convexities) but the results obtained so far have shown the potential of the approach. As described with the illustrative example, the networks obtained by the proposed approach are expected to be consistent with what is finally achieved in terms of industrial hardware.

5. Acknowledgements

Financial support provided by CONACYT (Mexico) is gratefully acknowledged.

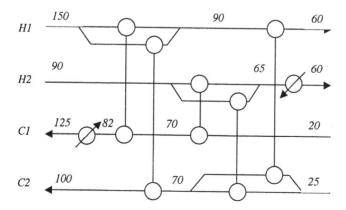

Figure 1 Resulting HEN for the Example

Table 1 Temperature Specifications for the Example

Stream	T_{in} (°C)	T_{out} (°C)	F CP (KW/°C)
H1	150	60	20
H2	90	60	80
C1	20	125	25
C2	25	100	30
Steam	180	180	
Cooling Water	10	15	

Table 2 Film Heat Transfer Coefficients for Two Designs of the Example

Stream	Design 1			Design 2		
	Contact Area (m^2)	h (W/m^2°C)	ΔP (Kpa)	Contact Area (m^2)	h (W/m^2°C)	ΔP (Kpa)
H1	139.42	734	20	154.71	721	20
H2	259.24	893	30	293.52	680	6.71
C1	308.91	572	10	325.39	565	10
C2	169.49	1049	60	200.08	814	23.17

References

Chen, J. J. J. (1987), Letter to Editors: Comments on Improvement on a Replacement for the Logarithmic Mean, *Chem. Eng. Sci.*, 42, 2488.

Linhoff, B. and J. R. Flower (1978), Synthesis of Heat Exchanger Networks, *AIChE Journal*, 24, 633.

Nie, X. R. and X. X. Zhu (1999), Heat Exchanger Network Retrofit Considering Pressure Drop and Heat-Transfer Enhancement, *AIChE Journal*, 45, 1239.

Papoulias S. A. and I. E. Grossmann (1983), A Structural Optimization Approach in Process Synthesis-II. Heat Recovery Networks. *Comp. Chem. Eng.* 7, 707.

Polley, G. T., Panjeh Shahi, M. H. and Jegede, F. O. (1990), Pressure Drop Considerations in the Retrofit of Heat Exchanger Networks, *Trans IChemE*, 68, Part A, 211.

Polley, G. T. and M. H. Panjeh Shahi (1991), Interfacing Heat Exchanger Network Synthesis and Detailed Heat Exchanger Design, *Trans IchemE*, 69, 445.

Shenoy, U. V., (1995), Heat Exchanger Network Synthesis. Process Optimization by Energy and Resource Analysis, Gulf Publishing Company.

Serna, G. M., (1999) Development of Rigorous Methods for Heat Integration of Chemical Processes, Ph.D. Thesis, Department of Chemical Engineering, Instituto Tecnológico de Celaya, Mexico.

Yee, T. F. and I. E. Grossmann (1990), Simultaneous Optimization Models for Heat Integration – II. Heat Exchanger Network Synthesis, *Comp. Chem. Eng.*, 14, 1165.

European Symposium on Computer Aided Process Engineering – 12
J. Grievink and J. van Schijndel (Editors)
© 2002 Elsevier Science B.V. All rights reserved.

Hierarchical Approach for Selecting Uncertain Parameters at the Conceptual Stage of Design

H. D. Goel[a]. P. M. Herder[a], M. P. C. Weijnen[a] and J. Grievink[b]

[a]Faculty of Technology, Policy and Management, [b]Faculty of Applied Sciences,
Delft University of Technology, 2600 GA, Delft, The Netherlands.
Email: H.D.Goel@tbm.tudelft.nl

Abstract

The problem of "process design and operations under uncertainty" has recently attracted the attention of many researchers from industry and academia. Consequently, many articles have been published about formulating design problems with uncertain parameters and solving the resulting complex stochastic optimization problems. However, relatively minor attention has been paid to the selection of uncertain parameters and variables in a particular design problem. In this paper, we introduce a systematic multilevel hierarchical procedure that uses preliminary qualitative information about the product and process to narrow down the search space for selecting "dominant" uncertain parameters. The application of the proposed procedure is presented via an illustrative process synthesis example of the HDA process.

1. Introduction

Whether technical or strategic, short term or long term, decisions often have to be made based on incomplete or imperfect information. In the context of the chemical industry, uncertainty effects the decisions made at the design (process synthesis) and operational stages (operational planning, scheduling) of processes. The uncertainty associated with these decisions often results in costly iterations at later stages of design or in an extreme case results in major revamps/shutdowns at the operational stage.

In recent years, significant progress has been made on developing rigorous optimization based methods to address uncertainty in design formulation at the conceptual stage (Pistikopoulos, 1995). These efforts focus primarily on the development and/or improvement of the algorithms to solve the complex stochastic optimization problems without giving much attention to the problem of identifying dominant uncertain parameters at the problem formulation stage. At first it may look trivial to select uncertain parameters for a particular design problem based on a designers' own experience. However, in most cases (unless the problem owner himself gives a concise list of uncertain parameters), the number of parameters that could be considered (to be) uncertain can be quite large and the selection of the most important ones to be difficult.

In this paper, we introduce a systematic multilevel hierarchical procedure that uses preliminary qualitative information about the product and the process to narrow down the search space for selecting "dominant" uncertain parameters. The purpose of this hierarchical procedure is to complement the existing rigorous optimization approaches by providing a subset of key uncertain parameters as input to those approaches.

2.Background

In an optimization-based method for process synthesis (without considering uncertainty), the process synthesis problem can be posed as MINLP problem (P1)

$$\max_{y,d} \quad P(x,y,d,z)$$
$$s.t. \quad h(x,y,d,z) = 0 \tag{P1}$$
$$g(x,y,d,z) \leq 0$$
$$x \in X, d \in D, z \in Z$$
$$y \in Y \subseteq [0,1]^p$$

where, d, x, z, y are the vectors of design, state, control, and binary variables, respectively. P(x,y,d,z) is a scalar economic objective function. The constraints h(x,y,d,z) correspond to mass and energy balances, equilibrium relationships and g(x,y,d,z) correspond to process design and operational specifications.

Unfortunately, at the conceptual stage, it is not often possible to assign a point value for most of the parameters (required in P1). Accurate information, for example, about the rate constants or transfer co-efficient may not be available. Time and budget constraints limit the experiments needed to get more accurate information, and, therefore, perfect information is not attainable. As a consequence, quite a number of parameters remain that could be considered uncertain at the design stage.

Table 1: Uncertainties classification based on source

Source	Uncertain variable/parameter
Supply market	Availability of raw material and utilities
Raw material	Feedstock, quality variations and prices
Utilities	Quality and price variations
Physical system	
- Operational	Flowrate, temperature, pressure and capacity variations (shutdown and expansion)
- Equipment related	Availability (reliability and maintainability), purchase cost and operating cost
- Model inherent	Kinetic constants, physical properties, transfer coefficients
Products	Product specifications, price, demand
Demand market	Demand of products and by products
Emission	Uncertain environmental index and regulatory policies

The most commonly used method to integrate uncertainty in different parameters (table 1) is to extend problem P1 to an augmented stochastic optimization problem (AP1) with θ as a vector for uncertain parameters. Mathematically, AP1 can be formulated as

$$\max_{x,y,d,z,\theta} \quad EP(x,y,d,z,\theta)$$
$$s.t. \quad h(x,y,d,z,\theta) = 0 \tag{AP1}$$
$$g(x,y,d,z,\theta) \leq 0$$
$$x \in X, d \in D, z \in Z$$
$$y \in Y \subseteq [0,1]^p$$
$$\theta = \text{vector of uncertain parameters}$$

In problem AP1, EP (x,y,d,z,θ) is an expected profit objective function. The formulation AP1 is generic in nature and a plethora of algorithms/approaches exists to solve problem AP1. However, due to computational limitations and scarcity of recourses (time and capital), it is required to limit the elements of the θ vector.

3. Hierarchical Procedure

In this section, we introduce a systematic hierarchical approach for selecting uncertain parameters at the conceptual stage of design. The basic idea behind the proposed step-by-step procedure is to use qualitative information about the process and product to systematically reduce the number of uncertain parameters (θ vector) that are required in problem AP1. In opposition to the existing approaches, where the selection of elements of θ vector is done on an *ad hoc* basis, our proposed procedure starts with a complete set of uncertain parameters (table 1) and progressively at each step reduces the dimensionality of the θ vector.

Generally speaking, the need to consider uncertainty in different parameters may be related to a particular kind of flexibility that has been sought for design. For example, in order to integrate market flexibility in a design we need to consider the uncertainty in product demand and price parameters. In our proposed procedure, we first identify the different kind of flexibilities that are dominant and then formally include them in the Basis of Design (BoD). Based on a specified flexibility in the BoD, it is relatively simple to select the members of the θ vector. In the following sections, we describe in detail the tasks that are required at each step.

3.1 Step 1 Identifying the flexibility requirements at the design stage

At this step, we identify the flexibilities required for a given project at the design stage. The identification of different flexibilities requires us to predict the expected innovations in the process and products. Here we can use the results of an empirical study by Hutcheson et al. (1995) in which they describe three-stages of process and product innovation that is observed in different process industries (as shown in figure 1). It is quite clear from the figure that as the industry matures, the focus of innovation gradually shifts from product to process. The information needed to identify the stage for a particular project is generally available at the design stage. For example, most of the commodity chemical projects are in the systemic stage, where the product and its market are matured.

Once the stage has been identified, we could use the characteristics of each stage (as described in Hutcheson et al. (1995)) to identify the flexibilities required for a

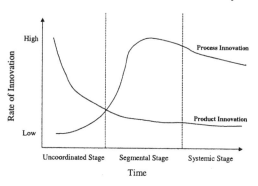

Figure 1: Profiles of product and process innovation (D'Souza and Williams, 2000).

particular project. The characteristics of the different stages allow us to assign flexibility requirements to a given stage. In table 2, we can see the different flexibilities that related to each stage.

Table 2: Characteristics of project in each phase

Uncoordinated	Segmental	Systemic
Innovation		
High product innovation	Low product innovation	Low Product innovation
Low process innovation	High process Innovation	Low process innovation
Characteristics		
Limited market	High demand	Saturated market
Limited knowledge about its applications	Focus on the production process	Competition is based on manufacturing costs reductions
High incentive to being first in the market	Much focus on the scale of process	Focus mainly on supply chain integration
Less attention is paid to equipment innovations		
Flexibility requirements		
Ability to accommodate different products	Easy to integrate selectivity and equipment improvements	Low process variability
Easy to expand or shutdown	Ability to accommodate variations in different raw materials	High availability requirements

3.2 Step 2 Formalizing the flexibility in the BoD

In the first step, we identified the flexibility requirements based on the preliminary information about process and product. In this step, we set to formally define them as design goals in the BoD. This step is important, as it will allow different parties, such as the project owner, designer and contractors, to have a clear consensus of the flexibility required in the project.

At present in the process system literature, the term "flexibility" is used to define the ability of a process system to accommodate continuous uncertainty (Swaney and Grossmann, 1983). This definition is broad and often used at the design stage for accommodating different continuous uncertainties with various time and physical scales. Unless stated explicitly in the BoD, a designer uses his experience and current process information in deciding the parameters that need to be considered to be uncertain in the model. However, in operational research, manufacturing flexibility is not considered to be one-dimensional and considerable efforts have been made to develop a standard taxonomy for flexibility that captures the different dimensions of flexibility (D'Souza and Williams, 2000). The taxonomy developed in their work (shown in table 3), although not developed specifically for the process industry, could, with minor modification, easily be adapted for the process industry.

Table 3: The dimensions of manufacturing flexibility (D'Souza and Williams, 2000)

Category 1: externally-driven flexibility dimensions	
1. Volume flexibility	This dimension of flexibility represents the ability to change the level of output of a manufacturing process
2. Variety flexibility	This dimension represents the ability of the manufacturing system to produce a number of different products and to introduce new products. Researchers have suggested the use of product mix and product modification as components of this dimension of manufacturing flexibility
Category 2 internally-driven flexibility dimensions	
Process Flexibility	This dimension represents the ability of the manufacturing system to adjust to and accommodate changes/disruptions in the manufacturing process. Examples of these changes/disruptions found in literature are, machine breakdowns, changes in the production schedules, or job sequencing
Materials handling flexibility	This dimension represents the ability of materials handling process to effectively deliver materials to the appropriate stages of the manufacturing process and position the part or the material in such a manner as to permit value adding operations

3.3 Step 3 Identifying uncertain parameter(s) based on defined flexibility goal(s)

So far in step 1 and step 2, we have identified and stated clearly the type of flexibility required in a particular design. In this step, we map out the set of dominant uncertain parameters from a complete set (table 1) based on the flexibility goal(s) defined in the BoD.

3.4 Step 4 Formulating problem with uncertain parameters

In this step we focus mainly on the computational aspects of the design problem under uncertainty. The aim of this step is to reduce the computational burden of the problem AP1 by using an alternative ways to represent particular uncertain parameters. For example, uncertainty in equipment breakdown is usually represented as probability distribution function (Pistikopoulos, 1995), which can be (assuming exponential failure distribution) alternatively represented in problem AP1 by an aggregate measures such as MTBF (mean time between failure) and MTTR (mean time to repair).

4. Illustrative Case Study: Hydrodealkylation (HDA) of Toluene

The hierarchical procedure developed in the previous section is applied to the synthesis of HDA process for benzene production. For a comprehensive description refer to Douglas (1988). The main purpose of this example is to concisely illustrate the basic ideas behind the method and its effectiveness as a formal procedure.

Step 1 Identifying the flexibility requirements at the design stage

From literature, it can be established that the HDA process and the main product benzene are quite mature (Douglas, 1988). Therefore, a systemic stage is applicable to this design project. Based on the characteristics cited in table 2 for the systemic stage, we can choose to have the goals of minimizing the process variability and maximizing the availability of the plant.

Step 2 Formalizing the flexibility in the BoD

Based on the descriptions of different flexibilities in table 3, we can choose to include process flexibility as a design goal in the BoD of this project.

Step 3 Identifying uncertain parameter(s) based on defined flexibility goal(s)

For the specified process flexibility goal in the BoD and this example, we determine the following key uncertain parameters: toluene composition, benzene specification (purity requirement), raw material and product prices, and the availabilities of key equipment in HDA plant.

Step 4 Formulating problem with uncertain parameters

For the cited uncertain parameters above, we can represent them all by probability distribution functions. However, increasing the number of parameters represented as distribution functions increases the computationally burden of the problem. In this step, we propose to use alternative ways (if possible) to represent the uncertainty in parameter identified in step 3. For this example, the uncertainty in breakdown of key equipment such as the compressor, the stabilizing column and the reactor, which is usually represented by failure rate distribution curves, can be represented by measures such as MTTR and MTBF values of the equipments.

In addition to illustrating the different steps, it is interesting to compare the set of uncertain parameters obtained by applying our procedure with the set of parameters that were considered uncertain for the HDA process synthesis example in the work of Chaudhari and Diwekar (1997). Both HDA examples share common uncertainty parameters, but not completely all of them. In particular, Chaudhari and Diwekar defined the initial cost of key equipments as an uncertain parameter, which we assume is not valid as the costs of key equipments does not fluctuates in future and are known at the initial stages. Another major difference is that they did not consider uncertainty in equipment reliability, which in turn affects the total availability of the HDA plant and hence the profitability (Goel et al., 2001). Other uncertainties such as fluctuations in the price of raw materials and the products that are identified with our procedure were considered in their work.

In order to demonstrate the accuracy of the proposed procedure, we would be required to undertake a detailed sensitivity analysis on the HDA and other process synthesis models. Nevertheless, this paper shows with the help of the simple illustrative HDA case study that the application of this procedure to a process will result in a systematic analysis of its uncertainty and associated parameters.

5. Concluding Remarks

A hierarchical procedure is presented for selecting uncertain parameters for a synthesis problem at the conceptual stage of design. The method uses only the preliminary process and product information to identify the kind of flexibility required. A standard taxonomy for manufacturing flexibility is used to define different kinds of flexibilities. We illustrated the method on a HDA process. It was found that it provides a designer with a formal framework for identifying dominant uncertain process parameters at the problem formulation stage.

The success of the developed procedure hinges on a proper definition of flexibility and the underlying subset of parameters that contribute to it. Therefore, more research is required in the area of flexibility taxonomy in the process industries and in mapping parameters to flexibility goal. Further, the effectiveness of the developed method needs to be established by means of additional case studies and comparisons such as a detailed sensitivity analysis.

6. References

Pistikopoulos, E.N., Uncertainty in process design and operations. *Computers and Chemical Engineering*, 1995, 19, S553-S563.

D'Souza, D.E. and F.P. Williams, Towards a taxonomy of manufacturing flexibility dimensions. *Journal of Operations Management*, 2000, 18, 577-593.

Douglas, J.M., Conceptual Design of Chemical Processes. 1988, McGraw Hill, NY.

Hutcheson, P., A.W. Pearson, and D.F. Ball, Innovation in process plant: A case study of ethylene. *Journal of Production Innovation Management*, 1995, 12, 415-430.

Swaney, R.E. and I.E. Grossmann, Optimization strategies for flexible chemical processes. *Computers and Chemical Engineering*, 1983, 7, 439-462.

Chaudhari, P.D. and U.M. Diwekar, Synthesis under uncertainty with simulators. Computers and Chemical Engineering, 1997, 21, 7, 733-738.

Goel, H.D., Herder, P.M. and Weijnen M.P.C., Application of life cycle costing at the conceptual design stage: A case study, *Proceedings of 6th World congress of Chemical Engineering*, 2001, 23-27 September, Melbourne, Australia.

European Symposium on Computer Aided Process Engineering – 12
J. Grievink and J. van Schijndel (Editors)
205
© 2002 Published by Elsevier Science B.V.

Optimal design of a continuous crystallizer with guaranteed parametric robust stability

R. Grosch, M. Mönnigmann, H. Briesen and W. Marquardt
Lehrstuhl für Prozesstechnik
RWTH-Aachen

Abstract

A recently developed approach that addresses the economically optimal design under uncertainty is adopted to the design of continuous crystallizers. A key property of the approach is the simultaneous treatment of the economics and the parametric robust stability of the operating point (Mönnigmann et al.; 2000, 2001). An optimization problem is formulated, which comprises of an economical objective function as well as equality and inequality constraints. In addition to the steady state model equations and usual feasibility constraints, another set of constraints is added to guarantee stability for the optimal operating point.

1. Introduction

Continuous crystallization plays a key role in the production of many particulate chemicals ranging from table salt to fertilizers. The production of bulk chemicals with decreasing economical margins necessitates the design of economically optimized plants. However, an economically favorable operation is often hindered by the high sensitivity to parametric variations of the process. For some industrial crystallizers, drift off during operation to an oscillatory state has been reported. As a consequence, the produced crystal size distribution (CSD) as well as the mean particle size vary over time. This undesired behavior negatively affects downstream process units such as filtration and drying. The product also often fails to meet customer quality specifications. At best, costly blending strategies are required to regain an acceptable product quality. The physical reasons for the oscillatory behavior are only understood qualitatively. In principle, oscillatory behavior can be counteracted by modifications of the process, the equipment or the control system. However, due to the unavoidable uncertainties in the model, such a retrofit design is a difficult and time-consuming task.

2. Crystallization process

The well known approach to modeling of crystallization processes is the use of the population balance equation (Randolph and Larson; 1988). In the case study presented here, a CMSMPR (continuous mixed suspension mixed product removal) crystallizer model is discretized with the method of moments to solve the population balance equation approximately. The implemented equations have been given by Chiu and Christofides (1999) using simple crystallization kinetics (Volmer's model for nucleation and McCabe's law for crystal growth). We choose this model for illustrative purposes.

$$\frac{\mathrm{d}m_0}{\mathrm{d}t} = -\frac{m_0}{\tau} + \left(1 - \frac{4}{3}\pi m_3\right)k_2 \exp\left\{-k_3\left(\frac{c}{c_s} - 1\right)^{-2}\right\} \tag{1}$$

$$\frac{\mathrm{d}m_j}{\mathrm{d}t} = -\frac{\mathrm{d}m_j}{\tau} + jk_1(c - c_0)m_{j-1}, \quad j = 1,\ldots,3 \tag{2}$$

$$\frac{\mathrm{d}c}{\mathrm{d}t} = \frac{(c_0 - c) - 4\pi k_1 \tau (c - c_s) m_s (\rho - c)}{\tau\left(1 - \frac{4}{3}\pi m_3\right)} \tag{3}$$

After initialization, this ODE system can be integrated for the state variables m_j and c, representing the j-th moment of the crystal size distribution and the concentration of solute in the liquid. c_0 and τ denote the concentration of solute in the feed and the residence time respectively, and are the main parameters of the process model. For explanation of the other variables please see Chiu and Christofides (1999).

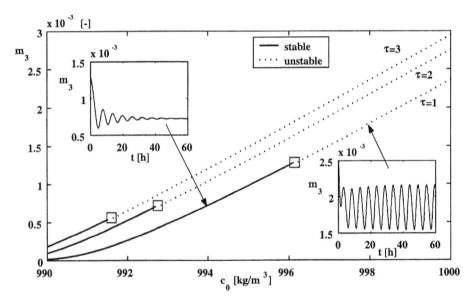

Figure 1: Stable and unstable points of operation depending on the process parameters.

For the example treated, it is known that stability can be lost through parameter variations resulting in the occurrence of sustained oscillations. This behavior is illustrated in Figure 1 with some results for the above model. For fixed τ the increase of feed concentration of solute leads to a loss of stability (square-marked points). The time trajectories of the third moment at two selected points of operation show the stable and the oscillatory behavior, respectively.

3. Optimization with constraints for parametric robust stability

In the optimization of the crystallization process, the existence of unstable solutions has to be addressed. An optimal but unstable point of operation cannot be accepted from a practical point of view without further measures for stabilization. It may, however, not be sufficient to guarantee that the optimal point is stable, but parameter uncertainties have to be accounted for. Process parameters such as the residence time τ and feed concentration c_0 cannot be fixed to precise nominal values, but are subject to uncertainty due to, e.g., measurement errors, drift or disturbances. Therefore any point of operation which is stable but too close to the stability boundary cannot be accepted, since a deviation from the nominal parameters may result in a loss of stability. This section outlines a recent approach (Mönnigmann and Marquardt, 2001) which is able to guarantee stability in process optimization in the presence of parametric uncertainty. Assuming that the uncertainty can be described by lower and upper bounds

$$p_i \in [p_i^{(0)} - \Delta p_i, \, p_i^{(0)} + \Delta p_i], \quad p_1 = c_0, \quad p_2 = \tau, \tag{4}$$

only points of operation $(p_1^{(0)}, p_2^{(0)}) = (c_0^{(0)}, \tau^{(0)})$ can be accepted, for which the process remains stable despite variations within (4). In Figure 2 the robustness square defined by (4) is overestimated by a circle. By requiring the circle to touch the stability boundary (bold line in Fig. 2) tangentially or to stay off this boundary, the process will remain stable for all parameter values within (4).

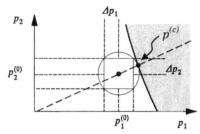

Figure 2: Robustness square at the stability boundary. For the example treated, the stability boundary results if squares in Fig. 1 are traced as τ is varied.

In Figure 2 the shortest distance from $(p_1^{(0)}, p_2^{(0)})$ to the stability boundary occurs along the bold dashed line which is normal to the stability boundary. The constraint for parametric robustness can be implemented by forcing the distance along this normal direction to be equal to or larger than the radius of the circle. Denoting the normal vector of appropriate length by r, the optimization problem with constraints for parametric robustness can be stated as

$$\max_{x, \, p^{(0)}, \, p^{(c)}, \, r} \Phi \tag{5a}$$

$$\text{s.t.} \quad f(x, p^{(0)}) = 0 \tag{5b}$$

$$g(x, p^{(0)}, p^{(c)}, r) = 0 \tag{5c}$$

$$p^{(0)} \le p^{(c)} - r \tag{5d}$$

where Φ is a profit function, x is the vector of state variables of a process model $dx/dt=f(x,p)$ like (1)-(3), g is a system defining the appropriate normal vector, and $p^{(0)}$ and $p^{(c)}$ refer to parameter values at the point of operation and the closest point on the stability boundary respectively. If more than one stability boundary exists in the parameter space close to $p^{(0)}$, constraints (5c-d) have to be stated repeatedly. For details on (5) and in particular on g, refer to Mönnigmann and Marquardt (2001). We stress that the method is applicable to an arbitrary number of uncertain parameters and to stability and feasibility boundaries of other types than treated here.

4. Optimization results

Here the results of the solution of problem formulation (5) applied to the crystallization process (1)-(3) are presented and compared to the optimization of the process without enforcing parametric stability.

The objective (5a) of the optimization is to maximize the solids production per residence time: $\Phi=\rho m_3/\tau$. For the sake of clarity we assume that only the feed concentration and the feed rate are subject to uncertainty: $\Delta c_0=1$ kg/m^3 and $\Delta\tau=0.1$ h. However, a growth or nucleation constant which cannot be measured accurately, could also be considered through this approach. The presented values are chosen arbitrarily. However, if knowledge on the uncertainty is available from experiments, it can be readily taken into account by the approach. Due to a limited pump capacity and the requirement of a crystal free feed, the process is subject to operability constraints, $\tau>0.6$ h and $c_0<1010$ kg/m^3, respectively.

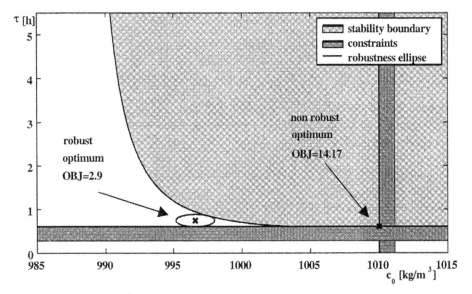

Figure 3: Robust (left) and non robust (right) optimum.

For reference the process is optimized without enforcing the robustness of the steady state (5c)-(5d). The result is presented in Figure 3, where the optimum is located in the corner of the operability constraints ($c_0=1010$ kg/m^3, $\tau=0.6$ h). Since this point is situated inside the unstable regime, the crystal mass production rate cycles within [13.4 kg/m^3/h, 15.5 kg/m^3/h] around the unstable steady state production rate of 14.7 kg/m^3/h. Also, the mean crystal size cycles within [0.31mm, 0.61mm] with a cycling time of about three hours, cf. Figure 4.

Since cycling operation is usually undesired, the optimization problem including the robustness constraints is solved next. For this case, the optimum solution is located at ($c_0=996.6$ kg/m^3, $\tau=0.74$ h) as can be seen in Figure 3. At this point the robustness ellipse is tangent to the stability boundary and the operability constraint $\tau>0.6$ h. The solids production rate and mean particle size are 2.9 kg/m^3/h and 0.47 mm, respectively. Since this operating point is situated at a required minimum parametric distance from the stability boundary, stable operation is guaranteed.

Due to the simplicity of the cost function, a rigorous comparison of the robust and unstable operating points is not possible. As the mean particle size oscillates at the unstable optimal point of operation, costly treatment of the product such as blending or sieving might be necessary to meet customer demands. Therefore, information on the product quality, and thus on the impact of an oscillating operation on the profit, has to be included in the cost function for a more realistic comparison of the robust steady state operation to cycling operation. Nevertheless, the two results provide a rough measure of the cost of stability. If robust stability is enforced in our case study, the production rate drops to about twenty percent of the production rate at the unstable optimum. This number indicates to which extend investments into further measures for process stabilization are economically reasonable. Such a retrofit design measure could be the implementation of a suitable controller. In fact, the underlying process can be stabilized by the implementation of a PI controller, which regulates the feed concentration, c_0, depending on the deviation of the second moment, m_2, from its set point, cf. Figure 4 which is an enlargement of Figure 3. The stabilization of the process by tuning of the controller gain, manifests in a shift of the stability boundary away from the operating point. The stabilization is also demonstrated by the subplots in Figure 4 which show the trajectory of the third moment of the CSD for the uncontrolled case and for the case with controller setting ($Kr=-0.7$, $Tn=1$) after a step variation of the feed set point of about 1 kg/m^3.

A comparison of the robust and the stabilized process is not directly possible. An economic comparison would have to include the cost of the controller and the more accurate actuators required. Tuning of the control parameters for the highly nonlinear model is also a difficult task. In the presented case, a nonlinear analysis of the model with a controller was necessary in order to find some arbitrary controller settings for which the model is stable. In the future, we suggest to apply formulation (5) to the model including the controller instead. For the presented example this can be interpreted illustratively as giving the optimizer a means to influence the size of the stable operating regime. Thus, the controller settings, which allow a minimum parametric distance from the instability boundary could be determined.

210

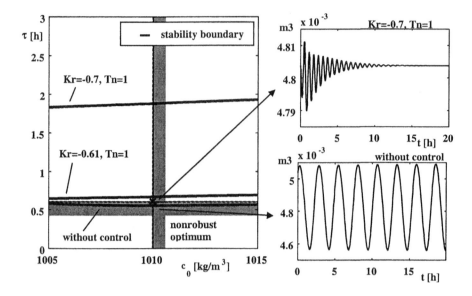

Figure 4: Effect of PI control on stability boundary and trajectory of third moment.

5. Discussion

The presented results which are obtained for a simple crystallization process model, demonstrate that the application of the adopted approach to crystallization processes is promising and that the approach provides the engineer with an additional tool useful to the design of continuous crystallization processes. In the future the adopted approach has to be evaluated by means of examples of industrial scale, building on more realistic models of the crystallization process. More rigorous problem formulations can also help to make the various results more comparable. For example, the product quality could be included in the objective or as constraints.

References

Mönnigmann, M., V. Gehrke, and W. Marquardt, 2000, Nonlinear dynamics in process design. CHISA 2000, 14th International Congress on Chemical and Process Engineering, 27-31.8.2000.

Chiu, T. and P.D. Christofides, 1999, Nonlinear control of particulate processes, AIChE J., 45(6), 1279-1297.

Randolph, A. and M. Larson, 1988, Theory of Particulate Processes. Academic Press, San Diego, 2nd edition.

Mönnigmann M. and W. Marquardt, 2001, Normal vectors on manifolds of critical points for parametric robustness of ODE systems, accepted for publication in J. Nonlinear Science.

European Symposium on Computer Aided Process Engineering – 12
J. Grievink and J. van Schijndel (Editors)
© 2002 Published by Elsevier Science B.V.

Modeling Coupled Distillation Column Sections Using Profile Maps

Holland, S.T. ; Tapp, M ; Hildebrandt, D*; Glasser, D
*Correspondence to:
Centre of Material and Process Synthesis
School of Process and Materials Engineering
University of The Witwatersrand
Private Bag 3, P.O. Wits 2050
Telephone (wk): +27 11 717 7557   Fax +27 11 717 7557
E-mail: dihil@chemeng.chmt.wits.ac.za

Introduction:

Differential equations (D.E.'s) have been employed in the preliminary design of distillation columns. The properties of D.E.'s have also been used to gain an understanding of the general behaviour of distillation systems. These approaches have, until now, only been applied to traditional distillation column configurations. There are however, potentially very exciting gains to be made in separation, by exploring alternative distillation configurations. This paper will discuss one such configuration, a coupled column system. A coupled column system is one where the products at the top and bottom of one column are fed directly into another column. These columns will be treated simply as column sections. The effects of one column on another, as well as the possible products will be analysed.

Background and Procedure:

The behaviour of distillation column sections can be approximated using differential equations analogous to those derived by Doherty (1977). The generalised differential equation describing the change of liquid composition in a column section is:

$$\frac{dx}{dn} = \frac{V}{L}\left[x - Y(x)\right] + \frac{\Delta}{L}\left[X_\Delta - x\right]$$

where $\Delta = [V - L]$ $X_\Delta = \left[VY^T - LX^T\right]/\Delta$

For a single column section, finding the composition profile is simply a matter of specifying Y^T and X^T (top vapour and liquid composition) and integrating the differential equation. If, however, the column is now coupled to a second column, such that both columns have the same X^T and Y^B (bottom vapour composition), but different reflux ratios and stages, the values of X^T and Y^T are no longer arbitrary. In addition, the values of X^B and Y^B are also linked. (See figure 1 below)

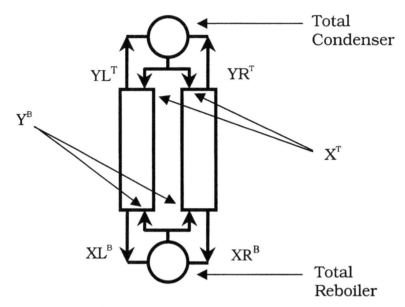

Figure 1: Coupled Column Sections

The differential equations for both columns must now be solved simultaneously with the additional constraints mentioned above ($XL^T = XR^T$ and $YL^B = YR^B$). The resulting system is quite unstable and numerical solutions quite difficult to achieve from an arbitrary initialisation point. A simple method for the determination of feasibility would be desirable.

In order to simplify the system and achieve an understanding of the dynamics involved, the infinite reflux case, (where L=V in both columns) and set values of X^T and Y^T are maintained, will be analysed. The resulting differential equations simplify to:

$$dx\big/_{dn} = [x - Y(x)] + [Y^T - X^T]$$

or

$$dx\big/_{dn} = [x - Y(x)] + X'_\Delta$$

For simplicity, the term $[Y^T - X^T]$ will be referred to as the difference vector.

As a result of the equal liquid and vapour flows on either side of the coupled system, simple mass balance dictates that the difference vector remain constant throughout each column. The difference vectors of the columns have equal magnitude but opposite direction.

These differential equations can be integrated in the same way as residue curve equations (as n→∞ and n→ - ∞) and produce two pinch points (Tapp et al.). If an entire map of profiles is produced, again in the same way as that of the residue curve map, (keeping the difference vector constant throughout the space), all possible profiles for any difference vector can be found. Regions outside the mass balance triangle are also mapped in order to track the pinch points. (See figure 2 below)

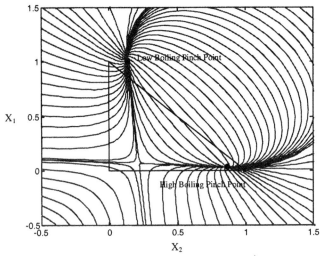

Figure 2: Liquid Profile Map for difference vector $X'_\Delta = [-0.0653 +0.0271]$

This profile map represents all possible solutions for one side of the coupled system. If the corresponding map is produced for the other side (negative deviation vector of first column) all possible solutions for this column can also be found. (See figure 3 below)

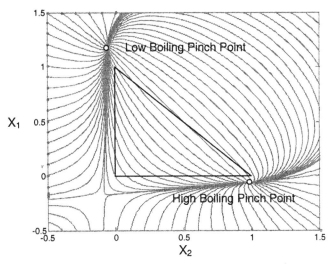

Figure 3: Liquid Profile Map for difference vector $(-X'_\Delta) = [-0.0653 -0.0271]$

Through mass balance and/or simulations, it can be shown that the feasible pairs of profiles for the coupled systems are of the form shown below (figure 4). i.e. the liquid profiles must intersect each other twice.

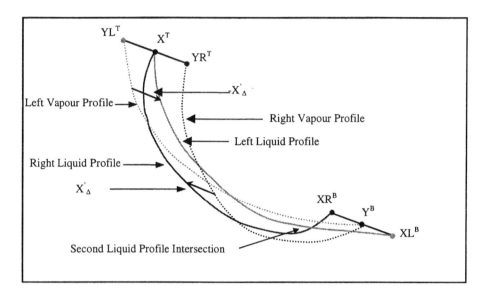

Figure 4: Coupled Column Section Profiles

Using this logic, if the positive and negative vector profile maps, of figures 2 and 3, are superimposed, it is possible to find all possible solutions for the system. They occur where the same liquid profiles intersect twice within the mass balance triangle.

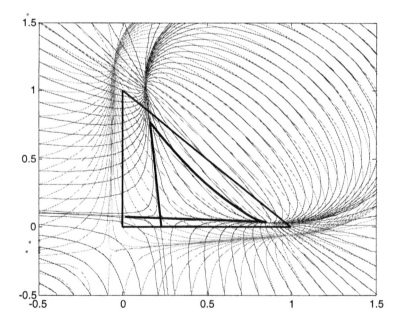

Figure 5: Superimposed Liquid Coupled Column Section Profiles

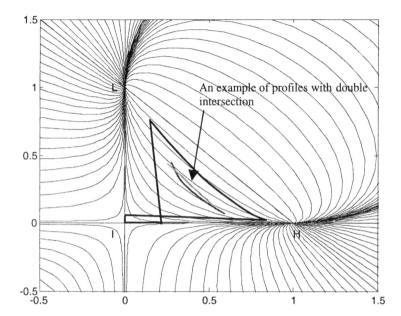

Figure 6: The feasible "Bow-Tie" Region shown on the residue curve map.

Conclusion

A profile map of solutions can be generated for the column section D.E. (for any X'_Δ). These solutions represent all possible composition profiles for a column section operating with the specified X'_Δ. The positive and negative X'_Δ maps can be superimposed and the entire region enclosing all possible double intersections found. This region has a "Bow-Tie" shape and represents all possible solutions of the coupled column system. These solutions can be used for the initialisation of the simultaneous coupled column section differential equations.

This technique is currently being investigated as a tool for the modelling of this and other novel column configurations. The ultimate aim is to break away from the constraints of the traditional distillation column and explore the potential of alternative designs.

Nomenclature:

X^T ....Liquid composition at the top of each column

YL^T .... Vapour composition at the top of left column

YR^T .... Vapour composition at the top of right column

XL^B ....Liquid composition at the bottom of the left column

XR^B ....Liquid composition at the bottom of the right column

Y^B ....Vapour composition at the bottom of each column

L...Liquid flowrate

V .... Vapour flowrate

$X^{\cdot}_\Delta$..Difference Vector $= [YL^T - X^T]$ or $[YR^T - X^T]$

X_i...Composition of component i

Mass balance Triangle ......Triangle bounded by mass balance points such that:
...... $X_1 + X_2 + X_3 = 1$

References:

Doherty, M.F., Perkins, J.D., *"On the Dynamics of Distillation Processes 1-3"*, Chem. Eng. Sci., 33, 1978.

Tapp, M; Holland, S.T.; Hildebrandt, D; Glasser, D, *"Column Profile Maps Part A: Derivation and Interpretation of Column Profile Maps "*. Paper in progress.

European Symposium on Computer Aided Process Engineering – 12
J. Grievink and J. van Schijndel (Editors)
© 2002 Elsevier Science B.V. All rights reserved.

217

Automating Reactor Network Synthesis: Finding a Candidate Attainable Region for Water-Gas Shift(WGS) Reaction

S. Kauchali, B. Hausberger, E. Mitova, D. Hildebrandt* & D. Glasser
*Centre of Materials and Process Synthesis
School of Process and Materials Engineering
University of the Witwatersrand
P/Bag x3, PO Wits, Johannesburg, 2050
South Africa

Abstract

We use the Attainable Region (AR) technique to generate reactor network synthesis solutions to the WGS system. We first do this using the conventional method as described by Nicol et al. (1999) to generate the AR for exothermic reversible reactions. We then generate the AR using the new *iso-state* algorithm and show the answers are essentially the same. We further use a *linear programming model* to show that no substantial extension to the candidate AR is possible at the level of resolution of the grid. These latter two methods are shown to be simple enough such that they could, in principle, be incorporated in software and be implemented by users who have a fair understanding of reactor system design. Generating candidate regions will provide the users with benchmarks for what can be attained for the WGS system.

1. Introduction

The Water-Gas shift reaction is industrially important. For example, in the Haber ammonia synthesis process, the WGS process has provided an economical means of producing hydrogen while removing carbon monoxide. There exist a number of reactor designs and flowsheet configurations for the WGS reactor in the production of hydrogen. It is common to carry out this operation using a simple series configuration of adiabatic beds using catalysts operating at high and low temperature ranges with the objective of minimizing catalyst mass and energy consumption for a specified conversion target. Up until now, there has not been a systematic study of the optimisation of reactor networks for the WGS reaction. The construction of the Attainable Region (AR) for a Reactor Network Synthesis problem provides important opportunities for the initial design and retrofitting of chemical processes. Furthermore, the AR provides useful design benchmarks that can be used to optimise existing reactor structures. In this paper we provide procedures to automate the construction of an AR so that the end users are not required to be familiar with the complex mathematics involved.

This paper begins by introducing the system we are looking at followed by defining the fundamental physical processes that are occurring and the state space we are going to be working in. The conventional AR techniques are going to be used to generate a candidate region with the corresponding reactor network required to achieve the region. We shall introduce two new algorithmic methods that are capable of the automatic generation of the candidate AR region and discuss some computational aspects associated with each method. The results obtained from the three techniques are presented and discussed. Finally, we conclude the paper with recommendations for possible research areas in the development of a technique to automate the AR generation.

2. The System

We will consider the case of a water gas shift reaction using standard Langmuir-Hinshelwood kinetics model for a high temperature shift catalyst, where the kinetic parameters were taken from Podolski and Kim (1974). This system was constrained to the standard operation conditions of 31bar, and a catalyst operating temperature range of 450 K to 823 K. A feed, representative of an industrial type, at 31 bar and 300 K, comprised 6.5 % Carbon Monoxide, 3.8 % Carbon Dioxide, and 28.3 % Hydrogen with the remainder being Water, was used. The Water to Dry Gas ratio in the feed was set to 1.6 in line with industrial practice. Furthermore, we have not explicitly included the cost for preheating the feed and have not considered both inter-stage and cooling processes.

2.1. The Fundamental Processes
The fundamental processes considered in this case are reaction and mixing. For this example, we allow free pre-heating of the feed up to the maximum allowable temperature for the catalyst.

2.2. The State Variables
The variables chosen were *CO Conversion (X)*, *Temperature (T)* and *Residence Time* (*τ*). These are all linearly independent and they completely describe the state of the system. Most objective functions for common optimisation problems are likely to be a function of these variables.

2.3. The Process Vectors
The process vectors are those for reaction and mixing, denoted R and v, respectively. Given the state vector C describing the X, T and τ at that point, the process vectors obey the following equations:

$$\mathbf{R} = \left[\frac{r_{f,co}}{N_{co}^0} \quad \frac{T_{ad}(X,T)r_{f,co}}{N_T \overline{C_p}} \quad 1 \right] \qquad \mathbf{v} = \begin{bmatrix} X^* - X & T^* - T & \tau^* - \tau \end{bmatrix} \qquad \dots(2)$$

3. Conventional Attainable Region Method

The classical Attainable Region (AR) approach as used by Nicol (1999) was used to obtain the following volume of possible states in Conversion-Temperature-Residence Time space. This volume is known as the Candidate Attainable Region and is represented in Figure 1. in the full 3-dimension, as well as the 2-dimensional projection into Temperature-Conversion space.

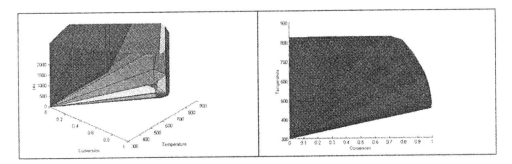

Figure 1. Representation of the Candidate Attainable Region for the High Temperature Water Gas Shift Catalyst in Conversion-Temperature-Residence Time State Space.

The general reactor network that can be used to achieve all the points in the candidate AR is:

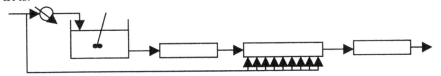

This consists of a preheater followed by a Continuously Stirred Tank Reactor (CSTR), followed by a Plug Flow Reactor (PFR), followed by a Differential Sidestream Reactor (DSR) taking the bypass feed from the original cold feed, followed by a further PFR. Even though this method gives the candidate AR and the optimal reactor structure, it requires manual intervention.

4. Algorithmic Methods

4.1. Isostate Algorithm

The Isostate Algorithm constructs the n-dimensional Attainable Regions from a set of $\binom{n}{2}$ 2 dimensional orthogonal subplanes. These planes have to date been oriented at regular intervals along the axes of the space, however this is not essential to the method. The method has three main stages of operation. The first or initialisation stage considers the full state space and generates the simple PFR and CSTR structures from the feed point(s). Should the processes considered have other simple well-defined processes such as batch boiling or flash operations, these can also be performed at this stage.

The second or growth stage, considers the convexified set of data points. The latest set of points is used in the construction of regularly spaced planes that span the range of the space considered. The construction of these subspaces is based on the selection and mixing of appropriate points from the full space dataset. These subspaces are extended through the local construction of Isostate DSR's within the plane, and where feasible Isostate CSTR's occur. Rooney et al., 2000, describe these structures and their control policies in detail.

The resulting extended subspaces are then convexified, and the subspace recombined into the full space dataset. The planes spacing are then recalculated based on the new

220

range, and the process repeated until the rate of volume or hyper-volume growth falls below a specified limit. For simpler analysis, it is possible to examine the growth in terms of the range of the space covered and use that as the termination criteria. The third is a Polishing stage, which makes use of full dimensional PFR's from the extreme points of the boundary to complete the surface.

4.1.1. Isostate Computational Run Times
The following computation times are reported for a Pentium III 866 MHz Intel Computer with 128 MB Ram.

- Initialisation stage: Initial Generation of Space   2.1 sec
- Growth Stage
  - Calculation of One Plane   0.05 – 0.07 sec
  - One Complete Iteration   70.4 sec
- Polishing Stage   3.4 sec

For the construction of the Candidate AR (one that satisfies the necessary conditions) we made use of the Isostate algorithm. The results from this calculation are shown in Figure 2 for the 2-dimensional and 3-dimensional spaces respectively.

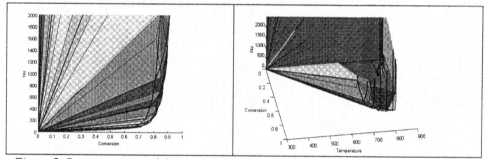

Figure 2. Representation of the candidate Attainable Region arrived at from the Isostate Algorithm (shaded region) comparing with traditional approach (solid black lines)

Having constructed the candidate Attainable Region, the structure was tested with a linear programming formulation for AR analysis. It was necessary to examine the space for possible extensions arising from isolated CSTR's that may have been omitted.

From Figure 2 as well as numerical comparisons we can see that the numerical methods have reproduced the earlier results, and can be used for the simple non-interactive generation of candidate Attainable Regions for optimisation and benchmark evaluations.

4.2. Linear Programming Models for AR Analysis
Kauchali et. al. (2000) outline some of the developments and properties that a candidate AR should obey in terms of Linear Programming (LP) models. Here, they consider the nonnegative concentration space constrained by mass balances on the components. Over this region, an arbitrarily large set of grid points are defined. At each grid point, the rate vector, based on a kinetic model is calculated. This set of points can be made as large as needed, but the total number of points was considered finite in this work. Now, let a CSTR operate at each grid point. The feed to each CSTR is a linear combination of the outputs from all other CSTR's and the feed to the system. For a constant density

system, isothermal operation and for any dimensional problem, we can write the CSTR design equation and the total connectivity equations at each grid point, which are essentially mass balance equations about CSTRs, mixing and splitting points. This leads to the derivation of to two linear programming algorithms. In one model an algorithm is developed that can be used to generate a candidate AR, while in the second model the algorithm is used to test a candidate region generated from any arbitrary technique for possible extensions. For this paper we consider the model that uses the LP formulation to test a given candidate AR for extension in the complement of the region.

4.2.1. AR Extension Test Using a Linear Programming Model

Consider a convex candidate AR that is constructed by any method and that may also satisfy the necessary conditions. We define a set of compositions points, k, from the boundary of the candidate AR. In addition to the process feed, each composition point from the boundary can feed the CSTR's operating in the complement. Here, an unambiguous measure of extending the candidate is to find a set of CSTR's that operates in the complement. To find such CSTR's we pose the following optimization problem:

$$Max_{\substack{F_i, FC_i^0, Q_{ij}, Q_{0,i} \in I^C \\ Q_{i,out}, Q_{0,out}, V_i}} \sum_{i \in I^C} V_i$$

s.t.

$$FC_i - FC_i^0 = V_i R_i$$

$$F_i = \sum_{j \in I^C} Q_{ji} + \sum_{k \in K} Q_{k,i}$$

$$F_i = \sum_{j \in I^C} Q_{ij} + Q_{i,out}$$

$$FC_i^0 = \sum_{j \in I^C} Q_{ji} c_j + \sum_{k \in K} Q_{k,i} c_k$$

$$1 = \sum_{k \in K} Q_{k,out} + Q$$

$$0 = Q - \sum_{i \in I^C} \sum_{k \in K} Q_{k,i}$$

$$0 = Q - \sum_{i \in I^C} Q_{i,out}$$

$$F_i, FC_i^0, Q_{ij}, Q_{k,i}, Q_{i,out}, Q, Q_{k,out}, V_i \geq 0$$

Furthermore, the design equations in the optimization problem are a consequence of mass balance equations about CSTRs (operating at point i in the composition space), mixing and splitting of streams entering or leaving a CSTR. Here, the overall system volumetric flowrate, Q_{feed}, is normalized to unity with a specified composition c_0, F_i is the total feed to the ith CSTR, FC_i^0 is the component inlet flowrate, $Q_{k,i}$ is the amount of overall feed that enters the CSTR, $Q_{k,out}$ is the amount of feed that completely bypasses all CSTR's, Q is the portion of feed that feeds to the CSTRs in the complement, Q_{ij} is the flowrate from the outlet of CSTR i to the inlet of CSTR j, $Q_{i,out}$ is the amount of material from the CSTR going to the outlet of the network, V_i is the volume of the CSTR, c_i is the outlet composition from CSTR i.

4.2.2. Linear Programming Results & Computational Run Times

The LP formulation was coded in GAMS and run on a PIII 800MHZ system with 128 MB ram. For a single rate vector in the complement and for 630 points representing the boundary of the candidate region, there were 1270 variables and 14 equations generated for the LP. The OSL solver took an average CPU time of 0.1 seconds to solve the program to optimality. For a 2,000,000 variable problem and 85,000 equations such as would be representative of a consideration of 1400 CSTR points, an average of 1200 CPU sec would be required. Several points in the complement of the AR as generated by the isostate algorithm were checked and no extensions were found.

5   Discussion

The candidate AR for the WGS system generated using the isostate algorithm compares very favourably to the region obtained by the conventional AR methods. The algorithm is robust in that it can handle multiple processes in any dimension with minimal user intervention. Moreover, the isostate algorithm successfully finds a candidate AR using reasonable computer resources. The LP algorithm, to test a candidate region, can potentially be incorporated within the isostate algorithm to provide a single algorithm that could help quicker convergence of the method, while simultaneously checking for possible extensions. However, the size of the LP formulation and the corresponding time to solve the LP is severely limited by the number of CSTR points required in the complement of the candidate region. For example, testing as little as 1400 CSTRs in the complement leads to having to solve a linear program with 2,000,000 variables and 85,000 equations.

6. Conclusions

We conclude that we have been able to produce a candidate attainable region through fully automated procedure. This has eliminated the need for a process synthesis expert to interact in order to produce the result. The method is robust, and has been able to rapidly and reliably generate the volume of the candidate attainable region. This technique has great potential as a benchmark for the evaluation and analysis of prospective designs. Currently, it is not possible to algorithmically interpret the boundary of the AR from this construction method as a reactor network structure. However work is in progress to remedy this problem.

The optimal reactor structure, as found by the conventional AR method, was found to be a pre-heater prior to a CSTR, followed by a PFR, followed by DSR taking the bypass feed from the original cold feed, followed by a final PFR.

7. References

Kauchali S., Rooney W.C., Biegler L.T., Glasser D., Hildebrandt D. , 2001, Linear Programming Formulations for Attainable Region Analysis, submitted as a conference paper at the AIChE Annual Meeting, Reno, USA.

Nicol W., 1998, Extending the Attainable Region Technique by Including Heat Exchange & Addressing Four Dimensional Problems, thesis, University of the Witwatersrand.

Podolski W.F. & Kim Y.G., 1974, Modeling the Water-Gas Shift Reaction, Ind.Eng.Chem., Process Des. Develop., Vol. 13, No 4, 415.

Rooney W.C., Hausberger B.P., Biegler L.T., Glasser D., 2000, Convex Attainable Region Projections for Reactor Network Synthesis, Comp. Chem. Eng. 24,2-6, 225.

European Symposium on Computer Aided Process Engineering – 12
J. Grievink and J. van Schijndel (Editors)

Integrated Design of Cooling Water Systems

Jin-Kuk Kim and Robin Smith
Department of Process Integration, UMIST
United Kingdom

Abstract

Overloading on cooling systems is one of major concerns to the petrochemical and chemical industries. The solution for overloading of cooling systems can be obtained by modifying the cooling water network with re-use design between coolers. This allows better cooling tower performance and increased cooling tower capacity. A novel methodology has been developed for the design of cooling networks to satisfy any supply conditions for the cooling tower and allows systematic exploration on interactions between the performance of the cooling tower and the design of cooling water networks. The new design method for cooling water networks provides the retrofit solutions for overloading of cooling systems.

1. Introduction

Cooling water is widely used in the petrochemical and refinery industry in order to satisfy a variety of cooling needs. As ambient air conditions change throughout the year, fluctuation of the tower performance is inevitable. Although the change of ambient air conditions is site-specific, tower overloading often occurs, especially when air conditions have a high temperature and humidity. Ambient temperature swing results in the oscillation of the load on the tower.

Aside from temperature swing, cooling systems become also bottlenecked when the cooling load of the process itself is increased. The overloading problems affect not only the cooling system itself but also the process performance. The later is a more serious problem, because a reliable process performance or desired product quality cannot be achieved when cooling systems cannot service the cooling demand of processes.

Therefore, many studies have been focused on solving cooling tower overloading. Simply adjusting operating conditions of cooling systems often solves cooling tower overloading problems for example, increasing air flowrate. The strategic use of taking hot blowdown from return cooling water and the instalment of air cooler have been taken as a practical way for debottlenecking of cooling systems. But the previous approaches have been based on the individual screening of components of cooling systems, rather than the system as a whole. Also, little attention has been given to the interactions of cooling water systems, although the changes of operating conditions of cooling water systems frequently happens. Cooling water systems have interactions between cooling water networks and the cooling tower performance. So cooling water systems should be designed and operated with consideration of economics and constraints of the cooling system components.

In this research, the process integration aspects will be highlighted, based on studies of systems interactions in cooling water systems. A novel and systematic design methodology will be presented to provide design guidelines for solving the overloading of the tower. Then, the retrofit analysis of the cooling water systems will be illustrated with example study.

2. New design concept of Cooling Water Systems

From cooling tower modelling (Kim, 2001), when the inlet cooling water has conditions of high temperature and low flowrate, the effectiveness of the cooling tower is high. In other words, the cooling tower removes more heat from the water when inlet cooling water has high temperature and low flowrate.

The current practice for cooling water network design most often uses parallel configurations. Under a parallel arrangement, return cooling water flowrate is maximised but the return temperature is minimised. These conditions will lead a poor cooling tower performance. The traditional parallel design method is not flexible when dealing with various process restrictions.

All cooling duties do not require cooling water at the cooling water supply temperature. This allows us, if appropriate, to change the cooling water network from a parallel to a series design. A series arrangement, in which cooling water is re-used in the network, will return the cooling water with a higher temperature and lower flowrate. From the predictions of cooling tower models, if the design configuration is converted from parallel to series arrangements, the cooling tower can service a higher heat load for the coolers.

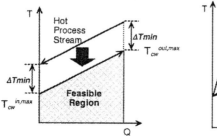
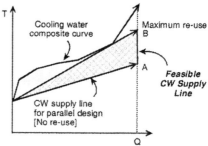

Figure 1. Limiting cooling water profile   Figure 2. Cooling water composite curve

In cooling water network analysis, it is assumed that any cooling-water-using operation can be represented as a counter-current heat exchange operation with a minimum temperature difference. The concept of the limiting water profile (Wang and Smith, 1994) is taken from water pinch analysis and shown in Figure 1 as a "limiting cooling water profile". This is defined here to be the maximum inlet and outlet temperatures for the cooling water stream. These allowable temperatures are limited by the "minimum temperature difference" (ΔT_{min}). In retrofit the temperature difference could be chosen to comply with the performance limitations of an existing heat exchanger under revised operating conditions of reduced temperature differences and increased flowrate. As the limiting profile represents water and energy characteristics simultaneously, the cooling water network can be manipulated on common basis throughout all coolers.

A cooling water composite curve is constructed by combining all individual profiles into a single curve within temperature intervals (Figure 2) and shows overall limiting conditions of the whole network. The overall cooling water flowrate and conditions for network are determined by matching the composite curve with a straight line: the cooling water supply line. Following the method given by Kuo and Smith (1998) allows the design of the cooling water network to achieve the target predicted by the supply line.

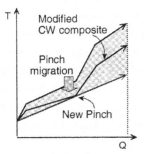

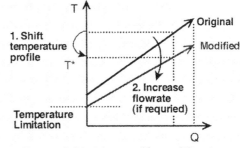

Figure 3. Pinch migration   Figure 4. Limiting profile modifications

If the cooling water supply line does not correspond with minimum flowrate (either because of system interactions or temperature constraints), then a pinch point is not created with the limiting cooling water composite curve. The setting could be between minimum flowrate (maximum re-use) and no re-use (parallel arrangement) as shown in Figure 2. The water main method is based on the concept of the pinch point and cannot be applied to problems without a pinch. The new design methodology should provide for cooling water networks without a pinch. To deal with network design without a pinch, the cooling water composite curve is modified within the feasible region and the problem changed into a problem with a pinch (Figure 3) by "pinch migration" method whilst maintaining the desired supply line. A "temperature shift" method is introduced to modify individual cooling water profiles. The cooling water network design can now be carried out using the cooling water mains method and the pinch migration and temperature shift method enables design with any target temperature.

3. Automated design of Cooling Water Networks

The conceptual method explained in previous section deals with cooling water systems by graphical representation of heat exchangers and therefore it is difficult to consider system constraints to be included (for example, forbidden/compulsory matches between units, complexity of networks, etc.). Also, it is a troublesome task to shift individual profiles according to "temperature shift" algorithms. The limitation of the conceptual method provides incentives to develop an automated method.

The automated method relies on the optimisation of a superstructure which can be set up to consider all possible connections between: i) fresh cooling water source and cooler ii) cooler and cooling water sink iii) coolers as shown in Figure 5.

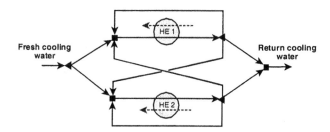

Figure 5. Cooling water network superstructure

The optimisation would give a network topology and assign the flowrate and temperatures of the streams. The above formulation results in the MINLP (Mixed-Integer Nonlinear Programming) problem. But the difficulty with nonconvexities would be arisen. Nonconvex problems are overcome by re-formulating the integration problems as a MILP model. From inspection of the characteristics in the outlet temperatures for parallel and maximum re-use designs, it can be deduced that the individual outlet temperatures are at their maximum values for any target conditions.

The design task is converted into the MILP problem when the outlet temperatures are treated as fixed parameters. The objective function is to minimise overall cooling water flowrate. When the overall outlet temperature is satisfied with the target temperature, the minimum overall flowrate of the network should be equal to the target flowrate. Regardless of the target temperature, the cooling water network can be designed retaining outlet temperatures at their maximum levels.

4. Debottlenecking of Cooling Water Systems

The cooling water composite curve can first be constructed from the limiting cooling water data. The cooling water network performance can be changed within a feasible region that is bounded by the maximum re-use supply line and the parallel design supply line (Figure 2). In Figure 2, the feasible cooling water supply line (line AB) represents the attainable outlet conditions by changing design configurations.

When the cooling tower is bottlenecked, the process cooling load becomes higher than tower heat removal and therefore, the inlet temperature of cooling water to process is higher than desired temperature. The heat removal of cooling water systems increases as the design configuration changes from parallel to maximum re-use (A to B in Figure 2). The target conditions for debottlenecking, where the inlet temperature to the cooling water network is satisfied, can be found using the cooling water system model.

The final stage is to design the cooling water network for the target conditions. The cooling water network design can be carried out using the cooling water mains method or automated design method.

The proposed debottlenecking procedure enables the cooling tower to manage the increased heat load by changing the network design from parallel to series arrangements. The design method targets the cooling tower conditions and then designs the cooling water network for the new target conditions.

The retrofit objective is to find a robust design of cooling water system that can cope with tower overloading and temperature swing. When the load on the tower is reduced from changes of network design and/or process changes, a cooling system may shut down some of the tower cells if tower consists of several cells. This means that the cooling systems can manage the increased cooling load. If the tower is not divided into separate cells, the air flowrate can be reduced, which results in a reduction of operating cost. When the load on the tower is increased from the changes of cooling water demand and/or ambient air conditions, the existing cooling system can remove the increased heat load of the tower without investment in new cooling equipment.

The concept of re-use design for cooling systems will now be considered for the Example study. The base case for cooling water systems is shown in Figure 6.

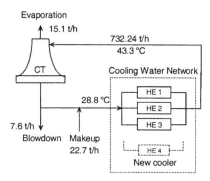

Figure 6. Base case of cooling water systems

With the new cooling water network design (Figure 7), the cooling tower manages the increased heat load by changing the network design. Efficient re-use of cooling water increases the operating range of the tower and also reduces the operating cost for water pumping and air supply, which is another benefit to tower operation.

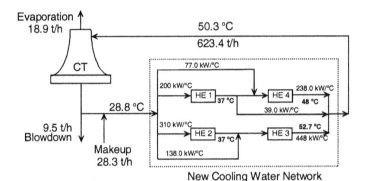

Figure 7. New cooling water network design

5. Conclusion

A new methodology for the design of cooling water networks has been developed to satisfy any supply conditions for the cooling tower. Design can be carried out with any target temperature with automated design method. From the interactions between the cooling tower performance and the design of the coolers, the proposed debottlenecking procedures allow increased capacity without investment in new cooling tower equipment when the cooling tower capacity is limiting.

References

Kim J. 2001, Cooling Water System Design. *PhD Thesis, UMIST*, UK

Kuo W. J. and Smith R., 1998, Designing for the Interactions between Water-use and Effluent Treatment. *Trans. IChemE*. 76, Part A, 287-301.

Wang, Y. P. and Smith, R., 1994, Wastewater Minimisation. *Chem. Engng Sci.*, 49, 981-1006.

European Symposium on Computer Aided Process Engineering – 12
J. Grievink and J. van Schijndel (Editors)
229

Non-linear behaviour of PFR-separator-recycle polymerization systems

Anton A. Kiss, Costin S. Bildea[*], Alexandre C. Dimian and Piet D. Iedema

University of Amsterdam, Department of Chemical Engineering
Nieuwe Achtergracht 166, 1018 WV, Amsterdam, The Netherlands

Abstract

This work investigates the non-linear behaviour of PFR-Separator-Recycle systems with reference to radical polymerization. In Reactor-Separator-Recycle systems, feasible operation is possible only if the reactor volume exceeds a critical value (i.e. $Da>Da^{cr}$). Generic reaction models of different complexities are considered. A Low-density poly-ethylene (LDPE) process is presented as an example.

1. Introduction

Reaction systems involving material recycles are commonplace in chemical industry (Figure 1). The problem of polymerization reactors coping with variations in production rate, as well as product specifications is addressed. This flexibility must be ensured by the reactor design and its operation policy.

In this context it is important to evaluate the effect of recycle on reactor stability. Pushpavanam and Kienle (2001) showed that state multiplicity, isolas, instability and limit cycles arise in CSTR – Separation – Recycle systems. Because these phenomena also occur in a stand-alone CSTR, the effect of recycle is not obvious. In contrast, the stand-alone PFR always has only a single stable steady-state. We show that a PFR involved in a recycle exhibits multiple steady-states because of the recycle effect.

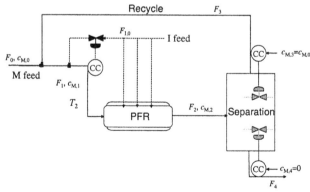

Figure 1. General structure of PFR-Separator-Recycle systems. Only reactant M is recycled. Product and recycle streams have fixed composition.

[*]C.S Bildea is currently at Delft University of Technology, Process Systems Engineering, Julianalaan 136, 2628 BL Delft, The Netherlands. Email: c.s.bildea@tnw.tudelft.nl

2. Model equations

The effect of reaction kinetics is systematically investigated by considering systems of increasing complexity: second order reaction, consecutive reactions and chain growth polymerisation (Table 1). The feed flow rate of reactant I is adjusted in order to keep its concentration at reactor inlet at a prescribed value ($z_{I,1}$).

Table 1. Investigated reaction mechanisms

Second order reaction	Consecutive reactions	Chain-growth polymerisation
$M + I \xrightarrow{k_1} P$	$I \xrightarrow{k_1} R$ $M + R \xrightarrow{k_2} P$	$I_2 \xrightarrow{k_d} 2I \bullet$ $M + I \bullet \xrightarrow{k_1} R_1 \bullet$ $R_n \bullet + M \xrightarrow{k_p} R_{n+1} \bullet$ $R_n \bullet + R_m \bullet \xrightarrow{k_t} P_n + P_m$

We consider dimensionless variables and parameters using feed flow rate F_0, feed concentration c_0 and reactor inlet temperature T_1 as reference values. These are: axial coordinate $0 \le \xi \le 1$, concentration $z_k(\xi)$, temperature $\theta(\xi)$, recycle flow rate f_2 and reactor-inlet concentration z_1; Damkohler number Da, dimensionless Arrhenius temperature α, adiabatic temperature rise B, heat-transfer capacity H, coolant temperature θ_c. The model equations are:

Component balances: $\dfrac{dz_k}{d\xi} = \dfrac{Da}{f_2} \cdot r_k$; k = 1 ... number of components $\qquad$ (1)

Energy balance: $\dfrac{d\theta}{d\xi} = \dfrac{Da}{f_2}\left(\sum_j B_j \cdot r_j - H \cdot (\theta - \theta_c) \right)$; j = 1 ... number of reactions $\qquad$ (2)

with the boundary conditions:

$$z_M(0) = 1 - \frac{z_{I1}}{z_{I0}}; \; z_I(0) = z_{I1}; \; \theta(0) = 0 \qquad (3)$$

$$f_2 = \frac{1}{1 - z_M(1) - \dfrac{z_{I1}}{z_{I0}}} \qquad (4)$$

By fixing the reactor-inlet temperature, energy feedback effects are excluded. Moreover, the plug-flow model of the stand-alone reactor has a unique solution. Hence, our analysis will identify the nonlinear phenomena caused by the material recycle.

The model equations can be solved by a shooting technique: start with an initial guess $z_M = z_M(1)$, calculate the reactor outlet flow rate f_2 (Eq. 4), integrate the PFR equations 1-2, check and update the guessed z_M. This implies that it is theoretically possible to reduce the model to one equation with one variable; therefore, the results of singularity theory with a single intrinsic variable can be applied.

3. Second order reaction

3.1. Isothermal

The second-order reaction $M + I \rightarrow P$ may be regarded as equivalent to the global stoichiometry of chain-growth polymerization. In Eq. 1 we consider:

$$Da = k_1 \frac{V}{F_0/c_{M,0}} c_{M,0} \; ; \; r_I = r_M = -I \cdot M \; ; \; z_{I,0}=1 \tag{5}$$

A parametric solution is possible:

$$Da = \frac{1}{\left(1-2z_{I,1}\right)\left(1-M-z_{I,1}\right)} \cdot \ln \frac{M \cdot z_{I,1}}{(1-z_{I,1})(2z_{I,1}+M-1)} \tag{6}$$

where $M \in \left[1-2z_{I,1}, 1-z_{I,1}\right]$ is the concentration of the reactant M at the reactor outlet.

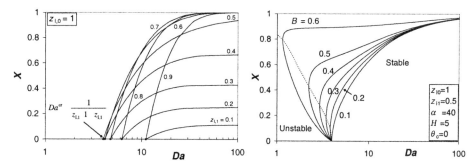

Figure 2. Two-reactants, second order reaction, isothermal reactor.

Figures 3. Two-reactants, second order reaction, non-isothermal reactor.

Figure 2 presents the bifurcation diagram (conversion $X = 1-M / (1-z_{I,1})$ vs. plant Damkohler number), for $z_{I,0}=1$ and different values of $z_{I,1}$. A feasible solution exists only if $Da>Da^{cr}$. The critical value is obtained by setting $M = 1-z_{I,1}$ in Eq. 6:

$$Da^{cr} = \frac{1}{z_{I,1}\left(1-z_{I,1}\right)} \tag{7}$$

Moreover, the minimum value of Da^{cr} is 4 when $z_{I,1}=1/2$ and $z_{I,0}=1$.

3.2. Non-Isothermal

Figure 3 presents the bifurcation diagram for the non-isothermal case. Multiple steady-states appear as a result of thermal effects, although the transcritical bifurcation remains intact. A higher value of B has a severe effect on state multiplicity, shifting the fold point to higher conversions. On the low conversion branch, increasing Da leads to decreasing conversion. This state is unstable and we will demonstrate this later.

4. Consecutive reactions

This system extends the previous one, by including an intermediate component. In general, the parameters of a dimensionless model can be defined in different ways. It is desirable, however, that the dimensionless parameters have physical significance.

One manifestation of chain reactions is that the total concentration of active intermediates at all times is very small. A quasi steady state exists (QSSA), in which the initiation and termination rates of intermediates are practically equal (Biesenberger, 1993). In the following, we will use the quasi steady-state approximation to obtain a reaction rate constant related to the consumption rate of the recycled reactant (M). This will be used to define the Damkohler number (Kiss *et al.*, 2002). The dimensionless model for this system includes the following parameters and variables:

$$Da = k_1 \cdot \frac{V}{F_0/c_{M,O}} \; ; \; r_I = -I \; ; \; r_M = -\beta \cdot M \cdot R \; ; \; r_R = I - \beta \cdot M \cdot R \; ; \; \beta = \frac{k_2 c_{M,0}}{k_1} \qquad (8)$$

Figure 4 presents the dependence of reaction conversion versus plant Damkohler number, for different values of β and $z_{I,0} = 1$, $z_{I,1} = 0.5$. There is a considerable change of the qualitative behaviour as compared to Figure 2: the transcritical bifurcation point disappears and in addition, all curves exhibit a turning (fold) point. Moreover, if QSSA is valid ($\beta \to \infty$), the conversion at the fold point is very small.

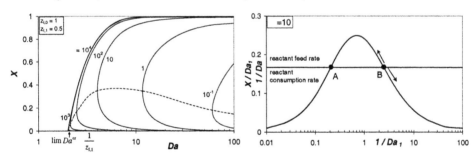

Figure 4. Consecutive reactions: X vs. Da bifurcation diagram.   *Figure 5. Instability of the low-conversion steady state.*

We can prove the instability of the low-conversion steady state using only the steady state model. Therefore, this result is independent of the dynamic separation model. For a stand-alone reactor, the amount of reactant consumed (F_1X), depends on the reactor's feed flow rate, F_1. This dependence is presented in Figure 5, using the dimensionless variables $F_1/(k_1c_{M,0}V) = 1/Da_1$ and $F_1X/(k_1c_{M,0}V) = X/Da_1$. In a reactor – separator – recycle system, the steady state values of the reactor-inlet flow rate are given by the intersections of this curve with the dimensionless amount of reactant fed in the process, $F_0/(k_1c_{M,0}V) = 1/Da$. Two steady states exist for $\beta = 10$ and $Da = 6$. To analyse the stability of the low-conversion state B, let us consider a positive deviation of the reactor inlet flow rate. At the right of point B, the amount of reactant fed in the process is larger than the amount of reactant consumed. Reactant accumulation occurs, leading to a further increase of the recycle and reactor-inlet flow rates; hence the steady state B is unstable. Note that these arguments, independent of the dynamic separation model, give a necessary but not a sufficient stability condition. To prove the stability of the high-conversion steady state A, the dynamic model is necessary.

5. Radical polymerisation

In this section we consider a simple radical polymerisation system which includes the main steps: initiation, propagation and termination. The concentration of initiator radicals I is subjected to the quasi steady state approximation, which is usually valid for polymerisation. Removal of QSSA does not change the results. The dimensionless model for this system includes the following parameters and variables:

$$Da = k_p \sqrt{\frac{k_d c_{M,0}}{k_t}} \frac{V}{F_0/c_{M,0}} \; ; \; r_I = -\frac{\gamma}{2\beta} I_2 \; ; \; r_M = -\beta \cdot M \cdot R - \frac{\gamma}{\beta} I_2 \; ; \; r_R = \frac{\gamma}{\beta} \cdot I_2 - \gamma \cdot \beta \cdot R^2 \; ;$$

$$\beta = \sqrt{\frac{k_t c_{M,0}}{k_d}} \; ; \; \gamma = \frac{2k_t}{k_p} \tag{9}$$

Figure 6 shows the bifurcation diagram (X vs. Da) for different values of γ and $z_{I2,1}=10^{-4}$, $\beta=10^3$. Each bifurcation diagram exhibits one fold (turning) point. No steady states exist for $Da < Da_F$, and two steady states exist for $Da > Da_F$. The conversion at the fold point is high for small values of γ (slow termination) therefore the multiplicity of states is important. In contrast, for large values of γ (for example $\gamma > 10^2$), the conversion of the lower branch has very small values. Hence, only the upper state existing for $Da > Da^F$ has practical significance.

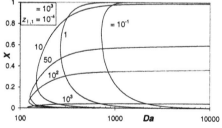
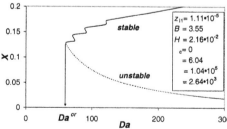

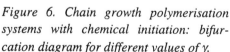

Figure 6. Chain growth polymerisation systems with chemical initiation: bifurcation diagram for different values of γ.

Figure 7. LDPE – bifurcation diagram. Chain transfer reactions do not affect the qualitative behaviour.

We have studied the LDPE process in order to investigate the effect of the chain-transfer to monomer and solvent reactions. Some additional kinetic mechanisms have been taken into account, like chain transfer to monomer or polymer and random-scission, but these do not affect the conversion, only the polymer characteristics. The stoichiometric model and kinetic parameters correspond to low-density polyethylene process conditions (Iedema *et al.*, 2000). Figure 7 shows multiple steady-states and several fold points, caused by the initiator fed along the reactor at three additional positions. Because parameter γ is high we expect a low conversion for the fold points. However the fold points are situated at a higher conversion, as a result of the thermal effects (Figure 3).

6. Discussion

In contrast to stand-alone reactors, Reactor-Separator-Recycle systems always exhibit a steady state at zero-conversion, infinite-recycle. This state is stable if the reactor volume is below a critical value, for a given feed flow rate and reaction kinetics. Feasible states can occur by two mechanisms:

- A transcritical bifurcation, at which the infinite-recycle state loses stability and a non-trivial state gets meaningful values and gains stability.
- A fold bifurcation (turning point), at which two steady states are born. This behaviour is generic for consecutive-reactions systems involving two reactants.

In practice, a design near the bifurcation points is dangerous, since changing operating conditions or uncertainty in design parameters can lead to a behaviour that is critically different from the expected one. A reactor design close to the fold point can suffer from serious operability problems (Bildea, 2000). Figures 4 and 6 show such unstable low-conversion states. The cases when the fold point is situated at higher conversion have practical importance from the design point of view.

It is important also to note the similarities between the non-linear behaviour of PFR and CSTR reactors in a Reactor-Separator-Recycle System (Kiss *et al.*, 2002).

7. Conclusions

1. A minimum reactor volume is necessary for a feasible operating point in a recycle system.

2. Multiple steady states are possible, even in case of simple reactions and isothermal operation.

3. The instability of the low-conversion branch sets a lower limit on the achievable conversion. For polymerisation systems, this has practical significance when the radicals' quasi steady state approximation is not valid (slow termination, gel-effect).

4. The thermal effects introduce new multiple steady states and shift the fold point to higher conversions.

8. References

Bildea, C.S., Dimian, A.C. and Iedema, P.D., 2000, Comp. Chem. Eng., 24, 209-214.

Biesenberger, J.A. and Sebastian, D.H., 1993, Principles of polymerisation engineering, Malabar-Florida: Krieger Publishing Company.

Iedema, P.D., Wulkow, M. & Hoefsloot, H.J.C., 2000. Macromolecules, 33, 7173-7184.

Pushpavanam, S. and Kienle, A., 2001, Chem. Eng. Sci., 56, 2837-2849.

Kiss, A.A., Bildea, C.S., Dimian, A.C. and Iedema, P.D., 2002, Chem. Eng. Sci., 57, 535-546.

European Symposium on Computer Aided Process Engineering – 12
J. Grievink and J. van Schijndel (Editors)
© 2002 Elsevier Science B.V. All rights reserved.

Application of Transport Equations for Heat and Mass Transfer to Distillation of Ethanol and Water

Gelein M. de Koeijer and Signe Kjelstrup
Norwegian University of Science and Technology, Department of Chemistry
N-7491 Trondheim, Norway

Abstract

The purpose of this work is to apply equations for heat and mass transport that are derived from irreversible thermodynamics to distillation. They contain six resistances based on experimental data, correlations, and kinetic theory. Data from an experimental rectifying column that separated ethanol from water were used. Overall resistances were obtained by fitting theoretical expressions for forces and resistances to the experimental entropy production rates. The entropy production in the interface was negligible with resistances from kinetic theory. In this case, we can therefore assume that there is local equilibrium across the interface. In our formulation the resistances have a contribution up to 11 % from the coupling between heat and mass transport in the liquid phase, i.e. thermal diffusion. In contrast to normal practise, this coupling effect can thus not be neglected.

1. Introduction

Non-equilibrium models of distillation are not only necessary to understand the phenomena that take place, but they are also feasible, due to the growth in computer calculation capacity. Taylor and Krishna(1993) and Krishna and Wesselingh(1997) have developed such models, using a Maxwell-Stefan type formulation of the transport problem. This work uses a new set of transport equations for heat and mass transport, recently derived from the entropy production rate in the system by Kjelstrup and De Koeijer(2002). In current non-equilibrium descriptions, the interface resistance is not included. Also, the heat-mass coupling effects (i.e. thermal diffusion or Soret effect) have not been taken into account. We have reasons to believe that the coupling of heat and mass transports may be important in distillation, see De Koeijer and Rivero(2002). The derived set includes those phenomena. The purpose of this work is to apply the new set and check the assumptions made for the interface and for the heat-mass coupling resistances. Experimental data from a rectifying column that separates ethanol from water in 10 trays shall be used, see De Koeijer and Rivero(2002).

2. Coupled transport processes in distillation

We consider the separation of two components on a tray in a distillation column. The tray in Fig. 1 shows vapour bubbles rising in a liquid. Vapour and liquid mole fractions are indicated, y and x, together with vapour and liquid flows, V and L. A close-up of the interface between the vapour and the liquid in a bubble is shown in Fig. 2. The bubble curvature is not visible on the scale chosen. The major parts of the bulk vapour and liquid are assumed to be totally mixed, while the films close to the interface are not. The gradients in temperature, T, and chemical potential, μ, across films and interface are illustrated. There is entropy production in the liquid film, the vapour film, and the interface. The interface may have an excess resistance, leading to jumps in intensive variables across it. The mass transfer rates are denoted J_1 and J_2 (mol/s), while the measurable heat transfer is J_q (J/s). Superscripts L, V or i indicate the phase in question. Positive direction of transport is from the liquid to the vapour. The rates of mass and heat are constant in time in a stationary column.

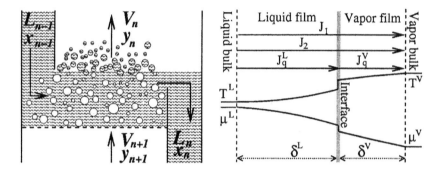

Figure1 Tray in a distillation column   *Figure 2 Close-up around interface*

The entropy production rate dS_n^{irr}/dt (W/K) on tray n for the system was derived in Kjelstrup and De Koeijer(2002), using the three transfer rates J_1 and J_2 and J_q^V in all parts. With an interface frame of reference, the result was:

$$\frac{dS_n^{irr}}{dt} = \frac{dS_n^{irr,L}}{dt} + \frac{dS_n^{irr,i}}{dt} + \frac{dS_n^{irr,V}}{dt} = J_{q,n}^V X_{q,n} + J_{1,n} X_{1,n} + J_{1,n} X_{1,n}$$
$$X_{q,n} = \bar{r}_{qq,n} J_{q,n}^V + \bar{r}_{q1,n} J_{1,n} + \bar{r}_{q2,n} J_{2,n}$$
$$X_{1,n} = \bar{r}_{1q,n} J_{q,n}^V + \bar{r}_{11,n} J_{1,n} + \bar{r}_{12,n} J_{2,n}$$
$$X_{2,n} = \bar{r}_{2q,n} J_{q,n}^V + \bar{r}_{21,n} J_{1,n} + \bar{r}_{22,n} J_{2,n}$$

(1)

Here X_n are the average forces conjugate to the transfer rates, and $\bar{r}_n$ are overall resistances for the vapour film, interface, and liquid film on tray n. The overall film resistance $\bar{r}_n$ is equal to the resistivity r times the film thickness δ divided by the transfer area A. Constant transfer area and film thickness ratio are used on all trays. The resistances of the interface include its thickness by definition, see Bedeaux et al(1992). The resistances of the layers in Fig. 2, $\bar{r}_n^V$, $\bar{r}_n^i$, and $\bar{r}_n^L$, are functions of the average temperature and composition on the tray n, i.e. $\bar{r}_n^i = r_n^i(T^{av}, y^{av})/A$, $\bar{r}_n^L = \delta^L r_n^L(T^{av}, x^{av})/A$, and $\bar{r}_n^V = \delta^V r_n^V(T^{av}, y^{av})/A$. The overall resistances become combinations of resistances for the two films and interface. Except for $\bar{r}_{qq,n}$, the overall resistance is not just a sum of the corresponding film and interface resistances. By applying Onsager's reciprocal relations, the following is obtained:

$$\bar{r}_{qq,n} = \bar{r}_{qq,n}^L + \bar{r}_{qq,n}^i + \bar{r}_{qq,n}^V$$

$$\bar{r}_{q1,n} = \bar{r}_{q1,n}^L + \bar{r}_{q1,n}^i + \bar{r}_{q1,n}^V + \bar{r}_{qq,n}^L \Delta_{vap}H_1$$

$$\bar{r}_{q2,n} = \bar{r}_{q2,n}^L + \bar{r}_{q2,n}^i + \bar{r}_{q2,n}^V + \bar{r}_{qq,n}^L \Delta_{vap}H_2$$

$$\bar{r}_{11,n} = \bar{r}_{11,n}^L + \bar{r}_{11,n}^i + \bar{r}_{11,n}^V + 2\bar{r}_{q1,n}^L \Delta_{vap}H_1 + \bar{r}_{qq,n}^L \Delta_{vap}H_1^2$$

$$\bar{r}_{22,n} = \bar{r}_{22,n}^L + \bar{r}_{22,n}^i + \bar{r}_{22,n}^V + 2\bar{r}_{q2,n}^L \Delta_{vap}H_2 + \bar{r}_{qq,n}^L \Delta_{vap}H_2^2$$

$$\bar{r}_{12,n} = \bar{r}_{12,n}^L + \bar{r}_{12,n}^i + \bar{r}_{12,n}^V + \bar{r}_{q1,n}^L \Delta_{vap}H_2 + \bar{r}_{q2,n}^L \Delta_{vap}H_1 + \bar{r}_{qq,n}^L \Delta_{vap}H_1 \Delta_{vap}H_2$$

(2)

Resistances contain terms with the heat of vapourisation $\Delta_{vap}H$, since we do not use J_q^L as a variable. The assumption of equilibrium across the interface is good when $\bar{r}_n^i$ are negligible. The transfer area A is the bubble interface area. The ratio of the liquid and vapour film thicknesses is estimated from the divergence of the constant energy flux with negligible Soret coefficients. For those variables we obtain:

$$A = \frac{8\pi\varepsilon}{d} V^{mix} \qquad\qquad \frac{\delta^L}{\delta^V} = \frac{\lambda^L}{\lambda^V} \frac{J_1 C_{P,1}^V + J_2 C_{P,2}^V}{J_1 C_{P,1}^L + J_2 C_{P,2}^L}$$

(3)

Here d is the average diameter of a bubble. The void fraction of bubbles on a tray is ε, and the volume of the mixture on a tray is V^{mix}. The thermal conductivity is λ, and C_P is the heat capacity.

3. Resistivities

Equation 4 gives the six resistivities from Kjelstrup and De Koeijer(2002) for the vapour and liquid films in terms of properties that can be found in the literature. The Gibbs-Duhem equation is applied.

$$r_{11}^L = \frac{Rx_2}{x_1 c^L \mathcal{D}_{12}^L} + \frac{r_{1q}^{L\,2}}{r_{qq}^L} \qquad r_{11}^V = \frac{Ry_2}{y_1 c^V \mathcal{D}_{12}^V} \qquad r_{12}^L = -\frac{x_2}{x_1} r_{11}^L \qquad r_{12}^V = -\frac{y_2}{y_1} r_{11}^V$$

$$r_{22}^L = -\frac{x_2}{x_1} r_{12}^L \qquad r_{22}^V = -\frac{y_2}{y_1} r_{12}^V \qquad r_{qq}^L = \frac{1}{\lambda^L T^2} \qquad r_{qq}^V = \frac{1}{\lambda^V T^2}$$

$$r_{1q}^L = -r_{qq}^L S_T RT^2 x_2\left(\frac{\partial \ln\gamma_1}{\partial \ln x_1} + 1\right) \qquad r_{1q}^V = 0 \qquad r_{2q}^L = -\frac{x_2}{x_1} r_{q1}^L \qquad r_{2q}^V = 0 \quad (4)$$

The density, heat of vapourisation and boiling temperatures were taken from Sinnot et al.(1983), Margules parameters for the activity coefficient γ from Gmehling and Onken(1977) page 153, heat capacities C_P from Smith and Van Ness(1987), thermal conductivities λ from Reid et al.(1987) using Filippov's equation in the liquid, Maxwell-

238

Stefan diffusion coefficients in the vapour $Đ_{12}^V$ by the Fuller correlation in Taylor and Krishna(1993), liquid diffusion coefficients from Tyn and Calus(1975), and the Soret coefficients S_T from Kolodner et al.(1988). R is the gas constant, 8.314 J/mol.K. The Soret coefficient in the vapour phase was neglected, see Hafskjold et al.(1993). The interface resistivities were taken from kinetic theory Bedeaux et al.(1992)(corrected in Kjelstrup and De Koeijer(2002)) with a condensation coefficient of 0.8.

4. Results and discussion

With $V^{mix}=2.0 \ 10^{-3} \ m^3$ (De Koeijer and Rivero(2002)), a void fraction of 0.7 and a bubble diameter of 10^{-2} m, we obtain A=3 m^2. Equation 3 gave on average a film thickness ratio of 10, and subsequently a liquid film thickness of $2.0 \ 10^{-3}$ m. The latter value was obtained by minimising the squared difference between the experimental entropy production rates from De Koeijer and Rivero(2002) and the the theoretical ones from Eq. 1. A lower, but not a higher, entropy production rate than the experimental one can be explained by effects that are not included in our model, Eq. 1, like pressure drops, turbulence and mixing. This means that the maximum film thickness is obtained by the method. These values seem reasonable, as Taylor and Krishna(1993) recommend film thicknesses between 10^{-3} and 10^{-5} m. De Koeijer and Rivero(2002) estimated that 90 % of the entropy production rate was due to heat and mass transport.

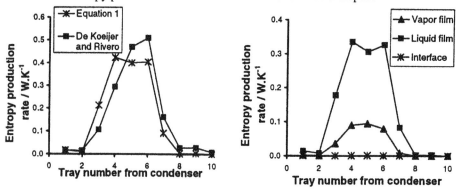

Figure 3 Entropy production rates from and De Koeijer and Rivero(2002) and Eq. 1

Figure 4 Entropy production rates from vapour film, interface and liquid film

Figure 3 gives the entropy production rate along the column trays, as given from experimental data De Koeijer and Rivero(2002), and from Eq. 1. We see that the last results follow the results from the experiments in a qualitative manner. One of the reasons for the discrepancies lies in the calculation method of the resistances. Figure 4 shows the contributions of the liquid film, vapour film, and the interface to the total entropy production rate, cf. Eq. 1. The contributions are 80 %, 20 % and 0.1 %, respectively. It is not a good assumption to neglect the resistance in the vapour film.

However, the assumption of equilibrium across the interface during evaporation or condensation in the two-component mixture is good. It is interesting that the entropy production per length of film is larger in the vapour than in the liquid.

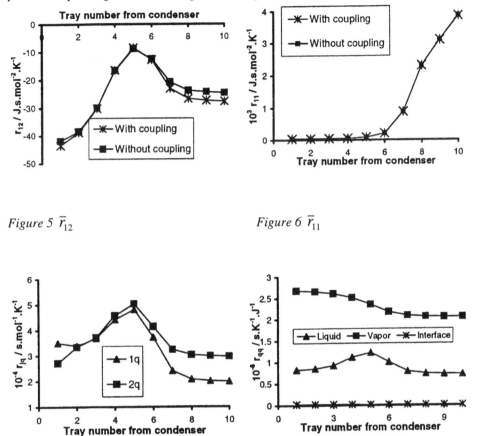

Figure 5 $\bar{r}_{12}$ Figure 6 $\bar{r}_{11}$

Figure 7 $\bar{r}_{1q}$ and $\bar{r}_{2q}$ Figure 8 $\bar{r}_{qq}$

Figures 5 and 6 show the main mass resistance $\bar{r}_{11}$ and mass-mass coupling resistance $\bar{r}_{12}$, when $\bar{r}_{1q}$ is neglected and not. The neglect causes a contribution up to 11 % for $\bar{r}_{12}$, 7 % for $\bar{r}_{11}$, and 4 % for $\bar{r}_{22}$. The Soret effect in the liquid may thus be important in dynamic models for distillation, which is in agreement with earlier observations in De Koeijer and Rivero(2002). This conclusion does not depend on the values of the transfer area or film thicknesses, because these variables are common to all resistances. The ratio between the thicknesses is important, however, because we neglect Soret effects in the vapour. When the vapour dominates the resistance at a ratio of 0.1, the contribution decreases to 0.7 %. From Eq. 4 we know that the mass-heat coupling resistivities $\bar{r}_{1q}$

and $\bar{r}_{2q}$ have an opposite sign, but in Fig. 7 the resistances $\bar{r}_{1q}$ and $\bar{r}_{2q}$ are both positive. So, terms with the heat of vapourisation are the main contributors to $\bar{r}_{1q}$ and $\bar{r}_{2q}$. Figure 8, which gives the three contributions to the main heat resistance $\bar{r}_{qq}$, indicates that the vapour is the largest contribution. Again the resistance of the interface is negligible. There are some differences with the Maxwell-Stefan approach in Taylor and Krishna(1993). The main ones are that they set all interface and heat-mass coupling resistances zero, include the latent heat in their formulations for heat transfer rate, and use the molar average velocity as frame of reference. A more extended discussion of the differences can be found in Kjelstrup and De Koeijer(2002).

5. Conclusions

By application of a set of transport equations for dynamic modelling of heat and mass transfer in distillation, we have obtained overall resistances for an ethanol-water distillation column. The assumption of equilibrium across the interface has been proven valid by using resistances given by kinetic theory. The Soret effect (or thermal diffusion) in the liquid gave a maximum contribution to the overall resistances of 4-11 %, and can therefore not be neglected for this column.

Acknowledgements

Discussions with Dick Bedeaux are appreciated. He suggested to use Eq. 3. The Research Council of Norway is thanked for a grant to Gelein de Koeijer.

References

Bedeaux, D., Smit, J., Hermans, L., and T. Ytrehus, 1992, Physica A 182, 388.

De Koeijer, G., and R. Rivero, 2002, Entropy Production and Exergy Loss in Experimental Distillation Columns, in preparation.

Gmehling, J. and U. Onken, Eds, 1977, Vapor-Liquid Equilibrium Data Collection – Chemistry Data Series. Vol.1, part 1. DECHEMA.

Hafskjold, B., Ikeshoji, T. and S. Ratkje, 1993, Mol. Phys. 80, 1389.

Kjelstrup, S. and G. de Koeijer, 2002, Transport equations for Distillation of Ethanol and Water from the Entropy Production Rate, in preparation.

Kolodner, P., Williams, H. and C. Moe, 1988, J. Chem. Phys 88(10), 6512.

Krishna, R. and J. Wesselingh, 1997, Chem. Eng. Sc. 52(6), 861.

Reid, R., and Prausnitz, J. and B. Poling, 1987, The Properties of Gases and Liquids. 4th edn. McGraw-Hill.

Sinnot, R., Coulson, J., and J. Richardson, 1983, An Introduction to Chemical Engineering Design. Vol. 6 of Chemical Engineering, Pergamon press.

Smith, J., and H. Van Ness, 1987, Introduction to Chemical Engieering Thermodynamics. 4th edsn. McGraw-Hill.

Taylor, R. and R. Kirshna, 1993, Multicomponent Mass Transfer. Wiley.

Tyn, M. and W. Calus, 1975, J. Chem. Eng Data 20(3), 310.

European Symposium on Computer Aided Process Engineering – 12
J. Grievink and J. van Schijndel (Editors)

Synthesis of Reactor/Separator Networks by the Conflict-based Analysis Approach

Xiao-Ning Li, Ben-Guang Rong and Andrzej Kraslawski*

Department of Chemical Technology, Lappeenranta University of Technology,
P.O.Box 20, FIN-53851, Lappeenranta, Finland. *E-mail: andrzej.kraslawski@lut.fi

Abstract

This paper presents a conflict-based analysis approach for the synthesis of reactor/separator superstructure towards the improvement of the quality and efficiency of the mathematical programming optimisation. A three-level hierarchical procedure is proposed for carrying out a combined method from the qualitative stage to the quantitative optimisation stage. In the first stage, there is the systematic formulation of the superstructure on the basis of multi objective conflicts analysis and phenomena-relationship structure abstraction. It has an important potential for gaining a more efficient solution in the next stage of mathematical optimisation. The proposed approach is illustrated with the synthesis of the hydrodealkylation of toluene (HDA) process.

1.Introduction

The process synthesis of reactor/separator (RS) systems is one of the important tasks of chemical process design. The prevailing methods for process synthesis are the mathematical programming and the hierarchical decomposition. The mathematical programming is mainly composed of three steps: the generation of alternatives superstructure, the formulation of mathematical program and the optimisation solution. How to evolve a highly-representative superstructure is a critical issue in this approach. Much work has been done for addressing this problem. Floquet et al. (1985a) proposed a tree searching algorithm for the reactor/separator sequence synthesis. Fredler et al. (1993) introduced a graph theory approach that has polynomial complexity to find all the interconnection in process networks. Nisoli et al. (1997) combined the attainalbe region approach for reactor synthesis with geometric concepts of feasibility of separation. While the issue of process synthesis for the fulfilment of the multi objective requirement is not addressed by the existing methods in the generation of the superstructure. It is observed that very often there occur different conflicts and contradictions when addressing the multi objective issue in the generation of the superstructure for the synthesis problems. The emerged problems are usually handled by the trade-offs. However it does not eliminate the fundamental conflicts and masks them by hard-to-fulfil and very subjective compromises. It may result in missing of the promising alternatives, unsatisfactory fulfilment of the design objectives and finally in the formulation of hard-to-solve MINLP problems. Therefore it is an important issue to develop an approach allowing for the generation of highly representative superstructures with regard to multi objective nature of the process design.

The objective of this work is applying the conflict-based analysis approach for the synthesis of RS networks. A combined method is proposed by carrying out the conflict-based analysis and mathematical programming. The efficient superstructure is systematically generated for the further mathematical optimisation. It is based on the multi objective conflicts analysis and the abstraction of the phenomena-relationship structure. HDA case study is presented for illustrating the proposed analysis approach.

2. The Combined Approach

The combined approach is performed in the three-level hierarchical procedure (Fig. 1). In the first step, the phenomena-relationship structure is extracted from the proposed general structure of all potential interconnections among the anticipated phenomena. The basic configuration is formulated under the consideration of the specific process streams. At the second level, the design objectives are identified and conflicts between them are analysed. The design principles are selected for removing those conflicts and grouped into phenomena blocks via so-called RS (reactor/separator) contradiction matrix. Applying the selected principles with the specific information, flowsheet superstructure is generated on the basis of the phenomena-relationship structure. Then mathematical optimisation is carried out for searching the optimal synthesis solution.

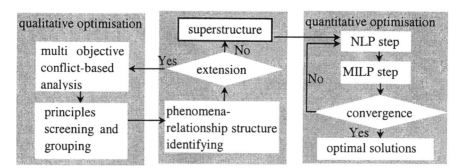

Fig. 1. The combined approach

2.1 Conflict-based methodology

Conflict-based analysis is derived from the TRIZ methodology (Theory for solving inventive problems) (Altshuller, 1998). TRIZ is a methodology to identify the system's conflicts and contradictions for solving the inventive problems. The main idea of TRIZ consists in the modification of the technical system by overcoming its internal contradictions. Therefore, it could be a promising method for improving superstructure by the conflict-based analysis based on the multi objective requirements.

The RS contradiction matrix is composed of 8 design objectives, such as capital cost, product quality, environmental impact, and 86 design principles for the synthesis of RS system. The design objectives form the rows and columns of the matrix and the design principles P_k (I_i, S_j) constitute the matrix elements (Table 1). The design principles P_k, k= 1-86 are extracted based on the available literature (Douglas, 1988 and Smith, 1995). Every principle is characterised by so-called influence coefficient I_i, i=1- 4 and flowsheet phenomena indicator S_j, j= 1-6. The influence coefficient I_i represents the

Table 1. A fragment of the rector/separator contradiction table

	1.captial cost	2.operation cost	3.product quality	4.environ. impact	5.safety	6.complex-ity	7.flexi-bility	8.controlla-bility
1	**	p2 (1, S1)	p1(1,s3)	p1(4,s3)	p60(1,s6)	p23(4,s3)	P32(3, s4)	p7(1, s2)
		...	...	...	...	...		...
2		**	p4(2,s6)	p4(4,s6)	p8(3,s3)	p23(4,s3)	p44(4,s4)	p13(3,s3)
			...	...	...	...	...	...

character of the influence on the two concerned objectives when applying the principle; e.g. If the application of the given principle will improve both objectives then use I_1. If the application of the given principle will worsen both objectives then use I_2. And finally, if the application of the given principle will improve one objective but worsen the other then use I_3 or I_4. The flowsheet phenomena corresponds to the region of the flowsheet structure in which the given principle should be applied; e.g. S_1= feed, S_4= purge, S_6 = recycle, etc.

The RS contradiction matrix reorganize the available design principles based on their possible influence on the design objectives. The design principles are activated if influence coefficient is I_1, I_3 or I_4 as well as the correlated phenomena. The selected principles are screened using RS matrix. The design principles are used for evolving the superstructure and analysing the potential optimal structure for the particular objectives.

2.2 Phenomena-relationship structure

Since the flowsheet superstructure may comprise many reactor and separator elements, it may become a non-trivial task for an efficient MINLP search. Therefore the phenomena-relationship structure is used here for eliminating the surplus phenomena and interconnectivities before generating the specific superstructure. The presented structure symbolizes the RS systems in a general form of phenomena relationship. All different configurations involving single or multiple RS can be obtained by matching the flow pattern of the process streams. The phenomena-relationship graph (Fig. 2) consists of two parts: one is the phenomena interconnection graph that shows all the possible connection among those phenomena; the other one is phenomena integration points which illustrate the possible locations of the process integration, such as combination of react-separate phenomena; complex distillation system; multifunctional, new reactor system. For the given specific problem, the specific phenomena-relationship structure can be identified.

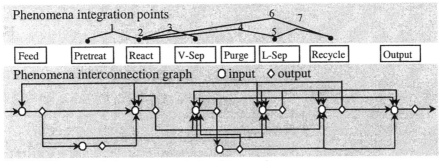

Fig. 2. The general phenomena-relationship graph

3. HDA Case Study

The HDA process has been extensively studied by Douglas (1988) by a hierarchical heuristic synthesis approach. The presented problems consist in the systematic generation of the highly-representative flowsheet superstructure, the identification of the flowsheet configuration and some of the operating conditions based on multi objective requirements.

The primary reaction of the HDA process is:

$$\text{Toluene} + H_2 \longrightarrow \text{Benzene} + CH_4$$

In addition to this desired reaction, there occurs an undesired reaction:

$$2\,\text{Benzene} \Leftrightarrow \text{Diphenyl} + H_2$$

A pure toluene stream is available at ambient conditions and the raw hydrogen is a purity of 95% with 5% methane. The homogenous gas phase reacts in the range of 895K and 978 K in order to assure the reasonable reaction rate and to prevent hydrocracking reactions. The hydrogen-to aromatics ration at the reactor inlet must exceed 5 and the reactor effluent must be rapidly quenched to 895 K to prevent coking.

3.1 Phenomena-relationship structure identification

For the HDA process, there is only one reaction so that no complex interconnection exists between reactor and separator networks. The pre-treatment block is not needed since the methane impurity in the feed stream is one of the products for the main reaction. Then the feed block is directly connected to the react block. There is performed the condensation of the aromatics from non-condensable hydrogen and methane. Therefore the output of reactor block is connected to the vapour and liquid separation blocks. The gas stream consists of the hydrogen feed material and methane impurities. Then vapour block connects to the recycle and purge blocks. The output of the recycle block connects to the feed, reaction blocks and separation blocks. Since the feed needs to be preheated before feeding to the reactor, the recycle block connects to the feed block. Then the specific phenomena-relationship structure is formulated (Fig. 3). The reactor effluent consists of five process streams: unreacted hydrogen and toluene, desired product benzene, and undesired diphenyl and methane. The structure configuration is identified by considering the processing of the specific streams, e.g.

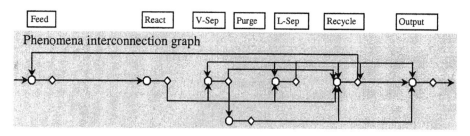

Fig. 3. The phenomena-relationship structure for HDA process

how to treat purge stream methane, by product diphenyl, and light end of the liquid separation. The phenomena integration offers the possible alternatives for this process: consider the different reactor types adiabatic or isothermal; use complex distillation systems such as the thermally coupled side-stream complex distillation column. Therefore the basic configuration for the superstructure is formulated through eliminating surplus interconnectivity of the phenomena and considering the specific information (Fig. 4). It will be modified by the selected principles on the basis of the multi objective conflicts analysis in the next step.

3.2 Multi objective conflicts analysis and principles selection

In this step, the synthesis work starts the analysis of the objectives via RS contradiction matrix. From the process description, it is clear that not only the economic criteria are important, but also the other objectives that are critical for maintaining product quality: the environmental impact for treating the by product, and the controllability for preventing coking. Finally five objectives are identified: capital cost (1), operational cost (2), product quality (3), environmental impact (4), and controllability (8). Sequentially 10 conflicts are formulated among them, such as conflicts between capital cost and operation cost (1 x 2), capital cost and product quality (1 x 3), environmental impact and controllability (3 x 8), etc.

Based on the value of the influence coefficients for the conflicting objectives, the principles are classified into three groups: positive principles which improve both objectives, negotiable ones which are improving one objective at the cost of decreasing another one, and finally, the negative principles that make both objectives getting worse. Under the consideration of the specific problem information, 36 principles are selected from the first two groups. They are grouped using the phenomena indicators and introduced into the blocks of phenomena-relationship structure (Table 2).

Table 2. The selected principles into phenomena groups

Feed S1	React s3	Liquid separate s5	Purge+ Vapour separation s4+s5	Recycle s6
p2, p7	p1,p8,p11,p 12,p13,p17	p26,p27,p32,p39, p40,p41,p44	p66, p68,p70,p73, 74,p75,p76, p78,p79,p80,p81,p82,p83,p85,p86	p4,p55,p56, p58,p59, p61

3.3 Flowsheet superstructure generation

Then the selected principles are applied for the developing of the flowsheet superstructure. Meanwhile they suggest the rational value range of some operating conditions and corresponding influences on the concerned objectives. For example, principle 2 (using an excess of one of the reactant) and its influence coefficients suggest that the use of one reactant could improve raw material quality but simultaneously increase the capital cost. From the point of view of the environmental concern, the principle 58 (recycling the by product for inhibiting its formation at the source) proposes recycling diphenyl by product instead of the recovery; and the principle 79 suggests that there should be vapour recovery system arranged on the purge stream. The principles 55, 56, 58, 59 show the detailed rules for processing product, by product, and unreacted feed by the different separation/recycle structure. The principles in the purge

246

and vapour separation blocks propose the new alternative for treating the light end, such as using membrane separation and further reaction for recovering the hydrogen. The purifying section can be added for purifying the products as it is suggested by the principle 73. The principles 39,40,41,42 are the rules concerning the sequence and operating condition of distillation separation when taking into account of the trade-off between capital and operation costs. In this way a specific superstructure of HDA process is generated (Fig. 5). Moreover the potential, optimal structures from the point of view of the particular objectives are indicated by the respective objective analysis. Those aspects are very important for the search of the optimal solutions in MINLP optimisation step.

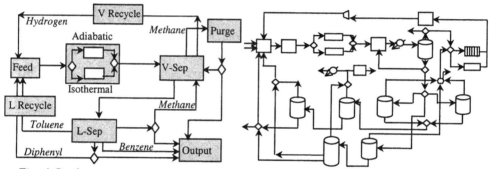

Fig. 4.Configuration for HDA superstructure   Fig. 5. Superstructure for HDA process

4. Conclusions

A conflict-based analysis approach is presented for the synthesis RS networks. A three-level hierarchical procedure is illustrated with the HDA case study. The abstraction of the phenomena-relationship structure can screen out a large number of surplus phenomena interconnectivities and produce promising alternatives. The principle selection, which is based on the multi objective conflicts analysis using RS matrix, offers all available principles for problems at hand. They are not only used for evolving and modifying the superstructure, but also for determining the rational value range of some operating conditions and indicating the potential optimal structures for the particular objectives. The superstructure is systematically generated by taking into account of the multi objective nature of the design process. It has an important potential for gaining a more efficient solution in the subsequent MINLP optimisation stage.

5. References

Altshuller, G., 1998, 40 principles: TRIZ keys to technical innovation, Inc. MA, USA.

Douglas, J. M., 1988, Conceptual design of chemical process, McGraw Hill, New York.

Smith, R., 1995, Chemical process design, McGraw Hill, New York.

Floquet, P., Pibouleau, L., Domenench, S., 1985a, Process Syst. Engng, Symp. Ser., No. 92, 415.

Friedler, F., Tarjan, K., Huang, Y. W., Fan, L. T, 1995, Computers Chem. Engng, 1993, 17, 929.

Nisoli, A., Malone, M. F., Doherty, M. F., 1997, AICHE J, 43, 374.

European Symposium on Computer Aided Process Engineering – 12
J. Grievink and J. van Schijndel (Editors)
© 2002 Published by Elsevier Science B.V.

Systematic decision-making technology for optimal multiphase reaction and reaction/reactive separation system design

Patrick Linke and Antonis Kokossis

Centre for Process and Information Systems Engineering, University of Surrey, Guildford, Surrey, GU2 7XH, U.K.

Abstract

This work introduces a general framework for the selection of process designs through simultaneous exploitation of reaction and separation options. The synthesis scheme exploits rich superstructures comprised of two types of generic synthesis units. A reactor/mass exchanger unit enables a detailed representation of the reaction and mass exchange phenomena. Conceptual representations of separation systems is facilitated through separation task units. The synthesis scheme supports the decision making process in both, early and late process design stages. A screening stage reveals design insights into the performance of complex reaction-separation systems early in design. This enables the inclusion of the relevant design information into the superstructure formulations of the subsequent design stage. The design options are systematically explored using stochastic optimisation. An example is presented to illustrate the approach.

1. Introduction

It is common practice to decompose the overall reaction/separation process synthesis problem into smaller sub-problems that can be handled using available design approaches. Progress in the development of process synthesis tools has been confined to the individual sub-problems and to relatively simple combinations of sub-systems, taking into account only pre-postulated phenomena. As a result, the complex interactions amongst reaction and separation systems are generally not fully identified and exploited. The lack of generality is a tribute to the complex process models involved as well as to the vast number of feasible process design candidates.

The most general synthesis framework presented to date is that of Papalexandri and Pistikopoulos (1996), who propose the use of generic heat/mass exchange modules to generate superstructures that are subsequently formulated as MINLPs. The general representation has taken up applications in reactive and reactor/distillation system synthesis (Ismail et al., 1999). On the downside, the approach carries the shortcomings of the MINLP technology in that the resulting design performances are likely not to be located in the globally optimal domain. Moreover, the approach is developed to address pre-postulated systems and does not offer ventures to support the identification of beneficial systems for a given problem. We present a general framework for the synthesis and optimisation of processes involving reaction and separation. The flexible representation framework allows to embed all novel as well as conventional process configurations into compact superstructure formulations that are useful for the generation of conceptual design insights and the synthesis of optimal processes based upon this information.

2. Screening and design

The design strategy accounts for two synthesis stages: screening and design. The screening stage aims at replacing the early problem decomposition based on intuition. This is realised by the introduction of simple separation models that enable the investigation of major trade-offs in the context of reactor design, facilitated through structural optimisation of relaxed reactor/separation task network superstructure models. Allowing any separation tasks to be present in the networks, the most important composition manipulations in the process network can be identified regardless of separation feasibility. By introducing constraints on specific separation tasks, sensitivity studies allow to investigate their impact on the performance targets and process layouts. Rather than final process layouts, the screening stage generates insights into optimal component separation, removal and recycle policies along with reactor mixing and contacting patterns. The obtained process layouts feature reactor and separation task combinations, and only those important tasks are subsequently assessed for separation feasibility. The identified relevant process design information is processed in a *design stage* by introducing the additional modelling information required for candidate reactive separation and mass exchange operations into a refined superstructure network model and by excluding infeasible separation tasks. Thus, more modelling rigour is added to give a more detailed account of the relevant reactive separation and separation options whilst the combinatorial size of the problem is reduced by the irrelevant design options. In both synthesis stages, superstructures of generic units are formulated and the performance targets as well as a set of design candidates are obtained subsequently via robust stochastic optimisation techniques.

3. Process representation schemes

The basic elements of the synthesis representation are the generic reactor/mass exchanger (RMX) units and the separation task units (STU). Figure 1 illustrates the units. A detailed description of the synthesis units and their functionalities is given in Linke (2001) and will be the subject of a future publication.

The underlying phenomena exploited in chemical process design are reaction, heat and mass exchange. This work employs a generic reactor / mass exchanger unit for a flexible and compact synthesis representation of all possibilities of phenomena exploitation. The RMX unit follows the *shadow compartment* concept developed for non-isothermal multiphase reactor network synthesis (Mehta and Kokossis: 1997, 2000). The unit consists of compartments in each phase or state present in the system under investigation and the streams processed in the different compartments of the generic unit can by definition exchange mass across a physical boundary, which can either be a phase boundary or a diffusion barrier. Each compartment features a superset of mixing patterns through which a compact representation of all possible contacting and mixing pattern combinations between streams of different phases can be realised within a single generic unit. The mutually exclusive mixing patternsconsidered for the compartments include well-mixed and segregated flow. Along with decisions on the existence of mass transfer links between compartments of the same generic unit and decisions on the consideration of reaction phenomena in the different compartments, a single RMX unit enables the representation of a reactor, a mass exchanger, a reactive mass exchanger or a combination of the above.

In contrast to the rigorous representation of reaction and mass transfer phenomena by RMX units, the separation task units (STU) represent venues for composition manipulations of streams without the need for detailed physical models. In accordance with the purpose of any separation system, the separation task units generate a number of outlet streams of different compositions by distributing the components present in the inlet stream amongst the outlet streams. In the *screening stage*, the aim is to identify those separation tasks, which have a positive impact on the performance of the reaction-separation system regardless of separation feasibility issues. In the most general case, any separation tasks can be performed by the units. However, the model accounts for venues to introduce biases and constraints. In the *design stage*, the STU enables the representation of separation tasks that can be performed using particular separation processes. In this case, the possible distribution policies of components to the outlet streams are constraint by the separation orders associated with the separation process. The STU performs a set of feasible separation tasks according to the separation order to define a set of outlet streams. Depending upon the order in which the tasks are performed, a variety of processing alternatives can be captured by a single unit.

The generic synthesis units described above allow for representation of all sections of general processes involving reaction and separation, *i.e.* reaction sections, reactive separation sections, mass exchange sections and sections performing separation tasks. The superstructures feature a number of generic RMX and separation task units as well as raw material sources and product sinks, the interconnections amongst which are realised by two types of stream networks: *Intra-phase streams* establish connections between the synthesis units, products and raw materials of the same state, whereas *Inter-phase streams* establish those connections across the state boundaries, *i.e.* the source

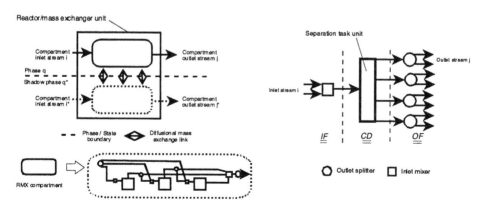

Figure 1.   *Illustration of generic synthesis units: RMX and STU.*

and sink of such a stream belong to adaptable states. The superstructures formulations evolve in the different synthesis stages according to the insights into the process obtained at the previous stages, which are incorporated by extending or constraining the synthesis unit functionalities as well as unit connectivities. Novelty is accounted for in the superstructures as the representations are not constraint to conventional process configurations but instead include all possible novel combinations of the synthesis units.

4. Network optimisation

In order to establish a basis for optimisation, the reaction and separation superstructure is formulated as a mathematical model that involves the balances for the RMX units, the mixers prior to the separation task units and the product mixers. The mathematical formulation incorporates models for reaction kinetics, physical property and mass transfer models, short-cut models and regression expressions for equipment sizing and costing, and the objective function. These relationships will introduce non-linearities and possibly discontinuities into the general superstructure network model. The network optimisation needs to exploit the structural as well as the continuous decisions within the network in order to arrive at optimal solutions. That is why the reaction and separation superstructures are optimised using stochastic search techniques in the form of Simulated Annealing and Tabu Search. Stochastic optimisation technology has been proven to converge statistically to the globally optimal domain (Aarts and van Laarhoven: 1985, Marcoulaki and Kokossis: 1999), *i.e.* to the performance limits.

5. Illustrative example

Novel designs for the Williams-Otto problem (Ray and Szekely, 1973) are sought. The raw materials A and B are fed to a reactor where the following reactions occur:

(1) $A + B \rightarrow C$
(2) $B + C \rightarrow P + E$
(3) $P + C \rightarrow G$

Component P is the desired product. Heavy waste G and component E are unwanted byproducts. The objective of the synthesis exercise is to find the designs that maximise the annual profit of the process for a minimum production rate of 400kg/hr of component P. The profit function includes the product value and raw material costs, waste treatment cost, reactor capital and operating cost as well as the cost incurred by separation. A summary of the problem data is given in Linke (2001).

The process is first synthesised to *screen* the most significant separation options. Any component present in the feed stream to separation task units can be assigned to any of the outlet streams, *i.e.* sharp splits are assumed and up to six outlet streams can leave a single distribution unit. The stochastic optimisation yields a profit target of around 618k$/yr. A variety of mostly complex design alternatives exist that can achieve the targets, featuring one, two, or three reactors and STUs. To evaluate the gain from complexity in a second screening stage, the separation tasks are associated with a lower cost bound, *i.e.* with the cost of the task of decanting component G. The stochastic search yields a slightly reduced new profit target of around 600k$/yr. As a result of penalising the existence of separation tasks, the complexity of the process structures is significantly reduced, containing only the performance enhancing separation tasks. The process design requirements from this stage are as follows: separation and removal of by-products E and G, separation and removal of desired product P, separation of components A, B, and C from the reacting mixture and distribution amongst the reactor units, excess of component B in the reaction zones to minimise by-product formation (low concentration of product P), and plug-flow behaviour in all reaction zones.

Based on the insights gained from screening, the appropriate separation venues can be identified and included in the *design stage*. Distillation enables separations of mixtures according to the order of volatilities P/PE/C/B/A (P: light, A: heavy) and hence allows separation in support of the raw material and intermediate recovery. Heavy G needs to be decanted prior to the operation, which can be achieved at a low cost in a decanter. However, component P forms an azeotropic mixture with component E resulting in a loss of desired product to the low value fuel. To provide a possible solution to this problem, a solvent is made available that allows selective extraction of desired product P from the mixture. The equilibrium relationship and solvent loading constraints are taken from Lakshmanan and Biegler (1996). The RMX unit can now take the functionality of a homogeneous reactor, a reactive extractor and an extractor. Stochastic optimisation yields a target performance of around 433k$/yr for this system. Designs with performances close to the target are grouped into two main categories according to their use of the solvent: reactive extraction designs and reactor-hybrid separation designs. Designs of the first category employ only RMX units featuring plug-flow compartments in the form of homogeneous reactors and reactive extractors. No separation tasks are utilised in the reactive phase, *i.e.* extraction of the desired product P in the course of reaction is the only separation required to achieve the target. A typical design is shown in Figures 2a. Designs of the second category feature RMX units in the form of homogeneous plug-flow reactors and counter-current extractors as well as a single separation task unit. Typical designs are illustrated in Figure 2b. The separation task unit facilitates a decanter for removal of solid waste G and a distillation column for separation of product P and by-product E from the raw materials and intermediates A, B and C. The novel designs utilise a distillation-extraction hybrid separation system for a complete recovery of valuable components P, A, B, and C. Additionally, a number of flow schemes employing reactive extractors and hybrid separation systems have been observed that closely achieve the performance limits of the system.

6. Conclusions

A synthesis framework to support the screening and design stages in integrated reaction and separation process design is presented. The processing alternatives are effectively captured by superstructure formulations of the proposed synthesis units that provide a venue for optimisation. The representation potential is not constrained to conventional reactor-separator or reactive separation designs but instead accounts for any novel combination of the synthesis units and their functionalities. Different features of the representation framework are utilised in the different synthesis stages. Conceptual

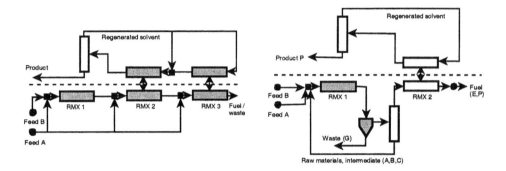

Figure 2.   *Process designs obtained from the design stage.*

screening stages allow for the extraction of optimal reaction-separation task interactions to guide the inclusion of the relevant design information in terms of the reactive separation and mass exchange options as well as the unit operations to be utilised in the rigorous design stages. In each synthesis stage, the performance targets and a set of optimal designs are extracted from superstructures using stochastic search techniques. Insights into the synthesis problem gained during a synthesis stage give rise to preferences, constraints or refined modelling information that is easily included into the flexible representation framework.

The functionalities and flexibility of the generic synthesis units and superstructure generations enable applications to general reaction and separation design problems, including the sub-problems multiphase reactor design, reactive separation systems design and (hybrid) separation systems synthesis. Future publications will address these issues in detail. A study of the activated sludge system can be found in Rigopoulos and Linke (2002).

References

Aarts, E. and van Laarhoven, P. (1985). *Philips J. Res.*, 40, 193-226.

Di Bella, C.W., and Stevens, W.F. (1965). *Ind. Eng. Chem. Proc. Des. Dev.*,4, 16-20.

Ismail, R.S., Pistikopoulos, E.N., and Papalexandri, K.P. (1999). *Chem. Eng. Sci.*, 54, 2721-2729.

Lakshmanan, L., and Biegler, L.T. (1996b). *Ind. Eng. Chem. Res.*, 35, 4523.

Linke, P. (2001). PhD thesis, UMIST, U.K.

Marcoulaki, E.C., and Kokossis, A.C. (1999). *AIChE J.*, 45, 1977-1991.

Mehta, V.L., and Kokossis, A.C., (1997). *Comp. Chem. Eng.*, 21, S325.

Mehta, V.L., and A.C. Kokossis (2000). *AIChE J.*, 46, 2256.

Papalexandri, K.P., and Pistikopoulos, E.N. (1996). *AIChE J.*, 42, 1010.

Ray, H., and Szekely, J. (1973). *Process Optimization*. New York: John Wiley and Sons.

Rigopoulos, S. and Linke, P (2002). *Comp. Chem. Eng.*, in press

European Symposium on Computer Aided Process Engineering – 12
J. Grievink and J. van Schijndel (Editors)
© 2002 Elsevier Science B.V. All rights reserved.

Alternative Processes For Production Of Anhydrous Ethanol Through Salt Extractive Distillation With Reconstitution Of The Separation Agent

Eliana L. Ligero, Teresa M. K. Ravagnani*

Faculdade de Engenharia Química - Universidade Estadual de Campinas – Cx.P. 6066
13083-970 Campinas, São Paulo, Brasil

ABSTRACT

Two process flowsheets were proposed for the production of anhydrous ethanol from a diluted aqueous solution of ethanol via extractive distillation with potassium acetate. In the first process, the diluted ethanol is directly fed in a salt distillation column and the salt is recovered in an evaporation system and in a spray dryer. In the second one, the diluted ethanol is pre-concentrated in a pre-concentration column and subsequently submitted to the salt distillation. In this case, the salt is recovered only in a "spray" dryer. In both the processes, the salt is recovered as a solid and is recycled in the extractive column. The equipment used in the processes was modeled separately and rigorously. Afterwards, they were connected for the global simulation of the plants. The results showed that the process with pre-concentration of diluted ethanol is more feasible in terms of energy consumption than the process without pre-concentration.

1. Introduction

The extractive distillation with liquid separation agents is a highly used technique in industries to separate azeotropic ethanol-water mixture is. The most employed liquid is benzene, which carcinogenic properties are well known and there is a trend to progressively eliminate its use.

One alternative process is the extractive distillation that uses soluble salts as separation agents. This distillation is similar to the extractive distillation with liquid agents Besides, the lower toxicity of certain salts, one of advantages of the use of salts is the production of a distillate completely free from the separation agent. Another favourable aspect is the high level of energy savings.

Theoretical and experimental studies about salt extractive distillation of ethanol-water (Cook and Furter, 1968; Schmitt and Vogelpohl, 1983; Barba et al., 1985; Martinez and Serrano, 1991; Vercher et al., 1991, Jiménez and Ravagnani, 1995 and Reis and Ravagnani, 2000), have demonstrated that it is possible to obtain high purity ethanol by using proper salts. However, most studies consider only the ethanol production and do not consider the salt recovery. Although the knowledge of the characteristics of salt extractive column is very important to obtain anhydrous ethanol, the study of the salt recovery is equally important, since this step is responsible for the economic feasibility of the process.

Two distinct processes are proposed and analysed in this study. Their basic difference is related to ethanol production: with or without a pre-concentrating distillation column. In this way, the processes are named according to the aqueous ethanol solution concentration fed in the salt column as follows: Process with Diluted Ethanol Feed and Process with Concentrated Ethanol Feed. The salt recovery consists of an evaporation system to pre-concentrate the salt solution, in case it is necessary, and a "spray" dryer. The focus of this work is to determinate the difference in energy consumption between the processes.

2. Proposed Processes

The potassium acetate was the salt used in both processes, which eliminates completely the azeotrope ethanol-water when present in small concentrations in the liquid phase of the column (Vercher et al., 1991). The ethanol-water solution to be separated has 2.4% molar of ethanol, 2,000 kmol/h and 30°C. The composition of the anhydrous ethanol is fixed in 98.9% molar.

2.1 Process with Diluted Ethanol Feed
The ethanol production is performed in a single extractive column. The salt recovery is executed in a multiple effect evaporation system and a spray dryer, figure 1. The diluted aqueous solution of potassium acetate from the bottom column is sent to the evaporation system, which purpose is to remove most part of water from the solution, before it is introduced in the spray dryer where the complete drying takes place.

2.2 Process with Concentrated Ethanol Feed
The second process consists of two distillation columns in the ethanol production, figure 2. The pre-concentrating column, which operates with no salt, has as function to produce a large amount of water as bottom product and an ethanol-water distillate with a composition close to the azeotrope. As a result, the salt column produces an aqueous concentrated potassium acetate solution as a bottom product. For this reason, there is no need to use an evaporation system since the salt is already concentrated.

3. Processes Simulation

The equipment in the processes are individually modeled and afterwards they are connected among themselves for the global simulation of the plants. As all the salt used in the processes is totally recovered and reintroduced in the extractive column, it is not necessary a sequential iterative method to simulate the global processes.

3.1 Salt Extractive Column
The salt column is simulated through the rigorous Sum-Rates method (Naphtali and Sandholm, 1971) with modifications in order to consider the salt in the liquid phase. Since the salt does not participate of the mass transfer process, the mass and energy balance equations remain equal to the equations without salt (Fredenslund et al. 1977). Due to the presence of salt, The required change is in the equilibrium relationships (Jiménez and Ravagnani, 1995):

$$E_{e,n} \; K_{e,n} \; V_e \; l_{e,n} \Bigg/ \left(L_e + \sum_{s=1}^{N_{sl}} v_s \; vs_{e,s} \right) + \left(1 - E_{e,n}\right) v_{e-1,n} \frac{V_e}{V_{e-1}} - v_{e,n} = 0 \qquad (1)$$

The prediction of the liquid-vapor equilibrium of the ethanol-water-potassium acetate is made by Sander et al. (1986) model. In the pre-concentrating column, figure 2, the expressions for the equilibrium relationships are identical those presented by Fredenslund et al. (1977).

3.2 The Evaporation System
The method used in the multiple effect evaporation system project is that one proposed by Holland (1975). The equations of a generic effect (n) are the energy balance, heat exchange rate, phase equilibrium and material balance. An exception is associated to the last effect, which has only the two first equations, because its pressure and composition are specified values.

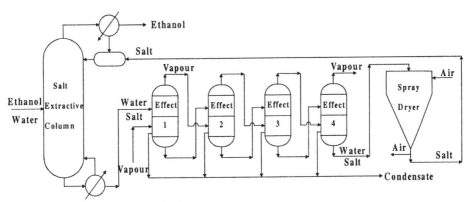

Figure 1 - Process with diluted ethanol feed

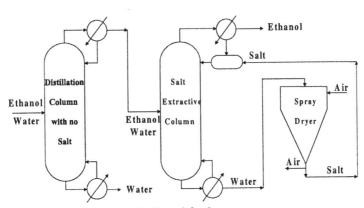

Figure 2 - Process with concentrated ethanol feed

3.3 Spray Drying
The model used to design the spray dryer is that proposed by Gauvin and Katta (1976). The three-dimensional equations of motion of a droplet in centrifugal and gravitational

fields are resolved simultaneously with the equations expressing the instantaneous heat and mass transfer rate to the droplets, the three-dimensional drying air flow pattern and the instantaneous properties of the drying gas.

4. Results

The design results of the units and some of their operating parameters are shown at Tables 1 and 2. More details are available in Ligero and Ravagnani (2000) and Ligero and Ravagnani (2001).

Table 1 – Process of Figure 1

Salt Distillation Column	
Operational Conditions	
Distillate	48.5 kmol/h (2.2t/h)
	98.9 mol% of ethanol
Design Results	
Number of stages	37
Optimum reflux	2.5
Salt flow rate	13.5 kmol/h (1.3 t/h)
Evaporation System	
Operational Conditions	
Feed components	Water/Salt (3.6wt%)
Feed flow rate	36.4 t/h
Feed Temperature	100.3°C
Steam pressure	1.01 10³ kPa
Vacuum	10.7 kPa
Final solution	60 wt% of salt
Design Results	
Number of effects	4
Heat area of each effect	76 m²
Steam flow rate	12.1 t/h
Spray Dryer	
Operational Conditions	
Feed flow rate	2.2 t/h
Feed temperature	62.4°C
Air humidity	0.005
Air initial temperature	377°C
Design Results	
Total height	6.6m
Air flow rate	9.5t/h
Product composition	100%wt of Salt

Table 2 – Process of Figure 2

Pre-Concentrating Column	
Operational Conditions	
Distillate	80 kmol/h (2.8t /h)
	60 mol% of ethanol
Efficiency	70%
Design Results	
Number of stages	22
Optimum reflux	1.1
Salt flow rate	---
Salt Distillation Column	
Operational Conditions	
Distillate	48.5 kmol/h (2.2 t/h)
	98.9 mol% of ethanol
Efficiency	60%
Design Results	
Number of stages	35
Reflux	1.2
Salt flow rate	0.63 t/h
Spray Dryer	
Operational Conditions	
Feed components	Water/Salt
Feed flow rate	1.2 t/h
Feed composition	52.8 wt% of salt
Feed temperature	115.4°C
Air humidity	0.005
Air initial temperature	377°C
Design Results	
Total height	7.8m
Air flow rate	5.7t/h
Product composition	100%wt of Salt

The design results of the processes are focused to the energy requirements and illustrated on Table 3, Process with Diluted Ethanol Feed and Table 4, Process with Concentrate Ethanol Feed.

Table 3 - Process of Figure 1

Energy Requirement	(MW)
Salt Extractive Column	4.8567
Evaporation System	21.6772
Spray Dryer	0.4540
Total	26.9879

Table 4 - Process of Figure 2		Pre-Concentrating Column	4.8439
		Salt Extractive Column	0.3073
Energy Requirement	**(MW)**	Spray Dryer	0.5563
		Total	5.7075

The basic difference between the processes is the diluted aqueous ethanol feed being or not concentrated before the salt distillation. The comparisons of energy requirement in the processes are divided in two parts: ethanol production and salt recovery.

In the Process with Diluted Ethanol Feed (Table 3), the salt extractive distillation column is responsible for the whole energy consumption in the ethanol production step. However, in the Process with Concentrated Ethanol Feed (Table 4), only 6% of the supplied energy in the ethanol production is consumed in the salt distillation, since the pre-concentrating column is responsible by most of the energy consumption. The comparison between the two processes shows the Process with Concentrated Ethanol Feed consumes almost 6% more energy in the ethanol production step. Also, the presence or not of the pre-concentrating column reflects directly in the salt recovery energy requirement. In the Process with Diluted Ethanol Feed, the multiple evaporation system is responsible by most of the energy consumption and only approximately 2% are consumed in the spray dryer. Otherwise, in the Process with Concentrated Ethanol Feed, the spray dryer is the unique responsible by the energy consumption in the salt recovery step. For this reason, the high level of energy consumption in the evaporation system makes the absence of a pre-concentrating distillation not viable. In fact, the energy consumption to recuperate salt in the Process of Diluted Ethanol Feed accounted for 82% of total energy consumption in the process, and in the Process of Concentrated Ethanol Feed only 9.7% of the total energy was used to recuperate the salt.

4. Conclusion

The two studied alternative processes of anhydrous ethanol production presented a small difference of 6% in the energy consumption during the anhydrous ethanol production step. Regarding energy consumption in the potassium acetate recovery, the comparison between the two processes has shown a high level of energy consumption in the evaporation system against the spray dryer. Therefore, the pre-concentration of the salt solution in the evaporation system has proven to being not feasible under the energetic point of view. Considering the energy requirements in the whole process to produce anhydrous ethanol from a diluted solution, the most feasible alternative is the of use a pre-concentrating column that eliminates the need of a multiple effect evaporation system in the salt recovery step.

Nomenclature

e   Arbitrary stage

$E_{e,n}$ Murphree Efficiency N_{sal} Total number of salts

$K_{e,n}$ Equilibrium rate V_e Total molar flow rate - vapor phase

L_e Total molar flow rate - liquid phase $v_{e,n}$ Molar flow rate of n - vapor phase

$l_{e,n}$ Molar flow rate of n -liquid phase $vs_{e,n}$ Molar flow rate of the salt s

| n | Arbitrary volatile component | v_s | Sum of the stoichiometric coefficients of the salt ions |

References

Barba, D., Brandini, V., and Di Giacomo, G. (1985), Hyperazeotropic Ethanol Salted-out by Extractive Distillation. Theoretical Evaluation and Experimental Check, Chemical Engineering Science, 40(12), 2287-2292

Cook, R. A. and Furter, W. F. (1968). Extractive Distillation Employing a Dissolved Salt as Separating Agent, The Canadian Journal of Chemical Engineering, 46, 119-123

Fredenslund, A., Gmehling, J. and Rasmussen, P. (1977), Vapor-Liquid Equilibria Using UNIFAC a Group Contribuition Method, Elsevier Science Publishing Company, Amsterdam.

Gauvin, W. H. and Katta, S. (1976), Basic Concepts of Spray Dryer Design., AIChE Journal, 22(4), 713-724

Holland, C. D. (1975), Fundamentals and Modeling of Separation Processes - Absorption, Distillation, Evaporation and Extraction, Prentice Hall Inc, Englewood Cliffs, New Jersey

Jiménez, A. P. C. and Ravagnani, S. P. (1995), Modelado y Simulacion del Proceso de Destilacion Extractiva Salina del Etanol, Informacion Tecnológica, Vol. 6, N0 5, pag. 17-20

Ligero, E. L., Ravagnani, T. M. K. (2001), Simulation of Salt Extractive Distillation with Spray Dryer Salt Recovery for Anhydrous Ethanol Production, Journal of Chemical Engineering of Japan, to be published

Ligero, E. L., Ravagnani, T. M. K. (2000), Proceso Alternativo para la Producción de Etanol Anhidro usando una Sal como Agente de Separación, Información Tecnológica, 11(4), 31-39

Martínez de la Ossa, E., Galán Serrano, M. A. (1991),Salt Effect on the Composition of Alcohols Obtained from Wine by Extractive Distillation, American Journal of Enology and Viticulture, 42(3), 252-254

Naphtali, L. M. and Sandholm, D. P. (1971), Multicomponent Separation Calculations by Linearization, AIChE Journal, 17(1), 148-153

Sander, B., Fredenslund A. and Rasmussen, P. (1986), Calculation of Vapour-Liquid Equilibria in Mixed Solvent/Salt Systems using an Extended Uniquac Equation, Chemical Engineering Science, 41(5), 1171-1183

Schmitt, D. and Vogelpohl, A. (1983), Distillation of Ethanol-Water Solutions in the Presence of Potassium Acetate, Separation Science and Technology, 18(6), 547-554

Vercher, E., Muñoz, A., and Martinez-Andreu, A., (1991), Isobaric Vapor-Liquid Equilibrium Data for the Ethanol-Water-Potassium Acetate and Ethanol-Water-(Potassium Acetate/Sodium Acetate) Systems, Journal of Chemical and Engineering, 36(3), 274-277

European Symposium on Computer Aided Process Engineering – 12
J. Grievink and J. van Schijndel (Editors)
© 2002 Elsevier Science B.V. All rights reserved.

Optimum Controllability of Distributed Systems based on Non-equilibrium Thermodynamics

F. Michiel Meeuse and Johan Grievink
Delft University of Technology, Department of Chemical Technology,
Julianalaan 136, 2628 BL Delft The Netherlands

Abstract

Integration of process design and control is currently an active research area within the Process Systems Engineering community. Most of the existing approaches only analyse the controllability of a given set of design alternatives. Meeuse et al. (2000,2001) have shown how non-equilibrium thermodynamics can be used to include controllability in the synthesis phase. Their approach addressed lumped systems only. In this paper we extend their approach to distributed systems, illustrating it with an application to heat exchanger example.

1 Introduction

Nowadays there is considerable research effort in the field of integration of process design and control. Several different approaches are available that can assist the process designer in designing processes with improved control performance. The first class of approaches relies on the use of controllability indicators. Skogestad and Postlethwaite (1996) present a brief overview of available indicators. An alternative method is to simultaneously design the process and the control system leading to Mixed Integer Dynamic Optimisation (MIDO) problems (e.g. Mohideen et al., 1996, Bansal, 2000).

Meeuse and Tousain (2002) presented a hybrid approach where design alternatives are compared based on the optimal closed-loop performance. In the sequential approaches the controllability assessment is a posteriori event in the *analysis* phase of a design.

Meeuse et al. (2000, 2001) started investigating the possibilities of including controllability in the *synthesis* phase by using a non-equilibrium thermodynamic description. They focus on the disturbance sensitivity aspects of controllability. Their work can be seen as an extension of the stability theory of process systems based on thermodynamics by Ydstie and co-workers (e.g. Ydstie and Alonso, 1997). Meeuse et al. (2000, 2001) only consider lumped systems. Here we extend this work to include distributed systems.

2 A thermodynamic system description

In non-equilibrium thermodynamics, rate processes are described by a combination of fluxes and forces. Close to equilibrium a flux is linear depended on the driving forces:

$$
\begin{aligned}
J_1 &= L_{11}X_1 + \cdots + L_{1n}X_n \\
&\vdots \qquad\qquad \ddots \\
J_n &= L_{n1}X_1 + \cdots + L_{nn}X_n
\end{aligned}
\tag{1}
$$

where J_I are the fluxes, L_{ij} are the phenomenological constants and X_I are the driving forces. Typical combinations of fluxes and driving forces are:

- energy flux: $J_u = L\nabla \dfrac{1}{T}$

- diffusion flux: $J_k = L\nabla \dfrac{\mu_k}{T}$

- chemical reaction: $J_{chem} = L\dfrac{\sum \nu_i \mu_i}{T}$

The entropy production in the system per volume, $\bar{p}_s$ is given as a simple function of the driving forces and fluxes:

$$\bar{p}_S = \sum_i J_i X_i \tag{2}$$

The design variable of a rate process involve the selection of a driving force or a flux and a spatial dimension: transfer areas for the fluxes or a volume for reactions. Meeuse et al. (2001) introduced the thermodynamic design factor, K, which is the product of the phenomenological constant and the area. This allows process design problems to be described by non-equilibrium thermodynamics. When the off-diagonal elements of (1) are neglected, the relation between a flow, F, and the driving force is given by:

$$F_i = aL_i X_i = K_i X_i \tag{3}$$

where a is the area. The total entropy production in the system, p_s, becomes:

$$p_S = \sum_i F_i X_i \tag{4}$$

From a thermodynamic point of view Meeuse et al. (2001) discriminate two different types of design situations: flow specified processes and force specified processes.
- In flow specified processes all design alternatives have the same flow F, but different driving forces, X. Examples are heat transfer and chemical reaction.
- In force specified processes. all design alternatives have the same driving force, X, but different flows, F. An example is distillation where all design alternatives have the same top and bottom composition and hence the same driving force for separation. However, the flow, related with the reflux rate, differs.

3   Non-equilibrium thermodynamics and process control

Ydstie and Alonso (1997) have showed how non-equilibrium thermodynamics can be used to describe stability properties of process systems. Their analysis is based on the passivity theory. Their main result is that for process systems a storage function can be defined such that process systems are passive and can hence be stabilised. This storage function is given by:

$$A = E - T_0 S + A_0 \tag{5}$$

where A is the storage function, E is the internal energy, T_0 the temperature of the surroundings, S the entropy and A_0 a constant. The balance equation for A is now given by:

$$\frac{dA}{dt} = \phi_A + p_A = \phi_A - T_0 p_S \qquad (6)$$

where ϕ_A is the net flow of A and p represent the production terms. When two steady-states are compared, an increase in the flow of A to the system can lead either to an increased entropy production, or an increase of the flow out of the system. The increase in entropy production is defined by:

$$\alpha = \frac{\Delta \phi_A - \Delta p_S}{\Delta \phi_A} \qquad (7)$$

where α is the dissipation coefficient. Meeuse et al. (2001) have shown that a small value of α is beneficial for control. Therefore, this implies that a large influence on the entropy production rate is favourable for controllability. A first order approximation of equation (4) leads to:

$$\Delta p_S = \sum F \Delta X + \sum \Delta F X = 2 \sum K X \Delta X \qquad (8)$$

For controllability it is desirable the the left hand side of equation (8) is large. This has the following implications:
- For a flow specified process (KX is fixed) a small driving force is beneficial for controllability since then ΔX is large.
- For a force specified process (X is fixed) a large flow is beneficial for controllability since the K is large.

4   Distributed systems

Meeuse et al. (2000, 2001) only considered ideal mixed systems. However a large number of process systems are distributed systems, e.g. tubular reactors, heat exchangers and packed distillation and absorption beds. The local driving forces can be integrated to obtain average driving forces. These average driving forces can then be used in the analyses. Let us consider a heat transfer process in the system shown in Figure 1.

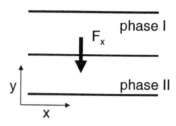

Figure 1. Distributed system.

The system consists of two phases with a heat flow from phase I, to phase II. The phases have a temperature gradient in the x-dimension. In the y-dimension the temperature within the phases are assumed constant. A heat flow will evolve due to the temperature difference between the two phases. The flux at position x is given by:

$$J_x = LX_x \qquad (9)$$

The product of the flux and the area gives the flow at this point:

$$F_x = J_x \Delta x a_i \qquad (10)$$

The total flow can be found by integrating the local heat flow over the x direction:

$$F_{total} = \int_0^1 F_x dx = K \int_0^1 X_x dx \qquad (11)$$

We now define the average driving force, $\overline{X}$, as:

$$\overline{X} = \int_0^1 X dx \qquad (12)$$

So the total flow in the system is given by:

$$F_{total} = K\overline{X} \qquad (13)$$

For a distributed system equation (3) is now replaced by equation (13). The controllability implication for distributed systems is the same as for lumped systems: a small driving force is favourable for controllability.

5   Example, heat exchanger design

We now consider four different designs of a heat exchanger, both co-current and counter-current. All designs have the same duty. Table 1 shows the design parameters of the different designs.

Table 1. Different heat exchanger designs

Design	flow mode	$T_{hot,in}$	$T_{hot,out}$	$T_{cold,in}$	$T_{cold,out}$
A	counter-current	50	40	20	27
B	counter-current	50	40	20	32
C	co current	50	40	20	27
D	co-current	50	40	20	32

Figure 2 shows the driving force profiles. Based on the theory presented above, a decreasing driving force should have a positive effect on the controllability. This would imply that the 'most controllable' design is D, followed by B, C and finally A. In order to verify this outcome we will apply an index that quantifies the controllability aspect considered in this work: static disturbance rejection capacities of process alternatives. This disturbance static sensitivity index is defined as (Seferlis and Grievink, 1999):

$$\min_{u} \quad \theta = (u - u_0)^T W_u (u - u_0) + (y - y_0)^T W_y (y - y_0)$$

st: process model   (14)

constraints on inputs and outputs

where $(u-u_0)$ is the manipulated variable deviation, $(y-y_0)$ is the controlled variable deviation and W_u and W_y are weighting matrices. This index is the static version of the LQG objective function.

A disturbance of 1°C in the cold temperature inlet was considered. The output and manipulated variables were scaled such that a difference of 1°C in the hot stream output temperature is penalised equally as a 20 % difference in manipulated variable action.

Table 2, controllability results

Design	normalised average driving force	disturbance sensitivity index
A	1.00	10.9
B	0.88	9.1
C	0.94	10.6
D	0.77	7.9

Table 2 shows for the various designs the disturbance sensitivity index and the average driving force. This Table shows that the controllability is closely correlated with the average driving force. The result is also inline with the statement by Luyben et al. (1999) that a small temperature gradient is beneficial for controllability. It is an interesting observation that this is independent of the flow configuration (co-current or counter current).

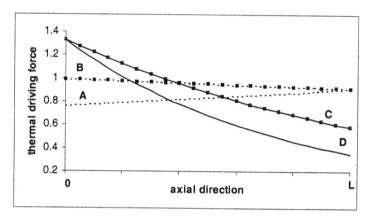

Figure 2. Driving force distribution in the different heat exchanger designs

6   Conclusions

In earlier work it was shown how non-equilibrium thermodynamics can be used to include controllability in the synthesis phase of chemical process design. In this work we have shown that their approach can be extended to include distributed systems. The heat exchanger example shows that there is a consistent relation between the average driving force and the disturbance sensitivity, independent of the flow configuration. Further validation of the method on more complex systems is in progress (e.g. reactor systems).

Literature

Bansal, V. (2000) Analysis, design and control optimization of process systems under uncertainty, PhD Thesis, University of London

Luyben, W.L., Tyreus, B.D. and Luyben, M.L. (1999) Plantwide process control, McGraw-Hill, New-York

Meeuse, F.M., Samyudia, Y. and Grievink, J. (2000) Design of controllable processes using irreversible thermodynamic structures, paper presented at the AIChE's 2000 Annual meeting , Los Angeles, U.S.A., paper 244d.

Meeuse, F.M., Samyudia, Y. and Grievink, J. (2001) Process design for controllability, a thermodynamic approach, *submitted to Ind & Eng. Chem. Res.*

Meeuse, F.M. and Tousain, R.L. (2002) Closed loop controllability analysis of process designs: Application to distillation column design, *Comp. Chem. Eng, in press*

Mohideen, M.J., Perkins, J.D., and Pistikopoulos, E.N. (1996) Optimal design of dynamic systems under uncertainty. *AIChE Journal* 42(8):2251-2272

Morari, M. (1983) Design of resilient processing plants III A general framework for the assessment of dynamic resilience. *Chem. Eng. Science* 38:1881-1891

Seferlis, P. and Grievink, J. (1999) Process control structure screening by steady-state multiple disturbance sensitivity analysis, *Comp. Chem. Eng*, vol 23, S309 - S312

Skogestad, S. and Postlethwaite I. (1996) Multivariable feedback control, Wiley, Chichester

Ydstie, B.E. and Alonso, A.A. (1997) Process systems and passivity via the Clausius-Planck inequality. *Systems & Control Letters* 30:253-264

European Symposium on Computer Aided Process Engineering – 12
J. Grievink and J. van Schijndel (Editors)

Internet-based equipment and model gallery

Marja Nappa[1], Tommi Karhela[2], Markku Hurme[1]

[1] Helsinki University of Technology,
Plant Design, P.O.Box 6100,FIN-02015 HUT, Finland

[2] VTT Industrial Systems, P.O.Box 1300, FIN-02044 VTT, Finland

Abstract

The paper presents an Internet-based integrated design environment, which aims to enhance the efficiency and quality of process and control design in process industry. The main idea is to improve the utilisation and distribution of simulation models and vendor-based equipment information to the designers and other customers in a ready available form. The information can be directly utilised by equipment selection tools and various process simulators both steady state and dynamic. The system is based on the use of Internet as a distribution channel and on the presentation form of information, which makes it compatible to different end user software.

1. Introduction

Process simulation is used more and more in process industry for process and control design, automation testing, operator training and support purposes. Various types of computer aided design tools are being used through the life cycle of the project; e.g. spreadsheet tools, steady state simulators, equipment design programs and dynamic simulators. Many aspects of successful and efficient process design require integration of design tools allowing data transfer between these systems. An existing integrated computer aided design system is ICAS (Gani et al, 1997).

An obstacle to the efficient use of various – especially the more sophisticated - simulation tools in process design is the time and work needed for building the simulation models. For example building a dynamic model of a process requires a lot of detailed equipment information, which is not easily available even in case the process includes standard equipment only. On the other hand there are many simulators, which are dedicated to special branches of process engineering; such as energy or pulp processes. Utilisation of these is time consuming in a total project context, if there is not a data transfer mechanism available to integrate them with the other design tools and simulators with wider context. Still the building of efficient integrated plants requires the use of both dedicated steady state and dynamic simulators.

Integration of control design with process design requires that dynamic simulation is used from early phases of process design. However, there are some obstacles in the way to large scale use of dynamic simulation in design. Modelling with current dynamic tools is often laborious and time consuming. The equipment parameters needed for a proper dynamic simulation are required from the beginning, even before the equipment details have been designed. In case standard commercial equipment is used, this information is available from the supplier. If the equipment is tailor made as in case of vessels and conventional heat exchangers, the preliminary information of detailed design is received from equipment dimensioning tools. One aim of the presented design environment is to enhance the use of dynamic simulation in the early phases of process design by providing means to easily transfer detailed equipment data from equipment manufacturers and vendors to process designers through the Internet-based information gallery.

2. Description of Internet based design environment

In the Gallery project a new kind of an infrastructure is specified and the required tools are built for storing information on process equipment and simulation models into an Internet-based database from where it can be easily used by simulators and process designers. In the Advice (ADVanced Integrated Component based Simulation Environment) project the same architecture is used to convert model configurations and parametrisation between various dynamic and steady state tools, and the necessary changes are implemented in the tools to enable their use under the same user interface. The whole system, integrated simulation environment, aims to sophisticated and modern process design. Figure 1 shows an example sketch of an integrated simulation environment. The integrated tools are not limited to those shown in Figure 1 but it is also possible to add other databases and programs such as company's project database and optimisation tools or design tools to the environment.

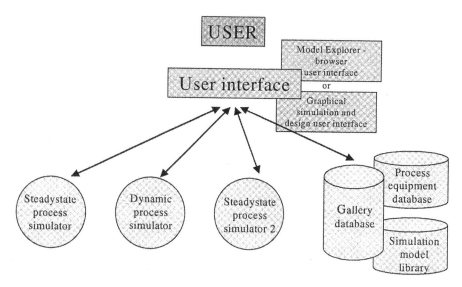

Figure 1. Internet based integrated design environment from the perspective of process designer.

By using the Gallery, the process designer can survey a whole range of components such as pumps, valves, piping, heat exchangers, disc filters and centrifugal cleaners. The database includes information on commercial components, including their technical specifications, performance and pressure drop characteristic curves. The designer can retrieve parameters through the Internet for equipment items used in the simulation model. All manual data transfer is minimised. For example the characteristic curve or dimensions of a pump or a valve can be retrieved for the simulation. The data provided in the database is kept up-to-date by the respective equipment suppliers. For suppliers the Gallery provides tools to make the updating process easier.

The Gallery allows the user to add his/her own simulation models and equipment information to the database for reuse and distribution. Simulation model configurations and parameters can be transferred from one simulator to another. While transferring from steady state simulator to dynamic simulator more specific data for process components is needed. From the database existing equipment information can be retrieved. In case of non-standard equipment (e.g. shell and tube heat exchangers), the Gallery provides integrated dimensioning tools that form a shortcut to more efficient design.

The Gallery includes tools which assist the use of the database and simulation programs. These can be, for example, dimensioning tools, data transformation tools, search engines and external models for simulators. All these are called extensions. The user can create an extension for a specific case and easily add this functionality to the system. For example when a new equipment supplier joins the Gallery the transformation tool specialised to convert supplier's equipment data can be added.

The users are divided into user groups and the data owner can give access rights for reading, writing and executing his information. When transferring the data over the Internet it is encrypted. In the Gallery, the SSL (Secure Socket Layer) is used for preventing leakage of information to an exterior party.

The architecture of integrated simulation environment is outlined in Figure 2. Model servers such as simulators and databases and client applications such as user interfaces communicate using an XML (eXtensible Markup Language) –based Gallery Query Language (GQL). The data model of Gallery is also based on XML and called Gallery Markup Language (GML). Due to the openness of the Gallery architecture, more model servers, client servers and extensions can easily be added in the environment. A more exact description of the architecture and the used data model can be found in Karhela (2002).

At the moment, a prototype of Gallery is available including example programs to test the features of the architecture and the data model. Some of the characters described above are under construction to be included in the Gallery version 1.0, which is scheduled at the end of the year 2002. The system is planned to be continuously extending tool, and once the groundwork has been done, more and more tools, simulators and data will be added.

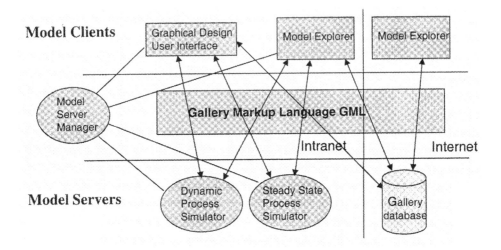

Model Clients

Model Servers

Figure 2. A typical setup of programs in the Gallery architecture.

3. Benefits of the Internet-based design environment

An integrated design environment is a heterogeneous system, in which design software components exchange data and configurations and share functional models. The integration involves following aspects.

The integration of *simulators* facilitates process designers work. This makes the integrated use of various dedicated (pulp, energy, dynamic etc.) simulators flexible without a need of manual information transfer from one program to another. Process models developed in steady state simulators can reused in dynamic simulators. Simulators can also share functional models, for example the same physical property databank can be used by several simulators.

An integrated design environment can be used trough one *user interface*, which is common to all simulators (steady state and dynamic) in the system. Since all programs in the system look and feel alike and the user can operate the tools in a similar manner, their use is more straight forward from the process designer's point of view.

The Gallery database enables the huge amount of *information* to be easily used in design and simulation. Common *equipment vendor data* is shared by all simulation programs in the system. The use Gallery based equipment data empowers especially the build-up of dynamic simulation models, since the detailed equipment data required can be directly retrieved from the database. On the other hand the process designers can share and distribute *simulation models* by filing them into Gallery model libraries and making them available through the Internet. The ready made simulation models, which include both equipment and control system models, would be of a great help in defining complex dynamic models for process units such as evaporator systems or sections of paper machines. Using the Gallery as source of information manual data transformation is minimised, which saves time and reduce the faults caused by human factors. The Gallery can be used in different locations as it is based on the Internet. This makes delivering of information efficient.

4. Example case

As an example of Gallery utilisation in design, we discuss a small process, which includes pump, heat exchanger, filter and control valve (Figure 3).

First a simulation of the material balance of the process is done with a steady state simulator. The choice of the simulator depends on the type of process designed; pulp, paper, energy, chemical etc. All simulators run under a common user interface, so the usage is straight forward to the designer. From this simulation a material balance is received with temperatures and pressures.

Next phase would be the selection of the commercial equipment types. This detailed information is also required by the subsequent dynamic simulation. The Gallery includes commercial data for pumps, control valves and filters. Also a simple dimensioning tool for pipe and shell heat exchangers is available.

Pump selection is done by starting extension called centrifugal pump search engine. Process designer is able to select the pump directly from the same graphical user interface, which is used for simulators. The selection is based on the simulated (or given) values on design flow rate, head and NPSH. The five most appropriate pumps are shown to designer. After the selection is done by the user, the pump equipment and process parameters are transferred automatically to the simulation model.

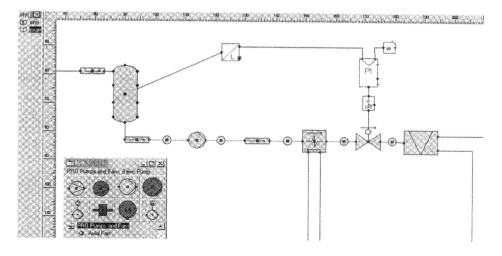

Figure 3. Diagram of the discussed design process.

A detailed dimensioning of the heat exchanger is also required, if dynamic simulation will be done. The Gallery includes a simple dimensioning tool for shell and tube heat exchangers, which selects an appropriate TEMA type exchanger and makes a preliminary mechanical dimensioning based on the steady state simulation data. Also a commercial control valve and a filter can be selected from the Gallery.

After this dynamic simulation can start. The same user interface is still used but now for APROS (Silvennoinen et al., 1989) dynamic simulator. APROS is a rigorous dynamic simulator for power industry and its pulp and paper equivalent is called APMS (Tuuri et al., 1995). By the dynamic simulation the behaviour of control loop is studied and the parameters tuned. From the dynamic simulation also the sizing of the pump and the valve can be checked. It can be seen, for instance, if the sizing is too tight for the pump & control valve system. If required, a new pump or valve can be selected from the Gallery.

In the presented way the data integration between simulators and equipment Gallery makes the early application of dynamic simulation possible and therefore results to better design quality of processes especially in dynamic sense.

5. Conclusion

The Internet-based equipment and model Gallery with the integrated simulation environment aims at more sophisticated process modelling and planning. In the environment, a collection of design and simulation tools can be used under the same user interface, which is expected to significantly speed up the design process and change engineering working methods. Especially the use of dynamic simulation will be far more feasible than today, as equipment data can be retrieved from the Gallery database, and as the dynamic model can be built using the results of a steady state solution.

6. References

Gani, R., Hytoft, G., Jaksland, C., Jensen, A., 1997, An integrated computer aided system for integrated design of chemical processes, Computers Chem Engng. 21, 1135.

Karhela, T. A Software Architecture for Configuration and Usage of Process Simulation Models: Component Technology and XML based Approach. To be published in VTT Publications http://www.inf.vtt.fi/pdf/. Otamedia 2002.

Silvennoinen, E., Juslin, K., Hänninen, M., Tiihonen, O., Kurki, J., Porkholm, K., 1989, The APROS software for process simulation and model development, VTT Research reports no. 618, Technical Research Centre of Finland, Espoo.

Tuuri, S., Niemenmaa, A.,Laukkanen, I., Lappalainen, J., Juslin, K., 1995, A multipurpose tool for dynamic simulation of paper and board mills, Proceedings of Eurosim 1995, European Simulation Society, Vienna.

Interval Analysis for identification and bounding of potential process structures in early design

A. R. F. O'Grady, I. D. L. Bogle & E. S. Fraga[1]
Centre for Process Systems Engineering
Department of Chemical Engineering
UCL (University College London)
Torrington Place, London WC1E 7JE, U.K.

Abstract

An automated design system, based on simultaneous graph generation and traversal, is extended through the use of interval analysis techniques to provide insight into the range of structures that can form the basis for good process designs. Interval analysis is used to bound the effects of discretization procedures used to generate the search graph. In particular, this paper investigates the effect of discrete component flows in structure selection and the bounding procedures can identify structures which can potentially achieve the global minimum cost solution. The extended search algorithm has been implemented in the Jacaranda object oriented automated design system. Short-cut models for distillation columns and physical property estimation methods have been extended to use interval analysis. Results for a distillation sequence example are presented. The example is a good test of the new procedures due to the wide range of component flows and purity specifications. Combined with the search procedure's ability to generate multiple ranked solutions, a range of alternative solution structures is identified, making the new system particularly well suited to early conceptual design.

1. Introduction

Automated design tools are intended for use at the early, conceptual stage of process design. At this stage, little is known about the structure of potentially good designs and one should minimize the number of a priori decisions about these structures. Automated design tools for process design are often based on the definition of a mixed integer nonlinear programme (MINLP) from a given superstructure. In defining the superstructure, decisions are made about possible alternatives, decisions that can limit the quality of subsequent solutions identified by the MINLP solution procedure. In this paper, we describe an algorithmic procedure, based on simultaneous graph generation and traversal, which attempts to identify the range of structures which may lead to good designs.

The core of the design procedure is an implicit enumeration (IE) method for generating and traversing a search graph that incorporates all possible solutions. The search graph consists of nodes, corresponding to processing unit alternatives, and edges, representing flow of material from unit to unit. The IE method has been implemented in Java as the Jacaranda generic framework for automated design using object oriented design principles (Fraga et al., 2000). In this work, a new search procedure is implemented

[1] Corresponding author; e.fraga@ucl.ac.uk; +44 (0) 20 7679 3817

within this framework through the use of subclassing to maximise the re-use of existing models and algorithms. The new search procedure is based on the use of interval analysis (Moore, 1966) to generate bounding information.

The basis of the graph generation procedure in Jacaranda is the use of discretization to convert all continuous quantities to sets of discrete values. When continuous quantities, such as stream component flows or unit operating pressures, are encountered, the nearest value from the discrete set for that variable is used. The level of coarseness for discretization is under user control and each variable can have a different level. An example of how discretization can encourage the use of automated design for exploration in early design is given by Laing & Fraga (1997).

The advantage of the use of discretization is the ability to tackle problems with complex mathematical models without concern for linearity or convexity issues. The disadvantage is the possibility of missing good solutions that fall between the discrete values considered by the search procedure. A preliminary attempt at understanding and exploiting the effects of discretization in Jacaranda through the use of interval analysis techniques was applied to discrete operating pressure levels in distillation sequence synthesis (O'Grady et al., 2001). This successful preliminary investigation motivated the further use of interval analysis techniques in Jacaranda.

This paper addresses the effect of the discretization of key component recovery fractions in distillation on flowsheet costs. The discretization of component flowrates is especially significant as they have a direct effect on the structure alternatives embedded in the implicit search graph. Interval analysis is applied in the distillation design procedures, generating designs spanning a range of key recovery fractions and providing a measure of the effect of discretization on structure choice. Furthermore, the analysis shows the range of structures that are appropriate for the problem, independent of the discretizations used.

1.1 Interval analysis

Interval analysis was first introduced by Moore (1966). An interval consists of a lower and an upper bound. Interval arithmetic ensures that the interval result of an operation on two intervals contains all achievable real values. This is useful as continuous real variables can be divided into discrete interval sections. If mathematical operations are carried out on these intervals, according to the rules of interval arithmetic, the result will contain the range of all possible values. As a result, the global optimum can be bounded. Hansen (1992) provides a detailed explanation of some interval methods and their application to global optimization.

Interval analysis has been previously applied to other process engineering problems. Schnepper et al. (1996) used interval analysis to find all the roots of several chemical engineering problems with mathematical certainty. Byrne et al (2000) used interval methods to globally optimize given flowsheet structures.

2. Methodology

Interval analysis is used to bound the effects of component flowrate discretization. There is a fundamental difference between the potential effect of component discretization and that of the discretization of other variables. The specification for product streams is often based on the molar (or mass) fraction of one or more of its components. Thus, the way in which components are discretized and how finely this discretization is performed affects the choice of final process structure.

2.1 Continuous flow discretization

Streams contain components, present within upper and lower bounds on their flow rate. Each component has a user-defined *base* level. After a unit design, the flow of each component in each product stream is checked against the component's base level. Below this threshold, the component flow is mapped to a *trace* interval, defined as [0,*base*]. This discretization at the lower end of flow rate range allows the dynamic programming aspect of Jacaranda to re-use solutions to sub-problems previously solved. The use of dynamic programming enables an efficient implementation of the algorithm and helps moderate the effect of the combinatorial growth of the search graph. Component flows above the base level are treated as continuous interval quantities.

The amount of each component in any stream in the flowsheets presented to the user is bounded. The contribution of a component does not disappear due to discretization as it did previously in Jacaranda. Even if the amount present falls below the base level, its presence is taken into account. The effect of trace amounts are incorporated into the search procedure to ensure that the search graph includes a fully representative set of process structures.

2.2 Distillation unit design

The case study presented below is based on distillation sequence synthesis. Therefore, for this case study, we have developed an *intervalised* distillation unit model based on the Fenske-Underwood-Gilliland equations. The model performs a semi-sharp separation in which the heavy and light keys distribute amongst the products of the unit. All other components distribute sharply to the appropriate product stream. The amount of distribution of key components is controlled by the key component recovery design parameter (in principle, both keys can have different recoveries but for the case study below it is assumed that both use the same value). The effect of discretization of the recovery parameter is investigated in the context of component flow discretization.

Component discretization is necessary because it is possible to specify a range of key component recovery fractions. The user specifies how many intervals within the range of recovery fractions are to be attempted each time a distillation unit is designed for a given feed stream. A unit design defines the stream outputs and the costs, all as intervals. For example, if distillation units are to be designed with key recovery fractions between 0.95 and 0.99, the user could specify that four intervals are to be used. For each stream that is to be processed, column designs are carried out with recovery value intervals [0.95,0.96], [0.96,0.97], [0.97,0.98] and [0.98,0.99]. Each of these four designs yields a top and bottom stream containing intervals of component flows. Jacaranda will subsequently generate appropriate search graphs corresponding to each of the interval output streams. These streams may require further processing or they may meet one of the product specifications. In order for a stream to be designated as a product, the whole interval must be within the product specification.

In designing a unit, the feed stream is checked for the presence of the light key in greater than trace amounts. If not present, the feed is checked for a more volatile component to be the light key. Any components that are present in trace amounts in the feed are passed through as trace amounts in the output streams of the unit. The presence of small amounts of a component are taken into account in the costing of a unit design but do not lead to unnecessary processing steps, reducing the computational effort without affecting the generation of suitable structural alternatives. Trace amounts are, however, fully considered in identifying valid product streams, as discussed above. The *intervalised* distillation design procedure allows a range of recovery fractions to be investigated whilst preventing potentially optimal structures from being missed between discrete levels.

2.3 Solution ranking

Jacaranda has two features that enhance its use in early design. Multiple criteria can be considered simultaneously and a ranked list of solutions may be generated for each criterion (Steffens et al., 1999). The number of solutions generated for each criterion is a user settable parameter. To ensure sufficient diversity in the solutions presented to the user, the ranking procedure, by default, ensures that only solutions that differ in structure will appear in the list of best solutions.

3. Case study: Benzene recycle separation problem

A stream, defined in Table 1, must be purified to achieve 98% purity for benzene. This stream appears in the separation section of a chlorobenzene process and the benzene is to be recycled back to the reactor. The other components are to be removed as waste, with the requirement that any waste stream contain less than 10 mol% benzene and less than 10 mol% chlorobenzene. Any output stream which consists of >90% chlorobenzene will also be accepted as a valid product stream. The flowsheet structure with the lowest capital cost for this separation is required.

3.1 Initial run

Distillation units were used with two key recovery intervals together spanning the range 97% to 99.9%. The base level for benzene was to 0.1% of its flow in the feed stream; all other components had base values of 10% of initial flow rates. The discrepancy is due to the difference in purity specifications for the benzene recycle stream when compared with the waste and product streams. The values of the threshold levels must be chosen to be consistent with the problem goals. If excessively large values are chosen, the required purity may not be attained as components that fall below this level are mapped to their *trace* interval values. If the values are too small, the efficiency of the procedure suffers as solutions are not reused due to small differences in flow rates.

The results from the initial run of Jacaranda is a list of solution structures ranked

Table 1. Composition and conditions of feed stream.

Component	Flowrate (kmol/s)
Benzene	0.97
Chlorobenzene	0.01
Di-chlorobenzene	0.01
Tri-chlorobenzene	0.01
Pressure	1 atm
Temperature	313 K

according to the lower bound of capital cost. The cost of the *best* solution bounds the global minimum with respect to the recovery fraction, noting that some other variables are fixed at one operating value. Figure 1 presents (the three cost bounds, indicated by "2" on the *x*-axis) both the cost bounds and the different structures identified as the best three solutions to this problem. From this initial run, we see that structures with both 2 and 3 units are equally as good, with respect to the bounds on their costs. There is little to distinguish between them and so further investigation is required.

3.2 Improvement of bounds

Increasing the number of key recovery intervals reduces the size of each interval but has the effect of increasing the size of the search graph and, hence, the computational effort required to solve the problem. Several further runs were performed, doubling the number of intervals each time. Figure 1, on the right , shows the capital cost bounds and the structures identified for the best 3 solutions using 16 intervals over the same range.

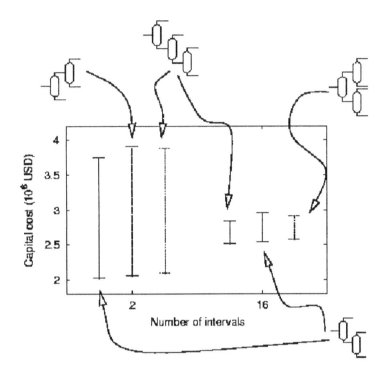

Figure 1. Cost bounds and process structures for initial and final runs.

When compared with the initial run, we see that the bounds have been tightened by increasing the number of intervals, as we would expect. Interestingly, two of the initial structures are still present, albeit in different ranking order, and a new structure has been identified. The "best" structure initially was based on 2 units but is now based on three units. Furthermore, the bounds on the costs are still similar and it is difficult to distinguish between them. Therefore, although a structure based on three units could have been dismissed by the initial run, the finer discretization leads us to avoid making this decision at this stage. The solutions obtained are all equivalent in terms of performance, based on the models used at this stage and

Table 2. Computational and search statistics.

Statistics	Number of Intervals	
	2	16
Problems	210	218838
Re-use (%)	60	92
Time (s)	6	3151

further discrimination is not possible without the use of more detailed models. However, these initial runs have enabled us to reduce the number of structures that should be considered with detailed models to two or three.

Statistics for the computational runs discussed are presented in Table 2 using a 850 MHz Pentium III computer with Java 1.3.

3.3 Discussion

The use of interval analysis to represent component flows provides the user with bounds on the costs of the solutions obtained. These bounds are not as tight as they possibly could be but they assure that the cost of the solution cannot be outside their limits. Optimization based on the lower bound of capital cost assures that the global optimum is bounded. In addition, the dismissal of less obvious structures can be prevented by using intervals to span the full range of a design variable.

Initially wide recovery fraction intervals yield solutions with wide capital cost bounds as runs with more, and hence finer, intervals are attempted, the bounds on the best solutions become tighter. This property may be useful in the future in scheme that adaptively changes discretization parameters in response to the bounds on solutions generated.

4. Conclusions

An automated design procedure based on the use of discretization to define a search graph can be an effective tool for early, conceptual design. However, to improve the effectiveness of this approach, the use of interval analysis has been implemented. Interval analysis generates bounds on the quality of the solutions generated. These bounds can provide insight into the effects of different design parameters on the potential solutions. It also prevents the dismissal of less obvious structures, enabling the use of automated design as a sieve in early design.

Acknowledgements.

The authors gratefully acknowledge funding from the EPSRC.

References

Byrne, R. P. and I.D.L Bogle, 2000, Global optimization of modular process flowsheets, Ind. Eng. Chem. Res. 39(11), 4296-4301.

Fraga, E. S., M.A. Steffens, I.D.L. Bogle and A.K. Hind, 2000, An object oriented framework for process synthesis and optimization, Foundations of Computer-Aided Process Design, Eds. Malone, M. F., Trainham, J. A., and Carnahan, B., AIChE Symposium Series 323(96), 446-449.

Hansen, E., 1992, Global Optimization Using Interval Analysis, Marcel Dekker, New York.

Laing, D. M. and E.S. Fraga, 1997, A case study on synthesis in preliminary design, Computers chem. Engng. 21(15), S53-S58.

Moore, R. E., 1966, Interval Analysis, Prentice Hall, Englewood Cliffs, New Jersey.

O'Grady, A. R. F., I.D.L. Bogle and E.S. Fraga, 2001, Interval analysis in automated design for bounded solutions, Chemical Papers, in press.

Schnepper, C. A. and M.A. Stadtherr, 1996, Robust process simulation using interval methods, Computers chem. Engng. 20(2), 187-199.

Steffens, M. A., E.S. Fraga and I.D.L Bogle, 1999, Multicriteria process synthesis for generating sustainable and economic bioprocesses, Computers chem. Engng. 23(10), 1455-1467.

European Symposium on Computer Aided Process Engineering – 12
J. Grievink and J. van Schijndel (Editors)

Hydrodesulphurization of Gasoils: Advantages of Counter-Current Gas-Liquid Contacting

J.A. Ojeda[1,2], J. Ramirez[2] and R. Krishna[1,*]

[1]Department of Chemical Engineering, University of Amsterdam, Nieuwe Achtergracht 166, 1018 WV Amsterdam, The Netherlands.
[2]Unidad de Investigación en Catálisis (UNICAT), Facultad de Química, Universidad Nacional Autónoma de México, Cd. Universitaria, México D.F.
*Correspondence to R. Krishna, fax +31 20 5255604; Email: krishna@science.uva.nl

Abstract

Hydrodesulphurization (HDS) of gasoils is conventionally carried out in the petroleum industry in *co*-current trickle bed reactors. The important disadvantage of co-current operation is that the build-up of H_2S along the reactor height is detrimental to the desulphurisation reaction; this is because of inhibition effects. In this paper we examine the benefits of carrying out the HDS reaction in a *counter*-current mode. We carried out simulations of the co- and counter-current configurations using an equilibrium stage model and taking dibenzothiophene (DBT) as a model sulphur compound. Our simulations show that counter-current operation can significantly reduce the requirements in the total amount of catalyst when ultra low levels of S are demanded in the product steam. For the same amount of catalyst, the conversion of DBT is significantly higher.

1. Introduction

Present regulations in the European Union (EU) and USA, among others countries, on the maximum levels of undesirable compounds in transport fuels has triggered an intensive quest for new catalytic systems and reactor technologies (Gates and Topsøe, 1997; Schulz et al., 1999; Sie, 1999). For diesel, the EU has the commitment to reduce the sulphur content below 50 ppm and aromatics till 0.0 % vol., as well as to improve the cetane number to a minimum of 51. The US specifications are not so strict because of its lower its diesel demand (as compared to the EU) and higher gasoline consumption. The fulfilment of this new diesel legislation represents a serious challenge for the refinery industry especially in view of the increasing diesel demand.

With the use of conventional catalytic systems and one-step reactor technology, typical HDS plants are able to reduce the S content in their diesel feedstocks to levels around 700-500 ppm (Froment et al., 1994; Gates and Topsøe, 1997). Deeper conversion levels are only possible with the use of two-stage technology and catalyst profiling (Xiaoliang Ma et al., 1994; Sie, 1999) for instance the SynSat[TM] and Arosat[TM] processes. The key aspect in these processes relies on the fact that removal of H_2S after the first stage has proved to be beneficial in order to reach ultra low levels of both S and aromatics in a second one. It is well recognized that HDS reactions are largely affected by the presence of H_2S (Broderick and Gates, 1981; Vrinat, 1983; Edvinsson and Irandoust, 1993; Sie

1999). Another improvement used in the two-stage technology is counter-current operation in the second reactor. This specification was an unavoidable requirement due to the use of a noble catalyst with partial resistance to poisoning , here counter-current gives a more favourable H_2S profile making most part of the second reactor to operate in a low H_2S concentration mode.

Some authors have pointed out that counter-current operation can also by applied for HDS (Xiaoliang et al., 1994; Van Hasselt et al., 1999; Sie, 1999). Here, we must say that the prime limitation for counter-current is the hardware implementation. Taking into account the catalyst dimension and the large gas and liquid flows usually found in HDS, the counter-current operation can not be directly applied in the conventional trickle-bed reactor because it leads to a prohibited pressure drop and flooding risks. Fortunately, new concepts on catalyst packing (Ellenberger and Krishna, 1994; and Van Hasselt et al., 1997) are very promising and currently the design of a HDS reactor using counter-current seems feasible from a hardware view point.

The aim of this work is to present a comparison of counter-current vs. co-current operation for the HDS of diesel. From the literature, we found that the most tackled approach for numerical simulations in the trickle-bed reactor is that of the non-equilibrium type, which makes use of Henry coefficients only on few key compounds (Froment et al., 1994; Korsten and Hoffmann, 1996; and Van Hasselt et al., 1999), and where a semi-rigorous description of mass and energy transport in the L/V interface is considered. In our mathematical description, the equilibrium-stage model was adopted to simulate both flow arrangements. We believe that this first comparison should lead us to elucidate how beneficial counter-current operation in HDS could be and what major challenges would be encountered. With this model a direct and clear comparison can be done. In this regard and from the best knowledge of the authors, this is the first time that the equilibrium stage model is used to compare counter-current and co-current in HDS.

2. Model development

The "diesel" liquid feed flow rate is 97.8 mol/s and is assumed to consist of nC16 with a dibenzothiophene (DBT) mole fraction of 0.1245 (corresponding to 1.82 wt%). Temperature and pressure for the liquid and gas feeds were kept constant (590 K, 50 bar). The entering gas stream (flow rate of 148.6 mol/s) consists of H_2, H_2S and CH_4. The H_2 mole fraction in the gas stream was fixed at 0.618. The mole fraction of H_2S in the entering gas stream was taken to be either 0, 0.1 or 0.2. The balance is made up of methane. Total flow and composition of the liquid and gas feed were taken from literature (Froment et al., 1994). Pressure drop within the reactor was not considered. Kinetic expressions of the LHHW type were adopted and validated with experimental data taken from the literature (Vanrysselberghe and Froment, 1996). The complex nature of these expressions account not only for adsorption of reactive species but also for the most interesting effect due to H_2S and other co-products like biphenyl and cyclohexylbenzene. In the modelling of the reactive equilibrium stages, the PRSV EOS was used for prediction of LVE (Stryjek and Vera, 1986). The accuracy of the LVE predictions was compared with experimental data in a wide range of temperature and pressure with satisfactory results for most simulation purposes. The enthalpy content of liquid and vapor was also calculated with the same EOS.

Table 1. MESH equations

Molar balance	$M_{ij}=L_jx_{ij}+V_jy_{ij}-(L_jx_{i,j-1}+V_{j+1}y_{i,j+1})+R_{ij}=0$
Equilibrium condition	$E_{ij}=K_{ij}x_{ij}-y_{ij}=0$
Summation restriction	$S^L_j=\Sigma x_{ij}-1=0$ and $S^V_j=\Sigma y_{ij}-1=0$
Entalphy balance	$H_j=L_jH^L_j+V_jH^V_j-(L_{j-1}H^L_{j-1}+V_{j+1}H^V_{j+1})+Q_j=0$

For co-current simulations we used an additional first equilibrium stage without catalyst, so that, the L/V mixture was in equilibrium before entering the reaction zone. This configuration resembles the industrial condition on the trickle-bed reactor. For counter-current, liquid and vapour were fed to the reactor without any additional preparation.

The set of MESH equations shown in Table 1 (Taylor and Krishna, 1993), extended to include chemical reactions, was solved with the support of the multi-purpose math library by IMSL^{TM}. The initial guess for iteration was the solution of an isothermal flash with a global composition given by the feeds. Convergence of these equations was done with a gradual increment in the catalyst loading. This method gave a reliable route in order to find a solution. For the numerical point of view, a total of 30 equilibrium stages shown to be sufficient and no distinctions in profiles were observed by using a higher subdivision. The catalyst loading was evenly distributed through all the stages. The reactor was simulated using both, the adiabatic and isothermal mode. With these two approaches, a separation of effects due to concentration of the reactive species (mainly as a result of the flow pattern) and those due to the temperature with a source in the heat of reaction (which has impact on several variables), was achieved.

3. Simulation results

For all the cases of industrial interest, counter-current achieves either higher conversion of DBT or lower content of S than co-current with a significant saving on the required amount of catalyst (Fig. 1). As expected, H_2S is a key parameter that determines how easily the HDS reactions proceed. The effect of H_2S on S consumption was important for both flow arrangements, being the negative effect more accentuated for co-current. Separation of H_2S seems to be a crucial and an essential step in order to avoid inhibition effects when ultra deep conversion levels are pursued. The H_2 and H_2S profiles (Fig. 2) in the liquid phase show that the higher efficiency of counter-current occurs in the second half of the reactor, where the concentration of H_2 reaches its maximum and the concentration of H_2S its minimum, allowing counter-current to speed the reaction despite the prevailing low concentration of DBT.

The increase in temperature due to chemical reaction has a positive effect on both configurations. In counter-current, a significant drop in the temperature was found in the final section of the reactor, where a partial evaporation of liquid is carried out by the initial contact of the outlet liquid and the inlet gas feed. In our simulations, we have only used few reactions to describe HDS, and probably, a more complete set of reactive species involving a higher heat of reaction could report thermal instability in counter-current. This is derived from its large parametric sensitivity associated with the existence of multiple steady states (Froment and Bischoff, 1990, p. 436).

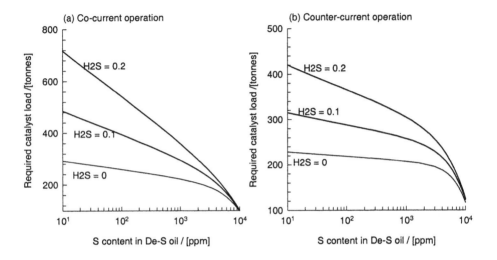

Fig. 1. Catalyst requirements for (a) co-current and (b) counter-current when different levels of contamination of H_2S, (mole fraction) in the gas feed, are used.

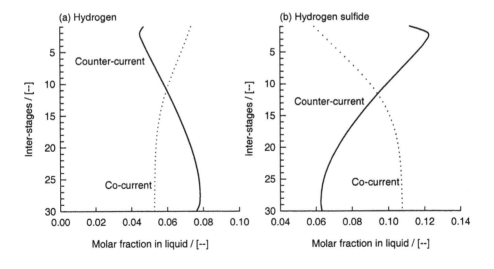

Fig. 2. Typical molar fraction profiles over the inter-stages.

Counter-current produces an internal liquid and vapor flow higher than co-current that provokes a dilution of the catalyst loading (Fig. 3). Implications of these results have important impact on counter-current reactor design due to a larger reactor diameter should be considered in order to avoid flooding risks. One drawback of counter-current is that it gives a lower liquid output than co-current due to the evaporation produced by the initial contact with the gas feed. On the other hand, this characteristic of counter-current could be capitalized by the possibility of flashing the liquid product at low temperature without risk of condensation of sulfur compounds giving extra advantages over the co-current operation.

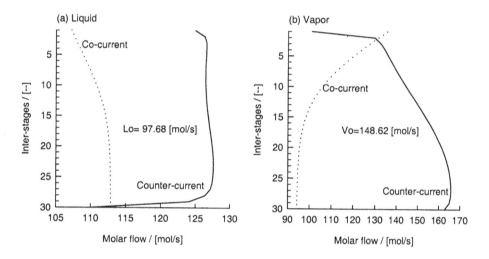

Fig. 3. Molar flow profiles over the inter-stages. Lo and Vo are the inlet molar flows.

4. Concluding remarks

Counter-current flow was compared vs co-current in the HDS of DBT using numerical simulation. In this work we have not only shown the advantages of counter-current over co-current but also some issues that deserve a more detailed examination. Two steps for further works can be to consider the use of the non-equilibrium model and to extend the treatment of chemical reaction to include intra-particle diffusion. Finally we believe that even though this comparison was performed by the use of equilibrium stages, which can be considered a very simplified method, the results from this study are useful for a preliminary insight in the design of the counter-current HDS reactor.

5. Acknowledgments

The authors want to thank the financial support granted to the PhD student Ojeda Nava by the Conacyt in México as well as by the Department of Chemical Engineering at the University of Amsterdam.

282

References

Broderick D.H. and Gates B.C., 1981, Hydrogenolysis and Hydrogenation of Dibenzothiophene Catalyzed by Sulfide CoO-MoO$_3$/γ-Al$_2$O$_3$: The Reaction Kinetics. AIChE Journal, vol. 27, no. 4, 663-672.

Gates B.C. and Topsøe H., 1997, Reactivities in deep catalytic hydrodesulfurization: chanllenges, opportunities, and the importance of 1-methyldibenzothiophene and 4,6-dimethyldibenzothiohene. Polyhedron, vol. 16, no. 18, 3213-3217.

Edvinsson R. and Irandoust S. 1993, Hydrodesulfurization of Dibenzothiophene in a Monolithic Catalyst reactor. Ind. Eng. Chem. Res. 32, 391-395.

Ellenberger J. and Krishna R. 1994, Counter-current operation of structured catalytically packed distillation columns: pressure drop, holdup and mixing. Chem. Eng. Science, 54, 1339-1345.

Froment G. F., Depauw G.A. and Vanrysselberghe V. 1994, Kinetic Modeling and Reactor Simulation in Hydrodesulfurization of Oil Fractions. Ind. Eng. Chem. Res., 33, 2975-2988.

Froment G. F., Bischoff K.B., 1990, Chemical Reactor Analysis and Design 2ed. Wiley series in chemical engineering.

Korsten H. and Hoffmann U. 1996, Three-phase reactor model for Hydrotreatment in Pilot trickle-bed reactors. AIChEJ, vol. 42, no. 5, 1350-1360.

Stryjek R. and Vera J. H. 1986, PRSV: An Improved Peng-Robinson Equation of State for Pure Compunds and Mixtures. The Can. J. Chem. Eng., vol. 64, 323-333.

Schulz H., Böhringer W., Waller P., Ousmanov F. 1999, Refractory sulfur compounds in gas oil. Fuel Processing Technology, 61, 5-41.

Sie S. T. 1999, Reaction order and role of hydrogen sulfide in deep hydrodesulfurization of gasoil: consequences for industrial reactor configurations. Fuel Processing Technology, 61, 149-177.

Taylor R. and Krishna R.. 1993, Multicomponent Mass Transfer. John Wiley, New York.

Van Hasselt B.W., Lebens P.J.M., Calis H.P.A., Kapteijn F., Sie S.T., Moulijn J.A., Van den Bleek C.M.. 1999, A numerical comparison of alternative three-phase reactors with a conventional trickle-bed reactor. The advantages of countercurrent flow for hydrodesulfurization. Chem. Eng. Science, 54, 4791-4799.

Van Hasselt B. W., Lindenbergh D. J., Calis H. P., Sie S. T. and Van Den Bleek C. M.. 1997, The three-level-of-porosity reactor. A novel reactor for countercurrent trickle-flow processes. Chem. Eng. Science, vol. 52, no. 21/22, 3901-3907.

Vanrysselberghe V. and Froment G. F., 1996, Hydrodesulfurization of Dibenzotiophene on a CoMo/Al$_2$O$_3$ Catalyst: Reaction Network and Kinetics. Ind. Eng. Chem. Res, 35, 3311-3318.

Vrinat M.L., 1983, The kinetics of the Hydrodesulfurization process. A review. Applied Catalysis, 6, 137-158.

Xiaoliang Ma, Kinya Sakanishi and Isao Mochida. 1994, Three-stage deep hydrodesulfurization and decolorization of diesel fuel with CoMo and NiMo catalyst at relatively low pressure. Fuel , vol. 73, no. 10, 1667-1671.

European Symposium on Computer Aided Process Engineering – 12
J. Grievink and J. van Schijndel (Editors)

Mixed-logic Dynamic Optimization Applied to Configuration and Sequencing of Batch Distillation Processes

J. Oldenburg[*], W. Marquardt[*,+], D. Heinz[**], D. Leineweber[**]

[*]Lehrstuhl für Prozesstechnik, RWTH Aachen, D-52056 Aachen, Germany.
[**]Bayer AG, D-51368 Leverkusen, Germany.

Abstract

This paper treats an approach for configuring equipment and sequencing of complex batch distillation processes based on mixed-integer dynamic optimization. We propose a logic-based modeling technique and an algorithm tailored to this class of problems and apply the approach to the design of a simple multistage batch distillation process.

1. Introduction

Batch distillation is frequently encountered in the specialty chemicals and pharmaceutical industries due to its high flexibility in design and operation. Methods for a simultaneous optimization of the design *and* the operational strategies have been proposed by several authors. Allgor and Barton (1997) and Sharif et al. (1998) address batch process design by formulating a *mixed-integer dynamic optimization* (MIDO) problem which incorporates both (time-)continuous and discrete decision variables in a dynamic *superstructure* model. The arising MIDO problems are solved by subdividing the original problem into two (ore more) subproblems which are, however, generated in different ways. In this work we show that the equipment configuration and sequencing problem of complex batch distillation can be formulated as a *mixed-logic dynamic optimization* (MLDO) problem in a rather natural way. We furthermore propose a tailored solution method to solve the arising MLDO problem which reveals to be efficient and robust as a preliminary case study indicates.

2. Problem statement

A batch process, being inherently transient, comprises various continuous time-varying degrees of freedom, such as the reflux ratio or heat duty of a batch distillation column. These *control* variables can be used to minimize a performance index specific to the batch process considered. The optimal trajectories of the control variables can be determined effectively using dynamic optimization techniques. If, however, the batch process structure or parts thereof are not predetermined, additional degrees of freedom can be used to further improve the process performance. These degrees of freedom are of a discrete nature since they usually represent the existence or non-existence of a batch process unit, a part thereof, or a connecting stream. Discrete decisions, which we here

---

[+] Corresponding author. E-mail address: marquardt@lfpt.rwth-aachen.de.

assume to be time-invariant, can be incorporated into the mathematical model by introducing binary variables $y \in \{0,1\}^{n_y}$ to form a dynamic superstructure model. Batch process units can be partly interconnected through streams in case of parallel operation in time, as e.g. a distillate accumulator. An optional stream between batch process units can easily be incorporated into the mathematical superstructure by an algebraic equation. Whereas stream connections are also known from continuously operated plants, batch processes are additionally often composed of a sequence of distinct operations. Therefore, the dynamic model is subdivided into n_s stages. The connection between process stages is expressed mathematically in terms of stage transition conditions.

The development of mathematical superstructure models is by no means a trivial task and the model formulation reveals a major impact on robustness and efficiency of the solution method. For these reasons, considerable effort has been spent in defining ways to build superstructure models that are generic to the extent possible and provide favorable properties in conjunction with especially designed numerical solution algorithms. Türkay and Grossmann (1996) employ disjunctive superstructure models for stationary problems arising in process synthesis of continuously operated plants. Inspired by these results, we transfer the basic ideas of disjunctive programming to the dynamic case and state the following (multistage) mixed-logic dynamic optimization problem:

$$\min_{u_k, p, Y} \quad \sum_{k=0}^{n_s-1} \Phi_k \left(z_k(t_{k+1}), p, t_{k+1} \right) + \sum_{i=1}^{n_Y} c^i \tag{1}$$

$$s.t. \quad f_k\left(\dot{z}_k, z_k, u_k, p, t \right) = 0, \quad t \in [t_k, t_{k+1}], \quad k \in K = \{0, ..., n_s - 1\}, \tag{2}$$

$$g_k\left(\dot{z}_k, z_k, u_k, p, t \right) \leq 0, \quad t \in [t_k, t_{k+1}], \quad k \in K = \{0, ..., n_s - 1\}, \tag{3}$$

$$l\left(z_0(t_0), p \right) = 0, \quad e_k\left(z_k(t_{k+1}), p \right), \quad k \in K = \{0, ..., n_s - 1\}, \tag{4}$$

$$z_{k+1}(t_{k+1}) - m_k\left(z_k(t_{k+1}), p \right) = 0, \quad k \in K = \{0, ..., n_s - 2\}, \tag{5}$$

$$
\begin{bmatrix}
Y_i \\
r_k^i\left(\dot{z}_k, z_k, u_k, p, t \right) \leq 0, \\
t \in [t_k, t_{k+1}], \quad k \in K = \{0, ..., n_s - 1\}, \\
s_k^i\left(z_k(t_{k+1}), p \right) = 0, \quad k \in K = \{0, ..., n_s - 1\}, \\
z_{k+1}(t_{k+1}) - v_k^i\left(z_k(t_{k+1}), p \right) = 0, \\
k \in K = \{0, ..., n_s - 2\}, \quad c^i = \gamma_i,
\end{bmatrix}
\vee
\begin{bmatrix}
\neg Y_i \\
\\
B^i [u_k^T, p^T]^T = 0, \\
\\
k \in K = \{0, ..., n_s - 1\}, \\
\\
c^i = 0,
\end{bmatrix}
\tag{6}
$$

$$\Omega(Y) = True. \tag{7}$$

We here assume that the number of process stages n_s is known a priori. $z_k \in \mathfrak{R}^{n_{zk}}$ denote differential and algebraic state variables. Time-invariant parameters and Boolean variables are represented by $p \in \mathfrak{R}^{n_p}$ and $Y \in \{True, False\}^{n_Y}$, respectively, while control variables are denoted by $u_k \in \mathfrak{R}^{n_{uk}}$. $f_k \in \mathfrak{R}^{n_{fk}}$ represents the set of differential-algebraic equations (DAEs) with a differential index of at most 1 for which consistent initial conditions are given in (4). Stage transition conditions (5) are used to map differential state variables values across the stage boundaries. Inequalities and equalities comprised in g_k and e_k are used to enforce path or end point constraints, respectively. $\Phi_k \in \Re$ denotes the objective function. Whereas Eqs. (1)–(5) hold globally, there are further inequalities, equalities and stage transition conditions included in disjunctions (6) that are only enforced if the corresponding Boolean variable Y_i is True. Otherwise, for each $Y_i = False$, these constraints are not enforced and a subset of the inputs and time-invariant parameters as well as the fixed cost contributions c^i may be set to zero. Without loss of generality, we only regard two-term disjunctions as stated in (6). Commonly, the Boolean variables themselves are constrained by logical conditions (7).

3. Logic-based solution approach

There are several options to solve the optimization problem (1)–(7). Solution methods reported in literature (e.g. Allgor and Barton, 1997, Sharif et al., 1998) rely on the reformulation of the mixed-logic into a mixed-integer dynamic optimization (MIDO) problem. This is achieved by replacing the Boolean variables with binary variables and by representing the disjunctions either by means of Big-M constraints or a convex-hull formulation. These two reformulation techniques are well known from mixed-integer nonlinear programming (MINLP) and process synthesis of continuously operated plants. A MIDO problem can be transformed into an MINLP problem by discretization of the control and state variables or the control variables only. The arising MINLP problems are often solved using decomposition strategies, such as e.g. Outer Approximation (OA) or Generalized Benders Decomposition (GBD) (Floudas, 1995). These algorithms rely on the decomposition of the original problem into a continuous nonlinear programming (NLP) and mixed-integer linear programming (MILP) problem, which are termed primal and master subproblems.

We employ a logic-based decomposition strategy to directly solve the MLDO problem (1)–(7) rather than applying a MIDO solution algorithm to solve a reformulated problem. Therefore, the primal problem is directly defined by Eqs. (1)–(7) with fixed Boolean variables Y. It forms an ordinary dynamic optimization problems including the constraints in those disjunctions for which the corresponding Boolean variable is $True$. Control variables and parameters set to zero in disjunctions where $Y_i = False$ can be removed from the primal problem, a fact that helps to improve its efficiency to a significant extent. Moreover, nonlinear constraints r_k^i can simply be neglected if $Y_i = False$.

An increased efficiency of the primal problems is especially important when considering that the solution to dynamic optimization problems involving large-scale process models is a computationally demanding task whereas, in contrast, the solution time for the master problems can be neglected.

In order to construct the master problems, each disjunction, for which $Y_i = True$, is linearized at the optimal solution of the primal problem with respect to the discretized degrees of freedom $\bar{u}_k$ and p. In fact, considering disjunctions (6) for which $Y_i = True$ ensures that linearizations for the master problems are generated for "active" process units around representative operating conditions only. We therefore avoid the accumulation of process information at physically irrelevant operating conditions and the corresponding non-convex constraints. Hence, the resulting linear disjunctive program is transformed into a mixed-integer linear programming (MILP) problem by a reformulation using Big-M constraints (Yeomans and Grossmann, 2000). MILP problems can be solved very efficiently using solution methods based on branch & bound (CPLEX, 1999). To ensure that the first master problem of the proposed algorithm contains sufficient information about the process superstructure, all nonlinear terms contained in disjunctions have to be linearized at least once. This is achieved by solving a number of initializing primal problems with fixed Boolean variables and by accumulating the corresponding linearizations in the first master problem. The minimum number of initializing primal problems to be solved can be determined from the solution of a set covering problem (Türkay and Grossmann, 1996).

4. Illustrative example problem

The example presented here is taken from the patent literature (Patent, 1997). The task is to separate a ternary mixture consisting of 100 kg n-Hexane, 700 kg n-Heptane and 200 kg n-Octane into components with a predefined purity of at least $\varphi_i = 0.99\ kg_i/kg_{total}$ (or $\chi_i = \varphi_i M_i/M_{total}$ in terms of mole fractions) with minimal energy demand in a sequence of two batch distillation phases. The multistage batch distillation process is operated in one single batch column that can be operated either inversely or regularly in each of the two phases. The column consists of 10 trays plus reboiler and condenser with a constant pressure drop of 1 mbar per tray and a constant reboiler duty of 50 kW. Since each batch distillation stage can be operated either inversely or regularly and either the residue or distillate of the first stage can be fed to the second we eventually have a total number of 8 structural alternatives which can be used together with the time-varying reflux ratios of the batch stages to minimize the time (which is proportional to energy demand) required for the entire batch process. The dynamic process model consists of two main parts: disjunctions as stated in (6) and model equations that hold independently of any discrete decision, i.e. a dynamic batch column model for stages S_1 and S_2 based on common simplifying assumptions. Due to space limitations we cannot state the complete model in this paper and exemplarily focus on two characteristic disjunctions instead. The first disjunction represents the discrete decision whether the columns are operated inversely or regularly. This is modeled by a direct relationship between the Boolean variable Y_1 and the top and bottom streams D^{S_1}, B^{S_1} as well as the top and bottom holdups at t_{0,S_1}:

$$
\begin{bmatrix}
Y_1 \\
D^{S_1} \leq D_{UB}^{S_1},\quad B^{S_1} = 0\ kmol/h, \\
H_B^{S_1}(t_{0,S_1}) = 9.897\ kmol, \\
H_C^{S_1}(t_{0,S_1}) = 0.01\ kmol,
\end{bmatrix}
\vee
\begin{bmatrix}
\neg Y_1 \\
B^{S_1} \leq B_{UB}^{S_1},\quad D^{S_1} = 0\ kmol/h, \\
H_B^{S_1}(t_{0,S_1}) = 0.01\ kmol, \\
H_C^{S_1}(t_{0,S_1}) = 9.897\ kmol.
\end{bmatrix}
$$

Another 4 two-term disjunctions are required to model the decision which component is to be removed during the first stage and which intermediate product is fed to the second stage. As an example, we state a disjunction in which the bottom residue is considered as product and the distillate accumulated at the top is fed to the second batch column:

$$
\begin{bmatrix}
Y_2 \\
x_{B,3}^{S_1}(t_{f,S_1}) \geq \chi_3, \\
F^{S_2} = Dist_C^{S_1}(t_{f,S_1}), \; x_{B,i}^{S_2}(t_{0,S_2}) = x_{C,i,acc.}^{S_1}(t_{f,S_1}),
\end{bmatrix}
\vee
\begin{bmatrix}
\neg Y_2 \\
x_{B,3}^{S_1}(t_{f,S_1}) \leq 1.0, \\
F^{S_2} \leq F_{UB}^{S_1}, \; x_{B,i}^{S_2}(t_{0,S_2}) \leq 1.0.
\end{bmatrix}
$$

Similar disjunctions are stated for the second batch stage. Altogether, the dynamic superstructure model contains 4 time-varying continuous degrees of freedom, i.e. D^{S_1}, B^{S_1}, D^{S_2}, B^{S_2} and 7 free initial values of the second batch stage. Furthermore, the problem comprises 10 discrete variables in terms of the Boolean variables which are, however, partly related to each other by logic constraints (7) in the form of propositional logic.

This MLDO is solved using the logic-based solution approach discussed in the previous section and a standard MIDO solution technique. With only 8 discrete alternatives one would usually prefer to solve the mixed-logic dynamic optimization problem by simple enumeration rather than spending the effort required when applying a sophisticated solution technique. We, nevertheless, apply the proposed MLDO technique to the simple problem to study the properties of the method. The proposed approach is compared to a MIDO solution algorithm based on Outer Approximation with the Augmented Penalty extension proposed by Viswanathan and Grossmann (1990) to account for the non-convexity of the problem under consideration.

The optimal solution is found in the fourth iteration (Table 1). In this configuration, n-Octane is accumulated in the bottom residue of stage 1 in a regular mode of operation. The second stage is operated inversely with the distillate of stage 1 charged at the top. The termination criterion for the non-convex problem is based on comparing the objective values of the primal subproblems of two subsequent major iterations, i.e. the algorithm is terminated when the primal solution values increase from one iterate to another (primal problems **6**: 21.36 h, **8**: 11.66 h). In order to ensure that the master subproblems contain sufficient information about the process superstructure in terms of accumulated linearizations, we impose a lower bound on the number of major iterations $L_{min} = 3$.

Table 1: Results obtained with MIDO (left) and MLDO algorithms (right).

Solut. method	MIDO				MLDO				
Iteration	1	2	3	4	0	1	2	3	4
Primal					Init.				
Structure	1	3	2	4	1,4,6,7	1	3	2	4
Batch time [h]	12.31	22.89	12.12	11.63	-	12.31	22.89	12.12	11.63
Master									
Structure	3	2	4	6	1	3	2	4	8
Batch time [h]	10.10	11.12	18.75	16.49	22.28	22.88	22.31	20.24	24.17
# Act. slacks	-	5	20	30	3	3	2	4	2

Although the saving of 40 min. in terms of batch time for the overall process is relatively small compared to the standard process structure **1** (two successive batches with regular operation, bottom residue of S_1 is fed to S_2), significant differences to some of the other design alternatives can be identified, such as structure **7** (inverse operation of S_1, regular operation of S_2, top residue of S_1 is fed to S_2), with a difference of 11.89 hours. The same result is obtained with the proposed algorithm for which 4 primal sub-problems have to be solved as initialization. Compared to 8 alternatives in total, this is a large number. This ratio will, however, decrease for problems with a higher combinatorial complexity. The number of active slack variables used for the augmented penalty terms is a measure for the non-convexity of the problem constraints. Obviously, the non-convexity plays a less dominant role for the MLDO method. This is explained by the fact that linearizations are generated for disjunctions only where $Y_i = True$.

5. Conclusions and future directions

In this contribution we address equipment configuration and sequencing of batch distillation processes by formulating a mixed-logic dynamic optimization problem. We furthermore propose a tailored solution approach for solving MLDO problems which is based on a sequential solution strategy and a logic-based decomposition strategy. On the one hand, the sequential approach enables us to use detailed dynamic models of (multistage) batch distillation processes for equipment configuration and sequencing. On the other hand, a logic-based dynamic superstructure model and a tailored solution method based on a decomposition strategy improves both the efficiency and robustness of the overall algorithm. A simple design problem of a multistage batch distillation process is used to illustrate the modeling technique and the solution. Future work will focus on investigating design problems involving more discrete alternatives and more complex dynamic process models as well as on a refinement of the optimization algorithm.

Acknowledgements
Financial support from Bayer AG, Leverkusen is gratefully acknowledged.

References
Allgor, R.J. and P.I. Barton, 1997, Mixed-integer dynamic optimization. Comput. Chem. Eng., S21, 451 – 456.

CPLEX, 1999, User's Manual, ILOG, France.

Floudas, C.A., 1995, Nonlinear and mixed-integer optimization, Oxford Univ. Press NY

Patent, 1997, Verfahren zur Durchführung von destillativen Trennungen in diskontinuierlicher Betriebsweise, Europäisches Patentamt, Patentblatt 1997/46.

Sharif, M., N. Shah and C.C. Pantelides, 1998, On the design of multicomponent batch distillation columns. Comput. Chem. Eng., S22, 69 – 76.

Türkay, M. and I.E. Grossmann, 1996, Logic-based MINLP algorithms for the optimal synthesis of process networks. Comput. Chem. Eng., 20, 959 – 978.

Viswanathan, J. and I.E. Grossmann, 1990, A combined penalty function and outer-approximation method for MINLP optimization, Comput. Chem. Eng., 14,769 –782.

Yeomans, H. and I.E Grossmann, 2000, Disjunctive programming models for the optimal design of distillation columns and separation sequences. Ind. Eng. Chem. Res., 39, 1637 – 1648.

European Symposium on Computer Aided Process Engineering – 12
J. Grievink and J. van Schijndel (Editors)

A General Framework for the Synthesis and Operational Design of Batch Processes

Irene Papaeconomou, Rafiqul Gani, Sten Bay Jørgensen
Computer Aided Process Engineering Center
Department of Chemical Engineering
Technical University of Denmark
DK-2800, Lyngby, Denmark

Abstract

The objective of this paper is to present a general problem formulation and a general methodology for the synthesis of batch operations and the operational design of individual batch processes, such as mixing, reaction and separation. The general methodology described supplies the batch routes, which is the sequence of batch operations performed in order to achieve a specific objective. Important features of the methodology are a set of rule-based algorithms that provide the operational model of the units. Such an algorithm is highlighted, together with the associated rules, for the operational design of batch reactors. A case study involving the feasible operation of a batch reactor with multiple desirable and undesirable reactions and operational constraints is presented. Application results including verification of the generated operational sequences (alternatives) through dynamic simulation are presented.

1. Introduction

A batch operation can be characterized in terms of the task (reaction, separation, mixing or any combination) that needs to be performed for a period of time. A batch route is a sequence of batch operations performed to achieve a specific objective. The synthesis of batch operations, therefore, entails the identification of the necessary tasks and their sequence. Most algorithms dealing with batch process development, however, start with a known (specified) batch route. An important question that has not been addressed yet is, how does one generate a set of batch routes from which the selection can be made? Also, how does one find the optimal batch route? In order to address these questions, it is necessary to define the tasks in terms of sub-tasks such as heating, cooling, charging and so on. What one needs to keep in mind is that the final state of one task will be limited by certain constraints, in order to remain feasible and is closely related to the desirable starting state of the next task. It is also essential to identify the impact of the individual sub-tasks on the process. This is done in many cases, by identifying the relationship between properties and operational principles, which also helps to define the operational constraints and corresponding actions.

The objective of this paper is to develop a general, systematic framework that will determine near optimal operational routes, with respect to safety and productivity, and feasible alternatives from thermodynamic and operational insights. The feasibility of an operational sequence is defined through state variables or supplementary variables that do not violate the constraints that are imposed on them. The methodology also addresses operational issues related to sub-task "start" and sub-task "end". Rule-based algorithms, which allow the generation of a set of feasible sequences of operations and cover different types of batch operations, have been developed. In these algorithms the decisions to be taken in order to achieve a feasible (or near optimal) task performance are limited and perceived through thermodynamic insights, the constitutive constraints of the system and the operational constraints imposed by the process. The algorithms can tackle the problem in batch processing related to variation in task performance, which has received little attention, as reported by Rippin (1993).

The operational sequence information is generated in terms of commands and actions such that they can easily be interpreted by a dynamic simulation engine or for actual operation of the equipment The completed methodology provides the operational route (model) within a batch process and the route from one process to the other. This information is important for planning and scheduling, since it actually provides the sequence of tasks and the corresponding batch times along with the operational model.

2. Problem formulation

The objective for the synthesis of batch operational sequences is to minimise the operating time and/or cost of operation. Some may be interested more in the optimal time, while others in the minimisation of energy. In general, an optimisation problem is formulated as described below.

Minimise $F_{obj}(x, y^*, u, p)$ (1)

where,   x: state variables

       y^*:   supplementary variables, which are given through a relation between state variables and/or other supplementary variables, $y^* = f(x, y^*)$

       u: optimisation variables, which can be the manipulated variables from a control point of view (control optimisation) or the design variables from a design point of view (process optimisation).

       p: parameters that are specified.

Minimisation of the objective function is subject to:

The process model $dx/dt = f(x, y^*, u, p, ...)$ (2)

and the model (equality) constraints $f_1(x, y^*) = 0$ (3)

Also, the operational constraints: $g_1(x) \leq 0$ (4)

                                    $g_2(y^*) \leq 0$ (5)

Therefore, given the identity of the compounds present in the system, their initial state, the identity of the products and their specifications, the thermodynamic models of the system and the reactions taking place together with their kinetic data, the objective is to determine a set of feasible operational sequences. By satisfying only the constraints, feasible alternatives can be generated while ordering these alternatives in terms of their calculated F_{obj} values, the optimal (at least local) can be identified.

Dynamic simulation is used to analyse and validate the algorithms. The key point here is to identify the relationships between physical and chemical properties and reaction, separation and mixing principles. The process design, the safety/hazard risk and the environmental impact of the processes are included as a set of constraints g_1 and g_2 for the problem, which are used for the rejection of infeasible operational routes.

3. Methodology

The general methodology consists of a set of rule-based algorithms and of a set of rules for the synthesis and operational design of batch processes. There are interconnecting rules to direct the user from one task to another (or from one algorithm to another). The individual algorithms include modelling the operation in each task (operational model of the unit) and generate the feasible operational sequence for the unit, while the general framework rules determine the overall batch route (synthesis). The interconnecting rules between the different sub-algorithms are represented in Figure 1 with the help of horizontal double arrows. According to these rules, decisions are taken for the sequencing of tasks, based on thermodynamic insights, the presence of separation boundaries, the identity of intermediates, etc. The end boxes (labelled as "one phase" or "crystallization") symbolise the sub-algorithms for each batch operation. For illustrative purposes, the algorithmic steps related to the path starting from "batch operation" and ending with "one phase" is highlighted in Figure 2 for a batch reactor.

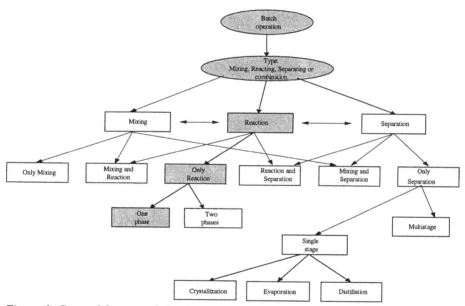

Figure 1. General framework for the synthesis of feasible operational sequences of batch operations

3.1 Algorithm

The generation of a feasible operational sequence for a batch reactor with multiple reactions is illustrated in Figure 2. The feasibility of the operation is defined through constraints in reaction selectivity and specified operational limits on temperature or pressure. Note that the selectivity S_{ij} of a reaction i over a competing reaction j is defined as r_i/r_j where r_i and r_j are the reaction rates of reactions i and j, respectively.

292

The algorithm consists of 5 main steps with step 1 being the definition of the problem. In step 2, dynamic simulations are performed for various types of operation (isothermal, adiabatic, etc.) until specified end criteria are reached. The purpose of this step is to generate operational data, in case they are not available. In step 3, the feasible operations (from step 2), if any, are identified. Alternatively or if all the simulated options are infeasible, either at any point or only at the end, the algorithm moves to step 4 where the intention is then to start the generation of a sequence of feasible sub-tasks (for corresponding periods of time). In step 4, the operation for period 1 is selected according to rule 1, among the simulated options from step 2, as the one where $S_{function}$ (figure 2) is highest and no constraints on T or P are violated. Step 5 is employed to identify the end of period 1 (when one of the constraints is violated) and the action that needs to be taken for the next period, which corresponds to a different operation. This step is repeated until the end criterion is satisfied. The rules 2-5 in the step are checks of constraints, namely the lower and upper temperature limits T_{min} and T_{max}, respectively, and the approximation of the selectivity at the end criterion reached time S'_{end}. If a violation of a constraint is detected by any of the rules, a corresponding action is taken which defines the next sub-task in the sequence. The new sub-task is assigned in terms of actions that a dynamic simulation engine can easily interpret (heating, cooling, etc.). For example, if the S'_{end} check is violated then the action to be taken is heating if the reaction of interest is more endothermic, endothermic or less exothermic compared to its competing reaction. A detailed description of the rules and actions can be obtained from the author. The operational sequence and the corresponding actions are obtained from the order of the sub-tasks determined by the rules by repeating step 5.

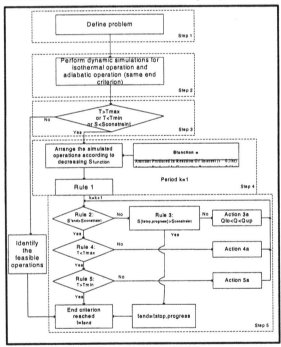

Figure 2. Algorithm for the generation of feasible operational sequences for a batch reactor

4. Results – A case study

This case study has been adapted from a paper by Allgor et al. (1996). A set of reactions is taking place in a batch reactor. Some of them are desired, since they lead to the product, while others are competing side reactions, which are unwanted:

Reaction 1: (R1) + (R2) → (I1)

Reaction 2: (R1) + (I1)→ (A)

Reaction 3: 2 (I1)→ (I2) + (B) + (H$_2$)

Reaction 4: (I2) + (I1)←→ (C)

The product is I2 and since it is produced from I1, we are very interested to promote reactions 1 and 3. This means that we are interested that reactant R1 is consumed in reaction 1 instead of 2 and also that reaction 2 is minimized, since I1 is consumed there to a byproduct. Therefore, the selectivity S_{12} must be high and above a certain value $S_{constraint}$. For the same reasons, it is desired that $S_{34} > S_{constraint}$. The value of selectivity constraint $S_{constraint}$, can be chosen arbitrarily but a high value is obviously desirable [for the case problem: $S_{constraint} = 15$]. Another operational constraint is the reactor temperature [$300K = T_{min} < T < T_{max} = 360K$].

In this paper, a brief description of the results is given. Applying the algorithm described above, seven feasible alternative operational sequences have been generated by applying in sub-task 2 different amounts of heat from a range of available heat input. Together with the generated sequences, the corresponding information related to the number, type and sequence of sub-tasks, their period (time) of operation, etc., are also generated. This information is then used to verify the feasibility of the alternatives and to compute the performance criteria through dynamic simulation. The simulation engine of ICAS (Gani et al. 1997) and BRIC (ICAS user's Manual, 2001) has been employed for dynamic simulation. It can be noted from Figure 3 that all the generated alternatives are feasible since none of the constraints are violated. Note that at the starting point the mole fraction of R2 is 0.225 and at the end point, the mole fraction of R2 is close to zero for all alternative sequences. For each sequence, the total operational time is also listed in Figure 3.

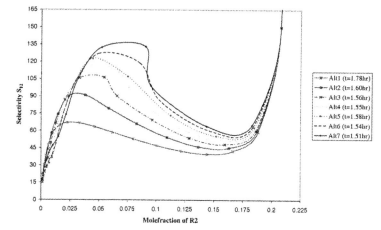

Figure 3: Simulated selectivity versus composition paths for alternative sequences

Figures 4a and 4b show the cumulative operational times for each alternative sequence and the correlation between the cost of operation and time of operation, respectively. Alternative 5 appears to be an exception and the exact cause needs to be further investigated.

294

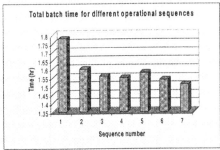

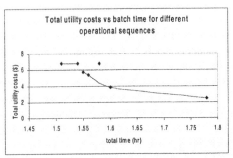

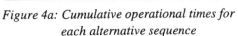

Figure 4a: Cumulative operational times for each alternative sequence

Figure 4b: Correlation of operational costs versus operational time

5. Conclusion

A general methodology that allows the generation of operational sequences/routes for batch operations has been developed. The operational design of batch processes has been tested for the special case of a batch reactor. All the alternatives generated by the algorithm have been verified through dynamic simulation and found to be feasible. Together with the generation of feasible alternatives, the algorithm also generates the necessary operational information that can be directly used for purposes of verification (experimental and/or simulation). The optimal sequence (at least local) of operations can be found among the generated feasible alternatives. Within the solution space considered in the case study, with respect to minimization of operational time (without taking into consideration the associated utility costs) alternative 7 appears to be the optimal, while with respect to minimization of utility costs (without taking into consideration the operational time) alternative 1 appear to be the optimal. However, if one's aim is to find the minimum time and the minimum utility costs, then since there is obviously a trade off between these competing objectives, a criterion needs to be established before an appropriate selection can be made. The methodology, the algorithm and the associated computer aided tools provide a systematic and integrated approach to the solution of problems involving batch operation sequence and verification. Current work is investigating batch operations involving mixing, reactions and separations in the same sequence in order to achieve a desired product.

References

Allgor R.J., M.D. Barrera, P.I. Barton and L.B. Evans, 1996, Optimal Batch Process Development. *Comp. & Chem. Eng.,* **20**, p.885.

Allgor R.J., L.B. Evans and P.I. Barton, 1999, Screening Models for Batch Process Development. Part I. Design Targets for Reaction/Distillation Networks. *Chem. Eng. Sci.,* **54**, p.4145.

Gani, R., G. Hytoft, C. Jaksland and A.K. Jensen, 1997, An integrated computer aided system for integrated design of chemical processes, *Comp. & Chem. Eng.,* **21**, p.1135.

ICAS Users manual, 2000, CAPEC report, CAPEC-DTU, Lyngby, Denmark.

Rippin D.W.T., 1993, Batch Process Systems Engineering: A Retrospective and Prospective Review. *Comp. & Chem. Eng.,* 17, p.S1.

Safe Process Plant Layout using Mathematical Programming

Dimitrios I. Patsiatzis and Lazaros G. Papageorgiou[1]

Centre for Process Systems Engineering, Department of Chemical Engineering, UCL (University College London), Torrington Place, London WC1E 7JE, U.K.

Abstract

This paper presents a general mathematical programming formulation, considering simultaneously process plant layout and safety. The proposed model determines the detailed process plant layout (coordinates and orientation of each equipment item), the number and type of protection devices in order to reduce possible accidents and the financial risk. The problem is formulated as a mixed integer non-linear programming (MINLP) model and its applicability is demonstrated by a literature example.

1. Introduction

The process plant layout problem involves decisions concerning the spatial allocation of equipment items and the required connections among them (Mecklenburgh, 1985). Increased competition has led contractors and chemical companies to consider these decisions during the design or retrofit of chemical plants. In general, the process plant layout problem may be characterised by a number of cost drivers such as connection, land area, construction costs, as well as management/engineering issues (for example production organisation). So far, safety aspects are considered in a rather simplified way by introducing constraints with respect to the minimum allowable distance between specific equipment items. However, it is evident that there is a need for considering safety aspects in more detail within process plant layout and design frameworks.

A number of different approaches have been considered for the process plant layout problem including graph partitioning for the allocation of units to sections (Jayakumar and Reklaitis, 1994) and a grid-based, mixed integer linear programming (MILP) model based on rectangular shapes (Georgiadis et al., 1997). Continuous domain MILP models have recently been suggested, determining simultaneously orientation and allocation of equipment items (Papageorgiou and Rotstein, 1998), utilising a piecewise-linear function representation for absolute value functionals (Ozyruth and Realff, 1999) and considering irregular equipment shapes and different equipment connectivity inputs and outputs (Barbosa-Povoa et al., 2001).

Particular attention to safety aspects of the process plant layout problem was given by (Penteado and Ciric, 1996) proposing an MINLP model. A heuristics method has been described by Fuchino et al. (1997) where the equipment modules are divided into subgroups and then sub-arranged within groups according to safety. Finally, the plant

---

[1] Author to whom correspondence should be addressed: Fax: +44 20 7383 2348; Phone + 44 20 7679 2563; email l.papageorgiou@ucl.ac.uk

layout problem utilising safety distances, has been solved efficiently by using genetic algorithms (Castel et al., 1998).

This work aims at extending previous continuous domain process plant layout models to include safety aspects using mathematical programming.

2. Problem Statement

In the formulation presented here, rectangular shapes are assumed for equipment items following current industrial practices. Rectilinear distances between the equipment items are used for a more realistic estimate of piping costs (Penteado and Ciric, 1996; Papageorgiou and Rotstein, 1998). Equipment items, which are allowed to rotate 90°, are assumed to be connected through their geometrical centres.

The single-floor process plant layout problem with safety aspects can be stated as follows:

Given:

- A set of equipment items (*i* or *j=1..N*) and their dimensions (α_i, β_i)
- Connectivity network
- Minimum safety distances between equipment items
- A collection of protection devices (*k=1..K*)
- Cost data (connection, equipment purchase, protection device purchase)
- A list of potential events at each unit
- An estimate of the probability of an accident propagating from one unit to the other as suggested in Penteado and Ciric (1996)

Determine:

The detailed layout (orientation, coordinates), the safety devices to be installed at each unit and the financial risk (expected losses if major accident happens)

So as to minimise the total plant layout cost.

3. Mathematical Formulation

3.1 Equipment orientation constraints

The length (l_i), and the depth (d_i) of equipment item *i* are determined by:

$$l_i = \alpha_i \cdot O_i + \beta_i \cdot (1 - O_i) \qquad \forall i \tag{1}$$

$$d_i = \alpha_i + \beta_i - l_i \qquad \forall i \tag{2}$$

where O_i is equal to *1* if $l_i = \alpha_i$; *0* otherwise.

3.2 Non-overlapping constraints

In order to avoid situations where two equipment items allocate the same physical location, non-overlapping constraints presented in Papageorgiou and Rotstein (1998) are included in the model. All non-overlapping constraints are written for *i=1..N-1*, *j=i+1..N*.

$$x_i - x_j + M \cdot (E1_{ij} + E2_{ij}) \geq \frac{l_i + l_j}{2} \tag{3}$$

$$x_j - x_i + M \cdot (1 - E1_{ij} + E2_{ij}) \geq \frac{l_i + l_j}{2} \tag{4}$$

$$y_i - y_j + M \cdot \left(1 + E1_{ij} - E2_{ij}\right) \geq \frac{d_i + d_j}{2} \tag{5}$$

$$y_j - y_i + M \cdot \left(2 - E1_{ij} - E2_{ij}\right) \geq \frac{d_i + d_j}{2} \tag{6}$$

where M is an appropriate upper bound, x_i, y_i. are the coordinates of the geometrical centre of item i and $E1_{ij}$, $E2_{ij}$ are binary variables (as used in Papageorgiou and Rotstein, 1998) required for the above disjunctions.

3.3 Distance constraints
The following distance constraints are included in the model in order to calculate the absolute distances between two equipment items in the x- and y- plane. All distance constraints are written for $i=1..N-1$, $j=i+1..N$.

$$R_{ij} - L_{ij} = x_i - x_j \tag{7}$$

$$A_{ij} - B_{ij} = y_i - y_j \tag{8}$$

$$R_{ij} \leq M \cdot W_{ij}^x \tag{9}$$

$$L_{ij} \leq M \cdot (1 - W_{ij}^x) \tag{10}$$

$$A_{ij} \leq M \cdot W_{ij}^y \tag{11}$$

$$B_{ij} \leq M \cdot (1 - W_{ij}^y) \tag{12}$$

where R_{ij} is the relative distance in x coordinates between items i and j if i is to the right of j, or L_{ij} if i is to the left of j. The relative distance in y coordinates between items i and j, is A_{ij} if i is above j or B_{ij} if i is below j. W_{ij}^x and W_{ij}^y are binary variables introduced to ensure that only one variable *at most* of each pair (R_{ij}, L_{ij}) and (A_{ij}, B_{ij}) is guaranteed to be non-zero. Thus, the rectilinear total distance, D_{ij}, between items i and j is given by:

$$D_{ij} = R_{ij} + L_{ij} + A_{ij} + B_{ij} \tag{13}$$

3.4 Additional layout constraints
Intersection with the origin of axis should be avoided:

$$x_i \geq \frac{l_i}{2} \quad \forall i \quad and \quad y_i \geq \frac{d_i}{2} \quad \forall i \tag{14}$$

3.5 Objective function
The objective function used is the minimisation of the total process plant layout cost associated with the connection, protection devices and financial risk:

$$min \sum_{i} \sum_{i \neq j, f_{ij}=1} C_{ij}^c \cdot D_{ij} + \sum_{i} \sum_{k} C_{ik}^{pd} \cdot Z_{jk} + \sum_{i} R_i^0 \cdot \prod_{k} (1 - RF_{ik} \cdot Z_{ik})$$
$$+ \sum_{j} \sum_{i \neq j} RR_{ij}^0 \cdot \prod_{k} (1 - RF_{jk} \cdot Z_{jk}) \qquad \text{(P)}$$

The connection cost is the first term in the objective function, where C_{ij}^c is the unit connection cost between units i and j and f_{ij} is a zero-one connection matrix.

The second term represents the cost of protection devices which are assumed to be installed only at the source of accidents. These devices can diminish the probability or the severity of accidents like relief valves or explosion suppression systems. C_{ik}^{pd} is the cost of protection device k installed at unit i which is the origin of the accident and Z_{jk} is a new binary variable to be determined (1 if a protection device is installed at unit i; 0 otherwise).

The financial risk is captured by the last two terms of the objective function. Similarly to the work presented in Penteado and Ciric (1996), the annual risk is expressed as a function of severity and probability of the accident and can be compared to the connection and protection devices cost (one-time costs) by computing the net present financial risk for a given lifetime of the plant and annual interest rate. In particular, the third term corresponds to the risk reduction, when a protection device k is installed at equipment item i, in the case of an accident at i. R_i^0 is the initial risk of accidents at item i without any protection device available and RF_{ik} is the risk reduction factor when protection device k is installed at item i. Note that for every protection device placed at unit i, the initial risk is reduced by a factor of $(1-RF_{ik})$. The last term is associated with the propagation of an accident from item j (origin) to unit i (target). RR_{ij}^0 is the financial risk related to accidents propagating from item j to item i *without* any protection device available at unit j:

$$RR_{ij}^0 = P_j \cdot e^{-\Phi_{ij}^1 (D_{ij} - D_{ij}^{min})} \cdot (\Phi_{ij}^2 \cdot (D_{ij} - D_{ij}^{min}) + \Phi_{ij}^3) \qquad \forall j, i \neq j$$

where P_j is the probability of an accident at unit j, D_{ij}^{min} is the minimum distance between units i and j and $\Phi_{ij}^1, \Phi_{ij}^2, \Phi_{ij}^3$ are parameters as described in Penteado and Ciric (1996) determined by applying the equivalent TNT method (Lees, 1980).

The overall problem is formulated as a non-convex MINLP model with non-linear objective function and linear constraints. Next, the applicability of this model is demonstrated by an illustrative example.

4. Example

Consider the ethylene oxide plant (see Figure 1) derived from the case study presented in Penteado and Ciric (1996) which considers three possible accidents: an explosion in the reactor (unit 1), the ethylene oxide absorber (unit 3) and the CO_2 absorber (unit 5). The probability of the accident to occur is 0.008 yr^{-1}, while the initial risk (R_i^0)

associated with the destruction of unit 1, 2 and 3 is $202000, $64400 and $49200, respectively. The available protection devices for installation at equipment items are additional cooling water (k-1), additional overpressure relief devices (k-2), additional fire relief devices (k-3), second skin on reactor (k-4), explosion protection system on reactor (k-5), duplicate control shutdown system on absorption towers (k-6) and duplicate control system with interlocking flow control on reactor (k-7). The cost and risk reduction factor for every protection device are shown in Table 1. The values for parameters Φ_{ij}^1, Φ_{ij}^3 are presented in Tables 2 and 3, respectively, while $\Phi_{ij}^2 = \Phi_{ij}^1 \cdot \Phi_{ij}^3$.
The connection cost is 98.4 ($/m).
The above MINLP example was modelled in the GAMS system (Brooke et al., 1998) using the DICOPT solver. Equipment dimensions and optimal solution details are given in Table 4. As expected, the high risk equipment items (units 1,3 and 5) are located far enough from each other so as to minimise the propagation risk. The total plant layout cost is $244851 with the following breakdown: 6.8% for connection, 20.4% for protection devices, 71.8% for financial risk from individual units and 1% for propagation financial risk cost. The protection devices selected are k-1, k-3 and k-7 for

Table 1. Protection device data

Device	Cost			RF_{ik}		
	Unit 1	Unit 2	Unit 3	Unit 1	Unit 2	Unit 3
k-1	5000	5000	5000	0.10	0.10	0.10
k-2	30000	20000	20000	0.24	0.24	0.24
k-3	15000	25000	25000	0.25	0.25	0.25
k-4	65000			0.60		
k-5	20000			0.20		
k-6		30000	30000		0.32	0.32
k-7	20000			0.46		

Table 2. Parameter Φ_{ij}^1

Origin	Target						
	Unit 1	Unit 2	Unit 3	Unit 4	Unit 5	Unit 6	Unit 7
Unit 1		0.231	0.485	0.205	0.485	0.442	0.489
Unit 3	0.246	0.323		0.345	0.286	0.314	0.223
Unit 5	0.273	0.327	0.225	0.354		0.470	0.279

Table 3. Parameter Φ_{ij}^3

Origin	Target						
	Unit 1	Unit 2	Unit 3	Unit 4	Unit 5	Unit 6	Unit 7
Unit 1		82984	807528	30176	613480	37720	2828
Unit 3	16427020	414920		181056	4601084	377200	113160
Unit 5	16427020	539396	6054060	196144		377200	113160

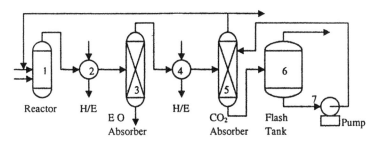

Reactor   H/E   H/E

E O CO_2 Flash   Pump

Absorber   Absorber   Tank

Figure1. Flowsheet for ethylene oxide plant

Table 4. Equipment item dimensions and optimal solutions

| Unit | Dimensions | | Optimal solution | | | |
| | | | Device | | No device | |
	α_i (m)	β_i (m)	x_i (m)	y_i (m)	x_i (m)	y_i (m)
1	5.22	5.22	15.66	2.61	21.06	2.61
2	11.42	11.42	5.71	18.82	21.06	10.93
3	7.68	7.68	22.11	34.72	4.24	22.47
4	8.48	8.48	15.66	18.82	4.24	42.74
5	7.68	7.68	39.62	13.68	21.06	42.74
6	2.60	2.60	32.11	18.82	9.78	47.88
7	2.40	2.40	34.61	18.82	4.24	50.38

unit 1 and *k-1* for both units 3 and 5 which are actually the same as the ones selected by Penteado and Ciric (1996). Table 4 also shows the optimal solution for the case where no protection devices can be installed with a total plant layout cost of $334289.

5. Concluding Remarks

In this work, an MINLP model has been proposed for the single floor process plant layout problem with safety considerations. The model determines simultaneously the optimal location (*i.e.* coordinates and orientation), the number and type of equipment protection devices in order to reduce possible accidents and the financial risk associated with individual units and propagation. Although the above model considers various trade-offs, there are a number of issues such as plant area and construction costs, flexibility on maintenance as well as use of piperacks that can further be investigated.

Current work focuses on enhancing solution efficiency of the resulting mathematical model by investigating model reformulation and approximation techniques.

6. References

Mecklenburgh J.C., 1985, Process Plant Layout. Institution of Chemical Engineers, London.

Jayakumar S. and G.V. Reklaitis, 1994, Comp. Chem. Eng. 18, 441.

Georgiadis M.C., G.E. Rotstein, and S. Macchietto, 1997, Ind. Eng. Chem. Res. 36, 4852.

Papageorgiou L.G. and G.E. Rotstein, 1998, Ind. Eng. Chem. Res. 37, 3631.

Ozyruth D.B. and M. J. Realff, 1999, AICHE J. 45, 2161.

Barbosa-Povoa A.P., R. Mateus and A.Q. Novais, 2001, Int. J. Prod. Res. 39, 2567.

Penteado F.D. and A.R. Ciric, 1996, Ind. Eng. Chem. Res. 35, 1354.

Fuchino T., T. Itoh and M. Muraki, 1997, J Chem. Eng. Jpn. 30, 896.

Castel C.M.L., R. Lakshmanan, J.M. Skilling and R. Banares-Alcantara, 1998, Comp. Chem. Eng. S22, S993.

Lees F. P., 1980, Loss Prevention Industries. Butterworths, Bsoston.

Brooke A., D. Kendrick, A. Meeraus and R. Ramam, 1998, GAMS: A User's Guide. The Scientific Press.

European Symposium on Computer Aided Process Engineering – 12
J. Grievink and J. van Schijndel (Editors)
© 2002 Published by Elsevier Science B.V.

Design of a Reactive Distillation Process For Ultra-Low Sulfur Diesel Production

Eduardo S. Perez-Cisneros[a,c]* , Salvador A. Granados-Aguilar[b], Pedro Huitzil-Melendez[c] and Tomas Viveros-Garcia[a]

(a) Universidad Autonoma Metropolitana-Iztapalapa.
Departamento de Ingenieria de Procesos e Hidraulica UAM-Iztapalapa.
Avenida San Rafael Atlixco 186, Iztapalapa, 09340, Mexico D. F.
Tel: 52 58.04.46.44   Fax: 52 58.04.49.00 E-mail: espc@xanum.uam.mx
(b) Facultad de Quimica UNAM. Ciudad Universitaria, México, D.F.
(c) Instituto Mexicano del Petróleo. Eje Central Lázaro Cárdenas 152. México, D.F.

Abstract

The analysis of the applicability of the reactive separation concepts to the deep-hydrodesulfurization process has been performed. Through the computation of the reactive residue curve maps a basic conceptual design of a reactive distillation column was obtained. This preliminary design considers two reactive zones, each one packed with a different catalyst, a Ni-Mo based catalyst for the bottom reactive section and a Co-Mo for the top reactive section. Six different column configurations were validated through numerical simulations. The results show that the complete elimination of the organo-sulfur compounds from a HC mixture is possible with appropriate design specifications for the reactive distillation column, i.e., reflux ratio, feed location and condenser duty.

1. Introduction

The worldwide regulations concerned to improve the quality of diesel transport fuel are increasing the pressure on the oil industry. The European Union agreed on specifications for the future quality of diesel. These specifications consider that a maximum permissible sulfur content of diesel will be 350 wppm from the year 2000, and 50 wppm from the year 2005. This required reduction is promoting changes in the oil refineries in terms of modifying the catalyst used and/or in the technology involved in the hydrodesulfurization process. That is, a higher activity of the commercial catalyst, and structural changes in the reactor configuration to increase the sulfur-compounds conversion, are needed.

Recently, Van Hasselt et al. (1999) have pointed out that a countercurrent operation of a trickle-bed reactor could lead to a higher reduction of sulfur content than the conventional cocurrent operation. They also show that the flooding limits (i.e., commercial flow rates) are attained by using a finned monolith catalyst packing. The analysis of these two characteristics suggests that reactive distillation could be an interesting technological alternative for the deep hydrodesulfurization of diesel. In a reactive distillation process the countercurrent operation is the natural operation mode and the flooding requirement can be obtained through, both, the catalyst packing arrangement or by manipulating the reflux and/or boiling ratio variable.

The objective of the present work is to analyze the applicability of the reactive separation concepts to the hydrodesulfurization process and based on this, to develop a conceptual design and perform numerical simulations of a reactive distillation process for the production of ultra-low sulfur diesel.

2. The Hydrodesulfurization Process and The *Element* Approach

For the analysis of the reactive system, the desulfurization of the various species is lumped into one single reaction of sulfur to H_2S:

$$v_A Org - S(1) + v_B H_2(2) \longrightarrow v_C Org - H_2(3) + v_D H_2 S(4) \tag{1}$$

Where the *Org-S* components represent the benzothiophene series compounds, with $v_A=1$, $v_B=1$, $v_C=1$, $v_D=1$. The graphical representation of the reactive system can be done using the *element* composition concept proposed by Perez-Cisneros et al. (1997). If we define the following *element* representation: $A=Org$, $B=S$ and $C=H_2$, the above reaction can be written in terms of elements, and specifically for the dibenzothiophene, as:

DBT (*AB*)+ H₂ (*C*)→ BiPh (*AC*)+ H₂S (*BC*) $\tag{2}$

3. The Reactive Residue Curve Map and The Conceptual Design

The computation of reactive residue curve maps at different operation pressures was performed. As an example, the residue curve map for the elimination of the dibenzothiophene is presented (see Figure 1). The reaction kinetics is the one obtained by Broderick and Gates (1981). The phase equilibrium calculations must be carried out carefully, because of the presence of H_2 and H_2S in the reacting mixture and a modified Peng-Robinson Equation was used (Stryjek and Vera, 1986). The equations used to compute the reactive residue curves are:

$$\frac{dx_i}{d\tau} = x_i - y_i + \frac{Da}{k_{ref}} \frac{kK_1K_2x_1x_2/\rho_L^2}{(1+K_2x_2/\rho_L)(1+K_1x_1/\rho_L+K_4x_4/\rho_L)^2} \tag{3}$$

Where

$$Da = \frac{M_{cat}}{V_0}k_{ref}; \quad d\tau = \frac{H_0}{V_0}dt \tag{4}$$

The reactive residue curve map diagram shows clearly three operation regions in terms of hydrogen concentration. One region of residue curves connect the Org-S - H_2S pure components stable nodes, indicating that this reaction-separation zone should be avoided in order to eliminate the H_2S compound from the liquid reacting mixture,

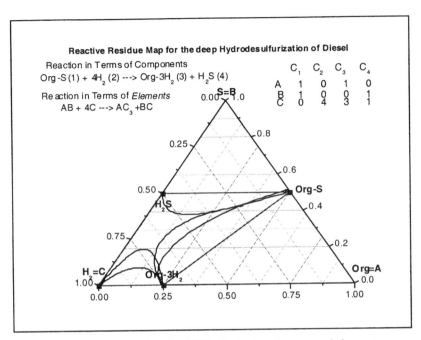

Fig. 1 *Reactive Residue curve Map for DBT elimination in terms of elements*

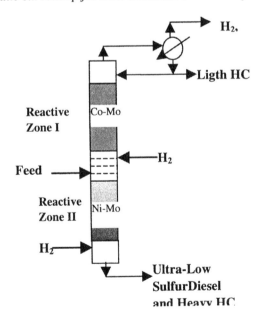

Fig. 2   *Conceptual basic design of a reactive distillation column for ultra-low sulfur diesel production*

considering that its presence reduces the activity of the catalyst. The other two regions of residue curves show a very interesting behavior of the reactive system. There are residue curves connecting the Org-S - Org-3H_2, and Org-3H_2 - H_2 stable nodes. These residue curves indicate that it could be possible to design a reactive distillation column with a reactive zone, where the elimination of sulfur from the organic compound is carried out simultaneously with the production of H_2S in the liquid phase and its subsequent vaporization, and another reactive zone where the most difficult sulfur compounds for hydrotreating (i.e., 4,6-dimethyl dibenzothiophene), could be handled with higher hydrogen concentrations. The analysis of these results has lead to a preliminary design of a reactive distillation column consisting of two reactive zones (see Figure 2), each one packed with a different catalyst, a Ni-Mo based catalyst for the bottom reactive section and a Co-Mo for the top reactive section.

4. Simulation Results and Discussion

After a basic conceptual design has been obtained, that is, a reactive distillation column comprising two reactive zones and non-reactive stages, six different column configurations were proposed and validated through numerical simulations. The mixture fed to the column considered was a paraffinic mixture with the following composition (mole fraction): DBT=0.0174022, C_{11}=0.4966322, C_{12}=0.31662451, C_{13}=0.10887454, C_{14}=0.00153543, C_{16}=0.05893112. The operation pressure of the reactive column is set at 30 atm and a H_2/HC feed relation of 3. The kinetic expression of Broderick and Gates (1981) was used. All configurations of the column considered 10 stages, with 6 reactive stages. Table 1 shows the details of the column configurations used for the simulations.

Table 1. Reactive distillation column configurations

	COLUMN 1	COLUMN 2	COLUMN 3	COLUMN 4	COLUMN 5	COLUMN 6
Reactive Stages (1)	3-4,6-8	3-4,6-8	3-4,6-8	3-4,6-8	3,5-8	4-8
Feed Stage)	5	5	5	5	4	3
Distillate Flow (kmol/sec)	320	330	330	350	330	330
Reflux Ratio	0.1	0.5	0.6	0.7	0.6	0.6
Top Temp. °C	200	200	200	226	200	200

(1) Stage 1 is a non reactive partial condenser, and stages 2, 9 and 10 are non-reactive stages.

In Figure 3 it can be observed a lower temperature profile for configuration 1. This is because the reflux ratio is 5 times lower than the other configurations, and less energy is required at the bottom of the column. The DBT conversion achieved with this column configuration was 85.3 %. Figure 4 shows the DBT composition profiles in the liquid phase along the reactive distillation column. It should be noticed that the higher composition corresponds to the feed stage and it decreases sharply through the reactive zones for configurations 2-6. However, for configuration 1, it can be observed that a complete elimination of DBT is not achieved, rather the DBT composition passes through a minimum at stage 8 and it increases at the non reactive stages. It is important to see that the column configuration 4 renders the complete elimination of DBT, since the DBT compositions at the ends of the reactive zones are close to zero. The column configurations 5 and 6 allow higher DBT concentration at the top of the column, while configurations 1, 2 and 3 allow higher DBT concentration at the bottom of the column.

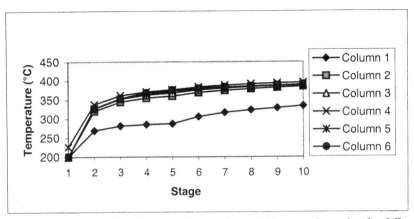

Fig. 3   Temperature profile through the reactive distillation column for the different column configurations

The conversion of DBT for the different column configurations was 85.3, 99.0, 99.2, 99.6, 99.5, 98.9, respectively. Figure 5 shows the H_2S liquid composition profile along the reactive column. It can be observed that for configurations 2-6 there is not great difference in the composition values, which suggest that in order to avoid catalyst activity inhibition, the operating conditions at the non-reactive stages, such as temperature could be increased or a vapor side stream could be located at the top reactive zone. Column configuration 1 shows higher concentration of H_2S because its lower temperature profile, reducing the vaporization of this component.

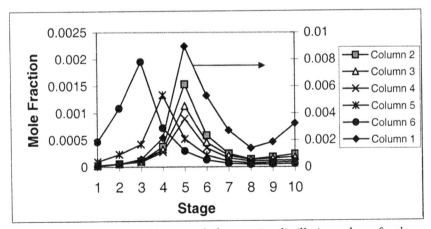

Fig. 4   DBT composition profile through the reactive distillation column for the different column configurations

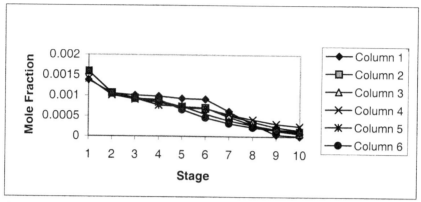

Fig. 5   H₂S composition profile in the liquid phase through the reactive distillation column for the different column configurations

5. Conclusions

An analysis in terms of reaction-separation feasibility of the deep hydrodesulfurization of diesel through reactive distillation has been performed. The basis of the analysis is the computation of reactive residue curve maps (*element* based) at different operation pressures. It was found that the deep-remotion of the difficult organo-sulfur compounds (i.e., DBT, 4-6 DMDBT) could be carried out at moderate pressures, i.e., 30 atm. Also, the analysis of the reactive residue curve map showed that it is possible to develop a basic conceptual design of a reactive distillation column for this purpose. Six different reactive distillation column configurations were validated through numerical simulations. The results show that the reflux ratio and feed location are very important variables to obtain optimal conversion of the organo-sulfur compounds in a specific configuration. Also, this conversion is high producing, in some cases, very low organo-sulfur compound exit streams (configuration 4). There is not doubt about the aplicability of reactive distillation to the deep-hydrodesulfurization process with important savings in hydrogen consumption and energy requirements.

6. References

Broderick D.H. and B.C. Gates (1981). Hydrogenolysis and Hydrogenation of Dibenzothiophene Catalyzed by Sulfide CoO-MoO₃/γ-Al₂O₃. The Reaction Kineticks. AIChE Journal, Vol 27, No. 4, pp. 663-672.

Perez-Cisneros E.S., R. Gani and M.L. Michelsen (1997). Reactive Separation Systems I. Computation of Physical and Chemical Equilibrium. Chem. Eng. Sci., Vol. 52, No. 4, pp. 527-543

Stryjek R. and J.H. Vera (1986), PRSV: An Improved Peng-Robinson Equation of State for Pure Compounds and Mixtures. The Canadian Lournal of Chemical Engineering, Vol. 64, April 1986, pp. 323-333

Van Hasselt B.W., P.J.M. Lebens, H.P.A. Calis, F. Kapteijn, S.T. Sie, J.A. Moulijn and C.M. van den Bleek (1999). A Numerical comparison of alternative three-phase reactors with a conventional trickle-bed reactor. The advantages of countercurrent flow for hydrodesulfurization. Chem. Eng. Sci. Vol. 54, pp. 4791-4799

European Symposium on Computer Aided Process Engineering – 12
J. Grievink and J. van Schijndel (Editors)

A Generalised Program for Fuel Cell Systems Simulation

S. Pierucci, D. Manca, S. Galli[1], M. Mangione[1]

CIIC Politecnico di Milano, Milano, Italy

[1] ENEA, Casaccia, Italy

Abstract

A computer program simulating the performance of a fuel cell system has been developed for the analysis of small-scale power plants and uotomotive technologies. FC2000 (Fuel Cells 2000) is based on a Sequential Modular Approach, is coded in Fortran and is user-interfaced by a preprocessor written in Visual Basic. All the system is compatible with a Window environment.

Models describing fuel cell equipments were either home developed or acquired from already existing codes desdcribed in literature (Pierucci et alt., 1985), while the preprocessor was specifically created for this application.

FC2000 has been succesfully applied to the study and design of transportation technologies and power-plant stationary systems.

The paper will describe FC2000 structure, will provide the content of specific fuel cell models and finally will discuss the results of a typical design where an innovative solution is proposed to increase process efficiencies.

1. Introduction

Considering the actual and estimated Natural Gas reserves, conversion of NG to hydrogen looks to be a way to introduce advanced electric generation technology (e.g. fuel cell) for automotive and stationary applications in a medium-term transition phase. Moreover environmental issues dictate the need of utilisation of non-fossil fuel sources and introduction of hydrogen as novel fuel. As the present cost of hydrogen from renewable energy sources is not competitive, at first it could be economically introduced via fossil fuel conversion.

In the framework of the Italian Scientific Research Ministry program aimed to promote R&D on solid polymer fuel cells for electric vehicles, ENEA and CIIC Department have been invited to carry out research activities with the aim to create a simulation program for fuel cell technology oriented to vehicle and small size stationary applications. A creation of a new program was compelled by contractual reasons with the Ministry.

The main objective was to create a versatile program capable to simulate typical fuel-cell based schemes which might provide prompt responses to production targets. Due to the need of utilising already developed software for the modelling of cells and partial oxidation equipments, the approach was necessarily oriented to a sequential modular one. Moreover both the difficulty of solving models based on ODE mixed with non linear algebraic equations and the need of providing a solid converging procedure capable to assist a wide range of potential users, confirmed the necessity to adopt SM approach although an EO might appear, in principle, more attractive.

As a resulting product, FC2000 demonstrates itself as a versatile, powerful and rigorous simulation program which can handle comprehensive fuel cell schemes in a good agreement with experimental data derived from pilot scale sites.

2. Models

FC2000 provides a convenient, flexible framework for integrating various component models, in Fortran and any Fortan-linkable language, into simple system configurations. A library of models for subcomponents and property tables common to many different systems are available, and users can easily add their own models as needed.

FC2000 includes ready-to-use component models: heat exchangers, fluid devices (splitter, nozzles, diffuser, gas turbine, pump, etc.), chemical reactors, fuel cells and separation units.

The program's mathematical utilities include a non-linear equation solver, a constrained non-linear optimiser (which handles both linear and non-linear constraints), an integrator, and a solver for ordinary differential equations.

The FC2000 environment is highly user-friendly and system configurations are set up with point-and-click features, using on-screen graphics; model parameters are easily changed; and pop-up windows are used to edit system configurations and for line and surface plots.

The software achieves rapid turnaround times in interfacing with component models. In present work models with various level of complexity (from lumped to three-dimensional ones) created by Industrial and University centres were included into the code. For the scope of the present paper two major models worth some details: the fuel cell and the PAOX models.

2.1 Fuel Cell Model

Two models for PEM (polymeric electrolyte membrane) fuel cells are currently available: a lumped semi-empirical model and a rigorous 2-dimensional one.

In the lumped semi-empirical model, the fuel cell performance is calculated as analytical function of the overall operating conditions: temperature, pressure and entering gas compositions (Amphlett et alt., 1995). The influence of the above conditions on thermodynamic cell voltage and overvoltages (activation, ohmic, diffusion losses) is evaluated with regressions of experimental data (Mangione, 1999). The resulting lumped semi-empirical model appears as adequate for the aim of an overall system simulation and when experimental current-voltage data at various operating conditions are available.

The rigorous 2-dimensional model (Recupero, 1996), specifically developed for PEM fuel cell design and diagnostic purposes, needs as input data the physico-chemical properties of mass transfer in the membrane and all the geometrical parameters of a single cell.

The model integrates along the cell the equations of mass transport, energy and momentum, providing, as a result, the distribution of physico-chemical parameters on each single cell surface. In other words the model describes the surface distribution of gas compositions, gas and cooling water temperatures, current densities and membrane hydration and temperatures, so that not well-distributed flows and high temperature hotspots can be easily checked. Mass transfer of water in polymeric membrane and its local conductivity as function of water content were analysed, detecting critical conditions for cathode flooding, membrane drying and high temperature membrane degradation.

Although the use of the rigorous model is limited to alike cell geometry and fluid flows (gas and cooling water flow patterns, mainly), its integration in FC2000 allows the evaluation of the performance of peculiar specific fuel cell technologies in off-design conditions (mainly different input gas compositions) without the need of troublesome experimental tests.

In both the models the operating voltage and output streams data and power are calculated as function of a requested current. Cell stack analysis is simply scaled up by the cell number.

2.2 PAOX Model

The mathematical one-dimensional model of methane partial oxidation reactor, derives from the work described by (Recupero, 1996) where the model was set up as a tool to support experiments on partial oxidation reactors. The model provides both temperature and concentration profiles of gases along the catalytic bed reactor. Modelling is based on the assumption, confirmed experimentally by lab-scale microreactor tests, that the methane partial oxidation reaction is controlled by mass and thermal transfers, being the reaction rate as high as it is not controlling the transfer phenomena. Some improvements were introduced to better evaluate the mass transfer coefficients of various gases along the reactor length, the radial temperature and the concentration profiles (Mangione, 1999). Model confirms the experimental behaviour of an early almost complete methane combustion, with an high temperature peek, followed by a decreasing temperature profile due to the endothermic steam reforming reaction and the slightly exothermic water shift gas reaction.

3. Applications

FC2000 has been applied to simulate fuel cell systems including both the different fuel cell technologies (PEM, molten carbonate, and phosphoric acid) and the related reforming processes (steam reforming, partial oxidation, autothermal) of traditional fuels such as methanol, methane and higher hydrocarbons, for production of an hydrogen rich anodic stream (Recupero et alt., 1998).

The aim was to evaluate the performance of various process configurations in different operating conditions in terms of optimal energy requirements and the definition of major design parameters. Off-design operations and transient performance and efficiency, as during start-up from cold conditions or during ramp-up or rump-down, were studied as well.

FC2000 has been implemented and successfully used for the design of a natural gas partial oxidation fuel processor for PEM fuel cell systems (Galli et alt., 2000), and currently is validated by experiments as illustrated in the next chapter.

4. Simulation of an integrated fuel cell system

A typical integrated fuel cell power system with natural gas (NG) as feedstock is illustrated in fig. 1. The process includes a PEM fuel cell, a "fuel processor" section that converts the fuel into a hydrogen rich gas mixture and finally the utility units such as turbo-compressor, burner, pumps an heat exchangers.

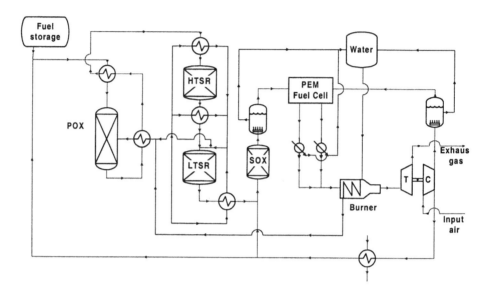

Figure. 1 : Integrated fuel cell system

NG, mixed with air, reacts in the high temperature partial oxidation reactor POX. In the next two-stage reactors (LTSR and HTSR) carbon monoxide reacts with steam according to the water gas shift reaction which increases the hydrogen content. A selective oxidation (SOX) reactor reduces with air the carbon monoxide content to the minimum concentration permitted by the fuel cell system. The stream flowing from fuel processing unit is humidified and then supplied, together with wet air, to the fuel cell. Due to the high percentage of heat produced by the fuel cell stack (about 50-60% of input energy), the energy recovery from output streams is the only mean to increase the energy efficiency off the system. As a consequence air/fuel mixture is pre-heated before entering the POX reactor while anodic and cathodic exhausts are burnt in a catalytic burner to produce steam for internal services into shift reactors and POX. Exhaust gas from burner is then expanded in a turbo-expander to produce further residual electric energy.

Analysis of such a fuel cell system was performed by FC2000. Preliminary test runs showed that the overall system performance was dependent on CH_4 conversion and on the related hydrogen yield occurring in POX reactor. Moreover it arose that an adjustment of the O_2/CH_4 molar ratio of POX input stream together with a controlled supply of superheated water steam along the reactor (favouring the steam reforming reaction) would have produced a 95.5% methane conversion and an increase in the reformer and overall system efficiencies. Steam injection would have had also the additional benefit of avoiding further supply of steam in successive high temperature water shift gas reactor.

Tables 1 and 2 report the major calculated operating conditions and efficiencies of fuel cell system and molar concentrations of main streams in the process, following the above mentioned modifications.

Table 1: Nominal operating conditions and efficiencies for fuel cell system of fig.1

Fuel Cell		Fuel Processor	
Cathodic and anodic pressure [bar]	3	POX input temp. [°C]	350
Operat. Temp. [°C]	78	O_2/CH_4 molar ratio	0.53
Current density [A/cm^2]	0.4	H_2O/CH_4 molar ratio	1
Gross fuel cell power [kW]	18.7	Efficiency [%]	92.9
Fuel cell Efficiency [%]	32.8		
System		**Utilities**	
Net system power [kW]	18	Water pump [kW]	0.25
System efficiency [%]	29.4	Turbine power [kW]	4
		Compressor power [kW]	4.5

Table 2 : Main stream molar concentrations (in percent) for fuel cell system of fig.1

	Input POX	Input HTSR	Input LTSR	Input SOX	Input fuel cell
O_2	15.	-	-	-	3.1
CO	-	10.6	6.8	0.8	-
CO_2	-	4.7	8.5	12.6	10.7
H_2O	-	10.9	7	13.3	14.5
H_2	-	38.1	42	42	33.6
CH_4	28.3	0.8	0.7	0.6	0.6

Fuel cell and Fuel Processor efficiencies are defined as ratio of the output energy to input energy in term of anodic gas high heating value (HHV). Overall efficiency considers as output the produced direct current (DC) power and the net power from utilities.

The above estimates were fulfilled using the simplified PEM fuel cell model and the rigorous POX reactor previously discussed.

5. Conclusions

FC2000 was thought and developed to provide researchers and process engineers with a flexible tool which could assist them and their activity in the field of fuel cells processes design and simulation.

FC2000 includes the most recent concepts in the field of modelling cells and partial oxidation reactors at the scale of small stationary stations and for automotive applications.

Sensitivity studies and detailed analysis of a complete fuel cell system were performed, demonstrating FC2000 ability to indicate innovative alternative configurations, which increased the overall process performance. The rational of the innovation was in a controlled supply of superheated water steam along the reactor. All this in favour of assisting chemical engineers in quantifying their conceptual improvements and ideas during design and experimental activity.

6. Acknowledgement

The development of FC2000 at CIIC Politecnico di Milano is supported by funding from the Ministry of University for research in the framework of a National Research program for fuel cell technology development. The implementation of new models was conducted with the co-operation and work performed previously by CNR-TAE of Messina (Italy), DIAM of University of Genoa (Italy) and ENEA (Italy).

7. References

Amphlett J.C., Baumert R., Mann R.,Peppley A. and Roberge P.,1995,
    J. Electrochemical Soc.,142, 1
Costamagna P., 2001, Chem. Eng. Sci. 56, 323
Galli S., Bernini E, Mangione M. and Recupero V, XIII WHEC, Beijing (China), June
    2000: Experimental Plant for Hydrogen Production via Natural Gas Catalytic Partial
    Oxidation.
Mangione M., 1999, Thesis in Chemical Engineering, University of Rome,1999-00:
    Studio sperimentale di un processo per la produzione di idrogeno per autotrazione
    (in Italian)
Pierucci S., Galli S. and Sogaro A.,1985, E.I.I.R Commission E2 "Heat Pump and
    System Recovery", Trondheim (Norway) June 1985: A Generalized Program for the
    Simulation of Absorption Heat Pump System.
Recupero V.,1996, Contract JOU2-CT93-0290, Final Report, April 1996 : Theoretical
    and experimental studies for a compact H_2 generator, via catalytic selective partial
    oxidation of CH_4, integrated with second generation of fuel cells.
Recupero V.,Pino L., Laganà M., Ciancia A. and A. Iacobazzi, 1998, 12th World
    Hydrogen Energy Conference, Buenos Aires, 1998: Fuel Cell Vehicle. An Italian
    Project to Develop a Hydrogen Generator.

European Symposium on Computer Aided Process Engineering – 12
J. Grievink and J. van Schijndel (Editors)
© 2002 Elsevier Science B.V. All rights reserved.

Reactive Extraction Process for Concentrating Citric Acid using an Electrolyte Mixed Solvent

Pinto,R.T.P.[1]; Lintomen, L.[1]; Broglio, M. I.[3], Luz Jr., L.F.L.[4] Meirelles,A.J.A.[2],
Maciel Filho, R.[1] and Wolf-Maciel,M.R.[1] - e-mail – renata@feq.unicamp.br

(1) Chemical Engineering Faculty / Separation Process Development Laboratory
(2) Food Engineering Faculty / Physical Separation Laboratory
(3) CPP - Rhodia Brazil S/A / (4) Federal University of Paraná
State University of Campinas – CP 6066 - CEP 13081-970 – Campinas – SP – Brazil

Abstract

A process for the separation and the concentration of citric acid was developed within the context of clean technology. It is based on the liquid-liquid phase equilibrium thermodynamic of a system containing salt, which indicates the liquid-liquid extraction process using electrolyte mixed solvent, as potentially attractive to recover citric acid from its fermentation broth. Experimental ternary and quaternary liquid-liquid equilibrium data were obtained in this work and the electrolyte activity coefficient model parameters were calculated from them. The ASPEN PLUS process commercial simulator was used for calculating the thermodynamic parameters, for simulating and for optimizing the process developed.

1. Introduction

In order to develop a solvent extraction process as an alternative to the classical precipitation process for recovering organic acids, a large number of liquid-liquid systems has been investigated. In the literature, several possible solvents can be found, such as mixed solvents constituted by ternary amines and long chain alcohol (Juang and Chang 1995; Procházka *et al.*, 1997). The principal disadvantage of using mixed solvents with amines is its high toxicity, which could mean high cost in the downstream processing. Grinberg *et al.* (1991) have studied the ternary system 2-butanol – citric acid – water, where 2-butanol is a simple solvent effective to recover citric acid. They have reported that the distribution of citric acid between the aqueous and alcoholic phases can be represented by a complex formation reaction. The principal disadvantage of using 2-butanol is that it is partially soluble in water and, in the presence of citric acid, the heterogeneous region is relatively narrow, reducing its practical importance.

Lintomen *et al.* (2001) report experimental liquid-liquid equilibrium data for water – 1-butanol – citric acid and for water - 2-butanol – citric acid systems. They have verified that the liquid-liquid equilibrium diagram for the system constituted by 1-butanol presents larger heterogeneous region and higher selectivity when compared with the system using 2-butanol, but with a smaller distribution coefficient. On the other hand, some researchers have, recently, shown that the salting-out effect has a potential application for solvent extraction process, since it enables good recovery of organic compounds from aqueous solutions since it increases the heterogeneous region, using an electrolyte in the solution (Marcilla *et al.* 1995, Al-Sahhaf and Kapetanovic, 1997; Lintomen *et al.*, 2000). Al-Sahhaf and Kapetanovic related that the liquid–liquid equilibrium in aqueous systems are determined by inter-molecular forces, predominantly, hydrogen bonds. The addition of a salt in such systems introduces ionic

314

forces that change the structure of the liquids at equilibrium. So, the aim of this work is to develop a liquid-liquid extraction process based on electrolyte effect to recover citric acid. An extensive experimental work on the salting-out effect was carried out. Binary interactions parameters of the activity model were regressed and the simulation and the optimization of the separation process were performed.

2. Thermodynamic Representation of the Studied Systems

The development of any liquid–liquid extraction process and its design requires the knowledge of the experimental phase equilibrium data (Pinto *et al.*, 2000). In this work, the electrolyte NRTL thermodynamic model and the parameter estimation procedure based on the Maximum Likelihood Principle presented in Aspen Plus Software Simulation version 10.2, were used. Water, 1-butanol and citric acid were used as molecular species in the model. NaCl and $MgCl_2$ were considered as totally dissociated, since the experimental liquid-liquid equilibrium data were obtained below the saturated point. Concerning the electrolyte NRTL model (Aspen Plus Process Software by Aspen Technology – Version 10.2), the parameters NRTL1 and NRTL2 represent the molecule-molecule binary interactions and NRTL3 represents the molecule-molecule nonrandonness factor. The parameters GMELCC, GMELCD and GMELCE represent the ion-molecule binary interactions and GMELCN represents the ion-molecule nonrandonness factor. Only this model was studied in this work, since this is the most common procedure found, although others models could be used.

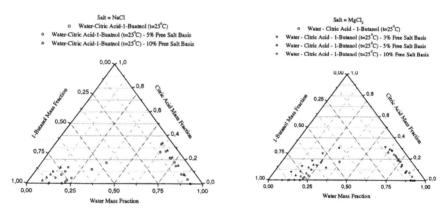

Figure 1: Salting-Out effect on the   Figure 2: Salting-Out effect on the
Liquid-Liquid Equilibrium Data   Liquid- Liquid Equilibrium Data

Figure 1 and 2 present the salting-out effect on the experimental liquid-liquid equilibrium data to the systems: water - citric acid - 1 butanol - NaCl and water - citric acid - 1 butanol - $MgCl_2$. In both, it can be verified that increasing the salt concentration in the overall phase, the heterogeneous region increases. In Tables 1 and 2, K is the citric acid distribution coefficient (mole fraction of citric acid in the organic phase / mole fraction of citric acid in the aqueous phase) and X is the medium citric acid concentration in the overall phase (% mass fraction). Tables 1 and 2 show the salting-out effect on the experimental distribution coefficient of citric acid for the systems studied by Lintomen *et al.* (2001) and for several salts obtained in this work, respectively. It can be observed that, depending on the salt used, the distribution coefficient of the citric acid in relation to the original ternary system (water – citric acid

– 1 butanol) can be improved or not, and that $MgCl_2$ has a larger effect on the coefficient distribution of the citric acid, when 1 butanol was used as solvent. On the other hand, the selectivity profile follows almost the same performance observed for the distribution coefficient, being $MgCl_2$ and $CaCl_2$ the most selective salts. The same distribution coefficient profile was obtained when 10% in mass of the studied salts was used. The systems water – citric acid – 1 butanol – NaCl and water – citric acid – 1 butanol – $MgCl_2$ were chosen to proceed with the thermodynamic evaluation. The first one was chosen to be possible to do a further comparison with the system that contains 2-butanol as solvent (Pinto et al., 2000), and the second one was chosen because it was the system that presented higher distribution coefficient and selectivity concerning to the desired compound (citric acid).

Table 1 - Citric Acid Distribution Coefficient at $t = 25^0$ C (Lintomen et. al. 2001)

System: water – citric acid – 1 butanol		System: water – citric acid – 2 butanol		System: water – citric acid – 2 butanol - NaCl		
X	K	X	K	X	K	K
					5% NaCl	10% NaCl
2.55%	0.8479	1.10%	1.1819	3.00%	1.1702	1.1489
5.20%	0.7490	2.00%	1.1276	6.00%	0.9091	1.0404
7.40%	0.8492	3.60%	1.0718	9.00%	0.8699	0.9677
15.00%	0.8942	-	-	12.00%	0.8824	0.8495
20.10%	0.9100	-	-	-	-	-

Table 2 - Citric Acid Distribution Coefficient at $t = 25^0$ C

System: water – citric acid – 1 butanol - salt								
X	K	K	K	K	K	K	K	K
	10% $MgCl_2$	5% $MgCl_2$	5% Na_2SO_4	5% KI	5% $CaCl_2$	10% NaCl	5% NaCl	5% KCl
6.00%	2.6186	1.3637	0.9726	0.8471	0.8536	0.9026	0.7947	0.5673
12.00%	2.2661	1.3383	0.9431	0.8669	0.7283	0.8861	0.7583	0.5585
18.00%	2.0315	1.3087	1.0244	0.8993	0.8347	0.8477	0.7088	0.5492
24.00%	1.7516	1.1724	0.9245	-	0.8974	0.8359	0.7222	0.5662
30.00%	1.4957	1.0334	1.0243	-	1.1576	0.7892	0.7493	0.5747

Figures 3 show the correlation results for the liquid-liquid equilibrium for the system water – citric acid – 1 butanol – NaCl (5 and 10% mass fraction). The intention here is to show the performance of the data calculated by the electrolyte NRTL. The agreement between the experimental and calculated data is representative, and larger deviations are obtained for the salt and for 1-butanol concentrations in the organic phase.

Figures 4 (a, b, and c) show the comparison between the experimental and calculated data for the systems water – citric acid – 1 butanol, and water – citric acid – 1 butanol – $MgCl_2$ (5 and 10% mass fraction). It can be verified that the calculated data of the ternary system (Figure 4 a) agrees with the experimental ones, except near to the plait point. Analyzing the calculated data of the quaternary systems (water – citric acid – 1 butanol - $MgCl_2$ (5 and 10% mass fractions) (Figures 4 b and c), it can be verified that increasing the salt concentration in the overall phase, the deviations between calculated and experimental data increase too. This effect can be expected, since the electrolyte NRTL model was developed to describe multicomponent electrolyte solutions for low salt concentration (Bochove et al., 2000).

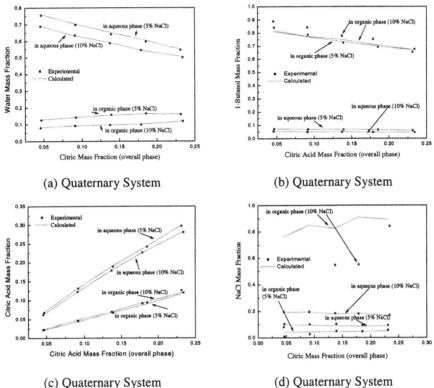

(a) Quaternary System   (b) Quaternary System

(c) Quaternary System   (d) Quaternary System

Figure 3: Composition Profiles of the water – citric acid – 1 butanol – NaCl (5 and 10% mass fractions) – t = 25⁰ C

The better correlation obtained with NaCl (compared with $MgCl_2$) can be justified by the fact that salts presenting larger salting-out effect, as $MgCl_2$, may introduce larger ionic forces, causing larger instability in the calculated data with the models developed so far and/or presented in the Aspen Plus. So, if the regressed parameters were used to carry out the simulations of the system containing $MgCl_2$, a poor representation of the process could be expected. Therefore, only the ternary system (water – citric acid – 1 butanol) and the quaternary system (water – citric acid – 1 butanol – NaCl) were considered for simulating the liquid-liquid extraction process. Taking this fact into account, we can expect, in the process simulation using NaCl, some deviation and we need to consider the qualitative results of the process.

3. Simulation and Optimization

The ASPEN PLUS process simulator uses a rigorous model for simulating liquid – liquid extractor. The inside-out approach algorithm for the column convergence was used. The electrolyte NRTL model was used to calculate the distribution coefficient and the thermodynamic parameters were obtained in the *regression case*. For the simulation, the design features of the extractor used were the default of the Aspen Plus Simulator. In terms of convergence criteria, no product stream specifications were made. They, endeed, were the results of the problem. However, only the number of stages were specified to carried out the simulations.

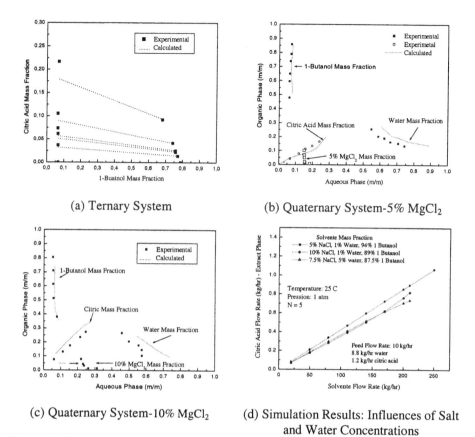

(a) Ternary System

(b) Quaternary System-5% MgCl$_2$

(c) Quaternary System-10% MgCl$_2$

(d) Simulation Results: Influences of Salt and Water Concentrations

Figure 4: Composition Profiles of the water – citric acid – 1 butanol and water – citric acid – 1 butanol – MgCl$_2$ (5 and 10% mass fractions) (t = 25^0 C) and Simulation Results

The optimization of the process was carried out aiming to maximize the citric acid concentration in the extract phase. The parameters that present larger influence on the process are: the solvent flow rate and the composition of the solvent. Table 3 shows the citric acid flow rate of the simulated process. For the other components, not represented in the table, the mass balance is also satisfied. A ratio of solvent to feed around 20 was found, when a mixed solvent containing NaCl was used. It can be observed that the citric acid split fraction (extract flow/ feed flow) decreases using NaCl, although the heterogeneous phase increases (Figure 1). These results are in accordance with the ones presented in Table 2, in which the addition of NaCl to the system has caused, normally, a small decrease in the distribution coefficient. The advantage in this case, however, is that it is possible to simulate the process with the feed concentration of the citric acid near to the one from the fermentation broth (about 12% mass fraction), since the heterogeneous region was increased with the addition of NaCl. Figure 4 d shows the influence of solvent mass flow rate on the citric acid concentration. Also, it can be verified that increasing the salt concentration in the solvent flow rate, the citric acid concentration increases, but this last effect is limited by the water quantity in the solvent feed flow, which is used for facilitating the salt dissociation. It is convenient to mention that it was necessary a large quantity of solvent for getting high recovery of citric acid.

This may be due to the thermodynamic data representation in relation to the experimental data.

Table 3: Mass Balance of the Citric Acid (kg/h)

Quaternary System: Water – Citric Acid – 1 Butanol – 10% NaCl				
Total Feed Flow Rate: 10 kg/h				
Component	Feed Flow	Solvent Flow	Raffinate Flow	Extract Flow
Citric Acid	1.2000	0.0000	0.138000	1.061200
Split Fraction	-	-	0.1150	0.8850
Ternary System: Water – Citric Acid – 1 Butanol				
Total Feed Flow Rate: 20 kg/h				
Component	Feed Flow	Solvent Flow	Raffinate Flow	Extract Flow
Citric Acid	0.8000	0.0000	0.003862	0.7961
Split Fraction	-	-	0.0099	0.9901

4. Concluding Remarks

The experimental liquid-liquid equilibrium data show that the use of an electrolyte system for recovering citric acid is feasible. On the other hand, the simulation results using NaCl show that it is qualitatively possible to have high recovery of citric acid exploring the salting-out effect, but high solvent flow rate was necessary. In order to minimize this problem, we can use salts that simultaneously increase the heterogeneous phase and the distribution coefficient, such as $MgCl_2$ and Na_2SO_4. However, it is clear, from the results, that improvements on the activity coefficient model of electrolyte systems are necessary, for improving the simulation results.

5. Acknowledgment

The authors are indebted to CNPq (Processes Numbers 141523/98-0 and 571683/1997-5, and Fapesp (Process Number: 00/00406-7) for the scholarships.

6. References

Al-Sahhaf, T.A. and Kapetanovic, E. (1997), J. Chem. Eng. Data, 42, 74.

Aspen Plus Manual version 10.2, Federal University of Paraná, Brazil.

Bochove, G.H., Krooshof, G.J.P., Loos, T. W. (2000), Fluid Phase Equilibria, 171, 45.

Grinberg, A., Povimonski, D. and Apelblat, A. (1991); Solvent extraction and ion Exchange, 9(1), 127.

Juang, R. and Chang, H. (1995); Ind & Eng. Chem. Res., 34 (4), 1294.

Lintomen, L., Pinto, R.T.P., Batista, E., Meirelles, A.J.A., and Wolf-Maciel, M.R. (2001), J. Chem. Eng. Data, 46, 546.

Lintomen, L., Pinto, R.T.P., Batista, E., Meirelles, A.J.A., and Wolf-Maciel, M.R. (2000), J. Chem. Eng. Data, 45, 1211.

Marcilla, A., Ruíz, F. and Olaya, M.M. (1995), Fluid Phase Equilibria, 105, 71.

Pinto, R.T.P., Lintomen,L, Meirelles, A.J.A., and Wolf-Maciel, M.R. (2000), Computer Aided Chemical Engineering 8 (ESCAPE 10) – Elsevier Science.

Procházka, J., Heyberger, A. and Volaufová, E. (1997), Ind. & Eng. Chem. Res., 36 (7), 2799.

European Symposium on Computer Aided Process Engineering – 12
J. Grievink and J. van Schijndel (Editors)

Synthesis of Thermodynamically Efficient Distillation Schemes for Multicomponent Separations

Ben-Guang Rong* and Andrzej Kraslawski

Department of Chemical Technology, Lappeenranta University of Technology, P.O. Box 20, FIN-53851, Lappeenranta, Finland. *Tel: +358 5 6216113, E-mail: benguang.rong@lut.fi

Abstract

In this paper, the synthesis of all the functionally distinct thermally coupled distillation schemes for an n-component mixture has been studied. The concept of the intended individual splits has been proposed with which all the inherent distinct separation sequences can be generated for a multicomponent distillation. The first splits of the feed mixture have been used to calculate the total number of the functionally distinct thermally coupled schemes for an n-component mixture. 22 functionally distinct thermally coupled schemes for a four-component mixture, and 719 for a five-component mixture, have been identified.

A simple and easy-to-use procedure is presented to draw all the functionally distinct thermally coupled distillation schemes for separating an n-component mixture into its pure components. The procedure can generate all the thermally coupled schemes included in the available superstructures for an n-component mixture. Moreover, some new feasible thermally coupled distillation configurations have been found which are not included in the known superstructures. This guarantees that the global optimum thermally coupled configuration for a multicomponent distillation can be obtained by the optimisation methods.

1. Introduction

Thermal coupling technique has been used to design multicomponent distillation systems which can reduce both energy consumption and capital cost compared with conventional simple column configurations. Petlyuk *et al.* (1965) first introduced the thermally coupled technique for ternary distillation and presented a fully thermally coupled configuration (FC). For ternary mixtures, there are also a side-stripper (SS) and a side-rectifier (SR) thermally coupled configurations. There were considerable works on the analysis of the relative advantages of the thermally coupled schemes for ternary separations (see literature in Agrawal, 1998a).

Recently, the efforts have been made for finding of new thermally coupled configurations in terms of its energy savings as well as operability for ternary mixtures (Agrawal, 1998b, 1999). As the extending of the alternatives space for ternary mixtures, it is realized that there is a critical issue to construct a complete search space to guarantee the optimal design for a specific separation problem (Agrawal, 1999).

While a lot of works have been contributed to the completeness of the possible distillation configurations for ternary mixtures, there were only a few papers for thermally coupled schemes for four or more component mixtures. Sargent and Gaminibandara (1976) presented a Petlyuk type fully thermally coupled scheme for a four-component mixture. Kaibel (1987) and Christiansen *et al.* (1997) illustrated some distillation columns with vertical partitions for multicomponent separations. Agrawal (1996) presented some satellite column arrangements of fully coupled schemes for four or more component mixtures. We have developed a shortcut synthesis procedure for thermally coupled schemes for four or more component mixtures (Rong et al. 2000). Recently, Agrawal (2000a) also illustrated how to draw thermodynamic equivalent fully thermally coupled schemes for four and five component mixtures. Since FC configuration is only one possibility among all the possible thermally coupled schemes for multicomponent distillations, some other configurations may have advantages in terms of operating and capital costs, as well as operability. Thus, for four or more component distillation, there definitely exists a problem that has not yet been solved – that can we draw all possible thermally coupled configurations for four or more components mixtures? The answer to this question is undoubtedly the most important step towards synthesis of the optimal thermally coupled distillation systems for multicomponent distillations. In this paper, a systematic procedure is presented to generate all the functionally distinct thermally coupled schemes for four or more component mixtures.

2. A Procedure to Draw All Distinct Thermally Coupled Schemes for a Multicomponent Distillation

We understand that the number of functionally distinct thermally coupled distillation schemes for a multicomponent distillation is determined by the number of distinct separation sequences which can be generated for a multicomponent mixture. For generation of all the possible separation sequences for a multicomponent distillation, we need to identify the number of the distinct sets of the intended individual splits which represent the different separation sequences for a multicomponent distillation. *A set of intended individual splits* is defined as a rank listed individual splits introduced into a multicomponent mixture with which a feasible separation sequence is determined for a multicomponent separation task. For example, for a ternary mixture, there are three distinct sets of intended individual splits, i.e. (A/BC, B/C), (AB/C, A/B), and (AB/BC, A/B, B/C). Thus, it is the number of the distinct sets of the intended individual splits generated for a multicomponent mixture that finally determines all the functionally distinct thermally coupled distillation schemes for the multicomponent mixture. By *a functionally distinct thermally coupled distillation scheme* for a multicomponent separation here we mean that at least one of the individual splits in the separation sequence is different from other possible thermally coupled schemes, while all the column units in the distillation configuration are thermally coupled interconnected, this guarantees that no simple columns are included in a functionally distinct thermally coupled distillation scheme. Here, we do not include the thermodynamically equivalent arrangements of a thermally coupled scheme into the space of the functionally distinct thermally coupled schemes for a multicomponent separation since they have the same intended individual splits. Obviously, for ternary mixtures there are three functionally distinct thermally coupled schemes which are the well-known side-stripper, side-rectifier and the Petlyuk column.

In this work, for the generation of all the possible thermally coupled distillation schemes for four or more component mixtures, we have the flexibility to conduct the separation of a submixture (including feed) with ternary or more components in all its feasible ways at any steps in determining a set of feasible individual splits for a multicomponent separation task. Then, the number of all the functionally distinct thermally coupled distillation schemes for an *n*-component mixture can be easily counted from the numbers of the functionally distinct thermally coupled distillation schemes of all its predecessor multicomponent mixtures with 3 to n-1 components ($n{\geq}4$). This is done by determining how many first splits can be generated for the feed mixture. The number of the first splits for an n-component mixture is calculated by $n(n-1)/2$ (Hu et al. 1991). Each first split will determine a branch for the generation of the distinct thermally coupled distillation schemes for the feed mixture.

Tables 1 and 2 illustrate the generation of all functionally distinct thermally coupled distillation configurations for a four-component and a five-component mixture respectively. There is an explosive combination of the number of functionally distinct thermally coupled schemes for a mixture with six or more components (e.g. 556456 schemes for six component mixtures).

Table1.All functionally distinct thermally coupled schemes for a four-component mixture

First Splits	Distributed Middle Component(s)	Number of Schemes
A/BCD	None	3
AB/CD	None	1
ABC/D	None	3
AB/BCD	B	3
ABC/CD	C	3
ABC/BCD	BC	9
		Total: 22

Table2.All functionally distinct thermally coupled schemes for a five-component mixture

First Splits	Distributed Middle Component(s)	Number of Schemes
A/BCDE	None	22
AB/CDE	None	3
ABC/DE	None	3
ABCD/E	None	22
AB/BCDE	B	22
ABC/CDE	C	9
ABCD/DE	D	22
ABC/BCDE	BC	66
ABCD/CDE	CD	66
ABCD/BCDE	BCD	484
		Total: 719

Network representations of multicomponent distillation configurations have been presented by Hu et al.(1991) and Agrawal (1996). Actually, we found that such a network representation to a feasible separation sequence uniquely designates a functionally distinct thermally coupled configuration for a multicomponent distillation. Table 3 presents all the 22 functionally distinct thermally coupled schemes represented in the distinct separation sequences for a four-component mixture. Figure 1 illustrates some typical functionally distinct thermally coupled schemes in Table 3 for a four-component mixture with a revised network representation. Some new configurations in Table 3 are drawn in Figure 2.

Among all the schemes in Table 3, there are five schemes of (a), (b), (d), (e) and (f) with all sharp splits in the separation sequences. These five schemes are produced from the well-known conventional simple column configurations. They constitute a unique subspace of thermally coupled configurations with only side-stripper(s) and/or side-rectifier(s) for any n-component mixtures (Rong and Kraslawski, 2001). There are four schemes of (j), (m), (p), and (r) with only one condenser and one reboiler in each scheme. The scheme (r) is the well-known Petlyuk type fully thermally coupled scheme as in Sargent and Gaminibandara(1976), while the other three are called satellite column fully coupled configurations(Agrawal, 1996). The schemes (c), (g), (h), (i), (k), (l), (n) and (q) can be distinguished by the fact that the total number of condenser(s) and reboiler(s) in each one equals three. It is interesting to note that there is a unique feature in schemes of (s), (t), (u) and (v), namely each of them can produce another intermediate volatility product with one more column unit in the scheme. Schemes (s) and (t) produce another intermediate volatility product B, while schemes (u) and (v) produce another C. They are feasible and useful configurations, especially in the situations where there might need the products with different purities (Agrawal, 2000b). Moreover, these schemes cannot be derived from the prior superstructures of Sargent and Gaminibandara(1976) and Agrawal(1996). They are shown in Figure 2a-d. Another interesting scheme is (o) shown in Figure 2e, though it has totally four heat exchangers, it might be attractive in some specific cases of the feed mixture. For example, if the splits A/B and C/D are easy ones and with approximately the same relative volatilities, and the split B/C is a difficult one with a significant amount of BC in the feed, then, the scheme might be more economically attractive than the schemes with only side-stripper(s) and/or side-rectifier(s) which also have totally four heat exchangers. Especially, it could be more promising for divided-wall equipment implementation as in Figure 2f.

Table 3. The 22 distinct thermally coupled schemes for a four-component mixture

a	A/BCD→B/CD→C/D	b	A/BCD→BC/D→B/C	c	A/BCD→B$\underline{C}$D→B/C→C/D
d	AB/CD→A/B→C/D	e	ABC/D→AB/C→A/B	f	ABC/D→A/BC→B/C
g	ABC/D→A$\underline{B}$C→A/B→B/C	h	A$\underline{B}$CD→B/CD→A/B→C/D	i	A$\underline{B}$CD→BC/D→A/B→B/C
j	A$\underline{B}$CD→B$\underline{C}$D→A/B→B/C→C/D	k	A$\underline{B}$CD→A/BC→B/C→C/D	l	A$\underline{B}$CD→AB/C→A/B→C/D
m	AB$\underline{C}$D→A$\underline{B}$C→A/B→B/C→C/D	n	A$\underline{B}$CD→A/BC→B$\underline{C}$D→B/C→C/D	o	A$\underline{B}$CD→A/BC→BC/D→B/C
p	AB$\underline{C}$D→AB/C→B/CD→A/B→C/D	q	AB$\underline{C}$D→A$\underline{B}$C→BC/D→A/B→B/C	r	AB$\underline{C}$D→A$\underline{B}$C→B$\underline{C}$D→A/B→B/C→C/L
s	AB$\underline{C}$D→A/BC→B/CD→B/C→C/D	t	AB$\underline{C}$D→A$\underline{B}$C→B/CD→A/B→B/C→C/D	u	AB$\underline{C}$D→AB/C→BC/D→A/B→B/C
v	AB$\underline{C}$D→AB/C→B$\underline{C}$D→A/B→B/C→C/D				

The following easy-to-use procedure can make it very simple to generate all functionally distinct thermally coupled distillation configurations for an n-component mixture.

1. Characterize the feed mixture by its relative volatilities and the feed composition.
2. Generate all the feasible first splits by considering all feasible individual splits for a multicomponent mixture. For example, there are 10 first splits for a five-component mixture as shown in Table 2.
3. Calculate the number of the feasible separation sequences for each first split generated in step 2. After this step, the total number of feasible separation sequences for an n-component mixture is determined. Note that all the feasible separation sequences generated here are inherent different separation sequences, since at least one of the individual splits in a separation sequence is different from all the other sequences.
4. Identify a first split as the intended first individual split and as the start point for the generation of a specific separation sequence for the feed mixture. For example, choose ABC/BCDE as the first split for a five-component mixture ABCDE.
5. For each of the two subgroups generated in step 4, if the number of its components is ternary or more, then introducing a subsequent intended individual split. For example, we introduce two sharp splits of A/BC and B/CDE for the two subgroups generated in step 4.
6. For the new subgroups generated in step 5, further introduce the intended individual splits for the new subgroups with binary or more components, until all the subgroups generated from present intended individual splits are the single component subgroups, i.e. the final products. For example, totally 7 intended individual splits are introduced for the separation of a five-component mixture as the illustrated example in the above steps, which determined the following separation sequence for the separation task. ABC/BCDE→A/BC→B/CDE→B/C→CD/DE→C/D→D/E.

7. Network representation of the separation sequence by all its intended individual splits. For example, the above five-component separation sequence is network represented in Figure 3a.
8. Draw the thermally coupled scheme based on the network representation of step 7. For example, the thermally coupled scheme of Figure 3a is drawn in Figure 3b.
9. Repeat steps from 5 to 8 for other interesting separation sequences for the selected first split in step 4. In this way, all the schemes in the selected first split branch can be drawn. For example, 66 schemes for the selected first split of ABC/BCDE.
10. Repeat steps from 4 to 9 to draw all the functionally distinct thermally coupled schemes for an *n*-component mixture. 719 schemes for a five-component mixture.

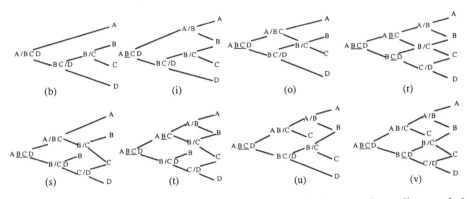

Figure 1. Network representations of some functionally distinct thermally coupled schemes for a four-component mixture in Table 3

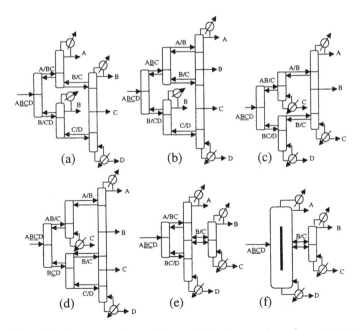

Figure 2. Some new thermally coupled schemes for a four-component mixture

324

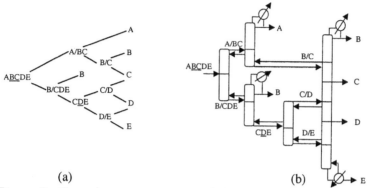

(a)   (b)

Figure 3. Network representation and the thermally coupled scheme for a five-component mixture

3. Conclusions

The synthesis of all the functionally distinct thermally coupled distillation schemes for an *n*-component mixture has been studied in this work. By identifying the possible individual splits for a multicomponent mixture, all the inherent distinct separation sequences can be generated for a multicomponent distillation. We found that a set of intended individual splits of a feasible separation sequence uniquely designates a functionally distinct thermally coupled scheme for a multicomponent distillation. The number of all the functionally distinct thermally coupled schemes for a multicomponent mixture can be calculated from all its feasible first splits of the feed mixture.

A simple and easy-to-use procedure is presented to draw all the functionally distinct thermally coupled distillation schemes for separating an *n*-component mixture into its pure components. The procedure can generate all the thermally coupled schemes included in the available superstructures for an *n*-component mixture. Moreover, some new feasible thermally coupled distillation configurations have been found which are not included in the known superstructures. This guarantees that the global optimum thermally coupled configuration for a multicomponent distillation can be obtained by the optimisation methods.

References

Agrawal, R. *Ind. Eng. Chem. Res.* **1996**, *35*, 1059.
Agrawal, R. *Trans. IChemE.* Vol. 78, Part A. **2000a**, 454.
Agrawal, R. *AIChE J.* **2000b**, *46*, 2198.
Agrawal, R.; Fidkowski, Z.T. *Ind. Eng. Chem. Res.* **1998a**, *37*, 3444.
Agrawal, R.; Fidkowski, Z.T. *AIChE J.* **1998b**, *44*, 2565.
Agrawal, R.; Fidkowski, Z.T. *Ind. Eng. Chem. Res.* **1999**, *38*, 2065.
Christiansen, A.C.; Skogestad, S.; Lien, K. *Comput. Chem. Eng.* **1997**, *21*, S237.
Hu, Z., Chen, B. and Rippin D.W.T. *AIChE Annual Meeting*, Los Angeles, CA, **1991**, Paper 155b.
Kaibel, G. *Chem. Eng. Technol.* **1987**, *10*, 92.
Petlyuk, F.B.; Platonov, V.M.; Slavinskii, D.M. *Inter. Chem. Eng.* **1965**, *5*, 555.
Rong, B.-G., Kraslawski, A.; Nyström, L., *Computer-Aided Chem. Eng.* 8, ESCAPE 10, **2000**, 595.
Rong, B.-G.; Kraslawski, A. *Submitted to AIChE J.*, March, **2001**.
Sargent, R.W.M.; Gaminibandara, K. *Optimization in Action;* Dixon. L.W.C., Ed.; Academic Press: London, **1976**, p. 267.

European Symposium on Computer Aided Process Engineering – 12
J. Grievink and J. van Schijndel (Editors)
© 2002 Elsevier Science B.V. All rights reserved.

Modelling of Multicomponent Reactive Absorption using Continuum Mechanics

D. Roquet, P. Cézac and M. Roques

Laboratoire de Génie des Procédés de Pau (L.G.P.P.)

5, rue Jules Ferry, 64000 Pau, France

email : damien.roquet@univ-pau.fr

A new model of multicomponent reactive absorption based on continuum mechanics is presented. It includes the equations of conservation of mass for species, momentum and energy. The diffusion is described by the Maxwell-Stefan model. The steady state resolution is performed in an equivalent two dimensional geometry. Simulation results are compared with experiments carried out on a laboratory scale column.

1. INTRODUCTION

Wet scrubbing is a widely used technique to remove heavy metal and acid compounds from flue gases. Because of the limitations of atmosphere emissions, optimisation of such processes is essential. Thus, the model taking into account the description of all physical and chemical phenomena is the most accurate tool available to provide a maximum of information on the process. However the theoretical description is quite complex: the different phases involved are multicomponent systems and the pollutants may react in the aqueous phase. The competitive chemical reactions are instantaneous or kinetically controlled. Furthermore the description of transport phenomena should take into account the interactions between species in multicomponent diffusion and the non ideality of the fluids thanks to a specific model of thermodynamics. The aim of this article is to present a general model of multicomponent reactive absorption based on the considerations above using empirical parameters. Firstly we will present a brief summary of the differents approaches that will lead us to the description of the model. Then we will apply it to the absorption in packed columns of acid gases in a aqueous solvent and we will compare the results with experiments carried out on a laboratory scale pilot.

2. STATE OF THE ART

Two kinds of model are presented in the literature. Firstly, the theoretical stage model, which describes the column as a succession of stages at thermodynamic equilibrium. Mass and heat transfers are driven by this equilibrium. It is very suitable in order to take into account each competitive chemical reaction. However it is an ideal case of mass transfer description and the simulated results could only be compared with experiments using an empirical coefficient called efficiency. Secondly, the transfer models which use empirical coefficients to predict the transfer of mass between phases. In packed columns, the two-film model (Whitman, 1923) is the most widely used. This model allows us to define mass transfer coefficients k_L and k_G for binary systems applying Fick's law of diffusion. In this case the coefficients could be obtained by experiments or by correlations (Lee and Tsui, 1999). Taylor and Krishna (1993) studied the case of multicomponent systems without reaction using the Maxwell-Stefan law of diffusion. When reactions occur, the mass transfer coefficient is modified using an enhancement factor. This factor is difficult to obtain in competitive chemical reactions (Versteeg and al., 1990). That's why recently, new models of non equilibrium have been proposed (Rascol and al., 1999 ; Schneider and al., 1999). Given an estimation of the diffusion film thickness, the mass balance relations are numerically solved in each film using an appropriate law of diffusion. This approach allows us to take directly into account the enhancement due to reaction without providing the enhancement factor. This is good progress in mass transfer description but in multicomponent mass transfer, the film thickness should be different for each component. Thus, we have developed a general model of multicomponent reactive absorption in a ionic system using continuum mechanics (Truesdell, 1969 ; Slattery, 1981) to describe local phenomena in each phase. This description does not require to use empirical mass transfer coefficient or film thickness.

3. GENERAL MODEL

Let us consider two phases (φ_1 and φ_2) separated by an interface I and ξ the unit normal. Let us consider the phases as a multicomponent body in which all quantities are continuous and differentiable as many times as desired. On this system two kinds of relation can be written thanks to the continuum mechanics theory in multicomponent system (Truesdell, 1969 ; Slattery, 1981) : conservation of properties in each phase and boundary conditions at the interface.

3.1 Equations of conservation
Using the transport theorem, the balances of each conservative quantity could be written as an equation of conservation applied to each point of the continua.
In order to describe the evolution of the various components we require to solve the conservation of mass for species $i \in \{1, ..., n_c - 1\}$. It takes into account the n_{rc} kinetically controlled chemical reactions and the n_{re} instantaneously balanced chemical reactions, as dissociation equilibrium in aqueous phase. The enhancements of these equilibria are obtained implicitly thanks to chemical equilibrium equations.

$$\frac{\partial}{\partial t}(\rho_i) + \vec{\nabla}(\rho_i \, \vec{v}_i) = \sum_{j=1}^{n_{rc}} v_{ij}^c \, r_j + \sum_{k=1}^{n_{re}} v_{ik}^e \, \xi_k \,, \; i \in \{1, ..., n_c - 1\} \tag{1}$$

$$\sum_{i=1}^{n_c} v_{ik} \cdot \mu_i = 0 \,, \; k \in \{1, ..., n_{re}\} \tag{2}$$

As we use only n_c - 1 equations of continuity for species, we have to solve the equation for overall conservation of mass.

$$\frac{\partial}{\partial t}(\rho_t) + \vec{\nabla}(\rho_t \, \vec{v}) = 0 \tag{3}$$

Equation (3) has been added to the system to prevent the total mass balance from numerical approximations.

Then, in order to describe the movement of the various constituents, we need to solve equations of momentum conservation. But as we don't know any fluid behaviour that can express the stress tensor for species, we use an approximate equation based on the Maxwell-Stefan approach (Taylor and Krishna, 1993 ; Slattery, 1981).

$$c_t R T \vec{d}_i = c_i \vec{\nabla} \mu_i + c_i s_i \cdot \vec{\nabla} T - \frac{\rho_i}{\rho_t} \vec{\nabla} P - \rho_i \left(\vec{f}_i - \sum_{j=1}^{n_c} \frac{\rho_j}{\rho_t} \vec{f}_j \right), \ i \in \{1, \ldots, n_c - 1\}$$ (4)

Neglecting the effect of heat transfer on the mass transfer (Soret effect), the driving force d_i is defined as follows :

$$\vec{d}_i = -\sum_{j=1}^{n_c} \frac{c_i c_j (\vec{v}_i - \vec{v}_j)}{c_t^2 D_{ij}}$$ (5)

Three main reasons have led us to this approach. Firstly, this description allows for the effect of the thermodynamic activity in ionic systems to be taken into account. Secondly, the effects of interactions between chemical species on rates of diffusion can be expressed. Finally, the external forces, such as the electrostatic ones occurring within an ionic system, are included. These forces are expressed thanks to the Nernst-Planck equation (Taylor and Krishna, 1993) and the electrical potential is implicitly obtained by solving the local electroneutrality.

$$\vec{f}_i = -\frac{z_i}{M_i} \cdot F^0 \cdot \vec{\nabla} \varphi, \ i \in \{1, \ldots, n_c\}$$ (6)

$$\sum_{i=1}^{n_c} z_i \cdot \frac{\rho_i}{M_i} = 0$$ (7)

In order to describe the behaviour of the whole phase, we use the conservation of the overall momentum. The stress tensor in laminar flow is achieved from the Newtonian model.

$$\frac{\partial}{\partial t} (\rho_t \vec{v}) + \vec{\nabla} (\rho_t \vec{v} \otimes \vec{v}) - \vec{\nabla} \overline{\overline{T}} - \rho_t \vec{f} = \vec{0}$$ (8)

$$\overline{\overline{T}} = \left(-P + \lambda \vec{\nabla} \cdot \vec{v} \right) \overline{\overline{I}} + 2\eta \overline{\overline{D}}$$ (9)

Finally, in order to describe the evolution of temperature, the conservation of energy is required. Thus, neglecting the viscous and the diffusive dissipations and the effect of mass transfer on heat transfer (Dufour effect), the energy balance is

$$0 = \frac{\partial}{\partial t} \left(\rho_t \left(\tilde{u} + \frac{1}{2} \vec{v}^2 \right) \right) + \vec{\nabla} \left(\rho_t \left(\tilde{u} + \frac{1}{2} \vec{v}^2 \right) \vec{v} \right) - \vec{\nabla} \cdot \left(\kappa \vec{\nabla} T \right) - \rho_t Q$$ (10)

Thus the global system is composed of the conservation of mass for species and of the overall mass, of Maxwell-Stefan's law of diffusion, of conservation of overall momentum and of the conservation of energy. The thermodynamic properties in the aqueous phase, such as enthalpy, internal energy, density or activity coefficients, are calculated from thermodynamic models adapted for electrolytes. The gas phase is considered as a perfect gas. Assuming the summation equations for mass concentration and for mass flow, the system can be solved. However we should present the boundary conditions used at interface.

3.2 Boundary conditions

As the interface is assumed to be immobile, the normal of the mass flows is conserved. It is applied to n_c-1 species and to the overall mass to prevent numerical approximations.

$$\left(\rho_i^{\varphi_2} \vec{v}_i^{\varphi_2} - \rho_i^{\varphi_1} \vec{v}_i^{\varphi_1} \right) \cdot \vec{\xi} = 0 \,, \ i \in \left\{ 1, \ldots, n_c -1 \right\} \tag{11}$$

$$\left(\rho_t^{\varphi_2} \vec{v}^{\varphi_2} - \rho_t^{\varphi_1} \vec{v}^{\varphi_1} \right) \cdot \vec{\xi} = 0 \tag{12}$$

As we assume that the thermodynamic equilibrium is reached at the interface, the chemical potentials for species, the temperatures and the pressures are equal.

$$\tilde{\mu}_i^{\varphi_2 \, I} - \tilde{\mu}_i^{\varphi_1 \, I} = 0 \,, \ i \in \left\{ 1, \ldots, n_c \right\} \tag{13}$$

$$P^{\varphi_2 \, I} - P^{\varphi_1 \, I} = 0 \tag{14}$$

$$T^{\varphi_2 \, I} - T^{\varphi_1 \, I} = 0 \tag{15}$$

4. NUMERICAL RESOLUTION

An equivalent two dimensional geometry (figure 1) is generated from gas and liquid fractions, and from the interfacial area. These three characteristics could be obtained by experiments or by correlations (Lee & Tsui, 1999). The height of packing is conserved.

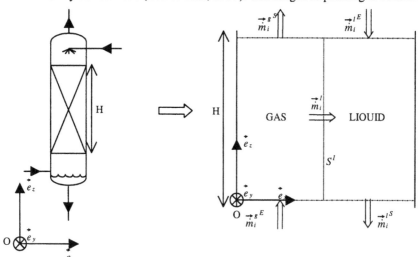

Figure 1: Two dimensional geometry, height of packing (z) by equivalent length (x)

The overall two dimensional system of differential and algebraic equations is solved in steady state using a finite volume method of discretisation. This method generates a non linear algebraic system solved by Newton-Raphson's method taking into account the sparse structure of the Jacobian matrix for minimising the storage space and the CPU time. Furthermore, the two dimensional grid uses a variable step to minimise the end effects. Thus, the equations are solved simultaneously and we obtain the profiles of concentration, temperature, pressure and velocity at each point of the geometry.

5. EXPERIMENT AND SIMULATION RESULTS

We investigated the CO2 absorption in soda inside a packed column at pH 10,1. The steady state simulations are validated by experiments performed on a laboratory scale packed column. The experiment measurements are reported in figure 2 and compared with the continuum mechanics model and with a rate based model taking into account the enhancement of the reactions (Versteeg and al. 1990). The phase's fractions and the wet interfacial area that permit us to define the two dimensional geometry are obtained with correlations (Lee and Tsui, 1999). The simulated pH profile with the continuum mechanics model is in accordance with the experiments, and is better than the rated-based model. Figure 3 and 4 are concentration profiles obtained thanks to the continuum mechanics model. These concentration profiles are suitable and in agreement with the theory.

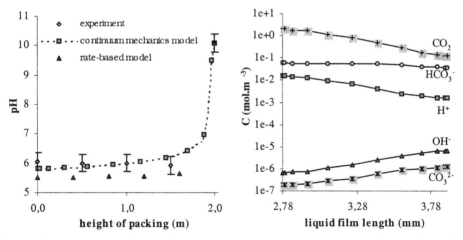

Figure 2: comparison of pH evolution   *Figure 3: concentration profiles at the outlet*

6. CONCLUSION

A new model of multicomponent reactive absorption has been presented. The comparison of the experiment and its simulation allows us to validate the model. Furthermore, global information, such as concentration or pH profiles, could be obtained thanks to a local description of transfer phenomena. One of the most interesting aspects of this model is that it requires only two parameters i.e. the wet interfacial area and the phase's fractions.

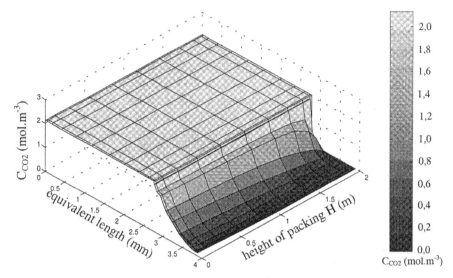

Figure 4 : Evolution of CO_2 concentration $(mol.m^{-3})$ in the two-dimensional geometry

NOTATIONS

c   molar concentration, $mol.m^{-3}$
d   driving force, m^{-1}
Đ   Maxwell-Stefan diffusivity, $m^2.s^{-1}$
D   deformation tensor, s^{-1}
f   external force, N
H   height of packing, m
m   mass flowrate, $kg.s^{-1}$
P   pressure, Pa
r   rate of chemical reaction, $kg.m^{-3}.s^{-1}$
R   gas constant, 8,314 $J.mol^{-1}.K^{-1}$
s   entropy, $J.kg^{-1}.K^{-1}$
S   area, m^2
T   temperature, K

T   stress tensor, $N.m^{-2}$
u   mass internal energy, $J.kg^{-1}$
v   velocity, $m.s^{-1}$
κ   thermal conductivity, $W.m^{-1}.K^{-1}$
λ   bulk viscosity, Pa.s
η   dynamic viscosity, Pa.s
μ   mass chemical potential, $J.kg^{-1}$
ν   stoechiometric coefficient of chemical reaction
ρ   mass concentration, $kg.m^{-3}$
ξ   enhancement of chemical equilibrium, $kg.m^{-3}.s^{-1}$

REFERENCES

Whitman, W.G., 1923, Chem. Met. Sci., 29, 147

Lee, S.-Y. and Y.P. Tsui, 1999, Chem, Eng. Prog., July, 23-49

Taylor, R. and R. Krishna, 1993, Multicomponent mass transfer, Wiley, New-York

Versteeg, G.F., J.A.M. Kuipers, F.P.H. Van Beckum and W.P.M. Van Swaaij, 1990, Chem. Eng. Sci., 45, 183-197

Rascol, E., M. Meyer and M. Prévost, 1999, ECCE2, Montpellier

Schneider, R., E.Y. Kenig, A. Gorak, 1999, Chem. Eng. Res. Des., 77, 633-638

Truesdell, C., 1969, Rational thermodynamics, McGraw-Hill, New-York

Slattery, J.C., 1981, Momentum, energy, and mass transfer in continua, McGraw-Hill, 2nd edition, New-York

European Symposium on Computer Aided Process Engineering – 12
J. Grievink and J. van Schijndel (Editors)
© 2002 Elsevier Science B.V. All rights reserved.

Optimization-Based Methodologies for Integrating Design and Control in Cryogenic Plants

M. Schenk[*], V. Sakizlis, J.D. Perkins and E.N. Pistikopoulos[1]
Centre for Process Systems Engineering, Department of Chemical Engineering,
Imperial College, Prince Consort Road, London SW7 2BY
United Kingdom

Abstract

The aim of this work has been to investigate the potential of advanced mixed-integer dynamic optimization (MIDO) strategies[2] to identify optimal design and control schemes for an industrial cryogenic air separation plant (ASP). The MIDO framework is applied to include, along with the control structure choice, selection and sizing of process components. From an economic point of view, this technique has shown that utility consumption and capital cost can be reduced, giving an improvement in the overall profitability. By considering discrete decisions within the optimization framework (instead of a sequential approach) great benefits can be achieved. The MIDO framework has shown to work efficiently for such a large problem, reaching a solution (optimal design and control structure) in four iterations.

1. Introduction

Cryogenic air separation is a highly competitive business where finding the proper balance between operating and capital costs is the key to commercial success. In the last few years, dynamic modelling has become a popular tool, especially in the industry, in order to improve process design and operation. Many simulations are run for different values of process parameters, however this is an exhaustive enumeration, and even when a feasible operation can be found, it does not guarantee an optimal solution. One method of overcoming this type of problem is to apply a systematic optimization approach. A mathematical algorithm can be employed which can automatically select a set of process parameters, optimizing a performance index while guaranteeing satisfaction of desired constraints (usually time-dependent).

Air separation takes place at extremely low temperatures, because only at low temperatures (around 100K) the predominant air components, namely oxygen, nitrogen and argon become liquid and can be separated by distillation. Commonly in ASPs there are two distillation columns, however, one of the the columns (used to separate oxygen and nitrogen) is split into two columns, one operating at low pressure (LP column) and the other at high pressure (HP column). This is done in order to integrate the reboiler and condenser, due to the high cost of condensing nitrogen. A third column is used to

---

[*] Now with Air Products Plc. Europe. United Kingdom.
[1] Author to whom all correspondence should be addressed, e-mail:e.pistikopoulos@ic.ac.uk
[2] Under developtment at the Centre for Process Systems Engineering, Imperial College (E.N. Pistikopoulos, J.D. Perkins and co-workers).

recover a very valuable argon (AR column) and this column is therefore, interconnected to the LP column. ASPs pose control problems of high complexity, and for this reason, it is necessary to address the underlying issue of the interactions of design optimization and plant control. When alternative schemes can be implemented, a cost-to-benefit analysis of the competing structures becomes an influencing factor. The aim of this study is to investigate the potential of advanced (dynamic mixed-integer) optimization strategies to identify optimal design and control schemes for ASPs. The objective of this work is then to include, along with the control structure, the selection and sizing of process components. The process components will include column diameters. Structural equipment decisions will include the number of trays in each column and the selection of the best control structure to improve the operation and control of the plant.

2. Problem Description

The problem addressed in this work can be stated as follows. Given,
- a fixed flow-sheet of the air separation plant (Figure 1),
- a set of manipulated variables and controlled outputs,
- a set and description of the process disturbances,
- a set of constraints defining feasible operation,
- a detailed process model, and
- an economic performance criterion (maximization objective)

The objective then is to determine the best multi-loop PI control strategy (i.e. control structure configuration and controller tuning parameters), equipment design and operating strategy which optimizes the expected performance criterion (net profit) for the ASP in the presence of the given process disturbances.

This leads to a mixed-integer dynamic optimization problem (MIDO). In this work, the problem is solved using control vector parameterisation techniques over a fixed time horizon of 24 hours. The framework was developed along similar lines to that used by Bansal (2000). The approach is based on Generalized Benders Decomposition (GBD) principles, iterating between a series of primal dynamic optimization problems and master, mixed-integer linear programming problem, but does not require the solution of an intermediate adjoint problem in order to build the master problem. Instead, in the primal dynamic optimization problem, basically a new set of variables, y_{bar}, are incorporated (fixed as: $y_{bar}^j = y^j$ at the jth iteration). The primal optimization problem now includes the following set of equality constraints:

$$h_y = y^j - y_{bar}^j = 0 \qquad (1)$$

where y is a set of continuous time-invariant search variables and y_{bar} is the set of complicating (binary) variables. Then a master problem, which is a mixed-integer linear problem (MILP), is posed as follows,

$$\min_{y_{bar},\eta} \eta \qquad (2)$$

subject to

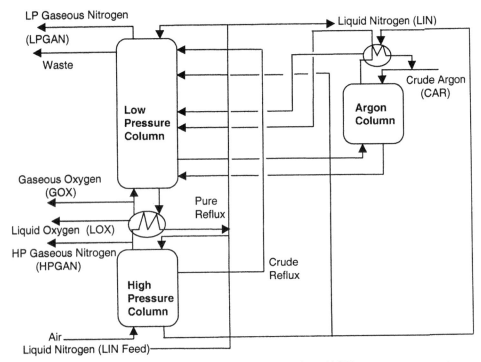

Figure 1. Schematic representation of an air separation plant (ASP)

$$\eta \geq J(x_d(t_f), x_a(t_f), u(t_f), d, y_{bar}) + \varpi^{jT}(y^j - y_{bar}), \qquad k\varepsilon K_{feas} \qquad (3)$$

$$0 \geq \varpi^{jT}(y^j - y_{bar}), \qquad k\varepsilon K_{\inf\ eas}$$

The solution of these master problems gives a new set of binary values for subsequent primal problems. The basic steps for the solution of the MIDO problem can be summarised as follows (for a minimization problem) (Bansal (2000) for more detailed explanation).

- Step 1. For fixed values of the integer variables, solve the jth primal problem to obtain a solution J^j. For computational reasons, omit the search variables and constraints that are not needed for the current choice of integer variables. Set Upper Bound (UB) as J^j.

- Step 2. Solve the primal problem at the solution found in Step 1 with the full set of search variables and constraints included (convergence achieved in one iteration). Obtain the Lagrange multipliers, ω^{jT} needed for constructing the master problem.

- Step 3. Solve the jth master problem with integer cuts (excluding the already obtained integer realization) to obtain the Lower Bound (LB).

- Step 4. If UB-LB $< \varepsilon$ (where is ε a termination tolerance) or the master problem is infeasible then stop. The optimal solution corresponds to the best solution obtained in Step 1. Otherwise, set the new iteration, j=j+1, and make y^{j+1} equal to the integer solution of the jth master problem and return to Step 1.

During this work, gPROMS (Process Systems Ltd., 1999) for the primal dynamic (and steady-state) optimization problems, and GAMS/CPLEX (Brooke *et.al.*, 1992) for the master MILP problems, were used to solve the full MIDO problem. The statistics of the model solved are given in Table 1. The dynamic model has been adapted from an Air Products proprietary model, which is a rigorous (steady-state and dynamic) detailed

Table 1. Model Statistics

Time Horizon	1440 min.
DAE Variables	10450 (without eqs. of the thermo model)
End-Point Inequalities	88
End-Point Equalities	297
Path Constraints	13
Pure Binary Constraints	253
Continuous Search Variables	19
Binary Search Variables	209

model of the plant. The objective function considered for this problem has been set as follows, **Economic Potential = Revenues – Utilities – Capital Cost.** Revenues included all the product streams (with the corresponding averaged market prices). The utilities cost was determined by considering the cost of running the main air compressor, and the capital cost, based on public-domain cost correlations, included the column shell and trays as well as adjacent tanks. The constraints considered include the hydraulic state of trays, the product purity, the areas of the columns, the feed to the AR column, the argon product, the air feed, the controller tuning parameters and the process operating point during the time interval of 24 hours considered (see Table 2). The design variables are distinguished between continuous, such as areas of the three columns, and non-continuous such as the number of trays per column and control scheme selection. The given disturbances are on the inlet air temperature and flowrate; the liquid nitrogen feed flowrate and the ramping specification. Ramping, i.e. switching from one operating point to another is a usual practice in the operation of ASPs. Generally these operational switching are due to different costs in electricity during the weekends in the countries were the plants are operated.

3. Results

As a first step, the MIDO algorithm was applied to generate a design in a pseudo steady-state mode, i.e. the plant was optimized for a given steady-state point only, to better compare and analyse the benefits of applying a dynamic approach. The pseudo-steady-state design converged in five iterations of the MIDO, while the dynamic MIDO only took four iterations to the optimal solution. In average, a steady-state optimization was obtained in approximately 3 hours of CPU time while a dynamic optimization took around 1.5 days of CPU time. This was the most time consuming part of the algorithm,

Table 2: Model constraints

Hydraulic/ State of Trays	Height of clear liquid, Liquid overflow, Pressure drop
Product Purity	Oxygen content in product streams
Areas of Columns	Argon column, HP column and LP column
Feed to Argon Column	Nitrogen content
Argon product	Argon flow-rate
Air Feed	Air flow-rate
Controller tuning parameters	Gains, Set-points, Reset times, Biases
Process operating point	Ramping strategy

Table 3. Results comparison (the steady-state optimal design = 1.00)

Objective function (maximization of performace objective)	0.85
HP column	
No.Trays	1.18
Diameter	1.03
LP column	
No. Trays	1.10
Diameter	0.95
AR Column	
No. Trays	1.12
Diameter	1.07

since the master problem solution took almost no time. Though this might seems lengthy, what is obtained from the solution of the MIDO algorithm is a full plant design and control scheme. Further advances in the optimization front end, would reduce the time spent in the dynamic optimizations. This work intent to demonstrate the capability of the MIDO formulation to handle real industrial problems and no effort has been put into speeding up the process.

The application of the MIDO approach for a time horizon of 24 hours gave the results presented in Table 3 (compared to the base-case, optimal steady-state design). The results from the dynamic approach leads to a more expensive design, *i.e.* only 85% of the profits, given by the steady-state optimal design, can be achieved for a dynamic structure, but this design in contrast to the steady-state design, guarantees the operability of the plant over the time subject to the given disturbances. The performance objective function is reduced in the dynamic approach in all the component terms, i.e. fewer revenues and more expenditure in capital costs as well as utilities. This is mainly attributed to the fact that the plant has to accommodate the ramping strategy and constraints enforced during the operational time. Column diameters are increased as well as the number of trays for the dynamic optimal solution, because the plant needs to satisfy the hydraulic constraints according to the process disturbances. It should again be emphasised that the steady-state optimal design, generated by the MIDO approach could not accommodate a pre-specified ramping requirements for this plant using the degrees of freedom available in the model. The optimal steady-state solution only guarantees the operation of the plant for a single-point operation, while the optimal dynamic solution gives a feasible operation of the plant over time. Figure 2 (a and b) shows some of the process constraints over time, for the optimal solution. A very interesting feature of the optimization framework is that, it not only gives the optimal plant structure, but also gives a 'set' of feasible schemes, since the solution of each iteration is a feasible (sub-optimal) solution. If, for any reason, the optimal solution cannot be used, then a set of possible solutions is available for the process design/control engineers.

Moreover, the control structure selection was obtained by a given set of 11 potential controlled variables (e.g., tray temperatures in the LP column, oxygen and nitrogen content in the different product streams, etc.) and 4 potential manipulated variables (such as argon, oxygen and nitrogen flowrate). In order to get an estimate of the

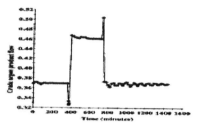 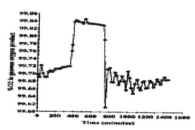

Figure 2. Operation variables during the time

controller tuning parameters (gains, reset times and set-points/biases) for each possible scheme, an open loop sensitivity analysis was previously carried out. The optimal solution found benefits also from the fact that both plant design and plant control are obtained simultaneously. The control scheme selected for this plant through the optimization framework is the best option possible and the tuning of the parameters is simultaneously optimized, so that no extra work is required in tuning, as is usually done in sequential-type approaches. By using this type of simultaneous optimization framework operational problems that may appear in a later stage due to the fact that the control structure is 'added' to an optimal design, are completely avoided.

Since the methodology has proven successful for ASPs, it holds the potential for providing a common frame of reference for engineers from process engineering, process control and process operations areas in discussing design and control strategies for general flowsheet structures. It has also shown the applicability of the MIDO technology to an industrial-size type of problem and opens the door to many more types of processes that can benefit from the technology, such as chemical, pharmaceutical, batch and periodically-operated processes.

4. Conclusions

This study demonstrates the economic and operability benefits of simultaneously considering process design and process control (including discrete decisions) within an optimization framework. This study has also shown that the state-of-the-art algorithms for solving mixed-integer dynamic optimization problems have progressed to a point where industrial problems can be readily tackled. The results from this work offer significant scope for other type of processes to benefit from the MIDO technology.

Acknowledgements

The authors would like to thank Air Products and Chemicals, Inc. for sponsoring this work through the Air Products / Imperial College Strategic Alliance. We would also like to thank the Air Products project team members for providing feedback and contributing ideas: David R. Vinson, David M. Espie, Jorge A. Mandler, Vincent White, Cristian A. Muhrer, and Mark M. Daichendt.

References

Bansal, V., 2000, PhD Thesis, Imperial College, University of London, U.K.
Brooke, A., Kendrick, D., and Meeraus, A., 1992, GAMS Release 2.25: A User's Guide. The Scientific Press, San Francisco, U.S.A.
gPROMS Advanced User's Guide, 1999, Process Systems Enterprise, London, U.K.

European Symposium on Computer Aided Process Engineering – 12
J. Grievink and J. van Schijndel (Editors)
© 2002 Elsevier Science B.V. All rights reserved.

Modeling Safety in a Distributed Technology Management Environment for More Cost-effective Conceptual Design of Chemical Process Plants

Bastiaan A. Schupp[ab], Saul M. Lemkowitz[b], L.H.J Goossens[c], A.R. Hale[c], H.J. Pasman[bd]

[a] *Corresponding Author*: Safety Science Group, Delft University of Technology, P.O. Box 5015, 2600 GA Delft, the Netherlands. e-mail: b.a.schupp@tbm.tudelft.nl
[b] Chemical Risk Management Group, Delft University of Technology
[c] Safety Science Group, Delft University of Technology
[d] TNO Defense Research, Delft, the Netherlands

Abstract

Profitability of the CPI can improve by better integrating safety into the design process. At present, conceptual designers lack means to design safety. This paper discusses a methodology, Design for Safety (DFS), that strives to provide these. It consists of two major concepts. A technology management environment, that is a distributed computer based collaborative design system, and a Safety Modeling Language.

1. Introduction

A number of major industrial accidents and environmental scandals in the early seventies led governments to impose increasingly stricter regulations. The Chemical Process Industry (CPI) complied with this new legislation, and its processes are now among the safest human activities. Though public concern about industrial safety has indeed become less pressing, a new challenge for safety management is emerging: profitability. The business environment is changing, and profitability is increasingly dependent on the ability of a market player to efficiently manage its technology (Herder 1999). Consequently, the motive for building a safe plant may well be shifting from ethical and legislative considerations to more direct economic incentives.

Today, Loss Prevention (LP) is very successful. It helps to prevent large accidents, and has reduced occupational hazards, and lowered environmental impacts. This is achieved by a number of formal methods (e.g. Hazop, QRA, IEC 61508, and LOPA), indexes (e.g. the Dow F&EI), and databases used to design protective equipment and ensure safe operation. Many advanced software tools also exist that aid design. During operation, increasingly effective Safety Management Systems help to keep risks constantly low.

Nevertheless, lacking proper means to achieve integration, Loss Prevention is still not adequately integrated at the early design stages (R&D, process development). Though not absolutely essential to safety itself, such means are required to increase profitability. In developing the Design For Safety (DFS) approach we strive to facilitate technology development such that both safety and profitability will benefit. To design and operate a technology competitively, other factors besides major accident prevention are also important. Though the potential costs of such accidents are huge, the CPI should be well able to prevent these using existing means. Thus, major accident prevention is not the first concern in developing our methodology. Facilitating the design of the optimal prevention strategy is, however.

1.1 Safety and profitability

Safety firstly influences *Capital Expenditure (CAPEX), and Operating Expenditure (OPEX)*. The investments to add protective equipment and extra investments in regular equipment, such as in more expensive materials, a different layout, redundant equipment, and similar, will increase CAPEX. The resulting complexity of a plant subsequently increases OPEX because of extra maintenance, workers, and training, and potentially more energy use and waste.

Though, more importantly, safety also adversely influences profitability because it can reduce *revenues*. Safety related problems may cause *Business Interruption (BI)* and *Market Loss (ML)*. BI may cause considerable financial losses, even without significant losses to company assets (Dreux 2001). For example, a safety trip does prevent all damage, but does not prevent (actually causes) BI. ML can potentially be worse. It can happen following an incident, but more likely because of a Time-To-Market (TTM) longer than that of a competitor. For instance following unforeseen issues that cause problems in late development stages or with regulating authorities. If a competitor takes over the market, not only years of R&D are wasted, but also all potential future revenues are lost.

1.2 Safety in process design

A designer has two basic means to reduce process risks, *prevention*, and *protection*:

- Prevention or Inherently Safer Design requires design of *alternative* technology in which hazards are absent. This means that the technology itself is changed, which must be achieved early during design, because otherwise the opportunity is lost.
- The second possibility is to design layers of protection, which requires *additional* technology to protect against the hazards still present in the primary technology. Though traditionally the most common approach, the drawbacks are complexity, and remaining hazards. Protection is added at a late (detailed) design stage.

The CPI currently focuses largely on protection, instead of at prevention. Prevention has potential benefits over protection, though. For instance, protective systems cause an increase in expenditure, and the risk of BI. Also, such systems can only be added at a late design stage, which may result in surprise. Prevention does not have these drawbacks.

Though we will not fully discuss why protection is still preferred over prevention, we will shortly explain one of the most fundamental to our approach we are developing, that also strongly relates to conceptual process design.

This can best be explained by an example. Though *hazards* can certainly be identified at the conceptual stage, e.g. flammability, toxicity or reactivity, it is difficult to tell whether this will result in a *safe plant*. The choice between two moderately hazardous solvents, a flammable one or a toxic one, or between an unstable but efficient reactor system versus a stable but less efficient reactor system, is not clear. Operational aspects, that only become known during or after detailed engineering, are relevant for this choice. Thus, the decision is taken on other grounds, and the decision on how to deal with safety is postponed till a stage that only allows for protection.

More in general, designers need means to find design alternatives (or knowledge) to synthesize this to produce a design, and to evaluate this design against the initial criteria. The problem is more complex however; designers may **not** have adequate means to:

- DESIGN safety at an early stage, thus not being provided with adequate means to find options and to decide
- Communicate amongst relevant actors (e.g. R&D, engineering)
- Anticipate on problems arising during later design phases
- Store and retrieve design information easily (e.g. intent, rationale, past experience, performance, experts)
- Apply tools continually during the design process. Currently most tools are applied only once during the design process, and are not integrated.

2. Design For Safety (DFS)

Design for safety strives to overcome these problems by facilitating reuse of past experience and communication amongst actors in the design process. Thus, to facilitate early prediction of how the proposed design is expected to perform once it becomes operational, and allowing designers of protection systems to concurrently start analyzing such systems and report problems without having to wait for the conceptual design to finish.

To achieve this, DFS uses the fact that designs are never entirely new, and therefore often reasonably well understood. For instance even if a process is chemically entirely new, the control system will be well understood, the chemistry of individual substances, and so on. Thus, also the associated risks are often better understood then from the perspective of the conceptual designer may seem the case.

This information is highly complex, and difficult to retrieve. DFS is for this reason facilitated by a knowledge-based system that has two main components, the technology management environment, and the safety modeling language. Basic to the methodology are object oriented design theory, concurrent engineering, and collaborative design systems. As these are all well-known concepts, they are not further discussed.

2.1 The technology management environment (TME)

The enabling technology for DFS is a technology management environment (TME). This is essentially a collaborative design system combined with a knowledge management system. Each actor participating in the design process is connected via a terminal and an Intranet system to a central database (Figure 1). The actor can thus find information, such as the design itself, intent and rationale, and the actors responsible for that in-

Figure 1: Technology Management Environment. The various actors collaborate using a common tool helping them to gain better insight and thus work more efficiently.

formation. The designer can also find information about previously known solutions, costs and performance. The database is not design specific. It acts as central knowledge base for all technology. The TME automatically searches for potential safety conflicts in the database, and subsequently warns the actor in charge to take action.

2.2 The Safety Modeling Language (SML)

The TME cannot function without an efficient manner to enter and use the knowledge stored in it. Creating a TME by simply connecting a set of computers to a database does not work. To facilitate this we are developing the Safety Modeling Language (SML). It is based on the Unified Modeling Language used in software engineering (Booch *et. al.* 1999). A Modeling Language obviously contains two elements. Modeling helps to visualize, document, and specify, but it also aids design activities, by offering templates (i.e. previously created models that only have to be adapted), the means for simulation and calculation, and to make problems explicit. A language offers a standardized vocabulary and rules (grammar) to communicate without confusion.

The basic philosophy is to keep the SML simple, as many must be able to apply it. The language itself must be easy to understand and use, and not produce complex diagrams (like fault trees), that cause much work, little overview, and confused decisions.

The Hazard, Barrier, Target model is the main building block of the language. Once a hazard has been identified (e.g. a toxic substance), it is relatively easy to find a related target, or vice versa. The next step is to ask whether to eliminate the hazard, or to place barriers between the hazard and the target. Thinking this way should trigger the creativity of designers, provided that ample information is available.

The language is constructed of the following elements:

- At the core of SML is the recursive Hazard, Barrier & Target model (rHBT)
- It is Object Oriented and Relation Oriented.
- It is implemented using a computer system (the TME)

Objects literally are all imaginable kinds of things that can be distinguished from other things, and thus can be given a name. Examples of objects are a table, chair, distillation column, reactor, pipe, and an operator. The latter of course are specific to chemical plants. Chemical reactions, and Hazards, Barriers and Targets are also objects.

In the SML objects are divided in two categories, design objects and HBT objects. Hazard and Target objects are associated with design objects. For instance, the substance toluene (an object present in a design) is associated with a hazard object, called flammability. Other objects present in the design are for, instance, human operators, which are target to flammable substances. Therefore, they are associated with a target object called flammability_target. This means that an Hazard-Target relation exists between humans and toluene. This relation can be analyzed, and subsequently barrier objects can be selected to protect the humans against the targets.

Barrier objects are more complex because these are associated with HT relations for which these are relevant. Barriers are, however, also targets themselves, which means that barriers can be vulnerable to functional hazards. For instance, a containment system can be a barrier to protect humans from an hazardous substance. However, this substance can also be corrosive. The containment system, is a target to this hazard. This is what we call recursion. The rHBT model is further illustrated in figure 2.

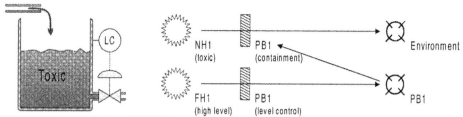

No.	Description	Comments
NH1	ToxicSubstance	The vessel contains a substance that is normally recognized as hazardous, hence normal hazard (NH)
NT1	Environment	The environment may include targest such as humans, technology or nature inside or outside battery limits. Normally these targets will be separately defined, as they may require different Barriers. In this simplified example they are lumped.
PB1	Containment System	Often, the containment system is the main barrier. The vessel is open, a high liquid level in the vessel thus is an hazard to the containment.
FH1	High Level	Though normally not recognized being a hazard, high level is functionally hazardous (FH) to PB1. *This is called recursion of the model to a lower level*
PB2	Level Control	Level control protects against high level. Functional hazards to PB2 are easily imaginable, e.g. valve failure

Figure 2: Simple example of how safety modeling in SML. Note the recursion principle.

2.3 Applications of DFS in the TME

The SML can be applied in various ways in the TME:

- For storage of information about objects, and for measurement of performance. Once an object has become part of a technology, operators can monitor for instance barriers and record their performance (e.g. failure on demand, resulting BI)

- As a design tool. Designers can identify Hazard Target pairs, and discuss with other actors how to prevent these, or about which barriers to use.

- For communication about decisions. Other actors can see what hazards and targets are present, why these are present, and what barriers are chosen for protection

- For concurrent engineering. Many Hazard Target pairs are the responsibility of multiple actors, e.g. a runaway can be prevented during *development* by adding a mass transfer limitation as a barrier, but also by adding a good control system, or a procedure during *engineering*.

3. Illustrative example: The Tosco-accident case

This example illustrates the use of SML in accident analysis, and how seemingly unrelated design stages are in fact closely related. We analyzed the January 21, 1997 accident at the Tosco Avon Refinery in Martinez, CA, USA. For details see the original EPA report (EPA 1998). The actual course of events is highly simplified here.

In this accident Loss of Containment of hot pressurized oil occurred, because a temperature excursion (runaway) in the upstream hydro cracker caused a pipe carrying the reactor product to down-stream processing units to rupture. The excursion occurred because of the formation of a hot spot in the catalyst bed. In the rHBT diagram presented in figure 3 this accident is modeled in more detail.

The prevention of such hot spots in the bed was one of the barriers that should have prevented this temperature excursion. Other barriers were a quench system, and operator intervention. This particular barrier is an example that demonstrates how multiple actors can participate in the design process. Hot spots can be prevented by a well designed catalyst bed (R&D), a well designed flow distribution system (engineering), and by regularly inspecting the catalyst bed (operation). In this case also an management of change process failed. Thus four seemingly unrelated design stages (R&D, engineering, engineering of the change, design of the inspection procedures) appear to have had a common responsibility. DFS strives to facilitate design by explicitly making such decisions starting from the design onset.

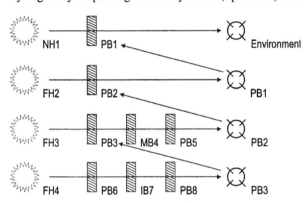

No.	Description	Comments
NH1	Hot Pressurized Oil	Hot oil from the cracker unit. Will cause an explosion when released
FH2	High Temp Oil	Above a certain temperature (and pressure) the pipe will burst.
FH3	High Temp. Cracker	Temperature excursion in cracker will disable temperature control
FH4	Hot Spot Occurrence	A Hot Spot in the bed will result in a temperature excursion.
PB1	Pipe Wall	Contains the fluid. Failure is catastrophic.
PB2	UpStream(P, T) control	The only barrier that protects PB1 from FH2
PB3	Hot Spot Prevention	This is one of the barriers to protect PB2 from FH3
MB4	Quench	Offers limited possibilities to control temperature excursion
PB5	Operator Intervention	When procedures followed very effective. Failed because of troublesome procedures.
PB6	Even flow Distribution	The system above the bed. Failed because of modification
IB7	Catalyst bed design	FH4 was overlooked during design→ not present
PB8	Catalyst bed inspection	Was not carried out
Env.	Environment	Below the unit an operator was present (to check for temperature excursion). He was killed. Also severe damage to surroundings.

Figure 3: The Tosco accident

References

P.M. Herder, Proc. Design in a changing environment., 1999, Ph.D. thesis, Delft Univeristy of Technology, Delft, the Netherlands

Mark S. Dreux, James F. Laboe, 2001, The Cost of failure, in proceedings of Making Process Safety Pay, Center of Chemical Process Safety (AIChE)

G. Booch, J. Rumbaugh, I. Jacobson, 1999 The Unified Modeling Language User Guide, Addison Wesley Longman, Reading MA

EPA Chemical Accident Investigation Report, 11/1998, Tosco Avon Refinery, Martinez, Cf.; U.S. EPA, Chemical Emergency Preparedness and Prevention Office; Washington D.C.; EPA 550-R-98-009; http://www.epa.gov/swercepp/pubs/tosco.pdf

European Symposium on Computer Aided Process Engineering – 12
J. Grievink and J. van Schijndel (Editors)
© 2002 Published by Elsevier Science B.V.

Synthesis of Azeotropic Separation Systems by Case-Based Reasoning

Timo Seuranen[1], Elina Pajula[2], Markku Hurme[1]

[1] Helsinki University of Technology, Laboratory of Plant Design,
P.O. Box 6100, FIN-02015 HUT, Finland

[2] KCL Science and Consulting, P.O.Box 70, FIN-02150 Espoo, Finland

Abstract

This paper describes a case-based reasoning (CBR) approach that supports the design of azeotropic separation process structures. CBR finds the most similar existing separation processes and applies the knowledge of their concept for solving new problems. The method is applicable especially in the early phases of process design.

1. Introduction

The objective of this paper is to introduce a method for finding feasible azeotropic separation processes by case-based reasoning (CBR). CBR is a method of reusing information of existing design cases for new designs. This means finding most alike existing processes and applying the knowledge of their separation capacity and design for solving new problems in process design. This is especially important in the early phases of process design when many alternatives should be quickly screened before a more detailed study. There is a great need for these kinds of screening tools in practice (Cornider, 2001).

The problem of synthesis of a simple separation sequences from the selection of single separations is studied in earlier papers (Pajula et al., 2001a and 2001b). However the use of CBR for synthesis of azeotropic separations is not yet discussed in detail. When dealing with multicomponent mixtures, the number of possible separation methods, their combinations and process structures to be screened is huge as well as the work involved. The synthesis method studies the physical and chemical properties of the species to be separated and uses the properties presenting most favourable possibilities for successful separation for retrieving the nearest cases to the current problem.

2. CBR in Process Synthesis

The main benefit of CBR approach is that readily available existing knowledge can be utilised systematically also in very large and complex problems like process synthesis and design. In this way the time-consuming conceptual screening phase of a design project can be fastened. Because generalisations are not needed in CBR, no data is lost. CBR gives answers to design problems in a straightforward way, but the results are dependent on the retrieval parameters and the adaptation applied. The strong interaction with the user makes the flexible and interactive use of existing data and design experience possible. The CBR search can be focused on different aspects by defining new search criteria and in this way the same case base can be used for several types of tasks. The system learns by updating the information in the database.

3. Complex Separation System

The presence of azeotropes adds some difficulties to separations and also the synthesis problem becomes much more complex. In order to separate azeotropic mixtures several technologies may be used: (Hilmen, 2000).

1. Pressure-swing distillation where a series of column operating at different pressures are used to separate binary azeotropes, which change appreciably in composition over a moderate pressure range or where a separating agent which forms a pressure-sensitive azeotrope is added to separate a pressure-insensitive azeotrope.
2. In homogeneous azeotropic distillation, a third component is added to modify the components relative volatility.
3. Heterogeneous azeotropic distillation is based on the same principle as homogeneous azeotropic distillation, but the added third component is partially miscible with one of the components, Solvent reprocessing is easy by means of a liquid-liquid separation system.
4. Reactive distillation is based on the transformation of one of the components into a component, which does not form an azeotrope with the other components.
5. Salted distillation consists in adding a ionic salt that dissociates in the liquid mixture and changes the azeotrope composition.

4. The General CBR Synthesis Algorithm

Distillation is the most feasible way to separate components in the majority of cases. Therefore the distillation related properties are studied first in the methodology, step 1. The strategy is to find first a feasible distillation system for the separations where ordinary distillation is possible. The distillation sequence is adapted from the strategy of the nearest existing design found. The remaining separation problems are solved with further reasoning which applies separation methods other than ordinary distillation. For this relative properties are calculated and the values that show potential for separation are used as retrieval parameters. The main steps in the approach are following (Pajula et al., 2001a):

Step 1: Distillation is applied whenever the relative volatility (α) is large enough. This decision to prefer distillation is up to the user. The first search for the solution is made using α's and reactivities as retrieval parameters. To make the search simpler, α's are classified as easy ($\alpha \geq 1.2$), possible, where mass separating agent (MSA) could be useful ($1.1 < \alpha < 1.2$) and difficult ($\alpha \leq 1.1$). A more accurate search is made (capacity and component types also as retrieval parameters) if several alternatives are found. The nearest strategy found is then applied in all the separations where ordinary distillation is applicable. If ordinary distillation is not feasible for all separations, the method continues to step 2.

Step 2: A suitable MSA is searched for each binary component pair that cannot be separated by conventional distillation. The retrieval parameters used are; types of components, concentrations, relative solubility parameter, dipole moment and dielectric constant. The found MSA is used for defining solubilities and other separation related properties for step 3.

Step 3: Relative physical property parameters (Jaksland et al., 1995) are calculated for each component pair that can't be separated by ordinary distillation. The parameter values are compared to the feasibility limits of different separation methods to find the feasible separation methods and to limit the search space.

Step 4. Separation strategy is searched using the relative parameters (min and max values) that are within the feasibility limits as retrieval parameters. For example crystallisation is considered very feasible if the relative melting point is greater or equal to 1.2. A more detailed search (concentration, capacity and component types also as retrieval parameters) is defined if several alternatives are found. If there are still several alternatives left, an economical comparison is needed. The separation strategy of the nearest found case is applied to the components that can't be separated by distillation.

5. Synthesis of Azeotropic Separations by CBR

The synthesis of azeotropic separations in the previous algorithm can be done by using the CBR approach. The hierarchy of CBR searches is done in the following order:
1. Separation in single column in atmospheric or non-atmospheric pressure
3. Separation in multiple columns in non-isobaric pressure
4. Separation by using MSA
5. Separation by using MSA and non-isobaric pressure
6. Separation by other means; reactive, membrane, extraction etc.
7. Separation by hybrid separations

When the separation problem is complicated, as in the case when azeotropes are present in the mixture, the definition of case description and retrieval parameters is more complex. One idea is to use the relative similarity based on the similarity of feed, product and azeotropic points. To be able to compare separation methods, where mass separating agent (MSA) is needed, a suitable MSA is searched for each binary component pair that cannot be separated by conventional distillation. The retrieval parameters used are types of components to be separated. A more accurate search is defined (concentrations, relative solubility parameter, polarity and dielectric constant also as retrieval parameters), if several alternatives are found. The found MSA is used for defining solubilities and other separation related properties. If the MSA has not been used earlier for same components, more rigorous studies, simulations and/or experiments are needed to confirm the suitability.

6. Examples for THF/Water Separation

6.1 Searching for a feasible non-isobaric distillation system

Task: Separate tetrahydrofuran (15 wt-%) from water. Purity requirement for tetrahydrofuran product is 99 wt-%.

This cannot be reached with ordinary distillation, because tetrahydrofuran and water form an azeotrope at 64 °C with 96 wt-% THF (Smallwood, 1993).

Search criteria:
1) Feed composition: 15 wt-% THF
2) Product compositions: 99 wt-% THF
3) Use pressure swing distillation
4) More than one column
5) No MSA

Using these parameters the nearest case found is:

Components: Tetrahydrofuran (THF) and water
Feed: 10 wt-% THF
Products: 99 and 3 wt-% THF
Separation: Distillation in a two column system operating at different pressure
             (p_1=1 bar, p_2=7,6 bar)

THF/water mixture contains minimum azeotrope, whose position can be shifted by changing system pressure in a two columns system (Figure 1.).

6.2: Selection of mass transfer agent

Task: Find suitable MSA for THF/water separation.

The search is made using following retrieval parameters: component type, solubility parameter, dipole moment and dielectric constant (these three describe solvent's separation capability). The nearest cases are shown in Table 1.

Table 1. Query and nearest cases in Example 6.2

	Query	Found 1	Found 2
Component 1 type	Water	Water	Water
Component 2 type	Ether	Ether	Acetate
Component 1	Water	Water	Water
Component 2	THF	Diethyl ether	Ethyl acetate
Solubility parameter	9.9	7.4	9.1
Dipole moment / D	1.75	1.3	1.7
Dielectric constant	7.6	4.34	6.02
MSA's		n-Hexane, Benzene, Toluene	n-Pentane, 2,2-Dimethyl-butane, Di-chloromethane
Similarity		0.92	0.85

The found MSA is used for defining solubilities and other separation related properties for step 3. If the MSA has not been used earlier for exactly the same components, more rigorous studies, simulations and/or experiments are needed to confirm the suitability. In this case n-pentane has been reported for THF/water separation (Smallwood, 1993).

6.3: Finding alternative separation methods

Task: Dehydrating of THF to containing less than100 ppm water.

The search is made for finding single separation method to THF dehydration. Adsorption and pervaporation seem possible separation methods based on the search.

Table 2. Query and nearest cases in Example 6.3

	Query	Found 1	Found 2	Found 3
Component types	water, ether	water, ether	water, ether	ether
Component 1	water	water	water	methanol
Component 2	THF	THF	THF	MTBE
Feed composition	5 wt-% THF	3 wt-%	15 wt-%	-
Product composition	100 ppm water	200 ppm	250 ppm	-
Separation method		Adsorption	Pervaporation	Reactive distillation
Comments		Type 5Å molecular sieves	Easily dehydrated to a few hundred ppm water	
Similarity		0.90	0.88	0.2

6.4: Finding hybrid operations

Possibility of further combined or hybrid operations need also to be taken into account. After reasoning a feasible separation system, the user should consider combining the unit operations one by one. This kind of combined operations may also be found in the retrieved cases. The query case is similar to example 6.3. The nearest case found is shown in Figure 2.

Table 3. Query and nearest cases in Example 6.4

	Query	Found 1	Found 2	Found 3
Component 1	water	water	water	water
Component 2	THF	THF	THF	THF
Feed comp.	5 wt-%	10 wt-%	15 wt-%	20 wt-%
Product comp. water	100 ppm	100 ppm	> 1 wt-%	> 1 wt-%
Separation methods	distillation/ extraction, distillation/ decant.	distillation/ extraction	distillation/ extraction	distillation/ decantation
Nr. of columns		3	3	2
MSA/solvent		NaOH	DMF	n-pentane
Similarity		0.90	0.82	0.80

348

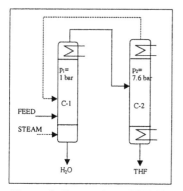

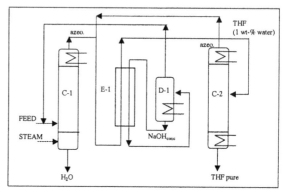

Figure 1. Separation of THF/Water mixture in two columns system.

Figure 2. Separation of THF/water mixture by distillation and extraction (salting-out).

7. Creativity and learning aspects

When new separation techniques emerge, for example hybrid membrane/distillation processes, the case base needs to be updated. When new cases are added to database process maturity factors (Pajula et al. 2001b) and feasibility limits need to be considered. Creativity of separation concepts can be enhanced by applying analogies in CBR queries. The queries can be made e.g. based on relative physical properties.

8. Conclusions

CBR separation synthesis algorithm for finding feasible separation processes and separation process structures utilising earlier design cases was extended to azeotropic separations. The main phases are CBR searches for; single column separations, multiple column separations in variable pressures, use of MSA in isobaric of non-isobaric conditions, the use of other separation methods and use of hybrid separations.

The method is useful especially in the early stages of a design, since it fasten conceptual design. The advantage compared to rule-based methods is that all the existing knowledge is available as cases and can be utilised in a non-reduced form. The method is flexible, since the user can focus the search by defining more accurate search parameters, if several nearly similar solution possibilities are available.

References

Barnicki, S. D., Fair, J.R., 1990, Ind. Eng. Chem. Res. 29(3), 421.

Cordiner, J.L., Use of prediction and modelling in early evaluation of process options, ESCAPE- 11, Elsevier 2001, 27.

Hilmen, 2000, E., Separation of Azeotropic Mixtures: Tools for Analysis and Studies on Batch Distillation Operation, Dr. Thesis, NUST, Trondheim.

Jaksland, C., Gani, R., Lien, K.M., 1995, Chem.Eng.Sci. 50 511.

Pajula, E., Seuranen, T., Hurme, M., 2001a, Selection of separation sequences by case-based reasoning, ESCAPE- 11, Elsevier, 469.

Pajula, E., Seuranen, T., Koiranen, T., Hurme, M., 2001b, Comp. Chem. Eng. 25, 775.

Smallwood, I., 1993, Solvent Recovery Handbook, McGraw-Hill Inc., New York.

European Symposium on Computer Aided Process Engineering – 12
J. Grievink and J. van Schijndel (Editors)

MINLP Retrofit of Heat Exchanger Networks Comprising Different Exchanger Types

Aleksander Soršak and Zdravko Kravanja

Faculty of Chemistry and Chemical Engineering, University of Maribor, Slovenia,

P.O. Box 219,

fax: +386 2 2527 774

Abstract

This paper describes the simultaneous MINLP retrofit of heat integrated heat exchanger networks. The stage-wise superstructure and the simultaneous model for HEN synthesis by Yee and Grossmann (1990), extended for different types of heat exchangers (Soršak and Kravanja 2001), have been now adapted to incorporate the existing exchangers for HEN retrofiting. The selection of different types of new exchangers and the (re)placement of the existing exchangers are modeled by disjunctions based on operating limitations and the required heat transfer area.

1. Introduction

Like the grass-roots design of HEN, the retrofit of an existing HEN can be traditionally carried out by two main approaches: the pinch analysis approach, e.g. Tjoe and Linhoff (1986), and the mathematical programming approach, e.g. Ciric and Floudas (1990). In order to determine the optimal retrofit topology of HEN simultaneously, the stage-wise superstructure model for HEN (Yee and Grossmann, 1990) was extended for the retrofit (Yee and Grossmann, 1991).

In general two problems can be observed in the synthesis and retrofit of HEN: (i) The first problem arises if only the pure counter flow heat exchangers are considered. In this case the solution may be suboptimal and the temperature distribution within the HEN can often be infeasible when other heat exchanger types, e.g. shell & tube heat exchangers with U-tubes, are preferred. (ii) The second problem arises if the Ft correction of the logarithmic mean is either assumed to be constant or neglected when the arrangement of streams in the exchanger is not a pure counter flow. In this case the proposed solutions can often be suboptimal or even infeasible.

In order to overcome the mentioned deficiencies, the simultaneous model by Yee and Grossmann (1990) was first extended to incorporate different exchanger types in the synthesis (Soršak in Kravanja, 2001) and now to consider different exchanger types in the retrofit. Additional energy balances for utility streams are added, constraint to limit the number of relocations and the objective function has been updated by the additional cost terms for the relocation, repiping and reconstruction of the existing exchanges.

Two examples are presented to illustrate the advantages and drawbacks of the retrofit using the proposed disjunctive MINLP model.

350

2. Modelling

2.1 Superstructure of HEN

In order to enable the selection of different exchanger types when the utility streams are used for the heat transfer, external heaters and coolers, as used in the original superstructure (Yee and Grossmann, 1990), are omitted and replaced by two additional stages within the superstructure (*Fig. 1a*). The heat transfer between the cold process and hot utility stream takes place within the first stage of the superstructure, while the heat transfer between the hot process and cold utility streams takes palce within the last stage of the superstructure. Each hot stream (*i*) can be potentially matched with each cold stream (*j*) witihin several stages (*k*). Each match is now represented by a match superstructure (SM) (*Fig.1b*) comprising new exchangers of different types, a bypass and existing exchangers of different types (*l*). Each existing heat exchanger is defined by the heat transfer area (A_l), the heat exchanger type (HET_l) and the location (POS_{ijkl}) of the exchanger within the HEN structure. So far, the double pipe (pure counter flow arrangement), shell & tube with U-tubes (combination of counter and cocurrent flow arrangement) and plate & frame (pure counter flow arrangement) exchanger types have been considered.

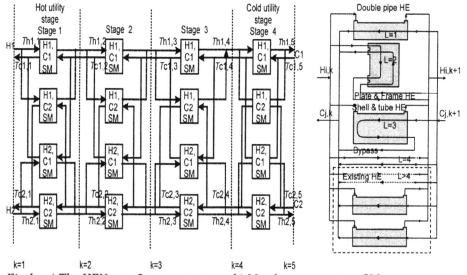

Fig.1: a) The HEN retrofit superstructure, b) Match superstructure SM

2.2 Disjunctive model for the retrofit of HEN

The model for the retrofit is based on the simultaneous model for the synthesis of HEN comprising different exchanger types (Soršak and Kravanja, 2001). The selection of types is modeled by convex hull formulations of disjunctions for different operating limitations and different required heat transfer areas. Since different types of heat exchangers involve different design geometries, additional constraints are specified to provide a feasible temperature distribution in HEN. Only new constraints and the most important modifications will now be described.

2.2.1 Energy balances of utility streams

Since the inlet and oulet temperatures of utility streams are constants, the heat capacity flowrates (F_i, F_j) for utility streams have been modelled as variables. Therefore, additional constaints have been specified:

$$T_{i2} = T_i^{OUT}; \quad HUT_i = 1$$

$$T_{jK-1} = T_j^{OUT}; \quad CUT_j = 1 \tag{1}$$

$$\left(T_i^{IN} - T_i^{OUT}\right) \cdot F_i = \sum_j \sum_k q_{ijk} \; ; CUT_j = 1$$

$$\left(T_j^{OUT} - T_j^{IN}\right) \cdot F_j = \sum_i \sum_k q_{ijk} \; ; HUT_i = 1$$

where, T_j^{IN} and T_i^{IN} are inlet temperatures for cold and hot streams, T_j^{OUT} and T_i^{OUT} are outlet temperatures for cold and hot streams, K defines the number of HEN superstructure stages, CUT is a cold utility assigning parameter that equals 1 when the cold stream is a utility stream, and HUT is a hot utility parameter that equals 1 when a hot stream is a utility stream. Thus, the linearity of the feasible space of the HEN retrofit model has been preserved.

2.2.2 Constraints to limit the number of relocations

The maximum number of relocations M_{rel} for the exisiting heat exchangers can be formulated by the following constraint:

$$\sum_i \sum_j \sum_k \sum_l y_{ijkl} \leq M_{rel} \; ; \forall l \in \{l : l > BID\} \tag{2}$$

where BID defines the index value of the bypass that is just preceeding the existing exchangers. When M_{rel} is small, the combinatorics is significantly simplified.

2.3 Objective function

The objective function contains the utility consumption costs (3a), the investment cost for the existing and new heat exchangers (3b) and the relocation costs for the existing heat exchangers (3c):

$$\sum_i \sum_j \sum_k \sum_l \left[\left(C^{cu} \cdot q_{ijkl}\right)_{j \in CUT} + \left(C^{hu} \cdot q_{ijkl}\right)_{i \in HUT} \right] + \tag{3a}$$

$$\sum_i \sum_j \sum_k \sum_l \left[Cf^l \cdot y_{ijkl} + Max\left(0, Cv^l \cdot \left(\frac{q_{ijkl}}{U_{ij} \cdot Ft_{ijkl} \cdot \Delta_{\ln} T_{ijkl}}\right) - A_l \right) \right] + \tag{3b}$$

$$\sum_i \sum_j \sum_k \sum_l \left[Cr^{ijkl} \cdot y_{ijkl} \right]_{\substack{l \in \{l:l>BID\} \\ POS_{ijkl}=0}} \tag{3c}$$

where the POS is the location parameter. It is equal to 1 for an occupied location within the HEN structure and has to be specified only for the existing heat exchangers since only the existing heat exchangers can be relocated. The temperature driving force is calculated by Chen's approximation (Chen, 1987). The parabolic smooth approximation of *Max* function is applied uniformly to estimate the investment costs of the extended area of existing exchangers and the area of new exchangers. In the latter case A_l are set to zero values. Charge coefficients are calculated according to the feasible area ranges for different exchanger types, based on the linearization of Guthrie's cost estimation function (Guthrie, 1969; Soršak and Kravanja, 2001). Originally, the Ft correction factor for the ST exchangers is defined by Underwood's equation (Hewitt et al., 1994). However, the form of the equation is inconvenient for optimization and has been replaced by the Ft approximation (Soršak and Kravanja, 1999).

3. Examples

The first example involves three hot and four cold streams (*Table 1*) within a four-stage superstructure of HEN. The HEN structure contains five heat exchangers with the annual operating costs of 283400 k$/a. Three different cases were carried out: a) the retrofit with fixed Ft, b) the one with variable Ft and c) the one where extensions of existing areas are prohibited.

Table 1: The first example data

Hot streams	FC [kW/K]	α [kW/(m²K)]	T_{in} [K]	T_{out} [K]	p [MPa]	C [$/(kW a)]
H1	250	0.6	470	400	2.0	
H2	25	1.0	450	390	1.0	
H3 (Utility)	--	5.0	510	505	1.0	250.0
Cold streams						
C1	240	0.7	330	390	1.0	
C2	13	1.0	410	500	1.0	
C3 (Utility)	--	1.0	300	320	1.0	21.0
C4 (Utility)	--	1.0	330	350	1.0	14.0

Existing heat exchangers:			
Match (i-j-k)	Type	Area [m²]	SM index (l)
1-1-2	ST	621.5	5
1-4-4	DP	125.0	6
2-2-3	PF	163.8	7
2-4-4	DP	28.7	8
3-2-1	DP	30.1	9

Table 2: Solutions of the first example.

a) Solution with the constant value of Ft ($Ft = 0.8$)				
Match (i,j,k,l)	Type	Ft	Area [m²]	Area^{Add}[m²]
1-1-3-5	ST	0.80	630.5	73.0
1-2-2-1	DP	1.00	79.7	0.0
1-4-4-6	DP	1.00	116.1	-150.0
2-2-3-7	PF	1.00	163.8	0.0
2-4-4-8	DP	1.00	28.7	17.2
3-2-1-9	DP	1.00	24.4	-5.5
b) Solution with the variable value of Ft.				
Match (i,j,k,l)	Type	Ft	Area[m²]	Area^{Add}[m²]
1-1-3-5	ST	0.88	630.5	9.0
1-2-2-1	DP	1.00	79.6	0.0
1-4-4-6	DP	1.00	116.1	-150.9
2-2-3-7	PF	1.00	163.8	0.0
2-4-4-8	DP	1.00	28.7	17.2
3-2-1-9	DP	1.00	24.4	-5.5
c) Solution with no allowed adaptation of existing heat exchangers.				
Match (i,j,k,l)	Type	Ft	Area[m²]	Area^{Add}[m²]
1-1-3-3	ST	0.88	630.5	0.0
1-2-2-1	DP	1.00	80.0	0.0
1-4-4-6	DP	1.00	116.1	-150.9
2-2-3-7	PF	1.00	163.8	0.0
2-4-4-5	ST	0.95	30.3	0.0
3-2-1-9	DP	1.00	24.4	-5.5

When Ft was fixed to 0.8 the optimal annual HEN cost was reduced to 244880 $/a (operating cost 159179 $/a plus annualized investment cost 85170 $/a). The optimal network comprises all existing heat exchangers, one new DP heat exchanger and additional 73.0 m^2 area of the ST heat exchanger (*l*=5) (*Table 2a*). The model contains 432 binary variables and was solved by MIPSYN. In the second case with variable *Ft* the optimal topology remains the same, but the annual HEN cost is reduced to 235070 $/a (*Table 2b*). Finally, when extensions of the areas are prohibited, the annual HEN cost is increased up to 265400 $/a (operating cost 159179 $/a plus annualized investment cost 75890 $/a) and a different topology is obtained: the ST heat exchanger (*l*=5) is relocated to a different position, two new heat exchangers are required (*Table 2c*) and one existing heat exchanger (*l*=8) is omitted because of the insufficient heat transfer area.

The second example comprises 5 hot and 6 cold streams (*Table 3*) with a six-stage superstructure of HEN. The stream data were taken from the HDA process case study. The HEN structure contains 12 exchangers with the annual HEN operating cost of 647060 $/a. Two retrofit cases were carried out: a) only one exchanger type (DP) was assumed, and b) different exchanger types can be selected.

Table 3: The second example data.

Hot streams	FC [kW/K]	α [kW/(m^{2}K)]	T_{in} [K]	T_{out} [K]	p [MPa]	C [$/(kW a)]
H1	49.27	0.15	823.20	299.89	3.5	
H2	27.54	0.90	330.85	329.85	3.5	
H3	1088.67	0.90	352.32	349.32	3.5	
H4	229.17	0.90	379.90	376.9	3.5	
H5 (Utility)	--	5.0	855.00	850.0	1.0	250.0
Cold streams						
C1	38.92	0.12	330.19	713.70	3.5	
C2	14.58	1.00	362.95	463.00	3.5	
C3	511.33	1.00	462.30	465.30	3.5	
C4	252.60	1.00	376.90	379.60	3.5	
C5	236.13	1.00	550.60	553.60	3.5	
C6 (Utility)	--	1.0	282.00	290.00	1.0	25.0
Existing heat exchangers						
Match (i-j-k)		Type		Area [m^2]		SM index (*l*)
1-1-2		ST		774.8		5
1-1-3		ST		736.8		6
1-1-4		DP		53.9		7
1-2-4		DP		156.0		8
1-3-3		DP		103.6		9
1-4-4		DP		47.1		10
1-5-2		DP		28.6		11
1-6-6		DP		919.8		12
2-6-6		DP		1.3		13
3-6-6		DP		106.4		14
4-6-6		DP		15.7		15
5-1-1		DP		76.4		16

The first case with 2100 binary variables was solved by MIPSYN using the Reduced integer strategy (RIS) by Soršak and Kravanja (in print). The annual HEN cost was reduced to 379640 $/a (operating cost 262060 $/a plus annualized investment cost 117580 $/a). The optimal topology comprises 11 existing heat exchangers, while one existing heat exchanger (*l*=9) is omitted (*Table 4a*). It is interesting to note that the obtained temperature distribution is not feasible for exchangers *l*=5 and *l*=6 which are actually ST exchangers. Therefore, the proposed solution, where only DT exchanger type is assumed, cannot be used for the retrofit of the exisiting network.

The example was solved by MIPSYN again, now considering different exchanger types. The annual HEN cost was reduced to 430020 $/a (operating cost 262060 $/a plus annualized investment cost 167960 $/a). The optimal topology differs from the previous one (*Table 4b*). Now 11 existing heat exchangers are involed in the HEN structure and two new ST exchangers are required. The entire family of solutions now contains a feasible temperature distribution. However, the number of binary variables is increased to 2700.

Table 4: Solutions of the second example.

a) Solution with the DP heat exchanger type.

Match (i,j,k,l)	Type	Ft	Area [m^2]	Area^{Add}[m^2]
1-1-2-5	DP	1.00	1013.9	239.1
1-1-3-6	DP	1.00	736.9	0.0
1-1-4-7	DP	1.00	254.5	200.3
1-2-4-8	DP	1.00	186.3	30.3
1-3-2-13	DP	1.00	46.6	45.3
1-4-4-10	DP	1.00	56.5	9.4
1-5-2-11	DP	1.00	34.6	6.0
1-6-6-12	DP	1.00	849.4	-70.3
2-6-6-16	DP	1.00	1.3	-75.1
3-6-6-14	DP	1.00	106.4	0.0
4-6-6-15	DP	1.00	15.7	0.0

b) Solution with all heat exchanger types.

Match (i,j,k,l)	Type	Ft	Area [m^2]	Area^{Add}[m^2]
1-1-2-5	ST	0.80	758.4	-16.5
1-1-3-6	ST	0.84	704.7	-32.2
1-1-4-3	ST	0.95	156.1	156.1
1-1-5-3	ST	0.81	649.2	0.0
1-2-5-8	DP	1.00	156.0	0.0
1-3-4-9	DP	1.00	111.6	8.0
1-4-4-10	DP	1.00	26.1	-21.0
1-5-2-11	DP	1.00	25.6	-3.0
1-6-6-12	DP	1.00	849.4	-70.3
2-6-6-13	DP	1.00	1.3	0.0
3-6-6-14	DP	1.00	106.4	0.0
4-6-6-15	DP	1.00	15.7	0.0

4. Conclusions

Both examples clearly demonstrate the advanteges of the proposed model. The proposed model allows a simultaneous heat integrated retrofit of HEN and enables the selection of optimal and feasible heat exchanger types during the optimization.

5. References

1. Ciric A.R. and C.A. Floudas, Comprehensive Optimization Model of the Heat Exchanger Network Retrofit Problem, Heat Recovery Systems & CHP, 10, 407-422 (1990)
2. Chen J.J.J., Letter to the editors: comments on improvement on a replacement for the logarithmic mean, *Chem.Engng.Sci.*, 42, 2488-2489 (1987)
3. Guthrie, K. M., Capital cost estimating, *Chem.Engng.*, 76, 114 (1969)
4. Hewitt G.F.,Shires, G.L. and Bott, T.R., Process heat transfer, *CRC Press*, 155-194 (1994)
5. Soršak, A. and Z. Kravanja, Simultaneous MINLP synthesis of heat exchanger networks comprising different exchanger types, Computers and Chem. Engineering Supplement, (2001)
6. Tjoe T.N. and B. Linhoff, Using pinch technology for process retrofit, Chemical Eng., 28,47-60 (1986)
7. Yee, T.F. and I.E. Grossmann, Optimization models for heat integration-II, Heat exchanger network synthesis, Computers and Chem.Engng., 14, 1165-1184 (1990)
8. Yee, T.F. and I.E. Grossmann, A Screening and Optimization Approach for the Optimal Retrofit of Heat Exchanger Networks, Ind. Eng. Chem. Res., 30, 146 (1991)

European Symposium on Computer Aided Process Engineering – 12
J. Grievink and J. van Schijndel (Editors)
© 2002 Elsevier Science B.V. All rights reserved.

Boundary Crossing during Azeotropic Distillation of Water-Ethanol-Methanol at Total Reflux: Influence of Interphase Mass Transfer

P.A.M. Springer and R. Krishna
Department of Chemical Engineering, University of Amsterdam
Nieuwe Achtergracht 166, 1018 WV Amsterdam, The Netherlands

Abstract

Experiments were carried out in a bubble cap distillation column operated at total reflux with the system: water (1) – ethanol (2) – methanol (3). This system has a binary azeotrope for the water-ethanol mixture, which leads to a slightly curved simple distillation boundary between the azeotrope and pure methanol. For certain starting compositions the measured distillation composition trajectory clearly demonstrate that crossing the distillation boundary is possible. In order to rationalize our experimental results, we develop a rigorous nonequilibrium (NEQ) stage model, incorporating the Maxwell-Stefan diffusion equations to describe transfer in either fluid phase. The developed NEQ model anticipates the boundary crossing effects and is in excellent agreement with a series of experiments carried out in different composition regions. In sharp contrast, an equilibrium (EQ) stage model fails even at the qualitative level to model the experiments. It is concluded that for reliable design of azeotropic distillation columns we must take interphase mass transfer effects into account in a rigorous manner.

1. Introduction

Residue curve maps have proven to be very useful as tools for the developing and design of column sequences in the field of distillation (e.g. Doherty and Malone, 2001). As such they can be used to predict the composition trajectories for packed and trayed distillation columns, provided that vapor and liquid phases are in thermodynamic equilibrium and the column is operating at total reflux. In the case of ternary azeotropic distillations, the residue curve map will be divided into two, or more, regions by the distillation boundaries. These boundaries cause more restrictions to the distillation operation considering the possible trajectories to be followed, in particular concerning possible "boundary-crossing". The existence, location and curvature of the distillation boundaries play a very important role in this. However, the constraints on possibilities of crossing any distillation boundary are based on the use of the equilibrium (EQ) stage model. There is evidence in the published literature (Pelkonen et al., 1997) that experimentally measured composition profiles are better simulated with nonequilibrium (NEQ) stage models, in which proper account is taken of mass transfer in either fluid phase by use of the Maxwell-Stefan diffusion equations (Taylor and Krishna, 1993). The Maxwell-Stefan formulation, based on the thermodynamics of irreversible processes, takes proper account of diffusional "coupling" between the species transfers i.e. the flux of any species depends on the driving forces of all the species present in the mixture. In a distillation column, the influence of species coupling manifests itself in significant differences in the component mass transfer efficiencies E_i^{MV}. The experimental results of Pelkonen raise the question whether the observed dramatic differences between EQ and NEQ model predictions are also obtained when the starting composition are not located precisely on the distillation boundary but on either side of it. The major objective of our work is to demonstrate that distillation boundaries can be crossed provided that the starting compositions are located within a finite region of composition on one side of the distillation boundary. Furthermore we aim to show that such boundary crossing phenomena can be predicted by the NEQ models and can be attributed to differences in component Murphree efficiencies.

2. Experiments

Experimental set-up
The experiments were carried out in a laboratory-scale distillation column supplied by Scott Nederland B.V. A complete description of the hardware and pictures of the column and bubble cap trays can be found on our web-site: http://ct-cr4.chem.uva.nl/distillation/. Also the experimental executions and analyze methods used for receiving the experimental data are extensively explained on this web-site.

Experimental results
The experimentally determined composition trajectories for a set of two experiments are shown in Fig. 1(a-b), along with the residue curve map for the system water-ethanol-methanol. At total reflux the composition of the vapor leaving any given stage equals the composition of the liquid arriving at that stage from above. Therefore, the vapor and liquid composition samples can be combined when plotting the experimental composition trajectories.

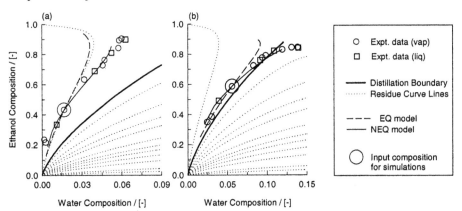

Fig 1. Experimental results (circles for vapor and squares for liquid samples) showing the column composition trajectories for the Water-Ethanol-Methanol system. Also shown are the simulation results showing the trajectories calculated by the EQ stage model (short dashed) and the NEQ stage model (solid) along with the residue curve map.

In the first experiment, Fig. 1(a), the column trajectory is located completely on the left side of the distillation boundary (indicated by a thick line). The second experiment exhibits boundary-crossing phenomena, Fig. 1(b). We also note that the experimental data points cut across the residue curves to the right at a sharp angle. Clearly, boundary-crossing phenomena is not in conformity with the assumption of thermodynamic phase equilibrium, which underlies the residue curve maps; this is evidenced by the fact that the experimental trajectories do not follow the residue curve lines. In order to understand, and rationalize, the boundary-crossing phenomena we develop a rigorous nonequilibrium (NEQ) stage model.

3. Nonequilibrium stage model development

The development of the NEQ stage model follows the ideas and concepts described in earlier publications (Taylor and Krishna, 1993; Krishnamurthy and Taylor, 1985). All our experiments were carried out in the bubble froth regime and visual observations showed that the bubbles were roughly of uniform size and shape. Therefore, we assume that the bubbles rise in plug flow and that the liquid phase is well mixed. The steady state component mass balance for for a total reflux operation can be written in terms of the overall number of transfer units for the vapor phase $[NTU_{OY}]$:

$$\frac{dy}{d\xi} = [NTU_{OY}](y^* - y),$$ (1)

where $\xi = h/h_f$ is the dimensional distance along the froth. Assuming that the matrix of overall mass transfer coefficients $[K_{OY}]$ does not vary along the froth height, $[NTU_{OY}]$ can be expressed by:

$$[NTU_{OY}] \equiv \int_0^{h_f} [[K_{OY}]a'/V_b] dh \equiv [K_{OY}]a'\tau_V$$ (2)

This implies that $[NTU_{OY}]$ can be calculated from knowledge of $[K_{OY}]$, the interfacial area per unit volume of vapor a' ($a'=6/d_b$; we assumed the bubbles to be spherical in shape with a diameter d_b) and the vapor residence time ($\tau_V=h_f/V_b$, where V_b is the bubble rise velocity estimated using the equation of Mendelson (1967)). The overall mass transfer coefficient, $[K_{OY}]$, is given by the addition of resistances formula:

$$[K_{OY}]^{-1} = [k_y]^{-1} + \frac{c_t^V}{c_t^L}[K_{eq}][k_x]^{-1},$$ (3)

in which $[K_{eq}]$ represents the diagonal matrix of K-values and $[k_y]$ and $[k_x]$ are the partial transfer coefficient matrices for the vapor and liquid phases respectively. For the vapor phase, the elements $k_{y,ij}$ can be estimated from the mass transfer coefficients of the constituent binary pairs, $\kappa_{y,ij}$, by making use of the Maxwell-Stefan formulation (Taylor and Krishna, 1993):

$$k_{y,ii} = \frac{z_i}{\kappa_{y,in}} + \sum_{k=1}^{n} \frac{z_k}{\kappa_{y,ik}}, \qquad k_{y,ij} = -z_i \left(\frac{1}{\kappa_{y,ij}} - \frac{1}{\kappa_{y,in}} \right) \qquad (4)$$

The partial transfer coefficient matrix for the liquid phase $[k_x]$ is calculated analogously. Hence, the definition of these binary mass transfer coefficients becomes essential for the description of the complete mass transfer model. Assuming that the contact time of the spherical bubbles is long enough so that the Fourier numbers are larger than about 0.06 ($Fo_{ij} = 4 Ð_{y,ij} \tau_V / d_b^2$), the binary vapor mass transfer coefficients $\kappa_{y,ij}$ can be estimated from the following reduced equation (Taylor and Krishna, 1993):

$$Sh = \frac{\kappa_{y,ij} d_b}{Ð_{y,ij}} = \frac{2}{3} \pi^2 \quad resulting \ in: \quad \kappa_{y,ij} = \frac{2\pi^2}{3} \frac{Ð_{y,ij}}{d_b} \qquad (5)$$

Eq. (5) leads to the important conclusion that $\kappa_{y,ij}$ would have an unity power dependence on the vapor diffusivity $Ð_{y,ij}$, which is in sharp contrast with the square-root dependence for small values of Fo; small vapor residence times.

The binary liquid mass transfer coefficients $\kappa_{x,ij}$ can be obtained from the penetration model (Taylor and Krishna, 1993). Evaluating the individual contributions of the liquid and vapor phase in eq (3) indicates that the mass transfer resistance is predominantly in the vapor phase.

Once the bubble diameter is set, eq. (1) can be easily integrated and solved for the composition leaving a distillation stage. Hence, the only parameter needed to solve the system of equations is the bubble diameter.

The operation pressure for all experiments was 101.3 kPa and the ideal gas law was used. Activity coefficients were calculated using NRTL parameters (Gmehling and Onken, 1977) and the vapor pressures using the Antoine equations. The vapor phase was assumed to be thermodynamically ideal. The reflux flow rate (0.006 mol/s) and bottom product flowrate (0.0 mol/s) were used for specifying the column operations. Since the column is operated at total reflux, the reflux flow rate determined the inner flowrates of vapor and liquid phases on each stage. Simulations of total reflux operations are "complicated" by the fact that there's no feed to the column at steady state. To overcome this problem we specify one of the experimental determined composition of the streams leaving or entering a stage as input parameter. The simulated profile of the total reflux run is forced to pass through this specified composition (vapor leaving stage 4; indicated by the large open circle in Fig. 1). The entire set of equations system was solved numerically by using the Newton's method (Krishnamurthy and Taylor, 1985).

4. Comparison of EQ and NEQ simulations with experiments

Spread over a wide region of the residue curve map, several experiments were carried out and simulated with the EQ stage model and the rigorous NEQ stage model, Fig. 2(a). Two of the whole set of experiments will be pointed out here and considered further in some detail. Let us first consider the experiment shown in Fig. 1(a). We note that while the experimental points cross the distillation boundary, the EQ model does not and remains on the left side, following closely the residue curves. Moreover, the experiments show that proceeding down the column, the compositions get richer in water, whereas the EQ model predicts that these trays should get richer in ethanol. The NEQ model requires specification of the bubble diameter. Decreasing the bubble diameter has the effect of increasing the mass transfer coefficient and makes the NEQ model tend towards the EQ model, see eq. (5). A range of bubble diameters has been tested from 1.5mm to 5.5mm. The best agreement with the experiments is obtained with $d_b = 5.0$mm. Fig. 1(b) shows that the NEQ model successfully anticipates the boundary crossing phenomena. Especially on the qualitative level, the NEQ model does a good job by predicting that the compositions proceeding down the column get progressively richer in water just like the experiments are showing. Even if the experiment does not depict any boundary crossing, the NEQ model still predicts better results than the EQ model, Fig. 1(a). Both the experiments and the NEQ model have the same tendency to cut across to the right of the residue curve, whereas the EQ model follows the trajectory dictated by the residue curve map.

In order to show that the bubble diameter $d_b = 5.0$mm is not merely a "convenient fit" of our ternary experiment, we also carried out several experiments with four different binary systems. Also for these experiments the NEQ simulations (with $d_b = 5.0$mm) describe the column trajectories very well for all the experimental results generated in our laboratory. We may conclude from the foregoing that boundary crossing is caused by multicomponent mass transfer effects. To explain this in some detail we consider the boundary crossing run, Fig. 1(b). From calculations, we confirmed that the *Fo* values exceed 0.06 in all cases, justifying the use of eq. (5).

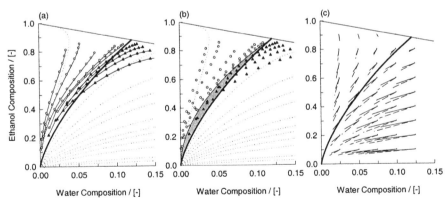

Fig 2. (a) NEQ simulation results over a wide range (b) All trajectories from (a) having compositions within the gray shaded "crossing" region will cross the distillation boundary to the right of this boundary (c) Calculated direction vectors using the EQ stage model (100% eff, dashed lines) and the NEQ stage model (solid lines) using a bubble-diameter of 5.0 mm.

To understand the phenomena of boundary crossing, we consider the component Murphree efficiencies. For the EQ model, the component efficiencies are all equal to unity. For the NEQ model the component efficiencies will, in general, differ from one another. The origin of the differences in E_i can be traced to the differences in the binary pair vapor diffusivities $Ð_{y,ij}$. Differences in the component efficiencies cause the actual composition trajectory followed on any given stage ($y_{i,L}$-$y_{i,E}$) to differ from the trajectory dictated by the EQ vector (y^{*}_{i}-$y_{i,E}$), see Fig. 2(c). Also in Fig. 1(a), it shows that the NEQ model trajectory cuts across to the right of the residue curves unlike the EQ model trajectory. It is this tendency to cut towards the right of the composition space that causes boundary crossing. By performing several NEQ simulations with various "starting compositions", we can determine the region within which the column trajectory will cross the distillation boundary from the left and end up with reboiler compositions in the right region of the distillation boundary, Fig. 2(b).

5. Conclusions

The following major conclusions can be drawn from the work presented above.

- The measured composition trajectories during distillation of water-ethanol-methanol under total reflux conditions in a bubble cap distillation column clearly demonstrate that crossing of a distillation boundary is possible
- An NEQ stage model is able to model the experimental results. The experimental results agree very well with the developed model in which a bubble size of 5.0mm is chosen. The NEQ model correctly anticipates boundary crossing
- An EQ model fails to anticipate boundary crossing in any experiment. The EQ model provides a much poorer representation of the column composition trajectories and do not even agree qualitatively with the experimental results.
- The differences in the NEQ and EQ trajectories emanate from differences in the component Murphree efficiencies, which in turn can be traced to differences in the binary pair vapor phase diffusivities $Ð_{y,ij}$.

References

Doherty, M.F. & Malone, M.F. (2001) *Conceptual design of distillation systems*, McGraw-Hill, New York.

Gmehling, J.L. & Onken, U. (1977) *Vapor-liquid equilibrium data collection*, Dechema: Frankfurt, Germany.

Krishnamurthy, R. & Taylor, R. (1985) Nonequilibrium stage model of multicomponent separation processes, *American Institute of Chemical Engineers Journal*, **32**, 449-465.

Mendelson, H.D. (1967) The prediction of bubble terminal velocities from wave theory, *American Institute of Chemical Engineers Journal*, **13**, 250-253.

Pelkonen, S., Kaesemann, R. & Gorak, A. (1997) Distillation lines for multicomponent separation in packed columns: theory and comparison with experiment, *Industrial and Engineering Chemistry Research*, **36**, 5392-5398.

Taylor, R. & Krishna, R. (1993) *Multicomponent mass transfer*, John Wiley: New York.

European Symposium on Computer Aided Process Engineering – 12
J. Grievink and J. van Schijndel (Editors)
© 2002 Elsevier Science B.V. All rights reserved.

Comparison of different mathematical programming techniques for mass exchange network synthesis

Z. Szitkai*, A.K. Msiza, D.M. Fraser, E. Rev*, Z. Lelkes*, Z. Fonyo*

Chemical Engineering Department, University of Cape Town,
Private Bag, Rondebosh, 7701, South Africa

*Chemical Engineering Department, Budapest University of
Technology and Economics, H-1521 Budapest, Hungary

Abstract

In this paper two design techniques, based on mathematical programming, for the synthesis of mass exchange networks (MENs) are compared. Problems not generally dealt with in the literature, and associated with these techniques, are highlighted. A method is presented for generating several feasible initial solutions to avoid accepting poor local solutions as final designs. A method of handling the discontinuity of the Kremser equation used for determination of the number of stages is also discussed. The new method does not rely on the use of integer variables. In addition, a method of generating MINLP solutions that feature integer stage-numbers is also presented. It is shown that insight-based superstructures assuming vertical mass transfer may fail to include the optimal structure of the MEN. Solutions of several MEN synthesis example problems are presented. Advanced pinch and mathematical programming-based solutions are compared. Both simple and advanced capital costing functions are used for the estimation of the total annual cost (TAC) of the MENs.

1. Introduction

Mass exchange networks (MENs) are systems of interconnected direct-contact mass-transfer units that use process lean streams or external mass separating agents to selectively remove certain components (often pollutants) from rich process streams.

The notion of mass exchange network synthesis (MENS) and a pinch-based solution methodology for the problem were presented first by El-Halwagi and Manousiouthakis (1989). Subsequently, Papalexandri et al. (1994) presented a mixed integer non-linear programming (MINLP) design technique for MENS. Later, Hallale and Fraser (1998) extended the pinch design method by setting up capital cost targets ahead of any design. Recently, Comeaux (2000) presented an optimisation based design methodology where the notion of vertical mass transfer is used to develop a superstructure optimised by non-linear programming (NLP). Using optimisation-based techniques, capital and variable costs of the network can be optimised simultaneously even in cases of large and multi-component design problems.

Our objective is to compare the different solution methodologies that can be used for MENS. In the first part of the paper the NLP design method of Comeaux is compared to the MINLP method of Papalexandri. After dealing with practical problems of the mathematical programming methods, the last part of this paper presents MINLP solutions for thirteen example problems solved with pinch methodology by Hallale (1998).

2. Comparison of two optimisation-based techniques

The MINLP method of Papalexandri and the insight-based NLP solution of Comeaux are compared here. The example problem 5.1 of Hallale (1998) is solved using both of the methods. Two different techniques for capital costing are used. At first, a simple capital costing procedure is applied, where it is assumed that the capital cost of an exchanger (in USD/yr) can be calculated by multiplying its theoretical number of stages by 4552 (see Papalexandri, 1994). Secondly, the advanced, exchanger volume-based capital costing of Hallale (1998) is used. Total annual costs (TAC) of the designed MENs can be found in Table 1. As a basis for the comparison, TACs of the supertargeted pinch solutions of Hallale and Fraser (1998) are also shown.

Table 1: Total annual costs of the different solutions of Hallale's example problem 5.1

Solution method	Simple capital costing	Advanced capital costing
Advanced pinch method of Hallale and Fraser	* 320,000 USD/yr. (target)	* 228,000 USD/yr.
NLP method of Comeaux	* 332,000 USD/yr.	249,150 USD/yr.
MINLP method of Papalexandri	306,108 USD/yr.	221,357 USD/yr.

* Original solutions of the referred authors are marked by an asterisk

The network obtained by the insight based NLP method using advanced capital costing is presented in Figure 1. Our MINLP designs based on Papalexandri's method can be seen in Figures 2 and 3 with simple and advanced capital costing structures, respectively.

Table 1 shows that the TAC of the MINLP designs and the advanced pinch solutions are approximately the same. This is seen more clearly when the structure of these solutions is compared. The costs also become closer when rounding up the stage numbers of the MINLP solutions (to $322 000 and $226 000 respectively). Table 1 also shows that Comeaux's method gives more expensive solutions. Hallale's design (see Figure 5.4 of Hallale (1998)) cannot be reproduced by using the insight-based NLP method of Comeaux. This is because the insight based superstructure for the example (see Figure 4.2 of Comeaux (2001)) does not enclose the structure of Hallale's solution. In order to be able to find Hallale's solution, additional possible matches should be added to the superstructure generated using the principle of vertical mass transfer. Although the NLP-based design method is computationally less expensive than the method of using MINLP, it lacks the advantages of the latter. Logical conditions for fixed costs, integer stage numbers, etc. cannot be handled in an NLP model.

Throughout this paper, optimisation problems are solved using the GAMS program package (Brook *et al.*, 1992). CONOPT-1 was used to solve the NLP models and DICOPT++ (running OSL and CONOPT-1) was used for the MINLP solutions.

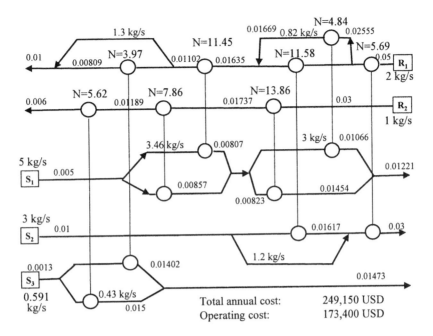

Figure 1: Insight-based NLP solution of Hallale's example problem 5.1. The capital cost of the network is calculated by Hallale's advanced capital costing method.

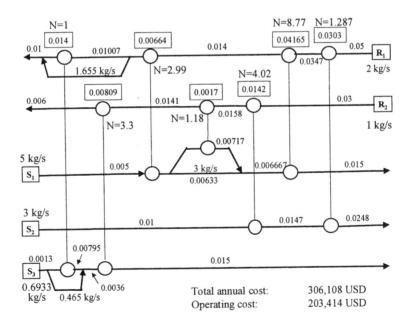

Figure 2: Our MINLP solution of Hallale's example problem 5.1. The solution is obtained by using a simple capital cost correlation.

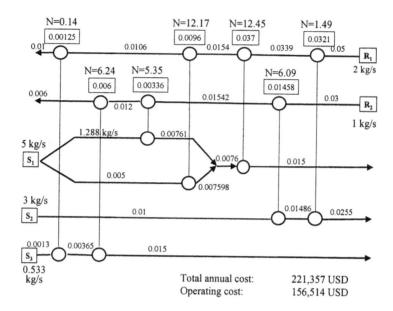

Figure 3: Our MINLP solution of Hallale's example problem 5.1. The capital cost of the network is calculated by the advanced capital costing method of Hallale and Fraser.

3. Three practical problems of using optimisation-based design techniques for mass exchange network synthesis

3.1 Generation of feasible initial solutions

Since the feasible space and often also the objective function of NLP and MINLP models for MENS are non-convex, global optimality of the solution cannot be guaranteed. Solutions obtained from different initial values have to be compared to prevent the selection of poor local solutions as final designs. Once a feasible solution is found starting from a particular initial solution, several other feasible initial values can be generated, by changing the objective function of the original problem. A solution featuring minimum MSA flowrates can, for example, be used as the initial solution for the original problem. Simple pinch solutions can also serve as intial solutions. If the problem is not severely non-convex, the initial feasible solution can be obtained from any nonzero set of initial values for the variables in the model.

3.2 NLP solution for the discontinuity problem of the Kremser equation

The capital cost of multistage exchangers is usually calculated using the Kremser equation. The Kremser equation determines the required number of equilibrium stages for a specified separation when the phase equilibrium relation is considered linear. Numerical problems arising from the discontinuity of the Kremser equation can be easily overcome in the case of MINLP formulations. When the MENS problem is formulated as a pure NLP problem, the binary variable for choosing between the two calculated stage-numbers (see Szitkai *et al.*, 2001) can be approximated by a continuous variable. Equation 1 can be introduced:

$$A - 1 = (A - 1 + \varepsilon) \cdot w \tag{1}$$

where A is the absorption factor of the mass exchange unit, and ε is a small positive number (eg. 0.001). The variable w takes the value of zero when $A=1$, and is close to one when $A \neq 1$. Using this method, the total cost of the network must be recalculated after optimization. Experience has shown that the difference is usually small. An obvious solution for the Kremser discontinuity problem could be the introduction of Equation 2, which prevents A to take the value of unity:

$$(A - 1)^2 \geq \varepsilon \tag{2}$$

According to our computational experience however, this method leads to severe numerical problems during the solution of the NLP, and therefore cannot be applied.

3.3 Integer stage numbers

The designer may often want to get solutions whereby the number of stages (N) are reported as integers. Buliding up the integer-values by according to Equation 3 requires the introduction of additional binary variables (y_i) in the MINLP model.

$$N = 2^0 \cdot y_1 + 2^1 \cdot y_2 + 2^2 \cdot y_3 + \ldots \tag{3}$$

When the expected number of stages for the exchangers in the superstrucure is large, the additional binary variables may extend the MINLP problem size over the solvability limit. The same method however, can be used in a two-level optimisation approach. After obtaining the solution with non-integer stage numbers, a second optimisation can be carried out, where the structure of the network is now fixed to that of the first solution (with non integer stage numbers). In the second level only the operating parameters of the network and the stage numbers around the first solution are optimised. For example, if an exchanger in the first level optimum has 16.3 stages then in the second level Equation 4 is introduced, allowing $15 \leq N \leq 18$:

$$N = 15 + 2^0 \cdot y_1 + 2^1 \cdot y_2 \tag{4}$$

In this way fewer binary variables are needed. However, the two-level method fails to find the optimum when the rounding affects the optimal structure of the network.

4. Solved MENS example problems

Without going into details, our MINLP solutions for most of the MENS example problems of the thesis of Hallale (1998) are presented in Table 2. Optimising for TAC, in most cases equally good solutions are obtained, with the exception of task 6.2 where the concentration range extends over six orders of magnitude hampering the scaling. Optimising just for CAP, usually better solutions are obtained, except in the case of example 3.1. Why the pinch solutions are not always reached with MINLP can be accounted for the inherent nonconvexity of the problem.

366

Table 2: Comparison of our MINLP solutions and the advanced pinch solutions of Hallale (1998). CAP indicates that the network was optimized only for its capital cost at fixed lean stream flow rates (operating costs).

Example	Objective function	Nick Hallale Target / Design (US$/yr.)	Our MINLP Solutions (US$/yr.)	MINLP vs. pinch
3.1	CAP	830 000 / 860 000	987 354	+14.8 %
3.2	CAP	448 000 / 455 000	453 302	-0.4 %
3.3	CAP	819 000 / 751 000	615 287	-18.1 %
3.4	CAP	591 760 / 637 000	637 280	0.0 %
4.1	CAP	296 000 / 298 000	255 068	-14.4 %
5.1	TAC	226 000 / 228 000	226 000	-0.9 %
5.2	TAC	226 000 / 228 000	226 000	-0.9 %
5.3	TAC	226 000 / 228 000	226 000	-0.9 %
5.4	TAC	49 000 / 49 000	50 279	+2.6 %
5.5	TAC	524 000 / 526 000	527 000	+0.2 %
6.1	TAC	692 000 / 706 000	720 000	+2.0 %
6.2	TAC	28 000 / 28 000	32 000	+14.3 %
6.3	CAP	591 000 / 539 000	536 000	-0.6 %

Conclusions

MINLP optimisation seems to be superior to the NLP optimisation of insight-based superstructures. The practical problems which were identified in optimisation-based MENS, namely generation of feasible initial solutions, discontinuity of the Kremser equation, and integer stage numbers can, however, be effectively handled, as described in Section 3. The solutions of MENS problems by the advanced pinch methods of Hallale and Fraser can be replaced by equally good solutions obtained using the MINLP methodology, presented in this paper.

References

Brook, Kendrick, Maereus, GAMS A User's Guide, Release 2.25, Scientific Press, 1992
Comeaux, R.G., (2000), Synthesis of MENs With Minimum Total Cost, MSc Thesis, Dept.of Process Integration, UMIST, Manchester, England.
El-Halwagi, M.M, and Manouthiousakis, V., (1989), Synthesis of Mass Exchange Networks, AIChEJ, **8**, 1233-1244.
Hallale N. and Fraser D.M., (1998), Capital Cost Targets for Mass Exchange Networks, Part I-II., Chem. Eng. Science, **53**(2), 293-313.
Hallale N. and Fraser D.M., (2000), Supertargeting for Mass Exchange Networks, Part I-II., Trans I Chem E, **78**, Part A, p.202-216
Hallale N., (1998), Capital Cost Targets for the Optimum Synthesis of Mass Exchange Networks, PhD thesis, University of Cape Town, Dept. of Chemical Engineering
Papalexandri K.P., Pistikopoulos, E.N., and Floudas, C.A. (1994), Mass Exchange Networks for Waste Minimization, Trans IChemE, **72**, Part A, 279-293.
Szitkai Z., Lelkes Z.,Rev E., Fonyo Z., Solution of MEN synthesis problems using MINLP: Formulations of the Kremser equation, ESCAPE-11, Elsevier, 2001.

European Symposium on Computer Aided Process Engineering – 12
J. Grievink and J. van Schijndel (Editors)
© 2002 Elsevier Science B.V. All rights reserved.

Optimal design of gas permeation membrane & membrane adsorption hybrid systems

Ramagopal Uppaluri[+], Robin Smith[+], Patrick Linke[#], Antonis Kokossis[#]

[+] Dept. of Process Integration, UMIST, Manchester, M60 1QD, UK
[#] Dept. of Chemical and Process Engineering, University of Surrey, Guildford, GU2 7XH, UK

Abstract

This work presents systematic synthesis procedures able to develop robustly optimal process designs for gas permeation membrane and hybrid membrane systems. A superstructure representation is proposed for membrane permeator compartments (stages), pressure equipment such as feed compressors, permeate recycle compressors, product compressors and vacuum pumps and separators (adsorbers) to capture alternative separation options. Allocation of vacuum pumps is followed using generic methodology. Various flow patterns such as cross-flow, counter-current and co-current can be simultaneously considered in the synthesis framework. The proposed representation builds upon previous efforts in reaction/separation process synthesis and can consider a variety of structural and operational options. Such options emerge from a number of possible module and process layouts. This work employs stochastic optimisation techniques in the form of simulated annealing. Applications of the proposed methodology to different case studies will be presented. Examples include the separation of air, up-gradation of lean hydrogen hydro-cracking refinery stream and acid gas removal from natural gas streams using membrane and fixed bed adsorption networks.

Introduction

Design of gas permeation membrane networks received considerable attention from various researchers (Lababidi et al [1], Qi and Henson [2], Qi and Henson [3]) whose procedures suggest deterministic techniques in the form of non-linear programming (NLP) and mixed-integer non-linear programming (MINLP) techniques. The objective behind this work is to develop an efficient and robust modelling and optimisation framework for the synthesis of gas permeation membrane and hybrid membrane systems so as to offer possible extensions to other membrane and hybrid membrane processes.

Synthesis representation

A superstructure representation is proposed for gas permeation membrane and hybrid membrane networks and allows two states for high pressure (retentate) and low pressure (permeate) side. Modelling of membrane stage (MEM) from the concept of shadow reactor compartment as proposed by Mehta and Kokossis [4] and is modified to represent membrane compartment (stage) as a plug flow reactor (PFR) without side streams and reaction with diffusional mass transfer between permeate and retentate compartments. Pressure drops along the compartment are neglected both on retentate and permeate side. Separator (SEP) modelling follows from the state task unit (STU) representation of Mehta, Linke & Kokossis [5] and allows sharp splits between the components.

Stream distribution on high pressure (retentate) side and low pressure (retentate) side follows from Linke, Mehta and Kokossis [6]. Stream distribution on retentate side allows all possible distribution options for feed, compartment (retentate compartment/separator), compartment product and product retentate streams. Stream distribution on permeate side allows recycle of compartment permeate and product permeate streams to compartments on retentate side using recycle compressors (RCOM). Vacuum application can be substantiated by the generic allotment of vacuum pumps (VAC). Feed compressor (FCOM) can be included using cost expressions if necessary.

The proposed representation can consider a variety of structural and operational options. Such options emerge from a number of possible module, separator and process layouts. The considered flow patterns include cases of cross flow, counter-current and co-current flow, and the various combinations of these. Additional layouts are produced with recycles and the removal of products from intermediate stages. The unified generic process representation offers the ability to yield both, the known, conventional as well as possible novel process configurations.

Optimisation framework

The objective function for membrane network and hybrid membrane network superstructure optimisation is taken as the total annualised cost comprising of cost of membrane compartment, recycle compressors, vacuum pumps, feed compressors and the cost of separation (adsorption). The membrane network representation outlined above is formulated as a mathematical model for optimisation. Non-linearities are involved mainly in the mass transfer expressions and cost correlations of the membrane systems and make the synthesis problem difficult to be handled using deterministic optimisation approaches. This work employs stochastic optimisation techniques in the form of simulated annealing. Basic advantages for stochastic optimisation exist in terms of providing multitude of solutions close to global optimal domain and ability to

Table 1. Problem specifications for enriched oxygen production			
P_H (bar)	1.07	C_{HP}^{RCOM}, C_{HP}^{FCOM} (HP/(kmol/s))	54927
P_L (bar)	0.2675	C_F^{RCOM}, C_F^{FCOM} ($ / HP)	1264
(Per/δ) O_2 (kmol/m^2.s.bar)	2.0491×10^{-4}	η^{RCOM}, η^{FCOM}	0.75
(Per/δ) N_2 (kmol/m^2.s.bar)	9.509×10^{-5}	C_{HP}^{VAC} ($ HP/(kmol/s))	1718213
C_{ann}^{mem} ($/m^2)	23.4	η^{VAC}	0.5

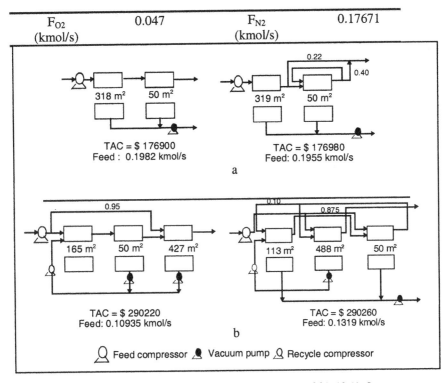

Figure 1: Optimised membrane networks for a) 30 % O_2 and b) 40 % O_2 permeate products

provide robustness in solving non-linear network synthesis problems (Mehta & Kokossis [4] and Linke [5], Effie & Kokossis [6]).

Illustrative examples:

1. Air separation using vacuum pumps:

Membrane network optimisation for enriched oxygen production is targeted using stochastic optimisation techniques. Vacuum application is considered on permeate side to offer additional network design complexity. Feed, membrane, network specifications are provided from Bhide and Stern [7] and are summarised in Table 1. Targets for network performance are provided as 10 tons of EPO_2 with 30 and 40 % O_2. Network optimisation considers simultaneous optimisation of feed flow rate and network design for required product specifications. Allocation of vacuum pump follows from a generic allocation methodology developed in this work. A single vacuum pump is allocated for each permeate compartment stream undergoing partial or complete permeate to retentate recycle and a single vacuum pump is allocated for all permeate streams undergoing no recycle and entering the product stream. Only cross flow is considered for the case study.

Results obtained for both the cases are summarised in Figure 1. As shown, optimised objective (TAC) value corresponds to a value of about $176,000 with an optimal feed rate value of about 0.191 kmol/s for 10 tons of EPO_2 at 30 % purity. Optimised objective value corresponds to about $ 289,000 with an optimal feed rate of about

370

0.1093 kmol/s for 10 tons of EPO$_2$ at 40 % purity. All the structures generated after stochastic optimisation presented good confidence, providing a standard deviation value of about 5 % for markov chain length values of above 40.

2. Hydrogen recovery from lean refinery stream

Stochastic optimisation is targeted for recovering hydrogen from multi-component lean hydro-cracking refinery stream at high purity and high recovery. Feed, membrane and network specifications are provided from Kaldis, Kapantaidikis and Sakellaropoulos [8] and Douglas [9]. Optimisation constraints are provided in the form of hydrogen recovery and hydrogen purity in the permeate stream. Two product specifications are considered, the first for low purity hydrogen and the second for high purity hydrogen. Further, additional cost is included in the form of product recompression.

Results obtained from stochastic optimisation are summarised in Figure 2. As shown, optimal network cost is about $ 660,000 for 95 % H$_2$ case and $ 890,000 for 99 % H$_2$ case. Optimised networks refer to the recycle of permeate stream of very high flow rate to a membrane stage with low membrane area using a recycle compressor for both the cases.Design targets obtained using stochastic optimisation provided a good confidence in the optimised network structures whose standard deviation was below a value of about 5 % at a markov chain length value of about 40. The following conclusions can be arrived for the present case study. Considerable contribution arises due to very high membrane areas for network cost. Improvements at existing selectivity to the membrane permeability values can provide industrial sustanance for illustrated membrane (polyimide). Further, about 20 % cost contribution arises due to permeate product recompression. Hence, there is a significant need to develop membranes that can reject hydrogen on the retentate side and provide lower network costs.

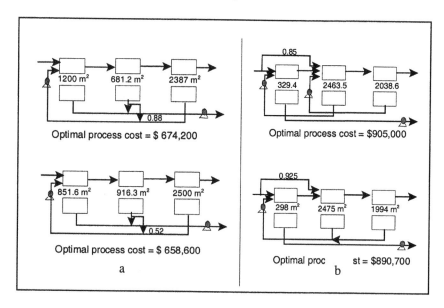

Figure 2: Optimal membrane networks for producing a) 95 % H$_2$ and b) 99 % H$_2$

Table 2. Problem specifications for natural gas sweetening			
P_H (bar)	45.0	C_{HP}^{RCOM} ($/HP)	6000
		F (kmol/s)	0.44683
P_L (bar)	1.5	η^{RCOM}	0.70
(Per/δ) $_{CH4}$	3.3482 x 10^{-7}	X_{CH4}	0.79
(Per/δ) $_{N2}$	3.3482 x 10^{-7}	X_{N2}	0.01
(Per/δ) $_{H2S}$	34.152 x 10^{-7}	X_{H2S}	0 – 0.1
(Per/δ) $_{CO2}$	7.3661 x 10^{-7}	X_{CO2}	0 – 0.1
C_{ann}^{mem} ($/m^2$)	60.0	C^{sep}	$ 50 / (kg H$_2$S adsorbed)

3. Natural gas sweetening using hybrid membrane – adsorption networks

Membrane-fixed bed adsorption systems are synthesised for the removal of acid gases (CO_2, H_2S) from natural gas streams. The problem data are summarised in Table 2. The fixed bed adsorbent considered is capable of complete separation of H_2S from other components in natural gas streams [10]. The study comprises network optimisations performed at different H_2S feed concentrations (0.5 – 10 % H_2S) whilst maintaining the total acid gas feed content at 20 %. The product specifications include the composition of H_2S in sweet gas set to zero. Natural gas throughput to the system is 0.44 kmol/s.

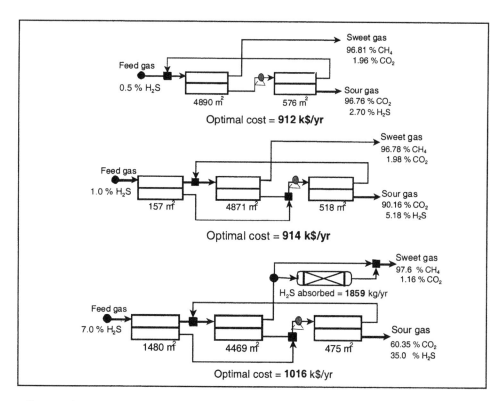

Figure 3: Optimal membrane and hybrid membrane networks for natural gas sweetening.

The objective value increases by about 10 % from a value of about $ 917,000 at very low H_2S feed concentrations (0.5 % vol) to a value of about $ 1,062, 000 at high H_2S feed concentration (10 % vol). Typical membrane and hybrid membrane configurations obtained after stochastic optimisation for natural gas sweetening problem are summarised in Figure 3. The optimal configuration consist of only membrane permeators below 4 % vol. H_2S feed concentration. Hybrid membrane configurations are obtained for higher H_2S feed concentrations. The objective value deterioated with an increasing H_2S content in natural gas as high membrane area is demanded for higher feed H_2S concentrations. For higher feed H_2S concentrations, fixed bed adsorption process is deployed for polishing the sweet gas product stream obtained from the membrane network.

Conclusions

A superstructure representation is developed for the optimisation of gas permeation membrane and hybrid membrane networks. The superstructure is rich in representation and can consider various flow patterns and can generate conventional and novel networks on optimisation. Optimisation methodology developed refers to stochastic optimisation in the form of simulated annealing to provide optimised membrane networks for desired product specifications. A very high degree of confidence is assured for performance targets using the proposed optimisation methodology.

Nomenclature

Per	Permeability, $kmol/(m^2.s.m.bar)$	C_{HP}	Power cost parameter
δ	Membrane thickness, m	C_F	Fixed cost parameter
P_H	Retentate pressure, bar	$C_{ann\,m}^{me}$	Annualised membrane cost, $\$/m^2$
P_P	Permeate pressure, bar	C^{sep}	Separator cost, $ / (kg H_2S adsorbed)
η	Compressor/pump efficiency		

References

1. Qi R, Henson M.A., *Sep. and Pur. Tech.*, **1997**, 13, 209-225
2. Qi R., Henson M.A., *Comp. Chem. Eng.*, **2000**, 24(12), 2719 – 2737
3. Lababidi, H., Al-Enezi, G. A., Ettonuey, H. M., *J. Mem. Sci.*, **1996**, 112, 185.
4. Mehta V.L., Kokossis A.C., *Comp. Chem. Eng.*, **1997**, 21, S 325
5. Linke P , Mehta V. L., Kokossis A.C., *ESCAPE – 10*, **2000**, 1165 – 1170
6. Marco ulaki E., Kokossis A.C., *Comp. Chem. Eng.*, **1996**, 20, S231
7. Bhide B. D and Stern S.A., *J. Mem. Sci.*, **1991**, 62:13-25.
8. Kaldis S.P., Kapantaidikis G.C., Sakellaropoulos G.P., *J. Mem. Sci.* **2000**, 173(2), 61 – 71.
9. Douglas J, Conceptual Design of Chemical processes, **1989**, Mc Graw Hill, NewYork
10. Carnell P.J.H., Joslin K.W., Woodham P.R., *Oil & Gas Journal*, **1995**, 93 (23), 52 – 55.

European Symposium on Computer Aided Process Engineering – 12
J. Grievink and J. van Schijndel (Editors)
© 2002 Published by Elsevier Science B.V.

Mixed integer optimization models for the Synthesis of Protein Purification Processes with product loss

Elsa Vasquez-Alvarez, Jose M. Pinto[†]
Department of Chemical Engineering, University of Sao Paulo
Av. Prof. Luciano Gualberto t. 3 n. 380, Sao Paulo, SP, 05508-900 Brazil

Abstract

The objective of this work is to develop a mixed integer linear programming (MILP) model for the synthesis of protein purification processes. Mathematical models for each chromatographic technique rely on physicochemical data on the protein mixture, which contains the desired product and provide information on its potential purification. In previous works, MILP models assumed the complete recovery of the desired protein. The present model incorporates losses in the target protein along the purification process, in order to evaluate the trade-off between product by purity and quantity. A formulation that is based on a *convex hull* representation is proposed to calculate the minimum number of steps from a set of chromatographic techniques that must achieve a specified purity level as well as the amount of product recovered. Model linearity is achieved by assuming that the product is recovered in discrete percentages. The methodology is validated in examples with experimental data.

1. Introduction

Many pharmaceutical products are proteins or polypeptides. These biotechnological products can be obtained from nature by extraction or produced by microorganisms genetically modified, namely recombinant proteins. In both cases, separation and purification of the desired protein are usually the most difficult stages in the whole process, and such stages may account for up to 60% of total cost (Lienqueo and Asenjo, 2000).

Depending on the degree of complexity of the mixtures that result from bioreactions, several recovery and purification operations may be necessary to isolate the desired product. The most important operations include chromatographic techniques that are critical for therapeutical products such as vaccines and antibiotics that require very high purity levels (98 - 99.9%). One of the main challenges in the synthesis of downstream purification stages is the appropriate selection and sequencing of chromatographic steps (Larsson *et al.*, 1997). Therefore, optimization methods (Vasquez-Alvarez et al., 2001) as well as expert systems (Bryant and Rowe, 1998) are useful tools for the design and synthesis of protein purification processes. Steffens *et al.* (2000) developed a synthesis technique for generating optimal downstream processing flowsheets for

[†]Author to whom all correspondence should be addressed. E-mail: jompinto@usp.br. Financial support from PADCT/CNPq under grant 62.0239/97 – QEQ, from ANTORCHAS and VITAE (Coop. Programs Argentina - Brasil - Chile) under grants A-13668/1-9 and B-11487/10B006.

biotechnological processes. The technique integrates the idea of screening units via physical property information into an implicit enumeration synthesis tool.

Mathematical programming approaches for process synthesis rely on the representation of algebraic equations with discrete variables. In previous works, Vasquez-Alvarez et al. (2001) and Vasquez-Alvarez and Pinto (2001) developed mixed-integer linear optimization models that implicitly assume the complete recovery of the desired protein. In the present work, the objective is to incorporate losses in the target protein along the purification process, in order to evaluate the trade-off between product quality given by purity and quantity.

2. Problem Description

Consider a complex protein mixture that must be purified by chromatographic techniques. The degree of separation depends on the protein partition differential between the stationary and mobile phases. Information on physicochemical properties can be used for the target and contaminant proteins and each chromatographic technique is able to perform the separation of the mixture by exploiting a specific physicochemical property, such as surface charge as a function of pH, surface hydrophobicity, molecular weight. For instance, ion exchange chromatography separates proteins based on their difference in charge. The charge of a protein depends on pH according to the titration curve. Ion exchange can use small differences in charge that yield a very high resolution and hence it is an extremely efficient operation to separate proteins.

Losses of target protein along the purification process are possible, and therefore we must evaluate the trade-off between product quality (given by purity) and quantity. In other words, the higher the purity achieved within each step, the smaller the product yield. In this sense, decisions involve the selection of techniques and their order as well as the percentage of product recovered.

3. Mathematical Model

Mixed-integer linear optimization models for the synthesis of purification bioprocesses were developed in previous work (Vasquez-Alvarez, et al., 2001). In previous models, implicit was the assumption that product recovery was complete. Consequently, the major decisions concerned selection and ordering of chromatographic operations. In the present case, the model must account for target protein losses along the purification process. In order to keep the optimization model linear, it is assumed that product must be recovered in discrete percentages.

3.1 Model of a chromatographic technique

The approach for each chromatographic technique is to approximate chromatograms by isosceles triangles. Moreover, physicochemical property data for the target protein as well as for the major contaminants are required.

It is assumed that the peaks in chromatograms have constant shapes and that the one on the left refers to the product and the other to the contaminant protein. In Figure 1, the shaded areas represent product that is removed with the contaminants for a given discrete recovery level (given by the index l). On the other hand, the dark-shaded areas (with base $B_{i,p,l}$) represent the amount of contaminant p that remains in the mixture (with the product) after applying chromatographic technique i. It is important to note that five situations may arise, depending to the relative position of the triangles. The first case corresponds to an almost complete overlap between the triangles; the other extreme occurs when both triangles are completely apart; finally, the remaining cases are shown in Figures 1(a), 1(b) and 1(c). In such figures, the amount of lost product can be determined from the ratio of the light-shaded area and the total area of the product chromatogram. Similarly, the amount of contaminant that remains in the mixture is calculated from the ratio of the dark-shaded and the total area.

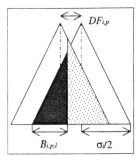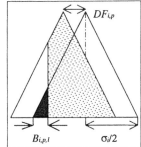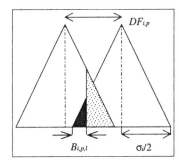

Figure 1. Peak representations in a chromatogram (a), (b) and (c)

The mathematical correlations applied for each chromatographic technique are given in Table 1 for protein p at chromatographic step i with discrete recovery level l. The ratio of proteins (p and dp) that remain in the mixture after and before chromatographic technique i at separation level l is denoted by $CF_{i,p,l}$.

The relationships expressed in Table 1 represent graphical approximations of the chromatograms for two different proteins. As a result, a fraction of proteins is admitted not to separate from the product (represented as 1.02 coefficients in Table 1). In Table 1, the first row indicates that purification is not carried out. In the following rows the purification degree increases (correspond to figures 1(a), 1(b) and (1c)), up to the case of total separation ($CF_{i,p} = 0.02$). The concentration factors $CF_{i,p}$ shown in Table 1 are introduced in the synthesis model that is described in the next section.

3.2 Synthesis model
An optimization model that minimizes the total number of chromatographic steps for a given purity level is proposed. This model relies on a convex hull representation that is derived from the following general disjunction:

$$
\bigvee_{i=1}^{I} \bigvee_{l=1}^{L} \begin{bmatrix} \lambda_{i,k,l} \\ m_{p,k+1} = CF_{i,p,l} \cdot m_{p,k} & \forall p \end{bmatrix} \vee \begin{bmatrix} \alpha_k \\ m_{p,k+1} & \forall p \end{bmatrix} \qquad k=1\ldots K-1 \qquad (1)
$$

Disjunction (1) contains $I.L+1$ elements for each order k. The first $I.L$ terms model the selection of step i in order k at level l, whereas the last term models no step selection.

Table 1. Mathematical relationships for chromatographic techniques

Deviation Factor	Base $B_{i,p,l}$	Mass reduction of protein p	
$0 \leq DF_{i,p} < \dfrac{\sigma_i}{10}$	$B_{i,p,l} = \sigma_i - DF_{i,p}$	$CF_{i,p,l} = 1$	$\forall p$
$\dfrac{\sigma_i}{10} \leq DF_{i,p} < \dfrac{\sigma_i}{2}$ (figs. 2a and 2b)	$\dfrac{\sigma_i}{2} - DF_{i,p} \leq B_{i,p,l} < \sigma_i - DF_{i,p}$	$CF_{i,p,l} = 1.02 \left[1-2\left(\dfrac{(\sigma_i - DF_{i,p} - B_{i,p,l})^2}{\sigma_i^2}\right)\right]$	$p=dp$
	$0 \leq B_{i,p,l} < \dfrac{\sigma_i}{2} - DF_{i,p}$ (fig 2a)	$CF_{i,p,l} = 1.02 \left[2\,\dfrac{(DF_{i,p} + B_{i,p,l})^2}{\sigma_i^2}\right]$	$p=dp$
	$\dfrac{\sigma_i}{2} \leq B_{i,p,l} < \sigma_i - DF_{i,p}$	$CF_{i,p,l} = 1.02 \left[1-\left(2\,\dfrac{(\sigma_i - B_{i,p,l})^2}{\sigma_i^2}\right)\right]$	$\forall p \neq dp$
	$0 \leq B_{ip,l} < \dfrac{\sigma_i}{2}$ (fig. 2b)	$CF_{i,p,l} = 1.02 \left(2\,\dfrac{B_{i,p,l}^2}{\sigma_i^2}\right)$	$\forall p \neq dp$
$\dfrac{\sigma_i}{2} \leq DF_{i,p} < \sigma_i$ (fig. 2b)	$0 \leq B_{i,p,l} < \dfrac{\sigma_i}{2}$	$CF_{i,p,l} = 1.02 \left[1-2\left(\dfrac{(\sigma_i - DF_{i,p} - B_{i,p,l})^2}{\sigma_i^2}\right)\right]$	$p=dp$
		$CF_{i,p,l} = 1.02 \left(2\,\dfrac{B_{i,p,l}^2}{\sigma_i^2}\right)$	$\forall p \neq dp$
$DF_{i,p} \geq \sigma_i$	$B_{i,p,l} = 0$	$CF_{i,p,l} = 0.02$	$\forall p$

$$\sum_i \sum_l \lambda_{i,k,l} + \alpha_k = 1 \qquad \forall k \qquad (2a)$$

$$\sum_k \sum_l \lambda_{i,k,l} \leq 1 \qquad \forall i \qquad (2b)$$

$$\sum_i \sum_l \lambda_{i,k+1,l} \leq \sum_i \sum_l \lambda_{i,k,l} \qquad \forall k \leq K\text{-}1 \qquad (3a)$$

$$Z_k \geq \sum_i \sum_l \lambda_{i,k,l} - \sum_i \sum_l \lambda_{i,k+1,l} \qquad \forall k \leq K\text{-}1 \qquad (3b)$$

$$\sum_i \sum_l \lambda_{i,k',l} \geq Z_k \qquad \forall k, k' \leq k \qquad (3c)$$

$$\sum_i \sum_l \lambda_{i,k',l} + Z_k \leq 1 \qquad \forall k, k' > k \qquad (3d)$$

$$\sum_k Z_k = 1 \qquad (3e)$$

$$m_{p,2} = \sum_i \sum_l CF_{i,p,l}.\lambda_{i,1,l}.m_{p,1} \qquad \forall p \qquad (4a)$$

$$m_{p,k+1} = \sum_i \sum_l CF_{i,p,l}.m^1_{i,p,k,l} + m^2_{p,k} \qquad \forall p,\ k=2...K\text{-}1 \qquad (4b)$$

$$m_{p,k} = \sum_i \sum_l m^1_{i,p,k,l} + m^2_{p,k} \qquad \forall p,\ k=2...K\text{-}1 \qquad (4c)$$

$$m^1_{i,p,k,l} \leq U\lambda_{i,k,l} \qquad \forall i, p, l,\ k=2...K\text{-}1 \qquad (4d)$$

$$m^2_{p,k} \leq U.\alpha_k \qquad \forall p,\ k=2...K\text{-}1 \qquad (4e)$$

$$m_{dp,k+1} \geq fp.\sum_{p'} m_{p',k+1} - U.(1-Z_k) \qquad k=1...K\text{-}1 \qquad (5)$$

$$m_{dp,k+1} \geq fr.\sum_p m_{p',1} - U.(1-Z_k) \qquad\qquad k=1...K-1 \qquad\qquad (6)$$

$$\lambda_{i,k,l} \in \{0,1\} \quad \forall i,k,l \qquad\qquad m_{p,k}, Z_k, m^1_{i,p,k,l}, m^2_{p,k}, \alpha_k \geq 0 \quad \forall i,p,k,l \qquad (7)$$

Constraint (2a) indicates that at most one step i may be chosen in order k. Slack variable α_k is activated if no steps are selected in order k. Constraint (2b) imposes that step i is selected at most once in the sequence and (3a) states that steps are assigned in increasing order. Constraints (3b)-(3e) define the last step of the sequence, denoted by Z_k. Constraint set (4) relates subsequent steps and is generated from disjunction (1). Constraints (5) and (6) enforce purity and yield specifications, respectively.

Objective function (8) that selects a minimum sequence for a given purity as well as yield is as follows:

$$Min\ S = \sum_i \sum_k \sum_l \lambda_{i,k,l} = \sum_k k.Z_k \qquad\qquad (8)$$

Alternatively, profit can be maximized by simply taking into account the revenue from product sales and operating costs of the chromatographic columns with available data.

4. Computational Results

The software GAMS/CPLEX 7.0 (Brooke et al., 1998) was used to implement the MILP model and to generate its solution. Two different examples of increasing size are solved, which correspond to the first two presented in Vasquez-Alvarez et al. (2001).

In *Example1*, taken from Lienqueo *et al.* (1998), we consider the purification of a mixture containing four proteins, all in equal concentration. Their physicochemical properties as well as the initial protein concentration of the mixture are shown in Vasquez-Alvarez et al. (2001). The required purity level for *p1* is 98%. If no product loss is considered, results are the same as those of model (M_{1a}) by Vasquez-Alvarez et al. (2001), which comprise three steps and 99.8% final purity. Nevertheless, if 4% loss of product is accepted, only two steps are necessary and 99.9% final purity is achieved.

In *Example 2*, we consider the purification of β-1,3 glucanase (8.3% initial concentration) that must be separated from eight contaminants; twenty-two chromatographic techniques are available (Lienqueo *et al.*, 1999). Consider *Case a* that corresponds to 94% purity and 100% recovery of β-1,3 glucanase and *Case b* given by 99% purity with 6% of product losses. In *Case a*, results are the same as in Vasquez-Alvarez et al (2001), and is shown in Fig. 2 (6 steps and 94.8% for final purity). For *Case b*, three techniques are employed and 99.7% final maximum purity is achieved, as is shown in Fig. 3. Statistical data for both examples are given in Table 2.

Table 2. Summary of statistical data for examples 1 and 2

Example	L	fr (%)	Integer variables	Continuous variables	Constraints	Nodes	CPU times (s)
1	1	100.0	168	618	874	20	1.46
	3	96.0	456	1674	1930	0	1.87
2	1	100.0	288	2378	2699	623	138.6
	4	94.0	1080	8912	9233	972	1333

378

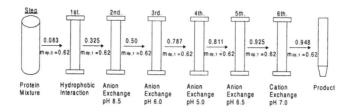

Fig 2. *Optimal results for example 2 – Case a*

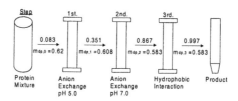

Fig 3. *Optimal results for example 2 – Case b*

5. Conclusions

This paper presented the development of an MILP for the synthesis of chromatographic steps for the purification of protein mixtures considering product losses. Results indicate that a systematic selection and sequencing of chromatographic steps may be obtained by the appropriate balance between yield and purity level.

Notation

$B_{i,k,l}$	width of contaminant peak, remains in mixture	$m_{p,k}$	mass of p after technique in order k
$CF_{i,p,l}$	conc. factor of cont. p after step i in level l	p	protein (product + contaminants)
$DF_{i,p}$	deviation factor for protein p in chrom. step i	S	objective function variable
Dp	desired protein (product)	U	upper bound on protein mass
Fp	specified purity level of dp	Z_k	binary variable that indicates if order k is last
Fr	specified yield level of dp	α_k	slack variable relative to order k
k	order in the sequence ($k = 1,...K$)	$\lambda_{i,k,l}$	boolean variable for selecting technique i
$Kd_{i,p}$	retention time of protein p in technique i		in order k at level l of product loss
l	level of target protein losses	σ_i	Width of chromatographic peak in step i

References

Brooke A., D. Kendrick, A. Meeraus, and R. Raman, 1998, GAMS -A user's guide, The Scientific Press, Redwood City, USA.

Bryant, C.H., and R.C. Rowe, 1998, Trac - Trends in Analytical Chemistry, 17, 18.

Larsson, G., S.B. Jorgensen, M.N. Pons, B. Sonnleitner, A. Tijsterman and N. Tichener–Hooker, 1997, J. Biotechnol., 59, 3.

Lienqueo, M.E., J.C. Salgado and J.A. Asenjo, 1998, Comp. Appl. Biot. - CAB7, 321, Osaka, Japan.

Lienqueo, M.E., J.C Salgado and J.A. Asenjo, 1999, J. Chem. Technol. Biot., 74, 293.

Lienqueo, M.E., and J.A. Asenjo, 2000, Comput. Chem. Eng., 24, 2339.

Steffens M.A., E.S. Fraga and I.D.L. Bogle, 2000, Biotechnol. and Bioeng., 68, 2, 218.

Vasquez-Alvarez E., M.E. Lienqueo and J.M. Pinto, 2001, Biotechnol Prog., 17, 685.

Vasquez-Alvarez E. and J.M. Pinto, 2001, 579, European Symposium on Computer Aided Process Engng - 11, Eds. R. Gani and S.B. Jorgensen, Elsevier, Amsterdam.

European Symposium on Computer Aided Process Engineering – 12
J. Grievink and J. van Schijndel (Editors)
© 2002 Elsevier Science B.V. All rights reserved.

379

Fresh Water by Reverse Osmosis Based Desalination Process: Optimisation of Design and Operating Parameters

A. Villafafila[1], I. M. Mujtaba[2*]

[1]Department of Chemical Engineering, University of Valladolid, (Spain)
[2]Department of Chemical Engineering, University of Bradford, BD7 1DP (U.K.)

Abstract

An optimisation framework for a Reverse Osmosis (RO) based desalination process is developed so as to maximize a profit function using different energy recovery devices, subject to general constraints. The optimal operating parameters (feed flux, feed pressure) and design parameters (internal diameter, total number of tubes) are determined by solving the optimisation problem using an efficient successive quadratic programming (SQP) based method. The optimal values for the decision variables depend on the constrains introduced, and are also sensitive to variations in water and energy prices, as well as feed concentration. The use of the emerging energy recovery devices is widely justified, reporting much higher reductions in operating costs than the traditional technology used for this purpose. Using pressure exchanger device it is possible to reduce energy consumption by up to 50%.

1. Introduction

Potable water supply is becoming a limited resource nowadays due to the increase in size and improved standard of living of communities. Desalination processes are used to produce fresh water from sea or brackish water. The processes can be divided in two groups: Power consuming processes (membrane processes such as electrodyalisis (ED) and RO, and freezing processes), and heat consuming processes (multiple effect distillation (MED), multiple stage flash (MSF) and mechanical vapour compression (MVC)).

For many years multistage flash distillation (MSF) was the predominant process for treating large amounts of seawater. Since 1960 due to the development of new RO modules and membranes, RO is becoming an attractive process for treating seawater in large scale. However, yield, energy consumption, energy recovery, capital cost will remain the important factors in choosing such technology.

Many mathematical models of transport phenomena across membranes have been developed (Sourirajan 1970, Lonsdale et al. 1965). Meares (1976) developed a computer program to predict the performance of hyperfiltration plants employing tubular modules in a tapered-flow cross section. Harris (1994) and Mandil et al. (1998) have focused on energy recovery of RO based desalination process.

---

* Author to whom all correspondence should be addressed. Email: i.m.mujtaba@bradford.ac.uk

380

In this work, RO based desalination process is considered to make fresh water from seawater. Loeb and Sourirajan type membrane in a tubular configuration is chosen for the process. This type of membrane has low clogging tendency, is easy to clean and reduces one of the most important problems such as polarisation (Meares, 1976). The process consists of a membrane module where a number of RO tubes are connected in series (Figure 1). Each tube is modelled by a set of algebraic equations (AEs) developed by Meares (1976). In this work we propose to solve the tube models sequentially. The output of tube j becomes the input for tube j+1. This approach allows evaluation of some of the process parameters (inlet velocity, bulk concentration, wall concentration and pressure) locally rather than using constant parameter value throughout, introducing a great error due to the considerable variation of those parameters in the axial direction. An optimisation problem using the process model is formulated to optimise design and operating parameters in order to maximize the profitability of the system. Different energy recovery devices (hydrodinamic turbines, pressure exchanger) are analysed and their influence on the economy of the system is studied. Given the input stream concentration, permeate pressure, temperature, membrane properties, and tube length, the optimisation problem determines the optimal feed pressure, internal diameter, number of tubes and feed flux so as to maximize the objective function, subject to equality (process model) and inequality constraints (linear bounds on the optimisation variables).

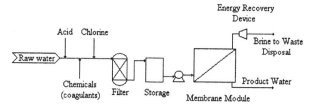

Figure 1: RO Process Using Energy Recovery Device.

2. Membrane Model

The equations used to model the behaviour of the membrane in the RO module consist of mass balances, pressure energy balance, and mass transport across the membrane. The model assumes turbulent flow, osmotic pressure represented by van't Hoff equation, tubular configuration, isothermal conditions, reflection coefficient σ approximately equal to the intrinsic salt rejection R_j', and Nerst film model to describe polarization phenomenon. The main equations used are shown below:

Hyperfiltrate flux at a fixed location: $\quad J_{v,i} = L_p[(p_{B,i} - p_{P,i}) - (Rj')^2 vRTc_{w,i}$ $\qquad$ (1)

Mass transport across the membrane: $\quad P_{j,i} = J_{v,i} x_{j,i} \pi d_i \Delta x$ $\qquad$ (2)

Mass balance: $\qquad\qquad\qquad F_{j,i} = P_{j,i} + B_{j,i}$ $\qquad$ (3)

Salt concentration in the brine: $\quad C_{B,i+1} = \left(\dfrac{u_i}{u_{i+1}}\right)c_{B,i} - \left[(1-R_j')\dfrac{4J_{v,i}}{u_{i+1}}\dfrac{\Delta x}{d_i}\right]C_{w,i}$ $\qquad$ (4)

Salt concentration in the permeate: $\quad C_{P,i} = C_{p,i-1}\dfrac{P_{i-1}}{P_i} + \dfrac{(1-R_j')C_{w,i}J_{v,i}\pi d_i \Delta x}{P_i}$ $\qquad$ (5)

Pressure balance:
$$p_{B,i+1} = \left(\frac{u_i}{u_{i+1}}\right)\left(p_{B,i} + 5\cdot10^{-8}\rho_{B,i}u_i^2 - 2\cdot10^{-7}f_F\Delta x u_i^2\frac{\rho_{B,i}}{d_i}\right)$$
$$-\left(\frac{4\Delta x J_{v,i}}{d_i u_{i+1}}\right)\left(p_{B,i} + 5\cdot10^{-8}\rho_{P,i}J_{v,i}^2\right) - 5\cdot10^{-8}\rho_{B,i}u_{i+1}^2 \tag{6}$$

In these equations i refers to the tube number, and goes from 1 to N; and j refers to the component (salt or water).

For a given feed concentration, if we specify the kind of membrane and its curing temperature (variables that fix the membrane permeability, L_p, and rejection coefficient, R_j'), the membrane configuration, tube length (Δx), permeate pressure (p_p) and operating temperature (T), and choose feed pressure to the first tube (p_F), feed flux (F), total number of tubes (N), and internal diameter (di) as decision variables to be optimised, we can calculate the brine pressure (eq. 6), the salt and water brine fluxes (eq. 3), the water and salt permeate fluxes (eq. 2), and brine and permeate salt concentrations (eq. 4 and 5). It is to be noted that in this work specifications on feed flux and pressure are relaxed and are optimised (as can be seen in the following section).

3. Formulation Of Optimization Problem

The optimization problem can be stated as follows:

Given: Feed concentration, membrane properties, temperature, permeate pressure, tube length, and energy recovery device

determine: Optimal feed pressure, feed flux, internal diameter and number of tubes

so as to: Maximize the profit of the system

subject to: Equality and inequality constraints

Mathematically, the optimization problem (**OP**) can be written as follows:

Maximize   Profit
p_F, F, d_i, N

Subject to   Equality constraints: Process model

Inequality constraints:
$N^{Lower} \le N \le N^{Upper}$
$p_F^{\ Lower} \le p_F \le p_F^{\ Upper}$
$d_i^{Lower} \le d_i \le d_i^{Upper}$
$F^{Lower} \le F \le F^{Upper}$
$Re \ge 5000$
$O(x)/p_B(x) \bullet 0.9$

where $O(x)$ is the osmotic pressure at x, and p_B is the brine pressure at x. The four inequality constraints are simple linear bounds (lower and upper) on N, p_F, d_i and F. Re is the Reynolds number (to be above 5000 to ensure turbulent regime). The last inequality is written to prevent the driving force to drop to limits that cause water permeation across the membrane to be too small or even to stop (Mandil et al., 1998).

Three simple profit functions with or without energy recovery are considered:

$$Profit = P\ C_{water} - E\ C_{energy} \tag{7}$$

Here P refers to the total flux of permeate, E to the amount of energy used, C_{water} is the price of fresh water, C_{energy} is the cost of energy used., with $E = \dfrac{F \cdot p_F}{\eta_{pump}}$ for the case of no energy recovery used; $E = \dfrac{F \cdot p_F}{\eta_{pump}} - B \cdot p_B \cdot \eta_{turbine}$ when using a turbine for energy recovery (Harris, 1994); $E = \dfrac{P \cdot p_F}{\eta_{pump}}$ when using a pressure exchanger for energy recovery (Geisler et al., 2001).

B refers to the brine flux leaving the system, η to the efficiency of the energy recovery device used, and p_B to the brine pressure.

Figure 1 shows a typical RO process with energy recovery device.

4. Solution Method

The problem presented above is a non-linear optimisation problem solved using the SQP based optimisation method (Chen, 1988).

For a given decision variables (F, p_F, d_i, N) and other specification mentioned in the previous sections the model equations for N tubes are solved sequentially using that the input for tube j+1 is equal to the output from tube j (Figure 2). This allows evaluation of the objective function and constraints.

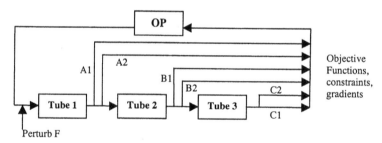

Figure 2. Function and gradient evaluation

The gradients of the objective function and constraints with respect to the decision variables are obtained by finite difference technique. At the function evaluation step the solution (objective function and constraints, permeate and brine fluxes of salt and water, and brine pressure) of each tube is stored as A1, B1, etc, respectively. To evaluate the gradient with respect to F, the variable is perturbed and the model equations of N tubes are solved again as shown in figure 2. The new solutions A2, B2, etc. are used to calculate the gradient of the objective function and constraints with respect to F. The same procedure is used for the rest of the decision variables.

5. Results

In this work we used tubular configuration, Sourirajan type of membrane, with $T_c = 92°C$, T = 16°C, $c_F = 35000$ ppm, and $\Delta x = 304,8$ cm. The pump, turbine and pressure exchange efficiencies are 0.75, 0.8 and 0.95, respectively. For the base case the water and energy prices are 1$/m³ and 0.06$/kWh, respectively.

The optimal values for the decision variables and the profit and energy required for the system for the 3 different energy recovery cases studied are shown in table 1. As the fresh water price is much higher than the energy price, N, d_i and p_F hit the upper bounds to give larger amount of permeate and to maximise the profit. So the production of more permeate compensates the additional energy costs. Pressure exchanger seems to be the most advantageous alternative as the operating costs are decreased up to 50%, but it has to be studied for each plant, having into account the capital costs. With a turbine, the cost decrease is lower (around 30-40 %) but the advantage over the no energy option recovery is also noted. When energy recovery is used the additional energy costs related to the operation with higher feed flux are compensated with the more permeate obtained.

Table 1. Energy Recovery Alternatives.

Cases		N	p_F (atm)	d_i (cm)	F (cm³/s)	E_{pump} (kW)	$E_{recovered}$ (kW)	Permeate (mol/s)	Profit ($/d)
1	No ER	200	100	2.54	400.0	5.3	0.0	10.8	10.0
2	Turbine	200	100	2.54	422.2	5.6	1.7	11.0	12.2
3	PE	200	100	2.54	598.4	3.3	3.5	11.6	14.1

The values for the feed pressure and feed flux to maximize the profit are very sensitive to the water and energy prices. If they are changed to 0.5 $/m³ and 0.09 $/m³ respectively, the new optimum decision variables will be different (Table 2). It is seen that the pressure and feed flux decreases and the profit is lower. Table 2 shows that the optimal pressure for the feed decreases, because the additional cost to apply that pressure is not compensated by the higher amount of permeate obtained. The same behavior is observed for the feed, and in all cases F hits the lower bound (it is more economical to get less permeate because its price is not so high). N and d_i hit the upper bound because they ensure more permeate without increasing the energy costs (the pressure drop is almost negligible). In case 1, operating without energy recovery reports a negative profit.

Table 2. Energy and Water Prices Influence on the Profit.

Cases		N	p_F (atm)	d_i (cm)	F (cm³/s)	E_{pump} (kW)	$E_{recovered}$ (kW)	Permeate (mol/s)	Profit ($/d)
1	No ER	200	87.8	2.54	400	4.7	0.0	9.2	-2.6
2	Turbine	200	86.1	2.54	400	4.6	1.5	8.9	0.7
3	PE	200	86.2	2.54	400	1.9	2.0	8.9	2.7

It is also observed that the operating cost reduces with increasing permeate with energy recovery systems in place. Also for a given fixed parameters (F, p_F, d_i, N) increasing feed salinity leads to the increased cost savings using pressure exchangers. The detailed results will be presented elsewhere.

6. Conclusions

RO based desalination process has been studied here. A number of RO membrane units (tubes) are connected in series to imrpove the recovery of fresh water and model equations for such tubes are developed. We propose to solve these models sequentially, output of tube j being the input of tube j+1. This allows taking into consideration the axial variations of inlet velocity, concentrations and pressure.

Optimisation problem formulation is presented to maximise a profit function while optimising design and operating parameters of the process. The optimisation problem is solved using an efficient SQP based technique. Options of energy recovery with different devices are included in the optimisation problem. The advantage of the use of this kind of machines has been tested. Sensitivity of water and energy prices on the design and operating parameters have been presented.

Alternative objective function and cost models can be easily introduced, without changing the main program structure developed in this work. Also, different model equations for alternative membrane configuration can be incorporated in the framework.

Nomenclature

B_j Flux of j in the brine stream, g-mol/s
C_B Salt concentration in brine, g-mol/cm^3
C_{energy} Energy price, $/kWh
C_P Salt concentration in permeate, g-mol/cm^3
C_w Salt concentration at the wall, g-mol/cm^3
Cwater Water price, $/m^3
d_i Internal diameter, cm
E Energy flux, kW
f_F Fanning friction factor
F_j Flux of j in the feed stream, g-mol/s
J_v Hyperfiltrate flux, cm/s
L_p Hydraulic permeability, cm/satm
N Total number of tubes
P_j Flux of j in the permeate stream, g-mol/s
p_B Brine pressure, MN/m^2

p_P Permeate pressure, MN/m^2
R Universal gas constant, J/Kg-mol
Rj' Intrinsic salt rejection coefficient
T Operating temperature, °C
T_c, Curing temperature of the membrane, °C
u Velocity of brine, cm/s
x_j Molar fraction of j
Δx Tube length, cm
O Osmotic pressure, MN/m^2
v Number of ions produced on complete disociation of one molecule of electrolyte.
η Efficiency
$\rho_{B/P}$ Brine/Permeate density, g/cm^3

References

Chen, L.,1988, PhD Thesis, Imperial College, London.
Geisler, P., K., Wolfgang, and T.A., Peters, 2001, Desalination, 135, 205.
Harris,C.,1994, *Desalination*, 125, 173.
Lonsdale, H.K., U. Merten, and R.L. Riley, 1965, Appl. Polym. Sci., 9,1341.
Mandil, M.A., H.A. Farag, M.N. Naim, and M.K. Attia, 1998, Desalination, 120, 89.
Meares, P., 1976, Membrane Separation Processes, Elsevier Scientific Publishing Company, Oxford.
Sourirajan, S., 1970, Reverse Osmosis, Logos Press, London.

European Symposium on Computer Aided Process Engineering – 12
J. Grievink and J. van Schijndel (Editors)
© 2002 Elsevier Science B.V. All rights reserved.

Batch Distillation of Zeotropic Mixtures in a Column with a Middle Vessel

M. Warter[*], D. Demicoli[**] and J. Stichlmair[**]

[*] Verfahrenstechnischer Anlagenbau, Linde AG, Höllriegelskreuth, Germany
[**] Lehrstuhl für Fluidverfahrenstechnik, Technische Universität München, Germany

Abstract

For the synthesis of batch distillation processes with a middle vessel column it is necessary to understand the influences and limits of important column parameters. The paper explains the geometrical and operational parameters of this type of column by computer-based simulations of a ternary zeotropic system. Moreover, it presents a comparison between experimental and simulation results for the separation of the ternary system in a cyclic operation with constant holdup.

1 Introduction

Batch distillation is a very efficient and advantageous unit operation for the separation of multicomponent mixtures into pure components. Traditionally, a so-called regular batch column is used for the process (*Fig. 1 a*). Unfortunately, this type of column configuration has some big disadvantages as, for instance, high temperatures in the feed vessel. One possibility to overcome some of these difficulties is the use of a middle vessel column (*Fig. 1 c*) which is a combination of a regular and inverted (*Fig. 1 b*) batch column. This type of column was originally proposed by Robinson and Gilliland (1950) [6].

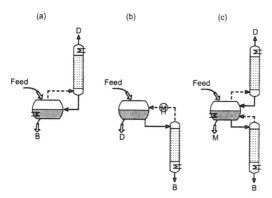

Fig. 1: Different column types – (a) regular, (b) inverted and (c) middle vessel column.

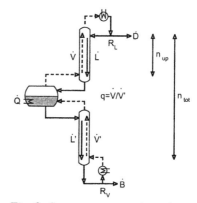

Fig. 2: Streams, geometric and operational parameters of a middle vessel column.

With a middle vessel column it is possible to carry out two separations simultaneously in the upper and in the lower column section. *Fig. 2* shows the streams and parameters of this type of column. It has two geometric parameters. One is defining the total number of stages, n_{tot} and the other, the location of the middle vessel, n_{up}. Additionally, the column has three operational parameters. The reflux and reboil ratios, R_L and R_V, define the ratio of the vapor to the liquid streams in each column section. The third parameter represents the ratio between the streams in the upper and the lower column sections. The ratio of the vapor streams, $q = \dot{V}/\dot{V}'$, is used for this parameter. It can be manipulated by adding or removing heat to the middle vessel.

To carry out a specified separation it is necessary to choose the column parameters in an appropriate way. Although there exist a lot of publications about the middle vessel column (e.g. [1-11]), there is nearly no work showing the influence of the geometric and operational parameters by computer-based simulations for a ternary zeotropic system. Moreover, there is no published experimental work on the separation of a ternary system.

2 Operational Parameters

2.1 Reflux and reboil ratio

By adjusting the reflux and the reboil ratios, R_L and R_V, products of specified concentrations can be simultaneously drawn off at the top and the bottom of the column. The regions of feasible top and bottom products are limited by very high and very low reflux and reboil ratios. *Fig. 3* shows these regions for two instances of time, with x_{F1} and x_{F2} being the corresponding liquid concentrations in the middle vessel. The limitation for very high reflux and reboil ratios can be easily calculated by a distillation line through the concentration of the liquid in the feed vessel, x_F. For very low reflux and reboil ratios, the top and the bottom products lie on the line of the preferred separation. This line goes through the liquid concentration x_F and the corresponding equilibrium vapor concentration y_F^*.

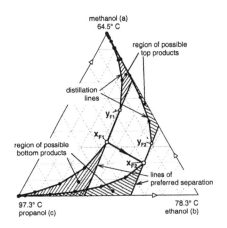

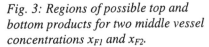

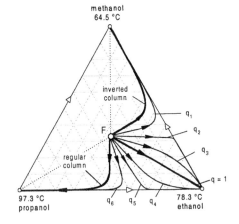

Fig. 3: Regions of possible top and bottom products for two middle vessel concentrations x_{F1} and x_{F2}.

Fig. 4: Influence of the vapor ratio q on the concentration path of the liquid in the middle vessel.

A comparison of the product regions at two different times with the concentrations x_{F1}

and x_{F2} in the middle vessel shows which products can be drawn off in constant concentration during a whole process. These are only products which lie near or on the binary edge between the low and the intermediate boiler and between the high and the intermediate boiler, *and* which are above the line of the preferred separation of x_{F1}.

2.2 Vapor ratio

While the reflux and reboil ratios are used for getting top and bottom products in a specified concentration, the third parameter, the vapor ratio q, can be used for steering the composition in the middle vessel. *Fig. 4* shows the concentration path of the liquid in the middle vessel for different vapor ratios q. For $q=1$, the liquid concentration tends directly to the intermediate boiler ethanol. For increasing vapor ratios, the concentration path becomes more and more similar to that of a regular batch column. Analogously, for decreasing vapor ratios, the concentration path gets more and more similar to that of an inverted batch column. Therefore, it is possible, by manipulating the vapor ratio q, to operate a middle vessel column in every condition between a regular and an inverted batch column.

3 Geometrical Parameters

3.1 Total number of stages

The total number of stages, n_{tot}, has a big influence on the energy demand of the process. For a decreasing number of total stages the energy demand increases steadily. Below a certain number, n_{min}, however, a specified separation can no longer be carried out. For a very high number of stages, the energy demand tends asymptotically to a minimum energy demand, E_{min} (*Fig. 5*).

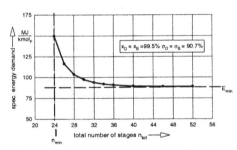

Fig. 5: Influence of the total number of stages on the process.

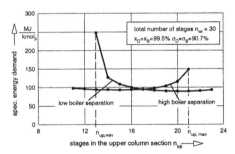

Fig. 6: Influence of the location of the middle vessel on the process.

3.2 Location of the middle vessel

Beside the total number of stages, the number of stages in the upper column section, n_{up}, has to be properly chosen (*Fig. 6*). For very low number of stages in the upper column section, the energy demand for the recovery of the low boiling product determines the energy demand of the whole process. Below a certain minimum value (here $n_{up,min} = 13$), the specified separation can no longer be carried out. For very high number of stages in the upper column section, the lower column section becomes very small. In this case, the energy demand for the recovery of the high boiler dominates the total energy demand. An upper bound, $n_{up,\ max}$, limits the possible number of stages, n_{up}. So, there exists an optimal location for the middle vessel (here $n_{up} = 17$) for which the energy demand of the process is minimal.

4 The Process

Once the operational and geometric parameters have been chosen properly, the whole process can be carried out. *Fig. 7* shows the simulation of a complete process using a constant reflux and reboil ratio operation policy. The feed was an equimolar mixture of methanol/ethanol/propanol. The simulation results show the concentration path of the liquid in the middle vessel and a certain number of liquid concentration profiles over the column height at different times of the process.

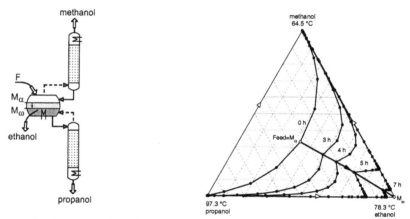

Fig. 7: Process for the separation of a ternary zeotropic mixture - concentration path of the liquid in the feed vessel and liquid concentration profiles of the column for different times.

During the process the concentration profiles in the upper column section lead towards the low boiler methanol which is the top product of the column. The concentration profiles in the lower column section lead to the high boiler propanol which is the bottom product. While methanol and propanol are withdrawn from the top and the bottom of the column respectively, the middle vessel enriches in the intermediate boiling ethanol, which can be removed as product from the middle vessel at the end of the process.

5 Experimental Investigation of a Cyclic Operation with Constant Holdup

In the experimental investigations, a cyclic operation with constant hold-up was chosen. The separation task was to remove low and high boiling impurities from an intermediate boiling product, which is a very common task in practice. *Fig. 8 left* shows the apparatus for the experimental investigations. A preexisting regular batch column was modified to a middle vessel column by installing an additional column section *K4* and a vessel *B3*. In this section the liquid from the column is withdrawn by a valve *V2*. By manipulating this valve, it is possible to operate the column either as a middle vessel column or as a regular column. In the first case the liquid is fed to the middle vessel *B3* and in the second case the liquid is recycled right back to the column. The upcoming vapor can pass the column section *K4* unhindered.

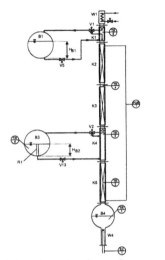

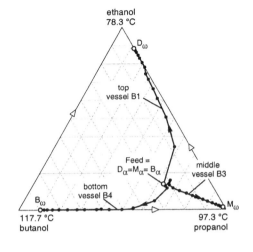

Fig. 8: Pilot plant for the cyclic operation with constant holdup.

Fig. 9: Concentration paths of the liquid in top, middle and bottom vessels during the cyclic operation with constant holdup.

At the beginning of the process the charge was fed to the vessels at the top, middle and the bottom. The feed consisted of ethanol, propanol and butanol. *Fig. 9* shows the liquid concentration path in each vessel during the process. The low and the high boiling impurities, ethanol and butanol, accumulate in the top and the bottom vessels respectively. Therefore, the concentration of the intermediate boiling product, propanol, increases in the middle vessel. At the end of the process, propanol is removed in the specified concentration from the feed vessel.

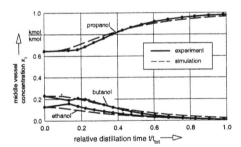

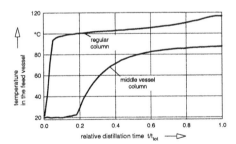

Fig. 10: Concentration in the middle vessel - experiment and simulation.

Fig. 11: Experimental comparison of the temperature in the feed vessel.

Additionally to the results of the experiment, the results of the simulation are plotted in the *Fig. 10*. Both curves show a good agreement, even though the modeling of the column is based on several assumptions like neglecting the fluid dynamics, constant molar hold-up in each vessel, fast energy dynamics, etc. *Fig. 11* shows a comparison of the temperature in the feed vessel with that of a regular column operated at constant reflux. By using a middle vessel column, the temperature can be reduced significantly.

This can be very important for substances which tend to decompose at high temperatures.

6 Summary

The paper explains the influences and limits of the geometrical and operational parameters of a middle vessel column. The results are based on computer-aided simulations of a ternary zeotropic separation. Furthermore, the paper shows an experimental investigation for removing low and high boiling impurities from an intermediate boiling product in a cyclic operation with constant holdup. The experimental results show a good agreement with the simulations.

7 Notation

B	Bottom fraction		q	Vapor ratio
D	Distillate		t	Time
F	Feed		x	Liquid concentration
M	Intermediate fraction		y	Vapor concentration
N	Number of stages		α, ω	Start/end of a process step
Q	Heat			

8 References

1. Barolo, M., G.B. Guarise, S.A. Rienzi and A. Trotta, 1998, *Understanding the Dynamics of a Batch Distillation Column with a Middle Vessel*, Computers chem. Engng Vol. 22, S37-S44
2. Farschman, C.A. and U. Diwekar, 1998, *Dual Composition Control in a Novel Batch Distillation Column*, Ind. Eng. Chem. Vol. 37, 1, 89-96
3. Hilmen, E.K., S. Skogestad, M.F. Doherty and M.F. Malone, 1997, *Integrated Design, Operation and Control of Batch Extractive Distillation with a Middle Vessel*, AIChE Annual Meeting 1997, Los Angeles, paper No. 201h
4. Meski, G.A. and M. Morari, 1995, *Design and Operation of a Batch Distillation Column with a Middle Vessel*, Computers chem. Engng Vol. 19 Suppl., S597-S602
5. Mujtaba, I.M. and S. Macchietto, 1994, *Optimal Operation of Multicomponent Batch Distillation - A Comparative Study Using Conventional and Unconventional Columns*, Proc. IFAC Symposium ADCHEM'94, Kyoto, Japan, 25-27, 401-406
6. Robinson, C.S. and E.R. Gilliland, 1950, *Elements of Fractional Distillation*, 4th ed. McGraw Hill, New York
7. Stichlmair, J. and J.R. Fair, 1998, *Distillation - Principles and Practices*, J. Wiley, New York
8. Warter, M. and J. Stichlmair, 1999a, *Batchwise Extractive Distillation in a Column with a Middle Vessel*, Computers chem. Engng., 23, Suppl., S915-S918
9. Warter, M. and J. Stichlmair, 1999b, *Batchwise Extractive Distillation in a Novel Modification of a Middle Vessel Column*, ECCE 2, 5.-7.10, CD-ROM Montpellier, France
10. Warter, M., J. Stichlmair, 2000a, *Batch Distillation of Azeotropic Mixtures in a Column with a Middle Vessel*, Computer-Aided Chemical Engineering, Vol. 8, Elsevier, p. 691-696
11. Warter, M., J. Stichlmair, 2000b, *Batch Distillation with a Middle Vessel Column*, CHISA 2000, Prag, 27.-31. August 2000, CD-ROM of full texts

European Symposium on Computer Aided Process Engineering – 12
J. Grievink and J. van Schijndel (Editors)

Wastewater Treatment Management Using Combined Water-Oxygen-Thermal Pinch Analysis

T K Zhelev

University of Durban-Westville

South Africa

Abstract

Presented paper address the problems of better management of resources (electrical energy, heat, water and other) when wastewater treatment is concerned. It intents to make one further step towards more cost-attractive environment protection activities. More precisely it focuses on centralised, semi-centralised and distributed wastewater treatment systems and attempts to provide a tool for decision making when a balance between on-site treatment and centralised treatment service is to be reached. The analysis considers benefits related to electrical energy saving, water and wastewater minimisation, and heat recovery. It is based on combination of Pinch principles application for heat, water and oxygen (Thermal Pinch, Water Pinch and Oxygen Pinch).

1. Introduction

Thermal Pinch technology was born during late 70s responding to problems related to the energy crisis. This technology was successfully used for energy management in large number of industrial applications world-wide. Water Pinch concept emerged during late 80s when it was recognised that the fresh water will become the most valuable aced of the new millennium. Both techniques are addressing problems of industrial technology from different point of view. Both suggest structural changes and integration actions that can lead to contradictions, overlaps and trade-offs.

From the other side, the analysis of heat and water relation shows that in average, around 60% of the water used in industrial processes is lost by evaporation. This fact gives enough indication for poor energy management. The call for general methodology considering simultaneously heat and water resources management and optimisation is apparent. The problem is complex because of the need for parallel consideration of heat and mass transfer and the big variety of mass transfer processes normally accompanied by heat transfer.

Pinch technology gives considerable hope to lay down the bridge between both processes because its principles appeared to be suitable for both heat and water. Here we should mention the work reported by El-Halwagi & Manousiouthakis (1989), Savulescu and R. Smith, (1998), the EC project Contract JOE_3-CT950036 (1997), recent works of Savin and Bashevich (2001), D. Koufos and T. Retsina (2001) and the study published by Bosnjakovic & Knoche (1999), where the authors are addressing the problems of heat and water integration in cooling towers - typical simultaneous heat and mass transfer unit operations, blamed for the major water loss in industry. We should also mention our first steps in this interesting direction trying to solve the problems of balance between humidification and dehumidification loads in contact economiser systems used for flue gas energy recovery of industrial boilers as it can be found in Zhelev & Semkov (2001).

2. Simultaneous Heat and Mass Transfer

As it was mentioned earlier, addressing heat and mass transfer processes in one single methodology targeting the optimal resources management requires systematisation of the great variety of simultaneous heat and mass transfer cases and optimisation goals (objectives). Heat and mass transfer are usually mutually supporting processes but they also can act in different direction. The general classification of these processes is according to the limiting process – (a) processes limited by the heat transfer and (b) these limited by the mass transfer. Examples of the first type are boiling, condensation, drying, distillation, rectification, etc., when to the second type can be allocated absorption, extraction, particle separation (settling, sedimentation), solution, concentration and many others. The problem of simultaneous consideration of possible energy and mass recovery increases the degree of complexity in both cases. In many of the above mentioned processes the heat and mass transfer can support each other when there are number of them where they can go in different directions. Because of the importance of water conservation the mass transfer consideration deals in first instance with water evaporation/condensation.

There are two general attempts to come up with a method solving the problem of management of energy and water: (i) Methods for simultaneous consideration of heat and water conservation in case of mass transfer limited processes. These methods are represented by the series of papers addressing the application of Pinch principles for water and heat minimisation in wastewater treatment processes, and (ii) Methods for simultaneous management of heat and mass transfer where the heat transfer is limiting. Obviously in this case it is quite important where the limiting process takes place – at the interface or in one of the phases.

3. The Pinch Approach

Classical thermal Pinch analysis incorporates two stages: (i) Targets setting (energy) and (ii) Formulation design guidelines to reach the targets. Water Pinch addresses similar issues: (i) Targets definition (minimum fresh water demand) and (ii) Design guidelines to optimise the structure of the wastewater treatment system. Coming to Oxygen Pinch, the analysis is broader, it sets targets in growth rate (the micro-organisms' health), oxygen solubility, residence time & oxidation energy load as suggested by Zhelev & Ntlahana (1999). Finally combining the heat and mass transfer

Pinch the expectations are even broader – targets in process balance, optimal operation and process control are expected.

The combined effort to solve simultaneously the problems of energy and water management led to the idea to construct combined composite curves matching both resources achieved utilising the streams mixing opportunities and using a new tool called Two dimensional grid diagram as suggested by Savulescu and Smith (1998). .

Zhelev & Bhaw (2000) reported the combination of oxygen and energy targeting applied to biodegradation operated centralised wastewater treatment plants. There the proposed oxygen management leads to benefits in saving electrical energy because of the aeration/agitation driven oxygen supply. The simultaneous water and oxygen management consideration represents the attempt to create a decision making tool helping factory managers to judge when and what part of their wastewater should be treated on site or send to centralised wastewater work (external service).

The Water Pinch concept addresses the quantity of effluents when the Oxygen Pinch considers the quality. In case the charges for wastewater treatment are based on quality of the effluents sent for treatment then the combined analysis helps to find the most cost-effective solution.

4. Combined Water-Thermal Pinch Analysis

Assumptions: In first instance the study is restricted to simultaneous heat and mass transfer processes without chemical reaction. The generalisation of methodology is very much desired.

The hypothesis is that the assumption of heat transfer direction from the stream with higher temperature to a stream with lower temperature can be related unequivocally to a direction of mass transfer from a "reach" to a "lean" stream, i.e. from a stream with higher concentration of particular component (components) to a stream with lower concentration of these components. Actually the mass transfer can be redirected in reverse direction if certain conditions are in place. These conditions usually are related to energy characteristics. For example, gas dehumidification using water can be realised if the water temperature is lower than the dew point temperature. In this case the direction of mass transfer depends on the partial pressure of the component into consideration. Assuming the partial pressure as measure of "concentration" - the reach streams are those with higher partial pressure and "lean" streams are those with lower partial pressure. Another general case to mention is the absorption, where higher temperature lowers the rate of absorption. The higher the solute temperature - the higher the solubility capacity. The conclusion is that the reach and lean streams (donors and acceptors of contamination) are unequivocally situated above or under the equilibrium conditions as much as the hot streams are on the top of the cold streams in their composite representation through the classical T/H diagram.

5. The Cost Model (Wastewater treatment cost model)

Data obtained from Hammarsdale wastewater treatment plant was filtered and tabulated. The wastewater treatment equation assumes that the two variables affecting the wastewater treatment cost are the flowrate and the quality of the wastewater. Thus an equation of the following form will be suitable:

$$C = a_0 . F^a{}_1 . Q^a{}_2 \qquad (1)$$

where C represents wastewater treatment cost in \$/day, F - wastewater flowrate in kl/day, Q - wastewater quality (COD) in mg COD/l, a_0, a_1 and a_2 are constants. Linearising the equation:

$$ln\ (C)\ = ln(a_0) + a_1.ln(F) + a_2.ln\ (Q) \qquad (2)$$

With C, F and Q data, multivariable linear regression was performed using Polymath to determine the constants a_0, a_1 and a_2.

5.1 Wastewater treatment cost with and without pinch applied

The stream data obtained for the poultry abattoir was used to calculate the wastewater characteristics. The wastewater characteristics, were the flowrate and the COD content. The calculations were performed firstly with no pinch analysis and then with combined water oxygen pinch analysis (CWOPA).

Using the collected data, the wastewater characteristics were calculated:

Flowrate (F)	=	*Process F + Scalding F + Washdown F*
	=	*473.39 kl/day + 131.93 kl/day + 170.73kl/day*
	=	*776.05 kl/day*
Mass flowrate (MF)	=	*Process MF + Scalding MF + Washdown MF*
	=	*712 kg COD/day + 132 kg COD/day + 457 kgCOD/day*
	=	*1301 kg COD/day*
Quality (Q)	=	*Mass flowrate (MF) / Flowrate (F) * 1000*
	=	*1301 kg COD/day / 776.05 kl/day * 1000*
	=	*1676.44 mg COD/l*

Regression of the data provided the following results:

ln(a₀)	=	*7.945*
a₁	=	*0.217615*
a₂	=	*0.0323635*

Thus, the wastewater treatment costs equation will be:

$$C = e^{(7.945 + 0.217615.ln(F) + 0.0323635.ln(Q))} \qquad (3)$$

To yield the wastewater treatment cost of 15265 \$/day. This was the cost of treating the wastewater on-site and represented treatment cost only. Similarly, the wastewater treatment cost for the brewery and the tannery was calculated. The results obtained were tabulated in Table 1 and 2.

Table 1. Wastewater treatment cost with no pinch analysis applied.

Plant	Flowrate (F) Kl/day	Quality (Q) mg COD/l	Treatment cost (C) $/day
Abattoir	776	1677	15265
Brewery	1153	1564	16600
Tannery	621	4203	14982

Table 2. Wastewater treatment costs with application of CWOPA

Plant	Distributed treatment plant			Centralised treatment plant		
	Flowrate kl/day	Quality mg COD/l	Cost $/day	Flowrate kKl/day	Quality mg COD/l	Cost $/day
Abattoir	473	2748	13930			3913
Brewery	694	2599	15110	1531	3953	5422
Tannery	364	8103	13620			8860

The savings are apparent and support the centralised wastewater treatment.

6. Water-Oxygen-Energy

As it was stated earlier, the oxygen pinch addresses problems of management of this resource in centralised waterworks when water pinch focuses on water and wastewater management at production sites. Energy integration worthy streams can only be found in those sites but not in centralised wastewater treatment sites where waters are sent form long distances and heat cannot be expected to be on industrial benefit. Moreover, extreme temperatures are not desirable for bio treatment conditions.

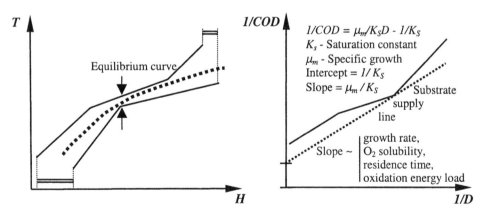

Figure 1. Targeting heat and mass   Figure 2. Targeting oxygen & energy

So, the trade-off of water reuse/recycle in production factories and at central wastewater servicing station will depend on cost analysis of particular case. This analysis can be performed using the cost model presented in previous chapter. Which water saving approach to be used on production sites depends on the type of mass transfer processes in place. If a mass transfer controlled processes are predominantly in place then an approach similar with the one suggested by Savulescu and Smith (1998) can be used. In the general case when a mixture of processes controlled by heat or/and mass transfer are in place then the T/H diagram including equilibrium presentation must be used.

6.1 Guidelines

It should be noted that the process streams subject to strictly recuperative heating or cooling should be separated from these allowing for regenerative heat exchange. Another words, in the general case there should be two composite curves - one for streams potentially allowed to exchange heat strictly trough a wall (like final products, explosive and flammable fluids, etc.), and second composite for streams to be heated or cooled in flexible way (eider through direct contact or through a wall). The same should apply to a complex production site. One needs to make use of the following tools: *T/H* Composite curves, *Concentration/mass* Composite curves, *1/COD* over *1/D* Composite curve, Two dimensional grid diagram and Combined composite curves with equilibrium conditions. For the particular case of wastewater treatment the trade-off between the heat utilisation and water minimisation will be addresses by Three dimensional grids if the water using processes are mass transfer limited and the combined composite curve with equilibrium presentation - if the processes involved are heat transfer limited. Part of wastewater streams to be sent for centralised treatment accompanied by the targets of energy to be used for aeration will be addressed by Oxygen Pinch composite curves.

7. Conclusion

Process integration guided simultaneous management of energy and water resources is proposed for classes simultaneous heat and mass transfer processes. The impact of extended Pinch theory addressing resources of mass and energy in industrial applications from system prospective will assist better decision making at both new design and retrofit level making these solutions more trustful and both – cost and environmentally acceptable.

8. References

El-Halwagi, M.M. and V. Manousiouthakis, 1989, Synthesis of Mass Exchange Networks, AIChE J., 8, 1233.

Savulescu, L. E. and R. Smith, 1998, Simultaneous Energy and Water Minimisation, AIChE Annual Meeting, Paper 13c.

Simultaneous Energy and Water Minimisation, 1997, Contract JOE_3-CT950036, EC JOULE III, Publishable report.

Bosnjakovic, F. & Knoche, K.-F., 1999, Pinch Analysis for Cooling Towers, *http//www.ltt.rwth-aahen.de/forschung/bilanzopti/florenz/index.html*, 1-12.

Zhelev, T.K. and K.A.Semkov, Analysis Combined Heat and Mass Pinch Analysis for More Efficient Flue Gas Energy Recovery, 2001, 6th World Congress in Chemical Engineering, Melbourne, Australia, September 23-27.

Zhelev, T.K. and N. Bhaw, Combined Water-Oxygen Pinch Analysis for Better Wastewater Treatment Management, Waste Management, 20, 2000, 665-670.

Zhelev, T.K. and L. Ntlhakana, 1999, Energy - Environment Closed Loop through Oxygen Pinch, Comp.& Chem.Eng. 23, s79.

Integrated Process Simulation and CFD for Improved Process Engineering

Stephen E. Zitney, Aspen Technology, Inc., Cambridge, MA 02141, U.S.A.
E-mail: steve.zitney@aspentech.com
Madhava Syamlal, Fluent Inc., Morgantown, WV 26505, U.S.A.
E-mail: mxs@fluent.com

Abstract

This paper describes recent efforts to seamlessly integrate process simulation and computational fluid dynamics (CFD) using open standard interfaces for computer-aided process engineering. A reaction-separation-recycle flowsheet coupled with a CFD stirred tank reactor model is presented as an example to demonstrate the applicability of the integration approach and its potential to improve process engineering. The results show that the combined simulation offers new opportunities to analyze and optimize overall plant performance with respect to mixing and fluid flow behavior.

1. Introduction

Process simulation and CFD are widely used by many operating companies in the process industries. Process simulators are used primarily in process engineering, but provide significant added value in modeling applications across the plant lifecycle, from conceptual design through process engineering and into plant operations. CFD is used primarily in the detailed engineering phase of the plant life cycle. Equipment designers and process engineers are increasingly using CFD to analyze the detailed flow and mixing processes within individual items of process equipment, such as chemical reactors, stirred tanks, fluidized beds, bubble columns, combustion systems, spray dryers, and other equipment (Bakker *et al.*, 2001).

Integrating process simulation and CFD is of much industrial interest because it improves the workflow between process engineering and detailed engineering by ensuring the use of consistent models and physical properties between these two phases of the plant lifecycle. In addition, the optimization of individual units using CFD is not done in isolation but within the context of the whole process, so that a global improvement is achieved, especially for cases in which plant operation depends on mixing and fluid dynamics. Strategies for combining process simulation and CFD vary in the degree of integration and the type of information exchanged:

1. The process simulator and CFD package model *different unit operations* and exchange stream information at flow boundaries.
2. The CFD and process simulation models are more tightly coupled and describe *distinct spatial domains* within the *same unit operation*. Mass and energy are

exchanged between the models not only at flow boundaries, but also across separating walls and membranes.

3. Representing an even tighter degree of integration, the process simulator and CFD package model the *same unit operation*, but *different physical phenomena* (e.g., Bezzo *et al.*, 2000). In this case, the process simulation model typically sends physical and transport property data (e.g., density, viscosity) to the CFD model, which in turn sends back flow-dependent information (e.g., turbulent kinetic energy). This strategy also includes the case where a CFD model of a single unit operation is represented by an interconnected network of models in the process simulator (e.g., Urban and Liberis, 1999).

Focusing on strategy one above, we describe here the coupling of a widely used commercial process simulator, Aspen Plus® (Aspen Technology, Inc.), and a popular CFD package, FLUENT™ (Fluent Inc.). The software integration is accomplished using a COM/CORBA bridge implementation of the open standard interfaces from Global CAPE-OPEN (e.g., Braunschweig *et al.*, 2000). To demonstrate the benefits of the integrated simulation environment, we use a FLUENT CFD model of a stirred tank reactor in an Aspen Plus steady-state simulation consisting of several unit operations (reactors, heat exchangers, and distillation columns).

2. Aspen Plus and FLUENT Integration

In this work, Aspen Plus 11.1 (Aspen Technology, Inc., 2001) and FLUENT 6.0 (Fluent Inc., 2001) are integrated using a CAPE-OPEN compliant COM/CORBA bridge, as shown in Figure 1. The bridge increases the flexibility of deployment by allowing Aspen Plus and FLUENT to run on different hardware platforms under different operating systems (i.e., Aspen Plus-Windows; FLUENT-Windows, UNIX).

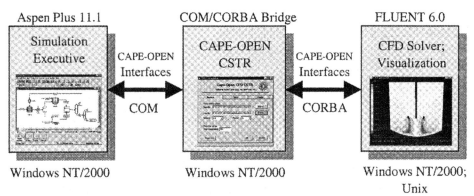

Figure 1. Software architecture for Aspen Plus and FLUENT integration

The CAPE-OPEN interfaces (e.g., Braunschweig *et al.*, 2000) are an established set of computer-aided process engineering standards for unit operations, reaction kinetics, thermodynamic and physical property calculations, and numerical solvers. We exploit the CAPE-OPEN unit operation interface to use (e.g., create, edit, solve) FLUENT CFD models in an Aspen Plus flowsheet. This interface also facilitates the bi-directional exchange of stream information between Aspen Plus and FLUENT. The CAPE-OPEN thermodynamics and reaction kinetics interfaces are used to send Aspen Plus physical property and reaction data to FLUENT.

3. Reaction-Separation-Recycle Example

To demonstrate the integration of process simulation and CFD, we use an Aspen Plus sequential-modular, steady-state simulation of a reaction-separation-recycle flowsheet (Figure 2). The first reactor, a stoichiometric reactor based on fractional conversion, is used to model a solvent-enhanced isomerization reaction, which produces feed A (see Table 1 below) for the second continuous-stirred tank reactor (CSTR).

An Aspen Plus duplicator block copies the outlet stream from the stoichiometric reactor to two duplicate input streams for the CSTR. One input stream goes to an Aspen Plus CSTR model, while the other goes to a FLUENT CSTR model. For any given run, we activate the Aspen Plus CSTR or the FLUENT CSTR. In the CSTR, feed A is mixed with feed B to generate the desired product P.

The downstream separation section consists of two distillation columns in series. The first column recovers solvent and recycles it back to the stoichiometric reactor. The second column separates the product P (bottoms) from the impurities (distillate).

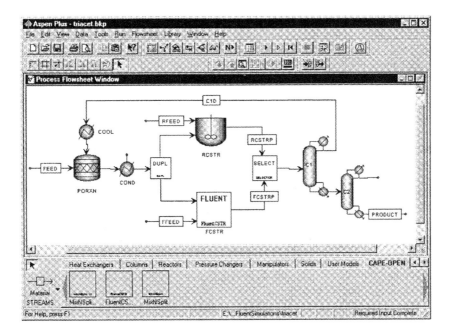

Figure 2. Aspen Plus flowsheet for reaction-separation-recycle example

3.1 CSTR Model
A detailed CFD model is required for the CSTR because mixing impacts the performance of the reactor. The FLUENT CSTR is registered as a CAPE-OPEN model

that appears in the CAPE-OPEN category of the Aspen Plus model library palette. A FLUENT CSTR model can be added to the flowsheet and configured in the same way as any other Aspen Plus model. Double-clicking on the FLUENT CSTR block in the process flowsheet window brings up a graphical user interface (GUI) via the CAPE-OPEN unit operation interface. This GUI is used for setting FLUENT start-up options (e.g., FLUENT case file name), solver options (e.g., maximum number of FLUENT iterations per Aspen Plus iteration), and basic CFD-related parameters (e.g., impeller speed) that are exposed as CAPE-OPEN parameters. Advanced CFD options can be set by launching the FLUENT GUI from the FLUENT CSTR block in Aspen Plus.

The CSTR models involve a homogeneous liquid-phase chemical reaction scheme with kinetic rates that vary over four orders of magnitude (Table 1). When mixing is rapid,

Table 1. Reaction Scheme (Rate)

A + B → R (medium-fast)
B + R → P (fast)
B + C → Q (slow)

formation of the desired product P is favored. When mixing is very slow, formation of Q is favored. At intermediate levels of mixing, formation of R is favored. All reactions compete for B, which enters the CSTR in a second feed stream.

The Aspen Plus CSTR model uses a built-in power law expression for calculating the Arrhenius rates for the rate-controlled reactions. The FLUENT CSTR model uses the finite-rate/eddy-dissipation model (for turbulent flows) that computes both the Arrhenius rate and the mixing rate and uses the smaller of the two. Since FLUENT incorporates the effect of mixing, its prediction of yield differs from the results predicted using the perfectly-mixed CSTR model in Aspen Plus.

The FLUENT CSTR model used here considers 2D axisymmetric swirling flow in the reactor. The standard k-ε model (Launder and Spalding, 1972) is used to model turbulence. This is a semi-empirical model based on model transport equations for the turbulent kinetic energy (k) and its dissipation rate (ε). In regions of the reactor where turbulence levels are high, the eddy lifetime k/ε is short and mixing is fast. As a result the reaction rate is not limited by small scale mixing of the reactants A and B. On the other hand, in regions with low turbulence levels, small scale mixing may be slow and limit the reaction rate. FLUENT solves the steady turbulent flow problem with its default segregated solution algorithm.

Given the tank style, feed location, and impeller geometry and location, we used MixSim 1.7 (Fluent Inc., 2000) to generate automatically the computational grid used by FLUENT. The CSTR used here consists of a high-efficiency axial impeller in a baffled, dish-bottomed tank with feed injection at the impeller. FLUENT solves the fluid flow problem using this structural information along with the inlet boundary conditions, equipment specifications (e.g., impeller speed), and solver settings.

3.2 Running the Integrated Simulation

The integrated Aspen Plus and FLUENT simulation involves an iterative sequential-modular solution process. To generate a good starting point, we first perform a steady-state simulation of the reaction-separation-recycle flowsheet using the Aspen Plus CSTR. We then deactivate the Aspen Plus CSTR, activate the FLUENT CSTR, and select the FLUENT stream as the inlet stream specification for the selector block. Next, we interactively run and monitor the combined simulation from within Aspen Plus.

Aspen Plus controls the integrated simulation and automatically executes the FLUENT CSTR model at each flowsheet iteration. The CFD results are saved at each Aspen Plus iteration so that subsequent FLUENT simulations converge more quickly. Stream information, physical properties, and reaction kinetic data are transferred automatically from Aspen Plus to FLUENT using CAPE-OPEN interfaces. For a given impeller speed, FLUENT computes the flow pattern and chemical species distribution. The weighted average of the field variables (e.g., species mass fractions) at the CSTR outlet are sent back to Aspen Plus also using the CAPE-OPEN interfaces.

Upon completion of the integrated simulation, we review results for streams, blocks, and overall convergence in Aspen Plus. From the CAPE-OPEN GUI for the FLUENT CSTR, we display basic CFD results such as contours of species mass fractions in the reactor. For advanced display options, we can automatically launch the FLUENT GUI.

3.3 Sensitivity Analysis

Figure 4 shows the results of a sensitivity analysis to determine how product purity and yield react to varying the CSTR impeller or shaft speed (i.e., mixing).

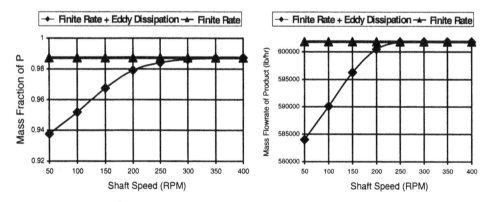

Figure 4. Sensitivity analysis of product purity and yield vs. impeller speed

When using the *finite-rate* reaction model in the FLUENT CSTR, the results for product mass fraction and flow rate are independent of impeller speed and match the Aspen Plus CSTR results. The results for the *hybrid finite-rate and eddy-dissipation* model show that faster mixing increases product purity and yield, approaching the finite-rate results at an impeller speed of about 300 rpm. It may also be possible to improve product quality and yield by modifying tank style, feed location, or impeller geometry.

Clearly, such integrated simulations can be used to analyze overall process performance with respect to important CFD design and operational parameters. The optimization of CFD parameters is not done in isolation but within the context of the whole process, so that a global improvement is achieved, especially for processes that depend on mixing and fluid dynamics.

4. Conclusions

This paper described the seamless integration of process simulation (Aspen Plus) and CFD (FLUENT) using standard CAPE-OPEN interfaces. The results for a reaction-separation-recycle flowsheet coupled with a detailed CFD model of a CSTR demonstrated the potential for improved process engineering. Continued focus on tighter integration between process simulation and CFD software will present many new opportunities to simulate entire plants based on the fundamental principles of flow-related phenomena, further enhancing process performance and efficiency.

5. Acknowledgements

This work was done with the support of the U.S. Department of Energy, under Award No. DE-FC26-00NT40954. However, any opinions, findings, conclusions, or recommendations expressed herein are those of the author(s) and do not necessarily reflect the views of the DOE. The authors gratefully acknowledge the contributions of Paul Felix, Maxwell Osawe, and Lanre Oshinowo (Fluent Inc.) and Joe Cleetus and Igor Lapshin (West Virginia Univ.) for software development and integration. We also thank Michael Halloran (AspenTech UK) for his support of the CAPE-OPEN interfaces.

6. References

Aspen Technology, Inc., 2001, Aspen Plus 11.1 User Guide, Cambridge, MA.

Bakker, A., A. H. Haidari, and L. M. Oshinowo, March 2001, Realize Greater Benefits from CFD, Chem. Eng. Progress, **97** (3), 45-53.

Bezzo, F., S. Macchietto and C. C. Pantelides, 2000, A General Framework for the Integration of Computational Fluid Dynamics and Process Simulation, Computers and Chemical Engineering, **24**, 653-658.

Braunschweig, B. L., C. C. Pantelides, H. I. Britt, and S. Sama, September 2000, Process Modeling: The Promise of Open Software Architectures, Chem. Eng. Progress, **96**(9), 65-76, http://www.global-cape-open.org.

Fluent Inc., 2000, MixSim 1.7 User's Guide, Lebanon, NH.

Fluent Inc., 2001, FLUENT 6.0 User's Guide, Lebanon, NH.

Launder, B. E. and D. B. Spalding, 1972, Lectures in Mathematical Models of Turbulence, Academic Press, London, England.

Urban, Z. and L. Liberis, 1999, Hybrid gPROMS-CFD Modeling of an Industrial Scale Crystalliser with Rigorous Crystal Nucleation and Growth Kinetics and a Full Population Balance, Presented at the Chemputers Europe 5 Conference, October 21-23, Dusseldorf, Germany.

European Symposium on Computer Aided Process Engineering – 12
J. Grievink and J. van Schijndel (Editors)
© 2002 Elsevier Science B.V. All rights reserved.

Frequency locking in a discontinuous periodically forced reactor

E. Mancusi[1], L.Russo[2], G. Continillo[1] and S. Crescitelli[2]

[1]Dipartimento di Ingegneria, Università del Sannio, Piazza Roma, 82100, Benevento

[2]Dipartimento di Ingegneria Chimica, Università degli Studi di Napoli "Federico II"
Piazzale Tecchio 80, 80125 Napoli

Abstract

The interaction between an external periodic forcing and natural frequencies of a reactor or system of reactors can give rise to an interesting dynamic behaviour. Particularly the system develops periodic responses where the periods are exact multiples of the period of the forcing. This behaviour is identified as creation and breaking of frequency locking and invariant tori. In this work, frequency locking and resonance regions are investigated for a reactor with discontinuously forced feed.

1. Introduction

In the last 30 years, the real advantages of forced unsteady-state operation over conventional steady-state regimes of catalytic fixed-bed reactors have been widely supported (Matros and Bunimovich, 1996; Eigenberger and Nieken, 1988; Neophytides and Froment, 1992). Special attention has been devoted to Reverse Flow Reactors (RFR) for catalytic combustion.

The optimal for these dynamically forced system is, usually, a periodic solution with the same period of the forcing. This means that in the case of RFR the period of the solution is twice the inversion time. However, depending on the operating or design parameters, those reactors can show a rich dynamical response as demonstrated by Řeháček et al. (1998) and Khinast et al. (1999). Indeed, it has been shown that multiperiodic, quasi-periodic or even chaotic regimes can be attained (Řeháček et al., 1998). Therefore, rational design and operation of a RFR requires efficient prediction of the prevailing regimes and accurate detection of multistability conditions.

In the present work we focus our attention on frequency locking, a typical phenomenon of time forced reactors. The model considers a homogeneous irreversible reaction conducted in non-isothermal conditions. These systems oscillate spontaneously, and the interaction between their natural frequencies and the forcing frequency gives rise to interesting dynamic behaviour, including frequency locking. Such phenomenon rather common in forced systems is related to the interaction between two frequencies (Kevrekidis et al., 1986a, 1986b; Schereiber et al.; 1988) e.g. the natural frequency of the system and the forcing frequency.

As a consequence of a Neimark-Sacker bifurcation 2-tori invariant quasi-periodic orbits emerge when the ratio between two frequencies (natural and forcing) is an irrational number. As the systems parameters are varied, the natural frequency varies and may "lock" on to the forcing frequency, when their ratio becomes rational. In the occurrence of Neimark-Sacker bifurcations, windows arise that contain periodic behaviour, where the periods are multiples of the forcing period. These windows, which in the solution diagram representation have the form of isolae delimited on both sides by saddle-node bifurcations, become wider as parameters move away from the N-S bifurcation locus (Kuznetsov, 1998). Within each of these windows, there are at least two periodic orbits (one stable and one unstable) that wind up over the torus surface.

By means of a purposely developed technique (Faraoni et al., 2001), this paper investigates on the existence of such kind of periodic behaviour in periodically forced reactors when the forcing is discontinuous.

2. The Mathematical Model

We consider a system of two CSTR with heat exchange between reactors and surroundings and with a periodically inverted feed (Żukowski and Berezowski, 2000). The model, essentially the same as in Faraoni et al. (2001), is represented by the following dimensionless equations:

$$\frac{d\alpha_1}{d\tau} = -\alpha_1 + IO(\tau)\alpha_2 + Da(1-\alpha_1)^n \exp\left(\frac{\gamma\beta\vartheta_1}{1+\beta\vartheta_1}\right)$$

$$\frac{d\vartheta_1}{d\tau} = -\vartheta_1 + IO(\tau)\vartheta_2 + Da(1-\alpha_1)^n \exp\left(\frac{\gamma\beta\vartheta_1}{1+\beta\vartheta_1}\right) + \delta(\vartheta_H - \vartheta_1)$$

$$\frac{d\alpha_2}{d\tau} = -\alpha_2 + (1-IO(\tau))\alpha_1 + Da(1-\alpha_2)^n \exp\left(\frac{\gamma\beta\vartheta_2}{1+\beta\vartheta_2}\right) \qquad (1)$$

$$\frac{d\vartheta_2}{d\tau} = -\vartheta_2 + (1-IO(\tau))\vartheta_1 + Da(1-\alpha_2)^n \exp\left(\frac{\gamma\beta\vartheta_2}{1+\beta\vartheta_2}\right) + \delta(\vartheta_H - \vartheta_2)$$

where α represents the conversion, ϑ the temperature, γ the activation energy, β the enthalpy of reaction, Da the Damköhler number, δ the heat exchange coefficient, and $IO(\tau) = \text{int}(\tau/\tau_p) - 2\text{int}(\tau/2\tau_p)$, τ_p being the inversion period.

The system described by Eq. (1) is a non-autonomous dynamical system made of four ordinary differential equations. As show in Faraoni et al. (2001), it is possible to define a Poincaré map that permits the study of the system in a global rather than in a local fashion. In fact, the continuous-time system can be replaced by a discrete-time system that is obtained by sampling all orbits every $T=2\tau_p$. In this way, all features of the underlying continuous-time system are preserved in the discrete-time system originating the map. Let $\mathbf{u}$ be the vector state of system (1), θ the vector parameter, and $\mathbf{f}$ the vector field, the associated map can be expressed as:

$$\mathbf{u}^{k+1} = P(\mathbf{u}^k) = \mathbf{u}^k + \int_0^T \mathbf{f}(\mathbf{u}, \theta, t) dt \qquad (2)$$

Hence, map P is a stroboscopic map obtained by sampling the system trajectory at times T. T-periodic orbits of system (1) correspond to fixed points of the map P. On the Poincaré section quasi-periodic orbits are reduced to invariant sets represented by one or more closed curves. A periodic orbit with period equal to a multiple p of T, where p is an integer, will correspond to fixed points of the p-iterate of P (i.e. P^p). If a pT-periodic solution is locked on the torus, the Poincaré section will be represented as a set of p points arranged along the "shadow" of the torus closed curves. The sequence by which these points are visited during the p iterations of the stroboscopic map defines the rotation number ρ of the orbit. For p-periodic orbits, ρ is a rational number

$$\rho = \frac{q}{p} \tag{3}$$

which represents the ratio between the number of revolutions along the meridian, q, that the orbit experiences prior to closing itself on to the pth point of the set, i.e. after having completed p revolutions along the parallel (Franceschini, 1983; Kuznetsov, 1998).

3. Resonance Zones

The bifurcational study of the system was conducted by picking the Damköhler number and the dimensionless heat transfer coefficient δ as bifurcation parameters. In this way the natural frequency of the system is changed while keeping constant the forcing frequency.

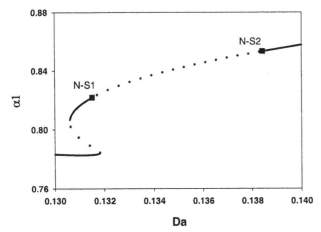

Figure 1 - Solution diagram for $\beta=2.7$, $\gamma=15$, $\delta=3$, $\vartheta_H=0$ and $n=1.5$. Solid lines represent stable periodic solutions, dotted lines the unstable periodic solutions; black squares indicate the N-S bifurcations.

As Da is varied, the solution diagram obtained for a given set of values of the other parameters is reported in Fig. 1. On the high-conversion branch two Neimark-Sacker (N-S1 and N-S2) bifurcations are found. These bifurcations are both supercritical, hence generate stable quasi-periodic behaviour. For Da in between the two N-S bifurcation values, chaotic, quasi-periodic as well as *p*-periodic solutions are found. In Figure 2 the Poincaré section relevant to two different (but very close to each other) values of *Da* is reported.

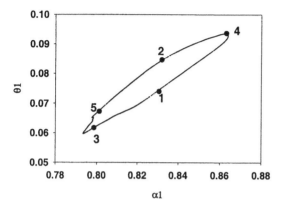

Figure 2 - Stroboscopic section illustrating frequency locking on a torus. In this figure a Poincaré section of stable torus (Da=0.1332) and a stable 2/5 resonant solution (Da=0.1334) are represented for two different values of the Damköhler number.

The points representing the 5-periodic solution in Fig. 2 are numbered according to the sequence by which they are visited. It appears that the orbit revolves twice before closing itself on to the 5th point, and this corresponds to the rotation number ρ=2/5.
By means of automatic parameter continuation (Faraoni et al., 2001), again using *Da* as the bifurcation parameter, the solution diagram shown in Fig. 3 has been constructed.

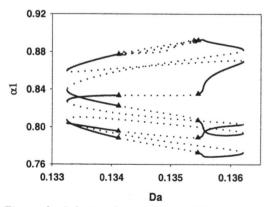

Figure 3 - Solution diagram for a 2/5 resonance solution. Filled triangles represent Pitchfork bifurcations and empty triangles Flip bifurcations.

It appears that, along with the stable 5-period cycle, an unstable 5-period cycle is also present. Both cycles are found on the same closed curve delimited by saddle-node bifurcations. The appearance of the solution diagrams as those shown in Fig. 3 is typical of time-periodic regime phenomena named frequency-locking. In fact, they arise and die in correspondence of saddle-node bifurcations (Kuznetsov, 1998). In Fig. 3 Flip and Pitchfork bifurcations are also reported. Pitchfork bifurcations generate non-symmetric solutions (an example is reported on the upper branch in Fig. 3). This solution generate in turn period-doubled solution by a Flip bifurcation. Via a Flip bifurcation cascade a chaotic scenario is attained (for $Da \in [0.1343; 0.1352]$).

Via parameter continuation it is also possible to obtain the bifurcation diagram that reports, in the (δ-Da) plane, the locus of Neimark-Sacker bifurcation and that of the fold bifurcation that delimit the frequency locking region.

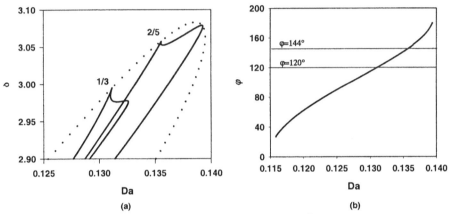

Figure 4 (a) - The bifurcation diagram in the plane δ-Da. Dotted lines: Locus of Neimark-Sacker; solid line: Locus of the Fold bifurcation. (b) Argument of the characteristic multipliers vs. Da in the Neimark-Sacker curve.

At any Neimark-Sacker bifurcation two Floquet multipliers are:

$$\mu_{1,2} = e^{\pm i\varphi} \qquad \varphi = 2\pi \frac{q}{p} \tag{5}$$

When q/p is a rational number, from the corresponding Neimark-Sacker bifurcation point a so-called Arnold tongue of a (q,p)-cycle (Arnold, 1983) emerges, which contains the (q,p) resonant solutions, and whose boundaries are loci of the saddle-node bifurcations. Resonant solutions are easily observable when q and p are small integers, and the corresponding Arnold tongues are rather large. To illustrate these facts, Figure 4a shows the locus of the Neimark-Sacker bifurcations, and the locus of the Fold bifurcations of the 2/5 and 1/3 resonant solutions, while Figure 4b represents the argument φ of the characteristic Floquet multiplier (the one with positive imaginary part) versus the Damköhler in the Neimark-Sacker bifurcation condition. The two values of the Damköhler number corresponding to the (1,3) and (2,5) bifurcation points (Fig. 4a) are the values at which φ is equal to 120° and 144° respectively (Fig. 4b).

4. Conclusions

This work analyses, for discontinuous forced reactors, resonant solutions with period equal to exact multiples of the period of the forcing action.

Continuation analysis, made possible by an *ad hoc* developed procedure, shows that such solutions arise in correspondence of Neimark-Sacker bifurcations on the torus, when the arguments of the Floquet multipliers are expressed by 2π times a rational number. Arnold tongues, formed by the loci of saddle-node bifurcations emerging from Neimark-Sacker bifurcations points, are easily found by the proposed methodology.

5 References

Arnold V.I., "Geometric methods in the theory of ordinary differential equations", Springer e Verlag, (1983).

Eigenberger G., and U. Nieken., "Catalytic Combustion with Periodical flow Reversal," *Chem. Eng. Sci.*, **43**, 2109 (1988).

Faraoni V., Mancusi E., Russo L. and G. Continillo, "Bifurcation analysis of periodically forced systems via continuation of a discrete map", European Symposium on Computer Aided Process Engineering -11, Kolding , Denmark, 27-30 May 2000.

Franceschini V., "Bifurcation of tori and phase locking in a dissipative system of differential equations", *Physica D*, **6**, 285 (1983).

Kevrekidis I. G., Aris R. and L. D. Schmidt, "The stirred tank forced", *Chem. Eng. Sci.*, **41**, 1549 (1986a).

Kevrekidis I. G, L. D. Schmidt and R. Aris, " On the dynamic of periodically forced chemical reactors", *Chem. Eng. Com.,* **30**, 323 (1986b).

Khinast J. Gong Y. O. and D. Luss, "Dependence of cooled Reverse-flow reactor dynamics on reactor model", *AIChE. J.*, **45**, 299, (1999).

Kuznetsov, Y. A., *Elements of applied bifurcation theory*, 2nd ed., Springer Verlag, New York, (1998).

Matros Y.S. and G. A. Bunimovich, "Reverse-flow operation in fixed bed catalytic reactors", *Catal. Rev Sci. Eng.*, **38**, (1996)

Neophytides, S.G., and G.G Froment., "A Bench Scale Study of Reversed Flow Methanol Synthesis," *Ind. Eng. Chem. Res.*, **31**, 1583 (1992).

Schreiber I., Dolník M., Choc P., Marek M., "Resonance behaviour in two parameters families of periodically forced oscillators", *Physics Letters A*, **128**, 66 (1988).

Řeháček J., Kubicek M. and M. Marek, "Periodic, quasi periodic and chaotic spatio-temporal patterns in a tubular catalytic reactor with periodic flow reversal", *Comp. Chem. Eng.,* **22**, (1998).

Żukowski W., and M. Berezowski, "Generation of chaotic oscillations in a system with flow reversal", *Chem. Eng. Sci.*, **55**, 339 (2000).

European Symposium on Computer Aided Process Engineering – 12
J. Grievink and J. van Schijndel (Editors)
© 2002 Elsevier Science B.V. All rights reserved.

Hybrid modelling: architecture for the solution of complex process systems

Ali Abbas[*], Victor Guevara and Jose Romagnoli
Laboratory for Process Systems Engineering
Chemical Engineering Department, University of Sydney NSW 2006 Australia

Abstract

Hybrid modelling is an important emerging field for chemical engineers and engineers in general. There are many engineering problems that require two or more unique and powerful software packages for a solution to be generated. Each program offers unique advantages and specialties, but there is to date no single program that incorporates all of the data and methods, which could possibly be used to solve every problem. A way of overcoming this shortfall is to develop a technique to connect different software packages, enabling the data communication across established links. This study is focused on such a development and presents a method for a novel connection between two powerful software packages, namely *gPROMS* and *HYSYS*. The connection demonstrates the advantages and flexibility of the new method. It is envisioned that an open system can be created whereby the different software packages can import and export each other's data and communicate with each other freely. The connection is demonstrated by a case study of a crystallisation pilot plant simulation.
Keywords: hybrid modelling, Software connectivity, crystallisation, *gPROMS*, *HYSYS*, *Excel*.

1. Introduction

In today's modern computer-lead world, more and more engineering problems are being solved with the help of modelling or simulation software packages. These tools are used to speed up the design process to enable the engineer to concentrate on problem definition and analysis. Examples of these software packages include spreadsheet-based software packages like *MS Excel*, analysis and programming languages such as *Matlab*, *Fortran*, *Speedup* or *gPROMS*, equation solvers such as *Polymath*, fluids packages and process modelling packages such as *CFX* and *HYSYS*. It is possible to use one or more of these tools to solve a given problem or to aid in the modelling of a given process. However, problems often need to be broken up into sections and solved sequentially or sometimes simultaneously. In some cases, the entire process model can be written in a single software package such as *gPROMS*. There are many problems however, which may need the concurrent use of two or more unique and powerful software packages for a solution to be generated. This is especially the case where a specific problem requires the use of say, both a physical properties package such as *SuperPro* and a programming

---

[*] Corresponding author: Tel: +61-2-9351 4337; Fax: +61-2-9351 2854
Email: aabbas@chem.eng.usyd.edu.au

language such as *gPROMS*. Each program offers unique advantages and specialties, but there is to date no single program that incorporates all of the data and methods, which may possibly be used to solve every problem. Nor will there ever be an all-encompassing process solver (Figure 1) that would be a solution to all chemical engineering modelling situations. As new hypotheses and theories are developed, process modelling tools will need to be restructured and improved. The field of engineering is already too broad for such a catch-all solver to be developed.

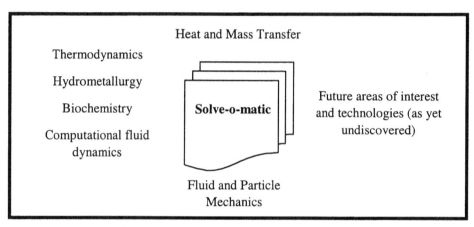

Figure 1. A diagram showing the capabilities of a theoretical 'catch-all' solver.

A way of overcoming this shortfall is to develop techniques to connect different software packages, enabling data communication across established links. This leads to increased accuracy and speed of calculation while increasing process modelling and problem solving capabilities. As Kakhu *et al.* (1998) point out, there are a number of reasons for developing a more open architectural approach to process modelling tools. Firstly, the users may wish for commercial software to interact with their internally generated or proprietary software. For example, the user may have patented their own design or process, which can be improved by its interaction with other software. Secondly, more specialised, third-party software may be more useful for modelling different parts of the process. Examples of such cases include specialist hydrodynamics modelling software and programs for designing heat exchangers. Finally, process modelling tools may be incorporated into an overall system. This is especially relevant for plant operations training, optimisation and control. The significance of this is that there is recognition that software interconnectivity is vital to modern chemical engineering. This interconnectivity will lead to the ability to create and maintain flexible and powerful hybrid models of processes and systems, essentially harnessing the powers of multiple tools. The hybrid architecture shown in Figure 2 facilitates the combining of the strengths of each particular program allowing powerful simulations to be developed. This architecture is an important, emerging field for chemical engineers. It is applicable for all situations and can be used to solve many complex problems and for modelling complex processes.

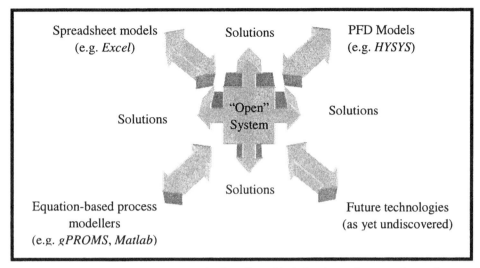

Figure 2. The open software vision which will enable hybrid simulations of complex processes to be modelled. It acts as a bridge or communicator between different specialist 3rd-party software, to generate hybrid models and solutions.

Two popular software packages currently in use by chemical engineers are *HYSYS* by Hyprotech Ltd and *gPROMS* by Process Systems Enterprise Ltd. *HYSYS* is a sequential-modular process modelling tool which enables the design of processes and the improvement of existing plant operations (Clark et.al., 2000). It has a number of advantages over equation-oriented modellers, including the ability to model a plant without prior programming skills and its employment of a user-friendly graphical user interface (GUI). *HYSYS* has models of entire unit operations pre-built into them. This leads to ease and speed of model generation and re-design. Its disadvantages include its 'black-box' approach to modelling. The user may not be fully aware of the equations and assumptions that underpin each unit model. Thus, it is useful for modelling complex processes constructed from less complex unit operations. *gPROMS* (the **general PRO**cess **M**odelling System), on the other hand, is an equation-oriented computational program designed for modelling complex processes. Its advantages include its flexibility to model a large range of simulations, including the ability to model both lumped and distributed parameter systems (Oh and Pantelides, 1996). As with all equation-oriented programs, the manual writing of code is time consuming and expensive, and it is easy to make errors (Clark et.al., 2000).

Typical processes that could benefit from the *gPROMS-HYSYS* link contain at least one complex unit operation but also include many other simple unit operations. The complex and distributed unit operation(s) (where properties vary in one or more spatial dimensions) such as solid-liquid separation (eg. in a crystalliser) can be modelled in an equation-based process simulator such as *gPROMS*. The rest of the process, containing standard unit operations can be modelled in *HYSYS*, for ease and speed of model generation. Further, 'hybrid systems' (Chaplee et. al, 2001) exhibiting discrete behaviour as well as continuous dynamics require a flexible modelling environment that takes into account the execution of operator events. This is particulary important in batch processes that are recipe driven such as crystallisation and polymerisation.

2. Developing *gPROMS-HYSYS* Hybrid Models

This section outlines the way in which a dynamic link was established between *gPROMS* and *HYSYS* through *MS Excel* to create a hybrid model of crystallisation pilot plant. The aims of this section are:

- To demonstrate the methods used to create a hybrid model between two modelling programs using *Excel* as a bridging program;
- To outline the way in which *HYSYS* and *gPROMS* can individually be linked with *Excel*;

MS Excel, developed by the Microsoft Corporation, is today being used by arguably all users in process engineering. Its strengths include its operability and the fact that it is used globally as the standard software for evaluating data. Thus, it is in a unique position to facilitate interconnectivity between different pieces of software.

Linking *gPROMS* with *Excel*: gFPI is the Foreign Process Interface capability of *gPROMS* (PSE Ltd(a), 2001) and is used to create links with *Excel*. gFPI is used to link *gPROMS* with foreign processes – external process modelling or properties programs. gFPI facilitates the real-time simultaneous importing and exporting of data from *gPROMS* to any foreign process. This enables *gPROMS* to interact with *Excel* through the in-built Dynamic Link Library (DLL) known as ExcelFP.dll. This DLL can be used to set up the required import/export links with *gPROMS*.

Linking *HYSYS* with *Excel*: Data in *HYSYS* can be exported via pasting a text link of the data to an *Excel* cell. This can be done to export time and any other process variable in Dynamic Mode. Importing and exporting variables to *HYSYS* is more readily achievable by using the *HYSYS* DDE capability.

Bridging with *Excel*: The *gPROMS-Excel* and *HYSYS-Excel* links can together be placed in a single *Excel* workbook, with a number of sheets. These include the compulsory gFPI sheet, a sheet to record the movement of *gPROMS* variables dynamically, a sheet used to call on a *HYSYS* time recording macro and a sheet used for data storage. Once this *Excel* file is set-up, a link between the desired variables from a *HYSYS* model and those variable from the *gPROMS model* is established dynamically. This completes the connection bridge and from here, the file can be used to link the two programs and extract process data dynamically from either source. The procedure for setting up the *Excel* software bridge is outlined in (PSE Ltd(b), 2001).

2.5 The case study: Industrial crystallisation pilot plant

This section describes the application of the hybrid modelling architecture to model a complex crystallisation pilot plant. This crystallisation pilot plant, which has been installed in the Department of Chemical Engineering at the University of Sydney, consists of a central unit operation, the crystalliser, and a utilities section for heating and cooling. The crystalliser has a volume of 80 litres, and can be operated in batch or semi-batch mode. It is heated and cooled by an oil stream flowing through the crystalliser jacket. The oil in the jacket is heated and cooled by way of two heat exchangers, one for heating and the other for cooling and is circulated through the heat exchangers by a pump. A boiler and a chiller are used to heat and cool the oil respectively. Cooling is used to generate the required supersaturation. As the solution is cooled, the solute will crystallise in accordance with the saturation concentration at that temperature. Another cooling circuit is used to cool the crystalliser contents and consits of a chilled water flow through a draft tube (a hollow circular anulus inside the crystalliser).

Figure 3. This figure depicts the software architecture of the model used.

The development of a crystalliser pilot plant model required the use of *gPROMS* to simulate the crystalliser unit operation, and the use of *HYSYS* to model the rest of the plant. Both *gPROMS* and *HYSYS* are capable of importing and exporting data into *Excel.* Figure 3 shows the overall software architecture of the model. *HYSYS* is responsible for the modelling of the pilot plant utilities section, including all heating and cooling flows, temperatures and pressures. *gPROMS* provides solutions for the crystalliser temperature and more significantly it provides the crystal size distribution (CSD) output from distributed population balance equations. *Excel,* while acting as a data link and used for data management and visualisation, is also utilised for the calculation of the crystaliser jacket and draft tube models. The temperature profiles and valve manipulations are shown in Figure 4. The effect of the manipulatin of a *HYSYS* valve in the form of reduction

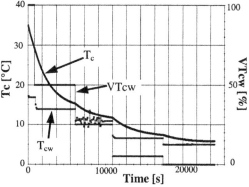

Figure 4. Profile of the draft tube by HYSYS and the crystalliser temperature by gPROMS.

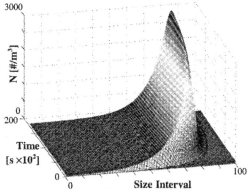

Figure 5. The crystal size distribution which generated by gPROMS using HYSYS data.

in valve opening (VTcw) of a hot stream from 100% to 0% is shown as steps. This effectivley reduces the inlet temperature of the draft tube (Tcw) in stepwise manner, which consequently cools the crystalliser contents. Tc is the crystalliser temperature and shows the step responses of the cooling profile across a crystallisation batch. The crystal size distribution shown in Figure 9 was generated using an algorithm in *gPROMS*, relying on *HYSYS* generated temperature and flow data.

2.6 Simulation issues

In both software packages, there is capability to set the simulation speed at multiples of real time. These capabilities in both process simulators enable the synchronous execution of the hybrid model at a much higher speed. The *HYSYS* simulation acts as a virtual plant with *gPROMS* simulation effectively acting as a CSD soft-sensor. Data transfer from *HYSYS* is set at a high rate allowing *gPROMS* to track changes of variables crossing the *Excel* interface almost continuously. This allows *gPROMS* to accurately predict the CSD and thus the overall error associated with the communication interval is minimised. Similarly, *gPROMS* data is communicated to *HYSYS* across the *Excel* interface at a high frequency. These data transfer settings are appropriate for a complex process with slow dynamics.

3. Conclusions

Architecture for the modelling of complex processes has been introduced. Separate sections of a process are built separately in each program. Each process modelling software is then linked interactively with *Excel*. The significance of the above result is that a model of a process can be built in parts, using a number of programs, and then connected to create a dynamic hybrid model that can be simulated under different operating conditions and schedules. This model can then be redesinged and maintained easily enabling modelling to be accomplished with increased speed, and conclusions reached with greater confidence. Extensions to the architecture include the implementation of other tools, particularly *Matlab* to perform advanced control and *CFX* to perform the fluid dynamics modelling. Another consequence of the software packages' real-time capabilities is that in future, such programs can be utilised in real plant situations. For example, instead of linking the *gPROMS* CSD model with *HYSYS*'s virtual plant via *Excel*, *gPROMS* can be connected with a real plant situation, to enable the sensing and monitoring of real plant conditions, in real time. Additionally, the *gPROMS* model can be linked with a Distributed Control System, to provide monitoring and control in real time. The possibilities of utilising such a hybrid model are endless.

References

Chaplee M., A. Sargousse, J. Lelann and X. Joulia, May 2001, Escape-11, An object-oriented environement for the modelling and simulation of hybrid processes, Kolding, Denmark, 29-34.

Clark G., D. Rossiter. D. and P. Chung, 2000, Trans IChemE, Intelligent Modelling Interface for Dynamic Process Simulators, 78 (Part A), 823-839.

Kakhu A., B. Keeping, Y. Lu, and C. Pantelides, 1998, AIChE Symposium Series, "An Open Software Architecture For Process Modelling And Model-Based Applications", 518-523.

Oh M. C. Pantelides, 1996, A Modelling and Simulation Language for Combined Lumped and Distributed Parameter Systems, *Computers and Chemical Engineering*, 20, 6-7, 611-633.

PSE Ltd(a/b)., May 2001, *gPROMS* Introductory User Guide/*gPROMS* Advanced User Guide – Release 2.0, Process Systems Enterprise Ltd, London.

Optimal Sensor Location and Reduced Order Observer Design for Distributed Process Systems

Antonio A. Alonso[1*], Ioannis Kevrekidis[2], Julio R. Banga[1] and Christos Frouzakis[2]

1. Process Engineering Lab, IIM-CSIC, Eduardo Cabello 6, 36208 Vigo, Spain
e-mail:antonio@iim.csic.es

2. Department of Chemical Engineering, Princeton University, Princeton, NJ, 08544-5263 USA

Abstract

This paper presents a systematic approach to efficiently reconstruct the infinite dimensional field in distributed process systems from a limited, and usually reduced, number of sensors. To that purpose, two basic tools are employed: on the one hand, a reduced order representation of the system which captures the most relevant dynamic features of the solution. On the other hand, the selection of the most appropriate type (and number) of measurements by the solution of a max-min optimization problem. These ideas will be illustrated on the problem of field reconstruction for unstable tubular reactors.

1. Introduction

The operation and control of distributed process systems such as chemical reactor or flow units usually requires precise information on the spatial distribution of the dynamic variables of interest (e.g. temperature, concentrations or fluid velocity fields). Unfortunately, that information is only available through a limited number of possibly expensive sensors (van den Berger et al, 2000). Such limitation justified over the past years a considerable research effort in establishing methods for efficient sensor placement and field reconstruction. Although the need for systematic methods was soon recognized, most techniques relied on exhaustive search over a pre-defined set of candidates (Keller and Bonvin, 1992). However, this approach, valid for a small number of locations, becomes useless when the number of possible location candidates increases, as it is the case, for instance, in flow reconstruction problems (Podvin and Lumley, 1998). Exceptions to exhaustive search include sub-optimal sequential selection alternatives, succesfully applied to place sensors in large space structrures (Kammer, 1991). Other interesting approaches include those recently developed by Antoniadis and Christofides (2000) or van den Berger et al. (2000), although limited by the type of dynamic system representation and control scheme employed.

The methodology we propose in this paper has been developed to handle those situations demanding sensor placement neither depend on the particular control structure nor on a particular dynamic model, as this could be poorly understood or unavailable. With this intention, our approach only requires knowledge of a low dimensional linear subspace capturing the most relevant features of the process operation. On this linear subspace, the optimal sensor placement will be computed as that which minimizes a criterion directly connected with the quality of the estimation. An efficient guided search algorithm, with optimal convergence properties, was developed to that purpose. We also show that such selection criterion will also result into dynamic observation schemes with fast convergence rates. These ideas will be illustrated on a case study involving concentration and temperature reconstruction from measurements for a tubular reactor operating at an unstable regime.

2. Reduced Order Representations of Distributed Dissipative Process Systems

The class of systems we are considering in this work are those described by sets of quasi-linear partial differential equations of the form:

$$u_t = \ell(u) + F(u) \tag{1}$$

where $\ell(\bullet)$ represents a linear parabolic operator defined on a 3-dimensional spatial domain D with smooth boundary and $F(u)$ a non-linear vector field. The solution $u(t,x)$, with appropriate initial and boundary conditions, will be referred to as the field and contains as elements functions such as temperatures and concentration defined in time and space. The dissipative nature of systems of the form (1) allows the representation of the solution in terms of an infinite series expansion of the form:

$$u = \sum_{j=1}^{\infty} c_j(t)\varphi_j(x) \tag{2}$$

where each element of the set of basis functions $\{\varphi_j(x)\}_{j=1}^{\infty}$ can be considered as the solution of the following eigenvalue problem: $\varphi_j(x) = \mu_j \int_D K(x,x')\varphi_j(x')dx'$.

Depending on the structure of the Kernel $K(x,x')$, different sets will emerge. Representative examples include the *spectral basis* or the empirical basis, also known as PODs (Proper Orthogonal Decomposition), obtained from the two point correlation kernel (Holmes et al, 1996). In both cases, the eigenspectrum $\{\mu_j\}_{j=1}^{\infty}$, associated to the basis allows the selection of a finite low dimensional set of orthonormal functions $\{\varphi_j(x)\}_{j=1}^{k}$ capturing the most relevant features of the solution. In this way, solution (2) can be approximated as a truncated series of the for $\tilde{u} = \sum_{j=1}^{k}\left(\int_D u,\varphi_j dx\right)\varphi_j(x)$.

Equivalently, projection of (1) on the elements of $\{\varphi_j\}_{j=1}^{k}$ leads to a state space-like approximation of the distributed system:

$$c_t = Ac + f(c) \quad \text{and} \quad \tilde{u} = [\varphi_1,...,\varphi_k]c \tag{3}$$

This representation constitutes the starting point to reconstruct the field from a limited and usually reduced number of measurements, as it will be described next

3. Field Reconstruction from Measurements

Let us consider two linear subspaces $E_m = \{e^i_j\}_{j=1}^m$ and $S_k = \{\phi_j\}_{j=1}^k$. Elements $\phi_j \in S_k$ must be interpreted as the n-discrete version of functions φ_j in (2), along the spatial coordinates x, so that $\phi_j \in R^n$ and $\phi_j^T \phi_k = \delta_{jk}$ for every j and k, with δ_{jk} being the Kronecker delta. The elements $e^i_j \in E_m$ are n-dimensional i-unit vectors which will be employed to project any $u \in R^n$ along the i-coordinate so that $u^T e^i_i = u_i$. With these preliminaries, the estimation problem can be stated as follows:

Given a vector v_m with elements being measurements of the solution v at m locations, reconstruct the remaining components of the solution at the unmeasured $n - m$ locations

This problem is equivalent to finding the set of coefficients $\hat{c}$ in (3) that minimises the distance between measurements v_m and the estimates $\hat{v}_m$ which leads to the standard minimum least squares formulae $\hat{c} = (QQ^T)^{-1} Q \hat{v}_m$, with $Q = \Phi^T P_m^T$ and $P_m = [..e^i_j..]^T$, $\Phi = [..\phi_j..]$. As it is well known (Golub and Van Loan, 1983), the estimation error will be influenced by the angles between subspaces E_m and S_k so that, large angles will deteriorate the quality of the estimation. Such observation suggests the minimisation of the maximum angle between these subspaces as a criterion to select the appropriate measurement subspace (sensor placement and type of measurements) compatible with a given number of available sensors. Combining this criterion with the singular value decomposition theorem (Golub and Van Loan, 1996) leads to the following statement of the optimal sensor placement problem:

$$\max_{E_m} \min_i \lambda_i(QQ^T) \tag{4}$$

where $\lambda_i(QQ^T)$ is the i-eigenvalue of QQ^T. This problem, although involved, can be efficiently approximated to a much more simple statement, as it will be seen in the next section. In what follows, we briefly discuss the implications of such selection criterion in dynamic observer design

3.1 Dynamic implications for observer design

A dynamic observer for systems described through the reduced order representation (3) is constructed as a replica of the system plus an extra-term which accounts for the differences between the measurements and the estimations. The following result establishes connections between optimal sensor placement and observer rate of convergence.

Theorem. *Let the system (3) be such that* $\|A\| = \alpha_m$, $f(c)$ *Lipschitz with constant* β *and* $\lambda_u I \geq QQ^T \geq \lambda_\ell I$. *Then the observer* $\hat{c}_t = A\hat{c} + f(\hat{c}) + \omega(v_m - Q^T\hat{c})$ *with* $\omega = P^{-1}Q$ *and* P *being a symmetric positive definite matrix satisfying:*

$$(A + \alpha I)^T P + P(A + \alpha I) = QQ^T - \eta I$$

with $0 < \eta < \lambda_\ell$ *and* $\alpha > \alpha_m + \beta(-1 + \lambda_u/\eta)$ *will make* $\varepsilon = (c - \hat{c}) \to 0$ *at an exponential rate proportional to* λ_ℓ.

The proof makes use of standard systems theory arguments (Rugh, 1993). In this way, a suitable Lyapunov function candidate $V = \varepsilon^T P \varepsilon$ is constructed and its time derivative computed along the error evolution equation to obtain $V_t \leq \varepsilon^T QQ^T \varepsilon$. The result then follows by making use of the Gronwall-Bellaman lemma. For a detailed description the reader is referred to (Alonso et al, 2001).

4. Optimal Selection of Sensor Placement

The solution of problem (4) is, in general, quite involved except for the trivial case $m = n$, where $QQ^T = diag(\phi_1^T \phi_1, \ldots, \phi_k^T \phi_k)$ becomes the identity matrix. However, as the number of measurements decreases, the resulting sub-vectors $P_m \phi_j$ will no longer be orthonormal which translates into non-zero off-diagonal elements in QQ^T. This simple observation suggests an alternative to problem (4) in which a subspace E_m is sought so to minimise *orthonormality distortion*. The alternative optimization problem can be then formulated as:

$$\max_{E_m} \min_i (s_1, \ldots s_i \ldots s_k) \qquad (5)$$

where each s_i is the i-diagonal element of QQ^T. In order to solve this problem, we propose an efficient guided search algorithm that globally converges to the optimum placement in fairly small computation times. A brief outline of the algorithm is presented next (for a detailed description see Alonso et al, 2001). Firstly, we define a set of k n-dimensional vectors σ_j with elements $\sigma_j^i = (\phi_j^i)^2$ for $j = 1, \ldots, k$ and associate to each of them an index vector set $\{\eta_j^p\}$ with $\eta_j^p \in R^m$ and an operator $\Im[\sigma_i; \{\eta_j^p\}]$ which computes all summations of the η_j^p elements of σ_i. Search will proceed on a k-dimensional space of all possible m-summations by constructing ℓ-sequences $\{\eta_j^p\}^\ell$ for each $j = 1, \ldots, k$ ordered so that $\Im(\sigma_j; \eta_j^p) \leq \Im(\sigma_j; \eta_j^{p-1})$. The iteration step is summarized as follows:

For a given ℓ, set up positive numbers ε, L_ℓ, and define for each j, intervals $[L_j^{low}, L_j^{up}]$ with $L_j^{low} = L_j^{up} - \varepsilon$

1. For a given j construct a sequence $\{\eta_j^p\}^\ell$ such that

$$L_j^{low} \leq \Im\left[\sigma_j; \{\eta_j^p\}^\ell\right] \leq L_j^{up} \text{ and } L_\ell \leq \Im\left[\sigma_i; \{\eta_j^p\}^\ell\right] \text{ for } i \neq j$$

2. Compute $s_i = \Im\left[\sigma_i; \{\eta_j^p\}^\ell\right]$ for $i = 1,...,k$ and $L_j = \min_i(s_i)$

3. Repeat step 2 for $j = 1,...,k$ and compute $L_{new} = \max_j(L_j)$

4. If $L_{new} \geq L_j^{up}$ stop along j. If not set $L_{\ell+1} = \sup(L_{new}, L_\ell)$, $L_j^{up} = L_j^{low}$ and go to step 1.

5. Illustrative Example

The different aspects of the methodology presented so far will be employed in this section to reconstruct, from a limited number of measurements, the concentration and temperature field in a tubular reactor with axial dispersion, in which a first order irreversible exothermic reaction is taking place. As it is depicted in Figure 1, the reactor operates at an unstable (oscillatory) regime. A detailed description of the process and parameters can be found in (Antoniadis and Christofides, 2000).

Following the schematic presented, we first construct a low order approximation of the original distributed system (Eqns (2)-(3)) using as a basis the set of POD functions. These are obtained, by the so-called direct method (Holmes et al, 1996), from a series of snapshots for concentration and temperature, as the ones depicted for temperature in Figure 1. The first three PODs were chosen as representatives of the operation since they captured more than the 99.9% of the energy of the system. In a second step, the optimal selection algorithm proposed in Section 4 is applied to the set S_3 to compute optimal sensor placements for a given number of measurements, among 32 possible locations for concentration and temperature in the reactor. Figure 2, illustrates how the algorithm evolves for $m = 10$, from a given initial set of s_i-summations to the optimal ones satisfying (5). Note that even for this case, the number of possible alternatives is in the order of 10^8 what would make exhaustive search inadequate. Computations for m's ranging between 5 and 15 showed that 10 measurements were enough to provide a fair reconstruction of the concentration and temperature fields during the operation. Optimal sensor arrangement and field reconstruction capability is illustrated in Figures 3 and 4. Field reconstruction for arbitrary snapshots during the operation is shown is Figure 3. The evolution of the first real and estimated coefficient c_1 (see (2)-(3)) for the optimal and an arbitrary arrangement (including only temperature measurements) is presented in Figure 4.

6. Conclusions

In this paper, a systematic technique was presented to efficiently solve the optimal sensor location problem and thus provide reliable field reconstruction from a limited

420

and usually reduced number of measurements. This was accomplished through a two-steps approach. First, a low dimensional representation of the solution of the original distributed system is obtained. Second, the most appropriate sensor type and locations are chosen on this low dimensional subspace through an efficient guided search algorithm that minimises orthonormality distortion.

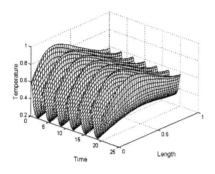

Figure 1. Typical snapshots for temperature

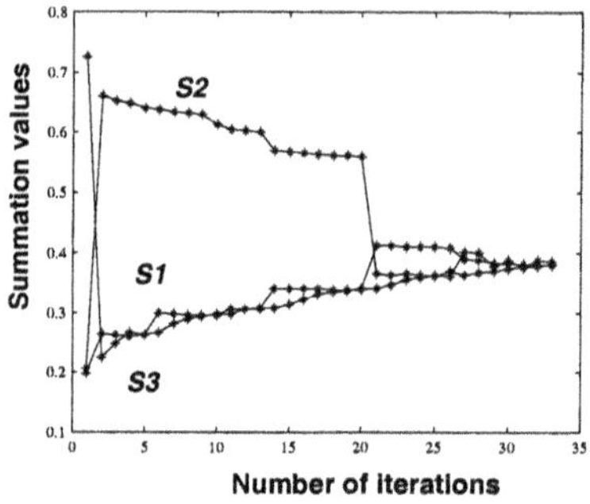

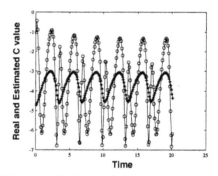

Figure 2. Evolution of the search algorithm

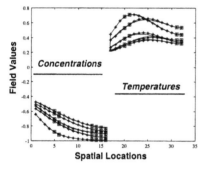

Figure 3. Field reconstruction from optimal sensor locations (squares). Crosses represent field estimations while continuous lines indicate the field profile in the reactor at different times

Figure 4. Evolution of the first c coefficient estimated from optimal (*) and arbitrary (o) sensor arrangements. Optimal estimation matches the real evolution

7. References

Antoniades, C. and P.D. Christofides, 2000, Comp. Chem. Eng. 24(2), 577.

Alonso, A., I. Kevrekidis and C. Frouzakis. Optimal field reconstruction of nonlinear distributed systems. Submitted AIChE J.

Golub G.H. and C. Van Loan, 1983, Matrix Computations. John Hopkins Univ. Press.

Holmes, P., J.L. Lumleyand G. Berkooz, 1996, Turbulence, Coherent Structures, Dynamical Systems and Symmetry. Cambridge Univ. Press.

Kammer, D.C., 1991, J. Guid. Contr., 19(3), 729

Keller, J.P. and D. Bonvin, 1992, Automatica, 28(1), 171

Podvin, B. and J. Lumley, 1998, Physics and Fluids, 10(5), 1182

Rugh, W.J., 1993, Linear System Theory. Prentice Hall, Englewood Cliffs, NJ

Van der Berg, F.W.J., H.C.J. Hoefsloot, H.F.M. Boelens, A.K. Smilde, 2000, Chem. Eng. Sci. 55, 827

European Symposium on Computer Aided Process Engineering – 12
J. Grievink and J. van Schijndel (Editors)
© 2002 Elsevier Science B.V. All rights reserved.

Dynamic Modeling of Catalytic Hydrogenation of Pyrolysis Gasoline in Trickle-Bed Reactor

A. Arpornwichanop[1], P. Kittisupakorn[1] and I.M. Mujtaba[2]*
[1] Department of Chemical Engineering, Chulalongkorn University
Bangkok, Thailand 10330
[2] Computational Process Engineering Group,
Department of Chemical Engineering, University of Bradford
West Yorkshire BD7 1DP, UK.

Abstract

Development of mathematical models for chemical process simulation has been the main focus of research for many years. In the past trickle bed reactor models for various reaction systems were validated with pilot plant data but the development of a dynamic model based on industrial data is rare. Kittisupakorn and Arpornwichanop (1999) have recently developed a steady state model of an industrial trickle-bed reactor for the above system. However, this model is only applicable to predict steady state conditions and inapplicable to design startup and shutdown as well as control studies. Therefore, in this work, a dynamic model of an industrial trickle-bed reactor for describing the hydrogenation reactions of a pyrolysis gasoline in an olefin plant has been developed based on actual plant data. The model results to a set of partial differential equations which were solved by method of lines with approximate spatial derivative terms by an orthogonal collocation method. The model was used to study the response of the process. The simulation results demonstrated that the dynamic response gave a reasonable path from a transient to steady state condition. This dynamic model will be used in future in the formulation of model based control strategies in order to control the reactor.

1. Introduction

A multiphase catalytic packed-bed reactor known as a trickle-bed reactor is one of several types of multiphase reactors widely used in industrial chemical processes. In trickle bed reactors, gas and liquid reactants flow through a fixed bed of catalyst (Satterfield, 1975) and many applications of this type of reactors are found in hydrogenation and hydrodesulfurization of petroleum fraction processes (Gianetto and Specchia, 1992).

During the past years many researches have focused on the simulation of trickle bed reactors using steady state models to describe a reactor behavior with different reaction systems. However, these models do not provide any information on the dynamics of the

---

* Correspondence concerning this paper should be addressed to I.M.Mujtaba
  (E-mail: I.M.Mujtaba@bradford.ac.uk)

reactor which is an important issue for controlling the reactor during startup and shutdown period and between steady state conditions. Therefore a dynamic model of the reactor which can give a good prediction of the dynamic behavior is needed.

Warna and Salmi (1996) developed rigorous dynamic models of the trickle-bed reactor based on three-film theory. The model equation for gas, liquid and solid phases comprises of a system of partial differential equations and ordinary differential equations. The solution of these equations was obtained by method of lines using finite differences to discretize the spatial derivative term. In spite of that the numerical approach gave a reliable solution, the proposed model was very complicated with too many unknown/uncertain parameters involved.

This work presents a simplified dynamic model of an industrial adiabatic trickle-bed reactor in which hydrogenation reactions of a pyrolysis gasoline in an olefin plant occur. The model assumes a pseudo-homogeneous system for liquid and solid phases in order to simplify the complexity involving multi-phases in the trickle bed reactor. Furthermore, components in the system are lumped into three pseudo-components: diolefins, olefins, and parraffins. Unknown/uncertain parameters are determined based on actual plant data.

2. Pyrolysis gasoline hydrogenation process

Pyrolysis gasoline is a by-product from a steam cracking process of an olefin production. Since the process is rarely selective, a large number of unsaturated compounds such as diolefins, olefins, alkenylaromatics and aromatics are produced. The high content of the unsaturated compounds (olefins and aromatics) gives the pyrolysis gasoline to be a valuable product either used as motor fuel with a high octane number or feedstock for an aromatic extraction. However, the presence of diolefins and alkenylaromatics makes it unstable and fail to meet stability specifications.

Hence, a hydrogenation process is designed to stabilize the raw pyrolysis gasoline. With the proper type of the process and its operating condition as well as catalyst, hydrogenation reactions of unstable compounds (diolefins and alkenylaromatics) occur without the hydrogenation of other unsaturated hydrocarbons (olefins and aromatics). The hydrogenation process consists of two fixed-bed reactors, the first and second stage reactor (Fig 1). The purpose of the first stage reactor in which selective hydrogenation reactions of diolefins and alkenylaromatics occur, is to stabilize the raw pyrolysis gasoline. The outlet from the first stage reactor is then sent to the second one for further hydrogenation of the remaining diolefins and olefins and desulfurization to meet aromatic sulfur specification. The quality of the obtained gasoline from the process is suitable for further processing to extract aromatics (Derrien, 1986).

Here, we aim to develop dynamic models for the first stage reactor which involves the gas (hydrogen), liquid (raw pyrolysis gasoline), and solid (catalyst) reactions. The first stage reactor contains approximately 60 m^3 of Ni/Al_2O_3 catalyst which is divided into two beds in order to control the reactor temperature by an addition of liquid quench between the bed. The reactor output goes through a separator unit and the low temperature liquid bottom product of the separator is used as the quench feed to the reactor. The reactor is designed with a diameter of 1.8 m. and height of the catalyst bed is 7.9 and 15.75 m. for the top and the bottom bed, respectively.

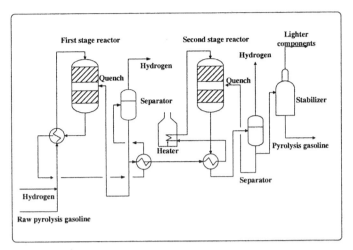

Fig. 1. A simplified diagram of gasoline hydrogenation process

3. Dynamic model of reactor

To develop a simplified dynamic model of a trickle-bed reactor for describing the hydrogenation reactions of a pyrolysis gasoline in an olefin plant, it is assumed that the reactor operates at unsteady state under adiabatic and isobaric condition and no heat loss occurs within the reactor. Furthermore, all hydrocarbon compounds are lumped into three groups: diolefins, olefins, and paraffins and the kinetic models involved are:

$$diolefins \xrightarrow{\ k_1\ } olefins \xrightarrow{\ k_2\ } paraffins$$

$$-R_1 = 4.27e^{\left(\frac{-224.96}{RT}\right)}C_D C_{H_2}$$

$$-R_2 = 13.91e^{\left(\frac{-142.93}{RT}\right)}C_O C_{H_2}$$

In addition, we also assume that the mixing process between the outlet from the top catalyst bed and the quench feed reaches steady state instantaneously. Hence, steady state model for the mixing at the quenching point is held. Other assumptions made are given in Kittisupakorn and Arpornwichanop (1999).

Using the assumptions mentioned above, the mass and energy balances can be described by a set of equations based on the series of steps: mass transfer of gaseous reactant from gas phase to bulk liquid, and then the reactions occur in the liquid phase.

Mass balances:

$$\frac{dC_{H_2,g}}{dt} = -\frac{u_g}{\varepsilon_g}\frac{dC_{H_2,g}}{dz} - \frac{k_{l,H_2}a_L}{\varepsilon_g}\left(C_{H_2,g} - C_{H_2,l}\right) \tag{1}$$

$$\frac{dC_{H_2,l}}{dt} = -\frac{u_l}{\varepsilon_l}\frac{dC_{H_2,l}}{dz} + \frac{k_{l,H_2}a_L}{\varepsilon_g}\left(C_{H_2,g} - C_{H_2,l}\right) - \frac{(1-\varepsilon)}{\varepsilon_l}\left[(-R_1) + (-R_2)\right] \tag{2}$$

$$\frac{dC_D}{dt} = -\frac{u_l}{\varepsilon_l}\frac{dC_D}{dz} - \frac{(1-\varepsilon)}{\varepsilon_l}\left[-R_1\right] \tag{3}$$

$$\frac{dC_O}{dt} = -\frac{u_l}{\varepsilon_l}\frac{dC_O}{dz} - \frac{(1-\varepsilon)}{\varepsilon_l}[(-R_2) - (-R_1)] \tag{4}$$

$$\frac{dC_P}{dt} = -\frac{u_l}{\varepsilon_l}\frac{dC_P}{dz} + \frac{(1-\varepsilon)}{\varepsilon_l}[-R_2] \tag{5}$$

where ε, ε_g, and ε_l are void fraction ($\varepsilon_g + \varepsilon_l$), gas fraction, and liquid fraction respectively. The ε_l can be calculated according to Eq. 6.

$$\varepsilon_l = 9.9\left(\frac{G_l d_p}{\mu_l}\right)^{\frac{1}{3}}\left(\frac{d_p^3 g \rho_l^2}{\mu_l^2}\right)^{-\frac{1}{3}} \tag{6}$$

For computing the gas-liquid mass transfer coefficient of Hydrogen, $k_{l,H2}$, the correlation in Eq. 7 is used.

$$\frac{k_{l,H_2} al}{D_{l,H_2}} = \alpha_1\left(\frac{G_l}{\mu_l}\right)^{\alpha_2}\left(\frac{\mu_l}{\rho_l D_{l,H_2}}\right)^{\frac{1}{2}} \tag{7}$$

where the coefficients α_1 and α_2 are function of a particle diameter (Korsten and Hoffmann, 1996). In this work, the diameter of catalyst pellet is about 3 mm; thus the values of these coefficients are 0.4 and 7 $(cm)^{-1.6}$ respectively.

Energy balance:

$$\frac{dT}{dt} = \frac{-(u_g C_{p,g}\rho_g + u_l C_{p,l}\rho_l)\frac{dT}{dz} - (\Delta H)(1-\varepsilon)[(-R_2) - (-R_1)]}{(\varepsilon_g \rho_g C_{p,g} + \varepsilon_l \rho_l C_{p,l})} \tag{8}$$

where ΔH is a heat of hydrogenation which is approximated to a value of -30 kcal/mole for each double bond reaction.

The dynamic model equations of a trickle-bed reactor developed above result to a set of partial differential equations (PDEs). These are solved numerically by a method of lines technique. The spatial derivative term in Eqs (1)–(5) and (8) is discretized by a orthogonal collocation method on finite elements. Each packed-bed of catalyst in the reactor is divided into 10 elements with an equal space and 2 internal collocation points are used for each element. An approximation of the spatial derivative terms make the PDEs reduce to a set of ODEs. Here, standard DASSL solver written in Fortran Program (Petzold, 1982) is used to determine the solution of the obtained ODEs.

4. Reactor simulation

The trickle bed reactor model proposed here needs the knowledge of parameters and kinetic data as well as an operating condition. For kinetic data of gasoline hydrogenation reaction, the optimization technique as given by Kittisupakorn and Arpornwichanop (1999) is used to estimate the kinetic parameters based on the actual plant data. The process parameters and the operating condition in a nominal case for the simulation can be seen in Kittisupakorn and Arpornwichanop (1999). It should be noted that in this work no disturbance is introduced into the reactor. This will be the subject of future work.

Figures 2 to 4 show the concentration profile of diolefins and olefins and the temperature profile respectively. It can be seen that the temperature is increased along the length of reactor due to exothermic hydrogenation reactions. Then, a quench stream is introduced to reduce the output temperature before feeding into the second catalyst bed. The flow rate of the quench stream is adjusted to match the feeding temperature of the second bed. This means that the output temperature of the first and second beds can be regulated to control product specifications. The steady state temperature profile in trickle bed reactor is given in Fig. 5. Simulation results show that the system takes approximately 700 s from initial conditions to reach the steady state conditions.

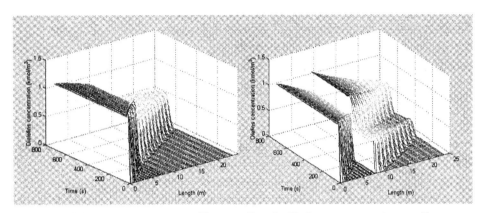

Fig. 2. Diolefins concentration profile.   Fig. 3. Olefins concentration profile.

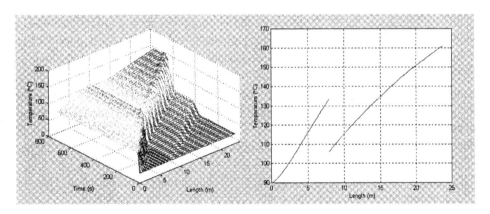

Fig. 4. Temperature profile.   Fig. 5. Temperature profile at steady
                                        state condition.

5. Conclusions

A simplified dynamic model for an industrial adiabatic trickle bed reactor in which hydrogenation reactions of a pyrolysis gasoline in an olefin plant occur, has been developed here. The complexity involving multiphases in the trickle bed reactor is simplified by assuming a pseudo-homogeneous system for liquid and solid phases. All components in the system are lumped into three pseudo-component: diolefins, olefins, and parraffins. The dynamic model results to a system of partial differential equations which is solved numerically by the method of lines. The orthogonal collocation method is used to discretize the spatial derivative term. Unknown/uncertain parameters are determined based on the actual plant data using optimization technique. Simulation results demonstrate that the dynamic response gives a reasonable path from a transient to a steady state condition. The future work will be the design of a model based controller based on this developed model to control product specifications subject to constraints and operating specifications.

Acknowledgement

Financial support from the National Science and Technology Development Agency to A. Arpornwichanop is gratefully acknowledged. The authors would like to thank Dr. Intarajang (Thai Olefin Company, Thailand) for his valuable advice and information.

Nomenclature

A	= reactor cross sectional area [m^2]
a_l	= ratios between gas-liquid interfacial area and reactor volume [m^{-1}]
C_i	= concentration of component "i" [kmol/m^3]
D	= diffusivity [m^2/s]
d_p	= catalyst diameter [m]
G_l	= mass velocity [kg/(hr.m^2)]
$K_{l,H2}$	= mass transfer coefficient in the liquid file [s^{-1}]
ΔH	= heat of reaction [J/kmol]
R	= rate of reaction [kmol/(m^3.hr)]
T	= temperature [K]
u	= superficial velocity [m/hr]
z	= reactor length [m]

Subscripts

D	= diolefins
G	= gas phase
H_2	= hydrogen gas
l	= liquid phase
O	= olefins
P	= paraffins

References

Derrien, M. L. (1986), Selective hydrogenation applied to the refining of petrochemical raw material produced by steam cracking, In catalytic hydrogenation. Cerveny, L. Ed., Elsevier: Amsterdam.

Gianetto, A., and V. Specchia (1992), Trickle-bed reactors: state of art and perspectives, Chemical Engineering Science, **47**(13/14), 3197-3213

Kittisupakorn, P., and A. Arpornwichanop (1999), A modeling of a trickle-bed reactor for gasoline hydrogenation process, AIChE Annual Meeting. Dallas, TX

Korsten, H., and U. Hoffmann (1996), Three-phase reactor model for hydrotreating in pilot trickle-bed reactors, AIChE Journal, **42**(5), 1350-1360

Petzold, L. R. (1982), A description of DASSL: a differential/algebraic system solver, SAND82-8637, Sandia national laboratories.

Satterfield, C. N. (1975), Trickle-bed reactors, AIChE Journal, **21**(2), 209-228

Villadsen, J., and M. L. Michelsen (1978), Solution of differential equation models by polynomial approximation, Prentice-Hall, Englewood Cliffs, New Jersey.

Warna, J., and T. Salmi (1996), Dynamic modelling of catalytic three phase reactors, Computers chem. Engng., **20**(1), 39-47

European Symposium on Computer Aided Process Engineering – 12
J. Grievink and J. van Schijndel (Editors)
© 2002 Elsevier Science B.V. All rights reserved.

An integrated dynamic modelling and simulation system for analysis of particulate processes

Nicoleta Balliu, Ian Cameron and Robert Newell*

CAPE Centre, Department of Chemical Engineering

University of Queensland, Brisbane, Queensland, Australia 4072

*Daesim Technologies Pty. Ltd.,

GPO Box 819, Brisbane, Qld 4001

Abstract

Particulate materials and their processing operations are an extremely important part of the worldwide manufacturing industry. In particular, granulation is a complex particle size enlargement process widely used in pharmaceutical, agricultural and fertilizer industries. The operation of continuous granulation plants can be very difficult. Problems with the design, control and operation of these systems can often be addressed through the use of computer aided modelling and simulation environments.

This work describes the development of a particulate library of dynamic models for particle processing which incorporate full particle distribution models. The major contribution of this work is to provide comprehensive dynamic modelling and simulation tools which capture the complexities of these systems in order to investigate the dynamic behaviour of a wide range of particulate processes. In particular, the development of a family of granulator models is presented which allows the user to tailor the models to specific applications through the selections of key mechanisms.

1. Introduction

Particle technology is a fertile area for research and is of great importance in a wide range of industries from pharmaceuticals to minerals, food and petrochemicals.

Various phenomena involving particle processing are still unclear and many design procedures are based more on past experiences. Therefore, a good knowledge of the mechanisms in particle processing is useful in product development, waste minimization and quality control.

Many practitioners are focused on better approaches to process analysis based on mathematical modelling and computer simulations. There are already well-established tools for steady state process design and investigations (e.g. Aspen Plus) (Aspentech, 2001). However, to understand complex dynamic systems, tools incorporating dynamic models that reflect the complex dynamics of the system are needed. Moreover, in the area of particle processing, there is a need for a comprehensive modelling and simulation environment, which allows process engineers to investigate the overall dynamic behaviour of a wide range of particulate systems from pharmaceuticals to fertilizers.

The objective of this work is to create and implement an integrated dynamic modelling and simulation system for analysis of major particle processing operations which include granulation, drying, particle reduction, particle separation and control structures. Models can be structured in a library and configured into a complex flow sheet using an object-based graphical configurator as done in Daesim Dynamics (Newell and Cameron, 2000). By choosing objects from component libraries we can simulate the time-dependent behaviour of systems whose underlying mechanisms can be defined by differential and algebraic equations (DAEs).

The integrated dynamic tool presents a combined package for understanding various dynamic systems that lead to improved product quality, economics and finally a better-designed process for optimal behaviour.

A case study for a granulation process based on the model library is presented in this paper. Due to the complexity of granulation processes we set out more fully the approach adopted in modelling these systems.

2. Description of the granulation circuit

In this section we describe the behaviour of a simulated granulation circuit for which the particulate library can be used. A particulate stream or slurry is piped into the granulation drum and binder can be sprayed on the feed and recycle. Granule growth occurs along the drum. Granules leaves the granulator into the dryer. After drying solids are screened to separate the product size. Over-size granules are crushed and recycled together with under-sized granules. Figure 1 shows a simple granulation circuit.

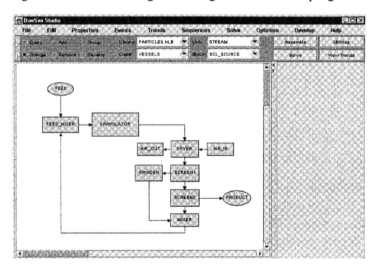

Figure 1: Schematic Diagram of the Granulation Circuit

3. Description of particulate library structure

The library structure is represented as flow sheet made up by using classes, which define a particular unit process, blocks, and connecting links. Models are composed of block and link objects. Block objects are model components whose behaviour is described by DAEs with parameters. Link objects connecting blocks are simply algebraic variables. Blocks and links codes are written in a version of structured text language.

LIBRARY CLASSES

Model class	Member
Particle size enlargement	➢ Granulator
	➢ High-shear mixer
	➢ Fluid bed
Particle drying	• Simple dryer
	• Complex dryer
Particle size separation	➢ Screens
	➢ Cyclone
Particle size reduction	• Ball mill
	• Hammer mill
Control & Instrumentation	➢ Valve
	➢ PI-Controller
	➢ Variable-Sensor

The most important link is the particulate stream which contains solid particles and fluid. The stream is defined in a matrix form, according to the number of particles size ranges – from 1 to M and the number of components present, 1 to NC. A temperature is also assigned to the particulate stream.

The above constitute the key models for simulating most particle processing flow sheets and in particular a generalized granulation circuit.

4. Modelling Hierarchies Using the Particulate Library

An important aspect of the particulate library is that it can be used to represent the model on various levels of granularity. As seen in Figure 3, at Level 1 we can represent the overall granulation circuit, with all the units involved. Simulations can be performed by choosing different units from the library and combine them together for particular applications. Sensitivity analyses of the influences on the granulation process of system variables can be easily evaluated and observed through simulations. At Level 2 we can decompose the system (Hangos and Cameron, 2001) to represent only the granulator, studying the dynamics of the whole unit. The granulator model can include mechanisms such as nucleation, layering, agglomeration and breakage or only nucleation and agglomeration or agglomeration and breakage, according to which assumptions were made. At Level 3 of decomposition , the model of granulation is defined in the library in such a way that we can perform simulations by switching between different mechanisms presented in the model. The same granulation model can be used in several interconnected granulator blocks – GRANULATOR 1, GRANULATOR 2, GRANULATOR 3 – and in each of them we can switch on/off different mechanisms. By employing this structure we are able to separate the various mechanisms/regimes present in the granulator. The concept of design and control of regime separated

430

granulation processes, based on granulation mechanisms has been applied by Wildeboer et al. (Wildeboer et al., 2001). The low level building blocks contain information regarding each mechanism or process present in the granulator. At this level we can define and characterize in detail nucleation, layering, agglomeration or breakage mechanisms as well as the chemical reactions in the granulator.

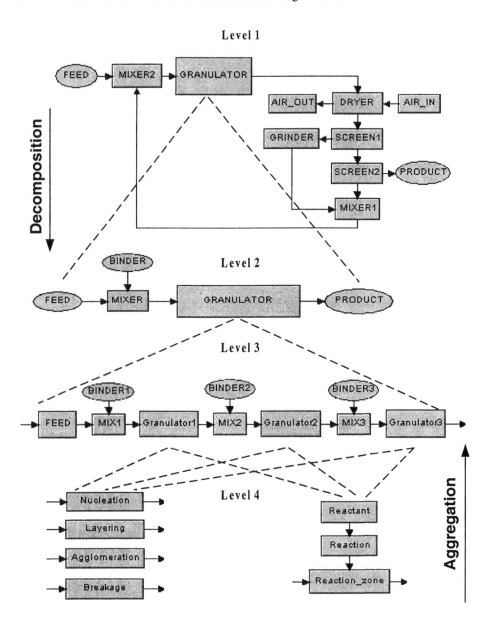

Figure 2: Hierarchical model representation

5. Model hierarchy application and key simulation studies

The above hierarchical system can be used to perform simulations for specific cases of interest. The influence of some parameters and mechanisms for Level 2 and Level 4 model representation are presented below.

5.1 Influence of increasing the amount of binder in the granulator inlet flow

We perform simulations by taking onto account the agglomeration mechanism into granulator. In figure 3a) we represent the product rates for different size distributions until the system reaches steady state. After a certain time, when the steady state is attained we increase the fluid content in the granulator. As seen in Figure 3b) the process responds fast and a deviation of the trends as we increase the inlet fluid flow can be noticed. Moreover, the strong correlation between the granule moisture and agglomeration process is also evidenced by the decrease in the proportion of smaller particles – size 003, size 006 and size 008 and the increase of larger particles – size 010, size 011 and size 012. These sizes referred to mean particle diameter of 0.4mm up to 3.17mm.

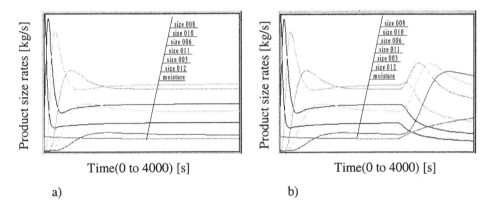

a)   b)

Figure 3: The influence of increasing the inlet flow moisture

5.1 Influence of granulator mechanisms

In the agglomeration model, granule growth by agglomeration is modelled with a sequential two stage kernel, a size independent one when the rate of collisions is assumed to be independent of particle size and a size dependent one, depends on particle size (Adetayo, 1993). This determines the shape of the granule size distribution. The variation of product rates with time for different size fractions assuming a size-independent kernel is presented in Figure 4a while the variation of the product rates taking into account a size-dependent one is presented in Figure 4b. In the case of the size-dependent kernel, the granule size distribution widens as a result of the production of large granules without any significant change to the small granule end of the distribution. There is in the model a possibility of choosing between the size-independent kernel and the size-dependent kernel.

432

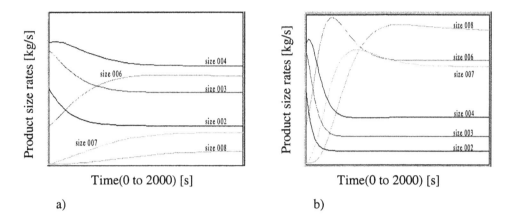

Figure 4: The influence of kernel structure in the agglomeration mechanism

These results can be used as the basis of key design changes in processing equipment to improve productivity. Control studies, structural optimization, optimal start-up, shut-down and grade changing can also be tested on such dynamic models.

6. Conclusions and recommendations

The particulate library presented in this chapter gives the user the choice of many different combinations of operations and mechanisms for various particulate processing flowsheets. Thus, the particulate library can be used to improve industrial operations following model validation and parameter estimation from experimental data. A sensitivity analysis can also be performed using the library models to determine the extent of dominance of various system parameters in any given industrial operation. Simulation results can be used as the basis for key design changes in the processing equipment to improve productivity. The library continues to be expanded to ensure that it meets the needs of industries where particulate systems are an important aspect of the operations.

The authors acknowledge part funding from Australian Research Council grant ARC 980026919.

7. References

Aspentech, 2001, http://www.aspentech.com.

Newell, R.B. and I.T. Cameron, 2000, Daesim Modelling and Simulation: User's guide, Daesim Technologies Pty. Ltd., http://www.daesim.com.

Hangos K. and I.T. Cameron, 2001, Process Modelling and Model Analysis, Academic Press, ISBN 0-12-156931-4.

Wildeboer, W.J., J. Litster and I.T. Cameron, 2001, Design of Regime Separated Continuous Granulators, University of Queensland P.P.S.D.C internal report.

Adetayo A.A., 1993, Modelling and Simulation of a Granulation Circuit, PhD Thesis, University of Queensland, Australia.

Computer Aided Neuro-fuzzy Control in Anodizing of Aluminium

A. F. Batzias and F. A. Batzias

Department of Industrial Management, University of Piraeus, Greece

Abstract

A combination of neuro-fuzzy networking and independent cluster analysis has been used to control alkaline etching in the sequence of aluminium anodizing processes. The objective was to avoid (a) failure or specific defects during a certain process and (b) creating an effect that might cause failure/defect during another downstream process. The fuzzy variables or input neurons used are the concentration of caustic soda, the bath temperature and the retention time. The defects under examination (namely, matness, etch staining, and inadequate cleaning) form the pattern classes of the neuro-fuzzy network. The independent clustering is based on (a) raw data and (b) specifications of the product, set by the client or the market. The algorithmic procedure applied herein can be used in other similar semi-continuous chemical processes, especially in the field of surface treatment of metals.

1. Introduction

Anodizing of aluminium articles is an electrochemical process for producing thin decorative and protective films. In the industrial terminology, anodizing is actually a series of processes, not always the same in their nature and/or order. E.g., for aluminium which will be used in the building construction industry, this series is: cleaning, etching, polishing, electrolytic brightening, sulphuric acid anodizing, dyeing, sealing, finishing; for aluminium which will be used in the aerospace industry, the pre-anodizing process should be changed according to the specifications set by the client, while anodizing is usually carried out in chromic acid bath and sealing precedes dyeing. In many cases, like in anodising of Aluminium, the reveal of a defect in the product may lead to the determination of a fault in an upstream process as the cause of the problem.

For analytical purposes, we can distinguish three artificial intelligence (AI) approaches for fault/defect diagnosis in the domain of Computer-Aided Process Engineering (CAPE). The first one of them is the expert system approach through knowledge acquisition/processing/structuring within a dynamic base, possibly incorporating both, swallow and deep knowledge (Kramer, 1987; Petti et el., 1990). The second approach is application of artificial neural networks, possibly incorporating fuzzy logic (Sorsa and Koivo, 1993; Schmitz and Aldrich, 1998). The third approach is a mixed one, trying to incorporate swallow/deep knowledge into a neural fuzzy system, i.e. to combine an expert with a neuro-fuzzy system either in an integrated form or in interactively cooperating subsystems (Ozyurt and Kandel, 1996; Ruiz et al., 2001).

434

The aim of this work is to present an original computer aided control procedure which guides the industrial user to avoid failure or specific defects during one of the processes described above. This control procedure is based on (i) neuro-fuzzy networking and (ii) independent data processing (shallow knowledge acquired by means of quality control measurements) mainly with cluster analysis. These two sub-procedures should run in parallel but in close interaction and under feedback corrective mechanisms.

2. Methodology

The algorithmic procedure, especially designed and implemented for the neuro-fuzzy and independent clustering combination, includes the following stages. Figure 1 illustrates the connection of stages, represented by the corresponding number or letter, in the case of activity or decision node, respectively.

1. Determination of variables
2. Collection of data
3. Selection (extraction) of data
4. Empirical determination of membership functions
5. Empirical design of initial fuzzy rules
6. Learning – production of new fuzzy rules and adapted membership functions
7. Prognosis of defects, based on new industrial data input
8. Comparison between prognostic results and real data
9. Independent cluster analysis
10. Determination of initial acceptable field of values of variables
11. Determination of final acceptable field of values of variables with fuzzy constraints
12. Formulation of practical rules
13. Industrial production under these rules

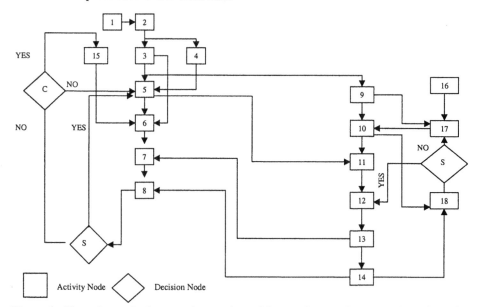

Figure 1: Flow chart of sub-procedures adopted for combining the neuro-fuzzy branch and the independent clustering branch

14. Quality control for revealing of defects
15. Choice/development of neuro – fuzzy network
16. Specifications of the product, set by the client or the market
17. Transformation of specifications into constraints of crisp values of variables
18. Comparison between real data and field-values determined in (10)
S.   Is the comparison satisfactory?
C.   Change of neuro-fuzzy network?

The initial membership functions of the input control variables are determined by
a) Partitioning the space of the input variable x_i (universe of discourse) into q_i partitions, allowing for overlapping of regions
b) Defining for each region [a,c] the characteristic value b of the fuzzy set that this membership function describes. Thus, triangular membership functions are obtained. The support set [a,c] and the characteristic value b are extracted from industrial data, mainly supplied by the anodizing department of the Hellenic Aerospace Industry S.A.

The construction of the initial fuzzy rules from numerical data used for classification consists of two phases:
a) Fuzzy partition of a pattern space – the projections of those partitions on the axes of our 3 variables form the partitioning referred in (a) of previous paragraph
b) Identification of a fuzzy rule for each fuzzy subspace. Depending from the density of the data points, a pattern space of high density is partitioned into fine fuzzy subspaces, while areas of low density are partitioned into coarse fuzzy subspaces (in the sense that the number of fuzzy subspaces is too large or too small, respectively) [Ishibuchi et al., 1992]. The rules corresponding to fuzzy subspaces are formulated by an operator of the plant, or an expert.

For the learning phase of the neuro-fuzzy system the NEFCLASS learning algorithm described by Nauck et al. (1997) was used. The learning patterns consist of an input pattern **p** (a vector with the control variables) and a target pattern **t**, which is a vector with the simple or combined defects as classes.
a) Rule Learning Algorithm.
   i) For each input pattern **p**: For each input variable x_i find the membership function that gives for p_i the maximum membership degree, and if there are no rules with these memberships functions, and the maximum number of rules k has not been reached yet, create a rule that incorporates them in the antecedent, and has class the one indicated by the target vector **t**.
   ii) Determination of the rule base, by following Simple Rule Learning, Best Rule Learning, or Best per Class Rule Learning procedure.
b) Fuzzy Set Learning Algorithm. The algorithm runs through the patterns, until a stop criterion is met. For each pattern (**p**,**t**), The fuzzy set that delivered the smallest membership degree for the current pattern and that is therefore responsible for the current rule activation is changed to the direction of minimizing the rule error.

Adaptive rule weights on rule units that may be applied to modify the output of the fuzzy rule by either changing the degree of fulfillment or the output fuzzy set are not used in order to keep the semantics of the NEFCLASS system; the rule weights prevent the linguistic interpretability of fuzzy systems used in fuzzy classification or fuzzy control, as it has been proved in (Nauck, 2000).

The production that takes place in Stage 13 provides real data to compare prognosis with measurements in Stage 8. If this comparison is satisfactory, both the neuro-fuzzy and the clustering sub-procedures are further improved.

3. Implementation and results

The algorithmic procedure depicted in Figure 1 was used to design and develop a computer program named "Aluminium Quality Control" or "AlQuCo". The raw data of alkaline etching, supplied by the anodizing department of the Hellenic Aerospace Industry S.A., was a file of learning patterns consisted of
a) a vector with three input variables (namely, time, temperature, NaOH concentration) as the input pattern **p**,
b) a target pattern **t**, as vector with the following eight classes, denoted by a number: 0 for no defect, 1 for etch staining, 2 for matness, 3 for inadequate cleaning, 4 for (1 AND 2), 5 for (2 AND 3), 6 for (1 AND 3), and 7 for (1 AND 2 AND 3); the elements of **t** were reduced to six since there were no patterns giving the classes 4 & 7.

Figure 2 shows a cross section at time 5 min of the 3D grid formed by the input rows. A distribution of classes is depicted in Figure 3. A specimen run based on this input gave:
a) the limiting coordinates X: (5.25, 7.25% w) for concentration, Y: (53, 67 °C) for temperature and Z: (4.75, 10 min] for time, as final results of clustering (see Figure 4); the corresponding compact parallelepiped represents the 3D area of permitted combinations for operating the alkaline etching bath under a double restriction of maximum allowable percentage of defects in both, the volume and the differential cross section for each control variable;
b) the adapted membership functions of Figure 5 and 2 x 4 x 3 = 24 rules according to all possible input combinations of time, concentration and temperature, respectively (the best per class learning was used, as the patterns were distributed at an equal number of clusters / class); the following two fuzzy rules is a representative sample:

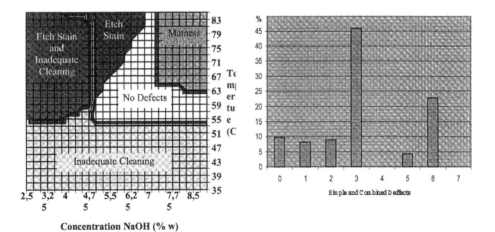

Concentration NaOH (% w)

Figure 2 (to the left): Cross section of the 3D grid depicting areas of defects
Figure 3 (to the right): Relative frequencies distribution (%) of defects

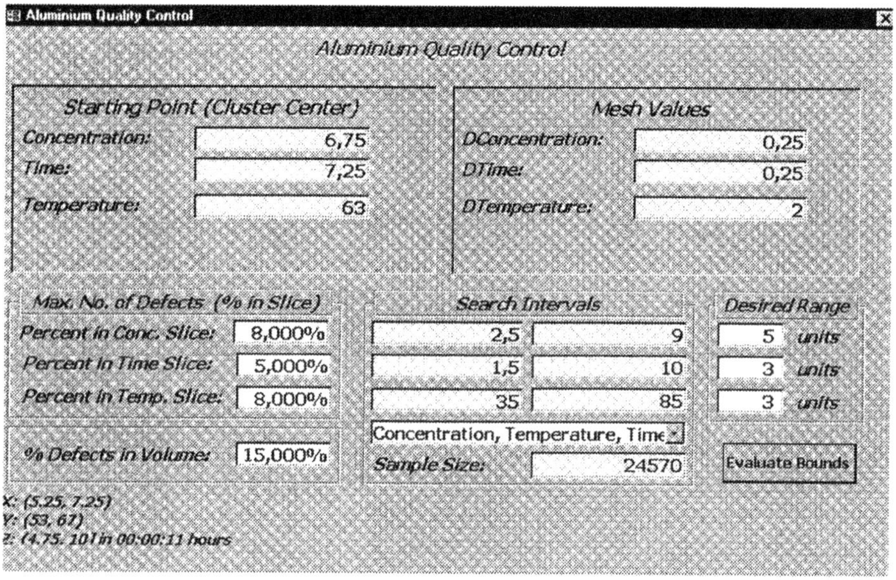

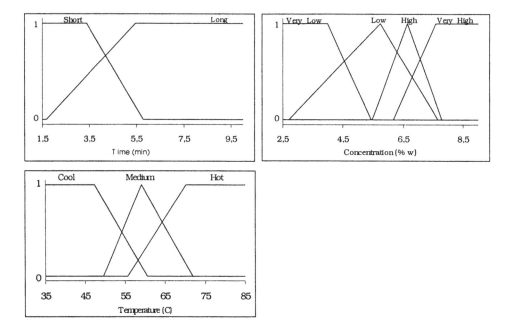

Figure 4. Screenshot of clustering results. The coordinates of cluster center is the starting point of an algorithmic determination of the 3D area of permitted combinations The desired range of each control variable in physical units is given by the corresponding mesh value multiplied by the units number quoted in the screen.

Figure 5: Adapted membership functions of the control variables after training (stage 6) obtained by using the learning set without noise.

I) If Time is Short and Concentration is Very_Low and Temperature is Cool then Class is Inadequate_Cleaning

II) If Time is Short and Concentration is Very_Low and Temperature is Medium then Class is Etch_Staining_Inadequate_Cleaning

c) Misclassification errors 8.1%, 8.4%, 8.9%, 9.6%, 10.4%, 11.2% for noise levels 0%, ±1%, ±2%, ±3%, ±4%, ±5%, respectively, in learning, resulting to testing errors within the range 8.70±0.15%, i.e. quite satisfactory for industrial application.

4. Conclusions

It has been proved that a combination of neuro-fuzzy networking and independent data processing (shallow knowledge acquired by means of quality control measurements) mainly with cluster analysis can be used in computer aided control of alkaline etching in the aluminium anodizing industry. The implementation of the corresponding algorithmic procedure within original proprietary software, which run with recent data supplied by the anodizing department of the Hellenic Aerospace Industry S.A., showed the ability of this combination to provide practical rules either for automatic control or for human supervision/conduction. The same software can be used in other similar semi-continuous chemical processes, especially in the field of surface treatment of metals/alloys, to decrease the response time, when a critical defect has been detected, and increase traceability according to ISO-standards of 9000 series.

Acknowledgements

Aluminium anodizing data supply from the Hellenic Aerospace Industry S.A. and financial support provided by the Research Center of the University of Piraeus are kindly acknowledged.

References

Ishibuchi, H., K. Nozaki and H. Tanaka, 1992, Distributed representation of fuzzy rules and its application to pattern classification, Fuzzy Sets and Systems, 52, 21

Kramer, M. A., 1987, Malfunction diagnosis using quantitative models with non-Boolean reasoning in expert systems, AIChE J. 33, 130.

Nauck, D., 2000, Adaptive Rule Weights in Neuro-Fuzzy Systems, Neural Comput. & Applic, 9, 60

Nauck, D., F. Klawonn, and R. Kruse, 1997, Foundations of neuro-fuzzy systems, John Willey & Sons, Chichester, England

Ozyurt B. and A. Kandel, 1996, A hybrid hierarchical neural network-fuzzy expert system approach to chemical process fault diagnosis, Fuzzy Sets & Systems 83, 11.

Petti, T. F., J. Klein, and P. S. Dhurjati, 1990, Diagnostic model processor: using deep knowledge for process built diagnosis, AIChE J. 36, 565.

Ruiz, D., J. M. Nougues, and L. Puigjaner, 2001, Fault diagnosis support system for complex chamical plants, Computers & Chem. Engng 25, 151.

Schmitz, G. P. J. and C. Aldrich, 1998, Neurofuzzy modelling of chemical process systems with ellipsoidal radial basis function neural networks and genetic algorithms, Computers & Chem. Engng 22, S1001.

Sorsa, T and H. N. Koivo, 1993, Applications of artificial neural networks in process fault diagnosis, Automatica 29, 843.

Design of Tubular Reactors in Recycle Systems

Costin S. Bildea[*], Susana C. Cruz, Alexandre C. Dimian and Piet D. Iedema
University of Amsterdam, Nieuwe Achtergracht 166, 1018 WV Amsterdam
[*]Delft University of Technology, Julianalaan 136, 2628 BL Delft,
The Netherlands

Abstract

The nonlinear behaviour of PFR – Separation – Recycle systems is analysed in terms of state multiplicity, instability and unfeasibility. These phenomena can occur for practical situations, as demonstrated by the HDA case study. They can be avoided by sufficient reactor cooling, high reactor-inlet temperature or fixed-recycle plantwide control.

1. Introduction

Reaction systems involving material recycle (Figure 1) are common to industrial practice. During a typical design, one specifies the performance of reactor (conversion, selectivity) and separation (quality of product and recycle streams), performs the mass balance and proceeds with detailed design of reactor and separation units. An optimisation procedure based on economic criteria can be applied. The final design can suffer, however, from serious operability problems caused by nonlinear phenomena. Pushpavanam and Kienle (2001) showed that state multiplicity, isolas, instability and limit cycles arise in CSTR – Separation – Recycle systems. Because these phenomena also occur in a stand-alone CSTR, the effect of recycle is not obvious. In contrast, the stand-alone PFR has a unique stable steady state. In the present article, the singularity theory (Golubitsky and Schaeffer, 1985) is used to classify the behaviour of PFR – Separation – Recycle systems. This includes multiplicity, instability or unfeasibility. By means of a case study, we show that these phenomena can be easily found in real plants.

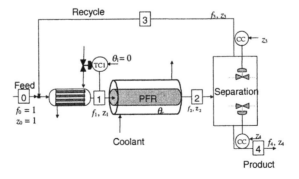

Figure 1. PFR – Separator – Recycle systems. Reactor effluent is processed by the separation section. The composition of product and recycle streams is controlled. An upstream heat exchanger fixes the reactor-inlet temperature.

2. Model equations

The dimensionless model assumes plug flow, constant physical properties and one-reactant, n-th order reaction. It contains the PFR model Eqs 1-3, the input-output mass balance Eq. 4 and the mass balance at the mixing point Eq. 5:

$$\frac{dX}{d\xi} = \frac{Da}{1+f_3} \cdot z_1^{n-1} \cdot (1-X) \cdot \exp\left(\frac{\gamma\theta}{1+\theta}\right) \tag{1}$$

$$\frac{d\theta}{d\xi} = \frac{Da}{1+f_3}\left(B \cdot z_1^n \cdot (1-X) \cdot \exp\left(\frac{\gamma\theta}{1+\theta}\right) - \beta \cdot (\theta - \theta_c)\right) \tag{2}$$

$$X(0) = 0; \ \theta(0) = 0 \tag{3}$$

$$(1 - z_4) = (1 + f_3 z_3) \cdot X(1) \tag{4}$$

$$1 + f_3 \cdot z_3 = z_1 \cdot (1 + f_3) \tag{5}$$

The dimensionless variables and parameters, defined using feed flow rate F_0, feed concentration c_0 and reactor inlet temperature T_1 as reference values, are: axial coordinate $0 \le \xi \le 1$, conversion $X(\xi)$, temperature $\theta(\xi)$, recycle flow rate f_3 and reactor-inlet concentration z_1; Damkohler number Da, activation energy γ, adiabatic temperature rise B, heat-transfer capacity β, coolant temperature θ_c, concentration of recycle and product streams z_3, z_4. For convenience, $X \equiv X(1)$ will stand for conversion at reactor outlet.

By fixing the reactor-inlet temperature, energy feedback effects are excluded. Moreover, the plug-flow model of the stand-alone reactor has a unique solution. Hence, our analysis will identify the nonlinear phenomena caused by the material recycle.

The model equations can be solved by a shooting technique: start with an initial guess $X = X(1)$, calculate the recycle flow rate f_3 (Eq. 4) and reactor-inlet concentration z_1 (Eq. 5), integrate the PFR equations 1 - 2, check and update the guess X. This implies that it is theoretically possible to reduce the model to one equation with one variable; therefore, the results of singularity theory with a single intrinsic variable can be applied. We also note that the feasible solutions, corresponding to positive flow rates, satisfy:

$$0 < X < 1 - z_4 \tag{6}$$

3. Nonlinear behaviour

Isothermal PFR

For a n-th order reaction in an isothermal PFR ($\theta = 0$), integration of Eq. 1 with boundary conditions Eqs 3 - 5 leads to:

$$\frac{Da \cdot X \cdot z_3^n (1 - z_4)^{n-1}}{\left((1 - z_4) - X \cdot (1 - z_3)\right)^n} = \frac{1}{n-1} \cdot \left(\left(\frac{1}{1-X}\right)^n - 1\right) \tag{7}$$

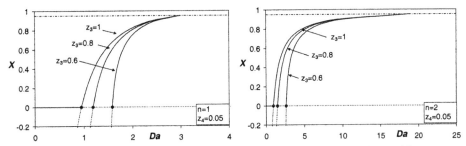

Figure 2. Isothermal PFR – Separator – Recycle Systems. Stable and unstable states are presented by continuous and dashed lines, respectively. Dots represent transcritical bifurcation points, where an exchange of stability takes place.

The X vs. Da dependence, which can be obtained from Eq. 7, is presented in Figure 2 for $n = 1$ and $n = 2$. For any Da, a trivial, unfeasible solution $(X, f_3) = (0, \infty)$ exists. The non-trivial solution is meaningful if Eq. 6 is satisfied. This is equivalent to the following feasibility condition:

$$\lim_{X \to 0} Da = \frac{1 - z_4}{z_3^n} = Da_T < Da < Da_{max} = \frac{z_4^{1-n} - 1}{n - 1} \tag{8}$$

Da_T represent a transcritical bifurcation point. This represents a feasibility boundary. A steady state can exist only if the reactor consumes the entire amount of reactant fed in the process. This is impossible for slow kinetics or small reactor volume, which lead to reactant accumulation and infinite recycle flow rate. The transcritical bifurcation does not exist in case of a stand-alone reactor. Interestingly, it occurs for the same value Da_T as in the case of CSTR – Separation – Recycle systems (Bildea et al., 2000).

Non-isothermal PFR

In this section we consider a first-order reaction, and non-isothermal PFR. In this case, the PFR – Separation – Recycle system can have multiple steady states. To achieve the steady state classification, we choose Da as distinguished (bifurcation) parameter. We fix the parameters related to reaction kinetics γ and thermodynamics B and to separation performance z_3, z_4. Then, by computing relevant singularities (Balakotaiah and Luss, 1984), we divide the space of the parameters concerning reactor design (β, θ_c) into regions with qualitatively different $Da - X$ bifurcation diagrams.

Figure 3 presents a typical phase diagram and four different bifurcation diagrams. In region I, one feasible steady state exists, if Da exceeds the critical value Da_T given by Eq. 8. Crossing the boundary-limit set BL_1 to region II, one fold point enters the feasibility region through the boundary $X = 0$, leading to state multiplicity. Let U and L signify steady states located on the upper and lower solution branches, respectively. Then, the multiplicity pattern in region II is $0 - 2 - 1(U) - 0$. This behaviour does not change if the PFR is replaced by a CSTR. Remarkably, the same simple equation gives the BL_1 set for both cases:

$$\left(\beta \cdot \theta_c \right)_{BL_1} = \frac{1}{\gamma} \cdot \left(\frac{z_3 - z_4}{1 - z_4} - B \cdot \gamma \cdot z_3 \right) \tag{9}$$

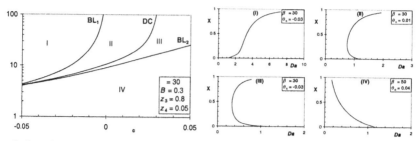

Figure 3. Steady state classification of non-isothermal PFR – Separation –Recycle systems.

Crossing the double-cross set DC to region III, the relative position of two solutions located at the feasibility boundaries changes. The multiplicity pattern becomes $0 - 2 - 1(L) - 0$. Finally, when the boundary-limit set BL_2 is crossed to region IV, the fold point exits the feasible region through the boundary $X = 1- z_4$. Hence, in region IV a unique steady state exists. This state is unstable, as will be demonstrated in the next paragraph. The amount of reactant consumed by the reactor is $F_1 c_1 X$, where F_1 and c_1 are dimensional variables. Its dependence on the reactor feed flow rate F_1 is presented in Figure 4 using the dimensionless variables $F_1/(kV) = 1/Da_1$ and $F_1 c_1 X/(c_0 kV) = z_1 X/Da_1$, where Da_1 uses the reactor-inlet flow rate as reference. The steady state values of the reactor-inlet flow rate F_1 are the intersections of this curve with the horizontal line representing the net amount of reactant fed in the process, given in a dimensionless form by $F_0 c_0 (1-z_4)/(c_0 kV) = (1-z_4)/Da$. Consider a small, positive deviation of the reactor inlet flow rate, from the steady state B. At the right of point B, the amount of reactant fed in the process is larger than the amount of reactant consumed. Reactant accumulation occurs, leading to a further increase of the recycle and reactor-inlet flow rates; hence the steady state B is unstable. This is independent on the dynamics, because this proof is based only on steady state considerations. The high-conversion branch can lose stability at Hopf bifurcation points, which can be identified only using a dynamic model. Moreover, additional nonlinear phenomena can occur if axial dispersion is included in the reactor model.

An operating point located on the unstable branch is more likely when low conversion is required. Instability can be avoided by low coolant temperature and large heat-transfer capacity (region I of Figure 3). This might be impossible, for example due to restricted heat-transfer area or heat-transfer coefficient. In this case, increasing the reactor-inlet temperature T_1 is an option. This decreases the dimensionless parameters γ and B, shifting the bifurcation varieties of Figure 3 in a favourable direction.

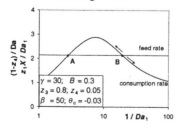

Figure 4. Instability of the low-conversion steady state

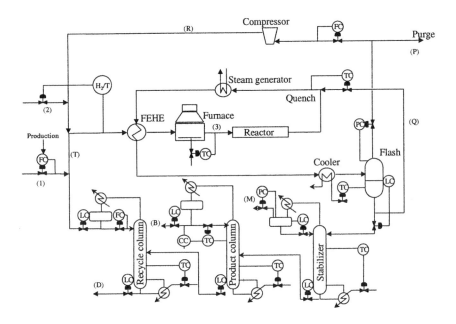

Figure 5. The toluene hydrodealkylation plant – flowsheet and plantwide control.

4. Case study

The HDA plant (Douglas, 1988) is used as a case study, in order to demonstrate that state multiplicity and instability can easily occur in recycle systems involving tubular reactors. In the HDA plant, benzene is produced by toluene hydrodealkylation, a mildly exothermic reaction with moderate activation energy:

$$C_6H_5 - CH_3 + H_2 \rightarrow C_6H_6 + CH_4 \qquad \Delta^r H_{950K} = -50,450 \text{ J/mol}; E_a = 217,600 \text{ J/mol}$$

The reaction takes place in an adiabatic tubular reactor. The reactor effluent is cooled and a gas-liquid split is performed. The gas stream is recycled, a purge being necessary to avoid methane accumulation. The liquid stream is processed by a train of distillation columns, which provides three product streams (lights, benzene, heavies) and the toluene recycle. The flowsheet and the control system are presented in Figure 5.

Two design variables must be specified, for example reactor conversion and gas recycle flow rate. Then, neglecting secondary reactions and considering perfect separation, a preliminary mass balance was performed and the reactor was designed. This yields Figure 6, presenting the conversion vs. reactor volume, for different values of the gas recycle flow rate. Two operating points were considered for further investigation: Design A – on the stable branch, far from the fold bifurcation; and Design B - on the unstable branch, close to the fold bifurcation. Then, the remaining units were designed and rigorous simulation was performed using AspenPlus. Finally, the simulation was exported to AspenDynamics, where control loops were provided and tuned.

For Design A, no operating problems were encountered. The plant was robust in face of various disturbances (e.g. ±25% production change) or uncertain design parameters.

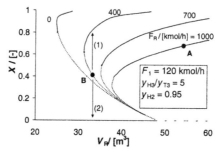

Figure 6. HDA plant – conversion vs. reactor volume bifurcation diagrams

In contrast, Design B suffers from serious operability problems. Starting from steady state, reaction ignition occurs (Figure 7a). If the reactor-inlet temperature deviates by only 2 °C from the design value, reaction extinction takes place (Figure 7b). The plant moves towards the trivial ($X=0$) steady state. All flow rates increase and one of the levels becomes uncontrollable. The plant can be stabilized, providing an additional control loop, which keeps the toluene recycle constant by changing the reactor-inlet temperature. However, extinction still occurs for some disturbances, suggesting a bifurcation close to the operating point. This will be the subject of future investigation.

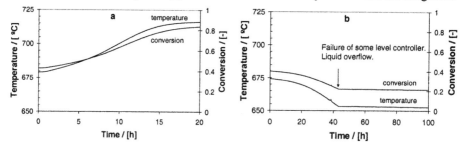

Figure 7. Dynamic simulation results

5. Conclusions

1. Interaction between reaction and separation through material recycle can lead to state multiplicity and instability, even if the stand-alone reactor has a unique stable state.
2. Instability is likely to occur for low conversions. It can be avoided by large cooling capacity, high reactor inlet temperature, or fixed-recycle plantwide control structure.
3. Considering only the nominal steady state operating point can lead to an un-operable plant. Nonlinear analysis is a way to identify and avoid such dangerous situations.

References

Balakotaiah, V. and Luss, D., 1984, *Chem. Eng. Sci.*, **39**, 865.
Bildea, C.S., Dimian, A.C. and Iedema, P.D., 2000, *Comp. Chem. Eng.*, **24**, 209.
Douglas, J.M., 1988, *Conceptual Design of Chemical Processes.* Mc-Graw Hill, New York.
Golubitsky, M. and Schaeffer, D., 1985, *Singularities and Groups in Bifurcation Theory.* Springer-Verlag , New York.
Pushpavanam, S. and Kienle, A., 2001, *Chem. Eng. Sci.*, **56**, 2837.

European Symposium on Computer Aided Process Engineering – 12
J. Grievink and J. van Schijndel (Editors)
© 2002 Elsevier Science B.V. All rights reserved.

445

Controllability of Reactive Batch Distillation Columns

Nicholas C.T. Biller and Eva Sørensen[1]

Centre for Process Systems Engineering,
Department of Chemical Engineering, University College London,
Torrington Place, London WC1E 7JE, United Kingdom

Abstract

Reactive batch distillation offers advantages by combining reaction and separation whilst improving yield through the removal of equilibrium limiting products. However, the process can be complex to control and in this paper, the controllability of reactive batch distillation columns is examined using rigorous models. A robust linearisation approach is presented which can be used for the investigation of controllability of both reactive and non-reactive systems. The synthesis of ethyl acetate is considered as a case study and it is concluded that, for this example, the reaction has minimal impact on the controllability.

1. Introduction

Reactive distillation has been applied successfully in industry where large capital and energy savings have been made through the integration of reaction and distillation into one system. It has therefore attracted much interest in the literature over the past few years (for a review, see Taylor and Krishna, 2000). Reactive distillation can improve yields through the removal of volatile products from the reaction zone, hence pushing the equilibrium towards the products. Operating reactive distillation in the batch mode is attractive to the fine chemical and pharmaceutical industries where high-value, low-volume products are manufactured and where flexibility is important. However, these industries operate to tight product specifications and slight variations in operating conditions can lead to the product being rejected. Sørensen and Skogestad (1994) concluded that some form of feedback control is essential and that open-loop policies for non-reactive distillation, such as constant reflux ratio or a predetermined reflux ratio profile, cannot be reliably applied to reactive batch distillation. Knowledge of how controllability of these processes is affected is therefore of great importance so that the unit, and its control system, can be designed to give the best possible performance. Previous studies into control of reactive batch distillation also include: Reuter et al. (1989) who mainly focussed on modelling, Sørensen et al. (1996) who considered optimal control and Monroy-Loperena and Alvarez-Ramirez (2000) who focussed on controller design.

A common approach to assessing controllability is through exhaustive simulation. This requires the control system to be designed, and then evaluated, against a finite set of

---

[1] Author to whom correspondance should be addressed.
Email: e.sorensen@ucl.ac.uk, fax: +44 (0)20 7383 2348

disturbances and set point changes. The drawback with this method is that it is impossible to guarantee that the conclusions drawn are due to the inherent properties of the process, which is what one is primarily interested in. The control system selected, or for that matter, the set of disturbances and set point changes chosen, could be responsible. Skogestad and Postlethwaite (1996) indicated that it is useful to assess controllability in the frequency domain rather than the time domain. In order to use frequency response controllability tools, it is necessary to use linearised versions of the non-linear process models with the assumption that the linear model accurately describes the non-linear model for a brief period after linearisation. It is, however, vital to confirm that this is indeed the case through comparison with non-linear simulation. It should be noted that, for highly non-linear processes such as distillation columns, the linear models will not accurately describe the process state at large deviations. When operating in batch mode, where the states are continuously changing, it is necessary to linearise the process model at multiple points to appreciate how controllability changes during the process in order to design the controllers.

2. Method

In this paper, an approach for the investigation of controllability in both reactive and non-reactive batch distillation columns is presented. The strategy is based on the use of rigorous dynamic models which are linearised at multiple points along the time trajectory. The linearised models are then employed to investigate the controllability of the process using frequency response tools. We will also investigate the impact of the reaction on the system by considering how controlled reacting and non-reacting systems respond to given disturbances.

2.1 Non-Linear Models
Previous studies on controllability of reactive batch distillation have used simple mathematical models to describe the process which were based on linear tray dynamics and an algebraic energy balance (Sørensen and Skogestad, 1994). In this paper, a rigorous dynamic model is used which includes the Francis Weir formula to describe liquid tray dynamics. Pressure dynamics, vapour phase holdup and a dynamic energy balance are also considered. A simplified version of this model, which assumes constant pressure and an algebraic energy balance, will also be used.

2.2 Controllability analysis approach
The approach proposed in this work for the investigation of controllability of both reactive and non-reactive batch distillation columns is as follows:
1. Using the rigorous model, simulate until the desired point of linearisation.
2. Initiate the simplified model with the process conditions of the rigorous model at the point of linearisation.
3. Linearise the simplified model at this point. (Linearising the rigorous model directly leads to numerical difficulties as will be shown later.)
4. Validate the linearised version of the simplified model by comparison with the rigorous model for typical open-loop process changes, such as step changes in inputs or disturbances.

5. Investigate the controllability of the process based on the linearised version of the simplified model.

6. Repeat steps 1-5 for the next point of linearisation.

It may be necessary to include controllers in the rigorous model to be able to reach the desired process conditions prior to linearisation. As long as these controllers are not included in the simple model, their effect will not influence the controllability. (The conditions at the point of linearisation are, nevertheless, a result of the controllers.)

The approach outlined above can be used to investigate the controllability of both reactive and non-reactive systems. However, in order to *compare* the controllability of the two systems, their process conditions must be as similar as possible for a comparison to be informative. Hence the approach needs to be modified:

1a. Using the rigorous model of the reactive system, follow the same procedure as before (steps 1-6). This is the *reactive system model*.

1b. Using the rigorous model of the reactive system, simulate until the desired point of linearisation. Now set the reaction terms in the mass and energy balances to zero and continue the approach for steps 2-6. This is the *non-reactive system model*.

In this work, linearisation of the process models were undertaken using the LINEARISE feature of the gPROMS modelling language (Process Systems Enterprise Ltd., 2001) which directly generates a linear state-space model from specified inputs and outputs.

3. Case Study

In this paper, we will demonstrate the approach outlined above using a case study of the synthesis of ethyl acetate from ethanol and acetic acid with water formed as a by-product. The reaction takes place in the liquid phase on each tray in the column section and in the reboiler. The kinetics and ideal physical properties were taken from Mujtaba and Macchietto (1997). The 8 tray reactive batch distillation column was initially charged with 22 *kmol* of feed (49% ethanol, 49% acetic acid and 2% water). Of this, 100 *mols* were placed in the reflux drum, 20 *mols* on each tray and the remainder was charged to the reboiler. As mentioned in Section 2, the system must first be brought to the desired operating point. Hence, the column shown in Figure 1 was operated with the following control scheme: The condenser cooling duty was used to maintain the condenser pressure at *0.5 atm*, the Distillate flowrate was employed to maintain the

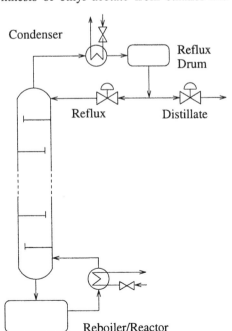

Figure 1: Reactive Batch Column

composition of ethyl acetate in the distillate product at *0.6* and the Reflux flow was used to maintain the level in the condenser at *100 mol*. All controllers were PI controllers. A constant reboiler heat duty of 88.5 kW was maintained throughout. There is a period of approximately 1.5 hours at the start of total reflux where no product is removed whilest the composition of ethyl acetate at the top of the column is building up. The batch is terminated when the distillate flow falls below 0.1 mol/s, after approximately 6 hrs.

3.1 Comparison between linear and non-linear models

In order to apply frequency response techniques to investigate the system controllability, the model was linearised using the approach detailed in this paper after 100 minutes of operation. The inputs to the linear model were selected as being: The Reflux and Distillate Flowrates and the Reboiler Heat Duty. The Outputs of interest were the Reflux Drum Level, Distillate Composition of ethyl acetate and the Temperature of the Reboiler.

The effect of a 1 *mol/s* increase in the reflux flowrate on the ethyl acetate distillate composition is shown in Figure 2. In this figure, the response of the linearised rigorous model is compared to the open loop response of the rigorous non-linear model. It is observed that the composition moves in the same direction as the non-linear although much quicker, but after approximately 25 seconds, it changes direction and actually

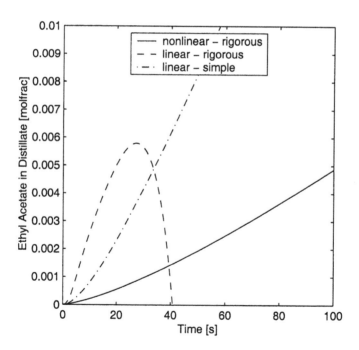

Figure 2: Open Loop models

becomes negative. This is believed to be due to the introduction of numerical noise during the linearisation. On the other hand, a closer response was given by linearising the simpler model with the same process conditions, also show in Figure 2. This trend,

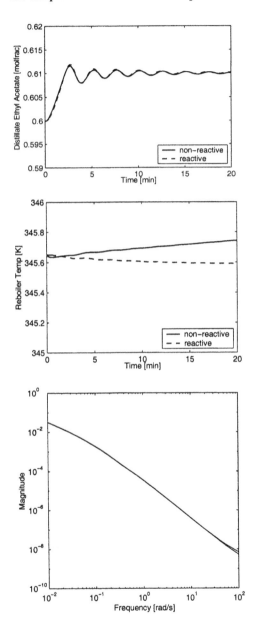

Figure 3: Reactive and non-reactive systems

i.e. large deviations between responses of the non-linear and linearised rigorous models, but similar responses of the non-linear rigorous and the simple linearised models, was found for a number of other step responses.

3.3 Comparison between reactive and non-reactive systems

In the following, the approach outlined earlier will be demonstrated when investigating the controllability of a reactive and a non-reactive system using the ethyl acetate synthesis example described above. Two linearised models were generated; one reactive system model and one non-reactive system model as described in Section 2. The *only* difference between the two is the inclusion of the reaction terms in the reactive system model. The two columns are both controlled using the control system outlined earlier until the point of linearisation.

Figure 3 (top and middle) shows the controlled responses of the column to a set point change of 0.01 in the setpoint for the composition controller. The top graph shows the response of the controlled ethyl acetate distillate composition and the middle graph shows the response of the uncontrolled reboiler temperature. Here, the system was simulated with reaction until 100 mins into the batch. The dashed line represents the response in the reactive system model and the solid line represents the response in the non-reactive system model. It can be seen that there is no significant difference between the reactive and non-reactive systems in terms of the distillate composition. There is a slight difference in the temperature

response which reflects the fact that ethyl acetate is not being produced in the non-reactive system model and therefore the temperature rises faster as the ethyl acetate concentration falls.

The process gain as a function of frequency for the reactive and the non-reactive model systems are shown in Figure 3 (bottom). There appears to very little difference between the reacting and non-reacting systems. The two curves separate slightly at high frequencies but the process gain is very small and therefore inconsequential. The phase is slightly different at low frequencies but otherwise similar (not shown). As already mentioned, the further one moves from the point of linearisation, the less reliable the linear model becomes. Therefore we cannot draw any meaningful conclusions from this area of the response.

4. Conclusion

In this paper, we have considered the controllability of rigorously modelled reactive batch distillation columns. In using the rigorous model, numerical difficulties were encountered whilst linearising which led to incorrect results. This was remedied by using a rigorous model to generate the process conditions, and linearising a simplified version of the model instead, which was then used for controllability analysis. For the case study of ethyl acetate synthesis, there is no appreciable difference in controllability between the reactive system and one where the reaction is not included, hence the chemical reaction in itself is not the reason why reactive distillation systems are harder to control than non-reactive ones, but rather the fact that reactive distillation systems have more degrees of freedom.

References

Monroy-Loperena, R and J Alvarez-Ramirez, Output-Feedback Control of Reactive Batch Distillation Columns, *Ind. Eng. Chem. Res.*, **39**, 378-386, 2000.

Mujtaba, I.M and S. Macchietto, Efficient optimization of batch distillation with chemical reaction using polynomial curve fitting techniques, *Ind. Eng. Chem. Res,* **36**(6), 2287-2295, 1997.

Process Systems Enterprise Ltd., *gPROMS Introductory User Guide,* 2001.

Reuter, E, G Wozny, and L Jeromin, Modelling of multicomponent batch distillation processes with chemical reaction and their control systems. *Comp. Chem. Eng.,* 13(4-5), 499-510, 1989.

Skogestad, S and I Postlethwaite, *Multivariable Feedback Control: Analysis and Design*, Wiley, Chichester, 1996.

Sørensen, E and S Skogestad, Control strategies for reactive batch distillation, *J. Proc Cont.,* **4**, 1994.

Sørensen, E, S Macchietto, G Stuart and S Skogestad, Optimal control and on-line operation of reactive batch distillation. *Comp. Chem. Eng.,* 20(12), 1491-1498, 1996.

Taylor R and R Krishna, Modelling reactive distillation, *Chem. Eng. Sci.,* 5, 5183-5229, 2000.

European Symposium on Computer Aided Process Engineering – 12
J. Grievink and J. van Schijndel (Editors)

Process Control Scheme for a 2-Bed Pressure Swing Adsorption Plant

M. Bitzer, M. Zeitz

Institut für Systemdynamik und Regelungstechnik, Universität Stuttgart,
Pfaffenwaldring 9, D-70550 Stuttgart, Germany

Abstract

Pressure swing adsorption (PSA) plants are used for the separation of gas mixtures and are operated as cyclic multi-step processes. Based on the rigorous distributed parameter model of the considered 2-bed PSA plant, a reduced controller design model is derived and a process control scheme is designed and validated by simulation.

1. Introduction

Pressure swing adsorption (PSA) is a common process technique for the separation of gas mixtures (Ruthven *et al.*, 1994). As example, a 2-bed PSA plant for the production of oxygen from air is considered (Bitzer and Zeitz, 2002). The plant whose flowsheet is depicted in Figure 1 is run in a 4-step cycle, i.e. the connections between the adsorbers are changed from one cycle step to the next according to the sketches in the same figure. Each fixed-bed adsorber is described by a set of nonlinear partial-differential-algebraic equations and changing boundary conditions (Unger, 1999). The whole PSA plant is therefore characterized as a hybrid and nonlinear distributed parameter system. It's inherent dynamics is influenced by the variable structure due to the cyclic switching of the valves as well as the distributed nature of the process.

The characteristic dynamics of PSA plants concerns the occurrence of nonlinear travelling concentration waves which are alternating their propagation direction as a consequence of the periodic process operation, see Figure 1. The duration of the cycle steps do considerably affect the product concentration, because they determine the extent of breakthrough of the concentration fronts at the product outlet of the adsorber beds. The duration of the cycle steps are therefore considered the manipulating variables of the process. If the adsorption plant is operated with fixed cycle step times, the plant approaches a cyclic steady state[1] (CSS), i.e. the conditions at the end of each cycle are identical to those at its start.

During the operation, the PSA plant is subject to disturbances influencing its performance. In order to control the purity of the product, i.e. the averaged concentration in the oxygen tank, see Figure 1, a periodic operating point has to be stabilized by a process control scheme which is manipulating the duration of the cycle steps. Other control tasks concern the realization of set-point changes or a start-up procedure. Up to now, there exists no model based process control strategy which is applicable to this type of

---

[1] A numerical approach for the determination and the optimization of the CSS of periodic adsorption processes was presented by Nilchan and Pantelides (1998).

plant and known to the authors[2]. The development of such process control strategies is also of interest concerning the use of PSA units within fuel cell systems (St-Pierre and Wilkinson, 2001). As first contributions to such a process control concept, a nonlinear observer with distributed parameters was designed (Bitzer and Zeitz, 2002) and a periodic control concept for a PSA plant driven by a simplified 2-step cycle was developed (Bitzer *et al.*, 2001).

In this paper, a process control scheme is presented which comprises a feedforward and a superimposed feedback controller. In the next section, the model of the 2-bed PSA plant is introduced in order to specify the control problem. Based on this, the controller design is presented and applied to the detailed PSA simulation model.

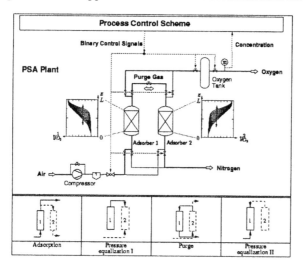

Figure 1: Flowsheet of a 2-bed PSA plant for the oxygen production from air with the visualization of the travelling oxygen concentration waves $y_{O_2}(z,t)$, $\iota \in \{1,2\}$ in the two adsorber beds and the realized coupling schemes for the adsorbers during the four cycle steps.

2. Two-bed pressure swing adsorption plant

The considered PSA plant depicted in Figure 1 is used for the production of oxygen from air for medical purposes. The produced oxygen is stored in a tank from which it is taken by the consumer. The considered adsorption model[3] treats air as a binary mixture of oxygen and nitrogen, and emanates from two phases, i.e. a gaseous phase and an adsorbed phase. Further, an isothermal process is assumed. The rigorous distributed parameter model with one space coordinate describing a single fixed-bed adsorber consists of a set of five quasilinear partial-differential-algebraic equations for the pressure $p(z,t)$, oxygen mole fraction $y_{o_2}(z,t)$ in the gaseous phase, the adsorbed amounts

---

[2] Christofides (2001) confirmes, that 'virtually no research has been done' concerning the control of hybrid distributed parameter systems.

[3] The derivation and discussion of the detailed model as well as simulation results can be found at full length in (Unger, 1999). The model equations together with simulation results are also given in (Bitzer and Zeitz, 2002).

$q_k(z,t)$, $k \in \{O_2, N_2\}$ of oxygen and nitrogen, and molar flux $\dot{n}(z,t)$ in dependence of space z and time t. By summarizing the states of each adsorber, i.e. $\mathbf{x}^i = [p^i, y_{O_2}^i, q_{O_2}^i, q_{N_2}^i, \dot{n}^i]^T$, $i \in \{1,2\}$, as well as the boundary input functions, i.e. $\mathbf{v}^i = [y_{O_2,in}^i, \dot{n}_{in}^i]^T$, the plant model can be written in the vector notation (Bitzer and Zeitz, 2002) of partial-differential-algebraic equations (PDAE), boundary conditions (BC), and initial conditions (IC) as follows:

$$\text{PDAE}: \quad B(\mathbf{x}^i)\frac{\partial \mathbf{x}^i}{\partial t} = A(\mathbf{x}^i)\frac{\partial \mathbf{x}^i}{\partial z} + \mathbf{f}(\mathbf{x}^i) \quad z \in (0,L), \; t > 0$$

$$\text{BC}: \qquad\qquad 0 = \varphi^j(\mathbf{x}^i, \mathbf{v}^i) \qquad\qquad z \in \{0,L\}, \; t \in \vartheta_k^j \qquad (1)$$

$$\text{IC}: \qquad \mathbf{x}^i(z,0) = \mathbf{x}_0^i(\mathbf{x}^i) \qquad\qquad z \in [0,L]$$

with the time interval $\vartheta_k^j = (t_k^j, t_k^{j+1}]$. The boundary conditions depend on the connections between the adsorbers during the jth cycle step of the kth cycle, see Figure 1. The model for the oxygen tank (index t) is given as a set of two ordinary differential equations

$$B^t(\mathbf{x}^t)\frac{d\mathbf{x}^t}{dt} = \mathbf{f}^t(\mathbf{x}^t, \mathbf{x}^i(L,t)) \qquad t > 0, \qquad \mathbf{x}^t(0) = \mathbf{x}_0^t \qquad (2)$$

with the state $\mathbf{x}^t = [p^t, y_{O_2}^t]^T$. The two adsorbers are run in a phase-shifted manner in order to attain a quasi-continuous production, as shown in Figure 1. Thereby, the duration Δt_k of the production/purge step is the manipulating variable of the process. The controlled variable is the purity P_k of the product, i.e. the oxygen mole fraction $y_{O_2}^t$ in the product tank, averaged over one cycle with the cycle time $T_k^c = T_k^c(\Delta t_k)$ i.e.

$$P_{k+1} := P(t_k + T_k^c) = \frac{1}{T_k^c}\int_{t_k}^{t_k + T_k^c} y_{O_2}^t(t)\,dt. \qquad (3)$$

The model of the PSA plant is implemented in the simulation environment DIVA (Köhler *et al*, 2001). Thereby, the model equations (1) are spatially discretized according to the Method of Lines. The simulation of the rigorous model (1)-(3) is used for the validation of the developed control strategy.

3. Controller design

The cycle step times are rather unconventional manipulating variables in controller design, especially as they occur implicitly in the model equations (1)-(3). Current approaches for the controller design for distributed process models are based on the derivation of a simplified model which captures the dominant system dynamics, see e.g. (Christofides, 2001). The derivation of a reduced order model for PSA plants is a challenge on its own since the structural changes and hybrid nature of the system have to be taken into account. In (Bitzer *et al.*, 2001), a reduced order model for the PSA plant was derived by approximating the travelling concentration waves by rectangularly shaped profiles and by introducing the front positions as new states of the reduced model. The controller has then been designed for this 'constant pattern wave' model.

454

This paper focuses directly on the time-discrete nature of the process which has to be taken as an intrinsic feature of the plant due to its cyclic operation. The duration Δt_k of the production/purge step, which is the manipulating variable, becomes apparent when the cyclic and time-discrete nature of the process is considered: the cyclic operation of the plant (1)-(3) defines a time-discrete state space model $\mathbf{x}_{k+1} = \mathbf{g}(\mathbf{x}_k, \Delta t_k)$ for the internal state $\mathbf{x}_k := \mathbf{x}(t_k)$. This is also valid for the purity $P_{k+1} = h(\mathbf{x}_k, \Delta t_k)$ of the PSA plant. Due to the complexity of the model (1)-(3), the iterative maps, which reflect also the fundamental solution of the plant model, cannot be calculated analytically. Therefore, a simplified model is required representing appropriately the input/output (I/O) behavior of the PSA plant. The approximated I/O model is based on the analysis of the cyclic steady state (CSS) as well as of the transient behavior of the simulation model (1)-(3).

3.1 Open-loop trajectory control
Simulation experiments showed that the plant operated in open-loop is stable. CSS is therefore reached when the duration of the production/purge step is kept constant, i.e.

$\Delta t_k = \Delta t^* = const.,$ then

$\mathbf{x}_k \to \mathbf{x}^* = \mathbf{g}(\mathbf{x}^*, \Delta t^*)$ and

$P_k \to P^* = h(\mathbf{x}^*, \Delta t^*) =: h_*(\Delta t_k)$ for

$t \to \infty.$ The rela-

tion $P_k = h_*(\Delta t_k)$ defines the stationary I/O behavior of the PSA plant and is determined by calculating the CSS using the detailed plant model (1)-(3), see Figure 2a. Inverting the function $h_*(\Delta t_k)$ enables the design of a feedforward con-

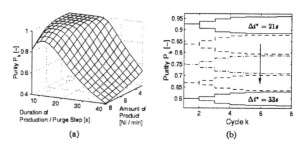

(a)   (b)

Figure 2: (a) Stationary I/O model depending on the duration Δt_k and the amount n_p of product drawn off the tank. (b) Step responses for $\Delta t_k = \pm 1.4s$ at the periodic operating points $\Delta t^* \in \{21s, 24s, 27s, 30s, 33s\}$ and $\dot{n}_- = 5Nl/\min.$

troller consisting of a static nonlinearity. Then, the duration Δt_k of the production/ purge step at a periodic operating point can be adjusted according to a desired purity P_k^d, i.e. $\Delta t_k^d = h_*^{-1}(P_k^d)$. In order to perform set-point changes, the global transient I/O behavior of the plant has to be taken into account also. As the inverse I/O behavior cannot be calculated analytically, it has to be approximated by a time discrete I/O model, e.g. $P_{k+1} = \tilde{h}(P_k, P_{k+1}, \dots, u_k)$ with u_k, representing the input variable, see also (Bitzer et al, 2001). Therefore, the transient I/O behavior of the plant is examined using the detailed simulation model (1)-(3). The simulation studies in Figure 2b illustrate that the I/O behavior can even be roughly approximated by a linear discrete model. In addition to the approximation of the I/O behavior, a desired trajectory P_k^d, for a set-point change has to be planned with respect to the transient plant dynamics. The desired trajectory P_k^d, is designed by a sufficiently smooth polynomial. The feedforward control-

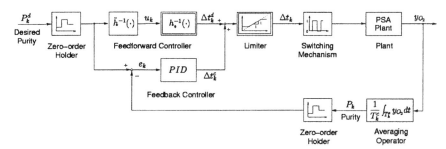

Figure 3: Block diagram of the feedforward controller and the superimposed feedback controller for the purity P_k of the PSA plant.

ler is then set up as a series connection of the approximated inverse transient I/O model and the static nonlinearity of the inverse stationary I/O model, i.e. $\Delta t_k^d = h_*^{-1}(u_k)$ with $u_k = \tilde{h}^{-1}(P_{k+1}^d, P_k^d, P_{k-1}^d, \ldots)$, see Figure 3.

3.2 Closed-loop control

The plant might be operated in open-loop, but due to model errors and disturbances, the purity P_k is influenced. Therefore, a superimposed feedback controller is necessary for the stabilization of periodic operating points as well as the desired trajectories during set-point changes. Transferring the process control scheme to the PSA plant leads to the control structure shown in Figure 3. Close to an operating point $(\mathbf{x}^*, \Delta t^*)$, the transient behavior of the plant can be described by the linearized equations

$$\Delta \mathbf{x}_{k+1} = \frac{\partial \mathbf{g}}{\partial \mathbf{x}}\bigg|_* \Delta \mathbf{x}_k + \frac{\partial \mathbf{g}}{\partial \Delta t_k}\bigg|_* \Delta \tilde{t}_k, \qquad \Delta P_{k+1} = \frac{\partial h}{\partial \mathbf{x}}\bigg|_* \Delta \mathbf{x}_k + \frac{\partial h}{\partial \Delta t_k}\bigg|_* \Delta \tilde{t}_k \tag{4}$$

with $\mathbf{x}_k = \mathbf{x}^* + \Delta \mathbf{x}_k$, $\Delta t_k = \Delta t^* + \Delta \tilde{t}_k$ and $P_k = P^* + \Delta P_k$. For the design of a feedback controller, the linearized discrete state space model (4) has to be identified according to the transient behavior close to a periodic operating point. For instance, simulated step responses at different operating points are given in Figure 2b. In a first step, a linear discrete controller can then be designed.

4. Simulation results

The discrete controller is set up as a PID-controller and designed using the simplified design model. Then, the controller is applied to the detailed model (1)-(3) of the PSA plant and the parameters of the PID-controller were readjusted by simulation studies. The simulation of the trajectory control of two consecutive set-point changes of the purity P_k is shown in Figure 4. The desired trajectories P_k^d for the set-point changes were set up using a 4th-order polynomial and the simulation results are looking very promising.

5. Conclusions and Outlook

A process control scheme for a 2-bed PSA plant has been presented. For the controller design, the stationary and transient I/O behavior of the plant has been examined. Even though the feedback controller has been designed in a first step with a rather coarse

456

linear discrete state space model, the performance of the control concept shows sufficiently good results. Future research will be focused on the derivation of more subtly reduced design models which consider the structural changes of the process and which provide a precise representation of the internal plant dynamics. Further issues are the extension of the process control concept to other cyclic mutli-step processes, e.g. a 3-bed PSA plant, as well as its experimental validation for a laboratory scale plant.

Acknowledgment

The authors gratefully acknowledge the cooperation with Prof. G. Eigenberger, W. Lengerer, and M. Stegmaier of the *Institut für Chemische Verfahrenstechnik* at the University of Stuttgart as well as the work of J. Gruber in course of his student thesis. The research is supported within SFB 412 by *Deutsche Forschungsgesellschaft (DFG)*.

References

Bitzer, M. and M. Zeitz (2002). Design of a nonlinear distributed parameter observer for a pressure swing adsorption plant. *Journal of Process Control.* In press.

Bitzer, M., F. J. Christophersen and M. Zeitz (2001). Periodic control of a pressure swing adsorption plant. In: *Preprints IFAC Workshop on Periodic Control Systems.* Como/Italy, August 27-28. pp. 85-90.

Christofides, P. D. (2001). Control of nonlinear distributed process systems: recent developments and challenges. *AIChE* **47**(3), 514-518.

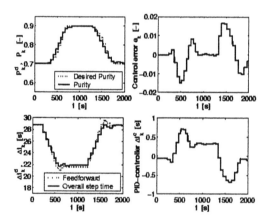

Figure 4: Simulation of the trajectory control for two consecutive set-point changes.

Köhler, R., K. D. Mohl, H. Schramm, M. Zeitz, A. Kienle, M. Mangold, E. Stein and E. D. Gilles (2001). Method of lines within the simulation environment DIVA for chemical processes. In: *Adaptive Method of Lines* (A. Vande Wouwer, P. Saucez and W. Schiesser, Eds.). pp. 367-402. CRC Press. Boca Raton/USA.

Nilchan, S. and C. C. Pantelides (1998). On the optimisation of periodic adsorption processes. *Adsorption* **4**, 113-147.

Ruthven, D. M., S. Farooq and K. S. Knaebel (1994). *Pressure Swing Adsorption.* VCH Publishers. New York, Weinheim, Cambridge.

St-Pierre, J. and D. P. Wilkinson (2001). Fuel cells: a new, efficient and cleaner power source. *AIChE* **47**(7), 1482-1486.

Unger, J. (1999). *Druckwechseladsorption zur Gastrennung - Modellierung, Simulation und Prozeßdynamik.* Fortschritt-Berichte Nr. 3/602. VDI-Verlag, Düsseldorf.

European Symposium on Computer Aided Process Engineering – 12
J. Grievink and J. van Schijndel (Editors)

Multiobjective Process Controllability Analysis[1]

Yi Cao[2][§], Zhijia Yang[*]

[§] School of Engineering, Cranfield University, Bedford MK43 0AL,UK

[*] Department of Energy Engineering, Zhejiang University, Hangzhou 310027,China

Abstract

A new approach for process controllability analysis by using multiobjective optimisation techniques is proposed. Within the approach, a set of performance specifications, such as minimum control error and input effort with closed-loop pole placement are represented as a set of linear matrix inequalities (LMI). The solution to the LMI conditions can be identified as feasible or infeasible. If the solution is feasible there is at least one controller that can make the closed-loop system satisfy all performance specifications simultaneously. Therefore, for the process plant, these performance specifications are achievable. Otherwise, they are unachievable. There is a Pareto-optimal set or a trade-off curve in the performance space to separate these two areas. The paper shows that such trade-off curves can be used for process controllability analysis, and therefore, can be applied to control structure selection problems.

1. Introduction

The issue of input-output controllability analysis has received increasing attention for a few decades. Input-output controllability is the ability of a plant to achieve acceptable control performance. Various tools and techniques have been developed and are available in the literature to quantify the inherent input-output controllability of a plant (Skogestad and Postlethwaite 1996). However most of these tools are mainly simple controllability indices. Each individual index only addresses one aspect of process controllability, which causes performance limitation, such as input constraints, unstable poles and zeros. It is still an open area to predict performance limitation jointly imposed by these factors. Nonlinear optimization has been used to predict performance limitation for a plant with input constraints and unstable zeros (Cao and Biss 1996a). The minimum input usage required to stable an unstable plant has been derived by Glover (1986) and recently by Havre and Skogestad (2001). Åström (2000) presented some results for performance limitations in SISO systems. For MIMO systems, the impact of unstable poles and zeros on closed-loop sensitivity and complementary sensitivity functions has been extensively studied by Chen (2000). However, general links between input constraints and unstable zeros have not yet been revealed by these studies.

In this paper, performance limitations are presented in trade-off curves, which are used in process controllability analysis. Boyd and Barratt (1991) have revealed that most

---

[1] Partially supported by the EPSRC grant GR/R57324.

[2] To whom corresponding should be addressed. Email: y.cao@cranfield.ac.uk

control system design specifications are affine and convex functions of the controller to be designed. The trade-off curves of these functions have been calculated as Pareto-optimal performance set. As summarised by Scherer *et al.* (1997), many control performance criteria can be represented as a set of linear matrix inequalities (LMI) (Boyd *et al.* 1994). These performance criteria include H_∞ and H_2 norms of certain closed-loop transfer functions and pole placement regions. Using LMI, various, even inconsistent performance requirements can be identified as feasible and infeasible in the performance space. These two areas are separated by the Pareto-optimal performance set. The Pareto-optimal performance set gives a clear picture about what is the achievable performance of a process control system and what performance trade-offs are necessary for control design.

The paper is organized as follows. Section 2 introduces the multiobjective control design specifications used in this paper. Two types of Pareto-optimal performance curves are designated for multiobjective controllability analysis. These curves are solvable by the off-shelf software with a small modification. Section 3 provides a complete case study to show the usage of these Pareto diagrams in control structure selection. The paper is concluded in Section 4.

2. Multiobjective Process Controllability and LMI

Consider a generalised control configuration shown in Figure 1. The block P represents a general plant, whilst block K is a controller. Signals, which link both blocks, are measured output, y and manipulated input, u. The signal w represents exogenous inputs, such as disturbances, references, noises and inputs from uncertainties and the exogenous output, z, is the control objective. Many control performance specifications can be expressed as the H_∞ norm of certain closed-loop transfer functions. Such functions could be the sensitivity function, S, complementary sensitivity function, T, the input sensitivity function, KS, or more generally, a closed-loop transfer function from w to z, i.e. $\|T_{zw}\|_\infty$. In H_∞ control design, a multiobjective performance specification is usually treated as a mixed sensitivity design problem, such as mixed S-KS, or mixed S-T objectives with suitable weighting functions. However, for the controllability analysis purpose, the multiple H_∞ norms are better to be considered simultaneously as a multiobjective optimisation problem. The multiobjective H_∞ optimisation can be solved by recently developed LMI techniques.

The H_∞ norm is a system norm where input and output signals both are in L_2 space. However, to consider the effect of input constraints, L_∞ space is more appropriate than L_2 in describing the control-input signal, u. If $w \in L_2$ and $u \in L_\infty$, then the induced norm

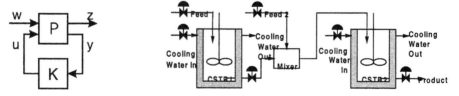

Figure 1 Generalised Control System   *Figure 2 Two-CSTR Plant*

from w to u is the generalised-H_2 norm (Rotea 1993, Scherer *et al.* 1997), denoted as $\|T_{uw}\|_g$. The LMI conditions for $\|T_{uw}\|_g < \beta$ (constant) are given by Scherer *et al.* (1997).

Transient response is always involved in performance requirements and can often be achieved by forcing the closed-loop poles into a suitable region. For example, the condition of all closed-loop poles on the left-half plane of $Re(p) \leq \alpha$ (a negative constant) is usually called α-stability condition. A more general pole region, equivalent to a set of LMI (Chilali and Gahinet 1996), is denoted as $S(\alpha, r, \theta)$:

$$S(\alpha, r, \theta) = \{p \mid Re(p) \leq \alpha, |p| \leq r, \tan(\theta)Re(p) \leq - |Im(p)| \} \tag{1}$$

The multiobjective optimisation problem to be considered in the paper is as follows:

$$\min \|T_{ew}\|_\infty, \quad \text{subject to: } \|T_{uw}\|_g < \beta \text{ and } p \in S(\alpha, r, \theta) \tag{2}$$

where e is the unit-feedback control error. This problem is ready to be solved by the off-shelf software (Gahinet *et al.* 1995). The MATLAB function, hinfmix in the LMI toolbox, originally designed for mixed H_2/H_∞ problem has been slightly modified for the generalised-H_2/H_∞ problem in (2). Two multiobjective Pareto diagrams, the minimum $\|T_{uw}\|_g$ against α-stability and the minimum $\|T_{uw}\|_g$ against the minimum $\|T_{ew}\|_\infty$ are to be produced by repeatedly calling the modified MATLAB function.

3. Case Study

The approach for multiobjective controllability analysis is applied to a two-CSTR process. The process is schematically shown in Figure 2. A full description of the system and an eight-state model can be found in Cao and Biss (1996b). To focus on the control structure selection problem discussed here, constant volume assumption is applied to the process, which leads to a six-state model to be used in the paper. The control problem is to maintain both tank temperatures at desired values in the presence of cooling-water temperature fluctuations within ±10 [K], i.e. $d=[T_{cw1} \ T_{cw2}]$. Three possible control configurations to be considered are:

S1: $u = [Q_{I1} \ Q_{I2}]$, two feed flowrates and $y = [T_{o1} \ T_{o2}]$, two tank outlet temperatures.

S2: $u = [Q_{cw1} \ Q_{cw2}]$, two cooling-water flowrates and y is the same as S1.

S3: u is the same as S2, but y has two extra secondary measurements, cooling-water outlet temperatures, i.e. $y = [T_{o1} \ T_{o2} \ T_{cwo1} \ T_{cwo2}]$.

The input constraints are $0.05 \leq Q_{I1}+Q_{I2} \leq 0.8$ [m^3/s] and $0.05 \leq Q_{cw1}, \ Q_{cw2} \leq 0.8$ [m^3/s]. To cope with the constraints, Q_{I1} and Q_{I2} are converted to total flowrate, $Q = Q_{I1}+Q_{I2}$ and flowrate ratio, $R = Q_{I2}/Q$. The new constraints are: $0.05 \leq Q \leq 0.8$ [m^3/s] and $0 \leq R \leq 1$.

For variable scaling, the manipulated variables are divided by the minimum distance from their steady state value to their boundary. The disturbance variables are divided by 10 [K], whilst the output variables is divided by 1 [K]. The final linearised model for configuration Si, i=1,2,3 is represented as:

$$dx/dt = Ax + B_iu + Ed, \quad y = C_ix, \quad B_2 = B_3, \quad C_1 = C_2 = \text{the first two rows of } C_3 \tag{3}$$

$$A = \begin{bmatrix} -17.9751 & -295.8655 & 0 & 0 & 0 & 0 \\ 0.0207 & 0.1889 & 0.0704 & 0 & 0 & 0 \\ 0 & 0.3879 & -0.8000 & 0 & 0 & 0 \\ 0.0977 & 0 & 0 & -18.0088 & -295.8655 & 0 \\ 0 & 0.0617 & 0 & 0.0131 & 0.0433 & 0.0589 \\ 0 & 0 & 0 & 0 & 0.3787 & -0.6220 \end{bmatrix}$$

$$[B_1 \ B_2 \ E] = \begin{bmatrix} 17.8996 & -13.7811 & 0 & 0 & 0 & 0 \\ -0.0131 & 0.0101 & 0 & 0 & 0 & 0 \\ 0 & 0 & -0.0294 & 0 & 0.0137 & 0 \\ 17.8636 & 17.8636 & 0 & 0 & 0 & 0 \\ -0.0082 & -0.0082 & 0 & 0 & 0 & 0 \\ 0 & 0 & 0 & -0.0235 & 0 & 0.0081 \end{bmatrix}$$

$$C_3 = \begin{bmatrix} 0 & 362.9950 & 0 & 0 & 0 & 0 \\ 0 & 0 & 0 & 0 & 362.9950 & 0 \\ 0 & 0 & 327.5600 & 0 & 0 & 0 \\ 0 & 0 & 0 & 0 & 0 & 335.4470 \end{bmatrix}$$

The effect of input constraints is normally assessed by the minimum singular value, which are 15.06 for S1 and 5.13 for S2 at steady state, i.e. S1 is better than S2 in terms of input constraints. S3 has two secondary measurements, thus this index cannot be directly applied. On other hand, S1 is the only configuration, which has two unstable zeros (10.33 and 10.31). Physically, this is because the effect of feed flowrate on tank temperature has two opposite directions – positive via reaction and negative due to lower feed temperature. Therefore, S2 and S3 are better than S1 in terms of unstable zeros. However, for overall performance, it is difficult to judge which configuration is the best only using these regular controllability measures. Therefore, the multiobjective controllability analysis approach described above is applied to this example.

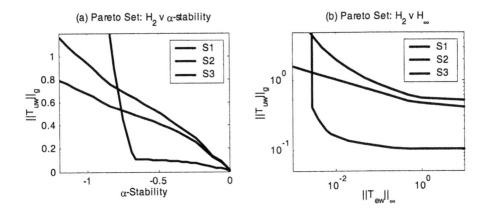

Figure 3 Pareto Diagrams with Pole Region r=20, θ=77.6° and α=-0.5 in (b)

The multiobjective problem is constructed by assigning $z = [e \ u]$ with $e \in L_2$ the control error and $u \in L_\infty$. The closed-loop poles region is defined as $r=20$ and $\theta=77.6°$ with α fixed to -0.5 or varying. To force zero error at steady-state, an integrator is inserted into each error channel and will be merged into the controller designed. Based on these conditions, the multiobjective Pareto diagrams are produced and shown in Figure 3. The results in Figure 3 show that the achievable performance of S1 ($\|T_{ew}\|_\infty$) is limited by its unstable zeros. However, S1 is still the best configuration when $-\alpha<0.6$ and $\|u\|_\infty< 1$. It is also shown that the nonsquare configuration, S3 does improve the controllability by introducing extra measurements into configuration S2. It can achieve almost the same performance as S1 within the input constraints. If the input constraints were permitted to increase slightly, S3 would even be better than S1. This observation is verified by the simulation results (see Figures 4). The controllers used for simulation are designed to achieve $\|T_{ew}\|_\infty$ at the values corresponding to $\|T_{uw}\|_g=1$ in Figure 3(b). The $\|T_{uw}\|_g$ values predicted in Figure 3(b) match the maximum input deviations observed in the simulation (see Figure 5).

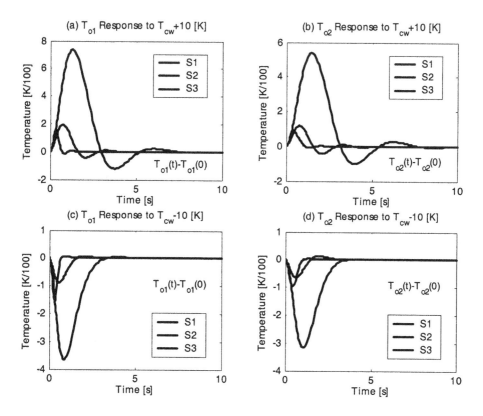

Figure 4 Output Response

4. Conclusion

The proposed approach for multiobjective controllability analysis is able to identify performance limitation imposed by multi-factors, such as unstable zeros and input constraints. It is also suitable for more sophisticated configurations, such as nonsquare, cascade and two degrees-of-freedom control. The produced Pareto diagrams can be directly used for control design trade-off. The generalised-H_2 norm is better than H_∞ norm to describe input with input constraints. The enforced closed-loop pole region makes the closed-loop time response more predictable.

462

References

Astrom, K.J., 2000, European Journal on Control, Vol.6, pp. 2—20.

Boyd, S. and Barratt, C., 1991, Linear Controller Design: Limits of Performance, Prentice-Hall, Englewood Cliffs.

Boyd, S., Ghaoui, L.E., Feron, E. and Balakrishnan, V., 1994, Linear Matrix Inequalities in System and Control Theory, SIAM, Philadelphia.

Cao, Y. and Biss, D., 1996a, Computers and Chem. Engng. Vol. 20, pp. 337—346.

Cao, Y. and Biss, D., 1996b, Journal of Process Control, Vol.6, No.1, pp. 37—48.

Chen, J, 2000, IEEE Transactions on Automatic Control, Vol.45, pp. 1098—1115.

Gahinet, P., Nemirovski, A., Laub, A. and Chilali, M., (1995), The LMI Control Toolbox, The MathWorks, Inc.

Chilali, M. and Gahinet, P., 1996, IEEE Transactions on Automatic Control, Vol. 41, pp. 358—367.

Glover, K., 1986, Int. Journal of Control, Vol. 43, pp. 741—766.

Havre, K. and Skogestad, S., 2001, Int. Journal of Control, Vol. 74, pp. 1131—1139

Rotea, M. A., 1993, Automatica, Vol. 29, pp. 373—385.

Scherer, C., Gahinet, P. and Chilali, M., 1997, IEEE Transactions on Automatic Control, Vol.42, pp. 896—911.

Skogestad, S. and Postlethwaite, I., 1996, Multivariable Feedback Control: Analysis and Design, John Wiley and Sons, Chichester.

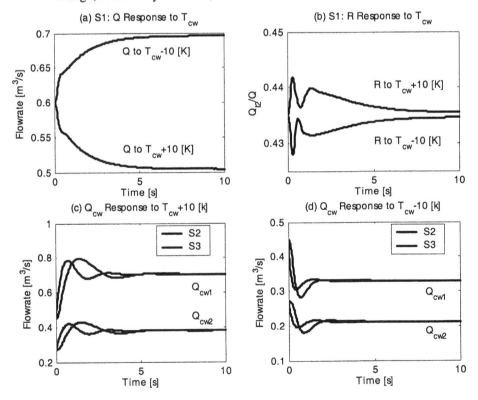

Figure 5 Input Response

European Symposium on Computer Aided Process Engineering – 12
J. Grievink and J. van Schijndel (Editors)

Control of the Rotary Calciner for Soda Ash Production

V.M. Cristea, P.S. Agachi

" Babeş-Bolyai " University of Cluj-Napoca, Faculty of Chemistry and Chemical
Engineering, 11 Arany Janos, 3400 Cluj-Napoca, Romania,
e-mail: mcristea@chem.ubbcluj.ro

Abstract

The paper presents the dynamic simulation results obtained by modelling and control of
the rotary calciner for sodium carbonate production by the endothermic decomposition
of sodium sesquicarbonate. The calciner model with distributed parameters consists of
three zones, each described by sets of PDE's: the combustion zone, the reaction zone
and the disengaging zone. Dynamic responses to typical disturbances are presented and
compared with industrial and literature data. As main target variable, the sodium
sesquicarbonate content control at the discharge end of the calciner is investigated using
different control approaches. The classical PID control and the Nonlinear Model
Predictive Control (NMPC) algorithms are compared in mixed feedforward and
feedback schemes. Incentives are revealed for particular control structures using
feedforward NMPC, becoming also a feasible approach for the control of similar
processes taking place in rotary calciners.

1. Introduction

Crude soda ash is one of the basic products in chemical industry that is used as raw
material for a large variety of end products. Either obtained from natural trona ore or
from decomposition of sodium bicarbonate obtained by carbonation of ammoniacal lye,
the rotary calciner is one of the main processing units. The dynamic simulator of the
rotary calciner proves to be a valuable tool for studying different construction design
approaches, operating strategies and control system design configurations. The present
work investigates different control approaches pointing out the benefits of particular
control schemes able to be directly implemented on the industrial unit.

2. Research Results

2.1 Mathematical model description

The rotary calciner consists of a cylindrical shell with its long axis having a small
inclination from the horizontal direction. The trona ore is introduced at the highest end
of the shell and moves towards the lower discharge end due to the rotation and the
inclination of the cylinder. Inside the shell, in the feed end zone, the combustion of the
natural gas flux is taking place, supplying the heat necessary for the endothermic
reaction. Three zones can be distinguished along the shell in the direction of the co-
current movement of the gases and solid material, figure 1.

The first (bare) zone is the combustion zone, where the combustion flame is developed
and raw material is heated up. The second (lifter) zone is the reaction zone, where solid

464

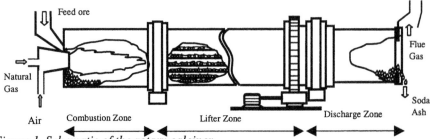

Figure 1. Schematic of the rotary calciner.

material showers due to lifters that elevate the solid and spill it over the cross section, increasing the mass and energy transfer. The last (bare) zone is the disengaging zone where solid and gaseous products separate and are evacuated. The main assumptions for the mathematical model development are: all parameters in a radial cross section are constant, both gas and solids velocity is considered constant, heat transfer by conduction and radiation are negligible in the axial direction and coefficients of convection, emissivities, latent heat and heat of reaction are temperature independent.

Elimination of the moisture entering the calciner with the trona ore is performed within the falling rate stage of the drying process, when the rate of drying is proportional to the free moisture content of the solids (moisture content less than the critical moisture content of 10%). The mass balance of the moisture flow is described by the equation (see Notation Section at the end of the paper for the list of variables):

$$\frac{1}{V_m}\frac{\partial}{\partial t}(Q_h) = -\frac{\partial}{\partial l}(Q_h) - \frac{h_t A(T_g - T_m)}{L_v(0.1Q_m)}Q_h.$$ (1)

The mass balance equation for the total amount of sodium sesquicarbonate contained in the solid is:

$$\frac{1}{V_m}\frac{\partial}{\partial t}(S_q Q_m) = -\frac{\partial}{\partial l}(S_q Q_m) - Q_m \frac{K_r}{V_m} S_q\, e^{-E/(RT_m)}.$$ (2)

The mass balance equations for the solids and gaseous products are:

$$\frac{1}{V_m}\frac{\partial}{\partial t}(Q_m) = -\frac{\partial}{\partial l}(Q_m) - \frac{Q_m}{100}\frac{K_r}{V_m}S_q\left(\frac{67}{226}\right)e^{-E/(RT_m)} - \frac{h_t A(T_g - T_m)}{L_v(0.1Q_m)}Q_h$$ (3)

$$\frac{1}{V_g}\frac{\partial}{\partial t}(Q_g) = -\frac{\partial}{\partial l}(Q_g) + \frac{Q_m}{100}\frac{K_r}{V_m}S_q\left(\frac{67}{226}\right)e^{-E/(RT_m)} + \frac{h_t A(T_g - T_m)}{L_v(0.1Q_m)}Q_h - \frac{\partial}{\partial l}(Q_c)$$ (4)

where the last term in equation (4) is only present in the first zone of the calciner.

The heat balance equations for the gas, solid, flame and calciner wall are presented in the following equations:

$$\frac{Q_g C_g}{V_g}\frac{\partial}{\partial t}(T_g) = -Q_g C_g \frac{\partial}{\partial l}(T_g) + H_{fg} + H_{vg} - H_{gm} - H_{gw}$$ (5)

$$\frac{Q_m C_m}{V_m}\frac{\partial}{\partial t}(T_m) = -Q_m C_m \frac{\partial}{\partial l}(T_m) + H_{gm} + H_{wm} + H_{fm} - H_{ev} - H_{re} \tag{6}$$

$$\frac{Q_c C_c}{V_g}\frac{\partial}{\partial t}(T_f) = -Q_c C_c \frac{\partial}{\partial l}(T_f) - H_{fw} - H_{fm} - H_{fg} \tag{7}$$

$$M_w C_w \frac{\partial}{\partial t}(T_w) = H_{gw} - H_{wm} - H_{w0}. \tag{8}$$

The heat fluxes considered in the above equations are: the heat flux for drying the solids H_{ev}, the heat flux due to the reaction H_{re}, the heat flux from vapors to the gas H_{vg}, the heat flux to the surroundings H_{wo}, the heat flux from flame to solid material H_{fm}, the heat fux from flame to wall H_{fw} and the heat flux from flame to calciner gas H_{fg}. The heat fluxes gas to solid material H_{gm}, gas to wall H_{gw} and wall to solid material H_{wm} are specified according to the different zones of the calciner.

The mathematical model is based on the model described by Ciftci and Kim (1999) and presents a good fit with industrial data. The Finite Element Method was used for solving the set of partial differential equations, based on 150 finite elements (FEMLAB, 2000).

2.2 Simulation results

First, the simulation reveals the steady state behaviour of some important process variables, for all of the three distinct zones of the calciner, as presented in figure 2.

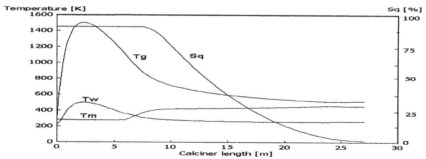

Figure 2. Steady state profiles of the sodium sesquicarbonate content S_q, solid T_m, gas T_g and wall T_w temperatures.

The gas temperature T_g has a maximum placed in the first zone where the flame is developed and for which, a linear decay of the flowrate is considered. The wall temperature T_w also presents a maximum in this zone due to the substantial radiation flux of the flame. The solid temperature T_m rises promptly at the beginning of the second zone where mass and energy transfer is intensified. Further, the solid temperature remains almost constant due to equilibrium between the heat transfer from the gas and the endothermic reaction sink. The sesquicarbonate content of the solid S_q decays only in the last two zones, reaching a value of about *1%* at the discharge end, (further considered as the setpoint value for sesquicarbonate control).

Second, the dynamic behaviour of the calciner was also investigated in the presence of some typical disturbances. Some of the representative results are presented in figure 3 and figure 4. Figure 3 presents the response of the solid flowrate Q_m to a step upset of $\Delta Q_m = +7000\ kg/h$ at the feed end. Figure 4 shows the solid temperature T_m response to a step upset of $\Delta T_m = +10\ ^0K$ incoming also from the feed end. The presented results show the profiles of the process variables along the calciner length at six time steps $(t_1=0\ h,\ t_2=0.05\ h,\ t_3=0.10\ h,\ t_4=0.15\ h,\ t_5=0.23\ h,\ t_6=2\ h,)$ until the steady state is restored.

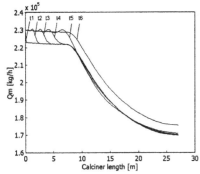 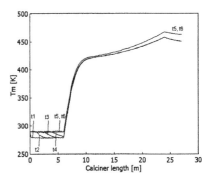

Fig. 3. Solid flowrate Qm dynamic response. *Fig. 4. Solid temp. T$_m$ dynamic response.*

Inverse response was noticed for some of the process variables related (mainly) to the solid. The reason for this behaviour may be the fact that disturbance effects travel along the calciner with different velocity: a rapid one due to the high velocity of the gas flowrate and a sluggish one due to the solid inertia (mass or heat), (Cristea et al. 2001). The main simulated process variables are varying in the range of values reported by the industrial unit. The dynamic results confirm, by the time lags, the industrial process retention time exhibited by the solids in the different zones of the calciner.

2.3 Control of the decomposition process

The main impediment in performing efficient control of the decomposition process is the lack of measured data obtained from different zones of the calciner, primarily needed for feedback control. This feature determines large time lags and dead time for variables measured at the discharge end, rising therefore an important challenge for the control system. As the concentration of the soda ash at the discharge end of the calciner or equivalent, the sodium sesquicarbonate content of the solid S_q, is the target variable, the control of the latter was investigated. Both PID control and Nonlinear Model Predictive Control (NMPC) approaches have been implemented and tested in the presence of typical disturbances, such as: step increase of the solid inlet flowrate $\Delta Q_m = +500 kg/h$, step decrease of the solid inlet temperature $\Delta T_m = -4\ ^0K$ and step increase of inlet sodium sesquicarbonate content of the solid $\Delta S_q = +3\%$; all disturbances applied at the time $t=0.1\ h$. Natural gas flowrate has been used as manipulated variable.

First, direct control of the sesquicarbonate content S_q at the discharge end has been investigated. Three control approaches have been considered: feedback PID control, feedback NMPC control and combined feedback-feedforward NMPC control. The simulation results in case of S_q control have presented incentives for the combined feedback-feedforward NMPC control. Although a sampling time of $0.05\ h$ has been considered in order to take into account for the delay of the required on line S_q analyser, this control approach is not feasible both from practical and economical point of view.

Second, different process variables have been inspected as possible candidates for indirect control of the sesquicarbonate content. Solid (at the discharge end) outlet, gases outlet and wall temperatures, as controlled (and measured) variables, have been considered and tested. The solid outlet temperature T_m control proved to be the most favourable control scheme, with direct influence on the target S_q variable. The solid temperature control in the presence of inlet (at the feed end) ΔQ_m and ΔT_m disturbances, for both solid temperature controlled variable and sesquicarbonate target variable, are comparatively presented in figure 5 and figure 6, for all of the three investigated control approaches.

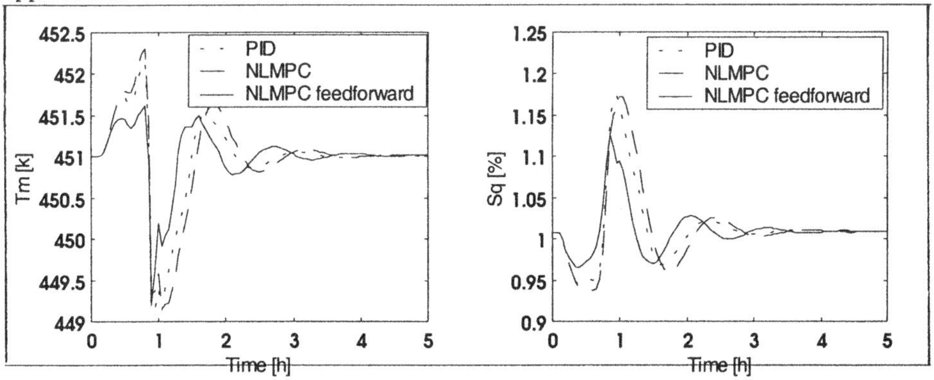

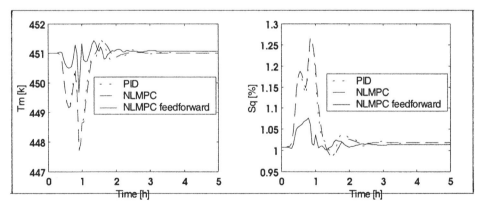

Figure 5. Solid temperature control and sesquicarbonate content dynamic response in the presence of the step decrease of the solid inlet temperature $\Delta T_m=-4\ ^0K$ disturbance.

Figure 6. Solid temperature control and sesquicarbonate content dynamic response in the presence of the step increase of the solid inlet flowrate $\Delta Q_m=+500kg/h$ disturbance.

The control results show superior performance for the NMPC feedforward-feedback control structure compared to the other control approaches. This conclusion is attested by the shorter settling time and the reduced overshoot (low integral square error) of the sesquicarbonate content. Similar behaviour has been noticed for other disturbances.

Although the gas temperature is affected by shorter response time, the control of this variable led to unacceptable offset in the sesquicarbonate control. The cascade control of the solid outlet temperature, as primary controlled variable, associated with the gas outlet temperature, as secondary controlled variable, failed to accomplish improved control performance.

3. Conclusions

The dynamic simulator with distributed parameters of the rotary calciner for soda ash production reflects the complex behaviour of the unit. Large time lags coupled with dead time and inverse response for some of the main process variables put severe restrictions on desirable control performance. Several process variables have been tested as possible controlled variables. The different control approaches investigated for the control of the sesquicarbonate content target variable, ranging from simple PID to NMPC control with mixed feedforward-feedback, proved the incentives for the solid temperature NMPC control in a combined feedforward-feedback control scheme. Further improvement of the control performance are expected to be obtained by the use of state observers based on spatial and temporal profiles of the main process variables.

Notation

A =specific area, [m^2/m]

C_c, C_g, C_m, C_v =specific heat for combustion products, gas, solid and vapors, [kcal/(kg K)]

e_f, e_g, e_m, e_w =emissivities

F_{fm}, F_{fw} =form factor for radiative heat transfer

h_f, h_g, h_0, h_t, h_w =heat-transfer coefficients, [kcal/(h m^2 K)]

K_r =frequency factor [h^{-1}]

L =length of the calciner, [m]

$L_1, L_2, L_3, L_4, L_1{'}, L_3{'}$ = arc lengths for bare and lifter zones, [m]

L_v =latent heat of water vaporization,[kcal/kg]

M_w =mass per unit length of the calciner wall, [kg/m]

$Q_c, Q_g, Q_m, Q_h, Q_{CH4}$ =flowrate of combustion products, calciner gas, solid, moisture, natural gas, [kg/h]

S_q =sodium sesquicarbonate content of the solid, [%]

T_f, T_g, T_m, T_0, T_w =temperature of the flame, gas, solid, surroundings, calciner wall, [K]

U = combined heat transfer coefficient [kcal/(h m^2 K)]

V_g, V_m =calciner gas and solid velocity, [m/h]

ΔH =heat of reaction, [kcal/kg]

References

Ciftci, S. and N.K. Kim, 1999, Control Schemes for an Industrial Rotary Calciner with a Heat Shield around the Combustion Zone, Ind. Eng. Chem. Res., 38, 1007.

Cristea, V.M., S.P. Agachi and S. Zafiu, 2001, Simulation af an Industrial Rotary Calciner for Soda Ash Production, IcheaP-5, Florence, 245.

* * * FEMLAB, Users Guide, COMSOL AB, 2000.

European Symposium on Computer Aided Process Engineering – 12
J. Grievink and J. van Schijndel (Editors)
469

A simulation tool for the wood drying process

Helge Didriksen and Jan Sandvig Nielsen

dk-TEKNIK ENERGY&ENVIRONMENT, Gladsaxe Møllevej 15, 2860 Søborg,
Denmark

Abstract

A new software tool for the simulation of drying of wood in drying kilns can be used to support the kiln operators in running the drying process efficiently. The simulation tool predicts the moisture in the wood boards during the drying batch. Furthermore, internal tensions in the boards are predicted indicating the risk of board cracking, a severe quality reduction of the wood. The simulation tool can help optimise the drying process with regards to product quality, process capacity and energy costs.

1. Introduction

The art of drying wood boards in a drying kiln is, in short, to dry the wood as quick as possible and as gentle as possible. A short drying time gives a high process capacity and low energy costs. A gentle drying gives a high product quality. The two overall objectives are, to some extent, contradictory. Checking of the wood, and other timber quality degradings, will occur if the wood is dried too quickly. On the other hand, an exaggerated caution will lead to a decrease in process capacity and an unnecessary high-energy consumption. The new simulation tool can simulate the drying batch and the tool can be used by operators and technical management to optimise the drying schemes with regards to product quality, process capacity and energy consumption.

2. Mathematical model

The core of the simulation tool is a detailed dynamic model of the drying kiln and of the wood boards. Many researchers have been dealing with modelling of the wood drying process in order to explain and predict how the wood dries and also the appearance of quality degradations such as cracking. Salin (1990), Stanish et. al. (1986) and Turner (1986) are some of the most important sources for the developing of the model in this work. The mathematical models consist of two distinct parts; a thermodynamical model and a tension/stress model.

Detailed model equations of the model are described in Didriksen et. al. (2001)

2.1 Model structure

The wood board is divided into a number of layers as shown in figure 1. There is mass- and energy exchange between the layers as shown in the figure. Deformations and tensions of each layer arises when the wood is dried down to a moisture content below the so called Fiber Saturation Point (FSP).

470

In each layer all process variables are assumed to be constant. In this way, the model is able to describe the moisture content profile and the tension profile of the board in the board depth direction. In other words, a one dimensional model structure is chosen. This is of course a simplification since the wood dries at a different speed at the corners and edges of the boards than in the middle, which also influences the tensions and deformations. However, the simplification is justified by the board's dimensions, they are long and flat, and the fact that the boards are laying closely side by side. A two dimensional of the wood board drying simulation is described e.g. in Turner (1986) and Perré and Turner (2001).

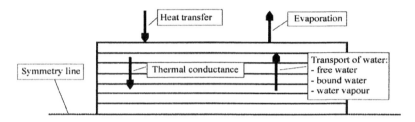

Figure 1. Model structure, wood board

2.2 Thermodynamic model

The thermodynamical model is based upon the conservation of mass and energy and calculates how the water is "transported" through the boards and is evaporated at the board surface throughout the drying batch. The water evaporation and heat transfer from drying air to the boards depends upon the climate in the kiln and upon the drying air velocity. The calculations account for the drying properties of the type of wood in question. Wood is a complicated cellular porous material consisting of a number of tubular like cells as shown schematically in figure 2.

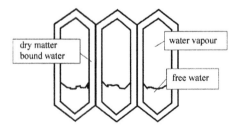

Figure 2. Wood structure, schematically

The water in the wood may exist in three different states; free water in the cell cavities, water vapour in the voids and bound water hygroscopically held in the cell walls. The transport mechanisms enabling moisture movement are capillarity liquid flow, η_f, water vapour diffusion, η_{vd}, and bound liquid diffusion, η_{bd}, respectively. Heat is transferred between the layers by thermal conduction, Q_{cond}.

Conservation of mass (water), m_w, in layer no. i:

$$\frac{1}{A}\frac{dm_{w,i}}{dt} = \left(n_{f,i+1} + n_{vd,i+1} + n_{bd,i+1} - n_{f,i} - n_{vd,i} - n_{bd,i}\right) \qquad (1)$$

Conservation of energy, U, in layer no. i:

$$\frac{1}{A}\frac{dU_i}{dt} = \left(n_{f,i+1} + n_{bd,i+1}\right)T_{i+1}c_{p,v} + \left(n_{vd,i+1}\right)h_{vapour,i+1}$$
$$\left(n_{f,i} + n_{bd,i}\right)T_i, c_p - \left(n_{vd,i}\right)h_{vapour,i} + Q_{cond,i+1} - Q_{cond,i} \qquad (2)$$

2.3 Tension model
Throughout the drying batch, the different layers of the board will have different moisture contents. When the moisture content reaches a certain value (the so-called Fiber Saturation Piont), the wood starts to shrink. Because the boards shrink uneven in the thickness direction, the board will be exposed to uneven shrinking and tensions in the board will arise. If the tensions exceed a certain limit, the board will check. Checking of the boards represent a serious degradation of the product quality.

The tension model describes the deformations and induced stresses in the drying wood. The tension model is put "on top" of the thermodynamical model. Four types of deformation are included in the tension model; shrinking ε_s , elastic strain ε_e , mechano-sorptive creep ε_{ms} and viscoelastic creep ε_{vs}.

$$\varepsilon = \varepsilon_s + \varepsilon_e + \varepsilon_{ms} + \varepsilon_{vs} \qquad (3)$$

When the values of the different deformations for each layer are calculated, an average deformation for the entire board, $\bar{\varepsilon}$, is calculated.

$$\bar{\varepsilon} = \frac{\sum_{j=1}^{n} E_j\left(\varepsilon_s + \varepsilon_{ms} + \varepsilon_{vs}\right)\Delta x_j}{\sum_{j=1}^{n} E_j \Delta x_j} \qquad (4)$$

Next, the tensions in the different layers of the boards, σ_j, can be calculated

$$\sigma_j = E_j\left(\bar{\varepsilon} - \varepsilon_{k,j} - \varepsilon_{ms} - \varepsilon_v\right) \qquad (5)$$

E is here the (temperature and moisture dependent) elasticity module and Δx is the thickness of each layer.

The estimated tensions are indications of the risk of board cracking.

472

3. Simulation case

The simulation tool is demonstrated by showing the drying of ash. Ash is a wood type, which is difficult to dry without quality degradations, and therefore, a very long drying time is required. The simulation case is based upon process data from a kiln at a Danish floor producer.

The current dimensional data and wood species data is typed in the simulation tool. Furthermore, a typical drying scheme; consisting of dry and wet temperature of the drying gas, is typed in. The drying scheme is shown in figure 3.

Notice the very sharp reduction in gas moisture at about t = 360 hours. When the operator measures moisture content below the FSP, he knows that the wood can stand a tougher drying because the wood is physically stronger at lower moisture contents.

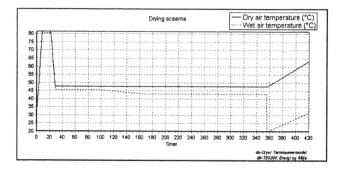

Figure 3. Original drying scheme (dry and wet air temperatures).

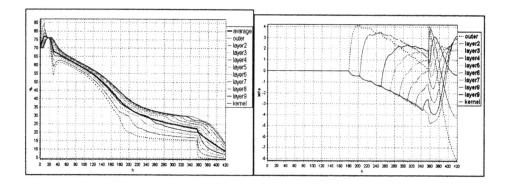

Figure 4. Moisture- and tension-profiles of the wood boards, original drying scheme.

The moisture contents and the tensions of the different layers of the boards are shown in figure 4. Internal tensions arise when the moisture content in the outer layers of the boards is below the FSP, in this case 28 % moisture. The outer layers will start shrinking at this moisture content. The other layers, however, have not started to shrink at this point. This is why a large positive tension appears in the outer layer and smaller negative tensions appear in the other layers. Later on, the moisture content in the second layer reaches the FSP, and a positive tension appears here also. This pattern repeats. When the inner layers starts shrinking, the outer layers have "stiffened" in a stretched position. This can make the tension profile turn, resulting in positive tension values for the inner layers and negative tension values for the outer layers

The values of the tension are indications of the risk of cracking of the boards.

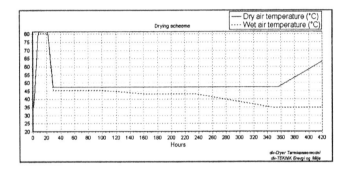

Figure 5. Modified drying scheme (dry and wet air temperatures).

On the basis of the simulation results from the simulation tool, a modified drying scheme is proposed. The dry and wet temperature of the drying gas is shown in figure 5. The resulting moisture and tension profiles of the boards are shown in figure 6.

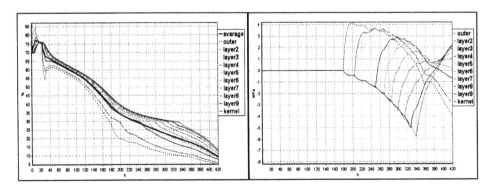

Figure 6. Moisture- and tension-profiles of the wood boards, modified drying scheme

The boards are dried to the same final moisture content value as with the original drying scheme (8.8 %) and within the same drying time (420 hours). The energy consumption (heat only) using the modified scheme is, however, reduced from approximately 35.1 MWh to 31.6 MWh. In other words, the modified drying scheme results in a heat energy reduction of about 10 %.

The maximum predicted tension in the boards is lower in the modified scheme than in the original. This indicates that the risk of board cracking will be reduced using the modified scheme and the product quality of the boards will be, on the average, increased.

Furthermore, the tensions in the boards are considerable lower at the end of the drying with the modified scheme. The final tension profile in the dried boards is a quality parameter. This is because cutting of boards with high internal tensions can result in problems like board bending. The tensions of the boards are reduced by a subsequent steaming. Nevertheless, a starting point with lower tensions before the steaming will be an advantage.

The simulation program can also be used in a similar way to achieve a faster drying scheme without higher risks of cracking, resulting in increased production capacity of the drying process.

Acknowledgement

The Danish Energy Agency and Junckers Industrier A/S have supported this work.

References

Didriksen H., Nielsen J.S., Hansen M.W. (2001). "Energibesparelser ved optimering af tørreprocesser gennem anvendelse af modeller og effektiv regulering (report in danish)", dk-TEKNIK ENERGY&ENVIRONMENT

Perré P. and. Turner W. (2000). An Efficient Two-Dimensional CV-FE Drying Model Developed for Heterogeneous and Anisotropic Materials. Proceedings of the 12th International Drying Symposium IDS2000

Salin J.G. (1990). Simulation of the timber drying process. Prediciton of moisture and quality changes. EKONO Oy, Helsinki Finland. (Doctor of Technology Thesis)

Stanish M.A., Schajer G.S., Kayihan Ferhan, (1986). A Mathematical Model of Drying for Hygroscopic Porous Media. AIChE Journal Vol. 32, No. 8

Turner I.W. (1986). A two-dimensional orthotopic model for simulation of wood drying processes. App. Math. Modelling 1996, Vol. 20, January.

European Symposium on Computer Aided Process Engineering – 12
J. Grievink and J. van Schijndel (Editors)
© 2002 Elsevier Science B.V. All rights reserved.

Optimisation of a Methyl Acetate Production Process by Reactive Batch Distillation

S. Elgue [a], L. Prat [a], M. Cabassud [a], J.M. Le Lann [a], J. Cézerac [b]

[a] Laboratoire de Génie Chimique, UMR 5503, CNRS/INPT(ENSIACET)/UPS,
118 Route de Narbonne, 31077 Toulouse Cedex, France
[b] Sanofi-Synthelabo, 45 Chemin de Météline, B.P. 15, 04201 Sisteron Cedex, France

Abstract

A general framework for the dynamic simulation and optimisation of global batch synthesis has been developed. In this paper, its application to the optimisation of a methyl acetate production process by reactive batch distillation is presented. Experiments performed on a batch pilot plant allow validating the dynamic model. Hence, optimal tuning of the operating parameters of the reactive batch distillation has been investigated by means of the dynamic optimisation procedure.

1. Introduction

The synthesis of fine chemical or pharmaceuticals, widely carried out in batch processes, imply many successive reaction and separation steps. Synthesis optimisation is often restricted to the determination of the optimal operating conditions of each step separately. Therefore, such an approach does not definitely lead to the optimal conditions for global synthesis. For example, optimising the conversion of a reaction for which separation between the desired product and the by-products is more difficult than between the reactants, will involve an important operating cost, due to further difficulties in the separation scheme. Thus, necessity to simultaneously integrate all the process steps in a single optimisation approach clearly appears. For this purpose, recent issue in the dynamic simulation associated with efficient optimisation method may be exploited to accomplish this goal (Wajge and Reklaitis, 1999, Elgue et al., 2001).

In reactive distillation process, the reaction and distillation steps simultaneously occur in the single unit. Unlike conventional configurations of reactors followed by separators, the reactants or products are continuously separated from the liquid reaction phase. This key feature allows an enhanced conversion in equilibrium limited reactions (Agreda et al., 1990). In this way, the application of our framework to a reactive batch distillation process appears obvious.

The present paper is divided into two parts. The first part details the simulation model and presents the associated validation carried out through a process of methyl acetate production. The second part focus on the optimisation of this synthesis and the resultant benefits, according to the proposed methodology.

2. Process modelling

A general modelling able to consider various configuration of reaction separation process has been developed (Elgue et al, 2001). In the purpose of the present study, the mathematical model results of the detailed and rigorous description of a batch reactor connected with a overhead distillation column. In this framework, a tray by tray approach has been adopted, the column condenser representing the first plate, the reactor the last one. Thus, the dynamic model is described by a set of differential and algebraic equations (DAE system) made around each plate.

Ordinary differential equations (ODE) are due to energy balances, total and component mass balances. With regard to the reactor, it has to be noted that thermal modelling of the vessel and the jacket is particularly detailed. Algebraic equations (AE) consist of vapour liquid equilibrium relationships, summation equations, physical property estimations. In order to reduce the complexity of the model, the following typical assumptions have been made, on each tray: perfect mixing between vapour bubbles and liquid, equilibrium between liquid and vapour bubbles and introduction of the Murphree efficiency, negligible vapour holdup compared to the liquid holdup and constant volume of the liquid holdup.

For models determination (enthalpy model, equilibrium constant, hydrodynamic relationship and bubble point temperature) mathematical model is connected to Prophy, a complete physical property estimation system with associated data bank. DISCo (Sargousse et al., 1999), a general solver of DAE systems based on the Gear method, obtains the solution of the mathematical model. Besides its accuracy and numerical robustness, DISCo offers substantial integration velocity thanks to the use of operator sparse and its automatic initialisation procedure.

3. Experimental

Bonnaillie et al. (2001) considered a methyl acetate production process by reactive distillation. The batch pilot plant and the experimental results of their study have been exploited in the present work. The batch pilot plant involves a glass reactor with an overhead distillation column. The bottom plant is composed of a stirred jacketed glass reactor of 5 litres volume. The jacket is provided with a heat transfer fluid circulating in a boiler at around 160°C. The overhead distillation column consists of a multiknit packed column (50 cm in length and 10 cm in diameter) with a condenser and a distillate tank. The condenser involves a spiral coil heat exchanger provided with cooling water. An adjustable timer regulates periodic switching between distillate tank and reflux to the column, with a constant reflux policy.

$$CH_3-OH \;+\; CH_3-C\overset{O}{\underset{OH}{\big\langle}} \;\overset{H_2SO_4}{\rightleftharpoons}\; H_2O \;+\; CH_3-C\overset{O}{\underset{O-CH_3}{\big\langle}} \tag{1}$$

Methanol	Acetic acid	Water	Methyl acetate
(65°C)	(118°C)	(100°C)	(57°C)

Production of methyl acetate is ensured by the addition of acetic acid to methanol with sulfuric acid as homogeneous catalyst, as can bee shown on the reaction scheme (1).

Comparison of boiling point temperatures show the higher volatility of methyl acetate relative to the other components. Thus, methyl acetate removals, by distillation, enable the reaction to reach higher conversion than theoretical equilibrium value. Three experiments have been exploited: experiment with reaction only, experiment with distillation only (following a reaction phase), experiment of reactive distillation, i.e. coupling of reaction and distillation. The experimental conditions are listed in table 1.

Table 1: experimental conditions

	Experiment	Initial molar ratio	Reflux ratio	Catalyst amount
a	Reaction	1.4	∞	5 ml
b	Distillation	1.4	2.5	5 ml
c	Coupling	1.4	∞ , 2.5	5 ml

4. Simulation validation

In order to verify the mathematical model accuracy, the experiments carried out by Bonnaillie et al. (2001) have been simulated. The same thermodynamic (NRTL) and kinetic models have been used. It has to be noted that reaction rate model is a simple kinetic model (2), in agreement with data reported in literature (Smith,1939).

$$r = k_{ester} \exp(\frac{-E_A}{RT})(C_{AAc}.C_{MeOH} - \frac{C_{H2O}.C_{MeAc}}{K_{eq}}) \tag{2}$$

with : k_{ester} pre-exponential factor = 3300 $l.mol^{-1}.mn^{-1}.ml^{-1}H_2SO_4$
   E_A activation energy = 41800 $J.mol^{-1}$
   K_{eq} equilibrium constant = 5

For each experiment (reaction, distillation and coupling) the simulation are compared to samples withdrawn into reactor (fig. 1.a to 1.c). Figure 1.b and 1.c show, during coupling and distillation experiments, light differences between experimental and simulated compositions. This differences are explainable, on the one hand by the strong thermodynamic non-ideality of the mixture and on the other hand by the kinetic model simplicity. In fact complex kinetic models, taking into account the non-ideality of the mixture and so integrating thermodynamic models (Pöpken et al., 2000) would probably provide more accuracy. Nevertheless, the good agreement between experiments and their mathematical representations allows validating the simulation tool. Simulations also show, through coupling results, the advantages linked to reactive distillation and emphasize the necessity of an optimal operation policy.

5. Dynamic optimisation

The purpose of the present study is to optimise the reactive distillation process of production of methyl acetate. Generally, operating time, reaction yield, energy consumption, safety and environmental constraints are the key points of reactive batch distillation processes. In the present process, the column is always operating with a total reflux policy during reaction phases and with a constant reflux policy during distillation

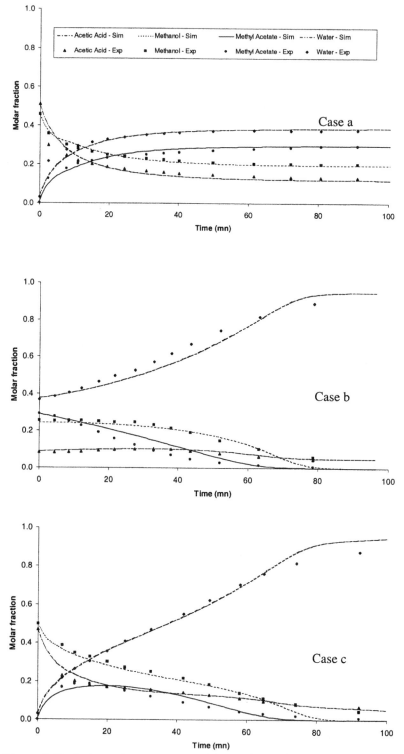

Figure 1: Variations of reactor composition with time for cases a,b and c

phases. Thus, the energy consumption (heat provided by the boiler, condenser cooling water) is only a function of the operating time. In this way, optimisation of the process only involves two criterions: operating time and conversion of reactants. Therefore, a function combination of operating time and conversion, with various weights, has been established in order to optimise the production process. Hence, a mono-objective optimisation method has been used, a successive quadratic programming method (SQP). All the different processes, reaction-distillation or coupling, have the same dynamic structure (fig. 2). In fact, according to the total reflux end time ($t_{R\infty}$), before or after the time of reaction equilibrium (t_{eq}), the process is respectively a coupling or a reaction-distillation process. Thus, the total reflux end time appears to be a main variable of the optimisation problems, as well as reflux ratio (R) and operating time(t_{op}).

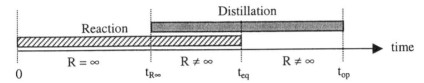

Figure 2: dynamic structure of batch reactive distillation process

The total reflux end time is a very sensitive parameter of reactive batch distillation. In fact, if total reflux policy ends too early, conversion will be reduced owing to the distillation of reactants. On the contrary, a too long total reflux policy will lead to additional operating time and so waste of time. Experimentally, the appropriate total reflux time is very difficult to correctly estimate. Thus, its determination by means of optimisation represents a very challenging objective.

Two main kind of optimisation problems have been studied (table 2). In the first one, the minimum operating time necessary to obtain the desired reactant conversion, is determined. In the second one, objective functions combination of operating time and conversion with various weighting are minimise.

Table 2: Optimisation problems

Kind	Objective function	Variables	Constraints
1.	Operating time	Operating time	Model equations
		Total reflux end time	Acid Conversion
		Reflux ratio	
2.	Operating time, acetic acid conversion	Operating time	Model equations
		Total reflux end time	
		Reflux ratio	

Several optimisation problems, of the two kinds previously presented, have been solved. The resultant solutions are detailed in table 3. The two first solutions present the optimal operating conditions for respectively 93 percent and 95 percent of acetic acid conversion (C_{onv}). The three other solutions present the operating conditions leading to optimal criterions, criterions for which the conversion weight respectively decrease.

Table 3: results of the optimisation problems

Objective function	Constraint	Operating time (mn)	Total reflux end (mn)	Reflux ratio	Acetic Acid conversion
t_{op}	$C_{onv} \geq 93\ \%$	80	15	1.8	93.0 %
t_{op}	$C_{onv} \geq 95\ \%$	118	20	3.0	95.0 %
$t_{op}+ (1-C_{onv})\ x\ 33$	/	141	25	2.3	95.9 %
$t_{op}+ (1-C_{onv})\ x\ 17$	/	105	23	2.5	94.5 %
$t_{op}+ (1-C_{onv})\ x\ 1.7$	/	48	9	1.0	85.8 %

Results of the first kind of optimisation problems show that a significant total reflux time (more than 15 mn) is required for high conversion of reactants. The second kind show that if conversion is privileged, total reflux time is almost invariable (around 23 mn) and only further operating time allows reaching higher conversion.

6. Conclusion

The developed framework is found to be useful in the determination of optimal operating conditions for the methyl acetate production by reactive batch distillation. For different production conditions, this framework allows, in particular, determining the appropriate value of the total reflux period. For the continuation of the operation, a constant reflux policy has been adopted, due to the experimental device. Nevertheless, this program shows to be able to determine more complex optimal reflux policies (piecewise constant reflux profile, piecewise linear reflux profile, complex function of reflux profile) as part of other applications. In fact, this framework is not restricted to reactive batch distillation processes, but is also able to represent and optimise various processes integrating batch reactor and batch distillation column.

References

Agreda V.H., L.R. Partin, W.H. Heise, 1990, High-purity methyl acetate via reactive distillation, Chem. Eng. Prog., Feb, pp. 40-46.

Bonnaillie L., X.M. Meyer, A.M. Wilhelm, 2001, Teaching reactive distillation : experimental results and dynamic simulation of pedagogical batch pilot-plant, ISMR2, Nuremberg.

Elgue S., M. Cabassud, L. Prat, J.M. Le Lann, G. Casamatta, J. Cézerac, 2001, Optimisation of global pharmaceutical syntheses integrating environmental aspects, Escape 11, Kolding, Denmark.

Pöpken T., L. Götze, J. Gmehling, 2000, Reaction kinetics and chemical equilibrium of homogeneously and heterogeneously catalyzed acetic acid esterification with methanol and methyl acetate hydrolysis, Ind. Eng. Chem. Res., 39, pp. 2601-2611

Sargousse A., J.M. Le Lann, X. Joulia, L. Jourda, 1999, DISCo: un nouvel environnement de simulation orienté objet, MOSIM'99, Annecy, France

Smith H.A., 1939, Kinetics of the catalyzed esterification of normal aliphatic acids in methyl alcohol, J. Am. Chem. Soc., 61, pp. 254-260

Wajge R.M., G.V. Reklaitis, 1999, RBDOPT: a general purpose object-oriented module for distributed campaign optimization of reaction batch distillation, Chem. Eng. J., 75, pp. 57-68

European Symposium on Computer Aided Process Engineering – 12
J. Grievink and J. van Schijndel (Editors)
© 2002 Elsevier Science B.V. All rights reserved.

Computer design of a new predictive adaptive controller coupling neural networks and kalman filter applied to siso and mimo control

L. Ender[a], R. Scheffer[b] and R. Maciel Filho[b]

[a]Chemical Engineering Department, Regional University of Blumenau, Rua Antônio da Veiga 140; CP 1507, Blumenau - SC, Brazil, CEP 89010-971
[b]LOPCA/DPQ, Faculty of Chemical Engineering, State University of Campinas (UNICAMP), Cidade Universitária Zeferino Vaz, CP 6066, Campinas - SP, Brazil, CEP 13081-970

Abstract

This work presents a predictive control algorithm based on constraint neural networks as internal non-linear model with a tuning algorithm based on the Kalman filter. The algorithm utilises a sequential quadratic programming algorithm to compute the next action of the manipulated process variables. The predictive control parameter, the suppression factor, is optimised on-line by a standard Kalman filter. The suppression factor is identified by a method based on the relative gain. The algorithm was tested on distinct chemical processes, a penicillin fermentation process (SISO) and a fixed bed catalytic reactor (MIMO). It shows that the suppression factor can be identified on-line, but a scaling factor has to be introduced because the process derivatives can become large. The proposed procedure still reduces the number of parameters to be adjusted in case of MIMO systems.

1. Introduction

Model Predictive Control (MPC) concept has been widely accepted in industry applications and extensively studied by academia. The main reasons for such popularity of the predictive control strategies are the intuitiveness and the explicit contraint handling. Several versions of MPC techniques are Model Algorithmic Control (MAC), Dynamic Matrix Control (DMC) and Internal Model Control (IMC). Although the above techniques differ from each other is some details, they are fundamentally the same. All of them are based on linear process modelling (Zhan and Ishida, 1997).

The most utilised model predictive control algorithm is the dynamic matrix control algorithm. Several modifications of the DMC algorithm were already proposed, such as identification by state space models and the use of non-linear MPC algorithms consisting of neural networks with a SQP optimisation algorithm (Economou et al. , 1986; Lee and Biegler, 1988; Piche et al. 2000; Ender and Maciel Filho, 2000). In this work the non-linear approach is used, as dealing with a variety of process conditions and disturbances demands a good process identification.

All model predictive control algorithms minimise the following objective function,

$$J = \min_{u} \arg \sum_{j=1}^{p} \sum_{k=1}^{N_P} \left(y_{ref,j,k} - y_{j,k} \right)^2 + \sum_{j=1}^{m} \lambda_j \sum_{k=1}^{N_C} \left(u_{j,k} - u_{j,k-1} \right)^2 \quad (1)$$

where p is the number of controlled outputs, m the number of manipulated input variables, N_p the prediction horizon, N_c the controller horizon and λ_j the suppression factor of the corresponding manipulated input variable u_j.

The manipulated inputs and controlled outputs are subjected to the following constraints:

$$y_{min} \leq \hat{y}(k+i) \leq y_{max} \qquad\qquad i = 1, \cdots, N_p \qquad\qquad (2)$$

$$u_{min} \leq u(k+i-1) \leq u_{max} \qquad\qquad i = 1, \cdots, N_u \qquad\qquad (3)$$

$$|u(k+i-1) - u(k+i-1)| \leq \Delta u_{max} \qquad i = 1, \cdots, N_u \qquad\qquad (4)$$

Adopting the receding horizon technique, only the first control action is implemented and all the calculations are repeated at each sampling time.

The only difference in the different type of model predictive control algorithms is the way in which the output of the process is predicted. In this case a neural network is used and thus the output is predicted by

$$y_{j,k} = f(y_{k-1}, u_{k-1}, d_{k-1}) + w_{k-1} \qquad\qquad (5)$$

To cope with process changes the neural network is trained on-line by a switching method as proposed in Ender and Maciel (2000). Due to the non-linear prediction model used, an sequential quadratic programming method has to be used.

The suppression factor λ assures that no exaggerated control action is calculated and influences the systems dynamics. A too small λ results in large control actions, which can result in a instable response, while a too large λ results in a sluggish response. Normally, the parameter is tuned manually until a desired process behaviour is obtained, which can be a very time-consuming procedure. Additionally, different values of λ might be needed for other operating conditions.

An automatic estimation procedure for λ was developed in Ender et al. (2001), which still showed a small overshoot for some set-point changes. Therefore it was searched for another identification scheme of the suppression factor, which is presented in the next paragraph. The new identification schemes are compared to the identification scheme derived from the optimisation criterion as described in Ender et al. (2001) and are applied to two distinct chemical processes: the penicillin fed-batch process (SISO case) and a fixed bed catalytic reactor (MIMO cases).

1.1 Estimation schemes of the suppression factor
The adjustment or tuning algorithm of the parameter λ is based on the standard Kalman filter. To be able to adjust λ a dynamical system has to be created which can observe the state of the parameter λ as in:

$$\lambda_{k+1} = \lambda_k(+ w_k)$$
$$z_k = C \lambda_k(+ v_k) \qquad\qquad (6)$$

where w_k and v_k are random variables with a normal distribution of N(0,Q) and N(0,R) respectively. z_k is the measurement related to the state λ_k. Normally the noise of a parameter state is zero, but a small process noise results in a more stable filter.

The observation equation of the λ was derived from the minimisation criterion J in equation 1 in Ender et al. (2001) by setting the derivative of J to the input change, Δu, equal to zero.

One of the shortcoming of this approach is that the adjustment of the suppression factor is based on the past and not on the present data. Thus the algorithm changes λ in a feedback way. It would be desirable that the suppression factor is raised, when the process is changing rapidly. This assures that the process will not show oscillatory behaviour.

It is used an intuitive approach, which is based on the definition of the relative gain.

$$\lambda_{k+1} = \lambda_k \left(+ w_k\right)$$
$$abs\left(\frac{y_{k+1} - y_k}{\Delta t_{sample}}\right) = abs\left(u_k - u_{k-1}\right)\lambda_k\left(+ v_k\right) \tag{7}$$

The absolute value is taken as the suppression factor has a positive value. In case of the fixed bed reactor, the derivatives are very large and it was necessary to put a velocity factor, ϕ, which results in the following identification system:

$$\lambda_{1,k+1} = \lambda_{1,k}\left(+ w_k\right)$$
$$\lambda_{2,k+1} = \lambda_{2,k}\left(+ w_k\right) \tag{8}$$
$$\phi\begin{bmatrix} abs\left(\dfrac{y_{1,k+1} - y_{1,k}}{\Delta t_{sample}}\right) \\ abs\left(\dfrac{y_{2,k+1} - y_{2,k}}{\Delta t_{sample}}\right) \end{bmatrix} = \begin{bmatrix} abs\left(u_{1,k} - u_{1,k-1}\right) & 0 \\ 0 & abs\left(u_{2,k} - u_{2,k-1}\right) \end{bmatrix}\begin{bmatrix} \lambda_{1,k+1} \\ \lambda_{2,k+1} \end{bmatrix}$$

This approach leads to the introduction of a new parameter which still has to be tuned manually. But the number of parameters to be tuned will be reduced.

2. Results

One of the case-studies considered is a large industrial fermentation process for the production of penicillin. The simulation is based on a model of Rodrigues (1999), which is validated with industrial data. The process is a fed-batch process and falls down in two parts, the growing phase and the production phase. In the growing phase a high sugar level is maintained, while in the production phase it has to be kept low as it inhibits the penicillin production.
The emphasis is put on the production phase where the feeding strategy of the sugar substrate has to be chosen carefully to maximise the penicillin production.

A single input, single output (SISO) system was set-up to verify the proposed process control scheme. The controlled variable was the dissolved oxygen concentration which was controlled by the rotation speed. Various constraints are applicable here as, maintaining the dissolved oxygen concentration above 30% and avoiding high rotation

484

speeds which destroy the fungi. A white noise was imposed on the dissolved oxygen concentration to simulate measurement noise.

The other case-study is a fixed bed catalytic reactor for the production of acetaldehyde by the ethyl alcohol oxidation over a Fe-Mo catalyst. The effective control of those reactors is fundamental to obtain a safe operation, especially when high performance is desired. The control problem of such reactors is not an easy task since they are non-linear, distributed and time-varying systems and show an inverse response due to differences in heat capacities from solids and fluid. The dynamic behaviour of the catalytic fixed bed reactor was modelled by the model proposed by Toledo (2000) and the control is done taking two temperatures at collocation points along the reactor tube.

For both cases the servo problem as well as the regulator problem was undertaken.

It is presented in figures 1 and 3 the servo and regulator control of the dissolved oxygen concentration. The evolution of the suppression factor is shown in figure 2 and 4. It can be seen that the dissolved oxygen concentration does not show any overshoot anymore in comparison with the identification method proposed in Ender et al. (2001). The suppression factor converges to about the same value when measure noise is present. Without noise the suppression factor does not converge to the same value in case of the servo problem. In case of the regulator problem the suppression factor converges to about the same value. In the servo case three perturbations are done, while in the regulator case 15 perturbations are done because of an optimal substrate feeding strategy applied (Rodrigues and Maciel Filho, 1999). This generates much more information content for the Kalman filter to adjust the suppression factor.

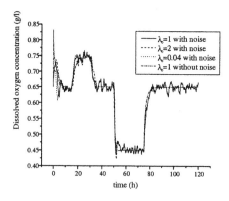
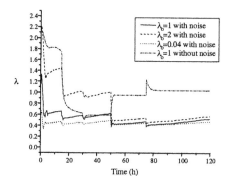

Figure 1: Dissolved oxygen concentration of the penicillin reactor under servo control

Figure 2: Evolution of the suppression factor under servo control for the penicillin reactor

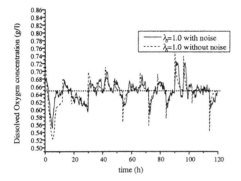

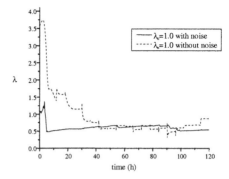

Figure 3: Dissolved oxygen concentration of the penicillin reactor under regulator control

Figure 4: Evolution of the suppression factor under regulator control for the penicillin reactor

In figure 5 and 7 the servo and regulator temperature control is shown of the fixed bed reactor. If no velocity factor, ϕ, is used, then the suppression factor becomes very large. This makes the SQP algorithm unable to find a solution as the manipulated input variable is too much restricted. Still, the value of the velocity factor is not restricted to a very small range, which would make the tuning of this factor a difficult task again. It can be seen from figure 5 and 7 that control is not that much affected when the factor differs a factor of 10 as the suppression factors do not differ a factor of such value (figure 6 and 8). But it should be noted that in the servo control case the second suppression factor (not shown) differed a lot, which explains the difference in control observed in the servo control case.

The suppression factor changes mainly when the process changes. From figures 6 and 8 it can be seen that large peeks occurs for the suppression factor when the set-point is changed. This is obvious from the applied identification scheme which is a function of the derivative of the process. In the penicillin process it was not needed a suppression factor as it has a slow dynamic behaviour. The fixed bed catalytic reactor has a very fast dynamics and thus it has high values of the derivative which result in a high suppression factor. Thus the velocity factor should be chosen in accordance with the gradient occurring under process changes.

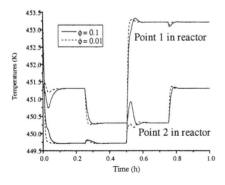

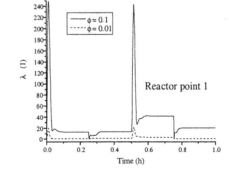

Figure 5: Temperatures at two points along the fixed bed reactor under servo control

Figure 6: evaluation of suppression factor of first manipulated variable under servo control of the fixed bed reactor

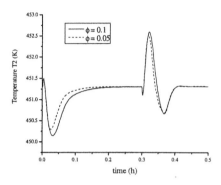

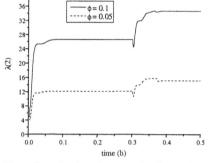

Figure 7: Temperature at second points along the fixed bed reactor under regulator control

Figure 8: evaluation of suppression factor of second manipulated variable under regulator control of the fixed bed reactor

The introduction of the velocity factor means introducing a new parameter to be tuned, while it is wanted to eliminate the tuning of the suppression factor. While the number of parameters to be tuned for MIMO systems is reduced in this way, it is still not the ideal case where no parameter has to be tuned. Therefore it has to be continued the search for other ways to identify the suppression factors, which result in no introduction of another parameter.

3. Conclusions

The proposed control scheme of a model predictive control algorithm with a constrained neural network internal model with auto-tuning capabilities results in a satisfactory control. The suppression factor is adjusted on-line by a standard Kalman filter through an identification method comparable to the relative gain. In rapid changing processes it was needed to introduce a new parameter called the velocity factor, otherwise the suppression can become too large prohibiting control of the process. It seems that this velocity factor is not restricted to a very narrow range and in case of MIMO processes the number of parameters to be tuned manually is lowered. Still, it has to be searched for better ways of identifying the suppression factor, which do not lead to introduction of new parameters.

References

Economou, G.G., M. Morari, B.O.Palsson, (1986) Ind. Eng. Chem Process Des. Dev., 25, 403-411

Ender, L. and R. Maciel Filho (2000), Comp. & Chem. Engineering, 24, 937-943

Ender, L., R. Scheffer and R. Maciel Filho (2001), Proceedings of ESCAPE 11, Denmark, 639-644

Li, W.C. and L.T. Biegler (1988), Ind. Eng. Chem. Res, 27, 1421-1433

Pinche, S., B.S. Rodsari, D.Johnson, M. Gerules (2000), IEEE control systems magazine, 53-61

Rodrigues, J.A.D and R. Maciel Filho (1999), Chemical Engineering Science, 54(13-14), 2745-2751

Toledo, E. de V. (2000), PhD-thesis, State University of Campinas, School of Chemical Engineering, "Modelling and control of fixed bed catalytic reactors"

Zhan, J. and M. Ishida, (1997) Computers and Chemical Engineering, 21, 2, 201-210

European Symposium on Computer Aided Process Engineering – 12
J. Grievink and J. van Schijndel (Editors)
© 2002 Elsevier Science B.V. All rights reserved.

Comparison of Dynamic Approaches for Batch Monitoring

N. M. Fletcher, A. J. Morris and E. B. Martin

Centre for Process Analytics and Control Technology,
University of Newcastle, Newcastle upon Tyne, NE1 7RU, UK

Abstract

The first stage in the development of a performance monitoring tool is a representative model of the process. Traditionally batch processes are associated with non-linear, dynamic behaviour. One method of developing a non-linear model is through a local modelling approach. These local models can be combined to provide a global non-linear representation of the process. The paper develops and compares local models formed using linear Partial Least Squares (PLS), dynamic linear PLS and ARX structures with global models based on dynamic linear and non-linear PLS, as well as non-linear PLS using a simulation of a fed-batch fermentation process. Finally the use of a local linear dynamic PLS model for process performance monitoring representation was evaluated.

1. Introduction

The use of multivariate statistical projection based techniques to model and monitor batch process have been recognised as one approach to enhancing process performance and contributing to an increased understanding of process behaviour. The key techniques include multi-way principal component analysis, multi-way partial least squares (Nomikos and MacGregor, 1994, 1995) and batch observation level analysis (Wold et al., 1998). However batch processes often exhibit non-linear, time variant behaviour with variables exhibiting serial and cross correlation. These characteristics challenge linear statistical multivariate batch monitoring techniques.

To address these limitations, PLS has been extended to dynamic PLS to capture the process dynamics through the augmentation of the input data matrix with lagged values of the past input and output observations (e.g. Kaspar and Ray, 1993, Baffi et al. 2000). two possible approaches to removing the process non-linearities are through the removal of the mean trajectory or the application of non-linear PLS. Alternatively the batch process trajectories can be sub-divided into a number of distinct operating regions with a linear model being fitted to each operating region. These individual models can then be pieced together, thereby providing an overall non-linear global model (e.g. Foss et al., 1995). Such a local model based structure provides the potential for a novel approach to batch process performance monitoring.

In the paper a number of techniques are used to develop the local model, including multi-way PLS, dynamic multi-way PLS and AutoRegressive with eXogenous (ARX) representations. The methodology is demonstrated by application to a simulation of a batch process and compared with the global modelling approaches of multi-way PLS,

ARX and non-linear PLS. Finally a process performance monitoring representation is developed from the local linear PLS models.

2. Local Models

By partitioning the batch trajectories into a number of operating regions where linear approximations apply, a non-linear global model can be approximated through the piecing together of local models. The model developed for the local operating regime is valid for the process operating under specific conditions and gradually becoming invalid outside of that region. A validity function vector, ρ_j, $(0 \le \rho \le 1)$ defines the weight for a specific operating regime at each time point, ϕ, throughout the entire batch. In this initial study linear interpolation is used to provide a smooth transition between the operating regimes (Foss $et\ al$, 1995). The interpolation function is calculated as:

$$\omega_j(\phi) = \frac{\rho_j(\phi)}{\sum\limits_{j=1}^{N} \rho_j(\phi)} \quad \text{where} \quad \sum\limits_{j=1}^{N} \omega_j(\phi) = 1 \quad \forall \ \phi, \tag{1}$$

where N is the number of operating regions, j, within the batch process. A nominal PLS model is then built for each local region using cross-validation for latent variable selection. The loadings ($\mathbf{p}_{jk}$) for each latent variable, k, and the weights ($\mathbf{w}_j$), define the coefficients of each local model that are then stitched together using the interpolation function, ω_j, for the prediction of unseen batches:

$$\mathbf{P}_k = \sum\limits_{j=1}^{N} \mathbf{p}_{jk}^T \omega_j \quad \text{for } k = 1, \ldots, \text{no. of latent variables} \tag{2}$$

where j = 1,..,N are the number of local models, $\mathbf{P}_k$ is the interpolated loading vector to be used in the prediction of the unseen batches for each latent variable and ω_j is the interpolation function for each operating regime calculated from Equation 1.

3. Simulation Example

The simulation used to investigate this methodology was the fermentation of glucose to gluconic acid (Foss $et\ al$, 1995). One hundred batches were simulated for the nominal batch set and thirty for the unseen test batches. The inputs to the model were cell concentration (x_1), glucose concentration (x_2), gluconolactone concentration (x_3), dissolved oxygen concentration (x_4) and oxygen uptake rate, OUR (x_5) and the output was gluconic acid concentration (y_1).

The variables selected to define the different operating regimes were x_1 and x_5. Three operating regimes were defined. Operating regime 1 was identified to be where OUR and cell concentration increase. Regime 2 was defined as the levelling off of the OUR gradient and its subsequent decrease. The third regime is defined to start when cell concentration begins to level off. Fig. 1 shows a plot of the five process variables. The operating range for each regime is also identified. From Fig. 1, the validity function ρ_i is calculated (Eq. 2). Fig. 2 shows the interpolation functions for the three operating regimes.

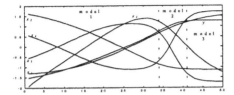 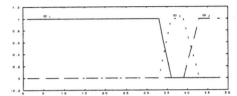

Fig. 1. Operating regimes.   Fig. 2. Interpolation functions for 3 local
                                                regimes

4. Results

The 100 nominal batches were stacked to give a data matrix **X** with each variable being normalised to zero mean and unit variance. For dynamic PLS, the variable lags were determined by first fitting an ARX model to the data. From the weight of the parameters of the ARX model (Fig. 3), a lag for each variable was identified. In Fig. 3 it can clearly be seen that for variable one, a lag of four is more significant than previous lags, consequently 4 lags were included in the final model. The number of lags included for each variable differed between each of the global and local models.

4.1 Results for the Global Model.

Cross-validation was applied to both the original data and the data set comprising the lagged variables. Four latent variables were selected for dynamic PLS (DPLS), two for PLS, three for non-linear PLS (NL-PLS) and two for PLS where the mean trajectory of each variable had been removed ($PLS_{mean-traj}$). This was confirmed by analysing the regression coefficients of the latent variables and their associated standard errors. Fig. 4 shows the regression coefficients ± the standard error for the global DPLS model. From Fig. 4 it can be concluded that from latent variables seven onwards are non-significant.

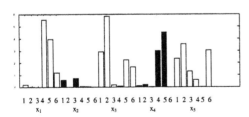 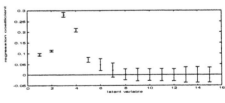

Fig. 3. ARX parameters for the   Fig. 4. Regression coefficients with
   determination of variable delays   standard error bars.
   (lags).

490

Table 1. Prediction of the output using global models

	RSS (x 10^{-2})	**BIC**(x 10^{-3})
ARX	8.1	3.7
PLS	354.5	6.3
PLS$_{\text{mean traj}}$	101.1	15
DPLS	11.9	4.0
NLPLS	180.4	5.8
NL-DPLS	8.1	3.7

The Bayesian Information Criterion (BIC) and the residual sum of squares (RSS) were used to determine how well each modelling technique predicted the output in terms of model parsimony and the ability of the representation to model unseen data. Table 1 summarises the results of the different modelling approaches. As expected the prediction of the output variable improves significantly when the batch dynamics are included in the analysis. The models that exhibited the best prediction capabilities were the ARX and non-linear dynamic PLS (NL-DPLS) representations. The following plots show the prediction of the output for one batch using PLS and NL-DPLS, Fig. 5 and 6.

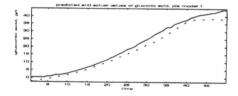

Fig. 5. Predicted and actual values for *Fig. 6. Predicted and actual values for*
PLS model   *NL-DPLS model*

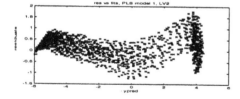 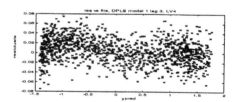

Fig. 7. Residuals versus fitted values for *Fig. 8. Residuals versus fitted values for*
PLS model   *DPLS model.*

An off-set can clearly be seen between the actual and predicted values when the model was based on linear PLS. In comparison, no off-set is seen for NL-DPLS. The residuals of the fitted model are plotted against the predicted values of the output to investigate whether structure remains in the model. The underlying structure of the data is not captured by PLS, Fig. 7. This is in contrast with the residuals from the DPLS model (Fig. 8). Some structure remains in the NLPLS model, whereas the NL-DPLS residuals are structureless (not shown). The residuals for the PLS model with the mean trajectory removed exhibit a high level of structure (not shown) similar to those shown in Fig. 7.

Table 2. Number of latent variables

Model	Local model 1	Local model 2	Local model 3
PLS	3	4	4
DPLS	4	4	5

Table 3 Prediction of the output using local models

Model	RSS (x 10^{-2})	BIC (x 10^{-3})
ARX	8.1	3.8
PLS	351.0	6.7
DPLS	9.9	3.9

4.2 Results for Local Modelling

PLS, DPLS and ARX models were selected for the local modelling stages. Table 2 summarises the number of latent variables selected for each local model using cross-validation. Table 3 shows that there is a significant improvement in the prediction capabilities when a dynamic local model approach is used. No improvement is seen in the prediction of the output when using local PLS models compared to that achieved with the global PLS model.

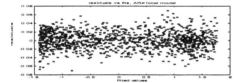   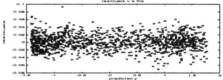

Fig. 9. Residuals versus fitted values for   Fig. 10. Residuals versus fitted values for
   local ARX models.   local DPLS models.

Fig. 9 shows the residual structure for the local model using an ARX model. It is evident that there is no structure remaining. The residuals of the local dynamic linear PLS model (Fig. 10) can be seen to exhibit less structure than those of the DPLS global model (Fig. 8). From the results it can be concluded that the use of local dynamic PLS models gave slightly improved prediction capabilities over those of the global dynamic PLS model. Once again it can be seen that inclusion of the dynamics of the process into the analysis resulted in a reduction in the structure in the residuals.

4.3 Monitoring Charts

The latent variable scores calculated from the nominal models were used to monitor the progression of a batch through time. The interpolation function, ω, was used to interpolate between the scores of each local model. The limits were calculated using ± 3 standard deviations of the average nominal score at each time point. Figures 11 to 14 show the monitoring charts for nominal batches for a DPLS global model and a DPLS local linear modelling approach. The scores can be seen to follow a similar trajectory

492

for latent variable 1 for both the DPLS global model and the DPLS local model (Fig 11 and 13). However, for latent variable 4, the DPLS local model approach shows a structured trajectory (Fig. 12), highlighting that information is retained in the lower order latent variables that is not detected in the global model (Fig. 14).

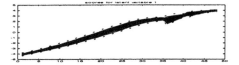

Fig. 11 Control chart for DPLS local
model scores for latent variable 1.

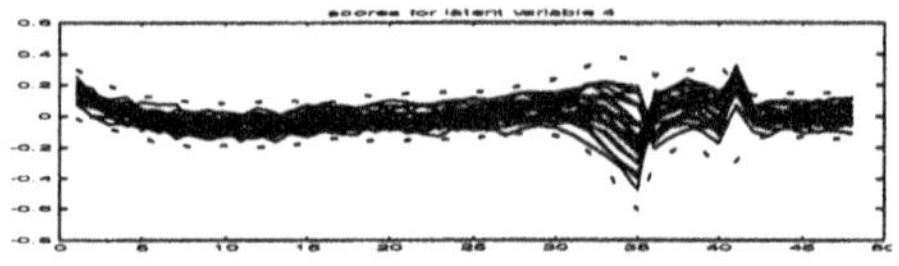

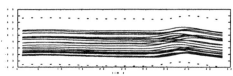

Fig. 12. Control chart for DPLS local
model scores for latent variable 4

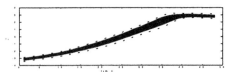

Fig. 13 Control chart for DPLS global
model scores for latent variable 4.

Fig. 14. Control chart for DPLS global
model scores for latent variable 1.

5. Conclusions

The use of a local modelling approach has been shown to result in a reduction in the structure of the residuals, leading to a more useful model compared to the overall global model when using dynamic PLS and non-linear DPLS approaches. Although the ARX model marginally exhibited the most accurate and parsimonious results, the PLS approaches have the potential to form the basis of a performance monitoring scheme using the latent variable scores and model residuals. An example is given of how the model can be used for through batch dynamic monitoring. This is an important area of research for the improved monitoring of dynamic, non-linear batch processes. Future research will address the modelling and dynamic performance monitoring of more complex non-linear processes, such as a large-scale industrial fermentation process.

Acknowledgements

Miss Fletcher acknowledges the EPSRC and CPACT for financial support of her PhD.

References

Baffi, G., E.B. Martin and A.J. Morris, 1999, Comput. Chem. Eng., 23, 395.

Foss, B.A., T.A. Johansen and A.V. Sørensen, 1995, CEP, 3, 389.

Kaspar, M.H. and W.H. Ray, 1993, Chem. Eng. Sci., 48(20), 3447.

Nomikos, P. and J. F. MacGregor, 1994, AIChE Journal 40 1361.

Nomikos, P. and J. F. MacGregor, 1995, Technometrics, 37, 41.

Wold, S., N. Kettaneh, H. Friden and A. Holmberg,. 1988, Chemom. Intell. Lab. Syst., 44, 331.

European Symposium on Computer Aided Process Engineering – 12
J. Grievink and J. van Schijndel (Editors)

493

Optimisation and Experimental Verification of Startup Policies for Distillation Columns

Günter Wozny and Pu Li

Institute of Process and Plant Technology, Technical University Berlin
KWT9, 10623 Berlin, Germany

Abstract

Startup of distillation columns is one of the most difficult operations of chemical processes. Since it often lasts a period of time, leads to off-products and costs much energy, optimisation of startup operating policies for distillation columns is of great interest in process industry. In the last few years we have accomplished both theoretical studies and experimental verifications with the purpose of developing optimal startup policies for distillation columns. As a result, significant reduction of startup time period can be achieved by implementing the developed optimal policies. This paper summarises our recent results from these studies.

1. Introduction

Due to its nature of phase transition, large time delay and strong interaction between variables, startup of distillation columns is one of the most difficult operations of chemical processes. Since the process is unproductive in the startup period, it is desired to shorten this period by optimising startup policies. The procedure of a startup from a cold, empty column to the required operating point consists of three phases (Ruiz et al., 1988): 1) heating the column by the rising vapour, 2) filling the trays by the reflux and 3) running the column to the defined steady state. The third phase requires the longest time and therefore has the potential to reduce the startup period by developing optimal policies. Despite its significance very few work has been done on optimisation of column startup policies. Total reflux (Ruiz et al., 1988) as well as zero reflux (Kruse et al., 1996) policies with a large reboiler duty have been proposed. The switching time from total or zero reflux to values at the steady state is determined by the criterion proposed by Yasuoka et al. (1987), i.e. at the time point when the difference between the temperature at the steady state on some trays and their measured value reaches the minimum. These are *empirical approaches* used for reducing the startup time of distillation columns. In the last few years we have carried out both theoretical and experimental studies with the purpose of developing optimal operating policies for distillation columns (Kruse et al., 1995; Flender et al., 1998; Löwe et al., 2000). *Model-based optimisation* has been adopted in these studies. Three different models are proposed and their parameters validated on the real plants. Two optimisation approaches are used and modified for column startup optimisation. The developed policies are verified on different pilot columns. As a result, by implementing the optimal policies significant reduction of startup time can be achieved. This paper summarises our recent results from these studies.

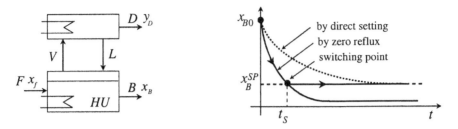

Fig. 1: A two-stage model (left) and its reboiler composition profiles (right)

2. Modelling Column Startup

Three different column models are formulated for column startup operations. A simple *two-stage model* is used to estimate the behaviours of startup and to gain a rough insight into the dynamics of distillation columns. As shown in Fig. 1 (left), the model consists of a total condenser and a reboiler. The component balance of the system is then

$$HU \frac{dx_B}{dt} = Fx_f - Dy_D - Bx_B \tag{1}$$

where $F = D + B$ and $y_D = Kx_B$. To study the trajectory of the reboiler composition x_B influenced by the reflux flow L, we assume HU, F, x_f, V are constant. Then the time constant of the light component composition in the reboiler x_B will be

$$\tau = \frac{HU}{F + (K-1)(V-L)} \tag{2}$$

It indicates that τ will be reduced if the reflux is decreased. Fig. 1 (right) illustrates the reboiler composition profiles caused by the direct setting strategy (setting the steady state value of reflux during whole startup period) and by the zero reflux strategy. To achieve an optimal startup, a proper switching point t_s from zero reflux to direct setting is needed, such that the composition from x_{B0} to x_B^{SP} along the arrow-pointed trajectory. The reboiler duty V has the same impact on x_B but in the opposite direction. In the same way, the influence of reflux and reboiler duty on the top composition y_D can be analysed. As a result, the heuristic for startup is to run the column with a period of a maximal reboiler duty and zero reflux and then switch to their steady state value.

The second model to describe startup behaviours is a detailed *tray-by-tray model* composed of dynamic component as well as energy balances, vapour-liquid equilibrium and tray hydraulics. Fig. 2 shows a general tray of this model, with the variables $x_{i,j}, y_{i,j}, L_j, V_j, M_j, P_j, T_j$ as component liquid as well as vapour composition, liquid as well as vapour flow, holdup, pressure and temperature on the tray. The whole model equations lead to a complicated large-scale DAE system.

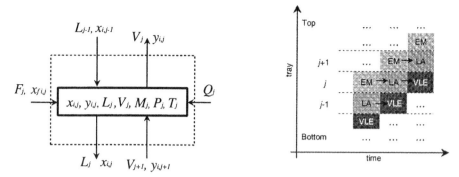

Fig. 3: A general tray of the 2nd model   Fig. 4: State transition of trays during startup

With this equilibrium model, the startup behaviour is described from the first time point at which equilibrium is reached on all trays. Model parameters like tray efficiencies can be validated by comparison of simulated and experimental results. Based on this model, the third model proposed is a *hybrid model* that depicts column startup from a cold empty state (Wang et al., 2001). Each tray will be described from a non-equilibrium phase in which only mass and energy transfer are taking place to an equilibrium phase in which vapour-liquid equilibrium is reached. The switching point between these two phases is determined by the bubble-point temperature at the operating pressure. Fig. 4 illustrates the state transition of the trays: from the empty cold state (EM) → liquid accumulation (LA) → vapour-liquid equilibrium (VLE). Using this model the simulation of startup procedures becomes more reliable.

3. Optimisation Approaches

In most cases, the aim of optimisation of distillation column startup is to minimise the startup period. It is a dynamic optimisation problem usually with reflux rate and reboiler duty as the decision variables. Our solution strategy to this problem is to use a model-based optimisation. A general dynamic optimisation problem can be described as

$$\min \quad t_{startup} \tag{3}$$
$$\text{s. t.} \quad g\,(\,dx/dt, x, u, t\,) = 0 \tag{4}$$
$$h\,(\,dx/dt, x, u, t\,) \geq 0 \tag{5}$$
$$x\,(0) = x_0 \tag{6}$$

where $t_{startup}$, g and h are the objective function, model equations and process constraints, respectively. x and u are state and decision variables. Here g in (4) can be any of the above startup models and h in (5) the predefined steady state specifications. As noted in (6), an initial state x_0 of the column should be given. We used two different approaches to solve such dynamic optimisation problems. The first one is a gradient-based *sequential approach* consisting two computation layers: an optimisation layer where the decision variables are solved by SQP and a simulation layer where the state variables and their sensitivities to the decision variables are computed by solving the model equations with the Newton method (Li et al., 1998). Collocation on finite elements is used for the discretisation of the dynamic system.

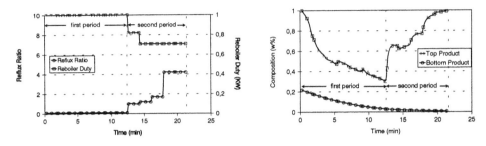

Fig. 5: Optimal policy (left) and composition profiles (right) for the packed column

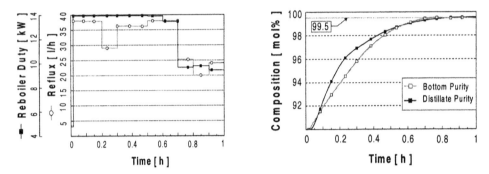

Fig. 6: Optimal policy (left) and purity profiles (right) for the bubble tray column

The approach is applied to start up a pilot packed column for separating a mixture of two fatty alcohols. The aim of optimisation is to run the column to the top and bottom product purity specification in a minimum time period. The results of optimisation (Fig. 5 (left)) show that the reboiler duty should be at a maximum value and there should be no reflux for a period of time, and after that both decision variables should slowly approach their steady state value. Under this policy, both product compositions reach their desired value in around 22 min. Note that the top composition decreases in the first period and then returns to the specified value. The second approach we have used is *simulated annealing* (SA) which is a stochastic search method. The advantage of this method is that it does not require sensitivity information and thus can be connected directly to an available simulator (Hanke and Li, 2000; Li et al. 2000). Since commercial simulation software is widely used in industry, using SA is an easy way to conduct optimisation. The shortcoming of this method is its low computation efficiency, i.e. many runs of simulation are needed to reach the solution. SA is applied to the startup study of a pilot column with 20 bubble-cap trays for separating a methanol-water mixture. The equilibrium model is used in the problem formulation. The model was implemented in the software SPEEDUP as a simulator which is called by a file of SA. The problem definition is the minimisation of the time period from an initial state of the column to the desired steady state. Fig. 6 shows the computed optimal operation policy and the product purity profiles. Different from Fig. 5, both the reflux rate and reboiler duty shown in Fig. 6 (left) for this column should be high in the first period and decreased to the steady state value in the second period. The reason is that one is a packed column with very small holdup and the other is a tray-by-tray column.

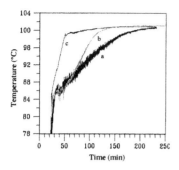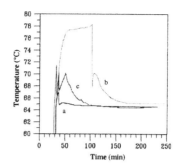

Fig. 7: Temperature profiles of bottom (left) and top (right) of the tray column

4. Experimental Verification on Different Pilot Plants

The *first pilot column* has a diameter of 100 mm and 20 bubble-cap trays for separating a water-methanol mixture under atmospheric pressure. The plant has an electrical reboiler and a total condenser. It is equipped with a control system and necessary measurements and electrical valves for flow control. Startup of the column to the steady state with a purity of 99.5mol% for methanol and water was studied. Fig. 7 shows the measured bottom and top temperature profiles by different startup policies: a – direct setting; b – zero reflux; c – optimal policy. All three experiments had the same feed condition (composition: 29 mol%, flow rate: 15 l/h, temperature: 60 °C). It can be seen that the time taken for reaching both bottom and top temperature at steady state was 220min, 170min and 120min by the three different policies, respectively.

The *second pilot plant* is a heat-integrated column system (Fig. 8) consisting of a high pressure (HP) and a low pressure (LP) column, with 28 and 20 bubble-cap trays, respectively. The vapour from HP is introduced as the heating medium to the reboiler of LP. The plant is so constructed that it is possible to operate the process in downstream, upstream and parallel arrangements. Due to the heat-integration startup of the plant becomes complicated. A fully rigorous optimisation was not chosen, because modelling the build-up of the pressure of HP during startup is difficult and under investigation. Startup policies in parallel operation for this plant are developed from analysing the startup heuristic (Fig. 1) and the optimal results of the previous column (Fig. 6). According to Fig. 6, both the reflux flow and reboiler duty should be at a maximum value (about 1.5 times of the steady state value) and switch to their steady state value at a suitable time point. For the heat-integrated column system, HP should be started with the maximum reboiler duty and both HP and LP should have the maximum reflux. When the bottom temperature of LP reaches nearly its desired value, the pressure build-up of HP will be shortly finished. This means the reboiler duty of HP and reflux flow of HP and LP should be switched to their steady state value at this time point. This near-optimal policy was applied to the plant. Fig. 9 shows the measured temperature profiles of the two heat-integrated columns in parallel operation. It took 3.2h for both columns to reach their desired steady state. The conventional direct setting strategy (both reboiler duty and reflux flow) was also tested for the same feed condition. The experimental result shows that the total startup time of the column system was 9h.

498

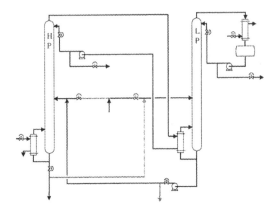

Fig. 8: The heat-integrated column system (left) and its flow-sheet (right)

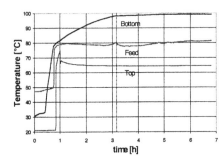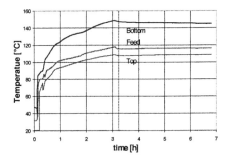

Fig. 9: Temperature profiles of LP (left) and HP (right) by the near-optimal policy

5. Conclusions and Acknowledgements

Model-based optimisation was used for developing time-optimal policies for distillation columns. The policies were verified on different pilot plants. Significant reduction of startup time was achieved by implementing the optimal policies in comparison to the conventional startup strategy. From these results some heuristic rules for column startup can be derived. These are now being further studied on startup of columns for reactive as well as three-phase distillation. We thank Deutsche Forschungsgemeinschaft (DFG) for the financial support in this work under the contract WO565/10-2-3.

References

Flender, M., G. Fieg, and G. Wozny, Proceedings DYCOPS-5, Corfu, June 8-10, 1998.

Hanke, M. and P. Li, 2000, Comp. Chem. Eng. 24, 1.

Kruse, Ch., G. Fieg, and G. Wozny, 1996, J. Proc. Cont. 6, 187.

Li, P., H. Arellano, G. Wozny and E. Reuter, 1998, Ind. Eng. Chem. Res. 37, 1341.

Li., P., K. Löwe, H. Arellano and G. Wozny, 2000, Chem. Eng. Proc. 39, 357.

Löwe, K., P. Li and G. Wozny, 2000, Chem. Eng. Technol. 23, 841.

Ruiz, A., I. Carmeron and R. Gani, 1988, Comp. Chem. Eng. 12, 1.

Wang, L., P. Li, G. Wozny and S.Q. Wang, 2001 AIChE Annual Meeting, paper 85h.

Yasuoka, H., E. Nakanisshi and E. Kunugita, 1987, Ind. Chem. Eng., 27, 466.

European Symposium on Computer Aided Process Engineering – 12
J. Grievink and J. van Schijndel (Editors)
© 2002 Published by Elsevier Science B.V.

Optimal Number of Stages in Distillation with respect to Controllability

Marius S. Govatsmark, and Sigurd Skogestad[1]
Department of Chemical Engineering, NTNU
N-7491 Trondheim, Norway

Abstract

The central question to be examined in this paper is if it is optimal to have a large or small number of stages in a distillation column with respect to controllability when the objective is to have dual composition control. With multivariable controllers and without considering model uncertainty few stages shows somewhat better controllability than many stages because (i) the available manipulated variables have larger effect on the outputs (allowing larger changes in manipulated variables for few stages) and (ii) it is not necessary to reject the disturbances as fast as for many stages. However, in reality there will always be model uncertainty, and with uncertainty included the conclusion is reversed: it is better to have many stages. The reason is that with more stages the system is less interactive and thus less sensitive to uncertainty. Physically, with many stages a pinch zone develops around the feed stage, which tends to decouple the two column ends from each other.

1. Introduction

We want to evaluate if it is optimal with a large or a small number of stages in a distillation column with respect to controllability. Economic objectives like design cost (connected to number of stages in the column and necessary dimensions for internal flows) and operation cost (energy cost connected to internal flows) are not considered here. The study is for dual composition where we have a given purity specification in the top and in the bottom of the column. The conventional LV-configuration is used for stabilizing the condenser and reboiler holdups. Skogestad (1997) claims that it is better to have many stages. He writes: *How should the column be designed to make feedback control easier? In terms of composition control, the best is probably to add extra stages. This has two potential advantages:*

1. *It makes it possible to over-purify the products with only a minor penalty in terms of energy cost; recall the expression for $V_{min}=1/(1-\alpha)F$ which is independent of the purity. The control will then be less sensitive to disturbances.*

2. *If we do not over-purify the products, then with "too many" stages a pinch zone will develop around the feed stage. This pinch zone will effectively stop composition changes to spread between the top and bottom part of the column, and will therefore lead to a decoupling of the two column ends, which is good for control.*

---

[1] e-mail: skoge@chembio.ntnu.no; phone: +47-7357-4154; fax: +47-7359-4080

However, this finding is disputed by Meeuse and Tousain (2001) who claim, based on optimal design of LQG controllers, that it is better to have few stages. The objective of this paper is to study this issue in more detail.

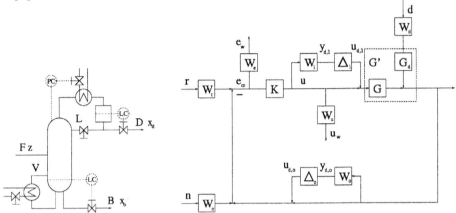

Figure 1: D istillation column
(One-feed two-product)

Figure 2. Block diagram for μ-analysis

2. Column data

A distillation column separating two-component feed is studied, see figure 1. The feed is saturated liquid. There are two remaining degrees of freedom when we assume given column pressure and liquid holdup in the reboiler and condenser. The model details are described in Skogestad (1997). The controlled variables are composition of light component in top product (x_d) and composition of light component in bottom product (x_b). With the indicated conventional control configuration for pressure and levels, the two remaining manipulated variables are reflux flow rate (L) and vapor boil-up flow rate (V). The disturbances are feed flow rate (F=1±0.2) and feed composition (z_F=0.5±0.1). The number of stages is varied between 25 and 51. The base case has 41 trays with the feed tray in the middle (tray 21). Column data is summarized in table 1.

Table 1: Column data

Controlled variables (y)	x_d	0.99	kmol/kmol
	x_d	0.01	kmol/kmol
Manipulated variables (u)	L	21.5-2.2	kmol/min
	V	22.0-2.7	kmol/min
Disturbances (d)	F	1	kmol/min
	z_F	0.1	kmol/kmol
Key hydraulic parameters	τ_l	0.02-0.2	min
	τ_d	0.02-0.2	min
	τ_b	0.02-0.2	min
	$\Sigma\tau_i$	0.5-10.2	min
Thermodynamic data	A	1.5	-
Number of stages	N_T	25-51	-
Feed stage number	N_F	13-26	-

3. Analysis of the controllers

Figure 2 shows the block diagram for the system where model uncertainty is included both as input and output uncertainty. This setup is based the setup previously used by Lundstrom and Skogestad (1995). G' is the plant model which consists of the disturbance gain G_d and the process gain G. G' has two outputs (x_d and x_b) and four inputs (L, V, F and z_F). The model is scaled with respect to acceptable control error ($y_{mad}= [0.01\ 0.01]^T$), allowed variation in manipulated variables ($u_{mad} = u_{nominal}$) and expected disturbances ($d_{mad}= 0.2d_{nominal}$). Maximum expected setpoint changes (r_{mad}) are $[0.01\ 0.01]^T$. K is the controller. W_r, W_d and W_n are weight matrices for setpoints r, disturbances d and measurement noise n. W_e and W_u are weights respectively on deviation from desired setpoints e and manipulated variables u. Model uncertainty is represented by $W_i\Delta_i$ which models input uncertainty, and $\Delta_o W_o$ which models output uncertainty. Δ_i and Δ_o are any diagonal matrices with H_∞-norm less than one. The weighting matrices are diagonal with elements:

$$w_r=r_{mad}/y_{mad}=1/(\tau_r s+1)=1/(30s+1) \tag{1}$$

$$w_d=1,\ w_n=10^{-4},\ w_u=0.1 \tag{2}$$

$$w_e=(\tau_I/\ M_s s+1)/(\tau_I s+A)=(0.5s+1)/(s+10^{-4}) \tag{3}$$

τ_I (=1min) is the closed-loop response time and M_s (=2) is the maximum allowed peak of the sensitivity function. In practice integral action is necessary when A is very small. We use $A=10^{-4}$, except when analyzing the controllers with no integral action, for which we use $A^*=0.5$. For the input uncertainty we use

$$w_i=(\tau_i s+M_{i,0})/(\tau_i/M_{i,\infty}s+1)=(s+0.2)/(0.5s+1) \tag{4}$$

$M_{i,0}$ (=0.2) is the relative gain uncertainty in the inputs with low frequencies, $M_{i,\infty}$ (=2) is the relative gain uncertainty in the inputs at high frequencies and the τ_i (=1.0 min) is the delay in inputs. For the output we use

$$w_o=(\tau_o s+M_{o,0})/(\tau_o/M_{o,\infty}s+1)= s/(0.5s+1) \tag{5}$$

The relative uncertainty in the measurements are at low frequencies ($M_{o,0}$) assumed equal 0 and at high frequencies ($M_{i,\infty}$) assumed equal 2. τ_o (= 1 min) corresponds to a delay up in each measurement.

For the system in Figure 2, μ_{NP} is the H_∞-norm of the transfer function from the scaled inputs [r d n] to the scaled outputs [e_w u_w], or equivalently tells us by which factor the performance weights must be reduced to have the scaled errors less than 1. μ_{RS} tells by which factor the uncertainty (the Δ-blocks) must be reduced to guarantee stability. μ_{RP} tells by which factor the uncertainty and performance weights must be reduced to give the worst-case scaled errors less than 1. In summary, μ_{RP}, μ_{RS} and μ_{NP} should be as small as possible, and preferably less than 1.

4. LQG-control

We first follow Meeuse and Tousain (2001) and design a quadratic optimal controller (LQG) where we only consider disturbances and measurement noise, with no model uncertainty included. The design of the LQG-controller is based on a scaled, linearized model of the plant:

$$dx/dt = Ax + Bu + w_d = Ax + Bu + B_d d, \quad y = Cx + n \tag{6}$$

The process noise (w_d) and measurement noise (n) are assumed to be white noise with respectively covariances W and V. The LQG-problem is to find the optimal controller u(t) which minimizes

$$J = E(\lim_{T \to \infty} \tfrac{1}{T} \int_0^T [x^T(t)Qx(t) + u^T(t)Ru(t)]dt) \tag{7}$$

where design parameters are $Q = Q^T \geq 0$ and $R = R^T > 0$. We design LQG-controllers for different number of trays. The inputs are weighted equal 0.1 ($r_{ii} = 0.01$). Top composition and bottom composition are weighted equal 1($q_{ii} = 1$). The remaining states have zero weights. The covariance matrix to the process noise (W) is selected as $B_d B_d^T$. The measurement noise is assumed small and the covariance matrix for the measurement noise is selected as [0.0001 0.0001]. The results when using no integral action, are summarized in table 2.

Table 2. LQG- and μ-optimal controller analysis for different number of stages

N_T/N_F	V/F	LQG - No integral action				LQG - Integral action			μ-optimal		
		$10^5 J^*$	μ_{NP}^*	μ_{RS}^*	μ_{RP}^*	μ_{NP}	μ_{RS}	μ_{RP}	μ_{NP}	μ_{RS}	μ_{RP}
25/13	22.0	0.054	2.001	1.158	2.004	0.384	6.544	6.549	0.691	1.230	1.258
27/14	9.69	0.201	2.028	1.169	2.034	0.159	6.862	6.875	0.714	1.118	1.122
29/15	6.63	0.346	2.074	1.126	2.088	0.103	6.645	6.656	0.780	1.124	1.130
31/16	5.25	0.474	2.139	1.114	2.154	0.078	6.306	6.327	0.834	1.133	1.152
33/17	4.48	0.581	2.226	1.096	2.241	0.064	6.326	6.344	0.832	1.105	1.128
35/18	3.98	0.676	2.332	1.060	2.350	0.055	6.171	6.178	0.874	1.118	1.144
37/19	3.64	0.760	2.457	1.064	2.478	0.049	6.067	6.075	0.865	1.099	1.112
39/20	3.39	0.836	2.600	1.050	2.624	0.044	6.152	6.160	0.876	1.095	1.106
41/21	3.21	0.905	2.759	1.020	2.786	0.041	6.240	6.248	0.878	1.086	1.108
43/22	3.06	0.970	2.931	1.024	2.963	0.039	6.302	6.311	0.869	1.084	1.086
45/23	2.95	1.031	3.114	1.016	3.150	0.037	6.289	6.297	0.871	1.079	1.086
47/24	2.86	1.089	3.303	0.996	3.345	0.035	6.289	6.297	0.883	1.087	1.098
49/25	2.79	1.144	3.495	0.988	3.543	0.036	6.365	6.374	0.856	1.065	1.071
51/26	2.73	1.196	3.684	0.993	3.739	0.037	6.439	6.447	0.845	1.061	1.063

μ_{NP}^*, μ_{RS}^* and μ_{RP}^* are computed with $A^* = 0.5$, i.e. $w_e^* = (0.5s+1)/(s+0.5)$. We see that the value of the objective function J (and μ_{RP}) is smallest for few number of stages, confirming the findings of Meeuse and Tousain (2001). Note that the internal vapor flow V in the column is larger with few stages. μ_{RS} shows that a large number of stages is best when we have model uncertainty. The uncertainties have small effect on the robust performance compared to other inputs (disturbances and reference tracking). If we increase the relative weights on the outputs (q=1,r=0.0001), the results change.

The results when using LQG-control with integral action included, are summarized in table 2. With increasing number of trays μ_{NP} is decreased and the nominal performance is improved. μ_{RP} and μ_{RS} are large and show that robust performance for the LQG-controller with integral action is far from acceptable when we have model uncertainties, including delays. The LQG-controllers do not clearly indicate if it is optimal with few or many trays.

5. μ-optimal control

The μ-optimal controller minimizes the structured singular value μ_{RP} for the system. The μ-optimal controller is designed by DK-iteration (Doyle et.al.,1982). The results are summarized in table 2. With increasing number of trays μ_{NP} is decreased and the nominal performance is reduced. With increasing number of trays both μ_{RS} and μ_{RP} is decreased and both robust stability and robust performance are improved. This confirms the claims of Skogestad (1997).

6. Discussion

In order to explain the contradictionary results, we will now look at the effect of the different number of stages with respect to some simple controllability measures: process gain (G), disturbance gain (G_d) and interactions (RGA). With few stages the manipulated inputs have larger effect on the outputs. One reason is that the internal flows, e.g. V, and thus the allowed changes in manipulated variables (=$u_{nominal}$) are larger for few stages. From disturbance gains we see that many stages require somewhat faster control to reject the disturbances than few stages, though the column holdup is assumed larger for many than few stages. This may be explained by that with few stages the internal flows are larger. Figure 4 shows the 1,1-element in the relative gain array for 25, 31 and 41 stages. The RGA-values, and thus the two-way interactions, are much higher with few stages, especially at low frequencies.

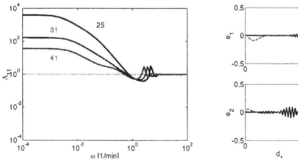

Figure 3: Λ_{11} for 25, 31 and 41 stages   Figure 4: Responses for μ-optimal (stable) and LQG-controller (unstable)

Interactions pose no problem when designing a multivariable controller with a perfect model, but may pose serious problems when uncertainty is included. A more detailed study is based on LQG-controller design. This study shows that nominally it is preferable with few trays when we only consider disturbances and many trays when we only consider reference tracking. If we consider both disturbances and reference tracking, the controller tuning decides if it is optimal to have few or many trays. Nominally it is preferable with few trays when the weights on the outputs are relatively small (q=1, r=0.01), because the reference tracking is no significant problem (w_r=1). Nominally it is preferable with many trays when the weights on the outputs are relatively large (q=1,r=0.0001) or include integral action, because the reference tracking has significant effect. When the controller output weights are relatively large, the uncertainties have large effect on the performance. LQG-controllers give bad performance, but do not clearly indicate if it is optimal to have few or many trays. In the μ-optimal controller design we consider uncertainty including delays, and the resulting controllers show better performance for many stages because of less interaction. Figure 4 shows simulations of step changes in the disturbances for 25 stages when using a LQG-controller with integral action included and when using a μ-optimal controller. We have taken into account some model input uncertainty including delay (e.g $G_p=GG_{extra}D$ where D=diag[0.8 1.2] and $G_{extra}=diag[1/(0.02s+1)^5]$). The LQG-controller is unstable, which is expected since $μ_{RP}$ is much larger than one (see table 3).

7. Conclusion

In conclusion, we find that a large number of trays gives somewhat better controllability than a small number of trays. The seemingly contradictory results of Meeuse and Tousain (2001) are correct, but only hold when having no model uncertainty including delay (and no reference tracking), which is of limited practical interest.

References

Doyle, J.C., J.E. Wall and G. Stein , 1982, Performance and robustness analysis for structured uncertainty. IEEE Conf. on decision and control.

Lundstrom, P. and S. Skogestad, 1995, Opportunities and difficulties with 5x5 distillation control. J. Proc. Cont. 5(4), 249-261.

Meeuse, F. M. and R. L. Tousain, 2001, Closed loop controllability analysis of process designs: Application to distillation column design. In: Escape-11.

Skogestad, S., 1997, Dynamics and control of distillation columns - a tutorial introduction. *Trans. IChemE* 75(Part A), 539-562. Also Plenary lecture at Distillation and Absorption '97, Maastricht, September 9-10, 1997, C97-5, V1 pp. 23-58.

European Symposium on Computer Aided Process Engineering – 12
J. Grievink and J. van Schijndel (Editors)

Dynamic Optimisation of Batch Distillation with a Middle Vessel using Neural Network Techniques

M. A. Greaves[1], I. M. Mujtaba[1*], M. Barolo[2], A. Trotta[2], and M. A. Hussain[3]

[1]Department of Chemical Engineering, University of Bradford,
West Yorkshire BD7 1DP (U.K.)

[2]Department of Chemical Engineering, University of Padova, via Marzolo 9,
35131 Padova PD (Italy)

[3]Department of Chemical Engineering, University of Malaya,
50603 Kuala Lumpur (Malaysia)

Abstract

A rigorous model (validated against experimental pilot plant data) of a Middle Vessel Batch Distillation Column (MVC) is used to generate a set of data, which is then used to develop a neural network (NN) based model of the MVC column. A very good match between the "plant" data and the data generated by the NN based model is eventually achieved. A dynamic optimisation problem incorporating the NN based model is then formulated to maximise the total amount of specified products while optimising the reflux and reboil ratios. The problem is solved using an efficient algorithm at the expense of few CPU seconds.

1. Introduction

Batch distillation is frequently used in chemicals and pharmaceutical industries for the manufacture of fine and specialty chemicals where the product demand frequently changes. In regular batch distillation columns, the products and slops are withdrawn sequentially from the top of the column according to their relative volatilities and the heavier cut is recovered from the reboiler at the end of the batch (Mujtaba and Macchietto, 1996; Barolo et al., 1998).

Bortolini and Guarise (1970) and Hasebe et al. (1992) suggested a different column configuration that has a middle vessel attached to the feed tray, to which the feed is charged, with continuous recycling of liquid feed between the feed tray and the middle vessel. This column configuration is defined as a middle-vessel column (MVC), and over the last few years a lot of attention has been given to such a column (Skogestad et al., 1997; Cheong and Barton, 1999). One of the advantages of the MVC process (Figure 1) is that it can produce three products (on spec or off-spec or a combination of both) from the top, middle and bottom of the column simultaneously. Barolo et al. (1996) provided the first comprehensive experimental results on a pilot-plant MVC column.

Batch distillation operation is a hard task to model because of the integrating behaviour of the system (Edgar, 1996). Optimising the performance of the column is even harder as the large system of differential and algebraic equations (DAEs) describing the process makes the solution of the dynamic optimisation problem very expensive from a computational point of view, even with the computational power currently available (Mujtaba and Macchietto, 1996).

Therefore, the aims of this study are to: (1) Model the process using neural network (NN) techniques rather than to base the model on first principles; (2) Evaluate the model against data generated by a rigorous MVC simulator (Barolo *et al.*, 1998); (3) Develop a dynamic optimisation framework using the NN based model to optimise operating parameters.

This will avoid the need for time consuming experiments (or computation using rigorous optimisation techniques) in order to find the optimum operating conditions to satisfy a specified separation task (product amount and purity).

2. Pilot Plant and Rigorous Modelling

Barolo *et al.* (1998) developed a rigorous dynamic model for a pilot plant MVC column separating a binary ethanol/water mixture. The model was validated against a set of experimental data. The column consisted of 30 sieve trays (0.3 m diameter and 9.9 m total height), a vertical steam heated thermosiphon reboiler (capacity 90 L), and a horizontal water cooled shell and tube condenser. The reflux drum (capacity 40 L) was open to atmosphere. The middle vessel had a maximum capacity of 500 L.

3. Neural Network Based Modelling

A feed forward network structure was selected as this structure is robust and performs well in approximating the complex behaviour of distillation processes (Greaves *et al.*, 2001).

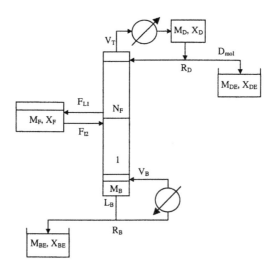

Figure 1 – Schematic of Batch Distillation Column with Middle Vessel

---

* Author to whom all corresspondence should be addressed E-mail: I.M.Mujtaba@Bradford.ac.uk

Table 1 – Data Set 20 with parameters: $R_D = 0.855$, $R_B = 0.727$, $F_{L2} = 0.1789$, $D_{MOL} =$ 0.00926 (kmol/min), $B_{MOL} = 0.03345$ (kmol/min), $R_{MOL} = 0.0546$ (kmol/min), $V_B = 0.089$ (kmol/min) and $F_{L1} = 0.14$ (kmol/min)

Time (mins)	M_{DE} (kmol)	X_{DE}	M_F (kmol)	X_F
0	0	0	4.921	0.322
10	0.093	0.865	4.532	0.331
20	0.185	0.862	4.143	0.343
:	:	:	:	:
120	1.111	0.856	0.251	0.672

Since the purpose of this work is to propose a general framework for the optimization of batch columns, rather than to consider the solution of a particular case study, a simple separation task is considered; we are not claiming that an MVC is necessarily the best column configuration for this separation task. We consider an ethanol/water mixture and the objective is to obtain two products with different ethanol compositions, while minimizing the ethanol losses from the bottom of the column. The lighter cut is obtained via the distillate stream from the top of the column, while the heavier cut is segregated in the middle vessel and recovered at the end of the batch. In the MVC column, R_D, R_B, F_{L1}, F_{L2} and V_B (defined in Figure 1 and in the nomenclature section) are the governing parameters that will affect the separation tasks. For a given vapour load (V_B) and liquid flow from the feed tray to the middle vessel (F_{L1}) sets of data were generated using the rigorous model. A "grid" of reflux and reboil ratios was considered, and the operation was carried out, regardless of what the objective of the distillation was. Table 1 shows a typical set (set 20) of data generated.

Four neural networks were used to predict the state variables M_{DE}, X_{DE}, M_F and X_F at any given batch time as a function of neural network input parameters as shown below:

$$\text{NN output } (P_i) = f(R_D, R_B, D_{MOL}, B_{MOL}, Z_0, t_1) \tag{1}$$

where $P_0 = \{M_{DE0}, X_{DE0}, M_{F0}, X_{F0}\}_{t=0}$ and $P_f = \{M_{DE}, X_{DE}, M_F, X_F\}_{t=tf}$

are the time dependant state variables defining the separation achieved at time t. It is important to note that at any time t, M_{BE} and X_{BE} (Bottom Product amount and composition) can be calculated by overall mass balance, assuming negligible tray holdup.

The multilayered feedforward network used is trained by a backpropagation method using a momentum term as well as an adaptive learning rate to speed up the rate of convergence (Greaves *et al.*, 2001). Two data sets were used to train the neural networks and one for validation.

Figures 2 and 3 show the results for data set 20. The figures compare the amount of product and composition for Overhead product tank (M_{DE}, X_{DE}) and the Middle Vessel (M_F, X_F) from the data set with those calculated by the NN based model. It can be seen that the results are an excellent match.

508

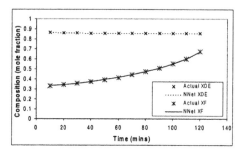
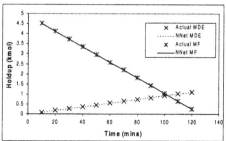

Figure 2 - Overhead Product for Set 20   *Figure 3 - Mid-Vessel Product for Set 20*

4. Optimisation Formulation and solution

The dynamic optimisation problem with an objective to maximise the amount of distillate and middle vessel products can be described as:

given: the column configuration, the feed mixture, separation tasks
determine: optimal reflux ratio, reboil ratio and batch time
so as to maximise: the amount of distillate and middle vessel products
subject to: equality and inequality constraints (e.g. bounds, etc.)

Mathematically the problem can be written as:

$$\textbf{(P1)} \qquad \text{Max} \qquad\qquad (M_{DE}+M_F)$$
$$R_D(t), R_B(t), t$$

subject to:

Process Model (DAE/Hybrid/NN)	(Equality constraint)
$X_{DE}^{*}-\varepsilon_1 \leq X_{DE} \leq X_{DE}^{*}+\varepsilon_1$	(Composition constraint)
$X_F^{*}-\varepsilon_2 \leq X_F \leq X_F^{*}+\varepsilon_2$	(Composition constraint)
$M_F \geq 0$	(Holdup constraint)
$M_{F0}X_{F0}-(M_{DE}X_{DE}+M_FX_F) \geq 0$	(Component balance constraint)
$R_D^{L} \leq R_D \leq R_D^{U}$, $R_B^{L} \leq R_B \leq R_B^{U}$	(Reflux and reboil ratio bounds)
$t^{L} \leq t \leq t^{U}$	(Time bounds)

where X_{DE}^{*} and X_F^{*} are the specified purity of the overhead and middle vessel products; ε_1 and ε_2 are small but finite values (both set to 0.001); R_D^{L} and R_D^{U} are the lower and upper bounds of R_D; R_B^{L} and R_B^{U} are the lower and upper bounds of R_B; t^{L} and t^{U} are the lower and upper bounds of t.

In the past, rigorous DAE model or hybrid (simple DAE coupled with NN) model based optimisation of regular batch distillation have been considered by Mujtaba and Macchietto (1996) and Mujtaba and Hussain (1998). However, these required the full integration of the DAE or Hybrid models several times during optimisation to evaluate objective function and gradients and were computationally intensive. In comparison, the proposed NN based model in this work is much faster and can evaluate the state variables at any given time without requiring stepwise evaluation as in DAEs integration.

The above optimisation problem was solved using an efficient Successive Quadratic Programming (SQP) based method (Chen, 1988). Forward Finite Difference method was used to calculate the gradients of the objective function and the constraints (required by the optimiser) with respect to the optimisation variables. The reason for choosing a finite difference method as opposed to the analytical based method is purely due to difficulty in evaluating an analytical gradients for the NN based process model.

Table 2 - Summary of Optimised and Simulated Results

			Simulated			Optimised				
Run	Data Set	t min	R_D	R_B	$M_{DE}+M_F$ kmol	t min	R_D	R_B	$M_{DE}+M_F$ kmol	%Improvement in $M_{DE}+M_F$
1	6	70	.639	.641	1.806	67.18	.639	.642	1.9476	7.84%
2	15	110	.787	.726	1.659	72.90	.802	.704	2.2313	34.49%
3	19	150	.849	.778	1.584	139.74	.843	.771	1.7030	7.51%
4	21	90	.852	.642	0.916	87.88	.847	.641	1.2689	38.53%
5	22	90	.884	.643	0.916	91.26	.865	.647	2.2326	143.73%
6	23	130	.885	.727	1.066	116.57	.876	.721	1.6752	57.15%
7	24	160	.885	.778	1.361	144.77	.872	.771	1.7245	26.71%
8	25	220	.869	.844	1.658	176.93	.879	.821	1.9532	17.80%
9	11	130	.661	.888	2.993	90.94	.758	.915	4.5509	52.05%
10	18	200	.845	.857	1.955	102.04	.834	.881	3.5122	79.65%
11	1	260	.880	.891	2.029	246.68	.883	.896	2.2911	12.92%
12	9	100	.701	.735	1.955	84.45	.729	.724	2.2343	14.28%
13	20	120	.855	.726	1.362	98.62	.842	.715	2.1108	54.98%

Table 2 shows the results obtained from the rigorous simulator and the neural network based optimised results in terms of batch time, reflux and reboil ratio, total distillate and middle vessel products. To get a rough idea of the effectiveness of the optimization routine, the %Improvement in the total amount of products obtained by optimisation over that obtained by the first-principle simulator are also reported in the table. In most cases significant reduction in batch time and increase in total products are observed.

It was also verified that a good match exists between the results obtained by off-line optimisation and those obtained by implementing the optimal values of R_D and R_B into the rigorous simulator. These results are not included here for the sake of conciseness, and will be the subject of a subsequent report.

5. Conclusions

A rigorous model of an MVC, validated against experimental data in a previous study, is used to generate a set of data, which is then used to develop a Neural Network based process model. An excellent fit between the first-principle model data and the data generated by NN was eventually achieved. A dynamic optimisation framework incorporating the NN based model has been developed to optimise operating parameters such as reflux and reboil ratios and batch time. It has been shown that the optimisation framework is able to improve on the values obtained through the simulator for Batch Time and Total amount of product collected. Performing experimental optimisation can take many hours whereas computational optimisation through the proposed approach takes few cpu seconds, therefore allowing for more efficient design of experiments.

6. Nomenclature

B_{MOL}, D_{MOL}	= Bottoms and distillate molar rate (kmol/min)
FL1	= Withdrawal rate from the column (kmol/min)
F_{L2}	= Feed rate to the column (kmol/min)
L_B	= Bottoms residue molar rate (kmol/min)
M_B, M_D	= Bottom and reflux drum holdup (kmol)
M_{BE}, M_{DE}	= Amount of bottom and distillateproduct (kmol)
M_F	= Holdup of the middle vessel (kmol)
N_F	= Feed plate location
N_T	= Total number of plates
R_B, R_D	= Internal reboil and reflux ratio [0,1]
R_{MOL}	= Reflux molar rate (kmol/min)
t_f	= Final Batch Time (min)
V_B, V_T	= Vapor boilup rate (bottom and top) (kmol/min)
X_{BE}, X_{DE}	= Ethanol mole fraction of the bottom and distillate product
X_F	= Ethanol mole fraction in the middle vessel

7. Acknowledgement

The University of Bradford Studentship to M.A. Greaves and the UK Royal Society supports to M.A. Hussain and I.M. Mujtaba are gratefully acknowledged.

M. Barolo gratefully acknowledges the financial support from the University of Padova (Progetto di Ateneo "VirSens"; # M1P012-1999).

8. References

Barolo, M., Guarise, G.B., Rienzi, S.A., Trotta, A. and Macchietto, S. (1996), *Ind. Eng. Chem. Res.*, **35**, pp. 4612.

Barolo, M., Guarise, G.B., Rienzi, S.A. and Trotta, A. (1998), *Comp. Chem. Eng.*, **22**, pp. S37.

Bortolini, P. and Guarise, G.B. (1970). *Quad. Ing. Chim. Ital.*, **6**, pp. 1.

Chen, C.L. (1988), PhD Thesis, Imperial College, London.

Cheong, W. Barton, P.I (1999), *Ind. Eng. Chem. Res.*, **38**, pp. 1504.

Edgar, T. F. (1996), *J. Proc. Contr.*, **6**, pp. 99.

Greaves, M.A., Mujtaba, I.M. and Hussain, M.A. (2001), In *Application of Neural Network and Other Learning Technologies in Process Engineering* (I.M. Mujtaba and M.A. Hussain, eds.), pp 149, Imperial College Press, London.

Hasebe, S., Aziz B.B. Abdul, Hashimoto, I. and Watanabe, T. (1992), In *IFAC Workshop on Interactions Between Process Design and Process Control* (J.Perkins Ed), London, Sept 7-8, Pergamon Press.

Mujtaba, I.M. and Macchietto, S.(1996), *J. Proc. Contr.*, **6**, pp. 27.

Mujtaba, I.M. and Hussain, M.A. (1998), *Comput. Chem. Engng.*, **22**, pp. S621.

Skogestad, S., Wittgens, B., Sørensen, E. and Litto, R. (1997), *AIChE J.*, **43**, pp. 971.

European Symposium on Computer Aided Process Engineering – 12
J. Grievink and J. van Schijndel (Editors)

511

A Two-Level Strategy of Integrated Dynamic Optimization and Control of Industrial Processes – a Case Study[*]

J. V. Kadam[1], M. Schlegel[1], W. Marquardt[1],

R. L. Tousain[2], D. H. van Hessem[2], J. van den Berg[2] and O. H. Bosgra[2]

[1]Lehrstuhl für Prozesstechnik, RWTH Aachen,
Turmstr. 46, 52064 Aachen, Germany

[2]Systems and Control Group, Mech. Eng., TU Delft,
Mekelweg 2, 2628 CD Delft, The Netherlands

Abstract

This paper discusses a two-level strategy integrating dynamic trajectory optimization and control for the operation of chemical processes. The benefit of an online dynamic re-optimization of operational trajectories in case of disturbances is illustrated by a case study on a semi-batch reactive distillation process producing methyl acetate.

1. Introduction

Increasing competition in the chemical industry requires a more agile plant operation in order to increase productivity under flexible operating conditions while decreasing the overall production cost (Backx et al., 2000). This demands economic optimization of the plant operation. However, existing techniques such as stationary real time optimization and linear model predictive control (MPC) generally use steady-state and/or linear representations of a plant model. They are limited with respect to the achievable flexibility and economic benefit, especially when considering intentionally dynamic processes such as continuous processes with grade transitions and batch processes.

There is an evident need for model based process operation strategies which support the dynamic nonlinear behavior of production plants. More recent techniques such as dynamic trajectory optimization and nonlinear model predictive control (NMPC) are still subject to research, and often the size of the applicable process model is still a limiting factor. Moreover, the integration of model predictive control and dynamic optimization for an optimal plant operation is an open field of research, which is e.g. studied in the EU-funded project INCOOP*. Various strategies have been suggested to implement such an integration. In the so-called direct approach (Helbig et al., 2000) the two main tasks, economic trajectory optimization and control, are solved simultaneously repetitively on each sample time of the process. This corresponds to a single-level optimal

---

[*] This work has been funded by the European Commision under grant G1RD-CT-1999-00146 in the "INCOOP" project (www.lfpt.RWTH-Aachen.de/INCOOP). The authors gratefully acknowledge the fruitful discussions with all the INCOOP team members.

control strategy. However, for large-scale and highly nonlinear processes this approach is intractable e.g. due to computational limitations.

In this paper, we employ a vertical decomposition approach: Here, the problem is decomposed into an upper level dynamic trajectory (re-)optimization, and a lower level (nonlinear) MPC which drives the process along the current optimal trajectory determined on the upper level. The interaction between the two levels is a key issue for the feasibility of such a decomposition. Data from the plant processed by a suitable estimation procedure enables nonlinear model-based feedback which can be utilized for several purposes: The dynamic optimization needs not to be performed each sample time but instead depending upon the nature of external disturbances. The feasibility of this approach is shown by means of a case study.

2. Problem definition

The goal of an optimal process operation is to maximize profit. In the ideal case, a perfect model of the process exists, the initial state x_0 at the beginning of the operation is known exactly and the process is not disturbed. Then the associated optimal trajectories for the operational degrees of freedom can be determined entirely off-line through the solution of an optimal control problem (P1):

$$\min_{u,t_f} \Phi(x,u,t_0,t_f) \tag{P1}$$

$$s.t. \quad 0 = f(\dot{x},x,u,d,t), \quad x(t_0) = x_0$$

$$y = g(x,u,d,t)$$

$$0 \geq h(x,u,d)$$

$$t \in [t_0,t_f]$$

In this formulation, x denotes the system states with initial conditions x_0, u free operational variables and d given parameters. f contains the differential-algebraic process model, g maps the system state to the outputs y. Constraints such as path and endpoint constraints to be enforced are collected in h. Φ denotes an economic objective function to be minimized on the time horizon $[t_0, t_f]$ of the process operation. In principle, problem (P1) can be solved by standard techniques for dynamic optimization (e.g. Betts, 2001) to determine an optimal u and optionally the final time of operation t_f, e.g. in the case of batch operation or for minimizing transition time in continuous processes.

3. Decomposition of dynamic optimization and control

The fact that the assumptions stated above are not fulfilled prevents an off-line solution of problem (P1) from being sufficient in any practical application. This is mainly due to model uncertainty and time-varying external disturbances $d(t)$ and unknown initial conditions x_0. To cope with this situation, a successive re-optimization of problem (P1) with updated models and initial conditions based on process measurements is required. However, the control relevant dynamics of typical processes will be too fast to enable real-time closed loop dynamic optimization. This is because current numerical techniques are not able to solve the problem (P1) for industrial-size applications involving large models sufficiently fast on the sample frequency and given small prediction horizon. High sampling frequency generally demands shorter prediction horizons and this might cause feasibility problems as well.

Alternatively, instead of solving (P1) directly, we consider a hierarchical decomposition. The two level strategy decomposes the problem into an upper level economic optimization problem (P2a) and a lower level control problem (P2a), as shown in Figure 1.

(P2a)

$$\min_{u^{ref},t_{fi}} \ \overline{\Phi}(\overline{x},u^{ref},t_0,t_f)$$

$$s.t. \ 0 = \overline{f}(\dot{\overline{x}},\overline{x},u^{ref},\overline{d},\overline{t}), \ \overline{x}(t_{0i}) = \overline{x}_{0i}$$

$$y^{ref} = \overline{g}(\overline{x},u^{ref},\overline{d},\overline{t})$$

$$0 \geq \overline{h}(\overline{x},u^{ref},\overline{d})$$

$$\overline{t} \in [\overline{t}_{0i},\overline{t}_{fi}]$$

$$\overline{t}_{0i+1} = \overline{t}_{0i} + \Delta\overline{t}, \quad \overline{t}_{fi+1} = \overline{t}_{fi} + \Delta\overline{t}$$

(P2b)

$$\min_{u} \sum_i (y_i - y^{ref})^T Q(y_i - y^{ref}) + (u_i - u^{ref})^T R(u_i - u^{ref})$$
$$+ (\tilde{x}_N - \overline{x}_N)^T P(\tilde{x}_N - \overline{x}_N)$$

$$s.t. \ 0 = \tilde{f}(\dot{\tilde{x}},\tilde{x},u,\tilde{d},\tilde{t}), \ \tilde{x}(t_{0i}) = \tilde{x}_{0i}$$

$$y = \tilde{g}(\tilde{x},u,\tilde{d},\tilde{t})$$

$$0 \geq \tilde{h}(\tilde{x},u,\tilde{d})$$

$$\tilde{t} \in [\tilde{t}_{0i},\tilde{t}_{fi}]$$

$$\tilde{t}_{0i+1} = \tilde{t}_{0i} + \Delta\tilde{t}, \quad \tilde{t}_{fi+1} = \tilde{t}_{fi} + \Delta\tilde{t}$$

The former is a dynamic real-time optimization (D-RTO) problem, which determines trajectories u^{ref}, y^{ref} for all relevant process variables such that an economical objective function $\overline{\Phi}$ is minimized and constraints $\overline{h}$ are satisfied. Only economic objectives such as maximization of production or minimization of process operation time are considered in $\overline{\Phi}$. The problem is repetitively solved on the rest of the entire time horizon on a sample time $\Delta\overline{t}$ for an update of the previous reference trajectories. The sample time has to be sufficiently large to capture the slow process dynamics, yet small enough to make flexible economic optimization possible. The re-optimization may not be necessary at each optimization sample time, instead it can be done based on the disturbance dynamics. The process model f used for the optimization has to have sufficient prediction quality and should cover a wide range of process dynamics.

On the lower level, an MPC problem (P2b) is solved in such a way that the process variables track the optimal reference trajectories in a strict operation envelope computed on the D-RTO level. The operation envelope, especially for controls u, is a small region around the reference trajectories; thus the MPC is referred to as delta mode MPC. The MPC sample time $\Delta\tilde{t}$ has to be significantly smaller than the D-RTO sample time $\Delta\overline{t}$, since it has to handle the fast, control relevant process dynamics. One requirement for the process model f used on the MPC level, which might be different from the model f used on the D-RTO level, is that it has to be simple enough, such that problem (P2b) can be solved sufficiently fast. A good prediction quality of f is required for the shorter time horizon $\lfloor t_{0i},t_{fi} \rfloor$ of (P2b). The initial conditions $\overline{x}_{0i},\tilde{x}_{0i}$ and disturbances $\overline{d},\tilde{d}$ for D-RTO and MPC are estimated from process measurements by a suitable estimation procedure such as an extended Kalman filter (EKF).

A proper separation of disturbances on different time scales is crucial for the decomposition of control and optimization, since the actions on both levels are induced by some kind of disturbance. Besides physical disturbances acting on the process directly,

514

changing external conditions such as market and environmental conditions also can be viewed as disturbances, because they require an update of reference trajectories for the optimal process operation. For example, it is conceivable that prices or product specification requirements change during the process operation. The production should then be adapted to the new situation, which can be done by a re-optimization of the process operation. The estimator (cf. Figure 1) estimates disturbances for which disturbance models have been added to the process model. The time-scale separation decomposes slowly varying or persistent from stochastic disturbances with time constants smaller than the prediction horizon of the MPC. The decision for a possible re-optimization is based on a disturbance sensitivity analysis of the optimal reference trajectories. A re-optimization is started only if persistent disturbances have been detected and have high sensitivities. On the lower level, both types of disturbances are used in the nonlinear prediction of the process. Persistent disturbances are taken up as a bias on the corresponding variables in the MPC, whereas stochastic disturbances are accounted for via the disturbance models added to the process dynamics. In this fashion the MPC problem (P2b) can be solved to obtain the updated control moves (cf. Lee and Ricker, 1994).

The structure in Figure 1 differs from the one suggested by Helbig et al. (2000): The sequence of estimation and time-scale separation is reversed. Both alternatives seem to have their merits and need to be investigated in the future.

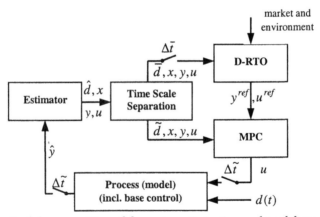

Figure 1. Vertical decomposition of dynamic optimization and model predictive control

4. Case study

The concept introduced above has been implemented in prototype software tools and applied to an industrial-size test example. Details on the numerical algorithms used in the different modules are beyond the scope of this paper. The process studied is a semi-batch reactive distillation process producing methyl acetate (MA) by esterification of acetic acid (AC) with methanol (MT) and byproduct water (W) (cf. continuous process described e.g. in Agreda et al., 1990). The process is started up with pure MT in the batch still and AC as a side feed stream. A gPROMS (gPROMS, 2001) model with 74 differential and 743 algebraic equations has been developed for this process. The dynamic optimization problems (P2a) have been solved using the optimizer ADOPT

(Schlegel et al., 2001). The objective is to maximize the amount of product (MA) for a fixed batch time of 4 hours (optimum, found by an off-line optimization with free final time) under strict enforcement of product purity of 0.95. The operational degrees of freedom are the reflux ratio and the reboiler vapor stream. Optimal profiles calculated by an off-line optimization run are shown in Figure 2 and 3 (solid lines) for the nominal case without disturbances.

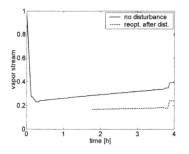 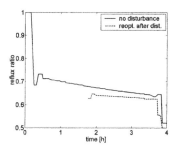

Figure 2. Control profiles for the nominal case and with re-optimization.

The disturbance scenario considered in our study is a drop of 50% in the feed rate of the side stream, which occurs before 1.75 hours. This is a persistent disturbance which is directly measurable and effects the product significantly. The analysis of the sensitivity of the optimal solution to the disturbance has shown that the nominal optimal trajectories need not to be updated for disturbance values less than 25% and these are handled at the MPC level. This decision making strategy for considering re-optimization or MPC at a current sample time subject to disturbances is proven to be suitable for this case study. However, further research is needed in this area.

The performance of the two level strategy is compared with using NMPC, delta mode MPC only and open loop operation. The product quality and the amount of product obtained using the above control strategies are depicted in Figure 3. If the original optimal trajectories would be followed (open-loop strategy) further, the disturbance prevents the required product quality of 0.95 to be met (* line in Figure 3) and leads to economic losses for this batch. A delta mode MPC that enforces a strict operation envelope around the reference trajectories and an NMPC without considering such an envelope are applied separately. The results depicted in Figure 3 (dash-dotted and dotted line resp.) show that these approaches are not economically viable (produces off-spec and less amount of product) for the given disturbance scenario.

The two level strategy of integrated dynamic optimization and control is then applied to the problem. A delta-mode MPC (constraints on control actions) is employed as a lower level MPC. The disturbance is recognized and a re-optimization of the trajectories is started (triggered by the sensitivity-based strategy). The new optimal operational trajectories are determined in order to meet the desired requirements. The re-optimization, which takes the changed state of the process due to the disturbance into account leads to changed optimal control profiles (Figure 2 -dashed lines). The profiles in Figure 3, left (dashed line) show that the product quality of 0.95 is met in the closed loop operation. Figure 3, right (dashed line) shows that more amount of on-spec product is produced. Thus the two level strategy guaranties an economical feasible operation

516

which is not guaranteed by the NMPC. Note that a rigorous nonlinear model is used at the MPC level which is the best option that can be considered.

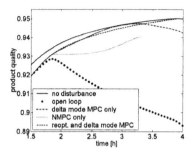 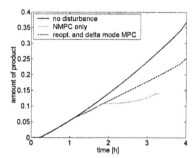

Figure 3. Product quality and amount of product

5. Conclusion

In this paper it has been explained that a more flexible plant operation to cope with changing market and operating conditions can be achieved by a systematic integration of dynamic optimization and model predictive control techniques. A vertical decomposition appears to be a viable strategy, which guaranties overall feasibility that might not be possible by an MPC only. With the help of a simulation study the benefit of the two level strategy, especially the dynamic re-optimization during the operation has been illustrated. Future research work in this area is required on a rigorous strategy for the separation of time scales, the relation of the process models used on the different levels, the choice of appropriate numerical algorithms on the different levels, etc.

References

Agreda V. H., L. R. Partin and W. H. Heise (1990). High-Purity Methyl Acetate via Reactive Distillation, *Chem. Eng. Prog.*, 86, 40-46.

Backx, T., O. Bosgra and W. Marquardt (2000). Integration of Model Predictive Control and Optimization of Processes. In: *IFAC Symposium on Advanced Control of Chemical Processes*. Vol. 1., 249-260.

Betts, J.T. (2001). *Practical methods for optimal control using nonlinear programming*. SIAM, Philadelphia.

gPROMS User Guide Version 2.0 (2001). Process Systems Enterprise Ltd., London.

Helbig, A., O. Abel and W. Marquardt (2000). Structural concepts for optimization based control of transient processes. In: Allgöwer, F. and A. Zheng (eds.), *Progress in Systems and Control Theory*, 295-311, Birkhäuser, Basel.

Lee, J. H. and N. L. Ricker (1994). Extended Kalman Filter Based Nonlinear Model Predictive Control. *Ind. Eng. Chem. Res.*, 33, 1530-541.

Schlegel, M., Th. Binder, A. Cruse, J. Oldenburg and W. Marquardt (2001). Dynamic Optimization Using a Wavelet Based Adaptive Control Vector Parameterization Strategy. In: Gani, R. and S. B. Jørgensen (eds.): *European Symposium on Computer Aided Process Engineering - 11*, 1071-1076, Elsevier.

European Symposium on Computer Aided Process Engineering – 12
J. Grievink and J. van Schijndel (Editors)
517

Bayesian Parameter Estimation in Batch Polymerisation

Zhen Lu, Elaine Martin and Julian Morris
Centre for Process Analytics and Control Technology
University of Newcastle, Newcastle upon Tyne, NE1 7RU, UK

Abstract

A Bayesian estimation framework is proposed for the tracking of time-varying parameters. The Bayesian approach is a statistical procedure that allows the systematic incorporation of prior knowledge about the model and model parameters, the appropriate weighting of experimental data, and the use of probabilistic models for the modelling of sources of experimental error. The interplay between these elements determines the best model parameter estimates. The proposed approach is evaluated by application to a dynamic simulation of a solution methyl methacrylate (MMA) batch polymerisation reactor. The Bayesian parameter adaptive filter is shown to be particularily successful in tracking time-varying model parameters.

1. Introduction

The operating objectives in many batch polymerisation processes require the satisfaction of complex property requirements for the final polymer whilst simultaneously reducing production costs. Most mechanical and rheological properties of polymer products are directly, or indirectly, linked to the molecular structural properties of the polymer chains (e.g. molecular weight distribution (MWD), copolymer composition distribution (CCD), chain sequence length distribution (CSD), etc.), which are difficult (sometimes impossible) to measure on-line.

Average polymer molecular weight properties (e.g. number and weight average molecular weights), that can be indirectly inferred from the on-line measurement of the solution viscosity or melt index of the polymer, are often selected as the major controlled variables that need to be maintained within well-determined limits so that desired product quality criteria can be satisfied.

Control strategies require that pre-determined trajectories for key process variables (e.g. reactor temperature) are implemented during batch operation (e.g. Thomas and Kiparissides, 1984). However, the operation of the batch polymerisation reactor is influenced by both process disturbances and model parameter variations due to the inherent process-model mismatch or changing operating conditions. Unless the time-varying model parameter values and any subsequent optimal control trajectory is updated regularly during batch operation, the control strategy will fail to meet the product quality specifications and the operating requirements (e.g. Ruppen et al. 1997, Crowley and Choi, 1997, 1998).

Crowley and Choi (1997, 1998) provided a comprehensive study of the optimal control of the molecular weight distribution in a batch free radical polymerisation experimental reactor. They proposed a control scheme incorporating state estimation and optimisation that could follow the desired molecular weight distribution by manipulating the temperature profile in the reactor. An Extended Kalman Filter (EKF) was used to retrieve the entire state vector based on process measurements (e.g. polymerisation temperature, and conversion) and product quality variables (e.g. molecular weight distribution). A model predictive control algorithm was then implemented to track the sub-optimal temperature profiles. However, they did not consider model parameter uncertainty and process disturbances in calculating the 'optimal set point sequence'. Such a policy may fail to meet the product quality specifications under model parameter variations (e.g. termination rate constant due to gel-effect). In this study, an alternative to the EKF is proposed for the tracking of time-varying parameters through the introduction of a Bayesian estimation framework.

2. Bayesian Parameter Estimation

Important information is lost when the model parameters are represented by a single point value, as in standard estimation and identification procedures, rather than by a full distribution as in Bayesian approaches. This is one of the main reasons for choosing a Bayesian approach to parameter estimation over conventional methods. It is assumed that the process has a state vector $\mathbf{x}(t)$ and is described by the continuous time process model:

$$\dot{\mathbf{x}}(t) = f(\mathbf{x}(t), \mathbf{u}(t), \theta, t) \tag{1}$$

where the measurement model takes the form:

$$\hat{\mathbf{y}}(t_k) = \mathbf{h}(\mathbf{x}(t_k), t_k) \tag{2}$$

The problem is to estimate a vector of parameters, θ, about which there may be some prior beliefs which can be expressed as a probability density function, $p(\theta)$. This prior distribution may be arrived at either from using previously monitored data, or by subjective process engineering judgement where the density expresses the users knowledge before the experimental data was collected, which in most applications is somewhat subjective. $\hat{\mathbf{y}}$ is a vector of model predictions, which depend on the parameters θ. After obtaining the process observations, $\mathbf{y}$, which have a probability distribution that is a function of the unknown parameters, the dependence of $\mathbf{y}$ on θ can be expressed as the conditional probability density function $p(\mathbf{y}|\theta)$, where the experimental error is denoted as $\varepsilon(\theta)$ such that the following error model applies:

$$\varepsilon(\theta) = \mathbf{y} - \hat{\mathbf{y}} \tag{3}$$

It is now assumed that the errors are modelled statistically such that there is a conditional probability density of ε for known θ, $p(\varepsilon \mid \theta)$. Then if θ is known, the conditional probability density $p(\mathbf{y} \mid \theta)$ can be represented as:

$$p(\mathbf{y} \mid \theta) = p(\mathbf{y} - \hat{\mathbf{y}} \mid \theta) = p(\varepsilon \mid \theta) \tag{4}$$

To update the probability density of the unknowns θ after new process data have been obtained, Bayes' theorem is used:

$$p(\theta \mid \mathbf{y}) = \frac{p(\mathbf{y} \mid \theta) p(\theta)}{p(\mathbf{y})} \tag{5}$$

where $p(\theta \mid \mathbf{y})$ is the Bayesian posterior distribution and contains all known information about the unknown parameter(s) after the on-line measurement information has been incorporated into the prior information. The denominator, $p(\mathbf{y})$ is the unconditional probability density of the observation data and from the law of total probability is given by:

$$p(\mathbf{y}) = \int p(\mathbf{y} \mid \theta) p(\theta) d\theta \tag{6}$$

This value is a constant and acts as a normalising constant. To construct the Bayesian parameter estimator, the posterior density is then expressed as:

$$p(\theta \mid \mathbf{y}) = \frac{1}{p(\mathbf{y})} e^{(\log p(\mathbf{y}\mid\theta) + \log p(\theta))} = Ke^{-J} \tag{7}$$

For a point estimate of the unknown parameter(s) θ, $\hat{\theta}$ is used as the value that maximises the a-posterior density. That is, $\hat{\theta}$ maximises the probability that the estimate is correct. To maximise the posterior probability density $p(\theta \mid \mathbf{y})$, the function J can be minimised:

$$J = -\log p(\mathbf{y} \mid \theta) - \log p(\theta) \tag{8}$$

The above function sums new measured process data and prior information and gives the appropriate weighting according to the knowledge of the statistical error. In this initial study of this new approach to parameter estimation, it is assumed that the errors are zero mean Gaussian errors. The error probability density then becomes:

$$p(\varepsilon \mid \theta) = \frac{1}{(2\pi \mid \sigma \mid)^{\frac{1}{2}}} \exp(-\frac{1}{2} \varepsilon^T \sigma^{-1} \varepsilon) \tag{9}$$

where $\sigma = E[\varepsilon\varepsilon^T]$, $E[\varepsilon] = 0$ and σ is the error covariance matrix. Thus, a working expression for J is:

$$J = \frac{1}{2}[\mathbf{y} - \hat{\mathbf{y}}]^T \sigma^{-1}[\mathbf{y} - \hat{\mathbf{y}}] + \frac{1}{2}\log(2\pi \mid \sigma \mid) - \log p(\theta) \tag{10}$$

and the Bayesian parameter estimation problem is defined as:

$$\min_{\theta} \quad J \quad \text{s.t.} \quad \theta \in \Omega_\theta \tag{11}$$

where Ω_θ is a set determined by lower and upper bounds on the elements of θ.

3. Results

The process studied is the free radical polymerisation reactor of methyl-methacrylate (MMA) (Mourikas *et al.* 2001). A mathematical model describes the dynamic behaviour of an experimental pilot scale system (Figure 1). Heating and cooling of the reaction mixture is achieved by controlling the flows of a hot and a cold water stream, through the reactor jacket. The polymerisation temperature is controlled by a cascade control system consisting of a primary PID and two secondary PI controllers. The polymerisation is highly exothermic and exhibits a strong acceleration in polymerisation rate due to gel-effects. Batch duration is 120 minutes.

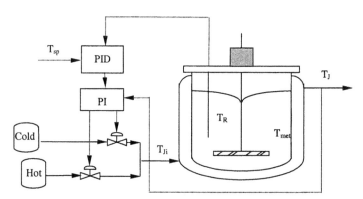

Figure 1. Plant Polymerisation Reactor

In practice important kinetic parameters such as k_p, the propagation rate constant and k_d, the initiator rate constant, cannot be determined accurately and may vary during the polymerisation. In this study, the propagation rate constant, k_p, is represented by $k_p = k_{p0}g^s_{p,corr}$ with the stochastic correction term $g^s_{p,corr}$, and the initiator decomposition rate constant k_d is represented by $k_d = k_{d0}g^s_{d,corr}$ with the stochastic correction term $g^s_{d,corr}$. In the EKF, a random walk modle is assumed for the behaviour of the stochastic state. In the process, the actual values of $g^s_{p,corr}$ and $g^s_{d,corr}$ are assumed to vary linearly from 0.9 at the nominal point to 0.78 and 0.66 for the propagation and initiator decomposition rates respectively.

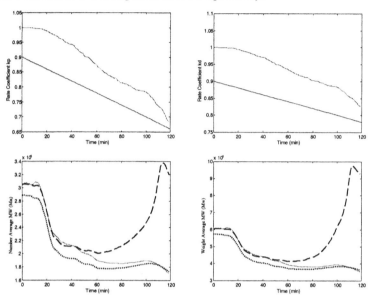

Figure 2. Performance of the adaptive EKF

The most common case in industrial practice is where the available on-line measurements are taken to be monomer conversion from an on-line densitometer, the reactor and the jacket inlet and outlet temperatures. In this study these three measurements are assumed to be available. In the simulations that follows, Gaussian zero mean white noise was added to the measurements to simulate measurement noise. In all the plots the large-dashed line (------) is the MMA process model, the dotted line represents the 'actual polymerisation process' (.......) and the solid line (———) represents the estimated values. The EKF performance is shown in Figure 2 with plots of Number Average Molecular Weight, *Mn*, and Weight Average Molecular Weight, *Mw*. Clearly the EKF based parameter adaptation cannot follow the change in rate constants leading to severe model-plant mismatch and hence errors between the predicted and actual values of *Mn* and *Mw*. This relatively simple example reflects the EKF parameter adaptation concerns expressed by Scali, *et al.* (1997).

Kozub and MacGregor (1992) indicated that the prediction of the molecular properties (number average molecular weight, weight average molecular weight and polydispersity) is vulnerable to model mismatch since these states are not observable. The main difficulty is the lack of any on-line measurements of the properties of a polymer molecular weight. The results shown above illustrate this problem. Instead of adapting the model and parameters off-line using the off-line measurements and analysis of the molecular properties from subsequent batch runs, Bayesian parameter estimation is used. The result is shown in Figure 3, demonstrates the significant

522

improvement that can be achieved using a Bayesian approach that involves the estimation of a distribution based parameter value rather than a spot value. The updated model can be observed to track the real process tightly, resulting in good estimates of the polymer properties.

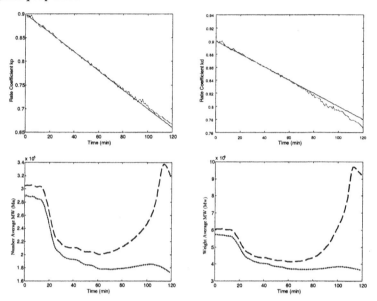

Figure 3. Performance of the Bayesian parameter estimator

4. Conclusions

An alternative approach to EKF based parameter estimation using Bayesian methods is proposed. In this initial study, only the impact of parameter variation is considered. Although only two critical parameters were chosen, the potential of the approach is clearly demonstrated. Future research will address the combined on-line identification of important model parameters, model states and reactor initial conditions. This will provide a firm base on which to build robust on-line optimal reactor control policies.

Acknowledgements

Mr Zhen Lu acknowledges CPACT and the University of Newcastle for the financial support of his PhD and Prof. Kiparissides, CPERI, Thessaloniki, for the simulation.

References

Thomas, I. M., and C. Kiparissides, 1984, Can. J. Chem. Eng., 62, 284.
Ruppen, D., D. Bonvin., and D. W. T. Rippin, 1997, Comput. Chem. Eng., 22, 185-199.
Crowley, T. J., and K. Y. Choi, 1997, Ind. Eng. Chem. Res., 36, 3676.
Crowley, T. J., and K. Y. Choi, 1998, Chem. Eng. Sci., 53, 2769.
Mourikas, G., P. Seferlis, J. Morris and C. Kiparissides, 2001, ESCAPE-11, Denmark.
Scali,C., M. Morretta, and C. Semino, 1997, J. Proc. Cont., 7, 5, 357.
De Valliere, P. and D. Bonvin, 1990, Comput. Chem. Eng., 14, 799.

European Symposium on Computer Aided Process Engineering – 12
J. Grievink and J. van Schijndel (Editors)
© 2002 Elsevier Science B.V. All rights reserved.

Super Model-Based Techniques for Batch Performance Monitoring

Lindsay McPherson, Julian Morris, Elaine Martin
Centre for Process Analytics and Control Technology
University of Newcastle, Newcastle Upon Tyne, NE1 7RU

Abstract

By combining mechanistic and empirical–based models, a process performance monitoring representation of a dynamic, non-linear process can be developed with the model-plant mismatch forming the basis of the monitoring scheme. In practice, the mechanistic model will not be perfect and therefore the residuals will contain structure. A modified model-based approach, Super Model-Based PCA (SMBPCA), is proposed which incorporates an additional residual modelling stage to remove structure from the residuals. The approach is evaluated on a simulation of a batch process using a number of residual modelling techniques including Partial Least Squares (PLS), dynamic PLS, ARX and dynamic Canonical Correlation Analysis (CCA). The out-of-control average run lengths for these techniques show that the SMBPCA approach gives improved process monitoring and fault detection compared to standard multivariate techniques.

1.Introduction

Standard multivariate methods for the monitoring of batch processes do not take into account the non-linear dynamic behaviour associated with batch processes. One approach that has been proposed to overcome the limitations of traditional multivariate batch techniques is that of model-based process performance monitoring. By combining a mechanistic model and an empirical–based model, a process performance monitoring representation of a dynamic, non-linear process can be developed with the residuals calculated from the model-plant mismatch forming the basis of the monitoring scheme.

Model-based Principal Component Analysis (MBPCA) has previously been applied to a simulation of an exothermic batch reactor and an ethylene compressor (Wachs and Lewin, 1998; Rotem, *et al.* 2000). The batch reactor was modelled using an exact first principles model of the process. However in an industrial environment, the assumption that a perfect first principles model can be built is generally not realisable. An evaluation of MBPCA was carried out by McPherson *et al.* (2001) on a simulation of an exothermic batch reactor (Wachs and Lewin, 1998) using a first principles model in which small parameter errors were introduced. The results showed that MBPCA did not perform significantly better than standard batch observation level PCA (Wold *et al,* 1998) with respect to fault detection. Returning to the original objective of removing the non-linear and dynamic components associated with process data, a reason for this can be established. Examination of the residuals, after the model-based technique had been applied, showed that serial correlation and non-normal behaviour was still present. A modified model-based approach, termed Super Model-based PCA, is proposed whereby an additional residual modelling stage is incorporated to remove the remaining structure and to obtain a set of unstructured residuals prior to the application of batch observation level PCA (Wold *et al.* 1998).

2. Benchmark simulation

The Super Model-based PCA approach was investigated on a simulation of an exothermic batch reactor (Wachs and Lewin, 1998). The two-stage simulation consists of five dynamic mass and energy balances representing three temperatures within the reactor and two reactant concentrations. Throughout the simulation four process variables were monitored, three reactor temperatures and the position of the coolant control valve. The reactant concentrations were not measured as it was assumed they would not be available on-line.

Using Monte Carlo simulation, a nominal data set of 40 batches was generated. The first principles model of the process was built using the same five equations and, as discussed previously, parameter errors were introduced to emulate plant-model mismatch. Three different faults (Wachs and Lewin, 1998) were introduced, (i) a gradual decrease in the activation energy of 1% from time point 150 to time point 180, (ii) a gradual decrease in the heat transfer coefficients of 15% from time point 150 to 180 and, (iii) an increase in the initial concentration of component A by 5%.

3. Super model-based PCA

In Model-Based PCA (MBPCA), unless a perfect mechanistic model of the process is available, serial correlation and non-normality will still be present in the residuals that are used as the basis of the monitoring representation. These attributes can mask the detection of abnormalities and deviations entering the process, thereby reducing the fault detection capability of MBPCA. Super Model-Based PCA (SMBPCA) includes an additional residual modelling stage prior to the application of PCA to remove the presence of serial correlation and non-normality (Figure 1). This results in a set of unstructured residuals to which linear statistical projection techniques can be applied. The SMBPCA algorithm is as follows:

1. Plant data is collected under normal operating conditions.
2. The model-predicted values are calculated and subtracted from plant-measured values, giving a set of structured residuals.
3. An error model is used to predict the values of the structured residuals from the original variables.
4. Predicted values of the structured values are subtracted from the original structured residuals, giving a set of unstructured residuals.
5. PCA or other monitoring methodologies are applied to the unstructured residuals and the confidence limits for the nominal representation are calculated.
6. The subsequent monitoring of the process is then based on the analysis performed on a nominal data set of unstructured residuals.

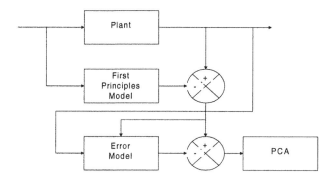

Fig. 1. Schematic of Super Model-based PCA

Five different types of error model have been investigated to model the structured residuals, Partial Least Squares (PLS), Autoregressive with eXogeneous input (ARX), dynamic PLS, dynamic non-linear PLS and dynamic Canonical Correlation Analysis (CCA).

3.1 Super model-based PCA with Partial Least Squares

Initially the error model was defined using a static PLS model, i.e. the residuals were modelled using the original variables. Comparing the normal probability plots of the original data and the residuals generated from the model-based approach (Figure 2) with the equivalent plot for the SMBPCA residuals that formed the basis of the PCA monitoring scheme (Figure 4a), it can be concluded that the SMBPCA approach reduces the non-linearity in the data. However, it is observed that the SMBPCA result is not significantly better than that obtained using the model-based PCA approach. To obtain an appreciation for the level of serial correlation in the data, plots of the partial autocorrelation function (PACF) were examined. Figure 4b shows the PACF plot for the residuals from the super model-based approach that can be compared with the equivalent plots for the original process data and the residuals generated from the model-based approach, Figure 3. For both MBPCA and SMBPCA with a PLS model, the level of serial correlation in the residuals is seen to approximately the same.

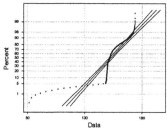
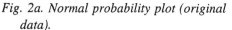
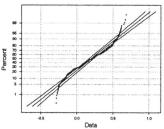

Fig. 2a. Normal probability plot (original data).

Fig. 2b. Normal probability plot of the residuals (MBPCA).

526

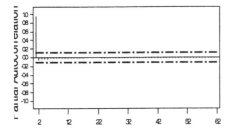

Fig. 3a. PACF plot (original data).

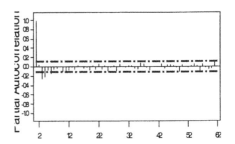

Fig. 3b. PACF plot of residuals (MBPCA)

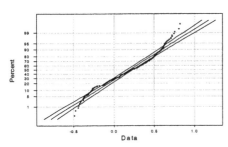

Fig. 4a. Normal probability plot of the
residuals (SMBPCA + PLS).

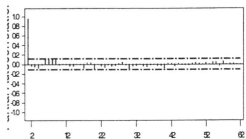

Fig 4b. PACF plot (SMBPCA + PLS)

3.2 SMBPCA with Dynamic Error Models

From the partial autocorrelation plots, it is evident that the process data contains serial correlation that can be modelled by an autoregressive model (Figure 3a). A number of different model structures were investigated to remove the serial correlation. Firstly the PLS residual model was replaced by an AutoRegressive with eXogeneous input (ARX) time series model. The ARX residual model was then replaced by a dynamic PLS representation (Qin, 1993). Dynamic PLS is a regression technique where PLS is applied to past values of the process variables and the residuals based on an ARX structure. By lagging the past values of the process variables and residuals, the impact of dynamic behaviour is addressed. Thirdly a dynamic non-linear PLS model (Baffi *et al.* 2000) was applied to the structured residuals. The non-linear PLS algorithm incorporates a polynomial (quadratic) function within the PLS algorithm through weight updating of the PLS inner and outer models, making it suitable for use in modelling non-linear systems. By integrating this non-linear PLS algorithm within an ARX framework, both the dynamics and non-linearity in the data are taken into account.

Finally Canonical Correlation Analysis (CCA), a latent variable method that maximises the correlation structure between variable sets X and Y, was used to model the residuals. A set of orthogonal latent variables in X and Y that are most highly correlated are calculated. In this study, CCA was extended into an ARX-structured framework, dynamic CCA, so as to allow the serial correlation in the data to be taken into account.

4. Fault Detection Analysis

The observation level scores for SMBPCA using a PLS residual model were used to develop a monitoring representation (Wold *et al.* 1998). The results are shown in Figure 5 for each fault type. Likewise the plots for SMBPCA with a dynamic non-linear PLS error model are shown, Figure 6. Based on such plots, the out-of-control Average Run Length (ARL) was calculated from one hundred batches. The out-of-control ARL is defined as the average number of samples between fault occurrence and fault detection. This measure was used to assess the fault detection ability of each technique. Figure 7 summarises the out-of-control ARL for each of the super model-based techniques and the ARL for standard batch observation level PCA and MBPCA.

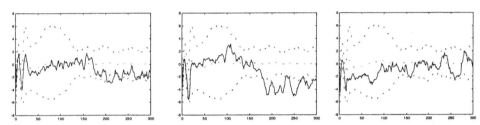

Figure 5: Observation level scores plots for faults 1,2 and 3 for SMBPCA + PLS.

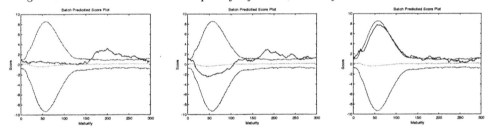

Figure 6: Observation level scores plots for faults 1,2 and 3 for SMBPCA + Dynamic Non-linear PLS.

For each of the three fault types the ARL has been reduced through the application of the super model-based techniques compared to standard observation level PCA. SMBPCA with PLS is the least effective of the super model-based methods. Its inadequacy in dealing with structured residuals is because a dynamic, non-linear element is not included in the modelling. For the first two fault types, the decrease in activation energy and the decrease in heat transfer coefficient, SMBPCA with dynamic non-linear PLS gives the best fault detection performance. Examination of the residuals (not shown) confirms that this is due to the removal of structure. For the third fault type, the increase in initial concentration of a reactant, SMBPCA + dynamic CCA gave the best performance, reducing the ARL by approximately 100 samples compared to standard PCA. Again this can be attributed to the removal of structure from the residuals. In general, all four super model-based techniques that involved dynamic components (ARX, dynamic PLS, dynamic non-linear PLS and dynamic CCA) gave approximately the same performance with respect to fault detection capability for faults

528

1 and 2. SMBPCA with Dynamic CCA and SMBPCA with ARX exhibited similar performance for the third fault type. Again the dynamic techniques showed significant improvement over the conventional methods.

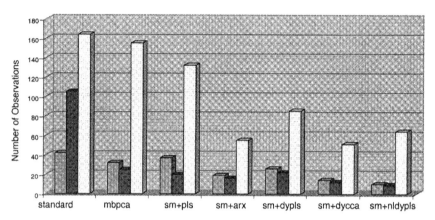

Figure 7: Average run length for faults 1 (▨), fault 2 (■), fault 3 (□).

5. Conclusions

Super model-based PCA has been shown to be an effective tool for the monitoring of batch processes through the reduction in ARL compared to standard batch observation PCA. The limitations of standard observation level PCA and model-based PCA in dealing with the structure present in the residuals has been overcome by incorporating an additional residual modelling stage prior to the application of PCA. Of the different residuals modelling techniques tested, SMBPCA + dynamic non-linear PLS and SMBPCA + dynamic CCA exhibit the best fault detection capability. Examination of the residuals confirmed this result was due to the removal of structure and non-linearity from the data. Future work will address more complex processes and the need to differentiate between model-plant mismatch and process disturbances.

Acknowledgements

LM acknowledges the EPSRC and the EU project PERFECT (Esprit Project 28870) for supporting her PhD studies.

References

Baffi, G., E. B. Martin and A. J. Morris, 2000, Chemo. Intell. Lab. Syst., 52, 5.
McPherson, L. A., Martin,E.B., Morris, A.J., 2001, IChemE, APC 6, York, UK, 23.
Qin, J, 1993, Proc. 32nd CDC, Texas, USA, 2617.
Rotem, Y., Wachs, A., Lewin, D.R. 2000, AIChE., 46(9), 1825.
Wachs, A. and D. R. Lewin, 1998, Proc. Proc. IFAC DYCOPS-5, Corfu, Greece, 86.
Wold, S., N. Kettaneh, H. Friden and A. Holmberg, 1998, Chem. Intell. Lab. Syst.44, 331.

European Symposium on Computer Aided Process Engineering – 12
J. Grievink and J. van Schijndel (Editors)

Closed loop indicators for controllability analysis

F. Michiel Meeuse[1], Yudi Samyudia[2] and Johan Grievink[1]

[1]Delft University of Technology, Department of Chemical Technology,
Julianalaan 136, 2628 BL Delft The Netherlands

[2]McMaster University, Department of Chemical Engineering,
1280 Main St West Hamilton, ON L8S 4L8

Abstract

This paper presents an approach to analyse the controllability of process alternatives based on the optimal *closed-loop* performance. The approach presented by Tousain and Meeuse (2001) is extended to include the robustness properties in terms of two different uncertainty models: parametric uncertainty and co-prime factor uncertainty. The simulation results on a distillation column demonstrate that the optimal design parameters using the nominal closed-loop controllability will not alter after introducing the robustness criterion.

1   Introduction

It is nowadays generally accepted that the traditional sequential design of a chemical process and the control system can lead to plants that are difficult or even impossible to control, or to sub-optimal combinations of the process and the control system. Morari (1983) used the concept of perfect control to identify the aspects that inherently limit the achievable control performance of process systems: time delays, right half plane zeros, input saturation and plant/model mismatch. A large number of alternative controllability indices have been proposed, including the Relative Gain Array (RGA), Singular Value decomposition based indices and the Closed-Loop Disturbance Gain (CLDG). The basic idea of all these indices is to screen alternative designs on potential control problems afterwards. However the main limitation of these indices is that the relation with the closed-loop behaviour is often unclear.

An alternative approach is the simultaneous design of the process and the control system by Mixed Integer Dynamic Optimisation (MIDO). In this approach first a superstructure is generated that contains a set of feasible process and controller options. Then optimisation techniques are employed in order to find the combination of process and controller that optimise the desired objective function under certain disturbance scenarios (Mohideen et al., 1996 and Bansal, 2000). The main advantage of these approaches is that in principle the optimum combination of the process and control system can be found. However detailed dynamic models are required and the solution is computational demanding. Moreover in all published cases the control structure was limited to multi-loop SISO PI controllers. In principle more general control algorithms could be considered, however this will increase the computational complexity even more. So we consider these methods as very useful in more detailed design phases, but too complicated for the earlier, screening phases.

Tousain and Meeuse (2001) have presented an alternative approach based on optimal control: **Closed Loop Controllability (CLC)**, which is a combination of the two approaches presented above. For a (linearized) model the optimal weighted closed loop variance is optimised by searching over the entire design space. For a linear model an exact solution of the optimal controller can be attained, eliminating the need for dynamic optimisation. The main advantage compared with the more traditional controllability indices is that the screening can be compared based on the best achievable *closed-loop* performances. However the computational effort required is considerable less than in the simultaneous approach. There is no need for dynamic optimisation because the optimal controller is given by an exact solution

Two limitations of this approach are that it relies on linear models and linear controllers and that no robustness properties can be guaranteed. Despite the fact that process systems are in general non-linear, a linear analysis often suffices e.g. by designing static nonlinear compensators (Morari, 1992). In this work we will extend this CLC approach to include robustness properties.

2 Basic modelling assumptions

We consider process systems that can be described by a set of Differential Algebraic Equations (DAE):

$$\dot{x}(t) = f\big(x(t), z(t), u(t), d(t), \theta(t), p\big),$$

$$0 = g\big(x(t), z(t), u(t), d(t), \theta(t), p\big), \tag{1}$$

$$y(t) = h\big(x(t), z(t), \theta(t), n(t)\big),$$

where $x(t) \in \mathbb{R}^{n_x}, z(t) \in \mathbb{R}^{n_z}, u(t) \in \mathbb{R}^{n_u}$ and $d(t) \in \mathbb{R}^{n_d}$ are respectively the state, algebraic, input and disturbance variables, $y(t) \in \mathbb{R}^{n_d}$ are selected output variables, $\theta(t) \in \mathbb{R}^{n_\theta}$ are the uncertain parameters, $n(t) \in \mathbb{R}^{n_y}$ is the measurement noise and $p \in P = P_1 \times P_2 \times \ldots \times P_{n_p}$ are the n_p design parameters, where P_i are the parameter sets. These sets can contain real-valued and integer-valued numbers. Real-valued design parameters are for example dimensions, pressures and temperatures. Integer-valued design parameters are related to structural design decisions e.g. the number of trays in a column or the type of reactor.

The behaviour of the plant in or close to its steady state operating point(s) can be described using linearised models. For all alternatives the non-linear model is linearised with respect to the states and input variables in it's operating point(s). We will use the standard state-space notation for these models:

$$\Delta \dot{x} = A(p)\Delta x + B(p)\Delta u + d,$$

$$\Delta y_m = C(p)\Delta x + n \tag{2}$$

where Δx, Δu and Δy_m are respectively the deviations of the states, the inputs and the measured outputs from their steady state values and n is the measurement noise. The system matrices $[A, B, C]$ are functions of p only. The system can be augmented with disturbance models. For stochastic disturbances and measurement noise the augmented system becomes:

$$\begin{bmatrix} \dot{x}_p \\ \dot{x}_d \end{bmatrix} = \begin{bmatrix} A_p & 0 \\ 0 & A_d \end{bmatrix} \begin{bmatrix} x_p \\ x_d \end{bmatrix} + \begin{bmatrix} B \\ 0 \end{bmatrix} u + \begin{bmatrix} 0 \\ B_d \end{bmatrix} [w_d],$$

$$y = \begin{bmatrix} C & 0 \end{bmatrix} \begin{bmatrix} x_p \\ x_d \end{bmatrix} + w_n, \tag{3}$$

where $[A_d, B_d]$ is the realisation filter for the disturbances, x_p and x_d are the states of the process and the filter, w_d and w_n are Gaussian white noise stochastic variables with respectively covariance matrices R_d and R_n and y_m are the measurements.

3   Closed loop optimal performance

Tousain and Meeuse (2001) have presented an approach for the CLC based on optimal performance. Their approach is based on the Linear Quadratic Gaussian (LQG) controller. The idea is to find the control $u(t)$ that minimises:

$$J_{clvar} = E\left\{ \lim_{T \to \infty} \frac{1}{T} \int_0^T \left(\Delta y^T Q \Delta y + \Delta u^T R \Delta u \right) dt \right\}, \tag{4}$$

where J is the objective function, E denotes expectation, and Q and R are weighting matrices of appropriate dimensions. For linear systems an exact solution exists for this controller, given by combining the optimal state observer (Kalman filter) with the optimal state feedback controller for the deterministic Linear Quadratic Regulator problem (Kwakernaak and Sivan, 1972). Tousain and Meeuse (2001) define the closed loop controllability as the minimum of (4), optimised over all controllers:

$$CLC = \min_{controllers} E\left\{ \lim_{T \to \infty} \frac{1}{T} \int_0^T \left(\Delta y^T Q \Delta y + \Delta u^T R \Delta u \right) dt \right\}, \tag{5}$$

Meeuse and Tousain (2002) have presented two alternative ways in which both the closed-loop controllability and the process economics can be considered. In the multi objective optimisation formulation both the closed-loop controllability and the economics are considered as objectives. An alternative formulation, closed-loop economic optimisation, includes the closed loop variance in the constraints. In this paper we will focus on a revised formulation of the closed loop controllability, ignoring the economics.

4   Closed loop robustness

The controller designed in the previous section achieves optimal performance when the real plant equals the model. In general there will always be plant model mismatch. Take for instance only the fact the controller with optimal performance was based on a linearised model. Therefore any controller that will be implemented must also meet the design specifications despite the uncertainty in the system. Hence the controllers must be robust. However the LQG controller used in the CLC analysis presented by Tousain and Meeuse (2001) has no robustness properties. We will now show how this method can be extended with robustness properties. We will consider two types of uncertainties: parametric uncertainties and coprime factor uncertainties.

4.1 Parametric uncertainty

Lets assume that the uncertain parameters in the system are indicated with θ. The parameter range is specified by:

$$\Theta = \left\{ \theta \mid \theta_L \leq \theta \leq \theta_U \right\}. \tag{6}$$

The worst case, close loop minimum variance for a given design is then given by:

$$CLC_{R1} = \max_{\theta \in \Theta} \; \min_{controller} \; E\left\{ \lim_{T \to \infty} \frac{1}{T} \int_0^T \left(\Delta y^T Q \Delta y + \Delta u^T R \Delta u \right) dt \right\}. \tag{7}$$

where CLC_{R1} is the closed loop controllability in case of parametric disturbances. Since the nominal values of θ are included in Θ, the following property always holds:

$$CLC_{R1} \geq CLC$$

4.2 Coprime factor uncertainty

An alternative uncertainty description, often used in the area of robust control is coprime factor uncertainty. Assume that the process transfer function, G(s), has a normalised left coprime factorisation (McFarlane and Glover, 1989):

$$G(s) = M_l^{-1}(s) N_l(s), \tag{8}$$

where both $M_l(s)$ and $N_l(s)$ are stable. The perturbed plant can then be written as:

$$G(s) = \left(M_l + \Delta_M \right)^{-1} \left(N_l + \Delta_N \right), \tag{9}$$

as also show in Figure 1. This uncertainty model is very general.

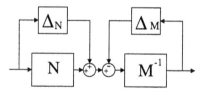

Figure 1. Coprime factor uncertainty.

The robust stability margin is the maximum perturbation for which the closed-loop system is stable. McFarlane and Glover (1989) have shown that there exists an exact solution for the controller that maximises this stability margin. The stability margin, ε, for a controller K is given by:

$$\frac{1}{\varepsilon} = \left\| \begin{bmatrix} K \\ I \end{bmatrix} (I - GK)^{-1} M^{-1} \right\|_\infty , \tag{10}$$

In case of coprime factor uncertainty the CLC is defined as:

$$CLC_{R2} = \min_{controller} \; E\left\{ \lim_{T \to \infty} \frac{1}{T} \int_0^T \left(\Delta y^T Q \Delta y + \Delta u^T R \Delta u \right) dt \right\},$$

$$st: \qquad \varepsilon \geq \varepsilon_{min}$$

Balas et al. (1998) state that values of ε > 0.2 – 0.3 are generally satisfactory.

5 Example

Tousain and Meeuse (2001) have applied their CLC indicator on the distillation column system presented by Skogestad (1997). We will now include the two robustness measures, CLC_{R1} and CLC_{R2} in their analysis.

System description

The system under study is a binary distillation column. The feed with $x_f = 0.5$ should be separated in a top and bottom product with respectively $x = 0.01$ and $x = 0.99$. The vapour/liquid equilibrium is described with a constant relative volatility. A simultaneous disturbance in the feed flow rate and feed composition are considered. The design parameters are the number of trays ($31 \leq NT \leq 51$) and the feed tray locations ($NF = centre \pm 5$).

Parametric uncertainty

The uncertain parameters considered are the feed rate and the feed composition:

$$\theta = \left\{ \begin{array}{l|l} F & 0.9 \leq F \leq 1.1 \\ x_F & 0.45 \leq x_F \leq 0.55 \end{array} \right\}$$

Figure 2a shows the objective function for the nominal values of F and x_f as defined by equation 4, whereas Figure 2b shows the robust CLC results as defined by equation 7. It can clearly be seen that the robust CLC results give the same trends as the results presented by Tousain and Meeuse (2001). The optimal design parameters using the CLC_{R1} are not altered, compared with the optimal design parameters using CLC. Obviously the values of the robust CLC indicator are higher than the nominal values since the values of maximised over the uncertainty domain.

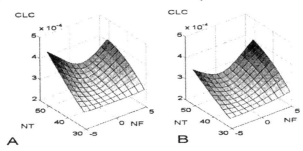

Figure 2. CLC results. A maximised over θ, B nominal value of θ.

Coprime factor uncertainty

Figure 3 presents the stability margins for the various designs related to the coprime factor uncertainty. When this stability margin is added as a constraint to the optimisation problem, the feasible area is reduced. In line with Balas et al. (1998) we add the constraint that ε > 0.2. The resulting feasible area is indicated in Figure 3. The values of CLC_{R2} in the feasible area are the same as the values of the CLC. Hence the optimum design parameters in case of coprime factor uncertainty have not changed, compared with the nominal design.

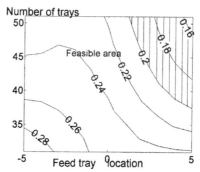

Figure 3. Co prime factor uncertainty stability margin

6   Conclusions

We have shown how the closed loop controllability as introduced by Tousain and Meeuse (2001) can be extended to include robustness properties. Uncertainties can explicitly be taken into account. For parametric uncertainty this leads to a min-max-min formulation. Other types of uncertainty can be modelled as co-prime factor uncertainty where the stability margin can be added as robustness constraint.

Literature

Balas, G.J., Doyle, J.C., Glover, K., Packard, A. and Smith, R. (1998) μ-Analysis and Synthesis Toolbox, The Mathworks Inc.

Bansal, V. (2000) Analysis, design and control optimization of process systems under uncertainty, PhD Thesis, University of London

Kwakernaak, H. and Sivan, R. (1972) Linear optimal control systems, Wiley, New-York

McFarlane, D.C. and Glover, K. (1989) Robust controller design using normalized coprime factor plant descriptions, Springer-Verlag, Berlin

Meeuse, F.M. and Tousain, R.L. (2002) Closed loop controllability analysis of process designs: application to distillation column design, accepted for publication in Comp & Chem. Eng.

Mohideen, M.J., Perkins, J.D., and Pistikopoulos, E.N. (1996) Optimal design of dynamic systems under uncertainty. *AIChE Journal* 42(8):2251-2272

Morari, M. (1983) Design of resilient processing plants III A general framework for the assessment of dynamic resilience. *Chem. Eng. Science* 38:1881-1891

Morari, M. (1992) Effect of design on the controllability of continuous plants, in: Perkins, J.D. (Ed) Interactions between process design and process control, IFAC Workshop, London

Tousain, R.L. and Meeuse, F.M. (2001) Closed loop controllability analysis of process designs: Application to distillation column design, in Gani, R. and Jorgensen, S.B. (Eds) Computer Aided Chemical Engineering, 9, 799 - 804

European Symposium on Computer Aided Process Engineering – 12
J. Grievink and J. van Schijndel (Editors)
© 2002 Published by Elsevier Science B.V.

Non linear dynamics of a network of reactors with periodical feed switching

L. Russo[1], E. Mancusi[2], P. L. Maffettone[3] and S. Crescitelli[1]

[1]Dipartimento di Ingegneria Chimica, Università degli Studi di Napoli "Federico II"
Piazzale Tecchio 80, 80125 Napoli, Italia
[2]Dipartimento di Ingegneria, Università del Sannio, Piazza Roma, 82100, Benevento, Italia
[3]Dipartimento di Scienza dei Materiali ed Ingegneria Chimica, Politecnico di Torino, Corso Duca degli Abruzzi 24, 10129 Torino, Italia

Abstract

In this paper we first assess the symmetry properties of a periodically forced network of reactors. Then, by making use of symmetry properties, the dynamic behaviour of the system is characterised. Bifurcation diagrams are derived with a continuation technique based on a suitable map, and symmetric and non-symmetric regimes are detected and described Possible bifurcation scenarios and, in particular, symmetry-breaking bifurcations are discussed.

1. Introduction

There is a growing interest in periodically forced reactors in the chemical engineering literature related to the possible improvement of heterogeneous catalysed processes.

Dynamic regimes can be achieved in different ways. For example, one (or more) input variable is forced to vary continuously in time. Alternatively, Matros (1985) proposed several methods to sustain a dynamic behaviour by periodically changing of the feed position while keeping the feed composition and temperature constant. In this context, proposed implementations are: The reverse flow reactors (RFRs), where the flow direction is periodically inverted, and the network of reactors (NTW), where the forcing is obtained with a periodical permutation of the reactor order.

Recently, for the treatment of lean waste gases, Brinkmann et al. (1999) have analysed a NTW of three catalytic burners operated with a valve system that allowed the cyclic permutation of their order. They have showed that the NTW is an efficient alternative to RFR to carry out autothermal VOC oxidation even at extremely low adiabatic temperature rise.

From a practical point of view, the optimal regime of the above mentioned dynamically forced system is a periodic regime with the same period of the forcing. However, depending on the operating or design parameters, those reactor configurations can show a rich dynamical response as demonstrated by Řeháček et al. (1992, 1998) and Khinast and Luss (2000) for RFRs. Indeed, it has been shown that multiperiodic, quasi-periodic or even chaotic regimes can be attained (Řeháček et al.; 1992, 1998). Řeháček et al. (1998) have also investigated the symmetry properties of the regime solutions of RFRs. To our knowledge, a similar analysis is still missing for the case of an NTWs.

It is well known that a complete dynamical characterization of an industrial process is useful (and sometime required) to develop adequate control strategies and to safely operate and design the reactors.

The discontinuous time forcing induces a spatio-temporal symmetry (Russo et al., 2002) into reactor model, and thus symmetric and asymmetric regime solutions are possible. The symmetry properties affect the bifurcation scenario: Some bifurcations cannot take place for the symmetry of the systems, independently of the intrinsic dynamics of the unforced system. It will be shown that solution symmetry cannot be lost through pitchforks as for RFRs, but through more complex phenomena such as resonance (frequency locking, e.g., Kuznetsov, 1998).

We have exploited the symmetry properties of such systems to implement a numerical technique based on standard pseudo-arclength continuation methods (Faraoni et al., 2001). The numerical approach here used to perform the bifurcation analysis is based on the numerical construction of a suitable map that once iterated gives the Poincarè map (Russo et al., 2002). To this end, public domain continuation software (AUTO97, Doedel et al., 1997) was used. With this approach it is possible to carry out an accurate bifurcation analysis and to characterize symmetric and non-symmetric regimes.

2. Reactor systems

In this work, we refer to simple reactor systems that are typically operated with a discontinuous forcing: A network of 3 identical CSTRs. In this case, every switch time the feed and the discharge positions are shifted according to a cyclic permutation.
In each CSTR the irreversible reaction A → B takes place following the kinetics:

$$r = k_0 C_A^q \exp\left(\frac{-E}{RT}\right) \tag{1}$$

The CSTRs are modelled with the two dimensionless ordinary differential equations, one for the mass balance and the other for the enthalpy balance, as follows:

$$\frac{d\alpha_i}{dt} = -\alpha_i + g(t)\alpha_{i-1} + Da(1-\alpha_i)^q \exp\left(\frac{\gamma\beta\theta_i}{1+\beta\theta_i}\right)$$

$$\frac{d\theta_i}{dt} = -\theta_i + g(t)\theta_{i-1} + Da(1-\alpha_i)^q \exp\left(\frac{\gamma\beta\theta_i}{1+\beta\theta_i}\right) + B(\theta_H - \theta_i) \tag{2}$$

$$g(t) = \begin{cases} 0 & \text{if } 0 \leq \frac{t}{\tau}(\text{mod }3) < 1 \\ 1 & \text{if } \frac{t}{\tau}(\text{mod }3) > 1 \end{cases}$$

In Eq. 2, the index i=1,2,3 identifies the reactor, while g is a square wave function introduced to account for the discontinuous feed scheme. It is apparent that the vector field is changed discontinuously in time, and after a time 3τ=T it recovers the initial configuration.

With some algebraic manipulation the system in Eq. (2) can be compactly represented as follows:

$$\frac{d\mathbf{u}}{dt} = \mathbf{F}(\mathbf{u}, \vartheta, t) \tag{3}$$

where $\mathbf{u} \in \mathbb{R}^m$ (in our case $\mathbf{u} = \{\alpha_1, \theta_1, \ldots, \alpha_i, , \theta_i \ldots \alpha_n, \theta_n\}$) is the state vector, $\vartheta \in \mathbb{R}^h$ (in our case $\vartheta = \{q, Da, B, \gamma, \beta, \tau, \theta_H\}$) is the parameter vector.
Discontinuous forcing introduces symmetries in the dynamical system. In the present case, the symmetry is a spatio-temporal one due to the discontinuous vector field that obeys to the following invariance property (Russo et al., 2002; Lamb, 1995):

$$GF(\mathbf{u}, \vartheta, t) = \mathbf{F}(G\mathbf{u}, \vartheta, t\text{-}\tau) \quad \text{with} \quad G = \begin{pmatrix} 0 & 1 & 0 \\ 0 & 0 & 1 \\ 1 & 0 & 0 \end{pmatrix} \tag{4}$$

The matrix G is the generator of the symmetry group Z_3, and thus this matrix has the properties that $G^3 = I$. The permutation is indeed:

$$\begin{pmatrix} v_1 \\ v_2 \\ v_3 \end{pmatrix} \xrightarrow{G} \begin{pmatrix} v_2 \\ v_3 \\ v_1 \end{pmatrix} \tag{5}$$

It is well known (Golubitsky et al., 1988) that symmetric systems can have symmetric solutions. Thus, periodic and symmetric solutions of system given by Eq. (3) are such that:

$$u(t) = Gu(t - \tau) \tag{6}$$

If a nonsymmetric solution exists, then the model symmetries imply the existence of other two nonsymmetric solutions. The property reported in Eq. 4 implies that if a nonsymmetric periodic solution exists, $\mathbf{u}(t)$, then the applications of G and G^2 onto $\mathbf{u}(t)$ give the other two solutions.
A peculiar property of systems with spatio-temporal symmetry is that the Poincaré map is the iterate of another map (Lamb, 1995). In the case at hand, if P is the Poincarè map, the following equation holds:

$$P = (G\varphi_\tau)^3 = H^3 \tag{7}$$

where φ_τ represents the evolution operator of system reported by Eq. (2) when it is not forced. Fixed points of P that are fixed points of H as well correspond to symmetric periodic solutions. On the other hand, fixed points of P that are not fixed points of H represent nonsymmetric periodic solutions.

3. Results

The bifurcation study was conducted by considering as bifurcation parameters the time switch (τ). The symmetry property reported in Eq. (7) allows the continuation of symmetric solution branches by continuing fixed points of map H. This approach reduces by a factor 2/3 the computation time of solution diagrams with respect to continuation of fixed points of the P map. In fact, for the case of map H, the ODE set can be integrated for a time τ whereas the map P not only requires an integration for a time T but also needs a permutations. The continuation has been implemented by using the software AUTO97 (Doedel et al., 1997) as reported by Faraoni et al. (2001).

In Fig. 1 the solution diagram for a network of 3 identical CSTRs is shown. The bifurcating parameter is the switch time τ

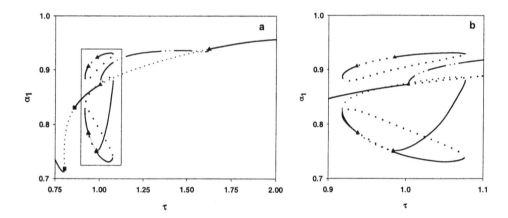

Figure 1 The solution diagrams report, as a representation of the regime solution, the dimensionless conversion (α_1) of the first reactor vs. one third of the dimensionless period of the forcing (τ). Solid lines represent stable T-periodic solutions; dashed lines unstable T-periodic solutions; dashed dot-dot lines stable 2T-periodic solutions; filled triangle Flip bifurcations; filled squares Neimark-Sacker bifurcations. Figure b shows details of the delimited rectangular region.

The solution diagram reported in Fig. 1 does not contain pitchfork bifurcations that are the bifurcations through which RFR symmetric solutions (Z_2 symmetric) lose symmetry (Russo et. al, 2002). Indeed, that kind of bifurcations cannot be encountered in Z_3 symmetric systems. Rather, loss of symmetry may take place through more complex phenomena as resonance (frequency locking). These phenomena can take place in the system under consideration since there are parameter regions in which it behaves as an oscillator when operated in unforced conditions. This means that the system is characterised by a natural frequency, which, in presence of an external periodic forcing, may couple with the external frequency giving rise to stable resonance regions (Arnold,

1983, Kevrekidis et al., 1986a, 1986b). It should be noted that the loss of symmetry implies that three different solutions are found, all sharing the same stability properties and bifurcations. By exploiting Eq. (4), once one solution is determined the others are easily calculated. In fact, they are calculated by applying G and G^2 to the fixed point from which the continuation was started.

For the sake of illustration, the symmetry properties of both the basic solution and that of the resonant isola are shown in Fig. 2.

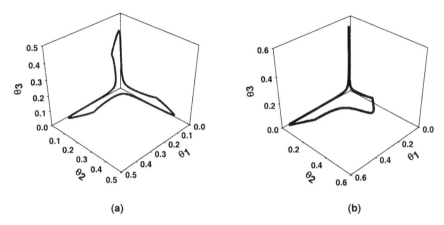

(a)   (b)

Figure 2 Orbit projections in the phase space. (a) and (b) are obtained for the same value of the switch time ($\tau=1$) and starting from two different initial conditions. (a) is symmetric; (b) is nonsymmetric

The solution reported in Fig. 2-(a) represents a symmetric solution (is a fixed point of map H). In fact, the projection is indifferent to any axis permutation. The resonant solutions relative to the frequency locking isola are nonsymmetric as it appears in Fig. 2-(b). In the case of periodic solution of the symmetric kind, the time series in the three reactors are identical but shifted in time of a time τ as shown in Fig. 3-(a). In the case of nonsymmetric solutions the time series of the three reactors are completely different from each other as visible in Fig. 3-(b).

4. Conclusions

This work has showed that reactor systems operated under periodic forcing posses symmetry properties effectively imposed by the forcing itself. The presence of such symmetry can be exploited to ease the dynamical analysis since it determines possible bifurcation scenarios and suggests multiplicity when nonsymmetric solutions are found.

540

(a)   (b)

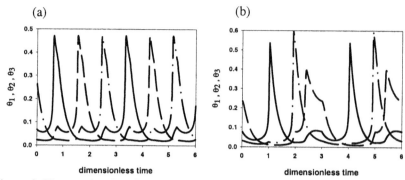

Figure 3. Time series data for a symmetric solution (a) and a nonsymmetric solution (b) of period T.

5. References

Arnold V.I., "Geometric methods in the theory of ordinary differential equations", Springer Verlag, (1983).

Brinkmann, M., A. A. Barresi, M. Vanni and Baldi G., " Unsteady state treatment of a very lean waste gases in a network of catalytic burners," *Catalysis Today*, **47**, 263 (1999).

Doedel E. J., Champneys A. R., Fairgrieve T. F., Kuznetsov Y. A., Sanstede B., and Wang X., "AUTO97: continuation and bifurcation software for ordinary differential equations", July (1997).

Faraoni V., Mancusi E., Russo L. and Continillo G., " Bifurcation analysis of periodically forced systems via continuation of a discrete map", ESCAPE-11, 27-30 May, (2001), Kolding, Denmark

Golubitsky, M., I. Stewart and D. G. Schaeffer, *Singularities and groups in bifurcation theory*, Vol II, Springer-Verlag, New York (1988).

Kevrekidis I.G., R. Aris and L.D. Schmidt, "The stirred tank forced", *Chem. Eng. Sci.*, **41**, 1549, (1986)a.

Kevrekidis I.G., R. Aris and Schmidt L. D., "some common features of periodically forced reacting systems", *Chem. Eng. Sci.*, **41**, 1263, (1986)b.

Khinast, J., and D. Luss, "Efficient bifurcation analysis of periodically-forced distributed parameters systems," *Comp. Chem. Eng.* , **24**, 139 (2000).

Kuznetsov, Y. A., Elements of applied bifurcation theory, 2nd ed., Springer Verlag, New York, (1998).

Lamb J.S.W., "Resonant driving and k-symmetry", Physics Letters A, 199, 55, (1995).

Matros Y.S, Unsteady Processes in Catalytic Reactor , Elsevier, Amsterdam, (1985).

Řeháček J., Kubíček M. and Marek M., "Modelling of a tubular catalytic reactor with flow-reversal", *Chem. Eng. Sci.*, **47**, 2897 (1992).

Řeháček, J., M. Kubíček and M. Marek, "Periodic, quasi-periodic and chaotic spatio-temporal patterns in a tubular catalytic reactor with periodic flow reversal," *Comp. Chem. Eng.*, **22**, 283 (1998).

Russo L., Mancusi E., P.L. Maffettone and S. Crescitelli, "Symmetry properties and bifurcation analysis of a class of periodically forced reactors", submitted to *Chem. Eng. Sci,* (2002).

European Symposium on Computer Aided Process Engineering – 12
J. Grievink and J. van Schijndel (Editors)
541

Robust Model-based Controllers via Parametric Programming

V. Sakizlis, N.M.P. Kakalis, V. Dua, J.D. Perkins and E.N. Pistikopoulos [*]
Centre for Process Systems Engineering,
Department of Chemical Engineering, Imperial College,
London SW7 2BY, U.K.

Abstract

In this paper a method is presented for deriving the explicit robust model-based optimal control law for constrained linear dynamic systems. The controller underlying structure is derived off-line via parametric programming before any actual process implementation takes place. The proposed control scheme guarantees feasibility under the presence of uncertainties by explicitly incorporating in the design stage a feasibility test and then by adapting on-line the origin according to the current disturbance realizations.

1. Introduction

Contrary to conventional control design methods, model predictive control (MPC) is particularly effective for dealing with a broad class of complex multivariable constrained processes. MPC determines the optimal future control profile according to a prediction of the system behavior over a receding time horizon. The control actions are computed by solving repetitively an on-line optimal control problem over a receding horizon every time a state measurement or estimation becomes available. The capabilities of MPC are limited mainly by the rigorous on-line calculations that make it applicable mostly to slowly varying processes. This shortcoming is surpassed by employing a different type of model-based controllers the so-called parametric controllers (Pistikopoulos et al., 2000, Bemporad *et al.*, 2002). These controllers are based on recently proposed novel parametric programming algorithms, developed in our research group at Imperial College, and succeed in obtaining the explicit mapping of the optimal control actions in the space of the current states. Thus, a model-based feedback control law for the system is derived off-line, hence, avoiding the restrictive on-line computations.

However, the inevitable presence of uncertainties, pertaining for instance to model inaccuracies, catalyst deactivation or heat-exchanger fouling, and the impact of persistent unmeasured disturbances, typically corresponding to slow variations in feed and utility conditions, have largely been ignored while designing the parametric controllers. Consequently, the performance of this novel control technique may lead to infeasibilities due to inaccurate forecasting of the process behaviour. These infeasibilities may result in situations such as off-spec production or hazardous plant

---

[*] To whom correspondence should be addressed. Tel.: (44) (0) 20 7594 6620, Fax: (44) (0) 20 7594 6606, E-mail: e.pistikopoulos@ic.ac.uk

operation. Hence, a modification of the explicit control law is necessary to ensure feasible and safe operation in the presence of disturbance and uncertainty.

In this work a novel methodology is presented for designing robust model based parametric controllers for general dynamic systems. The optimal control policy is derived off-line as a function of the process states via our parametric programming based theory and techniques (Dua *et al.*, 2002; Sakizlis *et al.*, 2001). The proposed control scheme manages to capture the effect of uncertainties and achieves satisfactory disturbance attenuation, thus avoiding the occurrence of infeasibilities. This is achieved (i) for the case of slowly varying disturbances, via suitably adjusting the state steady point based on disturbance estimates and implicitly adapting the control law to the current conditions; and (ii) for bounded uncertainties, by incorporating in the parametric controller design, an explicit feasibility constraint that ensures feasible operation for all possible uncertain parameter realizations.

2. Theoretical Developments

2.1 Parametric Controller

For deriving the explicit model – based control law for a process system, the following receding horizon optimal control problem is formulated:

$$\varphi(x_{t|0}) = \min_{v^N} x_{t|N}^T P x_{t|N} + \sum_{k=0}^{N-1} \left[y_{t|k}^T Q y_{t|k} + v_{t|k}^T R v_{t|k} \right]$$

$$\begin{aligned} \text{s.t. } & x_{t|k+1} = A_1 x_{t|k} + A_2 v_{t|k} \\ & y_{t|k} = B_1 x_{t|k} + B_2 v_{t|k} \\ & x_{t|0} = x^* \\ & 0 \geq g(y_{k|t}, x_{k|t}, v_{k|t}) = C_o y_{t|k} + C_1 x_{k|t} + C_2 v_{t|k} + C_4 \qquad k = 0,..N \end{aligned} \qquad (1)$$

where $x \in X \subseteq \Re^n$ are the states, $y \in \Re^m$, are the outputs and $v \in \Re^q$ are the controls; t is the time a measurement is available from the plant and k denotes the time element over the prediction horizon N. The outputs are the variables that we aim to control, i.e. to drive to their set-point, (temperatures, concentrations) whereas the states are the variables that fully characterize the current process conditions (enthalpies, specific volume). v^N denotes the sequence of the control vector over the receding horizon. The constraints g, which may pertain to product specifications or environmental and safety regulations, completely define the feasible operating region. By considering the current states x^* as parameters, problem (1) is recast as a multiparametric quadratic program (mp-QP). The solution of that problem (Dua *et al.*, 2002; Bemporad *et al.*, 2002) consists of a set of affine control functions in terms of the states and a set of regions where these functions are valid. This mapping of the manipulating inputs in the state space constitutes a control law for the system. The mathematical form of the parametric controller is as follows:

$$\left\{ v_{c\,t|0}(x^*) = a_c x^* + b_c; \quad \text{if } CR_c^1 x^* + CR_c^2 \leq 0 \right\} \qquad \text{for } c = 1,..., N_c \qquad (2)$$

where N_c is the number of regions in the state space, a_c, CR^1_c and b_c, CR^2_c are constant vector and matrices and the index c designates that each region admits a different control law. Vector $v_{ct|0}$ is the first element of the control sequence, whereas similar expressions are derived for the rest of the control elements.

2.2 Disturbance compensation

The model-based parametric controller described in the previous paragraph fails to address the impact of *persistent slowly varying unmeasured disturbances* on the process performance. The incorporation of a disturbance regulator in a predictive controller is usually performed by estimating the disturbance values and updating the model accordingly to ensure nonzero output target tracking (Muske and Rawlings, 1993; Loeblein and Perkins, 1999). The equivalent concept is applied here, generating a mechanism for updating the model on-line and modifying accordingly the derived control law. A distinction is done between two systems (i) one being the real process plant and (ii) the other comprising a model that represents an estimate of the process behavior, as shown in **Table 1**:

Table 1: Real vs. Prediction Model

Real System	$\hat{x}_{t|k+1} = A_1\hat{x}_{t|k} + A_2 v_{t|k} + W\theta_{t|k}$ $\hat{y}_{t|k} = B_1\hat{x}_{t|k} + B_2 v_{t|k} + F\theta_{t|k}$	Prediction model	$x_{t|k+1} = A_1 x_{t|k} + A_2 v_{t|k} + Ww_t$ $y_{t|k} = B_1 x_{t|k} + B_2 v_{t|k} + Fw_t$
θ:	real disturbance	w:	disturbance estimate

Vector w is modelled as a step disturbance. Once a disturbance enters into the system, at time k, there is an anticipated discrepancy between the measured output $\hat{y}$ and the predicted output y. Based on this discrepancy an estimate of the current disturbance is obtained via Kalman Filters or via a recursive or moving horizon least square estimator (Henson and Seborg, 1997). Then the disturbance estimate is employed for computing a new steady state point $[x_s\ v_s]$. If the dimension of output variables y is equal to the dimension of control inputs v and no input and state constraints are violated in the new steady state, then the new steady state is obtained by solving the linear system:

$$(I - A_1)x_s - A_2 v_s = W \cdot w; \qquad B_1 x_s + B_2 v_s = -F \cdot w \qquad (4)$$

Cases when q=dim $(v) \geq m$=dim (y) or when there are active control/state constraints are treated in a similar fashion. Based on $[x_s\ v_s]$, the control law (2) is shifted as follows:

$$\left\{ v_{ct|0}(\hat{x}^*) = a_c(\hat{x}^* - x_s) + v_s + b_c; \quad \text{if } CR_c^1(\hat{x}^* - x_s) + CR_c^2 \leq 0; \right\} \quad \text{for } c = 1,\dots, N_c \quad (5)$$

This technique implies the construction of a state-of-the art adaptive parametric controller. This controller achieves output feedback by readjusting automatically its tunings to account for the presence of disturbances.

2.3 Robust Parametric Controller

During the implementation of the parametric controller (5), constraint violations are likely to occur when the disturbance estimate is inaccurate or the process is subject to additional unknown *uncertain* variations. The traditional technique for designing a robust model based controller that avoids this shortcoming, relies on minimizing on-line the worst case cost, thus leading to a min-max optimal control problem (Campo and Morari, 1987). This can be (i) computationally prohibitive due to the complete exploration of the uncertainty extreme points and (ii) may also lead to a conservative controller since the worst-case cost is penalized in the objective. Recently, Bemporad *et al.*, (2001) based on the work of Pistikopoulos *et al.*, (2000) and Bemporad *et al.*, (2002) developed a technique for moving the computations off-line, but their approach merely minimizes the worst case ∞- norm of the output/input deviations that may cause deadbeat control and may result in a conservative control policy.

Here, feasibility analysis theory (Halemane and Grossmann, 1983; Bansal *et al.*, 2000) is adopted for deriving the robust parametric controller (as in Kakalis *et al.*, 2001). The system is considered to be perturbed by a vector of unknown input affine time-varying uncertainties ω that are assumed to belong to a compact polyhedral set $\omega \in \Omega \subseteq \mathfrak{R}$:

$$x_{t|k+1} = A_1 x_{t|k} + A_2 v_{t|k} + \omega_{t|k} \tag{6}$$

Note, that we distinguish between the input disturbance θ and the uncertainty ω in the sense that the latter has a higher frequency of variation and it lacks any reliable estimate. The *robust controller* can be defined as the controller that provides a single control action that ensures constrains' satisfaction for all the possible uncertainty realizations over the complete horizon. This is mathematically translated to the following constraint:

$$\forall \omega_{t|k} \in \Omega(\forall j \in J[g_j(x_{t|k}, v_{t|k}, \omega_{t|k}) \le 0]) \Leftrightarrow \max_{\omega} \min_{u}\{u \mid u \ge g_j, \forall j \in J\} \le 0 \tag{7}$$

This flexibility test condition (7) is incorporated explicitly as a constraint in the optimization problem (1), aiming to derive a robust parametric controller. The solution of the inner infinite max-min problem is accomplished by recasting problem (7) subject to the dynamics (6) as a linear parametric optimization problem (mp-LP). This results a set of piecewise affine expressions (Bansal *et al.*, 2000) for the feasibility function u in terms of the uncertain parameters, the controls and the current states: $u_i = d_i \cdot \omega_{t|k} + f_i \cdot x^* + p_i \cdot v_{t|k} + h_i$, $\forall i = 1,..I$. Next, by exploiting the linearity of these functions and the convexity of the uncertainty set Ω, it is proved that the critical uncertainties that bottleneck the plant operation lie on a finite subset ω^j of the uncertainty space vertices. Thus, the feasibility constraint (7) is equivalently, replaced by a finite set of constraints corresponding to the critical uncertain combinations. This reformulation gives rise to the following multiperiod receding horizon optimal control problem:

$$\varphi(x_{t|0}) = \min_{v^N} x_{t|N}^{l_n \; T} P x_{t|N}^{l_n} + \sum_{k=0}^{N-1}\left[y_{t|k}^{l_n \; T} Q y_{t|k}^{l_n} + v_{t|k}^T R v_{t|k} \right]$$
$$\text{s.t. } x_{t|k+1}^l = A_1 x_{t|k}^l + A_2 v_{t|k} + \omega_{t|k}^l$$
$$y_{t|k}^l = B_1 x_{t|k}^l + B_2 v_{t|k} \tag{8}$$
$$x_{t|0} = x^*$$
$$0 \ge g(y_{t|k}^l, x_{t|k}^l, v_{k|t}) \quad k = 0,.., N; \; 1 = 1,..., L;$$

where l_n corresponds to the nominal scenario of the uncertainty w. The solution of (8) (i) ensures feasible operation without exploring the complete space of the uncertainty realization as in Campo and Morari, (1987) and (ii) corresponds to a less conservative control action since it minimizes the nominal and not the worst case quadratic cost. By treating the current states x^* as parameters the optimal control problem (8) is recast as a parametric quadratic problem (mp-QP) and its solution results in a robust parametric controller of the form of (2). This controller apart for accounting for the presence of uncertainties is also capable of rejecting slowly varying disturbances by performing origin rearrangement as described in **section 2.2**.

3. Illustrative Example

The demonstrative example is concerned with deriving the explicit robust control law for an evaporator process studied in Kookos and Perkins, 2001. The controlled states are the pressure P_2 and the concentration C_2 of the product stream. Their nominal reference values are: P_2=50.57KPa; C_2=25%. The manipulating inputs are the pressure of the utility steam P_{100} that heats up the feed and the cooling water flow F_{200} that condenses the by-products. The active output constraint when operating around the nominal point is: $C_2 \geq 25\%$ whereas additional constraints are also imposed on the inputs.. The feed flowrate F_1 is considered as a disturbance that can be estimated only within accuracy of ±0.25kg/s but varies within a wider range.

For deriving the robust control law, first the feasibility problem (7), (6) is solved with a horizon N=2, considering F_1 as a deterministic uncertainty $\omega = F_1 - F_{1nom}$, where F_{1nom}= 10kg/sec is the nominal uncertainty value. The critical scenarios are: $\omega_{t|0}^{1,2,3}$=0.25,0.25,-0.25; $\omega_{t|1}^{1,2,3}$=0.25,-0.25,-0.25. A multiperiod MPC problem (8) based on those scenarios is formulated and solved as an mp-QP treating the current states as parameters. The solution features a robust parametric controller that is shown in Figure 1. The execution of the control law is shown in Figure 2. The system is initially perturbed and as it is driven back to the origin a sequence of step disturbances occur. The robust controller with origin readjustment generates a back-off from the nominal point to accommodate these variations, whereas the nominal controller although it does reject the disturbance at steady state exhibits severe constraint violations.

4 Conclusions

In this paper a novel framework is presented for designing robust model-based parametric controllers of linear dynamic systems that are subject to input disturbances and uncertainties. The controller consists of piecewise affine expressions for the control variables in terms of the states. The implementation of the control action is achieved by simple linear function evaluations, thus avoiding any expensive on-line computations. The controller guarantees robustness by means of satisfaction of constraints and disturbance attenuation.

5 References

Bansal V., J.D. Perkins and E.N. Pistikopoulos, 2000, AIChE, 46, 335.
Bemporad A., F. Borrelli and M. Morari, 2001, Proc. Eur. Cont. Conf.
Bemporad, M. Morari, V. Dua and E.N. Pistikopoulos, 2002, Automatica, 38, 3.
Dua V., N.A. Bozinis and E.N. Pistikopoulos, 2002, A multiparametric programming approach for mixed integer and quadratic process engineering problems. Accepted in Comp. Chem. Eng. (also appearing in ESCAPE-11 Proc. pp.979).
Halemane K.P. and I.E. Grossmann, 1983, AIChE J., 29, 428.

546

Henson, M.A. and D.E. Seborg , 1997, Nonlinear process control, Prentice Hall.

Kakalis N.M.P., V. Dua, V. Sakizlis, J.D. Perkins and E.N. Pistikopoulos. 2002 A parametric optimization approach for robust MPC. Accepted in IFAC Cong. Aut. Cont.

Kookos I.K. and J.D. Perkins, 2001, accepted in J. Proc. Contr.

Loeblein C. A and J.D. Perkins, 1999, AIChE J., 45, 1018.

Muske K.R. and J.B. Rawlings, 1993, AIChE J., 32, 262.

Pistikopoulos, E.N., V. Dua, N.A. Bozinis, A. Bemporad and M. Morari (2000). Comput. Chem. Eng. 24, 183.

Sakizlis V., J.D. Perkins and E.N. Pistikopoulos, 2001, Multiparametric Dynamic Optimization of Linear Quadratic Optimal Control Problems: Theory and Applications. Accepted in: Optimization and Control in Chemical Engineering. Editor: R. Luus.

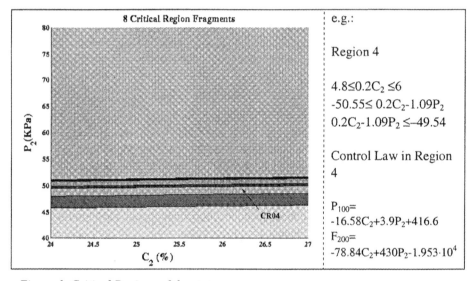

Figure 1. Critical Regions of the state space

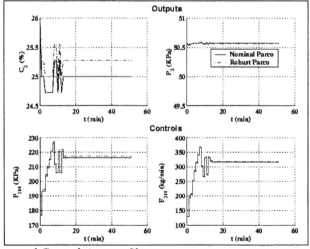

Figure 2. Output and Control time profiles.

European Symposium on Computer Aided Process Engineering – 12
J. Grievink and J. van Schijndel (Editors)
© 2002 Published by Elsevier Science B.V.

Dynamic Trajectory Optimization Between Unstable Steady-States of a Class of CSTRs

Andrea Silva B., Antonio Flores T[*]. and Juan José Arrieta C.
Universidad Iberoamericana, Prolongación Paseo de la Reforma 880 México D.F., 01210, México.

Abstract

In this work the computation of optimal dynamic transition trajectories between unstable steady-states was addressed. Using orthogonal collocation on finite elements, the optimal control problem was discretized and cast as a non-linear optimization program. Transition trajectories were computed for a CSTR operating around a non-linear multiplicity region. With the proposed optimization formulation optimal transition trajectories involving unstable steady-states were successfully computed.

1. INTRODUCTION

Design of transitions between steady-state operations is an important problem in the chemical process industry. For instance, in the polymerization industry transitions are carried out to switch from producing a certain polymer A to produce a different polymer B (A and B are the same polymer but with different characteristics such as molecular weight distribution). Another important problem for which transitions are sought occurs during plant start-up/shut-down. Since some of the economic attractive operating steady-states are open-loop unstable, it becomes important to be able to obtain optimal dynamic transition trajectories between such unstable operating points.

When a mathematical model of the process is available it becomes feasible to compute optimal transition trajectories leading to an open-loop optimal control problem (OCP). Roughly speaking three forms of solving the OCP have been proposed [1]: (i) direct application of the Pontryagin's maximum principle giving rise to a two-point boundary value problem, (ii) parameterization of the control trajectory leading to a non-linear optimization program where both the objective function and the constraints are evaluated by integrating the model equations and (iii) full discretization of the OCP leading to a non-linear optimization program. In principle any of these methods can be used for computing optimal transition trajectories between open-loop stable steady-states (where stability is defined in the Hurwitz sense). However, when it comes to compute optimal transition trajectories between: (a) stable-unstable, (b) unstable-stable and (c) unstable-unstable steady-states the full discretization approach of the OCP looks as the methodology that should be used to cope with this sort of OCPs [2]. In fact approaches based on the control trajectory parameterization strategy could be used for state transitions between unstable steady-states by first achieving closed-loop

---

[*] Author to whom correspondence should be addressed. E-mail: antonio.flores@uia.mx, phone/fax: +52 5 267 42 79, http://200.13.98.241/~antonio

stabilization of the system and then computing the transition trajectory; such approach has been used in [3] to initially solve unstable transition problems. Direct application of the control trajectory parameterization approach is totally unfeasible because during model system numerical integration the states will not be attracted to an unstable operating point and will converge to a stable one [2]. As far as we know unstable dynamic transition problems have not been previously addressed in the open academic literature; hence we trust to make a contribution to the OCP area. For solving the open-loop optimal transition trajectory problem between unstable steady-states, the full discretization approach was used in this work. To discretize the model the orthogonal collocation method on finite elements was used and the resulting OCP was solved as a non-linear optimization program.

2. PROBLEM DEFINITION

In order to compute transition trajectories, the CSTR model as proposed by Hicks and Ray [4] was used. Because the original parameters set used by these authors did not lead to multiple steady-states, some of the values were modified in order to end-up with a multiplicity map. In dimensionless form the model is given by:

$$\frac{dy_1}{dt} = \frac{1-y_1}{\theta} - k_{10} \exp\left(-N/y_2\right) y_1 = f_1 \tag{1}$$

$$\frac{dy_2}{dt} = \frac{y_f - y_2}{\theta} + k_{10} \exp\left(-N/y_2\right) y_1 - \alpha u (y_2 - y_c) = f_2 \tag{2}$$

where y_1 stands for dimensionless concentration (c/c_f), y_2 is the dimensionless temperature (T/Jc_f), y_c is the dimensionless coolant temperature (T_c/Jc_f), y_f is the dimensionless feed temperature (T_f/Jc_f), and u is the cooling flowrate. Table 1 contains the numerical values of the parameters used in this work; this set of parameter values lead to operate around the multiplicity region shown in figure 1.

Table 1. Parameter values.

θ	20	Residence time	T_f	300	Feed temperature
J	100	$(-\Delta H)/(\rho C_p)$	k_{10}	300	Preexponential factor
c_f	7.6	Feed Concentration	T_c	290	Coolant temperature
α	1.95×10^{-4}	Dimensionless heat transfer area	N	5	$E_1/(RJc_f)$

3. PROBLEM FORMULATION

3.1 Orthogonal collocation on finite elements

The dynamic mathematical model was fully discretized using orthogonal collocation on finite elements (OCFE). The approach consisted in forming the discretized version of the residual equation stated as:

$$r = y' - f(y, u) \tag{3}$$

where $r = [r_1\ r_2]^T$, $y' = [dy_1/dt\ dy_2/dt]^T$, $y = [y_1\ y_2]^T$ and $f = [f_1 f_2]^T$. In the collocation method the residual is minimized forcing it to pass through zero only in a finite number of discrete points [5]:

$$\int_0^1 \mathbf{r}\delta(x - x_i)dx = 0 \tag{4}$$

where $\delta(x-x_i)$ is the Dirac delta function. If the discrete points are chosen as the roots of an orthogonal polynomial then the discretization procedure is called the orthogonal collocation method. The way of discretizing the model consists in representing each unknown variable as a linear combination between a set of coefficients (γ_j)and a set of basis functions $\phi_j(\tau_i)$ (commonly taken as Lagrange polynomials):

$$y(\tau_i) = \sum_{j=0}^{N+1} y_j \phi_j(\tau_i) \tag{5}$$

$$\phi_j(\tau_i) = \prod_{\substack{k=0 \\ k \neq j}}^{N+1} \frac{\tau_i - \tau_k}{\tau_j - \tau_k} \tag{6}$$

where N is the number of internal collocation points and τ_i corresponds to the *ith*-orthogonal collocation root. In terms of the discretized model the residual equation is rewritten as:

$$\mathbf{r}(\tau_i) = \sum_{j=0}^{N+1} y_j \dot{\phi}_j(\tau_i) - h\mathbf{f}(\mathbf{y}, u) \tag{7}$$

where h stands for the element length. The first derivative of the Lagrange polynomial is computed from:

$$\dot{\phi}_j(\tau_i) = \left[\prod_{\substack{k=0 \\ k \neq j}}^{N+1} \frac{1}{\tau_j - \tau_k} \right] \left[\sum_{\substack{p=0 \\ p \neq j}}^{N+1} \left(\prod_{\substack{l=0 \\ l \neq j, l \neq p}}^{N+1} (\tau_i - \tau_l) \right) \right] \tag{8}$$

For each *kth*-element (except the last one) the discretized mathematical model is given by (notice that $A_{ij} = \dot{\phi}_j(\tau_i)$):

- Mass balance

$$\sum_{j=1}^{N+2} A_{ij} y_{1,j} - h_k f_1(\mathbf{y}) = 0, i = 2, N+1 \tag{9}$$

- Energy balance

$$\sum_{j=1}^{N+2} A_{ij} y_{2,j} - h_k f_2(\mathbf{y}, u) = 0, i = 2, N+1 \tag{10}$$

- Equations for solution continuity between elements (mass and energy)

$$\frac{1}{h_k} \sum_{j=1}^{N+2} A_{N+2,j} y_{1,j} - \frac{1}{h_{k+1}} \sum_{j=1}^{N+2} A_{1j} y_{1,j} = 0, i = 2, N+1 \tag{11}$$

$$\frac{1}{h_k} \sum_{j=1}^{N+2} A_{N+2,j} y_{2,j} - \frac{1}{h_{k+1}} \sum_{j=1}^{N+2} A_{1j} y_{2,j} = 0, i = 2, N+1 \tag{12}$$

- Total transition time

$$\sum_{k=1}^{NE} h_k \leq \theta \tag{13}$$

where *NE* is the number of finite elements. For the last element only the mass and energy discretized equations are taken into account, but collocation is done including the end-point of the solution space. This means that for the last element $i=2,N+2$.

550

3.2 Dynamic optimization

In terms of the discretized model the dynamic optimization problem can be formulated as:

$$\min_{\mathbf{y}_1,\mathbf{y}_2,\mathbf{u},\mathbf{h}} \int_0^\theta \left[\alpha_1(y_1-y_{1s})^2+\alpha_2(y_2-y_{2s})^2+\alpha_3(u-u_s)^2\right]dt \qquad (14)$$

s.t.

Eqs. 9 – 13 are met

$$y_{1,l}\le\mathbf{y}_1\le y_{1,u}$$

$$y_{2,l}\le\mathbf{y}_2\le y_{2,u}$$

$$u_l\le\mathbf{u}\le u_u$$

$$h_l\le\mathbf{h}\le h_u$$

in the above formulation $\alpha_1=10^6$, $\alpha_2=2\times10^3$ and $\alpha_3=10^{-3}$; such weighting values were suggested by Hicks and Ray. They tried different weighting factors and found that the optimal trajectories were independent of the used weight factors. In particular in this work these factors were selected because concentration was considered the main output variable; a small weight factor was used in the manipulated variable in order to avoid strong control actions. For each transition the subscript s represents the final steady-state operating conditions, while subscripts l and u stand for lower and upper bounds on the corresponding variables.

4. RESULTS AND DISCUSSION

Optimal transition trajectories were sought for the operating points shown in figure 1; in this figure s and u represent stable and unstable steady-states, respectively. Several transition cases were computed and they are summarized in table 2. The right number of both finite elements and internal collocation points were selected by trial and error; it was observed that due to the nature of the solution curves several discretization points were needed. So, it was found that 2 internal collocation points and 15 finite elements were sufficient.

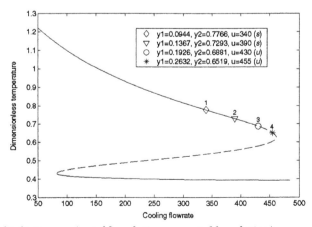

Figure 1. Multiplicity map (- stable solution, -- unstable solution).

Table 2. Definition of transition points.

Transition	Initial state		Final state	Transition	Initial state		Final state
A	2(s)	→	1(s)	C	3(u)	→	2(s)
B	2(s)	→	3(u)	D	3(u)	→	4(u)

In order to verify the correctness of our proposed optimization formulation a stable-stable transition was computed (see figure 2) and compared to that obtained using commercial optimal control software [6]; both solutions were practically the same. Next, transitions starting/ending on an unstable steady-state were computed; they correspond to B and C transitions which actually are the same transition but in opposite direction. The optimal transition trajectories are shown in figures 3 and 4. Finally, a transition starting and ending on unstable steady-states was obtained (transition D). The optimal transition trajectory is presented in figure 5. It can be observed that for all transition cases, the concentration profile is always soft whereas the temperature and cooling flowrate trajectories present fast dynamics. Also, in all optimal trajectories the total transition time is significantly smaller than the residence time.

Since transitions B, C and D involve unstable steady-states, it is unfeasible to test them on the open-loop system; they have to be tested under closed-loop control. It was observed that the proposed formulation was particularly sensitive to the finite element size; even though it was a decision variable it had to be strongly bounded to reduce the frequency of response discontinuities. Also it can be mentioned that in most cases the finite element length hit the upper bound, so at the end it was noticed that for the proposed formulation, the finite element size could have not been a decision variable. Nevertheless it was helpful in finding the correct finite element size.

Due to the steady-states unstable nature strong numerical problems were expected when the OCP was transformed into an equivalent set of non-linear algebraic equations (i.e. singularity).However, for the transitions addressed in this work, such expected problems did not arise.

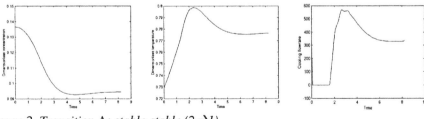

Figure 2. Transition A: *stable-stable (2→1)*

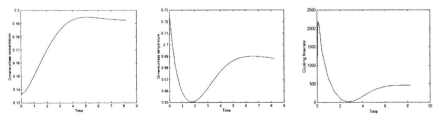

Figure 3. Transition B: *stable-unstable (2→3)*

5. CONCLUSIONS

The results of this work seem to indicate that the proposed optimization formulation is a feasible approach to compute dynamic transition trajectories that involve unstable steady-states. Contrary to what was expected, no significant numerical problems due to the steady-states unstable nature, emerged. We expect to apply a similar formulation to obtain optimal transition profiles in more complex systems (i.e. polymerization reactors).

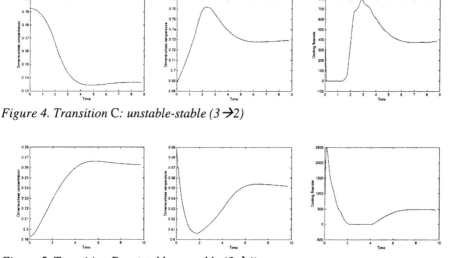

Figure 4. Transition C: unstable-stable (3→2)

Figure 5. Transition D: unstable-unstable (3→4)

6. BIBLIOGRAPHY

[1] Sargent,R.W.H., 2001, Manuscript in preparation.

[2] Biegler,L., 1984, Comp.Chem.Eng., 8,3/4,243-248.

[3] Flores,A., J.Alvarez,E.Saldivar and G.Oaxaca, 2001, Dycops 6, Korea.

[4] Hicks,G.A. and Ray,W.H., 1971, Can.J.Chem.Eng. 40-522-529.

[5] Celia,M.A. and Gray W.G., 1992, Numerical Methods for Differential Equations, Prentice-Hall.

[6] http://www.accescom.com/~adam/riots.

European Symposium on Computer Aided Process Engineering – 12
J. Grievink and J. van Schijndel (Editors)
© 2002 Elsevier Science B.V. All rights reserved.

MPC Control of a Predenitrification Plant Using Linear Subspace Models

Oscar A. Z. Sotomayor[1], Song W. Park[1] and Claudio Garcia[2]

(1) LSCP-Department of Chemical Engineering
(2) LAC-Department of Telecommunications and Control Engineering
Polytechnic School of the University of São Paulo
Av. Prof. Luciano Gualberto, trav.3, n.380, 05508-900 São Paulo-SP, Brazil
Fax: +55 11 3813.2380, E-mail: oscar@lscp.pqi.ep.usp.br

Abstract

In this paper, a model-based predictive control (MPC) technique is designed aiming to control the nitrogen (N)-removal from domestic sewage in an activated sludge wastewater treatment plant. The objective is to control the nitrate concentrations in both, anoxic and aerobic zones of the bioreactor and, therefore, to inferentially to control the effluent inorganic nitrogen concentration. The synthesis of the MPC controller is based on a linear subspace (state-space) model of the process, where an identification horizon is added to include a sequence of past inputs/outputs. This sequence can be used to estimate the model or the updated state of the process, thus eliminating the need for a state observer. Two different MPC control configurations are compared and the result shows the successful of the control application. The linear state-space model was obtained using subspace identification methods.

1. Introduction

Activated sludge (AS) is the most widespread biological process for wastewater treatment. In this process, N-removal is performed in two stages: nitrification and denitrification, under aerobic and anoxic conditions, respectively. In comparison with conventional AS plants (chemical oxygen demand - COD compounds removal and, often partly, ammonium), N-removal AS plants are more complex. The co-existence of nitrification and denitrification processes is accompanied by new operational problems (Yuan, 1999). Increasing the efficiency of one process will always have negative impacts on the efficiency of the other one. Hence, automatic control of N-removal AS plants is necessary to achieve the adequate performance of the overall system.

The control of N-removal AS plants is mainly focused on nitrate, ammonium and inorganic nitrogen (nitrate plus ammonium) removal. This paper is related to control of nitrate removal. In a pre-denitrification plant, the following approaches are possible: (1) control of nitrate concentration in the aerobic zone by manipulating the internal recycling flow rate, (2) control of nitrate concentration in the anoxic zone by manipulating the internal recycling flow rate, and (3) control of nitrate concentration in the anoxic zone by manipulating an external carbon source flow rate. Carlsson and Rehnström (2001) proposed two single-loop controllers to concurrently regulate both approach (1) and (3). Nevertheless, these approaches (1 to 3) are highly interrelated and for the optimal control of the process they should be simultaneously performed in a multivariable control philosophy.

Model-based predictive control (MPC) is currently the most widely implemented advanced process control technology for process plants (Qin and Badgwell, 1997), and they are commonly found in the medium level of a plant-wide control structure. The MPC formulation naturally handles multivariable interactions and constraints. On the other hand, an alternative to use polynomial models (e.g. ARMAX, CARIMA, Markov parameter models, etc), that can be quite cumbersome to obtain in the general multivariable case, are the state-space models (Viberg, 1995). In this case, a prediction model can be easily constructed making use of the extended observability matrix.

In this paper, a MPC controller, based on a linear subspace (state-space) model of the process, is designed aiming to control a predenitrifying AS plant, used for COD and N removal from domestic sewage, as shown in fig. 1.

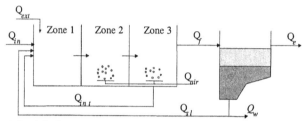

Fig. 1. Layout of the AS plant

The layout of the process is constituted by a bioreactor, composed of an anoxic zone (zone 1) and two aerobic zones (zones 2 and 3) coupled with a secondary settler. In the aerobic zones, the DO concentration is controlled in 2 mg O_2/l by simple PI controllers. The objective here proposed is to control the nitrate concentrations in both, the anoxic zone and the last aerobic zone ($S_{NO,1}$ and $S_{NO,3}$, respectively), by manipulating the internal recycle flow rate Q_{int}, and the external carbon source flow rate, Q_{ext}. Two control configurations are tested by simulation: one taking into account the influent flow rate Q_{in} as a manipulated variable (in a 3x2 system) and another one considering it as being constant (in a 2x2 system). The state-space model is obtained by using subspace identification methods, with input/output data obtained from the ASWWTP-USP benchmark (Sotomayor *et al.*, 2001a).

2. A Prediction Model (PM) of the Process

Consider that the process is suitably described by a discrete linear time-invariant (LTI) state-space model of the form:

$$x_{k+1} = Ax_k + Bu_k$$
$$y_k = Cx_k$$

(1)

where x is the state vector, u is the input vector, y is the output vector, A is the state transition matrix, B is the input matrix and C is the output matrix, and the time index k denotes the sampling instant. A PM is defined to be of the form:

$$y_{k+1/L} = p_L(k) + F_L \Delta u_{k/L} \tag{2}$$

where L is the prediction horizon, $y_{k+1/L}$ is a vector of future outputs, $\Delta u_{k/L}$ is a vector of future input changes, and $p_L(k)$ is a known vector that represents the information of the past, which is used to predict the future. This vector is a function of the number J (identification horizon) and the state-space model matrices. F_L is a constant lower triangular matrix, which is a function of the state-space model matrices. A simple algorithm to compute $p_L(k)$ and F_L is given by (Di Ruscio and Foss, 1998):

$$p_L(k) = O_L A^J O_J^\dagger y_{k-J+1/J} + P_L u_{k-J+1/J-1} \tag{3}$$

$$F_L = \begin{bmatrix} O_L B & H_L^d \end{bmatrix} \tag{4}$$

with the matrix P_L, which is related to past control inputs, defined as:

$$P_L = O_L A \Gamma_{J-1}^d - O_L A^J O_J^\dagger H_L^d \tag{5}$$

where O_L is the extended observability matrix for the pair (A,C), with L block rows, $O_J^\dagger = \left(O_J^T O_J \right)^{-1} O_J^T$ is the Moore-Penrose pseudo-inverse of the extended observability matrix O_J for the pair (A,C), with J block rows, Γ_{J-1}^d is the reverse extended controllability matrix for the pair (A,B), with J-1 block columns, and the H_L^d is the lower block triangular Toeplitz matrix for the triple (A,B,C), with L block rows and L-1 block columns.

3. Model Predictive Control

As can be seen, this PM (2) is independent of the state vector. Hence, there is no need for a state observer. The MPC law is found by minimizing the following discrete time LQ objective:

$$\Im_k = \left(y_{k+1/L} - r_{k+1/L} \right)^T Q \left(y_{k+1/L} - r_{k+1/L} \right) + \Delta u_{k/L}^T R \Delta u_{k/L} + u_{k/L}^T P u_{k/L} \tag{6}$$

where $r_{k+1/L}$ is a vector of future references and $u_{k/L}$ is a vector of future inputs. Q, R and P are block diagonal weighting matrices. The problem can be formulated as:

$$\min_{\Delta u_{k/L}} \Im_k \tag{7}$$

subject to linear constraints on u_k, Δu_k and y_k. The constraints can be written as an equivalent linear inequality of the form:

$$\alpha \cdot \Delta u_{k/L} \le \beta_k \tag{8}$$

It is convenient to find the relationship between $\Delta u_{k/L}$ and $u_{k/L}$ in order to formulate the constraints in terms of future deviation variables $\Delta u_{k/L}$ by using:

$$u_{k/L} = S \cdot \Delta u_{k/L} + c \cdot u_{k-1} \tag{9}$$

where S is a lower block triangular identity matrix and c is block rows identity matrix, of suitable sizes. The input constraints are of the form:

$$S\Delta u_{k/L} \leq u_{k/L}^{\max} - cu_{k-1}, \quad \Delta u_{k/L} \leq \Delta u_{k/L}^{\max}, \quad F_L \Delta u_{k/L} \leq y_{k/L}^{\max} - p_L(k)$$
$$-S\Delta u_{k/L} \leq -u_{k/L}^{\min} + cu_{k-1} \quad -\Delta u_{k/L} \leq -\Delta u_{k/L}^{\min} \quad -F_L \Delta u_{k/L} \leq -y_{k/L}^{\min} + p_L(k)$$
$$(10)$$

The objective functional $\Im_k$ may be written in terms of $\Delta u_{k/L}$. $y_{k+1/L}$ can be eliminated from $\Im_k$ by using the PM (2). $u_{k/L}$ can be eliminated from $\Im_k$ by using (9). Therefore, the LQ objective functional becomes:

$$\Im_k = \Delta u_{k/L}^T H \Delta u_{k/L} + 2f_k^T \Delta u_{k/L} + \Im_k^0 \tag{11}$$

where: $H = R + F_L^T Q F_L + S^T P S$, $f_k = F_L^T Q(p_L(k) - r_{k+1/L}) + S^T P c u_{k-1}$, and
$\Im_k^0 = (p_L(k) - r_{k+1/L})^T Q(p_L(k) - r_{k+1/L}) + u_{k-1}^T c^T P c u_{k-1}$.

H is the Hessian matrix, a constant positive definite matrix. f_k is a time-varying vector, independent of the unknown control deviation variable. $\Im_k^0$ is a known time-varying scalar, independent of the optimization problem. The problem can be solved by the following equivalent QP approach:

$$\min_{\Delta u_{k/L}} \left(\Delta u_{k/L}^T H \Delta u_{k/L} + 2f_k^T \Delta u_{k/L} \right) \tag{12}$$

subject to (8). When $\Delta u_{k/L}$ is computed, the control signal to be applied to the process is $u_k = u_{k-1} + \Delta u_k$. (Note that only the first change in $\Delta u_{k/L}$ is used)

4. A State-Space Model of the Process

The linear state-space model is obtained by using subspace identification methods, see Sotomayor *et al.* (2001b). This subspace model can be rewritten such that the measured disturbances d (Q_{in}, influent substrate concentration $S_{S,in}$ and influent ammonium concentration $S_{NH,in}$) are considered as inputs and included in the input vector as:

$$x_{k+1} = Ax_k + \begin{bmatrix} B_u & B_d \end{bmatrix} \cdot \begin{bmatrix} u_k \\ d_k \end{bmatrix}, \quad y_k = Cx_k \tag{13}$$

where, $A = \begin{bmatrix} 0.9763 & 0.0199 & 03263 \\ 0.0062 & 0.8818 & 0.0907 \\ -0.0024 & 0.0072 & 0.9758 \end{bmatrix}$, $C = \begin{bmatrix} 0.2259 & -0.4026 & -0.1810 \\ 0.2664 & 0.2876 & -0.4633 \end{bmatrix}$, and

$$B = \begin{bmatrix} B_u & B_d \end{bmatrix} = \begin{bmatrix} 0.0368 & -0.0434 & -0.1537 & -0.0431 & -0.0045 \\ -0.1505 & 0.0234 & 0.0357 & 0.0283 & -0.0044 \\ 0.0167 & -0.0100 & -0.0091 & -0.0003 & 0.0039 \end{bmatrix}$$

Aiming to incorporate feedforward/feedback action and integral error, these dynamics have be modeled and then be included in the state vector (13), as:

$$\begin{bmatrix} x_{k+1} \\ d_{k+1} \\ q_{k+1} \\ r_{K+1} \end{bmatrix} = \begin{bmatrix} A & B_d & 0 & 0 \\ 0 & A_d & 0 & 0 \\ -T_s K_i C & 0 & I & T_s K_i \\ 0 & 0 & 0 & A_r \end{bmatrix} \begin{bmatrix} x_k \\ d_k \\ q_k \\ r_K \end{bmatrix} + \begin{bmatrix} B_u \\ 0 \\ 0 \\ 0 \end{bmatrix} u_k, \quad y_k = \begin{bmatrix} C & 0 & 0 & 0 \end{bmatrix} \begin{bmatrix} x_k \\ d_k \\ q_k \\ r_K \end{bmatrix} \tag{14}$$

where A_d and A_r are diagonal matrix such that $d_{K+1} = A_d \cdot d_k$ and $r_{k+1} = A_r \cdot r_k$, respectively. K_i is a weighting matrix (diagonal positive definite), T_s is the sampling time and q_k is the integral error vector of the form $\dfrac{q_{k+1} - q_k}{T_s} = K_i(y_k - r_k)$. The augmented model (14) is used instead of (1) in the development of the MPC controller.

5. Simulation Results

In order to test the performance of the MPC controller, two configurations are tested, both of them only for set-point changes (servo case). It is necessary to state that depending on the use or not of Q_{in} as a manipulated variable, the dimensions of the matrices B_u and B_d change and, therefore, the objective function (12) and the constraints (10) are different for each control configuration. The tuning parameters of both configurations are not presented here.

5.1 3x2 system
In this configuration, the influent flow is considered as a manipulated variable. In predenitrifying processes, the raw sewage is used as carbon source. Therefore, this configuration is applied to make good use of the influent COD concentration. In fig. 2 the responses of the process to set-point changes are shown. It can be observed that the system responds quite well. The effluent inorganic nitrogen concentration ($S_{NO,3} + S_{NH,3}$) is maintained at low level.

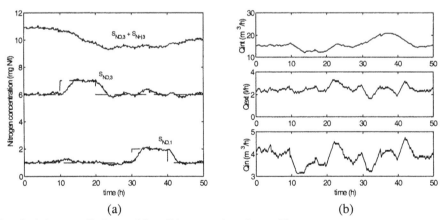

(a)   (b)

Fig. 2. (a) controlled variables, (b) manipulated variables.

5.2 2x2 system
In this case, the influent flow is kept constant (i.e. previously controlled). Fig. 3 show the response of the process to set-point changes. The variables are well-controlled. Nevertheless, a higher control effort is required.

Aiming to compare the performance of both configurations, here it is adopted the integrated squared error (ISE), which is based on the response system, defined as:

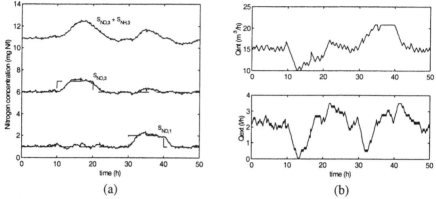

Fig. 3. (a) controlled variables, (b) manipulated variables.

$$ISE = \int \left(y_k - r_k\right)^2 dt \qquad (15)$$

In table 1, it can be observed that the 2x2 system presents a better performance.

Table 1. Numerical performance comparison for the two control configurations

System	ISE_1 ($S_{NO,1}$)	ISE_3 ($S_{NO,3}$)	$ISE_T = ISE_1 + ISE_3$
3x2	3.15	4.08	7.23
2x2	2.84	3.13	5.97

6. Conclusions

In this paper, a MPC controller, based on a linear state-space model of the process, was implemented to improve N-removal capability of a predenitrication AS plant. The control was successful for set-point changes in both control configurations presented. In the 2x2 system, a higher control effort was required in the manipulated variables than in the 3x2 system. The 2x2 system presented a better performance in controlling the nitrate concentrations (see table 1), but the 3x2 system was better successful in controlling (inferentially) the effluent inorganic nitrogen concentration (see fig. 2).

Acknowledgment: The authors gratefully thank the financial support from FAPESP under grant 98/12375-7. They are also grateful to Dr. David Di Ruscio from the Telemark Institute of Technology, Norway.

7. References

Carlsson, B. and A. Renhström, 2001. Proc. ICA 2001, Sweden, 229-236.
Di Ruscio, D. and B. Foss, 1998. Proc. DYCOPS 5, Greece, 304-309.
Qin, J. and T. Badgwell, 1997, *AIChE Symp. Series* **93**, 232-256.
Sotomayor, O.A.Z., S.W. Park and C. Garcia, 2001a, *Braz. J. Chem. Eng.* **18**, 81-101.
Sotomayor, O.A.Z., S.W. Park and C. Garcia, 2001b, *Contr. Eng. Prac.*, DYCOPS 6 special issue, (Submitted).
Viberg, M., 1995, *Automatica* **31**, 1835-1851.
Yuan, Z., 1999, Proc. ECB 9, Belgium, 2217.

European Symposium on Computer Aided Process Engineering – 12
J. Grievink and J. van Schijndel (Editors)

Relational modeling of chemical processes for control logic verification

Adam L. Turk[1]

Gary J. Powers[2]

[1]Delft University of Technology
The Netherlands

[2]Carnegie Mellon University
United States of America

Abstract

The area of formal methods offers improved fault detection for hybrid processes, such as chemical ones. Verification of a chemical process requires the construction of a finite state representation from the phenomena exhibited by the control system, physical process, operating procedures, and human behavior. The exhibited phenomena are modeled as a set of states, the transitions between these states, and their triggering events. In particular, the states and transitions are based upon landmarks and their relative position while the triggering events are based upon the actions in the system. This methodology capitalizes on the importance of relational properties instead of absolute ones in verifying a chemical process. The proposed methodology was applied to two industrial examples: a leak test procedure and a thermal oxidation process.

1. Introduction

Detection of faults within control logic can lead to a more reliable process with respect to safety. Based upon current industrial practices, fault analysis of a modest sized process is difficult due to its large number of events or state space. Fortunately, improved fault detection for chemical processes may be found in the hybrid systems area where formal logic and mathematical methods are being developed for verifying sequential behavior of processes. In order to verify a chemical process using these techniques, a finite state representation of the control system, physical process, operating procedures, and human behavior must be built. This paper presents a method for synthesizing of a relational model that captures the relevant behavioral states that a process exhibits during operation.

2. Background

Progress has been made in the area of verifying mechanical and chemical processes for safety. Several industrial processes, including an air traffic control system and an aircraft guidance system, have been verified using symbolic model checkers (Anderson *et al.*, 1996; Sreemani and Atlee, 1996). In these industrial examples, faults in the process were discovered and corrected, increasing reliability and safety.

560

Symbolic model checking has been used to verify the control logic of discrete control systems in chemical processes, including such complex systems such as a furnace standard and solids transport system (Moon, 1992; Probst, 1996). Formal methods, such as SMV, are used in the semiconductor and computer industries to uncover potential logic faults in microprocessors and software (Burch *et al.*, 1991; McMillan and Schwalbe, 1991; McMillan, 1993). One strength of SMV comes from its symbolic representation (binary decision diagrams) of the sequential and finite state space that describes the process. Another strength of this formal method is its efficient use of fixed point algorithms to search the state space for faults. These abilities allow SMV to completely verify large processes which can be typically on the order of 10^{20} states or larger cite (Burch *et al.*, 1990; Clark *et al.*, 1994). With these strengths, SMV has shown that it may be a valuable tool for efficiently verifying the logic for any type of process.

Verification with the symbolic model checker SMV requires the process to be represented as a finite state machine or non-timed discrete model. Chemical and mechanical processes are normally a combination of continuous and discrete systems. A key challenge to modeling these hybrid systems as a finite state machine is to represent the discrete domain with little modification while still capturing the dynamics and time-dependent phenomena of the continuous domain.

3. Modeling

In model checking, any process, such as a chemical one, must be represented as a finite state machine. A finite state machine is a set of states, triggering events, transitions, and initial conditions that characterize a process. The states within the representation describe the conditions of the process while the transitions identify the possible paths between these states. The events determine the time when transitions between states can occur. The key to an accurate and appropriate model is to capture the significant exhibited phenomena of the process with respect to the specifications as a set of states, transitions, and events.

The state variables that describe the process, including the continuous ones, must be defined by a set or range of discrete values such as Boolean, integer, or enumerated type. The particular discrete values of the state variables are defined by landmarks in the physical chemical process, control system, operating procedures, and specifications. The landmarks from these different sources form the natural breaks or boundaries between displayed phenomena of the process. The values identified by these landmarks are relative to the state of the system and not an absolute numeric value. The relative value of the variables helps to reduce the size and complexity of the state space while modeling the

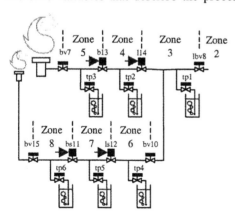

Figure 1: Piping and valve diagram for a typical combustion system for the leak testing procedure

appropriate phenomena of the physical process. The verification of a chemical process is concerned with the transition to critical or unsafe states rather than a change in the numeric value of a state variable. The landmarks determine the significant states that the discrete model needs to capture for verification. In addition to the states, the transitions between states and their triggers are also important for verification. These transitions and triggering events can be determined from observation, experimental data, theoretical mathematical equations, control logic, and operating procedures. These sources identify the possible sequence of states in a process and the actions that trigger the progression or movement along these paths.

4. Industrial Examples

The synthesis procedure for a relational model was applied to two industrial examples: a leak test procedure and a thermal oxidation process. The modeling of these industrial examples demonstrates the procedure's ability to represent the physical system, control systems, operating procedures, operator behavior and their interactions.

4.1 Leak Test Procedure

The leak test procedure is a structured approach for checking the seal across valves in a combustion system. In particular, the procedure tests for leaks across the valves located in a pipe network (fig. 1). The procedure pressurizes the pipe network by initiating the light up sequence of the combustion system which is followed by an emergency shutdown of the system. The pipe network is divided by the valves into eight pressure zones or pipe segments. The valves in each pipe section are tested by the presence of bubbling or no bubbling when a tap valve is opened. Once a leaking valve is detected, then the valve is repaired immediately. The procedure is begun anew after the repair of a leaking valve.

The key operational issue of the procedure is the reliable testing and detection of a leaking valve. If a leaking valve is not properly diagnosed by the procedure then it can result in a fatal explosion. On the other hand, unnecessary replacement of a non-leaking valve wastes time and money. Both concerns are critical for the proper operation of the leak test procedure and form the basis of the specifications used to verify the procedure. The specifications also drive the construction of the finite state machine of the process. In particular, they highlight the need to include the steps of the procedure along with the operator's behavior in the finite state machine. The representation of the leak test procedure and the operator behavior as a set of states, transitions, and events was developed as follows. A state variable was created that moved though the steps in the procedure. At each given step, the necessary tasks were performed such as opening a tap valve to check bubbling. The bubbling at a given tap valve was linked to that pipe segment having pressure. The pressure in the pipe segments was defined as a Boolean state variable where zero described no pressure and one pressure. The pressure for a given segment was conditioned upon the opening of its associated valves and the pressure values beyond the opened valves. The pressure in a given pipe segment was also dependent on a separate Boolean variables that described the potential leaking of associated variables. This failure mode along with others were added to the finite state representation of the procedure and verified.

The leak test procedure was first verified by Probst with respect to the proper diagnosis of valves (Probst 1996). Building upon this work, we verified the procedure for the proper diagnosis of valves with faults generated by incorrect operator behavior and from the dynamics of a leaking valve. The computational results for the verification of several different leak testing models are presented in table 1 (Turk, 1999). The failure mode of the operator not testing or skipping a pipe section was added to the discrete model of the procedure. This failure mode shares characteristics with the dynamics of a slow leaking valve. Both faults lead to a positive result where one should not exist. Therefore, only one model is needed to see if either event will lead to a misdiagnosis. In addition to the representation with these faults, table 1 lists several other models that change the manner in which the position of the valves is determined. Originally, the position of the valves was determined by the step that the procedure was currently performing. This modeling assumption opened and closed valves even though the particular step associated with that action was not executed. The finite state machine was modified so that a particular step had to be performed in order for given valves to open and close. The values for the reachable states and transition relations in table 1 show the moderate complexity and small size of the discrete model, respectively. The result from the verification demonstrated that the leak test procedure would misdiagnose leaking valves if the operator skipped steps or the leak in a valve was too slow. Possible solutions for these failures might be a checklist for the operator and using pressure gauges instead of a bubble test.

Table 1: Information on verified leak test procedure models by symbolic model checking (SMV) with dynamic variable ordering.

Model Name	Boolean Variables	Reachable States	Transition Relations (OBDD Nodes)	CPU Time (sec)[a]
Base Case	24	5,944	11,495	3
Segment Skip (1)	24	14,611	14,205	8
Valves: Tap	30	14,611	28,102	26
Valves: All	38	14,611	95,100	232
Valves	32	14,611	54,064	128
Macro-Pres	32	14,552	37,632	71
Expand-Proc	32	5,153	44,688	6
Segment Skip (2)	32	15,984	49,601	18

[a]Computations were performed on a HP 812/70 workstation.

4.2 Thermal Oxidation Process

A thermal oxidation process burns organic vapor in an effluent air stream from a metal casting pit before it is vented to the atmosphere (fig. 2). The casting pit is swept with air in order to remove any vaporized organic coolant before its concentration reaches the explosion limit and becomes a danger to operators. However, this air stream can not be released to the atmosphere due to environmental regulation. In order to vent the air stream to the atmosphere, the organic vapor is first burned off in a thermal oxidation unit.

The specifications were derived from the operating procedures of the thermal oxidation unit. In particular, the specifications described the existence of a flame from the main burner and the stability of the temperature in the unit. The existence of a flame is

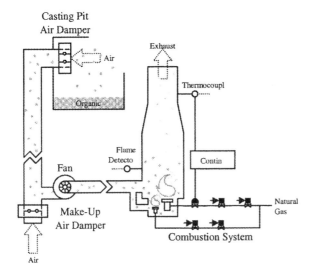

Casting Pit
Air Damper

Exhaust

Air

Organic

Thermocoupl

Flame
Detecto

Contin

Fan

Natural
Gas

Make-Up
Air Damper

Combustion System

Air

Figure 2: Diagram for the thermal oxidation

represented by a Boolean variable where the zero describes no flame present while the one indicates a flame. The transition between these values is dependent on the flow of fuel, flow of oxygen, and an ignition source. The temperature of the thermal oxidation was divided into three discrete values based upon the control landmarks of high temperature at 1700° F and low temperature at 1500° F. These two setpoints from the control system create three discrete temperature region: below operating temperature, at operating temperature, and above operating temperature. In the finite state machine, these temperature regions are represented as an integer range from zero to two. The triggering events for the transition between these temperature states are the presence of the flame and the amount of energy generated by the flame. In turn, the energy generated is dependent on the flow rate of the effluent air, the flow rate of the natural gas, and the caloric value of the gas. The energy contribution from the organic vapor is assumed negligible.

Several different models were built for the thermal oxidation unit. Each model represents the thermal oxidation process along with a failure mode that could potentially impact the stability of the system. The number of Boolean variables, reachable states, nodes in the transition relation, and computational time for these models are presented in table 2 (Turk, 1999). The value for reachable states indicates a simple system while the number of transition nodes shows its large model size with respect to the leak test procedure.

Table 2: Information on verified thermal oxidation process models by symbolic model checking (SMV) with conjunctive partitioning

Model Name	Boolean Variables	Reachable States	Transition Relations (OBDD Nodes)	CPU Time (sec)[a]
Base Case	66	23	865,280	28
Fail Pit Damper	66	57	1,342,126	57
Fail Gas Pres. (High)	66	59	1,376,861	112
Fail Over Temp.	66	69	1,558,914	72
Fail Under Temp.	66	58	1,869,591	112
Drift Temp.	66	34	1,477,692	49
Fail High Temp.	66	67	1,838,378	162
Fail Flame	66	81	1,682,027	185
Fail Flame Detect	66	82	1,499,612	181

[a]Computations were performed on a Sun Ultra 5 workstation.

The verification results from the different failure modes showed that the thermal oxidation system would appropriately extinguish the flame when it could not recover from the instability. However, the thermal oxidation model could recover from several of the failure modes but only if the gain of the disturbance was small. The only fault that the system failed to correct was the drifting of the temperature given by the thermocouple. The drifting of the temperature would cause the system to act incorrectly. This fault can be fixed by adding a secondary thermocouple and associated logic against which the temperature of the thermal oxidation unit can be checked. This discrete model of the thermal oxidation unit proved the safeness of the thermal oxidation process with respect to its specifications.

5. Concluding Remarks

The modeling methodology was used to represent two complex chemical processes. The significant exhibited phenomena of these systems with respect to the specifications were captured as a set of events, transition, and triggering events. This representation capitalized on the importance of relational properties instead of the absolute values in describing the chemical process. These discrete models were verified using SMV against a given set of specifications. The results from these industrial examples showed that each system had potential failure paths. The behavior of the operator in the leak test procedure could lead to misdiagnosed valves while the thermocouple in the second example would not function properly due to the drifting of the temperature value. Overall, the modeling methodology demonstrated its ability to create an accurate and appropriate representation that can be used to verify the safeness of a complex chemical process.

6. References

Anderson, R.J., P. Beame, S. Burns, W. Chan, F. Modugno, D. Notkin, and J. Reese, 1996, Proc. of the Fourth ACM SIGSOFT Symp. on the Found. of Software Eng., Assoc. of Comp. Mach.

Burch, J.R., E.M. Clarke, and K.L. McMillan, 1990, In the Proc. of the 27th ACM/IEEE Design Automation Conf.

Burch, J.R., E.M. Clarke, and D.E. Long, 1991, In the Proc. of the 28th ACM/IEEE Design Automation Conf.

Clarke, E.M., O. Grumberg, and D.E. Long, 1994, ACM Trans. on Prog. Lang. 16(5).

McMillan, K.L. and J. Schwalbe, 1991, In the Proceedings of the 1991 International Symp. on Shared Memory Multiprocessors.

McMillan, K.L., 1993, Symbolic Model Checking, Kluwer.

Moon, I., 1992, Autmatic Verification of Discrete Chemical Process Control System, Carnegie Mellon University.

Probst, S.T., 1996, Chemical Process Safety and Operability Analysis Using Symbolic Model Checking, Carnegie Mellon University.

Sreemani, T. and J. Atlee, 1996 Technical Report CS96-05, Department of Computer Science, University of Waterloo.

Turk, A.L., 1999, Event Modeling and Verification of Chemical Processes Using Symbolic Model Checking, Carnegie Mellon University.

European Symposium on Computer Aided Process Engineering – 12
J. Grievink and J. van Schijndel (Editors)
© 2002 Elsevier Science B.V. All rights reserved.

Development of Dynamic Models for
Fixed Bed Catalytic Reactors

E. C. Vasco de Toledo, E. R. de Morais and R. Maciel Filho
Faculty of Chemical Engineering, State University of Campinas (UNICAMP)
CP 6066 - CEP 13081-970 - Campinas, SP - Brazil.
email: urso@lopca.feq.unicamp.br - FAX +55-1937883910

Abstract

Heterogeneous and pseudo-homogeneous dynamic models, for fixed bed catalytic reactors are presented in this work. They allow to consider variations in the physical properties of the fluid and in the heat and mass transfer coefficients, as well as the heat exchange through the jacket of the reactor. The models permit a study on the dynamic behaviour of the system including the predictions of the inverse response phenomena. This work also allows to understand the model differences and their prediction capabilities.

1. Introduction

The catalytic chemical reactor exhibit complex dynamic behaviour resulting from the non-linear distributed features which, among other things, give rise to inverse response resulting in catastrophic instabilities such as temperature runaway. The non-linearities are a consequence of heat generation by chemical reaction, and the inverse response arises from the presence of different heat capacities of the fluid and solid as well as the bulk flow of fluid causing interactions between heat and mass transfer phases. This causes differential rates of propagation of heat and mass transfer which influence the heat generation through reaction on the solid catalyst, (McGreavy and Maciel, 1989).

Reliable models depend on the insight of how the dominant physic-chemical mechanisms and external factors which affect the overall performance. However, when on-line applications are required, simplified models have to be used which can keep the essential characteristics of the system. In this work, models for fixed bed catalytic reactors, based on heterogeneous and pseudo-homogeneous approach were developed. The proposed heterogeneous dynamic models for fixed bed catalytic reactors consist on mass and heat balance equations for the catalyst particles as well as for the gas phases, include the resistances to mass and heat transfer at the gas-solid interface and consider the resistances inside the catalyst particle. The heterogeneous dynamic models are used in applications where computational accuracy may be more emphasized than computational speed, for instance, reactor design, planning of startups, shutdowns and emergency procedure, (Martinez et al., 1985, Pellegrini et al., 1989). For real time implementation, as control and optimisation on-line, it is required to have the ability to overcome computational burden with a faster and easy numerical solution when compared to rigorous heterogeneous models. Bearing this in mind, reduced heterogeneous and pseudo-homogeneous models were developed. The reduced models was done through mathematical order reduction, which eliminates the spatial co-ordinate of the catalyst particle and promote radially lumped-differential formulations. The pseudo-homogeneous model were developed based on the approach which incorporates the thermal capacity of the fluid and solid, $(\rho C_p)f$ and $(\rho C_p)s$, respectively, (Vasco de Toledo, 1999).

2. Reduction Techniques

The solution of diffusion/reaction multidimensional problems present difficulties associated with a large analytic involvement and also request considerable computational effort. Thus it is convenient to propose simpler formulations for the original system of partial differential equations, through the reduction of the number of its independent variables. Therefore, one or more independent variables can be integrated, leading to approximate formulations that retain detailed local information in the remaining variable as well as medium information in the directions eliminated by the integration. This information comes from the boundary conditions related to the eliminated directions. Two different reduction approaches, to know Hermite and Classic, generating differentiates lumped formulations were investigated. These techniques generate models that describe the axial profiles as a function of the time for the convenient explicit elimination of the dependence in the radial variable in the case of the fixed bed catalytic reactor.

2.1 Hermite Reduction
Hermite developed a way to approach an integral based on the values of the integrating, y, and their derivatives, y', on the limits of the integration. The technique makes use of the approach $H_{1,1}$ and simultaneously of the theorem of the mean value for spherical co-ordinates, leading the generation of the radial medium variables, (Corrêa and Cotta, 1996).

$$[\ ]m = 3\int_0^1 r^2 [\ ]dr \tag{1}$$

where $[\ ]_m$ defines a mean radial value of the amount inside of the left bracket.

$$H_{1,1} = \int_0^1 y(x)dx \cong \frac{1}{2}[y(0) + y(1)] + \frac{1}{12}[y'(0) - y'(1)] \tag{2}$$

2.2 Classic Reduction
The technique makes uses of the theorem of the mean value, given by equation (1) to generate the reduced models.

3. Dynamic Models

The models of the reactor were generated under the following considerations: variation of physical properties, mass and heat transfer coefficients, along the reactor length; the plug flow profile of velocity; intraparticle gradient negligible (pseudo-homogeneous and heterogenou II models); mass transfer resistance between the gas and the catalyst surface is neglected (pseudo-homogeneous model); axial dispersion was neglected. In this equations the following notations are used: Fluid, F; Solid, S; Mass Balance, MB; Energy Balance, EB; Coolant Fluid Equation, CFE; Continuity Equation, CE; Momentun Equation, ME.

3.1 Heterogeneous Model I - Rigorous Model

$$\frac{\partial X_g}{\partial t} = \frac{D_{efi}}{R_t^2} \frac{1}{r} \frac{\partial}{\partial r}\left[r\frac{\partial X_g}{\partial r}\right] - \frac{G_i}{\rho_g L}\frac{\partial X_g}{\partial z} + \frac{k_g a_m \rho_B}{\varepsilon}(X_s^* - X_g) \quad \text{(FMB)} \tag{3}$$

$$\rho_g C_{pg}\varepsilon\frac{\partial T_g}{\partial t} = \frac{\lambda_{ef}}{R_t^2}\frac{1}{r}\frac{\partial}{\partial r}\left[r\frac{\partial T_g}{\partial r}\right] - \frac{\varepsilon G_i C_{pg}}{L}\frac{\partial T_g}{\partial z} + h_f a_m \rho_B (T_s^* - T_g) \quad \text{(FEB)} \tag{4}$$

$$\varepsilon_s \frac{\partial X_s}{\partial t} = \frac{D_s}{R_p^2}\frac{1}{r_p^2}\frac{\partial}{\partial r_p}\left(r_p^2\frac{\partial X_s}{\partial r_p}\right) - \frac{PM\,\rho_s R_w(X_s,T_s)}{\rho_s} \quad \text{(SMB)} \tag{5}$$

$$\rho_s C_{ps}\frac{\partial T_s}{\partial t} = \frac{\lambda_s}{R_p^2}\frac{1}{r_p^2}\frac{\partial}{\partial r_p}\left(r_p^2\frac{\partial T_s}{\partial r_p}\right) + \frac{\rho_s(-\Delta H_R)R_w(X_s,T_s)}{(R+1)} \quad \text{(SEB)} \tag{6}$$

$$\frac{\partial}{\partial z}(\rho_g V_g) = 0 \quad \text{(CE)} \tag{7}$$

$$\frac{\partial T_R}{\partial t} = -\frac{u_R}{L}\frac{\partial T_R}{\partial z} + \frac{2\,U}{R_t\,\rho_R\,C_{pR}}(T_g(1,z,t) - T_R) \quad \text{(CFE)} \tag{8}$$

$$\frac{\partial p}{\partial z} = -\frac{G^2 L}{\rho_g D_p g_c}f \quad \text{(ME)} \tag{9}$$

with the following boundary conditions:

$$r = 0 \quad \frac{\partial X_g}{\partial r} = \frac{\partial T_g}{\partial r} = 0, \quad r_p = 0 \quad \frac{\partial X_s}{\partial r_p} = \frac{\partial T_s}{\partial r_p} = 0 \quad \text{(symmetry)} \tag{10}$$

$$r = 1 \quad \frac{\partial X_g}{\partial r} = 0, \frac{\partial T_g}{\partial r} = \text{Bih}\,(T_R - T_g(1,z,t)) \quad \text{for all } z \tag{11}$$

$$r_p = 1 \quad \frac{\partial X_s}{\partial r_p} = \frac{k_g R_p}{D_s}\left(X_s - X_s^*\right), \frac{\partial T_s}{\partial r_p} = \frac{h_f R_p}{\lambda_s}\left(T_g - T_s^*\right) \quad \text{for all } z \tag{12}$$

$$z = 0 \quad X_g = X_s = 0, T_g = T_{go}, T_s = T_{so}, T_R = T_{Ro}, p = p_o \quad \text{for all } r \tag{13}$$

The Equations 7, 8, 9, 10, 11 and 13 are valid also for the remaining models. The following notations is used, a_m is the heat transfer area for catalyst (m^2/kg catalyst); Bih, number of Biot; C_p, calorific capacity (kcal/kg.K); D, radial effective diffusivity (m/h); D_p, particle diameter (m); f, friction factor; G, mass flow velocity (kg/m^2.h); g_c, conversion factor; h_f, heat transfer coefficient particle to fluid (kcal/m^2.h.K); h_W, convective heat transfer coefficient in the vicinity of the wall (kcal/m^2.h.K); k_g, mass transfer coefficient particle to fluid (m/s); L, length of the reactor (m); p, pressure of the reactor (atm); PM, the mean molecular weight (kg/kmol); r, dimensionless radial distance of the reactor; r_p, dimensionless radial distance of the reactor; R, air/ethanol ratio; R_t, reactor radius (m); R_p, particle radius (m); R_W, rate of the oxidation (kmol of reactant mixture/h.kgcat.); T, reactor temperature (K); T_{fo}, feed temperature (K); T_{go}, feed temperature (K); T_{so}, catalyst feed temperature (K); T(1,z,t); wall temperature of the reagent fluid (K); T_{ro}, coolant feed temperature (K); t, time (h); u,velocity (m/h); U, global heat transfer coefficient (kcal/m^2.h.K); V, velocity (m/h); X, conversion; z,dimensionless axial distance. λ, conductivity (kcal/m.h.K); ΔH_R, enthalpy of reaction molar (kcal/kmol); ρ, density (kg/m^3); ρ_B, catalyst density (kgcat/m^3); ρ_s, catalyst density (kgcat/m^3); ε, porosity; Subscripts: ef, effective; f, fluid; g, gas; i, interstitial; o, feed; R, refrigerant; s, solid; Superscripts: s, condition at external surface.

3.2 Heterogeneous Model II - Reduced Models

$$\frac{\partial X_g}{\partial t} = \frac{D_{efi}}{R_t^2}\frac{1}{r}\frac{\partial}{\partial r}\left[r\frac{\partial X_g}{\partial r}\right] - \frac{G_i}{\rho_g L}\frac{\partial X_g}{\partial z} + Bihm\frac{k_g\, a_m\, \rho_B}{\varepsilon}(X_s - X_g) \quad \text{(FMB)} \tag{14}$$

$$\rho_g C_{pg}\varepsilon\frac{\partial T_g}{\partial t} = \frac{\lambda_{ef}}{R_t^2}\frac{1}{r}\frac{\partial}{\partial r}\left[r\frac{\partial T_g}{\partial r}\right] - \frac{\varepsilon G_i C_{pg}}{L}\frac{\partial T_g}{\partial z} + Biht\, h_f\, a_m \rho_B\,(T_s - T_g) \quad \text{(FEB)} \tag{15}$$

$$\varepsilon_s (1-\varepsilon)\frac{\partial X_s}{\partial t} = -k_g\, a_m\, \rho_B\,(X_s - X_g) + Bihm\frac{PM\,(1-\varepsilon)\,\rho_s\, R_w(X_s,T_s)}{\rho_g} \quad \text{(SMB)} \tag{16}$$

$$(1-\varepsilon)\rho_s\, C_{ps}\frac{\partial T_s}{\partial t} = -h_f\, a_m\, \rho_B\,(T_s - T_g) + Biht\frac{(1-\varepsilon)\,\rho_s\,(-\Delta H_R)\,R_w(X_s,T_s)}{(R+1)} \quad \text{(SEB)} \tag{17}$$

Classic Reduction	$Bihm = 1$	$Biht = 1$
HermiteReduction	$Bihm = \dfrac{4}{4+K_g R_p/D_s}$	$Biht = \dfrac{4}{4+h_f R_p/\lambda_s}$

(18)

3.2 Pseudo-Homogeneous Model

$$\frac{\partial X}{\partial t} = \frac{D_{efi}}{R_t^2}\frac{1}{r}\frac{\partial}{\partial r}\left[r\frac{\partial X}{\partial r}\right] - \frac{Gi}{\rho_g L}\frac{\partial X}{\partial z} + \frac{(1-\varepsilon)\,PM\,\rho_s}{\varepsilon\,\rho_g}R_w(X,T) \quad \text{(MB)} \tag{19}$$

$$\left(\varepsilon\,(\rho_g\, C_{pg}) + (1-\varepsilon)\,(\rho_s\, C_{ps})\right)\frac{\partial T}{\partial t} = \frac{\lambda_{ef}}{R_t^2}\frac{1}{r}\frac{\partial}{\partial r}\left[r\frac{\partial T}{\partial r}\right] - \frac{\varepsilon Gi\, C_{pg}}{L}\frac{\partial T}{\partial z} + \frac{(1-\varepsilon)\,\rho_s\,(-\Delta H_R)\,R_w(X,T)}{(R+1)} \tag{20}$$

$$\text{(EB)}$$

As a case study, the catalytic oxidation of the ethanol to acetaldehyde over Fe-Mo, R_W, catalyst was considered, Vasco de Toledo (1999). It is a strongly exothermic reaction, representative of a important class of industrial processes.

$$R_w = \frac{2K_1 K_2 P_{O2} P_{ET}}{K_3 K_1 P_{ET} P_{AC} + K_1 P_{ET} + 2K_2 P_{O2} + K_3 K_4 P_{AC} P_{H2O}} \tag{21}$$

Where P_{O2}, P_{ET}, P_{H2O}, P_{AC} are partial pressure of oxygen, ethanol, water and acetaldehyde respectively, and the K_i are the kinetic constants in the Arrhenius form.

It is worthwhile mentioning that the rigorous heterogeneous model takes into account all the possibles resistances that may be important which are neglected in simpler mathematical models. Such models may be useful for different applications, depending upon the need of the prediction accuracy and required computer time.

4. Results and Discussions

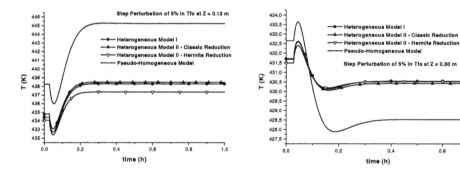

Figure 1 - Reactor temperature response for step perturbation in T_{fo}.

Figure 2 - Reactor temperature response for step perturbation in T_{fo}.

The numeric solution of the models was obtained using the method of the lines in conjunction with the orthogonal collocation which showed to be an effective procedure for the space discretization (radial and axial directions) in conjunction with the LSODAR algorithm for the integration in time. For the numerical solution are used 5 collocation points for the radial direction (fluid and particle) and 7 points for the axial direction. Figures 1 and 2 show the dynamic behaviour of the heterogeneous and pseudo-homogeneous models. In these figures the inverse response phenomena is observed in the profile of the reactor temperature due to a step perturbation in the reactant feed temperature, T_{fo}. This phenomenon is typical in fixed bed catalytic reactors. In Figure 3 is represented the temperature dynamic behaviour of the reactor due to a step perturbation in the mass flow velocity, G. It can be seen in these figures, that the dynamic behaviour of the heterogeneous and pseudo-homogeneous models is qualitatively similar, but quantitatively different. It is worthwhile mentioning that the reduced models require about half of the computer time to be solved when compared to the rigorous heterogeneous model. About the same relation is obtained when reduced models and pseudo-homogeneous model are compared, which indicates the potential of reduced models for real time implementations. These three figures show that the use of a certain model type (heterogeneous or pseudo homogeneous) to represent the reactor dynamic behaviour may lead to different reactor predictions. The temperature dynamic

570

behaviour, Figure 4, observed for a step perturbation in coolant feed temperature, T_{Ro}, allows a better understanding of the reactor behaviour along the reactor length, including the hot spot. The used model was the Heterogeneous Model I.

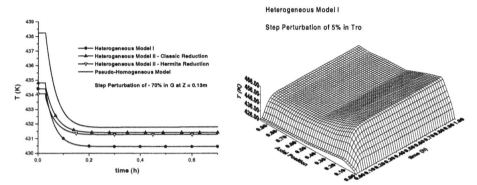

Figure 3 - Reactor temperature response for step perturbation in G. *Figure 4 - Reactor temperature response for step perturbation in T_{Ro}.*

5. Conclusions

The proposed models, specifically the heterogeneous, have shown to be able to predict the main characteristics of the dynamic behaviour of fixed bed catalytic reactors, including the inverse response phenomena. This knowledge is essential to design and control of the reactor. However, the computational time demanded for the solution of the Heterogeneous Model I is high in comparison to the simplified models, making possible the application of this models for applications where computational time is not limited. However, when on-line applications are required simplified models have to be used. The models based on reduction techniques and pseudo-homogeneous approach overcome computational burden with a faster and easy numerical solution, as well as other difficulties found in rigorous heterogeneous models, especially related to the large number of parameters and sophisticated numerical procedures required to the solution.

6. References

Corrêa, E. J. and Cotta, R. M., 1994, Improved Lumped-Differential Formulations of Transient Heat Conduction Problems, presented at III Congress of Mechanical Engineering on North-Northeast, PA, Brazil.

Martinez, O. M., Pereira Duarte, S. J., and Lemcoff N., O., 1985, Modeling of Fixed Bed Catalytic Reactors, Comput. Chem. Eng., 9, 5, 535-545.

McGreavy, C. and Maciel Filho, R., 1989, Dynamic Behaviour of Fixed Bed Catalytic Reactors, presented at IFAC Dynamics and Control of Chemical Reactors, Maastricht, The Netherlands.

Pellegrini, L., Biardi, G. and Ranzi, E., 1989, Dynamic Model of Packed-Bed Tubular Reactors, Computers Chem. Eng., 13, 4/5, 511-518.

Vasco de Toledo, E. C. V., 1999, Modelling, Simulation and Control of the Fixed Bed Catalytic Reactors, Ph.D. Thesis, University of Campinas, São Paulo, Brazil.

European Symposium on Computer Aided Process Engineering – 12
J. Grievink and J. van Schijndel (Editors)

Improving the Control of an Industrial Sugar Crystalliser: a Dynamic Optimisation Approach

T. T. L. Vu and P. A. Schneider
James Cook University, School of Engineering
Townsville Queensland 4811 Australia

Abstract

The controllability of an industrial sugar crystalliser is investigated. Using a simultaneous integration and optimisation approach, the optimal control problem is formulated in GAMS, based on a dynamic model of the vacuum pan validated against plant data. MINOS 2.50 solves this open-loop model, yielding: the minimum batch time, set-point trajectories and the optimal switching from high to low purity feed. These results are implemented within double-loop PI controllers. Alternative control variables are proposed, replacing traditional process outputs, which are not fundamentally process-relevant. Dynamic simulation results show that the proposed control schemes are satisfactory in the face of errors in growth rate correlations and variations in feed properties and initial batch conditions. Preliminary full-scale investigations have shown promise and will be extended into next year's sugar cane crushing season.

1. Introduction

The number 6 fed-batch evaporative crystalliser, also known as a vacuum pan, located at CSR's Macknade Sugar Mill (North Queensland, Australia) operates through two distinct phases, composed of many sub-steps. The existing conductivity-based control scheme, developed several decades ago, dictates the feed policy to this unit operation. Increasing production capacity, while maintaining process flexibility and quality, is of paramount importance to CSR and so they made this unit available for testing purposes. In the first phase of the batch Vu and Schneider (2001) propose two control schemes, using state-based control variables to dictate the feed and steam rates. Our investigation into the second phase of the batch is presented in this paper. The second phase of the batch is composed of three sub-steps and is more complex than the first phase, since this phase includes a switch from one feed material to another.

Previous authors (Frew, 1973 and Chew and Goh, 1989) concentrated on solving the optimal control problem, using an open-loop model. Their solutions could not handle uncertainties in initial conditions, variations in feed properties and plant-model mismatch. Their proposed optimal control solutions were never implemented.

This paper first briefly describes the process at hand. A brief overview of the solution technique is then given. The keynote is the comparison between two different control schemes to determine the feed and evaporation rate policies of the fed-batch process. Finally closed-loop responses within vacuum pan dynamic simulations and preliminary implementation data from the factory are presented for discussion.

2. Process Description

A schematic of a vacuum pan is presented below in Figure 1, showing the steam and various feed inlets. The starting and full levels of the vessel are indicated.

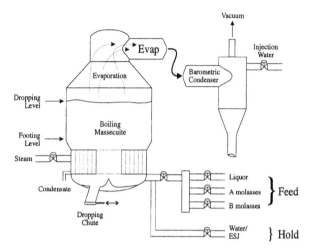

Figure 1: Schematic of a Vacuum Pan

The second phase of the crystallisation process includes three main sub-steps.

1. *Foundation*: steam heating is resumed with approximately 50 tonnes of footing material, known as massecuite (M), which is composed of sugar crystals (x_4) suspended in mother molasses (*Mol*). Molasses (*Mol*) is a solution comprising dissolved sucrose (x_1), dissolved non-sucrose impurities (x_2) and water (x_3). Liquor, a high purity feed, is simultaneously introduced to enrich the mother molasses, depleted due to sugar crystallisation.
2. *Boilback*: liquor feed is switched to A-molasses, a lower purity feed material, until the pan reaches its full volume.
3. *Heavy-up*: the pan is full. Feeding is stopped, but steam continues, leading to exhaustion of the mother molasses. After this step the massecuite is centrifuged, in order to recover the product sugar crystal.

Rigorously determining the schedules for feed and steam rates, as well as the feed switching point, requires a more sophisticated solution method than is presently available to industrialists. The essential steps leading to batch vacuum pan optimisation are described next.

3. Methods

The following sections present the problem formulation, including discussion on the constraints acting upon this system, followed by a brief description of the implementation.

3.1 Problem formulation

The process dynamic model of the vacuum pan can be fully described by mass, energy and population balances. However, the dynamics of the crystal size distribution are simplified, while the energy balance is written in the simplest form possible. In other words the evaporation rate (E) is assumed to be proportional to the steam rate to the vessel. The mass balances of water, impurities and sucrose in the pan are straightforward. Readers should consult Vu and Schneider (2000, 2001) for more details.

Three main constraints involve the crystal content (CC), the solution oversaturation (OS) and the target purity of the mother liquor at the end of Heavy-up.

The crystal content is defined as the mass fraction of crystal in the vessel, according to

$$CC = x_4/M \leq 0.55 \tag{1}$$

The solution sucrose oversaturation OS is the driving force for crystallisation. The higher the OS, the faster the crystal population will grow. However, the OS has an upper limit, termed the critical oversaturation or OS_{crit}. Beyond this point, nucleation of new unwanted crystals occurs, resulting in downstream processing inefficiencies. This critical level has been previously defined by (Broadfoot and Wright, 1972). At all times the fractional oversaturation, known as FOS, should be kept below unity.

$$FOS(= OS/OS_{crit}) \leq 1 \tag{2}$$

The last constraint determines the switching point from fresh liquor to A-molasses feed in order to achieve the desired final molasses purity, or $Pty(= x_1/x_1 + x_2)$, defined as mass ratio of sucrose to total dissolved solids in the molasses.

$$Pty \leq 0.75 \qquad \text{at Heavy - up} \tag{3}$$

The upper limits of massecuite, crystal content and purity also serve as termination conditions for the second phase. In order words, when M reaches 100 tonnes and Pty drops to 0.75, feeding is stopped, but steam continues until the crystal content reaches 0.55.

Challenges facing the batch pan problem include uncertainties in feed properties and initial conditions, such as purity, $Brix(= x_1 + x_2 + x_4/M)$, CC, footing seed diameter, etc. Due to upstream process fluctuations, the ranges of these variations can be large at the beginning of the batch. However, during the second phase, these variations become less important. After Boilback, the crystal content increases to a high level, reducing the natural circulation in the vessel and, consequently, the rate of heat transfer. This loss of circulation and the subsequent reduction in heat transfer are not presently modelled, which must be taken into account for factory implementations.

The optimal control problem formulated for the second phase is based on the nominal conditions. The objective function, OF, presented in (4) contains two different targets: minimum batch operating time and optimum trajectories of M, CC or FOS set points and, importantly, the switching point from the Foundation to Boilback sub-steps.

$$OF = \min \left\{ w_t \, time + w_m \sum_{i=1}^{i=nfe} \left[M(i) - M_{sp}(i) \right]^2 + w_c \sum_{i=1}^{i=nfe} \left[CC(i) - CC_{sp}(i) \right]^2 \right.$$

$$\left. + w_f \sum_{i=1}^{i=nfe} \left[FOS(i) - FOS_{sp}(i) \right]^2 \right\} \tag{4}$$

This type of dynamic optimisation problem can be solved using a simultaneous integration and optimisation technique, which approximates the state variables by interpolation polynomials (Vu and Schneider, 2000, 2001). These are differentiated and then back-substituted into the state equations, converting them to a set of algebraic equations. The scaling factors w_m, w_c or w_f in the above equation force M, CC or FOS to follow the desired optimum trajectories of M_{sp}, CC_{sp} or FOS_{sp}. The *time* variable in (4) is actually replaced by the summation of the element lengths. It is important to note that the number of finite elements (*nfe*) during Foundation, Boilback and Heavy-up are specified, but their lengths are decision variables in the optimisation problem. Thus the optimal switching points from one sub-step to another can be determined.

3.2 Implementation

Instead of using feed and steam flow profiles, set-point trajectories will be implemented within two PI controllers. Two different control schemes are proposed. Both schemes select the mass-controller as a primary loop because mass can be directly and accurately measured and it is one of the ending conditions stated above. Another advantage is that the mass set-point trajectory will not be affected by the presence of errors in the growth rate expressions. This dominant loop controls the mass in the pan by adjusting the feed rate.

First Scheme: M-F/CC-Steam - The second loop controls CC by adjusting steam. CC cannot directly be measured, but can be estimated using either a state estimation scheme or by combining other process outputs. Since CC is the second ending condition, this loop is simpler to be implemented. The greatest disadvantage of the first scheme is that the FOS remains uncontrolled. This might result in constraint (2) violation due to variations in feed properties and initial conditions. The gains and reset times of these two controllers therefore must be tuned against a worst-case scenario, requiring an iterative dual-level optimisation, as discussed in Vu and Schneider (2001).

Second Scheme: M-F/FOS-Steam - To avoid these tuning problems, the second loop could control FOS by adjusting the steam flow rate. The gains should be set at the highest values, without leading to oscillations in the controlled or manipulated variables. Therefore they must be tuned against the fastest situation and tested with other cases. At high values of CC, FOS control is safer and more robust but this loop pairing also has some disadvantages. First, the secondary loop in this case must include a logical operation to stop the batch, using the ending condition of CC. Second, FOS is derived from many other correlations, which might bring in a large error in FOS determinations.

Dynamic simulations obtained from two control schemes and preliminary implementation results are compared in the following section.

4. Results

The scheme *M*-Feed/*FOS*-Steam was selected as the option to test under factory conditions. Figure 2 shows *FOS*, *CC* and *M* profiles in the pan. Solid curves representing factory implementation are plotted against broken lines for simulation. A significant deviation between these two curves especially at the end of the batch is due to heat transfer limitations on evaporation rate from the pan. It should be noted that the mass variable, *M*, was indirectly controlled, but it nonetheless follows the path prescribed by the optimisation results. This factory trial was necessarily run at a low FOS set point. Future experiments at the factory will determine how much higher this set point can be, before the system suffers.

Dynamic simulation results using both control scheme options applied to the second phase of operation are shown in Figure 3: *CC*-Steam in the left and *FOS*-Steam on the right, showing the controlled (*M*, *CC*, *FOS*) and manipulated variables (Feed and Steam). In terms of batch time reductions, the *FOS*-Steam loop should be used over that of the *CC*-Steam, since it maintains a higher level of sucrose oversaturation throughout the batch, and therefore sustains crystal growth at a higher rate.

5. Conclusions

The controllability of an industrial sugar crystalliser has been investigated through the second phase of operation in a similar fashion used in the first phase investigation. New controlled variables: *M*, *CC* and *FOS* can replace the traditional output-based conductivity control. A double-loop control strategy is also proposed to replace the existing single loop controller. Based on the preliminary implementation, results from the factory indicate that the loop pairing of *M*-Feed/*FOS*-Steam functions well in the second batch phase. Future work will evaluate alternative implementations of both proposed control schemes across both phases of this fed-batch crystallisation. Further more the pan model will be augmented with some forms of correlations for the overall heat transfer coefficient across the heating section.

6. References

Broadfoot, R. and P.G. Wright, 1972, Nucleation Studies, Proceedings of The Australian Society of Sugar Cane Technologist 39th Conference, 353.

Chew, E. P. and Goh, C. J., 1989, On Minimum Time Optimal Control of Batch Crystallization of Sugar, Chem. Eng. Comm. 80, 225.

Frew, J. A., 1973, Optimal Control of Batch Sugar Crystallization, Ind. Eng. Chem. Process Des. Develop 12, 460.

Vu, T.T.L. and P.A. Schneider, 2000, Operability Analysis of an Industrial Crystalliser, CHEMECA Proceedings 2000.

Vu, T.T.L. and P.A. Schneider, 2001, Controllability Analysis of an Industrial Crystalliser, 6th World Congress Proceedings 2001.

7. Acknowledgements

Both authors would like to acknowledge CSR Ltd for their strong and enthusiastic support of this project. Dr Vu would like to acknowledge the SRDC for funding under their CP2002 project program.

576

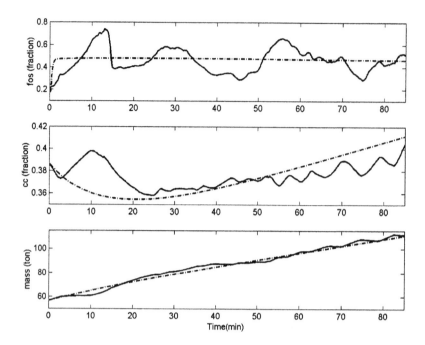

Figure 2: FOS, CC and M Profiles During Phase One of the Batch (M-F/FOS-Steam) (Solid line: implementation; broken line: simulation)

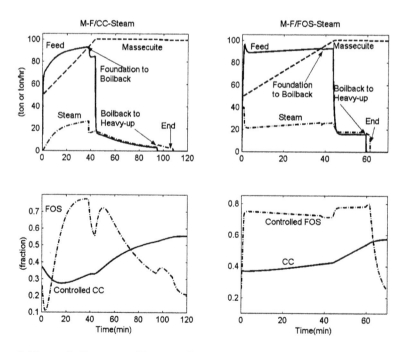

Figure 3: Dynamic Responses Using Different Control Schemes in Phase Two of the Batch.

European Symposium on Computer Aided Process Engineering – 12
J. Grievink and J. van Schijndel (Editors)

577

A Scheme for Whole Temperature Profile Control in Distributed Parameter Systems

T Wahl, WE Jones and JA Wilson
School of Chemical, Environmental & Mining Engineering
University of Nottingham, UK

Abstract

A control strategy is proposed for distributed process systems in which the whole measured temperature profile through the system is held as close as possible to a desired steady state profile defined externally via some global optimisation (not considered here). To do this the concept of a minimum error profile (MEP) is introduced which represents the closest feasible approach to the desired profile. The MEP then becomes the set-point or target for regulation. The implementation involves both state and parameter estimation, via extended Kalman filtering, and dynamic optimal control via LQR. Performance of the approach is illustrated with an example based upon a cascade of heated stirred tanks.

1. Introduction

Distributed process plant items are often instrumented with an array of measurement sensors. For example an array of temperature probes is often installed to measure bed temperatures along a fixed bed catalytic reactor or tray temperatures along a distillation column. However, control in many such situations is based upon use of only one or two sensors selected from the complete set, the remaining sensors being used to provide monitoring information for the operator. Here we propose a strategy for control of the whole measured profile with the objective of holding it (i.e. the measured profile) as close as possible to a desired profile (defined externally to meet some global optimum). Temperature profile control is mentioned by Edgar et al. (2001) but no more detailed explanation or specific mathematical solution is proposed. In the present work the minimum error temperature profile (MEP) is introduced. This combines Edgar et al.'s (2001) objective with standard mathematical derivations as performed in Sage (1968).

The MEP can be contained in a complete control scheme linking steady state optimization, system identification and dynamic optimal control. The supervisory-level global optimization of operating conditions is performed intermittently via general non-linear programming. The results of this calculation are introduced via the MEP calculation which iterates much faster at the sub-level and provides a real-time shortcut approach to the optimized solution. Identification, e.g. via extended Kalman filter (EKF), is performed to obtain estimates of system state and parameter changes. The controller, e.g. a linear quadratic regulator (LQR), provides control action to be undertaken. Calculations performed on the sub-level are executed every sample instant and results from the global optimization are input when they are available.

Disturbance compensation is performed via changes in MEP where controller error is its difference from the actual profile.

2. Control Scheme

An extended Kalman filter (EKF), driven with on-line measurements of the temperature profile, provides state and parameter estimates as input to the MEP control strategy. These estimates also feed the supervisory global steady state optimization, which delivers, on a slow intermittant timescale, the desired steady state temperature profile, Δx_{Sup} expressed here as a vector of temperature perturbations around a fixed steady operating point. The MEP control objective is to hold the system as close as possible to the desired profile in a squared error sense by minimizing

$$e = (\Delta x - \Delta x_{Sup})^T (\Delta x - \Delta x_{Sup}) \qquad (1)$$

where Δx is the corresponding perturbation in system steady state temperature profile. The dynamic behaviour of the temperature profile can be represented, as for example by Ackermann (1985), in discrete time as

$$\Delta x(k+1) = A_d \, \Delta x(k) + B_d \, \Delta u_{LQR}(k) + B_d \, \Delta u(k) + \Delta b(k) + \Delta w(k) \qquad (2)$$

where $\Delta x(k)$ is the actual current profile, A_d is the discrete state transition matrix, B_d is its associated driving matrix, Δu is the control action component needed to achieve a desired steady state profile while $\Delta u_{LQR}(k)$ is the additional transient action contributed by the LQR to drive optimally to the steady state. Unknown process disturbances are represented by $\Delta w(k)$, a set of white noise inputs, and $\Delta b(k)$, a set of biases representing unknown deterministic inputs. Values for $\Delta b(k)$ come as parameter estimates from the EKF. At any time instant k the steady state temperature profile predicted using this model will be

$$\Delta x = f_i \, \Delta u(k) + f_b(k) \qquad (3)$$

where $f_i = (I - A_d)^{-1} B_d$ and $f_b(k) = (I - A_d)^{-1} \Delta b(k)$. To minimize e the required control action Δu is given as

$$\Delta u = \Delta u_{MEP} = (f_i^T f_i) f_i^T (f_b(k) - \Delta x_{Sup}) \qquad (4)$$

which will take the system to steady state at the minimum error temperature profile Δx_{MEP}. $\Delta u_{LQR}(k)$ is then the current, transient action superimposed by the LQR in order to drive the current error $\Delta x(k) - \Delta x_{MEP}$ optimally towards zero.

3. A Case Study Example

To demonstrate the performance and application of the approach we present a simple distributed process case study involving a single liquid stream flowing through a long cascade of mixed stages each with an unknown, independently varying heat input, as shown in Figure 1. The result is formation of a temperature profile along the cascade. Both the flow rate and temperature of the feed stream can be manipulated to control the profile against disturbances in the heat inputs.

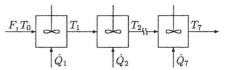

Figure 1: A simple distributed system with a temperature profile – a long cascade of heated vessels with unknown heat inputs.

Normal operating conditions are given as: Inlet temperature $T_0=30°C$; Flow rate $F=5000kg/h$; Mass of liquid per vessel $m_i=3000kg$; Number of simulated vessels $n=7$; Specific heat capacity $c=4kJ/(kg\ K)$. The overall heat input is chosen to be $10^6kJ/h$, raising the temperature by $50°C$. The heat is introduced as bell curve along the vessels with its centre at the middle vessel. The resulting temperature profile is sigmoidal.

Two test disturbances are considered to demonstrate the system response. First, the temperature in the middle vessel is perturbed by $10°C$, see the noise-free response in Figure 2(a). It can be seen that the effect of the temperature perturbation is flushed downstream through the system until all temperatures return to their initial steady state.

Second, the heat input, $\Delta\dot{Q}$, is increased in the middle vessel by 30% of its normal value, see Figure 2(b). This leads to a steady state perturbation from the 4th vessel on.

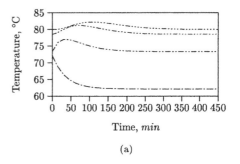

(a)

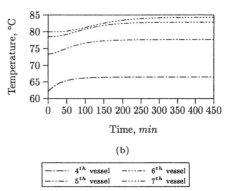

(b)

| ---·--- 4th vessel | ------ 6th vessel |
| ---·--- 5th vessel | ------ 7th vessel |

Figure 2: Open loop responses to disturbances in middle vessel temperature (a) and heat input (b).

4. Estimator, MEP and Controller Tuning for Case Study

Both process and measurement noise are included when considering closed loop control. Process noise, Δw, implemented as in Equation 2, and measurement noise, Δv, have standard deviations of $0.5°C$ in temperature and $50kg/h$ in flowrate.

Heat inputs $\Delta\dot{Q}$ are estimated as parameters in an augmented filter state (i.e. the bias term in Equation 2 is handled as $\Delta b = C_d \, \Delta\dot{Q}$). An error covariance of $(5\times10^{-4}\dot{Q})^2 I$ is assigned. It should be noted that this covariance matrix can also be viewed as a tuning parameter. If the factor 5×10^{-4} is increased then parameter estimates will converge faster. The drawback is that due to noise the estimates will fluctuate more. Under closed loop control this leads to stronger variations in in manipulated variables. Similar effects attach to assigning uncertainty to the initial guess of system state and paramteres Δx_f.

However, these assignments have no great influence to the observers performance since representative presentation are taken when error covariances converged to steady values. The LQR's tuning parameters are chosen to be $Q_r = 10^4 I$ and $R_r = I$, its objective function weighting matrices, Q_r for the state and R_r for the manipulated variables. When operating on its own the LQR's state is taken to be $\Delta x_r = \Delta x$. Here it is assumed to utilize the steady state gain which is to be re-calculated each sampling interval.

5. Steady State Profile Control

To demonstrate the response of the MEP to an extreme disturbance, the peak heat input is shifted from the middle to the first vessel in the cascade. The overall heat input is kept constant so that the same system outlet temperature is reached, see the steady state open loop response in Figure 3. Following this disturbance the LQR alone and, in a second simulation, the MEP plus LQR are applied for system control. The smaller final steady state deviations from the desired initial profile under MEP+LQR control are apparent, with a squared error advantage of 68% over the LQR in the table. The MEP-based profile crosses the desired profile twice instead of once, as with the purely LQR-based strategy. Final steady state values for manipulated variables are also given in the table in consistent units.

	Initial	Open Loop	LQR	MEP+ LQR
e_{MEP}	0	2230	511	164
T_0	30	30	12	1
F	5000	5000	4252	2360

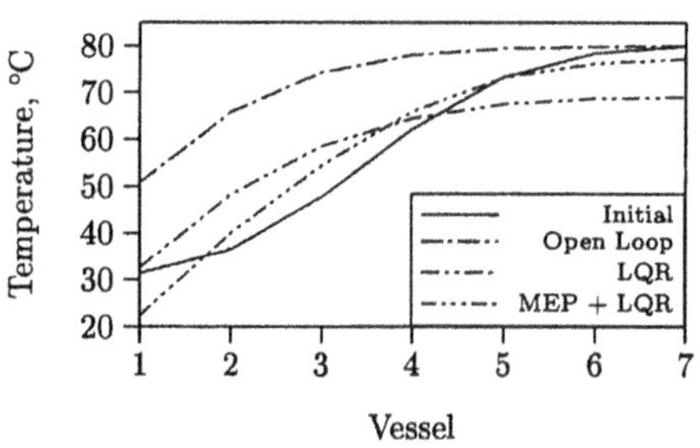

Figure 3: Steady state profiles reached under various control strategies after a large disturbance in heat input profile.

6. Dynamic Profile Control

Temperature Disturbance: The system response to the 10°C disturbance, as in Section 3, is presented in Figure 4(a). A moving target temperature profile, Δx_{MEP}, can be observed. This is due to the delay and inaccuracy of parameter estimation. Due to noise and the disturbance injection the EKF produces varying heat input estimates. This leads to re-calculation of a new target temperature profile which the controller tries to follow. This is predominantly seen in vessels three and four where a shift in target comes back into it's initial range of noise induced changes. The influence of noise is even more evident in the manipulated variable movements in Figures 4(b) and 4(c). The target for

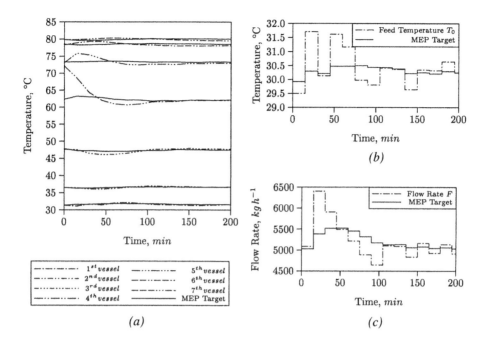

Figure 4: Closed loop response to temperature disturbance in middle vessel by 10K. Presented are vessel temperature and their MEP targets (a), inlet temperature (b) and feed flow rate (c).

the flow rate returns much earlier to its initial range than the inlet temperature does. However, the deviation is also more distinct than for the feed temperature case. In both subfigures it can be seen that their actual values approach their targets until they would meet in no-noise steady state.

Heat Input Disturbance: Response to disturbance in middle vessel heat input, as in Section 3, is presented in Figure 5(a). Due to the sustained change in heat input the target state does not go back to it's initial range. Again, most distinctive changes in target state can be observed in vessels three and four. Transient changes of state can be most clearly seen in vessels four and five. It takes the control scheme approximately 150min to bring the system to it's new target. The arithmetic averages of the MEP are given by $\text{avrg}(\Delta \mathbf{x}_{MEP})\big|_{195\min}^{435\min} = [0.48°C \quad -0.09°C \quad -1.18°C \quad 1.22°C \quad 0.17°C \quad -0.26°C \quad -0.34°C]^T$ and $\text{avrg}(\Delta \mathbf{u}_{MEP})\big|_{195\min}^{435\min} = [0.68°C \quad 570kg/h]^T$. Observing changes of inlet temperature in Figure 5(b) shows a distinct shift towards the approximate target given by 30.68°C. A similar pattern occurs for the feed flow rate as presented in Figure 5(c). It increases towards 5570kg/h.

582

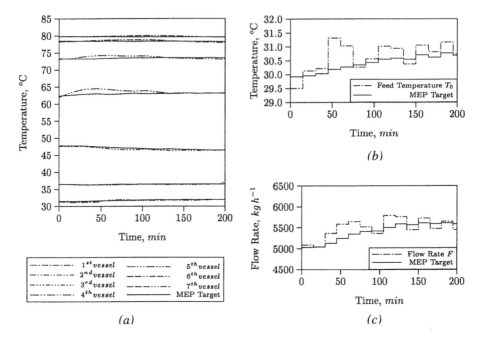

Figure 5: Closed loop response to heat input disturbance in middle vessel by 30%. Presented are vessel temperature and their MEP targets (a), inlet temperature (b) and feed flow rate (c).

7. Conclusions

The concept of the minimum error profile (MEP) is presented for control of whole temperature profile in distributed parameter systems faced with parameter drift. The approach minimizes in a least squares sense the error from a predefined desired profile and forms the target for application of the linear quadratic regulater (LQR).

Application of the proposed contol scheme to a simple example, a cascade of heated stirred tanks, has demonstrated the feasibility and performance relative to a conventional LQR implementation. The approach has the potential for quite direct application to control of fixed-bed catalytic reactors and counter-current mass transfer processes.

8. References

Ackermann, J. (1985), *Sampled data control systems: analysis and synthesis, robust system design*, Communication and Control Engineering Series, Springer Press

Edgar, T.F., Himmelblau, D.M. and Lasdon, L.S. (2001), *Optimization of chemical processes*, Chemical Engineering Series, second edn, McGraw-Hill

Sage, A.P. (1968), *Optimum systems control*, first edn, Prentice-Hall

European Symposium on Computer Aided Process Engineering – 12
J. Grievink and J. van Schijndel (Editors)
© 2002 Elsevier Science B.V. All rights reserved.

Dynamic Plantwide Modelling, Flowsheet Simulation and Nonlinear Analysis of an Industrial Production Plant

R. Waschler[1*], A.Kienle[1], A. Anoprienko[2], T. Osipova[3]

[1]Max-Planck-Institute of Dynamics of Complex Technical Systems
Sandtorstr. 1, 39106 Magdeburg
[2]Donetsk State Technical University, Artemstr. 58, 340000 Donetsk, Ukraine
[3]AZOT, Sewerodonetsk, Ukraine

Abstract

A detailed dynamic model of the Monsanto process for the production of acetic acid is presented. The nonlinear behaviour of the process is investigated and implications different basic control structures have on its dynamic behaviour are pointed out.

1. Introduction

Plantwide control issues have recently been attracting increasing interest within the process control community, as can be seen by the publication of several review articles during the last couple of years (see, e.g., Skogestad and Larsson, 1998, and references therein). While promising great potential for rendering processes more economic, it is widely recognized that the control of highly integrated plants must be regarded as a most challenging task.

It therefore seems appropriate first to gain insight into the behaviour of complex plants and also the associated control problems by considering suitable case studies. However, apart from the classic Tennessee Eastman benchmark problem (Downs and Vogel, 1993), and the case studies provided in Luyben etal. (1999), only few examples of industrial scale are found in the literature, a fact that is most likely to be attributed to the enormous efforts that come along with rigorously modelling an entire production plant.

The purpose of this contribution is to present such an industrial example, the so-called Monsanto process for the production of acetic acid. It captures features that are believed to be typical of many chemical plants and therefore of common interest beyond this particular system. As our point of view is absolutely consistent with Luyben's, who stated that "the primary mathematical tool in the solution of the plantwide control problem is a rigorous, nonlinear model of the entire plant" (Luyben etal. 1999), we regard the model presented in this article as a sound basis for our forthcoming investigations on plantwide control. Emphasis in this contribution, however, is on the model itself, the numerical analysis of the nonlinear behaviour of the process and the impact basic control structures have on its dynamic performance.

---

[*] Corresponding author, phone: ++49-391-6110-375; e-mail: waschler@mpi-magdeburg.mpg.de

2. The Process

Nowadays the most common and most economic way of producing acetic acid on an industrial scale is according to the Monsanto process, labeled by leading experts in the field as "one of the triumphs" of modern homogeneous catalysis (Maitlis etal 1996). Fig. 1 presents a simplified flowsheet that captures the main ingredients of the real plant. Peripheral equipment is left out for reasons of clarity.

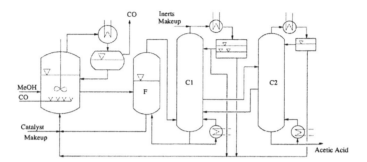

Fig.1: Simplified flowsheet of the Monsanto process

The liquid educt methanol and the gaseous educt carbon monooxide are fed to an adiabatic continuous reactor where the exothermic carbonylation reaction to acetic acid according to

$$CH_3OH + CO \rightarrow CH_3COOH, \quad \Delta H_R = -138,6 \text{ kJ/mol} \tag{1}$$

is carried out in homogeneous aqueous liquid phase at boiling point conditions of about 30 bar and 185°C. The reaction is homogeneously catalyzed by means of a rhodium (or iridium) carbonyl catalyst (denoted as Rh) which in turn has to be activated by a promotor methyl iodide (MeI). The heat of reaction is removed by evaporative cooling, with mainly the excess CO leaving the process. The liquid reaction product enters a flashdrum F operated at about 1.5 bar. This separator serves for the recovery of the extremely expensive catalyst, which – as essentially being nonvolatile – is recycled to the reactor along with a huge portion of the reactor effluent, while the catalyst-free vapour product is fed to a first distillation column C1. The main purpose of this column is the recovery of promotor MeI, accomplished by means of a liquid-liquid phase split in a decanter atop the column. While MeI and a water-rich stream are recycled to the reactor, the main product stream is withdrawn from an intermediate tray and fed to a second column C2 where the remaining inert component water is separated and recycled. Almost pure acetic acid is obtained as bottoms product of column C2.

3. The Model

A detailed dynamic model of the entire plant has been implemented in the simulation environment DIVA (Mangold etal. 2000). Here, we restrict ourselves to the key characteristics of the system that both distinguish it from other case studies and which are the origin of its most intriguing traits.

3.1 Reactor

At the heart of our overall model we have the reactor model which is of hybrid nature as it has to allow for switching between boiling and non-boiling conditions as consequence of a disturbance or during startup. Note that at standard operating conditions the temperature inside the reactor is the boiling point.

For a detailed calculation of the vapour-liquid equilibrium we apply the UNIQUAC method to obtain activity coefficients of the liquid phase and the Redlich-Kwong correlation for the prediction of vapour phase fugacities. One special characteristic of our system originates from CO being supercritical at the given standard operating conditions. To approximately describe its solubility we follow Prausnitz etal. (1986) and apply the concept of ideal CO solubility in the reaction mixture.

Dynamic component material balances and a liquid phase summation condition give the mole fractions of the five components we consider, i.e., methanol, CO, acetic acid, methyl iodide and water. The catalyst x_{Rh} is assumed to be a system parameter. We neglect side reactions and restrict ourselves to the brutto reaction (1). It has been intensely studied (e.g. Maitlis etal. 1996) and the underlying complex reaction mechanism giving rise to (1) seems fairly well understood. In particular, the reaction was distinguished as being of approximately first order in the catalyst and the promotor and zeroth order in the educts as long as the standing concentration of these reactants is sufficiently large. Yet, for small concentrations – and as there is almost 100% conversion of methanol this applies to the given process - the reaction order will shift to one in the limiting component, a fact we account for by introducing Michaelis-Menten-like dependencies for the reactants in the rate law, yielding

$$r = 0.4986 \cdot e^{-7830(\frac{1}{T}-\frac{1}{443})} \cdot c_{Rh}^{0.99} \cdot c_{Mel}^{1.05} \cdot \frac{K_{MeoH} \cdot c_{MeoH}}{1 + K_{MeoH} \cdot c_{MeoH}} \cdot \frac{K_{CO} \cdot c_{CO}}{1 + K_{CO} \cdot c_{CO}} \ . \tag{2}$$

Exact values for the constants K_{MeOH} and K_{CO} are not known, however this is no problem as we could show that the sensitivity with respect to the K_j is small as long as they are given physically reasonable values. For our simulation studies we chose $K_{MeOH} = K_{CO} = 100$.

3.2 Separation system and basic control structure

The flashdrum is operated adiabatically, and for all units of the separation system we use detailed dynamic vapour-liquid equilibrium models including component material and energy balances.

We assume that basic inventory control is guaranteed for all units and is incorporated in the models of all single units in terms of constant holdups and constant pressures.

3.3 Steady state simulation results

Let us mention briefly that agreement between steady state simulation results of our entire model and actual measurements from a real plant - located in Sewerodonetsk, Ukraine, with a capacity of about 150000 t/a – is very satisfactory, given the fact that our model is purely predictive.

4. Nonlinear Behaviour

We will now briefly investigate some nonlinear features of the process and try to emphasize the implications they have on possible control problems, always bearing in mind that our eventual goal is directed towards plantwide control.

4.1 Isolated Reactor

As the key to understanding the overall plant dynamics is a thorough understanding of the reactor dynamics we will set out looking at the stand-alone reactor.

Suppose all feeds and recycles to the reactor are fixed and we vary the amount of catalyst inside the reactor. Computationally, this can be done using the method for continuation of steady states in DIVA (Mangold etal. 2000). Fig. 2 illustrates that over a

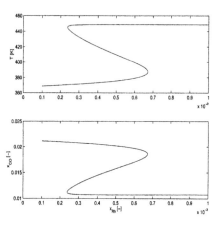

Fig. 2: Multiple steady states

certain range of the parameter x_{Rh} (abscissa) three steady states can occur. It can be shown that this type of behaviour is a general feature of any two-phase reactor where one of the reactants is by far more volatile than the other components (Waschler etal. 2001). In such cases the temperature will decrease with increasing reactant concentration, leading to a decrease of reaction rate. This can be interpreted as a (nonisothermal) self-inhibition of the reaction. It is a well known fact that any kind of self-inhibition is a potential source of multiplicity and instability. In the Monsanto process, CO (lower subplot) represents this by far most volatile component. Fig. 2 depicts the direct and inverse correlation between dissolved CO and reactor temperature (upper subplot). Note that these considerations are extremely important from an operating point of view as there is the problem of catalyst deactivation that could lead to a sudden extinction of the reaction once a lower threshold for catalyst concentration is crossed.

Another typical example exhibiting this type of characteristics is the ethylene glycol reactive distillation system (Gehrke and Marquardt, 1997).

4.2 Reactor-separator system

Let us now approach the realm of plantwide operation by closing the first recycle loop from the flashdrum back to the reactor. It has been widely recognized that control structure selection is the decisive task related to the plantwide control problem and this becomes evident already for the simple reactor-separator system (see also Pushpavanam and Kienle, 2001). Here, we are particularly interested to see how two different control structures cope with a disturbance as it may be encountered during the operation of a plant.

Let the setup where reactor holdup, or liquid level, respectively, is controlled by manipulating the reactor effluent be denoted by control structure 1 (CS1, see Fig. 3). In control structure 2 (CS2) the reactor effluent itself is flow-controlled and one recycle flowrate from a downstream separation unit is manipulated to control the reactor level.

Again assume all feeds and recycles to be fixed, except for the flash recycle, of course, and the manipulated recycle in CS2. Then assume that after one hour of standard operation there occurs a 20 K drop in methanol feed temperature $T_{F,MeOH}$ (subplot upper left in Fig. 4) lasting for two hours, e.g. due to disturbances in the utility system. As can be seen, the effect of this disturbance on CS1 (solid lines in Fig. 4) is drastic: the drop in energy supplied to the reactor decreases the reactor temperature T_R (upper right), which in turn causes the reaction rate to decrease. Thus, less CO is consumed, resulting in an increase in x_{CO} (lower right) and an additional decrease in temperature because of the mechanism described above. Due to the material

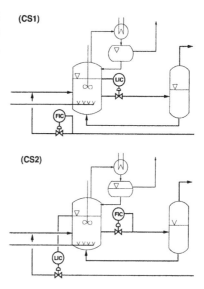

Fig.3: Different control structures

and energy integration with the flash, this effect is further amplified: the cold flash recycle increases as soon as the flash inlet temperature T_R drops. This larger and colder recycle decreases the reactor temperature even further and simultaneously increases the reactor effluent L_R (lower left), and is therefore identified as the origin of the accumulation of flowrates in the coupled system. In other words, we observe some sort of nonlinear snowball effect, similar to the one described by Luyben etal. (1999).

To avoid the obvious shortcomings of CS1 we follow Luyben's heuristic guidelines in terms of control structure selection and apply CS2 (dashed-dotted lines in Fig. 4), which turns out to be far superior in rejecting this disturbance. Fixing reactor effluent to some degree decouples the two subsystems. In particular, the flash will only be subject to a slight disturbance in composition and temperature, as opposed to the large flowrate disturbance encountered for CS1. As a consequence, the state of the flash will vary to a much smaller extent and the amplification of the self-inhibitory reactor characteristics is prevented. Thus, in CS2, the reactor can almost settle to a new steady state even within two hours after the disturbance.

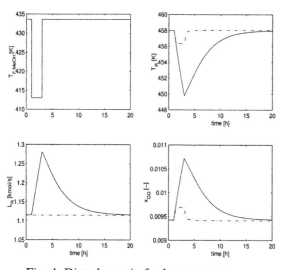

Fig. 4: Disturbance in feed temperature

4.3 Overall plant

As can be expected simulation of the entire plant shows many of the features commonly attributed to highly integrated processes with several recycles. Most noticeably, we observe an increase in the system's time constants and inverse response behaviour to load disturbances. In contrast to other reported case studies, where it is crucial to purge out inerts and avoid accumulation of both inerts and reactants, it turns out that for the Monsanto process keeping track of the inert water and promotor methyl iodide, i.e., controlling their material balances using makeup streams, is a key requirement in order to run the complete process in a stable way.

5. Conclusions

We have presented a dynamic nonlinear model of an industrial production plant and highlighted the nonlinear behaviour of the underlying process. Basic concepts for stable and robust operation at the nominal operating point were discussed. Future work will also focus on the design of supervisory control strategies for startup and the efficient performance of load changes, two problems representing big challenges in terms of plantwide control. Obviously, the efficiency of all new strategies will eventually have to be compared to well-established control techniques as, e.g., model predictive control.

Acknowledgement:
The authors gratefully acknowledge the financial support of the International Bureau of the German Ministry of Education and Research (BMBF) through grant UKR 00/004, supporting scientific technological cooperation.

References
Downs, J.J., and E.F. Vogel, 1993, A Plant-Wide Industrial Process Control Problem, Computers Chem. Engng. 17 (3), 245-255.
Gehrke, V., and W. Marquardt, 1997, A singularity approach to the study of reactive distillation, Comp. Chem. Eng. (21, Suppl.), 1001-1006.
Luyben, W.L., B.D. Tyreus and M.L. Luyben, 1999, Plantwide Process Control. McGraw-Hill, New York.
Pushpavanam, S., and A. Kienle, 2001, Nonlinear behavior of an ideal reactor separator network with mass recycle, Chem. Eng. Science (56), 2836-2849.
Maitlis, P.M., A. Haynes, G.J. Sunley, and M.J. Howard, 1996, Methanol carbonylation revisited: thirty years on, J. Chem. Soc., Dalton Trans. 2187-2196.
Mangold, M., A. Kienle, and K.D. Mohl, 2000, Nonlinear computation using DIVA – Methods and applications, Chem. Eng. Science (55), 441-454.
Prausnitz, J.M., R.N. Lichtenthaler, and E.G. de Azevedo, 1986, Molecular thermodynamics of fluid-phase equilibria, Prentice-Hall, Englewood Cliffs, N.J.
Skogestad, S., and T. Larsson, 1998, A review of plantwide control, Internal Report, Dept. of Chem. Eng., Norwegian Univ. of Science and Technology, Trondheim.
Waschler, R., S. Pushpavanam, and A. Kienle, 2001, Multiplicity features of two-phase reactors, in preparation.

European Symposium on Computer Aided Process Engineering – 12
J. Grievink and J. van Schijndel (Editors)
© 2002 Elsevier Science B.V. All rights reserved.

Flatness-based optimization of batch processes

M.E. van Wissen, S.Palanki* and J. Grievink,

Delft University of Technology, Department of Chemical Technology,
Julianalaan 136, 2628 BL Delft, The Netherlands.
* Florida A&M University- Florida State University, Department of Chemical
Engineering, Tallahassee,FL,32310-6046, USA.

Abstract

A flatness optimization framework is proposed, to deal with optimization of batch processes under uncertainty. Via the concept of differential flatness we transform the problem, such that the terminal-cost optimization problem can be solved in a cascade optimization scheme. The results have been tested on an example of a batch bioreactor.

1. Introduction

A wide variety of specialty chemicals are made in batch reactors. In batch process operations, the variables change considerably with time and thus there is no constant setpoint around which the process can be regulated. Because there is no steady state, the objective is to optimize some objective function, which expresses the system performance. The optimal operating policy for a given batch process is usually calculated under the assumption of a perfect model. However, realistic applications are subject to uncertainty in initial conditions, model mismatch, and process disturbances, all of which affect the optimal solution. This provides the economic drive for *on-line* calculation and implementation of the optimal operating policy.

Normally, the optimization of batch processes leads to a piecewise, discontinuous solution. A *cascade optimization scheme* is proposed by Visser et al. (2000) to implement such an optimal trajectory despite disturbances and uncertainty. It is very often the case, that the optimal solution of such a problem lies on the input bounds, or state constraints, or a combination of both. In the proposed method, we make explicitly use of this information. Also uncertainty is treated in the framework of cascade optimization. The classical approach is via the Hamilton-Jacobi-Bellman (HJB) formulation.

In batch reactor modeling, a lot of systems have been found, which have the so-called *flatness* property. For this class of systems, we propose a new theory which does not rely on the HJB-formulation. Via the concept of *differential flatness*, which is basically a method to dispose of the differential equations, we have developed an approach for which *flat* systems can be treated in the cascade optimization scheme. Moreover, the feedback controller, needed in the cascade optimization is now easy to obtain. The proposed scheme is illustrated on a simulation of an optimization problem in batch bioreactors.

2. Optimization Problem

Problem formulation without uncertainty
Batch optimization problems typically involve both dynamic and static constraints and fall under the class of dynamic optimization problems. It is true, that in most batch chemical processes, the inputs are flow-rates that enter the system equations in an affine manner (Srinivasan et al. (1997), Srinivasan et al. (2000), Visser et al. (2000)), but quite a few processes have been found for which this is not the case (Rouchon and Rudolph(2000)). The terminal-cost optimization of general dynamical systems can be stated as follows:

$$P: \quad \min_{u(t)} J(x(t_f), u(t_f)) \qquad (1a)$$

subject to

$$\dot{x}(t) = f(x(t), u(t)), t \in [t_0, t_f), \quad 0 = x(t_0) - x_0, \qquad (1b)$$

$$0 \le c(x(t), u(t)), t \in [t_0, t_f), \quad 0 \le c_p(x(t_f), u(t_f)). \qquad (1c)$$

where x is the n-vector of state variables with known initial conditions x_0, u the m-vector of control variables. We exploit the degree of freedom in $u(t)$ to minimize an (economical) objective function P subject to path and endpoint constraints in $x(t)$ and $u(t)$. The right-hand side of the nonlinear dynamical system is defined as the n-vector f. The constraints to be enforced during the process and at the final time t_f are given by c and c_p respectively. For a more detailed description of this problem, see Kumar and Daoutidis (1995).

The classical solution of (1) can be found by solving the well-known Hamilton-Jacobi-Bellman (HJB) formulation, which is easier to solve when the system is in control-affine form, (Palanki et al. (1991), Visser et al. (2000)).

Problem formulation with uncertainty
For many realistic applications, we can assume that the model structure is known, but the model parameters are unknown or only known within bounds. For this case, the terminal-cost optimization of general dynamical systems is as follows:

$$P: \quad \min_{u(t)} J(x(t_f, \theta), u(t_f, \theta)) \qquad (2a)$$

$$\text{s.t.} \quad \dot{x}(t) = f(x(t), u(t), \theta) + d(t), \ 0 = x(t_0) - x_0, \qquad (2b)$$

$$0 \le c(x(t), u(t), \theta), 0 \le c_p(x(t_f), u(t_f)), \qquad (2c)$$

in which θ is the p-vector θ of uncertain parameters and d is the function representing the unknown disturbances. Here, we choose to have probabilistic parametric uncertainty for θ, in which the objective function P is the expected value of a random variable, i.e.

$$P: \quad \min_{u(t)} E[J(x(t_f,\theta),u(t_f,\theta))]. \tag{3}$$

In realistic cases, one can think of the expected product quality or quantity to be maximized, or the expected loss to be minimized. The resulting optimization problem can be found in Srinivasan et al. (2000b), p.14.

3. Differential flatness

Differential flatness has been introduced by Fliess et al. (1995) in their studies of the feedback linearization problem in the context of differential algebra. A system is flat if we can find a set of outputs (equal in number to the number of inputs) such that all states and inputs can be determined from these outputs without integration. More to the point, if the system has states x (n-vector), and inputs u (m-vector), then the system is *flat* if we can find outputs y (m-vector) of the form

$$y = y(x,u,\dot{u},...,u^{(l)}), \tag{4}$$

such that,

$$x = x(y,\dot{y},...,y^{(q)}), \tag{5a}$$
$$u = u(x,u,\dot{u},...,u^{(q)}). \tag{5b}$$

The outputs y are called *flat* (or *linearizing*) *outputs*.

Differentially flat systems are useful in situations where explicit trajectory tracking generation is required. The reason for this, is that the behaviour of flat systems is determined by the flat outputs, and hence we can plan trajectories in output space, and then map these appropriate inputs. This property can be quite useful, when dealing with only a few flat outputs in comparison with the number of states and the number of inputs. For optimization problems the main advantage lies in the reduction of the computational effort, as there is no need to numerically solve the sensitivity differential equations in the nonlinear program (NLP).

For the dynamic optimization problem P, we replace the equations (1) (or (2)) with the expressions containing all the flat outputs and higher derivatives.

Another useful property of flat systems that we can use, called *endogenous dynamic feedback* (Fliess et al. (1995)) which is a dynamic feedback of the form (v is the m-vector of new inputs)

$$\dot{z} = a(x,z,v), \tag{6a}$$
$$u = b(x,z,v), \tag{6b}$$

such that z satisfying equation (6a) can be expressed as a function of x and u and a finite number of their derivatives:

$$z = \alpha(x, u, ..., u^1),$$ (7a)

The endogenous dynamic feedback can be used in the cascade optimization scheme as proposed by Visser et al. (2000), which combines notions of feedback control with notions from the field of optimization under uncertainty, and hence can be seen as a practical implementation strategy for problems containing uncertainty, see Visser et al. (1997). For a detailed description of trajectory planning of differentially flat systems and applications, see Faiz et al. (2001) and Veeraklaew and Agrawal (2001).

4. Cascade optimization scheme

In the cascade optimization framework proposed by Visser et al. (2000), the optimization problem is transformed into a tracking problem by use of so-called invariant signals obtained via the HJB-approach. In short, in the cascade optimization a 'low level' tracking controller is used, which guarantees that the system stays close to the optimal trajectory. After that, 'a high level' optimizer is used to guarantee optimality despite disturbances. We use the same framework, but instead of Visser et al. (2000), obtain the tracking controller for free, using the differential flatness approach.

5. Results

Illustrating example
The proposed methodology will be applied to a nonlinear model of fermentation of whey lactose to lactic acid by *Lactobacillum bulgaricus* in a batch bioreactor (Agrawal et al., (1989)). The mass balances are given by:

$$\dot{x}_1 = \mu(\mathbf{x})x_1 - ux_1 / x_4$$ (8a)

$$\dot{x}_2 = u(S_f - x_2) / x_4 - \mu(\mathbf{x})x_1 / Y_{X/S}$$ (8b)

$$\dot{x}_3 = (\alpha\mu(\mathbf{x}) + \beta)x_1 - ux_3 / x_4$$ (8c)

$$\dot{x}_4 = u$$ (8d)

$$\mu(\mathbf{x}) = \frac{\mu_m(1 - x_3 / P_m)x_2}{K_m + x_2 + x_2^2 / K_i},$$ (8e)

$$0 \le u \le 10 \ 1h^{-1}, \ x_4 \le 10 \ 1.$$ (8f)

All the symbols and initial states are explained in Table 1.

Table 1. Initial states and parameters

Symbol	State/parameters	Value
x_1	Biomass concentration	1 g/l
x_2	Substrate concentration	0 g/l
x_3	Product concentration	0.5 g/l
x_4	Volume	2 l
S_f	Substrate feed conc.	15 g/l
K_m	Substrate saturation const.	1.2 g/l
K_i	Substrate inhibition const.	22 g/l
P_m	Product inhibition const.	50 g/l
$Y_{X/S}$	Cell mass yield	0.4 g/g
α	Growth assoc. prod. yield	2.2 g/g
β	Non-growth assoc.pr. yield	0.2 h^{-1}
μ_m	Max. spec. growth rate	0.48 h^{-1}
t_f	Final time	11.6 h

The states in the model (8) are related by the following equation:

$$x_4(x_1 + Y_{X/S}(x_2 - S_f)) = \hat{C} = x_4^0(x_1^0 + Y_{X/S}(x_2^0 - S_f)) \tag{9}$$

In Mahadevan et al. (2001) the flat output for this problem has been derived, using (9). This has been used in the optimization problem of maximization the amount of product formed at the end of the batch, such that equations (8a)-(8e) are satisfied. In Figure 1 the trajectories and inputs of the optimal solution are plotted. Note that the input does not change too much.

Conclusions

We proposed an optimization approach of general batch processes using the concept of differential flatness. Especially, in the cascade optimization scheme, the construction of the feedback controller is easy, once the flat outputs are obtained. In comparison with classical methods (e.g., the JB-formulation), the proposed method could also lead to reductions in computing time on-line, because of the transformation to a NLP.

A disadvantage of the proposed method is that, beforehand, one does not know if a system is differentially flat. Moreover, even if one knows that a given system is flat, the flat ouputs are not always easy to obtain and might involve long algebraic manipulations. However, for a large group of reactor models for instance, it has been shown, see Rouchon and Rudolph (2000), that these are differentially flat and the flat outputs have been calculated. It is interesting to see if for all these models, the terminal-cost optimization problem under parametric uncertainty can be dealt with using differential flatness. Another subject under investigation is the robustness of the method against parametric uncertainty.

594

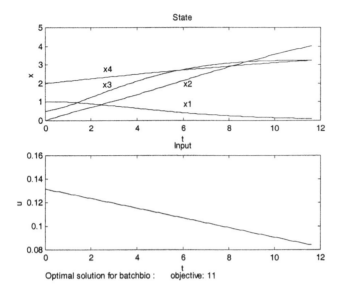

Figure 1. The states and inputs of the optimal solution of the batch bioreactor.

References

Agrawal,P.,G.Koshy and M.Ramseier, 1989, An algorithm for operating a fed-batch fermenter at optimum specific growth rate, Biotechn. Bioengg., 33, 115.

Faiz, N.,S.K. Agrawal and R.M. Murray, 2001, Trajectory planning of differentially flat systems with dynamics and inequalities, J. of Guid., Cont. Dyn., 24(2),219.

Fliess, M., J. Lévine, P. Martin and P. Rouchon, 1995, Flatness and defect of non-linear systems: introductory theory and examples, Int. J. Cont. 61(6), 1327.

Kumar, A. and P. Daoutidis, 1995, Feedback control of nonlinear differential-algebraic-equation systems, AIChE J. 41(3), 619.

Mahadevan, R., S.K. Agrawal and F.J. Doyle (2001), Differential flatness based nonlinear predictive control of fed batch bioreactors, Cont. Engg. Pract 9, 889

Palanki, S., C.Kravaris and H.Y. Wang, 1991, Synthesis of state feedback laws for end-point optimization in batch processes, Chem. Eng. Sc. 48(1), 135.

Rouchon, P. and J. Rudolph, 2000, Réacteurs chimiques différentiellement plats: planification et suivi de trajectories, *to be published.*

Srinivasan, B., E. Visser and D. Bonvin, 1997, Optimization-based control with imposed feedback structures, IFAC ADCHEM'97, 635.

Srinivasan, B., S. Palanki and D. Bonvin, 2000a, A tutorial on the optimization of batch processes: I. Characterization of the optimal solution, *to be published.*

Srinivasan, B., S. Palanki and D. Bonvin, 2000b, A tutorial on the optimization of batch processes: II. Handling uncertainty using measurements, *to be published.*

Veeraklaew, T. and S.K. Agrawal, 2001, New computional framework for trajectory optimization of higher-order dynamic systems, J. of Guid., Cont. Dyn., 24(2),228.

Visser, E., B. Srinivasan, S. Palanki and D. Bonvin, 2000, A feedback-based implementation scheme for batch process optimization, J. Proc. Cont. 10(5), 399.

European Symposium on Computer Aided Process Engineering – 12
J. Grievink and J. van Schijndel (Editors)

595

Modeling and optimization of a sugar plant using hybrid Batch Model Predictive Control

Michiel E. van Wissen, Joost F.C. Smeets[*], Ad Muller[**] and Peter J.T. Verheijen
Delft University of Technology, Department of Chemical Engineering
Julianalaan 136, 2628 BL Delft, The Netherlands.
*TNO-TPD, Department of Control Engineering
PO Box 155, 2600 AD Delft, The Netherlands
**COSUN Food Technology Centre
Oostelijke Havendijk 15, 4704 RA Roosendaal, The Netherlands

Abstract

This paper presents the modeling and optimization of a sugar plant regarding heat management. The optimization has been done using Batch Model Predictive Control, which takes into account the interactions between the batch and continuous units of the plant. On the basis of the results, the development of an advice system to assist operators in the plant has been started.

1. Introduction

There are two reasons why model-based optimization is of high importance in the sugar industry. Using model-based optimization, one is able to predict a regular sugar quality and one can handle the large amount of energy involved. Uncertainty reduction and process restrictions motivate the choice of Model Predictive Control (MPC), a control technique that calculates a sequence of control signals in such a way that it minimizes a cost function over a prediction horizon.

In this work, the interactions between both batch and continuous units play a role, hence an adapted version of MPC is proposed, the so-called hybrid Batch-MPC. This is to distinguish from Lee et al. (1999), who used the term Batch-MPC.

An existing mass-balance model of a beet sugar factory has been extended with energy balances, recycles and variable cycle times for batch units. It is a hybrid model, consisting of batch and continuous vacuum pans and intermediate storage buffers. The product flows and volumes are calculated in the mass components: water, sugar, non-sugar and crystal. Relevant pan-program stages such as filling, seeding, boiling-up and emptying have been included.

For the extended model, control variables and process restrictions have been identified. The model has also been validated with data from last year's campaign. The model was then extended with hybrid Batch-MPC in order to optimize the energy management system.

The objective was to reduce the fluctuations in the energy system which consists of the evaporators and steamsupply for the vacuum pans.

All of the preceding is going to be used in an operator-support system, which helps the operator in making decisions, i.e. the operator-support system is able to predict the consequence of an action taken by an operator (the system can deal with a what-if scenario).

2. Modeling

The sugar house and evaporation section of Suiker Unie in Groningen consists of the following units (see Figure 1). In the A, B and C-pans (also called the vacuum pans) the crystallization takes place, which are being fed by the S-1, S-A, S-B and S-C pans (also called the seed-station), in which the seed crystals used in the vacuum pans are grown. The purified juice stream passes through the evaporators (in which the solids content is typically increased from 15%-18% (thin juice) to 68%-74% (thick juice)), and subsequently through the vacuum pans. The thick juice is mixed with remelted B-and C-sugar to obtain so-called Standard Liquor which is processed in the A-pans (batch process) to end up with a solids content of typically 92-96 %, which after centrifugation leads to white sugar and A-syrup. The syrup is processed in the continuous B-pan to yield B-sugar and B-syrup which in turn is processed in the continuous C-pan and separated into C-sugar and molasses (C-syrup). Molasses, with a solids content of 80%-85%, is used in cattle-feed and in the production of alcohol. The B-and C-sugar are remelted in thin juice and returned to the A-pans.

Between the different pans, there are intermediate storage buffers, in order not to overflow a level in a certain pan. The B- and C-pans are mainly used to dispose of pollution in the juice stream. Most of the crystallization is done in the A-pans, and this happens in a batchwise manner, to have a better control of the crystal size and high supersaturations. The A- and S-pans are operated in a batchwise manner, which consists of the following steps: filling, concentrating, seeding, stabilising, boiling up, emptying steam cleaning and ready for the next strike. It should be added that the processing time of the A- and S-pans is variable, which is generally the case for batch operations (Nott and Lee, 1999).

The function of the evaporation section is twofold:
-It thickens the juice stream.
-It is used as a heat-generator for the sugarhouse and other parts of the sugar factory.

The process is modeled in Matlab/Simulink on basis of mass-balances, as in Bubnik et al. (1995). The product flows and volumes are calculated in mass components: water, sugar, non-sugar and crystal. This leads to a model with approximately 25-30 variables (e.g. levels, flows, program counters and dry substance content (brix)).

Heat management
In the original model, only mass-balances were taken into account. We introduced vapour streams between the different S-pans, A-pans and B- and C-pan, and with this we modeled the heat demand.

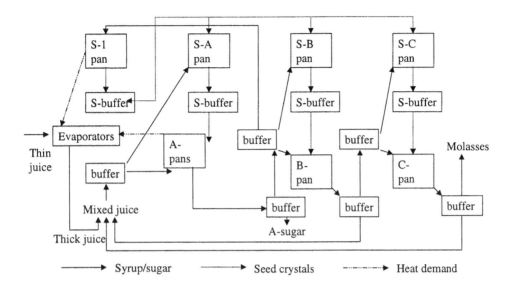

Figure 1. Units and streams of a sugarhouse and evaporation-section. Note that the heat demands of the B-, C- and S-pans are not indicated in the Figure (Bubnik et al. 1995)).

Urbaniec (1989) and Chung (2000) present a detailed description of how to model heat streams in a sugar plant.

3. Validation

For the model-validation we used three measurement-series (approximately with the length of one day, one week, and one month). With these data, we performed model-validation in two ways:

1. Model validation via mass balances.

For the modeling we used the 'one-week' measurement-data. We used checks on mass balances to verify that the 'one-day' measurement data were correct within experimental error, but not so good, when we compared it to the 'one-month' measurement-data. The main reason for this were shutdowns during that period, and it is unsure if there were losses during these shut downs

2. Model validation via measurement data.

For the three available measurements we compared the data with the model output. In order to this correctly one can compare the model results with the measurement data, just by plotting them in one plot. By doing so, one gets a fast overview if there are any discrepancies between the measurements and the model. First, we compared the data-sets, and if necessary, we used a second order filter to obtain an 'average' value of a certain variable. After that, we used the obtained data set for the validation.

A wide array of validation methods is available in literature, and most of them depend on the characteristics of the model and its intended use (Rao et al., 1998). One of the measures, that establishes that the simulation response and the data response match with a certain 'desired' tolerance (Law and Kelton, 1991) is given by

$$\left[\sum_{i=1}^{n} \frac{(y_i - z_i)^2}{\sigma_i^2} \right] \Bigg/ n \tag{1}$$

in which y is the data response, z is the simulation response, n is the number of variables, σ_i, $i=1...n$, is the desired tolerance for the i-th variable. The value of (1) should be close to 1 if the desired tolerance is chosen correctly. For our data we used σ in the range of 5% to 15% depending on the estimation of the actual error.

4. Model analysis

After analysing the model, we came to the conclusion, that the optimization and control problem for the sugar house is not trivial because of the amount of interactions within the process:

-The interaction between the heat demand of the sugarhouse and the evaporation section is reflected in the dry substance content (brix) of the thick juice.

-Dependent on the shapes of the heat-demand-brix curves for the evaporation section and sugarhouse, the development of the thick-juice brix is intrinsically stable or unstable. An illustration of this is given in Figure 2.

Assume we have a constant available heat in the sugarhouse and evaporation-section for varying values of the brix. Suppose at a given time in the process, there is a certain heat demand to the sugar and a certain brix, represented by point A. To this point corresponds a point B, which represents the amount of heat available to the evaporation section. Depending on the equilibrium curve of heat vs. brix in the evaporation-section, the brix of the thick juice will tend towards point C or point D. This tendency repeats until the amount of heat available to the evaporation section matches the amount needed to realise the actual thick juice brix (for example point E using the heat vs. brix-curve 1, leading to a stable situation). In practice, variations in brix can be controlled by adjusting the amount of of heat to the evaporation section, the brix of thin juice to the evaporation section, or the heat demand to the sugarhouse. Hence, one of the optimization variables should be the brix of the thin juice.

- The behaviour of the process in the B- and C-pans is to a great extent dependent on the operational frequency of the A-and S-pans.

- Brix and flow of the thin juice determine the operational frequency of the A-pans and S-pans. This is important for the optimization, because the brix and flow of the thin juice are a handle to schedule the A-pans and S-pans.

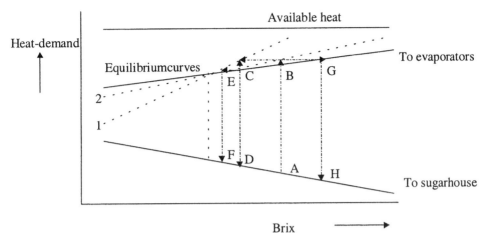

Figure 2. Brix vs. heat-demand

- A change in operation-frequencies of the A-pans leads to variations in brix of the mixed juice through. This also needs to be taken into account when developing an optimization scheme.

-The control of the flows to the B- and C-pans is not easy, because of the supply of these buffers to the S-B and S-C pans (see Figure 1). For an operator this is difficult to control manually, because every control-action taken might be good within 1-2 hours, but can lead to unforeseen actions later (say, after 8-10 hours). This is because of the long delay times in the buffers.

5. Proposed optimization using hybrid Batch MPC

The widespread use of Model Predictive Control (MPC) in chemical industries is because it has several nice features (see Garcia et al. (1989), Morari and Lee (1999)). MPC can be used for handling multivariable and/or constrained problems and its concepts are easy to understand for operators with only a limited knowledge of control (Bordons and Camacho, 2000) .To our knowledge no MPC-application combined with a scheduling-algorithm applied to a mixed batch/continuous system is applied. The case of scheduling a problem in a sugar factory, used to illustrate a hybrid scheduling approach, has been done by Nott and Lee (1999), but no MPC was involved here. Applications of MPC to continuous problems in a sugar plant have been reported by several authors, e.g. Bordons and Camacho (2000), Prada et al. (2000), to name but a few. In our approach, we have reduced the complex problem to a simpler problem (with less A-pans and only the relevant inputs and outputs), with all the characteristics of the more complex problem, such as its hybrid character. For this problem, the proposed objective function (OBJ) to be maximised is

OBJ= profit - batch costs - idle penalty - change flow penalty   (2)

in which 'profit' is the turnover for the production through the system, 'batch costs' are the costs associated with beginning a batch, 'idle penalty' is the penalty for scheduling idle batch units and 'change flow penalty' is the penalty for changes in the continuous flowrate. The constraints under which the objective function OBJ (2) is maximised are

1. The model is limited to stable domains (for the thin juice brix, see Figure 2).
2. Physical constraints, such as bounds on the flows and levels in the vacuum-pans and seed-station.
3. Bounds on the (variable) cycle times of the A- and S-pans.
4. The start-frequency of the batch-section (A-, S-1-, S-A-, S-B- and S-C- pans) is chosen in such a way that the waiting time between pans is constrained.

Using this objective function and constraints, the ratio between vapor input from the evaporators in the sugar mill and the steam demand of the sugarhouse is optimised.

The results of applying hybrid Batch MPC are satisfactory for the simple problem. For the more complex model we expect a significant reduction in fluctuations and heat peaks in the heat supply to the sugarhouse. Hence, we expect an increased on-spec time with tighter specs. But, more importantly, a much more stable process-operation is achieved within a time-horizon of 6-8 hours.

6. Conclusions

In the present paper we modeled the sugar house and evaporation-section of a sugar plant with the purpose of determing the heat streams in the process. The modeling has been done on basis of mass- and heat balances, which were checked on real plant data.
After that a reduced model has been used for hybrid Batch MPC, which is a novel methodology combining scheduling and predictive control. The results are now going to be applied to the more complex model.

References

Bordons, C. and E.F. Camacho, 2000, Applications of model predictive controls in a sugar factory, Proc. ADCHEM 2000, 329.

Bubnik, Z. P. Kadlec, D. Urban and M. Bruhns, 1995, Sugar Technologists Manual.

Chung, C.C. *eds.*, 2000, Handbook of Sugar Refining, John Wiley, New York.

Garcia, C.E., D.M. Prett and M. Morari, 1989, Model predictive control: theory and practice – a survey, Aut, 25(3), 335.

Lee, K.S., I.-S. Chin and H.J. Lee, 1999, Model predictive control technique combined with iterative learning for batch processes, AIChE J., 45(10), 2175.

Nott, P.H. and P.L.Lee, 1999, Sets formulation to schedule mixed batch/continuous process plants with variable cycle time, Comp.Chem.Eng. 23, 875.

Prada, C., C. Alonso, F. Morilla and M. Bollain, 2000, Supervision and advanced control in a beet sugar factory, Proc. ADCHEM 2000, 341.

Law, A.M., and W.D. Kelton, 1991, Simulation Modeling & Analysis, McGraw-Hill.

Rao, L., L. Owen and D. Goldsman, 1998, Development and application of a validation framework for traffic simulation models, Proc.1998 Winter Sim.Conf.1998, 1079.

Urbaniec, K., 1989, Modern energy economy in beet sugar factories, Elsevier, Amsterdam.

European Symposium on Computer Aided Process Engineering – 12
J. Grievink and J. van Schijndel (Editors)
© 2002 Elsevier Science B.V. All rights reserved.

Development of an Internet-based Process Control System

S. H. Yang[*], X. Chen[*], D. W. Edwards[+], and J. L. Alty[*]

[*]Department of Computer Science, Loughborough University, Loughborough, Leicestershire, LE11 3TU UK

[+]Department of Chemical Engineering, Loughborough University, Loughborough, Leicestershire, LE11 3TU UK

Abstract

This contribution presents an approach for the design of Internet-based process control. Uniquely, it addresses multi-user collaboration and Internet transmission latency. Video feedback, text based chatting, and a whiteboard are embedded in the system and shared by multiple communicating users. Virtual supervision parameter control is implemented to overcome dynamic delays caused by the Internet traffic. The experimental results from the test bed show that the Internet-based control system can have a similar behaviour to the local control system if properly designed.

1. Introduction

In the last decade, the most successful network developed has been the Internet and its users are worldwide. Internet technology offers unprecedented interconnection capability and ways of distributing collaborative work, and this has great potential for the high-level control of process plants. As a basis for the possible next generation of control systems, the concept of Internet-based process control has been introduced in recent years which would allow managers and/or operators remote access for monitoring and making adjustments to process plant operation. For example they could adapt to quick changes in the markets. It also allows collaboration between skilled plant managers situated in geographically diverse locations and the possibility of single group support for multiple installations.

To date, most research work on Internet-based process control has resulted in small-scale demonstrations (Atherton, 1998; Cushing, 2000). Some researchers in this area, from higher educational institutions, focus on developing web-based virtual control laboratories for distance learning purposes (Aktan, et al., 1996; Ko, et al., 2001). They allow a remotely-located user to conduct experiments in their control engineering laboratory via the Internet. Unfortunately none of them discusses the limitations caused by Internet environment features such as Internet transmission latency and user isolation. However, Internet time delay and multiple-user collaboration are two essential issues which must be addressed in the design of Internet-based control system for industries and for web-based control experiments.

This paper aims to develop an Internet-based process control system for a water tank in our control laboratory and use this system as a test bed for investigating the effect of

Internet time delays, concurrent user access and the nature of the communication between multiple users. The rest of this paper is organized as follows. In the next section the system architecture is described, including hardware structure and software structure. The Internet time delay and the virtual supervision parameter control approach (VSPC) are discussed in section 3. Section 4 provides a possible way of avoiding the conflict between multiple users. The system implementation and some experimental results are given in section 5. Section 6 provides the conclusions.

2. System Description

2.1 Hardware structure

As shown in Figure 1, the whole system consists of five parts, which are a water tank, a data acquisition (DAQ) instrument, a web server, a web camera, and several web clients. The tank is filled by the inlet flow controlled by a hand valve and is emptied into a drainage tank through a connection pipe and a pump. The outlet flow is controlled by a local control system located at the server to maintain the liquid level of the tank at a desired value. The local control system of the tank is located in the server machine. The server machine and the DAQ instrument are connected and wired by RS-232c serial cables. Through the serial cable, the real-time data is exchanged between the server machine and the instrument. A web camera connected to the server

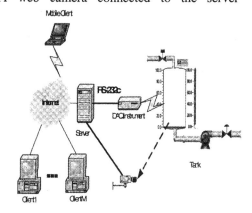

machine provides visual information to the users through a video server. Because the web camera is independent from the DAQ, it can be considered as an extra sensor. The server provides the standard control functions as well as the Internet services, and acts as the video server. The Internet service is implemented mainly based on the LabView G-Server (NI, 2000). In addition to the standard Internet service, the server also needs to establish the connections between the clients and the local controller. Using a web browser, several remotely-located users are allowed to simultaneously monitor and control the tank.

Figure 1. Hardware structure of the Internet-based controls

2.2 Software structure

The system software can be divided into two parts, the client side and the server side. Whilst the client side interacts with users, the server side is not only a web server, but also includes the control and data acquisition program to achieve the control task.

From a functional perspective, there are two programs in the client side as shown in Figure 2, for controlling and monitoring functions respectively, which are the control panel and the monitor panel. The control panel responds to interactions from users. The users can use it to issue commands and/or change the parameters of the controller. Through the TCP protocol, the control panel establishes the connection with the server. In addition to sending information to the server, it is also necessary to receive information from the server. If any client changes the parameters of the controller or issues a command, the server will broadcast the change to every registered user. The control panel deals with this information in order to synchronise the change and indicate the correct status of the controller. The monitor panel provides two functions, the dynamic image and the video & chatting system. The dynamic image consists of graphic information including the process flowchart and the dynamic trends of process variables, which provide the essential information of the current system status. The video & chatting system is designed to provide the visual information for monitoring the facility and the communication channels for multi-users. Multi-users can chat to each other by sending a text message and/or share a white board.

On the server side, the service can be divided into two parts, the command service and the data service. The command service handles incoming requests, and interprets the received information to parameters and commands for the controller and the data service. It also broadcasts the incoming information to every registered client in order to synchronise the client information. In addition, it also handles multiple client conditions such as concurrent user access. The data service is designed mainly to generate an image according to the client requirement, and to send the image embedded in an html page to the clients. The mediator establishes a bridge between the controller and the instrument. The controller deals with standard automatic set-point and manual control. Microsoft NetMeeting is used to support T.120/H323 video standard facilitates for the video & chatting function.

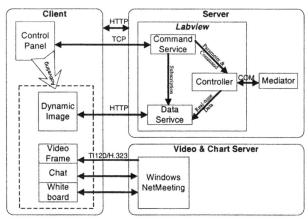

Figure 2. Software Structure of the Internet-based control system

3. Virtual Supervision Parameter Control

3.1 Internet time delay
One of the difficulties in Internet-based process control is the Internet transmission latency. Luo and Chen (2000) have repeatedly tested the transmitting efficiency of the

Internet by sending 64 bytes data every time from their Web server to different remote Web servers. The resulting statistics of the experiments show that the latency of the Internet contains the serious and uncertain time delays. A block diagram of the Internet-based control system is drawn in Figure 3. The total time of performing an operation (a control action) per cycle is $t_1 + t_2 + t_3 + t_4$ where the four types of time delay are:

t_1 time delay in making control decision by a remote operator.

t_2 time delay in transmitting a control command from the remote operator to the local system (the web server).

t_3 execution time of the local system to perform the control action.

t_4 time delay in transmitting the information from the local system to the remote operator.

If each of the four time delays is always a constant, the Internet-based control has a constant time delay. Unfortunately, as shown in Luo and Chen's experiments (2000) this is not the case. The Internet time delay, i.e. t_2 and t_4, increases with distance, but the delay depends also on the number of nodes traversed. Also the delay strongly depends on the Internet load. It is somewhat unreasonable to model the Internet time delay for accurate prediction at every instant. Therefore a control architecture, which is insensitive to the time delay, is needed for the Internet-based control system.

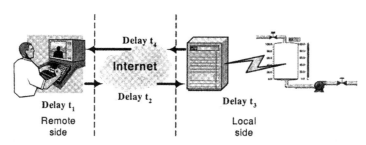

Figure 3. The block diagram of the Internet-based control.

3.2 Virtual supervision parameter control

The Virtual Supervision Parameter Control (VSPC) strategy is one practical approach for Internet-based control, which is insensitive to the time delay. As shown in Figure 4, the detailed control functions are implemented in the local control system. Internet-based control over VSPC is invoked only when the updated parameters like setpoints and Proportional-Integral-Derivative (PID) parameters are required to be sent to the local control system. The new set of VSPC parameters is used as input for the local control system until the next set of parameters is received. VSPC offers a high safety level because the local control system is working as its redundant system. Furthermore it is likely not to be greatly affected by the Internet time delay because the Internet time delays t_2 and t_4 are excluded from the close loop of the control system as shown in Figure 4.

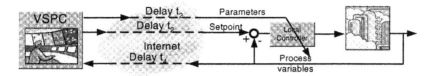

Figure 4. Virtual supervision parameter control.

4. Concurrent User Access

Compared with the traditional DCS system, the special feature of the internet-based control system is multiple users. Although the DCS system allows several operators and/or engineers to operate the system at the same time, they normally sit in the same operation room. Therefore, coordination amongst them is not a real problem. In the Internet-based control system, the operators cannot see each other, or even meet previously. It is likely that multiple users may try to concurrently control a particular process variable. For example, in VSPC, if multi- users try to change the set-point simultaneously, the system may be unable to operate. The set-point of the controller fluctuates from point to point. The coordination among multiple users becomes very important. One of the promising ways is to assign users with different priorities. The user with a high priority can immediately overwrite the commands issued by users with lower priorities. After a new command is accepted, the system will be blocked for a certain period of time and refuse to accept any further command from users with equal or lower priorities.

5. System Implementation And Experimental Results

The system is implemented using Java applets and $L_{AB}V_{IEW}$ virtual instruments (VI). Figure 5 illustrates the remotely-located users interface. The left-hand side column is the control panel, and the right-hand side column is the monitoring panel. The control panel is a java applet where Web users can issue the control command and/or change parameters of the controller. All the information in the control panel will be updated immediately once any other registered user has made any change on them in order to indicate the correct status of the controller. The monitoring panel is switched between the process flowchart, the process trends, and the video & chatting panel. The process flowchart indicates the current status of the process. The dynamic trends show the process responses under a control command. The experimental results show that by using VSPC, the Internet-based process control system can have a similar behaviour to the local control system even with some Internet traffic delay. Figure 5 illustrates that how the video provides the remote users with the visual information of the process. Text chatting and whiteboard pop windows are invoked by pressing a corresponding button below the video, which provide users with a communication channel for co-operation.

6. Conclusions

Internet technologies can provide web clients a platform not only for remotely monitoring the behaviour of the process plants, but also for remotely controlling the

606

plants as well. In this paper an Internet-based control system for a water tank in our process control laboratory has been developed. The issues in the design of the Internet-based control system, concerned with the Internet time delay, multi-user cooperation, and concurrent user access have been addressed. The concept differs from other approaches in that it provides a way for communication and conflict resolution between multiple users, and the VSPC control strategy excludes the Internet time delay from the close loop of the control system and is likely not to be greatly affected by the Internet traffic. The experiment results show that the Internet-based control system may have a similar behaviour to the local control system under the VSPC scheme.

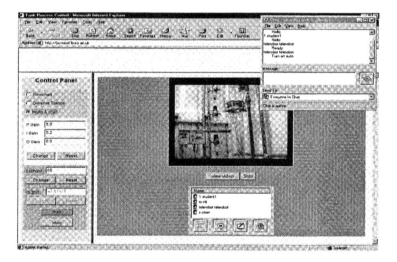

Figure 5. Video & chatting panel

Acknowledgement

The contribution is part of the work of the EPSRC (Grant No. GR/R13371/01) funded project "design of Internet-based process control".

References

Aktan, B., Bohus, C.A., Crowl, L.A., and Shor, M.H., 1996, Distance learning applied to control engineering laboratories. IEEE Trans. on Education, 39(3), pp. 320-326.

Atherton, R, 1998, Java Object Technology can be Next Process Control Wave. Control Engineering, 45(13), pp. 81-85.

Cushing, M., 2000, Process control across the Internet, Chemical Engineering. May, pp.80-82.

Ko, C.C., Chen, B.M., Chen, J., Zhuang, Y. and Tan, K.C., 2001, Development of a web-based laboratory for control experiments on a coupled tank apparatus. IEEE Trans. on education, 44(1), pp. 76-86.

Luo, R. C. and Chen, T.M., 2000, Development of a multibehaviour-based mobile robot for remote supervisory control through the Internet. IEEE Trans. on mechatronics, 5(4), 376-385.

European Symposium on Computer Aided Process Engineering – 12
J. Grievink and J. van Schijndel (Editors)
© 2002 Elsevier Science B.V. All rights reserved.

On-Line Optimal Control of Particle Size Distribution in Emulsion Polymerisation

J. Zeaiter[*], J.A Romagnoli, G.W. Barton and V.G. Gomes
Laboratory for Process Systems Engineering
Department of Chemical Engineering
University of Sydney NSW 2006 Australia

Abstract

This paper concerns itself with the question of the on-line optimal control of the particle size distribution (PSD) from an emulsion polymerisation reactor, using the monomer feed rate as a manipulated variable. A recently developed (and validated) reactor model was employed in the formulation of this control strategy. The model was implemented within the gPROMS software package and used as a "soft sensor" for the on-line estimation of the PSD within a model-based predictive control (MPC) formulation, so as to determine on-line the optimal trajectory for the monomer feed rate. For control purposes, the distributed nature of the PSD was represented by the leading moments, with the control objective being formulated as a function of both the breadth of the distribution and the average particle diameter. Off-line sample analysis, carried out using capillary hydrodynamic fractionation, was also incorporated into the control strategy (as irregular measurements), so as to update the on-line model predictions. Experimental studies on a laboratory scale styrene polymerisation reactor showed that such an approach was able to accurately predict the behaviour of the reactor, as well as improving its performance.

Introduction

Control of the PSD in emulsion polymerisation by means of closed-loop strategies is a challenging problem (Dimitratos, 1994). Difficulties associated with the on-line measurement of the PSD, together with the complex mechanisms involved in emulsion polymerisation systems in general, limit operational options and make the control problem a formidable task. In such cases, conventional control strategies fail to ensure a consistent product quality, with the result that industrial practitioners have to rely on traditional "recipes" and experience. In recent times, however, advances in process understanding, mathematical modelling, soft-sensor technology and model-based control techniques have been such as to offer polymerisation reactor operators the chance of achieving major improvements in process operation and product quality. A viable approach here would seem to involve combining these powerful tools into an effective model-based control strategy - and this is presented in detail in this paper.

---

[*] Corresponding author. Tel.: +61-2-9351-4337; fax: +61-2-93512854
Email: zeaiter@chem.eng.usyd.edu.au

A validated mathematical model (Zeaiter, 2001), based on the kinetic mechanisms of free-radical emulsion polymerisation, was used in this study. This model includes diffusion-controlled kinetics at high monomer conversion, and comprises a set of rigorously developed population balance equations. Using this model, the problem of achieving on-line optimal control of the PSD in the semi-batch emulsion polymerisation of styrene was investigated for the case where the final product has a predefined distribution.

To achieve this goal, an input/output model predictive controller (MPC) was developed to calculate the optimal trajectory on-line, with the full dynamic model being run in parallel and used as a "soft sensor" to provide an on-line estimate of the PSD. The motive behind choosing MPC is the fact that it is the only strategy that can incorporate a number of performance criteria and is capable of utilising any available dynamic process model. A comprehensive review of MPC theory and applications is available in the literature (Garcia, 1989).

Simulation Model

A detailed dynamic model for a perfectly mixed, semi-batch reactor has previously been developed (Zeaiter, 2001) for the styrene emulsion polymerisation system. This model is population balance based and solves a series of integro-partial differential equations coupled with a set of differential and algebraic equations. This equation set describes all known physical and chemical mechanisms that occur within both the particle phase and the aqueous continuous phase, including diffusion-controlled kinetics at high monomer conversion where the transition from the zero-one regime to the pseudo-bulk regime occurs. Within this model, both particle growth and nucleation are assumed to occur in discrete intervals of the particle size. Values for all physical constants and kinetic parameters were obtained from the open literature.

The resulting set of equations describing the polymerisation of styrene (initiated by potassium persulfate and stabilised by sodium lauryl sulfate) are numerically solved using the commercial gPROMS package (*Process Systems Enterprise Ltd*). Model validation against experimental data has been carried out over a wide range of operating conditions (principally combinations of reaction temperature and monomer feedrate) with model predictions able to adequately describe the entire polymerisation process.

Control Strategy

In this study, the objective of the control algorithm was to ensure the production of a polymer with a pre-specified PSD in the minimum reaction time. A particle size polydispersity index (*PSPI*) was chosen as the optimisation objective function, with particle concentration densities being set as constraints. The objective function to be maximised was, thus, defined as:

$$\underset{F_m}{Max}\left[PSPI(r, t_{final})\right] \qquad (1)$$

where t_{final} is the total processing time. For operational reasons, the manipulated variable (*ie* the monomer feedrate) was specified with the following lower and upper bounds:

$$5\times10^{-5} \, moles / \sec \le F_m \le 5\times10^{-4} \, moles / \sec \qquad (2)$$

The final PSD "shape" was included in this optimisation formulation in the form of end-point inequality constraints, given in terms of the final molar concentration density of particles:

$$n_{min} < n(r, t_{final}) < n_{max} \qquad (3)$$

In this equation, the limits n_{min} and n_{max} are specified so as to match the required PSD. Since we are dealing with a semi-batch (or fed-batch) process here, the maximisation of the *PSPI* is subject to additional constraints accounting for the total amount of monomer available in the "recipe", $N_{m,T}$, and the permissible total run time. These constraints are defined as follows:

$$N_{m,T} = 1.6 \, moles \quad and \quad t_{min} \le t \le t_{max} \qquad (4)$$

In this problem, the population balance equations were discretised with respect to the particle radius, r. The required *PSPI* profile was calculated off-line through an interface to the gOPT dynamic optimisation code (also from *Process Systems Enterprise Ltd*).

On-Line Optimal Control

In order to predict on-line the optimal trajectory for the monomer addition, a nonlinear model-based predictive controller (MPC) was used. The MPC algorithm uses an input-output model (Clarke, 1994; Camacho, 1999; Kanjilal, 1995) whose step response coefficients are determined as follow:

- Using the dynamic model, solve for the output using a constant monomer feedrate. The calculated output is designated by y^{ss}.

- Introduce a consecutive set of step changes into the monomer feedrate, and solve the dynamic model 'into the future'. The resultant output is designated by y^{step}.

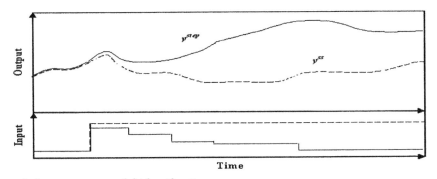

Figure 1: Input-output model identification

610

The step response coefficients are then calculated as follow:

$$a_i = \frac{y^{step}(k+i) - y^{ss}(k+i)}{\Delta u(k+i)} \qquad (5)$$

where u represents the model input (*ie* the monomer feedrate in this case). This input-output model identification technique is illustrated in Figure 1. Model predictions can then be calculated at every sampling time from the calculated step-response coefficients using the following:

$$y(k+1) = \sum_{i=1}^{N} a_i \Delta u(k+1-i) + y^{ss}(k+1) + d(k+1) \qquad (6)$$

Modelling error and the impact of unmeasured disturbances are included in the last term on the right hand side of Equation 6. Once the process output has been measured at a given point in time, $d(k)$ can be estimated by assuming all future d values are equal, as illustrated in Equation 7:

$$d(k) = y^{meas}(k) - \sum_{i=1}^{N} a_i \Delta u(k-i) - y^{ss}(k) = d(k+1) = \cdots\cdots = d(k+N) \qquad (7)$$

In this formulation, the controller has to calculate the set of control moves (Δu) into the future that allows the system to follow a pre-defined set-point trajectory. However, only the first control move is implemented on the process, with the entire optimisation being repeated at the next sampling time.

Experimental Validation

A major problem in attempting to implement any form of advanced control strategy is the lack of appropriate on-line sensors for the measurement of the PSD. In practice, such measurements require sampling, dilution and off-line analysis and data processing, typically using capillary hydrodynamic fractionation (CHDF).

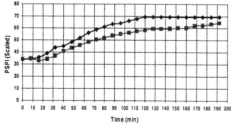

Figure 2: Control of the PSD via the PSPI.
◆ *Set-point,* ■ *Measurement*

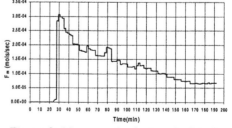

Figure 3: Monomer feedrate calculated online by MPC.

Typical analysis time by CHDF is of the order 10-15 minutes, making this method impractical for on-line monitoring and control applications. To overcome this problem, a "soft sensor" approach was employed, with the full mechanistic model (Zeaiter, 2001) being used to provide on-line estimates of the polymer PSD. This approach requires the use of real-time gPROMS execution. All relevant operating conditions (such as monomer flowrate and reactor temperature) are taken from the reactor at discrete time intervals as on-line measurements, and used by the dynamic model to estimate the PSD. Note that a commercial SCADA system interfaces the polymerisation reactor to the computer running the MPC package, the latter being written in MS-Excel and Microsoft Visual Basic. PSD measurements are obtained from this model-based "soft sensor" every 110 seconds and fed to the MPC which calculates the appropriate control action (*ie* the monomer feedrate for the next time interval). Off-line samples are also taken from the process every 20 minutes, and the CHDF measurements compared to the (model) predicted results. Any modelling error was handled within the MPC algorithm as an unmeasured disturbance, as previously described. In this MPC formulation, the *PSPI* was calculated using the step response method, also described previously.

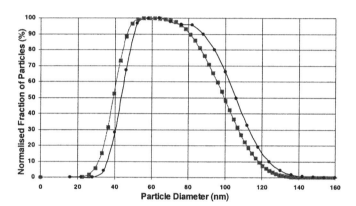

Figure 4: Optimal particle size distribution of the final product.
  ● Estimated (gPROMS model), ■ Measured (CHDF)

In this MPC application, the prediction horizon, the control moves, and the sampling time were set at 30, 5 and 470 seconds, respectively. The (relatively large) magnitude of the sampling time is related to the speed of the process dynamics. As the average volume growth rate for particles in emulsion polymerisation is very small (of the order 2×10^{-22} L/s), then a relatively long sampling time is needed for discernible changes to be observed.

Setpoint tracking of the *PSPI* by manipulating the monomer feedrate was investigated by first solving off-line the optimal control problem for the case where a specified (and quite broad) PSD was required for the final product. An optimal trajectory was thus generated for the monomer feedrate so as to give the required *PSPI* profile. This profile was then used on-line as the setpoint for the MPC. The ability of the MPC to hold the controlled output as close as possible to the setpoint is illustrated in Figure 2. Although the *PSPI* did increase throughout the run (and, consequently, the PSD continually broadened), an offset of variable magnitude was apparent at every sampling time. The reason for this behaviour lies in the tuning of the MPC, as the controller was set up to avoid large changes in the manipulated variable (so as to prevent the monomer flowrate hitting a bound). The weightings employed favoured this course of action at the expense of closely tracking the setpoint. As shown in Figure 3, the monomer feedrate profile was found to decrease with time after the batch pre-period (*ie* where no monomer was added) of 25 minutes, with no constraint violation observed for the manipulated variable during the entire run.

The final shape of the PSD was in good agreement with the estimated "soft sensor" result (see Figure 4), with a broad distribution being obtained, as originally defined as the objective of the control strategy.

Conclusions

This paper reports on the successful development and experimental implementation of an advanced control scheme for a styrene emulsion polymerisation reactor. This scheme was developed with a particular focus on achieving control over the full PSD of the polymer product (rather than the more simple option of controlling the mean particle size). Using a recently developed (and experimentally validated) detailed dynamic model of a semi-batch polymerisation reactor as an on-line "soft sensor" for the PSD, and as the means for calculating dynamic step response coefficients, it was possible to develop a model-based predictive control scheme whose objective was to provide a specified PSD for the final product. The success of the work reported indicates that this methodology could be readily extended to simultaneously provide tight control over a number of product attributes (*eg* in terms of both the particle and molecular weight distributions).

References

Camacho E.F. and Bordons C. (1999) , Model Predictive Control, Springer, NY.

Garcia C. E., Prett D. M., and Morari M. (1989), Automatica, 'Model Predictive Control: Theory and Practice-a Survey', 25, 335-348.

Clarke D. (1994), Advances in Model-Based Predictive Control, Oxford ; New York : Oxford University Press.

Dimitratos J., Elicabe G. and Georgakis C. (1994), AIChE Journal, 'Control of Emulsion Polymerization Reactors', 40, n12, 1993-2021.

Kanjilal P.P. (1995), Adaptive Prediction and Predictive Control, Stevenage, U.K.

Process Systems Enterprise Ltd. (PSE), gPROMS Advanced User Guide, Ver.2.0, United Kingdom.

Schork F.J., Deshpande P.B., Leffew K.W. and Nadkarni V.M. (1993), Control of Polymerization Reactors, M. Dekker , NY.

Zeaiter J., Romagnoli J.A., Gomes V.G., Barton G.W. and Gilbert R.G., 'Operation of Semibatch Emulsion Polymerisation Reactors: Modeling, Validation and Effect of Operating Conditions', submitted to Chemical Engineering Science Journal.

European Symposium on Computer Aided Process Engineering – 12
J. Grievink and J. van Schijndel (Editors)

Cyclic Production and Cleaning Scheduling of Multiproduct Continuous Plants

Alessandro Alle[1], Jose M. Pinto[1*] and Lazaros G. Papageorgiou[2]

(1) Department of Chemical Engineering, University of Sao Paulo, Av. Prof. Luciano Gualberto t. 3 n. 380, Sao Paulo, SP, 05508-900 Brazil.

(2) Centre for Process Systems Engineering, Department of Chemical Engineering, UCL (University College London), Torrington Place, London WC1E 7JE, U.K.

Abstract

The objective of this paper is to extend previous scheduling models (Pinto and Grossmann, 1994; Alle and Pinto, 2001) based on continuous time representation to include cleaning considerations. The proposed mixed integer non-linear programming (MINLP) model aims at simultaneously scheduling production and cleaning tasks of multiproduct multistage plants with performance decay. The resulting mathematical model has a linear objective function to be maximized over a convex solution space thus allowing global optimum to be obtained with an Outer Approximation algorithm. A case study demonstrates the applicability of the model and its potential benefits in comparison with a hierarchical approach.

1. Introduction

This paper addresses the problem of cyclic scheduling of multiproduct multistage continuous plants (one unit per stage) with performance decay. Performance decay is a serious problem in many chemical processes and has several causes such as catalyst deactivation, fouling *etc*. As performance decreases, processing units require cleaning to restore productivity to initial state. The decision of stopping a unit for maintenance interferes with the production schedule. Therefore, it is important to consider both production and cleaning scheduling decisions simultaneously.

Several analytical methods for determining the optimal cleaning schedules for equipment items have been proposed in the literature. The main drawback of these approaches is that they are restricted to a single equipment item. Only recently, Jain and Grossmann (1998) studied the scheduling of single stage with parallel units whose performance decreases with time and therefore shutdown for maintenance after regular intervals is required. Georgiadis *et al.* (2000) and Georgiadis and Papageorgiou (2000) have presented cleaning scheduling approaches applied to heat exchanger networks under fouling conditions based on discrete-time mathematical formulations.

This work aims at extending previous scheduling models (Pinto and Grossmann, 1994; Alle and Pinto, 2001) for multiproduct, multistage continuous plants based on continuous time representation to include cleaning considerations.

---

* To whom correspondence should be addressed. Fax: +55 11 3813 2380, E-mail: jompinto@usp.br

2. Problem Description

The plant is composed of M stages with one unit per stage, m, with no inventory limitations (see Fig. 1). The plant processes NP products in the same order in all stages (permutation flowshop plant). For every product, i, a fixed demand rate, d_i, should be satisfied.

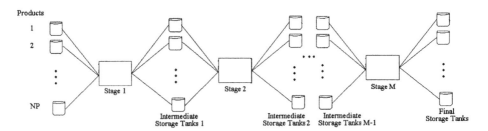

Fig. 1. Multiproduct multistage cyclic continuous plant.

Process yields at some stages may decrease with time. Therefore, units must shutdown for periodic cleaning. These cleaning tasks require both specific time duration and cost depending on the units and products processed. A typical production and cleaning schedule comprises product campaigns with subcycles, as shown in fig. 2.

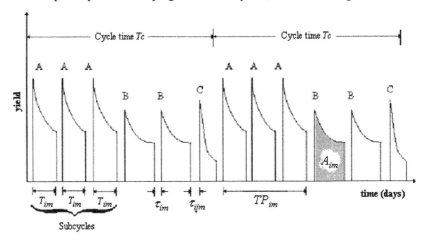

Fig 2. Cycle and subcycles in a unit with decay performance.

The yield decay for product i at stage m is assumed to be an exponential function of processing time, t, as in Jain and Grossmann (1998).

$$\eta_{im}(t) = c_{im} + a_{im} \exp\left(-b_{im}t\right) \tag{1}$$

The overall objective of the planning model is to maximize profit over a given cycle time. The profit is given by the difference between revenues (sales of the final products) and costs (raw material, cleaning and transition) divided by cycle time duration. Transition times and costs are incurred when the line is changed from one product to another. As transition times and costs are sequence dependent, there are trade-offs concerning the product sequence. A sequence that minimizes total transition costs may not be the one that maximizes the availability of the plant because transition times and costs are not necessarily proportional. There are two extremes in terms of cleaning policies. The units may be cleaned frequently and therefore units run at large yields. Consequently, the productivity per unit of processing time is larger and raw material consumption is lower. On the other hand, the cleaning cost increases and the availability of the units decreases due to the time spent for cleaning. At the other extreme, when units are not cleaned frequently, the cleaning cost gets lower and the availability higher. In addition, the raw material cost increases while performance falls due to increased production losses. Thus there are significant trade-offs among cleaning and raw material costs as well as availability and yields. From the above discussion, it is evident that production and cleaning scheduling decisions should be considered in the same framework.

Overall, the production and cleaning scheduling problem can be stated as follows:

Given are cycle time (Tc), performance decay functions, final product prices (P_i), raw material (Cf_i), cleaning (Cl_{im}) and transition costs (Ctr_{ij}), cleaning (τ_{im}) and transition times (τ_{ijm}), feeding rates (G_{im}), maximum number of subcycles (R_{im}^{max});

determine product sequence (Z_{ij}), start times (TS_{im}), processing times (TP_{im}), number of subcycles (R_{im}), amounts of raw material (F_i) consumed, final products (W_{im})

so as to maximize overall plant profit over the cycle time.

3. Mathematical Model

The mathematical formulation is based on a previous continuous-time scheduling model for continuous multiproduct plants (Alle and Pinto, 2001) that is extended here to capture cleaning aspects. First, the folowing binary variables are introduced:

Z_{ij} : *1* if product *i* precedes product *j*; *0* otherwise.

X_{imr} : *1* if *r* subcycles of product *i* take place at stage *m*; *0* otherwise.

As the plant is a flowshop, every product *j* must be preceded by the same product *i* at all stages. Only one product succeeds and precedes the other:

$$\sum_i Z_{ij} = 1 \quad \forall j \quad \text{and} \quad \sum_j Z_{ij} = 1 \quad \forall i \tag{2}$$

The number of subcycles required for product *i* at stage *m* is defined by:

$$\sum_{r=1}^{R_{im}^{max}} X_{imr} = 1 \qquad \forall i, m \tag{3}$$

The total amount of product *i* produced at stage *m* during one subcycle is as follows:

$$W_{im} = G_{im} RA_{im} \quad \forall i, m \tag{4}$$

where RA_{im} replaces the product between R_{im} (number of subcycles) and A_{im}. The latter is determined by integrating the yield decay function over the duration of one subcycle:

$$A_{im} = c_{im}T_{im} + a_{im}/b_{im}\left[1 - \exp\left(-b_{im}T_{im}\right)\right] \qquad \forall i, m \tag{5}$$

or alternatively the following convex-region constraint is used:

$$A_{im} \leq c_{im}T_{im} + a_{im}/b_{im}\left[1 - \exp\left(-b_{im}T_{im}\right)\right] \qquad \forall i, m \tag{6}$$

The amount produced at stage *m* must be completely consumed at stage *m+1* in order to avoid accumulation of material within cycles:

$$W_{im} = G_{i,m+1}RT_{i,m+1} \qquad \forall i, m = 1...M - 1 \tag{7}$$

where RT_{im}, which represents summed duration of subcycles, replaces the product of R_{im} and T_{im}. The following constraints are required in order to define the RA_{im} and RT_{im} variables:

$$\overline{AX}_{imr} \leq A_{im}^{up} X_{imr} \qquad \forall i, m, r \tag{8}$$

$$A_{im} = \sum_r \overline{AX}_{imr} \qquad \forall i, m \tag{9}$$

$$RA_{im} = \sum_r r \overline{AX}_{imr} \qquad \forall i, m \tag{10}$$

$$\overline{TX}_{imr} \leq T_{im}^{up} X_{imr} \qquad \forall i, m, r \tag{11}$$

$$T_{im} = \sum_r \overline{TX}_{imr} \qquad \forall i, m \tag{12}$$

$$RT_{im} = \sum_r r \overline{TX}_{imr} \qquad \forall i, m \tag{13}$$

The total demand on final products must be satisfied at the cycle end:

$$W_{iM} \geq d_i Tc \qquad \forall i \tag{14}$$

The total time that product i is processed at stage m, TP_{im}, is given by the summed duration of subcycles, RT_{im}, plus the time spent for cleaning:

$$TP_{im} = RT_{im} + \tau_{im} \sum_r (r-1) X_{imr} \qquad \forall i, m \tag{15}$$

At any stage, the sum of the total occupation time plus the transition times for all products must not exceed the cycle time, Tc.

$$Tc \geq \sum_i \left(TP_{im} + \sum_j \tau_{ijm} Z_{ij} \right) \qquad \forall m \tag{16}$$

As the schedule is cyclic, product 1 is arbitrarily chosen as the first to enter the production line to decrease solution degeneracy:

$$TS_{11} = \sum_i \tau_{i11} Z_{i1} \tag{17}$$

Constraint (18) states that product j starts immediately after the processing of the preceding product i plus the correspondent transition time.

$$-Tc\left(1 - Z_{ij}\right) \leq TS_{jm} - \left(TS_{im} + TP_{im} + \tau_{ijm} Z_{ij}\right) \leq Tc\left(1 - Z_{ij}\right) \qquad \forall i, j \geq 2, i \neq j, m \tag{18}$$

As the plant is a flowshop, the processing of product i at stage m must start (end) before the start (end) of processing of the same product at stage $m+1$:

$$TS_{im} \leq TS_{i,m+1} \text{ and } TS_{im} + TP_{im} \leq TS_{i,m+1} + TP_{i,m+1} \qquad \forall i, m = 1...M - 1 \tag{19}$$

There is a limit on the time that stage m can process product i before a cleaning task.

$$T_{im} \leq T_{im}^{up} \qquad \forall i, m \tag{20}$$

The objective function, Pro, to be maximized here is the difference between sales of the final products and costs (raw material, cleaning and transition) over a given cycle time:

$$Max \ \ Pro = \frac{1}{Tc} \sum_i \left(P_i W_i - Cf_i F_i - \sum_m Cl_{im} \sum_r (r-1) X_{imr} - \sum_j Ctr_{ij} Z_{ij} - Pm_i \sum_{m=2}^M TS_{im} \right) \tag{21}$$

Note that the last term in (21) represents a small penalty in order to prevent solution degeneracy as start times at the stages other than the first are unbounded.

The above mathematical optimization model corresponds to an MINLP model with linear objective function over a convex solution space. The resulting MINLP model can be solved by the OA/ER/AP method (Viswanathan and Grossmann, 1990) up to the global optimality using GAMS 2.5 (Brooke et al., 1998) with DICOPT-2 solver.

4. Case study

Given a cycle time of 360 days, 5 different products are processed in 3 consecutive stages (see fig. 3) with decaying first and second stages and third stages with constant performance. At most 4 subcycles are allowed for every product and the maximum time between stops in the units with decay is 60 days. Table 1 presents data involved.

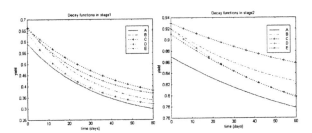

Fig 3. Performance decay in stages 1 and 2.

Table 1. Plant data for Case Study.

Pr.	P_i ($/ton)	G_{i1} (ton/d)	G_{i2} (ton/d)	G_{i3} (ton/d)	τ_{i1} (day)	τ_{i2} (day)		m	C_m ($/day)
A	56.2	100	53	51	1	1.5		1	6000
B	53.6	95	38	35	1.5	1		2	5500
C	52.6	90	33	33	1	1.5		3	4000
D	50.8	105	41	42	1.5	1.5			
E	45.0	102	55	49	1	1.5			

$$Cf_i = 0.15P_i; \ Cl_{im} = C_m\tau_{im}$$

$$Ctr_{ij} = 0.8\sum_m C_m\tau_{ijm}$$

	Transition times, τ_{ijm} (days)														
	Stage 1					Stage 2					Stage 3				
Pr.	A	B	C	D	E	A	B	C	D	E	A	B	C	D	E
A	0	1.5	3	2.5	2	0	3	1.5	2.5	2	0	1.5	1.5	3	3
B	2.5	0	1.5	2	3	2.5	0	3	2	1.5	2.5	0	1.5	2	1.5
C	2	3	0	1.5	3	2	1.5	0	3	3	2	1.5	0	1.5	3
D	3	3	3	0	1.5	1.5	3	3	0	3	1.5	3	2	0	1.5
E	1.5	2.5	3	3	0	3	2	2	1.5	0	1.5	2	3	1.5	0

To demonstrate the effectiveness of the proposed MINLP model, a hierarchical alternative approach is studied which comprises three steps. At the first step, the product sequence is obtained by minimizing the total transition cost. The second step determines subcycle duration for the first processing stage of each product by keeping the corresponding yield above 80% of the initial value. This step introduces new upper bounds for the maximum subcycle duration, that varies from 18 to 25 days, depending

618

on the product. Finally, at the third step, a reduced version of the original MINLP model is solved with fixed product sequence and tighter bounds on subcycle lengths.

Different product sequences and timings are obtained from this hierarchical approach and proposed MINLP model[1] as shown in fig. 5.

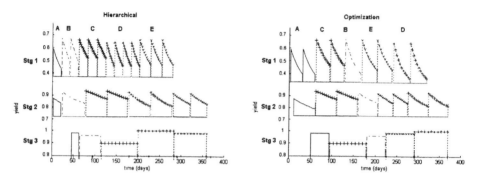

Fig. 5. Comparison between hierarchical and optimization solutions

Moreover, the optimal sequence, ACBED, has an objective function value of 570 $/day which represents a 11.5% improvement when compared with the hierarchical solution (511 $/day).

5. Conclusions

In this work, the problem of production and cleaning scheduling for multistage multiproduct continuous plants with performance decay has been studied. The overall problem has been formulated as a convex MINLP model. Finally, the advantages of applying optimization-based approach have also been emphasized when compared the proposed model with a hierarchical approach.

Acknowledgment

The authors would like to acknowledge support received from FAPESP (Fundacao de Amparo a Pesquisa do Estado de Sao Paulo) under grants 99/02657-8 and 98/14384-3.

6. References

Alle, A. and J.M. Pinto, 2001, Proc. DYCOPS 6, Korea, 213-218.
Brooke, A., D. Kendrick and A. Meeraus, 1998, *GAMS – A Users' Guide*. The Scientific Press, Redwood City.
Georgiadis, M.C. and L.G. Papageorgiou, 2000, *Chem. Eng. Res. Des.*, **78**, 168-179.
Georgiadis, M.C., L.G. Papageorgiou and S. Macchietto, 2000, *Ind. Eng. Chem. Res.*, **39**, 441-454.
Jain, V. and I.E. Grossmann, 1998, *AIChE J.*, **44**, 1623-1636.
Pinto, J.M. and. I.E. Grossmann, 1994, *Comp. Chem. Engng.*, **18**, 797-816.
Viswanathan, J. and I.E. Grossmann, 1990, *Comp. Chem. Engng.*, **14**, 769-782.

---

[1] Optimal solution required 356 CPUs in a Compaq ML-570 Workstation.

European Symposium on Computer Aided Process Engineering – 12
J. Grievink and J. van Schijndel (Editors)
© 2002 Elsevier Science B.V. All rights reserved.

Improving the Efficiency of Batch Distillation by a New Operation Mode

H. Arellano-Garcia, W. Martini, M. Wendt, P. Li, G. Wozny
Institute of Process and Plant Technology, Technische Universität Berlin
10623 Berlin, Germany

Abstract

Batch distillation processes are well-known for their high degree of flexibility. A feature of batch distillation is that it produces not only the desired products but also off-cuts. Conventionally, off-cuts are recycled to the reboiler of the column for the next batch. In this work, we propose a new operation mode for batch distillation, namely, the off-cuts will be recycled in form of a continuous feed flow into the column. The separation effect is promoted in this way and thus economical benefits can be achieved. Simulation and optimization based on a rigorous model are carried out to study the properties of this operation mode and develop optimal operating policies. Results of applying this mode to two industrial batch columns show significant improvements of operation efficiency in comparison to the conventional recycle strategy.

1. Introduction

Batch distillation is used in chemical industry for producing small amounts of products with high added value and for processes where flexibility is needed. For processes with a rather limited conversion or separation performance, off-cut recycles can be introduced to recover unused raw materials. Many authors studied circumstances under which slop recycling is worthwhile (Bonny, 1999; Wajge & Reklaitis, 1998). The main problem is to decide when and how much recycle can be profitable. In the conventional operation mode off-cuts are recharged to the column reboiler for the next batch. Several previous studies on treating off-cuts in this mode have been made (Bonny et al., 1994; Sorenson & Skogestad, 1994; Macchietto & Mujtaba, 1996; Wajge & Reklaitis, 1998), in which recycle strategies were developed through simulation and optimization.

However, mixing the off-cut with fresh feed for the next charge reduces the separation effect and thus leads to a low operation efficiency. A new operation mode for batch distillation is proposed to facilitate the separation effect by recycling the off-cut in form of a continuous feed flow into the column. Since the composition of the off-cuts lies between the initial composition in the reboiler and that of the distillate, feeding the off-cuts to the column shortens the way of separation. Thus the efficiency of batch distillation can be improved, e.g., less batch time and energy consumption and more product amount. The performance of this new operation mode is carried out through simulation and model-based optimization of two industrial batch distillation processes. As a result, using this operation mode significant improvement of the operation efficiency can be achieved in comparison to recycling them to the reboiler.

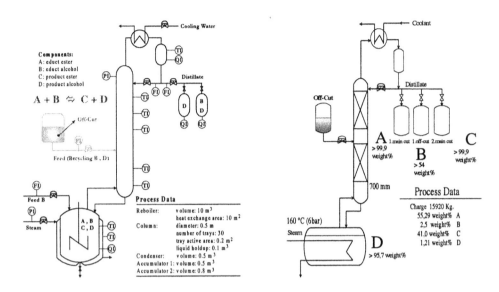

Fig. 1: Two industrial batch distillation processes with the proposed operation mode.

2. Modeling of two industrial processes

As shown in Fig. 1, we consider two industrial batch distillation processes to study the properties of the new operation mode. The first process is a semi-batch distillation with a chemical reaction (transesterification) taking place in the reboiler (Fig. 1, left). During the batch, a limited amount of educt alcohol will be fed to the reboiler to accelerate the reaction rate. With the conventional operation mode the total batch time was about 26h. We use a detailed dynamic model which is validated with the measured data from the experiment. Included are component balance, energy balance and vapor-liquid equilibrium on each tray in the model. The holdup, the pressure drop of each tray, and non-equimolar flow in the column are taken into consideration. The reaction kinetics are used to depict the chemical reaction in the reboiler. The second process is a batch distillation with a packed column for separating a 4-component mixture (Fig. 1 right). In this system there is an azeotropic point and the column pressure is used as a control variable as well. To describe the packed column we use a detailed dynamic tray-by-tray model. The number of the theoretical trays is calculated corresponding to the height of the packing. The holdup of each theoretical tray is computed with the correlation model proposed by Engel et al. (1997). The vapor load from the reboiler to the column is restricted by the F-factor (vapor load term) of the column as well as the heating capacity of the plant.

These two processes were studied for optimization with the purpose of minimization of the batch time and some improvement compared to the empirical operating strategy was achieved (Li et al., 1998; Wendt et al., 2000). However, they did not consider recycling of the off-cuts. In this work, we investigate the possibility to further improve the performance of these two processes by the new operating mode. The model of both processes is modified such that the off-cut of the previous batch will be fed to a tray of the column during the next batch.

3. Problem formulation and solution approach

In industry practice, it is often desired to minimize the batch time of such processes. Thus we consider the time-optimal problem to find optimal policies for the new operation mode for the two industrial processes, which can be described as follows:

$$\min \; t_f\left(F_R(t), F_C(t), R_v(t), t_u, t_f\right)$$

s.t. the model equation system and

$$x_{D,1}(t) \geq x_{D,1}^{SP} \qquad t_0 \leq t \leq t_u$$
$$x_{C,NST}(t_f) \geq x_{C,NSP}^{SP}$$
$$\int_{t0}^{t_f} F_R(t)dt \leq M_1$$
$$\int_{t0}^{t_f} F_C(t)dt \leq M_2$$
$$F_R^L \leq F_R(t) \leq F_R^U$$
$$F_C^L \leq F_C(t) \leq F_C^U$$
$$R_v^L \leq R_v(t) \leq R_v^U$$

$$\min \; t_f\left(F_R(t), P(t), R_v(t), t_{u1}, t_{u2}, t_f\right)$$

s.t. the model equation system and

$$x_A \geq x_A^{SP}$$
$$x_B \geq x_B^{SP}$$
$$x_C \geq x_C^{SP}$$
$$x_{D,NST}(t_f) \geq x_{D,NSP}^{SP}$$
$$\int_{t0}^{t_f} F_R(t)dt \leq M_1$$
$$F_R^L \leq F_R(t) \leq F_R^U$$
$$F_C^L \leq P(t) \leq F_C^U$$
$$R_v^L \leq R_v(t) \leq R_v^U$$

The recycle stream F_R, pure educt alcohol stream F_C and reflux ratio R_v are the control variables to be optimized for the first process. The switching time t_u from the main-cut to the off-cut should also be decided. The output constraints are the distillate purity during the main-cut period and the product ester purity in the reboiler at the end of the batch. The input constraints consist of the limited amount of both off-cut M_1 and fresh educt alcohol M_2 as well as their bound limitations.

In the second problem, beside the reflux ratio and the recycle stream the column pressure is also used as a control variable due to the available compressor. The process is operated in a packed column to separate a 4-component-mixture, with A, B, C, D representing from the lightest to the heaviest component (see Fig. 1). Three main cuts (fractions A and C from the top of the column and the fraction D remaining in the reboiler) will be obtained during the batch. An off-cut mainly containing B will be also received from the distillate. The heaviest component D has no vapor phase and remains in the reboiler during the whole operation. The output constraints are the specifications of the average composition in the three product vessels $x_A^{SP}, x_B^{SP}, x_C^{SP}$ and the composition in the reboiler $x_{D,NST}^{SP}$, respectively.

Both processes possess strong nonlinear behaviors and the model leads to a large-scale DAE system. To efficiently simulate and optimize the processes, we use collocation on finite elements to discretize the dynamic model equations. Through this discretization the dynamic optimization problem is transformed to a nonlinear programming (NLP) problem. It should be noted that the choice of the feed tray for the recycle stream should also be optimized. But this will lead to a dynamic MINLP problem. To prevent the problem becoming too complicated, simulation has been made to decide the feed tray.

622

To solve the optimization problem, we applied the sequential optimization method by Li et al. (1998), i.e. the entire algorithm is divided into one layer for optimization and one layer for simulation. The model equations are integrated in the simulation layer, so that the state variables and their sensitivities can be computed by given controls. The control variables are computed in the optimization layer by SQP as the only decision variables. A detailed derivation of the optimization approach can be found in Li et al. (1998).

4. Computation results

The two problems are solved by the sequential approach and the results compared with those of the conventional recycle mode. To have a base for comparison, the same amount of off-cut is either continuously fed to the column or charged to the reboiler at the beginning of the batch. Fig. 2, 3 and 4 show the optimal policies by the new operation mode for the first process. The recycle stream should be fed to the column in the main-cut period (Fig. 2) since it contains the two alcohols and thus it is favorable to separate it in this period. The educt alcohol feed flow has a similar profile (Fig. 3).

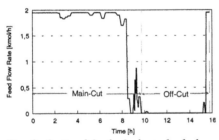

Fig. 2: Optimal feed of recycle stream.

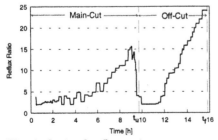

Fig. 3: Optimal feed of educt alcohol.

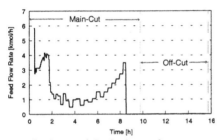

Fig. 4: Optimal reflux ratio.

This is because the temperature of the reboiler will be considerably raised in the off-cut period. The reflux ratio (Fig. 4) should be small at the beginning and increased gradually in the main-cut period in order to guarantee the purity constraint of the product alcohol, while it will be decreased for a period of time after the switching to the off-cut so as to quickly pull out the product alcohol remained in the column and condenser. Then the reflux can be increased so that the evaporated educt alcohol will be brought to the reboiler. Since the transesterification is a reversible reaction, the effect of the recycle stream to the column can be seen from the composition of the two alcohols in the reboiler. Fig. 5 shows their optimal profiles of the amount in the reboiler by both the new mode and the conventional mode. During the main-cut period, the amount of product alcohol is significantly smaller by the new mode than that by the conventional mode. This means that the reaction velocity in the direction of products will be increased. The total batch time by the new mode is 15.8h and it is 17.8h by the conventional mode.

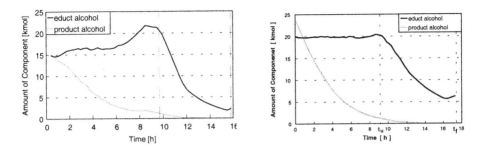

Fig. 5: Alcohol composition by the new mode (left) and conventional mode (right)

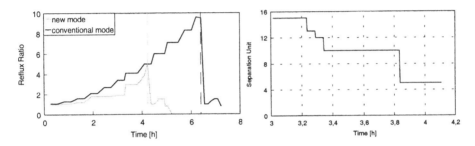

Fig. 6: Optimal reflux ratio for the conventional and the new operation mode (left)
and the optimal feed position of the recycle stream (right).

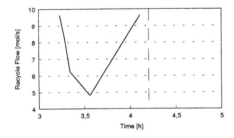

Fig. 7: Optimal recycle flow rate

For the second process, due to the amount of component A in the off-cut, the whole content of the off-cut should be pumped into the column by the end of the first main cut period. Thus only in this period, the differences between the conventional and the new operation mode can be seen. Fig. 6 (left) show the optimal trajectories of the reflux in this period for both the conventional and the new operation mode.

In Fig. 6 (right) the optimal feed position of the recycle stream corresponding to the theoretical tray number is illustrated. This is approximately corresponding to the position in the column where the composition is equal to the feed composition. Fig. 7 shows the optimal recycle flow rate during the first period. It has to be noted that the difference concerning the optimal policies of the column pressure are only marginal, since a higher pressure is favorable in case of a constant F-Factor during the first main cut period. But the pressure is restricted by an upper bound due to the temperature of the reboiler heating steam. Due to the fact that in the new operation mode the off-cut of the previous batch is kept separated from the liquid mixture in the reboiler from the beginning, it has only a little amount of component B compared to the composition in the conventional operation mode. This leads to the fact that at the beginning the VLE relation between component A and C is more dominating, which causes an increasing volatility and thus a much lower reflux rate is required for fulfilling the purity constraint of the first fraction. However, due to the decreasing amount of component A in the column during the batch, the supply of the recycle flow has to begin at a certain time as a compensation of this decrease. On the other hand, the content of the feed tank has to be depleted early enough before the first switch to the next fraction, in order to provide enough time for separating the remaining amount of component A, which originally comes from the recycle flow.

The computed trajectory of the recycle flow (Fig. 7) indicates two physical phenomenon. Actually with the proceeding time, with a decreasing amount of component A, a stronger compensation and thus a higher feed flow rate is desired. On the other hand high liquid flows in later time intervals also causes a longer period until the first switch to the next fraction can be done due to the higher amount of liquid in the column, which needs to be distilled. Thus the optimized curve indicates a trade-off between those two contradictory desires. However, it should be noted that the shape of the curve does not have a strong impact on the targets of this optimization problem.

5. Conclusions and Acknowledgement

We propose a new operation mode to improve the efficiency of batch distillation processes. For the minimization of the batch time two industrial processes have been considered. Model-based optimization was used for developing time-optimal policies for the batch distillation processes. The computation results show that the new operation mode is effective and workable. The effect of the recycle stream will be further studied concerning multiple batch campaigns. We thank the financial support from Deutsche Forschungsgemeinschaft (DFG) under the contract WO 565/12-1.

References

Betlem, B. H. L., H. C. Krijnsen and H. Huijnen, 1998, Chem. Eng. Journal, 71.
Bonny, L., 1999, Ind. Eng. Chem. Res., 38, 4759.
Bonny, L., S. Domenech, P. Floquet and L. Pibouleau, 1994, Chem. Eng. Process., 33.
Engel, V., J. Stichlmair and W. Geipel, 1997, IchemE Symposium No. 142, 939-948.
Kim, K. and U. M. Diwekar, 2000, AICHE Journal, Vol. 46, Nr. 12.
Li, P., H. Arellano-Garcia, G. Wozny and E. Reuter, 1998, Ind. Eng. Chem. Res., 37.
Mujtaba, I. M, and S. Macchietto, 1992, Comput. Chem. Eng., 16, 273-280.
Reid, R. C., J. M. Prausnitz and T. K. Sherwood, 1977, The Properties of Gases and Liquids, McGraw-Hill, New York.
Sørensen, E. and S. Skogestad, 1994, Journal of Process Control, 4, 205-217.
Wajge, R. M. and G. V. Reklaitis, 1998, Ind. Eng. Chem. Res., 37, 1910-1916.
Wendt, M.; P. Li and G. Wozny, 2000, ESCAPE-10, May 7-10, Florence.

European Symposium on Computer Aided Process Engineering – 12
J. Grievink and J. van Schijndel (Editors)

Planning and scheduling the value-added chain

M. Badell[1], M.Aguer[1], G.Santos[2] and L. Puigjaner[3]

[1]Dept.Enterp.Management, Esc.Univ.Ing.Tecn.Indust EUETIB., Urgell, 187, Barcelona

[2]Cimade S.L., Rambla Catalunya 17 3° 1ª , Barcelona 08007, Spain

[3]Chem.Eng. Dept., Univ. Politècnica de Catalunya, Diagonal 647, Barcelona 08028

This paper contains details of a computer aided decision-making tool under development for manufacturing and process industry that integrates financial/production trade-off planning, scheduling and optimisation in the enterprise value added chain.

Abstract

The enterprise-wide optimisation of its value-added chain is performed using as target key performance indicators, KPI, that consider its business operation influence in the value added. A value-added chain can be seen as the value view of the supply chain. Having this in mind, new schedule strategies can be devised reshaping current developments for the supply chain management. The new approach is to perform an integrated and simultaneous financial/supply chain trade-off planning and scheduling optimising its economical/shareholder value. Consequently the result of this work is a decision-making tool for the optimal management of a value-added chain in manufacturing and process industries. The systematic deployment of more qualified information and the substitution of intuitive sequential decision making by simultaneous optimisation of activities could be the key of new enterprise-wide optimisation systems (Badell, 2001). Managers could make better business and technology decisions if they can use accurate simulation tools and timely process data to evaluate plans. The challenge is to obtain trade off solutions with common maximum performance measures while satisfying customers. This system gives to the financial-managers (company economist side), a tailor-made version of the scheduling tool helping them do their work. The overall simulation allows to test different alternatives during planning through a supply chain schedule with all the operation and financial information online with absolute transparency of the limitations and interactions occurring at plant and business level within each alternative. With this help it is possible to keep the visibility of the cumbersome interactions between the plant floor in multi-site levels and to change the today slave/blind position of business level during the supply chain decision-making procedures. The benefits of the value added chain management system are shown through a case study.

1. Introduction

The aim of this work is to change the current slave/blind position of the financial manager in the supply chain decision-making procedure. With the system proposed their claimed problems can disappear, being able to improve enterprise competitiveness by knowing where the money is and where will it be due to production scheduling decisions; by knowing projections of financial and/or production decisions; by looking at a rough-cut activity/process/product costing; by knowing the company cumulative value added/profit graphically showing distance from the breakeven point; by looking at interactions of cash flow with production at the plant floor; by making simulations with business level scope using updated financial status of the company. During simultaneous financial-supply chain scheduling, financial operations are placed in dominant position. In the classical time variable is overridden also money supporting the well known *"time is money"*. This time-money variable is included in the trade-off solution with the multiobjective economic optimisation of the whole system. A friendly intuitive interface allows the financial decision selection from a set of possible actions, operations and indicators as discounts, investment in marketable securities, credit and loans, factoring, pledging, advancing, etc. needed to simulate/evaluate the convenience of the monetary uses or reproductions proposed. The status of value, cash and profit and the effects of this operations are visualized in the simulation tool environment.

2. Previous work

In the past the clerical role of the enterprise finance organization was centered on oversight and control consuming even 80% of working time on fiduciary duties. Dramatic changes in business scenario have driven finance organizations to reconfigure its role at a more prevalent position and active participation as value producer, strategist, high qualified adviser, annalist and business partner in order to improve the corporative value, and hence, its competitive advantages. The working time percent relation of the old financial profile has to be changed from a 80 to 20% relation to its inverse 20 to 80%, leaving the most time to fulfill the new tasks. The information demanded by decision makers to measure and manage performance requires greater quality, interoperability and time precision, so consequently a more complex computer-aided tool. The majority of finance applications are commercial off-the-shelf packages with individual analysis of items. This system modifies finance tools when uses the simultaneous analysis of actions to better assess the current role and organize finance to add value. The system is not only flexible in the type of enterprise and the type of competitive advantage, but also in the type of procedure and which performance measures and objective functions to use. Due to the system modular approach the applications are easily set in house establishing the adequate KPI to control and measure the performance of activities included in the objective functions to optimize. The traditional financial data, usually overaggregated and too late informed, is translated into a single set of meaningful information given on line in a partnership interoperable platform shared by management, finance and supply chain operations, including the value market diversified operations. Thus two systems must be shared in parallel in financial duties in the management space: the classical transactional accounting system and the simulation tool capable of giving optimal alternatives with in live overall information of activities interoperating to add value to corporations.

Almost all aspects of the cash management deterministic modeling problems have been discussed in the literature several decades ago. A review reveals that the majority of cash management deterministic models deal with a combination of three decision types: cash position management, short term investing and short-term borrowing.

Was Keynes in 1936 who introduced the transactions precautionary and speculative motives for holding cash. Lutz and Lutz, 1951, pointed that cash inflows/outflows are not normally synchronized so a positive cash balance is required to operate. Several deterministic inventory models were suggested by Gregory, 1976, Baumol, 1952, and Miller and Orr,1966 who consider the fluctuations in net cash flows as completely stochastic. The intertemporal features of cash management attracted attention (Beranek, 1963) to the uncertain possibility of using dynamic programming to solve the problem. In contrast succesful results with linear programming were first applied to finance by Charnes et al., 1963. Most of these studies concentrate either on capital budgeting and the problem of financial planning. Robichek, Teichroew, and Jones, 1965, focused the cash management with the intertemporal aspect of the problem for short-term considering several decision variables, but excluding securities transactions and cash balance. Orgler in 1969 used an unequal multiperiod linear programming model capturing the day-to-day aspect of the cash management problem in which includes the four major types of decision variables: payments, short-term financing, cash balance, and securities transactions with the amount and maturity that are defined and also derived by the model. Srinivasan, 1986, reviews his formulation about the cash management problem as a transshipment problem. A simpler network optimization approach gives advantage over conventional linear programs in computer performance. Lerner (1968) uses simulation by using the standard deviation of the elements in financial planning. Klein, 1998, created a knowledge based decision support system coupling financial planning and production planning for the financial analysis giving as output the balance sheets and income statements.

The research in cash management modeling focused more on the specific types of decisions paying less attention to a broader objective. Although the intertemporal aspects in the financial environment were rigorously studied, the whole sequence of interrelated problems in an enterprise were not considered, likely due to the lack of adequate software and computers. But thanks to the segmented focus now are available polished methodologies that can be applied in the adequate computer framework.

3. Key Performance Indicators, KPI

The selection of the set of operative and financial performance measures appropriate for the objectives to optimise in the integrated system is one of the most difficult and polemic aspects to decide. Further, for each enterprise the selection must capture its strategic aim to hold the competitive race that is also dependent on the kind of business. If response to market is the key, or if the product innovation is the advantage, or if it is service, then KPIs must have the appropriate scope to explore the different areas having this point of view. If several KPIs determine the exact advantage, its interrelations must be balanced in order to avoid the overlapping in detriment between them. In addition accounting systems are failing in the companies where the assets are increased via intangibles assets.

The KPI problem in software development is that usually management tools use profit as the primary indicator of corporate performance, viewing all aspects on those terms.

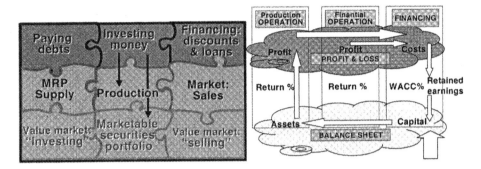

Figure 1. Value-added chain activities in interoperation and enhanced interoperable framework to be covered by in house meaningful KPIs.

The profit policy frequently gives more weight to the short term objectives and long term viability becomes threatened. As a result of this disadvantage good decisions to add value could be affected in an effort to diminish costs. Executives have a strong preference for this single indicator of performance which is well tested and gives unambiguous signals. The task of selecting KPIs is cumbersome: the accounting system based on the transactional principles and stated five centuries ago is in crisis. The financial and non-financial types of performance measurements has to be reconsidered in an effort to fill the gap or deficit of information about the enterprise intangible assets in the actual accounting and value market systems. Using multiple indicators is hard because they are difficult to design and to relate one to another. But the new trends emphasize on multiple indicators that help to draw the complexity of corporate activity and give more weight to treasury. What comes in and out – cash inflows/outflows – is unmistakable. Nevertheless, taking into account that net cash flow could only be post determined, the expected future treasury balance is the most reliable performance measure today available to set an objective to achieve. But the expected cash flows are mainly dependent on forecasts and non-financial (not quantitative) performance measures. Consequently, enterprises now have not well tested neither defined the adequate tools to provide reliable information for internal/external decision making. Dramatic changes in enterprises are now taking place. Main benefits are obtained through intangibles assets' values. Enterprises are now supported not by profit but by development expectative. While a new accountancy version is under construction or retrofitting, all enterprise system must have a flexible framework for easy KPI adaptation during this changing period. The tool designed is prepared to include traditional or shareholder value-added-based KPI objectives.

4. Case Study

One of the tasks of a financial officer involves the cash management and related financial instruments on a short-term basis so as to produce extra revenues for the corporation that otherwise would go to the banks. The goal is to determine the optimal cash management decisions for the firm. An example implemented via Excel illustrates the modeling framework of the system capable of making the simultaneous optimization of the supply-chain and financial operations. By means of a web-based agent customers orders are sequenced using the TicTacToe algorithm and a timing algorithm that optimizes the schedule giving as output a Gantt chart with the timing of supply chain operations and financial needs. A linear programming model with rolling horizon possibilities uses the set of cash inflows and outflows optimally scheduled taking into account customer satisfaction. A simplified objective function to maximize in time horizon T is the sum of payments $X_{g,j}$ and marketable securities $(y_{i,j} - z_{i,j})$ revenues discounting the costs of a credit line $w_{h,g}$. Technical coefficients C, D, E, and F adjust quantities depending on the timing of periods i, g, j, h (payment, maturity, sale periods, etc.) where actions incur.

$$Max \ Z = \sum_{j}^{T} C_{g,j} X_{g,j} + \sum_{j=2}^{T+1} \sum_{j=1}^{T} (D_{i,j} y_{i,j} - E_{i,j} z_{i,j}) - \sum_{g=1}^{T} F_{h,g} w_{h,g} \qquad (1)$$

It is chosen a time horizon of a week (128 h), divided unequally into twelve periods (see figure 2). The production plan considers sixteen batches to fulfill the orders received of eight products. The information of the monetary needs is included in the financial model. Salaries, utility costs, and fixed costs will not be taken into account. Production expenses during the week will consider initial zero stock and raw material needs. The economic situation of the case study is based on the information given in a financial balance sheet. Initial capital considered will be 1000 monetary units (mu), two times the minimal net cash flow 500 mu beneath which a short term loan must be requested. This value was assumed taking into account the variability of cash outflow, a random variable with normal distribution and mean value 372 mu per day for a week production cycle. The portfolio of marketable securities held by the firm at the beginning of the first period includes several sets of securities with known face values and maturity periods, only one maturing beyond the horizon. All marketable securities can be sold prior to maturity at a discount or loss for the firm. A short term financing source is represented by an open line of credit. Under the agreement with the bank, loans can be obtained at the beginning of any period and are due after one year at a monthly interest rate of 0.5 percent. Early repayments are not permitted. The payment decisions to be considered correspond to accounts payable with 2 percent 10 days, net 30 days terms of credit (2-10/N-30). All payments of raw materials are fulfilled within the horizon. It is assumed that all bills are received in the first half of the periods and that payments, including sales of final products, are made beginning the periods.

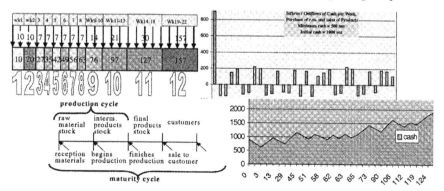

Figure 2. Time horizon divided in twelve unequal periods and inflows/outflows of cash during the production maturity cycle

Any part of the bills can be paid either at the first ten days with a 2% discount or at face value after 30 days. It remains to be decided upon what part of the bills to pay in which period. The net fixed cash flows, costs and revenues associated with the transactions in marketable securities are given. The company has heuristic rules: take all discounts, if are all possible; use fully the line of credit; if it is necessary to get a discount, sell not mature marketable securities. Several possibilities to "balance" the cash budget in periods are obtained introducing subjective management constraints. Revenues of 6% were obtained besides the not measured opportunity cost of assuring liquidity all time.

5. Conclusions and future work

Good manufacturing practices in financial management must begin with an appropriate logistic support to solve the present insuficiencies. The focus of this work has as target the design of adequate tools able to support in time optimal budgets. This work constitutes an advance since not only syncronizes cash inflows and outflows and assures a safety cash stock but solves the perishable property of money. making full use of the possibilities of the short term funding. The benefit of using the idle cash in financial operations gives revenues of more than 6%, enough to compensate the implicit monetary losses to its perishable character. This, together with other advantages already mentioned, leads to a competent, rigorous and consistent financial system prototype adequate for the competitive race. Financial support from European Community is gratefully acknowledged (VIPNET and GCO projects).

6. References

Badell, M. and L. Puigjaner, 2001, Discover a powerful tool for scheduling in ERM systems, Hydrocarbon Processing, 80, 3, 160.
Baumol, W. J., 1952, The Transactions Demand for Cash: An Inventory Theoretic Approach, Quarterly Journal of Economics, 66, 4 , 545.
Beranek, W., 1963, Analysis for Financial Decisions, R.D. Irwin, Homewood, Illinois.
Charnes, A., Cooper, and W. Miller , 1963, Breakeven Budgeting and Programming to Goals, Journal of Accounting Research, 1, 1, 16.
Gregory, G., 1976, Cash Flow Models: A Review. Omega 4, 6, 643.
Ijiri J., Levy, F. K., and Lyon, R. C., 1963, A Linear Programming Model for Budgeting and Financial Planning, Journal of Accounting Research, 1, 2, 198.
Keynes, J. M., 1936, The General Theory of Employment, Interest and Money, Harcourt Brace and Company, New York, 170, 194.
Klein, M.R., 1998, Coupling financial planning and production planning models. Xth Intern. Work. Sem. on Production Economics Proceed., V.3., Innsbruk, Igls, Austria.
Lerner, E., 1968, Simulating a cash budget. California Management Review. 11, 2, 78.
Lutz, F. A. and Lutz, V., 1951, Theory of Investment and the Firm, Princeton Univ., NJ.
Miller, M. H., and Orr, R., 1966, A Model of the Demand for Money by Firms, The Quarterly Journal of Economics, 80, 3, 413.
Orgler, Y. E., 1969, Unequal-period Model for Cash Management Decisions, Manag.Science, 14.
Robichek, A. A., Teichroew, D. and Jones, J. M., 1965, Optimal Short-term Financing Decision, Management Science, 12, 1, 1.
Srinivasan, V., 1986, Deterministic Cash Flow Management, Omega, 14, 2, 145.
Weingartner, H. M., 1963, Mathematical Programming and the Analysis of Capital Budgeting Problems, Prentice-Hall Inc., Englewood Cliffs, NJ.

European Symposium on Computer Aided Process Engineering – 12
J. Grievink and J. van Schijndel (Editors)
© 2002 Published by Elsevier Science B.V.

Optimisation of Oilfield Development Production Capacity

Richard J. Barnes, Patrick Linke and Antonis Kokossis
Centre for Process & Information Systems Engineering, School of Engineering,
University of Surrey, Guildford, GU2 7XH, U.K.

Abstract

The paper presents an optimisation method for the systematic evaluation of the economic production capacity of an offshore oil production platform. The problem is formulated as sequential mixed-integer linear programs to keep the mathematical comlexity at a low level. Continuous variables represent individual well, jacket and topsides costs and discrete variables are used to select or deselect individual wells within a defined field grid. The mathematical formulation is concise and efficient to enable future extensions to consider uncertainty in reservoir performance and actual development costs. The new method is illustrated with two hypothetical fields that are based on real-life examples in the North Sea.

Introduction

The production capacity is the most critical design decisions in the development of a new oil field as it defines the overall size of the facility and the rate of revenue generation. Particularly for an offshore installation, it can be expensive and difficult to change the capacity of an installation after it has been installed. If the facility has been oversized, it is not economic or practical to replace existing processing equipment with smaller units. If the facility were undersized, it is equally difficult to install an extra train of processing equipment or to install a larger processing train. It is possible to de-bottleneck installations, but this can be very costly. It is current practise to design the facility to produce between about 10% and 20% of the reserves each year. Economic analysis should ideally be used to explore the effect of different production capacities, but there are no guidelines or support systems to determine the most economic capacity from basic field data. The objective of this study is to identify the factors affecting the most economic capacity for the facilities and to develop a method of calculating the optimum design capacity. Such a method could either make savings in avoiding building an oversized installation, or produce extra revenue by ensuring that the installation was built with adequate capacity. Previous work in oilfield infrastructure planning has produced large scale MILP (Nygreen et al., 1998; Iyer et al., 1998) and MINLP (van den Heever and Grossmann, 2000) formulations. This work will draw on problem domain knowledge to reduce the overall problem into sequential, easy-to-solve but realistic MILPs.

Problem representation and model developments

This work aims at the development of a generic model for an offshore field that can be used to identify the criteria and capture the relationships between the different field parameters. In view of the large number of possible design aspects in oil field exploration, the model needs to strike a fine balance between its size and the accuracy of its predictions so that efficient optimisation techniques can be deployed for decision making. For the model to reflect the major system trade-offs, it needs to capture the reservoir areal extent, the productivity of individual wells, drilling costs, platform cost, processing and topsides cost, export pipeline cost, operating and maintenance cost, export pipeline cost, and oil revenue. These variables directly or indirectly affect the platform design capacity. We develop a model that captures these interactions from which the configuration giving the greatest financial return can be determined by optimisation. The complete model that has been developed is an MILP and will be the subject of a future publication. The main modelling aspects and assumptions are discussed in the remains of this section.

The reservoir is assumed to act as a single tank and no account of drawdown around a particular well is considered. However, in the fully developed model this will be an unrealistic assumption. The reservoir is divided up in a grid with each cell being selected to give a reasonable and typical well spacing assuming one downhole target in the centre of each cell. The peak production rate of each individual well is specified as an input parameter. The model is structured such that other depths and peak production rates can be incorporated. The field is assumed to be offshore. The model is capable of modelling fields in other locations and water depth with minimal alteration.

Well and drilling costs are based on the total length of the well drilled to the top of the reservoir. Two types of wells are considered. The first is a vertical well in which there is no well deviation from the centre. The second type of well geometry is known as build and hold. In this type of well, the initial section is drilled vertically, and then a deviation angle is built until a straight projection from this point intersects the target location. The cost of each well is a combination of a fixed cost for the wellhead, design and project management and a variable costs such as casing and actual drilling costs that depends on the well length.

The total cost of the production platform accounts for the costs of the jacket, topsides and pipeline. An in-house cost estimation package was used to estimate the cost of these three portions of the platform over a wide range of Design Production Rates (DPRs). Quadratic equations to describe the costs were fitted to enable cost calculations of platform depending on the production capacity.

For each DPR a target production profile was generated: 25% of the DPR is produced in Year 1, 50% in Year 2, 75% in Year 3 and 100% in the fourth and subsequent years until 85% of the recoverable reserves have been produced. After this point, the field is assumed to be in decline resulting in an exponential decline in production.

The actual production profile assumes that the full capacity of each selected well is produced and that there is sufficient capacity in the production facilities to process and export this flowrate. This assumption results in more production than the target profile, particularly when the field is in decline. This profile represents a similar practise to that which would be adopted in a real field.

Each well production rate is reduced by a function of the cumulative oil produced at the end of the previous year and the recoverable reserves (Individual Well Profiles). This reduction represents the general decline in productivity with field life and is described by the production reduction factor

$$F_{PR} = \frac{RR - 0.5CP}{RR}$$

Where RR are the recoverable reserves and CP is the cumulative production. For each field, an optimum location for the drilling centre for the development is initially calculated. This is determined from a simplified model of all possible downhole targets. For each surface location, represented by the grid locations, the lowest drilling cost wells are determined to meet or exceed the specified production rate. The optimum drilling centre was determined using a mean value of the DPRs to be investigated for that particular field. The minimum well cost was calculated for each grid location and the minimum cost of all locations was selected as the drilling centre. This optimisation problem gives rise to a MILP formulation to explore all possible drilling centre locations. Initially, the cost of drilling to each downhole target from each location is calculated. These parameters are used in the optimisation model with two scalars representing the coordinates of the drilling centre. The existence of each well is represented by a binary variable. For each drilling centre location, a minimum total cost is calculated by solving the MILP problem. This minimum cost corresponds to the optimum drilling centre location. This was then used as a fixed location for subsequent calculations involved in planning for field development.

The financial analysis is performed using a spreadsheet. Costs for drilling, platform and topsides, and export pipelines are annualised over a period of four years before the field comes into production. After this time, wells are assumed to be drilled in the year preceding the year they are required in operation. The total capital expenditure for the jacket and topsides is annualised over a period of four years in the proportions of 15%, 25%, 30% and 30%. The pipeline costs are assumed to be annualised over a two year period with expenditure being equal in both years whereas the drilling costs are not adjusted for inflation. The operating cost is assumed to be 4% of the capital expenditure and not adjusted for inflation. The revenue from the field is calculated for a range of oil prices. From this data the Net Revenue and Cash Flow is calculated for each year. The Net Present Value at a specified discount rate and the Internal Rate of Return, IRR, are then calculated for each case.

Implementation

The basic model to determine the drilling centre locations for each year of the project has been implemented in the General Algebraic Modelling System (GAMS) and solved using the CPLEX solver. The field model is implemented using a mathematical modelling package and a customised program written in C++. This structure was chosen to reduce the complexity of the overall problem when the full field life is modelled and also to more easily manage data transfer between the optimisations performed in the different years of the project. The customised program generates the GAMS input file for each year, executes the optimisation of the basic model, extracts the required information from the GAMS output file and then generates the new GAMS input file for the subsequent year. The program also calculates the target production rate and the Production Reduction Factor, F_{PR} after each year. At the end of each set of runs at a specified production rate, the data is transferred to a spreadsheet to calculate the economic performance of the field at that production rate. Results are reported as an Internal Rate of Return (IRR).

Once the optimum drilling centre has been determined, the development of the field is determined for each year of the exploration project in a subsequent optimisation stage. A DPR is specified for the particular case to be optimised. Annual production rates follow the profile described in the previous section. The objective is to minimise the cost of the wells. The binary variables are handled such that wells selected in the previous year are forced to exist in the current year and only those idle wells can be selected. Moreover, the Production Reduction Factor is incorporated into the production equation to capture general field and well productivity decline. The coordinates of the fixed, single drilling centre from the previous optimisation stage and the minimum production rate for the year are defined as input parameters.

A set of downhole well targets are defined by coordinates and potential production rate. This data is held in a separate field specific data file to enable simple change of field characteristics. Parameters are also defined to describe well geometry and well cost. Well length and drilling costs are then calculated from the fixed drilling centre and the downhole target for each well. The well cost remains constant for any given drilling centre so that the original problem is reduced from a MINLP to an easy-to-solve MILP with the existence of wells being associated with a binary variable. The objective function is to minimise the total drilling cost from the single drilling centre. An inequality constraint on the specified production rate is also defined. For each year, the MILP is implemented in the General Algebraic Modelling System (GAMS) and solved using the CPLEX solver. A C++ program is created to manage the optimisations and the data transfer between the optimisation problems solved for the individual project years.

Examples

Two hypothetical fields have been modelled to determine the effect of DPR. Each field has been based on a 500m by 500m grid. This grid is used to represent the downhole targets of each of the potential wells to be drilled from a single drilling centre. Two hypothetical reservoirs have been modelled, one with recoverable reserves of 500 MM bbl comparable to a large North sea field such as Nelson or Fulmar; the other with recoverable reserves of 50 MM bbl comparable to one of the smaller North sea fields such as Fife or Banff. The grid representations of the two fields is shown in Figure 1.

The West Field contains 224 potential well locations and has 500 MM bbl recoverable reserves. The model was run with DPRs of between 50,000 and 450,000BPD. IRRs were calculated using an oil price of between $5 and $30/bbl. These results are shown in Figure 2. Above a DPR of about 150,000 BPD, the IRR curve initially increases with DPR at a decreasing rate and is essentially flat over a wide range of DPR. A peak is not observed.

The North Field contains 59 potential well locations and has 50 MM bbl recoverable reserves. The model was run with DPRs of between 5,000 and 70,000BPD. The field requires a minimum oil price of $15/bbl to be economic and generate a positive IRR. IRRs were calculated using an oil price of between $15 and $30/bbl. These results are shown in Figure 3. Above a DPR of about 25,000 BPD, the IRR curve is essentially flat, indicating similar returns irrespective of the DPR. Again a distinct maximum did not occur.

Conclusions

The present model does not indicate any clear optimum in the Design Production Rate for these fields for any oil price considered. The results indicate that, provided the fields are developed with a DPR in excess of a minimum rate, the IRR will be within 5% of the maximum. These results appear to indicate that the decision on DPR is less important than perceived by the practitioners.

Future work will address the identification of the factors causing the stabilisation of IRR. It is also intended to enhance the model so that the Production Reduction Factor is applied to individual well production rather than to total field production. The assumption of the reservoir acting as a single tank is an over-simplification that could be contributing to the present results.

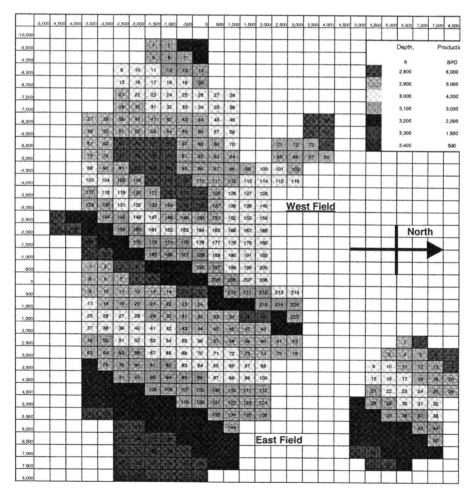

Figure 1. Layout of the hypothetical oil fields Recoverable reserves: 500 MM bbl (West Field), 50 MM bbl (North Field).

636

The current assumption of build up to full production capacity is also perhaps somewhat conservative and the effect of a delay in reaching full production will also be investigated. This change will have the effect of significantly increasing the discounted revenue from the field. Another area to be investigated further is the effect of uncertainty. Uncertainty arises in a number of different areas, including the estimation of recoverable reserves, the actual cost of the installation compared with the original estimate, and the well drilling cost and well productivity. It is intended to investigate these effects as the continuing work in this study.

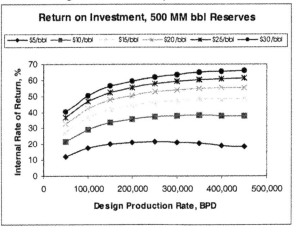

Figure 2. Variation in IRR with Design Production Rate for different oil prices (West field).

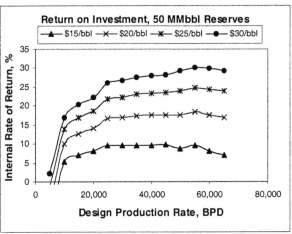

Figure 3. Variation in IRR with Design Production Rate for different oil prices (North field).

References

Iyer RR, I.E. Grossmann, S. Vasantharajan, and A.S. Cullick (1998). *Ind. Eng. Chem. Res.* 37 (4): 1380.

Nygreen, B., M. Christiansen, K. Haugen, T. Bjorkvoll and O. Kristiansen (1998). *Ann. Oper. Res.* 82, 251.

Van den Heever, S. and I.E. Grossmann (2000). *Ind. Eng. Chem. Res.* 39, 1955

European Symposium on Computer Aided Process Engineering – 12
J. Grievink and J. van Schijndel (Editors)

Optimal operation of fed-batch sugar crystallisation with recycle

Ben H.L.Betlem, Sander van Rosmalen, Peter B.Thonus, Brian Roffel

Dept. of Chemical Technology, University of Twente, The Netherlands

Abstract

In sugar batch crystallisation, multiple sequential crystallisers (pans) are necessary to obtain the required exhaustion of the feed. In this simulation study, the production rate is optimised by distributing the exhaustion effort between the first two pans. To describe the crystal size distribution (CSD), a dispersion based population balance has been used. An appropriate CSD of the seeds and a nucleation equation fulfil the initial and boundary conditions. In the growth rate kinetics, supersaturation and activation energy are formulated as a function of the impurities. The pan simulations as well as the interactions between the first two pans have been fitted by industrial data. Optimisation of the production rate implies maximisation of the exhaustion of the first pan.

1. Introduction

Many (fed-)batch processes can be optimised by partitioning the operation in a production step and one or more subsequent separate exhaustion steps. The intermediate products of the exhaustion steps are recycled to the feed of the first step (Figure 1). By this partitioning, the production step becomes easier as the exhaustion remains limited and the feed is enriched by recycle flow, which has a higher purity. Also, the exhaustion becomes easier since the product of the exhaustion steps satisfies a lower requirement. However, a larger recycle flow means a larger throughput in both steps. Therefore, a maximum in the production rate may arise for a certain distribution of the exhaustion between the production step and exhaustion step (Betlem and Roffel, 1997).

In the case of sugar fed-batch crystallisation, it is not possible to produce pure sugar crystals with the required size from impure syrup in one process step, while ensuring that the syrup is sufficiently exhausted. The sugar purity of the feed dry mass contents is about 93%. The first crystallisation step produces 720 μm crystals with a purity of nearly 100%. The subsequent exhaustion step produces 520 μm crystals, which are melted for recycling. The syrup is exhausted to a remainder with a purity of 75%. The

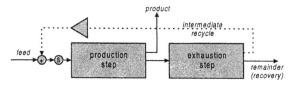

Figure 1   Batch operation consisting of production and exhaustion phase.

distribution of the exhaustion between the two steps will strongly depend on the influence of the impurities on the growth rate. When the influence is higher, the exhaustion of the production phase will be more restricted.

In this article, first, a suitable crystallisation model will be presented. Next, the model parameters will be fitted on three industrial pans with large differences in impurity levels. The fit is performed according to a sensitivity analysis. A number of simulations have been made to analyse the influence of the control variables on the production rate. Finally, conclusions will be drawn about the optimal recycle strategy. The simulations are performed with the program package gPROMS and the partial differential equations are solved numerically by the second order backwards finite difference method.

2. Sugar Crystallisation Model

The process concerns a fed-batch evaporation crystallisation of sucrose in a mixed suspension vessel at low constant pressure. The heat supply is condensing steam. The major assumptions for the process are the following.

- The vessel is sufficiently mixed to consider all variables to be lumped.
- The controllable heat supply can be set directly.
- The temperature of the boiling suspension is in equilibrium with the vapour. This means that the temperature directly depends on the pressure and composition.
- Only the net growth rate and nucleation are described.

One fed-batch step consists of several phases. First, a fixed part of the feed juice amount is initiated in a pan. This is thickened until a pre-defined supersaturation level has been reached. Then, a crystal seed slurry is introduced and the crystallisation phase starts. A constant feed juice flow and heat flow are added until all feed has been consumed. Next, the suspension is centrifuged to separate the crystals from the mother liquor. To wash away the impurities thoroughly, hot water is injected in the centrifuges. For the identification and simulation, only the crystallisation phase is considered. For the optimisation, also the thickening phase and the centrifuge separation are included.

The driving force of crystallisation is the supersaturation of the solution. The super-saturation ΔZ_S can be modelled according to Eq. 1. In this equation S, I, and W are the weight of the sucrose, impurities and water contents of the mother liquor. The sucrose content at equilibrium, E_{eq}, can be calculated according to the Wiklund equation. Some impurities, such as salts, will increase the supersaturation, whereas sucrose like substances, such as raffinose, have an opposite effect. For low impurity-water ratios, (I/W), the second term will result in a slightly decreased S_{eq}, whereas for higher ratios the third term is dominant and S_{eq} will increase linearly with the ratio.

$$\Delta Z_S = S/(S+I+W) - S_{eq}/(S_{eq}+I+W) \tag{1}$$

$$S_{eq}/W = Wiklund\{I/W\} \cdot S_{eq,pure}/W \tag{2}$$

$$Wiklund\{I/W\} = a + (1-a) \cdot \exp(-c \cdot I/W) + b \cdot I/W \tag{3}$$

It is generally accepted that the mechanism of crystal growth is determined by two steps: mass transfer by diffusion from the mother liquor bulk to the crystal surface followed by surface integration. Both rate constants are temperature dependent. The kinetic growth model of Eq. 4 is often used. In the presence of raffinose, sucrose crystals grow with approximately a second-order dependence on supersaturation (Liang et al., 1989). Eq. 5 takes the increase of the activation energy due the presence of impurities into account. It contains a Langmuir term considering the influence of the surface coverage by impurities.

$$\overline{G} = A\{I/W\}\cdot \exp(-\frac{E_c\{I/W\}}{R\cdot T})\cdot \Delta Z_s{}^k \tag{4}$$

$$E_c\{I/W\} = E_{c,pure}\left(1 + a_I(\theta_I)^\kappa\right) \quad , \text{with} \quad \theta_I = \frac{b_I\cdot I/W}{1 + b_I\cdot I/W} \tag{5}$$

Theories to describe crystallisation are based on constant growth rate, size-dependent growth rate, or growth rate dispersion (GRD). GRD means that crystals of the same size grow under the same conditions at different rates (Ulrich, 1989). This phenomenon should only be observed when the surface integration step is rate controlling. Although, sucrose crystallisation is mass transfer rate controlled, crystals grown exhibit GRD (Liang, 1987). So, the population density for growth rate dispersion can be given by:

$$\frac{\partial n^*(L_c,t)}{\partial t} + \overline{G}\cdot \frac{\partial n^*(L_c,t)}{\partial L_c} = D_g\cdot \frac{\partial^2 n^*(L_c,t)}{\partial L_c{}^2} \tag{6}$$

In Eq. 6, population density n^* [m^{-1}] is the amount of crystals with characteristic length L_c at time t. $\overline{G}$ and D_g describe the growth rate and the diffusivity. The initial condition is the CSD of the seed slurry. Its properties can be effectively described by:

$$n^*(L_c,0) = \alpha\cdot (L_c)^\beta\cdot \exp(-\gamma\cdot L_c), \quad \alpha,\beta,\gamma > 0 \tag{7}$$

The boundary condition is determined by the nucleation of new crystals. Crystal generation from solution is called primary nucleation and requires high levels of supersaturation. Secondary nucleation is the term used to describe the formation of new crystals from already present crystals. This nucleation is more likely to take place in industrial crystallisers, as it requires a supersaturation level which is considerably lower (Myerson, 1993). Empirical expressions are used to describe the nucleation.

$$n^*(0,t) = \frac{B\cdot (S+I+W)}{\overline{G}} \quad , \text{where} \quad B = c_1\cdot \overline{G}^{c_2} \tag{8}$$

From the population distribution the ith-moment can be derived according to Eq. 9. Subsequently, the mean aperture MA and the coefficient of variation CV of the CSD can be calculated.

640

$$m_i^* = \int\limits_{L_c=0}^{\infty} n^* \cdot L_c^i \cdot dL_c \quad , \text{with} \quad MA = \frac{m_4^*}{m_3^*} \quad , \text{and} \quad CV = \frac{\sigma}{MA} = \frac{\sqrt{m_5^*/m_3^*} - MA}{MA} \tag{9}$$

In addition to the Eqs. 1 to 9, which model the crystal formation, mass balances and an energy balance are needed for the accumulation of crystal sugar, dissolved sugar, impurities (non-sugars) and water. Nearly all balances are simple differential equations. The amount of crystal sucrose is determined from the third momentum by considering the density ρ_{SC}, and volumetric shape factor k_V, which is assumed to be constant.

The physical properties are taken from the Sugar Technologist Manual (Bubzik *et al.*, 1995). This concerns properties such as: (1) solubility of sugar, (2) viscosities of the mother liquor and the massecuite (overall pan contents), (3) specific heat capacities of crystal sugar, feed, mother liquor, and the massecuite, (4) densities of crystal sugar and pure or impure, under-saturated, saturated or supersaturated mother liquor, and (5) boiling point and boiling point elevation of the mother liquor, which depends on pressure, solid content (brix) and impurity level.

3. Model Identification and Simulation of Industrial Pans

For identification, limited information was available: the season average of the initial and final compositions of the pans, the batch times and some global indications about final MA, CV, viscosity, and supersaturation level. The average feed rate, average heat supply and seed slurry amount, which are the input variables, are determined by the

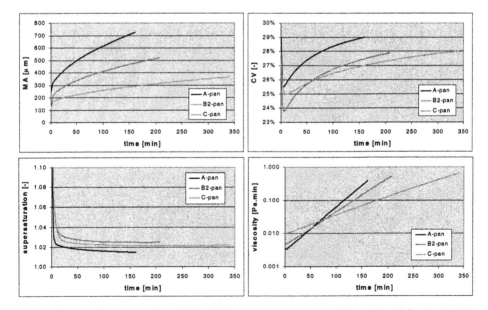

Figure 2   Simulation results of MA, CV, supersaturation, and viscosity of the A-, B-, C-pan during a batch.

final values of the massecuite, brix and MA. From a sensitivity analysis, it was found that neither the supersaturation exponent, κ, nor the frequency factor, A, exhibit any discernible effect on the key output variables. The average activation energy, E_c, has been identified by the final crystal mass while the diffusivity factor, D_g, can be derived from the final CV.

The simulations of the A-, B-, and C-pan (Figure 2) show that after thickening, supersaturation is high. Therefore, initially the MA increases fast and consequently the CV decreases at the same pace. During the batch the supersaturation level remains low, consequently the MA and CV increase slowly. However, simultaneously the viscosity increases exponentially and determines the possibilities to control the exhaustion.

4. Production Curve and Optimal Operation

The process requirements are the MA of the A-pan product and the exhaustion of the B-pan. The available manipulated variables are: the feed rate, the heat supply rate, the initial juice volume (relative to the feed amount), the initial supersaturation, and the seed slurry amount. The initial values are taken constant. The seed slurry amount is used to obtain the requirements. Therefore, the only remaining control is the feed to heat ratio. This determines directly the supersaturation. In sugar crystallisation, producing a narrow crystal size distribution (CSD) is required. To achieve this objective, the super-saturation level should be kept within the metastable zone without causing nucleation.

To determine the influence of the feed to heat ratio, *production curves* are drawn. These curves reflect the production for different batch times, while maintaining the product requirements (Rippin, 1983). The different batch times are obtained by varying the feed, while the heat supply is constant. Figure 3a shows the normalised production curves of the A-, B- and C-pan. The normalisation is obtained by plotting the relative production against the batch time corrected for differences in heat supply. For feed without impurities, the production curves of the pans coincide. From these curves and additional simulations, the influence of the impurities can be determined. The decrease in relative growth rate can amount up to 25% for the C-pan (Figure 3b).

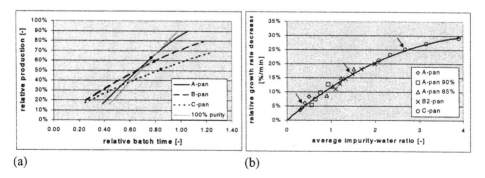

(a)   (b)

Figure 3   A-, B-, C-pan. (a) normalised production curves, (b) influence of impurities on the growth rate. (arrows indicate the three industrial pans.)

642

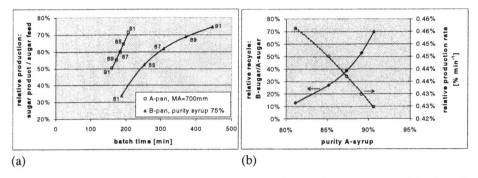

(a)   (b)

Figure 4   Determination of the optimal operation . (a) production curves of the A- and B-pan. (b) production rate inclusive thickening

At the lower limit of the production curve (Figure 3), at high feed rate no supersaturation is obtained, whereas at the upper limit nearly all water is evaporated and the viscosity is limiting. The tangent from the origin to the curve indicates the maximum production rate. The A-pan appears to have a constraint optimum, whereas the B-, and C-pan have a free optimum.

In Figure 4a the production curves of the A- and B-pan are shown. The requirement for the A-pan is a product MA of 720 μm , whereas for the B-pan the requirement is 75% syrup purity. The obtained distribution of the exhaustion effort between the pans is indicated by the purity of the A-syrup. The production rate is optimal when the A-pan is exhausted to 81%. A decrease of the impurity level does not justify a larger recycle.

5. Conclusions

A model for fed-batch evaporation crystallisation has been developed, which describes the dynamic behaviour and production of the different pans well. The influences of the impurities on the production are implemented through functions for the supersaturation and the activation energy. From the simulations it was found that the growth rate decrease due to impurities are approximately 5%, 15% and 25% for the A-, B-, and C-pan respectivily. As this influence of the impurities is restricted, the optimal recycle is a constraint optimum obtained by maximum exhaustion of the A-pan.

References

Betlem, B.H.L., Roffel, B. (1997). Integrated manufacturing of cyclic/batch processes. *ICSC-WMC 97*, Auckland, New Zeeland, 428-432.

Bubnik, Z., Kadkec, P. (1995). *Sugar Technologist manual.* 8th edition, Bartens.

Liang, B., Hartel, R.W., Berglund, K.A. (1989). Effects of raffinose on sucrose crystal growth kinetics and rate dispersion. *AIChE Journal*, 35, 2053-2057.

Myerson, A.S. (1993). *Handbook of Crystallisation*. Butterworth-Heinemann, Boston.

Rippin, D.W.T. (1983). Simulation of single- and multi-product batch chemical plants for optimal design and operation. *Comp. Chem. Engng*, 30, 137-156.

Ulrich, J. (1989). Growth rate dispersion - A review. *Cryst. Res. Technol.*, 24, 249-257.

European Symposium on Computer Aided Process Engineering – 12
J. Grievink and J. van Schijndel (Editors)

Economically Optimal Grade Change Trajectories: Application on a Dow Polystyrene Process Model

Wim Van Brempt[1], Peter Van Overschee[1] , Ton Backx[2], Jobert Ludlage[2],
Philippe Hayot[3], Louis Oostvogels[3], Shamsur Rahman[4]

[1]IPCOS, Technologielaan 11/0101, 3001 Heverlee, Belgium
Wim.VanBrempt@ipcos.be, Peter.VanOverschee@ipcos.be
[2]IPCOS, Bosscheweg 145a, 5282 WV Boxtel, The Netherlands
Ton.Backx@ipcos.nl, Jobert.Ludlage@ipcos.nl
[3]DOW Benelux, P.O. Box 48, 4530 AA Terneuzen, The Netherlands
phayot@dow.com, apoostvogels@dow.com
[4]The Dow Chemical Co., 1400 Building, Midland, Michigan 48640, USA,
srahman@dow.com

Abstract

A novel dynamic optimizer PathFinder has been applied to a dynamic model of a Dow polystyrene production facility at Tessenderlo, Belgium. PathFinder optimizes grade transitions subject to an economic cost function. Introduction of process constraints allows for a gradual migration from the currently used transition towards a more optimal transition. The results show a significant improvement in added value during a grade transition.

1. Introduction

The chemical process industry is facing a huge problem to increase their capital productivity. A solution to this problem is demand driven process operation such that exactly these products can be produced that have market demand and take price advantage of a scarce market. Flexible operation of production is therefore required [Backx, *et al.*, 1998]. A new integrated process control and transition optimization technology is needed for this purpose. A very important requirement for this technology is to enable the calculations of feasible and economically attractive grade transitions.

The idea of optimization of grade transitions in the polymer industry has been introduced by McAuley [McAuley and MacGregor, 1992]. Based on rigorous dynamic models optimal open-loop paths are calculated. The cost function has been improved into a more straightforward economical framework [Van der Schot et al, 1999]. In order to cope with the strong non-linear cost function, the PathFinder rigorous model based dynamic optimizer has been developed. An application on a Dow polystyrene production facility is discussed. The paper is organized as follows. In Section 2 the formulation of an economic optimization criterion is given. Subsequently, in Section 3 a framework for integration of trajectory control and trajectory optimization is explained. In Section 4 relevant aspects of the polystyrene solution process are described. Finally,

644

Section 5 describes the application of PathFinder on a Dow polystyrene production facility.

2. Economical optimization criterion

Since the incentive for the elaboration of optimal grade changes is merely economical, it is reflected in an economically driven optimization criterion (Eq. 1). The goal is to maximize added value (AV) during a time horizon T while making the transition from one grade to another grade.

$$AV(T) = \int_0^T price(t) throughput(t) dt - \int_0^T \sum_i feed_i(t) \cos t_i(t) dt \qquad (1)$$

The first term accounts for the benefits gained during the trajectory by producing the desired end-product. It depends on the production throughput and is a highly non-linear function with regard to product price. The high non-linearity arises from the fact that mostly a good price is paid for on-spec material, while the market price for wide-spec or off-spec material is significantly lower. Although integration softens the non-linearity, it remains nevertheless very severe. The specifications are typically expressed in terms of product properties, which are themselves non-linear (dynamic) functions of the process conditions. The second term in (Eq.1) accounts for the economic costs of all the feed stock materials and cost of plant operation. The optimizer searches for the optimal process manipulations, such that the resulting trajectory is economically optimal. The relation between optimization parameters, the process and the economic cost is shown in Figure 1. It is clear that a dynamic process model is needed to enable the calculation of the Added Value given the applied process manipulations.

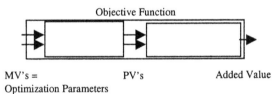

Figure 1. Relation between Manipulated Variables (MV's), Process Variables (PV's) (such as flows, product properties, holdups...) and Added Value

End-point constraints on CV's, path constraints on CV's and input constraints (absolute boundaries and rate of change constraints) on MV's are added to the optimizer restricting the optimization freedom. These constraints guarantee a safe and feasible operation during the transition and also guarantee that the desired product properties and production level are achieved after the transition. They also constrain the optimizer freedom such that the new trajectory doesn't differ too much from the initial trajectory. The last reason is important when one has no blindfolded confidence in the process model. Adding constraints will allow one to migrate slowly from a well-known recipe to a new recipe. PathFinder is a robust and fast solution for the above optimization problem. Though the objective function is strongly non-linear, typically 5 up to 10 trajectory simulations and model linearizations are needed for the cases that have been analyzed (compared to 500 up to 1000 model evaluations with a SQP optimization scheme). These model evaluations are the bottleneck for a faster calculation time. In [Van Brempt, *et al.*, 2001] relevant implementation topics are discussed.

3. Integrated Trajectory Control and Optimization Technology

A general framework has been set up in order to cope with the challenge to integrate trajectory optimization and trajectory control [Van Brempt, *et al.*, 2000]. The key idea is explained in Figure 2.

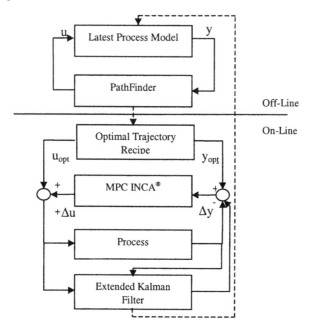

Figure 2. Integration of MPC control technology and optimization technology

PathFinder calculates off-line dynamic economically optimal grade change recipes. These manipulated and controlled variable trajectories are as such applied to the process. The controller only corrects for the deviations Δu and Δy ('delta mode') from the process input-output setpoints u_{opt} and y_{opt} that are given by the optimizer. The delta-mode guarantees a best of both worlds operation [Van Brempt, *et al.*, 2000]. Indeed, it would be a pity to have the trajectory, which has been carefully designed with non-linear knowledge, overridden by a linear model controller. Therefore this trajectory is applied as such to the process and the controller is only allowed to shift the deviations of the input-output trajectory (u_{opt}, y_{opt}) between the controller input and output. As explained in the previous section, the long trajectory simulation time determines the optimization calculation time. Therefore PathFinder is started several hours before the trajectory has to be initiated, with up-to-date market conditions and the latest instance of the rigorous model that is known, with the latest state updates in case an Extended

Kalman Filter is available. Once the optimal trajectory is calculated and acknowledged by a Product Engineer, it is sent to the controller environment. It will be graphically available to the panel operators, such that they can intervene depending on actual circumstances.

4. The Polystyrene Solution Process

Polystyrene is mostly produced using a solution process. A simplified layout of a typical polystyrene solution process can be found in Figure 3. The process consists of a combination of several plug flow reactors. Styrene, a solvent and in some cases an initiator are fed to the first reactor. Reactors are usually operated at sequentially higher temperatures with a final conversion at 60-90%. Unreacted monomer and the solvent are separated from the polymer under vacuum. The hot melt is then pelletized while the monomer and solvent are condensed and recycled. Additives may be added at different places in the process. Various polystyrene grades can be produced on the same process using a carefully chosen set of flow, temperature and pressure setpoints (SPi) that we will refer to as a "recipe". A given polystyrene grade will be characterized by a set of properties (PROPj). When changing from one polymer grade to another polymer grade, setpoints must be moved from one recipe to the other, driving the process through a zone where off-specification product is made. Typically the transition path for recipe setpoints will be selected to minimize production of low value off-spec product.

Dow developed a rigorous dynamic model for the specific process used for the production of polystyrene at Tessenderlo, Belgium. The model has been validated over the entire operation range. In order to minimize plant-model mismatch an on-line state estimator has been implemented.

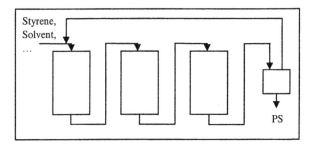

Figure 3. Schematic description of a typical solution polystyrene process

5. Application of PathFinder on the Dow Polystyrene Process

PathFinder's optimization technology is applied on a model of the Dow polystyrene production facility at Tessenderlo, Belgium. Fourteen variables were used as MV's for PathFinder including feed flows and composition setpoints and temperature setpoints for the reaction as well as for the separation sections of the process. For each variable 13 fixed move times were defined, resulting in 182 degrees of freedom for the optimizer. Path constraints are applied on 8 process variables, rate of change constraints on 10 MV's and absolute boundaries on all 14 MV's. Optimal trajectories were subsequently calculated for two different market situations depending on the price of off-grade product (Table 1). Indeed, off-grade product can be used in low-end applications and its price is therefore subject to offer and demand fluctuations. Both market situations are characterized by a considerable difference between the on-spec and off-spec material price. In the first case off-spec material represents a serious loss compared to the raw material, while in the second case off-spec material can be sold with a benefit compared to the monomer that is being used.

Table 1: Two different Market Conditions

Price	Case 1	Case 2
Onspec - Offspec	High	Low
Offspec - Styrene	Negative	Positive

In Figure 4 optimized trajectories are shown for both situations. PathFinder started from a trajectory that was given by Dow. The initial trajectory shows first a production decrease, followed by a transition from one grade to another grade. Notice that in both situations the off-spec time has considerably been shortened to about one quarter of the original off-spec time. The optimizer fully exploits the dynamical behavior of the process within the freedom of the specification range to optimize the transition. In both cases production has been increased in the original grade. The production has been increased more in the second case, since the off-spec is not penalized as much as in the first case. In Figure 4 also some MV trajectories are shown. Observe that the original trajectory was a rather quasi steady state transition, while the new trajectories are fully dynamic. The reactor feed flow (MV2) is obviously manipulated and related to the production increase. The rapid change of MV3 is responsible for the fast clipping of the PROP2 value. If this brusque change of the value of MV3 would be unacceptable for some reasons, it could be limited by introducing a proper rate of change constraint.

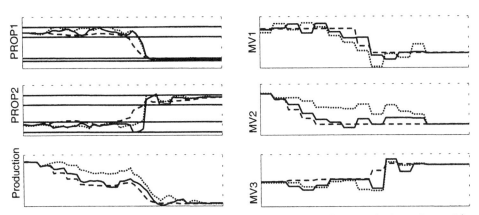

Figure 4 Trajectories for quality variables PROP1 and PROP2 (together with specification boundaries) and for production rate, as well as trajectories for three selected MV's for three cases: initial trajectory (---), Case 1 (solid) and Case 2 (...)

In Figure 5 a path constraint (PC1) is shown that was imposed on a specific process variable known as a constraint: operating beyond that point would lead to undesirable effects. The optimizer pushes the process against this boundary to increase profit.

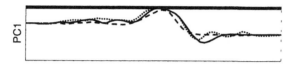

Figure 5. Trajectory for a constrained process variable for three cases: initial trajectory (---), Case 1 (solid) and Case 2 (...). The thick solid line corresponds to the upper boundary that is imposed on this variable.

6. Conclusion

A robust optimization technology PathFinder for the calculation of economical optimal grade change trajectories has been presented. It is seamlessly integrated with a model predictive control technology such that controlled (optimal) grade transitions are straightforward. PathFinder derives dynamic grade change-over trajectories that optimize added value. Several types of constraints can be entered, such that a safe operation can always be guaranteed and such that a gradual migration from a known recipe to a renewed recipe is obtained. PathFinder has been successfully applied on a dynamic model of a Dow polystyrene production facility at Tessenderlo, Belgium. The results showed considerable shortening of the off-spec time as well as a reduction of the overall cost of a grade transition for two different market conditions.

7. References

Backx, T., O. Bosgra and W. Marquardt, (1998), Towards intentional dynamics in supply chain conscious process operations, proc. FOCAPO 1998, 5-7 July 1998, Snowbird Resort, Utah, USA

Backx, T., O. Bosgra and W. Marquardt (2000), Integration of Model Predictive Control and Optimization of Processes, proc. AdChem 2000, June 2000, Pisa, Italy

McAuley, K.B. and MacGregor, J.F., 1992, Optimal Grade Change Transitions in a Gas Phase Polyethylene Reactor, AIChE Journal, October 1992, Vol. 38, No 10, pp 1564-1575

Van der Schot, J.J., R.L. Tousain, A.C.P.M. Backx and O.H. Bosgra (1999), SSQP for the solution of large scale dynamic-economic optimization problems, proc. ESCAPE 1999, June 1999, Budapest, Hungary

Van Brempt W., Backx T., Ludlage J., Van Overschee P., De Moor B., Tousain R. (2000), A high performance model predictive controller: application on a polyethylene gas phase reactor, proc. AdChem 2000, June 2000, Pisa, Italy

Van Brempt W., Backx T., Ludlage J., Van Overschee P.. 2001, Optimal Trajectories for Grade Change Control: application on a polyethylene gas phase reactor, Preprints DYCOPS6, June 2001, Cheju Island, South Korea

European Symposium on Computer Aided Process Engineering – 12
J. Grievink and J. van Schijndel (Editors)
© 2002 Elsevier Science B.V. All rights reserved.

Short-Term Scheduling of a Polymer Compounding Plant

P. Castro[†], A. P. F. D. Barbosa-Póvoa[*,‡] and H. Matos[†]

[†]Departamento de Engenharia Química and [‡]Centro de Estudos de Gestão
Instituto Superior Técnico, 1049-001, Lisboa, Portugal

Abstract

This paper addresses the optimal short-term scheduling of a three parallel production line polymer compounding plant, whose equipments require cleaning between product changeovers. A very effective user-friendly software tool was developed, which consists of a general scheduling model coupled with the capabilities of Microsoft Excel for data handling and analysis. The scheduling model is based on a Resource Task Network discrete time formulation and leads to Mixed Integer Linear Programming problems. As outputs the user can access the optimal schedules for a number of different objectives.

1. Introduction

The short-term schedule of a multiproduct plant is to be periodically made to satisfy a given set of production orders for different products within a fixed time horizon. The orders may come directly from customers or be generated to meet inventory requirements. In the multiproduct plant considered (a two-stage, three parallel line facility), the list of products to be manufactured changes every day. Hence, a cyclic production policy is inadequate to follow market demands and a non-regular production pattern is adopted. The scheduling of a given set of product orders involves important decisions regarding: *i*) the assignment of products to a given production line and *ii*) product sequencing. The aim is to minimise changeover times and meet due dates. To help with this decision making process and to improve customer satisfaction an efficient scheduling tool was developed.

2. Problem Analysis

The process can be viewed as consisting of two limiting stages: *i*) mixing and *ii*) extruding. Three batch mixers (CPE-049, CPE-055 and CPE-075) are suitable for the first stage while three semi-continuous extruders (CPE-001, CPE-002 and CPE-003) can handle the second stage. Each mixer has its own maximum and minimum capacities. Whenever the amount to be produced exceeds the maximum capacity of the mixer more than one batch is required. The extruding task can start as soon as the first mixing batch is completed.

---

[*] Corresponding author. Tel: +351-218419014. Fax: +351-218417638. E-mail: apovoa@ist.utl.pt

The plant is capable of producing several products resulting from the incorporation and dispersion of different colours into polymers, usually low-density polyethylene. The number of colours that can be produced at the plant exceeds one thousand. These can be grouped into 15 different families (set C) according to their main tint, ranging from white to black. Every time a product is changed, the mixers and the extruders require cleaning. Changeover times are dependent on three entities: i) the equipment handling the product (an element of set E); ii) the colour of the product that has been processed and iii) the colour of the product that is going to be processed immediately after. For a given equipment there are at most three distinct cleaning times for every colour tone. Also, cleaning from a colour tone to a different one lasts the same time regardless of the final colour tone. These two characteristics clearly suggest that, in certain situations, two or more colour tones can be grouped into a new family.

To reduce the number of colour tones to consider in each equipment and consequently the number of cleaning tasks required to model the process, a colour-grouping algorithm was developed. The algorithm identifies functional equivalences among subsets of the available resources, thereby allowing a more aggregate treatment of such resources (Dimitriadis et al.,1998). As a result, the RTN representation of the process becomes simpler and the size of the resulting formulation becomes smaller.

3. Mathematical Model

The general Resource Task Network (RTN) representation (Pantelides, 1994) is used to model the scheduling problem. The mathematical formulation is based on a discrete representation of time, where the horizon is divided into a number of intervals of equal and fixed duration, and gives rise to a MILP problem. All system events are forced to coincide with one of the interval boundaries.

3.1. RTN Process Representation

The RTN representation regards all processes as bipartite graphs comprising two types of nodes: resources and tasks. Each task is an operation that transforms a certain set of resources into another set. On the other hand, the relevant resources are classified into two different types: i) resources representing the relevant material states, set R; ii) resources representing the two possible equipment states, set S.

The adopted superstructure, where the three production lines are clearly identified, is given in Figure 1. For each product, five different resources are required ($R1$ represents the raw materials, $R2$ through $R4$ represent the intermediate mixing states, one for each line, and $R5$ the final product. Clean ($S1$) and dirty ($S2$) states are also considered. Besides the equipment index, these states need to be referred to a certain colour tone, omitted for simplicity. In Figure 1 it is assumed that the total amount of the mixed intermediate is produced at the end of the first mixing batch. These two situations are distinguished in the RTN: arrows denoting production of resources at the end of the first batch are connected to the block in an intermediate position while those denoting production at the end of the last batch are connected further to the right. Although the extruders operate semi-continuously, the extruding tasks are modelled as batch: the mixed product ($R2$, $R3$ or $R4$) is totally consumed at the beginning of the task, while the final product is totally produced at its end.

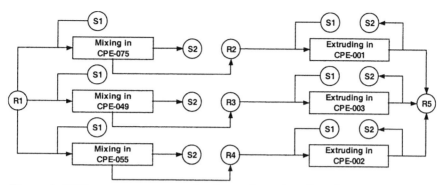

Figure 1- RTN process representation for each product. Part 1: processing tasks.

Between processing tasks the equipments require cleaning. Each cleaning task consumes a dirty state (*S2*) at its beginning and produces a clean state (*S1*) at its end.

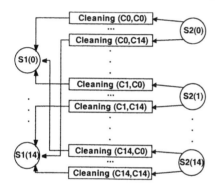

However, it is not enough to refer to a piece of equipment as clean. It is also necessary to state which colour it has been cleaned for. As the process superstructure must include all possible colour combinations (15×15) for each equipment, its representation is quite complex (see Figure 2) The first cleaning task (*C0,C0*) consumes the white dirty state *S2*(0) and produces the white clean state, *S1*(0). As the cleaning times vary between equipments, even for the same two colours, the number of cleaning tasks to consider is usually not the same in different equipments (a certain colour group will include more or less colours, see section 2).

Figure 2- RTN for each equipment.
Part 2: cleaning tasks.

3.2. Mathematical Formulation

The short-term scheduling of the plant is achieved by the following constraints.

Excess resource balances: the excess amount of a resource at a given time interval *t* is equal to that at the previous time interval adjusted by the amount consumed by all tasks starting at the beginning of *t* and by the amount produced by all tasks starting at a previous time interval and ending at *t*.

$$R_{r,p,t} = R_{r,p,t-1} + \sum_{e\in E}\sum_{\theta=0}^{\tau_{e,p}^{P-1B}} \mu_{e,r,p,\theta}^{P} N_{e,p,t-\theta}^{P} \quad \forall r\in R, p\in P, t\in T \tag{1}$$

$$S_{e,s,c,t} = S_{e,s,c,t-1} + \sum_{p\in P}\sum_{\theta=0}^{\tau_{e,p}^{P}} v_{e,p,s,c,\theta}^{P} N_{e,p,t-\theta}^{P} + \sum_{c'\in C}\sum_{\theta=0}^{\tau_{e,c,c'}^{L}} v_{e,c,s,c',\theta}^{L}\Big|_{s=2\wedge\theta=0} N_{e,c,c',t-\theta}^{L} + \gamma_{e,s,c,t}\Big|_{s=2}$$

$$- NB_{e,c,t}\Big|_{s=2\wedge e>3} + \sum_{c'\in C}\sum_{\theta=0}^{\tau_{e,c',c}^{L}} \bar{v}_{e,c',c,s,\theta}^{L}\Big|_{s=1\wedge\theta=\tau_{e,c,c'}^{L}} N_{e,c',c,t-\theta}^{L} \quad \forall e\in E, s\in S, c\in C, t\in T \tag{2}$$

In the above equations, the parameters $\mu^P_{e,r,p,\theta}$ and $v^P_{e,p,s,c,\theta}$ represent the amount of resource r (state s of colour c) of product p produced in equipment e at a time θ relative to the start of the processing task. Parameters $v^L_{e,c,s,c',\theta}$ and $\bar{v}^L_{e,c,c',s,\theta}$ are the equivalents of $v^P_{e,p,s,c,\theta}$ for the cleaning tasks, while parameter $\tau^{P-1B}_{e,p}$ represents the duration of the first mixing batch for mixers ($e \leq 3$) and the total processing time ($\tau^P_{e,p}$) for extruders.

Parameter $\tau^L_{e,c,c'}$ represents the required cleaning time in equipment e to change from colour c to colour c'. Finally, to account for equipment usage resulting from orders that began in the previous time horizon, parameter $\gamma_{e,s,c,t}$ is used. The excess variables $R_{r,p,t}$ are positive continuous variables while the excess variables $S_{e,s,c,t}$ as well as the extent variables $N^P_{e,p,t-\theta}$ (for processing tasks) and $N^L_{e,c,c',t-\theta}$ (for cleaning tasks) are of the binary type. Variables $NB_{e,c,t}$ are also binary and represent the consumption, at time t, of the dirty state of colour c in equipment e (if extruder). By using these variables, degenerate solutions near the end of the time horizon are avoided.

Excess resource constraints: In the beginning of the time horizon all raw materials are available in an amount equal to the order size (M_p). At that point, all intermediates and final products are unavailable and the state of all equipments is neutral:

$$R_{1,p,0} = M_p \quad \forall p \in P \tag{3}$$

$$R_{r,p,0} = 0 \quad \forall r \in R, r > 1, p \in P \tag{4}$$

$$S_{e,s,c,0} = 0 \quad \forall e \in E, s \in S, c \in C \tag{5}$$

To reduce the number of degenerate solutions, the beginning of the extruding tasks is made to coincide with the end of the corresponding mixing tasks. Also, the cleaning tasks are executed immediately after the end of the processing tasks.

$$R_{r,p,t} = 0 \quad \forall r \in R, 1 < r < 5, p \in P, t \in T \tag{6}$$

$$S_{e,2,c,t} = 0 \quad \forall e \in E, c \in C, t \in T \tag{7}$$

Objective function: Two alternative objective functions will be used. The first, production maximisation, will be used to select from the set of total orders, those that can be produced in a given time horizon. Then, the second objective will be used to minimize product delays among schedules with minimum cleaning times.

$$\max \sum_{\substack{t \in T \\ t=|T|}} \sum_{p \in P} R_{5,p,t} \tag{8}$$

$$\min \sum_{t \in T} \sum_{c \in C} \sum_{\substack{c' \in C \\ e > 3}} \sum_{e \in E} N^L_{e,c,c',t} \tau^L_{e,c,c'} TUNIT + \sum_{p \in P} \frac{TR_p}{|P|^3} \tag{9}$$

The tardiness of each order (TR_p, a positive continuous variable) is given by:

$$TR_p \geq \sum_{t \in T} \sum_{\substack{e \in E \\ e > 3}} N^P_{e,p,t}(T_t + \tau^P_{e,p} TUNIT) - d_p \quad \forall p \in P \tag{10}$$

where d_p represents the due date of product p, T_t the absolute starting time of interval t and $TUNIT$ the length of each time interval.

4. A User Friendly Interface

To handle the scheduling problem a user-friendly software tool was developed. By user-friendly we mean easy data input and generation of the problem's solution without going through its complex modelling aspects. The scheduling tool has two components: i) the general scheduling model (described in section 3.2), implemented in GAMS and ii) an interface to the GAMS software. Microsoft Excel was chosen as the interface since most users are familiarised with spreadsheets and because it has a programming tool associated with it, Visual Basic. Any problem is solved in four steps. In the first step, the problem data is provided in a worksheet of the Excel spreadsheet. In the second step, the data is analysed in order to reduce the number of colour tones to consider and converted to the GAMS format. The problem data is then included in the general GAMS input file, which is solved in the third step. Once the optimal solution is reached, the results are exported to Excel and displayed in the form of Gantt charts.

5. Model Solving

A real case study will be used to illustrate the behaviour of the model. It consists of nineteen products and a horizon of one week (the plant operates 5 days a week and 9 hours per day). Product data is provided in Table 1. Due to previous orders, the extruders only become available after 450 (CPE-001), 330 (CPE-002) and 90 minutes (CPE-003) and still require cleaning (from *preto*, *branco* and *amarelo*, respectively). The OSL solver solved the resulting models on a Pentium III-450 MHz machine.

Table 1-Product data (demand in kg, due date in days)

Product	Colour	Demand	Due Date	Product	Colour	Demand	Due Date
1	Verde	500	2	11	Vermelho	500	3
2	Verde	900	3	12	Preto	500	4
3	Verde	150	4	13	Verde	3000	3
4	Azul	250	5	14	Vermelho	400	2
5	Azul	150	2	15	Castanho	250	2
6	Azul	450	1	16	Castanho	150	5
7	Azul	100	4	17	Vermelho	100	4
8	Azul	150	3	18	Castanho	1000	4
9	Amarelo	3000	5	19	Cinzento	3000	3
10	Amarelo	1250	1	Total		15800	

Table 2- Computational statistics

Objective function	Eq. 8	Eq. 9
TUNIT (min)	15	5
Time Intervals	180	540
Integer variables	65639	119249
Continuous variables	99356	185900
Constraints	31317	62770
Obj. relaxed MILP	13462.5	258.49
Obj. MILP	13450	264.78
CPU (s)	7670	16732

When solving the problem using the first objective function (equation 8) it is found that it is impossible to produce all 19 products. A maximum of 13450 kg was obtained, corresponding to the sum of the demands of products 1, 2, 3, 9, 10, 13, 14, 15, 18, 19.

654

These products are then selected for production. Next, in order to minimise changeover times, the problem is solved using the second objective function (equation 9). A lower discretisation of the time interval is chosen (5 instead of 15 minutes) in order to reduce the errors involved in rounding the cleaning and processing times to a multiple of the interval length. The optimal schedule is given in Figure 3. As expected, products with the same colour tone are adjacent in the schedule. Product delays are the following: 2.94, 0.43, 1.86, 1.99, 0.92 and 0.71 days for products 1, 10, 13, 15, 18 and 19, respectively. The problem statistics are given in Table 2.

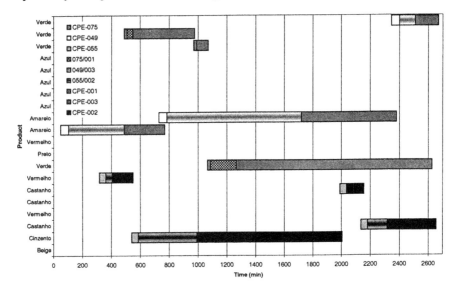

Figure 3- Optimal schedule

6. Conclusions

This paper addresses the short-term scheduling problem of a polymer compounding plant. A model, based on a RTN discrete-time formulation was developed to handle the problem. The process superstructure consists of three production lines for each product and all the cleaning possibilities for each equipment item. In order to reduce the number of cleaning possibilities, an algorithm was developed which, by looking at the cleaning times between colours involved in a given set of products, groups, if possible, two or more colours into one new group. The scheduling model was coupled with Microsoft Excel so as to provide a user-friendly interface. Using the production maximisation criterion, the developed software tool selects, from a given set of orders, those most suitable to be produced in a given time horizon. Then, for the selected orders, the optimal schedule that minimises product delays among schedules with minimal changeover times is generated.

References

Castro, P., 2001, Optimal Short-Term Scheduling of Industrial Batch Processes. PhD Thesis. Instituto Superior Técnico, Universidade Técnica de Lisboa, Portugal.

Dimitriadis, A., Shah, N. and Pantelides, C., 1998, Comp. Chem. Eng., 22, S563.

Pantelides, C. C., 1994, Unified Frameworks for the Optimal Process Planning and Scheduling. In Proc. 2nd Conference on FOCAPO. CACHE Publications, New York.

European Symposium on Computer Aided Process Engineering – 12
J. Grievink and J. van Schijndel (Editors)
© 2002 Elsevier Science B.V. All rights reserved.

Short-Term Site-Wide Maintenance Scheduling

Kwok-Yuen Cheung[1], Chi-Wai Hui[1],
Haruo Sakamoto[2], Kentaro Hirata[2] and Lionel, O'Young[3]

[1] Chemical Engineering Department, Hong Kong University of Science and
Technology, Clear Water Bay, Hong Kong.
[2] Safety Engineering and Environmental Integrity Lab., Process Systems Engineering
and Production Technologies Field, MCC-Group Science & Technology Research
Center, Mitsubishi Chemical Corporation, 1, Toho-cho,Yokkaichi Mie 510-8530, Japan.
[3] MC Research & Innovation Center Inc., 444 Castro St., Mt. View, CA 94041, USA.

Abstract

Preventive maintenance is essential for every chemical production site, as it can prevent equipment failure and accidents. In order to minimize the production loss caused by maintenance, the maintenance has to be carefully scheduled. To obtain an optimal maintenance strategy for a whole production site, the maintenance schedules of the production and utility plant have to be tackled simultaneously considering site-wide utility and material balances. However, the interconnections between production and utility system make the scheduling problem become very complex and difficult. In this paper, a multi-period mixed integer linear programming model, Site Model, is proposed as an aid to optimize short-term site-wide maintenance schedule. A special formulation is adopted to handle pre-set utility and material demand profiles during the shutdown, maintenance and start-up periods of plants.

1. Introduction

A chemical production site consists of variety production and utility plants. All of them require regular maintenance to enhance their reliability and to reduce losses caused by emergency shutdowns. Plant maintenance requires huge capital and human resources and causes losses due to the suspension of production. It also affects the material and utility balances between the production and the utility units. A good maintenance schedule should carefully take into account all these factors providing a feasible and economical solution. However, these considerations make the maintenance scheduling be a very complicated task.

In this paper, a multi-period mixed integer linear programming (MILP) model, called Site Model, is proposed for solving this type of maintenance scheduling problem. A Site Model includes all the major energy and material balances of the production and utility plants, and their interconnections and constraints.

Optimizing maintenance requires tackling both long-term scheduling and short-term schedules (Cheung and Hui, 2001). On a chemical production site, not every plant has to be maintained every year. There are always some plants being operated while others are down for maintenance. Due to the limits on the availability of materials, utilities and skilled workers, production plants and utility facilities are normally divided into groups and maintained separately in different maintenance periods. A long-term maintenance schedule provides the grouping and rough timing of plant maintenance over a two to five year time span. It takes into account the basic material and utility balances during each maintenance period to determine long-term production and inventory strategies, and policies on selling and buying material and utility services.

Superimposed on the long-term maintenance schedule, short-term scheduling determines the details, such as the exact timing of the main operations (e.g. shutdown, overhaul, inspection and startup) of all the maintained plants during a maintenance period. These operations are normally completed in a time span of four to ten weeks in which utility and labor demands could be significantly fluctuated. The main objective of short-term scheduling is to guarantee the feasibility during the main operations, smooth utility and labor demands, and minimize production losses.

Only small literature addresses scheduling maintenance of large-scale chemical production sites. Some of it focused on the short-term scheduling of multipurpose batch plants with integration of maintenance and production (Dedopoulos and Shah, 1995 and Pistikopoulos et al., 2001). Kim and Han (2001) have discussed maintenance scheduling of utility systems by assuming a fixed configuration of utility and production systems and utility demands. Hui and Natori (1996) included maintenance feasibility as a consideration in a case study of utility system expansion. Tan and Kramer (1997) scheduled maintenance for chemical processes. So far, maintenance schedules for a chemical production complex including the interaction between production and utility systems have not been investigated in any of this work.

The idea of applying a site model for long-term site-wide maintenance scheduling has previously been discussed by Cheung and Hui (2001). This paper will focus on the short-term maintenance scheduling of a chemical production complex that contains both production and utility plants.

2. Problem definitions

A simple example has been adopted from Hui (2000) to illustrate short-term scheduling in this paper (Figure 1). The example site consists of eight production plants (ETY, VCM, PVC, PE, PP, A, B and C) and a utility plant that consists of two boilers (B1 and B2) and three turbines (T1, T2 and T3).

Not all the processes and utility facilities need to be maintained in every maintenance period. In the following examples, only the ETY, VCM, PP, PE and PVC plants are maintained together with B2 and T2. Other units are operated during the maintenance period. The whole maintenance period lasts for a maximum of 31 days. Each day has three shifts (day, night and midnight). At each shift, prices and importation limits of electricity are different. Before a unit undergoes maintenance, a few days are required to shutdown all the process equipment. Overhaul is then carried in the coming days of weeks requiring intensive labor and heavy machinery involvements. After the overhaul, the plant requires a few days to resume (or startup) to normal operating conditions. The lengths of the shutdown and startup periods are fixed. The length of maintenance could either be fixed or given a lowest limit to allow the optimizer to determine the optimum length. Utility, material and manpower demands of the maintained plants vary during their shutdown, maintenance and startup. These demand profiles are given as an input for the optimization.

The objective of the site model is to maximize the overall profitability (or minimize operating cost) of the chemical production site during the shutdown maintenance period by simultaneously taking into account raw-material costs, product prices, inventory holding costs, fuel costs, electricity variable and fixed costs, and maintenance demand profiles.

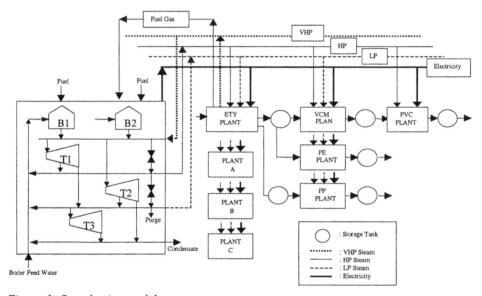

Figure 1: Sample site-model.

3. Definition of a Site Model

A site-model is a general mathematical programming model that contains the major material and energy balances of both production and utility system of a chemical production site. Some site-wide constraints, such as, seasonal production and utility demands, fuel and electricity supply contracts, are also included. To enable a site-model for production planning, the model has to be formulated in multi-period time frame such as days or months. Since the main applications of a site-model targets for large-scale planning or scheduling problems therefore using a linear model is the most suitable and sufficient. For problem like shutdown maintenance scheduling, integer variables are imposed into the model for determining equipment on/off and/or selection.

The number of parameters, equations and variables used in a site model increases rapidly with the number of plant units and time periods. Managing the model becomes

658

an important issue for industrial scale problems. In the proposed site-model, variables and equations are carefully defined to make the model manageable using a database system. Details of the model such as data, constraints, equations and GAMS (Brooke *et al.*, 1992) input files are given at ***totalsite.ust.hk***.

4. Case Studies

4.1 Base Case:

Currently, plant maintenance is scheduled by experience and with some simple heuristics. The main concern of the scheduling is to guarantee feasible material and utility balances. Payment for skilled labor is one major concern in scheduling process.

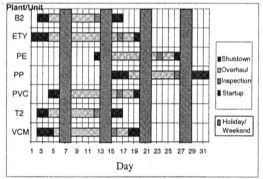

Figure 2: Schedule of Base Case.

To maintain steam balance, the boilers are normally the last one to shutdown and the first to be started up. With these considerations, a maintenance schedule can be manually created as in Figure 2. This schedule is then verified using the site model to make sure of its feasibility.

4.2 Case 1:

Figure 3 shows an optimum maintenance schedule generated using the site model directly. The model optimized the overall profit taking into account of the variation of electricity prices given in current electricity contract and other site-wide constraints.

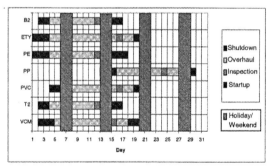

Figure 3: Schedule of Case 1.

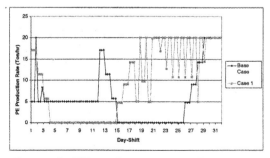

Figure 4: PE production rates of Base Case and Case 1.

Compared with the base case, this schedule increases the total labor cost. Since most of the plants are maintained in the early part of the maintenance period, increased the requirement of external skilled workers. Electricity cost has also increased due to increased production. In return, the overall profit increases, mainly due to the increase of PE production. The PE plant is shut down together with the ETY plant that reduces the need from ETY inventory allowing the downstream processes of ETY to increase production during and after the ETY plant shutdown. The PE production rates in the Base Case and Case 1 are shown in Figure 4.

4.3 Case 2:

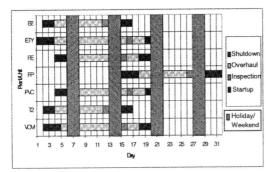

Figure 5: Schedule of Case 2.

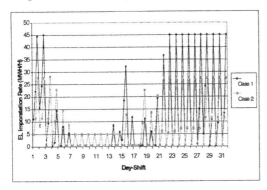

Figure 6: Electricity importation of Case 1 and 2.

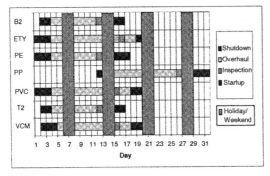

Figure 7: Schedule of Case 3.

Assuming that variation of current electricity contract is allowed providing an additional degree of freedom in the optimization. Electricity cost in a long-term purchasing contract is normally divided into two parts; fixed cost and variable cost. The fixed cost of electricity depends mainly on the amount of peak importation that is normally happen during the main utility facilities shutdown. The variable cost is calculated based upon the overall electricity consumption. In this study, the current electricity contract allows maximum importation of 45MW during the shutdown period. The optimum schedule with consideration of renewing the electricity contract is shown in Figure 5. Peak consumption of electricity is reduced to 28MW resulting a 43 million Yen decrease in electricity cost. Since the production rate remains high compared with the Base Case, the overall electricity price is still 2 million Yen higher than the Base Case. Electricity importation levels for Case 1 & 2 are shown in Figure 6.

4.4 Case 3:

In the previous cases, the length of the overhauling interval is fixed. In Case 3, the length of the overhaul is allowed to change providing an additional degree of freedom for increasing profit. The resulted schedule is shown in Figure 7. The PP plant has extended its overhauling interval from 11 days to 14 days. PP is the least profitable product. Extending the PP plant maintenance allows utilities and materials to be better utilized for other productions resulting a slightly better profit. A comparison of the cases is given in Table 1.

5. Conclusions

A MILP model, Site-Model, is evaluated in this paper for tackling a short-term site-wide maintenance scheduling problem. The case studies demonstrated the importance of considering utilities, materials and inventories, as well as variation of utility contracts simultaneously when scheduling the maintenance. A good maintenance schedule should not only guarantee feasibility, but also maximize overall profitability.

Table 1: Results Comparison.

Cost & Profit (Million Yen)	Base Case	Case 1	Case 2	Case 3
Product Revenue	2331	2707	2630	2600
Labor Cost	-18	-19	-19	-21
Production Cost	-1175	-1504	-1411	-1377
Total Elect. Cost	-65	-110	-67	-64
Total Profit	1137	1183	1200	1202

6. References

Brooke, A., D. Kendrick, and A. Meeraus, 1992, GAMS - A User's Guide (Release 2.25); The Scientific Press: San Francisco, CA.

Cheung, K.Y. and C.W. Hui, 2001, Total-Site Maintenance Scheduling. Proceedings of 4th Conference on Process Integration, Modeling, and Optimization for Energy Saving and Pollution Reduction (PRES'01), Florence, Italy, 20-23 May 2001.

Dedopoulos, I.T. and N. Shah, 1995, Optimal Short-term Scheduling of Maintenance and Production for Multipurpose Plants. Industrial and Engineering Chemistry Research, 34, 192-201.

Hui, C.W., 2000, Determining Marginal Values of Intermediate Materials and Utilities Using a Site Model, Computers and Chemical Engineering, 24, 1023-1029.

Hui, C.W. and Y. Natori, 1996, An Industrial Application Using Mixed-Integer Programming Technique: A Multi-Period Utility System Model, Computers and Chemical Engineering, 20, S1577-S1582.

Kim, J.H. and C. Han, 2001, Short-term Multiperiod Optimal Planning of Utility Systems Using Heuristics and Dynamic Programming. Industrial and Engineering Chemistry Research, 40, 1928-1938.

Pistikopoulos, E.N., C.G. Vassiliadis, J. Arvela, and L.G. Papageorgiou, 2001, Interactions of Maintenance and Production Planning for Multipurpose Process Plants — A System Effectiveness Approach. Industrial and Engineering Chemistry Research, 40, 3195-3207.

Tan, J.S. and M.A. Kramer, 1997, A General Framework for Preventive Maintenance Optimization in Chemical Process Operations, Computers and Chemical Engineering, 21 (12), 1451-1469.

7. Acknowledgments

The authors would like to acknowledge financial support from the Research Grant Council of Hong Kong (Grant No. 6014/99P), and financial and technical supports from Mitsubishi Chemical Corporation, Yokkaichi Plant.

European Symposium on Computer Aided Process Engineering – 12
J. Grievink and J. van Schijndel (Editors)
© 2002 Published by Elsevier Science B.V.

Modelling Multi-site Production to Optimise Customer Orders

D. Feord[1], C. Jakeman[2], N. Shah[3]

[1]Dow Deutschland GmbH & Co. OHG, D-77836 Rheinmunster, Germany
[2]Process Systems Enterprise Ltd. London W6 9DA, U.K.
[3]Centre for Process Systems Engineering, Imperial College of Science,
Technology and Medicine,London SW7 2BY, U.K.

Abstract

Short term planning of production networks across geographically linked sites is complex, especially during periods of severe or catastrophic system failure. This paper presents a method of modelling the production network, incorporating actual operational, demand and inventory data and predicting, based on the commercial strategy, which orders should be met, which delayed and which will not be delivered. The development and operation of this model will be discussed.

Introduction

The internal restructuring of large multinational chemical companies into smaller business units has meant that the emphasis of optimising production & supply has shifted away from single large integrated sites to the management of business production networks across several regional sites. Each of these sites may in turn be a stand-alone plant or part of an integrated site. Correct management of this geographical integrated production network is critical for business success. Enterprise management tools such as SAP allow for basic production planning but are not suitable for addressing problems that occur when the production network under-performs. These tools often have difficulty in representing the complexity of a chemical plant in sufficient detail to make them useful for optimisation. Examples of under-performance can include unplanned production outages, late arrival of raw materials, variable shift patterns etc.

A model has been developed to predict the performance of a specific business geographical production network. Its primary role is to identify, depending on the optimisation criteria, which orders can be fulfilled on time, those that will have to be delayed and those that cannot be fulfilled. The predictions have a time horizon of one to two weeks, depending on the sparseness of the available demand data. Optimisation criteria used are based upon heuristic customer priority levels set by the commercial organisation, which represent the business strategy and therefore optimal long-term returns.

Production Problem

The problem is composed of a network of three regional sites, which employ both continuous and batch processing. Figure 1 shows a simplified schematic of this problem.

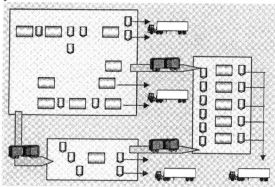

Figure 1 Simplified Production Network Description

From initial to final product there can be up to four different product formations, including combinations with different products also made in the chain. Each product produced in the chain may either be consumed downstream or sold directly to a customer. Only one product in the whole chain is made in more than one plant. Optimal profitability implies that production plants are run at 100% and inventories are held to a minimum. There is very little slack in the system to compensate for errors. Problems occur primarily because of,

- Unplanned production outage.
- Reduced production manpower levels.
- Delayed internal deliveries.
- Backlog of late orders that delay other orders in the system.

In order to maintain committed deliveries, on time, to strategic customers a tool is needed that represents the network mathematically, optimises existing orders against customer priority levels and indicates to the planners what actions should be taken.

Production Network Model

The production network has been represented and solved using the gBSS software package, N. Shah et al (1995) from PSE, an MILP programme. gBSS uses the State Task Network representation to model a process. A state corresponds to any material during the processing, e.g. raw material, intermediate or products. A task corresponds to any processing step, which can include standard physical and chemical processing but may also include packaging, loading, transportation etc. The flexible gBSS modelling language allows the production units in the network to be broken down into smaller units that represent the critical steps. Continuous units may be represented simply e.g. one task with a given hourly capacity. Batch units require a more detailed representation to determine each potential rate-limiting step, either volume e.g. reactor contents or time-wise e.g. a drumming line, Rickard (2000). The tasks are then modelled individually to form a network representing the total process. The flexibility to represent the chemical process close to its actual operation was the major reason for choosing the gBSS package.

The total multi-site network is modelled as one single network. Each individual site is modelled as one State Task Network. The sites are then linked together by modelling the transportation as tasks that hold the transported material making it unavailable for the given transportation time after which it becomes available in the new location. In this way one large network is formed. The complete production model is large, comprising about 100 equipment items, 180 materials and over 200 recipes.

gBSS also allows for utilities to be defined optionally with a given capacity profile. For this problem classical utilities such as steam and process water availability are not important. However the ability to include labour levels in gBSS enables some of the more important constraints in the batch processing to be accounted for. The labour required for tasks that have labour as a constraint are defined as part of the recipe. A matrix of recipes and resources representing the products made on the network is maintained. In this way the equipment on which each individual product may be made, process rates and utility requirements, as well as raw materials, unit ratios and material storage are defined.

Data Conditioning

The representation of the production system with the network model enables the customer order production sequence to be optimised over a given time horizon, which is normally 1-2 weeks. Meaningful optimisation is only possible with adequate and representative input data. Data must also be easily obtainable i.e. in an electronic form, to enable the manipulation and conditioning necessary to create the complete problem file to be solved by gBSS. This data necessary is shown in figure 2.

Process capability data represents the actual processing capability of the plant. In times of unplanned outages a database enables production personnel to enter the estimated outage time and the reduced production rate of the plant during this period. This results in a fast response to the supply chain on the potential impact of processing problems. This information is included in the simulation, thus a closer representation of the true state of the production network capacity is obtained. If no reduced capacity is entered the process capability is assumed to be maximum production rate.

Demand data. Obtained on a daily basis. Data for 1-2 weeks is taken, depending on the problem time horizon. In general, sparseness of data beyond 10 days can give misleading results. Included in this data is, order volume, delivery date, material ID, customer, and priority level.

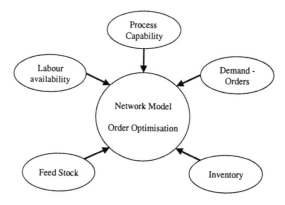

Figure 2 Data requirements to complete problem description

Inventory data. This data includes inventory volume and storage capacity. It is obtained as for the demand data. Filtered process control data, collected directly from the plant is used to obtain the intermediate and final product inventory held in plant tanks. In this way the best system data is made available.

Raw materials data is obtained initially as part of the inventory data. It is assumed that during the time horizon sufficient raw materials are available. A potential raw material shortage during the optimisation time horizon may be entered manually.

Labour availability via shift planning is obtained on a weekly basis, and can vary due to illness, holiday's etc. As labour is a constraint for some batch operating equipment this information is important to obtaining good results. Data entry is manual.

Optimisation Criteria

The primary objective of this work is to develop a tool which optimises orders against customer priority levels for unplanned production circumstances. This is a business decision and is made today by the commercial organisation, working with the supply chain and manufacturing. It is not intended at the moment that this tool should automate the process but rather to aid and improve the timing and accuracy of the decisions that need to be made. In order to do this the model optimisation must reflect the decision-making process, which is strategic rather than on a direct cost basis.

The network model may be run in two optimisation modes The first is "make in full" where all orders must be produced .An extended deadline is created and those that are made in this extended deadline are penalised. The cost function for this mode is given in equation 1. The optimisation attempts to minimise the value of the cost function. This method requires that the problem solution is feasible, which is not always the case. The second mode is "make on time" where all the products that are not made on time are neglected from the schedule. The cost function for this mode is given in equation 2. The optimisation attempts to maximise the objective function and therefore ensure that maximum number of high priority customers is served.

Make in Full   If (delivery date < Nominal date) **Cost function** = 0

If (nominal date < delivery date < Hard extended deadline)

$$\textbf{Cost function} = \Sigma_i(\text{Amount delayed (MT)} * \text{Weighting} *$$
$$(\text{Delivery date} - \text{Nominal date})) \tag{1}$$

Make on Time

$$\textbf{Cost function} = \Sigma_i(\text{Weighting} * \text{Amount made (MT)} +$$
$$\text{Inventory Created (MT)} * \text{Value}) \tag{2}$$

The Weightings for each order drive the cost function. Each customer is assigned a priority level by the business. The customer priorities are maintained in a matrix within the programme and are not changed unless a strategic change occurs

By scaling the priority levels correctly the model will be biased towards always delivering the highest priority customers. Weighting too biased towards high priority customers will tend to lead the model to ignore lower priority customers even if there is spare capacity for production. A weighting biased too evenly could mean that high volume low priority customers take preference over lower volume higher priority customers. Therefore care and attention are needed in finding the correct balance for the cost function weightings.

Results

The model is run at most on a daily basis. Running time is typically 10 - 30 minutes depending on the problem size and complexity as well as the method of optimisation. All operation is controlled via an Excel interface. It has been found that it is useful to run the model often (once a day) as even under "ideal" operating conditions the model will pinpoint errors in planning that had been overlooked. Figure 3 shows a typical results format. Each order is available in a spreadsheet, including order information such as amount, deadline, product ID, customer etc. Orders where the optimisation has indicated that the order should be delayed or cancelled are highlighted for easy identification. A Gantt chart is also available. This is particularly useful in identifying bottlenecks and reasons behind the delay of products if it is not obvious.

Benefits

The model benefits the business and planners as it enables a quick and accurate assessment of customer order fulfilment at any particular time, over a given working horizon. This foresight has often allowed corrective action to be taken to enable order fulfilment. Scenarios where this approach has already proved useful include planned and unplanned shutdowns, unplanned labour shortages and over demand (allocation). Other benefits have included the identification of processing bottlenecks and the early recognition of problem orders due to sub-optimal planning.

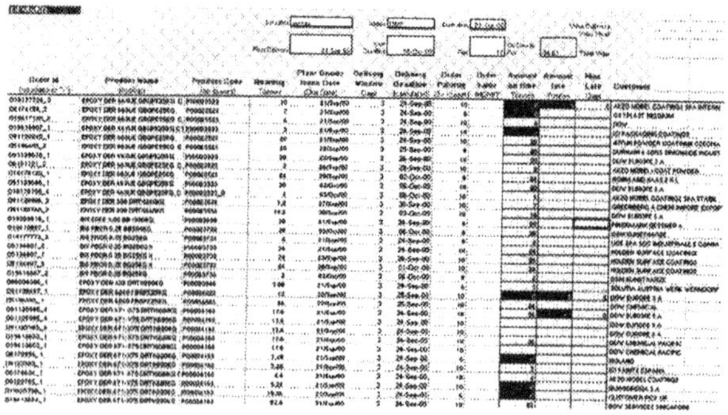

Figure 3 Typical results indication product availability

Conclusions

Through this approach it has been shown that it is possible to prioritise customer orders for geographically separate but linked production sites. Under abnormal operating conditions or catastrophic failures, the supply chain and commercial organisation can swiftly decide which orders will be met, delayed or not fulfilled in accordance with the strategy of the business.

Also consistent use of the model has enabled system problems to be identified such as inter-site transportation, manpower levels and over demand. In this way long term solutions for these problems are sought.

gBSS has proven capable of modelling the continuous and batch processing by breaking down these units into small units representing the critical steps. In addition the ability to represent transportation and utilities such as process manning levels has enabled an accurate representation of the total production network and its constraints.

References

J.Rickard, 2000, Combining Optimal Off-line Scheduling with On-line Supervisory
        Control, World Batch Forum – 2000 European Conference
N. Shah, C.C. Pantelides, L.G. Papageorgiou, L. Liberis, P. Riminucci and K. Kuriyan,
        1995, gBSS: An Integrated Software System for Multipurpose Plant
        Optimisation, 46-48, (Proc. IChemE Res. Event, Edinburgh, 1995)

European Symposium on Computer Aided Process Engineering – 12
J. Grievink and J. van Schijndel (Editors)
© 2002 Elsevier Science B.V. All rights reserved.

A New Event-Driven MILP Formulation for Short-Term Scheduling of Continuous Production Facilities

Nikolaos F. Giannelos[*] and Michael C. Georgiadis
Chemical Process Engineering Research Institute (CPERI)
PO Box 361, Thessaloniki 57001, GR, *Email: ngiannel@cperi.certh.gr

Abstract

A new mathematical formulation for scheduling multi-purpose continuous processes is presented, based on an event-driven representation of time, and resulting into a mixed-integer linear programming (MILP) model. The formulation is applicable to arbitrary process structures, variable unit processing rates, sequence-dependent changeovers, and flexible storage requirements. A medium-to-large scale manufacturing process is examined to illustrate the applicability and efficiency of the method. The formulation is shown to compare favorably with existing continuous-time models.

1. Introduction

A novel event-driven formulation is developed and applied to the short-term scheduling of continuous processes. Relevant existing approaches may be grossly divided into models applicable to mixed (batch/continuous) production facilities (Schilling and Pantelides, 1996; Zhang and Sargent, 1998; Mockus and Reklaitis, 1999) or specialized formulations for continuous processes of restricted topologies (Karimi and McDonald, 1997; Méndez and Cerdá, 2001). To manage problem sizes, schemes for treating continuous time rely on uniform (a single time scale for all process resources) or non-uniform (multiple time scales) representations. The proposed approach utilizes a non-uniform continuous time construct outlined in section 2 below. The resulting MILP model is presented in section 3. Applications are shown in section 4 and the relative strengths of the proposed approach conclude the discussion.

2. Methods

The formulation is based on the state-task network (Kondili et al., 1993). To avoid usage of tri-indexed variables (task/unit/time), any task that can be performed in k units (k>1) is treated as k distinct tasks of identical properties (states consumed/produced, stoichiometric coefficients of task recipe, etc). The time grid construction relies on the definition of n event points, $(t_1, t_2, ..., t_{n-1}, t_n)$, for every task in the process. Event points are determined by the end of task execution. Actual event times corresponding to the same event point are generally non-identical for different tasks in the process. The non-uniform time representation can therefore be thought of as $|I|$ distinct time axes/grids, where $|I|$ is the number of distinct tasks in the process. Coordination of the multiple time scales is driven by unit utilization considerations (for tasks performed in the same unit), as well as material balance considerations (for tasks processing the same material state). To ensure material balance feasibility, all continuous tasks processing the same state and terminating at the same event point (if any), are forced to start and end at the same real time, so that a rate-based balance is posed unambiguously.

3. Mathematical Formulation

3.1 Nomenclature

Sets: I (index i) is the set of all tasks. S (index s) is the set of material states. T (index t) is the set of event points. U (index u) is the set of units. I_u is the set of tasks performed in unit u, $I_u \subseteq I$. I^c is the set of continuous tasks, $I^c \subseteq I$. I^c_s is the set of continuous tasks consuming or producing state s, $I^c_s \subseteq I^c$. I^{st} is the set of flexible storage tasks, $I^{st} \subseteq I$. I^{st}_s is the set of storage tasks suitable for state s, $I^{st}_s \subseteq I^{st}$. $I^{ch}_{ii'}$ is the set of continuous tasks i, i' with changeover requirements. S^{fis} is the set of states with finite intermediate storage requirements, $S^{fis} \subseteq S$. S^p is the set of final product states, $S^p \subseteq S$. The HEAD(•) operator indexes the first element of set •.

Parameters: D^{min}_s (D^{max}_s) is the min (max) demand for state s. θ^{min}_i (θ^{max}_i) is the min (max) duration of task $i \in I^c$. R^{min}_i (R^{max}_i) is the min (max) processing rate of task $i \in I^c$. ST^{max}_s is the maximum dedicated storage limit for state s. ST^0_s is the initial amount of state s. v_s is the value of state s. λ_{si} is the fraction of state s engaged in task recipe i. $\theta_{ii'}$ is the changeover time from task i to i'. C^{max} is the time horizon, and C^{ch}_u the minimum cumulative changeover time in unit u.

Variables: θ_{it} is the duration of task $i \in I^c$ for event t. θ_{st} is the duration of all continuous tasks processing state s ($i \in I^c_s$) at event t. τ_{it} is the ending time of task $i \in I$ for event t. τ_{st} is the ending time of all tasks engaging state s ($i \in I_s$) at event t. ξ_{it} is the total extent (rate × duration) of task $i \in I^c$ for event t. x_{it} is 1 if task $i \in I$ terminates at t, 0 otherwise. ST_{st} is the amount of state s at the end of t. y_{st} is 1 if $ST_{st} \neq 0$, 0 otherwise.

3.2 Basic Model

$$x_{it} \, \theta^{min}_i \leq \theta_{it} \leq x_{it} \, \theta^{max}_i \qquad\qquad \forall \, i \in I^c, t \in T \qquad (1)$$

$$\theta_{it} \, R^{min}_i \leq \xi_{it} \leq \theta_{it} \, R^{max}_i \qquad\qquad \forall \, i \in I^c, t \in T \qquad (2)$$

$$\tau_{it} - \tau_{st} \leq C^{max} \, (1 - x_{it}) \qquad\qquad \forall \, s \in S, i \in I^c_s, t \in T \qquad (3)$$

$$\tau_{it} - \tau_{st} \geq -C^{max} \, (1 - x_{it}) \qquad\qquad \forall \, s \in S, i \in I^c_s, t \in T \qquad (4)$$

$$\theta_{it} - \theta_{st} \leq C^{max} \, (1 - x_{it}) \qquad\qquad \forall \, s \in S, i \in I^c_s, t \in T \qquad (5)$$

$$\theta_{it} - \theta_{st} \geq -C^{max} \, (1 - x_{it}) \qquad\qquad \forall \, s \in S, i \in I^c_s, t \in T \qquad (6)$$

$$\tau_{it} - \theta_{it} \geq \tau_{i,t-1} \qquad\qquad \forall \, i \in I^c, t \in T \qquad (7)$$

$$\tau_{st} - \theta_{st} \geq \tau_{s,t-1} \qquad\qquad \forall \, s \in S, t \in T \qquad (8)$$

$$ST_{st} = ST_{s,t-1} + \sum_{i \in I^c_s} \lambda_{si} \, \xi_{it} \qquad\qquad \forall \ s \in S, t \in T \qquad (9)$$

$$\sum_{i \in I_u} x_{it} \leq 1 \qquad\qquad \forall \ u \in U, t \in T \qquad (10)$$

$$\tau_{it} = \tau_{i't} \qquad\qquad \forall \ u \in U, i,i' \in I_u, i = HEAD(I_u), i' \neq i, t \in T \qquad (11)$$

$$\tau_{i't'} - \theta_{i't'} - \tau_{it} \geq \theta_{ii'} - C^{max} \left(2 - x_{it} - x_{i't'} + \sum_{i'' \in I_u} \sum_{t<t''<t'} x_{i''t''} \right) \quad \forall \ u \in U, i,i' \in I_u, I^{ch}_{ii'}, t<t' \qquad (12)$$

$$\sum_{t \in T} \sum_{i \in I^c, I_u} \theta_{it} \leq C^{max} - C^{ch}_u \qquad\qquad \forall \ u \in U \qquad (13)$$

$$ST_{st} \leq ST^{max}_s \qquad\qquad \forall \ s \in S^{fis}, t \in T \qquad (14)$$

$$D^{min}_s \leq ST_{s,tn} \leq D^{max}_s \qquad\qquad \forall \ s \in S^p, t \in T \qquad (15)$$

Constraint (1) ensures that continuous task durations lie within allowable processing times. Constraint (2) limits the total extent of any continuous task based on allowable rates. Constraints (3)-(6) coordinate the multiple time scales for continuous tasks engaging the same state s at the same event index t. When $x_{it} = 1$, constraints (3)-(4) enforce $\theta_{it} = \theta_{st}$; applied to all continuous tasks processing s, the ending times of these tasks are commutatively forced to equality. Similarly, constraints (5)-(6) equate the durations of all continuous tasks performed at the same event point and engaging the same state. Constraints (7)-(8) impose the required monotonicity on event times for the same task/state. Note that the upper bound of all $\tau_{i,tn}$, $\tau_{s,tn}$ variables is C^{max}. Constraint (9) is the typical multi-period material balance expression; in view of the multiple time scales, (10) is only meaningful by virtue of constraints (3)-(6). Constraint (10) ensures that no unit is assigned to multiple tasks concurrently. Because of the non-uniform time representation, (10) is sensible by virtue of constraint (11), which replaces multiple independent time grids with a single one for tasks executed in the same unit. Constraint (12) ensures that if task i is performed in unit u at event t, task i' is performed in unit u at event t' > t, and no other task is performed in u between i and i', then the starting time of task i' is sufficiently greater than the ending time of task i for the required changeover. Changeovers are used to provide improved linear relaxations of the model in constraint (13), where C^{ch}_u is the minimum cumulative changeover time required in unit u. Dedicated intermediate storage requirements are expressed in (14), and product demands are deterministic as in (15).

3.3 Flexible Storage Extensions

$$ST_{st} \leq \sum_{i \in I^{st}_s} x_{it} V^{max}_i \qquad\qquad \forall \ s \in S^{fis}, t \in T \qquad (16)$$

$$ST_{st} \leq y_{st} \sum_{i \in I^{st}_s} V^{max}_i \qquad\qquad \forall \ s \in S^{fis}, t \in T \qquad (17)$$

$$x_{i,t+1} \geq x_{it} + y_{st} - 1 \qquad\qquad \forall \ s \in S^{fis}, i \in I^{st}_s, t \in T, t<t_n \qquad (18)$$

$$\tau_{it} \geq \tau_{st} - C^{max} \left(1 - x_{it} \right) \qquad\qquad \forall \ s \in S^{fis}, i \in I^{st}_s, t \in T \qquad (19)$$

$$\tau_{i,t-1} \le \tau_{st} - \theta_{st} + C^{max} (1-x_{it}) \qquad\qquad \forall\, s \in S^{fis},\, i \in I^{st}_s,\, t \in T,\, t \ne t_1 \qquad (20)$$

$$\tau_{i,t-1} \le \tau_{s,t-1} + C^{max} (2-x_{it}-x_{i,t-1}) + C^{max} (1-y_{s,t-1}) \qquad \forall\, s \in S^{fis},\, i \in I^{st}_s,\, t \in T,\, t \ne t_1 \qquad (21)$$

$$\tau_{it} \ge \tau_{i,t-1} \qquad\qquad\qquad \forall\, i \in I^{st},\, t \in T,\, t \ne t_1 \qquad (22)$$

Constraint (16) activates suitable storage tasks when the amount of state s at the end of all t events processing the state is positive. Constraint (17) identifies all events t where there is a positive amount of state s by setting $y_{st} = 1$. By virtue of (16), there will be one or more suitable storage tasks active ($x_{it} = 1$). Equation (18) then ensures that all these storage tasks become active for the next event point, t+1, as well. The timings of storage tasks are determined in (19)-(21). Equation (19) ensures that the ending time of a storage task for state s at t is at least equal to the ending times of all continuous task events involving the state. Constraint (20), combined with (19), ensures that the duration of a storage task for state s at t is at least equal to the duration of all continuous task events involving the state. In fact, it can easily be verified that the combined effect of (19)-(20) is the condition $\tau_{it}-\tau_{i,t-1} \ge \theta_{st}$. For storage tasks active at consecutive events t−1 and t with $ST_{s,t-1} \ne 0$, a stronger condition is required as posed by (21). In this case, the storage task duration as determined by combined application of (19) and (21) becomes $\tau_{it}-\tau_{i,t-1} \ge \tau_{st}-\tau_{s,t-1}$. Constraint (22) simply imposes the time monotonicity on ending times of storage events.

3.4 Objective

$$\max z = \sum_{s \in S^P} v_s\, ST_{s,tn} \qquad\qquad\qquad (23)$$

subject to constraints (1)-(15) for dedicated intermediate storage, or, (1)-(13) and (15)-(22) in the flexible intermediate storage case.

4. Application

The continuous processing facility under investigation is shown in Figure 1. Process specifications are compiled in Table 1. Feed-stocks are mixed in three units to produce seven intermediates, which are packed into fifteen final products ('D' denoting drummed products, 'C' canned ones). There are three storage tanks for the intermediate states. Packing is the rate limiting step, and yield is affected by storage utilization. Two cases are investigated: (a) unlimited intermediate storage with min demands from Table 1, and, (b) flexible intermediate storage with the same demands. Analytical results for (b) are presented in Table 2 and Figure 2. The instance is solved to absolute optimality. The modest running time is due to the number of decision variables and the linear relaxation value. The schedule features maximum utilization of all packing lines in terms of rates and changeovers. Process yield is not critically limited by storage. In fact, additional runs indicate that only two storage tanks are sufficient for obtaining practically optimal solutions within 0.22% of the reported one in 2 cpu mins. The unlimited intermediate storage instance is also solved easily to optimality. Random instances are generally solvable within 1% of their relaxation in a few cpu mins. All results were obtained on a single 400 MHz processor via a GAMS/Cplex interface.

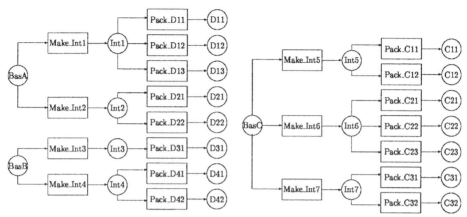

Figure 1. Simplified STN representation for case study.

Table 1. Summary of specifications for case study.

State	ST^0	D^{min}	State	ST^0	D^{min}
BasA, BasB, BasC	∞	-	D42	0	144
Int1-Int7	0	-	C11	0	42.5
D11	0	220	C12	0	114.5
D12	0	251	C21	0	53
D13	0	116	C22	0	2.5
D21	0	15	C23	0	16.5
D22	0	7	C31	0	13.5
D31	0	47	C32	0	17.5
D41	0	8.5			

Unit	Rate/ Capacity	Suitability	Changeovers	
Mixer 1	17	Int1-Int2	-	-
Mixers 2-3	12.24	Int5-Int7	-	-
	17	Int3-Int4	-	-
Tanks 1-3	60	Int1-Int7	-	-
Line 1	5.8333	D12, D21, D42	{D12,D21} ↔ D42	1h
Line 2	2.7083	D13, D22,C11,C31	{D13,D22} ↔ {C11,C13}	4h
Line 3	5.5714	D11, D31	PD11 ↔ PD31	1h
Line 4	2.2410	C23, D41		
	3.3333	C22, C32	{C23,D41} ↔ {C22,C32}	2h
Line 5	5.3571	C12, C21	-	-

Table 2. Comparison of results for case study.

	Events	Int/Cont Vars	Relaxation	Solution	CPU
This work	4	220/673	2695.32	2695.32	397
Schilling (1997)	30	1042/2746	2724	2604	3407
Zhang and Sargent (1998)	30	1318/3237	2724	2556	1085

672

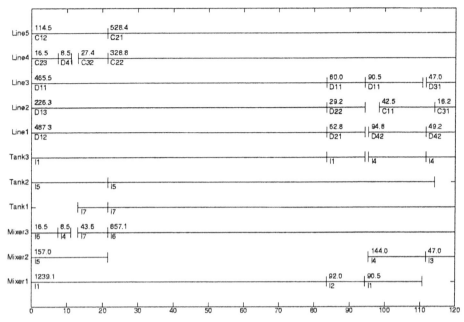

Figure 2. Finite intermediate storage schedule for case study.

5. Conclusions

This work introduced an event-based formulation to the short-term continuous process scheduling problem. The model was shown to produce optimal schedules on several variants of a benchmark case study. Due to the utilization of a non-uniform time grid, the proposed method reduces the number of event points by taking advantage of any independent sub-structures existent in the state-task incidence matrix. Depending upon the specific problem at hand, the handling of events in this work may be functionally equivalent to a uniform time scale employing an order of magnitude more points/slots. Work is in progress towards a unified formulation for mixed production facilities.

6. References

Karimi, I.A. and C.M. McDonald, 1997, Ind. Eng. Chem. Res. 36, 2701.

Kondili, E., C.C. Pantelides and R.W.H. Sargent, 1993, Comput. Chem. Eng.17, 211.

Méndez, C.A. and J. Cerdá, 2001, Proceedings of the European Symposium on CAPE 11, Eds. R. Gani and S.B. Jørgensen, Elsevier, New York, p. 693.

Mockus, L. and G.V. Reklaitis, 1999, Ind. Eng. Chem. Res. 38, 197.

Schilling, G., 1997, Algorithms for Short-Term and Periodic Process Scheduling and Rescheduling, PhD Thesis, Imperial College, London, UK.

Schilling, G. and C.C. Pantelides, 1996, Comput. Chem. Eng. 20, S1221.

Zhang, X. and R.W.H. Sargent, 1998, Comput. Chem. Eng. 22, 1287.

European Symposium on Computer Aided Process Engineering – 12
J. Grievink and J. van Schijndel (Editors)

A supporting tool for improvement of batch manufacturing

Petra Heijnen and Zofia Verwater-Lukszo
Delft University of Technology
The Netherlands

Abstract

Integration of the activities in a complex system as a batch plant will need extra support. The Management Decision Tool, as described in this paper, will give batch plant managers the opportunity to perform a quick scan of the plant focusing on objectives and activities, and their interactions. As a result the most promising activities for improvements are determined, an optimal way for performing product innovations is stipulated and new ideas for process innovations are generated.

1. Introduction and objectives

At the Delft University of Technology an interfaculty research project aimed at improving the efficiency of batch process operations was started in 1998. The industrial partners in this research endeavour represent the chemical and metallurgical process industry, the food industry and a number of agro-based companies. The close cooperation with the industrial partners and the identification of their needs for support on integrated plant management result among others in the development of a management decision tool. The objective of this tool is to support management in taking the following decisions:

- Which improvements at which activities in the plant will contribute to the overall objective of the enterprise?
- Which interactions between the activities should be improved to increase their effectiveness as well as their efficiency?
- How to organise product innovations in the existing plant in an optimal way?

To answer these questions a thorough analysis is needed of the company's structure, the goals, the desired performance, the way the activities are performed and the expected improvements. System engineering will be a suitable approach for such an analysis. An industrial site can be seen as a large integrated system, which is built from objects (sub-systems: departments, key activities etc) linked together by material and information flows in a complex structure.

In this paper the implemented ideas for the Management Decision Tool (MDT) will be discussed. The execution of the proposed method will give the plant manager the opportunity to get a quick scan of promising improvements in the batch plant and to manage his plant in an integrated way. Moreover, it will support the plant manager in the creative search for process innovations.

2. The Management Decision Tool

The MDT is applicable to each batch plant. The key model of MDT uses a so-called standard plant containing activities, which take place in almost every batch manufacturing plants, as well as activities that take place in all kind of production environments. To use the MDT in a dedicated plant, the present models need only to be adapted to the specific situation and do not have to be built from the ground.

The MDT consists of four different steps through which the user is guided. After performing all the steps the user has gained more insight in the organisation of the plant with respect to the plant objectives and their relative weights compared to the main objective of the plant, the activities in the plant and their mutual interactions and the most important possibilities for improvement.

3. Objectives of the plant

The first step in the MDT covers a thorough analysis of the objectives of the batch plant. For the MDT an objective tree is developed that will be applicable to most batch plants. In an objective tree the objectives are arranged in a hierarchical way starting from the main objective, which characterizes and defines the area of interest, i.e. the system involved (Keeney, 1992 & 1993). The main objective of the standard batch plant is formulated with respect to the long-term existence of the plant.

This strategic objective is made operational by dividing it into more concrete sub-objectives. The objective tree is split up in this way into three horizontal levels (Keuning, 2000)

1. The *strategic objectives* as regards the organization of the relations between the plant and its environment.
2. The *tactical objectives* as regards the choice of a good structure for the organization and management of the resources in the plant in such a way that the strategic objectives are achieved.
3. The *operational objectives* as regards the use of the resources that are available in the plant in such a way that the tactical objectives are achieved.

Figure 1 shows part of the objective tree as available in the MDT. The user of the MDT could adapt the objective tree to the specific situation in the plant considered. The adaptations will in general be carried out at the level of operational objectives.

4. Activities and their interactions

In the second step of the MDT the activities, which guarantee that the operational objectives of the plant are achieved in practice, are formulated. For the MDT an activity model is developed in which the activities of the standard batch plant have been modelled by using the so-called IDEF0 technique. IDEF0 is a method to model activities and their mutual relations in a hierarchical way (FIPS PUBS 183, 1994).

In the MDT firstly all activities are defined that may influence the measure in which the operational objectives are achieved. These activities have a high degree of detail and they can be combined into domed activities. After a number of these compositions the so-called context activity will be achieved. In the same way the interactions between the activities will be modelled and combined into interactions with less degree of detail.

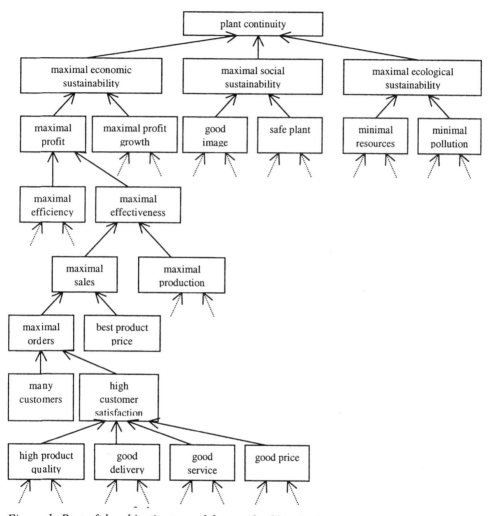

Figure 1: Part of the objective tree of the standard batch plant

The interactions that are distinguished in an IDEF0 model are
1. Input; what is transformed by the activity into the output,
2. Output; what is produced by the activity,
3. Control; what controls the activity,
4. Mechanism; what performs the activity.

676

In the activity model as used in the MDT a fifth flow is being introduced for the objectives to which an activity may contribute. An arrow coming from the activity and pointing upwards will denote this flow.

As an example the objective *high product quality* will be considered. This objective is decomposed into five sub-objectives. For every operational objective several activities are performed to guarantee that the objective is achieved. These activities are denoted by the grey boxes.

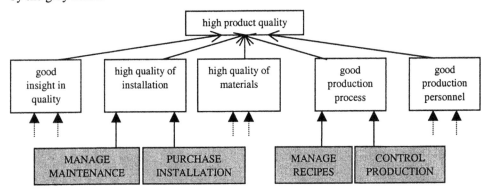

Figure 2: Activities added to the operational objectives

In Figure 3 a small part of the activity model for the standard batch plant is shown, where the interactions between the activities "manage maintenance" and "control production" are central.

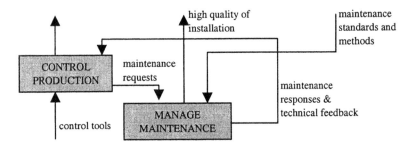

Figure 3: IDEF0 diagram of "Control Production" and "Manage Maintenance"

5. Contribution of the objectives to the overall objective

In the third step of the MDT the user will be asked to order the objectives in the objective tree to the degree in which they contribute to the main objective.

The pairwise comparison method (Saaty, 1980) will be used for the ranking of the objectives. In every group of objectives in the objective tree, i.e. all child objectives of one parent, every pair of two objectives is compared. The user decides for every two objectives which one will yield the highest contribution to the main objective. Doing this for every pair in the group a unique preference list will result. With the formula

$$w_i = \frac{k_i}{\sum\limits_{j=1}^{m} k_j}, \quad \text{where } k_i = \sum\limits_{r=i}^{m} \frac{1}{r} \tag{1}$$

the weight w_i of the ith objective in the preference list is calculated. The relative weight of an operational objective can now be calculated by multiplying its weight with the weights of all its parents.

As an example we look again at the five child objectives of *high product quality*.

    a.   *Good production process*
    b.   *High quality of installation*
    c.   *High quality of materials*
    d.   *Good insight in quality*
    e.   *Good production personnel*

The user is asked to decide for randomly chosen pairs, which of the two is more important to achieve a *high product quality*. Assume that the user decides to the following preference structure: $d \succ c \succ a \succ b \succ e$. With Equation (1) the relative weights of the objectives are:

    1.   *Good insight in quality*, weight: 0.46
    2.   *High quality of materials*, weight: 0.26
    3.   *Good production process*, weight: 0.16
    4.   *High quality of installation*, weight: 0.09
    5.   *Good production personnel*, weight: 0.04

When all weights of the operational objectives are known by multiplying the relative group weights with all the weights of their parents, the objectives with the highest contribution to the main objective of the plant are determined. With the results of the previous step the activities that are responsible for these objectives can be found. For improving the degree in which the main objective is being achieved, the user should focus on the improvement of these activities.

6. Efficiency of activities and possible improvements

In the last step of the MDT the possible improvements of the activities are being mapped. Here a difference is made between the effectiveness of the activity and the efficiency of the activity.

The effectiveness of the activity is defined as the degree in which the activity contributes to its objectives. Most operational objectives can be influenced by more than one activity, as shown in the activity model. To be able to measure the degree in which an objective is achieved, a performance indicator will be linked to the objective. To measure the effectiveness by activity the performance indicator of an objective should be split into several sub indicators that each can be influenced by just one activity.

The efficiency of the activity is defined as the costs or time needed to transform the input of the activity into its output.

As mentioned before the objective of the MDT is to find activities that are good candidates for improvements, i.e. which will yield a high contribution to the main objective of the plant. Measuring the effectiveness of the activity by its performance indicators can determine the actual contribution of the activity to the main objective.

Improvement of the effectiveness of the activity may not be performed at the expense of the efficiency of the activity. With the assistance of the activity model the user can determine which input the activity requires to deliver the right output. The right input on the right moment will improve in general the efficiency as well as the effectiveness of the activity.

With the results of the MDT the user may
1. determine which activities are good candidates for improvement,
2. determine the actual performance of the activities,
3. improve the activities by better tuning of their interactions.

7. Results and final remarks

The MDT is still under development. The modelling phases for objectives and activities are already evaluated and applied in the plants of the industrial partners. At the moment research concentrates on the definition of the contribution of the chosen operational objectives to the overall plant goal. This will be realised by discussions with a selected team of process people (plant managers, quality managers, schedulers and operators) from an industrial plant in the food and chemical industry. Next, several case studies will be performed for further evaluation.

References

IDEF0: FIPS PUBS 183, 1994, Standard for Integration Definition for Function Modelling.

Keeney, Ralph L., 1993, Decisions with multiple objectives; preferences and value tradeoffs. New York Cambridge University Press.

Keeney, Ralph L., 1992, Value-focused thinking; a path to creative decisionmaking. Cambridge, Mass. Harvard University Press.

Keuning, D., D.J. Eppink, 2000, Management en Organisatie, theorie en toepassing, Educatieve partners Nederland BV, (in Dutch).

Saaty, T.L., 1980, The Analytic Hierarchy Process, McGraw-Hill Book Co, New York.

European Symposium on Computer Aided Process Engineering – 12
J. Grievink and J. van Schijndel (Editors)
© 2002 Elsevier Science B.V. All rights reserved.

Comparison of three scheduling methods for a batch process with a recycle

J.Heinonen and F.Pettersson

jukka.heinonen@abo.fi, frank.pettersson@abo.fi

Heat Engineering Laboratory, Department of Chemical Engineering
Åbo Akademi University, Biskopsgatan 8, FIN-20500 Åbo, Finland

Abstract

A short term scheduling problem from an industrial case is modelled in three different ways. A discretized MILP-model, a heuristic approach and a genetic algorithm are tested. The scheduling task is complicated by a recycle connection. Storage capacity for raw materials and intermediates are restricted. Some productions steps are operated continuously and some in batch mode. A one-week schedule and a two-week schedule are constructed with the different models and compared with industrial results. The genetic algorithm outperforms the other approaches, though the computational efforts are larger.

1. Introduction

The significance of effective methods for production scheduling and planning has increased due to a general effort to improve profitability. The difficulties in scheduling tasks arise primarily from sharing of resources and equipment and the desire to operate the process cost-effectively. Most commonly the need for such methods occur in production operating in batch mode. Typically batch processes are used in production of pharmaceuticals, food and specialty chemicals. One of the most important advantages of batch processes is their flexibility. To be able to exploit this advantage, an effective production planning is of great importance. Some process steps (e.g. crystallization) are also better controlled in batch mode compared to their continuous counterparts. In spite of a common trend towards continuous processes, the batch process continues to be the best alternative in several cases.

Methods for short term scheduling using mathematical programming have been studied since the 1970s. The problems are described with either discretized time intervals or continuous time formulations. Although, the computational capacity increases continuously, one problem using mixed integer programming is the combinatorial character of scheduling problems, resulting in large problems with long execution times when problems of industrial relevance are considered. The planning horizon obtained in this work varies from 10 to 15 hours using a state task network formulation, presented by Kondili et al. (1993), when modest computational times are required.
In order to speed up the computation, a heuristic solution method was developed based on an exhaustive search of the most promising steps for each new decision to be made.

680

The drawback of this kind of methods is the need for major changes in the strategy when changes in the process setup occur.

Evolutionary programming, and especially genetic algorithms, has been quite successfully applied to i.e. job shop and travelling salesman problems thus encouraging the development of a genetic algorithm based planning tool for the case problem.

2. Problem description

The process consists of ten pairs of pre- and main reactors operating in batch mode. Two different feeds exist. Processing times in the reactors are considerably long, varying from 30 hours to 40 hours, depending on the selected raw materials. The reactors are fed through four parallel concentration units, operating continuously. After the reactors the product is separated. The remaining part is recycled through a continuously operating reactivation process step back to the concentration units.

The task is to find an optimal sequence of unit operations when the planning horizon is a week or more. The process is operated in campaigns, thus all available resources are dedicated to the current product. The objective is to maximize the product stream, while fulfilling a number of constraints imposed upon the process equipment, e.g. available units capacity limits, storage limits for all intermediates and amount of raw material available. Concentrators should also, if possible, be operated without interruptions. The cistern profiles must also be within certain safety limits. Unexpected changes or disturbances leads to situations where replanning has to be performed, setting conditions on a fast planning method.

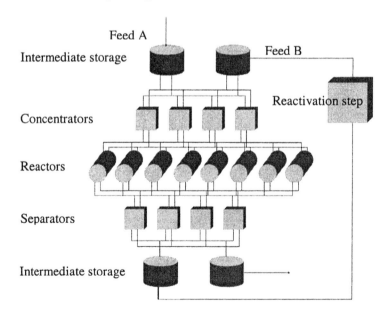

Figure 1. An overview of the studied process.

3. Model development

The concentrators are modelled with four distinct properties in mind. First, they should (if possible) be operated continuously, since interruptions in their performance are costly. Second, since there are a number of different concentrators to choose from, the number of changes in the concentration procedure, once started, should be restricted to maximum one. This is beneficial from an operators point of view since many changes take much of an operators time. Thirdly, the first property should not be allowed to compromise in any way the overall optimum of the solution, meaning that the concentration should still be efficiently sequenced but not by cutting down the number of starting reactors just to ensure continuous concentrator operation. And as fourth, the concentration event into a reactor should be continuous, and not contain any gaps in the resulting sequencing.

The separation step can only separate a single batch at any given time. The MILP-model does not necessary use a FIFO-queue for the separation, since a batch can be put on hold in favour of a batch that might prove to be more advanteguous. The heuristic method and the genetic algorithm use a strict FIFO-queue.

The reactivation step is modelled as a linear input/output flow with the reactivation yield determining the difference in their values.

3.1 The discretized MILP-model

The scheduling of the multistage process was formulated as a state-task network for the MILP-model inspired by Sahinidis and Grossmann (1991) as well as Voudoris and Grossmann (1996). The time interval for the discretization was set at 1 hour. This means that the resulting Gantt-charts contain some crudeness, but it is no real problem since the concentrator flow rates can be regulated accordingly. For detailed information of how the model works see Heinonen and Pettersson (2001). The solution times for different subproblem horizon lengths can be seen in table 1. It is clear that large problems must be divided into smaller subproblems and solved sequentially in order to make schedules with reasonable computational efforts. Schedules with longer timespans than 15 hours are required since the cycle times for the reactors are very long. It can be mentioned here that a continuous MILP-model was tried but due to the recycle step the problem was not easily translated into the formulation proposed by Pinto and Grossmann (1995).

Table 1. Solution times for different subproblem horizon lengths (discretized MILP-model).

horizon (h)	solution time (min)	# binary variables
10	0.18	1000
20	10.48	2000
30	128.87	3000

Solution procedure
The user inputs the current state of the process through a graphical user interface (GUI) (including flow rates, reactor stages and cistern profiles etc.) and a MILP-model is generated on basis of the input state. The model is sent to Cplex, which is a

commercially available solver for MILP-problems. Cplex solves the subproblem and returns the results to the GUI where the next subproblem is automatically generated. The system loops until the required time horizon for the entire schedule is reached and the entire schedule is presented to the user. Gantt-charts for the reactors, concentrators and the separation step are created as well as cistern profiles.

3.2 The heuristic approach

This model utilizes a decision based search-tree in determining the best way to use the concentrators. The unit operation times are floating point variables and the resulting charts are more accurate since no time discretization is needed.

Solution procedure

The basic input task is the same as described above, and one gets similar Gantt-charts upon completion of the run. No separate solver is needed since search trees are standard data structures and can be implemented with ease. The operation alternatives for the concentrator at time t are added to the search tree and the cistern profiles and such are simulated for those decisions. After the simulation step we are at time $t + t_{conc}$, where t_{conc} is the time it takes for the actual concentrator to fill the reactor, and new decisions have to be made. The alternatives are again inserted into the tree and the system loops until the desired schedule length is reached. At average there are three decisions at every decision point in the schedule for the concentrators. Each level in the search tree thus contains 3^n decisions, where n is the tree level, which have to be simulated. This effectively puts an upper bound on the subproblem size. 100 hours has been found as a reasonable length.

3.2 The genetic algorithm

The basic input procedure and Gantt-charts are similar as above. The genetic algorithm (GA) is capable of solving the entire time horizon of one month in one single system though it requires seemingly lengthy calculations. Shorter systems might save some valuable calculation time but may not give as good schedules. All unit operation times are floating point variables and result in a greater accuracy than the discretized MILP-model. No separate solver is used. A single gene is a decision of which reactor and which feed a concentrator should start with. The chromosomes thus use a 'permutation encoding'-type of representation. A chromosome consists of a sequence of genes (unit operations for a concentrator) and all the concentrators and their operations formed together allow the system to be simulated. A fitness value is calculated for each chromosome, which is based upon the amount of resulting products as well as how the cistern profiles behave in the simulated system.

Solution procedure

The algorithm is initialized by creating a set of chromosomes with random genes. Then the schedule implied by the chromosomes is simulated and fitness values calculated. The next generation of chromosomes is determined by means of roulette-wheel selection and single- or two-point crossover. A small mutation rate is kept, as well as elitism. Since the elitist chromosome has a tendency to often be a selected parent, a diversity check is made at regular intervals. If deemed necessary, the mutation rate can be increased by a user chosen amount during a single generation. Thus the GA can effectively break free from local optimums. The system can be set to loop for a specific amount of generations or the user can stop it at will and later restart it from the same point.

4. Illustrative problem

A common starting situation for the three approaches has been taken from an already implemented industrial schedule. A one-week schedule and a two-week schedule are constructed with the different models and compared to the industrial results. The amount of products manufactured while fulfilling all cistern profile- and other constraints is taken as a measure of the schedule's relative goodness. Table 2 shows the results for the scheduled horizon. A resulting two-week Gantt-chart for the process can be seen in figure 2.

Table 2. The amount of manufactured products for the different schedules and models.

	1 week schedule	2 week schedule
Implemented	48.41	98.50
MILP	49.98	97.35
Heuristic	47.83	95.37
GA	50.44	97.73

The GA outperforms both the discretized MILP and the heuristic approach. The relatively good results with the MILP-model, despite the time discretization, can be partially explained by a different approach to the separation step as explained before. The good implemented result for the two-week schedule is possible since in real life feeds are sometimes mixed if a cisternprofile is near its upper extreme. Mixing should be avoided, and the models thus operate under strict rules not to mix the feeds.

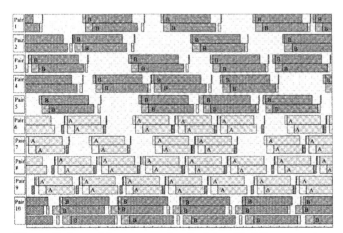

Figure 2. A resulting two-week schedule with full sequencing made with the GA. The letters stand for the different feeds, the sequence in front is the concentration part after which the reaction starts, and the sequence at the end is the cleaning part.

5. Computational experience

Table 2. The average solution times for the different approaches, on a 600MHz PC

	1 week schedule	2 week schedule
MILP	2.45 min	4.60 min
Heuristic	2.26 min	6.82 min
GA	6.77 min	38.25 min

The discretized MILP-model with problem subtasking performs best while the heuristic comes as close second. The GA requires the lengthiest calculations. Since the GA is stochastic by nature, the time required until convergence varies. A two-week schedule could take 20 minutes or it could last for as long as an hour. Two-point crossover was used during the runs for the GA. The times are average from three different runs.

6. Conclusion

The GA not only gives the best schedules, it also allows us to construct schedules with a large enough time horizon in a single step. When speed is an issue (e.g. during a mechanical breakdown when a modified schedule is needed fast) the MILP model in its present form could be preferable. Then again a shorter time horizon could be run with the GA.

The heuristic approach did not meet the expectations on either the computational speed or the obtained solutions. It does, however, cope with reasonable planning horizons without the separation into subproblems as in the MILP approach. It can be argued that longer horizons give better schedules since you can be relatively sure that the process is not driven into unwanted states. This was seen as a problem in the MILP approach.

Acknowledgment - The financial support from TEKES, the Finnish Technology Agency, is gratefully acknowledged.

7. References

Goldberg D.E. 1999, Genetic algorithms in search, optimization & machine learning. Addison Wesley, California.

Heinonen J. and Pettersson F. 2001, Scheduling a process with a recycle and units operating continuously and in batch mode, ICHEAP 5, Florence, Italy.

Kondili E., Pantelides, C.C., and R.W.H. Sargent. 1993, A general algorithm for scheduling batch operations. *Comput. Chem. Eng.* 17, 211.

Pinto J.M. and Grossmann I.E. 1995, A Continuous Time Mixed Integer Linear Programming Model for Short-Term Scheduling of Multistage Batch Plants. *Ind. Eng. Chem. Res.* 34, 3037.

Sahinidis N. V. and Grossmann I. E. 1991, Reformulation of Multiperiod MILP Models for Planning and Scheduling of Chem. Processes, *Computers chem. Eng.* 15, 4, 255.

Voudoris V. T. and Grossmann I. E. 1996, MILP Model for Scheduling and Design of a Special Class of Multipurpose Batch Plants, *Computers chem. Eng.* 20, 11, 1335.

European Symposium on Computer Aided Process Engineering – 12
J. Grievink and J. van Schijndel (Editors)

685

Computer-Aided Design of Redundant Sensor Networks

Georges Heyen[1] , Marie-Noëlle Dumont[1], Boris Kalitventzeff[2]

1 : Laboratoire d'Analyse et Synthèse des Systèmes Chimiques, Université de Liège,
Sart Tilman B6A, B-4000 Liège (Belgium), email G.Heyen@ulg.ac.be
2 : BELSIM s.a., rue Georges Berotte 29A, B 4470 Saint-Georges-sur-Meuse, Belgium

Abstract

A systematic method to design sensor networks able to identify key process parameters with a required precision at a minimal cost is presented. The procedure is based on a linearised model, derived automatically from a rigorous non-linear data reconciliation model. A genetic algorithm is used to select the sensor types and locations.

1. Problem position

The application of data reconciliation to plant monitoring in now considered as standard practice. Redundant measurements allow reducing the uncertainty due to random errors. Unmeasured parameters can be estimated safely from reconciled state variables.

However little has been published on the design of measurement systems allowing to achieve a prescribed accuracy for key process parameters, and to secure enough redundancy to ensure resilience with respect to sensor failures. Madron (1972) solves the linear mass balance case using a graph-oriented method. Bagajewicz (1997) analyses the problem for mass balance networks, where constraint equations are linear.

In this study, we propose a general mathematical formulation of the sensor selection and location problem, in order to reduce the cost of the measurement system while providing estimates of all specified key process parameters within a prescribed accuracy. The goal is to extend the capability of previously published algorithms, and to address a broader problem, not being restricted to flow measurements and linear constraints. For sure minimising the cost is not the only possible objective function: designing a sensor network should better be treated as a multi-objective optimisation problem, and address other features, such as redundancy and resiliency to equipment failure or capability to detect gross errors. This study is a first step toward that goal, and we plan to later refine the technique by selecting a more general objective function.

In the optimisation problem formulation, the major contribution to the objective function is the annualised operating cost of the measurement system. The set of constraint equations is obtained by linearising the process model at the nominal operating conditions, assuming steady state. The process model is complemented with link equations that relate the state variables to any accepted measurements, or to key process parameters whose values should be estimated from the set of measurements. In our case, the set of state variables for process streams comprises all stream temperatures, pressures and partial molar flow rates. In order to handle total flow rate measurements, the link equation describing the mass flow rate as the sum of all partial molar flow rates weighted by the component's molar mass has to be defined. Similarly, link equations relating the molar or mass fractions to the partial molar flow rates have also to be added for any stream where an analytical sensor can be located.

Link equations have also to be added to express key process parameters, such as heat transfer coefficients, reaction extents or compressor efficiencies.

In the present study, we assume that all variables are measured; those that are actually unmeasured will be handled as measured variables with a large standard deviation. Thus data reconciliation, taking linearised constraints into account, requires the solution of :

$$\min (X-X')^T W (X-X') \tag{1}$$
$$\text{s.t.} \quad A X + D = 0$$

where X is the array of process variables (size m), X' is the set of measured values and $W = \text{diag}(1/\sigma_i^2)$ is the weight matrix (the diagonal terms of the inverse of the measurement covariance matrix). The linear approximation of the constraints is easily obtained from the Jacobian matrix A of the non-linear model evaluated at the solution. Constraints are handled using Lagrange formulation:

$$\min_{X,\lambda} L = (X-X')^T W (X-X') + 2\lambda^T (A X + D) \tag{2}$$

Solving for stationarity conditions:

$$\begin{bmatrix} X \\ \lambda \end{bmatrix} = \begin{bmatrix} W & A^T \\ A & 0 \end{bmatrix}^{-1} \begin{bmatrix} WX' \\ -D \end{bmatrix} = M^{-1} \begin{bmatrix} WX' \\ -D \end{bmatrix} \tag{3}$$

Matrix M can easily be built, knowing the variance of measured variables appearing in sub matrix W, and the model Jacobian matrix A (which is constant). This matrix will be modified when assigning sensors to variables. Any diagonal element of matrix W will remain zero as long as a sensor is not assigned to the corresponding process variable; it will be computed from the sensor precision and the variable value when a sensor is assigned, as shown later in section 2.3. In fact we are not interested in the solution of system (3), since measured values X' are not known. The variance of reconciled values X is related to the variance of measurements X' as shown in Heyen et al. (1996):

$$\text{var}(X_i) = \sum_{j=1}^{m} \frac{([M^{-1}]_{ij})^2}{\text{var}(X'_j)} \tag{4}$$

The elements of M^{-1} are obtained by calculating a LU factorisation of matrix M. In case matrix M is singular, we can conclude that the measurement set has to be rejected, since it does not allow observing all the variables. Row i of M^{-1} is obtained by back substitution using the LU factors, using a right hand side vector whose components are δ_{ij} (Kronecker factor: $\delta_{ij}=1$ when i=j, $\delta_{ij}=0$ otherwise).

In the summation of equation (4), we only take into account variables X'_j that have been assigned a sensor, the variance of unmeasured variables being set to infinity.

2. Algorithm description

Solution of the sensor network problem is carried out in 7 steps:
1. Process model formulation and definition of link equations.
2. Model solution for the nominal operating conditions and model linearisation.
3. Specification of the sensor database and related costs.
4. Specification of the precision requirements for observed variables.
5. Verifications of problem feasibility.
6. Optimisation of the sensor network.

7. Report generation.

Each of the steps is described in details before presenting a test case.

2.1 Process model formulation and definition of link equations.

This is easily done in the Vali 3 data reconciliation software, which is used as the basis for this work (Belsim 2001). The model is formulated by drawing a flowsheet using icons representing the common unit operations, and linking them with material and energy streams. Any acceptable measurement of a quantity that is not a state variable (T, P, partial molar flow rate) requires the definition of an extra variable and the associated link equation, what is done automatically for standard measurement types (e.g. mass or volume flow rate, density, dew point, molar or mass fractions, etc). Similarly, extra variables and link equations must be defined for any process parameter to be assessed from the plant measurements. A proper choice of extra variables is important, since many state variables cannot be measured in practice (e.g. no device exists to directly measure a partial molar flow rate or an enthalpy flow).

To solve the model, enough variables need to be set by assigning them values corresponding to the nominal operating conditions. The set of specified variables must match at least the degrees of freedom of the model, but overspecifications are allowed, since a least square solution will be obtained by the data reconciliation algorithm.

2.2 Model solution for the nominal operating conditions and model linearisation.

The data reconciliation problem will be solved, either using a large-scale SQP solver, or the Lagrange multiplier approach. When the solution is found, the value of all state variables and extra variables is available, and the sensitivity analysis is carried out (Heyen et al, 1996). A dump file is generated, containing all variable values, and the non-zero coefficients of the Jacobian matrix of the model and link equations. All variables are identified by a unique name indicating its type (e.g. S32.T is the temperature of stream S32, E102.K is the overall heat transfer coefficient of heat exchanger E102, S32.MFH2O is the molar fraction of component H_2O in stream S32).

2.3 Specification of the sensor database and related costs.

A data file must be prepared that defines for each sensor type the following parameters:

- the sensor name
- the annualised cost of operating such a sensor
- parameters a_i and b_i of the equation allowing to estimate the sensor accuracy σ_i from the measured value x_i', according to the relation: $\sigma_i = a_i + b_i x_i'$
- a character string pattern to match the name of any process variable that can be measured by the given sensor (e.g. a chromatograph will match any mole fraction, thus will have the pattern MF*, while an oxygen analyser will be characterized by the pattern MFO2)

2.4 Specification of the precision requirements for observed variables.

A data file must be prepared that defines the following information for the selected key performance indicators or for any process variable to be assessed:

- the composite variable name (stream or unit name + parameter name)
- the required standard deviation σ_i^t, either as an absolute value, or as a percentage of the measured value.

2.5 Verification of problem feasibility.

Before attempting to optimise the sensor network, the programme first checks the existence of a solution. It solves the linearised data reconciliation problem assuming all possible sensors have been implemented. This provides also an upper limit C_{max} for the cost of the sensor network.

A feasible solution is found when two conditions are met:

- the problem matrix **M** is not singular;
- the standard deviation σ_i of all selected reconciled variables is lower than the specified value σ_i^t.

When the second condition is not met, several options can be examined:

- add more precise instruments in the sensor definition file;
- extend the choice of sensors by allowing measurement of other variable types;
- modify the process definition by adding extra variables and link equations, allowing more variables besides state variables to be measured.

2.6 Optimisation of the sensor network.

Knowing that a feasible solution exists, we can start a search for a lower cost configuration. The optimisation problem as posed involves a large number of binary variables (in the order of number of streams * number of sensor types). The objective function is multimodal for most problems. However identifying sets of suboptimal solutions is of interest, since other criteria besides cost might influence the selection process. Since the problem is highly combinatorial and not differentiable, we attempted to solve it using a genetic algorithm (Goldberg, 1986). The implementation we adopted is based on the freeware code developed by Carroll (1998). The selection scheme used involves tournament selection with a shuffling technique for choosing random pairs for mating. The evolution algorithm includes jump mutation, creep mutation, and the option for single-point or uniform crossover.

The sensor selection is represented by a long string (gene) of binary decision variables (chromosomes); in the problem analysis phase, all possible sensor allocations are identified by finding matches between variable names (see section 2.2) and sensor definition strings (see section 2.3). A decision variable is added each time that a match is found. Multiple sensors with different performance and cost can be assigned to the same process variable.

The initial gene population is generated randomly. Since we know from the number of variables and the number of constraint equations what is the number of degrees of freedom of the problem, we can bias the initial sensor population by fixing a rather high probability of selection (typically 80%) for each sensor. We found however that this parameter is not critical. The initial population count does not appear to be critical either. Problems with a few hundred binary variables were solved by following the evolution of populations of 10 to 40 genes, 20 being our most frequent choice.

Each time a population has been generated, the fitness of its members has to be evaluated. For each gene representing a sensor assignment, we can estimate the cost C of the network, by summing the individual costs of all selected sensors. We also have to build the corresponding matrix M (equation 3) and factorise it. In an initial version of the code, we used SUBROUTINE MA29AD from Harwell library (1984), aimed to factorise a general symmetric matrix. However later experience with larger test problems indicated that taking advantage of the sparsity of M matrix could save significant computer time. Indeed using Belsim's sparse matrix code reduced the computer time by a factor of 25 for a problem involving 312 variables.

The standard deviation σ_i of all process variables is then estimated using equation 4.

This allows calculating a penalty function P that takes into account the uncertainty affecting all observed variables:

$$P = \sum_{i=1}^{m} P_i \qquad \text{where} \quad P_i = \frac{\sigma_i}{\sigma_i^t} \quad \text{when} \ \sigma_i \leq \sigma_i^t \qquad (5)$$

$$P_i = 0.01 \min\left(10, \frac{\sigma_i}{\sigma_i^t} \right)^2 \quad \text{when} \ \sigma_i > \sigma_i^t \qquad (6)$$

The fitness function F of the population is then evaluated as follows:

- if matrix M is singular, return $F = - C_{max}$
- otherwise return $F = -(C + P)$

Penalty function (5) increases (slightly) the merit of sensor networks that perform better than specified. Penalty function (6) penalises genes that do not meet the specified accuracy, but it does not reject them totally, since some of their chromosomes might code interesting sensor sub networks.

The population is submitted to evolution according to the mating, crossover and mutation strategy. Care is taken that the current best gene is always kept in the population, and is duplicated in case it should be submitted to mutation. After a specified number of generations, the value of the best member of the population is monitored. When no improvement is detected for a number of generations, the current best gene is accepted as a solution. There is no guarantee that this solution is an optimal one, but it is feasible and (much) better than the initial one.

2.7 Report generation.

The program reports the best obtained configurations, as a list of sensors assigned to process variables to be measured. The predicted standard deviation for all process variables is also reported, as well as a comparison between the achieved and target accuracies for all key process parameters.

3. An example

As an example we tried to design a sensor network for an ammonia synthesis loop. The process involves a 5-component mixture (N_2, H_2, NH_3, CH_4, Ar), 13 units, 19 process streams, 10 utility streams (cooling water, refrigerant, boiler feed water and steam). The units are: 2-stage compressor with 2 intercoolers, recycle mixer, recycle compressor, reactor preheater, ammonia converter, waste heat boiler, water cooled condenser, ammonia cooled condenser, vapour-liquid separator, purge divider and flash drum for expanded ammonia condensate.

The model involves 181 variables and 131 constraint equations. Accuracy targets are specified for 45 variables. It includes extra measurable variables (molar fractions) and some unit parameters to be monitored (e.g. heat exchange transfer coefficient, compressor efficiency, and extent of reaction, departure from equilibrium),

The sensor database allows choosing between sensors of different accuracy and cost, namely 2 temperature sensors, 3 flow meters, 2 composition analysers and 2 types of pressure gauges. The program detects that up to 166 sensors could be installed. Thus the solution space involves $2^{166} = 9.3 \times 10^{49}$ solutions (most of them being unfeasible).

We let the search algorithm generate 4000 populations, which required 80000 function evaluations and 776 CPU seconds (PC with 1.33 GHz AMD Athlon processor, program compiled with Compaq Fortran compiler, only local optimisation). The initial cost function was $C_{max}=17340$ cost units with all sensors selected. Optimisation brought it down to 1110 cost units. A solution within 60 cost units of the final one was attained after 5700 evaluations (in less then one minute).

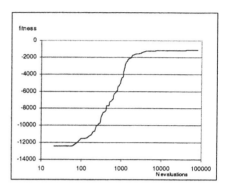

4. Conclusions and future work

The available software could be improved by allowing more flexibility in the sensor definition (e.g. defining acceptable application ranges for each sensor type), or by implementing different objective functions besides the cost. Possible objectives could address the resiliency of the sensor network to equipment failures, or the capability to detect gross errors, in the line proposed by Bagajewicz (2001).

There is no guarantee that this solution found with the proposed method is an optimal one, but it is feasible and (much) better than the initial one. Thus we claim that this algorithm contributes to the rational design of sensor networks.

References

Bagajewicz M.J., 1997, chapter 6 in "Process Plant Instrumentation: Design and Upgrade", Technomic Publishing Company.

Bagajewicz M.J., 2001, Design and Retrofit of Sensor Networks in Process Plants, AIChE J., 43(9), 2300-2306

Belsim., 2001, VALI 3 User's Guide, Belsim s.a, B 4470 Saint-Georges-sur-Meuse, Belgium

Carroll D.L., 1998, FORTRAN Genetic Algorithm Driver version 1.7, download from <http://www.staff.uiuc.edu/~carroll/ga.html>

Goldberg, D.E.,1989,"Genetic Algorithms in Search, Optimization and Machine Learning," Addison-Wesley

Heyen G. , Maréchal E. , Kalitventzeff B., 1996, Sensitivity Calculations and Variance Analysis in Plant Measurement Reconciliation, Computers and Chemical Engineering, vol. 20S, 539-544

Madron F., 1992, section 6.3 in "Process Plant Performance Measurement and Data Processing for Optimization and Retrofits", Ellis Horwood

European Symposium on Computer Aided Process Engineering – 12
J. Grievink and J. van Schijndel (Editors)
691

Improving Gelatin Plant Productivity By Modelling The Demineralization Process Of Animal Bones

D.A. Horneman[1], M. Ottens[1], M. Tesson[2], L.A.M. van der Wielen[1]

[1]Kluyver Laboratory for Biotechnology, Delft University of Technology, Julianalaan 67, 2628 BC, Delft, The Netherlands
[2]Delft Gelatin BV, P.O. Box 3, 2600 AA Delft, The Netherlands

Abstract

A dynamic model is developed that simulates a large size, full scale animal bone demineralization process, which is operated as a simulated moving bed system. The model at the particle level, is based on the unreacted shrinking core model. To solve the system of couples PDE's and ODE's the simulating package gPROMS is used. The model described is succesfully applied to find the optimal process conditions and to increase the productivity of the existing significantly.

1. Introduction

Gelatin is derived from collagen, the primary protein component of animal connective tissues like bone, skin and tendons (Croome and Clegg, 1965, Mark et al. 1987) Outside the United States, cattle bones are the most important source for collagen (Babel et al., 2000). Current industrial practice involves a first bone extraction step with hot water to reduce the fat content to approximately 1 %. After this step the bone particles contain on average 20-25 % of collagen. The principal non-collagen component of bone is the mineral salt tri-calcium phosphate $Ca_3(PO_4)_2$, which is extracted by a dilute acid stream (usually 5 % hydrochloric acid HCl (Makarewicz et al., 1980)). This extraction step is called the demineralization and is described by the following stoichiometric equations:

$$Ca_3(PO_4)_2^S + 4\,HCl^L \rightarrow Ca(H_2PO_4)_2^L + 2\,CaCl_2^L \qquad (1)$$
$$CaCO_3^S + 2\,HCl^L \rightarrow CO_2^G + H_2O^L + CaCl_2^L \qquad (2)$$

HCl diffuses into the bone particles and reacts with $Ca_3(PO_4)_2$ to form mono-calcium phosphate $Ca(H_2PO_4)_2$ and calcium chloride $CaCl_2$ that are soluble in the acid stream. This demineralization of bone is considered to be diffusion limited, with the reactions considered being almost instantaneous and complete (Croome and Clegg, 1965, Makarewicz et al., 1980).

Delft Gelatin BV, a Dutch gelatin producer, uses serial reactors for the demineralization of shredded animal bone "particles" (Figure 1). It is possible to have real counter current contact between the solid bone particles and the fluid by appropriately connecting and scheduling the reactors to the acid stream, like in a simulated moving bed system (Schulte and Strube, 2001). In these systems, the solid phase is *not* actually moving but the countercurrent motion is simulated. The acid stream flows through the reactors starting in the reactor with the most highly demineralized bone particles and ending in the reactor filled with fresh, non-demineralized bone particles. Upon

692

completion of the demineralization in the first reactor, this reactor is emptied and refilled with fresh bone particles and placed at the end of the series.

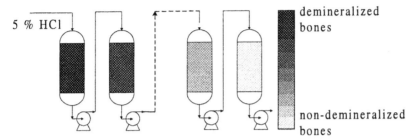

Figure 1. Schematic layout of the reactors system

The demineralization process has not changed much over the last 100 years. It is described in this study with the use of a dynamic mathematical model using the shrinking core model at the bone particle level. The purpose of this study is to show that this model is in good agreement with the real process and to use it for process optimization.

2. Theory

To describe the demineralization process, some general assumptions are made. Firstly, the demineralization is limited by internal diffusion with the reactions (1) and (2) being instantaneous and complete (Makarewicz *et al.*, 1980), (Croome and Clegg, 1965)). Secondly, the demineralization process does not influence the size and the shape of the bone particles. In this way, a growing demineralized layer is formed around an unreacted core separated by the sharp reaction front (Makarewicz *et al.*, 1980). This is described by the unreacted shrinking core model ((Levenspiel, 1972). Figure 2 gives a schematic representation of the unreacted shrinking core model.

The flux of HCl, J_{HCl}, in (moles/m^{2}s) diffusing into the bone particle is described by Fick's first law:

$$J_{HCl} = -\frac{dN_{HCl}}{dt} = 4\pi r^2 D_e \frac{dC_{HCl}}{dr} \tag{3}$$

where N_{HCl} is the amount of HCl in moles, D_e is the effective diffusion coefficient (m^2/s) and r is radial position in the bone particle (m) (Figure 2).

The change in mineral amount can be described in terms of radial position of the core boundary:

$$-dN_{min} = -\beta \, dN_{HCl} = -C_{min} d(\tfrac{4}{3}\pi r_c^3) = -4\pi \, C_{min} r_c^2 \, dr_c \tag{4}$$

where β is the stoichiometric coefficient, C_{min} is the concentration of $Ca_3(PO_4)_2$ in moles per m^3 and r_c the radius of the non-demineralized core (Figure 2).

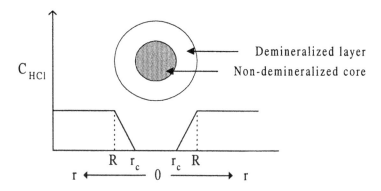

Figure 2. The unreacted shrinking core model

If the fraction of demineralized bone equals X than:

$$1-X = \frac{volume\,of\,non\,reacted\,core}{total\,volume\,of\,particle} = \frac{\frac{4}{3}\pi r_c^{\,3}}{\frac{4}{3}\pi R^3} = \left(\frac{r_c}{R}\right)^3 \tag{5}$$

The two balances (eq. 4 and 5) can now be written in terms of fraction demineralized bone:

$$-\frac{dN_{HCl}}{dt} = 4\pi D_e\,C_{HCl,bulk}\,R\,\frac{(1-X)^{\frac{1}{3}}}{1-(1-X)^{\frac{1}{3}}} \tag{6}$$

$$\frac{dX}{dt} = \frac{3b}{C_{min}R^2}\,D_e C_{HCl}\,\frac{(1-X)^{\frac{1}{3}}}{1-(1-X)^{\frac{1}{3}}} \tag{7}$$

2.1 Demineralization in packed bed reactors
The reactors used for demineralization are packed with the bone particles, which have a relatively broad size distribution, and also may differ in water and mineral content. To increase the accuracy of the calculations, the particles are divided into N fractions with different properties of the bone particles.
The balances for these packed reactors now become:

$$\frac{\partial C_{HCl}(t,z)}{\partial t} = -\frac{\Phi_v \partial C_{HCl}(t,z)}{\varepsilon A \partial z} - \sum_{n=1}^{N}\frac{4\pi R_n a_n D_e}{\varepsilon}\,C_{HCl}(t,l)\,\frac{(1-X_n(t,z))^{\frac{1}{3}}}{1-(1-X_n(t,z))^{\frac{1}{3}}} \tag{8}$$

$$\frac{dX_n(t,z)}{dt} = \frac{3\beta}{C_{min}R_n^2} D_e C_{HCl}(t,z) \frac{(1-X_n(t,z))^{\frac{1}{3}}}{1-(1-X_n(t,z))^{\frac{1}{3}}} \qquad z \in (0,L] \qquad (9)$$

where ε is the fraction of liquid in the reactor, a_n the number of bone particles of fraction n in the reactor and L the length of the reactor bed.

3. Modeling in gPROMS

For the calculations we have used the flexible scheduling and simulation package gPROMS. The first part of the model is a description of the demineralization in one reactor. This is done by the N partial differential equations (PDE's) of equation 7 and N ordinary differential equations (ODE's) of equation 8 where N is de number of bone fractions. The reactors are coupled by:

$$C_{HCl,in}^i = C_{HCl,out}^{i-1} \qquad\qquad i \in [2,\dots,NR] \qquad\qquad (12)$$

where NR is the number of reactors.

The reactors are switched mathematically, moving the content of reactors $i+1$ to i:

$$Q_z^i = Q_{z,old}^{i+1} \qquad\qquad i \in [1,2,\dots,NR-1] \qquad\qquad (13)$$

where Q_z^i, gives the value of variable or parameter Q in reactor i at length z after switching and $Q_{z,old}^i$ gives the value of variable or parameter Q in reactor i at length z before switching. After the switching procedure, new values are given to the parameters and variables in the new reactor (reactor NR).

The PDE's are solved with the methods of lines. In this method, the equations are rewritten into ODE's by using a discretization method. In this case, a first order forward discretization method is used.

4. Process characteristics

The dots in figure 3 show the fraction of undemineralized bones, Y, in top of the reactor as function of time. From this profile, it is possible to determine the demineralization time of the bone particles in the top of the reactor. The demineralization starts when Y decreases and is finished when $Y = 0$.

In the real process, the demineralization is followed by monitoring the density of the liquid flows in and out each reactor. Figure 3 shows an example for one of the reactors. The dashed line gives the density of the incoming flow and the solid line of the outcoming flow. The reactor is first placed at the end of the series of reactors. The incoming flow has a high density due to the high concentration of $Ca(H_2PO_4)_2$, which has been produced in the former reactors. During the process the reactor moves forward into the series. At a certain time the incoming flow of the reactor contains less

$Ca(H_2PO_4)_2$ resulting in a lower density. At this point, the demineralization starts in this reactor. The density of the flow out of the reactor is still high due to the production of $Ca(H_2PO_4)_2$ in the reactor, but decreases when the production of $Ca(H_2PO_4)_2$ decreases. At this point, the demineralization in the next reactor will start. In the end, the density of both flows will be the same, this is the end of the demineralization. The reactor is now the first one in the series, it will be emptied and refilled again and placed back at the end of the series.

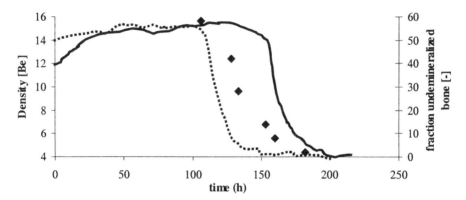

Figure 3. Density profiles of incoming (dashed line) and outcoming (solid line) flow and the ash content of the bone particles in top of the reactor (diamonds) as function of time.

The model calculates the fraction of undemineralized bones in the reactor but it is difficult to measure this continuously during the real process. Therefore Y is compared with the measured density profiles.

5. Results modeling

To simulate the real process the incoming flow and HCl concentration in this flow were measured as function of time and used as input variables in the model. All properties of the bone particles were given by Delft Gelatin. The particle diameter is between the 4 and 14 mm, the water content is about 9.6 wt % and the ash content (fraction non minerals) is about 34.5 wt %. The effective diffusivity was experimentally determined. Figure 4 shows the result of the model (line) with the measured density profile (squares and diamonds) of the real process. The left Y-axes gives a value of the fraction of undemineralized bone:

$$Y = 1 - \sum_{n=1}^{N} x_n \cdot X_n \qquad (14)$$

Figure 4 shows that the simulated profiles are in good agreement with the measured density. Therefore the model could be used for the optimization of the demineralization process. On the basis of the developed model it proved to be possible to increase the plant productivity substantially.

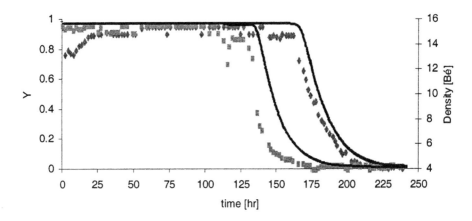

Figure 4. The measured density (squares and diamonds) of the flows in and out a reactor together with the model results (line).

6. Conclusion

The developed dynamical model, based on the unreacted shrinking core model, accurately describes the real large scale demineralization process. It has been applied succesfully for determination of the optimal process conditions to increase the productivity.

References

Babel, W., Schulz, D., Giesen-Wiese, M., Seybold, U., Gareis, H., Dick, E., Schrieber, R., Schott, A., Stein, W., 2000. Ullmann's Encyclopedia of Industrial Chemistry (6th ed.). 2000 Electronic Release.
Croome R.J., Clegg F.G., 1965, Photographic gelatin. The Focal Press, New York.
Levenspiel O., 1972, Chemical Reaction Engineering. John Wiley & Sons Inc., New York.
Makarewicz P.J., Harasta P.J., Webb S.L., 1980, The journal of Photographic Science. 28, 177.
Mark, H.F., Gaylord, N.G., Bikales, N.M., 1987, Encyclopedia of polymer Science and Engineering, vol 7, New York.
Schulte, M. and Strube J., 2001, J. Chromatogr. A. 906, 399.
Westerterp, K.R., Swaaij, W.P.M., van, Beenackers, A.A.C.M., 1987, Chemical Reactor Design and Operation. John Wiley & Sons, New York.

European Symposium on Computer Aided Process Engineering – 12
J. Grievink and J. van Schijndel (Editors)
© 2002 Elsevier Science B.V. All rights reserved.

A Mixed Integer Optimization Strategy for Integrated Gas/Oil Production

V. D. Kosmidis, J. D. Perkins and E. N. Pistikopoulos[1]
Centre for Process System Engineering, Department of Chemical Engineering ,
Imperial College, London SW7 2BY, U.K.

Abstract

The paper describes a mixed integer optimization formulation for the well allocation/operation of integrated gas/oil production systems, which takes into account the interactions between the reservoir, the pipeline network and the surface separation facility. To address the complexity and high dimensionality of the resulting optimization model, an efficient approximation solution strategy is proposed, which is based on outer approximation principles and illustrated with an example problem.

1. Introduction-problem statement

An integrated oil and gas production system comprises (i) the reservoir, which is defined as an accumulation of oil and gas in porous permeable rock, (ii) the wells, (iii) the headers, where the well streams are mixed, (iv) the flow lines which connect the headers to the separators, and (v) the separator facilities where the fluid is separated in oil, gas and water. Each well has a valve, the choke, which is used to control its flow rate; the region around the well inside the reservoir is called the well bore; finally, the wells, headers and flow lines define the pipeline network, as shown in Figure 1. The well allocation/operation problem of such an integrated oil and gas production system, as the one shown in figure 1 can be stated as follows: given (i) a set of wells, which could be either connected to headers or to separators, and (ii) a set of headers which could be connected to separators, the goal is to determine (i) the interconnections of wells to headers and flow lines to separators, and (ii) the corresponding well flow rates, which maximize oil production, at a particular time instant, while satisfying the underlying governing equations, such as mass, energy and momentum balances, and operational constraints, such as separator capacity constraints, well flow rate (upper and lower) bounds, maximum number of interconnection changes, etc. Previous attempts to address this problem include: (i) heuristic-based decomposition strategies which typically employ rules of thumb, such as 'shut in a well if it's water production violate an upper bound', 'allocate high gas producing wells to high pressure separator' etc (Fentor, 1984, Litvak et al, 1997), and (ii) mathematical programming approaches which only deal with simplified or special cases of the problem at hand (Fuji and Horne, 1994, Fang and Lo, 1996, Handley-Schachler et al, 2000). In this paper, we present a novel mixed integer optimization formulation and an efficient decomposition strategy for the solution of well allocation/optimization problems.

[1]Corresponding author. Tel: +44 20 75946620
Fax:+44 20 75946606. Email: e.pistikopoulos@ic.ac.uk

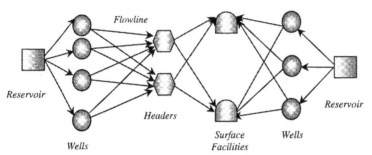

Figure 1. Components of an integrated oil and gas production system

2. Modelling issues

In order to develop a mathematical model to describe the discrete (infrastrucutre) and continuous (flow rate requirements) characteristics of the well allocation/operation problem, we firsrt introduce the following set of 0-1 binary variables:

$$y_k = \begin{cases} 1 \text{ if there is flow through well } k \ (k = 1,..., K) \\ 0 \text{ otherwise} \end{cases}$$

$$y_{k,n} = \begin{cases} 1 \text{ if well } k \ (k = 1,...,k) \text{ is connected to header } n \ (n = 1,..., N) \\ 0 \text{ otherwise} \end{cases}$$

$$y_{n,i} = \begin{cases} 1 \text{ if header } n \ (n = 1,..., N) \text{ is connected to separator } i \ (i = 1,..., I) \\ 0 \text{ otherwise} \end{cases}$$

The proposed mathematical model (see Kosmidis et al, for details) then includes (i) mass balances around the wells, headers and separators (ii) relationships regarding the flow rate of gas, water and liquid exiting the reservoir into each well (the well bore model, Litvak et al, 1997), (iii) relationships relating the flow through a well as a function of pressure and its design diameter (the choke model, Sachdeva, 1986), and (iv) logical, mixed-integer continuous constraints, such as for example, the requirement that each header be connected to one separator, etc. The mathematical model for the small pipeline network shown in figure 2 is given in Appendix I. Note that it involves (i) both algebraic and differential equations, and (ii) continuous and 0-1 binary variables, i.e. it corresponds to a mixed-integer dynamic optimization (MIDO) problem (Bansal et al 2000).

3. Solution Procedure

As the number of wells and headers increase the dimensionality of both continuous and integer space becomes prohibitly large for the application of recently proposed state-of the-art MIDO algorithms, especially for realistic oil fields with tens to hundreds of wells. Therefore, we propose instead an approximation solution strategy, which is based on the following ideas:

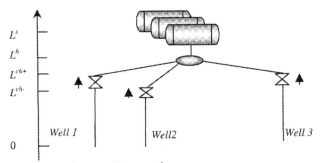

Figure 2. Three well network.

(i)   Projection of the well momentum balance and the choke valve equations to the space of header pressure and well flow rate (P, Q_o^U). The projection is achieved by setting the choke fully open and discretizing the header pressure. For each value of the header pressure, P, the well bore, well momentum balance and choke models comprise a square system of equations with its solution corresponding to the maximum well oil rate Q_o^U (fully open choke).

These pairs (P, Q_o^U) of discrete values are then approximated by a polynomial expression, which represents the well hydraulic constraints in the network and replaces the well momentum balance and choke models in the optimization model. Note that the well pressure constraints will be satisfied for any well flowrate below the maximum by reducing the choke diameter.

(ii)   Solution of the momentum balance equations of the header pressures for different values of the separator pressure, liquid flowrate, gas oil ratio and water cut, (see notation in Appendix) from which hydraulic look up tables can be developed (Litvak et al, 1997). Based on these tables, function and derivative evaluations can be performed simply by interpolation.

Considering the above approximations, the MIDO formulation of Appendix (i) can be recast as a mixed integer nonlinear optimization (MINLP) problem, as shown in Appendix (ii), which can be solved with an outer approximation algorithm (Grossmann and coworkers, 1990, 1996).

4. Illustrative example

The well characteristic of the three well pipeline network of Figure 2 is given in Table 1. The MINLP model of Appendix (ii) is then solved resulting in an optimal configuration for maximum oil production, which involves only two wells with the third one shut in, as shown in Table 2. It is interesting to compare the optimal solution with the one generated by the application of the heuristic rule, which states that for maximizing oil production the wells chokes ought to be fully open. As also shown in Table 2, the comparison clearly demonstrates that (i) heuristic rules may lead to suboptimal strategies, and (ii) an increase in oil production of 175 *barrels/day* is observed by applying the proposed optimization strategy. Furthermore, it must be noticed that (i) the optimal solution of the proposed strategy is always feasible, when it is applied to the exact system due to certain concavity properties (see Kosmidis et al for details) and (ii) the optimal solution of the proposed strategy compared to that of the exact problem is extremely closed and mainly depends on the discretization of the header pressure (see Kosmidis et al for details).

Table 1. Well characteristics of the three well network.

Reservoir / pipe parameters	Well 1	Well 2	Well3	Flowline
Reservoir Pressure (*psia*)	2370	4650	4250	
Productivity Index (*stb/psia day*)	3.0	9.0	3.3	
GOR (*scf/stb*)	5100	1900	1600	
WC	0.93	0.165	0.15	
Vertical length (*ft*)	8000	6000	7000	22000 ft
Horizontal length (*ft*)	6000	4000	3000	0
Diameter (*in*)	3 in	3 in	3 in	6 in
Roughness	0.0001	0.0001	0.001	0.0001
Flow rate upper bound (*stb/day*)	1600	10000	5300	
Flow rate lower bound (*stb/day*)	200	530	470	

Table 2. Maximum oil production.

Structure	Objective function (*barrels/day*)
(y_1, y_2, y_3)=(*1,1,1*)	11929.2 (Heuristic)
(y_1, y_2, y_3)=(*0,1,1*)	12104.2 (optimization)

5. References

Bansal, V., Perkins, J. D., Pistikopoulos, E.N., Ross, R., and van Schijndel, J. M.G. Simulatneous Design and Control Optimization under Uncertainty. *Comp. Chem. Eng.* 2000, 24, 261.

Dutta-Roy, K. and Kattapuram, J. A New Approach to Gas-Lift Allocation Optimization. *SPE 38333*, 1997.

Fang, W. Y. and Lo, K. K. A Generalized Well-Management Scheme for Reservoir Simulation. *SPERE*. 1996, 14, 116.

Fentor, D.J. A Multi-Level Well Management Program for Modeling Offshore Oil Facilities. *SPE 12964*, 1984.

Fuji, H. and Horne, R. N. Multivariate Optimization of Networked Production Systems. *SPE 27617*, 1994.

Handley-Schachler, S., McKie, and Quintero, N. New mathematical Techniques for Optimisation of Oil and Gas Production Systems. *SPE 65161*, 2000.

Kosmidis, V., Perkins, J., and Pistikopoulos, E. N. *IRC technical report* (manuscript in preparation), 2001.

Litvak, M., Clark, B., Farichild, J., Fossume, M., MacDonald, C. and Wood, A. Integration of Prudhoe Bay Surface Pipeline Network and Full Field Reservoir Models. *SPE 38895, 1997.*

Sachedeva C. Two Phase Flow Through Chokes. *SPE 15657, 1986.*

Türkay, M., and Grossmann, I. E. Logic –based MINLP algorithms for the optimal synthesis of process networks. *Comp. Chem. Eng*. 1996, 20, 959.

Viswanathan, J., and Grossmann, I. E. A combined penalty function and outer-approximation method for MINLP optimization. *Comp. Chem. Eng*. 1990, 14, 769.

Appendix. Mixed integer optimization formulation

(i) Original mixed integer dynamic optimization model

$$max\ q_o$$

$$s.t$$

$$
\left[
\begin{array}{l}
\left[
\begin{array}{l}
y_k \\
q_{o,k} = f_o(\,P_k^{wf}(0)\,) \\
GOR_k = f_g(\,q_{o,k}\,) \\
WC_k = f_w(\,q_{o,k}\,) \\
q_{g,k} = GOR_k q_{o,k} \\
q_{w,k} = WC_k q_{L,k} \\
q_{L,k} = q_{o,k} + q_{w,k}
\end{array}
\right\}well\ bore \\[1em]
\left[
\begin{array}{l}
\dfrac{dP_k^{wf}}{dL} = f(\,P_k^{wf},GOR_k,WC_k,q_{L,k}\,),\quad L\in[0,L_k^{ch-}] \\
q_{L,k} = f(\,d_k,P_k^{ch,in},P_k^{ch,out}\,) \\
P_k^{ch,in} = P_k^{wf}(L_k^{ch-}) \\
P_k^{ch,out} = P_k^{f}(L_k^{ch+}) \\
\dfrac{dP_k^{f}}{dL} = f(\,P_k^{f},GOR_k,WC_k,q_{L,k}\,),\quad L\in[L_k^{ch+},L_h] \\
P_k^{f}(L_n) = P_n(L_n)
\end{array}
\right\}well\ momentum\ \&\ choke
\end{array}
\right]
\vee
\left[
\begin{array}{c}
\neg y_k \\
Bx=0
\end{array}
\right],k=1,2,3
$$

$$
\left.
\begin{array}{l}
\dfrac{dP_n}{dL} = f(\,P_n,GOR_n,WC_n,q_L\,),\quad \forall L\in[L_n,L_i] \\[1em]
P_i = P(L_i)
\end{array}
\right\}flowline\quad momentum
$$

$$
\left.
\begin{array}{l}
GOR_n q_{o,n} = q_g \\
WC_n q_{L,n} = q_w \\
q_{g,n} = \displaystyle\sum_k q_{g,k} \\
q_{o,n} = \displaystyle\sum_k q_{o,k} \\
q_{w,n} = \displaystyle\sum_k q_{w,k} \\
q_{L,n} = \displaystyle\sum_k (q_{o,k} + q_{w,k})
\end{array}
\right\}header\quad mass\quad balance
$$

(ii) Approximate reduced mixed integer optimization model (MINLP).

$$max \; q_o$$

$$s.t$$

$$\begin{bmatrix} y_k \\ GOR_k = f_g(q_{o,k}) \\ WC_k = f_w(q_{o,k}) \\ q_{g,k} = GOR_k q_{o,k} \\ q_{w,k} = WC_k q_{L,k} \\ q_{L,k} = q_{o,k} + q_{w,k} \\ q_o^L \le q_{o,k} \le q_o^U \\ P_n = f(Q_k^U) \\ q_{o,k} \le Q_o^U \end{bmatrix} \vee \begin{bmatrix} \neg y_k \\ Bx = 0 \end{bmatrix}, \; k = 1,2,3$$

$$P_n = f(P_i, GOR_n, WC_n, q_L)$$

$$GOR_n q_{o,n} = q_g$$

$$WC_n q_{L,n} = q_w$$

$$q_{g,n} = \sum_k q_{g,k}$$

$$q_{o,n} = \sum_k q_{o,k}$$

$$q_{w,n} = \sum_k q_{w,k}$$

$$q_{L,n} = \sum_k (q_{o,k} + q_{w,k})$$

(iii)   Notation

GOR	= Gas to oil ratio.
L	= Length of the pipe.
$q_{p.k}$	= Flow rate in standard conditions of phase p from well k.
q_p	= Flow rate of phase p at header.
Q_k^U	= Oil flow rate upper bound of k well.
P_k^{wf}	= Pressure of k well along the well tubing.
P_k^f	= Pressure of pipe which connects the choke of k of well to header n.
P_n	= Pressure of header n.
$P_k^{ch,in}, P_k^{ch,out}$	= Pressure upstream and downstream of choke of k well.
P_i	= Pressure of separator i.
WC	= Water cut, namely the ratio of water to liquid flow rate.
p	= Phase (oil, gas, water, liquid).

European Symposium on Computer Aided Process Engineering – 12
J. Grievink and J. van Schijndel (Editors)
© 2002 Elsevier Science B.V. All rights reserved.

Towards the optimisation of logistic decisions and process parameters of multipurpose batch plants

Thomas Löhl and Sebastian Engell

Process Control Laboratory, Dept. of Chemical Engineering,
University of Dortmund,
D-44221 Dortmund, Germany.
Phone: +49 (0)231/ 755-5127, Fax: +49 (0)231/ 755-5129
e-mail: {t.loehl | s.engell}@ct.uni-dortmund.de

Abstract

This contribution presents a simulation and scheduling environment which enables the simultaneous optimisation of both scheduling decisions and process parameters based on the reference model described in (IEC, 1997). The modelling effort is reduced since the structure of a batch plant is defined by generic model elements. The scheduling problem is solved by a genetic algorithm based on an imprecise model, the logistic optimisation model. During the genetic search the simulation of a model with more accurate elements - the process model - is utilised to refine the parameters of the logistic optimisation model. Despite the use of models of different accuracy and detail, no redundant modelling effort is required.

1. Introduction

The profitability of multipurpose batch plants depends on the choice of the parameters of each processing step and on good logistic decisions. The development of life-cycle models is gaining increasing importance, since they reduce the effort to build and maintain different models for every stage of a life-cycle (Vankatasubramanian et al., 2001). Since a proper scheduling can reveal equipment savings in the design stage, reduce inventories in the operational stage and help to identify bottlenecks during plant retrofit of existing plants, we regard the optimal scheduling as a key problem to be solved. The remainder of this contribution explains the major elements of the optimisation model and their relation to the structural elements of the core model (section 2), a brief description of the solution approach including aspects concerning the overall architecture of the realised software environment (section 3) and the discussion of the results obtained for a benchmark example (section 4). The conclusion (section 5) closes this contribution.

2. Modelling

We propose to use the reference models for batch control in multipurpose plants (IEC, 1997) as the basis of both the simulation models and the logistic optimisation model.

This standard defines the elements of batch processes, their basic properties and their dependencies. In the paper this object oriented model is called the core model for short.

2.1 The core model

The key aspect of the core model is the division of the process into the plant and the recipes. A recipe describes the procedural steps which are required to produce substances or to execute services. The elements of the master recipe are formulated independently of the specific equipment items of the plant. The plant model represents the physical structure of the plant and, in particular, the basic technical functions which the elements of the plant can perform. The assignment of the steps of the master recipe to the equipment items yields the control recipe. The control recipe determines the flow of material through the plant unambiguously and can be used by a supervisory control program without major adjustments.

For simulation purposes, no additional structural information beyond the core model is necessary, only the interpretation of the dynamic behaviour of the elements (e.g. filling) according to the chosen level of detail (e.g. constant flow or hydrostatically driven flow) must be available. This concept has been realised in the Batch Simulation Package (BASiP), for which a thorough description of the modelling, the event detection and the numerical performance can be found in (Fritz et al., 1999). However, for the logistic optimisation the core model cannot be used directly for two major reasons:

1. The degrees of freedom, which the optimisation algorithm must be able to exploit are not stated explicitly. A search for alternatives would necessitate time consuming search through the entire object hierarchy.
2. The information about the current state of the plant, the arbitration strategy and the scheduling objectives, to name but a few, are not contained in the core model.

2.2 The optimisation model structure

The logistic optimisation model is structured as three sets of graphs, where each set is assigned to one of the layers of abstractions represented by the type and the elements of the recipes (see figure 1). The *substance graph* is used to map alternative master recipes for the production of the desired substances or services on the master recipe procedure level. The *activity graph* refines the substances described in the substance graph to the states of material (e.g. location of material). The activity graph replaces each master recipe procedure of the substance graph by a set of control recipes to consider alternative resource utilisation. Both graphs are directed bipartite graphs consisting of circle nodes, representing the involved materials/states and rectangle nodes, representing the master or control recipe procedure respectively. The representation of these graphs is similar to the STN framework for mathematical programming (Kondili et al., 1992). The *phase-graph* specifies the control recipe procedure by the sequence of the recipe phases, where the edges define precedence relations. Every state of the activity graph contains a reference to the associated unit. The recipe phases reference equipment items of units which can be used exclusively or shared. Each equipment model (unit) administers the inventory history and the list of allocation intervals (see for example unit *B31* in figure 1).

2.3 The solution procedure

The large number of combinatorial decisions suggests to apply genetic algorithms (GA) to determine the sequence and resource allocation of the batches (Corne and Ross, 1997). Therefore, a linear, indirect coding of the batch objects as decision variables is used. A batch object contains the interval where the batch run can be performed and a reference to the task of the activity graph, i.e. the control recipe. The interpretation of the sequence of batch objects to be scheduled is carried out for each batch in the following two step procedure. In the first step, the material balances are evaluated to determine the time intervals where the

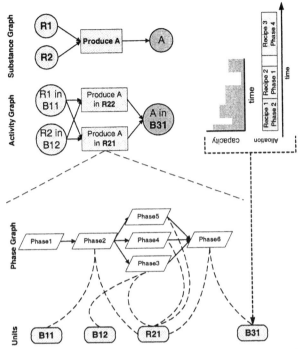

Figure 1: A part of the optimisation model which describes the production recipe of the example (see figure 4). The dashed lines are object references, the solid lines represents precedence relations. The dashed lines from the states to the units are omitted for clarity.

capacities of the states of the raw material and the products are within the feasible bounds. In the second step the starting time of the batch is adjusted to resolve all resource conflicts on the recipe phase level, using the concept of feasible time windows as proposed for example in (Rodrigues et al., 2000). The quality of the schedule is assessed using the logistic optimisation model for a specified objective. At predetermined number of generations the best schedule obtained so far is simulated by the BASiP Simulation module using the process model. The process model refines the elements of the core model. In this model, the dynamics are described by differential equations.

3. Realisation

In the realised software environment, the processes in a multipurpose batch plant are modelled graphically by specifying the plant topology as a flowsheet extended by the equipment phases, the master recipes using sequential-function-chart-like semantics (see figure 4) and the assignment of the recipe phases to the equipment phases for the generation of control recipes. The logistic optimisation model is generated automatically. The GA creates feasible schedules which are simulated using the process model. The results of the asynchronously running simulator are imported and the parameters of the optimisation model are updated. The figure 2 depicts the overall architecture focussing on the most important data flows. The local search strategy, the intermediate storage policy, the time representation and the configuration of the genetic algorithm (i.e. choice of the types of genetic operators, selection and mating strategy, constraint handling techniques, etc.) are determined by external configuration dialogs. Thus no modification on source code level is necessary to adjust performance critical

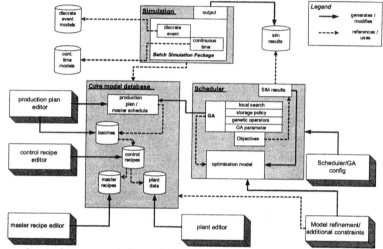

Figure 2: Architecture of the scheduling environment consisting of simulation, modelling and scheduling

parameters. The complete software is implemented in *C++* and uses the *GALib* (Wall, 1995) source code as the genetic algorithm class library.

4. Example and numerical results

The example illustrates the modelling approach, the scheduling performance and the influence of the detailed simulation on the course of the optimisation.

4.1 Problem description
The process under consideration is a batch process in which two liquid products (*D* and *E*) are produced from three liquid substances (*A,B* and *C*). Basically, the plant consists of three stages (see figure 3).

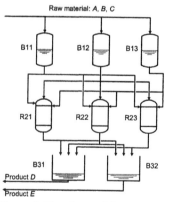

Figure 3: Topology of the benchmark plant

- Raw material buffer: The vessels B11, B12 and B13 buffer the raw materials *A,B* and *C*, respectively. Each tank is used exclusively for one raw material and may contain at most two batches of substance.

- Reaction: The three reactors R21, R22 and R23 may produce both products *D* or *E*. The master recipe of this stage consists of three parallel operations (see figure 4) which have to be synchronised.
- Product buffer: The tanks B31 and B32 buffer the products and are exclusively used for either *D* or *E*. Both of the tanks may contain at most three batches of product.

All equipment items of each stage are fully connected to the units of the next stage. The charge/ uncharge steps cannot be neglected and the flowrates are different for each control recipe and are only imprecisely known a-priori. The scheduling task is to produce an amount of 6 batches of each product. The algorithm has to determine the resource allocation on the recipe phase level, the sequence of batches and the starting times of every batch such that the makespan is minimised. The problem comprises 28 batches with 140 operations in total to be scheduled on a scheduling horizon of 600 equally spaced time units.

4.2 Numerical results

A GA was used for this example with a linear scaling, a roulette wheel selection and 20 % of the population are replaced by the offspring. The mutation operator changes the position and resource assignment of a batch with a probability of 0.3. The standard order crossover is applied with a probability of 0.9. The

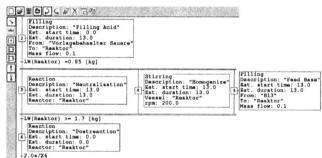

Figure 4: Screenshot of the production master recipe as a sequential function chart. The double bars indicate the parallel execution of the steps 3,4 and 5 until the mass in the vessel called Reaktor exceeds 1.7 kg.

population comprises 50 individuals and a constant duration of 100 generations was used as the termination criterion. This standard configuration is used without problem specific tailoring methods.

The convergence of the genetic algorithm and the influence of the detailed hybrid simulation is depicted in figure 5. The figure shows the rapid convergence of the best and the worst individuals towards better solutions. The duration of all process operations are updated after the 15th and then every 10 generations with the simulation results obtained by the hybrid simulation. The new parameter set had the greatest impact at the second time it occurred (see 25th generation). It is interesting to note that at this point the change of the durations of the operations has led to a broadening of the objective function value distribution of the population increasing the chance to find better solutions. The best solution of 319 time units is within the confidence interval obtained by a statistical t-test. An optimal solution for the problem without the more detailed simulation is 345 time units. The complete optimisation run took approximately 285 CPU seconds. Thus, 0.11 CPU seconds are required to create and evaluate one

708

schedule on the average. The algorithm was executed on a SUN UltraSparc-II (300 MHz) using 11 MB of memory.

For a more constraint problem (the starting times of the batches of two raw materials were fixed) of this process, the results were compared against a standard discrete time MILP/STN model using the GAMS/CPLEX solver. The proposed approach found an optimal solution after 11 seconds, whereas the mathematical programming approach required 5.1 minutes.

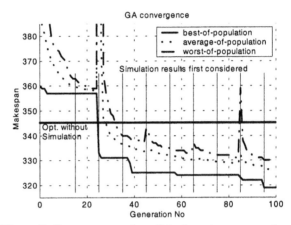

5. Conclusions

The validation of simulation models and the effective representation of the optimisation model

Figure 5: Convergence plot of the best, average and worst makespan.

with little additional modelling effort enables easy and consistent modelling as well as quick results. The proposed approach was able to improve the solution quality by 7 % compared to scheduling without simulation. Apart from the reduction of the modelling effort, the numerical experiments also demonstrate the efficiency of the optimisation method and the advantages of the combination of a detailed simulation with optimisation.

6. References

Corne, D. and P. Ross, 1997, Practical Issues and Recent Advances in Job- and Open-Shop Scheduling, in: Evolutionary Algorithms in Engineering Applications, Eds. D. Dasgupta and Z. Michalewicz, 531-546.

Fritz, M., A. Liefeldt and S. Engell, 1999, Recipe-Driven Batch Processes: Event Handling in Hybrid System Simulation, in: Proc. of 1999 IEEE int. Symposium on Computer Aided Control System Design (CACSD '99), Hawaii, 138-143.

IEC 61512-1, 1997, Batch Control, Part 1: Models and Terminology, International Electrotechnical Commission (IEC).

Kondili, E., C. C. Pantelides and R. W. H. Sargent, 1992, A general short algorithm for short-term scheduling of batch operations. Part I- MILP formulation, Computers and Chemical Engineering, 17 (2), 211-227.

Rodrigues, L. A., M. Graells, J. Canton, L. Gimeno, M. T. M Rodigues, A. Espuna and L. Puigjaner, 2000, Utilization of processing time windows to enhance planning and scheduling in short term multipurpose batch plants, Computers and Chemical Engineering, 24, 353-359.

Venkatasubramanian, V., J. Zhao, S. Viswanathan, C. Zhao, F. Mu, P. Harper and B. Russel, 2001, in: European Symposium on Computer Aided Process Engineering (ESCAPE) - 11, Eds. R. Gani and S. B. Jørgenson, 925-930.

Wall, M. B., 1996, GALib: A C++ Library of genetic algorithm components, Manual, Mechanical Engineering Department, MIT, http://lancet.mit.edu/ga/, (online: 22.10.01).

European Symposium on Computer Aided Process Engineering – 12
J. Grievink and J. van Schijndel (Editors)
© 2002 Elsevier Science B.V. All rights reserved.

Trilinear Models for Batch MSPC: Application to an Industrial Batch Pharmaceutical Process

Lopes J.A.[1] and Menezes J.C.[2]
Cemter for Biological & Chemical Engineering, Technical University of Lisbon
Av. Rovisco Pais, P-1049-001, Lisbon, Portugal
[1]joao.lopes@ist.utl.pt; [2]cardoso.menezes@ist.utl.pt
Phone: (+351) 218 417 347; Fax: (+351) 218 419 062

Abstract

In this paper, PARAFAC and Tucker3 models were compared with the commonly used multiway principal components analysis approach (MPCA) for multivariate process control of an industrial batch antibiotic production process. Two different approaches for on-line monitoring were used: sliding window (multiple models) and global window (single model) monitoring strategies. The later approach requires orthogonality for the time dimension scores. In this context, a modification of the Parafac algorithm was proposed. The Tucker3 and Parafac models as proposed here share an identical structure. Scores (D) and residuals (Q) statistics were used to on-line identify faults. We concluded that Parafac and Tucker3 models outperformed MPCA in terms of detection of faults specially when the statistic for scores is used. All models performed equally well in the residuals statistics. The sliding window strategy proved to be more appropriate to identify faults than the global window strategy. This is, to our best knowledge, the first time such study was performed for an industrial batch antibiotic process.

1. Introduction

Principal components analysis (PCA) has been extensively used in multivariate statistical process control (MSPC) in the past ten years (Martin and Morris, 1996; Chen and McAvoy, 1998; Kassidas et al., 1998). Multiway PCA (MPCA) is traditionally used to extend MSPC to batch processes (Nomikos and MacGregor, 1995; Albert and Kinley, 1996; Chen and McAvoy, 1998; Lopes and Menezes, 1998). Batch multivariate statistical process control (BMSPC) needs to account for three primary directions: the batch number, the monitored variables and the time dimension for each batch. Trilinear methods, such as Parafac or Tucker3 models, are more appropriate to model batch processes since they take into account the three-dimensional structure of the data (Bro, 1998). The purpose of this paper is to study the advantages of using trilinear models (Parafac and Tucker3 models) over standard MPCA models in terms of fault detection (Louwerse and Smilde, 2000). The Tucker3 and Parafac models are usually described with two different equations. In this paper, the same structure is used to describe both models. Two on-line fault detection strategies are compared. Equations to compute projected scores for each monitoring strategy are also presented here. A pharmaceutical process (antibiotic production by fermentation) was used as a case-study.

2. Methods

2.1 MPCA, Parafac and Tucker3 models

The MPCA model is equivalent to the PCA model. It uses a data decomposition commonly named unfolding to transform a data three-way array $\underline{X}(I \times J \times K)$ in a two-way array $X(I \times JK)$ where in general the batch dimension (I) is preserved (Kiers, 2000). Using SVD it is possible to extract an array $A(I \times R)$ of scores (variability among batches) and a loading matrix $C(JK \times R)$ containing information on variables and time dimension. R is the model number of components. $E(I \times JK)$ is the residuals array that depends on the chosen R..

$$X = AC^T + E \tag{1}$$

Parafac and Tucker3 are trilinear models because they preserve the trilinear structure of the data ($B(J \times R)$ and $C(K \times R)$ are the loadings for variables and time modes). They share the structure given by equation 2 (the operator $\otimes$ denotes the Kronecker product) (Bro, 1998).

$$X = AG(C \otimes B)^T + E \tag{2}$$

In a Parafac model, the core array $\underline{G}(R \times R \times R)$ is as shown in equation 3.

$$
\underline{G}_{parafac} =
\begin{bmatrix}
1 & 0 & \cdots & 0 & 0 & 0 & \cdots & 0 & & 0 & 0 & \cdots & 0 \\
0 & 0 & & 0 & 0 & 1 & & 0 & & 0 & 0 & & 0 \\
\vdots & & \ddots & \vdots & \vdots & & \ddots & \vdots & \cdots & \vdots & & \ddots & \vdots \\
0 & 0 & \cdots & 0 & 0 & 0 & \cdots & 0 & & 0 & 0 & \cdots & 1
\end{bmatrix} \tag{3}
$$

The Parafac model can also be written as $X = A(C \circ B)^T + E$, where $\circ$ denotes the Khatri-Rao operator (Bro, 1996). The Parafac model expressions are equivalent since for any two arrays B and C (with the same number of columns), it holds $(C \otimes B)G_{parafac}^T = C \circ B$. The model algorithms can be found elsewhere (Bro, 1998).

2.2 Monitoring Strategies

Two on-line monitoring strategies are presented here.

2.2.1 Sliding window strategy

The sliding window strategy consists on the model projection of a segment of the batch data as the process evolves. At each new batch time interval [k-Δk/2 , k+Δk/2] (where Δk is the window size), new B and C loading matrices are needed. The number of necessary loading matrices depends on the batch size (sampling frequency) and window size (Louwerse and Smilde, 2000). Equation 4 shows how to obtain the new scores.

$$a_{new} = x_{new} ZG^T (GZ^T ZG^T)^{-1} \tag{4}$$

In this equation, $Z = C \otimes B$. For a MPCA model, $a_{new} = x_{new} B$.

2.2.2 Global strategy

The global window approach involves using the full B and C loading matrices obtained from the collection of history nominal batches. At each new time interval the scores for the entire batch are estimated (Nomikos and MacGregor, 1995). The point here is to restrict the available batch information (say at time point k) to be consistent with the correlation structure up to that time point k. The scores are obtained by projecting the known data at time k onto the model. Equation 5 is used to obtain the scores for a new batch at time k for Tucker3 and Parafac models (Lopes, 2001).

$$a_{new,k} = x_{new,k} Z_k G^T \left(G Z_k^T Z_k G^T \right)^{-1} \tag{5}$$

In equation 5, $x_{new,k}$ is the vector containing the batch data up to time point k and the array Z_k is given by equation 6.

$$Z_k = \left[C_k \left(C_k^T C_k \right)^{-1} \right] \otimes B \tag{6}$$

The loadings matrix, C_k, contains the first k lines of C. Equation 6 holds if $C^T C = I$. This is only true for the Tucker3 model. The Parafac algorithm must be changed in order to account for this constraint (Lopes, 2001). In the iterative Parafac algorithm, we need to impose $C = XZ\left(Z^T X^T XZ \right)^{-0.5}$ each time the C matrix is estimated (where $Z = (B \circ A)\left[(B \circ A)^T (B \circ A) \right]^{-1}$). For the MPCA model, the scores are obatined with equation 7.

$$a_{new} = x_{new,k} C_k \left(C_k^T C_k \right)^T \tag{7}$$

2.3 Statistics

Multivariate statistical process control is based on two statistics: one for the scores (statistic D or Hotelling T^2) and one for the residuals (statistic Q). The D statistic measures the variability explained by the model, while the Q statistic measures the residuals. For each new batch i the statistic D can be obtained with equation 8 (Wise and Gallagher, 1996).

$$D_i = \left(a_i - \overline{A} \right) S_a^{-1} \left(a_i - \overline{A} \right)^T \tag{8}$$

The residual statistic for batch i is obtained with equation 9.

$$Q_i = \sum_{j=1}^{J} \sum_{k=1}^{K} e_{ijk}^2 \tag{9}$$

The confidence limits for these statistics were computed as explained in Nomikos and MacGregor (1995). Westerhuis et al. (1999) explain how to compute robust limits for these statistics. Because the Q_α limit is variable in time it is better to plot relative Q values in control charts ($Q_k/Q_{95\%}$).

Table 1. Non-nominal batches used to test models and monitoring strategies.

Case	Fault(s) description
A	Abrupt change of the air inlet flow at 49 hours
B	Low pH batch (since batch beginning)
C	Step tests applied periodically on substrates feeds

3. Case-Study/Experimental

An industrial pharmaceutical process (clavulanic acid production by fermentation) is presented here as a case-study (Neves et al., 2001). Fed-batch cultivation of a *streptomycete* strain was carried out using non-defined (complex) medium. The conditions used were typical of those employed routinely in industry for aerobic microbial growth. A fully instrumented bioreactor with an operating volume of 200 dm^3 was employed throughout this study. A total of twenty variables were measured. Some were obtained on-line while others were measured off-line with a frequency of about 4 hours except for the viscosity which was measured only once per day. Data pre-processing included outliers detection and noise reduction. Interpolation for missing/unavailable values was used due to the process slow dynamics. A 1 hour interval was chosen to synchronize the fermentation data.

3.1 Nominal Data
20 batches were operated under normal operation conditions (NOC). The 20 nominal batches were arranged as a three-way array with dimensions $\underline{X}(20 \times 20 \times 140)$.

3.2 Non-Nominal Data
Three non-nominal batches (with known faults) were used to test each model and each monitoring strategy, thereafter named cases A to C. Table 1 indicates the faults occurred in each non-nominal batch.

4. Results and Discussion

A major difference between the models is the captured variance for the same number of components. Because the number of parameters of the MPCA model is greater it also captures more variance for the same number of components. Note that a two-component MPCA model captures 47.2% of variance while a four-component Parafac model captures 42.9% and a four-component Tucker3 model captures 45.8%. The captured variances for Parafac models considering orthogonal C scores (as required for the global window strategy) are slightly lower due to the imposed constraint. For a four-component Parafac model with imposed orthogonality of C loadings, only 40.5% of variance is captured. To compare the error detection performance for the three models, two-component MPCA and four-component Tucker3 and Parafac models were selected. For the sliding window monitoring strategy 129 models ($\Delta t = 12$ hours) were built (note that the first model can only be built after 12 hours). One global model was built for the global window strategy. D and Q statistics were computed and probability values (p-values) were determined. It was considered that a fault is detected when the p-value is lower than 0.05 (95% confidence limit). Table 2 summarizes the times were an error was detected for the first time in the D statistic for each non-nominal case.

From Table 2 it is possible to conclude that Parafac and Tucker3 models performs better in terms of statistic D than MPCA models. This is specially true for the global monitoring strategy where a single model is used. However, errors are detected earlier when the sliding window strategy is used. The sliding window size must be selected in order to avoid false alarms. In this case (12 hours size) no false alarms were detected.

Table 2. Fault detection times for the three tested cases (a fault is detected when p-value<0.05).

	MPCA		Parafac		Tucker3	
	Sliding Window	Global	Sliding Window	Global	Sliding Window	Global
Case A	53 h	75 h	52 h	63 h	50 h	57 h
Case B	40 h	82 h	38 h	78 h	37 h	71 h
Case C	59 h	125 h	52 h	82 h	51 h	74 h

With respect to the residuals statistic (Q) small differences were observed between the three models tested. In general, errors are detected earlier in the Q statistic. In case A, all models detected an error in the Q statistic at 49 hours (where the abrupt change in the inlet air feed ocurred).

In case B, a fault (approximately at 40 hours) is detected by all models in both statistics. However, after a short period of time the control chart for statistic D returns to an in-control region. This is not verified in the Q statistic where an out-of-control signal persisted even after the 40 hours region (see Figure 1).

If an alarm is started, a comparison (difference) between the projected batch and the average of nominal batches variables contributions provides an indication of which variable(s) are causing the fault (Louwerse et al., 1999). These charts are called contribution plots. Figure 2 shows the contribution plots for residuals and scores, for case A, at the time the fault actually ocurred (49 hours). At 49 hours the fault was observed only on the residuals statistic (p-value<0.05). Clearly, variables 13, 14 and 17 (oxygen and carbon dioxide outlet-gas fractions and inlet-air flow) are causing the deviation.

5. Conclusions

It was found that trilinear models perform better than MPCA in terms of the statistic for scores. In the three tested cases the faults were detected earlier in the D statistic when trilinear models are used. A marginal gain was observed for the Tucker3 model.

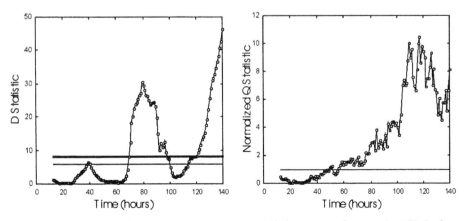

Figure 1. Sliding window monitoring strategy (12 hours window size) MSPC charts based on MPCA models (charts for case B).

714

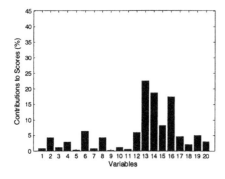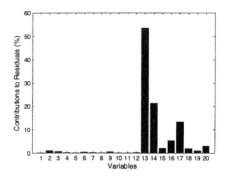

Figure 2. Fault identification contribution plots for scores and residuals, for case A at 49 hours (MPCA model/sliding window monitoring strategy).

All three models performed equally well when residuals statistic was used. In these cases the faults were primarily detected in the residuals statistic.

This happens when the models are unable to explain the variation observed in the projected data. The sliding window strategy is more appropriate than the global window strategy for fault detection. However, the later has the advantage to use only one set of loadings, thus, being less memory demanding in computer processing. Nevertheless, we found that the sliding window strategy is more sensitive to detect process deviations from normal operating conditions. The choice of the correct time window for fault detection would depend upon the process. Batches where faults occurred can be used to adjust this parameter. Future work is being directed, as new industrial data become available, to address these problems.

6. References

Albert, S. and R. Kinley, 2001, Trends Biotechnol. 19(2), 53-62.

Bro, R., 1996, IEEE T. Sig. Proces. 6795.

Bro, R., 1998, PhD Thesis, 286, University of Amsterdam.

Chen, G. and J. McAvoy , 1998, J. Proc. Cont. 8(5), 409-420.

Kassidas, A., P. Taylor and J. MacGregor, 1998 J. Proc. Cont., 8(5), 381-393.

Kiers, H., 2000, J. Chemometrics 14, 105-122.

Lopes, J., 2001, PhD thesis, 296, Technical University of Lisbon (in Portuguese).

Lopes, J. and J. Menezes, 1998, AIChE Symp. Series 94(320), 391-396.

Louwerse, D. and A. Smilde, 2000, Chem. Eng. Sci. 55, 1225-1235.

Louwerse, D., A. Tates, A.Smilde, G. Koot and H. Berndt, 1999, Chemometrics Intell. Lab. Syst. 46, 197-206.

Martin, E. and J. Morris, 1996, Trans. Inst. MC 18(1), 51-60.

Neves, A., L. Vieira and J. Menezes, 2001, Biotechnol. Bioeng. 72(6), 628-633.

Nomikos, P. and J. MacGregor, 1995, Technometrics 37(1), 41-59.

Westerhuis, J., S. Gurden and A. Smilde, 1999, J. Chemometrics 14, 335-349.

Wise, B. and N. Gallagher, 1996, J. Proc. Cont. 8(6), 329-348.

A Mixed Integer Programming Approach for Scheduling Commodities in a Pipeline

L. Magatão*, L.V.R. Arruda[†], F. Neves-Jr.[♦]

CEFET-PR, CPGEI

Av. Sete de Setembro, 3165, 80230-901 Curitiba, PR, Brazil

Tel.: +55 41 310-4707 - Fax: +55 41 310-4683

{*magatao, [†]arruda, [♦]neves}@cpgei.cefetpr.br

Abstract

This paper addresses the problem of developing an optimisation model to aid the operational decision-making process on pipeline systems. The model is applied on a real world pipeline, which connects an inland refinery to a harbour, conveying different types of commodities. The optimisation model was developed based on mixed integer linear programming (MILP) with uniform time discretisation. The MILP well-known computational burden was avoided by the problem domain decomposition. Simulation examples have demonstrated that the optimisation model was able to define new operating points to the pipeline system, providing significant cost saving.

1. Introduction

The oil industry has a strong influence upon the economic market. Research in this area may provide highly profit solutions and also avoid environmental damages. The oil distribution-planning problem is within this context. A wide net with trains, tankers, and pipelines are used to link harbours, refineries and consumers. According to Kennedy (1993), pipelines provide an efficient way to transport oil and gas. The maximum utilisation efficiency of this transportation medium becomes interesting to the oil industry. However, the operational decision-making on pipeline systems is still based on experience, with aid of manual calculation. According to Lee *et al.* (1996), mathematical programming techniques for long-term planning have been extensively studied and implemented, but much less work has been devoted to short-term scheduling, which in fact reproduces the operational decision-making process. The short-term scheduling requires the explicit modelling of discrete decisions. The approach to solve this problem is manifold. A general one is to use a mixed integer linear programming formulation. A complete survey in mixed integer programming and techniques for several application problems is presented in (Wolsey, 1998). The great concern of a real-word MILP formulation is related to the difficulty of finding solutions in a reasonable computational time. According to Applequist *et al.* (1997), a MILP feature of a practical problem requires a large number of integer variables, thus the computational expense has to be concerned. Subrahmanyam *et al.* (1995) demonstrate that decomposition strategies are a valid approach to avoid the combinatorial explosion introduced by integer variables.

2. Problem Definition

This work focuses on the short-term scheduling of activities in a specific pipeline system. It connects a harbour to an inland refinery. The pipeline is 93.5 km length, it can store a total volume of 7,314 m^3, and it connects a refinery tank farm to a harbour tank farm going along regions with 900-meter-altitude difference (Δh). The pipe conveys multiple types of commodities. It is possible to pump products either from the refinery to the harbour (this is called *flow* operation) or from the harbour to the refinery (this is called *reflow* operation). There is no physical separation between different products as they move in the pipe. Consequently, there is a contamination area between products: the interface. Some interfaces are operationally not recommended, and a *plug* (small volume of product) can be used to avoid a specific interface. However, plug inclusions increase the operating cost. The tank farm infrastructure, an up-to-date storage scenario, the pipeline flow rate details, and the demand requirements are known a priori. The scheduling process must take into account product availability, tankage constraints, pumping sequencing, flow rate determination, and a wide variety of operational requirements. The task is to specify the pipeline operation during a limited time horizon (*T*), providing low cost operational procedures. Figure 1 illustrates the pipeline physical structure overview.

3. Methodology

The methodology applied on this work is the mixed integer linear programming with uniform time discretisation. The computational complexity is concerned, and an optimisation structure is proposed to decompose the problem in blocks, providing a framework that aims to reduce the computational expense (Figure 2). The optimisation structure is based upon a MILP main model (*Main Model*), one auxiliary MILP model (*Tank Bound*), a time computation procedure (*Auxiliary Routine*), and a *Data Base*. The tank bound task involves the appropriate selection of some resources (tanks) for a given activity (pumping the demanded product). Its main inputs are demand requirements, product availability, and tankage constraints. As an output, it specifies the tanks to be used on operational procedures. The auxiliary routine takes into account the available time horizon, the product flow rate range, and demand requirements. It specifies temporal constraints, which must be respected by the main model. The main model determines the product pumping sequence and the flow rate details. It establishes the initial and the final time of each pumping activity. The final scheduling is attained by first solving the tank bound and the auxiliary routine, and, at last, the main model.

The modelling and optimisation tool *Extended LINGO/PC Release 6.0* (LINDO, 1999) was used to implement and solve the optimisation structure. LINGO is a commercial tool, which allows formulating linear and non-linear large problems, solving them, and analysing the solution. It has four internal solvers: a direct solver, a linear solver, a non-linear solver, and a branch and bound manager. The LINGO's solvers are all part of the same program, which is directly linked to its modelling language. This allows the data exchange directly through memory, rather than through intermediate files.

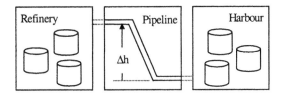

Figure 1. Pipeline physical structure overview.

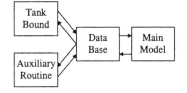

Figure 2. Optimisation structure.

4. Model Framework

Basically, the modelling process takes into account the following conditions: (i) pipeline can fill or empty only one tank at a time; (ii) tanks being emptied can not be filled, and tanks being filled cannot be emptied; (iii) a tank always stores the same product; (iv) the tank farm infrastructure limits must be respected; (v) the product flow rate range must be respected; (vi) the product demand has to be within an operational range; (vii) every product must be pumped continually; (viii) it is possible to use a plug between incompatible products, but plug inclusion increases the operating cost; (ix) the plug volume is significantly smaller than any demanded batch, so that its pumping time is neglected; (x) changeover times are neglected; (xi) use of plugs should be minimised; (xii) it is required a minimum time horizon (T_{min}) to pump the entire demand. In case $T = T_{min}$, every product is pumped at its maximum flow rate; (xiii) the system starts pumping at the initial time ($t = 1$). In case $T > T_{min}$, the pumping procedure can be finished before T, but the pipeline must remain pressurised. There is also a cost to maintain the pipe pressurised.

The mathematical approach, as stated, is based on MILP with uniform time discretisation. Space restrictions preclude a detailed problem formulation. Such information can be obtained in (Magatão, 2001). It is presented the main model objective function (1), exploiting its characteristics.

minimize COST =

$$
= CR_{pump} \cdot \sum_p (TFB_p^r - TIB_p^r) + CP_{pump} \cdot \sum_p (TFB_p^p - TIB_p^p) + \sum_{t=1}^{T-1} [Ce_t \cdot (PP_t + PR_t)] +
$$
$$
+ C_{plug} \cdot \left[\sum_p \sum_{pa} \sum_{t=TIR+1}^{TSR-1} (I_{p,pa} \cdot TR_{p,pa,t}) + \sum_p \sum_{pa} \sum_{t=TIP+1}^{TSP-1} (I_{p,pa} \cdot TP_{p,pa,t}) \right] + CS \cdot TS \tag{1}
$$

where p and pa are different products; t is the discretised time (h); T is the available time horizon (h); TIR, TSR, TIP, and TSP are temporal constraints determined by the auxiliary routine (h); C_{plug} is the average cost to pump a plug (\$); Ce_t is the average electric cost per flow rate unit at a time t (\$·h/m³); CP_{pump} is the average cost to pump a product from the harbour to the refinery (\$); CR_{pump} is the average cost to pump a product from the refinery to the harbour (\$); CS is the average cost to maintain the pipe pressurised (\$/h); $I_{p,pa}$ is a dimensionless parameter that assumes one if pumping p

followed by pa requires a plug between then, zero otherwise; PP_t is the flow rate (m³/h) at a time t (reflow procedure); PR_t is the flow rate (m³/h) at a time t (flow procedure); TFB_p^p is the end pumping time (h) of p (reflow procedure); TFB_p^r is the end pumping time (h) of p (flow procedure); TIB_p^p is the start pumping time (h) of p (reflow procedure); TIB_p^r is the start pumping time (h) of p (flow procedure); $TP_{p,pa,t}$ is a dimensionless variable that assumes one if the transition between p and pa occurs at a time t, zero otherwise (reflow procedure); TS is the time period that the pipe remains pressurised (h); $TR_{p,pa,t}$ is a dimensionless variable that assumes one if the transition between p and pa occurs at a time t, zero otherwise (flow procedure).

Expression (1) demonstrates that plug inclusions ($I_{p,pa} \cdot TR_{p,pa,t}$, $I_{p,pa} \cdot TP_{p,pa,t}$) increase the operating cost. So that, the optimisation solution method seeks scheduling solutions that minimise the plug usage. The product pumping time (TFB_p^r, TIB_p^r, TFB_p^p, TIB_p^p) increases the operating cost. This time is related to the flow rate (PR_t, PP_t) by an inverse ratio: if the flow rate increases, the product pumping time decreases. On the other hand, the operating cost is also directly influenced by flow rate variations - see the factor $Ce_t \cdot (PP_t + PR_t)$. Thus, the optimisation structure must determine the ideal flow rate policy during a limited time horizon (T). Maintaining the pipe pressurised also influences the operating cost ($CS \cdot TS$).

5. Results

This section considers an example involving the pumping of four products from the harbour to the refinery followed by another four pumped from the refinery to the harbour. Each product has two tanks enabled to sending operations. For simplicity, units were standardised and omitted. The normalisation is based on the pipeline volume. The entire pipe has 7,314 m³. It is admitted a NF (normalisation factor) that equally divides the pipe volume. The product demand is expressed based upon NF. As an example, $NF = 4$ determines batches of 1,828.5 m³ (7,314÷4). A normalised demand of two units represents a total demanded volume of 3,657 m³ (1,828.5x2). The system pumps, at most, one normalised volume per time unit. A normalised flow rate of one at a time t indicates that a volume of 1,828.5 m³ is pumped between times t and $t+1$. The time length selection of each discretised time span involves a trade-off between accurate operation and computational effort. The problem data was rounded, so that the time quantum could be increased and, thus, the number of decision variables decreased. It was adopted a uniform time discretisation of six hours, and $NF = 4$. Simulation covers since the minimum normalised time horizon ($T_{\min} = 20$) up to twenty-five normalised time units ($T = 25$). The pumping process starts from the harbour to the refinery; CP_{pump}, CR_{pump}, C_{plug}, and CS were considered unitary. Table 1 demonstrates the normalised electric cost value (Ce_t) at each time unit (t). The cost variation is due to on-peak demand hours. Pumping start time is at 6 a.m. ($t = 1$).

Table 1. Electric cost variation.

t	1	2	3	4	5	6	7	8	9	10	11	12	13	14	15	16	17	18	19	20	21	22	23	24	25
Ce_t	1	1	5	1	1	1	5	1	1	1	5	1	1	1	5	1	1	1	5	1	1	1	5	1	1

Table 2 is a system information sketch for the problem main features. It presents a priori information about demand requirements (*Demanded Amount*), flow rate range, and plug necessity. As an example, the sequence P1 followed by P2 demands the use of a plug.

Table 2. System information - main features.

Operation	Product	Demanded Amount	Flow Rate Range	Plug Necessity
	P1	1	0,5 - 1	P2, P4, P6, P8
Reflow	P2	1	0,5 - 1	P1, P5
	P3	2	0,5 - 1	P4, P8
	P4	1	0,25 - 0,5	P1, P3, P5, P8
	P5	1	0,5 - 1	P2, P4, P6, P8
Flow	P6	2	0,5 - 1	P1, P5
	P7	1	0,5 - 1	P4, P8
	P8	1	0,5 - 1	P1, P3, P5, P8

Table 3 provides information about the optimisation structure simulation on a Pentium III, 933MHz, 256 MB RAM. For each time horizon (T), the optimisation structure is run, and a specific normalised cost is attained - expression (1) value. It was not applied any optimality margin (Shah *et al.*, 1993). The auxiliary routine and the tank bound simulation data were neglected. These structures required a computational time lower than one second, for all simulation instances ($20 \leq T \leq 25$).

Table 3. Main model simulation data.

Time Horizon (T)	Total Number of Variables	Total Number of Binary Variables	Total Number of Constraints	Computational Time (s)	Normalised Cost ($)
20	468	112	1,371	4	63
21	559	134	1,559	24	60
22	650	156	1,747	189	59
23	741	178	1,935	1,346	58
24	832	200	2,123	2,633	59
25	923	222	2,311	19,140	60

In order to pump the entire demand, it is required a minimum time horizon ($T_{min} = 20$). In such a horizon, every product is pumped at its maximum flow rate. However, in case $T > T_{min}$ the optimisation structure determines the ideal flow rate policy. This flow rate is established based on both the available time horizon (T) and the electric cost variations (Table 1). Considering $T = 23$, Figure 3 shows the normalised flow rate determined by the optimisation structure. Figure 4 shows the normalised cost - expression (1) - as a time horizon function. It demonstrates the existence of a specific T that yields the minimum operating cost ($T = 23$). The cost versus time horizon function clearly demonstrates that a correct pipeline timing policy provides significant cost saving. For $T = 23$ the pumping sequencing determined by the optimisation structure is P4, P2, P3, P1, P5, P7, P6, and P8, which implies no use of plugs (Table 2).

720

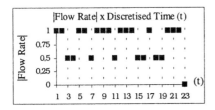
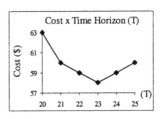

Figure 3. Flow rate versus discretised time.   *Figure 4. Cost versus time horizon.*

6. Conclusions

It was presented a mathematical programming approach to the economically important problem of oil distribution through pipelines. The task was to predict the pipeline operation during a limited time horizon, providing low cost operational procedures. It was applied the scheduling approach based on mixed integer linear programming with uniform time discretisation. The computational expense was concerned and an optimisation structure was proposed (Figure 2). The large-scale mixed integer linear problem was implemented and solved by using the commercial tool *Extended LINGO/PC Release 6.0*. Currently pipeline operation is based on experience, and no computer algorithm is used; plug product usage and energy consumption are not rigorously taken on account. Simulation examples indicate that economic improvements of 8% are feasible (Figure 4).

References

Applequist, G., O. Samikoglu, J.F. Pekny and G.V. Reklaitis, 1997, Issues in the use design and evolution of process scheduling and planning systems, ISA Transactions, 36, 2, 81-121.

Kennedy, J.L., 1993, Oil and Gas Pipeline Fundamentals, Penn Well Publishing Company.

Lee, H., J.M. Pinto, I.E. Grossmann and P. Sunwon, 1996, Mixed-Integer Linear Programming Model for Refinery Short-Term Scheduling of Crude Oil Unloading with Inventory Management, Ind. & Eng. Chem. Res., 35, 1630-1641.

LINDO, 1999, LINGO: The Modelling Language and Optimizer – User's Guide, LINDO Systems Inc, Chicago, Illinois.

Magatão, L., 2001, A Methodology for Sequencing Commodities in a Multi-Product Pipeline, Master Thesis, CPGEI/CEFET-PR, 170 pages (in Portuguese).

Shah, N., C.C. Pantelides and R.W.K. Sargent, 1993, A General Algorithm for Short-Term Scheduling of Batch Operations – II. Computational Issues, Comp. Chem. Engng., 17, 229-244.

Subrahmanyam, S., M.H. Bassett, J.F. Pekny and G.V. Reklaitis, 1995. Issues in Solving Large Scale Planning, Design and Scheduling Problems in Batch Chemical Plants, Comp. Chem. Engng., 19, suppl., S577-S582.

Wolsey A.L., 1998, Integer Programming, John Wiley & Sons Inc.

Acknowledgements

The authors acknowledge financial support from the Brazilian National Agency of Petroleum (PRH-ANP/MCT PRH10 CEFET-PR) and the CNPq (under grant 467311/00-5).

European Symposium on Computer Aided Process Engineering – 12
J. Grievink and J. van Schijndel (Editors)
© 2002 Elsevier Science B.V. All rights reserved.

An MILP Framework for Short-Term Scheduling of Single-Stage Batch Plants with Limited Discrete Resources

Carlos A. Méndez and Jaime Cerdá[*]
INTEC (UNL - CONICET)
Güemes 3450 - 3000 Santa Fe - ARGENTINA
E-mail: jcerda@intec.unl.edu.ar

Abstract

Dealing with limited discrete resources in batch scheduling problems usually produce a sharp increase in model size and computational requirements. This work introduces a novel MILP formulation where all discrete resources including processing units are treated uniformly. Moreover, the ordering of batches at any resource item is handled by a common set of sequencing variables so as to achieve an important saving in 0-1 variables. Pre-ordering rules significantly reducing the problem size can be easily embedded in the MILP framework. In addition, discrete resources could even be sequentially assigned when real world resource-constrained scheduling problems are tackled. Two examples involving the scheduling of up to 29 batches in a single-stage batch plant under severe manpower restraints were successfully solved. Comparison with prior work shows a notable reduction in CPU time of at least two orders of magnitude.

1. Introduction

In multiproduct batch plants, the processing tasks to be accomplished generally share manufacturing resources. Plant resources are usually classified as renewable and non-renewable ones. Renewable resources like processing units, manpower and utilities become again available for use after ending the task to which is currently assigned. A renewable resource is said to be discrete if it is consumed at a constant level throughout the entire processing task. Such resources are usually available by limited amounts that cannot be exceeded at any time of the scheduling period. To meet such constraints, it is necessary to monitor the resource usage level over the scheduling horizon. Schedules involving simultaneous tasks with a total resource requirement larger than the maximum supply are to be excluded from the problem feasible region. Discrete resource constraints are computationally costly when continuous-time domain representations are used. Reklaitis (1992) presented a comprehensive review of resource-constrained scheduling problems. Continuous time formulations based on both the resource-task network (RTN) and the partitioning of the time horizon into intervals of unknown duration usually account for resource constraints but at the expense of a sizable increase in the number of 0-1 variables (Pantelides, 1994; Schilling et al., 1996). In turn, Pinto and Grossmann (1997) introduced a logic-based approach treating the resource constraints as disjunctions to reduce the number of enumerated nodes by orders of magnitudes.

This paper introduces a novel continuous-time MILP framework for short-term scheduling of single-stage multiproduct batch plants under severe limitations in required discrete resources. Similarly to the RTN notion, all the discrete resources (processing units, manpower, utilities, etc.) are equally treated. However, the use of a common set of 0-1 variables to sequence the batches at any available resource item significantly bounds the model size growth when resource constraints must be considered. Compared with previous formulations, a notable saving in binary variables and CPU time is achieved and larger resource-constrained batch scheduling problems can be tackled.

2. Problem Statement

The problem of short-term batch scheduling under severe limitations in discrete resource supplies can be stated as follows. Given: (a) a single-stage multiproduct batch plant with several units $j \in J$ working in parallel; (b) a set of single-batch orders $i \in I$ with specified unit-dependent resource requirements, release-times and due-dates; (c) a set of discrete resources $r \in R$ with known limited supplies; (d) sequence-dependent changeover times at any resource item and (e) a specified time horizon. The problem goal is to determine a production schedule that, in addition to meeting resource allocation and batch sequencing constraints and optimising a particular problem objective like a weighted combination of batch earliness and tardiness, also satisfies the limitations on the total supplies of plant resources at any point in time.

3. The Mathematical Model

3.1 Timing constraints. The starting time of batch i (S_i) can be computed from its completion time by subtracting the processing time at the assigned unit j. Obviously, S_i must never be lower than the release-time of batch i.

$$S_i = C_i - \sum_{j \in J_i} pt_{ij} Y_{ij} \quad \forall i \in I \tag{1}$$

3.2 Discrete resource allocation constraints. Let RR_r be the available discrete resources items of type $r \in R$ and v_{irj} be the fixed amount of resource r required to process a batch $i \in I$ in unit $j \in J_i$. Resource type r may be referred to processing units, manpower, electricity, etc. As stated by Eqn (2.1), every batch $i \in I$ must be allocated to just a single unit $j \in J_i$. In turn, constraint (2.2) ensures that enough amount of each resource $r \in R_i'$ be allocated to meet the requirement of batch i,

$$\sum_{j \in J_i} Y_{ij} = 1 \quad \forall i \in I \tag{2.1}$$

$$\sum_{z \in RR_{ir}} q_{rz} Y_{iz} = \sum_{j \in J_i} v_{irj} Y_{ij} \quad r \in R'_i , \forall i \in I \tag{2.2}$$

where R'_i is the set of resources required by batch i other than processing units. Moreover, q_{rz} is the amount of resource r available at the resource item $z \in RR_r$.

3.3 Sequencing constraints. Assuming that a pair of batches $\{i, i'\}$ requires a common resource $r \in (R_i \cap R_{i'})$ and both have been assigned to the same resource item $z \in RR_r$ ($Y_{iz} = Y_{i'z} = 1$), then the completion time of batch i acts as a lower bound on the starting time of i' only if batch i is processed before. In such a case, the sequencing variable $X_{ii'}$ will be equal to one. Consequently, the sequencing constraint (3) is enforced and eqn. (4) becomes redundant. If instead the assignment variables Y_{iz} and $Y_{i'z}$ are both still equal to 1 but batch i' is first processed, then $X_{ii'} = 0$ and constraint (4) is now enforced. Otherwise, one or both assignment variables are zero and the value of $X_{ii'}$ is meaningless. Therefore, a single variable $X_{ii'}$ is required to control the relative ordering of any pair of batches $\{i,i'\}$ sharing a resource item. Sequencing constraints (3) and (4) explicitly account for sequence-dependent setup times at each common resource item z.

$$C_i + \hat{o}_{ii'z} \leq S_{i'} + M\left(1 - X_{ii'}\right) + M\left(2 - Y_{iz} - Y_{i'z}\right) \quad \forall i,i' \in I,\ i < i', r \in (R_i \cap R_{i'}), z \in RR_r \tag{3}$$

$$C_{i'} + \hat{o}_{i'iz} \leq S_i + M\ X_{ii'} + M\left(2 - Y_{iz} - Y_{i'z}\right) \quad \forall i,i' \in I,\ i < i', r \in (R_i \cap R_{i'}), z \in RR_r \tag{4}$$

Constraints (3) and (4) can be applied to sequence any pair of batches $\{i,i'\}$ at every resource item $z \in RR_r$ of any type $r \in (R_i \cap R_{i'})$ allocated to both ($Y_{iz} = Y_{i'z} = 1$). Moreover, the same binary variable $X_{ii'}$ can be used to denote the sequencing of batches $\{i, i'\}$ at any common resource item z. This implies an important saving of sequencing variables 0-1.

3.4 Sequencing constraints at any already allocated resource item z. Let us assume that resources of type $r \in R^{\#} \subset R$ have already been assigned. Suppose that I_z stand for the set of batches to which the resource item z has been allocated over the time horizon. For any pair of consecutive batches $\{i, i'\} \in I_z$ on the processing line of resource item z, the constraints (3)-(4) reduce to eqn. (5).

$$C_i + \tau_{ii'z} \leq S_{i'} \qquad \forall i,i' \in I_z (n_{iz} + 1 = n_{i'z}), r \in R^{\#}, z \in RR_r \tag{5}$$

Therefore, sequencing constraints (3)-(4) apply to non-allocated resources while constraint (5) controls the batch timing at already assigned resources. In this way, it is possible to sequentially allocate, for instance, first the processing units and then the manpower to batches to be processed. Batch sequencing at already assigned resources are assumed to remain unchanged during the allocation of other resource types. In a next paper, such a batch sequencing will be allowed to change too.

3.5 Order tardiness and earliness. The earliness E_i or tardiness T_i of batch i takes a positive value only if it is completed either earlier or later than its specified due date d_i.

$$T_i \geq C_i - d_i \qquad \forall i \in I \tag{6}$$

$$E_i \geq d_i - C_i \qquad \forall i \in I \tag{7}$$

3.6 Problem objective function. The problem goal is to complete the batches just in time by minimizing the overall weighted earliness and tardiness.

$$Min \quad \sum_{i \in I} (\alpha_i E_i + \beta_i T_i) \tag{8}$$

In this way, the batch scheduling problem with limited discrete resources has been modelled as an MILP involving the set of constraints (1)-(7) and the objective function (8).

3.7 Using preordering rules to get a near-optimal schedule. Preordering rules arranging the batches at any resource item by decreasing due dates or slack times can be easily embedded in the proposed formulation to attain a further reduction in the number of sequencing variables and constraints. For example, if the batches are to be sequenced by the EDD rule and $d_i < d_{i'}$, then $X_{ii'}$ can be eliminated from the model. Moreover, constraint (4) for the pair of batches $\{i,i'\}$ can also be removed while the corresponding constraint (3) should be rewritten without the term involving $X_{ii'}$. Frequently, preordering rules allow one to discover a very good solution to real world scheduling problems that otherwise it would never be found.

4. Results and Discussion

The proposed MILP approach to the batch scheduling problem under severe discrete resource constraints will be illustrated by tackling a couple of examples. Example 1 previously studied by Pinto and Grossmann (1997) involves the scheduling of 12 batches in a single-stage batch plant over a one-month period. Though four parallel units (U_1, U_2, U_3, U_4) can be run in parallel, limited manpower supplies (two or three operators crews) prevents from running all the units at the same time. A single operator crew per unit is required. Data for Example 1 are those related to the first twelve orders in Table 1. Three cases were studied: (a) non-limited manpower; (b) three operator crews available; (c) two operator groups available. Similarly to prior work, the problem objective was to minimize batch earliness, assuming that due dates are imposed as hard constraints on the completion times ($T_i = 0$, for any batch i). Since processing units and manpower are allocated to batches at the same time, constraints (5) are ignored. Gantt charts describing the optimal solutions to the three cases are depicted in Figures 1a, 1b and 1c, respectively. Model sizes and computational requirements for both (i) the logic-based approach of Pinto and Grossmann (1997) with preordering constraints and (ii) the proposed MILP formulation with/without applying the minimum-slack-time rule, are shown in Table 2. It can be observed that our approach discovers a better solution to any of the three case studies. For Examples (1b) and (1c), the CPU time is reduced by a factor of 450 and 1100, respectively, when compared with the highly efficient logic-based approach. Example 2 deals with the scheduling of the twenty-nine orders in the same single-stage batch plant (see Table 1). This example was studied before by Méndez & Cerdá (2001) who reported, for the unconstrained case, the production schedule included in Figure 1d. Assuming that the equipment items have already been

Table 1. Order data

order	due date (day)	U_1	U_2	U_3	U_4	order	due date (day)	U_1	U_2	U_3	U_4
O_1	15	1.538			1.194	O_{16}	30	1.250			0.783
O_2	30	1.500			0.789	O_{17}	30	4.474			3.036
O_3	22	1.607			0.818	O_{18}	30		1.429		
O_4	25			1.564	2.143	O_{19}	13		3.130		2.687
O_5	20			0.736	1.017	O_{20}	19	2.424		1.074	1.600
O_6	30	5.263			3.200	O_{21}	30	7.317		3.614	
O_7	21	4.865		3.025	3.214	O_{22}	20			0.864	
O_8	26			1.500	1.440	O_{23}	12			3.624	
O_9	30			1.869	2.459	O_{24}	30			2.667	4.000
O_{10}	29		1.282			O_{25}	17	5.952		3.448	4.902
O_{11}	30		3.750		3.000	O_{26}	20	3.824			1.757
O_{12}	21		6.796	7.000	5.600	O_{27}	11	6.410			3.937
O_{13}	30	11.25			6.716	O_{28}	30	5.500			3.235
O_{14}	25	2.632			1.527	O_{29}	25				4.286
O_{15}	24	5.000			2.985						
setup time		0.180	0.175		0.237			0.180	0.175		0.237

allocated as shown in Figure 1d, the proposed MILP formulation, including constraints (5) for the units, was applied to also allocate the required manpower. It is supposed that three operator crews are available to run a similar number of units. In this manner, the schedule depicted in Figure 1e has been found. Gantt charts describing the assignment of units and manpower over the time horizon are shown. Model sizes and CPU requirements are also included in Table 2.

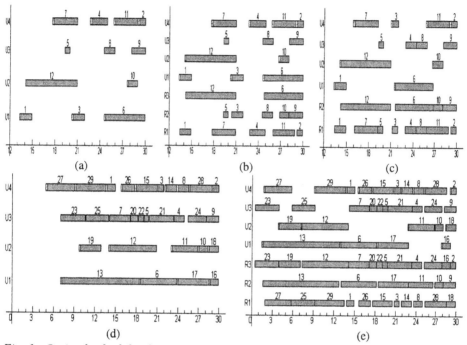

Fig. 1. *Optimal schedules for Example 1 (a) without resource constraints (b) with three operator crews (c) with two operator crews. Schedules for Example 2 (d) without resource constraints (e) with three operator crews.*

Table 2. *Model sizes and computational requirements*

Example	binary vars, cont. vars, rows	objective function	CPU time	nodes
#1. Logic-based approach with preordering constraints (Pinto and Grossmann, 1997)				
No resource constraints	1.581	63.56[a]	283	
Resource constrained (at most 3 units)	2.424	125.42[a]	673	
Resource constrained (at most 2 units)	8.323	927.16[a]	7341	
#1. This approach without preordering rules				
No resource constraints	82 , 12 , 202	1.026	0.11[b]	64
Resource constrained (at most 3 units)	127 , 12 , 610	1.895	7.91[b]	3071
Resource constrained (at most 2 units)	115 , 12 , 478	7.334	35.87[b]	19853
-This approach with preordering rules				
No resource constraints	25 , 12 , 119	1.026	0.05[b]	12
Resource constrained (at most 3 units)	61 , 12 , 329	1.895	0.28[b]	127
Resource constrained (at most 2 units)	49 , 12 , 263	7.334	0.82[b]	708
#2. MILP formulation (Méndez and Cerdá, 2001)				
No resource constraints		62.479		
#2. This approach with preordering rules				
Resource constrained (at most 3 units)	87 , 29 , 1352	110.57	89.97[b]	21907

[a] Seconds on IBM 6000-530 with GAMS/OSL. [b] Seconds on Pentium III PC (933 MHz) with ILOG/CPLEX

5. Conclusions

A highly efficient MILP formulation for the resource-constrained batch scheduling problem equally treating different types of discrete resources including equipment items has been developed. When compared with previous approaches, the proposed model shows a remarkable saving in 0-1 variables and a two-order-of-magnitude reduction in CPU time.

Nomenclature

(a) Sets

I orders to be scheduled

I_z orders assigned to the resource item z at the current schedule ($I_z \subseteq I$)

J_i available processing units to process order $i \in I$ ($J_i = RR_{ij}$)

R renewable resources (processing unit, manpower, utility, etc.)

$R^\#$ renewable resources that have already been assigned ($R^\# \subseteq R$)

R_i renewable resources required by order i ($R_i \subseteq R$)

R'_i renewable resources required by order i except processing units ($R'_i \subseteq R_i$)

RR_r available resource items of type $r \in R$

(b) Parameters

d_i due date of order $i \in I$

M a very large number

n_{ir} position of the order $i \in I$ on the current sequence of resource item r

pt_{ij} processing time of order $i \in I$ in unit j

q_{rz} amount of resource r assigned to resource item z

sl_i slack time of order i, $sl_i = d_i - Min\{pt_{ij}, j \in J_i\}$

su_{ij} setup time of order $i \in I$ in unit j

$\tau_{ii'r}$ sequence-dependent setup time between orders $i \in I$ and $i' \in I$ in resource item r

v_{irj} amount of renewable resource $r \in R$ required to process order i when order i is allocated to unit j

α_i weighting coefficient for earliness of order $i \in I$

β_i weighting coefficient for tardiness of order $i \in I$

(c) Variables

C_i completion time for order i

E_i earliness for order i

S_i starting time for order i

T_i tardiness for order i

$X_{ii'}$ binary variable denoting that order $i \in I$ is allocated before ($X_{ii'} = 1$) or after ($X_{ii'} = 0$) order $i' \in I$ in some common resource item r

Y_{ir} binary variable denoting the allocation of order i to resource item r

References

ILOG OPL Studio 2.1 User's Manual, 1999, ILOG S.A. France.

Méndez, C.A., Cerdá, J., 2001, Dynamic Scheduling in Multiproduct Batch Plants. Proceedings of 2nd Pan American Workshop on Process Systems Engineering, Guarujá, Sao Paulo, Brazil.

Pantelides, C.C., 1994, Unified Frameworks for Optimal Process Planning and Scheduling. In Foundations of Computer Aided Process Operations, Austin, TX, 253.

Pinto, J. M., Grossmann, I. E., 1997, A Logic-based Approach to Scheduling Problems with Resource Constraints. Comput. Chem. Eng., 21, 801.

Reklaitis, G.V., 1992, Overview of Scheduling and Planning of Batch Process Operations. NATO Advanced Study Institute-Batch Process Systems Engineering, Antalya, Turkey.

Schilling G., Pantelides C. C., 1996, A Simple Continuous-Time Process Scheduling Formulation and a Novel Solution Algorithm. Comput. Chem. Eng., 20, S1221.

European Symposium on Computer Aided Process Engineering – 12
J. Grievink and J. van Schijndel (Editors)
© 2002 Elsevier Science B.V. All rights reserved.

State-Space Residual Based Monitoring

E. Mercer, E. B. Martin and A. J. Morris
Centre for Process Analytics and Control Technology
University of Newcastle, Newcastle upon Tyne, NE1 7RU, England

Abstract

Although the process performance monitoring tools of dynamic Principal Component Analysis (PCA) and Canonical Variate Analysis (CVA) take into account process dynamics, the monitoring statistics still contain serial correlation. Consequently the traditional statistical basis for the calculation of control limits will be invalid resulting in either missed out-of-control signals or an excess of false alarms. A methodology is proposed whereby a CVA state-space model is first developed and then a PCA based monitoring scheme is formed using the model mismatch. In this case, the residuals will be independent and identically distributed and the standard control limits will be valid. The methodology is demonstrated on the benchmark Tennessee Eastman problem.

1. Introduction

The traditional approach to developing process representations for the monitoring of continuous processes have been the multivariate statistical projection techniques of Principal Component Analysis (PCA) and Partial Least Squares (PLS). These approaches do not take into account the serial correlation in the data. This has led to the development of dynamic monitoring tools including dynamic PCA (Ku *et al.*1995) and Canonical Variate Analysis (CVA) (Larimore, 1997). As the latent variables/states do not necessarily represent a specific physical measurement on the plant, the control limits are determined statistically. Calculation of the limits is based on the assumption that the data underlying the metrics is independent and identically distributed (i.i.d). This assumption tends to be invalid for the majority of chemical processes. If serial correlation is present in the data, then calculating the limits based on the assumption that the data is i.i.d. will result in an increase in the number of false alarms or in the fault detection time. One approach to removing the serial correlation is to increase the sampling period. However adopting this approach can delay the detection of process changes.

An alternative approach that has been proposed to address the problem of serial correlation in the resulting metrics has been to base the monitoring statistics on the CVA state space equation residuals, **w** and **e** (Simoglou *et al*, 1999), calculated from:

$$\mathbf{x}(t+1) = \mathbf{Cx}(t) + \mathbf{Gu}(t) + \mathbf{w}(t) \tag{1}$$

$$\mathbf{y}(t) = \mathbf{Hx}(t) + \mathbf{Au}(t) + \mathbf{Bw}(t) + \mathbf{e}(t) \tag{2}$$

where **x** is a state vector, and **u** and **y** are the process inputs and outputs respectively. The noise terms, **w** and **e**, are assumed to be i.i.d. with covariance matrices **Q** and **R**.

Simoglou *et al.* (1999) proposed two metrics based on the state-space residuals, T^2_w and T^2_e. If the state space equations adequately define the process, the residuals will be i.i.d. and the control limits can be calculated using the appropriate statistical distribution. Simoglou *et al.* also compared the statistical distributional approach to calculating the limits with the empirical reference distribution (Willemain and Runger, 1996) in terms of the false alarm rate and the time to fault detection. A number of monitoring statistics have also been developed based on the CVA states. The first of these was the T^2 statistic based on the k significant states, T^2_s, (Negiz and Cinar, 1997). This idea was further developed by using those states that are not considered significant as the basis of another metric, T^2_q, (Russell *et al.* 2000). Both statistics, T^2_s and T^2_q, have control limits based on the F-distribution. However the data is not i.i.d. and thus the calculated limits are not valid and will result in spurious alarms. This paper proposes an extension to the work of Simoglou *et al.* where PCA is performed on the state space output residuals, **e**. This allows the use of the PCA T^2 and SPE metrics to monitor the residuals, i.e. T^2_{PCA} and SPE_{PCA}. The limits can be calculated using an F-distribution as the residuals will be independent identically distributed.

2 The Tennessee Eastman Process

The Tennessee Eastman process simulation forms the basis of the subsequent study. The process consists of a reactor/separator/recycle arrangement and includes two gas/liquid reactions and two-side reactions that produce the same by-product. The simulation comprises 12 manipulated variables and 41 measured variables (see Downs and Vogel, 1993 for details). In this study, one manipulated variable, agitator speed, is constant so is excluded from the subsequent analysis. The Lyman control scheme for the 50/50 product ratio was used (Lyman and Georgakis, 1995).

Twenty-three data sets were generated of which three were associated with nominal data and twenty were based on pre-programmed faults (Table 1). Each run lasted 48 hours with samples taken every three minutes, with faults occurring after 8 hours (160 observations). A CVA state-space model was built from nominal data set one. This is in contrast to the approach of Russell *et al.* (2000) where a separate model was developed for each fault. A lag of three was selected for all variables and 29 states were included in the CVA model. These values were selected using the small sample corrected Akaike Information Criterion (AIC) (Hurvich *et al.* 1990). All data was scaled to zero mean, unit variance based on the first nominal data set.

The CVA based state space model built on the first nominal data set, resulted in residuals that were artificially low in magnitude. Thus a second nominal data set was used to calculate the residuals. These formed the basis of the PCA model and the control limits. The residuals, **e**, were obtained by subtracting the output model estimates from the process simulation outputs, whilst subtracting the state step-ahead predictions from the CVA calculated states gave the state errors, **w**. These two sets of residuals were then used to calculate the respective covariance matrices and the T^2_w and T^2_e statistics. The output prediction residuals, **e**, were normalised to mean zero and unit variance and PCA was applied. The number of principal components retained were those with an eigenvalue greater than one.

The false alarm rate for a 99% action limit was then investigated using the third nominal data set. From the results in Table 2, it can be concluded that T^2_{PCA} gave the value closest to that expected, i.e. 0.005. The poorest performance was exhibited by T^2_q and T^2_w which had false alarm rates in excess of 10%. The next step was to examine the serial correlation retained by each metric. This was done using the autocorrelation function (ACF) (Fig. 1) and the partial autocorrelation function (PACF) (Fig. 2). It can be seen that the false alarm rate is typically proportional to the level of serial correlation, i.e. the higher the level of serial correlation retained in the metric, the greater the false alarm rate.

The first measure used to compare the statistics' ability to detect faults was their sensitivity to the presence of faults, illustrated by the proportion of in-control signals once an initial fault had occurred. Faults 3, 9, and 15 were excluded due to them having no apparent effect on process operation. Table 3 shows the proportion of observations that showed an in-statistical-control reading once the prescribed fault had occurred. These results indicate that T^2_e gives the lowest missed detection rate for the majority of the faults with SPE_{PCA} giving marginally poorer results. However from the ACF and PACF, Figs. 1 and 2, SPE_{PCA} appears to contain less serial correlation. This suggests that the T^2_e limits may be artificially low, giving a high false missed detection rate.

Table 1. Tennessee Eastman Process Faults

Fault Number	Fault Description	Type
IDV(1)	A/C feed ratio, B composition constant (stream 4)	Step
IDV(2)	B composition, A/C ratio constant (stream 4)	Step
IDV(3)	D feed temperature (stream2)	Step
IDV(4)	Reactor cooling inlet temperature	Step
IDV(5)	Condenser cooling inlet temperature	Step
IDV(6)	A feed loss (stream 1)	Step
IDV(7)	C header pressure loss- reduced availability (stream 4)	Step
IDV(8)	A, B, C feed composition (stream4)	Random Variation
IDV(9)	D feed temperature (stream 2)	Random Variation
IDV(10)	C feed temperature (stream 4)	Random Variation
IDV(11)	Reactor cooling water inlet temperature	Random Variation
IDV(12)	Condenser cooling water inlet temperature	Random Variation
IDV(13)	Reaction kinetics	Slow Drift
IDV(14)	Reactor cooling water valve	Sticking
IDV(15)	Condenser cooling water valve	Sticking
IDV(16)	Unknown	Unknown
IDV(17)	Unknown	Unknown
IDV(18)	Unknown	Unknown
IDV(19)	Unknown	Unknown
IDV(20)	Unknown	Unknown

Table 2. False Alarm Rate

Statistic	False Alarm Rate	Statistic	False Alarm Rate
T^2_s	0.1076	T^2_e	0.0327
T^2_q	0.0549	T^2_{PCA}	0.0052
T^2_w	0.1014	SPE_{PCA}	0.0378

Table 3. Missed Fault Detection Rates (bold denotes best result)

Fault	T^2_s	T^2_q	T^2_w	T^2_e	T^2_{PCA}	SPE_{PCA}
IDV(1)	0.0038	0.0025	**0.0000**	0.0013	0.0025	0.0013
IDV(2)	0.0126	**0.0088**	0.0100	0.0125	0.0138	0.0176
IDV(4)	0.8952	0.0013	0.7215	**0.0000**	0.6939	0.0188
IDV(5)	0.6604	0.4811	0.6625	**0.0000**	0.0013	**0.0000**
IDV(6)	0.0013	0.0013	0.0000	**0.0000**	**0.0000**	**0.0000**
IDV(7)	0.3586	0.0013	0.3864	**0.0000**	0.0088	**0.0000**
IDV(8)	0.0189	0.0164	0.0151	**0.0088**	0.0176	0.0113
IDV(10)	0.1604	**0.0821**	0.1581	0.0828	0.0979	0.1380
IDV(11)	0.7854	**0.0909**	0.5069	0.2271	0.4881	0.3099
IDV(12)	0.0076	0.0063	0.0289	**0.0013**	0.0025	0.0038
IDV(13)	0.0467	0.0442	0.0351	**0.0402**	0.0439	**0.0402**
IDV(14)	0.0278	0.0013	**0.0000**	0.0013	0.1343	0.0100
IDV(16)	0.2551	0.0543	0.1280	**0.0703**	0.1205	0.1092
IDV(17)	0.1048	0.0240	0.0402	**0.0221**	0.0263	0.0226
IDV(18)	0.0922	0.0896	**0.0853**	0.0878	0.0954	0.0966
IDV(19)	0.6995	0.0025	0.2120	**0.0088**	0.1882	0.0452
IDV(20)	0.3333	0.1982	0.3526	**0.0765**	0.2509	0.0790

The next metric examined was the time to the initial out-of-control signal once a fault had occurred. Examining Table 4, the time between the fault occurring and an out-of-control value being registered, the best performers were T^2_e, T^2_w, and SPE_{PCA}. This is because although the statistics containing serial correlation, T^2_s and T^2_q, have a higher false detection rate (see Table 2), their ability to detect a true out-of-control signal more quickly is compromised by the effect of the previous in-control measurements.

In terms of sensitivity to faults and speed of detection once a fault has occurred, the above results show that the monitoring metrics based on **w** and **e**, the residuals from the state space equations, gave better results compared to those statistics based on the calculated states. The residual based statistics also gave fewer false out-of-control signals. These results are due to the reduced level of serial correlation in the residuals compared to the states, resulting in more appropriate statistical limits as the i.i.d. assumptions are upheld.

The statistic with the lowest level of serial correlation, T^2_{PCA}, did not give the best results but were close to those of the T^2_e and SPE_{PCA} results, although the level of serial correlation and false alarm rate in T^2_{PCA} suggests that it provided the best statistical control limits. The performance of the residual based techniques are comparable to those presented by Russell *et al.* (2000) who used a library of faults and limits based on the 99th percentile of a fault free validation data set.

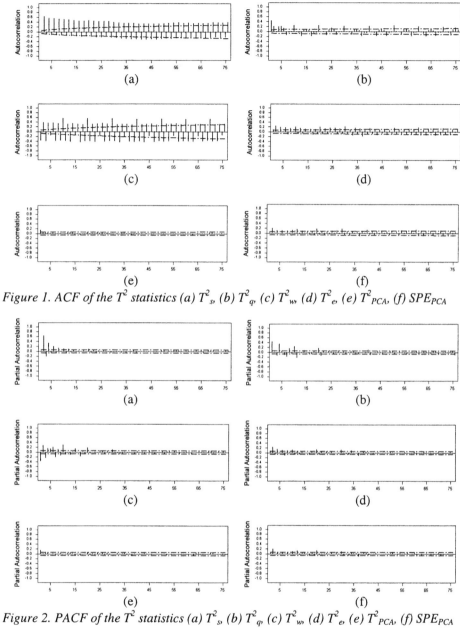

Figure 1. ACF of the T^2 statistics (a) T^2_s, (b) T^2_q, (c) T^2_w, (d) T^2_e, (e) T^2_{PCA}, (f) SPE_{PCA}

Figure 2. PACF of the T^2 statistics (a) T^2_s, (b) T^2_q, (c) T^2_w, (d) T^2_e, (e) T^2_{PCA}, (f) SPE_{PCA}

3 Conclusions

The Tennessee Eastman simulation was used to compare dynamic monitoring statistics based on process states and their residuals. The metrics examined were T^2_s based on the used CVA states, T^2_q based on the unused states, T^2_w and T^2_e based on the state space equation residuals, and two proposed statistics, T^2_{PCA} and SPE_{PCA} calculated by performing PCA on the output prediction residuals, **e**.

Table 4. Time to Fault Detection (mins) (bold denotes best result)

Fault	T^2_s	T^2_q	T^2_w	T^2_e	T^2_{PCA}	SPE_{PCA}
IDV(1)	9	6	3	**0**	6	3
IDV(2)	21	18	21	15	30	**12**
IDV(4)	6	3	**0**	**0**	**0**	**0**
IDV(5)	3	6	**0**	**0**	3	**0**
IDV(6)	3	3	**0**	**0**	**0**	**0**
IDV(7)	3	3	**0**	**0**	**0**	**0**
IDV(8)	**15**	30	18	18	33	18
IDV(10)	63	63	63	**60**	63	69
IDV(11)	33	**18**	**18**	**18**	21	**18**
IDV(12)	9	12	3	**0**	6	3
IDV(13)	111	105	**102**	**102**	108	**102**
IDV(14)	3	3	3	**0**	3	3
IDV(16)	33	24	**18**	**18**	21	**18**
IDV(17)	63	57	54	54	57	**51**
IDV(18)	237	228	222	**210**	225	240
IDV(19)	45	6	**0**	**0**	6	**0**
IDV(20)	225	195	**186**	222	219	**186**

The metrics based on the state space equation residuals, **w** and **e**, were shown to contain less serial correlation than those using the system states themselves, as the majority of the structure is contained within the states. The ability to detect faults both accurately and quickly was best shown by the T^2_e and SPE_{PCA} statistics, followed T^2_{PCA}, T^2_w, and T^2_q. The validity of the performance of the T^2_w, and T^2_q statistics was questionable due to the inherent level of serial correlation. The serial correlation contained within, and the false alarm rate of, T^2_e and SPE_{PCA} raises questions about the results. Based on these initial studies using monitoring statistics formulated from the output prediction residuals from the state space equations seem to give improved fault detection without the need to modify the statistically calculated control limits.

Acknowledgements

Mr Mercer acknowledges the EPSRC, BP and CPACT for financial support of his PhD.

References

Downs, J. J. and E. F. Vogel, 1993, Computers Chem. Engng., 17, 245.

Hurvich, C. M., R. Shumway and C. L. Tsai, 1990, Biometrika, 77, 709.

Ku, W.F., R.H. Storer and C. Georgakis, 1995, Chemo. & Intell. Sys. 30, 179

Larimore, W. E., 1997, In Statistical Methods in Control and Signal Processing (Eds, Katayama, T. and Sugimoto, S.) Marcel Dekker, New York, 83.

Lyman, P. R. and C. Georgakis, 1995, Computers Chem. Engng., 19, 321.

Negiz, A. and A. Cinar, 1997, AIChE Journal, 43, 2002-2020.

Russell, E. L., L. H. Chiangand, R. D. Braatz, 2000, Chemo. & Intell. Sys. 51, 81.

Simoglou, A. E. B. Martin and A. J. Morris, 1999, Computers Chem. Engng., S277.

Willemain, T. R. and G. C. Runger, 1996, Journal of Quality Technology, 28(1), 31.

European Symposium on Computer Aided Process Engineering – 12
J. Grievink and J. van Schijndel (Editors)
© 2002 Published by Elsevier Science B.V.

Performance Monitoring for Process Control and Optimisation

Arnoud Nougues, Pierre Vadnais, Rob Snoeren
Shell Global Solutions
The Netherlands

Abstract

Over the last two decades many oil and petrochemical companies have installed Advanced Process Control (APC) and Closed Loop Optimisers in their plants. Within Shell for example there have been about 550 APC projects and 30 Closed Loop Optimisers installed that add 430 million Euro's per annum to the bottom line. Potentially this can grow to 700 million Euro's per annum.

For the coming years the challenge will be to maintain the optimum performance of the existing applications while at the same time implementing new projects. With skilled resources remaining limited, it means that more innovative steps have to be taken.

With this background, Shell has developed a number of methodologies to monitor the performance of controllers and optimisers. The key objective is to benchmark the performance against 'best in class' performance, identify non-compliances, and diagnose possible problems (tuning, modelling, etc.) so that the appropriate corrective actions can be taken.

Economic Incentive of Application Performance Monitoring

APC applications essentially consist of multivariable controllers, varying in size from small local applications, e.g. with 2 or 3 Manipulated Variables (MV's), to unit-wide optimising multivariable applications where an economic variable is explicitly included in the control strategy for LP type optimisation.

In a refinery where all major APC applications have been implemented, reported benefits are typically in the range of 10-15 US cents/BBL overall. For an average site processing 150 KBBL/D, this translates into benefits in the order of 5 to 8 MMUS$/Y. Similar numbers apply to petrochemical plants.

For refinery optimisers the tangible benefits range from 5 to 10 US cents/BBL throughput. Achieved benefits come from three main sources:

- Feed increase from more stable operation, closer to the limiting operating constraints and products specifications,
- Yield improvement of the more valuable products,
- Energy savings.

Besides the tangible benefits, extensive lists of non-tangible benefits have been acknowledged as important spin-offs from APC /optimisation projects.

The benefits at stake are therefore considerable. If the applications performance is not sufficient, then only a fraction of the expected benefit will be achieved, implying an economic loss from non-compliance of the applications, and a lower than expected return on investment of the APC and optimisation implementation projects.

The importance of Performance Monitoring for Applications Maintenance

Effective applications maintenance nowadays implies putting in place a performance monitoring system with the following general characteristics:

- Provision of the right amount of information (%uptime, performance index, information to allow diagnosing the root cause of potential problems) in a systematic and concise way.
- The information should be readily available, at the engineer's desktop, and available in real time with no effort, preferably using web-enabled technology.
- Remote maintenance option: as more and more applications are put on-line, and under the general competitive pressure to cut costs, control and optimisation manpower at sites is generally scarce and overloaded. Remote maintenance then becomes an attractive option, to allow support from specialised staff in a central location (technical head office, or contractor's office).
- Application performance should be measured against pre-defined targets, and corrective action taken as required when non-compliances occur.

APC and Base Layer Control Monitoring

Shell has developed and is in the process of further developing a complete suite of software packages, called MD (Monitoring and Diagnosis) for monitoring the performance of control loops and to assist in troubleshooting loops that fail to meet their performance target. The tools apply to multivariable controls, of Shell technology (SMOC) as well as multivariable controls from any APC vendor, and they apply to traditional Single Input-Single Output loops (e.g. PID controller).

The central element of MD is a client-server information system for control loop performance tracking. MD is linked to various commercial plant data historians (e.g. Yokogawa's Exaquantum, OSI PI), where the basic real-time control loop status and performance information resides. Each day performance statistics are automatically calculated and stored in a dedicated Relational Data Base. Control engineers are notified if control loops are performing below predefined targets by daily email summary reports. Next to this, the user can enter the report mode where the statistical information can be browsed.

For every control loop and Controlled Variable (CV), MD provides the following statistical information:

- % in Service: optional controller and unit availability tags are monitored to determine whether the controller is in service.
- % Uptime: loop uptime is determined from the controller mode status.
- % in Compliance: this statistic indicates if a CV deviates significantly from either a setpoint or Min./Max. limits. The bound to indicate a significant deviation is determined from a user specified tolerance (CL) for each CV. If the CV is within the $\pm$ CL bound about the control limits (set range), the CV is considered to be in

compliance. The information is reported as daily and monthly averages based on calculations carried out using typically one-minute data.

%in Service, %Uptime, %in Compliance together with a user defined cost factor are used to derive a cost incentive which is reported on a daily and monthly basis to the user (PONC : Price of Non-Conformance).

Monitoring loop performance is not sufficient. Additional tools are required to help analyse loop performance related problems and troubleshoot under-performing loops efficiently. A number of innovative proprietary loop performance diagnosis techniques have been developed by Shell and are part of the MD suite of packages:

- Average closed loop response curves: both CV error and MV response curves are calculated and plotted. Average response curves provide a visual summary of the shape and response time of SISO as well as multivariable control loops, in response to the actual disturbances affecting the process and in response to setpoint changes. The average response curves are derived from fitting an ARMA (Auto-Regressive Moving Average) model to the loop time series data, typically over several hours or days of normal closed loop operation.

- Plot of sliding window CV error standard deviation; comparison with best achievable performance from Minimum Variance Controller (MVC): the CV error standard deviation is calculated over a representative time span, and then the calculation is repeated by sliding the window from the start to the end of the data time range. The standard deviation that would have been achieved by the fastest possible feedback controller (MVC) is shown on a parallel plot. The CV error standard deviation plots are useful in assessing the loop performance in relative terms (achieved CV error standard deviation, and how it evolves in time) as well as in absolute terms (comparison with reference MVC controller).

- Degrees of freedom and constraint analysis: this technique applies to multivariable control applications. The idea is to track and report which controlled variables in a closed loop multivariable system are active (i.e. driven to their upper or lower specification limit) and how often these are active. Correspondingly, the activity status of the manipulated variables is reported, i.e. which MV's are limited or unavailable and how often. This information, presented in the form of bar plots and trends, provides insight into the activity and performance of a complex multivariable controller, and helps diagnose control structure problems (e.g. insufficient degrees of freedom to achieve the required control objectives).

Real Time Optimisation Monitoring and diagnosis techniques

Recent monitoring and diagnosis developments have been focussed on the business objective of an optimiser.

An optimiser is totally and strictly driven by an economic objective function. This objective function is based on the best calculation available for the unit's margin using current market values for feeds, products and utilities. With such an objective function, the optimiser is clearly focused on the bottom line, on maximising profit.

In an attempt to help the understanding of the optimised solutions and to facilitate the communication between all involved, a few graphs were developed to show how the optimiser is improving the unit's margin and to pinpoint all the relevant constraints.

Product upgrading

The first graph, in Figure 1, shows the cumulative relative contribution of each element of the objective function to the increase in the unit's margin from the base case to the optimised case. A positive contribution is shown in green (light) and a negative contribution is shown in red (dark).

The cost elements (feeds, imports, utilities) have a positive contribution when they are decreased. The revenue elements (products) have a positive contribution when they are increased.

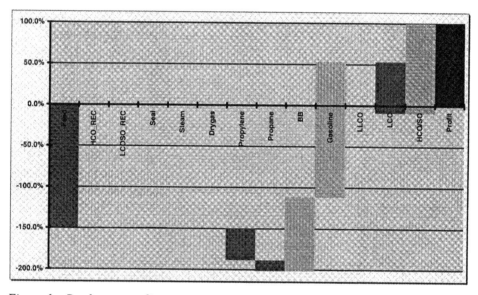

Figure 1 – Product upgrading

The graph gives a quick overview of the changes to the mass balance. For someone who knows the unit and has an idea of the current economic drivers, it is easy to evaluate the appropriateness of the changes. In some cases, further analysis is required to fully understand how the changes are actually implemented in the unit.

Independent moves

The second graph, in Figure 2, displays the cumulative relative contribution of each independent variable to the objective function. A positive contribution is shown in green (light) and a negative contribution is shown in red (dark).

The individual contributions are an approximate estimation of the true contribution of each independent variable based on the actual move and the average effect of the independent on the objective function (dO/dX). The purpose is mainly to single out the independents with the largest contributions.

First, it is important to know the main positive contributors represented by the longest green (light) bars on the right hand side. They can generally explain the changes to the mass balance that generated the profit seen in the first graph.

Second, it is important to understand the main negative contributors represented by the red (dark) bars on the left hand side. Why would an independent move "knowingly" in the direction that reduces its contribution to the objective function, unless it is to help some other independent(s) to provide an even greater contribution.

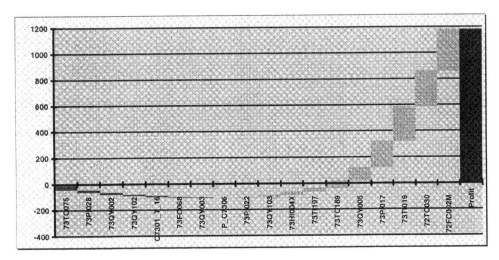

Figure 2 – Contribution of independent variables

Would you expect operators to predict that a move in the "wrong" direction for one variable will be favourably compensated by relaxing a constraint on another variable? Would you consider using a slogan going against the known profitable direction? Would you recommend to use a linear multi-input constraint pusher to reach the tenuous balance of a crude preheat train integration with the column pressure? The fact that an online optimiser can move some independents in the "wrong" direction to make more money with the others largely explain why the results of online optimisers are sometimes difficult to understand for all involved. It should also prove beyond any doubts the superiority of the technique for continuous optimisation of any operation.

Finally, it is possible to identify the independents with little or no effect on the objective function and investigate if they are over constrained, redundant or temporarily ineffective.

Constraints

Having seen what is happening in the first graph and how it is produced in the second graph, it is normal to wonder "Why not more?". The third graph, Figure 3, shows how far each independent has moved relative to its range of freedom. The range of freedom is shown by a green (light) frame in the direction (up or down) of a positive contribution to the objective function and by a red (dark) frame in the direction of a negative contribution to the objective function. If the independent moved in the "right" direction, the green frame is partially or totally filled by a green (light) bar proportional to the

738

extent of the move relative to the potential move. If the independent moved in the "wrong" direction, a similar red (dark) bar fills the red frame.

To maintain the business focus, the frames are relatively sized on their potential contribution to the objective function with 100% assigned to the largest potential contribution.

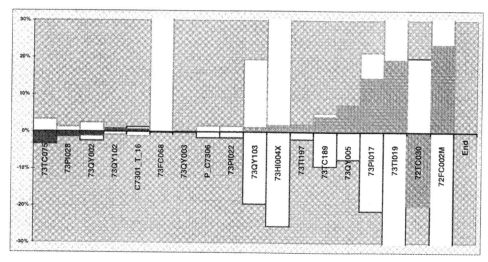

Figure 3 – Freedom of independent variables

But why are some frames only partially filled? Because some independents are limited by constraints on resulting variables. The independents can be self limited by their own direct constraints. They can also be limited by a number of dependent variables, which have reached the limit in the direction where the specific independent is trying to push.

Optimiser maintenance and performance monitoring services
Online optimisation is highly technical, but it is also a collaborative tool, a synergistic tool, which brings together the sum of the knowledge of all the team members about the process technology, the unit configuration, the economic incentives, the operating constraints, and the mathematical algorithms.

It appears that the commitment of the entire team (optimisation engineer, process engineer, operations, and economist) with full support of management is the most significant constant behind a successful optimisation project. Too often, we have seen a successful optimiser loose momentum after the transfer of the champion who could keep the spirit alive by stimulating the communication of the right information within and around the team.

Based on this experience, the continues pressure on operating costs, the lack of sufficient skilled supporting staff, Shell Global Solutions International has extended their services with APC / optimiser maintenance and performance monitoring making use of developed monitoring tools and highly skilled professionals.

Via web enabled technology day-to-day, worldwide support can be given from our head-offices in Amsterdam, Singapore, Kuala Lumpur and Houston.

European Symposium on Computer Aided Process Engineering – 12
J. Grievink and J. van Schijndel (Editors)
© 2002 Elsevier Science B.V. All rights reserved.

Optimization of Naphtha Feedstock Blending for Integrated Olefins-Aromatics Plant Production Scheduling

Y. Ota, K. Namatame, H. Hamataka, K. Nakagawa and H. Abe
Mitsubishi Chemical Corp. Kashima Plant, Ibaraki, Japan
A. Cervantes, I.B. Tjoa[†] and F. Valli,
MC Research and Innovation Center, Mountain View, CA 94041, USA.

Abstract

An optimization system for the planning and scheduling of a petrochemical complex is presented in this work. The complex uses naphtha, with a wide range of properties and costs, as its main feedstock. Depending on the plants production constraints and market conditions, the optimization of the feedstock blending has a significant economical impact. We describe the production planning and scheduling system for the integrated petrochemical complex. This application covers scheduling decisions in the naphtha and gasoline tank yards, two olefins plants and two aromatics plants. The optimization problem poses challenges due to the nonlinearities of the process model and the combinatorial part of the naphtha delivery and tank yard operation. A model in the form of a mixed integer nonlinear programming (MINLP) problem is used. In this work, we discuss the practical project implementation. The benefits of using the scheduling system are reflected on the plant operation. Two case studies are presented. The optimal naphtha scheduling on the tank yard section gives more flexibility to the naphtha blends that are fed to the plants. This flexibility is reflected on a better plant utilization that makes the whole operation more profitable.

1. Introduction

Production scheduling is an important operational activity in a chemical industry. The main focus of scheduling has been on batch processes. A comprehensive review on scheduling and planning of batch plant operations can be found in Reklaitis (1991, 1992), and a summary of basic scheduling techniques can be found in Pekny and Zentner (1994). Although there are many reported linear programming approaches for the planning problem of continuous processes, there are few studies reported on the optimization of the scheduling problem because it requires a large nonlinear model with many degrees of freedom. A reliable solution method for this type of problem is not widely available. The first reported MINLP based ethylene production scheduling model was presented by Tjoa et. al. (1997). Pinto et. al. (2000) addressed a nonlinear model for the a refinery plannning and scheduling problem where the blending problem is also a very important issue. In this paper, we extend the methodology of Tjoa et. al. (1997) to solve a large petrochemical complex model that includes a tank yard, two ethylene plants and two aromatic plants.

[†] Corresponding author: tjoa@mcric.com.

2. Problem Description

The production scheduling covers operational decisions for a tank yard section, two ethylene and two aromatics plants. Given a monthly production target of key products from the planning system, the main task for the scheduling system is to decide daily operational decisions such as transfer rates of feedstocks in the tank yard section, production target of key products, etc. The main challenge in this scheduling problem is to allocate the right composition of naphtha in the charging tank that maximizes the utilization of the main resources: furnaces and separation units. Since naphtha prices vary with its composition, the optimal allocation of the naphtha feedstock has a significant financial impact. In the following subsections, we describe the problem and constraints without presenting the mathematical formulation. The main purpose is to understand the economic opportunity and complexity of the model.

2.1 Tank Yard Section

Naphtha feedstock poses a fairly complex problem for the ethylene producer due to the wide variety of supplied naphtha qualities. The complexity of the tank yard management is related to naphtha delivery, storage, transfer, and mixing. In this problem, most of the naphtha is delivered by ship, and the rest is delivered through pipes on fixed dates. On a typical month, we need to schedule about 10 vessels that carry naphtha of various grades. There are few tanks at several stages before the feedstock reaches the charging tanks. Availability of extreme naphtha compositions at the charging tanks is very crucial for providing flexible online blending for meeting the targeted compositions and amounts required for each ethylene plant. Thus the objective of the tank yard section is to satisfy naphtha demand, maintain inventory volume and provide the ethylene plants with the optimal naphtha mixture.

In the model formulation, besides total and component mass balances on the naphtha tanks, the following constraints are formulated as a mixed integer nonlinear model:
- Naphtha storage tanks are divided into 3 stages that consist of 2 tanks each.
- A limited number of berths are available at stage 1 and 2.
- Other tank yard operational constraints due to piping network.

2.2 Ethylene plants

The main purpose of the ethylene plants model is to predict production rates for various feedstocks under an attainable plant operating condition. Here is the list of constraints for the ethylene plant:
- Nonlinear yield model for furnaces to represent 9 major products, from hydrogen to heavy ends. There are several furnace types for each plant. Furnace operational variables such as severity are considered in the model.
- Material flow between the two plants is allowed for certain products.
- Separation constraints that reflect the capacity of the separation columns.
- Utilities consumption and production in each plant.

A major economic trade off is the production of gasoline that is fed to the aromatic plants. Light feedstock produces insignificant amount of gasoline; on the other hand, heavy naphthas produce a significant amount of gasoline.

2.3 Aromatic Plants

The heavier products coming out from the separation units in the ethylene plants (gasoline and others) are the main input of the aromatic plants. We also have additional supplies of crude gasoline. Here, the compositions also vary. The objective of the aromatic plants is to transform the heavier products coming from upstream into aromatics. The most valuable product from the aromatics plants is benzene.

3. Modeling and Solution Approach

An accurate representation of the production scheduling requires a fairly detail model. The resulting model is often very large and complex. In order to reduce the modeling effort, we have developed an in house process planning and scheduling system that allows us to implement a project efficiently. Here we describe briefly the basic approach of our system.

3.1. Modeling Approach.

Our process planning and scheduling system uses a unit operation modeling approach, a similar approach as in the familiar process simulator systems. Here, we can handle both, process as well as operation constraints. Depending on the production operation flow diagram, the modeling environment will generate the appropriate mathematical programming formulation in GAMS modeling language, Brooke et. al. (1988). It can generate from a simple linear programming (LP) model to a more complex MINLP model. In this application, the system generates a MINLP model. The main advantage of this modeling approach is that we can modify the model representation efficiently.

3.2. Solution Approach.

The main advantage of our modeling approach is that the entire problem can be solved and optimized simultaneously using a mathematical programming method. Thus we can avoid a 'sequential' solution approach which can be very expensive for finding just a feasible solution given so many degrees of freedom in this model. The main challenge, however, is to develop a reliable solution methodology for solving a large scale MINLP model that has over a thousand binary variables and several thousands of continuous variables with over a thousand degrees of freedom. To illustrate the size of this application, this model has approximately 1,500 binary variables and 60,000 continuous variables.

Here we developed a proprietary solution strategy for solving a large scale MINLP model. Without disclosing the details, the basic idea of this strategy lies in decomposition of the full problem into two subproblems, a Mixed Integer Linear Programming (MILP) subproblem and a Nonlinear Programming (NLP) subproblems. The MILP model contains reliable linearized constraints based on process knowledge as well as mathematical programming techniques. The binary variables are decided from the MILP subproblem and the continuous variables are obtained from the NLP subproblem. Since we use GAMS modeling environment, there are several choices of solvers for solving each subproblem. Here, we use OSL (IBM, 1991) and CONOPT (Drud, 1992) for solving the MILP and NLP subproblems respectively.

Due to complexity of the process model, we use a similar decomposition strategy as described in Tjoa et. al. (1997). First, we solve the ethylene and aromatics sections based on two average naphtha compositions (light and heavy) in order to get an approximation to the amount of required feedstocks. This is an NLP problem, the nonlinearities appear in the yield models and the non-sharp split separation columns. Once we have an approximation of feedstock requirement, we solve the tank yard section as an MINLP model, trying to maintain the classification on the feedstocks into light and heavy. The objective is to keep one charging tank with heavy naphtha and the other one with light naphtha. The solution of the tank yard section satisfies the approximation of the feedstock requirements and also achieves a good classification of the naphtha. After solving these problems we resolved the ethylene and aromatics plants model with the fixed input to the charging tanks. This is again an NLP model. It takes approximately 20 min to solve the whole problem in a Pentium III, 800 MHz computer.

4. Case studies.

In order to demonstrate the importance of having an automated system for the plant scheduling we present two case studies. Both represent a one month schedule where the arrival of the naphtha shipments has been fixed. For a comparison purpose, the initial composition of the charging tanks is set the same, in this way the composition profiles only reflect the current schedule and not previous scheduling policies. In the first case we solved the problem with an open demand for ethylene and propylene, for the second case we set the demand equal to 85% of the production obtained for the open demand. The benzene demand for both cases is open.

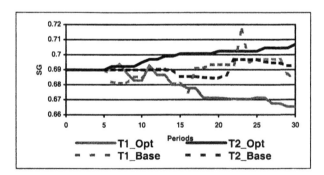

Figure 1. Charging Tanks Composition

The solutions of the optimization problem (Opt) are compared with those obtained by just getting a feasible solution (Base) for the tank yard section and an optimal solution for the ethylene and aromatics sections. The charging tanks (T1 and T2) compositions for these two solutions can be seen in Figure 1. For the optimized solutions (opt) there is a clear classification of the naphthas fed to the plant, the light naphthas go to T1 and the heavy naphthas go to T2. The naphthas are stored and transferred in the tank yard in an optimal way. In the base solution, there is no classification at all, the naphthas are stored and transferred just to satisfy the piping and inventory constraints.

Table 1. Cases Comparison

Case	Naphtha Avg SG	Ethylene Units	Propylene Units	Benzene Units	Hydrogen Units
Fix Demand Base	0.679	85%	85%	92%	92%
Fix Demand Opt	0.690	85%	85%	95%	92%
Open Demand Base	0.689	97%	100%	100%	97%
Open Demand Opt	0.686	100%	99%	100%	100%

The effect of the classification can be seen in Table 1 where the production value has been normalized based on the maximum value of each product for all cases. For the fix demand case the ethylene and propylene productions are obviously the same, but the production of benzene is larger for the optimal solution. Furthermore, the average SG of the naphtha feed in the optimized case is higher; this is reflected as a decrease in the operating costs because heavier naphtha is cheaper than lighter naphtha. The proper classification of the naphtha allows a wider range of naphtha compositions that can be fed to the plants, increasing the plant utilization and efficiency. In the second case the wider range of compositions is reflected as an increase in the production of ethylene, which is the most valuable product. The propylene and benzene productions are lower in the optimal case because their prices are lower than that for the ethylene. The efficiency can also be seen on the average SG as the plants are producing more ethylene with almost the same grade of naphtha.

For both cases, the production of other byproducts is very similar for both solutions, and the economical impact is smaller than that of ethylene, propylene and benzene. We also looked at the naphtha consumption by analyzing the naphtha final inventory at the end of the schedule. This can be done because the naphtha feed is fixed for all cases to the same value. Figure 2 and 3 show the results. Here we classified the naphthas in inventory according to their SG into 3 groups: light (< 0.675), heavy (>0.69) and medium (in between).

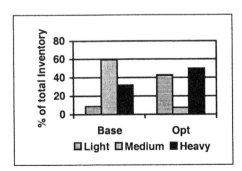

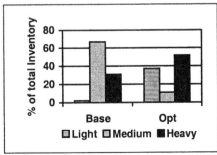

Figure 2. Final Naphtha Inventory Fixed Demand

Figure 3. Final Naphtha Inventory Open Demand

The decrease of medium naphtha in the optimal solutions is obviously the effect of the better classification of the naphtha feedstock. More importantly, the increase of light naphtha in the final inventory for both cases is larger than the increase of heavy naphtha for the optimal solutions. For example, for the fixed demand case the increment is 34 % for the light naphtha, but only 18 % for the heavy.

5. Actual Implementation

The current system was implemented under a Microsoft Windows environment. The general user interface (GUI) provides a modular, plug and play framework for connecting to the solution engine as well as accessing information on a database. The information is displayed in excel kind cells for easy editing. It also includes graph capabilities, which provides a better image of the process. The GUI allows easy browsing and editing of the information in the database. It can easily be reconfigured to the changing needs of the user, by simply creating, modifying or deleting queries in SQL language.

6. Conclusions

We have developed a system that allows us to improve our plant operation. The entire model that consists of a complex feedstock management, ethylene and aromatic plants is solved simultaneously using the proposed decomposition strategy. The robustness and utility of the system have been demonstrated on a real industrial application. The optimized solutions consistently demonstrate the benefits of the system. The economic impact can easily be identified, less light naphtha utilization and improve capacity utilization.

7. References

Brooke, A.; Kendrick, D. and Meeraus, A, GAMS - A user's guide (release 2.25). The Scientific Press, San Francisco (CA), 1988.

Drud, A.S., CONOPT - A GRG code for large scale nonlinear optimization - Reference manual. ARKI Consulting and Development A/S, Bagsvaerd, Denmark, 1992.

IBM, OSL (Optimization Subroutine Library) Guide and reference, release 2, Kingston, NY, 1991.

Pekny, J.F. and Zentner, M.G., 1994, Learning to solve process scheduling problems: the role of rigorous knowledge acquisition frameworks. In Foundation of Computer Aided Process Operations; (Rippin, D.W.T., Hale J.C. and Davis, J.F. eds.). CACHE, Austin, pp. 275-309.

Reklaitis, G.V., 1991, Perspectives on scheduling and planning of process operations. Presented at the Fourth International Symposium on Process Systems Engineering, Montebello (Canada).

Reklaitis, G.V., 1992, Overview of scheduling and planning of batch process operations. Presented at NATO Advanced Study Institute- Batch Process Systems Engineering, Antalya (Turkey).

I.B. Tjoa, Y. Ota, H. Matsuo and Y. Natori, 'Ethylene Plant Scheduling System Based on a MINLP Formulation,' a special edition of *Computers Chem. Eng.*, 1997.

European Symposium on Computer Aided Process Engineering – 12
J. Grievink and J. van Schijndel (Editors)

A Systematic Approach for the Optimal Operation and Maintenance of Heat Exchanger Networks

Hernán Rodera and Hiren K. Shethna

Hyprotech Ltd, Suite 800, 707 – 8th Avenue SW, Calgary, Alberta, T2P 1H5, Canada

e-mail: Hernan.Rodera@hyprotech.com; Hiren.Shethna@hyprotech.com.

Abstract

An automatic approach to study the behaviour of an existing heat exchanger network design when the operating conditions vary is presented. The purpose of this methodology is to facilitate the involvement of the engineer in the daily operation and maintenance of the network and its optimal performance prediction. The approach also assists in proposing improvements in the design that could lead to retrofit implementations. The final aim of the proposed methodology is to provide a tool capable of monitoring the performance of the individual heat exchanger units. Specification of the fouling factor is the first step in this direction. An example showing the application of this approach to a crude distillation column preheat train is presented.

1. Introduction

Synthesis of industrial heat exchanger networks is carried out usually by considering all possible set of operating conditions and by solving the problem to a final stage in which a practical design is obtained. The conceptual design phase evaluates the capital vs. operating cost trade-off and sets targets that are used to construct and optimise the network. Then, the final design is the result of a compromise between the optimal design and the designs capable of performing under different operating conditions. Therefore, limitations in the flexibility of the final design (when changes in these operating conditions occur) and in the capacity of the network to cope with fouling of heat exchanger units are usually encountered. Moreover, opportunities for retrofit improvements and their impact in the existing network are very difficult to evaluate. By these reasons, the operating engineer is faced with the challenge of adjusting the given design to the current operations. Because of the lack of tools to assess the effects that these changes have in the design, keeping the optimal conditions becomes a trial-and-error procedure. As a result the optimal operation of the network is compromised, the savings in energy consumption are reduced, and additional capital investment often is required. The need for a tool that allows the performance monitoring of the heat exchanger network is evident.

In this paper, an automatic approach to study the behaviour of heat exchanger network designs when the operating conditions vary is proposed. Its purpose is to facilitate the involvement of the engineer in the daily operation and maintenance of the network. Results showing the impact in the design are automatically provided when changes in the operating variables are observed. The tool is therefore available for online optimisation and the simultaneous evaluation of retrofit opportunities. Fouling of heat exchanger units (that produces a reduction in the area of transfer) and the evaluation of its impact in the whole network are easily conducted. An example of the preheat train of a crude distillation column shows the different aspects and advantages of the approach are presented.

2. The Heat Exchanger Network Calculator Approach

The basic equations for a single shell and tube heat exchanger unit that consider process and utility stream segmentation are:

$$Q_i = m_{ih}(Cp_{ih}^{in}T_{ih}^{in} - Cp_{ih}^{out}T_{ih}^{out} + C_{ih}^*) \tag{1}$$

$$Q_i = m_{ic}(Cp_{ic}^{out}T_{ic}^{out} - Cp_{ic}^{in}T_{ic}^{in} + C_{ic}^*) \tag{2}$$

$$Q_i = F_{ti}(UA)_i \Delta T_{ilm} \tag{3}$$

where C_{ih}^* / C_{ic}^* is a constant that accounts for the segmentation of the temperature vs. enthalpy curve of the corresponding hot/cold process stream in order to approximate the specific heat and consider a linear dependency of the enthalpy with temperature within the segments. In equation (3), F_{ti} refers to the F_t correction factor for the tube and shell exchanger unit i to account for noncountercurrent flow. The relations used in calculating this factor that requires values for all four extreme temperatures are summarised in Shenoy (1995). These temperatures also are used in calculating the logarithmic mean temperature difference ΔT_{ilm}. Once the mass balances are solved for the entire heat exchanger network, the set of variables for the single unit is comprise by the heat load Q_i, the heat exchanger area A_i and the four extreme temperatures.

2.1 Sequential vs. Simultaneous approach
Consider the simplest possible heat exchanger network shown in Figure 1. It comprises two heat exchanger forming a loop between the same hot and cold process streams. A set of equations (1) to (3) is written for each unit and two additional equations assure that the outlet temperatures of the first unit are the same as the inlet temperatures of the

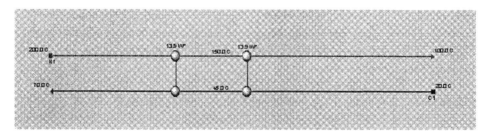

Figure 1. The Simplest Possible Heat Exchanger Network

second unit. The heat balances form a system of eight equations with twelve unknowns.

Four variables should be specified based on the degrees of freedom available. An iterative sequential approach that considers a single unit at a time can solve the system of equations provided that the two heat exchanger areas are not specified simultaneously. For the purpose of monitoring the performance of an existing heat exchanger network, it is required that the areas are kept constant. This is not possible with a sequential approach. Solving the whole set of heat balances simultaneously for the entire heat exchanger network has the advantage of allowing the specification of the both heat exchanger areas. The heat exchanger network calculator approach presented in this paper solves the entire system of equations for the heat exchanger network when this system is completely defined. In the case of partially defined systems, the approach determines and solves the largest completely define system. To assure global optimality for the solution, the nonlinear equation that has to be solved when area is specified is linearised as explained in the next section.

2.2 Linear Approximation when the Area Is Specified

When the area of a heat exchanger unit is specified, it becomes fixed and the system of equations (1) to (3) can be solved by iteration if the value of two of the extreme temperatures or one of these temperatures and the duty are know. The system is, however, nonlinear and difficult to solve. The partition of the entire area of the unit into equal area portions is proposed. The mean logarithmic temperature difference within each area section is then approximated by the corresponding linear arithmetic mean temperature difference.

2.3 Calculation of Utility Mass Flowrates

In the particular case of utility streams, loads are calculated using the Grand Composite Curve (GCC) based allocation method. For a single utility exchanger extreme temperatures are fixed. Therefore, the utility mass flowrate is calculated using equation (1) or (2) depending on utility side. In the case of multiple heat exchangers for a utility an iterative procedure is employed to calculate simultaneously the intermediate temperatures and the utility mass flowrate.

3. Heat Exchanger Fouling Analysis

In order to consider heat exchanger fouling, specification of the fouling factor for each particular unit is allowed. The new heat transfer coefficient for the heat exchanger is calculated by adding the additional resistance due to fouling. The capability of the Heat Exchanger Network Calculator approach to solve the system simultaneously makes possible the consideration of the different operating conditions through which an existing heat exchanger network undergoes when the fouling increases. Calculation of fouling factors based on measured extreme temperatures also is possible by using the Heat Exchanger Network Calculator approach. An step further is the analysis of alternative cleaning schedules and their influence in the total annual cost applying techniques existing in the literature (O'Donnell et al., 2001). Moreover, designs that avoid fouling or consider fouling mitigation can be investigated (Polley et al., 2000). All these new features are currently under development, and they will be part of future implementations.

4. Example

In this example, the preheat train of an atmospheric crude distillation column is considered. Process and utility stream data is presented in Table 1. The low-temperature crude stream is preheated prior to enter the pre-flash drum. The liquid from the pre-flash (high-temperature crude) is preheated before entering the furnace and mixed with the vapour stream coming from the pre-flash. The resulting stream enters the column that produces four products (naphtha, kerosene, diesel, and AGO) and bottoms residue. The condenser stream, waste H20, and three pumparounds are also available for integration.

The heat exchanger network corresponding to the preheat train of the atmospheric crude distillation column is shown in the grid diagram of Figure 2. The low temperature crude stream is heated up in five heat exchanger units prior to enter the pre-flash. The last unit uses high-pressure steam in order to reach the pre-flash temperature. The liquid exit

Table 1. Stream and Utility Data for Example

Stream	Segment	Tin (°C)	Tout (°C)	MCP(kW/°C)
Low Temperature Crude	1	30.0	108.1	333.6
	2	108.1	211.3	381.3
	3	211.3	232.2	481.5
High Temperature Crude	-	232.2	343.3	488.2
Naphtha	-	73.2	40.0	57.69
Waste H2O	-	73.2	30.0	6.842
Condenser	1	146.7	133.3	233.9
	2	133.3	120.0	202.2
	3	120.0	99.9	170.1
	4	99.9	73.2	338.5
Pumparound 1	1	167.1	116.1	172.4
	2	116.1	69.6	157.5
Kerosene	1	231.8	176.0	50.73
	2	176.0	120.0	46.31
Diesel	1	248.0	147.3	67.75
	2	147.3	50.0	58.44
Pumparound 2	-	263.5	180.2	123.1
AGO	1	297.4	203.2	21.99
	2	203.2	110.0	19.44
Pumparound 3	-	319.4	244.1	136.2
Residue	1	347.3	202.7	217.1
	2	202.7	75.0	179.6
Cooling Water	-	20.00	25.00	10800
LP Steam Generation	-	124.0	125.0	10258
MP Steam Generation	-	174.0	175.0	10258
HP Steam	-	250.0	249.0	33177
Fired Heat	-	1000	400.0	63.00

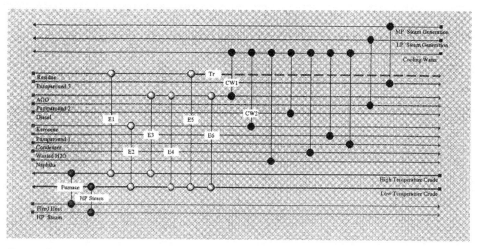

Figure 2. Crude Distillation Column Preheat Train

of the pre-flash is preheated in two units prior to enter the furnace. A series of coolers are used to reach the target temperatures for the products. Medium and low-pressure steam are generated using the two top pumparound streams. Notice the presence of loops in the network. With all process heat exchanger areas specified, a sequential iterative approach fails to calculate the entire network. By using the simultaneous Heat Exchanger Network Calculator approach, solution of the entire system of heat balance equations is possible. Process to process heat exchanger data is presented in Table 2.

In order to consider fouling of the process heat exchangers during the period of a year, a series of designs is obtained in which the fouling factor is specified. A first order degradation in the clean overall heat exchanger heat transfer coefficient (U_o) is used (O'Donnell et al., 2001) with $k = 0.35$ year^{-1}. Therefore, the fouling factor (R_d) is calculated by the equation:

$$R_d = \frac{1}{U_o}\left(e^{kt} - 1\right) \tag{4}$$

Table 3 shows the results obtained considering a design for each month starting from the clean-based design. As expected, all utility loads increase due to the degradation of the process to process heat exchanger performance. An increase in the Residue target temperature is also observed. Finally, the furnace inlet temperature degrades at a rate of

Table 2. Process Heat Exchanger Data for Example

Exchanger	Load (MW)	Area (m^2)	Shells	HTC (kJ/h.m^2°C)	FtFactor
E1	15.22	685.7	2	1505	0.8858
E2	0.67	20.03	1	1604	0.9993
E3	0.12	82.25	1	1765	0.9718
E4	0.73	21.45	1	1566	0.9983
E5	39.11	2029	6	1098	0.8037
E6	0.18	39.98	1	1268	0.9948

Table 3. Designs Considering Increasing Fouling Factor

Time (months)	Furnace Load (MW)	HP Steam Load (MW)	CW1 Load (kW)	CW2 Load (kW)	Tr (°C)
0	37.84	33.11	0.18	4.75	75.0
1	38.03	33.20	0.23	4.77	75.8
2	38.21	33.30	0.27	4.78	77.1
3	38.41	33.40	0.32	4.80	78.4
4	38.60	33.52	0.36	4.81	79.8
5	38.80	33.64	0.41	4.83	81.2
6	39.00	33.77	0.46	4.84	82.7
7	39.20	33.90	0.51	4.86	84.2
8	39.40	33.95	0.56	4.87	85.2
9	39.60	34.11	0.60	4.88	86.9
10	39.81	34.26	0.65	4.89	88.6
11	40.02	34.45	0.69	4.91	90.5
12	40.22	34.63	0.74	4.92	92.3

near 5°C per year due to fouling. This represents an increase in furnace duty of 2.4 MW or 6% of the total heat load. Considering a crude oil priced at 21$/barrel, the cost of heat exchanger fouling would be of around $250,000 per year.

5. Conclusions

An approach to simultaneously calculate a given heat exchanger network has been developed. The Heat Exchanger Network Calculator approach solves the entire set of equations for the network and allows area specification. The system is linearise by partitioning the total area in portions where the logarithmic mean temperature difference can be approximated by the corresponding arithmetic mean. Fouling factors are included to consider the degradation in the heat transfer capacity of the heat exchanger units. The approach is the base for the development of a performance-monitoring tool.

Acknowledgements
Dr. H. Rodera and Dr. H.K. Shethna thank Hyprotech Ltd. for allowing the use of HXNet software to produce the results of this paper.

References
HX-Net, Development Version, Hyprotech Ltd, Calgary, Alberta, Canada.
O'Donnell, B.R., B.A. Barna and C.D. Gosling, 2001, Optimize Heat Exchanger Cleaning Schedules, Chem. Eng. Prog. Vol. 97, No. 11, 56.
Polley, G.T., D.I. Wilson and S.J. Pugh, 2000, Designing Crude Oil Pre-Heat Trains with Fouling Mitigation, AIChE Spring Meeting, Atlanta.
Shenoy, U.V., 1995, Heat Exchanger Network Synthesis. Gulf Publishing Company, Houston, Texas.

European Symposium on Computer Aided Process Engineering – 12
J. Grievink and J. van Schijndel (Editors)
© 2002 Elsevier Science B.V. All rights reserved.

Scheduling of Continuous Processes Using Constraint-Based Search: An Application to Branch and Bound

L. C. A. Rodrigues, R. Carnieri, F. Neves Jr.
Centro Federal de Educação Tecnológica do Paraná (CEFET-PR)
Programa de Pós Graduação em Engenharia Elétrica e Informática Industrial (CPGEI)
Rua Sete de Setembro, 3165, 80230-901, Curitiba PR, Brasil
(lcar, carnieri, neves)@cpgei.cefetpr.br

Abstract

In this work a Branch and Bound approach based on constraint-based search (CBS) is proposed to the scheduling of continuous processes. The purpose of this work is to extend to continuous processes a CBS approach proposed previously to batch production (Rodrigues *et al.*, 2000). Tasks' time-windows are submitted to a constraint propagation procedure (CBS) that identifies existing orderings among tasks. Linear programming is used to determine the optimal flow rate for each bucket whenever all buckets are ordered in the branch and bound.

1. Introduction

Several MILP procedures have been recently proposed to the scheduling of continuous processes (Pinto *et al.*, 2000; Pinto and Moro, 2000; Stebel,2001; Magatão, 2001). But these authors are unanimous about the difficulties in solving real world problems due to its dimension. In this work a Branch and Bound approach based on CBS is proposed to the scheduling of continuous processes with the minimization of earliness. Production is identified in terms of buckets of tasks. The size of product buckets is made equal to the highest common factor of the storage capacity among available tanks. Therefore it takes an integer number of buckets to fill a tank. This assumption is necessary if the scheduling is posed as an ordering of tasks, as when CBS is used. As in previous works (Rodrigues *et al.*, 2000; Gimeno *et al.*, 2000; Rodrigues, 2000), the proposed approach uses a time-window to impose a time interval for the execution of each bucket. This approach has two phases: planning and scheduling. Time-windows are initially determined during planning based on the availability of raw materials and on the demand of final products. These time-windows are submitted to a constraint propagation procedure that identifies existing orderings among buckets of different tasks. Constraint propagation also enables the identification of disjunctions (undefined orderings) among buckets of different tasks. During scheduling the branch and bound procedure is used to set an ordering to these disjunctions. When all the orderings are set during branch and bound, linear programming is used to determine the flow rates of the continuous processes.

2. Problem definition

A problem proposed by Pinto and Moro (2000) is used to test the proposed branch and bound approach. The proposed problem is the optimization of production and inventory management of a liquefied petroleum gas (LPG) plant. The STN representation (Kondili et al., 1993) of this plant is presented in figure 1. A petroleum refinery yields a mixture of propane and butane (C3C4) to the LPG plant. C3C4 is sent to the depropanizer where two different tasks may be performed: i) production of separate yields of propane (C3) and butane (C4); ii) production of separate yields of propane for petrochemical purposes, also known as intermediate propane or *propint* (C3i), and butane (C4). An amount of C3C4 is bypassed whenever *propint* is produced. In this problem, any mix of C3 and C4 may be delivered to consumers as LPG, provided that the amount of C3 must be greater or equal to that of C4. It is also possible to deliver C3 as LPG. C4 may also be marketed as bottled gas or used to produce methyl-tert-butyl-ether (MTBE). The production of MTBE also results in the production of *raffinate* (Raff) as a byproduct. *Raffinate* is also used to the production of LPG. The storage farm comprises 8 spheres suitable for LPG, C3, and C3i. There are also 4 spheres suitable for C4 and *raffinate*. All buckets in this problem have the size of the storage sphere

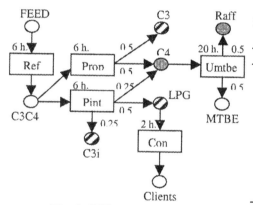

Fig. 1. STN representation

Table 1. Assignment and storage.

Task	Units	State	Storage
Ref	Prod	C3C4	ZW
Prop	Dep	C3	FIS(LPG)
		C4	FIS(C4)
Pint	Dep	C3i	FIS(LPG)
		C4	FIS(C4)
		LPG	FIS(LPG)
Umtbe	Umt	MTBE	UIS
		Raff	FIS(C4)
Cons	Pipe	Clients	UIS

3. Production planning

The amount of buckets of each product that will be produced is defined from final product demands. Latest finishing times (LFT) of the time-windows are obtained from a backward propagation of final products demands and its due dates. Earliest beginning times (EBT) are obtained from a forward propagation of raw materials availability. Time-windows are calculated based on the optimistic assumption that production will be accomplished with maximum flow rates. This is done because any time-window obtained from smaller flow rates fits within the time intervals defined by the optimistic time-windows. These initial time-windows are then submitted to constraint propagation that identifies existing orderings among buckets of tasks (Caseau and Laburthe, 1995; Baptiste and Le Pape, 1995). The problem becomes infeasible if these orderings are not respected. Therefore planning is in fact a pruning procedure that reduces the dimension of scheduling. The identification of orderings may impose time-windows reductions. A time-window is reduced during constraint propagation if it is identified that it is infeasible to process a task during a certain interval of time that belongs to the time-window of this task.

4. Constraint-based search

Constraint propagation takes into account equipment contention, storage restrictions, and pegging considerations due to mass balance. Analysis of equipment contention is performed by two procedures known as *explosion* and *edge-finding* (Caseau and Laburthe, 1995; Baptiste and Le Pape, 1995). *Explosion* and *edge-finding* are based on the analysis of time-windows feasibility. These two procedures are responsible for the identification of existing orderings between buckets of tasks that are processed in the same processing unit. Mass balance propagation is performed whenever a time-window is reduced. An example of pegging is that if the EBT of a bucket α is postponed, the EBT of the buckets that depend of α to be produced will also have its EBT postponed. Storage restrictions (as ZW, NIS, and FIS policies)[1] are also considered during constraint propagation (Rodrigues, 2000). ZW and NIS restrictions impose time-windows reductions in a similar way. Consider a task α that produces a product consumed by task β, and that this product has ZW storage restrictions. Any time-window reduction in task α will result in a similar change in task β. If storage demand is greater than the storage capacity, FIS restriction may impose orderings or result in disjunctions between tasks that share the same set of tanks. When scheduling is performed it is necessary to set an ordering for all disjunctions (imposed by processing and storage units) identified during planning. Therefore equipment contention and FIS storage must not be neglected, especially in continuous processes where FIS storage may be the main restriction to production.

5. Branch and bound

The branch and bound procedure is based on the definition of orderings to the disjunctions (among pairs of buckets) identified during planning. The proposed branch and bound uses constraint propagation to identify feasible orderings among buckets with disjunctions. New orderings may be identified by constraint propagation whenever an ordering is imposed during branch and bound execution. When an ordering is defined to all planning phase disjunctions at a solution node of the branch and bound, it can be stated that the solution is feasible. But the optimal flow rates must be determined since the optimistic assumption of maximum flow rates only leads to the lower bound of the analyzed solution. Linear programming is used to determine the optimal flow rate for each bucket of this solution. The branch and bound procedure minimizing earliness that has been implemented is presented bellow.

---

[1] ZW means "zero wait"; NIS means "no intermediate storage"; FIS means "finite intermediate storage".

Branch and bound procedure minimizing earliness:

i. Create a search tree G consisting solely of the start node s. Put s on a list called *OPEN* (indicating the nodes to be expanded). Set *LOWERBOUND* (the lower bound of the search procedure) to infinite ($+ \infty$). The search tree G contains all the nodes identified during branch and bound. Node s is the output of the planning phase.

ii. If *OPEN* is empty, end of search. In this case, if *LOWERBOUND* value is infinite ($+ \infty$), there is no feasible solution. Otherwise, optimal solution has been identified with an earliness value equal to *LOWERBOUND*.

iii. Select the first node on *OPEN* and remove it from this list. Call this node n. Expand node n, generating the set of its successors. Each expansion imposes an ordering to a pair of disjunctive buckets. Therefore each expansion generates two successor nodes. Constraint propagation is performed to all expanded nodes.

iv. Include successor nodes are that feasible but aren't solution nodes (where an ordering has been set to all planning disjunctions) in *OPEN*.

v. Use linear programming to define start time, processing time, and flow rates to all buckets of continuous processes to each successor node that is a feasible solution node. If there is a solution node with earliness value smaller than *LOWERBOUND*, set this earliness value as the new value of *LOWERBOUND*.

vi. Eliminate from *OPEN* all nodes with an earliness bound value greater than *LOWERBOUND*. Latest finishing times (LFT) of the time-windows impose an earliness bound to every node of the search tree. In a minimization problem, the monotone condition of the branch and bound search guaranties that a successor node always has an earliness bound value greater or equal to its ancestors earliness bound.

vii. Reorder *OPEN* according to the earliness bound value of its nodes. The smallest earliness bound is placed first in the list.

viii. Return to step *ii*.

The linear programming (LP) model used within the proposed branch and bound is presented bellow. Equation 1 presents the objective function, minimizing earliness. $DD_{i,b}$ represents the due date of bucket b of task i. $F_{i,b}$ is a variable representing the end of processing of bucket b of task i. B_i identifies the set of buckets of task i that are produced.

$$\text{Earliness} = \sum_i \sum_{b \in B_i} (DD_{i,b} - F_{i,b}) \tag{1}$$

Although not presented in this paper, the variables representing the start and end of processing, $S_{i,b}$ and $F_{i,b}$, are constrained by its time-windows. The most important restrictions in the LP model are those concerning precedence relations (equations 2 and 3). Equation 2 is used whenever there is a ZW restriction among two buckets. This equation is used for tasks linked by states with ZW storage and also to avoid processing gaps, assuring continuous processing among buckets of the same task. Equation 2 is generated if bucket b of task i is preceded by another task with zero wait connection ($ZW_{i,b}$ = True). The task that precedes bucket b of task i with zero wait connection is identified by $i\_zw_{i,b}$ and its bucket is identified by $b\_zw_{i,b}$. Equation 3 only imposes that when there is a precedence relation among tasks ($Prec_{i,b}$ = True) the processing end of bucket $b\_pr_{i,b}$ of task $i\_pr_{i,b}$ is smaller or equal to the processing start of bucket b of task i. There are no flow rate variables in this LP model. Since buckets have a fixed size, flow rates are indirectly obtained from its processing times ($TP_{i,b}$), as shown in equation 4. Maximal processing times are taken from minimal flow rates and minimal processing times are obtained from maximal flow rates.

$$S_{i,b} = F_{(i\_zw(i,b),\, b\_zw(i,b))} \qquad \forall\ i, b \in B_i\,/\,ZW_{i,b} = \text{True} \qquad (2)$$

$$S_{i,b} \geq F_{(i\_pr(i,b),\, b\_pr(i,b))} \qquad \forall\ i, b \in B_i\,/\,Prec_{i,b} = \text{True} \qquad (3)$$

$$TPmin_i \leq TP_{i,b} \leq TPmax_i \qquad \forall\ i, b \in B_i \qquad (4)$$

6. Results

Part of the proposed approach had been implemented previously to batch production (Rodrigues 2000). The most important differences between the two implementations are in the branch and bound, that had to be included. The constraint propagation mechanism was only changed to perform a deeper analysis of storage constraints. The LP procedure was implemented using LINGO (1999). A DLL (Dynamic Link Library) was developed to allow the solution of the LP within Visual Basic. Figure 2 is the result of preliminary tests with the proposed approach. It presents the time windows obtained from planning.

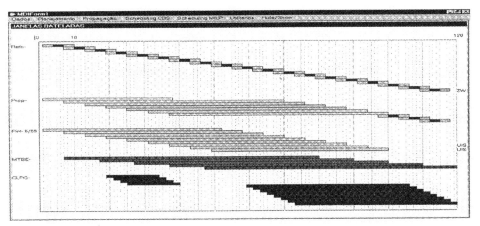

Fig 2. Time windows.

7. Conclusions

In this work we present an approach to the scheduling of continuous processes. The branch and bound that is proposed relies heavily on the pruning capacity of constraint-based search techniques. We claim that the definition of an ordering among tasks may reduce the dimension of the problem when compared to the existing MILP approaches. Since MILP procedures involve the definition of orderings and allocation of tasks at the same time. We believe that the smaller amount of decisions to be made in the branch and bound will allow us to solve problems of higher complexity than those solved by MILP.

8. Acknowledgements

The authors acknowledge the financial support of the Brazilian National Petroleum Agency (PRH-ANP/10 CEFET-PR) and of CNPq (Kit Enxoval Recém-Doutor).

9. References

Baptiste P. and Le Pape C., 1995, A Theoretical and Experimental Comparison of Constraint Propagation Techniques for Disjunctive Scheduling, Proceedings of 14th International Joint Conference on Artificial Intelligence, Montreal, Canada.

Caseau Y. And Laburthe F., 1995, Improving Branch and Bound for Job Shop Scheduling with Constraint Propagation, Proceedings of the 8th Franco-Japanese 4th Franco-Chinese Conference.

Gimeno, L., Rodrigues, M.T.M., Rodrigues, L.A., 2000, Constraint Propagation Tools in Multipurpose Batch Plants Short Term Planning, Proceedings of 2nd Conference on Management and Control and Production and Logistics, Grenoble, France.

Kondili, E., Pantelides, C.C., Sargent, R.W.H., 1993, A General Algorithm for Short Term Scheduling of Batch Operations – I. MILP Formulation, Computers and Chemical Engineering, 17, Pp. 211-227. Elsevier, Amsterdam.

LINDO Systems Inc., 1999, LINGO: The Modeling Language and Optimizer – User's Guide, Chicago, USA.

Magatão, L., 2001, Mathematical Programming Applied to the Optimization of Pipeline Operation, M.Sc. thesis, CPGEI, CEFET-PR, Brazil (in portuguese).

Pinto, J. M., Joly, M., Moro, L. F. L., 2000, Planning and Scheduling Models for Refinery Operations, Computers and Chemical Engineering, 24, Pp. 2259-2276. Elsevier, Amsterdam.

Pinto, J.M., Moro, L.F.L., 2000, A Mixed Integer Model for LPG Scheduling, In: European Symposium on Computer Aided Process Engineering-10, Comp. Aided Chemical Engineering (8) (S. Pierucci(Ed)), 1141-1146, Elsevier, Amsterdam.

Rodrigues, L.C.A., 2000, Planning and Scheduling of Multipurpose Batch Plants: A Decomposition Strategy Using Time-Windows, Phd Thesis, UNICAMP, Campinas, Brazil (in portuguese).

Rodrigues, L.C.A, Graells, M., Cantón, J., Gimeno, L., Rodrigues, M.T.M., Espuña, A., Puigjaner, L., 2000, Utilization of Processing Time Windows to Enhance Planning and Scheduling in Short Term Multipurpose Batch Plants, Computers and Chemical Engineering, 24, pp. 353-359. Elsevier, Amsterdam.

Stebel, S.L., 2001, Modeling of Liquefied Petroleum Gas Storage and Distribution in an Oil Refinery, M.Sc. thesis, CPGEI, CEFET-PR, Brazil (in portuguese).

European Symposium on Computer Aided Process Engineering – 12
J. Grievink and J. van Schijndel (Editors)
© 2002 Elsevier Science B.V. All rights reserved.

Real Time Batch Process Optimization within the environment of the Flexible Recipe.

Javier Romero, Antonio Espuña and Luis Puigjaner
Chemical Engineering Department, U. P. C.
E.T.S.E.I.B., Diagonal 647, E-08028 Barcelona, Spain.
e-mails: [romero, aec, lpc]@eq.upc.es

Abstract

Batch processes are normally thought to operate using a traditional fixed recipe. However, the fixed recipe has to be approximately adapted in a rather unsystematic way depending on experience and intuition of operators. Therefore, the concept of flexible recipe seems to be the adequate way to rationalize and systematize the adjustment procedure. In the flexible recipe context, the term recipe is used in a more abstract way by referring to a selected set of adjustable recipe items that control the process output.

In the present work, a framework for batch process real time optimisation considering the flexible recipe concept is presented.

As soon as a deviation is detected, the control recipe can be readjusted. Deviations are only detected at the process state assessment. The process state assessment compares the operating conditions at the sample time with the expected operating conditions that will depend on the initial conditions. For this reason and in order to perform this comparison, a predictive model is necessary. Then, a corrective model is required to adjust the control recipe. This second model describes the ultimate effect of the values measured at the time of the process state assessment as well as any run-time corrections applied during the remainder of the processing time.

1. Introduction

The traditional fixed recipe does not allow for adjustment to plant resource availability or to variations in both quality of raw materials and the actual process conditions. However, the industrial process is often subjected to various disturbances. For this reason, the fixed recipe is in fact approximately adapted but in a rather unsystematic way. Therefore, in order to systematize these adaptations the concept of flexible recipe might be appropriate, as the flexible recipe is systematically adapted at the moment (Rijnsdorp, 1991).

According to the standard ISA-SP88, the flexible recipe might be derived from a master recipe and subsequently used for generating and updating a control recipe. Verwater *et al.* (1995) introduced the concept of different levels between these two stages defined at ISA-SP88. With these levels, a better description of the different possible functionality of the flexible recipe is obtained. These levels are:

a) Master control recipe: a master recipe valid for a number of batches, but adjusted to the actual conditions, from which the individual control recipes per batch are derived.

b) Initialized control recipe: the adjustment of the still-adjustable process conditions of a master control recipe to the actual process conditions at the beginning of the batch.

c) Corrected control recipe: the result of adjusting the initialized control recipe to process deviations during the batch.

d) Finally, for monitoring and archiving purposes, the accomplished control recipe.

Figure 1, describes the interaction of these four flexible recipes in a real-plant environment.

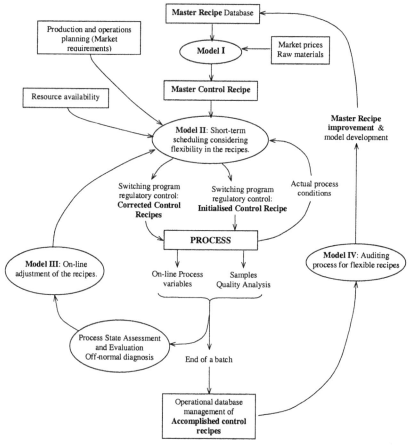

Figure 1. Real time flow data among different possible flexible recipes.

These models may be developed in laboratory experiments, during normal production by a systematic introduction of small changes in certain inputs, or by adjusting white models and simulating them under different conditions.

In this work, a new framework for the recipe correction is introduced. The aim of this approach is to optimize the entire batch process in front of disturbances. It is proposed a framework for generating the *corrected control recipe* from the *initialized (master) control recipe*. In Romero *et al.* (2001) is presented a framework for recipe initialization based on the same basic framework presented here.

2. Recipe correction interacting with a Rescheduling tool.

The recipe initialization is performed at the beginning of the batch phase only taking into account known initial deviations. But other run-time deviations may arise. However, under certain circumstances, it is possible to compensate the effects of these unknown disturbances during the batch run, provided that continuous or discrete measurements are available.

2.1 Flexible Recipe model for recipe correction.

Within the flexible recipe context, the term recipe is used in a more abstract way by referring to a selected set of adjustable items that control the process output. The Flexible Recipe Model is the relationship that correlates a batch process output as a function of the selected input items of the recipe. This model is regarded as a set of constraints on quality requirements and on production cost (Romero *et al*, 2001).

Recipe items are classified into; i) The vector of process operation conditions, poc_i, of stage i of a recipe. It includes items like temperature, pressure, type of catalyst, etc. ii) The product specification vector, ps_i, at the end of each process stage i of a recipe. It may consider items like conversion of a reactant, quality aspects, etc. iii) Processing time, TOP_i, at each stage i of a recipe. iv) And, waiting time, TW_i, the time between a stage finish and the next stage start time. From this, the vector of product variables of one batch stage may depend, according to a function • , on the processing time, on waiting time, on process set-points, and on product specifications at different stages $i*$, where the different inputs to stage i are produced. Then, within this model, product specifications, ps, and process operation condition, poc, are subject to be corrected within a flexibility region, • and • respectively.

While a batch process takes place, different on-line continuous process variables and discrete variables sampled at different moments are taken (see Figure 1). From this information, a *process state assessment* is performed. From this assessment, information about how the batch-process is being carried out is given to Model III (Flexible recipe model for recipe correction). The time at which process state assessment is performed, and so, at which Model III takes different actions might be different from the moment at which a deviation is detected. Finally, Model III will interact or will have integrated some scheduling algorithms. In order to perform the *process state assessment* comparison a prediction model is necessary. In order to adjust the control recipe, a correction model is necessary. Finally, in order to adjust the actual schedule, a rescheduling strategy may be necessary.

Prediction Model

This model, Eq.1, estimates the continuous and discrete sampled product variables in function of the actual control recipe, established in Model II, at the sample moment.

$$\mathbf{pvs}_{i}^{w} = \Psi^{pred} \left(TPA_{i}^{w}, \mathbf{ps}_{i*}, \frac{1}{TPA_{i}^{w}} \int^{TPA_{i}^{w}} \mathbf{poc}_{i} dt \right) \qquad (1)$$

where TPA_{w}^{i} is the *wth* moment at which stage i of a batch is assessed. $\mathbf{pvs}_{w}^{i}$ is the expected vector of product specifications at TPA_{i}^{w} moment.

Process State Assessment

The Process State Assessment basically consists on the evaluation of the batch-process run. This assessment consists on comparing the predicted product specification i, $\mathbf{pvs}^w_i$, (expected by Model II) with the real variable observed at the *wth* process statement, $\mathbf{ps}^w_i$. If this comparison is greater than a fixed permitted error, • , some actions will be taken in order to compensate this effect.

Correction Model

Describes the ultimate effect of the values measured at the time of the process state assessment as well as any run-time corrections applied during the remainder of the processing time. This model adjusts process operation conditions to set the values, established at the recipe initialization (Model II), of product specifications. This model adjusts these values in function of the processing time, waiting time, product variables at the beginning of the batch stage and of the deviations detected at the *wth* process assessment moment, Eq.2. There must be one correction model for each process state assessment moment.

$$\frac{1}{TOP_i - TPA_i^w} \int_{TPA_i^w}^{TOP_i} \mathbf{poc}_i \, dt = \Psi_w^{correct}\left(TOP_i, TW_i, \mathbf{ps}_i, \mathbf{ps}_{i\bullet}, \mathbf{pvs}_i^w, \frac{1}{TPA_i^w} \int^{TPA_i^w} \mathbf{poc}_i \, dt \right) (2)$$

Rescheduling strategy

The output of the flexible recipe model for recipe correction might give variations in processing time or resource consumption, which would make the existing schedule suboptimal or even infeasible. Therefore, a rescheduling strategy will have to be used. There are two basic alternatives to update a schedule when it becomes obsolete: generating a new schedule or altering the initial schedule to adapt it to the new conditions. The first alternative might in principle be better for maintaining optimal solutions, but these solutions are rarely achievable in practice and require prohibitive computation times. Here, it is proposed a retiming strategy to be integrated into the flexible recipe correction framework. At each deviation detected, optimization will be necessary to find the best corrected control recipe of the process. From this, it is proposed to solve the LP in Eq.3 to adjust the plant schedule to each recipe correction. When dealing with multipurpose plants, this strategy might not be able to make feasible some infeasible schedules. If this would happen, further actions should be taken. However, this aspect is out of the scope of this work.

$\min(Performance\_Criterion)$ *subject to,*

$$TI_i \geq 0 \quad \forall i$$
$$TF_i = TI_i + TOP_i + TW_i \quad \forall i$$
$$TI_i = TF_{i'} \quad \forall i, i' / \exists s \in \{\bar{S}_i \cap S_{i'}\} \tag{3}$$
$$TW_i \leq TW_i^{max} \quad \forall i$$

and subject to the correction flexible recipe model constraints

where TI_i and TF_i are the initial and ending time of each stage i of the batch.

2.2 Batch correction procedure

Within each batch-run, the algorithm of Figure 2 will be applied. This algorithm first predicts the expected deviations in process variables from the nominal values. Then, verifies if there exist significant discrepancies between the observed variables and the predicted. If so, fixes process variables of all batch-stages already performed and of the batch-stages that are currently being performed and are not the actual batch stage being assessed and re-optimizes the actual recipe taking into account the effect on the schedule timing.

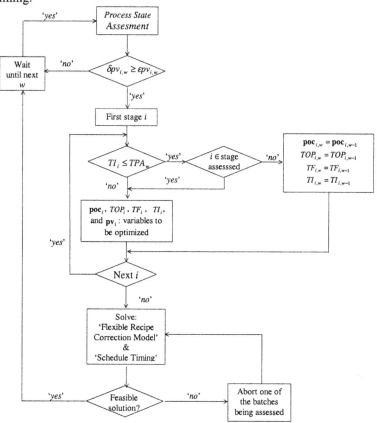

Figure 3. Batch correction procedure algorithm.

3. Case Study

To evaluate the potential use of this framework, a case study proposed by Cott and Macchietto (1989) is discussed. The plant produces three products (D, E and F). The production recipes of these four products are presented in Figure 3. The batches were scheduled into the plant using a minimum makespan scheduler with the nominal processing times and zero waiting time policy. The resulting proportion of batches scheduled were 6 batches of product E, 2 of D and 3 of F.

During actual operation, the processing times of all phases for all products were assumed to be subject to independent and normally distributed variations of a mean equal to zero and a standard deviation equal to 10 % of the nominal processing time.

In Figure 3 is presented the % equipment utilization for the nominal schedule and the resultant equipment utilization after introducing disturbances. It can be seen how disturbances introduce a waiting time in equipment unit 4 of 1.56 h., when readjusting the schedule with the retiming algorithm POMA (Projected Operation Modification Algorithm) proposed by Cott and Macchietto, 1989. When applying the batch correction procedure proposed, just without considering the possible flexible recipe, this waiting time is already reduced to 0.42 h. The nominal makespan of the case study is of 143 h which is increased to 145.6 h when applying the POMA but only to 145.1 h when applying the optimization procedure of the proposed batch correction procedure to react in front of the disturbances.

In order to consider flexibility in the recipe, black-box-linear models are assumed to be available for the different products. Process state assessments are performed only at the end of each recipe stage. Therefore, the prediction model itself is not necessary, as the expected product specifications at each process state assessment (pvs^1_i) are the ones given at the initialized control recipe. The correction model will be as,

$$\delta poc_{i+1} = d_i \, \delta ps_{i+1} - c_i \, \delta pvs^w_i - a_i \, \delta TOP_i - b_i \, \delta TW_i$$

where δ means the deviation from the nominal and expected values. It has been assumed an ideal process where all the parameters of the flexible recipe model are equal to 1. Finally, it has been assumed a performance criterion given different weights to production makespan, set-point modification and waiting time. With this, it is shown in Figure 3 how waiting time is removed from equipment unit 4 as well as production makespan is reduced. Notice that this improvement will probably imply a higher production cost due to some required set-point modifications.

Prod.	Equip.	TOP (h)	n° lots		Equipment unit	Nominal Recipe	POMA	No Recipe Flexibility	Flexible Recipe
						% Equipment utilization			
D	1	3	3		1	27	27.5	27.5	27.5
	4	5			2	51	51.0	51.0	51.0
	5	10			3	42	42.2	42.2	42.2
E	2	8	6		4	51	51.6	51.6	51.6
	3	4			5	135	134.3	134.3	133.8
	4	3					Waiting time (h)		
	5	10			1	0	0	0	0
F	1	6	3		2	0	0	0	0
	2	1			3	0	0	0	0
	3	6			4	0	1.56	0.42	0
	4	6			5	0	0	0	0
	5	13			Makespan (h)	143	145.6	145.1	144.8

Figure 3. Case study results. The proposed batch correction procedure has been solved using Lingo^{TM}.

4. References

Rijnsdorp, 1991, Integrated Process Control and Automation, Elsevier, Amsterdam.

Romero *et al.*, 2001, A new framework for batch process optimization using the flexible recipe. Ind. Engng. Chem. Res. Submitted.

Z. Verwater and K.J. Keesman, 1995, Computer-aided development of flexible batch production recipes, Production Planning and Control, 6, 320-330.

Cott and Macchietto, 1991, Minimizing the effects of batch process variability using online schedule modification. Comp. Chem. Engng, 13, 105-113.

European Symposium on Computer Aided Process Engineering – 12
J. Grievink and J. van Schijndel (Editors)
© 2002 Elsevier Science B.V. All rights reserved.

Smart Enterprise for Pulp & Paper Mills: Data Processing and Reconciliation

M. Sánchez[1], D. Leung[2], H. Konecsny[2], C. Bigaran[2], J.A. Romagnoli[3*]

[1]PLAPIQUI (UNS-CONICET), C. C. 717, 8000 – Bahía Blanca, ARGENTINA

[2] Visy Pulp & Paper Pty. Ltd. Tumut, NSW 2720, AUSTRALIA

[3] Chemical Engineering Department, University of Sydney, Sydney, NSW, 2006

Abstract

An ad-hoc data reconciliation procedure developed for the recausticizing section of a new pulp and paper industry is presented in this work. A comprehensive model was formulated to take into account different unit operation modes. It was also extended to incorporate specific knowledge of some pieces of equipment to increase redundancy, and consequently enhance estimate precision and gross error detectability.

1. Introduction

Visy Industries, one of a few major players in the Australian paper business, has recently started up a state of the art modern pulp and paper making facility. The mill produces 240 000 tonnes of Kraft pulp and brown packaging paper each year; 800 000 tonnes per annum of raw materials is sourced from local softwood plantations, that are supplemented by 50000 tonnes per annum of waste paper from domestic and commercial sources.

Smart Enterprise refers to a division at Visy Pulp and Paper Tumut that encompasses several areas of responsibility relating to data reconciliation, process modelling and simulation, process control and optimisation. Smart Enterprise´s goal is to improve plant performance by allowing operators to make more informed decisions in a shorter amount of time.

With this objective in mind, data processing and reconciliation arises as a key component of Smart Enterprise, as raw data are processed by this technique to produce a consistent set of data that constitutes a reliable input for other procedures. Furthermore, it will also allow the detection of faulty sensors, bias in measurements or other anomalies in the process operating data. A comprehensive analysis of data reconciliation strategies can be found in Romagnoli and Sánchez (1999).

In this work we present the distinctive features of the data reconciliation procedure especially developed for the recausticizing section of the plant. To our knowledge, it is the first strategy reported for a pulp and paper industry. The methodology succeeds in considering changes in operating conditions associated with global plant production strategies or simply with equipment changes for maintenance cleanings. Furthermore extra knowledge is incorporated in the process model, by using pump characteristic curves, to overcome the lack of redundancy of the system. The procedure is completed

---

* To whom correspondence should be addressed

with a gross error detection technique based on the individual residuals test and a graphical interface in which input may be entered and results may be displayed. The graphical interface was generated using Excel with the data imported/exported directly between the two programs.

2. Procedure development

Data Reconciliation is the process of adjusting or reconciling the process measurements to obtain more accurate estimates of flow rates, temperatures, compositions, etc... that are consistent with the material and energy balances. The classical procedure is based on the minimisation of the sum of the weighted difference between measured and estimated variables subject to a set of process constraints. It is valid under the assumption that no gross errors are present in the set of measurements. If it is not the case, gross error identification strategies are initially applied to determine and eliminate the set of suspect measurements.

Data Reconciliation strategies are supported by the location of a set of sensors in the plant, that allows the existence of redundant measurements. To know what type of information is available for a given process and set of instruments, an instrumentation analysis is performed. It allows classifying variables according to their feasibility of calculation. Measurements can be classified into redundant and non-redundant variables. The redundant ones are those whose value can be computed from the mathematical model that represents the plant and other measured variables. In turn, the unmeasured variables are called observable when they can be evaluated from the available measurements using model equations. Otherwise they are unobservable variables.

A data reconciliation procedure has been developed, following the classical approach, for the recausticizing area of the mill. This section plays two important roles in the production of white liquor. Firstly, it removes process impurities from the system in the green liquor filter. Secondly, the causticizing area increases the hydroxide (an important cooking chemical in the Kraft delignification process) content of the cooking liquor before it is fed into the digester area. The main features and motivations for each development stage of the procedure are presented below.

2.1 Model formulation

The model of the process is incorporated into the optimization problem as the set of constraints that should be satisfied at the solution point. But for the plant section under analysis, the model structure changes with time because the flow of some streams can be zero depending on the positions of 43 switches operated through the Distributed Control System. These streams are related to the green liquor feed (controlled from the recovery boiler) and to the cleaning cycles of the filters. Furthermore the presence of manual valves and streams connected to water locks can modify the model.

To tackle the problem of model variability with time, the broadest possible incidence matrix is first formulated. It has 146 rows and 306 variables associated with units and streams respectively. The model contains all streams interconnecting units (with the exception of those streams utilised only during plant maintenance) and fictitious streams that account for the net accumulation in the filters and tanks. It is assumed that only these units have an unsteady state operation mode.

A particular operation state is associated with specific positions for the switches, valves, and tank heights. Starting from this information and the all-inclusive model, a procedure was developed that determine the set of units and streams that participate in an operation state. It is based on row and column elimination of the general incidence matrix.

2.2 Instrumentation analysis

Considering only mass balance calculations, an instrumentation analysis is performed to know what information is available from the process with the existing instrumentation, that consists of 17 magnetic flowmeters and 15 level sensors. The analysis is based on the variable classification procedure that uses the Q-R matrix decomposition (Sánchez and Romagnoli, 1996).

Different case studies are conducted whose results are included in Table 1. It contains the number of units and streams involved in the corresponding model, and the amount of redundant (R), non-redundant (NR), observable (O) and unobservable (UO) variables.

Normal operation condition is first considered (Case 1), i.e. all switch positions, height of tanks and valves are as indicated by the P&I diagram as normal. In this case the number of measurements whose values can be adjusted through reconciliation, i.e. the redundant measurements, is zero. This is a serious drawback not only to enhance measurement's precision but to detect and identify gross errors.

Then a model that excludes vent tanks from the previous one is analysed (Case 2). Other vapour streams are also eliminated in Case 3, but the increment in system redundancy was not important. The incorporation of temperature measurements and energy balances in the model was discarded beforehand because its benefit was obviously worthless, due to the low number of temperature measurements that are present.

As any implementation of a data reconciliation procedure is settled on the existence of redundant variables, the next goal was to incorporate extra knowledge to the process model to increase system redundancy.

Table 1: Instrumentation Analysis Results

Case	Units	Streams	R	NR	O	UO
1	120	262	0	32	31	199
2	120	255	3	29	35	188
3	106	218	3	29	46	140

2.3 Extended model formulation

A study was undertaken to determine if measured variables related to pump operation could be used to formulate equations for flowrate adjustment, that is, relationships containing measured flowrates, constants and other measurements.

In the recausticizing area there are five centrifugal pumps of measured variable-speed. They are important pieces of equipment because they act as final elements of flowrate control loops that manage the input flow to filters and the slaker from tanks. The sensor configuration around each pump is shown in Figure 1. It consists of a magnetic flowmeter and two pressure gages installed on the pump output stream and a level sensor located on the tank, but there is no pressure sensors at pump inlet.

First a correlation was obtained using information from pump characteristic curves. It allows estimating the flow rate of liquor/slurry being processed by the pump in terms of the head and speed of the pump. This correlation is not useful for flowrate adjustment by itself because, the lack of pressure sensors at the pump suction avoids measuring directly the pressure rise of the liquid over the pump. Instead the energy balance of the fluid, from the liquid maximum height in the tank to the pressure sensor downstream of the pump, is used as an extra model equation because it can be formulated in terms of measured variables and parameters. It is expressed as follows

$$\frac{P_1}{\rho g} + \alpha_1 \frac{v_1^2}{2g} + h_1 + \frac{W_s}{g} + h_f + h_c = \frac{P_2}{\rho g} + \alpha_2 \frac{v_2^2}{2g} + h_2 \tag{1}$$

where: P = Pressure (kPa), ρ = Density (kg/m3), α = Velocity profile factor, g = Acceleration due to gravity (m/s^2), h = Height (m), W_s/g = Shaft Work (m), h_f = Head loss due to friction (m), h_c = Head loss due to contractions (m).

In equation (1), the velocity in the tank is assumed to be zero, the pressure at point 1 is given by the tank pressure (usually atmospheric), the pressure at point 2 is read from the pressure indicator, the height terms are derived from the level indicator on the tank and the geometry of the pipe work obtained from isometric drawings, the head loss due to friction and contractions is formulated in terms of pipe geometry and flow rate.

The process model is enlarged with seven non-linear semi-empirical equations, corresponding to balances indicated by formula (1). Also redundant equations that consider the equality of pressure due to hardware redundancy are included in the model. For the normal operation state, the results of variable classification are: R=25, NR=22, O=32, UO=197. Redundant measurements include flowrates, but also pressures and pump speeds that has not been considered in Case 1.

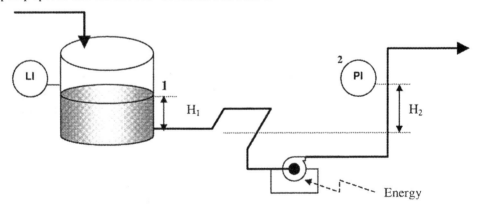

Figure 1: Tank and Pump Configuration

2.4 Data reconciliation (DR)

A current set of measurements and switch positions is provided by the DCS at intervals of 2.5minutes. The current operating status is compared with the previous one. If the operational mode had changed the model is modified accordingly and the DR is performed on the measured redundant variables. Its inputs are the vector of average measurements, obtained from sensor outputs corresponding to the same operational state, and the current sector model is made up of the mass balances associated to non-zero flow streams and the redundant equations obtained for pump operation. If there is no change in the operational state of the process, the DR procedure is performed after a prefixed interval of ten minutes. Two data reconciliation procedures were derived, linear and non-linear. The linear one, uses the linearized version of the redundant model, consequently estimates are calculated straightforward. The solution of the non-linear problem was achieved using a SQP technique. Measurement errors are estimated using the information provided by the vendors. Parameter values are updated using joint parameter and measured variable estimation at sensible fixed interval. A graphical interface was developed in Excel to allow the model's results to be presented in a recognizable and easy to read format. Figure 2 below shows a sample of the on-line results for reconciled and measured values in the case of the dregs filter pressures. These graphs are used as a monitoring tool for operators, to decide if process is running under normal conditions. Their judgement is based on confirmed information.

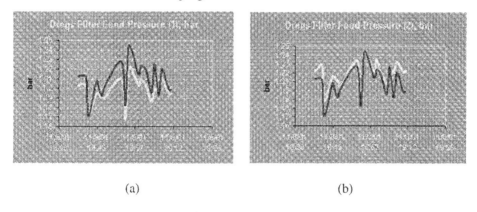

(a)   (b)

Figure 2: DR results for dregs filter pressures

2.5 Gross error detection

Data reconciliation deals with the problem of random errors. If gross errors are also present in the process data, they must be identified and removed before reconciliation. In our approach we have used the test on the estimates for gross error identification. Following Mah and Tamhane (1982), measurement residuals are defined as

$$e = y - \hat{y} \qquad (2)$$

where y and $\hat{y}$ are the vector of measurements and reconciled estimates respectively.

It is easy to show that $e \sim N(0, V)$ if there are no gross errors in the measurements, where V is the covariance of the residuals.

We conclude that there is a gross error present in the ith observation, if

$$|E_i| = |e_i| > E_{\alpha/2} \qquad (3)$$

where $E_{\alpha/2}$ is the upper $\alpha/2$ point of the standard normal distribution and α is the level of significance. In our case, $E_{\alpha/2}$ was obtained from the actual distribution of the plant data by approximating the distribution function of the residuals, based on historical information for normal operating conditions. Figure 3 illustrates the bar plot of the residuals (critical value based on statistics $E_{\alpha/2}$ and actual value e_i after reconciliation) for two operating conditions.

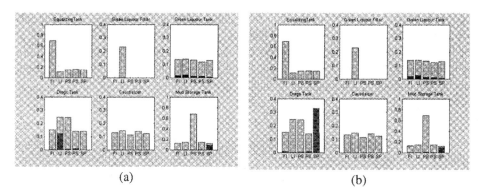

(a)   (b)

Figure 3: Residuals plot (a) Normal condition (b) Faulty condition

3. Conclusions

In this work, the study conducted for implementing a data reconciliation procedure at the recausticizing sector of the Visy Pulp and Paper Mill, Tumut is presented. The procedure points towards solving two main aspects. One of them is the variability with time of the model representing sector operation, a point not considered in previous data reconciliation procedures that work for only one operative mode. The second aspect involves the necessity of increasing system redundancy in order to detect and identify gross errors. In this sense semi-empirical relationships among measured variables are obtained using information from pump characteristic curves. This constitutes also an innovative way to overcome the difficulties associated with the lack of instrumentation.

4. References

Romagnoli, J.A. and M. Sánchez, 1999, Data Processing and Reconciliation for Chemical Process Operations. Academic Press, San Diego.

Sánchez, M. C. and J. Romagnoli, 1996, Use of Orthogonal Transformations in Classification/Data Reconciliation. Computers and Chemical Engineering, **20**, 483-493.

Mah R.S.H., Tamhane A.C., Detection of gross erros in Process Data, AIChE Hournal, Vol. 28, No. 5 pp. 828-830 (1982).

European Symposium on Computer Aided Process Engineering – 12
J. Grievink and J. van Schijndel (Editors)

A Framework for Consistency Analysis of Safety Properties and its Use in the Synthesis of Discrete-Event Controllers for Processes Systems

A. Sanchez[1], R. Gonzalez, A. Michel
Dept. of Elec. Eng. and Comp.
Centro de Investigacion y Estudios Avanzados (CINVESTAV)
Apdo. Postal 31-438, Guadalajara 45091, Jalisco, Mexico

Abstract

This paper proposes an "ad-hoc" formal framework for the analysis of three types of safety specifications and its use in the synthesis of a class of discrete event controllers. The notion of a specification set free of errors and redundancies is introduced as a *minimal set of consistent specifications* as well as procedures to establish it. The satisfiability verification of the specifications by the closed-loop behaviour model is also discussed. The use and advantages of the framework are illustrated with the synthesis of a class of discrete-event controller, termed procedural controller, for the operation of a batch reactor. Conflicts on the specification set were easily identified and corrected, reducing the synthesis effort. Satisfiability verification of the specifications by the closed-loop behaviour established to what extent the controller fulfils the specifications.

1. Introduction

Formal synthesis of controllers for event-driven operations using automata-based methods is frequently carried out employing a discrete event model of the process and a set of closed-loop behaviour specifications established by the designer. Both the model and the specification set are frequently assumed to be initially free of errors, which in few occasions is the case. Thus, the synthesis task becomes an iterative procedure in which the process model or specifications are modified in each iteration, until obtaining a satisfactory result. In most cases, not to mention large systems, it is very difficult to identify errors or inaccuracies either with incremental or monolithic specifications.

This paper proposes an "ad-hoc" formal framework to facilitate the synthesis of a class of discrete-event controllers, termed procedural controllers, for three types of safety specifications describing the conditional execution of sequences of controlled events. The framework helps the designer i) to capture this particular class of specifications, ii) to analyse the specification set and initiate the controller synthesis with a specification set free of inconsistencies and redundancies and iii) to determine what specifications are satisfiable by the obtained closed-loop behaviour. The paper first describes the modelling of a discrete-event process using standard automata as well as the notion of

---

[1] Corresponding author. e-mail address: arturo@gdl.cinvestav.mx. Fax: +52 (33) 3134 5579

semitrajectory. The definition of *procedural controller* is also presented together with an explanation of the type of process in which these controllers can be used. Section 3 introduces the definition of the safety properties realised as semitrajectories with specific semantics. Then, the notions of a *minimal set of consistent specifications* and *satisfiability* by the closed-loop behaviour are given. Computation procedures were implemented to carry out all required calculations. By using a minimal set of consistent specifications, the synthesis effort can be reduced substantially because this set is free of internal inconsistent behaviour and does not contain repeated specifications. Thus, if the resultant closed-loop behaviour does not satisfy some of the specifications, it is due only to controllability restrictions. Moreover, by verifying the satisfiability of the specifications by the closed-loop behaviour, it is established to what extent the controller fulfils the specifications. The paper then describes, using an example, how to incorporate these ideas in the synthesis of a procedural controller for the filling operation of a batch reactor. A closing section discusses what are the benefits of the proposed framework.

2. The modelling framework

A discrete-event process is modelled in the standard fashion by a finite state machine (FSM) $P = \{Q, V, \Sigma\ \delta, q_0, Q_m\}$, where Q is the state set; $V = \{v_1, v_2,..., v_n\}$ is the state-variable set with n, state variable number; a state-variable v_i takes values from a finite domain D_i plus the distinguished symbol ∞ meaning "any value". Σ is the transition set, divided into two disjoint sets: Σ_u (uncontrollable transitions) and Σ_c (controllable transitions); $\delta : \Sigma \times Q \to Q$, is the state-transition partial function; q_0 is the initial state; Q_m is the marked states set. A transition τ is enabled in state q if $\delta(q,\tau)$ is defined.

Definition. Assignment. An assignment over the variable set is a function $s:V \to D$ defined by the rule $v_i \to s(v_i)$ for i=1,...,n, where D is the union of the state-variable domains including the distinguished symbol ∞. The assignment s is represented by an n-tuple $(s(v_1), s(v_2),..., s(v_n))$ for i=1,...,n.

For each state $q \in Q$, an assignment can be associated by a function $\beta:Q \to A$ defined by $\beta(q)=s_q$, where A is the set of all possible assignments. Notice that $\beta(q)$ is not injective, thus it is possible to have two states with the same assignment.

Definition. Non-executable transition set. For an assignment s such that $\beta(q)=s$ for some $q \in Q$, the set of non-executable transitions of s is given by net(s)=$\{\tau \in \Sigma \mid \tau$ is an enabled transition in any q with assignment s which, by design, is not permitted to occur$\}$.

Definition. Semitrajectory. A semitrajectory of length m+1 is a finite mixed sequence of assignments and transitions $\pi = s_0{}^{\tau_1} s_1{}^{\tau_2}...{}^{\tau_m} s_m$. The relationship between assignments and transitions is established by $\gamma: A \times \Sigma \to A$ with the following rule: for $s_{i+1} = \gamma(s_i,\tau)$, there exists exactly one $j \in \{1,2,...,n\}$ such that $s_{i+1}(v_j) \neq s_i(v_j)$ and for all k=1,...,n such that $j \neq k$, $s_{i+1}(v_k) = s_i(v_k)$.

Definition. Covering (refinement). The assignment s covers assignment s' (equivalently, s' refines s) if and only if there exist at least one $j \in \{1,2,...,n\}$ such that $s(v_j) = \infty$, $s'(v_j) \neq \infty$ and for all $k= 1,2,...,n$ such that $j \neq k$, $s(v_k) = s'(v_k)$.

The control device used in this work is termed *procedural controller* (Sanchez *et al.* 99). This device is capable of forcing the execution of controllable transitions by pre-empting uncontrollable events. The controller is modelled as well as an FSM $C = \{X, \Sigma$ $\gamma, x_0, X_m\}$, where X is the state set; Σ is the same transition set as in the process model; $\eta : \Sigma \times X \rightarrow X$, is the state-transition partial function; x_0 is the initial state and X_m is the marked states set. In particular, for each $x \in X$, and $\sigma \in \Sigma$ such that $\eta(\sigma, x)$ is defined, either, 1) $\sigma \in \Sigma_u$ and for all $\sigma_c \in \Sigma_c$, $\eta(\sigma_c, x)$ is undefined, or 2) $\sigma \in \Sigma_c$ and for all $\sigma' \neq \sigma$ $\in \Sigma$, $\eta(\sigma', x)$ is undefined. That is, a procedural controller can either be in: 1) a state in which one of a set of uncontrollable transitions occurs or 2) a state in which the execution of the only controllable transition defined is enforced. Sanchez *et al.* (99) presented conditions of existence of a procedural controller, closed-loop invariant properties as well as procedures to calculate the controller.

3. Specification of safety properties

A safety property informally states that a "bad thing must not occur". When dealing with batch plants, we have found that most of the specifications for designing discrete-event controllers can be described in terms of three following types of safety properties. These specifications are modelled as semitrajectories.

Definition. Type 1 Safety Specification. Given an operation state, a set of control commands is executed sequentially. Any semitrajectory $\pi = s_0{}^{\tau 1} s_1{}^{\tau 2}... {}^{\tau m} s_m$ of length m+1, with m>0, models a type 1 specification. s_0 is called the *initial assignment* of π.

Definition. Type 2 Safety Specification. Given the occurrence of a process event, a set of control commands is executed sequentially. A semitrajectory $\pi = s_0{}^{\tau 1} s_1{}^{\tau 2}... {}^{\tau m} s_m$ of length m+1, with m>0, models a type 2 specification if and only if s_0 covers s_1. $\gamma(s_0, \tau)$ is used as the initial assignment s_1 from where the rest of π is executed. This semitrajectory is a compact way of representing the triggering of sequence by a transition. This type of semitrajectory will be written as $\pi = s_0{}^{\wedge \tau 1} s_1{}^{\tau 2}...s_m$, where expression $\wedge \tau_1$ indicates the emphasis over the exectuion of the event.

Definition. Type 3 Safety Specification. Given an operation state, a set of control commands is not permitted to occur. A specification of type 3 is modelled as a set of two-assignment semitrajectories of the form $\pi = s_0{}^{\tau} s_1$. s_0 and s_1 are fixed assignments whereas τ can be any transition such that $\gamma(s_0, \tau) = s_1$ and τ does not belong to set net(s_0). s_0 is the initial assignment. This type of semitrajectory will be written as $\pi = s_0{}^{\vee \tau} s_1$.

3. The minimal set of consistent semitrajectories

Semitrajectories π_1 and π_2 are *consistent* if and only if one of the following is true:
1. Initial assignments of π_1 and π_2 are not equal and do not cover each other.
2. If condition 1 is not true, and semitrajectories are not of type 3, then they must coincide after the initial assignments.
3. If condition 1 is not true and only one semitrajectory declares the execution of transition τ from its initial assignment, then the other semitrajectory must not forbid the execution of such transition.

Semitrajectories π_1 and π_2 are *duplicated* if and only if their initial assignments are
1. Either equal or one assignment covers the other and
2. Either coincide after the initial assignments or, for specifications of type 3, at least one transition of net(s_0) of π_1 belongs to net(s_0) of π_2.

A *set of consistent semitrajectories* is a set in which all semitrajectories are mutually consistent and a *minimal set of consistent semitrajectories* is a set of consistent semitrajectories without duplications. Computational procedures were implemented to obtain a minimal set of consistent semitrajectories.

4. Semitrajectory satisfiability in the closed-loop behaviour.

A semitrajectory $\pi = s_0{}^{\tau 1} s_1{}^{\tau 2} \dots {}^{\tau r} s_r$ of length $r+1$ is *accepted* by an FSM $M_\pi = \{X, V, \Sigma, \rho, x_0, X_m\}$. X is the state set; V is the state variable set of the process model with state assignments given by $\beta(x_i) = s_i$ for $i = 1, 2, \dots, r$; Σ is the transition set of the process model and $X_m = \{x_r\}$. The transition function $\rho: X \times \Sigma \to X$, such that if $\gamma(s_i, \tau)$ is defined in π, then $\rho(x_i, \tau)$ is defined in M_π. The initial state of M_π for specifications of type 1 and 3 is x_0. Otherwise, the initial state of M_π is x_1.

A semitrajectory π accepted by an FSM M_π is *satisfiable* in a closed-loop FSM model $M_{CL} = \{C, \Sigma, \varepsilon, c_0, C_m\}$, if and only if 1) each assignment $\beta(x_i)$ of state x_i in M_π is equal to or covers at least one assignment $\beta(c)$ of state c in of M_{CL} and 2) if $\rho(x_i, \tau)$ is defined, then $\varepsilon(q, \tau)$ is defined. That is, the FSM M_π "fits at least once into" the closed-loop FSM M_{CL}. Thus, semitrajectory satisfiability by the closed-loop model can be established by using a standard search strategy. The current computer procedure implementing the verification of specification satisfiability outputs whether the controller model satisfies a given specification. Otherwise, those specification states not existing in the closed-loop model are displayed as well as the closed-loop states in which $\varepsilon(q, \tau)$ is not defined for a given $\rho(x, \tau)$.

5. Example

The use of the framework is illustrated with the synthesis of the PLC control logic for the filling up phase of a batch reactor (shown in Figure 1) currently installed in a special lubricants factory. Table 1 includes the process and software components involved in the operation, together with FSMs models. Uncontrollable transitions are marked with an *. The rest of the transitions are commands that the PLC can use to control the

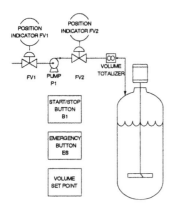

Figure 1. Diagram of batch

process. All initial and marked states are labelled with (0). The objective of the phase is to feed into the reactor a measured amount of a component from a warehouse location. During the filling, the operator can stop/restart the operation by pressing button B1 and the PLC must be able to stop and resume safely the filling. If the procedure does not succeed or the operator presses the emergency button, the controller blocks any possible action of the operator, issues the command to stop the pump and shuts any valve still open. The designer captured this phase using 12 semitrajectories shown in Table 2. Specification 1 and 7 are of type 2. The rest is of type 1. They are classified as normal, emergency and recovery procedures, which were exhaustively checked by hand. For each specification, a natural language statement is given first, followed by its associated semitrajectory. In order to present these semitrajectories in a concise manner, only the first assignment is shown. For subsequent assignments, it is indicated only what position and with what value it differs from the previous one. The process model was built using the method discussed in Sanchez (96) and the synthesis of the procedural controller was carried out using available tools (Sanchez *et al.* 99). The sizes of the process model and resultant procedural controller were 1024 and 124 respectively. Satisfaction of each semitrajectory by the closed-loop behaviour was then manually verified to the designer's satisfaction. Before translating the resulting FSM into programming code, a second verification round was carried out using the framework proposed here. It was found that the specification set was inconsistent and not minimal. Duplication was spotted in semitrajectories 2 and 11. Both specifications command to start the pump when the process is ready for the filling. The duplicate behaviour was attributed in this case to an oversight of the designer. More importantly, it was found that semitrajectories 8 and 9 were inconsistent. Both specifications used coverings in their respective initial assignments and were not comparable. That is, state-variable 3 was covered in specification 8 while it was refined in specification 9, and state-variables 5 and 7 were

Component		Transitions			
		Lbl	Description	From st.	To st.
1	B1 button	11*	switchOn	off (0)	on (1)
		12	switchOff	on (1)	off (0)
2	emerg. stop button	21*	switchOn	off (0)	on (1)
				on (1)	off (0)
3	vol. set point	31	FeedSP	NotInic (0)	Inic (1)
		32	NoSP	Inic (1)	NotInic (0)
4	vol. tot flag	41*	volOK	clear (0)	OK (1)
		42*	clear	OK (1)	clear (0)
5	FV2 valve	51	openFV2	closed (0)	open (1)
		52	closeFV2	open (1)	closed (0)
6	FV2 pos. ind.	61*	FV2opens	closed (0)	open (1)
		62*	FV2closes	open (1)	closed (0)
7	FV1 valve	71	openFV1	closed (0)	open (1)
		72	closeFV1	open (1)	closed (0)
8	FV1 pos. ind	81*	FV1opens	closed (0)	open (1)
		82*	FV1closes	open (1)	closed (0)
9	pump status	91	startPmp	off (0)	strtng (1)
		92*	pmpStarts	strtng (1)	on (3)
		93	stopPmp	on (3)	stppng (2)
		94*	pmpStops	stppng (2)	off (0)

Table 1. Elementary components of reactor and their associated FSM models (* = uncontrollable transition).

covered in specification 9 while they were refined in specification 8. This caused that there were 8 state assignments being shared by both states, which lead to the inconsistent behaviour. By verifying satisfiability of the specification set by the closed-loop model, it was found that the controller did not satisfy specifications 9 and 12. In the case of specification 9, this was because specification 8 was applied first in the synthesis procedure. For specification 12, it was found that it was not possible to reach in a controllable manner the state declared as initial in the specification. In a second exercise, specifications 11 and 12 were eliminated from the specification set. Specification 9 was modified, as shown in table 3, to avoid shared assignments. Using the new specification set (minimal and consistent), the newly synthesised controller was the same as the previous controller. All specifications were satisfied by the closed-loop model.

6. Conclusions

Obtaining the same results in both exercises indicates that, although the controller guarantees a safe (i.e. controllable) closed-loop behaviour, it was not known what the controller was actually controlling. From a practical point of view, this is not acceptable for any real process. Although it is not essential for the controller synthesis to debug the specification set to a minimal and consistent one, it was very useful for the better understanding of the controller role.

Bibliography

Sanchez A., G. Rotstein, N. Alsop and S. Macchietto.
Synthesis and implementation of procedural controllers for event-driven operations. AIChE J., 45, 8, 1753-1775, 1999

Sanchez A. Formal Specification and Synthesis of Procedural Controllers for Process Systems. LNCIS, v. 212, Springer-Verlag, 1996.

Acknowledgements. Partial financial support from CONACYT (grant 31108U) and the use of the application example from Interlub S.A. are kindly acknowledged.

Normal Operation
1. If button B120 is toggled, then PLC issues commands to open FV2 and FV1 $(0, 0, 0, 0, 0, 0, 0, 0, 0) \wedge^{11} (1:1)^{51} (5:1)^{71} (7:1)$
2. Once valves are open, then PLC issues command to start pump $(1, 0, 0, 0, 1, 1, 1, 1, 0)^{91} (9:1)$
3. If pump is successfully started, then the volume setpoint is feed to the totalizer $(1, 0, 0, 0, 1, 1, 1, 1, 3)^{31} (3:1)$
4. Once volume amount is reached, B120 is toggled, setpoint flag returns to not initialized and the PLC issues commands to close valves and stop pump $(1, 0, 1, 1, 1, 1, 1, 1, 3)^{12} (1:0)^{32} (3:0)^{52} (5:0)^{72} (7:0)^{93} (9:2)$
Emergency Operation
5. If FV2 shuts while filling up, then the PLC issues command to stop pump $(1, 0, \infty, 0, 1, 0, \infty, \infty, 3)^{93} (9:2)$
6. If FV1 shuts while filling up, then the PLC issues command to stop pump $(1, 0, \infty, 0, \infty, \infty, 1, 0, 3)^{93} (9:2)$
7. If emergency stop is activated, then button B1 is freed and operation cannot start again $(1, 1, \infty, \infty, \infty, \infty, \infty, \infty, \infty) \wedge^{12} (1:0)$
8. Once the emergency stop has been activated and the filling was under way, then the PLC must issue commands to close valves FV2 and FV1 $(0, 1, \infty, 0, 1, \infty, 1, \infty, 0)^{52} (5:0)^{62} (6:0)$
9. Once the pump is off, clear the volume set point $(0, 1, 1, 0, \infty, \infty, \infty, \infty, 0)^{32} (3:0)$
10. If emergency stop is activated and pump is on, then the PLC must issue a command to stop pump $(0, 1, \infty, 0, \infty, \infty, \infty, \infty, 3)^{93} (9: 2)$
Recovery procedures
11. After restarting normal operation, valves were opened successfully and pump is off, then the PLC must issue a command to start pump $(1, 0, \infty, 0, 1, 1, 1, 1, 0)^{91} (9:1)$
12. After restarting normal operation, if valves were closed, then the PLC issues commands to open them $(1, 0, 1, 0, 0, 0, 0, 0, 0)^{51} (5:1)^{61} (6:1)$

Table 2. Semitrajectories for filling phase

9. Once the pump is off, clear the volume set point $(0, 1, 1, 0, 0, \infty, 0, \infty, 0)^{32} (3:0)$

Table 3. Modified semitrajectory for specification 9

European Symposium on Computer Aided Process Engineering – 12
J. Grievink and J. van Schijndel (Editors)
© 2002 Elsevier Science B.V. All rights reserved.

Aggregated Batch Scheduling in a Feedback Structure

Guido Sand and Sebastian Engell

Process Control Laboratory, Department of Chemical Engineering,
University of Dortmund, Germany

Abstract

In this contribution a two-layer feedback structure to schedule batch chemical processes is specified. For a real-world benchmark the master scheduling problem is stated as a two-stage stochastic program and solved by a decomposition algorithm. Evaluation results are presented which prove the applicability of the approach.

1. Introduction

Scheduling operations in batch chemical plants constitutes large-scale combinatorial optimisation problems. Their modelling and solution has been addressed by a number of approaches from the mathematical programming domain, see e.g. Reklaitis (1996). In the majority of the publications the models are stated under the assumption of complete information, but in recent years researchers increasingly address the integration of uncertainties. Representative examples are the papers by Balasubramanian and Grossmann (2000), Honkomp et al. (1997), Ierapetritou et al. (1995), Petkov and Maranas (1997), and Sanmarti et al. (1997). However, almost all papers which consider the rescheduling of activities aim to maximise reliability, flexibility or robustness of a nominal schedule in the sense that the probability of having to perform extensive reactive modifications is minimized.

Our approach is different because we assume that reactive scheduling can be performed by solving detailed scheduling problems on an appropriate horizon online. Schulz (2001) showed for a real-world polymer production process that an MINLP-model for a 6 days horizon can be solved nearly optimally in one minute. Stimulated by this result, we propose to generate aggregated master schedules which explicitly reflect this considerable recourse capability. To this end, two-stage stochastic integer programs (2-SSIP) provide an appropriate modelling framework, see e.g. Birge and Louveaux (1997). We choose a formulation with a scenario based representation of uncertainty, which is, for instance, efficiently computational tractable by the algorithm of Carøe and Schultz (1999). Motivated by the mentioned process (chapter 2), we specify a two-layer feedback structure for real-time scheduling with recourse (chapter 3). The model on the upper layer is presented and evaluated (chapters 4 and 5), and an outlook on the future work concludes the contribution (chapter 6).

2. Problem Statement

Figure 1 schematically shows a multiproduct plant, which is used for the production of 2 types of polymer (expandable polystyrene - EPS) in 5 grain size fractions each. The preparation stage is less important for the problem and will not be considered here. Each polymerisation is performed batchwise according to a certain recipe (out of a set of 10 possible recipes), which specifies the EPS-type and the desired grain size distribution. Due to safety restrictions, a minimal offset between two batch starts has to be kept. The produced suspension is transferred into buffering mixers and continuously fed into separation units. The mixer levels have to be kept between certain thresholds while the corresponding finishing line is running; moreover, it may temporarily be shut down. The mixing process is described by a nonlinear equation.

Degrees of freedom are: 1. the starting times of the polymerisations, 2. the recipes for the polymerisations, 3. the operation modes of the finishing lines (on/ off), and 4. the feed rates into the separation units (within certain bounds). The process is influenced by significant operational and demand uncertainties, namely: 1. equipment breakdowns (polymerisation reactors), 2. not precisely reproducible grain size distributions, 3. variable processing times, and 4. vague demand forecasts. The production goal is to maximise expected profit, which is mainly determined by: 1. the revenue, which decreases with exceeding the due date, 2. fixed costs for each polymerisation, 3. fixed costs for each change in the operation modes of the finishing lines, 4. variable inventory costs, and 5. variable penalties for production shortfalls.

3. Solution Approach

Since the process is running continuously and the demand profiles reveal no periodical pattern, we propose a cascade-like feedback structure to optimise the process operations online (see figure 2). The aggregated layer works as a regularly master scheduler (MS), whereas the detailed layer takes the role of a quickly reacting scheduler (RS). Both parts are supposed to be based on stochastic integer programs with recourse. Their precise specifications are derived by applying these generic ideas to the specific process.

Considering the process dynamics, it appears reasonable to perform master scheduling updates in intervals of 2 days and to look 10 intervals ahead. This establishes a

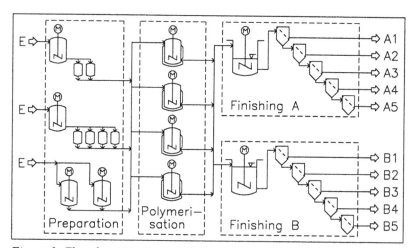

Figure 1: Flowsheet EPS-process

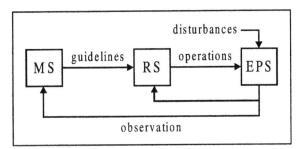

Figure 2: Cascade-like feedback structure

multiperiod time representation and an aggregation of decisions according to the intervals. The alternating sequence of observation and decision would naturally lead to a 10-stage stochastic program. However, to keep the problem computationally tractable it is approximated by a 2-stage stochastic program. In accordance with the specification of the detailed scheduler, the first stage covers the intervals 1 to 3. Decisions which are relevant to the indicated horizon and granularity are: 1. the number of polymerisation starts in each interval classified according to the recipes, and 2. the operation modes of the finishing lines. Relevant constraints are capacity restrictions imposed by the polymerisation stage and the finishing lines, whereas mixing effects are of less importance and do not affect feasibility. Since uncertainties are supposed to be modelled by scenarios with a priori known probabilities, it is favourable to consider the exogenuous disturbances based on fixed time intervals. The probability of the occurrence of an exogenuous event during an interval is only a function of its length, and for a point of time it is zero. Concerning the EPS process, the exogenuous uncertainties, namely equipment breakdowns and demand forecasts, are exactly those which have long-term effects, such that they should be considered on the aggregated layer. The generated production profiles and the operation modes of the first stage intervals serve as guidelines for the detailed scheduler.

Since the focus of this contribution is on the aggregated layer, the detailed scheduler is only sketched briefly. The model is intended to be based on an event driven time grid with 20 points. This corresponds to a horizon of 4 to 6 days, which is identical with the aggregated first stage. Those decisions which are associated with the first grid point form the first stage of a 2-SSIP. The process is modelled by detailed constraints, except for the nonlinear mixing effects: Simulations studies show, that the inaccuracy caused by a linear approximation is in the order of the inaccuracy caused by inexact data. The scenarios cover the endogenous (short-term) uncertainties, namely grain size distributions and processing times.

4. Aggregated Scheduling Model

The deterministic equivalent of a 2-SSIP with discrete scenarios is defined as follows (cf. Birge and Louveaux (1997)):

$$\max_{x, y_1, \dots, y_\Omega} \left\{ \sum_{\omega=1}^{\Omega} \pi_\omega \left(c^T x_\omega + q_\omega^T y_\omega\right) \text{s.t. } T_\omega x_\omega + W_\omega y_\omega = h_\omega, \, x_1 = \dots = x_\Omega, \atop x_\omega \in X, \, y_\omega \in Y, \, \omega = 1, \dots, \Omega \right\} \tag{1}$$

The first- and second-stage variable-vectors x and y belong to polyhedral sets X and Y with integer requirements. The parameter Ω denotes the number of scenarios with corresponding probabilities π. The constraints are formulated by means of the matrices T und W and the right hand side-vector h of suitable dimensions. The objective is to maximise a weighted sum of variables subject to the weighting-vectors c and q.

For the aggregated scheduling model of the EPS-process we define the following indices, parameters and variables:

1. Indices: intervals $i, j, k \in [1, \dots, I]$, delays $d \in [1, \dots, D]$, products $p \in [1, \dots, P]$, grain size fractions $f \in [1, \dots, F]$, and recipes $r \in [1, \dots, R]$.

2. Parameters: ratio ρ of a certain grain size fraction according to a certain recipe, customer demands B, bounds for polymerisation starts N^{max}, thresholds of mixer levels C^{min}, C^{max}, and bounds on feed rates F^{min}, F^{max}.

3. Variables: number of batches $N \in IN$, operation modes of finishing lines $z \in \{0,1\}$, mode change indicators $w \in IR_+$, amounts of delivered produkts $M \in IR_+$, amounts of inventories $M^+ \in IR_+$, and amounts of production shortfalls $B^- \in IR_+$.

The capacity constraints of the polymerisation stage (2) (see below) are stated dynamically, i.e. they represent the couplings between the intervals correctly, by bounding all interval sequences. The dynamic capacity constraints of the finishing lines (3) bound the polymerisation starts such that a) if the finishing line is operating, feed rates exist which keep the mixer level feasible or b) if the finishing line is not operating, the mixer is empty. The so-called non-anticipativity constraints (4), which say that the first stage variables do not depend on a particular uncertainty realisation, are straightforward.

$$\sum_{j=i}^{k} \sum_{p=1}^{P} \sum_{r_p=1}^{R_p} N_{jr_p\omega} \leq N_{ik\omega}^{max} \quad \forall i, k, \omega \,|\, i \leq k \leq I \tag{2}$$

$$x_{kp\omega} x_{(k+1)p\omega} C_{p\omega}^{min} - \left\{ {C_p^0 \text{ if } i = 1 \atop x_{(i-1)p\omega} x_{ip\omega} C_{p\omega}^{max} \text{ else}} \right\} + \sum_{j=i}^{k} x_{jp\omega} F_p^{min} \leq \sum_{j=i}^{k} \sum_{r_p=1}^{R_p} N_{jr_p\omega}$$

$$\leq x_{kp\omega} x_{(k+1)p\omega} C_{p\omega}^{max} - \left\{ {C_p^0 \text{ if } i = 1 \atop x_{(i-1)p\omega} x_{ip\omega} C_{p\omega}^{min} \text{ else}} \right\} + \sum_{j=i}^{k} x_{jp\omega} F_p^{max} \quad \forall i, k, p, \omega \,|\, i \leq k \leq I \tag{3}$$

$$N_{ir_p, \omega=1} = \dots = N_{ir_p, \omega=\Omega}, \quad x_{ip, \omega=1} = \dots = x_{ip, \omega=\Omega} \quad \forall i, p, f_p \tag{4}$$

To maximise the expected profit, demand and production profiles have to be balanced as well as possible. The demand constraints (5) (see below) offset the customer demands against the production shortfalls and (possibly delayed) product deliveries. The dynamic supply constraints (6) bound the amount of product which is delivered until a certain interval by the produced amount. The inventory is computed by the oversupply constraint (7), and finally, the mode changes of the finishing lines are detected by equation (8). The objective is to maximise a weighted sum of M, M^+, N, B^- and w subject to (1) – (8). This results in an MINLP which is tranformed into a large but structured MILP by common linearisation techniques (see e.g. Williams (1994)).

$$\sum_{d=1}^{D} \begin{Bmatrix} M_{d(i+d-1)f_p} & \text{if } i+d-1 \le I \\ 0 \text{ else} \end{Bmatrix} = B_{if_p\omega} - B^-_{if_p\omega} \quad \forall i, f_p, \omega \,|\, i \le I \tag{5}$$

$$\sum_{j=1}^{i} \sum_{d=1}^{D} \begin{Bmatrix} M_{djf_p\omega} & \text{if } j \ge d \\ 0 \text{ else} \end{Bmatrix} \le \sum_{j=1}^{i} \sum_{r_p=1}^{R_p} \rho_{f_p r_p} N_{jr_p\omega} \quad \forall i, f_p, \omega \,|\, i \le I \tag{6}$$

$$M^+_{if_p\omega} = \max\left(0; \sum_{j=1}^{i} \sum_{r_p=1}^{R_p} \rho_{f_p r_p} N_{jr_p\omega} - \sum_{j=1}^{i} B_{if_p\omega} \right) \quad \forall i, f_p, \omega \,|\, i \le I \tag{7}$$

$$w_{ip\omega} = \left| x_{(i-1)p\omega} - x_{ip\omega} \right| \quad \forall i, p, \omega \tag{8}$$

5. Model Evaluation

The 2-SSIP reveals a block-angular matrix structure, which is exploited by the dual decomposition algorithm of Carøe and Schultz (1999). It is based on the Lagrangian relaxation of the non-anticipativity constraints (4), without which the formulation decomposes into Ω independent subproblems, which differ in the right hand sides h only. CPLEX (2000) is used as a subproblem-solver and imbedded in a branch and bound-algorithm, which branches on the first-stage variables and generates bounds by solving the dual problems.

The model was implemented with the parameter values $I = 10$, $D = 3$, $P = 2$, $F = R = 5$ and randomly generated values for B and N^{max}, representing uncertain demand forecasts and equipment breakdowns, respectively. For $\Omega = 1$ the problem degenerates into a purely deterministic MILP with 725 variables (122 discrete) and 680 constraints; it is solved by CPLEX 7.0 within 20 s CPU-time on a SUN Ultra 2 300 with average optimality gaps of 2,8 % (all < 6 %). The problem size is nearly proportional to Ω so that 1000 scenarios yield much more than half a million variables and constraints. It is reasonable to limit the CPU-time to 8 h, which corresponds to up to 2 polymerisation starts in the running process. On a SUN Ultra Enterprise 450, average optimality gaps of 7,6 % (all < 10 %) could be achieved for the stochastic program.

By approximating the dynamic constraints statically the problem size decreases by 285 constraints ($\Omega = 1$) and the average optimality gap is approximately cut by half at the expense of reduced modelling precision. In fact, the static model leads to solution vectors which are structurally different from the above and of inferior quality. Furthermore, we substituted the model property that demands can be satisfied partially by the restriction that each demand must be met entirely, possibly later than the due date. The overall model size shrinks to 284 variables and 479 constraints, but the number of discrete variables rises. CPLEX is not capable to efficiently process this additional complexity such that the deterministic problems yield average optimality gaps of 63,8 %.

6. Conclusion And Perspectives

The results of this work demonstrate that for a real-world production process aggregated batch scheduling with recourse can be performed under real-time conditions. The explicit modelling of extensive recourse actions anticipates different reactions to feedback information and thereby expands the optimisation potential. The modelling of various sources of uncertainties is straightforward and allows for immediate utilization of empirical probability distributions. The large problem scale is successfully tackled by a well-posed model structure in combination with an efficient decomposition algorithm. Encouraged by the results for the master layer we will also apply the above methodology to the detailed layer based on the model from Schulz et al. (1998). Special challenges arise from the fact that the model will have stochastic matrices and continuous first stage variables.

7. Acknowledgements

We gratefully acknowledge financial support by the Deutsche Forschungsgemeinschaft within the priority programme "Real-Time Optimisation of Large-Scale Systems" under grant EN 152/17-1,2,3, and the very fruitful cooperation with the Chair of Discrete Mathematics, Gerhard-Mercator-University of Duisburg, Germany.

8. References

Balasubramanian, J. and I.E. Grossmann, 2000, Scheduling to minimize expected completion time in flowshop plants with uncertain processing times, Proc. ESCAPE-10, Ed. S. Pierucci, 79.

Birge, J.R. and F. Louveaux, 1997, Introduction to Stochastic Programming, Springer, New York.

Carøe, C.C. and R. Schultz, 1999, Dual decomposition in stochastic integer programming, Oper. Res. Lett. 24, 37.

CPLEX Optimisation Inc., 1989-2000, Using the CPLEX Callable Library, http://www.ilog.com.

Honkomp, S.J., L. Mockus and G.V. Reklaitis, 1997, Robust scheduling with processing time uncertainty, Comp. Chem. Engg. 21, 1055.

Ierapetritou, M.G., E.N. Pistikopoulos and C.A. Floudas, 1995, Operational planning under uncertainty, Comp. Chem. Engg. 20, 1499.

Petkov, S.B. and C.D. Maranas, 1997, Multiperiod planning and scheduling of multiproduct batch plants under demand uncertainty, Ind. Eng. Chem. Res. 36, 4864.

Reklaitis, G.V., 1996, Overview of scheduling and planning of batch operations, Batch Processing Systems Engineering, Eds. G.V. Reklaitis, A.K. Sunol and D.W.T. Rippin, Springer, Berlin, 660.

Sanmarti, E., A. Espunia and L. Puigjaner, 1997, Batch production and preventive maintenance scheduling under equipment failure uncertainty, Comp. Chem. Engg. 21, 1157.

Schulz, C., R. Rudolf and S. Engell, 1998, Scheduling of a Multiproduct Polymer Batch Plant, Proc. FOCAPO98 (Snowbird/ USA), 224.

Schulz, C., 2001, Modelling and Optimisation of a Multiproduct Batch Plant, PhD thesis, University of Dortmund, Germany (in German).

Williams, H.P., 1994, Model Building in Mathematical Programming, John Wiley & Sons, New York.

European Symposium on Computer Aided Process Engineering – 12
J. Grievink and J. van Schijndel (Editors)
781

Analysis of Parametric Sensibility of the Process of Production of Cyclohexanol

Santos, M.M. and Maciel Filho, R.

School of Chemical Engineering - State University of Campinas (UNICAMP)
CP 6066 – Campinas – SP – Brazil - 13081-970
e-mail: marcela@lopca.feq.unicamp.br

Abstract

The main objective of this paper is to carry out a parametric sensibility analysis for a cyclohexanol production multiphase reactor. Firstly, the mathematical steady-state model validated with industrial data, based on mass, energy and momentum balances, was submitted to a fractional factorial design. This method allows the understanding of interactions between variables with advantages over conventional methods, which involve changing one variable while fixing the others at certain levels. From this, some variables could be neglected and a complete factorial design could be developed. Using this procedure was possible to identify not only the main effects of the variables, but also the interaction effects among these variables. The behavior of the cyclohexanol reactor was extensively studied bearing in mind the determination of control strategies and optimization.

1. Introduction

The more economic operation form of a great scale production plant is the continuous one. Moreover the plant should satisfy demands of equipment operation, product quality, more rigorous environmental restrictions and also follow some economic order. A typical system of continuous processing is the cyclohexanol production plant, raw material for obtaining of several industrial products of high commercial value as, for example, nylon. Due to the interest of this substance, it is desirable to obtain models that reproduce qualitative and quantitatively the behavior of the system, beyond of the phenomenological understanding of the process. In this work a deterministic model for the steady state of the cyclohexanol multiphase reactor is considered and used as tool for process variable interactions. The reactor is composed by tubular modules immersed in a boiler. Some of the modules are constituted by concentric tubes through where the reagents and the refrigerant flow. Some modules, through where the only reagents flow, are used concentric tubes. The reagent flow from one module to another and the refrigerant is added to each of the modules. The refrigerant stream, that comes from the boiler and eventually of a new feeding of condensed (make-up), is divided among the tubular modules, with different flow rates. This arrangement allows the operator to play with individual coolant flow rates, which are added to the conventional manipulated variables (as reactant pressure, temperature and concentrations). This makes the problem to be highly multivariable so that the usual techniques to perform parametric sensibility tend to fail. Bearing this in mind, in this work is proposed the implementation of fractional and complete factorial design procedure in a hierarchical approach to solve the parametric sensibility analysis of a large scale multivariable process. The proposed is to apply the fractional factorial design to reduce the process dimensionality and then use the factorial design to the remaining variables. Although the factorial design be a known tool extensively used in experimental planning considering real data very few attention has been given to its use as sensibility analysis tool coupled with process modelling. In fact the application of statistical process control procedure do not cover the whole potential as factorial design coupled to deterministic model does.

2. Mathematical modeling and statistical analysis of the process

2.1 Mathematical Modeling

The deterministic mathematical model used to describe the reactor is based on the work by Santana (1999) and Toledo et al. (2001). This model considers the peculiarities of tubular modules that compose the reactor. The process model is constituted of conservation laws of mass, energy and momentum. The mass and energy balance are in the form, respectively:

$$\frac{dC}{dz} = f(C, T, P, Tr_n, \cdots) \tag{1}$$

$$\frac{dT}{dz} = f(C, T, P, Tr_n, \cdots) \tag{2}$$

where C is concentration, T is temperature, P is pressure and Tr_n is refrigerant temperature in the n-tubular module. Moreover, equations for predicting the heat coefficients must be present as well as a way to describe evaporation that may occur, depending upon the operating conditions. Each of these equations must be applied to each module for both regions, namely, the tubular and annular. Since the reactor is essentially a tubular one, axial dispersion is considered. Thus, the steady-state process model presents a set of ordinary differential equations if radial dispersion is neglected, which is, together with the hypothesis that the solid-liquid phase is a single pseudo-homogenized fluid, a reasonable simplification that can be made in order to reduce the complexity of the process model. The mathematical model was validated with industrial data.

2.2 Fractional factorial and complete factorial design

The univariable analysis involves changing one variable while fixing the others at constant levels and usually it does not provide good information on the problem since interactions among variables are not taken into account. To overcome such limitations the fractional factorial design (Box et al., 1978) appears to be a suitable procedure. To apply the method, it is necessary to plane the trials through of factorial design. This method is based on selection of a fixed level number to each variable and executes all possible combinations. When many variables or parameters are involved in the process, one can choose a fractional factorial design. The design is reduced and it is possible to evaluate the variable importance in the responses. In this proposed hierarchical approach the fractional factorial design is useful in the initial stages of process development, so, that technique can be used to identify the more important independent variables and select them to realize a complete factorial design.

Once the statistically significant variables were selected by means of fractional factorial design, the complete factorial design experiments were planned to obtain the principal and interaction effects of such variables.

3. Results and discussion

3.1 Fractional factorial design

Firstly, the influence of eight variables on mole fraction of phenol in the liquid phase, reactor temperature and pressure will be analyzed. The independent variables are: initial reagent temperature (To) and pressure (Po), flow rate of refrigerant introduced into some tubular modules (Vref1, Vref2, Vref3, Vref4, Vref5 and Vref6). Each variable is tested at two level, a superior (+) and an inferior (-), as shown in Table 1. The range of values is in agreement with real data and the temperature is normalized. Table 2 presents the simulations as well as the way they are to be conducted for the 2^{8-4} fractional factorial design with 16 trials. The statistically significant effects of the variables on each response for a 95% confidence level are presented in Figures 1 and 2 for mole fraction of phenol in the liquid phase and reactor temperature, respectively. For pressure only the initial reagent pressure have statistically significant effect, so the graphic is not be presented. In these figures, the flow rate of refrigerant is not statistically significant in any tubular module. Thus, they can be excluded of the complete factorial design.

These results make possible to define the variables, which have an effect on the responses of interest so that complete factorial design can be elaborated to study the reactor behavior. Taking into consideration the present results, the selected variable for the responses are initial reagent temperature and pressure.

Table 1. Superior and inferior levels to independent variables.

Level	To	Po (kg/cm^2)	Vref1 (kg/h)	Vref2 (kg/h)	Vref3 (kg/h)	Vref4 (kg/h)	Vref5 (kg/h)	Vref6 (kg/h)
-1	0.817	23.97	306.0	323.0	2329.0	561.0	1011.5	1190.0
+1	1.105	32.43	414.0	437.0	3151.0	759.0	1368.5	1610.0

3.2 Factorial design

In addition to the selected variable from fractional factorial design, the initial concentration of phenol (X_{f0}) and the quantity of catalyst (W_{cat}) will be used in the complete factorial design, since they are important manipulated variables.

Table 3 presents the superior and inferior levels and Table 4 outlines the complete factorial design, where the variables used are To, Po, X_{f0} and W_{cat}. The data are not experimental, but produced by simulation, as it consists in the Table 4 (trial 17), it is only possible to present a trial in the central point and the answer models are free from pure error.

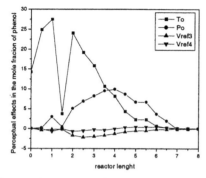

Figure 1. Perceptual effects in the mole fraction of phenol.

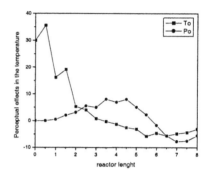

Figure 2. Perceptual effects in the temperature.

Table 2. Trials used in fractional factorial design.

Trial	To	Po	Vref1	Vref2	Vref3	Vref4	Vref5	Vref6
1	-1	-1	-1	+1	+1	+1	-1	+1
2	+1	-1	-1	-1	-1	+1	+1	+1
3	-1	+1	-1	-1	+1	-1	+1	+1
4	+1	+1	-1	+1	-1	-1	-1	+1
5	-1	-1	+1	+1	-1	-1	+1	+1
6	+1	-1	+1	-1	+1	-1	-1	+1
7	-1	+1	+1	-1	-1	+1	-1	+1
8	+1	+1	+1	+1	+1	+1	+1	+1
9	-1	-1	-1	-1	-1	-1	+1	-1
10	+1	-1	-1	+1	+1	-1	-1	-1
11	-1	+1	-1	+1	-1	+1	-1	-1
12	+1	+1	-1	-1	+1	+1	+1	-1
13	-1	-1	+1	-1	+1	+1	-1	-1
14	+1	-1	+1	+1	-1	+1	+1	-1
15	-1	+1	+1	+1	+1	-1	+1	-1
16	+1	+1	+1	-1	-1	-1	-1	-1

Figures 3 and 4 depict the effects for a 95% confidence level for mole fraction of phenol in the liquid phase and reactor temperature, respectively, and for pressure, only the initial reagent pressure have statistically significant effect, and the graphic is not presented.

For the mole fraction of phenol in the liquid phase, seen in Figure 3, only principal effects of the amount of catalyst are not statistically significant in any region of reactor. The increase in initial reagent temperature increases the mole fraction of phenol at the reactor exit that prejudices the reaction. The increase in initial reagent pressure and initial concentration of phenol decreases the mole fraction of phenol, and this is suitable to reaction, since phenol must present lower concentrations as possible because the rigorous environmental restrictions.

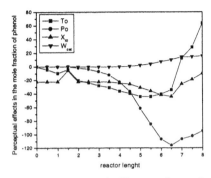

Figure 3. Perceptual effects in the mole fraction of phenol.

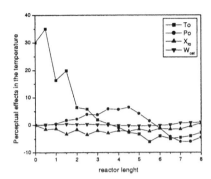

Figure 4. Perceptual effects in the temperature.

Table 3. Superior and inferior levels for independent variables for factorial design.

Level	To	Po (kg/cm^2)	X_{f0}	W_{cat}
-1	0.817	23.97	0.0	40.0
0	0.961	28.2	0.1	60.0
+1	1.105	32.43	0.2	80.0

Table 4. Trials used in complete factorial design.

Trial	To	Po	X_{f0}	W_{cat}
1	-1	-1	-1	-1
2	+1	-1	-1	-1
3	-1	+1	-1	-1
4	+1	+1	-1	-1
5	-1	-1	+1	-1
6	+1	-1	+1	-1
7	-1	+1	+1	-1
8	+1	+1	+1	-1
9	-1	-1	-1	+1
10	+1	-1	-1	+1
11	-1	+1	-1	+1
12	+1	+1	-1	+1
13	-1	-1	+1	+1
14	+1	-1	+1	+1
15	-1	+1	+1	+1
16	+1	+1	+1	+1
17	0	0	0	0

As Figure 4, where the principal effects in the reactor temperature are shown, again only principal effect of the quantity of catalyst are not statistically significant in any region of reactor. The increase in the other three independent variables generates decrease in the temperature.

786

The all interaction effects in the temperature are not statistically significant. In the mole fraction of phenol, shown in Figure 5, the effects that contain the amount of catalyst are not significant. The interaction among initial reagent pressure and initial concentration of phenol presents positive values in the last tubular modules of reactor and assumes null value at the reactor exit. The simultaneous increase of initial reagent temperature and initial concentration of phenol worsens the reaction, since greater concentration of phenol appears. On the other hand, the increase in the initial reagent temperature and pressure simultaneously is suitable to reaction (negative effects).

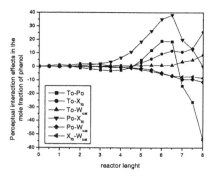

Figure 5. Perceptual interaction effects in the mole fraction of phenol.

4. Conclusions

The proposed hierarchical procedure based on fractional and complete factorial design appears to have a large potential to deal with large scale non-linear multivariable process. The methodology of fractional factorial design was shown to be very useful for determination of relevant variables. This makes possible to consider initially a large number of variables and then to obtain a smaller set of variables that are the most significant ones. For the case study, the use of this method reduced the number of variables from eight to two. With this, it was possible apply the complete factorial design, including more two variables.

Using the technique of factorial design it was possible study the reactor behavior through the principal and interaction effects, which are difficult to be identified in conventional parametric analysis procedures.

References

Box, G.E.P., Hunter, W.G., Hunter, J.S., 1978, Statistics for experimenters. An introduction to design, data analysis and model building, New York, Wiley.

Santana, P.L., 1999, Mathematical Modeling for three phase reactor: deterministic, neural and hybrid models, PhD Thesis, School of Chemical Engineering, Unicamp, São Paulo, Brazil (in Portuguese).

Toledo, E.C.V, Santana, P.L., Maciel, M.R.W., Maciel Filho, R., 2001, Dynamic Modelling of a Three-Phase Catalytic Slurry Reactor, presented at 5th International Conference on Gas-Liquid and Gas-Liquid-Solid Reactor Engineering - GLS 5, 6th World Congress of Chemical Engineering, Melbourne, Australia – To appear in Chemical Engineering Science, 2002.

European Symposium on Computer Aided Process Engineering – 12
J. Grievink and J. van Schijndel (Editors)
© 2002 Elsevier Science B.V. All rights reserved.

On line optimisation of maintenance tasks management using RTE approach

Sebastián Eloy Sequeira[1], Moisès Graells[2], Luis Puigjaner[1].
Chemical Engineering Department. Universitat Politènica de Catalunya
[1]ETSEIB, Av. Diagonal 647 Pav. G, 2° P Barcelona (08028) Spain
[2]EUETIB, Comte d'Urgell, 187, Barcelona (08036) Spain

Abstract

The efficiency of equipment units decreases depending on the time they have been operating due to fouling, formation of byproducts, catalysts deactivation etc. Therefore, periodical maintenance and/or cleaning is required to restore original conditions. This imposes a trade-off between shutdown and productivity improvement. When considering different production lines, the problem is usually addressed using mathematical programming (MINLP- Jain, and Grossmann, 1998, Georgiadis et al. 1999). The implementation of the decisions obtained in this way is unlikely to result in an optimal operation because of plant-model mismatch. Therefore, this work introduces both: simpler formulations (NLP and LP) and a procedure for the subsequent implementation of the discrete decisions involved in a real time environment. This methodology is an extension of the Real Time Evolution (RTE) approach for continuous processes (Sequeira et al., 2001b). The main advantages of the proposed approach are the robustness resulting from the use of on-line information and a scarce model dependency. An example illustrates the methodology from the planning stage to the on-line optimization including model mismatch and model uncertainty.

1. Introduction

The simplest case of maintenance is a single process whose performance (Instantaneous Objective Function, IOF) decreases with time illustrates. This will be the case for common semi-continuous (cyclic) operations as filtration, ionic exchange, catalyst with decreasing activity, etc. At some time t, the operation must finish in order to perform a maintenance task, which re-establishes the process initial conditions. Suppose that the maintenance task consumes a known time τ_m and has a cost C_m. Thus, the lower the time t, the higher the performance but at the expense of a cost and time occurrence because of the maintenance itself. Therefore, there is a trade off determining the time t_{opt} that maximizes the overall performance of the whole operating cycle. A possible formulation for this optimization problem consists of maximizing a Mean Objective Function (Buzzi et al., 1984; O'Donnel et al. 2001):

$$MOF\ (t) = (Benefits\ -\ Costs\ )\ /\ Cycle\ \_Time\ = \left(\int_0^t IOF\ (\tau)d\tau - C_m \right) \cdot (t + \tau_m)^{-1} \qquad (1)$$

Thus, the optimal maintenance time (t_{opt}) must satisfy:

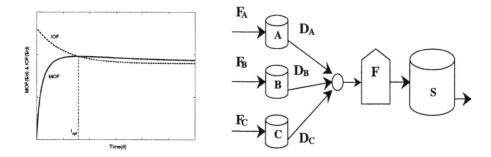

Figure1: IOF and MOF as functions of *Figure2: Motivating example scheme*
operation cycle duration

$$\frac{IOF(t_{opt})\cdot(t_{opt}+\tau_m)+C_m-\int_0^{t_{opt}}IOF(\tau)d\tau}{(t_{opt}+\tau_m)^2}=0 \quad hence, \quad IOF(t_{opt})=\frac{\int_0^{t_{opt}}IOF(\tau)d\tau-C_m}{(t_{opt}+\tau_m)}=MOF(t_{opt}) \quad (2)$$

Therefore, using an appropriate model of *IOF* the optimum policy can be found (Fig. 1).

2. Off -Line Approaches

Industrial practice poses more complex problems. In order to introduce such complexity, this section illustrates the effect of multiple feed presence trough a motivating example (Jain and Grossman, 1998). Three different raw materials (*A, B, C*) are available for arriving continuously to the corresponding storage tanks at a constant rate. These feed stocks are then processed sequentially in a reactor (the furnace), where because of cocking, the conversion decreases with time. The rate of arrival of different feed stocks is a decision variable (*Fm$_i$*) bounded by *Flo$_i$* and *Fup$_i$*. Every feedstock is processed in the furnace at a rate D_i. Whenever there is a product changeover, the furnace is cleaned, and the operating parameters are set so that the furnace operates at the best possible conversion for that particular feed. The changeover time for feed *i* is known and given by τm_i (sequence independent). The set up and cleaning cost for every feed is given by the constants Cm_i. Raw material at different grades (A, B, C) is processed in the furnace to obtain a final product *S* (Fig. 2) at a conversion depending, for a given set of conditions, on the grade *i* and that decreases with time:

$$X_i(t) = c_i + a_i \cdot e^{-b_i t} \tag{3}$$

where a_i, b_i, c_i are experimental parameters. Revenue is directly proportional to production through the price parameter P_i. The problem is to determine the policy that maximizes the profit. Specific data for the example considered are found in Table 1.

Table1: Data for the motivating example

Feed	τm_i (d)	D_i (ton/d)	a_i	b_i (1/d)	c_i	P_i ($/ton)	Cm_i ($)	Flo_i (ton/d)	Fup_i (ton/d)
A	2	1300	0.20	0.10	0.18	160	100	350	650
B	3	1000	0.18	0.13	0.10	90	90	300	600
C	3	1100	0.19	0.09	0.12	120	80	300	600

2.1 MINLP Formulation (F1)

Jain and Grossmann (1998) propose the following formulation. The objective function to maximize is the average profit during the cycle Z. There are seven decision variables: total processing time of feed i to the furnace (t_i), number of sub-cycles during the total cycle (n_i) and the total cycle duration (T_{cycle}). The total time devoted to feed i in furnace, including processing and cleanup time is given by Δt_i.

$$\max z = \frac{\sum_i \left\{ P_i \cdot D_i \cdot \left[c_i \cdot t_i + \frac{a_i}{b_i} \cdot n_i \cdot \left(1 - e^{-b_i t_i / n_i}\right) \right] - Cm_i \cdot n_i \right\}}{Tcycle} \tag{4}$$

$$ST: \qquad Fm_i \cdot T_{cycle} = D_i \cdot t_i, \qquad \Delta t_i = n_i \tau_i + t_i \qquad \forall i \tag{5}$$

$$\sum_i \Delta t_i \leq T_{cycle} \tag{6}$$

$$\Delta t_i \geq 0, T_{cycle} \geq 0, \qquad Flo_i \leq Fm_i \leq Fup_i, \qquad n_i \in Z^+ \tag{7}$$

The solution reported corresponds to that in Table 2 (F1):

Table 2: Results applying formulation F1, F2, F3 and F4.

Variable	F1 A	F1 B	F1 C	F2 A	F2 B	F2 C	F3 A	F3 B	F3 C	F4 A	F4 B	F4 C
n_i	4	1	2	-	-	-	-	-	-	-	-	-
t_i (d)	42.44	41.74	37.94	-	-	-	-	-	-	-	-	-
ts_i (d)	10.61	41.74	18.97	9.76	115	21.11	5.95	9.19	11.25	9.76	9.19	11.25
Pm_i ($/d)	53065	11168	24057	53109	9827	23694	51476	13905	24950	53109	13905	24950
Tf_i	0.363	0.322	0.316	0.380	0.308	0.311	0.668	0	0.332	0.600	0	0.390
Fm_i (t/d)	397	300	300	411	300	300	650	0	288	650.0	0	345.3
T_{cycle} (d)		139.1			-			-			-	
Z ($/d)		30430			30640			42670			41913	

This formulation has some disadvantages. The solution is strongly dependent of arbitrary upper bounds used for T_{cycle} and n_i. The reported solution corresponds to a maximum value for n_i of 4 and 140 for T_{cycle}. For higher values of T_{cycle} and n_i bounds, the dependence still remains (i. e. for $n_i \leq 6$ and $T_{cycle} \leq 200$, $Z = 30507$). Such fact suggests the possibility of another formulation, which does not consider the T_{cycle} concept. In the following, this formulation is introduced.

2.2 NLP formulation (F2 and F3)

Assume that the furnace is operated with the feed i during a time ts_i. The resulting mean conversion (Xm_i) during the sub-cycle time ts_i, can be easily computed, according to:

$$Xm_i(ts_i) = ts_i^{-1} \cdot \int_0^{ts_i} c_i + a_i \cdot e^{(-b_i \cdot ts_i)} \; dt = ts_i^{-1} \cdot \left[c_i \cdot ts_i + \frac{a_i}{b_i} \cdot (1 - e^{-b_i \cdot ts_i}) \right] \tag{8}$$

Therefore, for that feed, the contribution to profit Pm_i will be (Eq. 2):

$$Pm_i = (ts_i + \tau m_i)^{-1} \cdot (P_i \cdot D_i \cdot Xm_i \cdot ts_i - Cm_i) \tag{9}$$

Additionally, the average consumption of feed i (Fm_i) can be expressed as a function of the fraction of time (Tf_i) dedicated to producing with feed i in the furnace:

$$Fm_i = (ts_i + \tau m_i)^{-1} \cdot D_i \cdot ts_i \cdot Tf_i \qquad (10)$$

Based on those variables, an alternative mathematical programming model can be stated as follows:

$$\max z = \sum_i Pm_i \cdot Tf_i \qquad (11)$$

ST :

$$b_i \cdot Xm_i \cdot ts_i = c_i \cdot ts_i \cdot b_i + a_i - a_i \cdot e^{-b_i \cdot ts_i} \qquad \forall i \qquad (12)$$

$$Pm_i \cdot ts_i + Pm_i \cdot \tau_m = P_i \cdot D_i \cdot Xm_i \cdot ts_i - Cm_i \qquad \forall i \qquad (13)$$

$$Fm_i \cdot ts_i + Fm_i \cdot \tau_m = D_i \cdot ts_i \cdot Tf_i \qquad \forall i \qquad (14)$$

$$\sum_i Tf_i = 1 \qquad (15)$$

$$Flo_i \leq Fm_i \leq Fup_i$$

Under these circumstances, the decision variables are ts_i and Fm_i (the later in correspondence with Tf_i). The optimum is also found and the corresponding results are given in Table 2 (F2). Thus, this new formulation gives a better MOF value for the same problem. The main reason is that there is no constraint related to the cyclic operation (T_{cycle} variable and associated constraints). It should be noted that in both cases (F1 and F2), the contribution to profit (Pm_i's) obtained processing raw materials A and C is substantially higher than B. Additionally, the lower bound constraints over feed B and C are active. For the specific problem where the lower bounds on Fm_i are zero (which is likely the more common case), the results are shown also in Table 2 (F3). It can be seen that the improvement could reach about 35%, and that the upper bound in the most profitable feed is now the active constraint.

2.3 LP formulation (F4)
Using the concepts introduced previously, a useful solution approach is proposed as a practical alternative to the last case. According to the information for every feed, the individual optimal values for ts_i can be evaluated using the equation 2, and then the correspondent Pm_i's using the equation 1. This allows dealing a priori with the non-linear terms of the problem. After that, the second formulation F2 is simplified to an LP formulation F4:

$$\max z = \sum_i Pm_i \cdot Tf_i \qquad (16)$$

ST :

$$Tf_i \leq \frac{Fup_i \cdot (ts_i + \tau_m)}{D_i \cdot ts_i} \qquad \forall i \qquad (17)$$

$$\sum_i Tf_i = 1 \qquad (18)$$

Such an approximation reduces the number of decision variables to a half (only the Tf_i, or what is equivalent, Fm_i). Additionally, as a consequence of the constraint imposed over the Tf_i set (Eq. 18), is possible to eliminate another decision variable. The numerical solution for this example is shown in Table 2 (F4). The resulting z is only 1.8 % lower than the optimal (F3).

2. On-Line Approaches

Suppose that the on-line information needed to compute IOF at a given time interval k is available. Suppose again that $IOF(k)$ monotonally decreases. Under such circumstances, there will be an optimal period of time k_{opt} for performing the corresponding maintenance task (Fig. 1). According to that, a simple solution is given by answering at every period k: "Should we stop for maintenance at this period or the following?". The answer can be obtained just by comparing of $MOF(k)$ and $MOF(k+1)$, at every interval k. Naturally, if $MOF(k) < MOF(k+1)$ it is better to wait. Otherwise, when $MOF(k) \geq MOF(k+1)$ then we should stop at period k. It can be observed than this is equivalent to find when: $MOF(k) - MOF(k+1) = 0$, which is the discrete form of equation 2, and hence provides the optimal solution. This affirmation holds, even when C_m and τ_m are functions of t rather than constant values. This kind of approach, termed Real Time Evolution (RTE, Sequeira et. Al, 2001b), relies on an evolutionary basis rather than formal optimization and has been successfully applied for optimizing continuous process in real time. The implementation of the mathematical programming result (F4) using a dynamic simulation environment (ASPEN CUSTOMER MODELLER) is shown in Figure 3.

However, there are two main reasons for disagreement between the mathematical programming results and that obtained after its implementation over the plant. In the first place, a model mismatch. This will be the case for "bad" values of the model parameters a_i, b_i and c_i for the considered example (or even structural mismatch). Secondly, as these parameters are determined by statistical techniques, usually least squares fitting, they are just averaged values. Therefore, the instant plant behavior will vary according to the degree of deviation observed during the adjustment stage. As a consequence, the results will vary according to the implementation methodology, because the previous factors will have more or less influence on the global behavior. A Montecarlo simulation was performed in order to reflect the model mismatch and the instant uncertainty over, a_i, b_i and c_i parameters. The formulation F4 was then applied in two different ways: First, using the policy given by the fixed ts_i values resulting from formulation F4. Second, following the RTE procedure to on-line determinate these values. Figure 4 shows the difference between the objective function values (Z_{RTE} and Z_{F4}) obtained for different values of e (model mismatch) and σ (uncertainty), where can be seen the expected benefits obtained when implementing an RTE system.

3. Conclusions

This works presents both, a way for off-line calculation and a methodology for the on-line implementation of production scheduling problems relative to maintenance tasks. The advantages of the proposed approach are its simplicity and robustness, which make

792

it interesting for application to industrial cases where a DCS has been already installed. Certainly, this approach may not guaranty global optimum but, depending on the case, this will be largely compensated because its robustness achieved over the two key aspects always present: model uncertainty and model mismatch. In any case, future work includes investigating methodologies needed for reaching the global optimum on-line. However, the difference between the optimum and the proposed approach is likely to diminish with the problem complexity (in terms of number of feeds and furnaces) as well as when the parameter values of different feed-reactor pairs are similar.

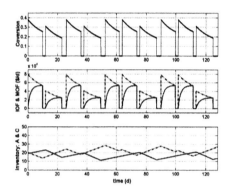

Figure 3: Simulation results when applying F4 over the plant

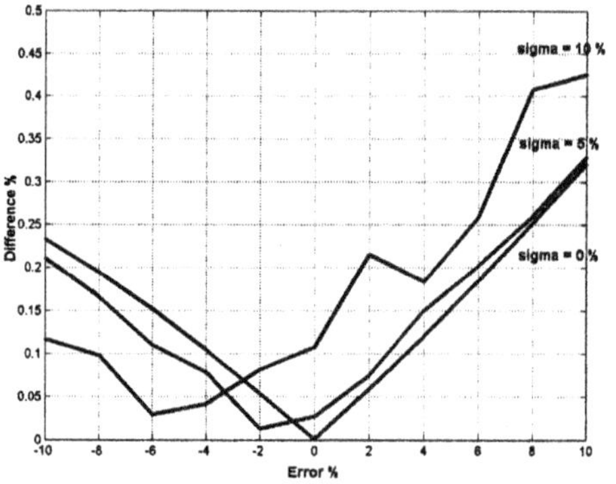

Figure4: Benefits obtained using RTE in the presence of model mismatch and uncertainty

4. Acknowledgements

One of the authors (S. E. S.) wishes to acknowledge to the Spanish "Ministerio de Ciencia y Tecnología" for the financial support (grant FPI).
Financial support from CICYT (MCyT, Spain) is gratefully acknowledged (project REALISSTICO, QUI-99-1091).

5. References

Buzzi, G., M. Morbidelli, P. Forzatti and S. Carra, 1984, *Int. Chem. Eng.* **24**, 441.

Georgiadis, M. C., L. Papageorgiou, S. Machietto, 1999, *Comp. Chem. Engng. Suppl.*, S203.

Jain, V. and I. Grossmann, 1998, *AIChE J.* **44**, 1623.

O'Donnell, B. R., B. A. Barna and C. D. Gosling, 2001, *Chem. Eng. Prog.* **97**, 56.

Sequeira S. E., M. Graells and L. Puigjaner, 2001a, Computer-Aided Chemical Engineering, Vol. 9: *ESCAPE-11*, Eds. R. Gani and S. B. Jørgensen, Elsevier, Amsterdam.

Sequeira, S. E., M. Graells and L. Puigjaner, 2001b, *Ind. Eng. Chem. Res.* (submitted).

European Symposium on Computer Aided Process Engineering – 12
J. Grievink and J. van Schijndel (Editors)
© 2002 Elsevier Science B.V. All rights reserved.

Operation Decision Support System using Plant Design Information

Yukiyasu Shimada[1], Hossam A.Gabbar[2] and Kazuhiko Suzuki[1]

[1]Department of Systems Engineering, Okayama University
[2]Frontier Collaborative Research Center, Tokyo Institute of Technology

Abstract

Operation support system (OSS) has been investigated to support the operator decision making in abnormal plant condition. The usefulness of OSS depends on the efficiency of the utilized operation decision method. This paper proposes an enhanced operation decision support system (ODSS), as an important function in OSS, using plant design information. At the plant design stage, the potential factors of dangerous situations are clarified and the causal relationships between causes and effects are investigated as a result of the risk assessment practices. Safety design and operating procedures of abnormal situations are examined based on such knowledge, which can also be useful to decide the appropriate operation at the plant operation stage. This means that plant design rationales can be reflected into the operation decision making. Accordingly, the system can show the reason of selecting the next operation. Prototype system of ODSS is developed based on the proposed operation decision method and applied to HDS process as a case study.

1. Introduction

Plant operation is automated by introducing computer-controlled system; this including chemical plants. Such automation needs minimum intervention from the operator, especially in normal operation. While the reliability of the devices and equipments, which composes chemical plant, are improving and many safeguards system are developed, most of the accidents occuring in actual plants are due to operator mistakes. Therefore, the development of operation support system (OSS) to support the operator decision making in abnormal plant conditions is essential. In case of process malfunction, expert operator infers the cause of malfunction and its effect on plant condition and decides the approproate operation based on these plant condition (SCEJ, 1999). The usefulness of OSS depends on how the operation decision making methods are efficient.

From the other side, most of the safety problems, related to plant operation, are discussed at the plant design stage, and are embodied through the risk assessment, instrumentation design, and safety facility design. However, it is pointed out that operators at the plant operation sites perform their daily work without having such safety management information, which includes design intent, objective and rationale. In order to overcome such limitations, this paper proposes the enhanced operation decision making method using the plant design information, which is implemented in a prototype system for operation decision support system (ODSS).

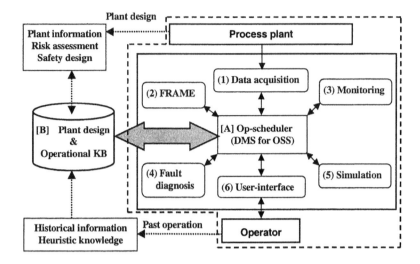

Fig.1 System architecture of operation support system.

2. Outline of Operation Support System

In case of process malfunction, expert operator infers the cause of malfunction and its effects on plant condition and decides the appropriate operation based on this plant condition. Vedam et al. (1999) proposed the framework of OSS highlighted by a dotted line in Fig.1. The roles of each module are summarized as follows;

(1) Data Acquisition: This module acquires on-line data from the plant and provides it to other modules.

(2) FRAME (Fault paRAmeter Magnitude Estimation): The magnitude and the rate of change of the root causes are estimated.

(3) Monitoring: The process data for the presence of abnormalities are monitored.

(4) Fault diagnosis: The root causes of detected process malfunction are identified.

(5) Simulation: The consequences of a detected abnormal situation are estimated.

(6) User-interface: The status of the process and the output results from each module are communicated to operator through this module (Shimada et al., 2000). And the plant condition and corresponding operation are indicated to operator.

[A] Op-scheduler (Data Management System, DMS includes ODSS in this paper): This module manages the exchange of information among modules. The appropriate operation is decided based on the output results of each module.

[B] Knowledgebase (KB): KB stores the plant design information, heuristic operation information and is searched to carry out each module.

The above system architecture will efficiently suppor operator's judgement. Vedam et al. focused to assist the operator in quantitative diagnosis and assessment of current and future consequences of abnormal situation. The main target of this paper is the realization of ODSS, which is an important function within OSS. The proposed ODSS decides the next operation by referring to plant design information, including safety management information, based on the output results from each module. This idea has been realized in prototype system of ODSS.

3. Operation Decision Making

At the plant emergency situation, operator has to judge whether the operation to be continued or stopped considering safety factors. In this proposed solution, operation decision making is performed by searching the information about the abnormal operation based on the plant condition.

3.1 Operation Category

Plant abnormal operation can be classified as three major categories: (a) recovery operation, (b) partial shutdown operation (PSD), and (c) total shutdown operation (TSD). Recovery operation should be considered first that returns the plant back to normal condition by taking a measure against root cause of malfunction or by using the any suitable prevention means. PSD stops part of the plant temporary to protect the fault propagation within certain area under consideration of effective restart. TSD stops the whole plant safely to avoid the crucial problem. The proposed ODSS decides the appropriate operation in these operation categories and instructs operators with detailed operation procedures.

3.2 Knowledgebase (KB) of plant design information

During the plant design stage, safety measures are considered and studied comprehensively to ensure the plant safety. These safety measures include information about plant abnormal operation and design rationale, which can be used to support operation decision making. In this proposed method, the results of risk assessment and the safety design information as well as the plant information are stored in the KB.

(1) Plant information (plant structure, process behavior and operation):
Plant information on plant structure, process behavior and operation is stored in the KB. Such information is needed to analyze how the process malfunction propagates within the plant. Abnormal operation procedures can be derived from normal operation procedure and used as the information on the operation support procedure.

(2) Results of risk assessment (process malfunction, causes and effects):
Information about the process malfunction's cause identification and its effect on plant condition can be acquired using risk assessment techniques such as HAZOP, FTA, etc.. Such information can be used to check the relationships between the malfunction and causes/effects. Also, the results of risk assessment, including information about the severity of process malfunction and fault propagation speed, can be used to analyze the severity of events/situation during the plant operation.

(3) Safety design information (prevention and protection means):
At the safety design, well-balanced safety facility design can be carried out based on information of fault propagation models and likelihood and severity of hazardous impact event. Recently this safety design is carried out based on IPL (Independent Protection Layer) concept (AIChE/CCPS, 1993). In this research, we have studied the design of the prevention means (Ex. stand-by pump) for high reliable plant operation, and protection means (Ex. Depressuring system) to protect against the fault propagation. The information collected and developed during the design of both the prevention and protection means can be used effectively to support operator deciding making.

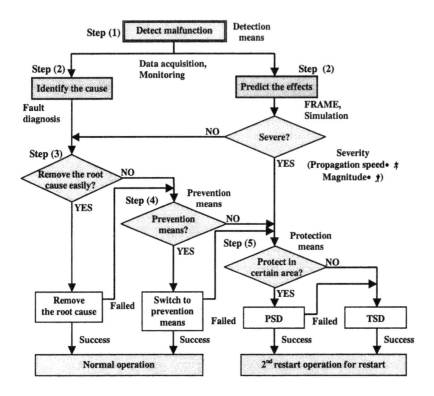

Fig.2 Basic procedure of operation decision making.

3.3 Operation Decision Making Procedure

Previously, it has been pointed out that OSS makes no sense if it cannot identify the cause of the process malfunction. In this research work, the algorithm of operation decision making to protect against fault propagation is proposed, even if the cause cannot be identified as, shown in Fig.2 (Shimada et al., 2001).

Steps (1) & (2): Detect malfunction and infer the cause and its effects

When process malfunction is detected, the cause and effects are identified by SDG-model-based and knowledge-based reasoning, or by using simulation techniques (Vedam et al., 1999). These methods are out of scope in this paper.

Steps (3) & (4): For recovery operation

If the effects are not severe and the root cause of the malfunction can be removed easily, the ODSS decides the removal operation to recover the plant back to normal condition. The information (KB) on recovery operation against the causes of the malfunction is searched. If the root cause of malfunction cannot be identified, or the measure for the root cause fails, the switch to prevention means is considered. If the prevention means such as stand-by pump can be available, then the ODSS can decide to switch to them as a recovery operation.

Step (5): For shutdown operation

If the recovery operation in steps (3) and (4) fails, or its effect on plant condition is severe, the shutdown operation is considered to protect against the fault propagation. First the ODSS tries to select the PSD to protect against the fault propagation.

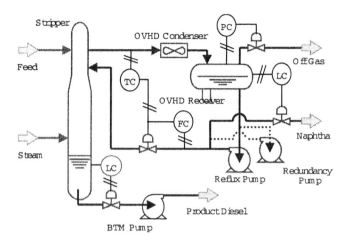

Fig.3 Stripping area of Hydrodesulfurization Plant.

If the speed of fault propagation is high and there is a possibility that the process malfunction may expand in the whole plant and lead to the accident, the ODSS selects the TSD such as addition of shortstop.

In this proposed method, OSS can suggest the operation according to the intents of plant safety design, because plant information and safety design information considerations at the plant design stage are included in the KB. The output results from ODSS is instructed to operator through the use-interface and operator can decide the appropriate operation finaly. The different features from the conventional methods are to systematize the design information, including safety management information, and to use them for operation decision making positively.

3.4 Prototype for Operation Decision Support System

A prototype ODSS is utilized as an experimental testbed for the proposed method. The environment is being implemented in Visual Basic (VB). The stripper area of HDS (Hydrodesulfurization) plant process is used as a case study, as shown in Fig.3. HDS plant is well-known process for removing sulfur from refinery distillates through a reaction with hydrogen. The stripper area removes H_2S, which is produced at previous reaction area and produces the high purity diesel as the product. Overhead vapor from stripper is partially condensed in stripper overhead receiver and a part of it is taken out as by-product, Naphtha. In this process area, a redundancy pump is designed as a prevention mean for reflux pump.

It is assumed that the temperature malfunction of stripper (Stripper-T(+)) in Fig.3 occurs. Fig.4 shows the candidate operations, which were displayed as output, within the developed ODSS. The left window shows the corresponding operation of: cause, temperature control of stripper as a recovery operation, and an emergency shutdown operation as a PSD. The right window shows the corresponding operation of: effect, the protection of cavitations as recovery operation, PSDs against stripping error and the prediction of pump cavitations. When operator selects one operation, the detailed operating procedures and the objectives (types) of such operation are displayed.

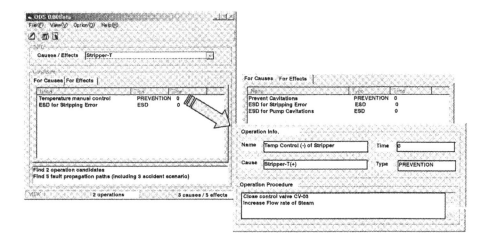

Fig.4 Output results from operation decision support system.

4. Conclusion

This paper proposes an enhanced operation decision making method using plant design information, which is considered as a positive step towards the development of an effective operation support system (OSS). The presented approach can be used in plant abnormal operation to decide the appropriate next operation. The ideas in this paper have been successfully tested in an example of HDS continuous plant. The proposed operation decision support system (ODSS) can suggest the appropriate operation and explain the reason of selecting such operation. This will enable plant operator to understand the safety management information in clearer way, which will have a positive impact on reducing the human errors during plant operation. Also the offered solution can help process designers to debate the different design rationales during the plant design stage.

Acknowledgement

This work was funded by Japan Society for the Promotion of Science, Japan.

References

AIChE/CCPS, 1993, Guidelines for Engineering Design for Process Safety

SCEJ, Society of Chemical Engineer Japan, 1999, Chemical Engineering Report, 38 (in Japanese).

Shimada,Y., H.A.Gabbar and K.Suzuki, 2000, Study on Designing the Operation Support System, Proc. of PSAM5 (Osaka), 4, 2637.

Shimada,Y., H.A.Gabbar, K.Suzuki and Y.Naka, 2001, Advanced Decision Method for Plant Operation Support System, Proc. of Loss Prevention 2001 (Stockholm), 619.

Vedam,H., S.Dash and V.Venkatasubramanian, 1999, An Intelligent Operator Decision Support System for Abnormal Situation Management, Computers & Chemical Engineering, Suppl., S577.

European Symposium on Computer Aided Process Engineering – 12
J. Grievink and J. van Schijndel (Editors)
© 2002 Elsevier Science B.V. All rights reserved.

Supply Chain Optimization Involving Long-Term Decision-Making

Jehoon Song[1], Jin-Kwang Bok[2], Hyungjin Park[1], and Sunwon Park[1]

[1]Department of Chemical Engineering, Korea Advanced Institute of Science and Technology, 373-1 Guseong-dong, Yuseong-gu, Daejeon 305-701, Korea

[2]Samsung SDS, 707-19 Yeoksam-dong, Gangnam-gu, Seoul 135-918, Korea

Abstract

This paper deals with the optimal design problem of multiproduct, multi-echelon supply chains for Supply Chain Management. The supply chains are composed of plants, warehouses, distribution centers, and customers. The locations of the plants and the customers are fixed. The locations and capacities of warehouses and distribution centers should be determined. The transportation and production of products should be determined, too. A mixed integer linear programming is used for the mathematical modeling. The objective is to minimize total cost of supply chains. We propose more detail modeling techniques in the cost calculations for the transportation of products and the installation of warehouses and distribution centers on the basis of the recent research on the supply network design by Tsiakis et al. (2001).

1. Introduction

Recently, many companies have tried to seek various methods to implement SCM. However, the basic formation can be explained with the Supply Chain Operations Reference-model (SCOR) proposed by Supply Chain Council (SCC). SCOR divides the activities of supply chains into 4 categories shown in Figure 1. It illustrates that the plan category has influence on the supply chains more comprehensively than remaining three categories. Among many activities in the plan category, supply chain optimization that means determination of optimal location of facilities like plants, warehouses (WHs), and distribution centers (DCs), supply route, and transportation method is one of key issues of SCM (Simchi-Levi et al., 2000). Typical supply chains are shown in Figure 2. Actually, IBM has experienced cost reduction effect by 20-30% through supply chain optimization (Quinn, 1997).

In this paper, we solve a supply chain optimization problem focusing on the points as follows on the basis of the work of Tsiakis et al. (2001) that reflects recent research trends.

1) The distance between echelons should be considered as well as the transportation volume of products when the transportation costs are calculated. The transportation cost should be expressed as a discontinuous piecewise linear function of the transportation volume, and a continuous piecewise linear function of the distance between echelons.

2) The installation costs of WHs and DCs should be linear functions of their capacities. The calculated initial installation costs are paid by installment considering the compound interest rate.

3) The single source assumption is applied only to the links between DCs and customers.

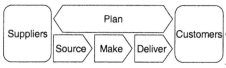

Figure 1. Supply Chain Operations Reference process.

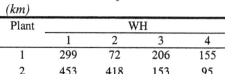

Plants
Distribution
Warehouses Centers Customers

Figure 2. Supply chains.

Table 1. Distance from Plant to WH (km)

Plant	WH			
	1	2	3	4
1	299	72	206	155
2	453	418	153	95
3	390	240	313	359

Table 2. Distance from WH to DC (km)

WH	DC			
	1	2	3	4
1	96	201	153	281
2	303	280	376	193
3	145	437	97	471
4	248	99	244	30

Table 3. Distance from DC to customer (km)

	C1	C2	C3	C4	C5	C6	C7	C8	C9	C10	C11
DC1	429	156	323	269	58	208	39	188	134	309	110
DC2	282	351	216	139	160	406	319	181	28	294	158
DC3	365	197	168	139	65	338	151	333	270	541	297
DC4	109	216	432	22	357	132	151	210	366	284	126

2. Problem Description

We deal with the supply chains including three products, three plants, four candidate WHs, four candidate DCs, and eleven customers. The location of plants and customers are fixed. To satisfy customers' demands for each product at the least expense in supply chain, the location of WHs and DCs, production in plants, and the capacities of WHs and DCs should be determined. The transportation links and transportation volume among plants, WHs, DCs, and Customers should be also determined.

3. Mathematical Model

3.1 Objective Function

The objective of optimization is to minimize total cost in supply chains. Total cost = {installation cost of WHs} + {installation cost of DCs} + {production cost in plants} + {operation cost in WHs} + {operation cost in DCs} + {transportation cost from plants to WHs} + {transportation cost from WHs to DCs} + {transportation cost from DCs to Customers}

3.2 Constraints

The constraints used in this problem are on the basis of Tsiakis et al. (2001). To avoid simple duplication, the categories of constraints are just referred. But in detail, we explain the cost calculation techniques considered in this paper.

Network composition
Transportation
Mass Balance
Capacity
Transportation cost

In general, the transportation cost of products between two echelons can be expressed as a function of transportation volume and the distance between them.

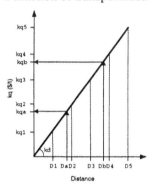

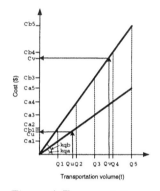

Figure 3. Transportation cost as a linear function of the distance.

Figure 4. Transportation cost as a linear function of the transportation volume.

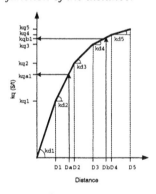

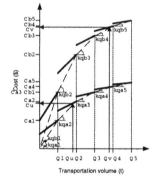

Figure 5. Transportation cost as a continuous piecewise linear function of the distance.

Figure 6. Transportation cost as a discontinuous piecewise linear function of the transportation volume.

$$CTR = f(Q, D) \qquad (1)$$

where *CTR*: Transportatin cost, *Q*: Transportation volume, *D*: Transportation Distance.
If we calculate roughly, the transportation cost is proportional to the transportation volume with constant transportation distances, and to the transportation distance with constant transportation volume.

$$CTR = kDQ \qquad (2)$$

where k: proportional constant.
Then, since Q is a decision variable, and D is a parameter, eqn. 2 can be expressed as a divided form.

$$CTR = kqQ \tag{3}$$

$$kq = kdD \tag{4}$$

where kq: transportation cost per unit transportation volume, kd: transportation cost per unit transportation volume and distance.

That is to say, if D is given as a parameter, constant kq can be obtained since kd is also constant. Therefore, if only decision variable Q is determined, the transportation cost is obtained. Figures 3-4 illustrate this relation graphically.

In Figure 3, if the transportation distance is Da, kqa is fixed by eqn. 4. This kqa means the slope in Figure 4. If transportation volume Qa is determined, transportation cost Cu is obtained by eqn. 3. In the same way, if distance is Db that is different from Da, different slope kqb is obtained. If transportation volume is Qv at this time, the transportation cost is Cv.

Realistically, however, the transportation cost is not linearly proportional to the transportation volume and the distance. Based on the economy of scale, the larger transportation volume, the lower transportation cost per unit volume at constant distance, and the farther transportation distance, the lower transportation cost per unit distance at constant transportation volume. Furthermore, the transportation cost is expressed as a continuous piecewise linear function of transportation distance, and as a discontinuous piecewise linear function of transportation volume. Figures 5-6 illustrate this relation graphically.

In Figure 5, the number of possible ranges of transportation distance is S, and the distance of the point where the slope changes is Ds. The slope in the interval $[D_{s-1}, D_s]$ is kds. Then, based on the economy of scale, the slopes have the features as follows.

$$kd_{s-1} \geq kd_s, \qquad \forall s \tag{5}$$

Generally, kd_s forms a sequence that has a certain rule such as geometric sequence.

If the sequence $\{kd_s\}$ is a geometric sequence, the general term of the sequence is like this.

$$kd_s = kd_1 CR^{s-1} \tag{6}$$

where CR: commom ratio.

If the distance Da ($D_{s-1} \leq Da \leq D_s$) is given, the transportation cost per unit transportation volume is like this.

$$kqa_1 = \frac{kq_s - kq_{s-1}}{D_s - D_{s-1}}(Da - D_{s-1}) + kq_{s-1} \tag{7}$$

kqa_1 obtained in this way is a basic slope of the transportation cost function of transportation volume in Figure 6.

In Figure 6, the number of possible range of transportation volume is R, and the volume of the point where the slope changes is Q_r. The continuous slope in the interval $[Q_{r-1}, Q_r]$ expressed as a dashed-line is kq_r. Then based on the economy of scale, the slopes have the features as follows.

$$kq_{r-1} \geq kq_r, \qquad \forall r \tag{8}$$

However, the real slopes of the cost for the interval $[Q_{r-1}, Q_r]$ are discontinuous. They are smaller than continuous slope by some ratio. So they make discontinuity. Let the ratio h. Generally, kq_r forms a sequence that has a certain rule such as geometric sequence. If the sequence $\{kq_r\}$ is a geometric sequence, the general term of the sequence is like this. The first term kqa1 is from eqn. 7.

$$kqa_r = kqa_1 CR^{r-1} \tag{9}$$

If the distance Qu ($Q_{r-1} \le Qu \le Q_r$) is given, the transportation cost per transportation volume is like this.

$$Cu = kqa_r h(Qu - Q_r) + Ca_r \tag{10}$$

However, since Qu is not a parameter but a decision variable, it is possible to exist in any interval of R intervals. Accordingly, a new set of binary variables Z_r that denote if Qu is in a certain interval $[Q_{r-1}, Q_r]$ is introduced.

Therefore, the transportation cost is determined by the equations as follows.

$$Q_{r-1} Z_r \le Qu_r \le Q_r Z_r, \qquad \forall r \tag{11}$$

$$\sum_{r=1}^{R} Z_r = 1 \tag{12}$$

$$Qu = \sum_{r=1}^{R} Qu_r \tag{13}$$

$$Cu_r = kqa_r h(Qu_r - Q_r Z_r) + Ca_r Z_r \tag{14}$$

$$kqa_r = kqa_1 CR^{r-1} \tag{15}$$

$$CTR = \sum_{r=1}^{R} Cu_r \tag{16}$$

Table 4. Variation range of transportation cost rate based on transportation volume and distance.

	Transportation Volume (ton/week)				Distance (km)
	Plant-WH	WH-DC	DC-C	Unit cost ($/ton)	
Range1	0-20	0-20	10	400	100
Range2	20-40	20-40	20	600	200
Range3	40-60	40-60	30	700	300
Range4	60-80	60-80	40	750	400
Range5	80-100	80-100	50	775	500

Table 5. Production cost at each plant and handling cost for each product at warehouses and distribution centers ($/ton).

Product	Production cost			Operation cost							
	P1	P2	P3	WH1	WH2	WH3	WH4	DC1	DC2	DC3	DC4
PD1	62.27	59.45	61.44	4.25	4.55	4.98	4.93	4.25	4.55	4.98	4.93
PD2	33.33	35.44	37.55	5.28	4.06	4.25	4.55	5.28	4.06	4.25	4.55
PD3	92.56	90.01	88.79	4.98	4.93	4.06	5.28	4.98	4.93	4.06	5.28

Installation cost

In general, the installation costs of WHs and DCs can be expressed as a function of the capacity.

$$CI = f(G) \tag{17}$$

where CI: installation cost G: capacity.

The installation cost is proportional to the capacity.

$$CI = kgG \tag{18}$$

where kg: installation cost per unit capacity.

However, this cost is initial investment cost. For calculation with operating costs, it is necessary to convert into the payment per unit time. Simply, the cost divided into usage time is used, but we use installment-paying method in this problem. If the initial installation cost of WHs or DCs is A, the period is n (520 weeks), and the rate of interest is r (weekly 0.001), then the installment paid every week is expressed as eqn 19.

$$a = \frac{r(1+r)^n}{(1+r)^n - 1} A \tag{19}$$

Therefore, the time-based installation cost of WHs and DCs is expressed as follows.

$$CJ = \frac{r(1+r)^n}{(1+r)^n - 1} kgG \tag{20}$$

where CJ: installation cost per unit time.

3.3 Results

The optimization model involves 1160 binary variables, 2622 continuous variables, and 4320 equations. This MILP model was solved with CPLEX in GAMS 2.50. 1416 seconds of CPU time was taken to solve the model. The optimal supply chain configuration is shown in Figure 7. Selected data are shown in Table 6.

4. Conclusions

Supply chain otimization problem is an essential part in SCM. We addressed supply chain optimization problem including multiproduct, multi-echelon facilities. A MILP model based on the work by Tsiakis et al. (2000) was used, but we focused on the realistic cost calculation. We considered the distance and transportation volume with continuous/discontinuous piecewise linear functions. The compound interest rate was considered for the calculation of installation costs. Considerations of Multi-period and uncertainty are needed for further study.

Acknowledgement

This work was partially surpported by the Brain Korea 21 Projects.

References

Quinn, F.J., 1997, The Payoff!, The Supply Chain Series Part Six, Logistics Management & Distribution Report, Dec. 1.

Simchi-Levi, D., Kaminsky, P., Simchi-Levi, E., 2000, Designing and Managing the Supply chain, McGraw-Hill, Singapore.

Tsiakis, P., Shah, N., Pantelides, C.C., 2001, Design of Multi-echelon Supply Chain Networks under Demand Uncertainty, Ind. Eng. Chem. Res. 40, 3585.

Table 6. Transportation volume (ton/week)

		PD1	PD2	PD3			PD1	PD2	PD3
P1		40	40	49		DC1	22	43	55
P2	WH4	40	78	20	WH4	DC2	75	83	0
P3		40	40	40		DC3	45	44	20

		PD1	PD2	PD3			PD1	PD2	PD3
DC1	C5		34	11	DC2	C2		21	5
	C7	7	7	7		C3	28	9	
	C8	5		27		C6		50	
	C11	10	34			C9	15		16
						C10		3	33
DC4	C1	43							
	C4	12		20					

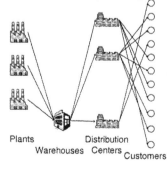

Plants Distribution
Warehouses Centers Customers

Figure 7. Optimal supply chains configuration.

European Symposium on Computer Aided Process Engineering – 12
J. Grievink and J. van Schijndel (Editors)
© 2002 Elsevier Science B.V. All rights reserved.

Modelling Liquefied Petroleum Gas Storage and Distribution

S.L. Stebel, F. Neves-Jr., L.V.R. Arruda, J.A. Fabro, L.C.A. Rodrigues
E-mails: {stebel, neves, arruda, joao, lcar}@cpgei.cefetpr.br
CEFET-PR, CPGEI
Av. Sete de Setembro, 3165, 80230-901 Curitiba, PR, BRASIL
Tel.: +55 41 310-4707   Fax.: +55 41 310-4683

Abstract

This paper presents two models for the Liquefied Petroleum Gas (LPG) transfer and storage operations in a refinery. First, a simulation model is proposed, based on Petri Nets, which integrates the continuous and discrete parts of the process. The second model uses mixed integer linear programming techniques for the optimisation and scheduling of the system. The models allow the visualization and simulation of the problem, helping the system operator to easily test and correct scheduling plans despite the complexity of the operations. Results from the simulation model, when applied on the optimisation model, can then be used to reduce the severity of the problem.

1. Introduction

The optimisation of operational processes is critical to the success of petrochemical organisations. The optimal allocation of productive features takes into account monetary, physical, and operational restrictions imposed by the structures and production processes. Optimal allocation is a priority for the organizations. In several activities of business, the classical problem of optimisation appears: how can critical features be used most efficiently? Oil refineries have used computational features directed at the activities of planning and scheduling since the 1950's. The techniques commonly used for these activities are based on linear programming (Williams, 1999). This methodology is still predominantly employed in the modelling of chemical plants, despite its modelling capacity limitations.

This paper focuses on the development of two approaches for LPG distribution and storage operations. The main goal is to provide an auxiliary tool for the decision making process. The Petri Nets model enables a diagnosis of critical points of scheduling. The objective of the second approach proposed is minimising operational costs. The operational and physical constraints, as well as product demands, are considered in both approaches. Since storage plays a major role in the scheduling of continuous processes, the proposed approaches are based on the different possible states of the spheres. These are: ready to receive, receiving, ready to deliver, delivering, and product analysis.

The developed models allow a better storage and distribution system understanding, providing a simulation tool. A graphic process simulation facilitates the scheduler's task, allowing the scheduler to easily experiment and correct the scheduling, despite the complexity of the operations in a real life scenario. Moreover, the modelling process allows the detection of bottlenecks and provides mathematical problem analysis.

2. Problem Definition

LPG is basically a mix of hydrocarbons with 3 (C3 – propane) and 4 (C4 – butane) carbon atoms, receiving its name from the fact that it can be liquefied by compression at room temperature. This product may be used as domestic fuel for cooking and heating. The fact that LPG can be liquefied at relatively low pressure facilitates the storage of large amounts using, in general, spherical tanks, known simply as spheres. In a typical refinery, the catalytic cracking process is the principle method of producing LPG with smaller amounts being produced by the crude distillation column, delayed coking, etc (Pinto and Moro, 2000).

The deficit of Liquefied Petroleum Gas (LPG) is a persistent problem in some Brazilian refineries. The Araucária refinery, located in southern Brazil, is an example. It only produces ¾ of its LPG demand. The final quarter is supplied by tankers, which unload in a harbour. The product is then conveyed to the inland refinery by a pipeline system. The refinery is responsible for the LPG distribution to small delivery bases as well as to a pipeline of finished products (Schechtman et al., 2000). In order to manage storage constraints some operational decisions must be taken by the scheduler. The decisions are currently taken based on personal experience, with the aid of manual calculations. Due to storage and distribution complexity, the operational decision making process becomes a difficult procedure (Stebel et al., 2001a). Some decisions to be taken by the scheduler are shown in figure 1. The overall scheme of the LPG storage and distribution area is shown in figure 2.

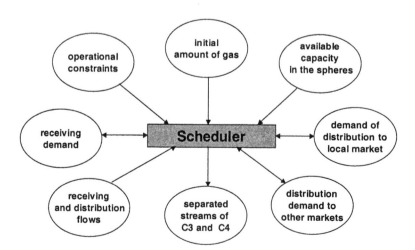

Figure 1: Scheduler made decisions

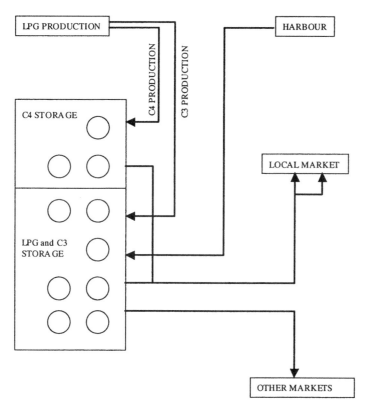

Figure 2: LPG storage and distribution

3. Simulation Model

The LPG storage and distribution process is continuous. The spheres are lined up with the pipeline and then loaded (or unloaded) continuously, in accordance with the flow of each product involved in the operations. However, the decision operations can be considered discrete events in time, for example, at the beginning of the receiving operation. Therefore, the LPG storage and distribution process modelling can be done using Petri Nets. To achieve this, t-temporised nets (Reisig, 1985) were used. Utilising this net, the time for the accomplishment of a transition is placed at the transition, determining a minimum time until a transition is qualified and can be triggered. The duration of each process task is codified by the time definition (receiving, period of product analysis, and delivery).

The Petri Nets allow dynamic simulations of problem instances. The model allows the visualization and simulation of a problem. It helps to test and correct scheduling plans despite the complexity of the operations, as well as making it possible to identify bottlenecks within the system.

The model is separated in two main parts, one for the receiving of gas and another for its delivery. The receiving model has two sources, that are (Stebel et al., 2001a):

- the constant gas flow (C3 and C4) from process units;
- the flow from external sources, represented by a harbour, which transfers LPG to refinery using a pipeline.

After the spheres receive any product it is necessary to wait some time until the product can be delivered to customers. This permits water separation and draining as well as product analysis (Pinto and Moro, 2000).

The delivery process is also carried through two pipelines that are in the LPG area. One of the pipes sends gas to the local market and the other to other markets (located in another state). Local companies can receive gas in three ways: C3, LPG, C3 + C4 (mixed in line). C3 and C4, when mixed in line, must be sent at the same time and the same quantity. The delivery to other markets must be of a minimum amount of gas. Due to the fact that the pipeline is used by a lot of products.

The Petri Nets model was divided into blocks, LPG receiving process, storage and sending. The processes of receiving and sending were modelled taking into account their continuous character (Drath et al., 1998). Therefore, the intermediate states of the processes can be visualized by these models. This was possible through discretisation of the gas flow and using timed transitions. Each token, in this part of the model, represents an amount of 50 m^3 of gas. This value represents the highest common divisor between the various flow rates of a pipeline system.

The simulation software used to implement the Petri Nets model was the Visual Object Net + + (Drath, 2001).

4. Optimisation Model

The optimisation model is based on mixed integer linear programming with a uniform discretisation of time (Stebel, 2001b). The representation of time and the model structure itself are the two main reasons for building the model (Pinto and Moro, 2000). When the problem has a large scheduling time horizon, it is not practical to use a uniform discretisation of time, which makes problem solution a difficult and time-consuming task. The problem in question has a scheduling horizon of between one and two days, allowing a simulation with discretisation units of one, two or four hours. The proposed approaches are based on the different possible states of the spheres causing an increase in the model size.

The considered model has, as its objective function, the minimisation of the costs involved in this area of the plant. These costs are: changeover (between two spheres), storage, and electricity. These costs were normalised to represent the qualitative aspects of the system. The model will be more efficient if the costs are derived from another model. The specialist information will be represented in a rule based system. Through this, the different costs will exert a distinguishing influence on the model. The constraints are derived from operational procedures, material balance, flow rate, and demand. The total number of constraints and variables depends on the input data of the problem.

Shown in (1) is an example of model constraint (Stebel, 2001b). Each sphere, at all times, should be either: receiving ($ER_{e,o,p,t}$), analysing a product ($EA_{e,t}$), delivering ($EE_{e,d,p,t}$), ready to receive ($LR_{e,t}$), or ready to deliver ($LE_{e,t}$). These variables are binaries and they represent the different possible states of the spheres:

$$\sum_{o \in O_1} \sum_{p \in P_1} ER_{e,o,p,t} + EA_{e,t} + \sum_{d \in D_1} \sum_{p \in P_1} EE_{e,d,p,t} + LR_{e,t} + LE_{e,t} = 1$$

$$\forall e \in EPROP, t \tag{1}$$

The LINGO (LINDO, 1999) solver was used to implement the MILP optimisation model. The simulations were done on a PC (Pentium III 850 MHz). In order to reduce the CPU time, the relative optimality criterion was set to a non-zero value and was tested to reduce the size of the search tree providing significant timesaving.

5. Model Integration And Results

It is possible to make an analysis of the feasibility problem with the simulation model. It is also possible to determine the minimum resources to be carried by all the operations. However this analyses does not determine an optimal point of operation. This is only possible through the optimisation model. The structure of the optimisation model enables the insertion of the amount of resources to be simulated. For each simulated resource the number of the binary variable can be increased. Therefore, using minimum resources, optimal scheduling can be identified without the necessity of generating binary variables, which are not used. Consequently allowing for a reduction in the size of the optimisation model.

An example is the number of spheres necessary to carry out all the operations. This number is used in the optimisation model to search for the best answer, however, only searching in a narrow search tree. Table 1 shows an example of this. From the simulation in Petri Nets model it was possible to perform all the operations with six spheres (4 propane and 2 butane), but this model cannot guarantee optimal solution. This is possible through the optimisation model. If all spheres are used (7 propane and 3 butane) the CPU time will be 1min 27s with simulation in contrast to 7min 12s without. This difference occurs because of the increase in binary variables.

Table 1: Computational Results

number of spheres	0–1 variab.	total variab.	nodes	objective function	constraints	iterations	CPU time
6	342	1768	22	106	4340	30476	1min27s
10	576	3656	26	106	9054	71259	7min12s

6. Conclusions

The LPG scheduling problem lies basically in the determination of the best policy for the utilisation of storage resources. Petri Nets checks the feasibility of the modelling of a continuous process through discrete events. This model integrates the continuous part of the process (load and unload) to the discrete part of operational control. It allows the visualization and simulation of some instances of the problem. This problem can be modelled as a MILP model with uniform discretisation of time. The user's interface is carried out through a spreadsheet, where the data input is kept and the Gantt chart is presented. With the change of some restrictions, these models can be applied to other refineries. The results obtained are better than those obtained currently by the operator with the use of manual calculations. Not only can the developed models be used to test receiving and sending plans to find the most suitable scheduling, but also to detect and correct possible problems arising. Furthermore, the modelling process allows for the detection of problem bottlenecks. These models can result in costs reduction or an increase in the plant profit. Integrating models can facilitate obtaining results more efficiently, but at the moment this is done manually.

Acknowledgments

The authors acknowledge the financial support from the Brazilian National Agency of Petroleum (PRH-ANP/MCT PRH10 CEFET PR).

References

Drath, R., U. Engmann and S. Schwuchow, 1998, Hybrid Aspects of Modelling Manufacturing Systems by Using Modified Petri Nets, International Federation of Automatic Control – Preprints of the 5th IFAC Workshop on Intelligent Manufacturing Systems, Gramado – RS, Brazil.

Drath, R. Short User's Guide for Visual Object Net + +. Obtained in September 2001 at http://www.systemtechnik.tu-ilmenau.de/~drath.

LINDO, 1999, LINDO: The Modelling Language and Optimizer – User's Guide, LINDO Systems Inc. Chicago, Illinois.

Pinto, J.M., Moro, L.F.L., 2000, A Mixed Integer Model for LPG Scheduling, In: European Symposium on Computer Aided Process Engineering-10, Comp. Aided Chemical Engineering (8) (S. Pierucci(Ed)), 1141-1146, Elsevier, Amsterdam.

Reisig, W., 1985, Petri Nets – An Introduction, Springer-Verlag.

Schechtman, R., Vieira, J.V.C., Moreira, J.G.S, Costa, L.S., Nascimento, D.L., 2000, LPG Demand Outlook: 1999-2004, Proceedings of the Rio Oil & Gas Expo and Conference (in Portuguese).

Stebel, S.L., Fabro, J.A., Neves Jr, F., Arruda, L.V.R., Tazza, M., 2001a, Simulation of the LPG Transfer and Storage Process by Petri Nets, Proceedings of the V Brazilian Symposium of Artificial Intelligence (in Portuguese).

Stebel, S.L., 2001b, Modelling of the Liquefied Petroleum Gas Storage and Distribution in an Oil Refinery, Master Thesis, CPGEI / CEFET-PR, 100 pages (in Portuguese).

Williams H.P., 1999, Model Building in Mathematical Programming, 4th ed. Chichester (England): John Wiley & Sons Ltd.

European Symposium on Computer Aided Process Engineering – 12
J. Grievink and J. van Schijndel (Editors)
© 2002 Elsevier Science B.V. All rights reserved.

A Successive Mixed Integer Linear Programming approach to the grade change optimization problem for continuous chemical processes

Rob Tousain and Okko Bosgra
Mechanical Engineering Systems and Control Group
Delft University of Technology, Mekelweg 2, 2628 CD Delft, The Netherlands
o.h.bosgra@wbmt.tudelft.nl

Abstract

To enable flexible, market-oriented operation of continuous chemical processes the control of transitions between different operating conditions (hereafter called: grades) should be enhanced. Dynamic optimization is believed to be a major enabling technology in this respect. In this paper, we formulate the grade change problem as an economic optimization problem using a finite horizon evaluation of the added value of the processing plant. The resulting objective function is discontinuous due to the transitions between grade-regions which makes standard, gradient-based optimization methods unsuited for solving the problem. A new, Successive Mixed Integer Linear Programming approach is proposed and its potential is demonstrated on an example: the optimization of transitions in a binary distillation column.

1. Introduction

The operation of many multi-product (or multi-grade) continuous chemical processes is subject to rapidly changing market conditions and strong competition. Ideally, the production should continuously track the most favorable market conditions to guarantee good margins. The moderate status of most current industrial process control systems is a main obstacle in doing so. The result is that many multi-product plants still are operated in a highly inflexible fashion, running through a predetermined and fixed sequence of product types, also called a *slate*, or *product wheel*, as described in e.g. (Sinclair, 1987). One way to improve the flexibility of the operation is by the use of model-based dynamic optimization technology. Trajectories resulting from such an optimization effort can be combined with Advanced Process Control to achieve automated and well predictable control of process. The actual implementation of the optimal trajectories will not be treated in this paper, instead we will focus on the definition and the solution of the grade change optimization problem.

The use of model-based optimization in switchability analysis for multi-grade processes was discussed by White et al., 1996, where the feasibility problem was reformulated as a minimization problem with arbitrary quadratic penalties and solved using a gradient-based algorithm. A similar approach was used by McAuley and MacGregor, 1992, for the optimization of grade transitions in a polyethylene plant.

In this paper, we consider the optimization of grade transitions with *economic* cost functions. The new approach that is described in this paper proceeds via a *mixed integer* description of the grade regions and the corresponding production rates. The feasibility of the proposed method is illustrated by means of an example, grade change optimization for a binary distillation column.

2. Grade change problem formulation

2.1 The plant
We assume the continuous chemical manufacturing plant to be described by the following system of differential algebraic equations (DAE):

$$\dot{x} = f(x,u,y), \qquad 0 = g(x,u,y), \qquad z = C_x x + C_u u \qquad (1)$$

where x are the states, y the algebraic variables, u the inputs and z the so-called performance variables, i.e. all variables that are required for the performance evaluation (objective and constraints) of the plant. The operating constraints are expressed as follows

$$h(z) < 0 \qquad (2)$$

Many continuous chemical plants can be operated in different production *grades*. A production grade can be identified by a set of specific characteristics of the end products or the mix of end products that is produced. Examples of product grades are grades of high-density poly-ethylene and certain purities of a distillation product. We define a production grade g by a corresponding set of inequality constraints

$$R_g(z) < 0 \qquad (3)$$

The operation of a multi-grade plant can be characterized in terms of finite-time quasi-stationary tasks and transitions between those.

2.2 Quasi-stationary tasks: static optimization
The operation of the plant during the quasi-stationary intervals is assumed to be realized according to a static optimization of the process economics, where the objective function to be minimized is given as follows

$$L(z) = -\sum_e p^{Y,e} Y_e(z) + \sum_r p^{C,r} C_r(z) \qquad (4)$$

$Y_e(z)$ is the yield of end product e. $C_r(z)$ is the consumption of raw material or utility r. The prices of the end products and the raw materials/utilities are given by respectively $p^{Y,e}$ and $p^{C,r}$.

2.3 Transition tasks: dynamic optimization
We consider two production grades g and h and a transition between those. Initial and target conditions are given by the stationary optimal operating conditions for the two production grades, respectively $(\bar{x}^g, \bar{u}^g)$ and $(\bar{x}^h, \bar{u}^h)$.

In literature, no consistency exists as to the general definition of the grade change optimization problem. From a theoretical viewpoint it might be argued that the optimal changeover strategy is given by the solution of a time-optimal control problem, see e.g. (Lewis and Syrmos, 1995). However, note that the classical time-optimal control problem forces all states to reach the desired end-point as soon as possible. This does not imply in general that the fastest transition between grade g and h is achieved because the grades are defined by their corresponding sub sets of the state space. Also, the time-optimal formulation does not honor the fact that valuable end products and often invaluable off-spec materials are produced *during* the transition. Finally, market situations may exist in which there is no incentive to implement the fastest transition possible. For example, during periods of low market demand it may be more advisable to minimize transition costs instead.

An alternative formulation, which can be seen as a generalization of the economic grade change optimization problem for different market situations is proposed here. The corresponding finite-time economic objective function is the following:

$$J = \int_0^T - \sum_e \alpha^e Y_e(z(t)) + \sum_r \beta^r C_r(z(t)) dt \qquad (5)$$

where T is a fixed end-time, which should of course be chosen larger than the minimum transition time. α^e and β^r are weighting parameters. Emphasis on transition time can be introduced by selecting all β^r's equal to zero and the α^e corresponding to the product that is being produced in the target grade equal to a large value. Emphasis on transition economics can be introduced by selecting $\alpha^e = p^{Y,e}$, and $\beta^r = p^{C,r}$.

In the most general setting, a number of different end products is produced in each grade. The yield of end product e then is given as follows:

$$Y_e(z) = \sum_g G^g(z) M_e^g P(z) \qquad (6)$$

where $P(z)$ is a vector function representing all material flows from the plant and M_e^g, a row vector with zeros everywhere except for a '1' at at most one location, assigns the material flows to the eth end product for grade g. G^g are so-called grade variables and are defined as follows

$$G^g(z) = \begin{cases} 1, & \text{if } R_g(z) < 0 \\ 0, & \text{otherwise} \end{cases} \qquad (7)$$

The grade change optimization problem is defined as to minimize the objective (5), subject to the model equations (1), the path constraints (2), the initial and end-conditions corresponding to the departure grade and the target grade and with the product flows given by (6) and (7). The choice of the economic objective function makes the optimization problem discontinuous. Next, we will describe how this optimization problem can be solved.

3. Solution approach

3.1 Exploration of possible approaches

In this work we focus on the sequential approach to dynamic optimization (see e.g. (Vassiliadis, 1993)). The standard sequential approach uses an outer loop gradient-based Nonlinear Programming tool (e.g. Sequential Quadratic Programming or Generalized Reduced Gradient) to solve the finite-dimension optimization problem that results after control parametrization. This approach assumes the objective as a function of the parameters to be twice continuously differentiable which does not hold for the economic grade change problem. To circumvent this, we proposed in an earlier publication to approximate (7) by smooth functions (e.g. 'arctan'-functions) and we presented a tailor-made sequential optimization routine for solving the resulting smooth but strongly nonlinear problem (van der Schot et al., 1999). This approach works well on many examples, however it may get stuck in poor local minima since it uses gradient-based algorithms on a problem that is strongly nonlinear by nature. The approach we present here treats the discontinuities through the introduction of a set of binary decision variables in the optimization problem.

3.2 A Successive Mixed Integer Linear Programming approach

The crucial step in this approach is the introduction of *grade variables* $G_i^g \in \{0,1\}$ at time instances iT_s, $i = 0...T/T_s$. The following constraints enforce that the grade variable for which the corresponding grade constraints (3) are satisfied, is set to one:

$$R_g(z_i) - (1 - G_i^g)Q^g < 0, \qquad \sum_g G_i^g = 1 \qquad (8)$$

where $z_i = z(iT_s)$. This is a standard technique for coupling real and binary variables, see e.g. (Bemporad and Morari, 1999). Note that G_i^g enters these equations linearly. For feasibility of (8) we require the fixed parameter Q^g to be larger than the maximum that is attained by $R_g(z_i)$ on the feasible set in which z_i lives and that all grade regions are adjacent. Next, we introduce new continuous decision variables $Y_{e,i}^g$ which represent the flow of end product e during operation of the plant in grade g. We require

$$Y_{e,i}^g \le G_i^g Y_{e,u}^g \qquad (9)$$

where the fixed parameter $Y_{e,u}^g$ should be chosen larger than the maximum end product flow. $Y_{e,i}^g$ relates to the material flow $P(z)$ as follows

$$\sum_g \sum_e N_e^g Y_{e,i}^g = P(z_i) \qquad (10)$$

where N_e^g maps the flow of end product e to the material flow in grade g. Finally, the total end product yield at time iT_s is given by $Y_{e,i} = \sum_g Y_{e,i}^g$. Using the integer description of the grade regions and after discretization of the objective (5) (using e.g. the Riemann sum) and the path constraints (2), a mixed-integer nonlinear progamming (MINLP) formulation of the transition problem can be derived. Several approaches towards solving such MINLP's exist; most popular methods are branch and bound and cutting plane methods, see e.g. (Floudas, 1995) for an overview, however their applicability is generally limited to small problems only. For most problems, h, R_g, and P will be linear functions, leaving the process model (f, g) the only remaining nonlinearity. Therefore, we propose a *successive linearization approach* for the problem at hand. The linearization of the MINLP can be obtained by substituting the nonlinear dynamics by the linearized-time-varying (LTV) dynamics that describe the behavior along the solution that results from the previous iteration. The LTV dynamics can be obtained for example by sampling the solutions of the sensitivity equations (Støren and Herzberg, 1994).

The linearized (inner loop) problem then is a Mixed Integer Linear Program (MILP) which is solved in each iteration of the outer loop. Well-established Branch&Bound techniques can be used to solve the MILP inner loop problem to a desired accuracy. Only local minima can be guaranteed.

4. Application to a binary distillation column grade change problem

As an example of the grade change optimization we consider a model of a 20-tray binary distillation column based on the CONSTILL example by Ingham et al. 1994. The inputs are the reflux ratio (u_1) and the reboiler duty (u_2). The performance variables are respectively the top purity, the bottom impurity, the distillate flow, the bottom outflow, the reflux ratio and the reboiler duty. We distinguish 3 different top-products, 2 different bottom-products and hence 6 product grades given in the following table.

g	Constraints ($R_g(z)$)		g	Constraints ($R_g(z)$)	
1	$0.00 < z^1 < 0.98$	$0.00 < z^2 < 0.05$	4	$0.98 < z^1 < 0.99$	$0.05 < z^2 < 1.00$
2	$0.00 < z^1 < 0.98$	$0.05 < z^2 < 1.00$	5	$0.99 < z^1 < 1.00$	$0.00 < z^2 < 0.05$
3	$0.98 < z^1 < 0.99$	$0.00 < z^2 < 0.05$	6	$0.99 < z^1 < 1.00$	$0.05 < z^2 < 1.00$

We consider a changeover from grade 3 to grade 5 with the economic objective (5) and prices of the top product chosen equal to [1,2,4] for rising purity and of the bottom product [1.5,0.8] for rising impurity. Reboiler duty costs 0.25 per unit. The discretization interval used in the piecewise constant control parametrization and in the Riemann sum approximation of the objective is of length 0.3 hr. The optimization horizon has length 6 hr. Additional constraints on the rate of change of the inputs are imposed, respectively 0.25 and 200. The end point constraint is relaxed by 0.001 in order to ensure feasibility of the optimization.

Implementation of the optimization algorithm is done using gPROMS (for function and gradient evaluations) and GAMS with CPLEX (for the MILP optimization). The optimization converges in 7 iterations. Progress in each iteration is measured using a specific merit function that is based on the smooth approximation of the grade regions proposed in (van der Schot, 1999). Details are omitted. The optimization results are plotted in Figure 1 (in between the vertical rulers). Due to the economic attractiveness of grade 5 there is a clear incentive to establish the grade change (in this case equivalent to a change of the top purity) as quickly as possible. The rate-of-change constraint on the reflux ratio is bounding the performance.

5. Conclusions

A formulation of the grade change problem for continuous chemical processes as an economic optimization problem is presented. This formulation leads to a discontinuous optimization problem which cannot be solved using the standard sequential or

816

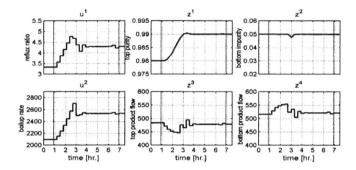

Figure 2: Optimal trajectories of the controls (left two images) and the performance variables for a transition from grade 3 to grade 5.

simultaneous dynamic optimization method. Using control parametrization, a Mixed Integer Linear description of the grade regions, and a Riemann sum approximation of the economic objective the problem can be transformed into a Mixed Integer Nonlinear Program (MINLP), which we can solve using a successive linearization approach. Search directions are computed from the Mixed Integer Linear Program (MILP) that results after linearization of the dynamics. The feasibility of the approach is demonstrated on an example: the optimization of grade changes in a binary distillation column.

References

Bemporad, A. and Morari, M. , 1999, Control of systems integrating logics, dynamics and constraints. Automatica, vol. 35, p. 407-427.

Floudas, C.A., ,1995, *Nonlinear and mixed-integer optimization: fundamentals and applications*, Topics in chemical engineering. Oxford University Press.

Lewis, F.L. and V.L. Syrmos, 1995, *Optimal Control*, Wiley, New York.

Ingham, J. and I.J. Dunn, and E. Heinzle, and J.E. Prenosil, 1994, *Chemical Engineering Dynamics*, VCH, New York.

McAuley, K.B. and J.F. MacGregor, 1992, Optimal Grade Transitions in a Gas Phase Polyethylene Reactor, *Aiche Journal*, vol. 38, p. 1564-1576.

van der Schot, J.J., R.L. Tousain, A.C.P.M. Backx, and O.H. Bosgra, 1999, *Computers and Chemical Engineering Supplement*, p. S507 - S510.

Sinclair, K.B., 1987, Grade change flexibility defined, determined, compared, *Proceedings fifth International Conference on Polyolefins*, Houston, Texas, USA.

Støren, S. and T. Hertzberg, 1994, The sequential linear quadratic programming algorithm for solving dynamic optimization problems - a review, *Computers and Chemical Engineering*, vol. 19, p. 495-500.

Vassiliadis, V.S., 1993, Computational Solution of Dynamic Optimisation Problems with General Differential-Algebraic Constraints, University of London.

White, V., J.D. Perkins, and D.M. Espie, 1996, Switchability analysis, *Computers and Chemical Engineering*, vol. 20, p. 469-474.

European Symposium on Computer Aided Process Engineering – 12
J. Grievink and J. van Schijndel (Editors)
© 2002 Elsevier Science B.V. All rights reserved.

An Integrated Framework for Multi-Objective Optimisation in Process Synthesis and Design

H. Alhammadi, G.W. Barton, J.A. Romagnoli and B. Alexander
Laboratory for Process Systems Engineering
Chemical Engineering Department, University of Sydney NSW 2006 Australia

Abstract

A multi-objective optimisation framework is used to identify trade-offs between various goals in flowsheet design/synthesis. Life Cycle Assessment (LCA) is used to handle the environmental considerations. The methodology is demonstrated using a vinyl chloride monomer (VCM) plant with varying degrees of heat integration as a case study.

Introduction

Industry is required to operate any process so as to satisfy economic, environmental and social objectives, while at the same time being readily operable. In general, the major challenge (both at the design and the operational stage) lies in resolving the conflicts between these objectives (Miettinen, 1999). A major task for process system engineers is to develop tools that assist in the trade-off scenarios arising in such multi-objective situations. Our ultimate goal is an integrated approach allowing all relevant objectives to be accounted for during the detailed design of a processing plant. To this end, we present here a framework for such a methodology that incorporates both economic and environmental objectives for assessing various levels of heat integration for a given process.

Life Cycle Assessment (LCA) is a methodology for estimating the environmental impacts associated with a given product, process or activity (Consoli *et al.*, 1993). It is a comprehensive technique that covers both "upstream" and "downstream" effects of the activity or product under examination, thus often being referred to as "cradle-to-grave" analysis. Being an accepted (and widely used) tool in this area (Azapagic and Clift, 1999), it was employed in this study to map the environmental impact potential of any given alternative.

Energy integration techniques are today an accepted means of both improving process economics and reducing environmental impacts (Rossiter, 1994; Linnhoff, 1994), with 'pinch analysis' being a simple means of determining the minimum heating and cooling requirements for a given process configuration. In this paper, thermal pinch analysis is included to demonstrate the incorporation of energy integration within a multi-objective optimisation framework. An obvious extension here is to include techniques for devising and assessing 'practical' energy integration networks within the methodology. However, it is regularly noted in the literature (Morari, 1983; Bildea and Dimian, 1999) that the inclusion of energy integration generally makes a process more difficult to control. Thus, another (on-going) extension to the methodology is the inclusion of process controllability as part of the multi-objective framework.

818

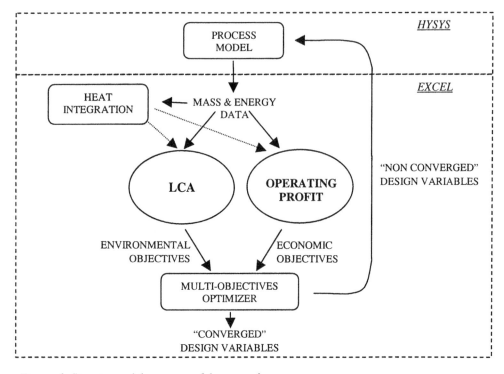

Figure 1. Structure of the proposed framework

In this paper, a vinyl chloride monomer (VCM) process is used to demonstrate our approach.

Framework and methodology

The proposed framework extends the work of Alexander *et al.* (2000), so as to include energy integration analysis within the multi-objective optimisation environment. Figure 1 is a schematic of the proposed methodology's structure, showing the inter-linking of the software tools used and the flow of data between them.

Process Model: The process is modelled within the Hysys© simulator (Hyprotech Ltd). This was selected as the simulation environment as it provides for both steady-state and dynamic modelling, while permitting the ready exchange of data with other software packages (including equation-oriented simulators, such as gPROMS). Mass and energy data from the Hysys© model are transferred to/from MS-Excel© using an object link and embed (OLE) communications protocol.

Environmental Model: In Excel, LCA is used (via impact potentials) to analyse the environmental impacts of the process and to formulate the environmental terms to be used in the multi-objective optimisation problem. All the upstream/external activities, from the extraction of raw materials to the provision of site utilities, are considered in this study. It should be noted that the LCA study was completed as a "cradle-to-gate" analysis, whereby the final usage and disposal phases of the various products are not considered. The environmental burdens of the upstream activities are determined using the SimaPro© commercial LCA database. The LCA analysis first performs an inventory analysis, followed by an environmental impact assessment. The impacts covered in this environmental model are global warming potential (GWP), ozone depletion potential (ODP), eutrophication potential (EP), acidification potential (AP), summer smog potential (SP), human toxicity potential (HTP) and process energy potential (PEP).

Economic Model: Also within Excel[©], calculations are performed (using the mass and energy data transferred from Hysys[©]) to formulate an economic objective based on the operating profit.

Heat Integration: Similarly, all heating and cooling requirements for the process are obtained from the data transferred from Hysys[©], and are tabulated in Excel[©]. Here, a pinch analysis is performed to determine the minimum utility requirements for any given operating point. These values are used in the economic and environmental models as 'best estimates' for an energy integrated version of the process.

Multi-Objective Optimisation Algorithm: The ε-constraint method is used to solve the multi-objective optimisation problem and to generate individual points on the Pareto curve. Note that a Pareto curve is a set of 'non-inferior' solutions defining a boundary beyond which none of the objectives can be improved without sacrificing at least one of the other objectives. The trade-off between objectives can, thus, be visualised through the Pareto curve over the set of design alternatives. The basic strategy for the ε-constraint method is to select one of the objectives (*eg* operating profit) to be optimised and to convert all other objective functions (*eg* the environmental impacts) into constraints by setting an upper bound ($\varepsilon_1, ..., \varepsilon_n$) for each of them (Miettinen, 1999). The algorithm for the ε-constraint approach then solves a problem posed as follows,

$$\text{Max Objective}_j$$

Subject to: 1. $\text{Objective}_i \le \varepsilon_i$; $i=1, ..., n : i \ne j$
                     2. Mass and energy balance constraints

where the Pareto curve is generated by parametrically varying the upper bound on the constrained objectives over each entire range, and solving the above for each case.

Description of Case Study

Figure 2 is a simplified block diagram of a typical vinyl chloride monomer (VCM) plant (McPherson *et al.*, 1979; Cowfer and Gorensek, 1997). This integrated process produces VCM from ethylene, chlorine, oxygen and a portion of the by-product hydrogen chloride (HCl). The major sections of this plant are as follows: (1) A direct chlorination of the ethylene to produce ethylene dichloride (EDC). (2) An oxy-chlorinator to produce EDC by reacting ethylene with oxygen and HCl. (3) The two crude EDC streams are mixed and purified in a pair of distillation columns (essentially to remove water and unwanted reaction by-products). (4) The pure EDC undergoes (partial) thermal cracking in a pyrolysis furnace to yield VCM and HCl. (5) VCM is separated from the HCl and EDC in another pair of distillation columns. Note that a portion of the HCl is recycled to the oxy-chlorination reactor to make EDC, while any unconverted EDC is recycled (via the purification) to the furnace.

820

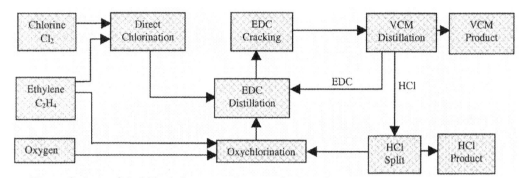

Figure 2. A simplified block diagram of a typical VCM plant

The process design variable selected here for (multi-objective) optimisation was the portion of HCl recycled to the oxy-chlorinator. It would be a straightforward extension to the framework to include multiple design variables - however, in this paper only a single variable was considered for ease of demonstration. Also, as the environmental potentials in this case all trend in the same direction, the impact potential most sensitive to this design variable (*ie* GWP) was chosen as an exemplar. The economic objective chosen was the operating profit, that is, the difference between the total value of the products and the total cost of the raw materials and utilities. Each objective function was normalised (over the specified range for the recycled HCl) and scaled so that 0 and 1 represents the best and worst value of the objective. The ε-constraint method was used to solve the multi-objective optimisation problem and obtain the Pareto curve. Here, the economic objective was optimised while the environmental objective was converted into a constraint with a specified upper bound.

This multi-objective optimisation problem was performed for three cases. In the first case, no heat integration was considered, while in the second the process was examined for optimal heat integration with the minimum heating and cooling requirements being determined using pinch analysis. These minimum utilities were then used in the environmental and economic models to formulate their respective objectives. In the final case, a single heat exchanger around the pyrolysis reactor was considered.

Results and Discussion

The Pareto curves for all three cases are shown in Figure 3. The curves for cases one and two provide the lower and upper bounds for all possible levels of heat integration at all operating points. The 'optimal heat integration' curve shows the maximum possible reduction achievable for the two objectives. However, as noted previously, this level of heat integration may well be impractical, as the heat exchanger network (HEN) required to realise it could make the resultant process extremely difficult to operate/control. For example, a HEN was realised for a single point on the optimal heat integration Pareto curve but this required nine exchangers and the splitting of two streams - not really a practical option. Thus, moving from the 'no heat integration' curve towards the 'optimal heat integration' curve involves other trade-offs to be considered (in addition to economic and environmental) - those of plant controllability and/or operability.

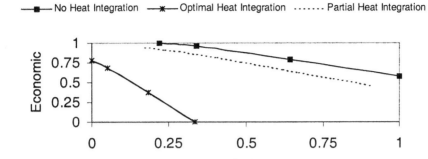

Figure 3. Pareto curves of the three cases

Table 1: Annual values of the objectives for various VCM plant designs

HCl Recycled %	No Heat Integration		Partial Heat Integration		Optimal Heat Integration	
	GWP 10^3 ton CO_2/yr	Profit M$/yr	GWP 10^3 ton CO_2/yr	Profit M$/yr	GWP 10^3 ton CO_2/yr	Profit M$/yr
20	2842	160	2762	164	2441	175
40	3010	163	2920	167	2515	181
60	3243	168	3140	173	2618	189
80	3548	175	3428	180	2750	201
100	3935	183	3790	190	2911	216

Also shown on Figure 3 is the Pareto curve for the case where a single heat exchanger is employed around the pyrolysis reactor (using the hot reactor effluent to heat the cold reactor feed). As expected, this example of partial heat integration resulted in a Pareto curve that lies between the two extremes.

The extent to which the Pareto curve has shifted by the inclusion of this one exchanger is less than a quarter of what is achievable, leaving scope for considerable improvement (remembering that the 'best' point on Figure 3 is the origin) but at the price of a (likely) reduction in process controllability and/or operability.

Table 1 summarises the numerical results given on the plots in Figure 3. Such a table quantifies the trade-offs possible between the economic, environmental and degree of heat integration objectives. In this table, the impact of employing different levels of heat integration is tracked as you progress along a row (*ie* for a given percentage of HCl recycled). Similarly, the trade-off between economic and environmental objectives as a function of the HCl recycled (for a given level of heat integration) is tracked as you proceed down a column.

822

Conclusions

In this paper, a methodology has been proposed that incorporates both economic and environmental considerations within a multi-objective optimisation framework that permits (for a given process) the inclusion of various levels of heat integration. The methodology as it stands enables us to draw 'boundary' Pareto curves corresponding to the maximum and minimum levels of heat integration for all operating points achievable by the process. It is also possible to use the proposed approach to draw the Pareto curve for any HEN between the calculated limits, and thus to quantify the trade-off between economic and environmental objectives.

It was noted that improved energy efficiency generally increases plant complexity and may well have significant impacts on plant operability and/or controllability. Thus, it will be necessary to explicitly incorporate the effects of increased energy efficiency as you move between the two 'boundary' Pareto curves. The inclusion of such operability and controllability analysis into the general framework is the next step in this work, so as to enable assessment of 'practical' HEN alternatives.

Acknowledgements
The authors wish to thank Professor J. Petrie for his LCA contribution, while Hasan Alhammadi acknowledges the financial support of The University of Bahrain.

References
Alexander, B., Barton, G.W., Petrie, J. and Romagnoli, J.A. (2000). Process Synthesis And Optimisation Tools For Environmental Design: Methodology And Structure, Computers & Chemical Engineering , 24, 1195-1200.

Azapagic, A. and Clift, R., 1999, The Application of Life Cycle Assessment to Process Optimization, Computers & Chemical Engineering, 23, 1509-1526.

Bildea, C. and Dimian, A., 1999, Interaction between Design and Control of a Heat-Integrated Distillation System with Prefractionator, Trans IChemE, 77, 597-608.

Consoli, F., Allen, D., Boustead, I., Fava, J., Franklin, W., Jensen, A., de Oude, N., Parrish, R., Perriman, R., Postlethwaite, D., Quay, B., Séguin, J. and Vigon, B. (eds.), 1993, Guidelines for Life-Cycle Assessment: A "Code of Practice". SETAC, USA.

Cowfer, J. and Gorensek, M., 1997, Vinyl Chloride: Encyclopaedia of Chemical Technology, 24, 851-882.

Linnhoff, B., 1994, Use Pinch Analysis to Knock Down Capital Costs and Emissions, Chemical Engineering Progress, August, 32-57.

McPherson, R., Starks, C. and Fryar, G., 1979, Vinyl Chloride Monomer – What You Should Know, Hydrocarbon Processing, March, 75-88.

Miettinen, K., 1999, Nonlinear Multi-objective Optimisation. Kluwer Int. Series.

Morari, M., 1983, Flexibility and Resiliency of Process Systems, Computers & Chemical Engineering, 7, 423-437.

Rossiter, A. P., 1994, Process Integration and Pollution Prevention. In: Pollution Prevention via Process and Product Modifications, AIChE, 90, 12-22.

European Symposium on Computer Aided Process Engineering – 12
J. Grievink and J. van Schijndel (Editors)
© 2002 Elsevier Science B.V. All rights reserved.

An approximate Optimal Moving Grid technique for the solution of Discretized Population Balances in Batch

Menwer M. Attarakih[1], Hans-Jörg Bart[1] and Naim M. Faqir[2]

[1]Kaiserslautern University, Faculty of Mechanical & Process Eng., Institute of Thermal Process Eng., POB 3049, D-67653 Kaiserslautern, Germany.

[2]University of Jordan, Faculty of Eng. & Technology, Chemical Eng. Department, 11942, Amman, Jordan.

Abstract

The numerical solution of droplet population balance equations by discretization is known to suffer from inherent finite domain errors (FDE). A new technique that minimizes the total FDE during the solution of discretized population balance equations (DPBE) using an approximate optimal moving grid for batch systems is established. This optimal technique is found very effective for tracking out steeply moving population density with a reasonable number of size intervals. The present technique exploits all the advantages of its fixed counterpart by preserving any two moments of the evolving population. The technique is found to improve the predictions of the number density, zero and first moments of the population.

1-Introduction

The population balance equation (PBE) for a well-stirred batch vessel could be written as (Ramkrishna, 2000):

$$\frac{\partial n(v,t)}{\partial t} + \frac{\partial [\dot{v}n(v,t)]}{\partial v} = \rho\{n(v,t),v,t\} \tag{1}$$

where $n(v,t)$ is the average number of droplets per unit volume of the vessel at time t. The first term on the left hand side denotes the rate of accumulation of droplets of size v, and the second term is the convective flux along the droplet volume coordinate. The term on the right hand side is the net rate of droplets generation by coalescence and breakage.

Despite the importance of Eq. (1) it rarely has an analytical solution. So, in general numerical solutions are sought where several methods are proposed in the literature. Kumar and Ramkrishna (1996a) critically reviewed the available methods where they concluded that the methods conserving both total number and volume of droplets are not only computationally efficient but are also accurate. These authors made great achievement in the discretization of the PBE by introducing a general framework of discretization. Their method preserves any two moments of the population, and converts the PBE into set of discrete partial differential equations. The developed methods are called the fixed and moving pivot techniques, where the latter is used in the present work due to its generality and accuracy.

Nevertheless, this discretization is by no means exact, and it is inherently associated with the so-called finite domain error (FDE) resulting from trying to use a finite droplet volume to approximate the infinite one. As so far, only Sovova and Prochazka (1981) tried to investigate rigorously the effect of the FDE on the accuracy of the solution of the DPBEs. They studied droplet breakage and coalescence in batch vessels at steady state and tried to estimate the FDE by extrapolating both ends of the droplet distribution. The main drawback of this technique is the general uncertainties associated with extrapolation and its lack of general relations to predict the time dependent FDE.

The objective of this work is to develop an approximate optimal moving grid technique for batch systems, based on the minimization of the total FDE. The proposed technique has the ability to conserve any two moments of the distribution. A general equation is also derived for the total FDE by approximate discretization of the general PBE.

2-The discretized PBE using the moving pivot technique

In the moving pivot technique, the droplet volume is discretized according to the partition (grid) $V_M \equiv \{v_{min}, v_2, \ldots v_M, v_{max}\}$ and the ith interval is denoted by $I_i = [v_i, v_{i+1})$. Kumar and Ramkrishna (1996b) derived the DPBEs, which conserve the total number and droplet volume for droplet breakage in batch vessel:

$$\frac{dN_i(t)}{dt} = \lambda_i N_i(t) + \sum_{k=i+1}^{M} \pi_{0,i,k} \Gamma_k N_k(t), \qquad i = 1, 2, \ldots M \tag{2}$$

$$\frac{dx_i(t)}{dt} = \eta_i + \frac{1}{N_i(t)} \sum_{k=i+1}^{M} (\pi_{1,i,k} - x_i \pi_{0,i,k}) \Gamma_k N_k(t), \qquad i = 1, 2, \ldots M \tag{3}$$

where x_i is the representative size of the population in the interval I_i and it is called the pivot, N_i is the total number of droplets associated with this pivot, $\eta_i = (\pi_{1,i,i} - x_i \pi_{0,i,i}) \Gamma_i$, $\lambda_i = (\pi_{0,i,i} - 1) \Gamma_i$, $\pi_{0,i,k}$ and $\pi_{1,i,k}$ are given by Kumar and Ramkrishna (1996b) and Γ_i is the breakage frequency.

3-The Finite Domain Error

In discretizing an equation defined over an infinite domain an inherent error is incurred due to the failure of taking into account the portion of the function lying outside the domain of discretization. This error is termed the total FDE and is represented by (Sovova and Prochazka, 1981):

$$\varepsilon_0(t) = \int_0^{v_{min}} n(v,t) dv + \int_{v_{max}}^{\infty} n(v,t) dv \tag{4}$$

Note that for a given number of intervals, M, and interval width, Δv, an optimal minimum droplet volume, v_{min}, exists and could be found by differentiating Eq. (4) with respect to v_{min} and set the result equal to zero:

$$n(v_{min}^*, t) - \frac{dv_{max}}{dv_{min}} n(v_{max}^*, t) = 0 \tag{5}$$

According to Eq. (5), the optimal minimum droplet volume must decrease as function of time to account for the increasing number density at the lower size range. This suggests the use of optimal moving grid for droplet breakage, which moves from the upper to the lower size ranges as function of time. Consequently, Eqs. (2) , (3) and (5) must be solved simultaneously at each instant of time to find such an optimal moving grid. Unfortunately, the solution is iterative by solving Eq. (5) at each integration step, and might mask the benefits gained by using the optimal moving grid. To compensate for this drawback, an approximate optimal moving grid technique is derived in the following section.

4-An approximate optimal moving grid technique

The total finite domain error, as defined above, will be close to the minimum value when both residuals are equal, which leads to an approximate optimal minimum droplet volume and hence optimal moving grid. This optimal moving grid should keep the number of intervals constant during grid movement, and hence redistribution of the population between the old and the newly formed grids is essential. This should be performed by conserving any two moments of the population in order to be consistent with Eqs. (2) and (3).

Now consider a typical geometric grid ($v_i(t) = \sigma^{i-1} v_{min}(t)$) at two instants of time: t and $t + \Delta t$ where the optimal minimum droplet volume moves from $v^*_{min}(t)$ to $v^*_{min}(t + \Delta t)$. Let $\gamma_i^{<i>}(t)$ be the fraction of droplets at the pivot $x_i(t)$ to be assigned to the pivot $x_i(t + \Delta t)$ and $\gamma_{i+1}^{<i>}(t)$ be the fraction of droplets at the pivot $x_i(t)$ to be assigned to the pivot $x_{i+1}(t + \Delta t)$. These fractions are found such that both number and volume of these droplets are conserved after redistribution. Accordingly, the discrete lower and upper residuals at this instant of time are given by:

$$FDE_0^L (t + \Delta t) = \gamma_1^{<0>} N_0(t) \tag{6}$$

$$FDE_0^U (t + \Delta t) = \sum_{i=M+1}^{\infty} \gamma_i^{<i-1>} N_{i-1}(t) + \gamma_i^{<i>} N_i(t) \tag{7}$$

where only the $(M+1)$th term in the summation above has a significant value for sufficiently large M or geometric factor σ.

The optimality condition implied by Eq. (5) could be approximately satisfied by forcing both sides of Eqs. (6) and (7) to be equal, which after some algebraic manipulation yields:

$$\frac{v^*_{min}(t + \Delta t)}{v^*_{min}(t)} = \frac{\dfrac{1}{\sigma - 1} N_M(t) + \dfrac{1}{\sigma} N_0(t)}{\dfrac{\sigma + 1}{\sigma} N_0(t) + \dfrac{1}{\sigma - 1} N_M(t) - N_{M+1}(t)} \tag{8}$$

5-Estimation of the lower and upper residuals

To estimate the lower residual, $N_0(t)$, it could be assumed that the first interval will only receive broken droplets from higher ones or from droplets within the interval itself with no droplets are lost through breakage from this interval (Laso et al., 1987; Hill and Ng, 1995). Consequently, an unsteady state number balance on this interval yields:

$$\frac{dN_0(t)}{dt} = \vartheta(v^*_{min}(t))\Gamma_0 N_0(t) + \sum_{k=1}^{M} \pi_{0,0,k}\Gamma_k N_k(t) \tag{9}$$

Since the width of the interval $[v_{min}, v^*_{min}(t)]$ is very small due to the geometric grid used in discretization, the pivot $x_0(t)$ is fixed at the middle of this interval. Similar arguments for the I_{M+1} interval leads to the following equation:

$$\frac{dN_{M+1}(t)}{dt} = \lambda_{M+1} N_{M+1}(t) \tag{10}$$

and the $(M+1)th$ pivot could be derived from Eq. (3). So Eqs. (6) and (7) along with Eqs. (9) and (10) define completely the lower and upper residuals at any instant of time for specified grid parameters σ and M.

6-Numerical results and discussion

By using a geometric grid, it should be mentioned that when the last interval is completely passed due to the grid movement, the new and the old interval boundaries completely coincide except for the first boundary. This suggests that the number densities could be updated only when $v^*_{min}(t)$ is less than or equal to $v_M(t)$ to exclude any numerical inaccuracies due to population redistribution. This strategy is adopted in the present solution algorithm using uniform daughter droplet distribution and linear breakage frequency over a relatively long period of time, $t=100$ (arbitrary time units) to illustrate the steepness of the number density. The solution algorithm is implemented using an exponential initial condition and a number of intervals, $M=15$, and $\sigma=2.0$. The analytical solution is given by by Ziff and McGrady (1985) for binary breakage. We start by comparing the exact and numerical FDE as well as the optimal minimum droplet volumes. Fig. 1-a shows these results, and it can be seen an excellent agreement between the numerical and exact FDE is obtained. As expected the optimal fixed FDE increases with time due to the failure to account for the increase in number density in the small size range as droplet breakage proceeds. This is actually equivalent to a loss of number of droplets from the system. To compensate for this, we let the grid move in an optimal manner as shown in Fig. 1-b, where the exact minimum droplet volume is depicted along with that predicted using the optimal moving grid algorithm. First the agreement between the optimal piecewise minimum droplet volume and the exact one is also excellent even when the grid moves so fast. Second the great influence of the optimal grid movement on the reduction of the total FDE is obvious when compared to the fixed grid (Fig. 1-a). Fig. 2-a shows the exact and numerical average number densities at the final time of simulation where, excellent agreement is perceptible. Also, one could see how the optimal moving grid leaves the approximately empty intervals (large sizes) to accommodate the increasing number densities in the small size range as expected.

Fig. 2-b shows the clear discrepancies between the discrete zero moment of the distribution using fixed and optimal moving grids respectively. As expected for a long time of droplet breakage, the number density becomes increasingly sharp. Failure to include the small size range of the population will induce appreciable errors in the total number density as a result of increasing the total FDE. The mean droplet volume is also over predicted, however to a small extent, when fixed grid is used because large number but small volume of droplets are lost at long times of breakage due to the increase in the total FDE. It should be mentioned that as the number of intervals decreases the sum of the residuals becomes the main source of the discretization error (relative to the integration error). Under these circumstances minimizations of these residuals (FDE) is the only way to reduce the discretization error if coarse grid is to be maintained.

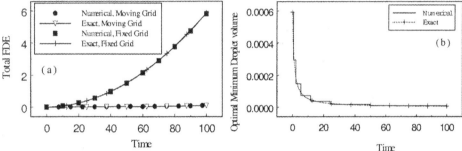

Figure 1: a-The effect of optimal grid movement on the finite domain error. b– Exact and numerical optimal minimum droplet volumes using geometric grid with factor σ = 2.0 and M = 15.

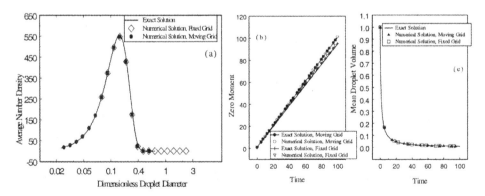

Figure 2: The effect of the optimal grid movement on: a- The average number density. b- The zero moment and c- The mean droplet volume using geometric grid with factor σ = 2.0 and M = 15

7-Conclusions

An optimal moving grid technique is developed for the solution of the DPBEs for droplet breakage in batch systems base on the minimization of the total FDE. The redistribution algorithm, on which this technique is based, is consistent with DPBEs by preserving any two moments of the distribution. Moreover, ordinary differential equations are derived to estimate the total FDE of the droplet distribution, which shows excellent agreement with the analytical solution studied in this work.

Nomenclature

FDE_0^L, FDE_0^U	lower and upper residuals based on zero moments of the distribution
M	total number of intervals used in droplet volume discretization
$N_i(t)$	total number of droplets in the ith interval, at time t
$n(v,t)dv$	number of droplets in size range v to $v+dv$, at time t per unit volume
v, v'	droplet volumes
v_{min}, $v_{max.}$	minimum and maximum droplet volumes
v_{min}^*, $v_{max.}^*$	optimal minimum and maximum droplet volumes
$\dot{v}$	droplet growth rate
$x_i(t)$	characteristic volume of the droplet population in the ith interval
t	time

Greek Symbols

$\Gamma(v)$	droplets breakage frequency
$\gamma_i^{<i>}$, $\gamma_i^{<i-1>}$	fractions of droplet assigned to the ith pivot
Δt	time increment
$\varepsilon_0(t)$	total finite domain error based on zero moment of the distribution
η_i	the ith eigenvalue of the pivot equations.
λ_i	the ith eigenvalue of the number density equations
σ	geometric grid factor
$\vartheta(v')$	number of droplets produced when droplet of volume, v', is broken

References

Hill, P. J. and Ng, K. M., 1995, New discretization procedure for the breakage equation. AIChE J., **41**, 1204-1216.

Kumar, S. and Ramkrishna, D., 1996a, On the solution of population balance equations by discretization-I. A fixed pivot technique. Chem. Engng. Sci., **51**, 1311-1332.

Kumar, S. and Ramkrishna, D., 1996b, On the solution of population balance equations by discretization-II. A moving pivot technique. Chem. Engng. Sci., **51**, 1333-1342.

Laso, M., Steiner, L. and Hartland, S., 1987, Dynamic simulation of liquid-liquid agitated dispersions-I. Derivation of a simplified model. Chem. Engng. Sci., **42**, 2429-2436.

Ramkrishna, D., 2000, Population Balances. San Diego: Academic Press.

Sovova, H. and Prochazka, J., 1981, Breakage and coalescence of drops in a batch stirred vessel-I Comparison of continuous and discrete models. Chem. Engng. Sci., **36**, 163-171.

Ziff, R. M. and McGrady, 1985, The kinetics of cluster fragmentation and depolymerisation. J. Phys. A: Math. Gen., **18**, 3027-3037

European Symposium on Computer Aided Process Engineering – 12
J. Grievink and J. van Schijndel (Editors)
© 2002 Elsevier Science B.V. All rights reserved.

Restructuring the Keywords Interface to Enhance CAPE Knowledge Acquisition via an Intelligent Agent

F. A. Batzias [a] and E. C. Marcoulaki [a, b]

[a] Department of Industrial Management, University of Piraeus, Greece
[b] Department of Physics, Faculty of Applied Sciences, NTUA, Greece

Abstract

This work proposes an improved KeyWord Interface (KWI) to enhance the efficiency of information retrieval when using an advanced search engine, as an intelligent agent. This can be achieved by restructuring the KWI into a new hierarchical structure based on an {n domains} by {3 levels} arrangement of keywords ($n \times 3$ KWI), forming a loose/adaptive semantic network. The hierarchical levels used in the suggested implementation are *set of species*, *logical category*, and *holistic entity*. As an illustration, the method is applied to an example of literature survey concerning a well-documented process engineering field. The results of the proposed technology are compared with the outcome of general-purpose search-engines built in common academic publication databases. The comparison reveals the advantage of intelligent searching in creating a local base according to the orders/interests of the researcher.

1. Introduction

Intelligent Agent (IA) technologies have been proposed recently, and partially been implemented, for the collective acquisition of information from large, unstructured, and heterogeneous information spaces like the Internet (O'Meara and Patel, 2001; Crestani and Lee, 2000; Zacharis and Panayiotopoulos, 1999; Etzioni and Weld, 1995; Maes, 1994). The quest for information is hereby assumed to support the initiation, progress and termination stages of a scientific research program, as demonstrated in Figure 1.

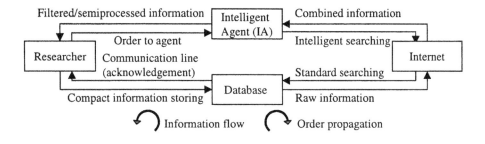

Figure 1: General scheme of information retrieval through an intelligent agent (IA)

At present, there is a vast amount of information on Computer – Aided Process Engineering (CAPE) available on Internet platform, in the form of relevant publications, data, applications, casestudies, etc. This information can be used to construct a CAPE knowledge base, to be interfaced with an available process engineering software tool. The base can be continually enriched and periodically restructured, either exclusively via the software designer, or inclusively via an IA and the designer. The former constitutes the usual present practice, while the inclusive scheme is mostly met on experimental stage in small-scale application. Regardless of the updating schemes applied, the KeyWord Interface (KWI) remains the main communication protocol and the most popular tool in the current database search engines.

This work considers the inadequacies of conventional KWI's to cooperate efficiently with the IA's. Evidence of this inadequacy is provided by searching the Internet for pre-specified targets. A solution is to define a more concrete communication protocol to improve the efficiency of the interaction between the researcher and the search engine. The proposed interface assumes a pyramid-like structure of keyword information levels, so the user can move uphill/downhill according to the abstraction of the search query.

2. Motivation and problem description

A scientific program is a problem describable in linguistic terms providing appropriate keywords and key-phrases for information retrieval. The main difficulty is that these terms are not unique and each of them frequently belongs to a class or category represented by a wider concept, possibly used as a substitute for the keyword.

Consider the quality control of the electrochemical anodization of aluminium, as a typical CAPE research program. There are over 40 low level keywords describing defects that might appear on the product surface (e.g. *burning*, *cracking*, *spotting*, *blooming*, *crawling*, *blistering, traffic marks*, *wood grain*). A survey is carried out in ScienceDirect ® (SD), a very reliable and extended database of relevant literature that uses the LEXIS®-NEXIS® search logic. Using the combination *anodi! AND alumin! AND (burn! OR bloom! OR blister! OR crawl!)* nothing is retrieved (until November 2001). The truncation is used to find the root word plus all the words made by adding letters to the end of it. Inefficiencies in returning results can be attributed to the use of: (i) a synonym, e.g. *fretting corrosion* instead of *traffic marks* or *polishing rings* instead of *wood grain*, and/or (ii) a roundabout description, e.g. "*catastrophic local dissolution of the anodic coating due to overheating*" instead of *burning*.

A commonsense way of extending the search domain is to conduct a more abstract search, i.e. by supplying the general keyword *defect*. However, when SD is called to process the combination *alumin! AND anodi! AND defect* only 6 articles are obtained, while *alumin! AND anodi! AND pit!* returns 53 and *alumin! AND anodi! AND crack!* returns 32 articles. Note that the intersection of the resulting lists for the two last queries contains 8 articles. As the search is carried out within the fields *Title-Abstract-Keywords*, it is safe to conclude there is no hierarchical structure in the *keywords* field for the SD database entries to conform to.

A solution for improving the communication protocol might be a hierarchical structuring of the keywords assigned to each article, either by the author or/and by a reviewer of the Base, according to a *controlled vocabulary*. This standardization seems unattainable in a huge central Base, partially due to lack of resources and partially to the complexity of categorization in many cases. It could though be realized (i) as an additional service offered by the Base in special areas, like Materials Technology, where standardization/ categorization is already at a satisfactory level due to increased interest from industry for quality control purposes, and/or (ii) when an ad hoc decentralized local base is structured to serve as an information aid for a research program. This work develops an algorithmic search procedure to support the semiautomatic parallel creation of the KWI and the local Base. The KWIs of the documents stored in this local Base form a network of interconnected terms with hierarchical structure, i.e. an ontology. CAPE-related knowledge fields are used in the implementation and illustration of the software tool.

3. Algorithmic procedure

The proposed hierarchical KWI structure is based on an $\{n$ topic domains$\}\times\{3$ levels$\}$ arrangement of keywords ($n\times3$ KWI), forming a loose/adaptable semantic network. The n domains considered here to be of interest to the process engineer are three, namely *materials, processes* and *management*. The levels of methodological approach assigned to each domain correspond to: *set of species, logical category* and *holistic entity*.

In the aluminium anodization example discussed previously, the *holistic entity* in the *materials* domain is *quality*, which includes the desired and undesired features of the products. The *logical categories* can be defined by the general keywords *defect* and *property*. The *set of species* is a collection of low-level word-descriptors (e.g. *cracking, blistering*). In the *process* domain, the *holistic entity* may be the collection of all the unit operations met in aluminium treatment (e.g. *anodization, dyeing, sealing*), a *logical category* is a member of this collection (e.g. *anodization*) and the set of species is the collection of alternative process schemes (e.g. *anodizing* with *sulfuric acid*). Similarly, the *management* domain includes design decision methods applied on the various abstraction levels of the *process* domain.

The developed algorithmic procedure for the creation of a $\{n\times3$ KWI$\}$-based Thesaurus is illustrated in Figure 2 and includes the following steps:
1. Description of the subject under investigation in proper linguistic terms
2. Creation of keyword structure according to width/depth of concepts, including the $n\times3$ novel KWI, and information storing (function of temporary/permanent memory)
3. Determination of the appropriate keyword combination within the *set of species*
4. Determination of the appropriate keyword combination in *logical category* terms
5. Searching in field, host, site, document
6. First evaluation of documents obtained as an output of searching
7. Second evaluation of documents – marking and sorting according to relevance
8. Presentation of dependence of marks assigned on the number of corresponding combinations found within the full text of each document (scatter diagram)

832

9. Choice of the appropriate function to correlate relevance with the number of keyword combinations retrieved in the full text – correlation parameters estimation
10. Cluster analysis of the new keywords extracted from the retrieved documents
11. Sorting of new keywords according to their relative frequency and elimination of the most frequent keyword in case of rejection via the feedback route
12. Identification of key phrases
13. Removal of stop and indifferent words (e.g. *and*, *or*, *with*, etc)
14. Determination and expansion of critical root words
15. Creation/enrichment of the n×3 KWI as a communication protocol
16. Local Base creation by storing the information and interconnecting the KWIs
R. What is the type of interface to be followed?
A. The number of relevant documents retrieved is: too big, too small, satisfactory?
P_j. Is the most frequent keyword accepted for incorporation in the structure of stage 2?
W. Is the desired change an elimination/substitution of keyword or an addition of field?
S. Is the fitting satisfactory?   Q. Is there another correlation function available?

The creation of the local knowledge Base under the semantic network of the structured

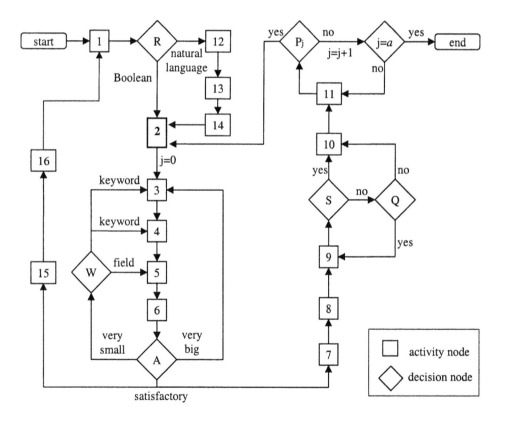

Figure 2: Flowchart of procedures adopted for restructuring the KWI and creating a local knowledge Base

keywords takes place in: (i) stage 2, via the route [...→11→P_j→2→...→11 →...], and
(ii) stages 15, 16. The procedure terminates when a proposed incorporation is rejected
a-times successively. Note that direct information retrieval via the route
[2→3→4→5→16] can be done by means of a simple search engine, possibly using a
tailored internal indexing system. The algorithm is hereby implemented for bottom-up
action, i.e. for extending the *set of species* assigned to a certain *logical category*, and the
logical categories assigned to a certain *holistic entity*.

4. Results and discussion

Example results of a bottom-up action are shown in the screenshot of Figure 3 based on
the initial query *alumin! AND anodi! AND crack!*. The choice shown here concerns the
inclusion of the new keyword *pit!* in the local Base of stage 16 as a member of the *set of
species*. Initially, *pit!* is proposed for incorporation, as it is the next most frequent term
encountered in stage 11. The decision for accepting it in node P_j is guided by
information concerning the origin of this entry, i.e. the journal/database name and the
frequency of appearance of this source (multiplicity). High multiplicity signifies
increased importance of this source with respect to the proposed keyword. Lower
multiplicity (as it is the case here) indicates a more widely diffused term, i.e. of higher
interdisciplinary value.

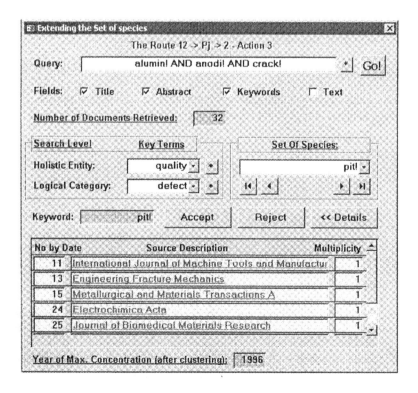

*Figure 3: Screenshot depicting a process of extending the keyword structure, by
incorporating a new keyword member in the set of species*

An example of a fervent field for the application of the proposed hierarchical KWI structure is that of materials design. The KWI could facilitate the management and assessment of information tabulated in property databases, and provided by computer-aided molecular design synthesis tools (e.g. Marcoulaki and Kokossis, 2000; Marcoulaki et al., 2000). The function (e.g. solvent) of the material may present the *holistic entity*. The *logical categories* could classify the molecular configurations performing the described function into aromatic, straight-chain, branched-chain etc. Each *logical category* contains a *set of* molecular configuration *species* (tabulated or designed) and anticipates the introduction of novel entries to extend the bank.

5. Conclusions and recommendations

The paper shows that several inadequacies appear when conventional KeyWord Interfaces (KWIs) are used in searching through a web-based database using an intelligent agent. The situation can be improved by using a structured communication protocol employing controlled vocabulary. A new hierarchical KWI configuration is proposed, assuming an {n domains} × {3 levels} arrangement of keywords, forming a loose/adaptive semantic network. The three hierarchical levels discussed here are *set of species*, *logical category* and *holistic entity*. An algorithmic procedure is developed for the domain of *materials*, which can be similarly applied to the domains of *processes* and *management*. The application of this procedure in an example search of the aluminium anodization literature, is found successful in enriching the hierarchical levels and creating a local knowledge Base in a continuous intelligent mode.

Acknowledgements

The authors kindly acknowledge the contribution of an anonymous reviewer, and the financial support provided by the Research Center of the University of Piraeus.

References

Crestani, F. and P. L. Lee, 2000, Searching the web by constrained spreading activation, Information Processing & Management 36 (4), 585

Etzioni, O. and D. Weld, 1995, Intelligent agents on the Internet: fact, fiction and forecast, IEEE Expert 10 (4), 44-49

Maes, P., 1994, Agents that reduce work and information overload, Commun. ACM 37 (7), 30

Marcoulaki, E. C. and A. C. Kokossis, 2000, On the development of novel chemicals using a systematic synthesis approach. I – Optimisation framework, Chem. Eng. Sci. 55 (13), 2529

Marcoulaki, E. C., A. C. Kokossis and F. A. Batzias, 2000, Novel chemicals for clean and efficient processes using stochastic optimisation, Computers Chem. Eng 24, 705

O'Meara, T. and A. Patel, 2001, A topic-specific web robot model based on restless bandits, IEEE Internet Computing, March-April 2001, 27

Zacharis, Z. N. and T. Panayiotopoulos, 1999, A learning personalized information agent for the WWW, Proc. ACAI-99 on Machine Learning and Intelligent Agents, Chania, Greece: EETN, 39

European Symposium on Computer Aided Process Engineering – 12
J. Grievink and J. van Schijndel (Editors)
© 2002 Elsevier Science B.V. All rights reserved.

835

Mining Textual Project Documentation
in Process Engineering

A. Becks [a], J.-C. Toebermann [b]

[a] Fraunhofer Institute for Applied Information Technology FIT,
D-53754 Sankt Augustin, Germany, andreas.becks@fit.fraunhofer.de

[b] RWTH Aachen - Computer Science V (Information Systems),
D-52056 Aachen, Germany, toebermann@cs.rwth-aachen.de

Abstract

The aim of Knowledge Management is to systematically create, maintain and distribute intellectual capital. To analyse numerical or structured data, techniques like "online analytical processing" and data mining methods are used. However, a lot of existing corporate information like manuals, guidelines, patents and project documentation is captured in textual, i.e. unstructured, form. Specific requirements for mining such textual data were analysed and reviewed and an according tool "DocMINER" was developed and evaluated. Using an example session with project documentation from the process engineering domain its features and usage are described and its potential to lower the necessary effort during typical work steps are demonstrated.

1. Introduction

High efforts in time and money are often spent to re-capture or even re-invent already existing knowledge. Knowledge Management can be seen as a way to systematically create, maintain and distribute intellectual capital. An important technological aspect is supporting companies in their efforts to analyse and learn from their existing proprietary data and documentation. Techniques for analysing numerical or structured data include "online analytical processing" (OLAP) and a broad spectrum of data mining methods. However, a lot of existing corporate information is captured in textual, i.e. unstructured, form. Examples are manuals, guidelines, patents and project documentation. Consequently, text mining is an emerging field of research.

Important aspects of text mining comprise document clustering and visualization. Graphically displaying complex information directly appeals to the powerful human visual perception, enabling users to easily identify patterns and trends in data collections. We developed the tool "DocMINER" (Document Maps for Information Elicitation and Retrieval) which uses a so-called document map for displaying the semantic similarity structure of document collections. Its motivation is to ease access to corporate text collections for analysis purposes. Its development was based on an empirical investigation of typical document analysis tasks in knowledge-intensive industries (Becks 2001).

In this paper we introduce document maps and then sketch specific requirements of text mining as well as the architecture and design of our tool. Using a detailed example session with project documentation from the process engineering domain, its features and usage are described and its potential to lower the necessary effort during typical work steps are demonstrated. By this example we show that the tool provides effective access to relevant parts of the documentation and enables the user to easily understand the relationships among the complex material.

2. Document Maps and the DocMiner System

Document Maps present the semantic structure of a document collection by using a suitable metaphor for intuitively visualizing 'document similarity'. The concept of document or group similarity is usually reflected by a notion of distance: the more similar two documents or document groups are, the closer they appear in the visualization. In literature different metaphors and granularities for visualization have been proposed, e.g. 2 or 3-dimensional scatter plots (Chalmers et al 1992, Wise et al. 1995) and 3D landscapes (Davidson et al. 1998) for presenting inter-document similarities, or category maps (Chen et al. 1996) for presenting the group structure. These visualizations enable the user to easily study the topical density or relatedness in non-trivial text collections.

In general, the application of document maps is useful when an explorative access to text collections is required. This is important especially in the context of knowledge management (Becks et al. 2001): here, the user is often interested in gaining insight into a text collection as a whole and in analysing aspects, e.g. figuring out relationships of single documents. Mostly, the user is not able to specify his information needs adequately and precisely in such a context. In the DocMINER project (Becks 2001) we have designed and evaluated a document map system for visually aiding text collection analysis tasks in knowledge management.

DocMINER supports an adaptable framework for generating a graphical overview, using a modular combination (fig. 1) of algorithms from classical information retrieval, spatial scaling, and self-organizing neural networks. The basic method allows a fine-granular (dis-)similarity analysis of specialized document collections. It can be tailored to domain-specific needs since the module for assessing the similarity of documents is exchangeable. For example, statistical methods (Salton 1971) to assess document

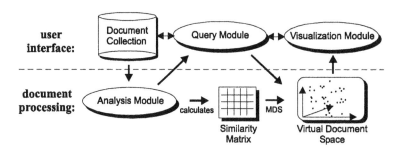

Figure 1: Modular approach for generating document map

similarity are usually well suited, but for some types of document collection improvements are possible using terminological (or description) logics (Meghini et al. 1996). Additionally, a semantic refinement extension based on fuzzy rules enables the analyst to incorporate a personal 'bias' into the map generation process.

DocMINER tightly integrates the graphical map display with explorative and goal-directed interaction methods. Documents are represented as points in a map that conveys a picture of the document collection structure, i.e. the grouping of similar documents and distribution of related document groups. Neighbouring points in bright shaded areas represent similar documents. Grey borders separate these areas: the darker the border, the stronger the separation, and thus the more dissimilar the single documents or document groups.

DocMINER comprises several interaction methods, which help to explore a given collection of documents step by step and to successively gain information on different levels of granularity. Its interface design was guided by following rules: overview first, zoom and filter, then details-on-demand (cf. Shneiderman 1996). Documents can be opened by point-and-click. System features include different zoom, scaling and sub-map functions, means to define and assign document symbols (e.g. to support the analysis of a priori given document classes), an annotation function, automatic map labelling and textual document group summaries, and a tight coupling with a query-driven retrieval interface.

3. Mining Project Documentation: An Example Session

We describe a detailed example session in which our tool is used to get familiar with a complex collection of project documentation, i.e. to understand relationships and dependencies of subprojects. Such a scenario typically arises when an employee gets involved in an existing long-term project. The example documentation is taken from a collaborative research project that aims at improving development processes in process engineering. The collection is made up of the documentation of 12 projects, each with various subtasks, so this example incorporates 120 documents.

Assuming that there is an accessible prepared pool of project documentation a new DocMINER project is defined. Then keywords are automatically extracted from the documents by an indexer; the inter-document similarity is assessed by keyword co-occurrence, and the document map is generated (for details cf. Becks 2001). Assume further, that one document is a 'starting' document, which was identified as clearly relevant to the employee's task within the project. That document is selected from the document list within DocMINER and is accordingly highlighted in the map. Additionally, its text is displayed for a quick orientation. From this map, fig. 2, a first document grouping is easily identified. Examples are:

- a group of relative 'similar' documents within the white to light grey 'plateau' at the centre of the map. The 'starting' document lies somewhat distinctly but still within this 'plateau'.
- the documents at the 'eastern' part are grouped internally, but more significantly they are all clearly separated from the other documents by an approximately vertical dark shaped boundary.

838

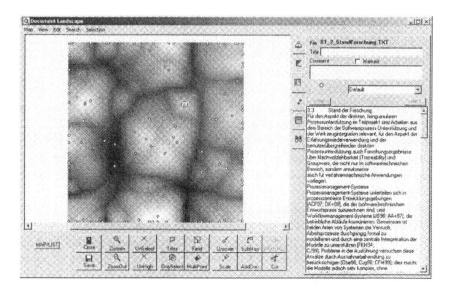

Figure 2: Generated document map with highlighted 'starting' document

- the few documents within the very dark area at the midst of the 'western' part are strongly separated from all other documents.

To get a first overall impression of the collection's topical structure, visually identified groups of documents are selected. The names of the documents and their most characteristic terms (in relation to the other documents) may be displayed directly in the map for an easier assessment. Additionally a term profile, e.g. the most significant terms of selected documents, may be displayed. It follows that the 'central plateau' with the 'starting' document is about supporting the flowsheet editing and evolution process. Figure 3 shows another step where it turns out, that the common theme of the documents in the 'north-eastern' corner is extrusion. Now it is possible to:
- colour-code documents according to discovered or predefined categories, e.g. red-mark all documents in the 'central plateau'
- make a more detailed investigation of document groups, e.g. zooming in the map at densely populated parts
- investigate 'bridging' documents between otherwise distinct document groups
- search for terms and highlight them in the map

For example a search for "extrusion OR extruder" reveals the highlighted documents in fig. 3. One of the retrieved documents lies in the 'central plateau'. Clicking on this document and scanning the displayed text shows that it considers the integration of the extrusion simulation within the flowsheet editing support.

Having gained a good understanding of the document collection a lot is already learned about the project and subproject themes as well as their relationships and inter-

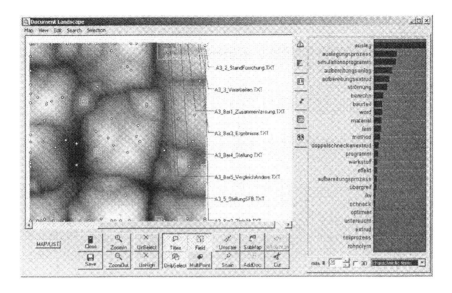

Figure 3: Document map with document title and characteristic terms of selected documents in the map as well as significant term profile in the right hand window. Highlighted documents found by search for "Extrusion OR Extruder".

dependencies. This now allows for a more goal-oriented and effective selection of further reading.

4. Conclusions

Data mining techniques are more and more frequently used on numerical or structured data to discover new knowledge and the benefit of such techniques is well proven. However, knowledge captured in textual documentation is also a very valuable information source for any organization, but methods and tools to explore and exploit such data are less mature.

We identified specific requirements and developed an appropriate tool with a powerful visualization and user-interaction for text mining. The tool was successfully used in many case studies, e.g. analysing usage scenarios for chemical process simulators (Becks et al. 1999) and for condensing and maintaining internal and external technical documentation of foundry simulator manuals (Becks et al. 2000). In the given example session we showed that the tool provides effective access to relevant parts of documentation typical for research in process engineering. The tool enabled the user to easily understand the relationships among the complex material. The benefit in reducing efforts needed during settling into a new project seems obvious, and was additionally confirmed in a laboratory study (Becks 2001).

First contacts are established for a broader industrial evaluation, among others in chemical engineering companies. The modular approach of the tool allows an easy adaptation to specific tasks. For example, modifications to enable an improved mining of "structured" textual data like simulation input files will be assessed. Furthermore, at

Fraunhofer FIT we are actually working on a tight integration of "DocMINER" with various other text access tools in conjunction with information brokering aspects. From this we expect a higher degree of flexibility and advanced adaptive features.

5. Acknowledgements

This work was supported by the DFG in its focused doctoral programme on Informatics and Engineering at RWTH Aachen and within the framework of the collaborative research center "SFB 476 IMPROVE".

6. References

Becks, A., J. Koeller, 1999, Automatically Structuring Textual Requirement Scenarios. Proc. 14th IEEE Int. Conf. on Automated Software Engineering, Cocoa Beach, Florida, USA

Becks, A., 2001, Visual Knowledge Management with Adaptable Document Maps, PhD-thesis, RWTH Aachen, Germany

Becks, A., M. Host, 2000, Visuell gestütztes Wissensmanagement mit Dokumenten-landkarten, Wissensmanagement July 2000 (in German)

Becks, A., C. Seeling, 2001, A Task-Model for Text Corpus Analysis in Knowledge Management. Proc. 8th Int. Conf. on User Modeling, Sonthofen, Germany

Chalmers, M., P. Chitson, 1992, Bead: Explorations in Information Visualization. Proc. 15th Int. ACM SIGIR Conf. on Research and Development in Information Retrieval, Copenhagen, Denmark

Chen, H., Ch. Schuffels, R. Orwig, 1996, Internet Categorization and Search: A Self-Organizing Approach. Journal of Visual Communication and Image Representation, Vol. 7, No. 1

Davidson, G.S., B. Hendrickson, D.K. Johnson, Ch.E. Meyers, B.N. Wylie, 1998, Knowledge Mining With VxInside: Discovery Through Interaction. Journal of Intelligent Information Systems, Vol. 11, No. 3

Meghini, C., U. Straccia, 1996, A Relevance Terminological Logic for Information Retrieval. Proc. 19th Int. ACM SIGIR Conf. on Research and Development in Information Retrieval, Zuerich, Switzerland

Salton, G. (Ed.), 1971, The SMART Retrieval System – Experiments in Automatic Document Processing. Prentice Hall, New Jersey

Shneiderman, B., 1996, The Eyes Have It: A Task by Data Type Taxonomy for Information Visualization. Technical Report 96–66, Institute for Systems Research, University of Maryland

Wise, J.A., J.J. Thomas, K. Pennock, D. Lantrip, M. Pottier, A. Schur, V. Crow, 1995, Visualizing the non-visual: Spatial analysis and interaction with information from text documents. Proc. IEEE Information Visualization 95

European Symposium on Computer Aided Process Engineering – 12
J. Grievink and J. van Schijndel (Editors)
© 2002 Elsevier Science B.V. All rights reserved.

Multilevel Dynamical Models for Polydisperse Systems: a Volume Averaging Approach

Béla G. Lakatos

Department of Process Engineering, University of Veszprém

H-8200 Veszprém, Hungary

Abstract

The paper presents the fundamental elements of multilevel dynamical models of polydisperse systems of chemical engineering. A three-level model is derived and generalised by means of the volume averaging method, modified appropriately for dispersed systems. The model errors and computational aspects are analysed and discussed.

1. Introduction

Solid-fluid heterogeneous systems often are modelled by pseudo-homogeneous models taking into account the effects of processes of the disperse elements by means of effectiveness factors (Datar et al.,1987, Ramkrishna and Arce,1989) When, however, the time scales of the disperse solid and continuous fluid phases are of similar order of magnitude then the pseudo-homogeneous models, especially in dynamic conditions, may lead to significant errors.

A characteristic feature of the solid-fluid systems is the solid and fluid phases exhibit significantly different length scales, mostly differing from each other by some orders of magnitude. This property provides a basis for application of the point sink approximation treating the dispersed elements as point sinks immersed in the environment formed by the continuous phase. Adsorption on porous adsorbents (Ruckenstein,1971, Bathia, 1997, Liu et al.,2000, Quintard and Whitaker,1993) and heterogeneous catalytic reactions in porous catalytic particles (Do and Rice,1982, Burghardt et al.,1988, Keil,1996, Möller and O'Connor,1996) have been modelled in this way, but only few works have been concerned with the special features of this modelling approach. Neogi and Ruckenstein,1980 examined the problem using ensemble averaging, Ramkrishna and Arce, 1989 studied the spectral properties of the operator representing the models, and recently, Lakatos (2001) analysed the problem in terms of multilevel dynamical models, formulated by means of the volume averaging technique.

The aim of the present work is to further develop the concept of multilevel dynamical models, and to discuss the computational aspects and model errors, arising in applying the multiscale volume averaging.

2. Multilevel Models: General Development

Consider the three-phase system shown schematically in Fig.1 in which the interfacial surfaces between the phases are complex and may vary in time. Let the characteristic length scales of the phases, termed α-, β- and γ-phases, respectively, differ from each

842

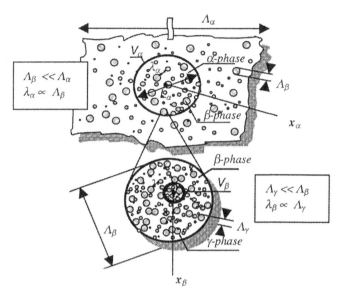

Figure 1. Three-phase system exhibiting three-level spatial hierarchy

other significantly. Then, in relation of the α- and β-phases, the α-phase is considered continuous and the β-phase is termed dispersed, but, in turn, in relation of the β- and γ-phases, the β-phase is the continuous one and the γ-phase is considered dispersed. Let ψ be a scalar quantity which in the phases is denoted by ψ_α, ψ_β and ψ_γ. The variation of ψ inside the phases is described by the balance equations

$$\rho_i \frac{\partial \psi_i}{\partial t} + \nabla \circ (j_i) = \pi_i, \quad i = \alpha, \beta, \gamma \tag{1}$$

where j_i is the flow density and π_i is the volumetric source density of ψ. The transport across the $\alpha\beta$- and $\beta\gamma$- interfaces is described by the boundary conditions

$$n_{ij} \circ (j_i - \rho_i \psi_i w_{ij}) + n_{ij} \circ (j_j - \rho_j \psi_j w_{ij}) = \sigma_{ij}, \quad i, j = \alpha, \beta \quad \text{and} \quad i, j = \beta, \gamma \tag{2}$$

where w_{ij} is the velocity of the ij-interface, σ_{ij} denotes surface source density of quantity ψ on the ij-interface, and n_{ij} is normal unit vector to the ij-interface.
Let us assume that it is possible to define such spatial averaging volumes

$$V_\alpha = \text{constant } L_\alpha^3 \quad \text{and} \quad V_\beta = \text{constant } L_\beta^3, \tag{3}$$

for the α- and β-phases, associated with coordinates x_α and x_β, that the conditions

$$\lambda_\alpha \ll L_\alpha \ll \Lambda_\alpha \quad \text{and} \quad \lambda_\beta \ll L_\beta \ll \Lambda_\beta \tag{4}$$

$$\Lambda_\beta \propto \lambda_\alpha \quad \text{and} \quad \Lambda_\gamma \propto \lambda_\beta \tag{5}$$

are satisfied. Then, following the procedure presented by Lakatos (2001) for a two-level model, the molecular (single) level mathematical model of the system can be converted into a three-level one by means of the modified volume averaging technique. In this case, the phase average $\langle .. \rangle_\alpha$ of the intensive quantity ψ in the α-phase is defined in the usual way (Whitaker,1967, Slattery,1967, Gray,1975)

$$\langle \psi_\alpha \rangle_\alpha (x_\alpha, t) = \frac{1}{V_\alpha} \int_{V_{\alpha\alpha}} \psi_\alpha dV \tag{6}$$

where $V_\alpha = V_{\alpha\alpha} + V_{\beta\alpha}$, $V_{\alpha\alpha}$ and $V_{\beta\alpha}$ are the partial volumes of the α- and β-phases in V_α, respectively. The phase average $\langle .. \rangle_\alpha$ of quantity ψ in the β-phase takes the form

$$\langle \psi_\beta \rangle_\alpha (x_\alpha, t) = \int_0^{V_{\beta\,max}} \langle \psi_\beta \rangle_P n_\beta (V_\beta, x_\alpha, t) V_\beta dV_\beta \tag{7}$$

where $\langle . \rangle_P$ denotes the average of ψ over a β-phase element (particle):

$$\langle \psi_\beta \rangle_P = \frac{1}{V_\beta} \int_{V_\beta} \psi_\beta dV \ . \tag{8}$$

In Eq.(7), the function $n_\beta : R_0^+ \times R^3 \times R_0^+ \to R_0^+$ is called the population density function of the β-particles which in the present case is determined in the following way: n_β is such a function that the equality

$$\int_0^{V_{\beta\,max}} g(V_\beta) n_\beta (V_\beta, t, x_\alpha) dV_\beta = \frac{1}{K} \sum_{k=1}^{K} g(V_{\beta k}) \tag{9}$$

is satisfied for each continuous and bounded function $g(.)$, where K is the number of β-particles. By means of this function $V_\alpha n_\beta (V_\beta, t, x_\alpha) dV_\beta$ expresses the number of particles having volume $(V_\beta, V_\beta + dV_\beta)$ at the moment of time t in the averaging volume V_α associated with coordinate x_α. The spatial averaging $\langle .. \rangle_\beta$ in relation of the β- and γ-phases is derived analogously.

Applying now, in turn, the averaging operators $\langle .. \rangle_\alpha$ and $\langle .. \rangle_\beta$ to Eqs (1)-(2), and taking into account that, because of the relations (3)-(5),

$$\langle \langle .. \rangle_\alpha \rangle_\beta = \langle .. \rangle_\alpha \tag{10}$$

as well as the appropriate volume averaging and general transport theorems, we obtain the following hierarchy of the model equations. The motion of ψ in the α-phase, i.e. on the α-level is described by the equation

$$\langle \rho_\alpha \rangle_\alpha \frac{\partial \langle \psi_\alpha \rangle_\alpha}{\partial t} + \langle \rho_\alpha \rangle_\alpha \nabla \circ \langle j_\alpha \rangle_\alpha - \langle \pi_\alpha \rangle_\alpha = - \int_0^{V_{\beta\,max}} \langle \psi_\beta \rangle_P \frac{dV_\beta}{dt} n_\beta dV_\beta +$$

$$+ \int_0^{V_{\beta\,max}} n_\beta \int_{A_\beta(V_\beta)} \langle j_\beta \rangle_\beta \circ n_\beta dA dV_\beta - \int_0^{V_{\beta\,max}} n_\beta \int_{A_\beta(V_\beta)} \langle \sigma_{\alpha\beta} \rangle_\beta n_\beta \circ dA dV_\beta \tag{11}$$

where the left hand side terms of Eq.(11) describe the variation of quantity ψ in the α-phase, while the right hand side terms describe the changes of ψ due to the variation of the volume of β-particles, the transfer of ψ through the $\alpha\beta$-interface, and the production of ψ by the $\sigma_{\alpha\beta}$ surface source density, respectively. Here, the population density function is determined by the population balance equation

$$\frac{\partial n_\beta}{\partial t} + \nabla \circ (\langle v_\beta \rangle_P n_\beta) + \frac{\partial}{\partial V_\beta} \left(\frac{dV_\beta}{dt} n_\beta \right) = \langle \pi_\beta \rangle_P n_\beta \tag{12}$$

describing the behaviour of β-particles, represented on the α-level as point sinks immersed and moving in the α-phase. Similarly, equations on the β-level are

$$\langle\rho_\beta\rangle_\beta \frac{\partial\langle\psi_\beta\rangle_\beta}{\partial t} + \langle\rho_\beta\rangle_\beta \nabla\circ\langle j_\beta\rangle_\beta - \langle\pi_\beta\rangle_\beta = -\int_0^{V_{\gamma\max}}\langle\psi_\gamma\rangle_P\frac{dV_\gamma}{dt}n_\gamma dV_\gamma +$$

$$+\int_0^{V_{\gamma\max}}n_\gamma\int_{A_\gamma(V_\gamma)}j_\lambda\circ n_\gamma dA dV_\gamma - \int_0^{V_{\gamma\max}}n_\gamma\int_{A_\gamma(V_\gamma)}\sigma_{\beta\gamma}n_\gamma\circ dA dV_\gamma \tag{13}$$

and

$$\frac{\partial n_\gamma}{\partial t} + \nabla\circ\left(\langle v_\gamma\rangle_P n_\gamma\right) + \frac{\partial}{\partial V_\gamma}\left(\frac{dV_\gamma}{dt}n_\gamma\right) = \langle\pi_\gamma\rangle_P n_\gamma. \tag{14}$$

Finally, equation on the γ-level

$$\rho_\gamma\frac{\partial\psi_\gamma}{\partial t} + \nabla\circ\left(v_\gamma\rho_\gamma\psi_\gamma + q_\gamma\right) = \pi_\gamma \tag{15}$$

describes the variation of quantity ψ inside the γ-particles. Here, q_γ denotes some non-convective component of the flow density that may be of complex nature, depending on the structure of particles. Eqs (11)-(15) are completed with the appropriate boundary and initial conditions. The boundary conditions for Eqs (13)-(14) describe the connection of the system with the environment while the boundary conditions for Eq.(15) describe the connection between the internal world of a γ-particle and its continuous phase environment.

3. Model Errors

In developing multilevel dynamical models by using the volume averaging technique some approximations have been made, yielding different errors. Namely, model errors, characteristic for that procedure, are produced by
- averaging nonlinear source functions,
- closing the averaged equations using approximating closure models,
- assuming that particles represented as point sinks are immersed in homogeneous environment given by the average value,
- intersecting particles by the surface of the averaging volume.

For the sake of illustration, Fig.2.a shows what is the difference between the approximated ψ_β and the real $\psi_{\beta real}$ profiles in a β-phase particle immersed an ideal and realistic environment. At the same time, Fig.2.b shows that from the particles intersected those, which have mass centre inside the averaging volume, are accounted for, but those that have mass centre outside do not. A detailed analysis and quantitative predictions of the model errors of this approach will be presented elsewhere.

4. Three-Level Model: Adsorption in Bidisperse Solids

A three level model describes a fixed bed adsorber with bidisperse adsorbents, i.e. in which the adsorbent pellets consist of small porous (micro) particles. Here, the α-level is formed by the gas phase, where the pellets are treated as point sinks, and is described

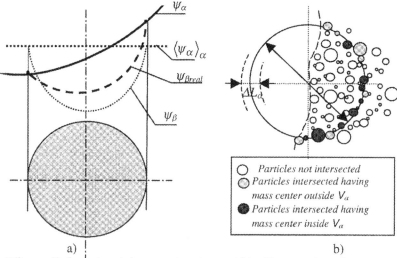

Figure 2. Effects of) the point sink approximation and b) of intersection of some particles by the averaging volume

by the axial dispersion model

$$\frac{\partial c_\alpha(x_\alpha,t)}{\partial t} = D_\alpha \frac{\partial^2 c_\alpha(x_\alpha,t)}{\partial x_\alpha^2} - v_\alpha \frac{\partial c_\alpha(x_\alpha,t)}{\partial x_\alpha} - \frac{3(1-\varepsilon)}{R_\beta \varepsilon} D_\beta \frac{\partial c_\beta(r_\beta,t)}{\partial r_\beta}\bigg|_{r_\beta=R_\beta} \qquad (16)$$

The spherical adsorbent pellets form the β-level, where the microparticles, often distributed in size, are considered point sinks

$$\frac{\partial c_\beta(r_\beta,t)}{\partial t} = \frac{D_\beta}{r_\beta^2} \frac{\partial}{\partial r_\beta}\left(r_\beta^2 \frac{\partial c_\beta(r_\beta,t)}{\partial r_\beta}\right) - \frac{3(1-\varepsilon_\beta)}{R_\gamma \varepsilon_\beta} D_\beta \int_0^{R_{\gamma\,max}} \frac{\partial c_\gamma(r_\gamma,r_\beta,t)}{\partial r_\gamma}\bigg|_{r_\gamma=R_\gamma} n(R_\gamma,t)dR_\gamma$$

$$(17)$$

Finally, the internal world of the microparticles forms the γ-level in this model. The complex structure of microparticles often yields intensive dynamic interactions between the adsorbate and the solid surface (Luikov,1980), so that adsorption proceeds under dynamic conditions, i.e. local relaxation phenomena should also be taken into consideration. Consequently, the transport and adsorption in the micropores is described by the equations involving also the local relaxation time scale τ_a of the adsorption process

$$\frac{\partial c_\gamma(r_\gamma,r_\beta,t)}{\partial t} = \frac{D_\gamma}{r_\gamma^2} \frac{\partial}{\partial r_\gamma}\left(r_\gamma^2 \frac{\partial c_\gamma(r_\gamma,r_\beta,t)}{\partial r_\gamma}\right) - \frac{\partial a(t,c_\gamma,\partial c_\gamma/\partial t)}{\partial t}, \quad a(t) = f(c_\gamma) + \tau_a \frac{\partial c_\gamma}{\partial t}$$

$$(18)$$

Here in Eqs (16)-(18), we used the following notation: c - concentration, x - axial coordinate, r - radial coordinate, R - radius, D - dispersion or diffusion coefficient, v - linear velocity, ε - porosity, a - adsorbed species, f - static adsorption equilibrium.)

The set of nonlinear partial differential equations (17)-(18), with the appropriate initial and boundary conditions, can be solved numerically by using multilevel computational schemes, determined in principle by the hierarchical structure of the model, shown schematically in Fig.3. A three-level orthogonal collocation scheme, has proved suitable for

846

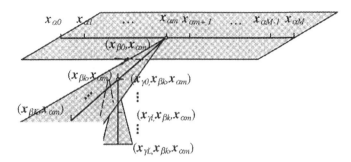

Figure 3. Three-level computational structure corresponding to the hierarchy of the system shown in Fig.1.

solving Eqs (16)-(18), similarly to that developed for catalytic reactors with biporous catalysts (Lakatos, 2001).

Also, this scheme illustrates that systems exhibiting such hierarchy usually are manipulated at the α-level. Then, accounting for the possible motion of the lower level elements and the detailed description of their internal world simultaneously produces contradictory requirements. Particles, due to their independent motion in the environment represented by the higher level, with significantly different internal histories may be present in the same averaging volume which provides a number of difficulties in computations. In such cases, often hybrid, i.e. combination of deterministic and stochastic, computations should be applied.

Acknowledgement
The author would like to thank the Hungarian Research Foundation for the financial supporting this work under Grant *T034406*.

5. References

Bathia, S.K., 1997, Che, Engng Sci., 52, 1377.
Burghardt, A., J. Rogut, and J. Gotkowska, 1988, Chem. Engng Sci. 43, 2463.
Carbonell R.G. and S. Whitaker, 1983, Chem. Engng Sci. 38, 1795.
Datar, A., B.D. Kulkarni and L.K. Doraiswamy, 1987, Chem. Engng Sci. 42, 1233.
Do, D.D. and R.G. Rice, 1982, Chem. Engng Sci. 37, 1471.
Gray, W.G., 1975, Chem. Engng Sci. 30, 229.
Hu, X. and D.D Do, 1995, AIChE Journal, 41, 1581.
Keil, F.J., 1996, Chem. Engng Sci. 51, 1543.
Möller, K.P. and C.T. O'Connor, C.T., 1996, Chem. Engng Sci. 51, 3403.
Lakatos, B.G., 2001, Chem. Engng Sci., 56, 659.
Liu, F., S.K. Bathia and I.I. Abarzhi, 2000, Computers chem. Engng, 24, 1981.
Luikov, A.V., 1980, Heat and Mass Transfer. Mir, Moscow.
Neogi, P. and E. Ruckenstein, 1980, AIChE Journal, 26, 787.
Quintard, M. S. and Whitaker, 1993, Chem. Engng Sci. 48, 2537.
Ramkrishna, D and P. Arce, 1989, Chem. Engng Sci. 44, 1949.
Ruckenstein, E., A.S. Vaidyanathan and G.R. Youngquist,1971,Chem. Engng Sci.,1305.
Slattery, J.C., 1967, AIChE Journal, 13, 1066.
Whitaker, S., 1967, AIChE Journal, 13, 420.

European Symposium on Computer Aided Process Engineering – 12
J. Grievink and J. van Schijndel (Editors)

Open Software Architecture For Process Simulation: The Current Status of CAPE-OPEN Standard

Jean-Pierre Belaud

Laboratoire de Génie Chimique (LGC, UMR CNRS 5503), Equipe Analyse
Fonctionnelle des Procédés, INPT-ENSIACET
118 route de Narbonne, 31077 Toulouse Cedex 4, France
JeanPierre.Belaud@ensiacet.fr

Michel Pons
ATOFINA Centre Technique de Lyon
BP 32 F-69492 Pierre-Bénite Cedex, France
Michel.PONS@atofina.com

Abstract

Traditionally simulation environments have been closed monolithic systems and the resulting bottlenecks in interoperability, reuse and innovation have led to the search for a more open and interoperable solution. The CAPE-OPEN (CO) effort, launched in January, 1997, is a standardisation process for achieving true plug and play of process industry simulation software components. The resulting CO standard is now being widely disseminated to the chemical engineering community. It relies on a technology that integrates up to date concepts from the software field such as a component-based approach. A number of software components based on this technology have been developed and are already available. Thanks to this new generation of CAPE tools, it is expected to reach cheaper, better and faster design, operation and control of processes. The CAPE-OPEN Laboratories Network (CO-LaN) consortium is in charge of managing the lifecycle of the CO standard.

1. Introduction and Objectives

The integration of external know-how in traditional simulation environments has to deal with proprietary code, and requires painful and expensive activities, while open software architectures are the way forward for the next generation of CAPE tools as demonstrated by (Braunschweig et al., 1999). The CO standard is the result of a world-wide collaboration between chemical and petroleum refining industries, academics, and CAPE software suppliers, all with a common goal of defining a standard for a component-based approach to process simulation. Thus any compliant component can be integrated instantly in any compliant environment. The standard is open, multi-platform, uniform and available free of charge.

Version 1.0 of the CO standard was released in February 2002. This paper explains how it will be subsequently managed by the CO-LaN. The associated technology and the scope of this standard are then described. Finally we present some CO-compliant commercial and non-commercial implementation of this technology.

2. CAPE-OPEN Laboratories Network

The (CO-LaN, 2001), a non-for-profit organisation, maintains the CO standard, disseminates information on the standard, releases software tools to test CO compliance and publishes interoperability test results. With more than 20 members in year 2001 (operating companies, vendors and academics), it provides a service to the CAPE community in all aspects of the CO standard. The bylaws of the CO-LaN ensure that it is open to the entire CAPE community for access and membership. Membership is definitely not required to use CO compliant software components or to develop such components. However the CO-LaN organises the CAPE community with respect to the CAPE-OPEN standard.

The CO-LaN operates through a web portal where visitors find or will soon find the CAPE-OPEN documentation, as well as additional resources for implementing and using CO compliant components (FAQ's, discussion board, how-to's, software migration support, etc.). Members of the CO-LaN find the services that help them to develop new standard interfaces or improve existing ones, as well as dedicated help for implementing CO compliant software components.
Special Interest Groups will be created and maintained by the CO-LaN for refining or extending the CO standard, following a careful analysis of the value creation brought to users and to vendors by each development and improvement. The SIGs will be managed in such a way as to bring these developments quickly to the market, updating as necessary the documentation and the tools delivered to the CAPE community.

3. CAPE-OPEN Technology

Key elements of CO technology are openness, interoperability, standardisation process and service. This technology integrates today's concepts in software domain, development tools and web enabled skills.
It embodies the *CO formal documentation set*, the *CO architecture*, and *the CO system model*. We focus below on the current potential for process simulation development engineers, so the *CO system model* is detailed while the *CO formal documentation set* and *the CO architecture* are only introduced.

3.1 CO formal documentation set
The CO standard follows a specific versioning system using a unique and global version number and is composed of a set of documents. These documents are organised according to a *CO formal documentation set*. This documentation set includes six blocks: *General Vision, Technical Architecture, Business Interfaces, COSE Interfaces, Common Interfaces and Implementation Specifications*.
- *General vision* contains documents that should be read first to get the standard general information, such as general requirements and needs.

- *Technical architecture* integrates the horizontal technical materials and defines an infrastructure for a process simulation based on the CO standard.
- *Business interfaces* contain all vertical interface specification documents. These interfaces are domain-specific interfaces for the CAPE application domain. They define CO components involved in a CO process simulation application.
- *COSE (CAPE-OPEN Simulator Executive) Interfaces* refer to horizontal interface specifications. They are interfaces for simulation environments such as simulator executives. Within this category, services of general use are defined such as *diagnostics* and *material template system* in order to be called by any CO components through a call back usage.
- *Common interfaces* enclose horizontal interface specification documents for handling concepts that may be required by any *Business* and *COSE interfaces*. This is a collection of interfaces that support basic functions and are always independent of *Business* and *COSE Interfaces*.
- *Implementation Specifications* contain the implementation of the *Business*, *COSE* and *Common Interfaces* specifications for a given distributed computing platform. All documents from *Business*, *COSE* and *Common Interfaces* are abstract specifications which create and document a conceptual model in an implementation neutral manner. Thus the design of CO is independent from any computing platform. It has the ability to be extended to any platform. The *Implementation Specifications* are available for (D)COM and CORBA through the Interface Definition Language libraries. In order to produce CO compliant software components, any software developer has to use these official libraries.

3.2 CO architecture

The *CO architecture* elements describe technical objectives and terminology and provide the infrastructure upon which supporting *Business*, *COSE* and *Common Interfaces* are based. This identifies the technologies associated with the CO standard, includes the object model which defines common semantics, and shows the reference model which embodies the CO interfaces categories, CO (compliant) software components and communication mode. That is based on the distributed component (heterogeneous) system and the object-oriented paradigm. The involved technologies are the UML notation (Rumbaugh et al., 1997), the OMG CORBA (OMG, 2001) and Microsoft (D)COM (Microsoft, 2001) middleware, as well as the Unified Processes and object-oriented programming languages (Meyer, 2001). The wide-scale industry adoption of this CO architecture provides application developers and end-users with the means to build web-enabled interoperable simulation software systems distributed across all major hardware, operating system and programming language environments.

3.3 CO system model

The *CO system model* represents the UML design model of standard. It defines the scope. The physical view of this model allows extraction of the CO software components and shows their dependency relationships. The logical view of this model organises the services, identifies the CO packages and CO interfaces, and designs the related structural organisation. The standard distinguishes two kinds of software components: *Process Modelling Components (PMCs)* and *Process Modelling*

Environments (*PMEs*), the latter making use of the services provided by the *PMCs*. Typically the *PMEs* are environments that support the construction of a process model and that allow the end-user to perform a variety of different tasks, such as process simulation or optimisation (Pantelides et al., 1995). Among the standardised *PMCs* there are: *Thermodynamic and Physical Properties*, *Physical Properties DataBases*, *Unit Operations*, *Numerical Solvers* and *Sequential Modular Tools*.

- *Thermodynamic and Physical Properties* component: In the area of physical properties, CO focuses on uniform fluids that are mixtures of pure components or pseudo-components, and whose quality can be described in terms of molar composition. The physical properties operations that have been provided with standardised interfaces are those required for the calculation of vapour-liquid or liquid-solid equilibria or subsets thereof, as well as other commonly used thermodynamic and transport properties. A key concept is that of a Material Object. Typically, each distinct material appearing in a process (in streams flowing between unit operations, as well as within individual unit operations) is characterised by one such object. Each unit operation module may interact with one or more Material Objects. To support the implementation of the above framework, the CO standard defines interfaces for Material Objects as well as for thermodynamic property packages, calculation routines and equilibrium servers.

- *Unit Operation* component: CO defines a comprehensive set of standard interfaces for unit operation modules being used within modular and steady-state *PMEs*. A unit operation module may have several ports that allow it to be connected to other modules and to exchange material, energy or information with them. In the material case (which is also the most common), the port is associated with a Material Object. Ports are given directions (input, output, or input-output). Unit operation modules also have sets of parameters. These represent information that is not associated with the ports but that the modules wish to expose to their clients. Typical examples include equipment design parameters (e.g. the geometry of a reactor) and important quantities computed by the module (e.g. the capital and operating cost of a reactor).

- *Numerical Solvers* component: As explained by (Belaud et al., 2001a) the CO standard focuses on the solution algorithms that are necessary for carrying out steady-state and dynamic simulation of lumped systems. In particular, this includes algorithms for the solution of large, sparse systems of non-linear algebraic equations (NLAEs) and mixed (ordinary) differential and algebraic equations (DAEs). Algorithms for the solution of the large sparse systems of linear algebraic equations (LAEs) that often arise as sub-problems in the solution of NLAEs and DAEs are also considered. The CO standard introduces new concepts, such as models and the equation set object (ESO), which is a software abstraction of a set of non-linear algebraic or mixed (ordinary) differential and algebraic equations. The standard ESO interface enables access to the structure of the system, as well as to information on the variables involved. The equations in any model may involve discontinuities. Discontinuous equations in a models are represented as state-transition networks (Avraam et al., 1998).

- *Sequential Modular Specific Tools* component: A key part of the operation of sequential modular simulation systems is the analysis of the process flowsheet in order to determine a suitable sequence of calculation of the unit operation modules (Westerberg et. al., 1979). Thus, typically the set of units in the flowsheet is partitioned into one or more disjoint subsets (maximal cyclic networks, MCNs) which may then be solved in sequence rather than simultaneously ("ordering"). The units within each MCN are linked by one or more recycle loops which are converged iteratively via the identification of appropriate "tear streams". The above tasks are typically carried out using a set of tools that operate on the directed graph representation of the flowsheet. The CO standard defines standard interfaces for the construction of these directed graphs, and for carrying out partitioning, ordering, tearing and sequencing operations on them.

- *Physical Properties Databases* component: CO defines how a database of recorded physical property values and model parameters can be connected to flowsheeting and other engineering programs. This interface deals with physical property data at discrete values of the variables of state (temperature, pressure, composition), as far as measured, correlated or estimated values are concerned.

4. Delivering Components

Software component developers are working either to bring new CO compliant products to the market place or to make existing software components CO compliant. In either case, these software components can be for commercial sale, for proprietary use within an organisation, or for proprietary delivery to a specific client. Commercial software such as detailed by (Belaud et al., 2001b) and non-commercial software based on the CO standard are already available. Only a few are listed in order to illustrate results and potentials of the CO standardisation effort.

4.2 Process modelling environment development
The CO technology is now delivered in commercial process simulation software: Hyprotech has developed the HYSYS Unit and Thermodynamic CO sockets, which make HYSYS.Process/Plant version 2.2 (and subsequent versions) a CO compliant *PME*. AspenTech has implemented a socket for CO *Thermodynamic and Physical Properties* components in Aspen Plus 10.2 and Aspen Properties 10.2 (and subsequent versions i.e. 11.1). Aspen Plus 10.2 also implements a socket for CO *Unit Operation* components. Process Systems Enterprise has released a new version of their gPROMS tool with a CO Thermodynamic socket. BELSIM SA has done the same for their VALI III data reconciliation tool.

4.3 Process modelling components development
Several *PMCs* are already implemented and many more are being developed. Some are only prototypes or for internal use only (for example ProSim SA delivered, exclusively to TotalFinaElf, a CO compliant thermodynamic server) while others are releases to the CAPE marketplace. In its CAPE-OPEN kit, for demonstration purposes, Hyprotech is distributing one Property Package and two Unit Operations which are CO compliant. AspenPlus can be used to create new CO physical property packages, which can be integrated in any CO compliant *PME*. Infochem has made its MultiFlash tool CO

compliant and has successfully tested it with Hysys.Process, AspenPlus and gProms. From the academic side, (Belaud et al., 2001a) from LGC-INP Toulouse institute have developed a CO-compliant *Numerical Solvers* component called Numerical Services Provider which supplies LAE, NLAE and DAE objects and acts as a real framework that makes up a reusable design for disseminating any solver algorithm through the CO standard. Furthermore, INPT has demonstrated a *Sequential Modular Specific Tools* PMC.

5. Conclusion

The CO standard gives process engineers more flexible process modelling tools by allowing simulation with software components from multiple sources, assembled easily in a simulation environment. Any CO compliant software component can be integrated in any CO compliant simulation environment by "plug and play". The CO standard benefits software component developers by increasing the usage of CAPE tools and reducing the development time thanks to the CO technology. This technology is based on the distributed component heterogeneous system and modern software development techniques. A non-for-profit organisation, CO-LaN, promotes and maintains the standard.

References
Avraam M., N. Shah and C.C. Pantelides, 1988, Modelling and. Optimisation of General Hybrid Systems in the Continuous Time Domain, Comput. Chem. Engng., 22S, S221-S228, 1998
Belaud J.P., K. Alloula, J.M. Le Lann and X. Joulia, Open software architecture for numerical solvers, 2001, Proceedings of European Symposium on Computer Aided Process Engineering-11, Elsevier, pp 967-972, 2001.
Belaud J.P., B. L. Braunschweig, M. Halloran, K. Irons and D. Piñol, 2001, New generation simulation environment: Technical vision of the CAPE-OPEN standard, Presentation at the AIChE Annual Meeting, Reno, NV, Nov 4-9, Modelling and Computations for Process Design, 2001.
Braunschweig B. L., C. C. Pantelides, H. I. Britt and S. Sama, Open software architectures for process modelling: current status and futures perspectives 1999, FOCAPD'99 Conference, Breckenridge, Colorado, 1999.
CO-LaN, 2001, CAPE-OPEN Laboratory Network web site: www.colan.org
Meyer B., 1997, Object-Oriented Software Construction, 2nd edition, Prentice Hall, 1997
Microsoft, 2001, Microsoft web site about COM: www.microsoft.com/com
OMG, 2001, Object Management Group web site about CORBA: www.corba.org
Pantelides C.C. and H. I. Britt, 1995, Multipurpose Process Modeling Environments. In L.T. Biegler and M.F. Doherty (Eds.), Proc. Conf. on Foundations of Computer-Aided Process Design '94. CACHE Publications, Austin, Texas. pp.128-141, 1995.
Rumbaugh J., I. Jacobsen and G. Booch, 1997, Unified Modeling Language Reference Manual, Addison Wesley, 1997.
Westerberg A.W, H.P. Hutchinson, R.L. Motard and P. Winter, 1979, Process Flowsheeting. Cambridge University Press, Cambridge, U.K, 1979.

European Symposium on Computer Aided Process Engineering – 12
J. Grievink and J. van Schijndel (Editors)
853

An Open Software Architecture for Steady-State Data Reconciliation and Parameter Estimation

Chouaib Benqlilou[1], Moisès Graells[2], Antonio Espuña[1] and Luis Puigjaner[1*]
Universitat Politècnica de Catalunya, Chemical Engineering Department,
[1]ETSEIB, Av. Diagonal 647, E-08028 – Barcelona (Spain)
[2]EUETIB, Comte d'Urgell, 187, E-08036 – Barcelona (Spain)
E-mails: [chouaib, graells, aec, lpc]@eq.upc.es

Abstract

In this paper a flexible and open architecture design for Parameter Estimation and Data Reconciliation (PEDR) software application is proposed by de-coupling it according to the functionalities involved. In the proposed approach the different components that are involved in this application and their interactions are specified and tested. The proposed architecture aims at an improved efficiency and upgrading of the PEDR application by allowing the exchangeability and connectivity of the present components in an easy and consistent way.

1. Introduction

Accurate process measurement is increasingly vital for modelling, monitoring and optimisation of chemical and related processes. Once gross errors are filtered, accurate process data may be obtained using Data Reconciliation (DR), a technique that "corrects" measured data by minimising the variance with respect to a process balance model. The same procedure may lead to model adjustment by determining the best model parameter values matching a set of reliable data. Thus, Parameter Estimation (PE) and DR are based on optimisation techniques that require the following types of elements: mathematical models of the process or processes, sets of process variable measurements and/or reconciled data and mathematical solvers.

Being an aggregation of different technologies, a PEDR tool should not be conceived as a monolithical application since it will not be possible to exchange one component without affecting the performance of others. This could be a serious drawback for taking full advantage of a growing number of tools, such as equation-based simulators, optimisation algorithms, databases, Distributed Control System (DCS), etc. Therefore, a flexible design requires all the PEDR components to be as de-coupled as possible.

This work introduces an open modular architecture for a PEDR software executive allowing the management of different sets of process data, process models and available mathematical solvers through standardised interfaces. This leads to the exchangeability of the corresponding software components. Furthermore, the communication between the corresponding objects has been established using Common Object Request Broker Architecture (CORBA) as middleware in order to handle data across the network.

Case studies are provided showing the great flexibility of the system for handling different plant structures and situations by matching various predefined or newly introduced process models with the corresponding available data sets. Moreover, the system could be also upgraded by plugging new optimisation solvers to it, when available.

2. The system architecture

2.1 Multi-module vs. monolithical applications

The performance of numerical solvers and process modeling tools has a direct impact on the results of Data Reconciliation and Parameter Estimation. Thus, the continuous improvement of such tools requires a flexible and modular structure for a PEDR system so that upgrading of this system may be easily achieved by replacing old modules. Since, these modules could be supplied from different providers, they may present a possible software and hardware incompatibility. Therefore, their integration and incorporation in a monolithical structure leads to high effort both, in cost and in implementation. These drawbacks may be overcome by using separated components that inter-operate through a well-specified interface. The use of standardised interfaces for open communication between software components has emerged as a promising challenge for the software application incompatibility (CAPE-OPEN, 2000). Currently, different vendors and industrial companies are already incorporating the CAPE-OPEN interfaces. By using the results of CAPE-OPEN, it is relatively easy to include and integrate modules from various suppliers and in a heterogeneous environment.

2.2 Architecture design

2.2.1 Modules

The monolithical application should be de-coupled as far as possible for an open and flexible architecture in such a way that each one of the present components could be replaced without affecting the application performance. The PEDR system can be functionally de-coupled into five main components: the PEDR Manager, process model, sets of process or reconciled measurements, mathematical solvers, and the PEDR client component as is shown in Fig. 1.

1. The main purpose of the PEDR Manager module is to gain access to the measured data and to the process model description in order to build up the optimization problem by interacting among the required modules (e.g. database and solvers). Furthermore, it is the responsible for variable classification, estimation of unmeasured observable variables and Gross Error detection (GED) tasks.
2. The main purpose of the process model component is to generate a model description of the process under consideration. It is important to notice that only steady-state processes are considered. The fundamental building block employed for this purpose is an equations-based model. However, any other available model (black box, simulation software package) may be used.

3. The process data module is responsible for acquiring measured process variable values for on-line as well as for off-line applications. These process variable data are often stored intermediately in a relational database or obtained directly from a DCS.

4. The mathematical solver module is responsible for the resolution of the DR or PE problems. The solver may be accessed by direct methods or via file (e.g. MPS, MathML...).

5. The PEDR Client component prepares the output generated by the PEDR Manager for easy use by the customer.

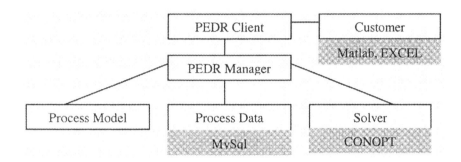

Figure 1: Architecture design for PEDR.

2.2.2   *Modules interaction*

In the first step, the database, the solver and the PEDR Manager are registered in order to make them accessible to their clients (there can be more than one client placed remotely). Once the servers are registered, the client component contacts and initiates the communication with the PEDR manager component through a specified interface. The client component initiates the PEDR Manager by asking it to reconcile/estimate a given set of process variables supplying the references of the following information: process measurements, the corresponding process models and the appropriate solvers. Then the PEDR manager accesses the process measurement and the process model, prepares the optimisation problem, and interacts with the solver to obtain the optimisation results that are finally offered to the client (Fig. 2).

3. Implementation

Component software and object-oriented approaches, which view each component in the above architecture as a separate object, were adopted. All the communications between objects are handled by CORBA and implemented in Java (Orfali *et al*, 1998). The sequence diagram (Fig. 2) represented in Unified Modeling Language (UML) shows the temporal sequence of steps to be followed in order to perform the DR or the PE.

Test prototype software has been developed to demonstrate the use and benefits of the proposed component architecture and the specification of the open interface proposed.

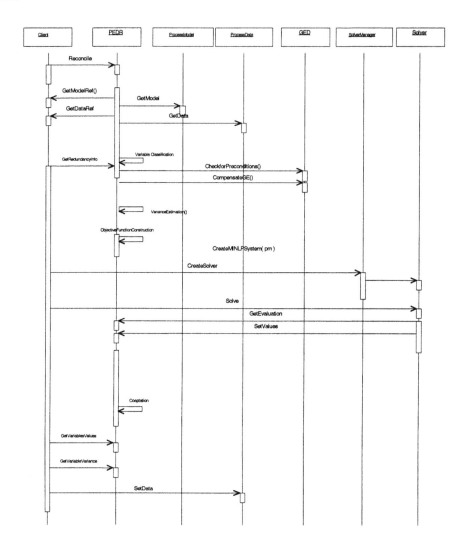

Figure 2: The sequence diagram.

4. Interfaces

4.1 The Model and Solver interfaces

For the sake of standardization, the interfaces to communicate the PEDR Manager with the solver and model modules should follow the CAPE-OPEN specifications (CAPE-OPEN, 1999). The process model interface specification assumes that the process model is described by a set of continuous equations, while the solver interface specification assumes a mathematical programming problem.

4.2 The database interface

This is not a general-purpose interface for process measurement data but only a specific one according to the requirement of a PEDR module. The proposed interface permits both to introduce and to retrieve data. This data could be measured, reconciled or estimated, and are grouped into sets of experiments (corresponding to experiments on the plant in different operating conditions).

4.3 The PEDR Manager interface

Finally, despite its internal modularity, PEDR manager had to expose a common interface to be invoked by any external client. Fig. 4 shows the interface in UML that is being proposed within the GLOBAL-CAPE-OPEN project.

Furthermore, the PEDR Manager provides a graphical and user-friendly interface (Fig. 3) designed according to the methods that the Manager exposes. First of all, a PEDR Client can choose to perform either a DR or a PE task. Then, it selects the measured data to be reconciled or used for parameter estimation, the required mathematical model to be used and the appropriate solver for solving the resulting optimization problem. Finally, the Client could ask the system to solve the problem.

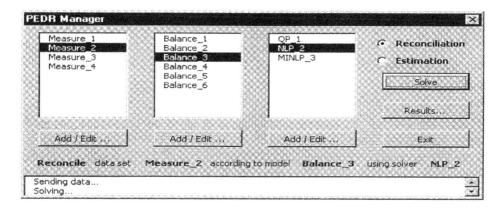

Figure 3: The Graphical User Interface for the PEDR Manager.

Figure 4: The PEDR Manager interface.

5. Conclusions

The main objective of the presented work has been the design and implementation of a software architecture for distributed Data Reconciliation and Parameter Estimation applications. The task of maintaining and supporting new process modelers, databases and/or solvers within the PEDR packages could be justified in cost and implementation using the proposed open software architecture. In this work the specification of the Data reconciliation and Parameter Estimation interfaces are conceived for steady state. However, the general approach adopted makes them likely to be of application for the dynamic case. Thus, the extension to the dynamic case can be implemented with relatively low effort without making any essential change to the other components.

Acknowledgment

Financial support received from European community (project Global-Cape-Open IMS 26691) and the Spanish "Ministerio de Educación, Cultura y Deporte" (project REALISSTICO QUI99-1091, CICYT) are gratefully acknowledged.

6. References

CAPE-OPEN 2000, Conceptual Design Document (CDD2) for Global-CAPE-OPEN Project. Available on the Word Wide Web at the URL http://www.global-cape-open.org/.

CAPE-OPEN 1999, Open interface specification (Numeric Solvers) for-CAPE-OPEN Project. Available on the Word Wide at the URL http://www.global-cape-open.org/CAPE-OPEN_standard.html.

Orfali,. R. and D. harkly. Client/Server programming with java and CORBA. John.Wiley & Sons, Inc., 1998.

European Symposium on Computer Aided Process Engineering – 12
J. Grievink and J. van Schijndel (Editors)
© 2002 Elsevier Science B.V. All rights reserved.

TRIZ and the evolution of CAPE tools
From FLOWTRAN® to CAPE-OPEN¬® and beyond

Bertrand Braunschweig, Institut Français du Pétrole, 1 & 4 avenue de Bois
Préau, 92852 Rueil Malmaison Cédex, France, bertrand.braunschweig@ifp.fr
Kerry Irons, The Dow Chemical Company, Engineering Sciences, 1400 Bldg, Midland
MI 48667, USA, ironsk@dow.com

Abstract

The paper looks at trends in evolution of software tools with a particular emphasis on
CAPE applications. Using results from TRIZ (Theory of Inventive Problem Solving), we
show why and how CAPE software follows a major trend towards distributed adaptive
heterogeneous components. Consequently, and thanks to the CAPE-OPEN
communication infrastructure, we show that CAPE models can be made of distributed
autonomous agents that we nickname "cogents".

1. Introduction

In the middle of the 20th century, Genrich Altshuller, a Russian engineer, analysed
hundreds of thousands of patents and scientific publications. From this analysis, he
developed TRIZ, the theory of inventive problem solving, together with a series of
practical tools for helping engineers solve technical problems. Among these tools and
theories, the substance-field theory gives a structured way of representing problems, the
patterns of evolution show the lifecycle of technical systems, the contradiction matrix
tells you how to resolve technical contradictions, using the forty principles that describe
common ways of improving technical systems. For example, if you want to increase the
strength of a device, without adding too much extra weight to it, the contradiction matrix
tells you that you can use principle 1, segmentation, or principle 8, counterweight, or
principle 15, dynamicity, or principle 40, composite materials. TRIZ is now used by a
wealth of Fortune 500 companies in support of their innovation processes.

When Atshuller developed TRIZ, he could not think of software components. These
objects just did not exist. But software components are technical objects too. They are in
fact some of the most complex technical objects produced by man. Many of the TRIZ
tools and theory elements only relate to concrete objects (e.g. principle 11, prior
counteraction, or principle 32, change the colour). But some of the principles can be
applied to software. Among these, a famous one is principle 26, copying : a simplified
and inexpensive copy should be used in place of a fragile original or an object that is
inconvenient to operate. Another one, principle 7, nesting or « matrioshka », introduces
the modular approaches to software development.
We really like two principles, 1: Segmentation, and 15: Dynamicity, which are also
considered to be patterns of evolution because of their importance. Segmentation shows

how systems evolve from an initial monolithic form into a set of independent parts, then eventually increasing the number of parts until each part becomes small enough that it cannot be identified anymore, such as in a powder. Further evolution based on this principle leads to similar functions obtained with liquids, gases or fields. Think of a bearing with balls suspension, replaced by microballs, then by gas suspension and finally by magnetic field.

Dynamicity introduces flexibility and adaptation by allowing the characteristics of an object, of an environment, or of a process, to be altered in order to find an optimal performance at each stage of an operation. Think of a traffic light that adapts its period depending on the traffic flow.

If you look around you, you will find examples of segmentation and dynamicity in technical objects. Here are a few in process engineering and in computer science:
- Segmentation : process operating companies are now experimenting with *microscale processes* that fit on a chip or on a PC board;
- Dynamicity : *feedback control loops* are everywhere in process plants; there are multilevel controllers too;
- Dynamicity and segmentation : what is now called "*mass customisation*" with end products being adapted towards the needs of their individual customers;
- Dynamicity : simulated counter-current processes operating in *non-equilibrium mode*;
- Segmentation: mainframe computers were replaced by *networks of smaller computers*, and then by clusters of PCs;
- Segmentation and dynamicity : in numerical modelling, *local adaptive grid methods* allow precise and efficient modelling and simulation of unit operations.

2. The evolution of CAPE tools

CAPE software, as other technical objects, follows the TRIZ trends of evolution by becoming segmented and dynamised. We present the first three stages of evolution.

- Stage 1: Monolithic to FORTRAN subroutines and modular architectures

The first CAPE software developed in the sixties and seventies were large monolithic systems. Developed in FORTRAN, they were designed as large multipurpose programs sharing data through COMMON declarations and using internal or external subroutines. Modular programming helped in facilitating maintenance and debugging, so it was quickly adopted. In-house code such as IFP's PGGC[1], BP's Genesis, or commercial software such as FLOWTRAN are examples of stage 1 of evolution.
They remained as such until recently when developers started to cut those systems into smaller pieces that would fit together.

- Stage 2: from object-oriented tools to component architectures

Modularity, object-oriented programming, component software, and n-tier architectures are the current paradigm for CAPE software development, and can be considered as the

---

[1] Programme Général de Génie Chimique

second stage of evolution. Although there are other component-based architectures, The CAPE-OPEN interoperability architecture, based on object orientation and middleware, is the best representative of this stage, as it appears to be the dominant one being adopted by a majority of players. CAPE-OPEN is now accepted as a standard for communication between simulation software components in process engineering. This leads to the availability of software components offered by leading vendors, research institutes, and specialized suppliers which enable the process industries to reach new quality and productivity levels in designing and operating their plants. See (Belaud et al., 2002) for more information.

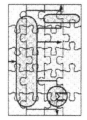 It is also important to mention the CAPE-OPEN Laboratories Network (CO-LaN) and its public catalog of components. CO-LaN, a non-profit society, was created by the Global CAPE-OPEN (GCO) project to become the internationally recognised testing and process management organisation for the CAPE-OPEN standard, and more generally to encourage the use of CAPE software tools in industry, administration and academia. More specifically, the main activities of the CO-LaN are to disseminate, to maintain and manage evolution of the specifications, to facilitate component and interoperability testing, and eventually to keep a public catalog of compliant components. Information on each CO software component is stored in a database which can be searched by visitors of the CO-LaN Web site (www.colan.org). The database can be searched, for example, by type of calculation performed by a piece of software, thus enabling a process engineer to find the right tool to perform any piece of process engineering work, while being sure that the found piece of software will plug easily in his/her CAPE-OPEN compliant simulation environment. Through the component directory, the CO-LaN will provide a mean for buyers and sellers to meet as well as an appropriate classification of software tools.

- Stage 3: from dynamic components to software agents
The third stage will be the one of dynamicity, as the needs for self-adaptation become increasingly important, in order to match the increasing diversity in usage. Self-adaptation can be obtained using current software technologies, such as Enterprise Java Beans and web services, which allow software components to discover their environment at runtime and to seamlessly integrate within these environments.

Current architectures, even though they allow distributed computing on heterogeneous hardware platforms, share the same paradigm for control and co-ordination: a central piece of software controls and co-ordinates execution of all software modules and components that together constitute the model and the solving mechanism of a system. One example is the central piece of software that is usually called the "simulation executive", or "COSE" in CAPE-OPEN architectures. Its tasks are numerous: it communicates with the user; it stores and retrieves data from files and databases; it manages simulation cases; it helps building the flowsheet and checks model topology; it attaches physical properties and thermodynamic systems to parts of the flowsheet; it manages the solving and optimisation algorithms; it launches and runs simulations; etc. All other modules (e.g. unit operations, thermodynamic calculation routines, physical property data sets, solvers and optimisers, data reconciliation algorithms, chemical kinetics, unit conversion systems etc.) are under control of the simulation environment and communicate with it in a hierarchical manner, as disciplined soldiers will execute their assignments and report to their superiors.

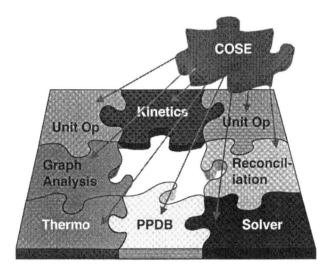

Figure 1: a COSE-centric architecture

3. Beyond CAPE-OPEN: COGENTS

- co.gent (-jɔnt): n. pl. [CSG. <CO-, abbrev. of CAPE-OPEN, + -gents, IST. agents] , new generation of simulation software based on cognitive and distributed heterogeneous software components.

Future CAPE tools will involve distributed architectures based on multi-agents technology where control and co-ordination are decentralised. Instead of the current approach, each piece of software, each module, each component (generically called "agents") lives its own life, able to negotiate and co-ordinate with other components in order to solve problems such as process design, fault diagnosis, or supply chain management. CAPE-OPEN interfaces can be extended towards decentralised architectures of adaptive process modelling agents nicknamed "COGENTS" (CAPE-OPEN Agents). These adaptive cogents will implement the third stage of CAPE tools, following the evolution process defined by Altshuller in TRIZ.

We describe interaction modes with cogents. An important piece of the system is the definition of an ontology of process modelling components. We refer the reader to publications on ontologies in CAPE, e.g. (Batres et al., 2001). We assume that such an ontology is defined in an XML Document Type Definition called CAPE-ML, such as proposed in (von Wedel, 2000).

A user will describe the simulation problem using the CAPE-ML language and interface, communicating with an enhanced COSE which understands CAPE-ML, knows the CO-LaN component catalogue, and is able to discuss with the individual cogents. The COSE will look for possible components in the library, then will start discussing with all potentially useful components - they will probably communicate with each other too - and establish one or several networks of cogents in order to solve the modelling problem. The COSE will then eventually run these networks, confront them with real process data, and possibly choose among the networks of cogents, following a quality criterion.

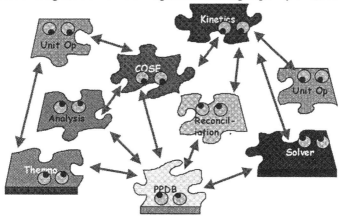

Figure 2: A distributed architecture

Such an agent-based distributed system will enable interoperability, inter-working, openness and integration of CAPE applications and services across platforms. This will enable businesses and organisations to deploy agile and integrated systems in support of the development of new value chains. Of course, the cogents concept demands significant evolution in general computer hardware and software capabilities. The requirements for added computing power to effectvely implement the cogents concepts are perhaps obvious. A (r)evolution in operating systems and/or software interface standards is necessary to allow the use of smart software as proposed for cogents.

4. Conclusion

The scenario that we draw will become real only in a few years time. But we can start working on it. Companies and research organisation who take this new step will gain a competitive advantage and be ready to offer Application Service Provider (ASP) CAPE services taking advantage of the current software technologies and of the many combinations offered by the internet.

Acknowledgements

Many thanks to Rafael Batres, Jean-Pierre Briot, Alexis Drogoul, Eric Fraga, Zahia Guessoum, Wolfgang Marquardt, Didier Paen, Pascal Roux, Sergi Sama, Philippe Vacher, Lars von Wedel, Aidong Yang, for their contributions to COGENTS ideas. Some of the ideas in this paper are developed in the new EC-funded IST project, "Cogents: Agent-Based Architecture For Numerical Simulation", contract IST-2001-34431.

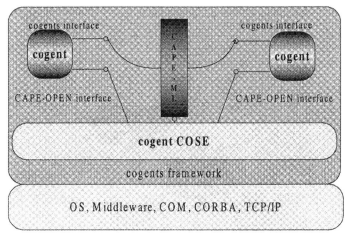

Figure 3: a possible cogents architecture;
note that cogents can be either CAPE-OPEN components wrapped with agent facilities,
or autonomous agents acting on behalf on CO components.

5. References

Altshuller G. (1954) 40 Principles : Triz Keys to Technical Innovation, Uri Fedoseev, Steven Rodman (Translator), Lev Shulyak (Translator)

Batres, R. et al (2001), A life-cycle approach for model reuse and exchange, ESCAPE-11 conference, Kolding, Denmark

Belaud J.-P., Braunschweig B., Pons M., (2002) Open Software Architecture For Process Simulation : The Current Status of CAPE-OPEN Standard, in this ESCAPE-12 conference.

Braunschweig B., (2000) Architectures ouvertes pour l'ingenierie de procedes: le standard CAPE-OPEN, rapport Arago 28, Observatoire Français des Technologies Avancées, Oct. 2000

Braunschweig B., Gani R. (editors), (2002) Software Architectures and Tools for Computer Aided Process Engineering, Elsevier (in print).

Briot J.-P., Demazeau Y., (2001) « Principes et Architecture des systèmes multi-agents», Hermès Science Publications, collection IC2, in print.

CO-LaN, CAPE-OPEN Laboratories Network web portal, www.colan.org

Guessoum Z., Briot J.P., (2002) From Active Objects to Autonomous Agents, Submitted to IEEE Concurrency - Special Series on Actors & Agents

Salamatov Y. (1999) TRIZ : The Right Solution at the Right Time. A Guide to Innovative Problem Solving., Insytec B.V.

Sycara K., Klusch M., Widoff S., Lu J., (1999) Dynamic Service Matchmaking Among Agents in Open Information Environments, SIGMOD Record (ACM Special Interests Group on Management of Data), Vol. 28, No. 1, March, 1999, pp. 47-53.

Von Wedel L., (2000) An Object Model for Chemical Process Models, WP 5: Advancing Open Process Engineering Modeling Concepts and Exchange Language, internal Global CAPE-OPEN document, December 2000

European Symposium on Computer Aided Process Engineering – 12
J. Grievink and J. van Schijndel (Editors)

Automatic Integration of High-Index Dynamic Systems

E.F.Costa Jr[1]., A.R.Secchi[2] and E.C.Biscaia Jr.[1*]

[1]Programa de Engenharia Química - COPPE/UFRJ - Rio de Janeiro – RJ - Brazil
[2]Departamento de Engenharia Química – UFRGS - Porto Alegre – RS - Brazil
*Author to whom correspondence should be addressed, evaristo@peq.coppe.ufrj.br

Abstract

A computational code for automatic characterisation, consistent initialisation and integration of generic DAE systems with index up to 3 is presented in this contribution. Two dynamic models are presented to illustrate the use and to characterise the performance of the code. The developed code has been tested in several dynamic simulations, the successful results and the minimal participation of the user during the simulation are encouraging characteristics to incorporate this code to processes dynamic simulators.

1. Introduction

When solving DAEs one must concern about the index of the system and about the consistency of the initial conditions. The index of a DAE system is the number of times that all or part of the system must be differentiated with respect to time (independent variable) in order to convert it into an explicit set of first order ODEs. Index 1 systems may be solved with modified ODE methods, while higher index systems (systems with index 2 or greater) require specific methods. Generally, higher index systems must be reduced to an index 1 problem and then solved using standard integration codes.

During the index reduction, some extra algebraic equations are obtained which generally correspond to derivatives of the original algebraic equations. Those hidden algebraic equations along with the original DAEs compose the extended system. Consistent initial values must satisfy not only the DAE system but also the resulting extended system.

A computational code for automatic characterisation of DAE system has been developed by Costa Jr. et al. (2001). In the present contribution, that code has been improved making possible the integration of DAE system with index up to 3 using the DASSLC computer code (Secchi, 1992). The Jacobian matrix has been calculated using automatic differentiation tools, resulting in a higher integration speed. It should be pointed out, that on this code the equations of the system could be presented on their original form avoiding any further manipulations.

2. Code Description

In order to use the automatic differentiation code ADOLC (Griewank et al., 1996), the code was written in C++. This computational language is also compatible with the code DASSLC, written in C, used to integrate the extended system. To use the present code, the user should write a program compatible with the syntax of DASSLC code as described in Secchi (1992).

The code accomplishes the following stages for the characterisation and resolution of the DAE systems:

(1) determination of the DAE system graph through numerical perturbation;

(2) execution of the algorithm proposed by Pantelides (1988) seeking the equation sets that should be differentiated;

(3) construction of the extended system by automatic differentiation;

(4) determination of the extended system graph;

(5) selection of the variables subset specified by the user;

(6) checking up consistency by structural analysis of the variables subset (If this subset is structurally inconsistent returns to stage 5.);

(7) checking up consistency by numerical analysis of the variables subset and determination of the initial conditions of the problem (If this subset is numerically inconsistent returns to stage 5.);

(8) construction of the Jacobian matrix and determination of the index 1 system.

3. Numerical Examples

3.1 The Start-up of a Batch Distillation Column

The model equations of this system, as presented by Costa Jr. et al. (2001), are:

$$\dot{n}_{ij} = V_{i+1} \ y_{i+1,j} + L_{i-1} \frac{n_{i-1,j}}{N_{i-1}} - L_i \frac{n_{i,j}}{N_i} - V_i \ y_{i,j} \tag{1a}$$

$$N_i = \sum_{j=1}^{nc} n_{i,j} \tag{1b}$$

$$N_i \ \dot{h}_i = V_{i+1} H_{i+1} + L_{i-1} h_{i-1} - L_i h_i - V_i H_i + Q_i \tag{1c}$$

$$y_{i,j} = \frac{n_{i,j}}{N_i} K(T_i) \tag{1d}$$

$$N_i h_i = \sum_{j=1}^{nc} n_{i,j} h_j(T_i) \tag{1e}$$

$$H_i = \sum_{j=1}^{nc} y_{i,j} H_j(T_i) \tag{1f}$$

$$\sum_{j=1}^{nc} y_{i,j} = 1 \tag{1g}$$

$$V_{i+1} = L_i \tag{1h}$$

where: np (=12) and nc (=5) are the numbers of trays and components of the column, respectively; i is the stage index (i is equal to 0 for the condenser and np+1 for the reboiler); j is the component index; y represents molar fraction in vapour phase; h and H are enthalpies in liquid and vapour phases; V_i and L_i are the vapour and liquid fluxes leaving the ith stage; $\underline{n}$ is the amount of a component in a stage; $\underline{N}$ is the stage holdup; $\underline{Q}$ is the heat removed from a stage and $\underline{T}$ represents temperature. The expressions to calculate the values of $\underline{h}$, $\underline{H}$ e $\underline{K}$ can be found in Holland and Liapis (1983).

This system has 223 equations and, as exposed in Costa Jr. et. al. (2001), the Equations (1b), (1d), (1e) and (1g) must be differentiated once (index 2). The resulting extended system has 335 equations and 405 variables. Thus, 70 variables should be assigned in the initial condition. There are $_{405}C_{70}$ (approximately 10^{80}) possible subsets to be assigned, and only some of them are feasible. The feasible initial subset has been specified as the amount of component $\underline{j}$ in every stage $\underline{i}$ ($\underline{n}_{i,j}$), according to Table 1. Figures 1a and 1b show some dynamic simulation results.

Table 1 - Specified initial conditions subset

Component	C_3H_8	$i\text{-}C_4H_{10}$	$n\text{-}C_4H_{10}$	$i\text{-}C_5H_{12}$	$n\text{-}C_5H_{12}$
$n_{0,j}$ (j = 0,...nc-1)	0.2	0.6	1.0	0.8	1.4
$n_{i,j}$ (i=1,...np and j=0,...nc-1)	0.05	0.15	0.25	0.2	0.35
$n_{13,j}$ (j=0, ... nc-1)	1.7	5.1	8.5	6.8	11.9

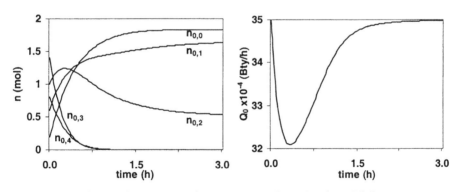

Figure 1 - Condenser dynamic simulation: (a) Number of moles of different components(left); (b) Heat duty(right).

Figure 1a reproduces the results of simulation reported in Holland and Liapis (1983). After 3 hours, the column is near to its steady state and the. heat removed from the condenser is the equal to the heat added in the reboiler, as presented in Figure 1b.

3.2 Product Purity Control of a Batch Distillation

The simplified mathematical model of this system, as presented by Logsdon and Biegler (1993), is presented below:

$$M_0 = -V/(R+1) \tag{2a}$$

$$M_0\dot{x}_{i,j} = V\big(x_{0,j} - y_{0,j} + R(x_{1,j} - x_{0,j})/(R+1)\big) \tag{2b}$$

$$M_i\dot{x}_{i,j} = V\big(y_{i-1,j} - y_{i,j} + R(x_{i+1,j} - x_{i,j})/(R+1)\big), \ i=1,...,np \tag{2c}$$

$$M_{n+1}\dot{x}_{n+1,j} = V(y_{n,j} - x_{n+1,j}) \tag{2d}$$

$$\sum_{j=1}^{nc} y_{i,j} = 1 \ , \ i=0,...,np+1 \tag{2e}$$

$$x_{n+1,1} = 0.998 \tag{2f}$$

where, $\underline{V}$ is the molar vapour flow rate; M_i is the molar holdup in the ith stage; $\underline{R}$ is the reflux ratio; $\underline{np}$ is the number of trays in the column; $\underline{i}$ is the stage index ($\underline{i}$ is equal to 0 to the reboiler and np+1 to the condenser); $\underline{j}$ is the component index; $\underline{x}$ and $\underline{y}$ represents molar fractions on liquid and vapor phases. The molar fraction of the last component, $\underline{x}_{i,nc}$, has been eliminated by using the Equation (3), and $y_{i,j}$ has been eliminated through the thermodynamic equilibrium relation, as presented in Equation (4).

$$x_{i,nc} = 1 - \sum_{i=1}^{nc-1} x_{i,j} \ , \ i=0,...,np+1 \tag{3}$$

$$y_{i,j} = K(T_i, x_{i,j})x_{i,j} \ \ , \ i=0,...,np+1 \tag{4}$$

The model has nc(np+2)+2 equations. There are (nc-1)(np+2)+1 differential equations and the remaining ones are algebraic. The (nc-1)(np+2)+1 differential variables are $x_{i,j}$ (j=1,...,nc-1; i=0,...,np+1) and M_0, and the (np+3) algebraic variables are T_i (i=0,..., np+1) and $\underline{R}$. The model is a high-index DAE system because the algebraic variable $\underline{R}$ does not appear in any algebraic equation.

The operating conditions, proposed by Logsdon and Biegler (1993) and used in this work, are presented in Table 2.

Operating Conditions (Logsdon and Biegler, 1993)		
V = 120 mol/h	M_0 = 100 mol	$x_{0,1}$ = 0.55, cyclohexane
np = 10 trays	M_i = 1 mol, i = 1,...,n+1	$x_{0,2}$ = 0.45, toluene

The characterisation of this system by the developed code has been the same reported by Logsdon and Biegler (1993). The equations that should be differentiated once are (2d) and (2e) for i=np (last tray). The Equation (2f) should be differentiated twice (index 3). Thus, in the extended system there are four new equations and two new variables: $\ddot{x}_{11,1}$ and $\dot{T}_{10}$. The number of degrees of freedom is 11 because there are 41 unknowns (variables and their time derivatives) and 30 equations.

The user must specify a subset of variables among the variables of the original system. Then, the user must choose 11 from the 39 original variables. There are more then 1.6 billions of possible subsets. In this stage, the user physical knowledge is very valuable because only few subsets are consistent. For example, it is not possible to choose, at the same time, the composition and the temperature of a stage. Besides, the specification of the purity in the Equation (2f) determines all variables of the two last stages of the column (last tray and the condenser). The values of the variables of these two stages are presented in Table 3. If a chosen subset does not follow these conditions, then it will be infeasible.

Table 3 – Variables values of the last two column stages		
$x_{10,1} = 0.9949$	$\dot{T}_{10} = 0\ °C/h$	$x_{11,1} = 0.9980$
$\dot{x}_{10,1} = 0\ h^{-1}$	$T_{11} = 80.77\ °C$	$\dot{x}_{11,1} = 0\ h^{-1}$
$T_{10} = 80.83\ °C$	$\dot{T}_{11} = 0\ °C/h$	

For any supplied initial condition, the values of the variables in the last two column stages will always be those supplied in Table 3. The consistent initial condition subset used in the present work is presented in Table 4.

Table 4 – Inital subset		
$H_0 = 100\ mol$	$T_1 = 89.80\ °C$	$T_5 = 82.60\ °C$
$x_{0,1} = 0.55$	$T_2 = 87.50\ °C$	$T_6 = 81.70\ °C$
$R = 1$	$T_3 = 85.40\ °C$	$T_7 = 81.10\ °C$
	$T_4 = 83.80\ °C$	$T_8 = 81.00\ °C$

This subset is structurally and numerically feasible. However, the variables of Table 4 cannot assume arbitrary values because the solution of the extended system can be out of the real domain. Another feasible subset can be obtained changing the variable $\underline{R}$ by the variable T_9. In this situation, meaningless results have been obtained, e.g. negative values of the reflux ratio. Other feasible subsets can be also specified.

The numerical solution of the extended system, using the initial subset of Table 4, is presented in Table 5.

Table 5 – Obtained initial conditions		
$\dot{x}_{0,1} = -0.2040\ h^{-1}$	$x_{4,1} = 0.8549$	$\dot{x}_{7,1} = -1.1423\ h^{-1}$
$x_{1,1} = 0.6070$	$\dot{x}_{4,1} = -1.0015\ h^{-1}$	$x_{8,1} = 0.9866$
$\dot{x}_{1,1} = 0.2927\ h^{-1}$	$x_{5,1} = 0.9100$	$\dot{x}_{8,1} = 0.0284\ h^{-1}$
$x_{2,1} = 0.6970$	$\dot{x}_{5,1} = -0.5111\ h^{-1}$	$x_{9,1} = 0.9911$
$\dot{x}_{2,1} = -1.8847\ h^{-1}$	$x_{6,1} = 0.9526$	$\dot{x}_{9,1} = 0.0210\ h^{-1}$
$x_{3,1} = 0.7845$	$\dot{x}_{6,1} = -0.4696\ h^{-1}$	$T_0 = 91.34\ °C$
$\dot{x}_{3,1} = -1.8111\ h^{-1}$	$x_{7,1} = 0.9817$	$T_9 = 80.91\ °C$

After the determination of the consistent initial conditions, the code automatically builds the index 1 extend system (without recompilation due to the use of automatic differentiation) to be integrated by DASSLC. The simulation results of the column production phase are presented in Figure 2.

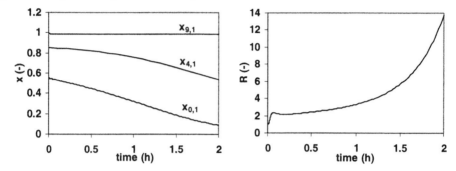

Figure 2 – Simulation results: (a) molar fraction of ciclohexane(left); (b) R (right).

As expected, the reflux ratio increases with the time. The small decrease between 0.08 and 0.18 is resulted of the initial conditions considered. The reflux ratio increases indefinitely with the time in order to assure the product purity. Its value is approximately 40,000 after 5 hours of production, resulting in a very low distillate production rate.

4 Conclusions

In this present contribution the code developed by Costa Jr. et al. (2001) has been improved making possible the direct numerical integration of DAE system with index up to 3 through the DASSLC computer code. Two dynamic models have been presented to illustrate the use and to characterise the performance of the code.

The new version of the code requires minimal intervention of the user during the simulation. This new feature shows the potentiality of incorporating the code to processes dynamic simulators.

5 References

Costa Jr., E.F., R.C. Vieira, A.R. Secchi and E.C. Biscaia, 2001, Automatic Structural Characterization of DAE Systems, Proc. of 11° European Symposium on Computer Aided Process Engineering - ESCAPE 11, 123.

Griewank, A., D. Juedes, H. Mitev, J. Utke, O. Vogel and A. Walther, 1996, ADOL-C: A Package for the Automatic Differentiation of Algorithms Written in C/C++, ACM TOMS 22 n° 2, 131-167, Algor. 755.

Holland C., A.I. Liapis, 1983, Computer Methods for Solving Dynamic Separation Problems, McGraw Hill Publishers.

Logsdon J.S. and L.T. Biegler, 1993, Accurate Determination of Optimal Reflux Policies for the Maximum Distillate Problem in Batch Distillation, Ind. Eng. Chem. Res. 32, 692-700.

Pantelides, C.C., 1988, The Consistent Initialisation of Differential-Algebraic Systems, SIAM J. Sci. Stat. Comp. 9, 213-231.

Secchi, A.R. (1992). "DASSLC: User's Manual, a Differential-Algebraic System Solver", Technical Report, http://www.enq.ufrgs.br/enqlib/numeric.DASSLC, UFRGS, Porto Alegre, RS/Brazil

European Symposium on Computer Aided Process Engineering – 12
J. Grievink and J. van Schijndel (Editors)

Modeling work processes in chemical engineering – from recording to supporting

M. Eggersmann, R. Schneider, W. Marquardt[*]
Lehrstuhl für Prozesstechnik, RWTH Aachen, D-52056 Aachen, Germany

Abstract

Design processes in chemical engineering are complex and highly creative. They show a large potential for improvement. In order to improve these work processes different steps including recording, formalization, analysis, and implementation are necessary. They will be described, together with their interactions, in this contribution.

1. Introduction

In a chemical engineering design project many software tools are used to support engineering activities. The focus of these tools is either on specific design tasks like simulation and optimization or on the management of information for example in a database. Not much attention has been paid to the work processes or workflow during process design (Subrahmanian et al., 1997). Our long term research objective is to improve the work processes by reengineering (similar to business process reengineering) and to support the individual designer during the design process by computer-based aids. For both objectives a thorough understanding of work processes in chemical engineering is necessary. In this paper, we present a conceptual approach towards workflow support.

In order to understand design processes it is essential to identify their characterizing features which are necessary to describe them in sufficient detail. In our opinion at least five elements are needed. Core elements are the *activity* to describe what has been done and the order of activities, which defines the *control flow*. Besides, the results of an activity and the data resources needed to perform it, have to be known. These are the input and output *information* of an activity resulting in an *information flow*. Another important element is the person performing the activity, or more precisely the *role* a person is performing in a design process.

In the next section we will summarize a procedure which can be followed to reach an improvement of work processes in chemical engineering. The consecutive sections describe the most important parts of such a procedure including recording of design processes, as well as their analysis and formalization.

2. An Approach towards Workflow Support

Figure 1 shows a simplified overview of the steps of the workflow support approach, under consideration of the elements introduced in Section 1. The first step towards

---

[*] Correspondence should be addressed to W. Marquardt, marquardt@lfpt.rwth-aachen.de

872

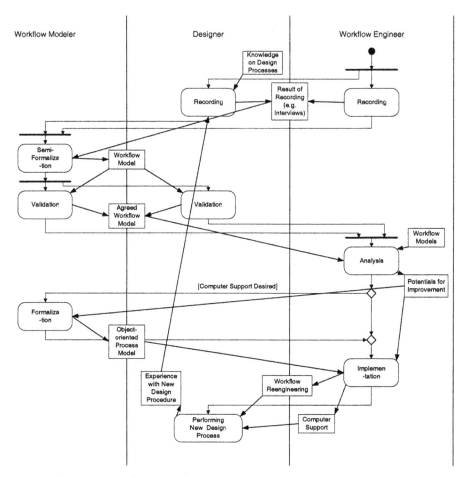

Figure. 1: An approach to workflow support

workflow support is to acquire knowledge about industrial work processes in process design. Whereas predefined design procedures may exist on a coarse level internally in companies or in the open literature (e.g. Douglas, 1988), not much knowledge exists about work processes on a finer level of granularity. This is mainly due to the creative character of design processes. Therefore *recording* of actual industrial work processes is necessary in the first place for understanding it (Bucciarelli, 1994). In principle, there are two different ways of recording: (i) The *designer* who has the best (tacit rather than explicit) knowledge about design processes protocols what he is doing during the design (self observation). (ii) The designer is interviewed after completing (part of) the design process by a person who has some background on interview techniques and the domain of interest, namely chemical engineering. We will call this person a *workflow engineer*. Besides the designer and workflow engineer, a *workflow modeler* is necessary who transforms the (usually informal) results of the recording or the interviews in a, what we call, *semiformal workflow model*. This model has to be easily understandable. Therefore a graphical representation is used, which facilitates the understanding. To eliminate misunderstandings this model is *validated* by the designer and the workflow modeler together.

On the basis of the agreed workflow model the workflow engineer *analyzes* the work process, to identify problems and to find possible ways of *reengineering* the work process or of formulating requirements for *computer support*. From the analysis of a certain number of design processes it should be possible to define standard workflows which have proven to be effective and can serve as some kind of template for further projects. These standard procedures could be directly *implemented* without the use of information technology: A project manager for example uses the standard workflows to organize his project or a designer can employ a certain design procedure to design a process unit. Obviously work processes can be even better supported by the application of computer tools. A software tool which captures information about various best practice workflows could guide a designer by suggesting appropriate activities and advising him how to perform those. A prerequisite for such a support system is a proper *formalization* of the work processes in a format which can be processed by a computer. This task has to be fulfilled by the workflow modeler. In contrast to a human-readable representation such a format must be unique and unambiguous and must not require or allow interpretations. After the application of new – hopefully better – design procedures their success has to be evaluated, again starting with the recording of the work process. In Figure 1 this is indicated by the feedback information flow from *performing of new design processes* to *recording*.

3. Recording and Representing Work Processes

In order to record and model work processes a method is needed which is easy to use and to understand as in most cases the designer is not an *expert* in work process modeling. When a project manager wants to record and represent the workflow in a particular design project he is responsible for, not too much *extra time* should be necessary to accomplish this. Also, when a workflow engineer is supposed to conduct interviews in order to elicit information about the work processes, the interviewee (the designer) needs to be able to *understand* the workflow representation so that he can give direct feedback for discussion and verification purposes.

These requirements are addressed by the modeling formalism C3 suggested by Killich et al. (1999). The name C3 refers to cooperation, coordination, and communication as the formalism is capable of representing these three features, which are typical for design processes. Although C3 is based on UML (Unified Modeling Language, Booch et al., 1998), a detailed knowledge of UML is not necessary to understand the models, as the reader may have noticed by intuitively understanding the content of Figure 1, which is modeled in C3. The elements of C3 are roles, activities, input and output information, control flows, information flows, synchronous communication, and tools. The last two are not used in the example of Figure 1.

All activities are assigned to a role within a so-called swim lane. The solid arrows represent control flows which indicate the order of the activities. Within each swim lane, the temporal order is indicated by the arrangement of the single activities and their connecting control flows. Information flow from an activity where information is produced to other activities which use this information is represented by a dashed line. Such an information flow is completed by the specification of the information transfered within an associated rectangular box. An information flow between two or more activities indicates which activities can only be performed after the completion of a previous one. The main elements for modelling the order of activities in C3 are, however, the control flows, whereas information flows provide additional information about the dependency between those activities. Tools used within an activity are indicated by an associated block arrow (not shown in Fig.1). The elements are predefined, but the user can fill them with arbitrary information.

Besides the ease of understanding its representation, C3 has the ability to be used directly as an interactive modeling technique during interviews (Scheele and Groeben, 1984). It allows a structured representation of the work process the interviewee has in his mind: the interviewer uses certain cards to represent the interviewee's answer; these cards represent the central elements of C3 like activities, information, and tools. The interviewee himself writes additional information on the cards. The cards are then pinned together with other informal, written comments on a large roll of wallpaper, which is then analyzed and (semi-)formalized by the workflow modeler. Finally the model is verified by the workflow modeler and the interviewee together.

4. Analysis and Implementation of Work Processes

Typically, knowledge about work processes in the domain of process design is only implicit. Writing down and modeling these processes has the benefit that the designers become more conscious about the way they solve problems as part of a design team. The C3 model of a particular design process can be employed to analyze the work processes, similar to the application of SADT models (Structured Analysis and Design Technique, Ross and Schoman, 1977). The C3 models may be used for example to identify potential ways for shortening the design process in the sense of concurrent engineering. By identifying the information flow it is possible to judge if one activity can be performed concurrently to others. Activities can be identified which lack good computer support. This way, necessary requirements on tool functionalities can be defined. The work process model clearly shows the interfaces between different roles in the design team or even between different departments and the information transfer between them. Hence, it can also be used to discover outsourcing opportunities. Outsourcing of specific design activities can be planned by assisting in defining interfaces between external and internal roles.

The benefits of such an analysis can be demonstrated by means of a case study. We modeled the workflow during conceptual design of a nylon6 process by means of the C3 formalism (Bayer et al., 2001a). The resulting work process model consists of more than 100 activities and involves seven roles. In a conventional design process the polymer processing subprocess is being developed after the polymer reaction subprocess has been specified. The C3 model of the case study helped to discover that the polymer processing part can be designed much earlier. The implementation of this insight significantly decreases the duration of the design process and hence the time to market. Additionally, concurrent engineering allows an earlier investigation of the effect the polymer processing unit has on the rest of the process. Problems and potentials for improvement can be identified earlier. In our case the knowledge about the polymer processing permitted a more economic design of the separation section.

5. Formalization

In the approach presented here the C3 notation is used for modeling past processes; its main feature is an easily understandable graphical representation. However, it is not unique nor unambiguous and therefore has to be interpreted by a human user. In order to use workflow models in a computer, they have to be unambiguous, and a more strict formalization is required. Standard work processes have to be modeled which are not related to a specific execution but rather can be used as templates for new work processes. These standard workflows are an abstraction from previous workflows. According to the object-oriented paradigm (Rumbaugh et al., 1991) the previous workflows can be seen as instances and the standard workflows as classes.

We adapted this view within CLiP (Conceptual Lifecycle Process Model) (Bayer et al., 2001b, Eggersmann et al., 2000), an object-oriented data model covering product data of the design and the work processes leading to the design. The work process model consists of three layers (instances, classes, and metaclasses). The instance layer contains those activities which have already been performed and are therefore well known. Standard workflows and standard activities, which can form templates for future activities are represented on the class layer. The metaclass layer includes the modeling concepts for the other two layers. However, it is not yet clear how certain object-oriented concepts like e.g. inheritance between classes can be applied to work processes.

The differences between C3 and CLiP can be explained by the example in figure 2. On the instance level the activity "Design CSTR for PA6 production" is modeled using the C3 formalism. Here it is represented what one specific person (the reaction expert) did and that he used the PA6 polymerization kinetics as input information and Pro II as a tool. By abstracting this activity, an activity "Design reactor" can be defined which is modeled within CLiP. We call this an activity class because the specific activity instance (and possibly further activities) do carry the same attributes and methods as this class. On the class level in CLiP no specific actor is modeled but the skill he is supposed to possess. This is necessary if activities should be assigned according to the skills of the designer. In the C3 model the skills are only implicitly represented by the role. Whereas the control flow is explicitly modeled in C3 it is only implicitly contained in the class level of CLiP by the definition of the information. The possibility to use alternative tools – Polymers Plus and Pro II – can also be modeled within CLiP.

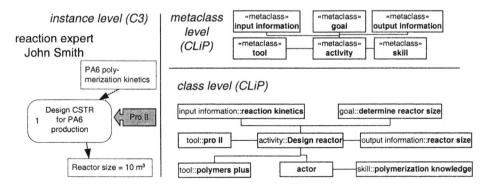

Figure. 2: C3 and CLiP model of an example activity

The abstraction and formalization must be done by a human and cannot be done automatically because it involves modeling and domain knowledge. Whereas C3 is easy and fast to use, modeling with CLiP requires both more time and more modeling knowledge. An approach to facilitate the transition from C3 models to CLiP is to extend the C3 graphics by a storage of links between e.g. activities and information.

6. Conclusions and Future Work

We have presented the steps necessary to support work processes in chemical engineering design and their interactions. A system is being implemented which currently supports the graphical representation and in the future shall be extended to facilitate the formalization. Open issues are still the transition of a textual description of the work process by a designer to a C3 model, and the abstraction of C3 models to classes of work processes in CLiP. Additionally, the analysis requires further systematization and the possibilities of computer support have to be evaluated.

Acknowledgements

This work is supported by the DFG, Deutsche Forschungsgemeinschaft, in the CRC 476 'IMPROVE'. The authors thank B. Bayer, C. Foltz, and M. Wolf for many fruitful discussions.

References

Bayer, B., M. Eggersmann, R. Gani, R. Schneider, 2001a, Case Studies for Process Design, In: Braunschweig, B., R. Gani, Software Architectures and Tools for Computer Aided Process Engineering, Elsevier Publishers, to be published.

Bayer, B., C. Krobb, W. Marquardt, 2001b, Technical report, Lehrstuhl für Prozesstechnik, RWTH Aachen, LPT-2001-15.

Booch, G., J. Rumbaugh, I. Jacobson, 1998, The Unified Modeling Language User Guide, Addison Wesley, Reading.

Bucciarelli, L.L., 1994, Designing Engineers, MIT Press, Cambridge, Massachusetts.

Douglas, J.M., 1988, Conceptual Design of Chemical Processes, McGraw-Hill, New York.

Eggersmann M., C. Krobb, W. Marquardt, 2000, A Modeling Language for Design Processes in Chemical Engineering. In: Laender, A.H.F., S.W. Liddle, V.S. Storey (Eds.): Lecture Notes in Computer Science 1920, Springer, Berlin, 369-382.

Killich, S., H. Luczak, C. Schlick, M. Weissenbach, S. Wiedenmaier, J. Ziegler, 1999, Behavior & Information Technology, 18, 325-338.

Ross, D.T., K.E. Schoman, 1977, IEEE T. Software Eng., SE-3, 1, 6-15.

Rumbaugh, J., M. Blaha, W. Premerlani, F. Eddy, W. Lorensen, 1991, Object-Oriented Modeling and Design, Prentice-Hall Inc., Englewood Cliffs, New Jersey.

Scheele, B., N. Groeben, 1984, Die Heidelberger Struktur-Legetechnik (SLT), Beltz, Weinheim.

Subrahmanian, E. S.L. Konda, Y. Reich,. A.W. Westerberg, the N-dim group, 1997, Comp. Chem. Engng., 21, Suppl., S1-S9.

European Symposium on Computer Aided Process Engineering – 12
J. Grievink and J. van Schijndel (Editors)
877
© 2002 Elsevier Science B.V. All rights reserved.

A Multi-Cellular Distributed Model for Nitric Oxide Transport in the Blood

Nael H. El-Farra, Panagiotis D. Christofides and James C. Liao
Department of Chemical Engineering
University of California, Los Angeles, CA 90095-1592

Abstract

In this work, a multi-cellular, spatially distributed model, that describes the production, transport and consumption of Nitric Oxide (NO) in blood vessels and the surrounding tissue, is developed. In contrast to previous modeling efforts, such as continuum and single-cell models, the current model accounts explicitly for the presence of, and interactions among, a population of red blood cells (RBCs) inside the blood vessel. Using mainly experimentally derived parameters, the model equations, subject to the appropriate boundary conditions, are solved for the NO concentration profile using an efficient finite-element algorithm.

1. Introduction

Despite the well-documented significance of NO, as a versatile reactive free radical that participates in a diverse array of vital biological functions in humans and animals, the transport of NO from its site of synthesis to its target remains a complex process that is not yet completely understood. For example, in order to exercise its regulatory effect in the circulatory system, NO must diffuse from the blood vessel wall to the external surrounding tissue. Due to its reactivity, however, NO may be consumed and degraded by numerous reactions before it reaches its target. In particular, NO diffuses into the lumen of the blood vessels where it is believed to react at very high rates with hemoglobin in red blood cells (Cassoly and Gibson, 1974) . The extremely fast kinetics of this reaction suggest, however, that most of the NO produced in the blood vessel wall would be consumed in the blood, thus severely limiting NO diffusion into surrounding tissue and compromising NO regulatory effects. This paradox has triggered a significant search, with the aid of coupled mathematical modeling and experiments, for the specific mechanisms that must exist to reduce NO consumption in the blood (see, e.g., (Lancaster, 1994; Vaughn et al., 1998; Vaughn et al., 2001)).

Despite these significant contributions to the modeling and analysis of NO transport, the particulate nature of the blood, namely the presence of, and interactions among, a population of RBCs inside the lumen of the blood vessel (as opposed to a homogeneous medium or just a single RBC where NO reacts with Hb), has not been investigated in any of the existing models so far. This factor contributes to further reduction of NO uptake in the blood by diminishing the actual exposure of NO to Hb further, compared to that predicted from the continuum model, and introducing additional diffusional barriers in the extracellular spaces between cells in the population. Motivated by these considerations, together with the recent advances in numerical methods and the development of efficient tools for the numerical simulation of complex systems, we develop in this study a detailed, multi-cellular, distributed model for NO transport, based on diffusion-reaction principles, that explicitly takes into account the presence of, and interactions among, a population of red blood cells inside the blood vessel. The model equations consist of several sets of PDEs that describe the production, diffusion, and consumption of NO in the tissue surrounding the blood vessel, the vessel wall, a cell-free zone near the vessel wall, the extracellular space between the population of RBCs, the membrane surrounding each cell, and, finally, the intracellular space inside the RBCs. The parameters for the model are obtained from experimental data reported in the literature (see, e.g., (Vaughn et al., 1998; Vaughn et al., 2001)) and the entire set of PDEs is solved for the NO concentration profile using efficient finite-element algorithms. The detailed model offers a more realistic understanding of the key mechanisms for NO transport and consumption, which, in turn, has crucial clinical implications for the development of practical technologies (e.g., artificial blood substitutes) that can be used to treat diseases attributed to imbalances in NO transport in humans.

2. Mathematical Modeling of NO transport

2.1 Governing equations

To provide a detailed, yet computationally tractable, model that captures the main features of NO transport in blood vessels and the surrounding tissue, we divide the system into three major compartments: the abluminal region, the endothelium, and the lumen. The lumen is further divided into a cell-free zone, extracellular spaces between the cell population in the vessel, the membrane surrounding each cell, and, finally, the intracellular spaces (see Figure 1). The system is modeled using polar coordinates.

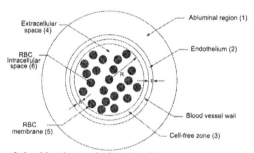

Figure 1: Geometry of the blood vessel showing the various compartments of the model and a distribution of RBCs.

The concentration of a diffusing reacting substance, such as NO, is described by the species mass balance. For NO, this balance, in its general form, can be written as

$$\frac{\partial C_{NO}}{\partial t} = D_{NO}\nabla^2 C_{NO} - \nabla C_{NO} \cdot v + R_{NO} \tag{1}$$

where ∇ is the vector gradient operator, ∇^2 is the Laplacian operator, C_{NO} is the concentration of NO, D_{NO} is the diffusion coefficient, and R_{NO} is the rate at which NO is produced or consumed by reaction. Two processes are involved in the transport of NO: the first term on the right hand side represents the diffusion of NO; the second represents the transport of NO by a molar averaged velocity. By focusing on the steady-state behavior and neglecting convective transport, we have that $\nabla C_{NO} \cdot v = 0$ and $\partial C_{NO}/\partial t = 0$, respectively. The system can therefore be treated as a two-dimensional problem, with NO concentration varying only in the radial (r) and azimuthal (θ) directions. The azimuthal variations in NO concentration arise from possible non-uniform consumption of NO inside the vessel due to a non-uniform (random) distribution of cells. The balance between NO diffusion and reaction in all compartments can be written for cylindrical coordinates as

$$D_{NO}\left(\frac{1}{r}\frac{\partial}{\partial r}\left(\frac{1}{r}\frac{\partial C_{NO}}{\partial r}\right)+\frac{1}{r^2}\frac{\partial^2 C_{NO}}{\partial r^2}\right)+R_{NO}=0 \qquad (2)$$

Eq.2 is used for all compartments, although the value of D_{NO} may differ depending on the intrinsic transport resistance in each region. Also, the expression for R_{NO} may differ in each region depending on whether NO is being produced or consumed and depending on the rate laws of the chemical reactions taking place in each region. Following (Vaughn et al., 1998), NO is assumed to be consumed by a second order reaction in the abluminal region, so R_{NO} takes the form $R_{NO}=-k_{ab}C_{NO}^2$, where k_{ab} is the reaction rate constant. In the endothelium, NO is also consumed by the above second order reaction but, in addition, it is produced by an enzyme that is partially bound to the membrane of endothelial cells. The expression for R_{NO} in this compartment therefore takes the form $R_{NO}=-k_{ab}C_{NO}^2+Q_{NO}$, where Q_{NO} is the total NO production rate per unit area of the endothelium. In the extracellular vascular lumen and inside the RBC membrane, NO is transported by diffusion (with a different diffusivity in each region) and R_{NO} is taken to be zero. Finally, inside each RBC, NO consumption by hemoglobin can be expressed by the rate equation given in (Vaughn et al., 1998) $R_{NO}=-k_{2,lu}C_{NO}C_{Hb}\approx-k_{1,lu}C_{NO}$, where $k_{2,lu}$ and $k_{1,lu}$ are the rate constants and C_{Hb} is the hemoglobin concentration in the red blood cell, which remains essentially constant so the reaction can be considered pseudo-first order in NO with the reaction rate constant $k_{1,lu}=k_{2,lu}C_{Hb}$.

In order to solve the resulting set of partial differential equations for NO concentration, we need to specify the boundary conditions. In the radial direction, one boundary condition is implied by the no-flux condition at the center while far from the vessel wall we have that the NO concentration changes slowly; therefore

$$\left.\frac{\partial C_{NO}}{\partial r}\right|_{r=0}=\left.\frac{\partial C_{NO}}{\partial r}\right|_{r\to\infty}=0 \qquad (3)$$

The rest of the boundary conditions in the radial direction are obtained by invoking continuity of the NO concentration profile and matching the fluxes at the interfaces between the various regions. For the azimuthal direction, periodic boundary conditions that express continuity of both the NO concentration and NO concentration gradients in the azimuthal direction are used.

2.2 Model parameters and numerical solution

The main parameters in the model include: 1) the diffusion coefficient of NO in the abluminal region, the endothelium, and the extracellular vascular lumen (all assumed to be the same), D_{ext}, 2) the diffusion coefficient of NO in the RBC membrane, D_{mem}, 3) the diffusion coefficient of NO inside the RBC D_{int}, 4) the NO production rate Q_{NO}, 5) the rate constants for NO degradation in each region k_{ab}, $k_{2,lu}$ and $k_{1,lu}$. The values of these parameters have been derived from experimental data reported in the literature (see (Vaughn et al., 1998; Vaughn et al., 2001) for details and references) and are given in Table 1. Other parameters used in the numerical simulations include the radius of the blood vessel, R, the thickness of the cell-depleted zone, δ, the thickness of the endothelium region producing NO, ε, the effective radius of the RBC, a (modeled as a sphere), the effective thickness of the RBC membrane, δ_{RBC}, and the concentration of hemoglobin inside the RBCs, C_{Hb}. The values of these parameters are given in Table 1.

Using the software FEMLAB, a finite-element algorithm was developed to discretize the spatial domain of the problem and numerically solve the model equations subject to the above boundary conditions and model parameters. Because the concentration of NO in the endothelium and close vicinity is of particular interest, an adaptive meshing (variable grid spacing) technique was used to allow a larger number of elements near the endothelial surfaces, where the concentration gradients are expected to be steep. In all simulations, the mesh was continuously refined to insure that the solution is grid-independent. Also, although the second boundary condition in Eq.6 applies to an infinite domain, it was implemented on the outermost elements of our finite mesh, typically 2,000-4,000 μm from the vessel axis.

Table 1. Model parameters

D_{ext}	3300 $\mu m^2 s^{-1}$
D_{int}	880 $\mu m^2 s^{-1}$
P_{mem}	450 $\mu m\, s^{-1}$
K_{ab}	0.05 $\mu M^{-1} s^{-1}$
$K_{1,lu}$	2.3×10^5 s^{-1}
Q_{NO}	10.6×10^{-14} $\mu mol.\mu m^2 s^{-1}$
R	50 μm
δ	2.5 μm
ε	2.5 μm
a	3.39 μm
δ_{RBC}	0.0078 μm
C_{Hb}	23 μM

3. Simulation results

To gain some insight, from the developed model, into the factors governing NO mass transport, we solve the model under the following four different scenarios:

3.1 Homogeneous case

In this case, the NO-hemoglobin interaction is assumed to take place uniformly everywhere in the vascular lumen. To simplify the presentation of our results, we will focus only on the mean NO concentration (averaged over θ) as a function of distance from the vessel axis. The resulting profile in this case is depicted by the solid line in Figure 2. It is clear from this profile that NO transported across the endothelium-lumen interface diffuses only very little into the lumen before it is scavenged by hemoglobin and completely depleted (note the zero concentration of NO inside the blood vessel). As expected, the NO concentration profile exhibits a maximum in the endothelium where NO is produced. However, by comparing the NO concentration gradients at this point, on the lumen side and the smooth muscle side, it is clear that the blood acts as a sink for NO, where the majority of NO produced in the endothelium flows into the lumen (note the steep gradient inside the lumen).

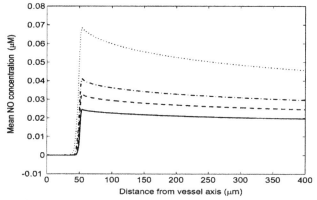

Figure 2: Mean NO concentration profiles as a function of distance from the vessel axis for continuum case (solid), with population of RBCs (dashed), with RBCs+cell-free zone (dashed-dotted), and with RBCs+cell-free zone+RBC membrane (dotted).

3.2 Particulate model

Since hemoglobin is packaged inside red blood cells, and is not free floating in the blood vessel, NO consumption by hemoglobin takes place only inside red blood cells, and not everywhere in the blood vessel. Furthermore, the diffusion of NO in the extracellular space between RBCs offers an additional barrier that could slow down NO uptake by the RBCs. To analyze these effects, we explicitly incorporate a population of RBCs in the vascular lumen compartment of our model. The RBCs are modeled as spheres, with an effective radius, a, that are distributed randomly throughout the lumen (see Table 1). The number of cells included depends on the hematocrit of the blood, which is the volume fraction of RBCs in the blood. Under normal physiological conditions, the hematocrit is around 50%. For our two-dimensional model, this number is equivalent to the fractional coverage, by cells, of the lumen compartment. For each cell, a mass balance of the form of Eq.2 describing the diffusion and consumption by hemoglobin of NO inside the cell is included in the model.. The resulting mean NO concentration profile for this case is shown by the dashed profile in Figure 2. As expected, the concentration profile exhibits a similar shape to the continuum case; however, the endothelial and abluminal NO concentrations increase, as more NO (not consumed by RBCs) is now available to diffuse into the smooth muscle tissue.

3.3 Cell-free zone

In the above calculations, the RBCs were assumed to be randomly distributed throughout the vascular lumen and, therefore, allowed to get arbitrarily close to the vessel wall. However, due to the hydrodynamic effects of blood flow, the RBCs tend to migrate towards the center of the vessel, thus creating a cell-depleted zone or layer near the vascular wall where NO has to travel across before it reaches Hb inside the RBCs. To account for this effect, we include a thin layer near the vascular wall of our model where no RBCs are present. The thickness of this layer depends on fluid mechanical considerations and is expected to be around 2.5µm for a 100µm -diameter vessel (Vaughn et al., 1998). Solving the model equations for this case, the resulting mean NO concentration profile is shown by the dashed-dotted profile in Figure 2. It is clear from this figure that under these conditions both the endothelial and abluminal NO concentrations increase almost twofold over that predicted form the continuum model.

3.4 RBC membrane permeability

So far in our model, we have neglected the potential role of the RBC membrane in reducing NO uptake in the blood by assuming that the RBC membrane is highly permeable to NO. Recently in (Vaughn et al., 2001), however, combination of an experimental technique, that overcomes experimental diffusional limitations, together with model analysis have been used to show that the RBC possesses an intrinsic mechanism that can slow down NO uptake. The results of this work point to the RBC membrane as a potential source for this mechanism. To investigate this effect in our model, we explicitly model the RBC membrane by enclosing each cell in the population with a thin layer whose thickness corresponds to the effective thickness (based on spherical geometry) of the RBC (see Table 1). Inside these layers, NO is transported by diffusion and is not consumed by hemoglobin. The value of the NO diffusion coefficient inside the membrane is computed directly from the permeability estimate obtained in (Vaughn et al., 2001) using the film theory approximation

$$D_{mem} \frac{\partial C_{NO}}{\partial r} \approx \frac{D_{mem}}{\delta_{RBC}} \left(C_{NO,ext} - C_{NO,cyt} \right) = P_{mem} \left(C_{NO,ext} - C_{NO,cyt} \right) \qquad (4)$$

where $C_{NO,ext}$ is the NO concentration at the interface between the membrane and extracellular space, $C_{NO,cyt}$ is the NO concentration at the interface between the membrane and the intracellular region (cytosol), and P_{mem} is the NO permeability of the membrane. By adding the membrane, the model now includes all six compartments depicted in Figure 1, and accounts for the host of transport barriers, analyzed above, combined. The mean NO concentration profile for the full model is shown by the dotted profile in Figure 2. As expected, when the full host of transport barriers are considered together, the increase in the endothelial and abluminal NO concentrations (over those obtained from the continuum model) is larger than the increase observed when fewer barriers are accounted for. Finally, we note that the results of this detailed model have been used to assess, quantitatively, the contribution of each mass transport barrier considered here to the overall resistance to NO uptake in the blood (see (El-Farra et al., 2002) for details).

4. References

Cassoly, R. and Q. H. Gibson, ``Conformation, co-operativity and ligand binding in human hemoglobin," *J. Mol. Biol.*, **91**, 3301-3313 (1974).

El-Farra, N. H., P. D. Christofides, and J. C. Liao, ``Analysis of nitric oxide transport barriers in blood vessels," *Annals. Biomed. Eng.*, submitted (2002).

Lancaster, J., ``Simulation of the diffusion and reaction of endogenously produced nitric oxide," *Proc. Natl. Acad. Sci. USA*, **91**, 8137-8141 (1994).

Vaughn, M. W., K.T. Huang, L. Kuo, and J. C. Liao, ``Erythrocytes consumption of nitric oxide: competition experiment and model analysis," *Nitric Oxide: Biology and Chemistry*, **5**, 18-31 (2001).

Vaughn, M. W., L. Kuo, and J. C. Liao, ``Effective diffusion distance of nitric oxide in the micocirculation," *Am. J. Physiol. Heart Circ. Physiol.*, **274**, H1705-H1714 (1998).

European Symposium on Computer Aided Process Engineering – 12
J. Grievink and J. van Schijndel (Editors)

Systems Engineering Approaches To Gene Expression Profiling

Sanjeev Garg and Luke E.K. Achenie

sgarg@engr.uconn.edu, achenie@engr.uconn.edu

Department of Chemical Engineering, University of Connecticut, Storrs, CT 06269, USA

Abstract

Several data analysis algorithms exist for the analysis of gene expression data resulting from cDNA microarray experiments. From a biological perspective, all of these have a number of disadvantages, some of which are addressed in this study. Dimensionality reduction, *a priori* specification of the number of classes and the need for a training set are a few of these disadvantages. To address these issues, we propose two novel approaches based on systems engineering principles. The proposed algorithms do not (1) require a training set, (2) require the *a priori* specification of the number of classes and (3) perform any dimensionality reduction. The first algorithm has been used on three gene expression data sets (yeast cell cycle data, human fibroblast response to serum data and the cutaneous melanoma data) from the open literature, while the second has been used on the fibroblast data set. The results found in general are at least in excellent agreement with studies in the open literature or they reveal further knowledge, which was not available previously. The present study, thus, establishes the viability and strength of the proposed algorithms for gene expression data analysis.

1. Introduction

Many different techniques are reported in the open literature for gene expression data analysis resulting from cDNA microarray experiments, namely data reduction techniques such as principal component analysis (Raychaudhuri et al., 2000), multidimensional scaling plots (Bittner et al., 2000), clustering techniques such as hierarchical clustering (Eisen et al., 1998), self-organizing maps (Tamayo et al., 12), knowledge-based support vector machines (Brown et al., 2000), "gene-shaving" (Hastie et al., 2000) and many more. These techniques have many advantages, but have some inherent disadvantages from a biological perspective. Any form of data reduction technique is generally biased and thus can result in the loss of meaningful biological information. Hierarchical clustering is best suited for data, which follow a hierarchical pattern; however it is not well suited for gene expression data analysis in general (Hastie et al., 2000). Self-organizing maps require the *a priori specification* of the number of cluster centers, which might not be known in many instances. Support vector machines need a training data set to *learn* the class information. The difficulty lies in the fact that the training data might not be a true representation of the whole data set. In this study, the proposed algorithms address most of the above issues associated with existing techniques in gene expression data analysis for *class discovery* and *class prediction as applied to gene expression data*. The proposed algorithms can be used efficiently in a unified strategy to perform genome wide classification and functional annotation of genes with previously unknown functionality.

2. Formulation

Gene expression matrix is an $N \times M$ matrix, where N is the number of genes and M is the number of attributes or time points. Typical numbers for N and M are in thousands (or hundreds) and hundreds (or tens), respectively. The goal of different data analysis algorithms is to group genes with similar gene expression patterns in different clusters. This is desired as genes with similar expression patterns may have similar functions based on homology.

The first of the proposed data analysis techniques is based on the *Growing Neural Gas* algorithm developed by Fritzke (1994a, 1994b and 1995). The growing neural gas technique employs a model that uses a growth mechanism of the *growing cell structures* (Hastie, 1994b) and the topology generation of *competitive Hebbian learning* (Martinetz, 1991). Starting with a small number of units (step 1), new units are inserted (step 8) successively in the network (subsequently referred to as the *net*) using the error measurements. These error measurements are gathered during the adaptation process (steps 4-7). In the proposed algorithm, the maximum number of cells in the net is fixed to be the total number of genes in the data set. In addition, the maximum number of connections for each cell is fixed to be one less than the number of cells in the net. A new performance criterion is defined and is used as a stopping criterion for the growth of the net (step 10). This is in contrast with strategies that employ the maximum number of cells. A step (step 11) has also been added to calculate the 'class-center' positions in the input space, based on the given gene expression data and the resulting net topology. An iteration counter is defined and is updated to keep track of the number of input data read by the algorithm. We define a normalized mean error as the ratio of mean error value to the number of experimental attributes in the gene expression data as a measure of 'how good the clustering is'. The steps of this algorithm can be more formally given as:

Step 0: Initialize the set A to contain two units a and b (with zero initial errors) at random positions w_a and w_b in R^M (M-dimensional feature space), that is $A = \{a, b\}$. Initialize the connection set C, $C \subset A \times A$ to the empty set as $C = \phi$.

Step 1: Get an input signal x from the given data.

Step 2: Find the nearest unit s_1 and the second nearest unit s_2 $(s_1, s_2 \in A)$ by:

$$s_1 = \arg \min_{x \in A} \|x - w_c\|$$

$$s_2 = \arg \min_{x \in A \setminus \{s_1\}} \|x - w_c\|$$

Step 3: Create a connection between s_1 and s_2 if it does not exist already: $C = C \cup \{(s_1, s_2)\}$. Set the age of the connection between s_1 and s_2 to zero.

Step 4: Add the squared distance between the input signal and the nearest unit in input space to the error variable:

$$E_{s_1} \leftarrow E_{s_1} + \left\| x - w_{s_1} \right\|^2 .$$

Step 5:   Adapt the reference vector of s_1 and its direct topological neighbors by fractions ε_b and ε_n, respectively, of the total distance to the input signal:

$w_{s_1} \leftarrow w_{s_1} + \varepsilon_b \left(x - w_{s_1} \right)$ and $w_n \leftarrow w_n + \varepsilon_n \left(x - w_n \right)$ for all direct topological neighbors n of s_1.

Step 6:   Increment the age of all connections emanating from s_1:

$$age(s_1, n) = \{ age(s_1, n) + 1 \} \text{ for all direct topological neighbors } n.$$

Step 7:   Remove connections with an age larger than a_{max}. If this results in points having no connections, remove the points as well.

Step 8:   If the number of input signals generated so far is an integer multiple of a parameter λ, insert a new unit as follows:

- Determine the unit q with the maximum accumulated error.
- Insert a new unit r halfway between q and its direct topological neighbor f with the largest error variable:

$$w_r = 0.5 \left(w_q + w_f \right)$$

- Insert connections connecting the new unit r with units q and f, and remove the original connection between q and f.
- Decrease the error variables of q and f by a factor α:

$$E_q \leftarrow E_q (1-\alpha) \quad E_f \leftarrow E_f (1-\alpha), \ 0 < \alpha < 1.$$

- Initialize the error variable of r from q and f:

$$E_r = (E_q + E_f).$$

Step 9:   Decrease the error variables of all units by β:

$$E_c \leftarrow E_c (1-\beta), \ 0 < \beta < 1$$

Step 10:   If the performance criterion ($E_{max} \leq E_{set}$) is not yet fulfilled go to Step 1 else continue.

Step 11:   Calculate the 'centers' by averaging the reference vectors of each unit c and its direct topological neighbors.

$$w_{center} = \left\{ (w_c + \sum_1^n w_i \ )/(n+1) \right\}.$$

Similar concepts are used in a second algorithm, the "*adaptive centroid algorithm*"(ACA), to group genes with similar expression patterns. An analogy is made and used from the center of mass calculation for a heterogeneously distributed mass. An overall centroid is calculated as the average expression pattern of all the unclassified genes at a particular stage. The gene nearest to this centroid, with distance equal to minimum distance, is located and a local search for other similar genes is performed. The centroid of all genes belonging to this local region is calculated and constantly updated. Local search is continued till the distance between this centroid and the initial gene located is less than the minimum distance (centroid condition) or the local distance for search is less than the minimum distance

(distance condition). When any of these conditions is violated a new cluster location is found by adapting the centroid of unclassified genes to a new location based on the average of unclassified genes at this stage and the gene nearest to it. The process is repeated and the centroid location is iteratively adapted till all the genes are classified. This algorithm is currently being reformulated as a mixed integer-programming model.

3. Results and Discussion

The data sets chosen for the first algorithm are: (a) a subset of yeast cell cycle data by Spellman et al., (1998); (b) a subset of human fibroblast response to serum data by Iyer et al., (1999); (c) data set for the study of cutaneous malignant melanoma by Bittner et al., (2000). The yeast cell cycle data corresponding to the genes for which all the experimental attributes are available (other than the alpha factor arrest and release) is chosen. The subset of the data for human fibroblast response to serum (available in the public domain) is selected for analysis in this paper. The first of the proposed algorithms is run with the following parameter values on these three data sets: $\varepsilon_b = 0.1$, $\varepsilon_n = 0.01$, $\alpha = 0.001$, $\beta = 0.001$, $\lambda = 1000$ and $a_{max} = 100$. It is observed that the algorithm performs equally well in the three data sets chosen and is able to extract the biologically meaningful class information correctly.

For the yeast data set, nine class-centers are observed. Based on the average gene expression, a few of these centers can be grouped together to give seven centers. The normalized mean error values range from 0.0508 to 0.0806. In addition, the low error values signify a close grouping of different class members around the class center. The algorithm results in fifteen class-centers when applied to the human fibroblast data. The higher number of class centers in the present study can be attributed to the further resolution or division of classes, into subclasses, reported in the previous study. The normalized mean error for fibroblast data ranges from 0.08 to 0.25. The error rates are again observed to be low and this proves that the class members are closely grouped together. Based on the average gene expression patterns these centers can be grouped together into nine different centers.

The gene expression patterns in the present study are in excellent agreement with the clusters "A-D, G-J" as reported in a previous study (Iyer et al., 1999). In the latter study, the superposition of the two clusters "E and F" results in a similar expression patterns found in the present study. We can therefore say that the two studies are in excellent agreement except that our approach led to a better resolution of the classes.

The cutaneous melanoma data was divided into two sets to avoid computer memory limitations during analysis. The even-numbered genes from the parent set were put into one group, while the odd-numbered genes were put into another group. Both groups were analyzed independently. The even-numbered gene group resulted in five class-centers. Based on these expression pattern we observe that we can differentiate the 19 tissue samples as in the published study. We can therefore classify the different cutaneous melanoma samples based on their gene expression patterns. This is in agreement with the findings of Bittner et al. (2000). The normalized mean error for these class centers lies in the range of 0.1395 – 0.7623. Three class-centers are observed in the odd-numbered gene group. The two sets have similar overall expression patterns. It should be noted that the two groups belong to one parent data set and should therefore, *ideally*, follow the same pattern. The difference in the resulting number of centers in the two groups analysis can be attributed to the fact that there is only a small fraction of genes that have different critical gene expression and are able to classify the samples in two classes. This fact is also in accordance with the findings in (Bittner et al., 2000). The normalized mean error for these class centers lies in the range of 0.1761 – 0.5716. The high error values seen in both of these groups are not a concern as the actual gene expression values for some samples greatly exceeded their fixed value of 50.00 in the analysis. Thus, the percentage error was quite low in these cases too.

The second algorithm ("ACA") was run with the fibroblast data sets. It is observed that 17 clusters are obtained by this method, as shown in Figure 1, but most of the genes are classified within first seven clusters and the remaining ten clusters have very low cardinalities. This can be seen as outliers being classified as individual clusters. Thus, this algorithm outperforms the previous algorithm in classifying the outliers.

Class prediction (the assigning of functionality to genes with previously unknown functions) can be done based on the assumption that genes with similar gene expression patterns may have similar functions (similarity by homology). Thus, genes with known function can define the function of other genes in the same class, whose function is currently unknown.

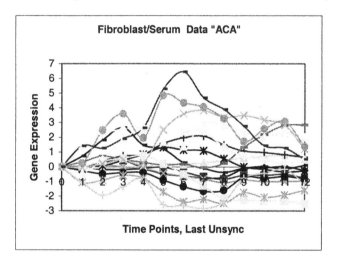

Figure 1. Average Gene Expression Patterns for Individual Classes

4. Conclusions

The proposed algorithms learn the true number of functional classes, which are also biologically significant, without requiring *a priori* information about the number of centers, as in self-organizing maps (Tamayo et al., 1999) or in support vector machines (Brown et al., 2000). The algorithms are unbiased as they use only the Euclidian distances and local error measures for the learning process and no assumption is made about the correlation of different genes. There is no dimensionality reduction and hence no knowledge loss, as in principal component analysis or multidimensional scaling. No assumption about the data distribution is made and, thus, the algorithms are expected to work for both hierarchical and non-hierarchical data sets. Therefore, the present study establishes the feasibility of the proposed algorithms for class discovery and shows that these algorithm excel over other algorithms by not having any of the disadvantages compared to reported algorithms for data analysis of gene expression data. Class prediction is done based on class membership, assignment by homology, genes with previously unknown functionality are assigned new functions similar to other genes in that class. This is envisaged to be a good starting basis for further analysis and performing experiments by molecular biologists.

5. References

Bittner, M., Meltzer, P., Chen, Y., Jiang, Y., Seftor, E., Hendrix, M., Radmacher M., Simon R., Yakhini, Z., Ben-Dor, A., Sampas, N., Dougherty, E., Wang E., Marincola, F., Gooden, C., Lueders, J., Glatfelter, A., Pollock, P., Carpten, J., Gillanders, E., Leja, D., Dietrich, K., Beaudry, C., Berens, M., Alberts, D., Sondak, V., Hayward, N. and Trent, J., 2000, Molecular classification of cutaneous malignant melanoma by gene expression profiling. *Nature* **406,** 536-540.

Brown, M.P., Grundy, W.N., Lin, D., Cristianini, N., Sugnet, C.W., Furey, T.S., Ares, M. Jr. and Haussler, D., 2000, Knowledge based analysis of microarray gene expression data by using support vector machines. *Proc. Natl. Acad. Sci. USA* **97,** 262-267.

Eisen, M.B., Spellman, P.T., Brown, P.O. and Botstein, D., 1998, Cluster analysis and display of genome-wide expression patterns. *Proc. Natl. Acad. Sci. USA* **95,** 14863-14868.

Fritzke, B., 1994a, Fast learning with incremental RBF networks. *Neural Processing Letters* **1(1),** 5.

Fritzke, B., 1994b, Growing cell structures – a self-organizing network for unsupervised and supervised learning. *Neural Networks* **7(9),** 1460.

Fritzke, B., 1995, A growing neural gas network learns topologies. *Advances in Neural Information Processing Systems* **7,** 625-632.

Hastie, T., Tibshirani, R., Eisen, M.B., Alizadeh, A., Levy, R., Staudt, L., Chan, W.C., Botstein, D. and Brown, P., 2000, 'Gene shaving' as a method for identifying distinct set of genes with similar expression patterns. *Genome Biology,* **1,** 1-31.

Iyer, V.R., Eisen, M.B., Ross, D.T., Schuler, G., Moore, T., Lee, J.C.F., Trent, J.M., Staudt, L.M., Hudson, J. Jr., Boguski, M.S., Lashkari, D., Shalon, D., Botstein, D. and Brown, P.O., 1999, The transcriptional program in the response of human fibroblasts to serum. *Science* **283,** 83-87.

Martinetz, T.M. and Schulten, K.J., 1991, A "neural gas" network learns topologies. In Kohonen, J., Makisara, K., Simula, O., and Kangas J., Editors, *Artificial Neural Networks,* 397-402, North-Holland, Amsterdam.

Raychaudhuri, S., Stuart, J.M. and Altman, R.B., 2000, Principal components analysis to summarize microarray experiments: application to sporulation time series. *Pac. Symp. Biocomput.* **5,** 455-466.

Spellman, P.T., Sherlock, G., Zang, M.Q., Iyer, V.R., Anders, K., Eisen, M.B., Brown, P.O., Botstein, D. and Futcher, B., 1998, Comprehensive identification of cell- cycle-regulated genes of the *Saccharomyces cerevisae* by microarray hybridization. *Mol. Biol. Cell* **9,** 3273-3297.

Tamayo, P., Slonim, D., Mesirov, J., Zhu, Q., Kitareewan, S., Dmitrovsky, E., Lander, E.S. and Golub, T.R., 1999, Interpreting patterns of gene expression with self organizing maps: methods and application to hematopoietic differentiation. *Proc. Natl. Acad. Sci. USA* **96,** 2907-2912.

European Symposium on Computer Aided Process Engineering – 12
J. Grievink and J. van Schijndel (Editors)
© 2002 Elsevier Science B.V. All rights reserved.

A Concurrent Engineering Approach for an Effective Process Design Support System

Solomon S. Gelete, René Bañares-Alcántara and Laureano Jiménez

University of Rovira i Virgili., Av. Països Catalans 26;. 43007 Tarragona, SPAIN. Tel.: +34-977-559673; Fax: x9667/21.
e-mail: rbanares, sgelete, ljimenez@etseq.urv.es

Abstract

The Concurrent Engineering (CE) methodology applied to chemical process design has the potential to improve the performance of chemical process design. This work presents a software prototype, CEPD/MODEL, that uses commercial software (e.g. HYSYS and AXSYS) and some tools developed in-house (e.g. CEPD-ART and CEPD-DOC) for the application of Concurrent Engineering to process design. Among the several requirements for a successful application of Concurrent Engineering in design we examine in some detail tool integration, organisational requirements related to the workflow and document management requirements. The paper presents the initial stages of the development of CEPD/MODEL, and how it can be used as a design support system.

1. Introduction and Objectives

The adaptation of a CE methodology to chemical process design has the potential to reduce design time and cost, and improve the quality of the designed chemical process, impact assessment and decision integration. These ideas have been applied successfully for more than 15 years in the aerospace, automotive, communications, and hardware/software industries with savings of up to 50% in development time and 30% of development costs (Rossenblat, 1991). Recent analyses of the impact of the application of new CAPE tools and techniques (which include the application of the CE approach) on the European process industries have estimated conservatively a reduction of 10% in investment costs and 5% in operation costs, i.e. around 9.8 and 10.95 bn euros/yr respectively (Perris & Bolton, 2001).

2. Requirements for the application of CE in process design

There are many definitions for the term "Concurrent Engineering" (CE). The definition adopted in our work is taken from the Concurrent Engineering Network of Excellence (CE-NET, 1998):

"Concurrent Engineering is a systematic approach for the integration and parallelisation of the design of a product and all its related processes (e.g. building, production, decommissioning) by concurrently taking into account upstream and downstream requirements".

The term *integration* relates to the incorporation of requirements that come from inter- and intra-organisational context, for example the alignment of the goals of the people and organisations involved in the development of a product. In turn, the term *parallelisation* refers to the concurrent execution of processes.

The CE approach is similar to the "tiger team" approach in small organisations, where a small group of people works closely for a common endeavour. However, for larger organisations the tiger team concept needs to be modified and restructured because team members can be at geographically different, networked locations, requiring far-reaching changes in the work culture and ethical values of the organisation (Tad, 1992). An example of these changes is to ensure that the hierarchical position of an individual is not a barrier to information exchange.

At this point it is also important to define the term Front End Engineering Design (FEED). FEED is a stage in the process life cycle encompassing the whole of its conceptual design and parts of its detailed design (including the specification of parts of the control, safety and environmental systems of the process). At FEED about 50% of process design task is completed and 80% of the cost of the design is committed (Battikha, 1996). FEED places a strong emphasis on an increased involvement of the client and on the revision of alternatives. Its output is a set of documents known as a *bid package*, typically consisting of a PFD, a P&ID, simulation cases and cost estimations, equipment data sheets, requests for quotation responses, equipment and piping layouts and 3D models.

The application of CE ideas and techniques involves a complex interaction of social and technological factors, see Figure 1. The current state of the art suggests that a generic information management system satisfying all the requirements neither exists nor will it be developed in the near future. However, the technology exists to implement substantial parts of CE and improve significantly current practice. Cultural, organisational and human factors needed for CE require new technological support. At the same time, new technologies will not be successful without taking into account their implications in the social factors (Kimura et al., 1992). Therefore, to improve business practices the application of new technology is needed, but improvements in technology will have a limited impact unless the underlying business processes also change to exploit these new capabilities (Bañares-Alcántara, 2000).

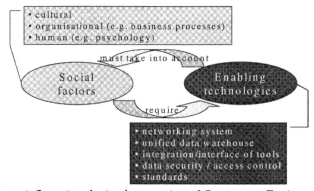

Figure 1. Factors influencing the implementation of Concurrent Engineering.

Among the several requirements for a successful application of Concurrent Engineering in design we examine in some detail the:

- Organisational requirements:
  Related to the workflow, i.e. an ordered sequence of activities that conform a project.
- Document handling:
  An analysis of the types of documents used and generated during the process design workflow, their possible handling mechanisms and the important features to consider when developing a document-handling tool.

3. CEPD/MODEL as a Prototype to Test the Application of CE

3.1. CEPD/MODEL

CEPD/MODEL is a software prototype designed to test the applicability of the Concurrent Engineering methodology during process design. As a precondition to fulfil the previous objective CEPD/MODEL must meet the requirements of a CE environment, in particular with regard to the integration of tools, transfer of information and document management. The workflow in CEPD/MODEL consists of the following steps:

1. Process simulation for conceptual design and as a tool in the final steps of process synthesis.
2. Generation of PFDs and P&IDs.
3. Integration with equipment detailed design applications.
4. Costing.
5. Generation of datasheets and reports.
6. Revision and work approval.
7. Document handling.

3.2. Requirements of CEPD/MODEL

One of the important targets of CE implementation is the availability of a document management system, i.e. a system to maintain access and manipulate documents quickly, securely and cost effectively. It has been reported that mismanaged files may account for up to half of the "unexpected" project delays (Linton & Hall, 1992). Manual procedures to manage paper documents are simply inadequate for electronic documents mainly because of poor security. Furthermore, just storing the files on a PC hard drive does not mean that they are being efficiently managed because electronic revisions, copies, backups and releases multiply each day. For this reason it is essential to have a document management system with features helpful to facilitate the automation of the CE process, these might include (Gelete, 2001):

- Document organisation and access (finding a document and allowing the examination of its contents).
- Flexibility in organisation (regrouping documents without physically moving them).
- Easy to access profile information.
- Document search capability and document security.
- Allowing a quick comparison for revision.

- Automation in the workflow.

The other basic aspect is the availability of the process design tools and their features related to integration and automation. The growth in technology and information management points to the availability of new process design tools and means of communication and interaction in the market. However, their application in the area of concurrent engineering for process design needs a substantial improvement in terms of interfacing the applications, document handling, management and automation. CEPD/MODEL encompasses a number of activities and tools (commercial and developed in-house) and their integration, see Table 1.

Table 1. Activities and tools in CEPD/MODEL.

Activity	Tools used by CEPD/MODEL
Process simulation	HYSYS$^{®}$
CAD	AUTOCAD-R14$^{®}$
PFD and P&ID generation	AXSYS$^{®}$
Detailed design of equipment	Various, e.g. HTFS$^{®}$ *(planned)*
Costing	ICARUS 2000$^{®}$ *(planned)*
Datasheet and report generation	AXSYS$^{®}$, Excel™, Access™
Work revision and approval	CEPD/ART
Document management	CEPD/DOC

Three of these packages warrant some further explanation:
- AXSYS$^{®}$ is commercial database software that provides storage and use of process stream, unit operation, equipment sizing and cost estimates data. It can also generate automatically PFD and P&ID diagrams from a HYSYS$^{®}$ output based on customisable rules.
- CEPD/ART is a software tool being developed for the revision of data extracted from AXSYS$^{®}$, e.g. lists the changes done to the approved data for design revision purposes. It is being implemented in Microsoft Access™, Visual Basic™ and Excel™.
- CEPD/DOC is a network-based document-handling tool for file storage and retrieval. It is being implemented in Microsoft Access™ and Visual Basic™. It is currently being expanded into a document management system compliant with the requirements listed in Section 3.

The components of CEPD/MODEL are organised around a database system as shown in Figure 2.

4. Case study

A full test of the workflow and tool integration capabilities of CEPD/MODEL is under development. In the meantime, and based on data from four sections of the refinery operated by REPSOL-YPF in Tarragona, a test of the integration and capabilities of the tools and of the potential of CE in the area has been completed as a case study. The test is done using data taken from a section of the Olefin (U 661) refinery plant of Repsol, in which the production of monomers such as ethylene and propylene is completed. It also

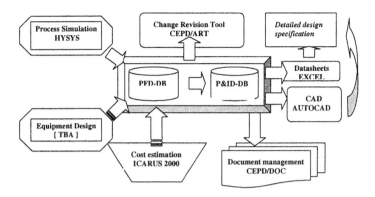

Figure 2. Structure of CEPD/MODEL.

produces sub-products such as ethane, propane, hydrogen, methane, C4s, aromatic components and fuel oil by pyrolysis.

This part of the plant takes naphtha as a feed and also propane, butane or residues of isomers generated in other parts of the site or imported from other sites.

The CEPD/MODEL testing simulation is done considering a section of the processes listed below.

1. Precooler , process gas drying and de-ethaniser.
2. Hydrogenation of acetylene.
3. Separation and purification of light gases.
4. Purification of ethylene.

The workflow of the test case in CEPD/MODEL includes:

1. Simulation of the whole section using HYSYS Plant.
2. Simulation data loading from HYSYS to AXSYS, and development of PFD and P&ID diagrams.
3. Detailed design work in AXSYS, including PFD and P&ID, completion and approval at the logout as a checking and approval of the original work.
4. Loading of the design data for the sake of later revision, and also its approval status information into CEPD-ART.
5. Design document handling using CEPD-DOC by manual data storage. This includes typical information about equipment from the company and registration and retrieval checking.
6. Data and information exchange with AUTOCAD R-14 for design specification format, using AXSYS customisation.
7. Data sheet and reports generation with the help of AXSYS

Significant improvement is expected by the application of CE for the improvement of process design workflow and by organising individual activities within the team work such as simulation, PFD and P&ID drawing, revision of design work and documentation and data sheet specification preparation. The considerations in the case study could also indicate that CE application can be automated if these activities are integrated and appropriately organised. Design revision time and cycle is the other concern that requires attention. CEPD- ART as a tool able to record and display the change in name, code, type (among others) in the approved file and the revision work, is thought to be helpful in reducing the revision time and work.

5. Conclusions

There are several advantages to be obtained from the development and application of a software model such as CEPD/MODEL, the main one being the increased understanding of alternative workflows for the FEED stage of process design. We also expect to obtain the same sort of advantages observed in other areas where CE has been applied, i.e. time reduction, improved workflow management, and an increased automation of parts of the design process. Valid conclusions could be drawn from the case study which show the future potential of CE application in process design:

- A Concurrent Engineering approach automation is a short-term possibility. Process design tool integration shows a potential in the area of process design workflow management and also in the improvement of organisation and work performance during process design.
- During the FEED stage of process design, considerations of other important process steps in the overall workflow can apparently be possible. This points to automation of the design process as an ideal target.
- The CEPD/MODEL approach is compliant with the CE approach as it permits task assignment and accomplishment through distant communication.

An important extension in the development is assisting the process design revision through transfer of relevant data and information from the records of the detailed design stage to the simulation/conceptual design stage. It is envisaged that this reverse flow of design information will increase the level of automation of the revision process.

6. References

Bañares-Alcántara, R., 2000, "Concurrent Process Engineering. State of the Art and Outstanding Research Issues", http://capenet.chemeng.ucl.ac.uk/ CAPE.Net website.

Battikha, N.E., 1996, "Applying Front-End Engineering to Control Projects", ICI Explosives, PE Magazine.

CE-NET, 1998, CE Network of Excellence, http://esoce.pl.ecp.fr/ce-nt/

Gelete, S.S., 2001, "A Concurrent Engineering Approach for an Effective Process Design Support System", Tesina, Dept. of Chem Eng, Univ. Rovira i Virgili.

Kimura, F., T. Kjellberg, F. Krause and W. Wozny, 1992, Conclusions of the First CIRP Int. Workshop on CE for Product Realisation, Tokyo, Japan.

Linton, L. and D. Hall, 1992, "First Principles of Concurrent Engineering – A Competitive Strategy for Product Development", CALS Technical Report 005.

Perris, T. and L. Bolton, 2001, "CAPE-21. Feasibility Study", http://CAPE-21.ucl.org.uk.

Rossenblatt, A., 1991, "Concurrent Engineering", IEEE Spectrum, July.

Tad, A.D., 1992, "Advancing CE using STEP", Concurrent Engineering Research, DARPA Project report, West Virginia.

Acknowledgements

We acknowledge the support of Hyprotech for the use of HYSYS® and AXSYS® under a university agreement, and to REPSOL-YPF for access to some of its process data.

European Symposium on Computer Aided Process Engineering – 12
J. Grievink and J. van Schijndel (Editors)
895

Agent-based Refinery Supply Chain Management

N. Julka, I. Karimi, R. Srinivasan
Department of Chemical and Environmental Engineering
National University of Singapore

Abstract

The refinery business involves tasks spanning across several departments and requiring close coordination and collaboration among them. Present-day business decision support systems are tuned to individual departments and their work processes and are incapable of exploiting the intricate interdependencies among them. There is a need for a decision support system that provides a structured way to make decisions at the enterprise level. In this paper, we address this critical need by developing an agent-based decision support system for refinery supply chain management. Software agents are used to emulate the internal departments of the refinery. These agents gather relevant information from across the enterprise, make decisions based on the embedded business policies and mimic the different business processes of the modeled enterprise. Uncertainties in decision-making are captured by stochastic elements embedded in the agents. Through this, the dynamics in the supply chain is emulated and the performance of the supply chain measured through department-specific key performance indicators.

1. Introduction

Decision-making is distributed across various departments in a refinery. Each department solves its own locally focused problems. But local improvements do not necessarily ensure that the overall process attains an optimum. Many times, the departmental objectives may conflict, and all decisions may not contribute positively to the overall performance of the refinery. Furthermore, the decision support systems available for these problems are disjoint and thus inadequate. They are unable to (1) integrate all the decision-making processes of a refinery (2) interface with the present systems in place (3) incorporate dynamic data from various sources and (4) assist various departments (e.g. operations, procurement, storage, logistics, etc.) at the same time. The need for integrated decision-making is obvious and is the new approach to manage business. The challenges associated with this approach include the difficulty of modeling the entire system and integration with present decision support systems for individual departments or tasks. In addition, the opportunity that the Internet offers in exchanging data seamlessly needs to be exploited in the B2B decision-making processes. In case of a refinery supply chain, information on the web includes the prices of crude, logistics, products, etc. The crude procurement process involves several departments and is critical to the refinery's bottom line. This paper describes a decision support system for refinery supply chain management.

Refinery supply chain management solutions have predominantly focused on various sub-problems of the complete refining business. Existing research addresses the optimization of refining operations, pooling and blending, planning and scheduling, and a limited integration of some aspects. Rigby et al. (1995), Amos et al. (1997) and Adhya et al. (1999) have addressed the pooling and blending problem for crudes and stored products in a refinery. Pinto et al. (2000) review the research on planning and scheduling. Zhang et al. (2001) propose a model to integrate the oil liquid flow system, the hydrogen system and the steam and power system for simultaneous optimization. Sullivan (1990) emphasizes the integration of the operations and process functions, with blend control and optimization strategies in a refinery, to achieve global optimization. Apart from manufacturing, business processes such as oil trading, logistics and product delivery are also important. Refineries need an integrated supply chain vision to achieve the desired competitive advantage in the present day dynamic business environment. An approach using the optimization of a model of all refinery supply chain entities, their relationships and associated processes is a challenging and computationally intensive problem. On the other hand, an approach using simulation allows only scenario modeling, and analysis is left to the user. A new approach is to blend optimization and simulation to create decision support systems (Padmos et al., 1999).

2. PRISMS

In an earlier paper (Julka et al., 2000), we proposed an agent-based framework for the modeling and analysis of a general supply chain. We classified the elements of supply chains as entities and flows. Entities include the operators in a supply chain, e.g. manufacturers and their internal departments, oil suppliers, internet exchanges, etc. These are modeled as emulation agents. Material and information are modeled as commodity and message objects. These objects are exchanged between agents to simulate the material and information flow in the supply chain network. The business processes of each entity in the supply chain are embedded into its agent in the form of grafcets. Each task that the entity performs is modeled as a thread in the grafcets. Threads are triggered by events such as the arrival of a message from another agent or the occurrence of a landmark event. Task threads have steps and transitions representing the various activities performed by the agent such as processing information, acquiring data, recording history and delegating tasks to other agents. Adding task threads to the grafcets chart increase the functionality of an agent. This agent-based supply chain modeling framework provides an environment where all business processes can be emulated in an integrated manner. Various supply chain scenarios can be configured; simulated and analyzed using this model based on user defined metrics or key performance indices (KPIs). The Petroleum Refinery Integrated Supply Chain Modeler and Simulator (PRISMS) utilizes this framework and is developed using Gensym's G2.. It models a refinery with procurement, sales, logistics, storage and operations departments. The refinery supply chain consists of oil suppliers, 3rd party logistics providers (3PL) and electronic exchanges for oil trading. The next section describes the modeled refinery and its crude procurement process.

2.1 Refinery Model

Our objective was to develop a system that allows the modeling and simulation of a refinery supply chain. The system should assist in evaluating the impact of different policy and planning parameters on the overall working of a refinery based on user-defined metrics. The entities in the refinery supply chain are:

1. Oil Exchange: Oil suppliers submit postings about crudes available and their prices.
2. Oil Suppliers: Sell crude oil to the refineries.
3. 3PLs: Transport crude from the oil supplier terminals to the refinery.

A brief overview of the internal departments of the refinery is as follows.
1. Procurement department coordinates the crude procurement process. It retrieves availability postings from the exchange and decides on crudes to consider for purchase by taking into account the crude properties, refinery targets, product data and logistics information.
2. Sales: The sales department provides the departments with the present and forecasted product prices and demands.
3. Storage: The storage department schedules the tanks and jetty. It also issues the requested amounts of crudes to the operations department.
4. Operations: This department decides the crudes to be run through the refinery every day. It also decides the various operating parameters for the refining process.
5. Logistics: The logistics department arranges 3PLs for the transport of crude from the oil supplier terminal to the refinery. It performs this through a bidding process.

For modeling purposes, the refinery business process is divided into three sub-processes crude selection and purchase, crude delivery and storage, and crude refining.

2.1.1 Crude Selection and Purchase

Crude procurement is a key business process in the refinery management. It is also very critical to refinery operations, as shutdown costs are huge and to be avoided under all circumstances. Most refineries purchase many crudes and process them in various mixes. Products with different quality can be obtained from the same crude mix by blending products and varying cut points. Volatility in crude/product prices, crude availabilities and product demands all impact the optimum crude mixes, purchased and refined. Jetty and storage tank scheduling, choice of logistics, uncertainty in ship arrivals, etc. impact the crude procurement and scheduling process. The crude procurement thus is a multi-dimensional process with a single objective of maximizing profit. Figure 1 maps the crude selection and purchase sub-process. The process is initiated based on the present stock of crudes and the estimated ship arrival schedule. Based on a forecast of product prices and demands, the procurement department evaluates the crudes available on the exchange and computes their netback values. The procurement team selects a set of crudes (crude basket) based on these netback values and sends the crude basket to the operations department. The operations department refines the basket based on any operational constraints or previous experience and returns it back to the procurement department. The procurement department compiles the locations and times of all crude pickups and forwards it to the logistics department.

2.1.2 Crude Delivery and Storage

The delivery and storage sub-process has three steps (Fig 2a) – dispatch of a ship for crude pickup, arrival of the ship at the pickup terminal and loading of the oil in the ship, and actual arrival of the ship at the refinery jetty and the unloading of oil from the ship.

2.1.3 Crude Refining

The crude refining process represents the actual processing in the refinery (Figure 2b).

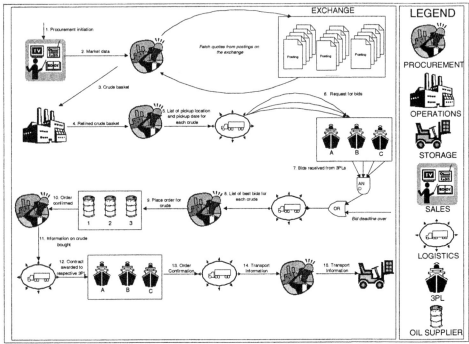

Figure 1. Modeled Refinery's Crude Purchasing Process

In PRISMS, the user creates a scenario by defining the refinery configuration, selecting the refinery supply chain entities, setting the planning parameters and specifying the simulation details. The prices of crudes and products and the costs of logistics are modeled stochastically and are also input to the system. Once a scenario is configured, the refinery operation is simulated for a specified number of days. The results of simulation are stored as attributes of the respective agents in PRISMS. The messages exchanged among the agents capture the communication between various entities of the refinery supply chain. PRISMS allows the user to add key performance indices (KPIs) such as production profiles, inventory profiles, crude quality profiles in storage tanks, supply demand curves, etc. by defining procedures to compute them.

Other assumptions of the modeled refinery are: (1) The refinery has five products: Gas to C4, Gasoline, Kerosene, Gas oil and Residue (2) Only one type of crude is procured in each cycle (3) There is only one jetty for the refinery (4) Crude is unloaded directly into storage tanks (5) The refining of crude is modeled as a single day batch process, i.e. crude is released to the operations in the morning and the products are produced on the same day (6) A tanker takes one day to unload all crude into storage tanks, and. (7) Crude mixtures have linear relationships with respect to the products.

3. Case Studies

PRISMS has been implemented to simulate the refinery supply chain described above. Here, we present three case studies to illustrate PRISMS' ability to support the following types of decisions:
1. Policy changes: Evaluate and compare refinery business policies
2. External changes: Understand refinery's response to business environment changes
3. Plant configuration: Evaluate impact of changes in plant configuration

Study 1: Impact of Procurement Policy

In each procurement cycle, crudes are purchased in discrete packets comprising one crude each. The amount of crude to be purchased in each packet is an important decision for the refinery. Two different practices are possible – fixed packet size and packet size based on the present inventory, schedule of crude arrivals and demands for products. In this study, we evaluated the impact of the fixed vs. variable packet size policies. For the former, a packet size of 125 kbbl was used. Six runs were performed for each scenario using a four-month horizon. The average throughput was calculated using the closing stocks on each day of production. Simulations revealed that the use of variable packet size reduced the average throughput by about 6% (106.9 kbbl/d to 100.3 kbbl/d) and decreased the average inventory level by about 6% (1.18 Mbbl to 1.12 Mbbl). Standard deviations for the throughput were 21.1 kbbl/d and 23.7 kbbl/d for the fixed and variable packet sizes respectively, while those for the inventory level were 0.17 Mbbl and 0.18 Mbbl. This reveals that the variable packet size policy lowers inventory levels while resulting in the same cycle-to-cycle variability as indicated by the similar values of the standard deviation.

Study 2: Effect of Demand Fluctuation

The ability to handle demand fluctuations is a key to a refinery's economic performance. Production targets are decided based on product demands estimated by the sales department. In this study, we evaluated the robustness of the refinery operation to volatile demand patterns and compared it against a relatively static demand. The average demand in both cases (volatile and static) was kept same, but the standard deviation changed. The refinery's ability to match the target production was quantified by the Root Mean Square of Percentage Shortfall (RMSPS) in daily production. It was observed from six runs of each scenario that the refinery business process could not effectively handle a volatile demand as the RMSPS was almost double (13.47 kbbl) that in the static demand scenario (6.56 kbbl).

Study 3: Benefit of Extra Storage Capacity

To enhance the refinery's inability to handle demand volatility, the option of increased capacity (additional 15% by adding another tank) in the tank farm was studied. It was observed that in this case, the RMSPS reduced to 9.49 kbbl and the additional inventory helped the refinery match the spikes in the demand.

4. Conclusions

Integrated supply chain management is crucial in today's business environment. The lack of decision support systems has led to business processes and policies being evaluated in isolation without consideration of their impact on the overall business performance. In this paper, we have proposed a system for emulating supply chains and illustrated it using the crude procurement process in a refinery. Individual departments of the refinery are modeled as intelligent agents that emulate the different business processes of the department. The system can be used to study the effects of internal policies of a refinery upon KPIs. A key application of the system is for comparing different business policies under a variety of business scenarios in order to identify the ones suitable for actual implementation in the enterprise. Future work will include embedding detailed solution techniques used by individual departments and studying

900

the effectiveness of these with reference to the whole refinery. Additional case studies dealing with oil trading decisions and long-term contracts are also planned.

References

Adhya, N., M. Tawarmalani and N.V. Sahinidis. A Lagrangian Approach to the Pooling Problem, Industrial Engineering and Chemistry Research, 38 pp. 1956-1972. 1999.

Amos, F., M. Ronnqvist and G. Gill. Modelling the pooling problem at the New Zealand Refining Company, Journal of the Operations Research Society, 48 pp. 767-778. 1997.

Julka, N., R. Srinivasan, I. Karimi, N. Viswanadham and A. Behl. Enabling Framework for Decision Support Systems in the e-Commerce Era. In AIChE, 12 -17 Nov. L.A., U.S.A.2000

Padmos, J., B. Hubbard, T. Duczmal and S. Saidi. How i2 integrates simulation in supply chain optimization. In Proceedings of the 1999 Winter Simulation Conference, Dec 5-8. Phoenix, Arizona, USA, pp. 1350-1355.

Pinto, J.M., M. Joly and L.F.L. Moro. Planning and scheduling models for refinery operations, Computers and Chemical Engineering, 24 pp. 2259-2276. 2000.

Rigby, B., L.S. Lasdon and A.D. Warren. The evolution of Texaco's blending systems: from OMEGA to StarBlend, Interfaces, 25 (5), pp. 64-83. 1995.

Sullivan, T.L. Refinery-wide blending control and optimization, Hydrocarbon Processing, May pp. 93-96. 1990.

Zhang, J., X.X. Zhu and G.P. Towler. A Simultaneous Optimization Strategy for Overall Integration in Refinery Planning, Industrial and Engineering Chemistry Research, 40 (12), pp. 2640-2653. 2001.

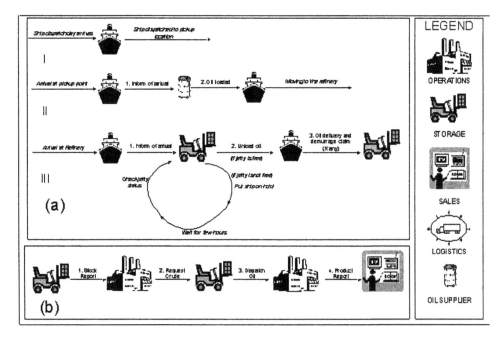

Figure 2. Refinery (a) Crude Delivery and Storage Process and (b) Crude Refining Process

European Symposium on Computer Aided Process Engineering – 12
J. Grievink and J. van Schijndel (Editors)
© 2002 Elsevier Science B.V. All rights reserved.

Using Continuous Time Stochastic Modelling and Nonparametric Statistics to Improve the Quality of First Principles Models

Niels Rode Kristensen[a], Henrik Madsen[b] and Sten Bay Jørgensen[a]

[a] Computer Aided Process Engineering Center, Department of Chemical Engineering
[b] Mathematical Statistics Section, Informatics and Mathematical Modelling
Technical University of Denmark, DTU, DK-2800 Lyngby, Denmark

Abstract

A methodology is presented that combines modelling based on first principles and data based modelling into a modelling cycle that facilitates fast decision-making based on statistical methods. A strong feature of this methodology is that given a first principles model along with process data, the corresponding modelling cycle can be used to easily, rapidly and in a statistically sound way produce a more reliable model of the given system for a given purpose. A computer-aided tool, which integrates the elements of the modelling cycle, is also presented, and an example is given of modelling a fed-batch bioreactor.

1. Introduction

The increasing use of computer simulations in analysis and design of process systems and recent advances in model based process control and process optimisation have made the development of rigorous dynamic process models increasingly important over the past couple of decades. Particularly in view of the increasing focus on batch and fed-batch operation in many areas of the process industry, the ability of such process models to describe nonlinear and time-varying behaviour has also become more important.

Altogether, these developments have necessitated faster development of new and improvement of existing first principles models, i.e. models based on physical insights and conservation balances. The purpose of this paper is to show how *continuous time stochastic modelling* and time series analysis tools based on *nonparametric statistics* can be used to facilitate this. Continuous time stochastic modelling is a grey-box approach to process modelling that combines deterministic and stochastic modelling through the use of stochastic differential equations (SDE's) and has previously been described in Kristensen et al. (2001a). Other previous contributions in the area of grey-box modelling include the work of Madsen and Melgaard (1991) and Bohlin and Graebe (1995) and references therein.

The outline of the paper is as follows: In Section 2 the overall methodology is described in terms of a modelling cycle, some details of the individual elements of this cycle are given, and a computer aided tool that facilitates the use of the overall methodology is briefly described. In Section 3 a case study is presented that shows how the methodology can be used to improve the quality of a first principles model of a simple fed-batch bioreactor. Conclusions are given in Section 4.

2. Methodology

The overall methodology can be described in terms of Figure 1, which shows the proposed continuous time stochastic modelling cycle described in the following.

2.1 Model construction

The first step in the modelling cycle deals with construction of the basic model, which is a *continuous-discrete stochastic state space model* consisting of a set of SDE's describing the dynamics of the system in continuous time and a set of algebraic equations describing measurements at discrete time instants, i.e.

$$dx_t = f(x_t, u_t, t, \theta)dt + \sigma(u_t, t, \theta)d\omega_t \tag{1}$$

$$y_k = h(x_k, u_k, t_k, \theta) + e_k \tag{2}$$

where t is time, x_t is a vector of state variables, u_t is a vector of input variables, y_k is a vector of measured output variables, θ is a vector of unknown parameters, $f(\cdot)$, $\sigma(\cdot)$ and $h(\cdot)$ are nonlinear functions, ω_t is a Wiener process and e_k is a $N(0, S(u_k, t_k, \theta))$ process. A detailed account of the advantages of using SDE's is given in Kristensen et al. (2001a).

2.2 Parameter estimation

The second step in the modelling cycle deals with estimation of the unknown parameters θ in (1) and (2) using data sets from one or more experiments. The properties of the basic model allow statistical methods to be applied for this purpose, e.g. *maximum likelihood* (ML) estimation, or *maximum a posteriori* (MAP) estimation if prior information about the parameters is available. More specifically, the unknown parameters can be determined by solving a variant of the optimisation problem

$$\hat{\theta} = \arg \min_{\theta \in \Theta} \left\{ -\ln \left(\left(\prod_{i=1}^{S} \prod_{k=1}^{N_i} \frac{\exp\left(-\frac{1}{2} \varepsilon_k^{i\,T} R_{k|k-1}^{i}{}^{-1} \varepsilon_k^{i}\right)}{\sqrt{\det R_{k|k-1}^{i}} \sqrt{2\pi}^l} \right) p\left(y_0^i \mid \theta\right) \right) \frac{\exp\left(-\frac{1}{2} \varepsilon_\theta^T \Sigma_\theta^{-1} \varepsilon_\theta\right)}{\sqrt{\det \Sigma_\theta} \sqrt{2\pi}^p} \right\} \tag{3}$$

by further conditioning on the initial conditions y_0^i in the individual experiments. ε_k^i and $R_{k|k-1}^i$ are the mean and covariance of the innovations from an extended Kalman filter at the k'th sample in the i'th experiment, and ε_θ and Σ_θ are the deviation from, and the covariance of a prior estimate of the parameters. A more detailed account of this formulation is given in Kristensen et al. (2001a), and details about the algorithms behind the corresponding estimation methods can be found in Kristensen et al. (2001b).

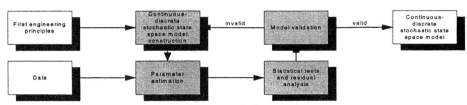

Figure 1. The continuous time stochastic modelling cycle.

2.3 Statistical tests and residual analysis

The third step in the modelling cycle deals with assessing the quality of the model once the unknown parameters have been estimated. The estimators described above are all approximately Gaussian, meaning that t-tests can be performed to test the hypothesis that a parameter is marginally insignificant. The test quantity is the value of the estimate of the parameter divided by the standard deviation of the estimate and is approximately t-distributed with a number of degrees of freedom that equals the number of data points minus the number of estimated parameters. To test the hypothesis that some parameters are simultaneously insignificant, several tests can be applied, e.g. a likelihood ratio test, a Lagrange multiplier test or a test based on Wald's W-statistic. These test quantities all have the same asymptotic χ^2-distribution with a number of degrees of freedom that equals the number of parameters to be tested for insignificance, but in the context of the proposed modelling cycle Wald's test has the advantage that no re-estimation is required. Details about the derivation of this statistic are given in Holst et al. (1992).

Another important aspect in assessing the quality of the model is to investigate its predictive capabilities by performing cross-validation and examining the corresponding residuals. Depending on the intended application of the model this can be done in a one-step-ahead prediction setting or in a pure simulation setting, and one of the most powerful methods is to compute and inspect the sample autocorrelation function (SACF) and the sample partial autocorrelation function (SPACF) of the residuals to detect if there are any significant lag dependencies, as this indicates that the model is incorrect. Nielsen and Madsen (2001) recently presented extensions of these linear tools to nonlinear systems, the lag-dependence function (LDF) and the partial lag-dependence function (PLDF), which are based on the close relation between correlation coefficients and values of the coefficients of determination for regression models and which extend to nonlinear systems by incorporating nonparametric regression in the form of additive models. In the context of the proposed modelling cycle the ability of the LDF and the PLDF to detect nonlinear lag-dependencies is particularly important.

2.4 Model validation

The last step in the modelling cycle deals with model validation or invalidation, or, more specifically, with whether, based on the information gathered in the previous step, the model is invalidated with respect to its intended application or not. If the model is invalidated, the modelling cycle is repeated by first changing the structure of the model in accordance with the information gathered in all steps of the previous cycle.

2.5 A computer aided tool for continuous time stochastic modelling

To facilitate the use of the proposed modelling cycle, a GUI-based computer-aided tool, called CTSM, has been developed, cf. Kristensen et al. (2001b). Within CTSM models of the kind (1)-(2) can be set up, unknown parameters can be estimated using a variant of (3), and statistical tests and residual analysis can be performed. CTSM is very flexible with respect to the data sets that can be used for estimation, as features for dealing with occasional outliers, irregular sample intervals and missing observations have been implemented. CTSM runs on Win32, Solaris and Linux platforms, and on Solaris platforms the program supports shared memory parallellization using OpenMP for improved performance.

3. Case study: Modelling a fed-batch bioreactor

To illustrate how the proposed modelling cycle can be used to improve the quality of a first principles model, a simple example is given. The process considered is a fed-batch bioreactor described by a simple unstructured model of biomass growth, i.e.

$$
d\begin{pmatrix} X \\ S \\ V \end{pmatrix} = \begin{pmatrix} \mu(S)X - \dfrac{XF}{V} \\ -\dfrac{\mu(S)X}{Y} + \dfrac{(S_F - S)F}{V} \\ F \end{pmatrix} dt + \begin{bmatrix} \sigma_{11} & 0 & 0 \\ 0 & \sigma_{22} & 0 \\ 0 & 0 & \sigma_{33} \end{bmatrix} d\omega_t ,
\tag{4}
$$

$$
\begin{pmatrix} y_1 \\ y_2 \\ y_3 \end{pmatrix}_k = \begin{pmatrix} X \\ S \\ V \end{pmatrix}_k + \begin{pmatrix} e_1 \\ e_2 \\ e_3 \end{pmatrix}_k , \qquad \begin{pmatrix} e_1 \\ e_2 \\ e_3 \end{pmatrix}_k \in \begin{pmatrix} N(0, S_{11}) \\ N(0, S_{22}) \\ N(0, S_{33}) \end{pmatrix}
\tag{5}
$$

where X is the biomass concentration, S is the substrate concentration, V is the volume of the fermenter, F is the feed flow rate, S_F (=10) is the feed concentration of substrate, Y (=0.5) is the yield coefficient of biomass and $\mu(S)$ is the growth rate. σ_{11}, σ_{22}, σ_{33}, S_{11}, S_{22} and S_{33} are stochastic parameters. Two different cases are considered for $\mu(s)$, corresponding to Monod kinetics with and without substrate inhibition, i.e.

$$
\mu(S) = \mu_{max} \frac{S}{K_2 S^2 + S + K_1}
\tag{6}
\qquad\qquad
\mu(S) = \mu'_{max} \frac{S}{S + K'_1}
\tag{7}
$$

In the following the model consisting of equations (4), (5) and (6), with $K_2=0.5$, is regarded as the true process to be modelled, and using the true parameter values in Table 2 two sets of data are generated by stochastic simulation. One data set is used for estimation and the other is used for validation. The model consisting of equations (4), (5) and (7) is regarded as an original first principles model, which in the context of the modelling cycle is the basic model. Using the estimation data set, the unknown parameters of the model are estimated with CTSM, giving the results in Table 1.

Table 1. Estimation results using the incorrect model structure.

Parameter	X_0	S_0	V_0	μ_{max}	K_1	σ_{11}	σ_{22}	σ_{33}	S_{11}	S_{22}	S_{33}
True value	1	0.245	1	-	-	0	0	0	0.01	0.001	0.01
Estimate	1.042	0.250	0.993	0.737	0.003	0.104	0.182	0.000	0.008	0.000	0.011
Std. Dev.	0.014	0.010	0.001	0.008	0.001	0.018	0.010	0.000	0.001	0.000	0.003
t-score	72.93	24.94	689.3	96.02	2.396	5.867	18.26	1.632	6.453	3.467	3.801
Significant	Yes	Yes	Yes	Yes	Yes	Yes	Yes	No	Yes	Yes	Yes

Results of marginal t-tests show that the only insignificant stochastic parameter is σ_{33}, whereas σ_{11} and σ_{22} are significant. This in turn indicates that the deterministic parts of the equations for X and S in (4) are incorrect in terms of describing the variations in the estimation data set. To investigate this further, residual analysis is performed. One-step-ahead prediction results on the validation data set are shown in Figure 2 and Figure 3 shows the SACF, SPACF, LDF and PLDF for the corresponding residuals. There are no significant lag dependencies in the residuals for y_1 and y_3, whereas in the residuals for y_2 there is a significant lag dependence at lag 1. This is an additional indication that the equation for S in (4) is incorrect. A final piece of evidence that something is wrong is gathered from the pure simulation results in Figure 2. The information now available clearly invalidates the model, particularly if its intended purpose is simulation, and the modelling cycle is repeated by modifying the structure of the model.

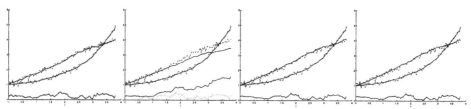

Figure 2. Cross-validation results. From left to right: One-step-ahead prediction and pure simulation using the incorrect model structure and one-step-ahead prediction and pure simulation using the correct model structure. (Solid: Predicted values, dashed: true y_1, dotted: true y_2, dash-dotted: true y_3).

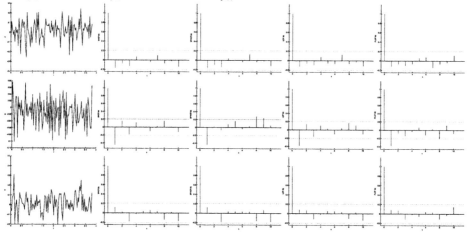

Figure 3. One-step-ahead prediction cross-validation residuals and corresponding SACF, SPACF, LDF and PLDF using the incorrect model structure. (Top: y_1, middle: y_2, bottom: y_3).

Table 2. Estimation results using the correct model structure.

Parameter	X_0	S_0	V_0	μ_{max}	K_1	σ_{11}	σ_{22}	σ_{33}	S_{11}	S_{22}	S_{33}
True value	1	0.245	1	1	0.03	0	0	0	0.01	0.001	0.01
Estimate	1.004	0.262	1.003	0.999	0.030	0.000	0.000	0.000	0.009	0.001	0.011
Std. Dev.	0.010	0.008	0.007	0.009	0.007	0.000	0.000	0.000	0.001	0.000	0.001
t-score	101.0	32.75	143.3	109.4	4.240	0.003	0.005	0.003	7.142	7.391	7.193
Significant	Yes	Yes	Yes	Yes	Yes	No	No	No	Yes	Yes	Yes

The information available suggests that the deterministic parts of the equations for X and S in (4) are incorrect, i.e. those parts of the model that depend on $\mu(S)$. Replacing (7) with the correct structure in (6) and re-estimating the unknown parameters with

CTSM, the results shown in Table 2 are obtained. Marginal t-tests indicate that all three stochastic parameters, σ_{11}, σ_{22} and σ_{33}, are now insignificant, and the hypothesis of simultaneous insignificance cannot be rejected when performing a test based on Wald's W-statistic. Additional evidence that the modified model is correct is gathered by performing residual analysis. One-step-ahead prediction results on the validation data set are shown in Figure 2, and the SACF, SPACF, LDF and PLDF (not shown) for the corresponding residuals show no significant lag dependencies. A final piece of evidence of the validity of the modified model is gathered from the pure simulation results in Figure 2. In summary, if the intended purpose of the original model was simulation or infinite-horizon prediction, e.g. for use in an MPC controller, it has been now been invalidated and a more reliable model has been developed. However, if the intended purpose of the original model was one-step-ahead prediction, it might still be suitable.

4. Conclusion

A methodology has been presented that combines modelling based on first principles and data based modelling through the use of stochastic differential equations and statistical methods for parameter estimation and model validation. The methodology features a modelling cycle that can be used to easily, rapidly and in a statistically sound way develop a reliable model of a given system. A computer-aided tool, called CTSM, which integrates the elements of the modelling cycle, has also been presented.

5. References

Bohlin, Torsten and Stefan F. Graebe (1995). Issues in Nonlinear Stochastic Grey-Box Identification. *International Journal of Adaptive Control and Signal Processing* 9, pp. 465-490.

Holst, Jan, Ulla Holst, Henrik Madsen and Henrik Melgaard (1992). Validation of Grey-Box Models. In: *Preprints of the IFAC Symposium on Adaptive Systems in Control and Signal Processing*, Grenoble, France, pp. 407-414.

Kristensen, Niels Rode, Henrik Madsen and Sten Bay Jørgensen (2001a). Computer Aided Continuous Time Stochastic Process Modelling. In: *European Symposium on Computer Aided Process Engineering – 11* (Rafiqul Gani and Sten Bay Jørgensen, Eds.), pp. 189-194.

Kristensen, Niels Rode, Henrik Melgaard and Henrik Madsen (2001b). *CTSM 2.0 – User's Guide*. DTU, Lyngby, Denmark.

Madsen, Henrik and Henrik Melgaard (1991). The Mathematical and Numerical Methods used in CTLSM. Technical Report 7/1991. IMM, DTU, Lyngby, Denmark.

Nielsen, Henrik Aalborg and Henrik Madsen (2001). A Generalization of Some Classical Time Series Tools. *Computational Statistics and Data Analysis* 37(1), pp. 13-31.

European Symposium on Computer Aided Process Engineering – 12
J. Grievink and J. van Schijndel (Editors)
© 2002 Elsevier Science B.V. All rights reserved.

Moving mesh generation with a sequential approach for solving PDEs

Y. I. Lim[1], J. M. Le Lann[2] and X. Joulia[3]

[1]CAPEC, DTU, 2800 Kgs. Lyngby, Denmark

[2,3]Laboratoire de Génie Chimique (LGC, UMR-CNRS 5503), INPT-ENSIACET
118 route de Narbonne, F-31400 Toulouse, France

Abstract

In moving mesh methods, physical PDEs and a mesh equation derived from equidistribution of an error metrics (so-called the monitor function) are simultaneously solved and meshes are dynamically concentrated on steep regions (Lim et al., 2001).

However, the simultaneous solution procedure of physical and mesh equations suffers typically from long computation time due to highly nonlinear coupling between the two equations. Moreover, the extended system (physical and mesh equations) may be sensitive to the tuning parameters such as a temporal relaxation factor. It is therefore useful to design a simple and robust moving mesh algorithm in one or multidimension.

In this study, we propose a sequential solution procedure including two separate parts: *prediction step* to obtain an approximate solution to a next time level (integration of physical PDEs) and *regriding step* at the next time level (mesh generation and solution interpolation). Convection terms, which appear in physical PDEs and a mesh equation, are discretized by a WENO (Weighted Essentially Non-Oscillatory) scheme under the conservative form.

This sequential approach is to keep the advantages of robustness and simplicity for the static adaptive grid method (local refinement by adding/deleting the meshes at a discrete time level) as well as of efficiency for the dynamic adaptive grid method (or moving mesh method) where the number of meshes is not changed. For illustration, a phase change problem is solved with the decomposition algorithm.

1. Introduction

For the numerical solution of dynamic systems involving steep moving fronts, a uniform fixed-grid structure is computationally inefficient, since a large number of meshes are required for an accurate solution. In such cases, methods which automatically adjust the size of the spatial step, namely moving mesh methods, are likely to be more successful in efficiently resolving critical regions of high spatial activity. Moving grid methods have important applications for a variety of physical and engineering problems (e.g., solid/fluid dynamics, combustion, heat transfer and phase changes) that require extremely fine meshes in a small portion of physical domain. Successful implementation of the adaptive strategy can increase the accuracy of the

---

[1] To whom correspondence should be addressed, e-mail: lim@kt.dtu.dk., Fax. +45 4593 2906

numerical approximation and also decrease the computational cost (Furzeland et al., 1990).

The moving grid methods (Miller & Miller, 1981; Dorfi & Drury, 1987; Huang & Russell, 1997), where the number of meshes is kept constant, could be very powerful due to the continuous grid adaptation with the solution evolution. However, intrinsic coupling between physical PDEs and mesh equations leads to a large system of nonlinear equations and much calculation time.

The simultaneous solution techniques have been well presented and tested in the literatures (Dorfi & Drury, 1987; Furzeland et al., 1990; Blom & Zegeling, 1994; Li & Petzold, 1997; Huang & Russell, 1997; Li et al., 1998), using a central or ENO discretization method for convection terms and well-known monitor functions as an error metric. In contrast, there are few articles on the sequential solution approaches where physical and mesh equations are solved separately or sequentially.

Mackenzie & Robertson (2000) proposed a two-step iterative method for fully discretized PDEs, using a Runge-Kutta ODE integrator. In the solution update step, the solutions are linearly interpolated. Tang & Tang (2000) used conservative-interpolation in the solution update step. The solutions of underlying PDEs and a mesh equation are obtained independently using a Runge-Kutta or BDF ODE integrator. For the two sequential approaches, physical PDEs considered do not contain a mesh convection term (dx/dt or $\dot{x}$). Therefore, the two static sequential methods may not be adequate for stiff systems.

Huang & Russell (1999) proposed a sequential approach based on the Lagrangian description involving mesh movement for two-dimensional mesh equations. However, using their algorithm, it is not easy to determine the next time stepsize for integration of physical and mesh equations. In next section, we present a new decomposition algorithm that allows us to easily implement on stiff systems in using a conventional BDF (Backward Differentiation Formula) ODE integrator.

2. Decomposition algorithm

The decomposition algorithm is to decouple the calculation of the mesh position from the solution. There are two advantages of the decomposition algorithm (i.e., sequential solution procedure). First the size of the discretized system (ODEs or DAEs) that arises at each time step is smaller. This is of great importance for the extension to multidimensional problems and for a reduction of the calculation time. The second advantage is that decomposition allows flexibility in the choice of iterative methods used to calculate the grid and the solution.

We propose a sequential solution procedure including two separate parts: *prediction step* to obtain an approximate solution to a next time level (integration of physical PDEs) and *regriding step* at the next time level (mesh generation and solution interpolation), as shown in Fig. 1. The theoretical background of this approach relies on the Sequential Regularization Method (SRM) proposed by Ascher & Lin (1996, 1997).

2.1 Prediction step

Given the physical solution (u^n), the mesh (x^n) and the time stepsize (Δt^n) at time $t=t^n$. Normally, initial meshes (x^0) are uniformly given. To obtain new initial mesh positions equi-distributed, it is recommended to update the mesh and the solution with a very small time stepsize (Δt) at the first time level (t=0).

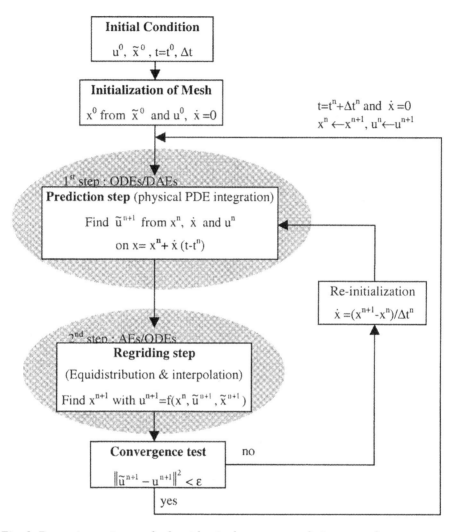

Fig. 1. Dynamic moving mesh algorithm in the two-step solution procedure.

A new solution ($\tilde{u}^{n+1}$) is computed by integrating time-dependent ODEs of Lagrangian description from $t=t^n$ to $t=t^n+\Delta t^n$, using x^n as the initial mesh.

$$\dot{u} - u_x \dot{x} = f(u, x, \dot{x}) \tag{1}$$

In the above equation, the flux correction term caused by mesh movement, $u_x \dot{x}$, may become a considerable value, when the mesh is updated after a relatively large time step (Δt^n). In this case, the convection term affects stability and accuracy of the whole equation (1). It is nature that the convection term (u_x) is discretized by a high resolution upwinding scheme to enhance accuracy as well as stability (Li & Petzold, 1997). In our study, the fifth-order WENO scheme (Jiang & Shu, 1996) is employed in the upwinding sense.

At the first iteration, mesh points are fixed with time, i.e., $\dot{x} = 0$. Hence, the mesh position x^n is not equidistributed any more with regard to the new solution $\tilde{u}^{n+1}$. The meshes will be redistributed in the regriding step.

2.2 Regriding step

In this step, the solution u^{n+1} is interpolated on new meshes x^{n+1} during mesh calculation, where a mesh equation derived from an error-metrics (i.e., monitor function) is required. To efficiently generate moving meshes, the monitor function must be appropriately selected according to problems considered. For the phase change problems, the integrable monitor function improves accuracy and temporal performance over the arclength or curvature monitor functions. To our knowledge, the integrable monitor function is useful for systems in which a value indicating discontinuous regions (e.g., phase change temperature or enthalpy) can be prescribed.

The mesh equation based on an integrable monitor function (Mackenzie & Robertson, 2000) is presented here. For the first and last mesh points (x_0 and x_N) fixed, each mesh x_i (i=1...N-1) is defined by:

$$(x_i - x_0) + \frac{\mu_1}{\mu_2}\sinh^{-1}(\mu_2(x_i - x^*)) - \frac{\mu_1}{\mu_2}(1 - i/N)\sinh^{-1}\mu_2(x_0 - x^*)) -$$

$$\frac{i}{N}\left((x_N - x_0) + \frac{\mu_1}{\mu_2}\sinh^{-1}\mu_2(x_N - x^*)\right) = 0 \tag{2}$$

where N is the total number of meshes and x^* is a point estimated (or given) as a phase change boundary. The parameters (μ_1 and μ_2) are positive constants that govern the smoothness or clustering of the grid around the point x^*. The regularization parameters are not sensitive to problems, to our knowledge. In our study, these values are set to $\mu_1=\mu_2=200$. Equation (2) can be solved easily by a Newton's iteration for nonlinear algebraic equations.

The solution u^{n+1} is updated by using a conservative piecewise cubic Hermite interpolant (E01BEF/E01BBF subroutines in the NAG numerical library). At the first iteration, the solution u^{n+1} on x^{n+1} is obtained from $\tilde{u}^{n+1}$ not considering mesh movement ($\dot{x} = 0$). Iteration must be performed between the two steps, after the convergence test, $\sum_i (\tilde{u}_i^{n+1} - u_i^{n+1})^2 < \varepsilon$.

2.3 Iteration between the prediction step and the regriding step

From the second iteration, mesh movement $\dot{x}$ can be calculated on the basis of x^{n+1} obtained in the previous iteration:

$$\dot{x}^n = \frac{x^{n+1} - x^n}{\Delta t^n} \tag{3}$$

In this iteration, the physical equation (1) is integrated with $\dot{x}^n$ that keeps constant during integration. The starting values of iteration, u^n and x^n remain constant. There is rarely any need to use a very strict tolerance for the convergence of the grid points. To our knowledge, a satisfactory solution can be obtained after 2-4 iterations. A numerical test follows in the next section.

3. Numerical study

We consider a heat conduction-diffusion PDE with phase change, where two phase change interfaces develop with time along a spatial direction (x) and enthalpy profiles (H) are discontinuous at the interfaces:

$$\frac{\partial H}{\partial t} = \frac{\partial^2 T}{\partial x^2} + \varphi(T) \tag{4}$$

with

$$\varphi(T) = \begin{cases} 0.336 + 3.457T, & T \le 0.6 \\ 1.708 + 1.220T, & T > 0.6 \end{cases} \tag{5}$$

$$H(T) = \begin{cases} 0.780T, & T < 1 \\ 0.780 \le H \le 0.780 + 0.331, & T = 1 \\ 0.780T + 0.331, & T > 1 \end{cases} \tag{6}$$

Refer to Mackenzie & Robertson (2000) for boundary, initial conditions and a differentiable enthalpy function.

Mesh evolution of 40 points with time and the enthalpy profiles are shown in Fig. 2. The phase change boundaries are well captured and near them meshes are concentrated, as shown in Fig. 2 (a). The enthalpy profiles agree well with the reference solutions (solid lines in Fig. 2 (b)) obtained by the 600-fixed grid system.

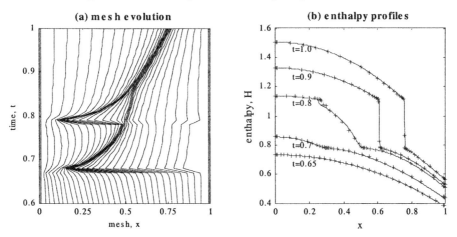

Fig. 2 Numerical results for the sequential moving mesh method based on the WENO scheme and the integrable monitor function.

4. Conclusion

With the same number of meshes, the moving mesh methods improve solution accuracy over the fixed mesh methods at the cost of the computational time. The fixed grid system often requires a large number of meshes and a long computational time in order to obtain numerical results with accuracy. Through the sequential moving mesh method with an appropriate monitor function, an accurate solution is obtained at a small number of meshes for a short computational time.

For the mesh convection term, the WENO (Weighted Essentially-Non Oscillatory) discretization scheme is used to enhance the moving mesh method for which the central discretization has been commonly used.

The sequential approach, where the physical equation and the mesh equation are solved separately, can reduce the computational time, since strong nonlinear coupling between the two equations is avoided.

References

Ascher, U. and P. Lin, 1996, Sequential regularization methods for higher index DAEs with constraint singularities: the linear index-2 case, SIAM J. Num. Anal., 33, 1921.

Ascher, U. and P. Lin, 1997, Sequential regularization methods for nonlinear higher index DAEs, SIAM J. Sci. Comput., 18, 160.

Blom, J. G. and P. A. Zegeling, 1994, Algorithm 731; a moving-grid interface for systems of one-dimensional time-dependent partial differential equations, ACM Trans. on Math. Software, 20, 194.

Dorfi, E. A. and L. O'C. Drury, 1987, Simple adaptive grids for 1-D initial value problems, J. Comput. Phys., 69, 175.

Furzeland, R. M., J. G. Verwer and P. A. Zegeling, 1990, A numerical study of three moving-grid methods for one-dimensional partial differential equations which are based on the method of lines, J. Comput. Phys., 89, 349.

Huang, W. and R. D. Russell, 1997, Analysis of moving mesh partial differential equations with spatial smoothing, SIAM J. Num. Anal., 34, 1106.

Huang, W. and R. D. Russell, 1999, Moving mesh strategy based on a gradient flow equation for two-dimensional problems, SIAM Sci. Comput., 20, 998.

Jiang, G. and C. W. Shu, 1996, Efficient implementation of weighted ENO schemes, J. Comp. Phy., 126, 202.

Li, S. and L. Petzold, 1997, Moving mesh methods with upwinding schemes for time-dependent PDEs, J. Comput. Phys., 131, 368.

Li, S., L. Petzold and Y. Ren, 1998, Stability of moving mesh systems of partial differential equations, SIAM J. Sci. Comput., 20, 719.

Lim, Y. I., J. M. Le Lann and X. Joulia, 2001, Moving mesh generation for tracking a shock or steep moving front, Comp. Chem. Eng., 25(4-6), 653.

Mackenzie, J. A. and M. L. Robertson, 2000, The numerical solution of one-dimensional phase change problems using an adaptive moving mesh method, J. Compt. Phys., 161(2), 537-557.

Miller, K. and R. N. Miller, 1981, Moving finite elements I, SIAM J. Numer. Anal., 18, 1019.

Tang, H. and T. Tang, 2000, Moving mesh methods for one and two-dimensional hyperbolic conservation laws, preprint, Department of Mathematics, The Hong Kong Baptist university, available in the website: http://www.math.hkbu.edu.hk/~ttang/.

European Symposium on Computer Aided Process Engineering – 12
J. Grievink and J. van Schijndel (Editors)
© 2002 Elsevier Science B.V. All rights reserved.

An Optimized Strategy for Equation-Oriented Global Optimization

Ricardo M. Lima and Romualdo L. Salcedo[*]

Departamento de Engenharia Química, Faculdade de Engenharia da Universidade do Porto, Rua Dr. Roberto Frias s/n, 4200-465 Porto, Portugal

Abstract

An optimized strategy for simulation and optimization of steady-state processes, under an equation-oriented environment, is presented. Equation-oriented environments apply a solution procedure to solve the entire system of non-linear algebraic equations arising from the mathematical model describing these processes. The difficulty in solving these systems may change drastically by specifying different independent variables for the degrees of freedom, as it has long being recognized. An algorithm that chooses the decision variables by minimizing the number and size of the subsystems of equations that need to be simultaneously solved for, while allowing for the inclusion of functional constraints, is used (Salcedo and Lima, 1999). With this algorithm, optimum sets of decision variables and the corresponding solution strategies are obtained.

This paper describes the implementation of this approach linked with a simulated annealing-based global optimizer. The proposed strategy was applied to the optimization of a reactor-extractor system and to a more difficult absorber-stripping system with heat integration. With these examples we pretend to compare different optimization procedures for each test case, respectively solving the entire system of equations, solving some smaller subsystems (a local optimum for the simulation step) or solving for the global optimum of the simulation step (which may correspond to a sequential solution). By optimizing the simulation step much more accurate results as well as significantly reduced CPU times are obtained, in comparison with simultaneous solution strategies, suggesting that this may be a powerful tool for global optimization.

1. Introduction

The simulation of steady-state chemical processes can be performed by three different approaches: sequential modular, simultaneous modular and equation-oriented. The ideas behind them, as well as the advantages and disadvantages of each approach are well known and described by several authors (Biegler et al.; 1997, Tolsma and Barton, 2000). The main idea behind the equation-oriented approach is the application of a solution strategy to all equations that describe the process. This set of equations is usually a non-linear system, very often with a sparse structure. The simultaneous solution of these systems is the key task of equation-oriented environments, which however require a proper initialisation (Zitney and Stadtherr, 1988). For the optimization of steady-state processes, the equation-oriented approach can be interfaced

---

[*] Author to whom correspondence should be addressed. E-mail: rsalcedo@fe.up.pt

with deterministic global search algorithms (Grossmann, 1994), or with stochastic algorithms (Salcedo, 1992; Cardoso et al., 1997, 2000). With stochastic algorithms, the solution of these optimization problems proceeds at two different levels: a) the simulation level that performs the solution of the system of equations and the calculation of a performance index; b) and the optimization level, which receives the performance index and is responsible to update the decision variables. This paper describes the implementation of an equation-oriented simulation approach integrated with an optimization framework suitable for the global optimization of non-linear programming (NLP) problems.

2. An equation-oriented environment for global optimization

The main idea behind the proposed approach is the optimization of the solution of the simulation step within a global optimization procedure. This is performed using the algorithm described by Salcedo and Lima (1999). Here the simulation step of an optimization is posed as an optimization problem itself, whereby a cost function, i.e. a performance index related to the difficulty of solving non-linear systems of equations, is associated with the choice of decision variables. The algorithm couples a combinatorial optimization algorithm with a tearing/partitioning algorithm (Ledet and Himmelblau, 1970). The global optimum for the simulation step may eventually correspond to a sequential solution, but it may also correspond to the solution of smaller subsystems (Salcedo and Lima, 1999). In any case, it is expected that choosing different decision variables will provide for a more efficient simulation procedure than the solution of the entire system of equations, no matter how efficient this might be, viz. even employing algorithms that exploit sparcity.

Simulated annealing-based optimizers are suitable algorithms for the global optimization of NLP problems due to their capability to escape local optima and to find solutions close to the global optimum. However, these optimizers need a high number of function evaluations (Grossmann and Biegler, 1995), which usually results in a great computational burden. Although the development of faster convergence is possible, when the simulation step involves the iterative solution of models the computational requirements can still be quite large. The present paper shows the implementation of an equation-oriented simulation strategy for the global optimization of NLP problems, which is able to: i) solve efficiently systems of non-linear equations; ii) minimize convergence problems; iii) minimize CPU times when couple with stochastic global optimizers. The proposed approach for the global optimization of NLP problems is shown in Fig.1, and is based in four steps: pre-processing of the data; optimization of the simulation step; symbolic manipulations; and numerical optimization. The next sections highlight the details of the implementation of this approach.

2.1 Data pre-processing
The equation-oriented environment proposed needs the user to specify the mathematical model that describes the chemical process in simple text files. One essential task of this step is the automatic build-up of the occurrence matrix, where the information about functional constraints is also stored.

2.2 Optimization of the simulation step

The choice of the decision variables is automatically performed by the combinatorial global optimizer and the corresponding output set is obtained. The combinatorial algorithm is responsible for the choice of the decision variables and the tearing/partitioning algorithm is responsible for obtaining the cost function of the simulation step. This cost function is based on the minimization of the number of subsystems to be simultaneously solved for, while allowing for the inclusion of functional constraints (Salcedo and Lima, 1999). The output set gives information about the sequence of solution and the occurrence of irreducible subsystems. In the proposed approach, whenever the global optimum for the simulation step corresponds to a sequential solution, there is no need of a numerical method to solve the system of equations (apart from possibly solving nonlinear functions in one variable). However, it may be impossible to find a sequential solution, and in these cases a numerical method is needed to solve the subsystems of equations. These may be solved either by a direct iterative process using tearing variables, or by a simultaneous approach using Newton-type algorithms. The latter approach was retained in the present work, since direct

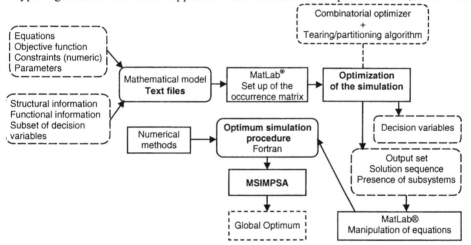

Figure 1. Flow diagram of information to perform an optimization using the simulation/optimization framework.

iteration often fails to converge (Lin and Mah, 1978). At the present we have implemented two numerical methods: a globally convergent Newton method from Press et al (1992) and a modified Newton algorithm combined with a trust region method from MINPACK, available from netlib. These were compiled into libraries interfaced with the MatLab® environment, where the model equations and constraints are initially written by the user.

2.3 Symbolic manipulations

This level receives the information contained in the output set. At this point, the Symbolic Toolbox from MatLab® (using the kernel from Maple®) is used to explicit variables whenever possible. If it turns out to be impossible to explicit one variable, a Newton-based method is applied to solve a non-linear function in a single variable. The automatic generation of several Fortran subroutines which comprise the entire mathematical model are also generated at this level.

2.4 Optimization

The optimization of the NLP problem is performed using MSIMPSA. This optimizer is a robust simulated annealing-based algorithm suitable for the global optimization of nonconvex NLP and MINLP constrained functions (Cardoso et al. 1997, 2001). Large-scale problems (large non-linear systems) may need the application of sparse-type solvers to improve the numerical solution of irreducible subsystems. The Fortran routine created in the previous level is built to be compliant with MSIMPSA, and all the compilations and links between the routines are automatically done and completely transparent to the user.

3. Case studies

The first case study represents a reactor-extractor-recycle system taken from Lee (1969). The second is an absorber-stripping system with heat integration, described by Umeda (1969) and more recently used as a test case by Ferreira and Salcedo (2001). For both cases the equations and a detailed description about these processes can be found in the cited literature. With these examples we pretend to compare different optimization procedures for each test case: i) solving the entire system of equations by a Newton-type method; ii) solving explicitly some equations followed by some smaller subsystems using Newton's method (a local optimum for the optimized simulation); iii) solving for the global optimum of the simulation procedure. A proper initialization of the variables to solve the system of equations was achieved by adopting the role of pseudo-decision variables for the initial estimates, leaving this function to the optimizer (Salcedo, 1989). Whenever the Newton-based methods were used, the error criterion was set to 10^{-6} both for the convergence of the function values and for the variables. The performance evaluation of the different solution procedures was carried out through the study of the results of 10 optimization runs for the simultaneous solution of all equations and 100 runs for the global optimum of the simulation step (each different run corresponds to a different starting point, which is randomly generated by MSIMPSA).

3.1 Case I

This case represents a reactor-extractor-recycle system composed by five isothermal continuous stirred tanks reactors followed by three crosscurrent extractors with recycle. The process is described by a system with 45 algebraic equations with 53 continuous variables, resulting in 8 degrees of freedom. The optimization of the solution procedure did not produce a sequential solution, but rather the solution of two subsystems, one with 2 equations and one with 12 equations. Figs. 1a-b show that the simultaneous solution of all equations presents much higher computational times and lower accuracies in arriving at the global optimum (24.77 kg/h).

3.2 Case II

This example represents an absorber-stripping system with heat integration. The process is described by 40 algebraic equations and 45 continuous variables, resulting in a NLP

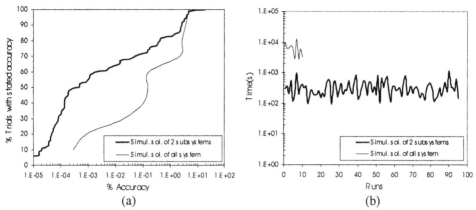

Figure 2. (a) Results for case I. (b) Results for case I, elapsed time for each run.

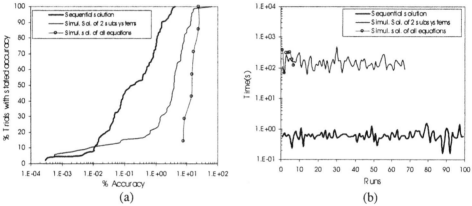

Figure 3. (a) Results for case II. (b) Results for case II, elapsed time for each run.

problem with 5 degrees of freedom. In this case, the global optimum for the solution procedure is a sequential solution (Salcedo and Lima, 1999). One local optimum for the solution procedure was obtained by choosing a different set of decision variables, resulting in an output set with one subsystem in seven equations, plus one subsystem with four equations. Figs. 2a-b show that the results are even more dramatic, where the optimized simulation corresponds to a sequential solution. The optimum value for this case corresponds to a profit of -$173,392.

4. Conclusions

The present paper shows that it is possible to increase the chance of convergence to the global optimum, while drastically decreasing the CPU times, by optimizing the simulation step of the optimization problem. This is performed by associating a cost function to the simulation step, and optimizing this step by choosing different sets of decision variables (a combinatorial problem). This combinatorial problem was solved by the same global optimizer that was used for the numerical optimization of the original NLP problem, viz. the MSIMPSA optimizer. A completely automated procedure has been build around this paradigm, using the MatLab® symbolic toolbox and Maple® kernel, with Fortran 77 coding to interface with the Fortran based MSIMPSA stochastic global optimizer.

The results of the proposed approach, for a reactor-extractor-recycle system and an absorber-stripping system with heat integration, show that much more accurate and faster results may be obtained rather than by using a simultaneous solution approach. This should be especially important when using stochastic algorithms for global optimization, due to their inherent large computational burden. Problems with a large number of decision variables could also benefit from the combinatorial optimization by simulated annealing, and this is currently under study.

Acknowledgement- This work was partially supported by FCT-MCT (Fundação para a Ciência e a Tecnologia - Ministério da Ciência e Tecnologia), under contracts PRAXIS XXI/BD/21481/99 and PRAXIS XXI/3/3.1/CEG/2641/95 employing the computational facilities of Institute for Systems and Robotics (ISR)-Porto.

5. References

Biegler, L.T., Grossman, I.E., Westerberg, A.W., 1997, Systematic Methods of Chemical Process Design, Prentice-Hall, NJ.

Cardoso, M.F., Salcedo R.L., Azevedo S.F. and D. Barbosa, 1997, A Simulated Annealing Approach to the Solution of MINLP Problems, Comput. Chem. Eng., 21, 12, 1349.

Cardoso, M.F., Salcedo R.L., Azevedo S.F. and D. Barbosa, 2000, Optimization of Reactive Distillation Process with Simulated Annealing, Chem. Eng. Sci., 55, 5059.

Ferreira, E.C. and R. Salcedo, 2001, A Tool to Optimize VOC Removal During Absorption and Stripping, Chem. Eng, 108, 1, 94.

Grossmann, I.E. and Biegler, L. T., 1995, Optimizing Chemical Processes, CHEMTECH, p. 27.

Grossmann, I.E., Daichent, M.M., 1994, New Trends in Optimization-Based Approaches to Process Synthesis, 5th Int. Symp. on PSE, En Sup Yoon, Ed., Kyongju, Korea, p.95.

Ledet, W.D. and D.M. Himmelblau, 1970, Decomposition Procedures for Solving Large Scale Systems, Adv. Chem. Eng., 8, 186.

Lee, E. S., Optimization of Complex Chemical Plants by a Gradient Technique, 1969, AIChE J., 15, 3, 393.

Lin, T.D. and R.S.H. Mah, 1978, A Sparse Computation System for Process Design and Simulation, Part I. Data Structures and Processing Techniques, AIChE J., 24, 830.

Motard, R.L., Shacham, M. and E.M. Rose, 1975, Steady State Chemical Process Simulation, AIChE Journal, 21, 3, 417.

Press, W.H., Flannery, B. P., Teukolsky, S. A., Veterling, W. T., 1992, Numerical Recipes in Fortran 77, The Art of Scientific Computing, Cambridge University Press, New York, vol.1.

Salcedo, R., 1989, Application of Random Search to the Optimization of Complex Systems, Chempor'89, *24c1-24c10.*

Salcedo, R.L. and R. Lima, On the Optimum Choice of Decision Variables for Equation-Oriented Global Optimization, 1999, Ind. Eng. Chem. Res., 38, 4742.

Salcedo, R.L., Solving Nonconvex Non-linear Programming and Mixed-Integer Non-linear Programming Problems with Adaptive Random Search, 1992, Ind. Eng. Chem. Res., 31, 1, 262.

Tolsma, J.E. and P.I. Barton, 2000, DAEPACK: An Open Modeling Environment for Legacy Models, Ind. Eng. Chem. Res., 39, 1826.

Umeda, T., 1969, Optimal Design of an Absorber-Stripper System, Ind. Eng. Chem. Proc. D.D., 8, 3, 308.

Zitney, S.E. and M.A. Stadtherr, 1988, Computational Experiments in Equation-Based Chemical Process Flowsheeting, Comput. Chem. Eng., 12, 12.

European Symposium on Computer Aided Process Engineering – 12
J. Grievink and J. van Schijndel (Editors)
© 2002 Elsevier Science B.V. All rights reserved.

919

Nonlinear Analysis of gPROMS Models Using DIVA via a CAPE ESO Interface

M. Mangold*, K.D. Mohl**, S. Grüner**, A. Kienle*, E.D. Gilles*,**

*Max-Planck-Institut für Dynamik komplexer technischer Systeme,
Sandtorstraße 1, D-39106 Magdeburg, Germany
Email: {mangold,kienle,gilles}@mpi-magdeburg.mpg.de
**Universität Stuttgart, Institut für Systemdynamik und Regelungstechnik,
Pfaffenwaldring 9, 70569 Stuttgart, Germany
Email:{mohl,gruener}@isr.uni-stuttgart.de

Abstract

The CAPE ESO interface of the process simulator gPROMS is used to pass model information to numerical methods contained in the simulation environment DIVA. By the interface, algorithms for the continuation of stable and unstable steady state and periodic solutions can be applied directly to gPROMS models. The use of the interface is illustrated by a detailed nonlinear model of an industrial reactive distillation column.

1. Introduction

In academia, numerical bifurcation analysis has been widely accepted as a useful tool for the investigation of chemical processes. However, although nonlinear effects are frequently encountered especially in reactive chemical processes, methods for bifurcation analysis are still rarely used in industry. One reason may be that in the past most packages for bifurcation analysis were tailored to small systems of ordinary differential equations and hardly applicable to realistic models of industrial processes. To overcome this difficulty, in the last years a set of methods for nonlinear analysis was implemented in the simulation environment DIVA (Kienle, 1995; Mangold, 2000). The methods can cope efficiently with high order differential algebraic models of chemical plants. The methods comprise algorithms for the one parameter continuation and stability analysis of steady states and periodic solutions, and for the continuation of co-dimension-zero singularities in two parameters. Usually, models in DIVA are formulated in a proprietary modelling language. In order to make the architecture more open, now an interface to CAPE equation set objects (ESOs) (CAPE-OPEN, 1999) has been implemented, as they are provided by other simulation tools like gPROMS (PSE, 2000). It is possible to apply the bifurcation analysis and the other numerical methods available in DIVA to gPROMS models via this interface. The primary motivation to realize the interface comes from a joint research project between several research institutes and an industrial partner on the subject of nonlinear dynamics in reactive systems. The interface is intended to facilitate the exchange of modern numerical methods on one hand and applications on the other hand. Experiences with the interface will be reported in this contribution, which is structured as follows. In the next section, a brief introduction to numerical bifurcation analysis will be given. The methods available in DIVA will be summarized. Then the communication between gPROMS and DIVA via the ESO interface will be discussed. Finally, the use of the interface will be illustrated by the example of an industrial reactive distillation column. First, the behaviour in the vicinity of the nominal operation point is investigated. The size of the operation window with respect to heat duty is analysed by one-parameter continuation. In the second step, a different set of operation conditions is used to demonstrate the nonlinear effects which may be encountered in a reactive distillation column.

2. Numerical Bifurcation Analysis in DIVA

The simulation environment DIVA is a software tool for dynamic flowsheet simulation of chemical processes which has been developed at the University of Stuttgart (Mohl et al., 1997). The plant model is formulated as a linearly implicit system of differential algebraic equations of the type

$$B(x, p)\dot{x} = f(x, p), \tag{1}$$

where x represents the state vector, p represents the parameter vector, B is a (possibly singular) left-hand-side matrix, and f is a right-hand-side vector.

In addition to methods for steady state and dynamic simulation, optimization, and parameter estimation, DIVA contains a package for numerical bifurcation analysis. The package comprises algorithms for the one-parameter continuation of steady states and periodic solutions as well as for the two-parameter continuation of saddle-node and Hopf bifurcations. The numerics have been tailored to systems of high dynamical order. For a detailed description of the numerical methods the interested reader is referred to (Kienle et al., 1995; Mangold et al., 2000). In the following, only a brief idea of the methods will be given. The central element of the bifurcation package is a continuation algorithm used to trace the solution curve of an under-determined system of algebraic equations

$$g(y) = 0, \quad g \in R^m, \quad y \in R^{m+1} \tag{2}$$

in an $(m+1)$ dimensional space. A predictor-corrector algorithm with local parameterisation and step-size control is used. A simple application of the continuation algorithm is the computation of stable and unstable steady state solutions as a function of some distinguished model parameter p. In this case, g is the right-hand side vector f of the model equations (1), and y consists of the state vector x and the parameter p. An eigenvalue monitor is used to determine the stability of the computed steady states and to detect singular points where one or several eigenvalues cross the imaginary axis. The singularities most frequently encountered in physical systems are saddle-node bifurcations (coincidence of two solutions) and Hopf bifurcations (stability change of steady state solutions and generation of periodic solutions). From bifurcation theory, necessary conditions for the singular points can be derived. Together with the steady state equations of the model they form an augmented equation system for the direct computation of the state vector x and the parameter p at a singular point. The augmented equation systems are generated automatically by DIVA. In the framework of the continuation algorithm, they are used to trace the curves of singular points in two parameters. The resulting curves form the boundary of regions of qualitatively different behaviour in the parameter space. A further application of the continuation algorithm is the continuation of stable and unstable periodic solutions in one parameter. For that purpose, the continuation algorithm is combined with a shooting method adapted to the special demands of high-order systems.

3. The CAPE ESO interface between gPROMS and DIVA

3.1 Architecture

The CAPE-OPEN project (CO-LaN, 2001) proposes a standardized interface between a flow sheet simulator and an external (open) solver. The flowsheet simulator is to provide model information in the form of a so-called Equation Set Object (ESO) whose specifications were defined in (CAPE-OPEN, 1999). gPROMS 1.8.0 (PSE, 2000) is able to create such an ESO and is used as the simulator in this study. DIVA serves as the external solver. It sets the variables of the gPROMS model and receives information on residuals and Jacobians by the ESO interface. DIVA and gPROMS run as separate processes. The inter-process communication is handled by the object request broker OmniORB 3 by AT&T (AT&T, 2000).

3.2 Comparison of Computation Times

The test example for the interface is the model of an industrial reactive distillation column which will be discussed in the next section. In the gPROMS formulation, the model consists of 250 differential and 5207 algebraic equations. The model size can be considered as typical for a realistic process model. In order to test the efficiency of the ESO interface, the model was also implemented in DIVA. Test runs were done on a 296 MHz Sun Enterprise under SunOS 5.6, and on a 1 GHz AMD Athlon PC under SuSE Linux 7.0. A computation for the bifurcation diagram in Fig. 2 takes 7.6 s of CPU time on the PC, if the model is implemented directly in DIVA, i.e. if the ESO interface is not used. The same computation takes 20.3 CPU s on the Sun. For the test of the ESO interface, gPROMS runs on the Sun and provides the residuals and Jacobians of the model under consideration. DIVA runs on the PC and carries out the nonlinear analysis. In this case, the same computation takes 10.5 CPU s on the PC plus 45 CPU s on the Sun. Obviously, the communication of the interface creates an overhead of computation time. However, in terms of time required to obtain computation results, the overhead seems still tolerable and is certainly preferable to the time consuming and error prone re-implementation of a complex model in a new software environment.

4. An Illustrative Example

The use of the interface between DIVA and gPROMS will be illustrated by the example of an industrial reactive distillation column schematically shown in Fig. 1 (Fernholz et al., 2001; Grüner et al., 2001). The column consists of 63 trays including the reboiler and the condenser. The liquid feed contains the components B and C from which the product E is produced. In the liquid phase, two reactions take place whose simplified reaction scheme is given in Fig. 1.

The lower boiling component A is removed as the top product. Its purity is specified in terms of a threshold for the top key component, the impurity C. The main product E is removed from the column as the bottom product. Its purity is determined by the bottom key component D. The model of the column consists of component material balances, quasi-static total material balances, and quasi-static energy balances for the trays. A constant volume hold-up and vapour–liquid equilibrium are assumed. Further details on the model can be found in (Grüner et al., 2001). As a first application example of continuation methods, the steady state behaviour of the column

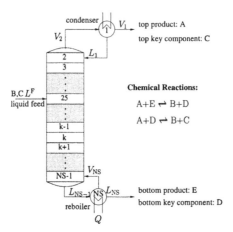

Fig. 1: Scheme of the reactive column

is studied in the vicinity of the nominal operation conditions. The bifurcation diagrams in Fig. 2 show steady state solutions as a function of the heat duty, which is one of the principal operating parameters. The left-hand-side diagram contains the steady state concentration of the key component C at the top of the column. The right-hand-side diagram shows concentration of the key component D at the bottom of the column. The horizontal dashed lines denote thresholds below which the specifications of the column are met. The vertical dashed lines indicate the nominal operation value of the heat duty. Obviously, in-spec operation of the plant is only possible in a very narrow window indicated by a grey band in Fig. 2. High performance controllers are required to keep the plant within this small operation window as shown in (Grüner et al., 2001; Fernholz et al., 2001).

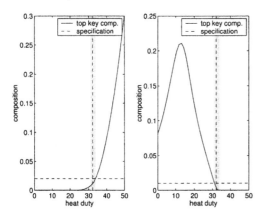

Fig. 2: Steady state solutions as a function of the heat duty. Left diagram: concentration of C at column top; right diagram: concentration of D at column bottom.; grey band: operation window.

The steady states in Fig. 2 are always unique and stable. However, instabilities and multiplicities of steady states are well-known patterns of behaviour in reactive distillation columns (e.g. (Mohl et al., 1999; Mohl et al., 2001). A second mode of operation is chosen to illustrate that the reactive distillation column studied here can also show more complicated nonlinear behaviour. In contrast to the nominal operation conditions, the components A and E are now fed to the column. The distillate mass flow is chosen as the bifurcation parameter. The resulting bifurcation diagram is shown in Fig. 3. Multiple steady states exist in a rather large parameter region. Multiplicities of three and five steady states are found. Since the column is operated close to reaction equilibrium, the nonlinearities may be attributed to the interactions between phase equilibrium and reaction equilibrium. In literature, also the nonlinear relationship between mass and molar flow rates has been reported as a potential source of multiplicities (Güttinger and Morari, 1999). However, further investigations show, that they are not relevant for the nonlinearities in Fig. 3. The same qualitative behaviour can also be reproduced, if a constant molar volume for the liquid mixture is assumed.

In addition to the multiple steady state solutions, a branch of periodic solutions is found. The periodic branch emanates from a subcritical Hopf bifurcation point, i.e. it is initially unstable. This means, that crossing the Hopf bifurcation by an increase of the distillate mass flow results in a sudden burst of periodic oscillations with large amplitudes. The branch of periodic solutions ends in a homoclinic orbit, which connects the stable and the unstable manifold of the saddle point marked "SP" in Fig. 3. Once the bifurcation diagram has been created, the existence of multiplicities can be easily verified by a dynamic simulation in gPROMS. For that purpose, the distillate mass flow is increased in a step-wise manner (see Fig. 4). Subsequently, it is reduced to the original value in the same steps. The different values of the mass flow applied in the dynamic simulation are indicated as vertical dotted lines in Fig. 3. The numbered labels in Fig. 3 and Fig. 4 denote corresponding solutions found by continuation (Fig. 3) and by dynamic simulation (Fig. 4). The dynamic simulation confirms for example the coexistence of the stable steady states "1" and "7" as well as "3" and "5", and the coexistence of the stable steady state "2" and the stable periodic solution "6".

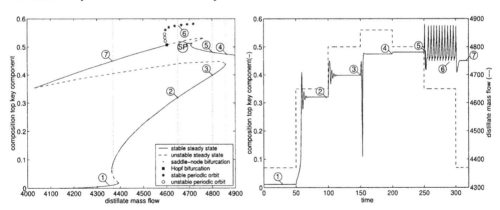

Fig. 3: Bifurcation diagram of the industrial reactive distillation column created in
DIVA

Fig. 4: Verification of the stable solutions
in Fig. 3 by dynamic simulation in
gPROMS

5. Conclusions

The nonlinear nature of many chemical engineering processes requires adequate numerical methods for its analysis. The simulation tool DIVA offers a package of continuation methods for that purpose. In order to make the methods available to other simulators, an interface to CAPE-ESOs has been implemented in DIVA. The interface

924

has been used to apply bifurcation analysis to the model of an industrial reactive distillation column formulated in gPROMS. The example shows that the interface is efficient enough to analyse realistic process models of high complexity within reasonable time, even if the DIVA and the gPROMS process run on different computers. The additional computational burden caused by the inter process communication is acceptable and certainly negligible compared to the effort necessary to implement a detailed chemical process model on a different simulation platform. Furthermore, the interface proves to be an efficient way to transfer new numerical tools from academia to applications in industry. In a common project, academic as well as industrial partners can draw benefits from such an interface. The industrial partner can test new methods with comparatively low efforts. The frequently met obstacle of having to re-implement a model and get familiar with a new software tool is minimised. The industrial partner can bring his application examples into the co-operation without having to reveal all the details about his model like physical property correlations. For the academic partner, on the other hand, it is easier to get interesting application examples for his tools and he gets some feedback on the usability of the developed methods.

6. Acknowledgements

The authors thank Bayer AG, Process Technology, Process Control, for providing the model of the reactive distillation column, and the group of Prof. Marquardt, RTWH Aachen, for their support during the implementation of the ESO interface in DIVA. The financial support by the German Bundesministerium für Bildung und Forschung (contract no. 03C0268B) is gratefully acknowledged. CAPE-OPEN was funded by the European Community under the Industrial and Materials Technologies Programme (Brite EuRam III), under contracts BRPR CT96-0293.

7. References

AT&T , 2000, OmniORB - Free High Performance CORBA 2 ORB. http://www.uk.research.att.com/omniORB

CAPE-OPEN, 1999, Open interface specification numerical solvers. Ref. Nr. CO-NUMR-EL-03, http://www.global-cape-open.org

CO-LaN, 2001, CAPE-OPEN Laboratory Network web site, http://www.colan.org

Fernholz, G., M. Friedrich, S. Grüner, K.D. Mohl, A. Kienle, E.D. Gilles, 2001, Linear MIMO controller design for an industrial reactive distillation column. Presented at DYCOPS'2001, June 3-6, 2001, Chejudo, Korea.

Grüner, S., K.D. Mohl, A. Kienle, E.D. Gilles, G. Fernholz, M. Friedrich, 2001, Nonlinear control of an industrial reactive distillation column. Accepted for publication in Control Engineering Practice

Güttinger, T.E., M. Morari, 1999, Predicting multiple steady states in distillation: singularity analysis and reactive systems, Ind. Eng. Chem. Res. , 38, 1633

Kienle, A., G. Lauschke, V. Gehrke, E.D. Gilles, 1995, On the dynamics of the circulation loop reactor – numerical methods and analysis. Chem. Eng. Sci. (50), 2361

Mangold, M., K.D. Mohl, A. Kienle, E.D. Gilles, 2000, Nonlinear computation DIVA -methods and applications. Chem. Eng. Sci. (55), 441

Mohl, K.D., A. Spieker, R. Köhler, E.D. Gilles, M. Zeitz, 1997, DIVA - a simulation environment for chemical engineering applications, in : Informatics, Cybernetics, and Computer Science (ICCS-97), Donetsk, Ukraine

Mohl, K.D., A. Kienle, E.D. Gilles, P. Rapmund, K. Sundmacher, U. Hoffmann, 1999, Steady state multiplicities in reactive distillation columns. Chem.Eng. Sci. (54), 1029

Mohl, K.D., A. Kienle, K. Sundmacher, E.D. Gilles, 2001, A theoretical study of kinetic instabilities in catalytic distillation. Chem. Engng. Sci. (56), 5239

PSE, 2000, gPROMS User Guide, Process Systems Enterprise Ltd., London

European Symposium on Computer Aided Process Engineering – 12
J. Grievink and J. van Schijndel (Editors)
© 2002 Elsevier Science B.V. All rights reserved.

Model Transformations in Multi Scale Modelling

S. McGahey and I. Cameron
CAPE Centre, Department of Chemical Engineering
The University of Queensland, Australia 4072

Abstract

Computer Aided Process Engineering (CAPE) requires computer based process models for most of its applications. Many of these applications require different attributes (accuracy, speed of solution, etc) of their models. For example, models that run in real-time sacrifice accuracy in exchange for speed, while offline applications do not need to be so concerned about the time spent in solving their models.

Therefore it comes as no surprise that a number of models may exist to describe a process, with each model having certain attributes that qualify it for a certain CAPE application. Such a collection of models is termed a Model Family

A formalised method for the semi-automated creation of new members of a model family from existing members has been created, and is outlined in this document as six operations that manipulate the boundary-volume / connection structure of a model. Two of these operations, Merging and Demerging are believed to be new additions to the set of previously non-formalised meta-modelling operations; Aggregation, Disaggregation, Addition and Neglection. The major contribution of these operations is that they generate certain modelling information automatically when multiple boundary volumes are merged into a new boundary volume and the original boundary volumes lose their identity, reducing the amount of information required from the modeller.

1. Introduction

Model development is an iterative process, in which many models are derived, tested and built upon until a model fitting the desired criteria is built. Subsequent modelling work may need to begin the search at the same place as the original model building began, rather than where it finished. This may be for a number of reasons, a common one being that the model currently in service is poorly suited for reuse because it is poorly understood (Foss et.al 1998).

Previous work has tried to make models more understandable in a number of ways. For example, MODKIT (Bogusch et al. 1996) creates a record of the decisions that were made during the modelling process and MODEL.LA (Stephanopoulos et al. 1990) utilises a Process Modelling Language (PML) (Marquardt 1996) that is rich in process engineering, rather than mathematical, terminology.

This work makes extensive use of a set of Process Modelling Objects (PMOs) from the PML "SCHEMA" (System for CHemical Engineering Model Adaptation) developed by Williams et al. (2001), to create an ordered object oriented representation, called a *model description* of a process model.

This contribution illustrates the use of the SCHEMA PMOs to develop two Meta Modelling concepts which give Process Engineering (PE) modellers added power in model building – especially for novel systems which will require a lot of model based work.

The first concept is that a group of models describing the same process system are related, and form a group called a model family, which is described in Section 2.

The second concept is that the relationships between members in such a model family can be described by and exploited with a set of model transformations, covered in Section 3.

2. Model Families

Section 1 introduced the idea that a number of PE models may exist to describe the same system. Such a collection of models is termed a *Model Family*. Work by Aris (1994) and Rice and Do (1995) notes the existence of such related model sets, however it is not until Foss et al. (1998) and Bogusch et al. (1996) that explicit references to the information contained in these groups are made. In extension to their ideas of retaining model version information, the model families described here relate all of the models describing a particular system, including those developed in different modelling projects. Ongoing work by Williams et al. (2001) is exploring the viability of these model families for recording the various modes contained in models of hybrid systems.

The families are constructed as follows. An initial model of a system, preferably one with a low level of detail, is presented as a starting point. New members of this model family are then derived from this model by performing appropriate model transformations (Section 3), as illustrated by the arrows between Models a, b, c and d in Figure 1. Each member of the model family needs to be a completely valid model, else the model family will quickly become crowded and meaningless due to the many models which cannot be used or related to by the user. In this figure, the model members are arranged in terms of the number of Boundary Volumes that each system contains, to show their relative structural complexity and provide some method of categorising the different family members. Models a, b and d have two Boundary Volumes while Model c has three Boundary Volumes.

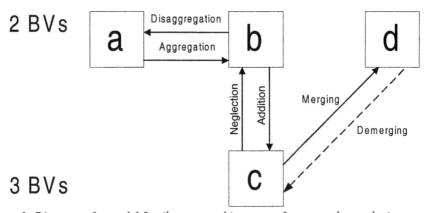

Figure 1: Diagram of a model family arranged in terms of structural complexity

The resulting arrangement of models and transformations provides an indication of how the various models are related to each other, supplementing any documentation present in the individual models. This extra information can assist a modeller to choose the most appropriate model for further development by detailing what models are already in existence, and how they differ.

3. Model Transformations

In most PE Computer Aided Modelling (CAM) systems, a simple PE modelling task such as assuming a Boundary Volume is negligible, or merging together a pair of Boundary Volumes, would require a long list of simpler tasks to be carried out. This is because the tools which have been designed for the various modelling tasks are built around existing CAM concepts, rather than being reinvented for the PE domain. Fischer and Lemke (1987) prove that by making a modelling package relevant to the domain in which it is used vastly improves the efficiency of the tasks performed with it. Therefore the introduction of the core modelling manipulations used in PE modelling as basic tools in a modelling package should drastically improve the efficiency of modellers using it.

Recent work by the authors has investigated the typical modifications that are made to PE models during model development, and described them as a set of Meta Modelling operations, termed Model Transformations. This work aims to give these tools an accepted identity and form, and then implement them in the SCHEMA system mentioned in Section 1. Two transformations that may not have been noticed previously, Merging and Demerging (Sections 3.5 and 3.6) are brought to light in addition to the more obvious manipulations of Aggregation, Disaggregation, Addition and Neglection.

The SCHEMA PMOs categorise model information in a very useful manner for these transformations, as will be seen in the following subsections. For the purposes of this discussion, they can be further divided up into two sets – the Structural Modelling Objects (SMOs, comprised of Models, Boundary Volumes and Connections) which describe the connectivity of the various elements in a described system, and the Internal Modelling Objects (IMOs, comprised of all that is not a SMO), which describe the internal elements of the SMOs. It is the introduction of these IMOs which sets this modelling system apart from other object oriented modelling languages which utilise concepts similar to the devices and connections of Marquardt (1996). Six set theory algorithms describe the SMO and IMO manipulations of these transformations, however space limitations mean that they can only be described briefly here, along with a few guidelines. The first two transformations require no explanation in terms of IMOs, and are nearly completely described by Figure 2. Addition and Neglection differ from these only in that they can also affect Assumption objects. Merging and Demerging involve intense IMO manipulations, and have a number of interesting issues which need to be considered when they are performed.

3.1 Aggregation

Aggregation groups a set of Boundary Volumes and/or Models together, and encapsulates them inside a new Model. This is a very simple operation which involves

928

manipulation of the SMOs of the selected group of objects, and the creation of a new Model object. Figure 2a illustrates a set of Boundary Volume (labelled Vessel and Jacket) and Connection (unlabelled) objects before aggregation, and Figure 2b shows the same objects after they have been aggregated into a Model object labelled "CSTR". All objects undergoing aggregation transformations retain their individual identities.

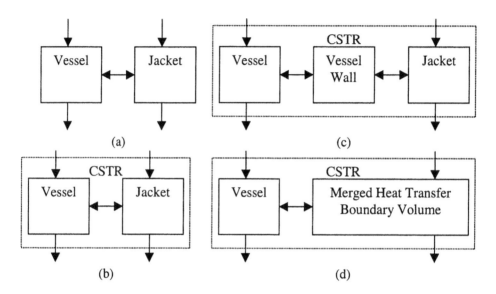

(a)

(c)

(b)

(d)

Figure 2 : Structural Modelling Objects as (a) individual entities, as (b) an aggregated group of entities, as (c) a group with an additional member, and (d) a group with a merged member

3.2 Disaggregation

Disaggregation is the reverse of aggregation. This involves removal of the encapsulating Model object and replacing it with its contents. This can be visualised by taking the set of objects illustrated in Figure 2b, and then removing them from the Model object "CSTR", which is then disposed of and then reintroducing them as individual objects into the object which originally contained the "CSTR" Model. The resultant set of model objects are represented by Figure 2a. Once again, all modelling objects retain their individual identities when they are removed from their Model group.

3.3 Addition

Addition is the operation of adding either a SMO or an assumption to a suitable object. An example of this operation on SMOs is the alteration of the object set shown in Figure 2b to that shown in Figure 2c. The modeller in this case has decided to enrich the model by adding information about a wall which separates the vessel and the jacket in the CSTR model. A second example of an additive operation is when a new assumption such as "vessel-is-isothermal" is applied to the vessel object of a CSTR. SCHEMA would then check that these additions to the model do not cause inconsistency problems, such as conflicting assumptions or violations of model limitations.

3.4 Neglection

Similar to Addition transformations, Neglection operations can involve SMOs or assumptions. A SMO neglection operation can be seen in the movement from Figure 2c to Figure 2b. In this example, a wall Boundary Volume is removed from the model, resulting in a model simplification. Neglection of assumptions is also possible. This illustrates the difference between transformation operations on Models, versus operations on Model Descriptions. If the operations were on a Model, then Neglection should always result in a simplified model, however neglecting a simplifying assumption, such as "vessel-is-isothermal" can result in a more complex model. The SCHEMA system deals with model descriptions.

3.5 Merging

In this operation, Boundary Volume and/or Model objects lose their individual identities, and are merged into a new Boundary Volume object which attempts to describe the original objects. This results in a reduction in information richness. As an example, consider the set of SMOs illustrated in Figure 2c. If the modeller is only interested in the events in the vessel, but wants to consider the effects of the heat transfer to the jacket through the vessel wall, then they might ignore all non-energy events outside the vessel Boundary Volume. Merging of the Wall and Jacket Boundary Volumes into one large pseudo-balance, and only considering the heat flows in this pseudo-balance's interactions with its environment will result in the object set shown in Figure 2d. This has a similar effect conceptually to assuming thermal equilibrium between the wall and the jacket, except that index problems are avoided. In this transformation operation, the objects involved in the merger lose their individual identities to create a new single Boundary Volume object. This operation is not always reversible due to information loss. The automated construction of the IMOs and SMOs of this object from the IMOs and SMOs of the original merging Boundary Volumes and/or Models is the subject of continuing work. Some typical, non algorithmic rules follow:

- Merging of Boundary Volumes which contain different phases will require the creation of a Pseudo-phase. All remaining connections will assume the new pseudo phase to be that phase they were connected to before merging for the purposes of determining appropriate transfer mechanisms.
- Connections which have one end external to the Merging process must be incorporated into the new Balance Volume, with appropriate provisions made for them in the IMOs of the new Boundary Volume.
- Connections which are anchored at both ends to Boundary Volumes or Models which are selected for Merging cease to exist after the Merging process.

3.6 Demerging

This operation changes a Boundary Volume into a Model Object which contains other newly-defined Boundary Volume objects, resulting in a general increase in information richness. This can be observed in changing the objects in Figure 2d into those in Figure 2c. Such a transformation can be compared to the addition transformation, with the consideration of a few algorithmic and conceptual guidelines, some of which follow:

- The SMOs of the new Boundary Volume set must account for the SMOs of the original Boundary Volume - all connections leaving the transformation area must be accounted for in the new set of Boundary Volumes.
- The overall effect of the new Boundary Volume set should have some similarity to the effects created by the IMOs of the original Boundary Volume.

4. Conclusions and Recommendations

This work has highlighted the concept of model transformations and illustrated their utility as basic modelling tools. Two novel transformations, merging and demerging have been defined to further empower these model transformations. When developing models of differing levels (scales) of detail or dimension, these transformations serve to save a lot of time and effort by carrying over and/or modifying relevant information from one model to another. The use of a highly conceptual system of modelling objects (in SCHEMA) leads to models that are information rich and useful in advanced modelling operations. The use of model families aids in the understanding and development of models and makes use of already existing information which is normally wasted. This provides a record of the model development as well as a collection of models which can be drawn upon should future development be required.

Bibliography

Aris, R., 1994, Mathematical Modelling Techniques, Dover Publications, Inc., New York

Bogusch, R., Lohmann, B. and Marquardt, W., 1996, Computer-Aided Process Modeling with ModKit, Technical Report #8, RWTH Aachen University of Technology

Fischer, G. and Lemke, A.C., 1987, Construction Kits and Design Environments: Steps Toward Human Problem-Domain Communication, Human-Computer Interaction, 3, 179-222

Foss, B.A., Lohmann, B. and Marquardt, W., 1998, A Field Study of the Industrial Modeling Process, Journal of Process Control , 8, 5-6, p325-338

Marquardt W., 1996, Trends in Computer Aided Process Modeling, Computers and Chemical Engineering, 20, 6/7, pp591-609

Rice, R.G. and Do, D.D., 1995, Applied Mathematics and Modeling for Chemical Engineers, John Wiley and Sons, Inc., New York.

Stephanopoulos, G., Henning, G. and Leone, H., 1990, Model.LA A Modeling Language for Process Engineering - I. The Formal Framework, Computers and Chemical Engineering, 14, 813-846

Williams, R., Hangos, K.M., McGahey, S. and Cameron, I.T., 2001, Assumption Based Modelling and Model Documentation, 6th World Congress of Chemical Engineering, Melbourne

European Symposium on Computer Aided Process Engineering – 12
J. Grievink and J. van Schijndel (Editors)
© 2002 Elsevier Science B.V. All rights reserved.

Application of Hybrid Models in Chemical Industry

G. Mogk, Th. Mrziglod, A. Schuppert

Bayer AG, Modelling & Methods, Leverkusen, Germany

{georg.mogk.gm, thomas.mrziglod.tm, andreas.schuppert.as}@bayer-ag.de

Abstract

The power of hybrid models as a combination of rigorous models and artificial neural networks (ANNs) was shown in several applications in different domains. This new technique is utilised in the area of chemical product development, process design and marketing applications for different demands. During a project in the last three years at Bayer hybrid modelling was advanced to a standard technique. This new modelling technique is completely integrated in the existing modelling software infrastructure for data based and rigorous models.

In this paper an overview of the theory of hybrid modelling and the software implementation is given. The capabilities of hybrid models will be demonstrated on industrial application examples.

1. Introduction

Model-based process control, simulation and optimisation are widely accepted to be key technologies for process improvement. In the wide area of fine chemicals as well as in biotechnology, however, the desired broad application of model-based process improvement, has not been performed yet. Although highly desirable closing the gaps in quantitative process knowledge for all subprocesses, and thereby allowing classical modelling techniques to be applied, is often not affordable.

For more than 10 years Artificial Neural Networks (ANN) have been successfully applied in various application areas at Bayer. Serious problems in the practical use of ANN are for instance the lack of extrapolation capability and the problem to integrate scientific information.

A significant improvement is achieved by the additional use of a-priori-known structural information about the process (see [2]). The central idea is that structural information is used to reduce the solution manifold of the system. This approach shall herein be called *Structured Hybrid Modelling*. A structured hybrid model (SHM) therefore consists of three components:

- rigorous submodels describing the i/o (input/output)-relation of those subprocesses which are well understood
- black-box submodels for those subprocesses where no rigorous model is available
- a model flowsheet describing the i/o-structure of all the submodels, joining them together by mapping the i/o-structure of the real process.

The new component with respect to the earlier, „simple" hybrid modelling methods (e.g [1]) is the explicit use of the structure of the process in the generation of the model flowsheet. In theoretical and practical investigations since 1995 (cited in [2]) it has been shown that by the use of hybrid models the extrapolation behaviour can be improved dramatically.

During a project in the last three years at Bayer hybrid modelling was advanced to a standard technique. This new modelling technique is completely integrated in the existing modelling software infrastructure for data based and rigorous models. Furthermore new numerical strategies for fitting hybrid models are developed.

As the mathematical foundations have been described elsewhere [2], in this paper we will focus our attention on a sketch of our software platform as well as on the demonstration of some real life application examples.

2. Theory of Hybrid Models

The complexity of black box models increases in principle exponentially with the number of input variables of the model (curse of dimensionality). The incorporation of the model flowsheet, however, allows a splitting of one black box model for the entire process into separate, small submodels for parts of the process reducing the number of input variables to each black-box submodel significantly. We will demonstrate this superior quality of SHMs on correlated input data with the following artificial example.

Let $z : \mathbf{R}^2 \to \mathbf{R}$ be a function with the known structure

$$z(x, y) = u(x) + v(y) + \alpha \cdot u(x) \cdot v(y) \tag{1}$$

where the functions $u(x), v(y)$ and the parameter α are unknown. For a concrete situation we generated data with the settings

$$u(x) = x^3, \quad v(y) = \sin(\frac{\pi}{2} y), \quad \alpha = 0.614 \quad \text{and} \quad x \in [-1;1], \ y = x + \gamma \cdot X,$$

where $X \in [-1;1]$ is a uniform distributed random number.

For the case $\gamma = 0.1$ and 100 data points (figure 1a) we identified the function z in two different ways. Using the hybrid structure we identified the function u, v as one-dimensional ANNs simultaneously with the parameter α. It is important to stress that we hereby used only the information on x, y and z. The model was identified with the Bayer internally developed hybrid modelling software described in the next section. Second we ignored the knowledge of the hybrid structure and identified z as ANN from x and y with a commercial neural network software. The resulting functions are shown together with the exact solution in figure 1b-d. Inside of the area with the training data (which is only the small diagonal area inside of the square $[-1,1]^2$ illustrated in figure 1a) all functions are identical. Outside of this region only the hybrid model is able to reproduce the original function. As can be seen, the values of the ANNs outside of the data region are arbitrary without any relation to reality. The error of the ANNs outside of the data region is more than a hundred times larger than the error of the hybrid model.

This simple example illustrates that hybrid models in contrast to ANN are able to extrapolate. In our example the extrapolation region is ten times larger than the data region! It is important to remark that it is only possible to identify u and v up to an arbitrary constant c.

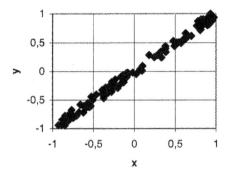

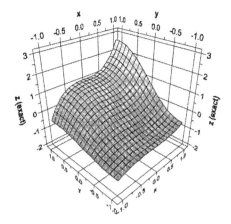

Figure 1a: Distribution of the input components of training data

Figure 1b: Exact solution

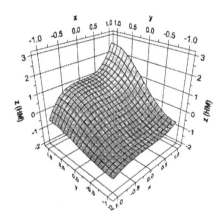

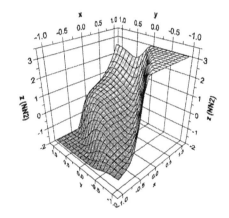

Figure 1c: Hybrid model

Figure 1d: Neural Network

3. The Bayer HybridTool

During an internal project since 1999 we have developed a specific software environment called *HybridTool* for the standardised generation and the identification of hybrid models. From a technical point of view a hybrid model is a network of databased submodels like ANNs and known rigorous submodels which have to be coded in MatLab. Several algorithms for the efficient identification of the databased submodels and of unknown parameters in the rigorous models are available. This allows a problem oriented choice of the best identification strategy as a combination of partial iterative and simultaneous model fitting. Thereby an automatic structure optimisation of all internal ANNs can be performed. We attached great importance to the efficient handling of incomplete datasets. Thereby the network topology is used automatically to improve the number of utilisable datasets.

HybridTool has a modular design as shown in Figure 2. The model generation step can be performed with the help of a graphical user interface. The user interface is connected

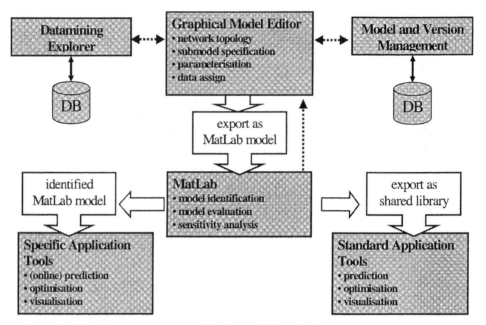

Figure 2: Modular organisation of the Bayer HybridTool

to our internal data preparation tool and a component for the model and version management. With the help of the data preparation tool arbitrary data sources are available. Actually, the model identification takes place within MatLab, which is triggered automatically from the user interface. The identified MatLab model can be exported in two different ways. For our internal standard application tool which can dynamically handle arbitrary ANNs and hybrid models, the model is converted into a shared library (DLL). Otherwise the MatLab model can be directly integrated into an application specific software environment, e.g. to use the model for quality (online) prediction or for optimisation purposes.

4. Applications

The potential of hybrid models and the features of HybridTool was demonstrated in several application at the chemical industry. Three of them are presented in the following.

4.1 Continuous polymerisation

As an example for an industrial application a continuous polymerisation plant will be described. The polymerisation of monomers and comonomers had been performed in an tube reactor with low residence time. A quantitative modelling of the reactor on a rigorous basis was not possible.

Using the chemical knowledge about the reactive part of the process a SHM for the reactor could be established. It can be described shortly as follows:

The melt index (MI), which had to be modelled, depends on the average chain length $<c>$ by a given power law. The chain length $<c>$ is controlled by yield with respect to the monomer feed and the number of chains produced per each mole of monomers:

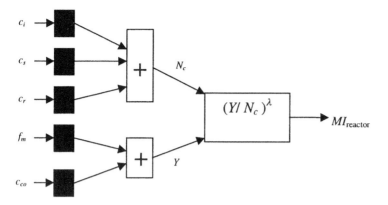

Figure 3: Structure of the hybrid model for the continuous polymerisation process

$$MI = \langle c \rangle^{\lambda} \sim \frac{\langle \text{yield} \rangle^{\lambda}}{\langle N_c \rangle^{\lambda}}$$

Using additional chemical information about the structure of the reaction network we arrived at a further reduction of the complexity of the SHM for the reactor subprocess as depicted in figure 3.

This model was used to fit the melt index of the polymerisation plant with respect to the input variables and was compared with a pure ANN. For the underlying real process, the ANN alone required about 2000 data sets to predict the melt index properly, whereas the SHM could be used for melt index-prediction after training on only 200 data sets. Moreover, the prediction quality of the SHM was about five percent higher than that of the neural network. Due to the extrapolation behaviour, the SHM allowed the analysis of the impact of monomer impurities on the melt index also in parameter region where no process data are available.

4.2 Quality management for polymer compounding

Among other things the quality of polymers depends on the formulation, the colorants and the quality of the raw materials which often shows strong variations. The aim of this application is to compensate the influence of quality variations with the help of formulation variations. Thereby the actual colour formulation has to be taken into account. Since the influence mechanism is partially unknown the situation is ideally suited to hybrid modelling strategies. Furthermore the number of input parameters is greater than 100. Hence the limited data information available makes it impossible to identify a pure ANN. Using the known network structure allows to reduce the number of inputs into the most complex ANN submodel to less than 20 and it is possible to generate accurate models for different quality parameters which are able to extrapolate.

4.3 Metal hydride process development

The most important quality parameters for a specific metal hydride product are the crystallite size and the median of the particle size distribution. Thereby each particle consists of many crystallites. Relevant influence parameters are the reactor temperature and the concentration of the ingredients. For the hybrid structure we use the fact that e.g. the crystallite size depends only on the total amount of solved metal and some metal

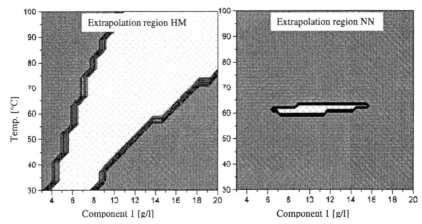

Figure 4: Comparison of the valid extrapolation region of the hybrid model and an ANN

complexes. Therefore the hybrid model allows the extrapolation into a wide temperature range and permits to locate economical process regions (see figure 4).

5. Conclusion

In the present paper SHMs with a given order for the interaction of black- and white-box submodels are presented. The model flowsheet describing the connections between the submodels maps the real structure of the process explicitly. With respect to black-box or unstructured hybrid models containing only one big black-box submodel lumping all unknown subprocesses, the functional structure of the SHM exhibits significant advantages over black-box and rigorous models.

The extrapolation capability of SHM allows its use in process analysis, on-line investigating for example the effect of impurities or other disturbances on the product quality. In case of any process disturbance, the operators will try to keep the product quality constant through changing appropriate control parameters. Therefore the values of the process disturbances and control parameters are closely correlated and a retrospective analysis of the process dependence on the disturbances is impossible using only black-box models. Appropriate SHMs, however, allow the extrapolation from correlated data, thereby rendering the analysis of the relation between product quality and process disturbances possible.

References

[1]   H.A.B. Te Braake, H.J.L. van Can, H.B. Verbruggen, Semi-mechanistic modelling of chemical processes with neural networks, Eng. Appl. Art. Intell. 11, (1998), pp. 507-515.

[2]   A. Schuppert, Extrapolability of structured hybrid models: a key to optimization of complex processes, in: Proceedings of EquaDiff '99, B.Fiedler, K.Gröger, J.Sprekels Eds., (2000), pp. 1135-1151

European Symposium on Computer Aided Process Engineering – 12
J. Grievink and J. van Schijndel (Editors)
© 2002 Elsevier Science B.V. All rights reserved.

Simulation of Two-Dimensional Dispersed Phase Systems

Stefan Motz, Natalie Bender and Ernst Dieter Gilles
Institute for System Dynamics and Control Technology
Pfaffenwaldring 9, University of Stuttgart, 70550 Stuttgart, Germany

Abstract

In this contribution, the recently published Method of Space-Time Conservation Element and Solution Element for the numerical solution of conservation laws is compared to state of the art Method of Lines based schemes. The influence of the numerical methods on the dynamic behaviour of one- and two-dimensional dispersed phase systems is investigated. Guidelines for a proper selection of numerical methods for the treatment of population balance models are given.

1. Introduction

In chemical engineering dispersed phase systems occur in a large variety. They play an important role in many industrial processes, such like e.g. crystallization, granulation, polymerisation or liquid-liquid extraction. A common characteristic of those processes is, that one or more dispersed phases of e.g. crystals, bubbles or drops, are embedded in a continuous medium. A suitable and commonly accepted concept for the modelling of dispersed phase systems is the population balance approach (Ramkrishna, 2000), which considers the dispersed phase as a population of individual particles that are distinguished from each other using some characteristic particle properties. The application of this approach leads in general to partial integro-differential equations.

In this contribution, numerical methods will be discussed for the simulation of multi-dimensional dispersed phase systems containing both continuous changes of the particle properties (e.g. particle growth or aging) and sources and sinks due to the breakage or aggregation of particles. In the following, the recently published Space-Time Conservation Element and Solution Element (CE/SE) Method (Chang et al., 1999), which was originally designed for CFD problems, will be extended for the treatment of partial integro-differential equations and compared with state of the art Method of Lines based schemes using standard finite volumes (upwind method) or flux limiters (Koren, 1993; Schiesser 1991). The numerical methods will be applied to one- and two-dimensional crystallization models considering one or two particle properties, respectively (Gerstlauer et al., 2001).

2. Application of the CE/SE Method for One-Dimensional Problems

For a brief introduction into the CE/SE method, a rather simple population balance accounting only for particle growth will be considered. Using a characteristic crystal length L as the property coordinate, this population balance can be formulated for a number density function F using the crystal growth rate G as

$$\frac{\partial F(L,t)}{\partial t} + \frac{\partial \big(G(L,t)\, F(L,t)\big)}{\partial L} = 0 \quad . \tag{2.1}$$

The application of the CE/SE method requires the discrete treatment of both the particle property coordinate L and the time t. As depicted in Fig. 1, the considered domain will be subdivided into Conservation Elements (CE), which are shifted against each other. Around those CEs, Solution Elements (SE) are arranged. They have to be considered as infinitesimal small elements or as connected lines.

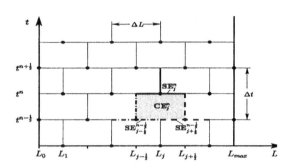

Figure 1: Subdivision of the (L,t)-domain.

Besides the differential form (2.1) of this conservation law, also the integral form

$$\int_{CE_j^n} \left(\frac{\partial F(L,t)}{\partial t} + \frac{\partial \big(G(L,t)\, F(L,t)\big)}{\partial L} \right) \; d\,CE_j^n \;=\; 0 \quad , \tag{2.2}$$

which has to be valid for each CE, is taken into consideration. On the SEs, the number density function F, and the flux term GF will be approximated by a first order Taylor expansion around the center points (j,n) of each SE, which are marked by dots in Fig. 1. By an integration over the boundary of each CE in a row, and applying Gauss' divergence theorem, an explicit time marching scheme on a fixed staggered mesh can be calculated (Chang et al., 1999). This results in a second order method, because there are two marching variables at each mesh point (j,n), one for the solution of F, another for the partial derivative of F with respect to L. The stability of this explicit method only depends on the local Courant number $\frac{|G|\Delta t}{\Delta L} \le 1$.

2.1 The treatment of integral terms using the CE/SE method
In order to include integral sink or source terms, which may account for breakage or aggregation phenomena, to the population balance (2.1), the CE/SE framework can be extended (Chang et al., 1999). The required integrations over the particle population can be carried out very accurately using the above introduced Taylor expansion.

3. Comparison of the CE/SE Method with Method of Lines Schemes

In this section, the CE/SE method will be compared to state of the art Method of Lines based schemes using finite volumes with simple upwind discretization or with a flux limiter (Koren, 1993; Schiesser 1991). The three methods will therefore first be applied to solve the simple population balance (2.1) as a test problem. Afterwards, the influence of the different numerical methods on the dynamic behaviour of a continuous crystallizer will be investigated.

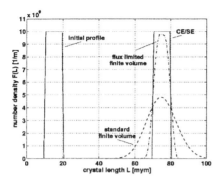

Figure 2: Simulation of Eq. (2.1) applying different numerical methods.

3.1 A one-dimensional test problem accounting for particle growth

For a reasonable comparison of the different numerical methods, a growing particle population with a constant growth rate of $G = 1\mu m / s$ is considered. Therefore, the numerical solution of the population balance (2.1) will be obtained by discretizing the particle length coordinate L in the range from 1 to $100\mu m$ into 100 finite volumes, respectively solution elements. The growth behaviour of an initially equally distributed number density F during $60s$ is then computed applying the different methods using MatLab[1] and can be seen in Fig. 2. Since it is possible to calculate an analytical solution of Eq. (2.1), which would be the movement of the initial profile without changing its shape, the quality and accuracy of the numerically calculated number densities can easily be estimated. Here, the application of the standard finite volume scheme using simple upwind discretization results in an extremely decreased and enlarged profile after $60s$ of constant growth. The reason for this behaviour is the numerical diffusion introduced by this method. With the flux limited finite volume scheme this high amount of numerical diffusion can be reduced considerably. This results in a less broadened, sharper profile. In contrast to the two Method of Lines based schemes, the profile computed using the CE/SE method does not differ from the exact solution. This follows from the fact, that this method doesn't introduce any numerical diffusion for the problem in Eq. (2.1) (Chang et al., 1999).

3.2 A one-dimensional population balance model for continuous crystallizers

In order to investigate the influence of the discussed numerical methods on the dynamic behaviour of a complex chemical process, a continuous crystallizer will be considered. The mathematical model describing this crystallization process includes very detailed microscopic models for crystal growth and attrition of crystals due to crystal-stirrer collisions (Gerstlauer et al., 2001). The simulation will be performed with an initially Gaussian distributed number density F, with an initial L_{50} of $520\mu m$. In order to compare the computed simulation results, the progression of the mass median crystal size L_{50} will be considered as a measure of the dynamic behaviour of the crystal population. As can be seen in Fig. 3, the computed dynamic behaviour of the crystallizer differs significantly depending on the chosen numerical method. The simulation using the CE/SE method results in a stable periodic behaviour that is

---

[1] MatLab 5.3, The Math Works Inc., 24 Prime Park Way, Natick, MA 01760-1500, USA

940

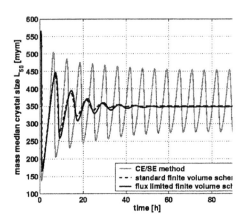

Figure 3: Simulation of a continuous crystallizer applying different numerical methods.

determined by growth and attrition of crystals, i.e. the production of fragments, whereas the application of both of the Method of Lines based schemes does lead to a steady state. These considerable differences in the qualitative process behaviour result from the rather high numerical diffusion introduced by the finite volume schemes, which ultimately cause the damping of the oscillations. Thus, the CE/SE method has been proven to be an applicable numerical method, compared to the other here discussed schemes, in order to simulate particulate processes in an accurate way.

4. Application of the CE/SE Method for Two-Dimensional Problems

In this section, the CE/SE method will be applied to simulate a two-dimensional problem. The considered population model accounts again for a crystallization process and uses two independent particle properties, the characteristic crystal length L and a molar energy w_P accounting for the lattice strain or the plastic deformation of the crystals (Gerstlauer et al., 2001). This two-dimensional crystallization model enables a more detailed simulation of the crystal growth, because the growth rate $G(L, w_P)$ depending on both particle properties allows a physical description of the experimentally often observed growth rate diffusion.

The derivation of the numerical scheme in case of this two-dimensional problem is straightforward, using the above described CE/SE methodology. The CEs now simply consist of three-dimensional volume elements in the (L, w_P, t)-domain and the SEs are two-dimensional plains that demarcate adjacent CEs. Again, an explicit time marching scheme can be derived by an integration over the surface areas of the CEs using Gauss' divergence theorem. The resulting marching variables at each mesh point are then the solution of F, and values for the partial derivatives of F with respect to L and w_P.

4.1 A two-dimensional test problem accounting for crystal growth

The efficiency of the two-dimensional CE/SE method is tested by solving a simple population balance with the two particle properties L and w_P

$$\frac{\partial F(L, w_P, t)}{\partial t} + \frac{\partial (G F(L, w_P, t))}{\partial L} + \frac{\partial (v_{w_P} F(L, w_P, t))}{\partial w_P} = 0 \quad . \tag{4.1}$$

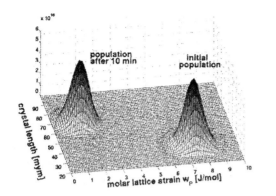

Figure 4: Simulation of Eq. (4.1) with the CE/SE method.

with constant internal velocities $G = 5 \cdot 10^{-8} \, \mu m / s$ (growth rate) and $v_{w_P} = \dfrac{dw_P}{dt} = -0.01 \, J /(mol \, s)$ (change of molar lattice strain). Starting with a narrow Gaussian distribution around $40 \mu m$ and $7.5 J / mol$, Fig. 4 shows the resulting particle population after $10 \, min$. For this calculation, the property domain (L, w_P) was discretized into 50×50 solution points. As can be seen from Fig. 4, this simulation can be performed very accurately using the CE/SE method, without any noticeable numerical diffusion. The application of this method is very efficient. The required CPU time for the calculation shown in Fig. 4 was only $40s$ on a state of the art Pentium 3 computer.

4.2 Simulation of the growth behaviour of attrition fragments

For the simulation of the growth behaviour of attrition fragments, a growth rate $G(L, w_P)$ depending on both particle properties will be applied. The relaxation of the stressed fragments that result from crystal-stirrer collisions will be described by an empirical approach. Therefore it will be assumed that the absolute stress energy stored in the crystal lattice is proportional to the crystal surface area ($\sim L^2$), for more details about this two-dimensional crystalliza-tion model the reader is referred to (Gerstlauer et al., 2001). The investigated growth behaviour starts with an initial fragment formation according to Rittinger's law. This hypothesis states the relationship

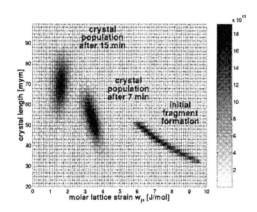

Figure 5: The simulated growth behaviour of attrition fragments.

$$w_{P, frag} = \frac{3 \cdot 10^{-4} \, J m / mol}{L_{frag}} \qquad (4.2)$$

between the length L_{frag} of an attrition fragment and its molar lattice strain $w_{P, frag}$. The changes of this fragment formation during $15 \, min$ of crystal growth can be seen in

Fig. 5. This simulation demonstrates the applicability of the CE/SE method to complex, high dimensional population models using more than one particle property. As in case of the above discussed two-dimensional test problem, the performance of this numerical method is also very efficient, since it provides very accurate numerical solutions, in terms of low numerical diffusion, with acceptable CPU times ($\sim 50s$ on a Pentium 3 computer for the simulation in Fig. 5).

5. Conclusions and Outlook

The main focus of the work presented here is the identification of proper numerical methods for the simulation of higher-dimensional dispersed phase systems that are mathematically described by population balances using more than one particle property. The presented comparison of an extended CE/SE method with state of the art Method of Lines based schemes for a one-dimensional test problem shows the superiority of the CE/SE method in terms of more accurate solutions with less numerical diffusion and higher computational efficiency. In case of a two-dimensional population model, the CE/SE method can also be applied successfully.

The presented simulation of a continuous crystallizer shows a strong dependence of the computed dynamic behaviour on the selected numerical methods. Due to the rather high amount on numerical diffusion introduced by the applied finite volume schemes, the simulations with the Method of Lines based schemes result in a steady state, instead of the periodic solution that is computed with the more accurate CE/SE method. The better computational efficiency and higher accuracy of the CE/SE method also allow the simulation of more complex crystal growth models accounting for the often observed growth rate diffusion. Another important point in this context is the model based identification of, especially, two-dimensional growth kinetics from measured data. In order to obtain kinetics that are not polluted by the applied numerical methods, the therefore required accurate simulations can be performed applying the CE/SE method.

6. References

Chang S.-C., X.-Y. Wang and C.-Y. Chow, 1999, The Space-Time Conservation Element and Solution Element Method: A New High-Resolution and Genuinely Multidimensional Paradigm for Solving Conservation Laws, *Journal of Computational Physics*, Vol. 156.

Gerstlauer A., A. Mitrović, S. Motz and E.-D. Gilles, 2001, A population model for crystallization processes using two independent particle properties, *Chemical Engineering Science*, Vol. 56.

Koren B., 1993, A robust upwind discretization method for advection, diffusion and source terms, In *Numerical Methods for Advection-Diffusion Problems*, Eds. C.B. Vreugdenhil and B. Koren.

Ramkrishna D., 2000, *Population Balances – Theory and Applications to Particulate Systems in Engineering*, Academic Press.

Schiesser W.E., 1991, *The Numerical Method of Lines*, Academic Press.

European Symposium on Computer Aided Process Engineering – 12
J. Grievink and J. van Schijndel (Editors)
© 2002 Elsevier Science B.V. All rights reserved.

Experimental Study and Advances in 3-D Simulation of Gas Flow in a Cyclone Using CFD

A.P. Peres[1], H.F. Meier[2], W.K. Huziwara[3], M. Mori[1]

[1] School of Chemical Engineering, UNICAMP, P.O. Box 6066, 13081-970, Campinas-SP, Brazil, E-mail: mori@feq.unicamp.br

[2] Department of Chemical Engineering, FURB, P.O. Box 1507, 89010-971, Blumenau-SC, Brazil. E-mail: meier@furb.br

[3] PETROBRAS, CENPES, Rio de Janeiro-RJ, Brazil.

Abstract

Experimental results and a 3-D simulation of gas flow in a cyclone are presented in this work. Inlet gas velocities of 11.0 m/s and 12.5 m/s and measurements of local pressures were used to determine radial distributions of the tangential velocity component at five axial positions throughout the equipment. The aim of this work was to analyze an anisotropic turbulence model, the Differential Stress Model (DSM). First and higher order interpolation schemes and a numerical strategy were used to assure stability and convergence of the numerical solutions carried out using the computational fluid-dynamics code CFX 4.4. The models showed a satisfactory capability to predict fluid dynamics behavior since the calculated distribution of velocity components match the experimental results very well.

1. Introduction

Cyclones, such as those in FCC units, have been used as solid particle separators in large-scale chemical processes, due to their low building and maintenance costs and the fact that they can be used under severe temperature and pressure conditions. The design of new cyclones and the analysis of the actual equipment can be achieved using computational fluid dynamics (CFD) techniques in order to obtain higher collection efficiency and a lower pressure drop. In our recent studies, Meier and Mori (1999) and Meier et al. (2000), the models analyzed were the standard k-ε, RNG k-ε and the Differential Stress Model (DSM) and it was observed that turbulence models based on the assumption of isotropy, such as standard k-ε and RNG k-ε, were inapplicable to the complex swirling flow in cyclones. On the other hand, whenever a turbulence model that considers the effect of the anisotropy of the Reynolds stress is used, such as DSM, an adequate interpolation scheme must also be considered for the prediction of flow in cyclones.

In this work, a 3-D simulation of the turbulent gas flow in the cyclone was carried out using the computational fluid-dynamics code CFX 4.4 by AEA Technology. The numerical solutions were obtained with a finite-volume method and body fitted grid generation aiming at the analysis of an anisotropic turbulence model (DSM) using first and higher order interpolation schemes. The numerical strategy adopted assured stability and convergence of the numerical solutions.

2. Mathematical Modeling

The time-averaged mathematical models along with the Reynolds decomposition governing mass and momentum transfers can be written as follows:

$$\frac{\partial \bar{\rho}}{\partial t} + \nabla \cdot \left(\overline{\rho \mathbf{v}} \right) = 0 \tag{1}$$

$$\frac{\partial \left(\overline{\rho \mathbf{v}} \right)}{\partial t} + \nabla \cdot \left(\overline{\rho \mathbf{v} \mathbf{v}} \right) = \overline{\rho \mathbf{g}} + \nabla \cdot \left(\overline{\sigma} - \overline{\rho \mathbf{v}' \mathbf{v}'} \right) \tag{2}$$

The last term of Equation (2), $\overline{\rho \mathbf{v}' \mathbf{v}'}$, is a time-averaged dyadic product of velocity fluctuations and is called Reynolds stress or turbulent stress. Some difficulties are faced in relating the dyadic product of velocity fluctuations with the time-averaged velocities. In the literature this kind of problem is known as "turbulence closure," and it is still considered an open problem in physics.

2.1 Turbulence Model

In engineering applications, there are two types of turbulence models. One is known as the eddy viscosity model, which assumes the Boussinesq hypothesis. Reynolds stress is related to time-averaged properties as strain tensor is related to laminar Newtonian flow. This model neglects all second-order correlations between fluctuating properties that appear during the application of the Reynolds decomposition procedure. The other is known as second-order closure, whereby Reynolds stress is assumed to have anisotropic behavior and also needs to predict the second-order correlation.

The model used in this work, the Differential Stress Model (DSM), is known as second-order closure and has one differential equation, or transport equation, for each component of Reynolds stress. These can generally be expressed by the following differential equation:

$$\frac{\partial \left(\rho \overline{\mathbf{v}' \mathbf{v}'} \right)}{\partial t} + \nabla \cdot \left(\rho \overline{\mathbf{v}' \mathbf{v}'} \mathbf{v} \right) = \nabla \cdot \left[\rho \frac{C_s}{\sigma_{DS}} \frac{k}{\varepsilon} \overline{\mathbf{v}' \mathbf{v}'} \left(\nabla \overline{\mathbf{v}' \mathbf{v}'} \right)^{T} \right] + \mathbf{P} - \phi - \frac{2}{3} \rho \varepsilon \mathbf{I} \tag{3}$$

in which **P** is a shear stress production tensor and is modeled as

$$\mathbf{P} = -\rho \left[\left[\overline{\mathbf{v}' \mathbf{v}'} \left(\nabla \mathbf{v} \right)^{T} + \left(\nabla \mathbf{v} \right) \overline{\mathbf{v}' \mathbf{v}'} \right] \right] \tag{4}$$

and ϕ is the pressure-strain correlation given for incompressible flow defined as

$$\phi = \phi_1 + \phi_2 \tag{5}$$

$$\phi_1 = -\rho C_{1S} \frac{\varepsilon}{k}\left(\overline{v'v'} - \frac{2}{3}kI \right) \tag{6}$$

$$\phi_2 = -C_{2S}\left(P - \frac{2}{3}P\,I \right) \tag{7}$$

P in this case is the trace of P tensor and C_S(0.22), σ_{DS}(1.0), C_{1S}(1.8) and C_{2S}(0.6) are the model's constants (Lauder et al., 1975).

It is also necessary to include an additional equation for the dissipation rate of turbulent kinetic energy that appears in Equation (3), and this is written:

$$\frac{\partial(\rho\varepsilon)}{\partial t} + \nabla.(\rho v \varepsilon) = \nabla.\left[\rho \frac{C_S}{\sigma^\varepsilon} \frac{k}{\varepsilon}\left(\overline{v'v'} \right)\nabla \varepsilon \right] + C_1 \frac{\varepsilon}{k}P - C_2\rho\frac{\varepsilon^2}{k} \tag{8}$$

in which C_1(1.44), C_2(1.92) and k are obtained directly from its definition ($k = 1/2\ \overline{v'^2}$).

2.2 Numerical Methods

In a general form, the numerical methods used to solve the models were the finite volume methods with a structured multiblock grid, generated by the body fitted on a generalized coordinate and collocated system. The pressure velocity couplings were the SIMPLEC (Simple Consistent) and PISO algorithms with interpolation schemes of first order, upwind and higher order, QUICK, Van Leer, CCCT and higher upwind.

The Rhie Chow algorithm with the AMG solver procedure was also used to improve the solution and to avoid numerical errors like check-boarding and zigzag due to the use of collocated grids and numerical errors caused by no generation of orthogonal cells during the construction of structured grids, for more details on methodologies see Maliska (1995).

The boundary conditions were uniform profiles at the inlet for all variables; no slip conditions at the walls; continuity conditions for all variables at the outlet, except for pressure where an open circuit condition with atmospheric pressure conditions were assumed. A laminar shear layer condition was also assumed for the wall with default models from the CFX 4.4 code.

3. Results

The experimental study was conducted in an acrylic cyclone settled in a pilot unit belonging to Six/Petrobras in São Mateus do Sul, Brazil. The inlet velocities of clean air were 11.0 m/s and 12.5 m/s and the measurements of local pressures obtained with a Pitot tube were used to determine the radial distributions of the tangential velocity component at five axial positions throughout the equipment (two in the cylindrical section, 0.90D and 1.35D from the cyclone roof, and three in the conical section, 2.39D, 3.36D and 4.32D from the cyclone roof). The peak of the tangential velocity like a

Rankine curve typical of flows in cyclones was obtained. The grid used in the numerical simulations had about 72,500 cells. The experimental cyclone configuration and a typical 3-D grid are shown in Figure 1.

Initially, the numerical solutions were obtained applying the upwind scheme for all variables to guarantee stability of the solution (one of the criteria adopted was a value of less than 10^{-8} for the euclidean norm of the source mass in the pressure-velocity coupling), but the results exhibited high numerical diffusion, as previously reported in the literature (Meier et al., 2000, Witt et al., 1999). The first solutions with upwind scheme were used as initial conditions. A higher order interpolation scheme was then introduced for the velocity components using a transient procedure. This has been found useful to overcome the difficulties of convergence presented by the DSM with higher order interpolation schemes. Nevertheless, it was observed that the steady state was achieved after about 1 second of real time.

Numerical solutions were obtained for the inlet gas velocities (11.0 m/s and 12.5m/s) at five different heights and the behavior of the higher order schemes were all similar (higher upwind, QUICK, CCCT and Van Leer). More details of the interpolation schemes can be seen in Guidelines of CFX 4.4 (2001). Figure 2 shows the numerical solutions obtained for the inlet gas velocities of 11.0 m/s and 12.5 m/s and using the higher upwind scheme to illustrate the experimental data and the numerical results.

Data obtained on the capability of the turbulence model to represent the radial distributions of tangential velocities throughout the cyclone was compared with the experimental data, and it was possible to verify a good agreement, especially for the velocity in the cylindrical section of the cyclone. In the conical section, the numerical results of tangential velocity were overpredicted, probably because of the imperfections in the acrylic surface of the conical section of the cyclone, which became evident during the experiments.

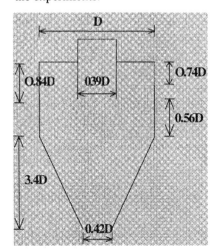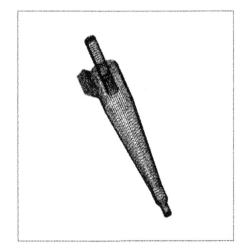

Figure 1 - 3-D grid and geometrical dimensions of the cyclone.

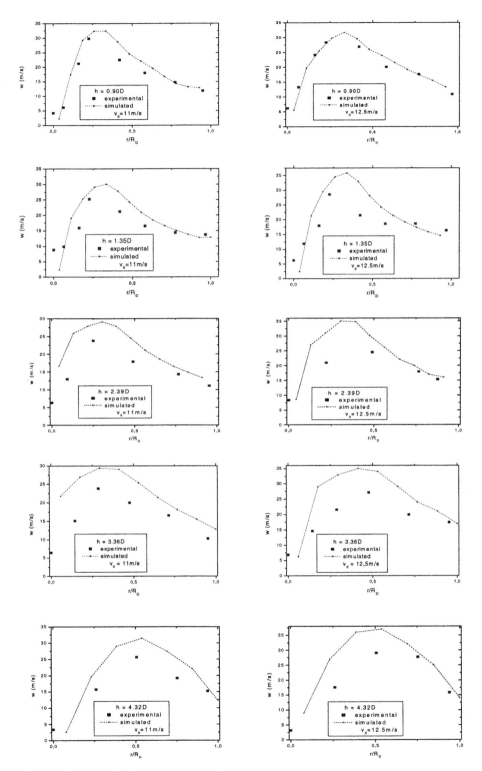

Figure 2 – Distributions of tangential velocity in the cyclone.

948

4. Conclusions

The Differential Stress Model (DSM) with higher order interpolation schemes (higher upwind, QUICK, CCCT and Van Leer) showed great capability to represent the swirling flow in the cyclone, and no significant difference was observed between the higher order schemes used. Higher order schemes avoid numerical diffusion but introduce instability and convergence difficulties that can be minimized by using appropriate solution procedures. Transient procedure used in this work had been found useful to overcome these difficulties.

5. References

Guidelines of CFX 4.4 User Guide, 2001, AEA Technology.

Lauder, D.E., Reece, G.J., Rodi, W., 1975, Progress in the Development of a Reynolds-Stress Turbulence Closure, J. Fluid Mech., 68, 537-566.

Maliska, C., 1995, Transferência de Calor e Mecânica dos Fluidos Computacional. LTC Editora, Rio de Janeiro, Brasil, 424p.

Meier H.F., Mori M., 1999, Anisotropic Behavior of the Reynolds Stress in Gas and Gas-Solid Flows in Cyclones, Powder Technology, 101, 108-119.

Meier H.F., Kasper, F.S., Peres, A.P., Huziwara, W.K., Mori, M., 2000, Comparison Between Turbulence Models for 3-D Turbulent Flows in Cyclones, Proceedings of XXI CILAMCE, 18p., Rio de Janeiro, Brasil.

Witt, P. J., Mittoni, L.J., Wu, J. and Shepherd, I.C. 1999, Validation of a CFD Model for Predicting Gas Flow in a Cyclone, Proceedings of CHEMECA99, Australia.

Acknowledgments

The authors are grateful to PETROBRAS for the financial support that makes this work possible and to Eng. Alexandre Trentin from Trentin Engenharia for his help in the experimental study in the cyclone at SIX/PETROBRAS.

Nomenclature

D — diameter of the cyclone
g — gravity acceleration
I — identity tensor
k — kinetic turbulent energy
p — pressure
P — shear stress production tensor
P — trace of the tensor **P**
t — time
v — velocity vector
w — tangential velocity component
ε — dissipation rate of turbulent kinetic energy
ϕ — pressure strain correlation
μ — viscosity
ρ — density
σ — stress tensor

Superscripts

- — mean time-averaged property
´ — fluctuation property
T — transpose tensor or matrix

European Symposium on Computer Aided Process Engineering – 12
J. Grievink and J. van Schijndel (Editors)
© 2002 Elsevier Science B.V. All rights reserved.

A New Algorithm for developing Dynamic Radial Basis Function Neural Network Models based on Genetic Algorithms

Haralambos Sarimveis, Alex Alexandridis, Stefanos Mazarakis and George Bafas
National Technical University of Athens
Department of Chemical Engineering
Greece

Abstract

A new algorithm for extracting valuable information from industrial data is presented in this paper. The proposed methodology produces dynamic Radial Basis Function (RBF) neural network models and uses Genetic Algorithms (GAs) to auto-configure the structure of the networks. The effectiveness of the method is illustrated through the development of a dynamical model for a chemical reactor, used in pulp and paper industry.

1. Introduction

Due to the always decreasing prices and increasing capacities of electronic data storage devices, today most industrial plants are collecting large volumes of process data in an every-day basis. However, in most cases the data remain unexploited, since the plant personnel rarely have the time and scientific background to work with the data and extract the important information. The desperate demand for new and more efficient algorithms to extract knowledge out of the plethora of available data was the motivation of this work. The proposed method is based on the powerful Radial Basis Functions (RBF) neural network architecture and employs Genetic Algorithms (GAs) to build dynamic models, using process input-output data.

RBF networks are continuously increasing their popularity due to a number of advantages compared to other types of Artificial Neural Networks (ANNs), which include better approximation capabilities, simple network structures and faster learning algorithms. The most important part in the development of an RBF network model is the selection of the structure of the model and the hidden node centers. Most of the popular training techniques determine only the network parameters, whereas the network structure is obtained by trial and error (Moody and Darken, 1989; Leonard and Kramer, 1991). Another family of algorithms uses various methods to determine the structure of the network as a first step that is separated from the actual objective, which is the minimization of the prediction error (Chen et al., 1990; Musavi et al., 1992). However, none of the above algorithms guarantees the selection of the optimum number of hidden nodes.

The inclusion of the structure selection in the formulation of the optimization problem is desirable, but results in a much more difficult problem, which cannot be easily solved by standard optimization methods. Genetic algorithms are stochastic methods, based on the principles of natural selection and evolution, and offer an interesting alternative optimization technique for such complicated problems (Michalewicz, 1996). Boozarjomehry and Svrcek (2001) proposed a method based on genetic algorithms, for the automatic design of feedforward neural network structures. Billings and Zheng (1995) used genetic algorithms for the training of RBF networks. However their method restricts the potential node centers only among the set of training data. The proposed methodology uses GAs to determine the optimum number of centers and the network parameters simultaneously. The method gives more freedom in the selection of the hidden node centers, since it defines a multidimensional grid in the input space and every knot in this grid represents a potential node center. The proposed training algorithm is illustrated through the development of a dynamic RBF network model for a Kamyr digester, based on real process data.

2. RBF Networks and Genetic Algorithms – An overview

RBF networks form a special class of neural networks, which consist of three layers. The input layer is used only to connect the network to its environment. The hidden layer contains a number of nodes, which apply a nonlinear transformation to the input variables, using a radial basis function, such as the Gaussian function, the thin plate spline function etc. The output layer is linear and serves as a summation unit. The typical structure of an RBF neural network can be seen in figure 1.

The standard training procedure of an RBF network selects the structure of the network by trial and error. Determination of the network parameters involves two phases: In the first one, the centers of the hidden layer nodes are obtained, based on the k-means clustering algorithm. In the second phase, the connection weights are calculated using simple linear regression.

On the other hand, GAs which will constitute the basis for developing the new method are iterative stochastic methodologies, that start with a random population of possible solutions. The individuals with the best characteristics are selected for reproduction and their "chromosomes" are transferred to the next generation. In order to emulate the way nature works, some genetic "operators" are added to the algorithms, such as mutation, where a chromosome of a single individual is altered randomly, and crossover, where new individuals are born from a random combination of the old ones.

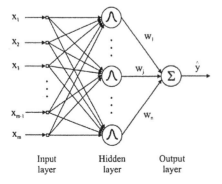

Figure 1. Typical structure of an RBF network

3. Configuration of RBF networks using Genetic Algorithms

The proposed methodology utilizes genetic algorithms to determine the optimum number of centers and the network parameters simultaneously, by minimizing an error function subject to the structure of the network, the hidden node centers and the parameters between the hidden and the output layer. The algorithm starts with an initial population of chromosomes, which represent possible network structures and contain the associated center locations. The centers are selected from a multi-dimensional grid that is defined on the input space. For each chromosome the weights between the hidden and the output layer are calculated using linear regression and the objective function is computed. New generations are produced by the standard genetic operators: crossover, mutation, deletion and addition. The algorithm stops after the specified number of generations has been completed. The chromosome which has produced the minimum value of the objective function is selected as the optimum neural network model.

Each chromosome (solution) is represented by a matrix, where the non zero rows correspond to the hidden node centers. The detailed description of the algorithm that follows assumes that M input-output examples are available, which are split into two sets, $\mathbf{X}$ and $\mathbf{Y}$. The number of input variables is represented by N, while only one output variable is used. The algorithm can be easily generalized for more than one output variables, since a different neural network can be developed for each output. $\mathbf{X}$ is an $M \bullet N$ matrix, where each row corresponds to an input vector, and $\mathbf{Y}$ is an $M \bullet 1$ vector, where each element corresponds to an output value. $\mathbf{X}$ and $\mathbf{Y}$ are scaled so that the values of all variables are positive. Additionally they are divided in three different subsets $(\mathbf{X}_1, \mathbf{Y}_1)$, $(\mathbf{X}_2, \mathbf{Y}_2)$ and $(\mathbf{X}_3, \mathbf{Y}_3)$ of size M_1, M_2 and M_3 correspondingly, namely the training, testing and validation set.

Using the first set of the input data, $\mathbf{X}_1$, the minimum and maximum values $x_{n,min}$ $x_{n,max}$ of every input variable x_n $(n=1, 2,..., N)$ are found and the space between the pair $(x_{n,min}, x_{n,max})$ is divided into m_n subspaces of range $\bullet$ $x_n = (x_{n,max}-x_{n,min})/ m_n$. The number of subspaces m_n is usually the same for each input variable. Then the following parameters of the genetic algorithm are selected: the maximum number of hidden nodes K, the size of the population L, the number of generations G and the probabilities of crossover (p_c), mutation (p_m), addition (p_a) and deletion (p_d). The genetic algorithm can be described as follows:

Step 1. Initialization: L matrices (chromosomes), $\mathbf{C}_1$, $\mathbf{C}_2$,..., $\mathbf{C}_L$, of size $K \bullet N$ are created with all zero elements. For every matrix $\mathbf{C}_l$, $(l=1, 2,..., L)$ a random integer number k_l from 1 to K is selected. The k_l first rows of the matrix $\mathbf{C}_l$ are replaced by r_l row vectors of size $1 \bullet N$ that are the centers of the corresponding network. The rows below k_l are left zero and do not correspond to a center. The elements of every row vector are given by the following equation:

$$c_{k,n} = x_{n,\min} + r_n \cdot \delta x_n \tag{1}$$

where $n=1, 2,..., N$, $k=1, 2,...,k_l$ and r_n a randomly selected integer number between 1 and m_n. At the end of this step, the matrices have the following form:

$$\begin{bmatrix} c_{1,1} & c_{1,2} & \cdots & c_{1,N} \\ c_{2,1} & c_{2,2} & \cdots & c_{2,N} \\ \vdots & \vdots & & \vdots \\ c_{k_l,1} & c_{k_l,2} & \cdots & c_{k_l,N} \\ 0 & 0 & \cdots & 0 \\ \vdots & \vdots & & \vdots \\ 0 & 0 & \cdots & 0 \end{bmatrix}$$

Step 2. Weight calculation: For every chromosome, C_l, the output weights of the network are calculated by the equation

$$\mathbf{w}_l = (\mathbf{A}_l^T \cdot \mathbf{A}_l)^{-1} \cdot (\mathbf{A}_l^T \cdot \mathbf{Y}_1) \tag{2}$$

where $\mathbf{A}_l$ is the $M_1 \times L$ matrix containing the responses of the hidden layer for the $\mathbf{X}_1$ subset of examples. It should be noted that the radial basis function used in the hidden nodes is the thin plate spline.

Step 3. Error calculation: In step 2, L networks have been developed, defined by the pairs $(\mathbf{C}_l, \mathbf{w}_l)$, $l=1, 2,\ldots, L$. The second subset of data $(\mathbf{X}_2, \mathbf{Y}_2)$ is now used as a testing set, in the following manner: The predictions of the L RBF networks $\hat{\mathbf{Y}}_{2,1}, \hat{\mathbf{Y}}_{2,2}, \ldots, \hat{\mathbf{Y}}_{2,L}$ are calculated given the input $\mathbf{X}_2$, and the corresponding error values, E_l, are obtained:

$$E_l = \left\| \hat{\mathbf{Y}}_{2,l} - \mathbf{Y}_2 \right\|_2 \tag{3}$$

The chromosome with the minimum error is kept in a separate matrix $\mathbf{B}$. The objective is to give more chances of surviving to the networks with the smaller error values. Therefore, the probability of selection p_l of every chromosome, C_l, is calculated by the following equation:

$$p_l = \left(\frac{1}{E_l} \right) \cdot \left(\sum_{l=1}^{L} \frac{1}{E_l} \right)^{-1} \tag{4}$$

and the cumulative probability by the equation:

$$q_l = \sum_{i=1}^{l} p_i \tag{5}$$

Step 4. New generation: In order to produce the new generation of L solutions, the following procedure is executed L times: A random number r in the range $0\ldots1$ is generated and the lth chromosome such that $q_{l-1} < r < q_l$ is selected. It is obvious that the chromosomes with a higher probability of selection will get more copies and others with lower probability of selection will most probably be deleted.

Step 5. Crossover: For every chromosome C_l in the new population a random number r_l between 0 and 1 is selected. If $p_c > r_l$ then chromosome C_l is selected for crossover. The number of chromosomes selected for crossover must be always greater or equal to 2 and even. If only one chromosome is selected then no crossover is executed. The chromosomes selected for crossover are grouped in pairs and for every pair all the data below a selected position are exchanged. The crossover position is selected randomly with the restriction that at least one of the two chromosomes in each pair has at least one non-zero row below it. The procedure is demonstrated with the example presented below, where the crossover operation is applied to chromosomes C_l, C_{l+1} and yields chromosomes C_l', C_{l+1}' in the new generation:

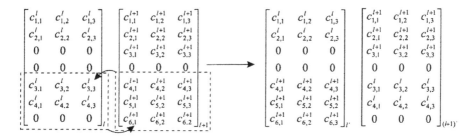

Step 6. Mutation: For every non-zero element $c_{k,n}^l$, $k=1,2,...,K$, $n=1, 2,...,N$ of each chromosome C_l, a random number r within the space $0...1$ is selected. If $p_m > r$, the element is replaced using Eq. 1.

Step 7. Addition and deletion: A binary value of 0 or 1 and a number r between 0 and 1 are generated randomly for each one of the chromosomes C_l. If the binary number is 0, and $p_d > r$ all the data below a randomly selected position are deleted. If 1 is selected and $p_a > r$, a random number of non-zero vectors created as described in step 1, replace an equal number of zero rows, so that the total number of rows will not exceed K.

Step 8. Replacement: The chromosome with the maximum error is replaced by chromosome **B** and the algorithm turns to step 2, unless the maximum number of iterations is reached. In the latter case, the algorithm stops and the network corresponding to the smallest prediction error is selected. Finally the model is validated using the third subset (X_3, Y_3), which has not been utilized throughout the entire training procedure.

4. Case study: Kappa number identification in a continuous digester

Kamyr digesters are complex tubular reactors, where the delignification of wood chips takes place through combined chemical treatment and thermal effects. The kappa number, which represents the amount of residual lignin in the pulp is the most important quality variable in the process. The proposed method was applied in order to build a dynamic model, that can predict the next value of the kappa number, using as inputs hourly past values of the three manipulated variables, which are three temperatures along the reactor.

For the particular application a set of noisy data from an industrial Kamyr digester was available. Because of very large retention times in the digester, past values of up to 12 hours were used, summing to a total of 36 input variables. The data set consisted of 400 data points, from which 185 were used for training, 100 for testing the model and 115

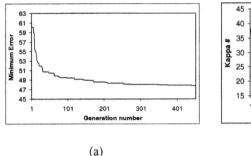
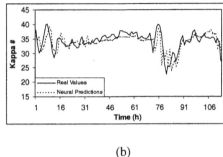

(a)

(b)

Figure 2. (a) Minimum SSE per generation (b) Kappa number prediction

as a validation set. The algorithm was applied using $K=40$, $L=20$, $p_c=0.25$, $p_m=0.01$, $p_a=0.005$ and $p_d=0.005$. The result after 450 generations, was an RBF network with 18 nodes. The training time in a Pentium IV 1400 Mhz processor was 18 mins. Figure 2a shows the reduction of the Sum of Squares Error (SSE) for the best network structure, as the algorithm proceeds. The neural network predcitions along with the real values, are shown in figure 2b for the validation set. Several runs of the algorithm with different parameter values resulted in similar results, showing that the method is insensitive to these parameters. Taking into account that the data are rather noisy, it can be seen that the model performs very well, thus rendering a set of collected data into an adequate model for the prediction of a crucial process parameter.

For comparison purposes we used the same data set to train an RBF network with the standard MacQueen k-means algorithm (Moody and Darken, 1989). The results showed that even by using the optimal structure with 18 nodes in the hidden layer, the SSE achieved by the standard method after 500 iterations, is 5% larger than the one obtained by the proposed algorithm.

5. Conclusions

The paper presents a new methodology for data mining, which produces RBF network models based on input – output data. The proposed algorithm uses genetic algorithms to determine the optimum structure of the network, in contrast to the standard RBF training methods, where the structure is selected by trial and error. In order to test the methodology, dynamic process data from an industrial reactor were used to build an RBF network model, which predicts successfully a crucial product quality parameter.

6. References

Billings, S.A. and G.L. Zheng, 1995, Neural Networks. 8, 877.
Boozarjomehry R.B. and W.Y. Zvrcek, 2001, Computers Chem. Engng. 25, 1075
Chen, S., S.A. Billings, C.F.N. Cowan and P.W. Grant, 1990, Int. J. Control. 52, 1327.
Leonard, J.A. and M.A. Kramer, 1991, IEEE Control Systems. 31.
Michalewicz, Z., 1996, Genetic algorithms + Data structures = Evolution Programs. Springer-Verlag, Berlin Heidelberg
Moody, J. and C. Darken, 1989, Neural Computation. 1, 281.
Musavi, M. T., W. Ahmed, K.H. Chan, K.B. Faris, and D.M. Hummels, 1992, Neural Networks. 5, 595.

European Symposium on Computer Aided Process Engineering – 12
J. Grievink and J. van Schijndel (Editors)
© 2002 Elsevier Science B.V. All rights reserved.

Synthesis of large-scale models: Theory and implementation in an industrial case

J. Pieter Schmal, Johan Grievink, Peter J.T. Verheijen
Technical University Delft, Department of Chemical Engineering
Julianalaan 136, 2628 BL Delft, The Netherlands

Abstract

Four different model synthesis approaches from different authors are discussed and a variant was developed based on experiences gained with the modelling of a petrochemical plant. A qualitative improvement in model synthesis has been achieved by balancing formalism with practical manageability. Some practical implementation issues are given.

1. Introduction

Over the years many people have contributed to development of the model building process (e.g. Aris 1994, Murthy et al. 1990, Marquardt 1995, Lohmann and Marquardt 1996, Hangos and Cameron 2001a). The development of models is similar to the design procedure of technical artefacts in general. Hence, the modelling procedure reflects the five steps of the generic design cycle: specifying the functional requirements, assessment of existing domain knowledge, synthesis of model structure, computation and analysis of behaviour, evaluation of model performance based on validation with plant data. The synthesis phase has received little attention in literature, as became apparent when the authors developed a model for a process that was part of a comprehensive cooperation project between industry and academia.

This project, INCOOP (INtegration process unit COntrol and plant-wide OPtimisation), researches the next generation of model based control and dynamic real time optimisation, techniques that rely on rigorous models.

In this article we will discuss our experiences in model synthesis based on an industrial case study. First we will describe the case study, before we discuss the model synthesis theory and compare it with existing approaches. After this we will give guidelines for large-scale model building and discuss implementation issues before we end with conclusions.

2. Case study

Within INCOOP a model of a petro-chemical plant has been built (figure 1). The plant contains heat integration, multiple recycles and multiple high purity separations, making it a complex non-linear plant. One single dynamic model was to be the base for the analysis of the plant behaviour for control, and optimisation. Furthermore, the model should function as plant replacement in the test phase of the new techniques. The model had to have high level of detail, sufficient accuracy, flexibility and robustness. The most complex version of the DAE model of the plant contains 2.000 differential and 23.000 algebraic equations.

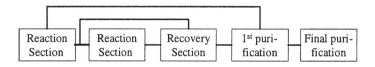

Figure 1: Block diagram of a petro-chemical plant

3. Model synthesis theory

We will assume that we have a properly defined modelling goal and we do allow for some iteration in both goal and model structure. Although Aris (1994) states that the formulation is nothing more than rational accounting, especially in the case of large-scale models some formalism is needed to minimize model errors. This formalism starts with defining the model building blocks used to set-up the model.

Two formal descriptions of model building blocks can be found in literature. Willems (2000) distinguishes between nodes that correspond with the different blocks; edges with connections between these blocks, terminals with connections between nodes or devices and devices contain laws for change. One level consists of edges, terminals and leafs (a terminal that is not connected). Another level consists of terminals and devices only. A terminal can only be associated with two real variables that reflect a force and the accompanying flux.

Marquardt's (1995) approach is also based on the idea of fluxes and forces. The fluxes are calculated in the connections and the forces in the devices. Marquardt splits the material entities in devices and connections, which in turn consist of composite or elementary elements. In Marquardt's approach a connection may well be a device on a lower level and an explicit distinction is made between (predominantly electrical) signals and phase flows.

Other non-formal building blocks can be found in the method of Lefkopoulos and Stadtherr (1993) that only contains equations and Cameron and Hangos (2001a) that contains different levels of procedures, i.e. procedures and sub-procedures.

The practical implementation led us to a more applicable approach. This approach consists of connections and devices with leaves to indicate one-sided connections. A connection can send multiple variables across the connection, not restricted to force flux combinations. A device may consist of devices coupled by connections, whereas a connection cannot be decomposed.

The model process is to a large extent driven by specific variables that are deduced from the model goal. In our case study the model has wide range of applications from plant replacement to optimisation and thus a wide range of variables needs to be modelled.

3.1 Abstraction

The abstraction is concerned with the translation of reality to a model and is thus one of the first steps in the synthesis. Every abstraction calls for a clear definition, place and character, of the system boundary. This character can be closed or open with respect to a phenomenon or variable. Reductions on the other hand are performed within the model. Abstractions and reductions are a direct result of the model goal and the capabilities of both modeller and (numerical) solvers. The abstractions and reductions can be of the following types:

- Space (e.g. minimum length scale considered 1 cm for instance)
- Time (e.g. fastest process considered dynamically 10 s)
- Phenomena (e.g. no liquid entrainment in distillation)
- Forced (e.g. lack of knowledge)
- Solver based (e.g. ignoring very small numbers)

Aris (1994) pays ample attention to abstractions and reductions, but in the sense of examples rather than an approach. Hangos and Cameron (2001a) point out that it is important to filter the set of all 'model controlling' mechanisms with respect to the modelling goal. In our case study we abstract our plant to figure 1 and define the system boundaries. For simplicity we start with boundaries at the walls of the equipment closed with respect to the surroundings except for explicit in- and outputs. The time scale we are interested in is in the range of seconds to days (time reduction). In a first step we considered two phases on our equilibrium tray and encountered an index problem later on that could not be handled by gPROMS, our modelling environment, forcing us to something different like describing the phases combined (solver based reduction).

3.2 Decomposition

Decomposition of the problem reduces the complexity and ensures the problem becomes manageable. The decomposition can be done on two levels as indicated by Marquardt (1994): structural and equational.

Structural decomposition

The structural decomposition concerns breaking down the original problem in smaller blocks, which contain equations that have something in common depending on how the system was decomposed. For every structural decomposition we need to define an abstraction. The way we decompose the problem can be (for both levels) the following classes:

- Locational: location (time, space or in process stream) is similar
- Tree: natural connection (a direct physical, causal or mathematical connection)
- Behavioural: similar behaviour
- Temporal: similar time scales
- Spatial: similar size
- Functional: similar functions (both operational and mathematical)

Orthogonal to these decompositions are the levels of detail and the levels of hierarchy.

Equational decomposition

The equational decomposition concerns the break down of equations in a block to sets of equations with a certain similarity. Hangos and Cameron (2001a) give a functional equational decomposition by splitting up the equations in one of the following classes: balance, transfer rate, property relation, balance volume relation, and equipment and control constraint.

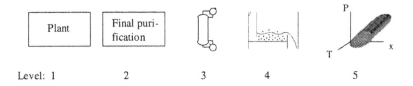

Level: 1   2   3   4   5

Figure 2: Locational and functional structural decomposition

In our case study we break down the plant into sections as in figure 1 and next into devices. A distillation column is further decomposed into trays, a tray into a physical property part and a device specific part. This is a five-level (figure 2) locational structural decomposition with a functional structural decomposition on the last level. In our case we had two different types of physical property relations that could easily be switched due to the functional structural decomposition. A tree equational decomposition was used to decompose the equations (figure 3).

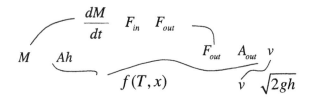

Figure 3: Tree equational aggregation

3.3 Aggregation
In contrast with the decomposition we start with equational aggregation since we can only do structural aggregation once the blocks are composed.

Equational aggregation
The equations in a block must be set-up in such a way that the block can function according to its specifications, i.e. it must fulfil the sub-task it received during the structural decomposition. The equations must be written down in accordance with the way they were decomposed, because it determines the logic needed to minimize errors. The composed blocks are the devices of our model.

Structural aggregation
The structural aggregation combines the composed blocks, the devices, with the help of connections and creates connections to the surroundings via leafs. The connections describe the information that is shared between devices. Connections and leafs arise from the character of the system boundary of the device that is open.
We choose to use a tree equational decomposition, therefore we start with the component balance (since we were among others interested in product purity). We write down the balance equation first and make sure all variables that occur in the balance are either specified or given by an additional equation (figure 3 for an example of a liquid vessel). Additional assumptions such as equilibrium on the tray could be incorporated at this stage. Next we construct the column by incorporating the other trays, condenser and

reboiler. To construct the purification section (figure 2) we couple all devices in this section before we hook it to the rest of the plant.

3.4 Level of detail

As stated before orthogonal to the decomposition and aggregation above we have the level of detail. Changing the level of detail can be caused by a need for more accuracy or detail, but also by a need for faster simulations or less complexity. Since the systems functions as a whole, the lowest level of detail usually determines the level of detail for the whole system and the highest level of detail usually determines the speed of the simulation, but certainly not always!

In our case study, we increased the level of detail by incorporating the heat-integration and decreased the level of accuracy in the physical properties to increase simulation speed, making the levels of detail more balanced over the system.

4 Implementation issues

The way the equations are set-up can have a great influence on the (re-)initialisation and numerical stability. Physically correct alterations due to discontinuities or unwanted limit behaviour are preferred, but may not always be available. In our case study we helped the initialisation by reverting to two directional flows, despite the discontinuities we had to introduce, because the operating region of the model was greatly enlarged.

5 Discussion

As stated by Foss et al (1998) the modelling process is poorly understood. This is caused by the fact that we are unable to predict the effect of some choices on the outcome of the simulation. It is the experiences of a modeller with respect to these sensitivities that play an important role during the model synthesis. Some experiences in the sense of pitfalls are given in Murthy et al. (1990).

Table 1: Main differences

	Willems	Marquardt	Hangos	Lefkopoulos	Presented
Complementary	Yes	No	Yes	No	Yes
Formality	High	High	Low	Low	Medium
Number of components	High	Medium	Low	High	Low
Strategy levels	2	2	>2	1	2

The main differences between the leading synthesis methods and our approach are given in table 1. Both Willems and Marquardt have defined a set of rules that should be obeyed for their devices and terminals or connections. In the case of Willems the terminals can only be associated with two real variables that are related in some sense, making the system unnecessary large (number of components). Marquardt's approach, in which a connection may become a device on a lower level, actually embodies multiple models and is therefore not complementary or has some overlap. Hangos' approach on the other hand is too unstructured (low formality) in the sense that there is no systematic way of setting up a consistent set of equations. Our approach is basically using the advantages of both Willems, Marquardt and Hangos approach. It balances the need to avoid errors by formality and the manageability to limit model building time.

The modelling tool heavily influences the modelling process (Foss et al. 1998), and thus usually determines the way we decompose our system. Since we used gPROMS in our case study, the model in the examples is well suited for equation-oriented solvers. Especially in the case of large-scale systems causality is usually difficult to determine and parallel equation-oriented solvers are then preferred.

During the aggregation of equations the choice for certain equations is made. It is well known that conservation laws increase numerical stability (Gershenfeld 1999). Whether making the model more linear really helps is questionable, the numerical solver has less problems, but larger inaccuracies in variables occur. For structural aggregation it is important that the streams between the blocks are clearly defined and connecting streams contain the same set of variables. A useful sub-class of the forced reductions are the connection breakers. A connection breaker simply fixes some variables in the connection, e.g. a temperature.

A proper decomposition in combination with modularity can help to change the level of detail from simple (at the start) to complex. The needed level of detail can best be obtained by investigating the levels below and above the current level of detail to see whether a step in either direction is useful.

Finally, model misuse poses a serious threat in large-scale models. Assumptions should therefore be translated into constraints where possible (Hangos and Cameron 2001b).

6. Conclusions

Using a model synthesis approach makes the synthesis qualitatively more efficient and robust to errors. The different model syntheses are comparable and the main differences are given in table 1. The key point of our approach compared to the others is that it balances the formality with practical manageability. Future work consists of investigation of the advantages of different decomposition strategies.

References

Aris, R., 1994, Mathematical modelling techniques, General Publishing Company Ltd.

Foss, B.A., B. Lohmann, and W. Marquardt, 1998, J.Proc.Cont. 8 (5-6), 325.

Gershenfeld, N., 1999, The nature of mathematical modelling, Cambridge University Press.

Hangos, K., and I. Cameron, 2001a, Process modelling and model analysis, Academic Press.

Hangos, K., and I. Cameron, 2001b, Comp. Chem. Engng. 25, 237

Lefkopoulos, A., and M.A. Stadtherr, 1993, Comp. Chem. Engng. 17, 399.

Lohmann, B., and W. Marquardt, 1996, Comp. Chem. Engng. 20, S213.

Marquardt, W., 1995, Methods of Model-Based Control, NATO-ASI Ser. E, Applied Sciences 293, 3, Kluwer Academic Publ.

Murthy, D.N.P., N.W. Page, and E.Y. Rodin, 1990, Mathematical modelling, Pergamon Press.

Willems, J.C., 2000, Math. Computers in Sim. 53, 227.

European Symposium on Computer Aided Process Engineering – 12
J. Grievink and J. van Schijndel (Editors)
© 2002 Published by Elsevier Science B.V.

Prediction of the Joint Molecular Weight-Long Chain Branching Distribution in Free-Radical Branched Polymerizations

P. Seferlis and C. Kiparissides
Department of Chemical Engineering and Chemical Process Engineering Research Institute, Aristotle University of Thessaloniki, P.O. Box 472, 540 06 Thessaloniki, Greece

Abstract

A novel approach based on orthogonal collocation on finite elements (OCFE) techniques is proposed for the prediction of the joint molecular weight – long chain branching distribution in free-radical polymerization. The OCFE formulation reduces the size of the model, preserves the nature of the balance equations and ensures the closure of the overall material balance regardless of the selected kinetic mechanism. The proposed approach is successfully applied in the free-radical branched polymerization of vinyl-acetate. The OCFE formulation shows high accuracy, when compared to the method of the moments, with improved predictive ability in systems characterized by strong diffusion controlled phenomena and long-chain branching.

1. Introduction

Most commonly employed polymer property indicators (e.g., mechanical strength, tear strength, rheological properties and so forth) are directly or indirectly linked with the molecular structural properties of the polymer chains (e.g., molecular weight distribution, MWD, long chain branching, LCB, and so forth), which are usually very difficult to measure on-line. Hence, the ability to accurately predict the molecular structural properties from mechanistic models becomes of significant importance.

The mathematical models dealing with the prediction of the molecular weight distribution in free-radical polymerization are based on kinetic lumping methods (Crowley and Choi, 1997), continuous variable approximations, Z-transforms, polynomial expansion methods (Tobita and Ito, 1993), variations of the method of the moments for branched polymer systems (Pladis and Kiparissides, 1998), Monte-Carlo simulations (Tobita, 1993), discrete weighted Galerkin formulation (Wulkow, 1992; Iedema *et al.*, 2000), and orthogonal collocation methods (Canu and Ray, 1991; Nele *et al.*, 1999). The proposed method employs orthogonal collocation on finite elements modeling techniques that preserve the nature of the balance eqautions and reduce significantly the size of the model, while providing accurate prediction of the joint weight chain length (WCL)-long chain branching (LCB) distribution. The effective computation of the WCL-LCB distribution becomes essential in the control of molecular weight properties via dynamic optimization of polymerization reactors.

2. Free-Radical Polymerization Kinetic Mechanism

A general kinetic mechanism that describes the free-radical polymerization of branched polymers includes the following elementary reactions:

Initiation: $\text{I} \xrightarrow{k_d} 2\text{R}^\bullet$

Chain initiation: $\text{R}^\bullet + \text{M} \xrightarrow{k_i} \text{R}_1$

Propagation: $\text{R}_x + \text{M} \xrightarrow{k_p} \text{R}_{x+1}$

Chain transfer to monomer: $\text{R}_x + \text{M} \xrightarrow{k_{fm}} \text{P}_x + \text{R}_1$

Chain transfer to solvent: $\text{R}_x + \text{S} \xrightarrow{k_{fs}} \text{P}_x + \text{R}_1$

Chain transfer to polymer: $\text{R}_x + \text{P}_y \xrightarrow{k_{fp}} \text{R}_y + \text{P}_x$

Reaction with terminal double bond: $\text{R}_x + \text{P}_y^= \xrightarrow{k_{db}} \text{R}_{x+y}$

Termination by disproportionation: $\text{R}_x + \text{R}_y \xrightarrow{k_{td}} \text{P}_x + \text{P}_y$

Termination by combination: $\text{R}_x + \text{R}_y \xrightarrow{k_{tc}} \text{P}_{x+y}$

The subscripts x and y denote the number of monomer units for the "live", R_x, and "dead", P_x, polymer chains. Symbols I, M, and S denote the initiator, monomer and solvent, respectively. The kinetic mechanism includes propagation and termination reactions, molecular weight control reactions by transfer to monomer and solvent, and long-chain branching formation by transfer to polymer and terminal double bond reactions. The present kinetic mechanism assumes that $\text{P}_x \approx \text{P}_x^=$, without significant loss of accuracy (Pladis and Kiparissides, 1998).

3. Molecular Weight Distribution in Free-Radical Polymerization

Orthogonal Collocation on Finite Elements Formulation
A key characteristic of the OCFE formulation is the treatment of the discrete polymer chain length domain, s_f, as a continuous one. Hence, the concentrations of the "live" and "dead" polymer chains, become continuous variables. The OCFE chain length domain was divided into a number, NE, of finite elements, with boundaries at $\zeta_0=1$, ζ_1, ζ_2, ..., ζ_{NE-1}, $\zeta_{NE}=s_f$. In every element a number of n interior collocation points, $[s_1, s_2,..., s_n]$, was specified from the roots of the Hahn family of discrete orthogonal polynomials.

The concentrations of the "live", $\tilde{R}(s)$, and "dead", $\tilde{P}(s)$, polymer chains within each element were approximated by continuous low-order Lagrange interpolation polynomials.

$$\tilde{R}(s) = \sum_{i=0}^{n} W_{i,j}^{R}(s)\tilde{R}(s_{i,j}) \qquad \tilde{P}(s) = \sum_{i=1}^{n} W_{i,j}^{P}(s)\tilde{P}(s_{i,j}) \qquad \zeta_{j-1} \leq s \leq \zeta_j \qquad j = 1,...,NE \qquad (1)$$

The tilde denotes approximation variables. The functions, $W_{i,j}^{R}(s)$ and $W_{i,j}^{P}(s)$ are Lagrange interpolation polynomials of order n+1 and n, respectively. The left boundary point of each element was also considered an interpolation point for the "live" polymer chains so that the boundary condition at chain length x=1 was transferred throughout the domain. Considering a batch polymerization reactor, the main requirement of the OCFE formulation forces the dynamic residual balances for the "live" and "dead" polymer chains to vanish at the selected collocation points, $s_{i,j}$.

Residual balance equation for "live" polymer chains at the collocation points:

$$\Re_{R_{i,j}} = -\frac{d\tilde{R}(s_{i,j})}{dt} + \left\{ k_I[R^\bullet][M] + (k_{fm}[M] + k_{fs}[S]) \sum_{s=1}^{s_f} \tilde{R}(s) \right\} \delta(s_{i,j} - 1)$$

$$+ k_p \left\{ \tilde{R}(s_{j,i} - 1) - \tilde{R}(s_j) \right\}[M] - \left\{ k_{fm}[M] + k_{fs}[S] \right\} \tilde{R}(s_{i,j}) + k_{fp} \left\{ s_{i,j} \tilde{P}(s_{i,j}) \right\} \sum_{s=1}^{s_f} \tilde{R}(s) \tag{2}$$

$$- k_{fp} \tilde{R}(s_{i,j}) \sum_{s=2}^{s_f} \left\{ s \tilde{P}(s) \right\} - k_{td} \tilde{R}(s_{i,j}) \sum_{s=1}^{s_f} \tilde{R}(s) - k_{tc} \tilde{R}(s_{i,j}) \sum_{s=1}^{s_f} \tilde{R}(s)$$

$$- k_{db} \tilde{R}(s_{i,j}) \sum_{s=2}^{s_f} \tilde{P}(s) + k_{db} \sum_{s=2}^{s_{i,j}-1} \left\{ \tilde{R}(s_{i,j} - s) \tilde{P}(s) \right\}$$

for i=0,...,n and j=1,...,NE

Residual balance equation for "dead" polymer chains at the collocation points:

$$\Re_{P_{i,j}} = -\frac{d\tilde{P}(s_{i,j})}{dt} + \left\{ k_{fm}[M] + k_{fs}[S] \right\} \tilde{R}(s_{i,j}) + k_{td} \tilde{R}(s_{i,j}) \sum_{s=1}^{s_f} \tilde{R}(s) - k_{db} \tilde{P}(s_{i,j}) \sum_{s=1}^{s_f} \tilde{R}(s) \tag{3}$$

$$- k_{fp} \left\{ s_{i,j} \tilde{P}(s_{i,j}) \right\} \sum_{s=1}^{s_f} \tilde{R}(s) + k_{fp} \tilde{R}(s_{i,j}) \sum_{s=2}^{s_f} \left\{ s \tilde{P}(s) \right\} + \frac{1}{2} k_{tc} \sum_{s=1}^{s_{i,j}-1} \left\{ \tilde{R}(s) \tilde{R}(s_{i,j} - s) \right\}$$

for i=1,...,n and j=1,...,NE.

The individual elements of the discrete summation terms that appear in eqs 2-3 were approximated using the Lagrange interpolation polynomials for the given element partition. Hence, the residual balances were solely expressed in terms of the "live" and "dead" polymer chain concentrations at the collocation points. The WCLD was then reconstructed from the following relationship applied at the collocation points:

$$WCLD(s_{i,j}) = s_{i,j} \tilde{P}(s_{i,j}) \bigg/ \sum_{s=2}^{s_f} \left\{ s \tilde{P}(s) \right\} \quad \text{for} \quad i = 1,...,n \quad \text{and} \quad j = 1,...,NE \tag{4}$$

The accuracy of the predicted WCLD mainly depends on the selected size of the chain length domain, the total number of finite elements, the partition of the domain in finite elements, and the total number of collocation points in the domain.

OCFE formulation for branch classes
The OCFE formulation was extended to the balance equations for the "live" and "dead" polymer chains of branch classes. Each branch class is defined as the fraction of the

entire polymer chain population with the same content of long chain branching (e.g., linear polymer chains, polymer chains with one LCB, two LCB etc.). The chain length domain for each branch class was partitioned in a number of finite elements, NE_c, with n interior collocation points specified from the roots of the Hahn polynomials. In a similar fashion as in the overall balances, the residual balance equations for each branch class were forced to vanish at the collocation points (Seferlis and Kiparissides, 2002). The branch class residual balances were solved in conjunction with the overall residual balance equations (eqs 2-3). Such a solution approach eliminates the error in the prediction of the WCLD for each branch class arising from a poor selection of the assumed degree of long-chain branching. This is however true, only for kinetic mechanisms that do not contain reactions that reduce the branch class.

4. Methyl-Methacrylate Bulk Polymerization

The outlined approach for the prediction of the WCLD was applied to the bulk polymerization of methyl-methacrylate (MMA) in a batch reactor. The polymerization is highly exothermic and exhibits a strong acceleration in polymerization rate due to gel-effect (e.g., the termination rate constant decreases with conversion). A total degree of polymerization equal to 656,840 monomer units was considered for the OCFE formulation. The chain length domain was partitioned into 42 finite elements with two collocation points in each finite element.

Two operating scenarios were considered: i) isothermal operation and ii) one step change in temperature during the batch duration. Figure 1 shows the WCLD achieved at the end of the batch for the two operating scenarios. The isothermal reactor produces a unimodal distribution, while the second scenario produces a bimodal distribution. The WCLD calculated using the OCFE formulation was compared to the one obtained from the summation of the instantaneous WCLD, when reconstructed from the corresponding leading moments, with excellent matching results. Usually high monomer conversions or strong gel-effect phenomena cause considerable broadening of the associated WCLD. An empirical rule that alternatively uses the Schulz-Flory (polydispersity PD<4) or the

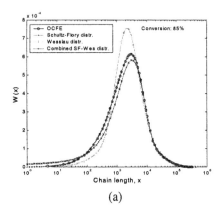 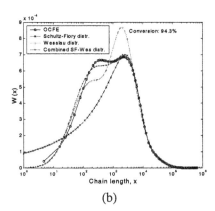

(a)   (b)

Figure 1. WCLD for MMA bulk polymerization (a. isothermal operation at 60°C, b. temperature step change of +10°C after 125 minutes of operation).

Wesslau (PD>4) distributions for the instantaneous WCLD was applied.

5. Vinyl-Acetate Solution Polymerization

The free-radical polymerization of vinyl-acetate was selected as a representative example of the production of branched polymers. Highly branched polymer chains are produced through the transfer to polymer and the terminal double bond polymerization reactions. Diffusion phenomena strongly affect the termination and propagation kinetic rate constants as monomer conversion increases. An OCFE scheme consisted of 72 finite elements and two collocation points per finite element was employed. A total degree of polymerization equal to 668,580 monomer units was selected, large enough to allow an accurate approximation of the WCLD for high monomer conversion rates. The residual balance system (overall and 41 branch classes) was consisted of 2878 differential equations.

Figure 2 compares the WCLD obtained from the OCFE formulation and the method of the moments at different monomer conversions. The overall WCLD with the method of the moments was calculated from the summation of the branch class WCLD reconstructed from their leading moments. A good agreement between the overall WCLD calculated from both methods is generally observed. Some small discrepancies are however present for short polymer chains. A closer analysis of the branch class distributions, suggests that the discrepancies are mainly attributed to the differences in the WCLD of the linear polymer chains. The reason is that the reconstruction of the WCLD for the linear polymer chains from its moments is prone to error due to the high polydispersity of the distribution. Such a behavior is more vividly observed at higher monomer conversion values (e.g., 90%). High conversion rates cause a dramatic reduction in the kinetic rate constants due to the gel-effect that subsequently results in the creation of a significant amount of linear oligomers (short polymer chains). The agreement between the two methods improves as the long-chain branching content increases. Figure 3 shows the joint WCL - LCB distribution for 41 branch classes.

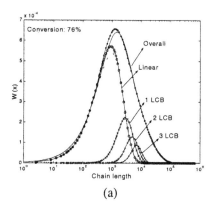
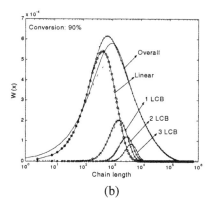

(a)   (b)

Figure 2. Overall and branch class WCLD for OCFE (lines with symbols) and method of moments (plain solid and broken lines).

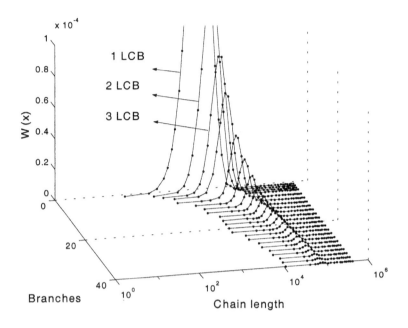

Figure 3. Joint WCL-LCB distribution.

6. Conclusions

A novel approach for the prediction of the weight chain length distribution based on the model reduction properties of OCFE techniques is presented. Comparison of the calculated WCLD using the OCFE modeling technique with the method of the moments verifies the accuracy of the proposed method. The power of the OCFE formulation as a prediction tool of the joint WCL-LCB distribution of complex polymer systems is evident from application to polymerization systems that are characterized by strong diffusion controlled reactions and high degree of long-chain branching.

7. References

Canu, P., and W. H. Ray, 1991, Comput, Chem. Eng. 15, 549.

Crowley, T. J., and K. Y. Choi, 1997, Ind. Eng. Chem. Res., 36, 1419.

Iedema, P. D., Wulkow M., and C. J. Hoefsloot, 2000, Macromol., 33, 7173.

Nele, M., Sayer, C., and J. C. Pinto, 1999, Macromol. Theory Simul., 8, 199.

Pladis, P., and C. Kiparissides, 1998, Chem. Eng. Sci., 53, 3315.

Seferlis, P., and C. Kiparissides, 2002, submitted Chem. Eng. Sci.

Tobita, H., 1993, J. Polym. Sci., 31, 1363.

Tobita, H., and K. Ito, 1993, Polym. Reac. Eng., 1, 407.

Wulkow, M., 1992, Impact of Computing in Sci. and Eng., 4, 153.

European Symposium on Computer Aided Process Engineering – 12
J. Grievink and J. van Schijndel (Editors)
© 2002 Elsevier Science B.V. All rights reserved.

Multiobjective Dynamic Optimization of Semi-Continuous Processes

C. M. Silva, E. C. Biscaia Jr.
PEQ/COPPE/UFRJ - Federal University of Rio de Janeiro - Brazil
evaristo@peq.coppe.ufrj.br

Abstract

Multiobjective dynamic optimization of Chemical Engineering systems is carried out using an improved genetic algorithm. A fed-batch bioreactor for foreign protein production in an inducible bacterial system has been optimized. The target of this process is to maximize the production of the foreign protein and minimize the inducer consumption. An improved genetic algorithm is proposed to generate the optimal operating policies of this system. A new concept of fitness function, based on the ranking procedure, was adopted. A fuzzy penalty function method is used to incorporate the constraints into the fitness function. A new class of operators is introduced to enhance the algorithm performance, and the standard ranking procedure is extended to solve multidimensional problems. The simulated results demonstrated the efficiency of the algorithm to find the Pareto optimal set, offering a viable strategy to solve complex dynamic optimization problems.

1. Introduction

Many chemical and biochemical processes are operated in semi-continuous mode. The dynamic behavior of such systems is usually highly complex and non-linear. In biotechnology processes, several phases can be distinguished during the operations, characterized by different substrate consumptions and metabolic production rates. Changes in the external environment often affect the internal composition of the cells as well as the cell morphology (Roubos et al., 1999).

In order to specify the optimal control strategies for these processes, multiobjective optimization methods that take dynamic behaviors into account are required. Such methods can deal with conflicting targets that although influence the reactor performance in opposing ways should be achieved simultaneously. Several studies have been reported on multiobjective dynamic optimization of batch and semi-batch processes (Butala et al., 1992, Wajge and Gupta, 1994, Sareen and Gupta, 1995, Bhaskar et al., 2001). Different methods have been proposed to determine optimal control policies for bioreactors (Lee and Ramirez, 1996, Tholudur et al., 2000, Canto et al., 2001). Roubos et al.(1999) pointed out the limits of these methods on number of control variables and model complexity and suggested the use of evolutionary techniques. In this contribution, an improved genetic algorithm (GA) is proposed to conduct multiobjective dynamic optimization. A challenging case study presenting different feeding strategies is considered to evaluate the performance of the algorithm.

2. Multiobjective Genetic Algorithms

The GA technique simulates a natural evolution process: the fittest species survive and propagate while the less successful tend to disappear. The multiobjective optimization procedure consists of a search for nondominated solutions. The concept of nondominance refers to the solutions for which no objective can be improved without worsening at least one of the other objectives. The nondominated solutions are superior to the others with respect to all targets, but comparatively good among themselves. Any of these solutions is an acceptable solution, as all are considered equivalent in dominance. The progress strategy is guided by the fitness evaluation, and consists of performing the population with genetic operators to generate the next population. Different adaptations of the original GA are presented in the literature (Cheng and Li, 1998, Toshinsky et al., 1999, Wang et al., 1998). A detailed background on the GA theory is reported in Goldberg (1989) and Busacca et al. (2001).

2.1. The proposed algorithm

The multiobjective optimization algorithm developed is an improved version of the GA proposed by Cheng and Li (1998). The standard ranking procedure is extended to treat multidimensional problems. A new class of operators - niche, Pareto-set filter and elitism - is introduced to reduce the necessary number of generations. A fitness function and a fuzzy penalty method are also adopted. The algorithm operates in a continuous variable space, which is computational fast and stable in converging to global optima.

The proposed GA procedure works through the following steps: a) creation of a random initial population; b) evaluation of the individuals and application of the penalty function method; c) ranking of the individuals, calculation of the fitness, registration of the best individuals; d) registration of all nondominated individuals in the Pareto set filter operator; e) selection of pairs of individuals as parents; f) crossover of the parents to generate the children; g) replacement of the individuals using the niche operator; h) genetic mutation; i) replacement of the individuals using the elitism operator.

Penalty function method
A fuzzy penalty function method has been adopted to treat constrained multiobjective optimization problems. This method incorporates the constraints into the objective functions by using a transferred function, which carries information on the point's position and feasibility (Cheng and Li, 1998). The penalization procedure consists of associating a finite value, established on the fuzzy logic theory, with the extent each constraint is violated. The largest amount violated of each point is used to determine the transferred function value of that point. As a result, all points in the feasible region present values between 0 and 1, while the infeasible ones are greater than the unity.

Ranking procedure
The ranking procedure consists of the classification of the individuals into categories according to the concept of dominance. First, all nondominated individuals of the population are identified and assigned rank 1. These individuals are virtually removed from the population and a new evaluation is conducted on the remaining individuals. The next set of nondominated points are identified and assigned rank 2. This procedure

continues until all the individuals are classified. Mathematically, the ranking method is conducted in four basic steps: a) the points are sorted according to the evaluation of an objective function, randomly chosen as a reference function; b) the sequence of points that simultaneously produces an increase on the reference function and a decrease on the other functions, or vice versa, are selected as potential candidates to the rank; c) repeat steps (a) and (b) until all the objective functions have been chosen as the reference function; d) all points selected at least (n-1) times in the step (b) will be assigned to the rank. These steps are repeated for each rank, until all the points are classified.

Fitness function

The fitness value represents a measure of each individual performance. Every individual belonging to the same rank class is considered equivalent and has the same probability of being selected for reproduction. The fitness function, F_k, is determined to each individual of the same rank k as follows (Cheng and Li, 1998):

$$F_k = \frac{N_r - k + 1}{SS} \tag{1}$$

$$SS = \frac{1}{P_s} \sum_{k=1}^{N_r} (N_r - k + 1) P_{sk} \tag{2}$$

where Nr is the highest rank of the population, P_s is the population size and P_{sk} is the population size of rank k. According to this fitness definition, the larger the population size at a rank is, the smaller the fitness of a point. Hence, the reproduction ratio of individuals at each rank depends on both the rank level and population size.

Niche operator

The niche operator determines which individuals will go to the next generation. The fitness of each child is calculated in the domain of its paternal population. The replacement of the parents only occurs if the child fitness exceeds the parents' inferior fitness. Otherwise, the parents will go to the next generation. This operator helps to avoid the genetic drift, which makes a population become clustered at certain regions, maintaining the appropriate diversity of the population.

Elitism Operator

The elitism procedure consists of the propagation of the best solutions to the next generation. For this purpose, it maintains an elitism file where the best solution of each individual objective function is registered. At each generation, this file is updated: if a better solution was generated it replaces the one stored. The individuals selected to the next population by the crossover and mutation procedures are submitted to the elitism operator. The points registered in the elitism file will randomly replace some of the candidates to the next generation. As a consequence, it increases the convergence of the optimization process as well as the robustness of the algorithm.

Pareto-set filter operator

All points assigned rank 1 are registered in the Pareto set filter at each generation. This file is dynamically updated by using a filter operator, in which the nondominated solutions of the current population are compared with those already stored in the file, from the previous generations. A new evaluation is conducted in the filter, according to the following rules: a) all points in the filter identified as nondominated, are recorded in the Pareto set file. The dominated ones are discarded; b) if the number of points in the file is inferior to the population size, the new nondominated points are stored. Otherwise, if the file is full, the most similar points in the Pareto set file are replaced.

At the end of the optimization process, the file itself comprises the Pareto optimal set and constitutes the result of the optimization.

The computational load associated to the proposed algorithm relies on the number of objective functions of the problem. At each generation, the objective functions are evaluated just once for each point. Therefore, the total number of the objective evaluations is a function of the population size, the number of generations and the number of objective optimized. The algorithm is fully presented in a previous work (Silva and Biscaia, 2001).

3. Formulation of the Multiobjective Optimization Problem

The case study deals with the production of induced foreign protein by recombinant bacteria. The mathematical formulation reported by Tholudur et al. (2000) has been modified to consider the reactor volume as a function of time:

$$\frac{dV}{dt} = q_G - q_I \tag{3}$$

$$\frac{d(VX)}{dt} = \mu XV \tag{4}$$

$$\frac{d(VG)}{dt} = q_G G_f - \frac{\mu XV}{Y} \tag{5}$$

$$\frac{d(VI)}{dt} = q_I I_f - r_I XV \tag{6}$$

$$\frac{d(VP)}{dt} = r_P XV \tag{7}$$

where X, G, I and P are the cell mass, glucose, inducer and protein concentrations respectively, and V, the reactor volume. G_f and I_f are the glucose and inducer feed concentrations, and q_G and q_I the feeding rates ($0 \leq q_G \leq 1$ and $0 \leq q_I \leq 1$). μ is the specific cell growth rate, r_I, the specific inducer inactivation rate, and r_P, the specific protein production rate. Y is the biomass yield on the substrate. The parameter function definitions and values are described in Tholudur et al. (2000). Aiming to increase the sensitivity of the model to the controls, a modified parameter function μ is presented:

$$\mu = \frac{\mu_{\max} G}{K_{CG} + G} \left(\frac{1 + \beta_1 I}{1 + \beta_2 I} \right) \tag{8}$$

where $\beta_1 = 8.33$ and $\beta_2 = 23.33$. The process is conducted in 15 hours. Glucose and inducer are fed simultaneously at different rates, during the whole process. The optimization problem involves the maximization of the foreign protein production and the minimization of the total inducer consumption. The decision variables are the inducer initial concentration, I_0, the inducer feed concentration, I_f, and the glucose and inducer feeding rates, $q_G(i)$ and $q_I(i)$ ($i = 1, 10$). A total of 22 variables are manipulated. Two constraints are identified: the amount of inducer available, I_{Total}, is 3 g, and the final reagent volume, $V(t_f)$, is required to be greater than 10 dm^3. This optimization problem is formulated as:

maximize
$$f_1 = P(t_f) \, V(t_f) \tag{9}$$

minimize
$$f_2 = I_f \int_0^{t_f} q_I(t) \, dt + V_0 I_0 \tag{10}$$

subject to
$$g_1: I_{Total}(t_f) \le 3 \text{ g}; \qquad g_2: V(t_f) \ge 10 \text{ dm}^3 \tag{11}$$

4. Results and Discussion

The proposed algorithm has been applied to determine the optimal operating policies for the formulated problem. The optimization has been carried out using the GA parameters shown in Table 1, established in a previous sensitivity study (Silva and Biscaia, 2001).

Table 1. Genetic algorithm parameters

Population size	20 individuals	Number of children /crossover	1
Crossover probability	75 %	Number of generations	230
Mutation probability	5 %		

Figure 1 (a) shows the Pareto optimal set obtained. It can be observed that the objective functions chosen are affected in opposing ways by changes in the decision variables. A desirable increase of the total protein production corresponds to an undesirable increase of the total inducer consumption. Thus, this case constitutes a multiobjective problem and is suitable to test the proposed algorithm. Figure 1 (b) depicts the optimal feeding strategy for the maximum protein production. The other optimal strategies were omitted for the sake of brevity. No specific tendency was identified in any of these curves. The complexity of the feeding strategies reinforces the use of an evolutionary technique.

Figure 2 presents the profiles of the protein production and inducer consumption, simulating the process under the optimal conditions for maximum protein production (max f_1), minimum inducer consumption (min f_2) and an intermediate result. The behavior of the profiles corroborates the optimization results: the optimal policy to maximize f_1 produces more protein and consumes more inducer; when minimizing f_2, less inducer is used and less protein is produced; the intermediate policy presents intermediate responses.

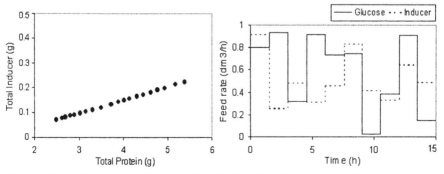

Figure 1. (a) Pareto optimal set; (b) Optimal feeding strategy for maximum protein production.

972

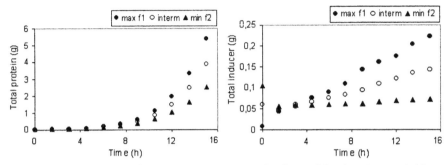

Figure 2. Simulation of the process using optimal values of the decision variables.

The computational time taken to generate the Pareto optimal set has been considerable low. Although this parameter depends on the desired accuracy of the optimum, the CPU time required for this problem was inferior to 2 min, on a 1.3 Gb Pentium 4 computer.

4. Conclusions

Our focus in this contribution has been to evaluate the performance of a proposed genetic algorithm in conducting multiobjective dynamic optimization. The algorithm has been applied to a fed-batch bioreactor. Different feeding strategies have been investigated in order to establish the optimal operating policies.

The results of the challenging case study confirm the efficiency of the proposed optimization approach to solve dynamic control problems. The ability of the algorithm to seek tradeoff surface regions has also been demonstrated.

It should be pointed out that the computational cost related to the proposed GA approach depends on the dimension of the problem and its level of complexity, regardless of the number of decision variables manipulated. It has been demonstrated that the proposed algorithm can successfully solve problems involving a large number of decision variables.

5. References

Bhaskar, V., S.K. Gupta and A.K. Ray, 2001, Comp. Chem. Engng. 25, 391.

Busacca, P.G., M. Marseguerra and E. Zio, 2001, Reliab. Engng Syst. Saf. 72, 59.

Butala, D., W. R. Liang and K. Y. Choi , 1992, J. Appl. Pol. Sci 44, 1759.

Canto, E.B, J. Banga, A. Alonso and V. Vassiliadis, 2001, Com. Chem. Engng. 25, 539.

Cheng, F.Y. and D. Li, 1998, AIAA J. 36, 1105.

Goldberg, D.E,. Genetic Algorithms in Search, Optimization, and Machine Learning, Addison-Wesley, Reading MA, 1989.

Lee, J. and W.F. Ramirez, 1996, Chem. Engng. Sci., 521, 51.

Roubos, J.A., G. van Straten and A.J.B. van Boxtel, 1999, J. Biotech. 67, 173.

Sareen, R. and S. Gupta, 1995, J. Appl. Pol. Sci. 58, 2357

Secchi, A.R., E.L. Lima and J.C. Pinto, 1990, Pol. Engng. Sci. 30, 1209.

Silva, C.M. and E.C. Biscaia, 2001, presented at 2nd Pan American Workshop on Process Systems Engn. (CEPAC/2001) and submitted to Comp. Chem. Engng.

Toshinsky, V.G., H. Sekimoto and G.I. Toshinsky 2000, Proc. Nucl. En. 27, 397.

Tholudur, A., W.F. Ramirez and J.D. McMillan, 2000, Comp. Chem. Engng. 24, 2545.

Wajge, R. and S. Gupta, 1994, Pol. Engng. Sci. 34, 1161.

Wang, K., Y. Qian, Y. Yuan and P. Yao, 1998, Comp. Chem. Engng. 23, 125.

European Symposium on Computer Aided Process Engineering – 12
J. Grievink and J. van Schijndel (Editors)
© 2002 Elsevier Science B.V. All rights reserved.

A MILP approach to the optimization of the operation policy of an emulsification process

M. Stork, R. L. Tousain and O. H. Bosgra
Mechanical Engineering Systems and Control Group
Delft University of Technology
The Netherlands

Abstract

This work addresses the model-based optimization of the operation policy of an emulsification process. In this paper the choice of the stirrer speed as function of the time, for reaching a certain predefined drop size distribution (DSD) in minimal time, is studied. It is argued that "standard" gradient based optimization techniques will fail to solve this optimization problem. An approach is suggested that approximates the original minimum time optimization problem as a Mixed Integer Linear Program (MILP). The MILP can be solved using well-proven, standard optimization codes and its solution is also a good solution of the original optimization problem. Applying it to the industrial process shows the feasibility of the approach and it illustrates the benefit of using model-based optimization for improving the operation policy of emulsification.

1. Introduction

Emulsification is an essential manufacturing technology in the food industry. Examples of emulsions[1] are mayonnaise and all kind of dressings. The equipment as typically used for the production of o/w (oil-in-water) emulsions is shown in figure 1. It consists of a stirred vessel in combination with a colloid mill and a recirculation loop. The vessel is equipped with a scraper stirrer: a device that consists of several blades rotating at low speed at a small distance from the vessel wall in order to achieve mixing and breaking of the oil drops. The colloid mill consists of a stator and a rotor. In the narrow gap between these the intensity of the (hydrodynamic) forces acting on the drops is very high, which causes the breakage of the oil drops. The process is operated fed-batch wise and typical production times are in the order of 10-15 minutes. For profit maximization it is desirable to decrease the production time while maintaining the product quality specifications. In this paper it is investigated if this can be established by using model-based optimization. To this end a model is developed that predicts the course of the DSD in time (the product quality is strongly influenced by the DSD). The model describes emulsification in a stirred vessel that is operated fed-batch wise. In this work it is used to calculate the stirrer speed as function of the time, such that a certain

---

[1] An emulsion is a dispersion of drops of one liquid in another one with which it is incompletely miscible (Israelachvili, 1993).

974

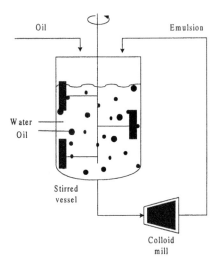

Oil   Emulsion

Water
Oil

Stirred
vessel

Colloid
mill

Figure 1: Equipment for the production of o/w-emulsions.

predefined DSD is reached in minimal time. To the authors knowledge this is not addressed in the literature. A short description of the model is given in Section 2. The model contains continuous and discontinuous dynamics and therefore belongs to the class of so called hybrid discrete/ continuous models (Barton et al., 98). In Section 3 the optimization problem is presented and it is argued that "standard" gradient based optimization techniques are not suited for the solution of it. An approach to solve general hybrid discrete/ continuous optimization problems is discussed in Barton et al. (2000) and in Avraam et al. (1998); it is called the MIDO (mixed-integer dynamic optimization) approach. In Avraam et al. (1998) it is applied to some toy examples and they ended up with a Mixed Integer Nonlinear Program (MINLP). MINLP's are generally hard to solve, severely limiting the applicability of the MIDO approach (Barton et al., 2000). We also use this approach (Section 4), however due to our specific model structure and by using results of Bemporad and Morari (1999) we are able to reformulate the original minimum time optimization problem as a MILP instead of a MINLP. This enables to solve the optimization problem. An application is presented in Section 5 and concluding remarks are made in Section 6.

2. Outline of the model

The hydrodynamic forces that are generated by the stirrer action cause the breakage of oil drops and because of this the DSD changes in time. The model describes the DSD(t) and it consists of two parts: a reactor model and a drop model. The reactor model describes the mixing in the vessel and the hydrodynamic forces that are generated by the stirrer action. This is modeled with compartment models and for each compartment a population balance equation PBE (Ramkrishna, D., 1985) is formulated to describe the DSD(t). It is assumed that drop breakage occurs in a small region round the stirrer with laminar flow and that in the bulk region the fluid is mixed. This is modeled with two

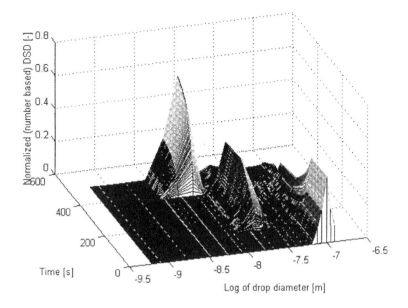

Figure 2: Course of the normalized (number based) in the bulk compartment.

compartments: the so-called laminar compartment and the bulk compartment. The drop model describes the phenomena occurring at the drop level. It describes the breakage condition, the breakup time and the number and sizes of the daughter drops. These relations are for the larger part based on the theory as currently available for the breakage of drops. Simulation results are included to illustrate that the DSD(t) is strongly affected by the stirrer speed. During the simulation the stirrer speed is kept constant for a period of 125 s, then its value is changed and kept constant for an other period of 125 s etc. The sequence as used is: 1, 3, 2 and 5 s^{-1}. The corresponding normalized DSD(t)[2] is shown in figure 2.

3. Features of the optimization problem

We want to choose the stirrer speed as function of the time such that a certain predefined DSD is reached in minimum time. Next it is argued why "standard" gradient based optimization techniques are not suited for the solution of this optimization problem. The breakage phenomena (the breakage condition, the number and the sizes of the daughter drops) depend heavily on the stirrer speed and exhibit discrete events. A very small increase of the stirrer speed may already lead to the breakage of certain drop sizes that would not break with a slightly lower value of the stirrer speed. Comparable behavior is observed for the formation of certain drop sizes; until some stirrer speed they are not formed whereas they are formed rapidly at a stirrer speed that is only slightly higher. A further increase of the stirrer speed may suddenly lead to the non-formation or even breakage of these drop sizes. Gradient based optimization methods will fail because of this behavior.

---

[2]The normalized DSD equals the DSD divided by the total number of drops.

4. MILP approach

Here an approach is suggested that approximates the original nonlinear optimization problem as a MILP. The method is derived as follows. First, by model analysis it can be shown that the strong non-linearity (discontinuity) is only in the dependence on the stirrer speed. Further, in small intervals of the stirrer speed the dynamics are bilinear (product of stirrer speed and states). These intervals form the *modes* of the system. At any given time, the system finds itself in exactly one mode; its mathematical behavior is then described by a given set of evolution ordinary differential equations. A transition from one mode to another is triggered when the stirrer speed passes a certain critical value. Hence, the stirrer speed determines completely in which mode the system is and when the transitions occur. This suggests that the model can be reformulated as a state-transition network (Avraam et al., 1998) where bilinear dynamics describe the behavior in a mode and where transitions between different modes are modeled using integer decision variables. However, this would result in a MINLP, which is undesirable. The dynamics are however linear if the stirrer speed is fixed at a constant value for the different modes. This enables to end up with a MILP. It is believed that this will not result in a loss of performance because the intervals are rather small (typically 0.05 s^{-1}). The objective, being to reach a certain end-point condition in minimum time, can be enforced through the introduction of another set of integer decision variables. The precise formulation of the optimization problem as a MILP is discussed next.

4.1 System behavior
We define a set of periods (k=1,...,N) of fixed duration. The following binary variable is introduced to characterize the system behavior:

$$\delta_k^s = \begin{cases} 1 & \text{if the system is in mode } s \text{ over period } k \\ 0 & \text{otherwise} \end{cases} \tag{1}$$

The system can be in only one mode at a point in time. This is expressed mathematically as $\sum_{s=1}^{n_s} \delta_k^s = 1$, with n_s the number of modes of the system. For the ease of implementation we use discrete time domain models. For each mode s we have the discrete time model:

$$x_{k+1} = A_k^s x_k + B_k^s \tag{2}$$

The states x_k are the number of drops, with a certain volume, in the laminar and bulk compartment. While the system is in mode s, the corresponding equations characterizing the behavior of the system must hold in that mode. In Bemporad and Morari (1999) an approach is suggested for this. We follow the main idea, some modifications are however made to facilitate the solution process (details are omitted).
A new continuous variable z_k^s is defined and with linear equality and inequality constraints it is enforced that this variable is equal to x_k as the system is in mode s and that it is zero as this is not the case. Mathematically:

$$x_{k+1} = \sum_{s=1}^{n_s} (A_k^s z_k^s + B_k^s \delta_k^s) \tag{3}$$

$$\sum_{s=1}^{n_s} z_k^s = x_k \tag{4}$$

$$z_k^s \le M1\delta_k^s \tag{5}$$

$$z_k^s \ge 0 \tag{6}$$

With $M1$ being a weighting matrix. Hence, if δ_k^s is zero then z_k^s is zero. Since δ_k^s is one for only one mode, $z_k^s = x_k$ for that mode.

4.2 End-point inequality constraints and the objective function

We now turn our attention to handling the inequality end-point constraints and the formulation of the objective function J. The following binary variable is introduced to characterize the inequality end-point constraints:

$$Y_k = \begin{cases} 1 & \text{if the inequality end - point constraints are met over period } k \\ 0 & \text{otherwise} \end{cases} \tag{7}$$

The constraints are now expressed as:

$$Y_k x_{\min} - x_k \le 0 \tag{8}$$

$$x_k - x_{\max} - (1 - Y_k) M2 \le 0 \tag{9}$$

With $M2$ being a weighting matrix. Hence, if x_k is less than its lower bound $x_{\min}$ then Y_k must be zero in order to satisfy inequality constraint (9). If x_k is larger than its upper bound $x_{\max}$ then Y_k is also set to zero. Y_k can be either one or zero if the bounds are met. Due to the formulation of the objective function J, Y_k will be set to one. The objective function is written as:

$$\max J = \sum_{k=1}^{N} Y_k \tag{10}$$

This way the number of periods, where the end-point constraints are satisfied, are maximized. Hence, the time needed to satisfy these constraints is minimized.

5. Optimization of the operation policy

Here the results of an optimization study are presented. A sample interval of 50 s is used for the derivation of the discrete time domain models and the time horizon is set to 500 s. Upper and lower bounds are formulated for all 38 states. The system has 55 modes, so the resulting optimization problem consists of 561 binary variables, 810 equality and 46816 inequality constraints. The MILP is solved using GAMS/CPLEX. It took 304 nodes (less than a hour) to find the optimum. Its value is 3, so after 400 s the

978

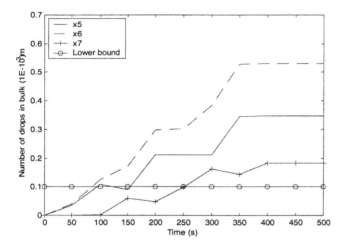

Figure 3: Optimal state trajectories in the bulk compartment.

desired DSD is reached. The corresponding trajectory for the stirrer speed is: 4.578, 4.578, 3.937, 4.578, 3.392, 3.837, 4.508, 3.392, 0.804 and 0.804 s^{-1}. Some of the optimal state trajectories are depicted in figure 3. The states $x5$, $x6$ and $x7$ are zero at the start of the process. With the stirrer speed of 4.578 s^{-1} they are formed from breakage of larger drops. The results of figure 3 might suggest that $x7$ is not formed during the first 100 s. These drops are however formed, but they rapidly break down to $x5$ and $x6$. After 100 s the stirrer speed is lowered: now it is too low to cause breakage of $x7$, hence its increase. These results are non-trivial illustrating the benefit of using model-based optimization for improving the operation policy of emulsification.

6. Concluding remarks

An approach is suggested that approximates the original minimum time optimization problem as a MILP. The solution of the MILP is also a good solution of the original optimization problem. Applying it to the industrial process shows the feasibility of the approach and it illustrates the benefit of using model-based optimization for improving the operation policy of emulsification. In order to establish what improvements can be reached the quality of the model has to be established; this is subject of current research.

References

Avraam, M. P., Shah, N. and Pantelides, C. C., 1998, Modelling and optimization general hybrid systems in the continuous time domain. Comp. Chem. Eng., 22: S221-S228.

Barton, W. F. , Allgor, R. J., Feehery, W. F. and Galán S., 1998, Dynamic optimization in a discontiuous world. Ind. Eng. Chem. Res., 37: 966-981.

Barton, W. F., Banga, J. R. and Galán, S., 2000, Comp. Chem. Eng., 24:2171-2182.

Bemporad, A. and Morari, M. , 1999, Control of systems integrating logics, dynamics and constraints. Automatica, 35: 407-427.

Israelachvili, J., 1993, The science and applications of emulsions- an overview, Colloids And Surfaces A: Physicochemical and Engineering Aspects, 91: 1-8.

Ramkrishna, D., 1985, The status of population balances. Rev. Chem. Eng., 3: 49-95.

European Symposium on Computer Aided Process Engineering – 12
J. Grievink and J. van Schijndel (Editors)

The Effect of Modelling Assumptions on the Differential Index of Lumped Process Models

Zs. Tuza, G. Szederkényi, K. M. Hangos

Computer and Automation Research Institute Hungarian Academy of Sciences
Department of Computer Science, University of Veszprém
Hungary

Abstract

The effect of model simplification assumptions on the differential index of DAE process models is investigated in this paper. Efficient incremental graph-theoretical algorithms are proposed to follow the changes in the variable-equation assignments during the modelling process. Case studies are used to demonstrate the operation of the algorithms and the effect of steady-state assumptions on the differential index of a simple process model.

1. Introduction

Lumped process models are in the form of differential-algebraic equations (Hangos, Cameron, 2001a) which are sometimes difficult to solve numerically, due to index and stiffness problems. The effect of some modelling decisions on the structural solvability has already been investigated (Moe, 1995) and it has been found that a change in the specification may transform an index-1 model to a higher-index one (Hangos, Cameron, 2001a). It is intuitively clear that simplification assumptions applied during the modelling process may also affect seriously the differential index of lumped process models.

The aim of this paper is to propose a polynomial-time incremental algorithms for analysing the changes in the differential index during the modelling process. The algorithm warns the modeller about decisions causing index problems and advises him/her to change the specification to meet solvability requirements.

2. Computational Structure of Lumped Process Models

For the purpose of the analysis, the structure of a lumped process model is described by a bipartite graph called the *equation-variable graph*. For this we use the fact that a lumped process model is given as a DAE system with the following special form.

$$
\begin{array}{lll}
(D) & u & = dx/dt = f(x, z_S, z) \\
(DS) & x & = \int u\, dt + x_S \\
(A) & 0 & = g(x, z_S, z) \\
(AS) & z_S & = spec \quad (x_S = spec)
\end{array}
$$

where x is the vector of differential variables, u contains their derivatives and x_S their initial values, z is the vector of algebraic variables, z_S is that of the specified algebraic variables and spec stands for a general given constant (different for each variable). Note that the variables in the specification $S = \{ z_S, x_S \}$ appear in the structure in a special way. The equation-variable graph of the above model is a bipartite graph, where one vertex class represents the set of equations, $((D), (DS), (A), (AS))$ and the other one contains the variables, (x, u, z, z_S, x_S). A variable-vertex is adjacent to an equation-vertex if and only if the variable in question appears in the corresponding equation.

Assume that the degrees-of-freedom requirement $(DOF=0)$ is satisfied. Then the DAE model above is of index 1 if and only if there exists a perfect matching in the equation-variable graph (Hangos, Cameron, 2001a).

2.1 The effect of model simplification transformations

A formal representation of assumptions in process is reported in the earlier paper (Hangos, Cameron, 2001b) where the modelling assumptions acted as formal transformations on process models. Similarly, model simplification transformations are described as graph transformations on the equation-variable graph. Assume that a full equation-variable assignment (a perfect matching) is given together with an equation-variable graph. Then a simplification transformation may be:

1. *edge-changing* transformation when only some non-matching edges are removed or added,
2. *assignment-changing* transformation when some of the matching edges are affected,
3. *vertex-changing* transformation when new equations and/or variables appear causing edge changes and change in the specification, too.

3. Algorithms for Finding Closest Maximum Assignment

Let us have an equation-variable graph with a full equation-variable assignment. Moreover, let us consider a model simplification transformation of the 2nd or 3rd type affecting the equation-variable assignment. *A closest maximum assignment is a (not necessarily full) assignment in the transformed equation-variable graph, which has the largest possible number of edges and under this requirement the largest number of matching edges in common with the original full assignment.*

3.1. The algorithms

Two different cases will be distinguished from the algorithmic point of view. The first case (covered by Algorithm 1) occurs when a model simplification transformation deletes just one matching edge from the model, while in the second case (Algorithm 2) more than one edges are deleted.

Algorithm 1. Let $B = (X, Y, E)$ be the equation-variable graph of the process model in question, with vertex classes X (variables) and Y (equations) and edge set E (dependence of equations on the variables). Let $F \subseteq E$ be the assignment given (i.e. a matching selected in B). We construct an auxiliary directed graph D as follows. The vertex set of D is $F \cup \{s, t\}$, where $s \in X$ and $t \in Y$ are the two vertices that have no matching edge. An ordered pair (s, e) – where $e = xy \in F$, $x \in X$, $y \in Y$ – is an edge in D if and only if $(s, y) \in E$. Similarly, a pair (e, t) with $e = xy \in F$ is an edge in D if and only if $(x, t) \in E$. Inside F, there is an edge from $f_1 = x_1 y_1$ to $f_2 = x_2 y_2$ if $(x_1, y_2) \in E - F$. Now, we run the *Breadth-First Search* algorithm (see e.g. Cormen, Leiserson and Rivest (1990), Section 23.2) on D, starting from s. If t is reachable from s along any directed path, then BFS also finds one shortest s–t path, say $P(s, t)$.

1. If there is no directed path from s to t then the given assignment F is of maximum size in B.

2. If t is reachable from s and BFS finds a shortest s–t path $P(s, t) = sf_1f_2...f_kt$ then an assignment closest to F and of size $|F| + 1$ is $(F - \{f_1, f_2, ..., f_k\}) \cup \{(s, y_1), (x_1, y_2), ..., (x_{k-1}, y_k), (x_k, t)\}$, where $f_i = x_i y_i$, $i = 1, ..., k$.

Since BFS can be implemented in linear time, the closest assignment is found in linear time, too.

Algorithm 2. Let again $B = (X, Y, E)$ be the equation-variable graph, and $F \subseteq E$ the assignment given. Assume $|X| = |Y| = n$. We define an edge-weight function w as

$$w(f) = n + 1 \quad \text{for all } f \in F, \qquad w(e) = n \quad \text{for all } e \in E - F.$$

We run a *Maximum-Weight Bipartite Matching* search algorithm on (B, w). If $M \subseteq E$ is the output of the algorithm, then the total weight of M is $w(M) = n|M| + |F \cap M| > (n + 1)(|M| - 1)$. Thus, the algorithm first maximizes the possible number of edges in a matching of B, and under this condition it maximizes the possible number of edges selected from F. This means precisely a closest maximum assignment.

Since a maximum-weight matching of a bipartite graph can be found in polynomial time (see e.g. Hopcroft and Karp (1973)), the same time bound is valid for the closest assignment, too.

3.2. The algorithmic complexity of the algorithms

The following theoretical results show the efficiency of our algorithms above.

Theorem 1. *There is a linear-time algorithm to decide whether an assignment is of maximum size. In particular, if just one edge is deleted from a perfect matching, it can be decided in linear time whether the modified system still has differential index 1. Moreover, if the assignment is below the maximum size by 1, then a closest assignment can be found in polynomial time.*

Theorem 2. *A closest assignment of maximum size can be found in polynomial time.*

There is a good reason for the different time complexities in the two theorems. For a closest maximum assignment, one cannot really expect a linear-time algorithm, because such an algorithm would solve the *Bipartite Matching* problem as a particular case where the initial „assignment" contains no edges. On the other hand, it remains an open question whether the removal of a *bounded* number k of edges admits an algorithm finding a closest maximum assignment more efficiently than Bipartite Matching. Theorem 1 yields a positive answer for $k = 1$.

4. Case Studies

The algorithm for finding the closest assignment is illustrated on the following example by examining the effect of steady state assumptions. The model used is a simplified version of the one described in (Hangos, Cameron, 2001b). The original model equations are the following (the labels of the equations are between parentheses):

$(d1)$ $\quad m_L = dM_L/dt = F - E - L$

$(a1)$ $\quad E = k_{LV} (P* - P^0)$

$(d2)$ $\quad u_L = dU_L/dt =$
$\qquad = F h_F - E h_{LV} - L h_L + Q + Q_E$

$(a2)$ $\quad Q_E = u_{LV} (T_L - T^0)$

$(ds1)$ $\quad M_L = \int m_L \, dt + M_{L0}$

$(a3)$ $\quad U_L = M_L c_{pL} T_L$

$(ds2)$ $\quad U_L = \int u_L \, dt + U_{L0}$

$(a4)$ $\quad P* = H T_L$

The specified variables are

$(as1)$ $\quad Q = $ **spec**

$(as4)$ $\quad M_{L0} = $ **spec**

$(as2)$ $\quad F = $ **spec**

$(as5)$ $\quad U_{L0} = $ **spec**

$(as3)$ $\quad L = $ **spec**

The constant physico-chemical parameters are: h_F, h_{LV}, h_L, c_{pL}, P^0, T^0, and **spec** denotes a given specified value for the variable in question (different for each variable). This model has differential index 1.

4.1 Test case 1

We examine the effect of a steady-state assumption on the liquid mass M_L. For this, let us introduce the following new equation in the modified model:

Add $(a*)$ $\qquad m_L = dM_L/dt = 0$

Since we have a new equation, we have to reduce the number of specified variables, i.e. the number of specification equations. This is done by changing the specifications $(as1)-(as4)$ in such a way that we delete equation $(as4)$ from the new model.

Del $(as4)$ $\qquad M_{L0} = $ **spec**

Using Algorithm 1 we find that the modified system still has differential index 1, and we can also determine the assignment which is closest to the original one. The algorithm is based on finding the shortest directed path in a directed graph assigned to the equation-variable graph. The results are shown in Figure 1 where the vertices s and t (see Algorithm 1) are (d_1) and M_{L0} respectively.

4.2 Test case 2

Now we investigate the effect of changing the assignment in a less fortunate way together with the above steady state assumption $(a*)$. For this, let us again introduce *(a*)* as a new equation in the modified model but now we delete the specification equation (as1) from the new model.

Del (as1) $Q = \textbf{spec}$

Using Algorithm 1 we find that the modified system is no longer of differential index 1.

4.3 Test case 3
Let us investigate what happens if we put two steady state assumptions to our original model. For this, let us introduce the steady state assumptions for both the mass M_L and the energy U_L in the original model, that is:

Add (a*) $m_L = dM_L/dt = 0$ *Add (b*)* $u_L = dU_L/dt = 0$

but now we need to delete two specification equations (as4) and (as5)

Del (as4) $M_{L0} = \textbf{spec}$ *Del (as5)* $U_{L0} = \textbf{spec}$

Using Algorithm 2 we find that the modified system is no longer of differential index 1.

4.4 Test case 4
Now we try to improve the situation in the previous test case with two steady state assumptions by selecting another, more fortunate specification equation to be deleted. For this, we again introduce the steady state assumptions *(a*)* and *(b*)* in the model, but now we delete two specification equations *(as4)* and *(as1)* from the new model.

Del (as4) $M_{L0} = \textbf{spec}$ *Del (as1)* $Q = \textbf{spec}$

Using Algorithm 2 we find that now the modified system has differential index 1, and we can determine the assignment which is closest to the original one shown in Figure 2. Note that in this case we obtain the same result by applying Algorithm 1 in two steps: the first step is the same as in Test case 1. In the second step we delete *(as1)* and add the specification equation $Q=\textbf{spec}$, and then find a matching in the resulting graph.

References

Cormen, T.H., Leiserson, C.E. and Rivest, R.L. (1990): *Introduction to Algorithms.* MIT Press.

Hangos, K.M. and I.T. Cameron, 2001a, *Process Modelling and Model Analysis.* Academic Press.

Hangos, K.M. and I.T. Cameron, 2001b, A Formal Representation of Assumptions in Process Modelling. *Comput. Chem. Engng.* **25**, 237-255.

Moe H.I., 1995, *Dynamic Process Simulation, Studies on Modeling and Index Reduction.* PhD Thesis, University of Trondheim

Hopcroft, J. and Karp, R.M. (1973): An $n^{5/2}$ algorithm for maximum matching in bipartite graphs. *SIAM J. Comput.* **2**, 225-231.

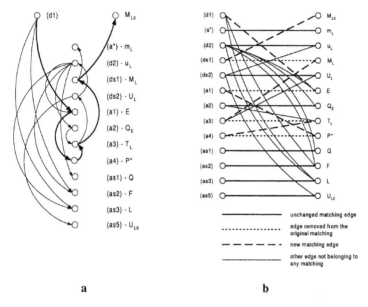

a   b

Figure 1. Test case 1. a - Determining the closest assignment by finding the shortest path in Test Case 1. b - The closest assignment in the equation-variable graph in Test Case 1. The edges belonging to the original matching are denoted by solid and dotted lines. The new matching consists of the solid (unchanged) and the dashed (new) lines.

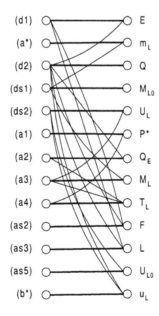

Figure 2. The resulting matching of Test Case 4. The edges belonging to the final matching are denoted by solid lines.

European Symposium on Computer Aided Process Engineering – 12
J. Grievink and J. van Schijndel (Editors)
© 2002 Elsevier Science B.V. All rights reserved.

985

Dynamic Simulation of Continuous Crystallizers by Moving Finite Elements with Piecewise Polynomials

Zsolt Ulbert and Béla G. Lakatos
Department of Process Engineering, University of Veszprém
H-8200 Veszprém, Hungary

Abstract

The moving finite element method is developed for numerical solution of the population balance equation of disperse systems. This approximation is obtained by using the Lagrange interpolation polynomials, spacing the interior nodes according to the orthogonal collocation technique. The method is used for solving the mixed set of nonlinear ordinary and partial integro-differential equations, forming a detailed dynamical model of the continuous cooling crystallizers.

1. Introduction

The population balance model is the adequate mathematical description of crystallization processes. This is a mixed set of ordinary and partial differential equations and the crucial point of using this modelling approach is the numerical solution of the population balance equation. A number of different methods have been presented for solution of this equation (Subramanian and Ramkrishna,1971, Gelbard and Seinfeld,1978, Chang and Wang,1984, Eyre *et al.*,1988, Hounslow *et al.,*1988, Steemson and White, 1988, Nicmanis and Hounslow,1998), but it rarely has been considered in coupling with the nonlinear ordinary integro-differential equations describing the variations of concentrations and temperature in time (Lakatos *et al.*,1984, Marchal *et al.*,1988, Rawlings *et al.*, 1992, Wulkow and Gerstlauer,1999, Mantzaris *et al.*,2001).

In dynamic conditions, however, the solutions of the population balance equation often exhibits sharp profiles and steep moving fronts of the particle size distribution in certain regions of the particle size. In such cases it seems to be desirable to move and place the grid points into those regions where they are most efficient in approximating the exact population density function. The moving finite element method, developed by Miller and Miller (1971) and Miller (1971) has proved satisfactory for that purpose in computing processes occurring in finite intervals (Sereno *et al*,1991, Coimbra *et al.*, 2001).

The aim of the paper is to apply this method for solving the population balance equation, allowing computations also for such environmental conditions that produce sharp profiles and steep moving fronts in the particle size distribution.

2. Population Balance Model

Consider a continuous cooling crystallizer, assuming that:
(1) The crystal suspension is well mixed;
(2) The volumetric feed and withdrawal rates of the crystallizer are constant and equal;

(3) There is a selective withdrawal of crystals characterised by the selection function $s(L)$;

(4) Crystal breakage and agglomeration are negligible;

(5) All new crystals are formed at a nominal size $L_n \cong 0$, so that we take $L_n = 0$.

Then the mathematical model of the crystallizer takes the following form:

the variation of the population density function $n(L,t)$ with time is described by the population balance equation

$$\frac{\partial n(L,t)}{\partial t} + \frac{\partial \left(G(L,c,c_{eq})n(L,t)\right)}{\partial L} - \frac{\partial^2 \left(D_G(L)n(L,t)\right)}{\partial L^2} = \frac{q}{V}\left(n_{in}(L,t) - s(L)n(L,t)\right) \qquad (1)$$

subject to the initial and boundary conditions

$$n(L,0) = n_0(L), \quad L \geq 0 \qquad (2)$$

$$\lim_{L \to 0}\left[G(L,c,c_{eq})n(L,t) - \frac{\partial\left(D_G(L)n(L,t)\right)}{\partial L}\right] = B(c,c_{eq},\mu_3), \quad t \geq 0 \qquad (3)$$

$$\lim_{L \to \infty} n(L,t) = 0 \quad \text{and} \quad \lim_{L \to \infty}\frac{\partial n(L,t)}{\partial L} = 0, \quad t \geq 0 \qquad (4)$$

where the rate of nucleation B is given by the power law relation

$$B(c,c_{eq},\mu_3) = k_b\left(c - c_{eq}(T)\right)^b \mu_3^j \qquad (5)$$

and overall linear growth rate of crystals G is expressed as

$$G(L,c,c_{eq}) = k_g(c - c_{eq})^g \varphi(L). \qquad (6)$$

Here: L - crystal size, D_G - growth rate dispersion coefficient, q - volumetric rate, V - volume, c -concentration of solute, c_{eq} - solubility, k_b, b, j, k_g, g - constants, φ - empirical function, T – temperature, and μ_3 is the third order moment of the population density function, given as

$$\mu_3 = \int_0^\infty L^3 n(L,t)dL \qquad (7)$$

The terms on the left hand side of eq.(1) describe, respectively, the accumulation, growth and size dispersion of crystals, while the two terms on the right hand side describe the feed and removal rates of crystals. Boundary condition (3) expresses that the flow density of crystals at the size of nuclei $L_n = 0$ equals to the overall rate of nucleation of new crystals.

In order to characterise the crystallizer entirely, we need also equations describing the variations with time of the concentrations of solute and solvent. The equation for the solute can be obtained from the overall mass balance

$$\frac{dc}{dt} = \frac{q}{V\varepsilon}(\varepsilon_{in}c_{in} - \varepsilon_{out}c) + \frac{q}{V\varepsilon}\left((1 - \varepsilon_{in}\rho) - (1 - \varepsilon_{out}\rho)\right) - \frac{k_V \rho}{\varepsilon}\frac{d\mu_3}{dt} \qquad (8)$$

while the mass balance of the solvent can be written in the form

$$\frac{dc_{sv}}{dt} = \frac{q}{V\varepsilon}(\varepsilon_{in}c_{svin} - \varepsilon_{out}c_{sv}) + \frac{k_V c_{sv}}{\varepsilon}\frac{d\mu_3}{dt}.$$ (9)

Finally, the heat balance can be written in the form

$$\frac{dT}{dt} = \frac{q(\Phi_{in} - \Phi_{out})}{V\varepsilon\Phi} + \frac{V}{\Phi}(-\Delta H_c)k_V\rho\frac{d\mu_3}{dt} - \frac{Ah}{\Phi}(T - T_Q).$$ (10)

where

$$\Phi = V\varepsilon(C_{sv}c_{sv} + Cc) + V(1-\varepsilon)C\rho, \quad \Phi_{in} = V\varepsilon_{in}(C_{sv}c_{svin} + Cc_{in}) + V(1-\varepsilon_{in})C_\rho\rho$$

$$\Phi_{out} = V\varepsilon_{out}(C_{sv}c_{sv} + Cc) + V(1-\varepsilon_{out})C_\rho\rho$$

(11)

(c_{sv} – concentration of solvent, C – heat capacity, ρ - density, k_V – volume sorm factor, ΔH_c – heat of crystallization, A – heat transfer surface, h – heat transfer coefficient, T_Q – temperature of the cooling medium). Here, because of the selective withdrawal, the voidage in the crystallizer and that in the outlet stream are not equal, i.e. in turn,

$$\varepsilon = 1 - k_V\mu_3, \quad \varepsilon_{in} = 1 - k_V\mu_{3in}, \quad \varepsilon_{out} = 1 - k_V\int_0^\infty s(L)L^3n(L,t)dL,$$ (12)

Therefore, the state of the crystallizer is given by the quaternion $[c(t),c_{sv}(t),T(t),n(.,t)]$ at time $t>0$, and its dynamics is described by the distributed parameter model formed by the mixed set of partial and ordinary integro-differential equations (1)-(12). The evolution in time of this system occurs in the state space $\boldsymbol{R}^3 \times \boldsymbol{N}$, which is the Descartes product of the vector space $\boldsymbol{R}^3$ of concentrations and temperature and of the function space $\boldsymbol{N}$ of the population density functions. Projecting $\boldsymbol{N}$ into some finite (K-)dimensional space spanned by appropriately chosen trial functions the problem is reduced to finding an approximating trajectory of the state of crystallizer in a $K+3$ dimensional state space.

3. The Moving Finite Element Method

In order to find a finite dimensional approximation of the system of equations (1)-(12), the moving finite element method is applied to the partial differential equation (1). Let us assume that the maximal size of crystals at time $t>0$ is $L_{max}(t)$, and let the interval $[0,L_{max}]$ be divided into M elements by $M+1$ separation nodes as it is shown in Fig.1. In the m-th element we define the local size variable l^m and Im local interpolation points, as well as the approximation $\hat{n}^m$ of the population density function n by using an $Im-1$ degree polynomial expressed through the Lagrange interpolation polynomials Λ_i^m as:

$$l^m \equiv L, \quad if \ L_m \leq L \leq L_{m+1}$$ (13)

$$\Lambda_i^m(l^m) = \prod_{\substack{j=1 \\ j\neq i}}^{Im} \frac{l^m - l_j^m}{l_i^m - l_j^m} \quad i = 1,2,...Im$$ (14)

$$\hat{n}^m(l^m,t) = \sum_{i=1}^{Im} n_i^m(t)\Lambda_i(l^m), \quad m = 1,2,...M.$$ (15)

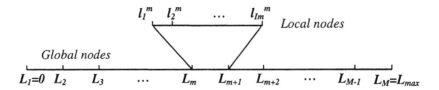

Figure 1. Finite elements on the interval [0,L_{max}] and the local nodes of the m-th element

If we now take all the K nodes, either the separation nodes or the local ones inside the finite elements, and use the notation $n_k(t) = \hat{n}(L_k,t) \approx n(L_k,t)$ at the k-th global node, $k=1,2,\ldots K$, then the approximating function can be written in the form

$$\hat{n}(L,t) = \sum_{k=1}^{K} n_k(t)\Psi_k(L) \approx n(L,t) \tag{16}$$

where Ψ_k, $k=1\rightarrow K$ denotes the global interpolation functions identical with the appropriate local interpolation polynomials on the finite elements, respectively.

In order to get the appropriate values of the time dependent coefficients $n_k(t)$ and global moving nodes $L_m(t)$ we minimise the functional

$$Q(t) = \int_{0}^{L_{max}} Res(L,t)^2 \, dL + \sum_{m=1}^{M} \left(\varepsilon_m \frac{d\big(L_{m+1}(t) - L_m(t)\big)}{dt} - S_m \right)^2 \tag{17}$$

with respect to all time derivatives dn_k/dt and dL_m/dt, $k=1\rightarrow K$, $m=1\rightarrow M$ as it was formulated by Miller (1981). Here, ε_m and S_m are the so called internodal viscosity function and internodal spring function, respectively, regularising the movement of the global nodes during the process, and $Res(L,t)$ stands for the residual of approximation of the population balance equation.

The solution of the minimising problem (17) generates a set of ordinary differential equations for the time dependent coefficients $n_k(t)$ and global nodes $L_m(t)$ which together with the differential equations (8)-(10) can be solved by an *ODE*-solver. In order to reduce the problems of stiffness often arising in solving the model equations of crystallizers, the set of eqs (1)-(12) had been scaled before applying the finite element method and solving the resulted set of ordinary differential equations by means of the *DASSL*-solver (Petzold,1983). The details of the numerical solution will be presented elsewhere.

4. Simulation Results and Discussion

In crystallizers, sharp profiles in population density function, describing the crystal size distribution, arise often when primary nucleation is the dominant mechanism of producing new crystals, described by the rate equation (1) with $j=0$ and $b>>1$. This occurs during the start up of continuous crystallizers or at the beginning of a batch crystallization process from empty vessel conditions, i.e. pure solvent conditions without any seed crystals (Ulbert and Lakatos,1999). In such cases, a large number of small crystals are produced in a very short interval of time, generating a large peak in the crystal size distribution which is moving and dispersing into the regions of larger crystal sizes slowly.

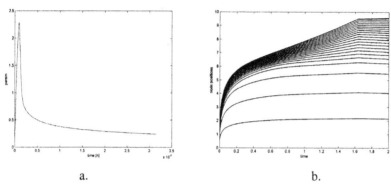

a.   b.

Figure 2. Variation a) of the nucleation rate, and b) of the positions of the separation nodes as a function of time

The variation of the nucleation rate in a start up process of a continuous cooling crystallizer without product classification is shown in Fig.2.a, and the variation of the positions of the separation nodes versus time t is presented in Fig.2.b. The evolution of the population density function in time is presented in Fig.3.a. In these runs, the size-dependency of the crystal growth was linear function of size, and the growth rate dispersion was assumed to be zero.

Steep moving fronts of the crystal size distribution along the size coordinate may arise also in continuous crystallizers under the conditions of autonomous, periodic, quasi-periodic or chaotic, oscillations. These phenomena have been examined mostly by using the moment equation model (Lakatos and Blickle,1995, Lakatos and Ulbert,2001), but

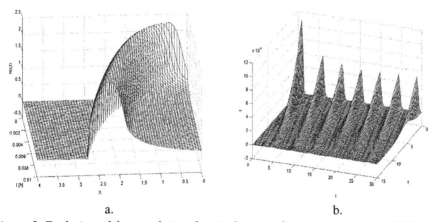

a.   b.

Figure 3. Evolution of the population density function during a) start up, and b) limit-cycle oscillations of crystallizer in the case of representative withdrawal of crystals

applying the present moving finite element solution, the predictions obtained by the moment equation model can be compared with the results given by the population balance model. In Fig.3.b, development of limit cycle oscillations is presented in isothermal case at parameter values for which the stability analysis based on the moment equation model predicted oscillatory instabilities.

6. Conclusions

The moving finite element method presented for numerical solution of the population balance equation appears to be a good technique for solving the usually highly nonlinear integro-differential equations of crystallisation models. Since, however, the crystal size distribution may vary some orders of magnitude in the dynamic crystallisation processes, this method needs some adaptation crossing the boundaries of the significantly differing size regions. Changing the parameters of the movement regularising functions (17) what however requires some experience can do such adaptation.

7. References

Chang, R.-Y. and M.-L. Wang, 1984, Computers chem. Engng, 8, 117.

Coimbra, M.C., C. Sereno and A. Rodrigues, 2001, Chem. Engng J., 84, 23.

Eyre, D., C.J. Wright and G. Reuter, 1988, J. Comp. Physics, 78, 288.

Gelbard, F. and J.H. Seinfeld, 1978, J. Comp. Phys., 28, 357.

Hounslow, M.J., R.L. Ryall and V.R. Marshall, 1988, AIChE J., 34, 1821.

Kumar, S, and Ramkrishna, 1997, Chem. Eng. Sci., 50, 4659.

Lakatos, B., E. Varga, S. Halász and T. Blickle, 1984, Simulation of batch crystallizers. In: Jancic, S.J. and E.J. de Jong (Eds), Industrial Crystallization'84. Elsevier, Amster-dam, 185.

Lakatos, B.G. and Zs. Ulbert, 2001, Proc. 8th Int. Workshop Ind.Crystallization. DUT, The Netherlands, 231.

Lakatos, G.B. and T. Blickle, 1995, Computers chem.Engng, 19, S501.

Mantzaris, N.V., P. Daoutidis and F. Srienc, 2001, Computers chem. Engng, 25, 1411.

Marchal, P., R. David, J.P. Klein and J. Villermaux, 1988, Chem. Eng. Sci., 43, 59.

Miller, K. and R.N. Miller, 1981, SIAM J. Numer. Anal., 18, 1019.

Miller, K., 1981, SIAM J. Numer. Anal., 18, 1033.

Nicmanis, M. and M.J. Hounslow, 1998, AIChE J., 44, 2258.

Petzold, L.R., 1983, A description of DASSL. In: Stepleman, R.S et al.(Eds), Scientific Computing, North Holland, Amsterdam, 65.

Ramkrishna, D., 2000, Population Balances: Theory and Applications to Particulate Systems in Engineering. Academic Press, San Diego.

Rawlings, J.B., W.R. Witkowski and J.W. Elton, 1992, Powder Technology., 69, 3.

Sereno, C., A. Rodrigues and J. Villadsen, 1991,Computers chem. Engng, 15, 25.

Steemson, M.L. and E.T. White, 1988, Computers chem. Engng, 12 81.

Subramanian, G. and D. Ramkrishna, 1971, Math.Biosciences, 10, 1.

Ulbert, Zs. and B.G. Lakatos, 1999, Computers chem.Engng, 23, S435.

Wulkow, M. and A. Gerstlauer, 1999, Proc. 14th Int. Symp. Ind. Cryst., IChemE, Rug-by, UK, no 7.

European Symposium on Computer Aided Process Engineering – 12
J. Grievink and J. van Schijndel (Editors)
© 2002 Elsevier Science B.V. All rights reserved.

An Object-Oriented Framework for Bill of Materials in Process Industries

Marcela Vegetti‡, Gabriela Henning†, Horacio Leone‡

†INTEC, Güemes 3450, 3000 - Santa Fe, Argentina. ghenning@intec.unl.edu.ar

‡INGAR/UTN, Avellaneda 3657, 3000 - Santa Fe, Argentina.
mvegetti@ceride.gov.ar, hleone@ceride.gov.ar

Abstract

Products that were formerly quite standard turn now into custom-made ones, leading to an enormous increase in product variants. High degrees of redundancy can occur in data management when closely related product structures are treated as independent bills of materials. Since the bill of materials (BOM) is one of the most fundamental data in industrial enterprises a lot of work has been devoted to it in last years. In this contribution, a new object-oriented BOM has been proposed. It easily manages crucial aspects that should be taken into account in a product representation, such as the efficient handling of product families and variants concepts, composition and decomposition structures and the possibility of describing restrictions. Moreover, the model easily accommodates the needs of various organizational units within a company.

1. Introduction

Product proliferation is uncontrollable nowadays. With customers demanding ever more customized products, manufacturers have responded with mass customization and even segment-of-one views of the market, as companies perceive each customer as an independent market segment. In addition to the enormous variety of product types, products life cycle has been shrinking dramatically. These phenomena, which have been referred as product flexibility by Van Veen and Wortmann (1992a), have become a new factor in competition. However, product flexibility has an enormous impact in supporting information systems, which are burdened by the growing number of products that must be managed.

Many problems arise in defining and maintaining big amounts of product data, which affect the performance of information systems with regards to aspects such as storage requirements, data entry support, retrieval support as well as correctness and timeliness of provided information. Indeed, many of the modules that comprise overall company information systems make use of product information either in the form of a recipe or a BOM. However, the actual way of presenting a BOM depends on factors such as the point of view from which products are seen. For instance, the materials requirement planning (MRP) function uses it intensively to explode the requirements of end products into the needs of intermediate ones and raw materials, thus launching work orders for the various production departments and purchase orders for the different suppliers. Therefore, the BOM supporting MRP must be consistent with the way a product is manufactured. On the contrary, the Master Planning Scheduling (MPS) function does not need those details and requires product information in terms of how products are sold instead of how they are manufactured, leading to different types of products models such as the modular and planning BOMs. Other functions within the production planning and control department like demand forecasting use different product information such as smoothing factors and previous forecast values. In addition, a variety of functional areas make use of the BOM (e.g., for calculating costs of final and intermediate products). These distinct needs have led in practice to a situation where many functional areas have developed their own product model, suitable for their particular requirements. As pointed out by Vollmann et al. (1997) a golden rule is that a company should have one, and only one BOM that should be maintained as an entity and be so designed that all legitimate company uses can be satisfied.

Another problem faced by many industries in the fact that conventional BOM representations are only suitable for discrete manufacturing industries where products are always fabricated by putting parts together (composition) in assembly processes. In other words, they do not handle BOM representations where products are obtained by decomposing raw materials, like in some food industries (milk and meat ones) and in the petrochemical business where hybrid structures (combining composition and decomposition types of operations) may be associated to products.

The problems described before reveal a demand for a new representation of BOMs, able to suit the needs of different functional areas, to efficiently deal with a growing number of product variants and to handle all types of production strategies. Associated with such representation, there is need for the corresponding bill of materials processor, the specific computer software that deals with data entry, maintenance and recovery. This contribution describes a conceptual representation of BOMs that tries to overcome the problems pointed out before. It has been organized as follows. First, a brief summary of former approaches is presented. Afterwards, the proposed model is introduced and then its application to a case-study is tackled. Finally, conclusions and future work are discussed.

2. Different types of bill of materials

At the simplest conceptual level, the BOM is a list of all the parts needed to produce a finished product. Thus, the BOM is generally represented by a **tree structure** whose root is a final product, and the descendants of each node represent the components or necessary materials to produce it (**multi-level BOM**). Nevertheless, there are other BOMs like the **single level** BOM and the **indented BOM**. A typical BOM has been defined by Scheer (1998) using two types of entities: **Part** and **BOM Relationship**. A **Part** entity can be a **finished product**, an **assembly**, a **component** or a **raw material**. Each **BOM Relationship** entity defines a link between a product P (parent) and a product Q (component) if Q is one of the direct components required to produce P.

In the traditional BOM, each variant is treated as an independent product. High degrees of redundancy can occur when closely related product structures are represented as independent bills of materials. This is true in the case of product variants, as well as in cases where the representation of a product is managed separately according to the different views of distinct functional areas of the organization. The creation of independent bills of materials for similar products can lead to an uncontrollable volume of data, associated with high storage and management costs, and potential inconsistency problems. Thus, special representations are to be developed to manage similar bills of materials. To achieve an efficient representation of variants, Scheer (1998) outlines some modifications to this model called **identical-part BOM**, **plus-minus BOM** and **multiple BOM**. Van Veen and Wortmann (1992a, b) describe other approaches, which are a little more complex but more efficient than the previous ones: **Modular BOM**, **Variant BOM** and **Generic BOM**, where the last two belong to the group of "generative BOM" systems. Chung and Fisher (1994), as well as Usher (1993), present a BOM model based on an object-oriented representation.

3. An object-oriented model for a hybrid BOM

The proposed model agrees with the "generative BOM" philosophy, where each variant structure is derived at the moment it is required by resorting to a valid specification, thus reducing the volume of stored data. Moreover, it is an object-oriented (OO) model that minimizes the coupling among the products and their variants, allowing to carry out changes in products´ structure with minimum consequences for the associated variants. The model, which is presented in Fig. 1 in a concise view (not showing specific attributes), considers the existence of similar products that have almost common structures and only differ in: (i) the presence (with quantity specification) or absence of the some components (at least one) in the structure, or (ii) the value of some of the characteristics that define the set of components. A group of similar products is referred as product family (named as **Product** in Fig. 1). The explicit representation of families allows the encapsulation of aggregated data usually stored in planning bills of materials. A particular member of this group is referred as a **variant**. A **Product** family represents a simple or a compound product. A **Compound Product** models at a generic level (i) an intermediate or a final product (intended for commercialization) resulting from the assembly or processing of raw materials and intermediates or (ii) a non-atomic raw material that can be decomposed into other products. These two situations are denoted in the model as **Composed-by** and **Decomposed-into** relationships and are described in the model by the **CStructure** and **DStructure** classes, which are subclasses of the **Structure** class. This class contains the number of units or the amount of the descendent product participating in the parent product, the classification of the structure and the restrictions that could exist for the required quantities of the component in its parent's composition. Regarding classification, a **Structure** relationship can be mandatory (when its links a component that is always present), optional (when the component might not participate) or mandatory-selective (when one out of a group of links should be present). It is also necessary to represent that in certain cases, either for technological or commercial reasons, not all the combinations of components are valid. Therefore, when defining a structure, a mechanism that allows expressing restrictions among parts is needed. It is possible to identify two groups of such constraints: **obligatory** (when a component family is included another is obliged to take part) and **incompatible** (when one component family is present another must not take part). These constraints are represented by the **P_Restriction** entity. According to Olsen et al. (1997), most restrictions should be included in generic structures in order to simplify the specification of product variants. Other constraints that may occur in a composition structure are quantity restrictions (**Q_Restriction**), restraining the "amount" in which a given component participates (it can a maximum or a minimum bound, a range, etc.). Apart from the previous constraints that apply to a generic level, there might be restrictions among specific instances of variants (**SV_Restriction**), where also obligatory and incompatible constraints can be identified.

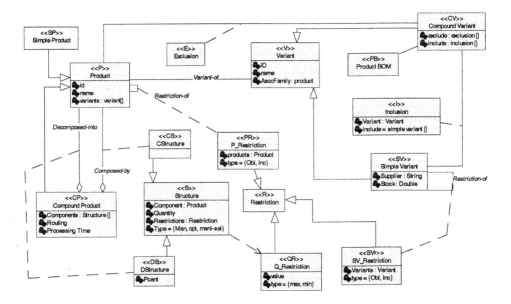

Figure 1: Class diagram of the proposed BOM model.

Since a **Variant** is a particular member of a product family (**Product**), to represent its membership to such family, the **Variant-of** relationship is included in the model. As there are simple and compound products, simple and compound product variants must also be introduced in the model. The first situation is represented by resorting to the **Simple Variant** class, while the second one by the **Compound Variant** class. For each **Compound Variant**, the associated **Compound Product** family is specified, from which it is possible to obtain the variant´s structure. Provided that certain components that are defined at the level of a family may not be present in a particular compound variant, the **Exclusion** relationship has been introduced in the model. On the other hand, a **Compound Variant** may be associated to a **Simple Variant** when there is a possibility of choosing among alternatives and a particular one is chosen; the selected variant is associated to the compound variant by means of an **Inclusion** relationship. This link may also contain information on the number of the simple components that are included in the compound variant.

4. Example

The proposed model has been tested by representing product data of different industrial enterprises that are characterized by having complex structures (like in the meat industry, where composition and decomposition operations exist), high number of variants, as well as needs for representing data both at the level of product families and instances. The case treated in this section and depicted in Fig. 2 corresponds to a candy production facility, described by Henning and Cerdá, 2000, where an enormous amount of variants needs to be tackled. For instance, from a particular unwrapped candy (such as the strawberry candy depicted in Fig. 2) many variants originate (**MPS Strawb. Price, MPS Straw Holid** and **MPS Strawb. Candy**) due to the adoption of different wrappings (**Strawb. candy, Holid.** and **Price**, which are **variants-of** the entity **wrapp**). Similarly, the same wrapped candy can be packed in different quantities and formats (in bags, tubes and jars of distinct capacities), thus leading to different end products. Due to space limitations Fig. 2 depicts just one end product (named **000XXX Strawb. Fil.**), but it shows different alternative packagings that lead to a greater variety of them. The fact of representing packagings as separate intermediate assemblies (**Packaging 000, 001**, etc.) allows their "reuse" in the definition of other products, since many products having different characteristics (flavor, shape, filling, etc.) are packed in the same way.

5. Conclusions and Future Work

Product data is used as the basis for many modules of an industrial information system. An object-oriented product model that tries to represent all the legitimate company uses of product information has been presented. It has been implemented using OODBMS technology that allows the creation of persistent objects, enabling the implementation of the object model as it was conceived, without transforming it into a relational outline. The chosen database administrator is VERSANT, Release 5.2 with the Java interface JVI (Java-Versant Interface, VERSANT Corporation, 1999). Future work involves intensive model testing to assess its suitability in relation to the distinct needs of the various functions of industrial organizations, easiness of data entry, modification, etc.

References

Chung, Y. and G. Fischer, 1994, A Conceptual Structure and Issues for an Object Oriented Bill of Materials (BOM) Data Model. *Computers Ind. Engng.*, 26, 321-339.

Henning, G.P. and J. Cerdá, 2000, Knowledge-based predictive and reactive scheduling in industrial environments, *Computers and Chemical Engineering*, 24, 2315-2338.

Olsen, K.A., Sætre, P. and A. Thorstenson, 1997, A Procedure-Oriented Generic Bill of Materials. *Computers Ind. Engng.*, 32, 29-45.

Scheer, A.W., 1998, *Business Process Engineering*. Springer-Verlag, Berlin- Heidelberg.

Usher, J.M., 1993, An Object-Oriented Approach to Product Modelling for Manufacturing Industries, *Computers Ind. Engng.*, 25, 557.

Van Veen, E.A. and J.C. Wortmann, 1992a, Generative bill of materials processing systems. *Production Planning & Control*, 3, 314-326.

Van Veen, E.A. and J.C. Wortmann, 1992b, New developments in generative BOM processing systems. *Production Planning & Control*, 3, 327-335.

Versant Corporation, 1999, J/VERSANT Interface Release 2.4.

Vollmann, T.E., Berry, W.L. and D.C. Whybark, 1997, Manufacturing Planning and Control Systems, Fourth Edition, Irwing McGraw-Hill, New York.

996

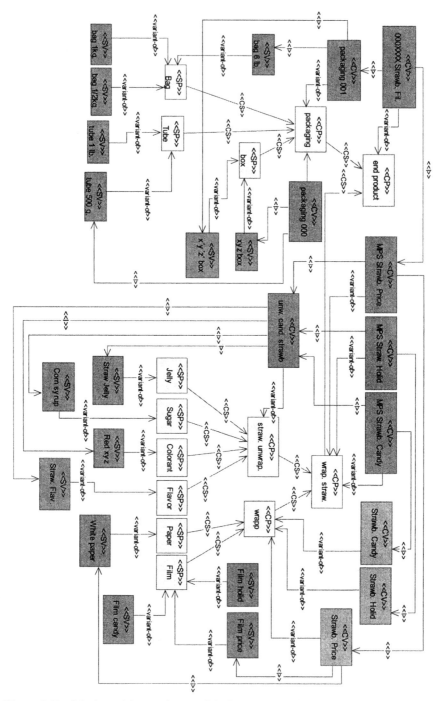

Figure 2: Partial view of the product model of a candy production facility

European Symposium on Computer Aided Process Engineering – 12
J. Grievink and J. van Schijndel (Editors)
© 2002 Elsevier Science B.V. All rights reserved.

997

A Hybrid Global Optimization Approach for Solving MINLP Models in Product Design

Yiping Wang and Luke E. K. Achenie*
Department of Chemical Engineering, University of Connecticut, Storrs, CT 06269
yiping@engr.uconn.edu and achenie@engr.uconn.edu
mailto:yiping@engr.uconn.edu   * Phone: (860) 486 2756. Fax: (860) 486 2959.

Abstract

This paper presents a hybrid global optimization approach for solving product design (specifically solvent design) problems modeled by mixed integer nonlinear programming (MINLP). The strategy incorporates a variant of the Outer Approximation mathematical programming algorithm and a soft computing global optimization approach (namely simulated annealing). The suggested approach is not provably globally optimal. However, computational experience with benchmark examples and solvent design MINLP models indicate strongly that the approach gives near globally optimal solutions.

Keywords: Global optimization; Simulated annealing; MINLP; Computer Aided Product Design (CAPD); Outer approximation (OA).

1. Introduction

In recent years, some computer-aided product design (CAPD) problems have been formulated as mixed integer nonlinear mathematical programming (MINLP) models in which performance requirements of the compounds are reflected in the objective and the constraints (Odele and Macchieto, 1993, Duvedi and Achenie, 1996, Churi and Achenie, 1996, Pistikopoulos and Stefanis, 1998, Sinha et al. 1999, Buxton et al. 1999, Wang and Achenie, 2001). Most of these CAPD models are multi-extremal and non-convex from an optimization point of view. As a result, care needs to be taken when employing locally optimal techniques such as generalized benders decomposition (GBD, Geoffrion, 1972) and outer approximation (OA/ER/AP, Viswanathan and Grossmann, 1990).

Developing efficient algorithms for identifying the global optima is not trivial and has been the target of research for some time. A recent book by Floudas (2000) presents an overview of the existing global algorithms for MINLP problems. Overall, the strategies in the open literature for global optimization of MINLP models can be classified as deterministic (Kocis and Grossmann, 1988; Floudas et al. 1989; Ryoo and Sahinidis, 1995), stochastic (Cardoso et al. 1997; Consta and Oliveria, 2001), or transformation techniques (Pörn et al., 1999). Deterministic approaches tend to use gradient-based techniques to solve sub-problems. Stochastic methods often do not use gradient information, and can find a global minimum with an asymptotic convergence guarantee in probability (i.e. with probability 1, Dekkers and Aarts, 1991), such as simulated annealing. Depending on the structure of the optimization model, Pörn et al. (1999) proposed some convexification techniques to reformulate posynominal and negative binominal terms in the constraints so that the nonconvex problem is transformed into a convex problem.

Some MINLP models can be reformulated or simplified to have the proper convexity assumptions such that existing algorithms can be employed. There are other types of MINLP models that do not lend themselves to such reformulations. Similarly, there are models for which approximation of physical property models may not be appropriate. For example, consider MINLP models that employ complicated activity models in phase equilibria. For these types of MINLP models we propose a hybrid method that incorporates stochastic and deterministic elements for global solutions. We initiate our investigation with the OA/ER/AP algorithm (Viswanathan and Grossmann, 1990). It should be emphasized that our aim is to provide a viable approach for finding a near global solution for nonconvex MINLP problems. The basic idea is to employ the simulated annealing method (Kirkpatrick et al. 1983) to solve the primal problem in OA/ER/AP algorithm to increase the possibility of getting a globally optimal solution. The paper is organized as follows. We start with an introduction to the simulated annealing method, followed by a discussion of the proposed hybrid global optimization method. To demonstrate the viability of the proposed approach, some benchmark MINLP examples and MINLP models of the solvent design problem are solved.

2. Simulated annealing (SA) approach

Simulated annealing is a stochastic method, which can avoid getting trapped in a local minimum (Dekkers and Aarts, 1991). The SA method originates from an analogy between a physical annealing process and the problem of finding solutions to optimization problems. The SA algorithm involves the following steps (Zhu and Xu, 1999):

Step 1: Initialization: set the initial and final values of the parameter T, namely T_0 and T_f. Also provide an initial guess x_0 for the vector of optimization variables. Set the iteration counter $k \leftarrow 0$.

Step 2: For the current value of $T = T_k$, perform a sequence (Markov chain loop with length L_0) of exploratory moves.

(2a) Generate a neighbor of x_k as x using a random number generator.

(2b) Let $\Delta = f(x) - f(x_k)$.

(2c) If $\Delta \leq 0$ or $exp(-\Delta/T) \geq random\ (0,1)$, set $x_k = x$.

where random (0,1) is a random number (between 0 and 1) from a given probability distribution. We employed both a normal distribution and a uniform distribution.

Step 3: if $T < T_f$, STOP, otherwise $T_{k+1} = fxn(T_k)$, and go to Step 2.

In the SA algorithm the following parameters need to be optimally specified (Wang and Achenie, 2001): T_0, T_f, and L_0. One also needs to identify a function $fxn(*)$ for decreasing the value of the control parameter. We chose $fxn(*)$ as follows: $T_{k+1} = \alpha T_k$, $0 < \alpha < 1$.

3. Proposed hybrid global optimization method (OA_Global)

The outer approximation (OA) algorithm was developed by Duran and Grossmann (1986) to solve a MINLP problem without equality constraints. To accommodate equality constraints, Kocis and Grossmann (1987) provided a variant, namely outer approximation with equality relaxation (OA/ER). Both OA and OA/ER can identify the global solution if certain convexity requirements are satisfied (Floudas, 1995). To soften the convexity assumptions, Viswanathan and Grossmann (1990) proposed a variant of the OA/ER algorithm by adding a penalty, namely OA/ER/AP (outer approximation with equality relaxation and augmented penalty). However, as shown in (Floudas, 1995), the OA/ER/AP may still end up with a locally optimal solution.

To obtain a global solution, we suggest a modification of OA/ER/AP, namely OA_Global. The premise for the proposed strategy is if a globally optimal solution for the nonconvex NLP primal problem is obtained in the OA/ER/AP, then a potentially global optimal solution will be obtained for the original MINLP problem. Simulated annealing is employed to solve the NLP primal problem to find a globally optimal solution.

The OA_Global algorithm addresses MINLP problems of the general form

$$
z = \min_{x,y} c^T y + f(x)
$$

$$
s.t
$$

$$
Ay + h(x) = 0 \tag{1}
$$

$$
By + g(x) \leq 0
$$

$$
Cy + Dx \leq d
$$

$$
x \in X = \{x \mid x \in R^n, x^L \leq x \leq x^U\}
$$

$$
y \in Y = \{0,1\}^m
$$

In problem (1), all the binary variables appear linearly. For the problem involving nonlinear binary variables, pseudo-continuous variables can be employed (Kocis and Grossmann, 1988). The OA_Global solves the NLP primal of problem (1) without making convexity assumptions. The algorithmic steps are similar to the OA/ER/AP algorithm (Viswanathan and Grossmann, 1990).

Since simulated annealing does not provide Lagrange multipliers, the formation of the relaxation matrix using Lagrange multipliers is not possible here. In this study, we have used the following heuristics to determine the elements of the relaxation matrix in the MILP master of problem (1).

$$
T_k = \{t_{ii}^k\} \tag{2}
$$

where

$$
t_{ii}^0 = \begin{cases} 1 & Ay^k + h_i(x^k) > 0 \\ 0 & Ay^k + h_i(x^k) = 0 \\ -1 & Ay^k + h_i(x^k) < 0 \end{cases} \tag{3}
$$

Theoretically at optimal solution x^k, the equality constraints should be satisfied, i.e.

$$Ay^k + h(x^k) = 0$$

Simulated annealing satisfies the equality constraints to within a given tolerance thus

$$|Ay^k + h(x^k)| \leq \varepsilon$$

To validate the OA_Global algorithm, a benchmark MINLP example from the open literature and a solvent design MINLP model are solved.

Benchmark example

$$\min f(x_1, x_2, x_3, y_1, y_2, y_3, y_4) = (y_1 - 1)^2 + (y_2 - 2)^2 + (y_3 - 1)^2$$
$$- \ln(y_4 + 1) + (x_1 - 1)^2 + (x_2 - 2)^2 + (x_3 - 3)^2$$

$s.t.$

$$y_1 + y_2 + y_3 + x_1 + x_2 + x_3 \leq 5$$
$$y_3^2 + x_1^2 + x_2^2 + x_3^2 \leq 5.5$$
$$y_1 + x_1 \leq 1.2$$
$$y_2 + x_2 \leq 1.8$$
$$y_3 + x_3 \leq 2.5$$
$$y_4 + x_1 \leq 1.2$$
$$y_1^2 + x_2^2 \leq 1.64$$
$$y_3^2 + x_3^2 \leq 4.25$$
$$y_2^2 + x_3^2 \leq 4.64$$
$$x_1, x_2, x_3 \geq 0$$
$$y_1, y_2, y_3, y_4 \in \{0,1\} \tag{4}$$

This nonconvex MINLP model was solved by Floudas and Aggarwal (1989), Ryoo and Sahinidis (1995), and Cardoso et al. (1997). The best solution from the open literature is $(x_1, x_2, x_3, y_1, y_2, y_3, y_4; f) = (0.2, 0.8, 1.908, 1, 1, 0, 1; 4.576)$ (Floudas and Aggarwal, 1989). OA_Global obtained $(x_1, x_2, x_3, y_1, y_2, y_3, y_4; f) = (0.206, 0.804, 1.910, 1, 1, 0, 1; 4.540)$ (see Table 1).

Table 1: Computational results of illustrative example by OA_Global

Simulated annealing parameters				Optimal solution	Objective function
T_0	α	ε	L_0		
1.0e6	0.98	1.0e-3	100	[0.206,0.804, 1.910,1,1,0,1]	4.540
Floudas et al. (1989)				[1,1,0,1,0.2,0.8,1.908]	4.576
Ryoo and Sahinidis (1995)				[1,1,0,1,0.2,0.8,1.907878]	4.579582
Cardoso et al. (1997)				[1,1,0,1,0.2,0.8,1.907878]	4.579582

4. Solvent Design Case Study

Solvent design for extraction of acetic acid from water has been studied in the open literature (Odele and Macchietto, 1993; Pretel et al., 1994; Hostrup et al. 1999). In these studies, tetramethyl hexane, 1-nonanol, and diisobutyl ketones were identified as potential solvents. These solvents however were obtained based on infinite dilution activity coefficient models. When the concentration of the solute is very low, this simplification is reasonable; otherwise it can lead to unsuitable solvent choices. In this paper, we employ the rigorous UNIFAC model to calculate liquid-liquid equilibrium. We note that the highly non-linear nature of this rigorous model is likely to cause problems for many MINLP global solvers. We hope researchers in global MINLP will use this problem as a benchmark problem for their algorithms. We are genuinely interested in whether any one can find a solution better than what is presented here.

Here we would like to design a globally optimal solvent for the extraction of acetic acid from water using the OA_Global algorithm. The objective is to maximize the distribution coefficient of acetic acid. Since one would like to extract as much acetic acid as possible, this is probably a better performance criterion than maximizing selectivity. For comparison purposes, we employ the same conditions as in the open literature (Hostrup et al. 1999). The selectivity of the candidate solvent should be greater than 7, and solvent loss should be less than 0.01. The boiling point of the solvent should be in the range 421K to 541K, and the melting point should be less than 310K. The results by OA_Global are given in Table 2.

Table 2: Computational results of acetic acid extraction

	OA_Global				Hostrup et al. (1999)
Simulated annealing parameters	T_0	α	ε	L_0	
	1.0e4	0.85	0.01	60	
Solvent	Propanedioic acid, methyl-dimethyl ester [609-02-9]				Diisobutyl ketone [108-83-8]
Optimal structure	(CH3COO)$_2$(CH$_3$)(CH)				(CH$_3$)$_4$(CH$_2$)$_2$(CH)$_2$CO
Molecular Weight	146				142
BP(K)	443.6 (449.7*)				462 (441*)
MP(K)	224.9				227 (227*)
-log(LC$_{50}$)	0.98				3.35
Selectivity of acetic acid	12.43				20.3
Distribution coefficient of acetic acid	0.66				0.28

*Experimental value

As shown in Table 2, propanedioic acid, methyl-dimethyl ester (abbreviated as PAMDE) is the potentially optimal solvent for extracting acetic acid from water under the specified conditions. The global solution was partly verified by partially enumerating the MINLP using the binary variables and solving the individual NLP's. Compared with Hostrup's results (Hostrup et al. 1999), PAMDE shows a higher distribution coefficient (as dictated by the performance objective function), but a lower selectivity than diisobutyl ketone. If selectivity is more important, then selectivity should be used in the objective function. In addition, PAMDE has a lower value of − log(LC$_{50}$) than diisobutyl ketone, which makes it environmentally benign.

5. Conclusions

This paper presents a hybrid global optimization strategy (for a given kind of MINLP model) based on the OA/ER/AP algorithm. The premise for OA_Global is if a global solution is obtained for NLP primal problem, then the final solution may be potentially global. Based on this, a near-global optimizer, namely simulated annealing, was employed to solve the NLP primal problem in OA/ER/AP.

The OA_Global algorithm is demonstrated through the solution of a benchamrk example and a realistic solvent design problem. Although the suggested approach has not been mathematically proven to be globally optimal, our experience so far indicates very strongly that a global solution of the problem is obtained.

References:

Buxton, A.; Linvingston, A. and Pistikopoulos, E. N., 1999, *AIChE J.* 45(4), 817.

Cardoso, M. F. Salcedo, R. L. de Azevedo, S. and Barbosa, D., 1997, *Computers Chen Engng*. 21 (12), 1349.

Churi, N.; Achenie, L. E. K., 1996, *Ind. Eng. Chem. Res.*, 35(10), 3788.

Costa, L. and Oliveira, P., 2001, *Computers Chem. Engng*. 25, 257.

Dekkers, A. and Aarts, E., 1991, *Mathematical Programming*, 50, 367.

Duran, M. A. and Grossmann, I. E., 1986, *Mathematical Programming* 36, 307.

Duvedi, A. P. and Achenie, L. E. K., 1996, *Chem. Eng. Science*, 51, 3727.

Floudas, C. A. Deterministic global optimisation: theory, algorithms and applications Kluwer Academic Publishers, 2000.

Floudas, C. A. Nonlinear and mixed-integer optimisation. Oxford University Press, NY 1995

Floudas, C. A.; Aggarwal, A. and Ciric, A. R., 1989, *Comp. Chem. Engng*. 13, 1117.

Geoffrion, A. M., 1972, *Journal Optimization and Theory Applications* 10, 237.

Hostrup M.; Harper, P. M.; and Gani, R., 1999, *Computers Chem. Engng*. 23, 1395.

Kirkpatrick, S.; Gelatt, C. D., Jr; Vecchi, M. P., 1983, *Science*, 220, 671.

Kocis, G. R. and Grossmann, I. E., 1987, *Ind. Engng. Chem. Res*. 26, 1869.

Kocis, G. R. and Grossmann, I. E., 1988, *Ind. Eng. Chem. Res*. 27, 1407.

Odele, O. and Machietto, S., 1993, *Fluid Phase Equilibria*. 82, 47.

Pistikopoulos, E. N. and Stefanis, S. K., 1998, *Computers Chem. Engng*. 22, 717.

Pörn, R. Harjunkoski, I. and Westerlund, T., 1999, Computers Chem. Engng.23, 439.

Pretel, E.J.; Lopez, P. A.; Bottini, S. B., 1994, *AIChE J.*, 40(8), 1349-1360

Ryoo, H. S. and Sahinidis, N. V. (1995). *Computers Chem. Engng*. 19(5), 551.

Sinha, M.; Achenie, L.K.; Ostrovsky G., 1999, *Computers Chem. Engng.*, 23, 1381.

Viswanathan, J. and Grossmann, I. E., 1990, *Computers Chem. Engng*. 14(7), 769.

Wang, Y. and Achenie, L. E. K., 2001, *ESCAPE-11*, 585.

Zhu, Y. and Xu, Z., 1999, *Fluid Phase Equilibria*, 154, 55.

European Symposium on Computer Aided Process Engineering – 12
J. Grievink and J. van Schijndel (Editors)
© 2002 Elsevier Science B.V. All rights reserved.

Cluster Identification using a Parallel Co-ordinate System for Knowledge Discovery and Nonlinear Optimization

K Wang[1], A Salhi[2] & E S Fraga[1] [*]

[1]Centre for Process Systems Engineering, Department of Chemical Engineering, UCL (University College London), London WC1E 7JE, U.K.

[2]Department of Mathematics, University of Essex, Colchester CO4 3SQ,U.K.

Abstract

The visualization of multi-dimensional data using a Parallel Co-ordinate System is applied to process optimization. A Scan Circle algorithm is proposed for identifying clusters of "close-to-a-line" points in an n-dimensional space by using the duality between the Cartesian co-ordinate and parallel co-ordinate systems. The lines identified by the algorithm are used to identify regions of interest in the domain of the constraints. These regions of interest can lead to the discovery of knowledge about the behaviour of the complex, nonlinear constraints and this knowledge is used as a basis for a genetic algorithm for efficient and robust nonlinear optimization.

1 Introduction

Data visualization plays an indispensable rôle in uncovering hidden structures or patterns for information extraction tools (Lee and Ong, 1996). Most graphical representations display data in two-dimensional Euclidean space. For data having more than three dimensions, other than drawing all the pair-wise relationship individually, dynamic elements such as colors, shapes and/or motions are normally used to enhance visual effects (Keim, 1997). However, chemical process design problems are so large that direct visualization of entire space through such methods is difficult. The Parallel Co-ordinate System (PCS) representation (Inselberg and Dimsdale, 1990) transforms the search for relations among design variables into a two-dimensional pattern recognition problem, and design points become amenable to visualization. This paper describes the use of PCS as a basis for multi-dimensional data visualization and subsequent knowledge discovery. Moreover, PCS is explored further and a Scan Circle Algorithm (SCA) is proposed to find clusters of "close-to-a-line" points, to identify such lines and hence to locate regions of interest in the problem domain.

Visualization and any subsequent knowledge discovery are useful in their own right. However, the knowledge discovered can provide useful information for optimization procedures, improving both reliability and efficiency. We therefore propose the combination of knowledge discovery with stochastic optimization for complex nonlinear process optimization. Genetic Algorithms (GA) (Goldberg, 1989) mimic natural evolution on an abstract level. In contrast to mathematical programming, the

---

[*] Author to whom correspondence should be addressed: e.fraga@ucl.ac.uk

evaluation of a candidate process can be done by simulation, making GAs attractive for optimization in simulation-based chemical process design. Furthermore, nonlinear, multiple and complex objectives can be included. For these reasons, genetic algorithms have attracted much interest in process optimization (Caldas & Norford, 2002). The Scan Circle and genetic algorithms are combined in this work to provide a robust and efficient optimization procedure.

2 Parallel Co-ordinate Systems

In a parallel co-ordinate system (PCS), n equally spaced vertical axes are used to represent n variables (Inselberg and Dimsdale, 1990). The distance between all the adjacent axes are assumed to equal one for simplicity of discussion. An n-dimensional point in PCS is represented by points on the n axes joined by a polygonal curve connecting the n points sequentially.

In the n-dimensional space, a line in the Cartesian co-ordinate system (CCS) can be represented by n-1 linearly independent equations (Inselberg and Dimsdale, 1990):

$$x_{i+1} = m_i x_i + b_i \qquad i = 1, 2, ..., n-1 \qquad (1)$$

whereas the same line is represented by a set of n-1 indexed points in PCS with co-ordinate values ($\frac{i}{1-m_i}, \frac{b_i}{1-m_i}$), $i = 1, 2, ..., n$-1.

3 Scan Circle Algorithm

The basis of the new procedure is the correspondence between CCS and PCS. Visualization by PCS is used to identify a particular type of pattern, namely clusters of "close-to-a-line" points, in the optimization search space. A novel Scan Circle algorithm (SCA) is proposed here to identify such clusters and lines which represent these clusters with the aim of providing insight into possible solutions to the optimization problem and also to provide support for the development of new optimization algorithms.

3.1 Rationale of SCA

As shown in Fig. 1(a), N lines in PCS (corresponding to N points in CCS) result in $N(N-1)/2$ intersections within two adjacent axes. SCA uses the intersections as centres of circles of radius r, called *scan-circles*. The core of the novel algorithm is the identification of a set of circles which include a large number of intersection points and the selection, from this set, of the *best circle*. The rationale is that if a point and a line are close to each other in CCS, their counterparts in PCS are even closer (Chou et. al., 1999). If the *best circle* exists, its centre point will correspond to a line in CCS which represents a "close-to-a-line" cluster of points. To determine whether such a cluster exists in an n-dimensional space, close proximity of the edges of polygonal lines within two adjacent axes being scanned alone is not sufficient. The entire set of polygonal lines containing these edges must also exhibit close proximity. The full polygonal lines spanning the n axes and which contain these edges must therefore be further analyzed.

3.2 SCA implementation

1. Initialize S to the set of N data points.
2. Each pair-wise combination of n variables is chosen to define a candidate region for analysis. In each region, e.g. the region ij shown as Fig. 1(a), the data points in S are represented by line segments with up to $N(N-1)/2$ intersections, where N is the size of S. Each intersection point defines a circle, of radius r, centred on the intersection point, and which contains a number of other intersection points. The value of r used is selected automatically as the minimum to ensure covering all feasible points. However, infeasible points are also covered, increasing the computational effort here and in the GA step described below. For each candidate region, a potentially *interesting circle*, the one with the most number of line segments, is selected. The circle with the largest number of line segments from the list of interesting circles is chosen and is known as the *best circle*. The region that contains the *best circle* is called the *best region*. The polygonal lines, whose line segments in the *best region* result in the intersections within the *best circle* (indicated by solid lines in Fig. 1(a)), are called the *interesting polygonal lines* and represent the first guess at a cluster of data points.
3. We now check to see if the *interesting polygonal lines* identified in step 2 represent clusters of points, with respect to the other variables, by analysing their situation in the other regions (i.e. all regions except the *best region*). Assuming that region ij was identified as the *best region* in step 2 (as indicated in Fig. 1(a)), variable i cannot be selected as the left axis and j as the right axis to define candidate regions. All other combinations are allowed.

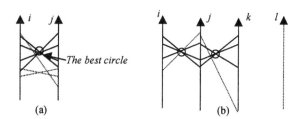

Fig. 1. SCA find the best circle automatically within candidate regions.

For each region, we identify a new scan circle considering only the *interesting polygonal lines* defined in step 2. If the number of the lines that intersect this new circle is greater than a specified number (10% of total feasible points in the results presented below), the region is added to the best region and the lines which define the cluster is the intersection of the original interesting polygonal lines and those which intersect this circle. In the list of remaining regions, region ji and all regions with the same left (i) or right (j) variables are removed before selecting the next region to analyse. The termination criterion for this whole step is when the number of lines in the cluster falls below the minimum allowed (e.g. 10% of original data set size) or when n-1 candidate regions have been analysed. At the end of this whole step, a cluster of points and associated pairs of variables will have been identified. Output this cluster.

4. Let S be S minus the lines in the cluster found in step 3. If S is sufficiently large (greater than 10% of original data set size), go back to step 2. Otherwise, terminate.

The result is a set of clusters. Each cluster is defined by a set of data points from which we derive linear relationships between the variables, as described next.

1006

3.3 Interesting Region Identification by SCA

As shown in Fig. 2, if three lines, L_A, L_B, L_O, in CCS are parallel, their corresponding points in PCS lie on a vertical line AOB (O is the center of the *best circle*). The converse is also true: a vertical line AOB in PCS represents a family of parallel lines in CCS. If the value of x_i is in [a,b], then the family of parallel lines represents the region CDEF. As line AB slides horizontally from point O to the edge of the *best circle*, the width of the region covering the parallel lines decreases gradually to 0. CDEF defines a family of different lines transformed from the vertical lines in the *best circle*.

If d represents the distance between point F and L_O in CCS, $1 \le \frac{d}{r} \le \sqrt{2}$ if $m < 0$ (Chou et. al., 1999). Therefore, CDEF is represented approximately by

$$a \le x_i \le b, \quad (m_i x_i + b_i) - r\sqrt{2(1+m_i^2)} \le x_j \le (m_i x_i + b_i) + r\sqrt{2(1+m_i^2)}. \quad (2)$$

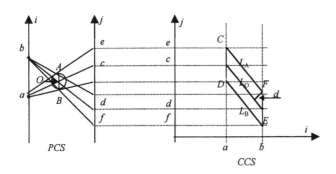

Fig. 2. The feasible region defined by such lines.

To meet with the assumption of $m<0$, the search of *the best circle* in each candidate region is restricted to the region between adjacent axes.

If the input points for SCA are feasible points, then the region CDEF is identified as being feasible. If, for a given optimization problem, several lines are found by SCA at the same time, each of these lines represents a *linear feasible region*. Their combination represents an approximation to the *nonlinear feasible region*.

4 Optimization By GA Using Feasible Region Knowledge

The task in applying a genetic algorithm to a new problem is the definition of the genetic operators, crossover and mutation. In this paper, we focus on the mutation operator. In mutation, a *linear feasible region* is selected randomly from all of the feasible regions identified by the SCA. We then carry out a Gaussian mutation (Goldberg, 1989) for a randomly selected solution, emulating a toin coss for each variable in the genome. For example, in the feasible region defined by Fig. 2, the variable of x_i is mutated within [a,b]; x_j is mutated within the corresponding dynamic feasible region: $[m_i x_i + b_i - r\sqrt{2(1+m_i^2)}, m_i x_i + b_i + r\sqrt{2(1+m_i^2)}]$. If the coin toss for a variable indicates that no mutation should be done, this is over-ridden if the variable is not within the feasible region. For the results presented below, the probability of mutation for individual variables, *pmut*, is 0.5. The total mutation rate (probability of mutation for members of the population) is 0.1. The crossover rate is 0.9. An overlapping population is used.

5 A Case Study

An oil stabilization process (OSP) (McCarthy et al., 2000) is used to demonstrate the new procedures. Fig. 3 shows the flow sheet structure in which, 1, 2, 3 and 4 represent flash vessels, M a mixer, and V a pressure valve. The feed consists of 12 hydrocarbons. There are $n=5$ continuous optimization variables: the flash temperature as a fraction relative to the bubble and dew points for the feed to each flash, x_1, x_2, x_3, x_4, with *initial search region* from 0 to 1, and the target pressure for the valve, x_5, with *initial search*

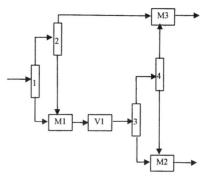

Fig. 3. OSP flowsheet structure.

region of 1 to 30 bar. The objective is to minimize the vapour pressure (Pv), at 310.8 K, of the liquid product such that $Pv<0.817$ bar (the liquid product specification).

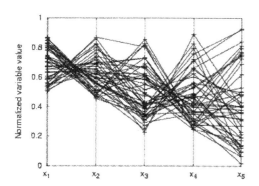

Fig. 4. Visualization of feasible points.

Jacaranda (Fraga et al., 2000) is used both for the generation of the initial data required by the SCA and visualization procedures and for the evaluation of the objective function.

10^n random points are generated initially within the initial search region. The feasible points are shown in Fig. 4 which leads us to define the *reduced search region*: $x_1 \in (0.5, 0.9)$, $x_2 \in (0.4, 0.9)$, $x_3 \in (0.2, 0.9)$, $x_4 \in (0.2, 0.9)$ and $x_5 \in (1, 30)$. SCA is subsequently used to define two *linear search regions*, using $r=0.08$ (Fig. 5).

To evaluate the effectiveness the linear feasible regions, we compare the results obtained by GAs with the mutation operator based on (A) the *initial search region*, (B) the *reduced search region* and (C) the *linear search region*. In all cases, the GA population size is 100 and a convergence termination criterion is used. The GA is run 10 times in each case and the distribution of the results and the standard deviation are shown in Fig. 6. The stopping criterion came into effect after approximately 20 generations in

$x_3 \in (0.27, 0.60)$	$x_3 \in (0.50, 0.80)$
$x_4 = -0.90\, x_3 + 0.94 \pm 0.20$	$x_4 = -0.94\, x_3 + 1.00 \pm 0.20$
$x_1 \in (0.61, 0.71)$	$x_1 \in (0.65, 0.80)$
$x_2 = -0.98\, x_1 + 1.30 \pm 0.20$	$x_2 = -1.08\, x_1 + 1.40 \pm 0.20$
$x'_5 = -2.42\, x_2 + 2.11 \pm 0.37$	$x'_5 = -3.37\, x_2 + 1.95 \pm 0.41$
$x_5 = x'_5 (30-1) +1$	$x_5 = x'_5 (30-1) +1$

Fig. 5. The linear search regions.

1008

all cases. The results indicate that the consistency of the GA is improved through the incorporation of knowledge of the feasible region.

6 Conclusions

A scan circle algorithm has been proposed for identifying interesting regions of feasibility based on parallel co-ordinate system multi-dimensional data visualization. The regions of interest provide some

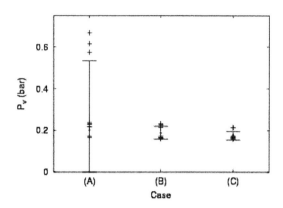

Fig. 6. Distribution of GA results for different cases.

knowledge about the behaviour of the complex, nonlinear constraints and can be used as a basis for the definition of genetic operators for a genetic algorithm. We have shown how the mutation operator may be adapted to exploit such knowledge. One industrial case has demonstrated that the combination of the scan circle and genetic algorithms leads to more effective optimization.

Acknowledgments

Funding provided by the EPSRC and valuable input from BP Amoco are gratefully acknowledged.

References

Caldas L. G. & L.K. Norford (2002), *A Design Optimization Tool Based on A Genetic Algorithm*, Automation In Construction, **11** 173-184.

Chou S.Y., S.W. Lin & C.S. Yeh (1999), *Cluster identification with parallel coordinates*, Pattern Recognition Letters, **20** 565-572.

Fraga, E. S., M. A. Steffens, I. D. L. Bogle & A. K. Hind (2000), *An object oriented framework for process synthesis and optimization*, in "Foundations of Computer-Aided Process Design," M. F. Malone, J. A. Trainham, & B. Carnahan (Editors), AIChE Symposium Series **323**(96) 446-449.

Goldberg, D.E. (1989), *Genetic algorithms in search optimization and machine Learning*, Addison Wesley, Reading.

Inselberg A. & B. Dimsdale (1990), *Parallel Coordinates: A tool for visualizing multidimensional geometry*, in Proceedings of the First IEEE Conference on Visualization, 361-378.

Keim D.A. (1997), *Visual database exploration Techniques*, in Proceedings of Tutorial Int. Conf. On Knowledge Discovery and Data Mining (KDD'97), Newport Beach, CA, http://www.informatik.uni-halle.de/~keim/PS/KDD97.pdf.

Lee H.Y. & H.L. Ong (1996). *Visualization support for data mining*, IEEE Expert **11**(5) 69-75.

McCarthy E. C., E. S. Fraga & J. W. Ponton (1998), *An automated procedure for multicomponent product synthesis*, Computers chem. Engng **22**(Suppl.) 877-884.

European Symposium on Computer Aided Process Engineering – 12
J. Grievink and J. van Schijndel (Editors)
© 2002 Elsevier Science B.V. All rights reserved.

Application of CFD on a Catalytic Rotating Basket Reactor

Warna J.[1], Ronnholm M.[1], Salmi T.[1], Keikko K.[2],

[1]Laboratory of Industrial Chemistry, Process Chemistry Group, Abo Akademi University,
Biskopsgatan 8, 20500 Turku Finland, Fax: +358-2-2154479, e-mail:
johan.warna@abo.fi
[2]Kemira Chemicals, Box 171, 90101 Oulu Finland

Keywords Catalytic Rotating Basket Reactor, CFD

Abstract

Using catalytic rotating basket reactors is often preferred instead of catalytic slurry reactors. This is especially true in the case of testing a specified catalyst particle size as the catalyst particles remain the same size during the experiment and they are not crushed as much as in a slurry reactor. However, in a rotating basket reactor, the liquid-solid mass transfer rate is poorer than in conventional slurry reactors

The use of rotating catalyst basket reactors in experimental work was introduced by Carberry (Carberry 1964). Different designs for spinning basket reactors for two- and three phase reactions have been presented. Turek and Winter (Turek and Winter 1990) made a comparison between two types of spinning basket reactors to determine the hydrogenation rate of butynediole with a nickel catalyst. Kawakami et. al. (Kawakami et. al.1976) studied a semibatch hydrogenation reaction of styrene in a basket reactor.

The goal with the different designs of rotating catalyst basket reactors is to achieve a sufficiently high flow rate of the fluid over the catalyst particles. To build and test different reactor designs is costly and time consuming. With the use of modern CFD tools it is possible to investigate different reactor designs with a computer model and finally build the reactor that shows the best performance.

In order to enhance the reactor performance, CFD calculations were used to estimate the liquid-solid mass transfer rates inside the rotating catalyst basket. Different catalyst basket designs were screened. The reactor design that showed the best performance was built and it was tested with catalytic esterification.

1. Rotating basket reactor

The reactor consists of the reactor vessel and a rotating catalyst basket connected to the shaft. The catalyst basket is made of a metal net. The size of the rotating basket reactor was 100 mm in diameter and 150 mm in height. The liquid amount loaded into the reactor was 0.5 l and the catalyst amount was 100 g. The rotating basket had four paddles on the side of the basket and four paddles at the bottom of the basket; the bottom and side paddles were tilted 45° (Fig. 1). The basket was rotating in anti-clockwise direction with a rotation speed of 320 rpm. The reactor operates in batch mode.

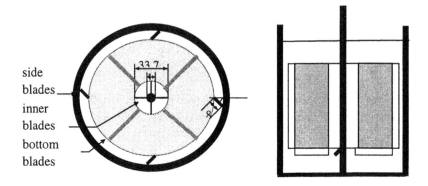

Figure 1. Top and side view of reactor 2.

Two reactors were used in this work. Reactor 1 was the original design of the laboratory test reactor, with no paddles in the middle at the rotation axis, but instead this space was filled with catalyst particles. Reactor 2 was constructed according to the CFD calculations that showed better liquid-solid mass transfer rates, as paddles were inserted in the middle of the reactor as shown in Fig 1.

2. Liquid-solid mass transfer

In order to compare the performance of the different reactor designs the liquid-solid mass transfer rates in the catalyst area were calculated. To estimate the limitations set by liquid-solid mass transfer, it is appropriate to use correlations for k_s of the form (Beenackers, 1993):

$$Sh = \frac{d_p k_s}{D} = 2 + a\,Re^{\alpha}\,Sc^{\beta} \tag{1}$$

where Sh, Re and Sc denote the Sherwood, Reynolds and Schmidt numbers, respectively. The diffusion coefficients were calculated from the Wilke-Chang equation. (Reid et al, 1988) The mass transfer coefficient k_s was obtained from eq. (1), Reynolds number, however, must be calculated with the actual velocity v' that the liquid passes the catalyst particles. The corrected velocity can be calculated from :

$$v' = \sqrt{v_x^2 + v_r^2 + \left(v_{rot} - v_{\omega}\right)^2} \tag{2}$$

where v_{rot} is the rotation velocity of the basket in the calculation point and v_x, v_r and v_{ω} are the velocities in the 3D coordinate system. The mass transfer rates k_s are shown in Fig (2), and it can be seen that the mass transfer rates are higher in the catalyst area in reactor 2.

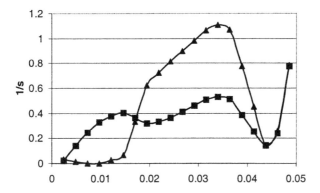

Figure 2. Mass transfer rates in the catalyst basket as a function of radial position. Reactor 1 (squares), reactor 2 (triangles)

3. Pressure drop and physical properties

The pressure drop in the catalyst area was calculated with a correlation equation. The friction factor was estimated from (Fried, 1989).

$$f = 1.5 \cdot \frac{2.2\left(\dfrac{64}{Re} + \dfrac{1.8}{Re^{0.1}}\right)(1 - \varepsilon)}{\varepsilon^3} \tag{3}$$

The flow resistance in porous areas (catalyst area) is accounted for with body forces in the CFD program code. The body force is obtained from:

$$B = \frac{f\rho v}{2d_p} \tag{4}$$

The body force caused by pressure drop is added to other body forces, i.e. centrifugal forces caused by rotation. The total body forces are included in the momentum transfer equation of the CFD program.

The influence of the viscosity and density on the flow field was investigated by running simulations with viscosity and density values in the range 0.3 10^{-3} Ns/m^2 - 2.22 10^{-3} Ns/m^2, 0.66 kg/dm^3 –1.0 kg/dm^3, respectively. In this viscosity and density range only very small differences in the flow fields were observed.

4. CFD model

The CFD calculations were performed with the software CFX-4.4. The turbulent k-ε method was applied in the area outside the rotating catalyst basket, while laminar and porous flow methods were used to describe the flow in the catalyst area. Reynolds value in the catalyst area was in the range 100-1500. The flow profiles in the reactor were modelled using a 90° sector (Fig 3) of the cylindrical reactor and assuming a four-fold flow periodicity. The rotation of the catalyst basket was calculated with both the sliding grid and the multiple frames of reference (MFR) methods (AEA Technology, 2001). The MFR method was found to be much more rapid than the sliding grid method. The number of calculation elements in the 90° sector was 51450.

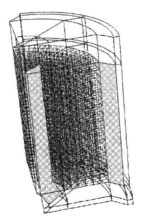

Figure 3 Calculation domain

Flow fields in a plane in the middle of the reactor show that the flow direction in reactor 2 is from the centre of the reactor through the catalyst towards the walls. In reactor 1 the flow direction for the liquid is mainly the same as the rotation direction as shown in Fig 4

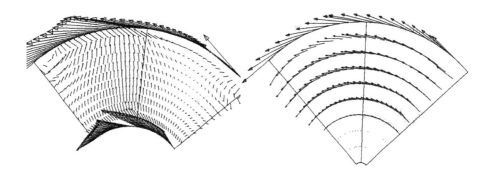

Figure 4 Flow directions in reactor 2 (left) and reactor 1 (right) in the centre of the reactor.

5. Chemical reaction

Chemical kinetics was also implemented in the CFD model. It showed out that, predicting the flow fields, catalyst basket rotation and the chemical reactions simultaneously required a lot of calculation time. Instead the flow profiles were firstly predicted in the absence of chemical reactions. Chemical reactions were then added and the obtained flow field was kept constant during the chemical reaction simulation. In this way, the calculation times became reasonable.

A catalytic reversible reaction: esterification of acetic acid (Mäki-Arvela et. al. 1999) over Amberlyst 15 catalyst, showed improved performance with the reactor 2 design shown in Fig. 6. By including the reaction rates in the CFD program the reaction in the reactor could also be simulated with the CFD program. The simulations showed a good agreement with experimental data presented in Fig. 5. Reaction zones could also be studied from the CFD calculations.

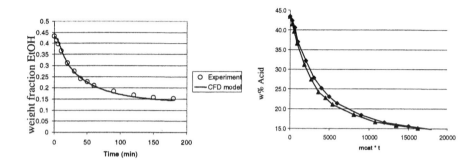

Figure 5 CFD calculation versus experimental data. Weight fraction as a function of time.

*Figure 6 Experiment with reactor 1 (lower line) and reactor 2 (upper line). Weight % as a function of mass of catalyst * time.*

6. Conclusions

A rotating catalytic basket reactor was modelled with CFD in order to improve the performance of the reactor. The CFD calculations showed that liquid-solid mass transfer rates could be improved by a change of the design of the rotating catalyst basket. The new and old reactor designs were compared with experiments with a catalytic esterification reaction and it was shown that the new reactor design gave an improved performance.

Acknowledgements

This work is part of the activities at the Åbo Akademi Process Chemistry Group within the Finnish Centre of Excellence Programme (2000-2005) by the Academy of Finland.

The financial support from TEKES (Technology Development Centre) is gratefully acknowledged.

Symbols

a_p	Mass transfer area / volume	m^{-1}
B	Body force	kg/m^3
d_p	Catalyst particle diameter	m
D	Diffusion coefficient	m^2/s
f	Friction factor (Eq. 3)	m^{-1}
k_s	Liquid-solid mass transfer coefficient	m/s
$k_s a_p$	Mass transfer coefficient, $k_s a_p = k_s 6/d_p$	s^{-1}
T	Temperature	K
v	Velocity	m/s

Greek letters
ε   Porosity
ρ   Density kg/m^3

Subscripts
x, r, ω   Indicating flow direction in axial, radial and rotation direction

References

Carberry, J. J. Designing Laboratory Catalytic Reactors. Ind. Eng. Chem. 1964, 56, 39

Turek F., Winter H., Effectiveness Factor in a Three-Phase Spinning Basket Reactor: Hydrogenation of Butynediol. Ind. Eng. Chem. Res., Vol. 29, No. 7,pp 1546-1549, 1990

Kawakami K., Ura S., Kusonki K., The Effectiveness Factor of a Catalyst Pellet in the Liquid-Phase Hydrogenation of Styrene. Chem. Eng. Jpn. 9, 392, 1976

Beenackers, A. A. C. M. & van Swaaij W. P. M., Mass Transfer in Gas-Liquid Slurry Reactors, Chem. Eng. Sci. Vol. 48, pp. 3109-3193, 1993

Reid, R. C., Prausnitz, J. M., Poling, B. E., , The Properties of Gases and Liquids, 4th Ed., McGraw-Hill Book Company, New York, 1988.

Fried, E. & Idelchik, I., Flow Resistance: A Design Guide For Engineers. Hemisphere Publishing Corporation., 1989

CFX-4.4: Solver, AEA Technology, 2001

Mäki-Arvela, P., Salmi, T., Sundell, M., Ekman, K., Peltonen, R., Lehtonen, J., Comparison of polyvinylbenzene an polyolefin supported sulphonic acid catalyst in the esterification of acetic acid, Applied Catalysis A: General 184 (1999) 25-32

Baldyga, J., & Bourne J., R,., Turbulent Mixing and Chemical Reactions, Wiley, 1999

Brucato, A. et. al. Numerical prediction of flow fields in baffled stirred vessels: A comparison of alternative modelling approaches, Chem. Eng. Sci. Vol 53, No. 21, pp. 3653-3684, 1998

Harris , C. K., Roeckaerts, D., Rosendal, F. J. J., Computational Fluid Dynamics for Chemical Reactor Engineering, Chem. Eng. Sci. Vol. 51, pp. 1569-1594, 1996

Kluytmans, J. H. J., Markusse A. P., Kuster B. F. M., Marin G. B., and Schouten J. C., Engineering Aspects of the Aqueous Noble Metal Catalysed Alcohol Oxidation, Catalysis Today, vol. 57, pp. 143-155, 2000

European Symposium on Computer Aided Process Engineering – 12
J. Grievink and J. van Schijndel (Editors)
© 2002 Elsevier Science B.V. All rights reserved.

A post-graduate study in Process Design: An Innovative Model in the Netherlands

Johan Grievink, Giljam Bierman, Cees P. Luteijn and Peter J.T. Verheijen
Faculty of Applied Sciences, Department of Chemical Technology
Delft University of Technology, 2600 GA, Delft, The Netherlands

Abstract

The design of sustainable, cost effective processes, with perfect control of product quality, calls for strongly enhanced capabilities of designers. A thorough training in both fundamentals of design methods and CAPE tools as well as in the associated work processes is required to become an expert designer. Manufacturing and engineering companies in the Netherlands have indicated that a more advanced training in design is needed to meet the upper range of design capabilities. The Dutch chemical engineering departments responded by developing specialised two-year postgraduate studies in process systems design. As an example the structure and contents of a post-graduate study in process design at Delft University of Technology is highlighted and evaluated.

1. Introduction

This paper describes and evaluates a higher level of engineering education in process design, as developed and practised by the chemical engineering departments of the technical universities in The Netherlands. It is a relatively new branch of the educational model for the engineering sciences with a clear focus on the integrative features of design. Furthermore, it offers a higher level of expertise to start a career in chemical engineering. This educational innovation must be placed against the background of the conventional academic engineering education on the European continent.

About a fifteen years ago the manufacturing industry in the Netherlands pointed out that in the future a higher level of expertise in design of complex technological systems was needed, for reasons outlined in section 2. The technical universities have responded by developing post-graduate studies in the design of technological artefacts, as a follow-up to the graduate MSc level. The objectives of these two-year, full time studies are reviewed in section 3. Taking the post-graduate study developed by the Chemical Technology Department of Delft University of Technology as an example, the specifics of this study are explained with particular attention to CAPE elements (section 4). The experiences with this study are predominantly positive (section 5). Yet, the rapid pace of change in CAPE technology in particular and the process industry in general calls for continued developments in order to remain effective (section 6).

2. Increased design capabilities: needs and means

2.1 Wider horizons in the Process Industry

The drive for a sustainable society and for products with a high customer satisfaction with respect to cost and quality challenges scientists and engineers to think and act over much wider horizons than before. For example, the design and operation of chemical processes must integrate at least four different perspectives.

- *The product life span*

A chemical product (e.g. a polymer) passes through three processes during its life span: (1) manufacturing, (2) application and (3) recycling or recovery (of spent product). The design of each of these processes must acknowledge the functional requirements of the other two processes and contribute to an optimisation over the entire product life span.

- *Integration in a supply chain*

Many plants are imbedded in a supply chain on a site or within an enterprise. The main product is often an intermediate, to serve as a feed for other plants or to be sold as a finished product. The supply chain will exert its influence in the design and operation of such a plant by imposing additional constraints and economic trade-offs along the chain.

- *Optimisation over process life span*

The physical life of a plant can be much longer than the economic life span of a product (grade), implying that the plant must be flexible enough to handle different product grades. Furthermore, minimisation of life cycle costs forces one to strike a balance between investments in equipment and the resulting contributions to operational costs.

- *Integration over multiple scales within and outside a process.*

The concept of a process as just a connection of unit operations is not sufficient anymore. At the upper scale (in time, space) integration between a process and the site (utilities, supply chain) is required, in both design and operations. However, it is at the lower scales that new opportunities arise for better control of product properties and for process innovations. Especially, processes for the manufacture of products with an internal spatial structure (e.g. polymers, crystals, emulsions, and catalysts) demand a multi-scale approach to the simultaneous designs of product and process.

In summary, the increasing interactions between a process and its business environment and the needs for a more finely tuned internal process structure call for enhanced design capabilities to integrate the multi-scale features.

2.2 CAPE related innovations

This growth in demands from industrial practice has stimulated new developments in process engineering. In addition, the autonomous developments in computing and information technology have given rise to a wide array of new methods and tools for process design, control and optimisation of operations. Grossmann and Westerberg (2000) present a list of some major accomplishments in the PSE/CAPE area. However, the level of abstraction of many methods and the complexity of the computational schemes of these tools are high. To become a reliable, expert user of these tools one requires a thorough training in the underlying fundamentals. The graduate education in chemical engineering (to the MSc level) does not offer enough room for such training.

3. Innovation in design education at Dutch universities.

3.1 Structure of university education in design

In late eighties the Netherlands industry were concerned about graduate education in engineering not being able to meet the upper range of the future demands for design capability. The main reason being the existing academic focus on developing capabilities for research rather than design. The education gives high priority to acquiring fundamental knowledge in the pertinent domains of natural sciences and engineering and to developing strong analytical skills. Although room is made for an introduction to design, it is not enough to confidently deal with the design complexity of modern technological systems. A doctorate in engineering research is not fully adequate for design either, although research on design methods and tools creates an excellent fundamental understanding in general. But a doctorate typically leads to highly trained specialists who can go in great detail, rather than pursuing the broader, integrative approach as needed for design. Furthermore, the organisational and social aspects of the work processes in design often fall beyond the scope of recognised research.

To close this gap the technical universities established a specialised post-graduate study in design of complex technological systems. It is a two-year, full time study, starting from a MSc level in engineering. This study is a generic complement to a Doctorate in engineering. The study is (partially) funded by the government, who gave equal status to the design and research students. The completion of an post-graduate design study is rewarded with a certified degree 'Master in Technological Design'. The design objects can be technical products, processes and systems for control and logistics. The study has a focus on the integrative aspects of systems engineering. *The key objective is to learn and master the design process, including its associated methods, tools and procedures.*

3.2 Objectives

The objectives of the advanced design study can be further specified by looking at aspects of the designer's conceptual environment and their mental outfit.

- *Interface between a designer and the business and society in general.*

To be capable of developing a consistent Basis of Design, including different performance criteria of economic, ecological and technological nature.

- *Design methods, tools and technical procedures.*

To get a thorough understanding and mastery of the technical aspects of design with a focus on the conceptual phase.

- *Work processes and project organisation.*

To understand and experiment with the social, organisational and management aspects of the work processes.

- *Knowledge of relevant disciplines of science and engineering.*

To extend and deepen knowledge of a relevant mix of disciplines.

- *Individual characteristics.*

To understand and train individual characteristics (creativity, social and management).

4. Post-graduate chemical process design in Delft

4.1 Entrance level

The preferred entrance level of expertise in design can be characterised by briefly reviewing the graduate education in Chemical Technology at Delft University as a typical example. Representative process engineering subjects in the core curriculum are:

- Chemical process technology
- Risk analysis and management;
- Process dynamics and control;
- Computer laboratory in process simulations;
- Process systems design (based on Douglas, 1988) or, chemical product design.

The integration and application of the acquired knowledge takes place at the end of the fourth year by means of the:

- Conceptual Process Design project   (12 weeks effort in a team of four students)

The courses and projects in process design are accredited by the British Institution of Chemical Engineering and ABET. While at this stage of education the students obtain the experience of making a design of a (conventional) continuous process with simple products (e.g. base chemicals), they do not yet master the design process in general. Students admitted to the post-graduate study end coming from other universities have a similar profile of initial design expertise.

4.2 Post-graduate study in Process and Equipment design

The post-graduate study aims for higher levels of abstraction, understanding and creativity, i.e. *to master the design process*. The focus remains on the conceptual design stage. A post-graduate design student learns:

(a) *To think in life span and supply chain dimensions (as outlined in section 2.1);*

(b) *To master methods and CAPE tools for the design and integration of a process, involving synthesis, analysis, evaluation and optimisation steps for the process and its control;*

(c) *Deal with the non-technical elements of the design process.*

These latter two points reflect the generic objectives outlines in section 3.2.

A more complete overview of the study and its organisation can be found at the web-site: http://www.dct.tudelft.nl, looking under Delft Ingenious Design.

4.3 Post-graduate study: first year courses and design project

The major part of the first year (28 weeks) is spent on knowledge expansion, both on engineering and non-technical topics. This expansion will be tailored to the needs of the individual student, depending on the profile acquired in the preceding education. The non-technical topics involve, among others, project organisation, economic evaluations in the process industry, presentation and writing skills. Some CAPE topics are:

- Advanced process design   (e.g. Biegler et.al., 1997; Seider et.al. 1999)
- Process & heat integration   (e.g. Smith, 1995)
- Loss prevention and sustainable technology.
- Thermodynamics for designers
- Process modelling and model analysis (e.g. Hangos & Cameron, 2001)

- Design of plant wide control   (e.g. Luyben et.al., 1998)
- Optimisation of Chemical Processes   (e.g. Edgar, et.al., 2001);

All teaching is in English, using international text books. For each course the students are also trained in using the associated CAPE tools, e.g. process dynamic simulators, heat integration design, control design. In addition to these topics students can take courses in catalysis and reactor engineering, and in separation technology.

The remainder of the first year (15 weeks effort) is devoted to a group design project, in which project management and teamwork can be practiced. A manufacturing or an engineering contracting firm usually commissions the design problem; e.g. design of a Gas-to-Liquid plant. The basis of design has to be extracted from the problem owners in a negotiation process. Design alternatives are to be generated, evaluated and compared. In addition to writing a design report, the results must be orally presented and defended to some experienced process designers of the commissioning firm.

4.4 Post-graduate study: second year design project

The full second year is spent on an individual design project (42 weeks) in an industrial setting on a contract basis. Here, the challenge is to demarcate a tractable design problem in a real life environment and solve it adequately in time and with limited, available means. The design is often combined with some experimental process development work to generate data for design or with optimisation of plant operations. This project can be carried out on the site of the commissioning firm or in a university laboratory. The design student is responsible for managing his own project. Consultants to the project advise the student, which will be academic staff and a process engineer of the commissioning firm. Some examples of recent design projects are:

- Degassing process for plastics;
- New process for IPA production;
- ETBE recovery using permeative distillation;
- Debottlenecking of a desulphurisation unit;
- Processes for food products (e.g., Yadhav et al., 2002).

5. Experiences and evaluation

5.1 Recruitment of students

Since the start in 1991 the number of applicants have exceeded the number of available places for many years. However, the number of admitted students with suitable qualifications was lower than the maximum capacity (~15 students per year). This screening for admittance has kept the attrition rate below the ten-percent. Since 1998 it is difficult to attract enough qualified students from the Netherlands:

- Low number of chemical engineering students (~ 60 % drop in 5 years);
- The short term prospect of making more money in industry as junior engineer;
- Concerns about international recognition of this type of study, since it deviates from the international education pattern BSc, MSc and PhD.

Yet, recruiting on an international scale appears to be effective; most design students are now from abroad. Although this study also offers a nice opportunity for process design engineers to take a break from an industrial job and upgrade their knowledge, this has hardly happened so far. It is an option to be developed for the future.

5.2 Starting positions of designers

About 40 % ends up in the process industry, 30 % with engineering-contracting firms, 7 % goes to consultancy firms and 7 % goes to work for (semi-) governmental institutes of technology. Less than 7 % stays on at a university (for a PhD degree). Industry recognises the added value of these designers by the fact that they are found to be effective almost immediately after their start. This is reflected in starting salaries and rate of rise.

5.3 Quality control with industrial feed back

Every five years a committee of the National Accreditation Board reviews the quality of the post-graduate design studies. Each committee includes design experts from the industry. For the Delft course these experts have given recommendations with respect to its contents; i.e. putting more emphasis on process dynamics and plant wide control design and on project management skills. The appreciation for these design study is also reflected by companies volunteering to offer design projects.

6. Future developments

This post-graduate design study has proven effective from an educational point of view. The process industry recognise the advanced skills of these designers. Attracting enough talented students and finding more international recognition remain challenges. The technical contents of the courses will be adapted to anticipate the developments in the process industry. More attention need to be given to processes for structured products and for multiple modes of production. Last but not least, the advances in computer aided product and process synthesis need to be captured in the teaching.

7. References

Biegler, L.T., I.E.Grossman, A.W.Westerberg, 1997, *Systematic methods of process design*, Prentice Hall

Douglas, J.M., 1988, *Conceptual Design of Chemical Processes*, McGraw-Hill.

Edgar, Th.F., D.M.Himmelblau, L.S.Lasdon, 2001, *Optimization of chemical processes,* McGraw-Hill

Grossmann, I.E., A.W.Westerberg, 2000, AIChE J., p. 1700, *Research challenges in process systems engineering*, Vol. 46.

Hangos, K., I.Cameron, 2001, *Process modelling and model analysis*, PSE Volume 4, Academic Press.

Yadhav, N., M.L.M.vander Stappen, R.Boom, G.Bierman and J.Grievink, 2002, *Conceptual design of processes for structured food products*, submitted to ESCAPE-12.

Luyben, W.L., B.D.Tyreus, M.L.Luyben, 1998, *Plantwide process control*, McGraw-Hill

Seider, W.D., J.D.Seader, D.R.Lewin, 1999, *Process design principles. Synthesis, analysis, and evaluation*, Wiley & Sons, Inc..

Smith, R., 1995, *Chemical process design*, McGraw-Hill.

European Symposium on Computer Aided Process Engineering – 12
J. Grievink and J. van Schijndel (Editors)
© 2002 Elsevier Science B.V. All rights reserved.

Preconceptions and Typical Problems of Teaching with Process Simulators

Laureano Jiménez and René Bañares-Alcántara

Chem. Eng. Dept, University Rovira i Virgili, Av. Països Catalans 26, 43007 Tarragona, Spain. Tel.: +34-977-559617; Fax: x9667/21; e-mail: ljimenez, rbanares@etseq.urv.es

Abstract

The basic input to any process simulator is the result of a set of decisions. For this reason, it is convenient to provide students with a *road map* to assist them in the solution of the typical dilemmas. In particular, continuous effort to foster off-line work (conceptual design, process synthesis) is required. Overall, the use of accurate physical properties and a correct selection of models for property estimation are key factors to succeed in process modelling, although they are frequently neglected.

1. What is Process Modelling?

Commercially available process simulators (HYSYS®, ASPENPLUS®, CHEMCAD®, PRO-II®...) help to solve, once all degrees of freedom are fixed, the mass and energy balance of a process with rigorous physical property calculations, kinetic considerations and detailed unit operation models. The simulator calculates the flowrate and compositions of the process streams, size/rates equipment, predict the operation variables and the dynamic behaviour.

The use of process simulators should result in the increase of the engineer's efficiency, and improve the opportunities to discern between configurations, to optimise variables and to explore the dynamic behaviour of the process. Traditionally, the faculty staff stress the emphasis in the design of individual units, perceived as the *core* characteristics of the design, while the considered *ancillary* issues (such as conceptual design and process synthesis) are left aside, despite the fact that the capital costs are fixed when the P&ID is established, prior to the final modelling.

2. Methodological Aspects

Computer aided process design plays a critical role in bridging the gap between theory and practice, as students face practical problems (operability, dynamic behaviour) and connect scatter units as a whole (separation units, reactors, and controllers). Technical competence is no longer sufficient to fulfil these objectives if it is not combined with *soft* skills (decision-making, problem solving, teamwork, communication and management abilities). To increase the retention and improve the integration of concepts, the design is organised as a '*stop, think and go*' procedure. This frame forces students to analyse, interpret and extract information to establish or re-direct decisions.

The traditional teacher and student roles change. Students assume increasing responsibility thus enhancing motivation. In turn, the faculty staff role is to guide students to prevent misconceptions, rather than to transmit formal knowledge to passive students (they freely operate the simulator without interference unless they fall in a dead-end situation). As the course advances, Socratic questions are posed to enhance critical thinking. In this way self-confidence is improved and autonomy is encouraged.

3. Decision-Making

The process simulator should not be used as a *black box*, because it requires a sequence of *decisions:* selection of a thermodynamic model, setting of the flowsheet, fixing the operating conditions, sizing and rating equipment or setting the control strategy. Conflict resolution has to be used, as often several criteria point in irreconcilable directions and their relative importance has to be balanced. In addition, it is not clear when/if real behaviour has to be introduced (e.g. subcooling, efficiency, leakage or entrainment).

For a beginner, a *too* agile mouse operation may foster *uncertainty* (students are used to solve close-ended problems and they are suddenly required to solve ill-defined problems), *insecurity* (students have to surf over an avalanche of results of varying relevance) and *helplessness* (students may choke with the huge amount of values). As a general rule, we can state that each mouse click is equivalent to a decision that must be supported by arguments.

4. What Does a Physical Property Model Imply?

Students perceive the physical property package selection as a *wild* territory. We should transmit students the idea that *absolutely* all properties (for pure components and mixtures) are computed based on these predictions. For this reason we suggest to validate each model in a two level approach, with respect to physical properties (pure component, VLE, LLE) and to plant data. In addition, and depending on the unit operation to model, the stress is switched to different properties (distillation: VLE; extraction: LLE; pumping: viscosity and density; reaction: kinetics and enthalpy; heat exchangers: specific heat and latent heat, etc.).

A list of the typical issues to take into account follows:

- *How to model the vapour phase?* Systems with volatile organic acids (e.g. acetic acid), aldehydes or HCN the Hayden-O'Connell or Nothnagel models must be selected due to vapour phase association. HF requires a specific model due to the formation of hexamers.
- *Number of phases.* Computing efficiency and convergence are improved if the presence of two/three phases is provided a priori.
- *Are there components at trace level?* In this case, special emphasis must be taken in the validation. In addition, tolerance can be relaxed to improve convergence.
- *What to do with supercritical components?* If supercritical gases are present, Henry's law in conjunction with an activity coefficient model should be used to calculate the solubility. Most equations of state can predict this behaviour without further consideration.
- *Modelling oil.* Petroleum and refinery products are systems with hundreds of non-polar components and thus it is convenient to estimate their properties in

terms of pseudo-components. The critical step is the oil characterisation from the assay data curves (TBP and ASTM distillation, density, viscosity, sulphur content), and it is imperative to focus on the discretisation methods.

- *Presence of electrolytes.* Ions do not participate directly in the phase equilibrium, but influence the activity coefficients of the other species because of their possibility to combine and precipitate. Typical electrolyte models are Pitzer (P<10 atm and $c_i < 6M$), NRTL-modified and Bromley-Pitzer (predictive).

- *Simulation with solids.* Each industrial sector describes the solids with specific properties (e.g. length of the cellulose fiber in the paper industry). Models are based on these attributes to characterise the solid and estimate any other property. Transport properties are important, as solid products require both the chemical composition and particle size distribution to be defined.

- *The component is not in the database!* Different alternatives are possible (e.g. use predictive models or simplify the mixture) depending on the importance of the missing component in the system. If a predictive model is used, it is strongly recommended to provide the simulator at least with accurate TB, MW and critical constants since all models are very sensitive to these parameters.

- *What can I do if there is a lack of properties?* We can not assume that all parameters are available just because no error message appeared. If additional properties are needed it is recommended to: a) search in databases revised by editors (DIPPR, DECHEMA); b) search in non-revised databanks (Reid, Perry, CRC); c) obtain experimental data, or d) use predictive models (Carlson, 1996).

4.1. Criteria to select physical property estimation methods

Due to the great casuistry (some methods are not recommended with certain components, while others are incompatible), selecting a method is a complex task (Ballinger et al, 1994; Carlson, 1996; Agarwall et at, 2001; Satyro et al, 2001). Methods can be clustered in equations of state and activity coefficient models, their applicability is compared in Table 1.

The commercial process simulators provide a vague selection procedure (HYSYS®, ASPENPLUS®, Pro-II®). Only CHEMCAD® has a selection algorithm implemented within the software. However there are several applications (5, 6) that can assist in these tasks:

1. Select the components and requests information about their importance.
2. Search for the nature of the properties of interest.
3. Assess the range of pressure, temperature and composition.
4. Check the available parameters and perform a model discrimination procedure. The parameters that are not available are estimated by a purely predictive method, e.g. group contribution methods.

4.2. When are purely predictive models recommended?

The purely predictive models are the *unique* option when qualitative results are sufficient or when no data is available. These models are based on the structural information of the components, and the interactions between functional groups are computed based on average values. They should be used with caution and their predictions must be validated.

Table 1. - Comparison between equations of state and activity coefficient models.

Equations of state	Activity coefficient models
Limited in ability in representing non-ideal liquids (polar components)	Valid for highly non-ideal mixtures (P<10 atm). If P > 10 atm use purely predictive models
Few binary parameters required	Many binary parameters required
Parameters extrapolate reasonably with temperature	Binary parameters are highly temperature dependent
Consistent in the critical region	Inconsistent in the critical region

4.3. Models for special systems

Some systems (e.g. absorption of acidic gases with amines, water and steam) present deviations so particular that it has been necessary develop specific models valid in a certain range of pressure, temperature and composition.

4.4. Why select different thermodynamic models in a case?

The operation of the chemical plants can be divided in different sections (e.g. reaction, separation and utilities), each one with certain particularities. Accordingly, it is frequent to use different thermodynamic models in each subsection, due to the disparity of objectives, components and operating conditions.

5. Convergence

For vendor companies the mathematical and thermodynamic methods are public, and the access to databases is obtained through commercial agreements. Thereby, one of the key factors in simulation is the *robustness* of the calculation methods. Some aspects to observe in order to improve convergence are to verify the information introduced, check the number of phases, provide estimated values, partition and stream tearing or modify the integration interval in a dynamic simulation. A common problem is to request for an unfeasible separation (e.g. disregard an azeotrope in a distillation column), usually pointing the finger to a convergence problem. These situations fosters the importance of the off-line work.

6. Preconceptions

It is common to start modelling just by sitting in front of a computer, without previously applying any systematic off-line work (e.g. search for information, planning and conceptual design). In this situation, students may feel that approaching process simulation relies purely on experience or is a question of luck, and a *trial and error* approach is reinforced. On the contrary, it is necessary to use a systematic procedure to detect preconceptions (P&ID ≠ PFD, recycle ≠ tear stream, unit operation ≠ icon).

Students often fail in their first approximation to process simulation because their notion of degrees of freedom is not clear, and they fail to conclude that the simulator requires *exactly* the same information than with the '*by hand*' approach. For instance, students have problems in identifying why each type of reactor (e.g. PFR, equilibrium) only accepts certain type of reactions (e.g kinetics, equilibrium). Counterexamples are used to help students to detect and correct erroneous preconceptions, in particular the implication of the flow-pressure relationship in dynamic modelling and that if a case had converged that means the results are feasible. In addition, the simulator is perceived as a labyrinth and a *road map* is provided (Figure 1).

6.1. Good practices in process simulation
- Build the PFD in several phases, make modifications step by step and approach problems by first isolating them.
- Classify components in high, medium and low priority; explore the properties prediction of pairs of high/high priority components and their combinations. Estimate the non-available parameters.
- Divide the process and product specifications into fixed, flexible and modifiable.
- Discern among different process alternatives (sensitivity analysis, optimisation).

7. Size Matters!

All units must be rated according to technical criteria (velocity for a pipe, flooding percentage for the column diameter, volume for a CSTR/PFR). Many problems that appear during dynamic simulation are due to incorrect sizing (flooding, reverse flow, uncontrollable valves).

8. Control Strategy

There is a growing recognition of the need to consider the controllability of a chemical process during its design. Setting the appropriate control strategy according to each objective (P, PI, PID, cascade, ratio) is performed off-line. It is a common mistake to consider all control schemes with the same importance (e.g. the reflux flowrate is controlled by the accumulator liquid level, thus indicating a lack of understanding of distillation; or reactor feeds that are controlled independently, without any ratio consideration).

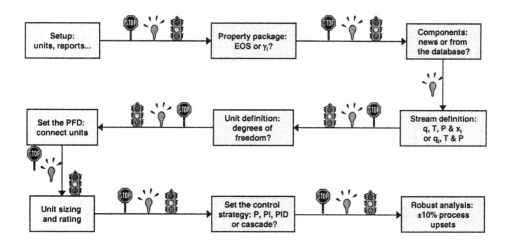

Figure 1. Road map for process simulation.

The controller parameters can be tuned with different models. The auto tuning value method has a good performance for simple controllers (level, pressure), while for complex configurations the open or closed Ziegler-Nichols strategy is commonly used.

9. Steady State is Not Enough

Anomalous, ambiguous and/or contradictory situations in steady state are usually detected during dynamic simulation, such as multiple stationary states (reactive distillation), or reverse flow. Also, dynamic analyses expose the need for extra equipment (for start-up or shut-down), exhibit typical asymmetric behaviour and help to select appropriate control variables. Among all the advantages of dynamic simulation, the main one is the possibility to analyse the system robustness for typical process upsets (±10%). In addition, the model can check if upstream or downstream units bottleneck the process.

10. The Analysis of Results

It is fundamental to foster in students the use of *graphical* analysis techniques (trends, behaviours and profiles). In this way, results are found more quickly, and in a fashion that may allow small problems to slip-in but big mistakes to be detected. In particular, beginners are treading a thin line separating fact from fiction, where the reality is the interpretation of the simulation results, based on knowledge and rigor.

11. Conclusions

Process simulation is not an objective in itself, but a tool that avoids the most *repetitive*, *routine* and *heavy tasks*. Its main problems in education are that the necessary computer requirements are expensive and that it is difficult to dedicate enough teaching effort for groups with many students. The use of a simulator is not exempt of risks, since there is the possibility to act as a video game player, without knowing the underlying assumptions, applying the necessary criteria and without carrying out critical analysis of the results (reality is the interpretation of the simulation results). To achieve good results in process simulation it is necessary to foster off-line work and emphasise the appropriate selection of physical property models.

12. References

Agarwall, R., Y.-K. Li, O. Santollani, M. A. Satyro and A. Vieler, 2001, Chem. Eng. Prog. 97(5), 42.
Ballinger, G.H., R. Bañares-Alcántara, D. Costello, E.S. Fraga, J. Krabbe, H. Labadibi, D.M. Laing, R.C. McKinnel, J.W. Ponton, N. Skilling and M.W. Spencely, 1994, Comp. Chem. Eng. 18.
Carlson, E.C., 1996, Chem. Eng. Prog. 10, 35.
http://www.ip-sol.com/TMS.htm. Accessed on Feb 2002.
http://www.virtualmaterials.com/products.html. Accessed on Feb 2002.
Satyro, M. A., R. Agarwall, Y.-K. Li, O. Santollani and A. Vieler, 2001, Chem. Eng. Prog. 97(6), 64.

European Symposium on Computer Aided Process Engineering – 12
J. Grievink and J. van Schijndel (Editors)

A Novel Course on Integrated Batch-Plant Management

Zofia Verwater-Lukszo and Petra Heijnen
Delft University of Technology
The Netherlands

Abstract

The course "Integrated Plant Management", developed at the Department of Technology, Policy and Management at the Delft University of Technology, is aimed to provide knowledge and understanding of the plant operation in such a way that the challenges imposed by the economic, environmental and social sustainability are made more transparent (Verwater-Lukszo e.a., 2001). The course focuses on the batch processing industry, but the most concepts are applicable to continuous and discrete industry as well.

The integration of the enterprise functions as strategic and tactical management, forecasting, planning, scheduling, recipe management, process execution, optimisation and control are central to the main course aim. To realise this integrated manner of plant management the modern concepts of manufacturing execution systems (MES), plant modelling according to the ISA-S88 and ISA-S95 standards, total quality management and system thinking are very useful. Those issues constitute the main focus of the course. The course is concluded by an emphasis on the importance of the integration programs for quality and environment, which can be realised and maintained according to the principles of the new ISO quality (ISO 9001:2000, 2000) and environmental (ISO14001: 1996, 1996) standards.

Modelling enterprise activities and production processes as well as experimental and model-based process optimisation are the enablers of the intended improvements. Monitoring business and process performance form the next step. The principles of Statistical Quality Control form on one hand a basis for a sound assurance of quality and enterprise performance. On the other hand, they create a framework for continual performance improvement.

1. Introduction

In the last few decades, developments in the global economic structure have changed the environment in which process industries operate. They have to cope with the following problems:

- more short-term dynamics in supply and end-products markets as well as more unpredictable and turbulent demand patterns
- more complicated processes which may be more difficult to operate
- short series of the manufactured products
- stricter requirements on product quality
- greater emphasis on shorter and more reliable production time
- a growing number of product grades and brands

- a need for improved customer service level.

A flexible batch-wise mode of operation is a good answer for these trends. Batch processing of higher added-value specialities has been a fast growing segment of the process industry (i.e. food & beverages, chemical, pharmaceutical, metal industry etc.) in most industrialised countries. However, the flexibility of a batch plant poses a difficult problem of the allocation to available equipment for producing the desired products and setting up a production plan to decide if, when, and in what amounts, products should be produced. Moreover, the dynamic character of processing steps, which do not operate in a steady-state mode, complicates further the operation and control of a batch plant. Despite their complexity, the attractiveness of batch processing plants should be mentioned, too. It is found in the flexibility they offer to produce different (types of) products with the same equipment and to use the same pieces of equipment for different processing operations. This feature makes batch plants eminently suitable for production situations with a large number of product grades and short series of (tailor made) products.

Chemical engineering research into methods to support an eco-efficient design of batch processes is gaining momentum since a few years. It is striking that, in comparison with batch processes design, the operation of batch processes is hardly explored as a chemical engineering field. The course "Integrated Plant Management", developed at the Department of Technology, Policy and Management at the Delft University of Technology, is aimed to provide knowledge and understanding of the batch-wise plant operation in such a way that the challenges imposed by the economic, environmental and social sustainability are made more transparent.

The course is delivered to the students of the Systems Engineering, Policy Analysis and Management (SEPA) educational program at the Delft University of Technology as part of the Master's curriculum. The students from other faculties, i.e. Chemical Engineering and Mechanical Engineering, participate in the course, too. The development of this course has gone through a number of stages. It started as part of the module Cleaner Production and Process Design in an educational European Socrates program ELCE (Environmental Life-Cycle Engineering) in which various universities developed four advanced modules in the area of environmental engineering (Verwater-Lukszo e.a., 1999). The original course on Cleaner Production and Process Design is still given to international students as a distance-learning module on the Internet, in particular via the World Wide Web. The two main application areas in this course are plant design and plant operation (Herder e.a., 2001). The course described in this paper, given for the regular SEPA students, is an extension of the ELCE module in the direction of more sustainable operation of an industrial plant, and in particular a batch plant. The students learn to see the "whole picture" of the plant operation with economic, ecological and technical aspects discussed together.

2. How to operate a plant in a changing environment?

Industrial companies operating in a rapidly changing world of global economy and more short-term dynamics are continuously searching for opportunities to improve their competitive position. This involves improvements on the one hand in the production processes that produce products more efficiently and on the other hand in the internal methods of operation that enable companies to be more effective. Surprisingly, improvements in internal methods related to integrated plant management paying attention to both economic as environmental issues have often received relatively little attention. Mostly, the companies concentrate on improving one task without taking sufficiently into account interactions with other activities and with the surroundings. The goal of the course it to create an understanding of bringing the batch plants operation in agreement with the strategic goals related to developments on the market. The leading thread running through the course is the integration of planning, scheduling, recipe, and quality management and process control according to operational objectives defined in compliance with strategic goals. The question that rises firstly, is:

How could strategic and tactical objectives (as those related to growth, profitability, sustainability) be translated to operational objectives (as those related to production results, personnel satisfaction, environmental impact of production activities etc.), which are measurable and achievable?

In the course the students learn to structure the enterprise objectives in a hierarchical way by using the so-called objective-tree technique. An objective tree gives a good structure for the hierarchy and connection between objectives. Firstly, the overall objective should be determined - it characterises the reason for interest and defines the breadth of concern. Next, other objectives, which contribute to this main objective, should be specified. The procedure continues till the main objectives (at the strategic level) are translated into the operational objectives. The general framework for the objective tree will be the same for all sites of the company.

Further, to support an integrated way of plant management a clear representation of activities performed in the company in relation to the specified objectives is desirable. In the course the students learn to model the key business activities by using so-called IDEF0 techniques, which represent the structure of activities in a hierarchical way (IDEF0, 1993). The activities in the system – in this case an industrial plant - are analysed independently of the objects (persons, departments) that perform them. Such a purely functional perspective allows for a clear separation of the issues of meaning from the issues of implementation.

The activity defined at the highest level is decomposed in a number of activities, which collectively should achieve the main objective of the plant. In the decomposed model an output from one activity can be an input to another activity. In this way activities can be combined in a chain or network. The decomposition stops at that level, which makes it possible to associate the objectives at the lower level of the objective tree with the decomposed activities.

Modelling enterprise activities and their interdependencies, and relating them to the operational objectives makes it possible to visualise how improvements or changes in one activity interact with other activities and which results could be expected. This visualisation supports the decision, which activities at the operational level contribute mostly to the overall objective of the plant, in other words which ones are most effective. Improvement of the efficiency of those activities will be most effective. To improve manufacturing processes, i.e. to improve quality, yields, environmental performance, statistical thinking plays an important role. In general, statistical thinking

can be used for the identification of problems or issues having common sources, in contract with those having unique causes. The following aspects could characterise it:
- the processes (activities) and not the products (outputs) are emphasised;
- identification, quantification, control and reduction of variation are the central issues;
- the problems are tackled on the basis of gathered data.

The learnt techniques for modelling objectives and activities as well as improving them are also useful in the implementation of quality and environmental management systems according to the ISO standards: respectively (ISO 9001:2000, 2000) and (ISO 14001:1996, 1996).

The quality standards, published in 2000, make the integration of a quality management system with an environmental one much easier. Both standards address:
- identification and satisfaction of the expectation of all involved parties (employees, owners, suppliers, society), aimed to achieve chosen objectives and to do this in an effective and efficient manner
- attain, maintain and improve overall organizational performance and capabilities.

The mentioned standards encourage the adoption of the so-called process approach (process model), in which continual improvements play a central role.

3. The program of the course

The course is given during two educational periods of six weeks each. Table 1 presents the main subject of each weekly part and the corresponding individual assignment. The course ends with an oral examination.

Table 1. The main subjects of the course "Integrated Plant Management"

NR.	Subjects and aims	Assignment
1.	Introduction to the course and to the process industry - Economic and ecological potential of the process industry in the Netherlands and in Europe - Various ways of processing (continue, batch, discrete) and various operating regimes (start-up, switch-over, shut down) - Mutually related enterprise functions - Concepts of activity modelling according to the SADT technique, including IDEF0 diagrams	Read the paper "Planning, scheduling and control systems: why can they not work together" of D.E. Shobry and D.C. White (Shobry e.a., 2000). Make a short description of forecasting, planning, scheduling and control activities in an industrial company Make an IDEF0 model of a manufacturing site with the software tool BPWin. Zoom in on the four functions: Forecast, Planning, Scheduling and Control.
2	Planning - Examples of planning problems - Short term planning, capacity planning - Customer Order Uncouple Point - Main technologies for planning (simulation, programming)	Extend the IDEF0 model with the planning activities

NR.	Subjects and aims	Assignment
3	Scheduling - Examples of scheduling problems - Industrial practice - Recent academic developments (disturbance management)	Extend the IDEF0 model with the scheduling activities
4	Quality control for process operation - Quality as reducing variations - Statistical thinking - Introduction Statistical Process Control (SPC)	In a plant there are three multi-purpose batch reactors. Monitor the manufacturing process with Cornerstone - a statistical package for industrial data analysis. The progress of the reaction is tested by continuous temperature measurements and by taking samples from the reactor.
5.	Quality improvement: Statistical process Control - Introduction to control charts - Development of control charts - Interpretation of control charts - Process judgement	Develop a control chart for the production process of epoxy resin. Calculate the process capability indices. Develop a cause-and -effect diagram for the cases with a wrong WPE-number (quality parameter)
6.	Experimental design for process optimisation - Introduction to the basis principles of design of experiments - Introduction to factorial design	Define designed experiments to find an optimised recipe for a fermentation process, for which a new recipe has to be found. Cornerstone is again the supporting tool.
7.	Experimental and model-based process optimisation - Discussion of various ways of process modelling - First principle models - Black-box models (regression and statistical test)	Use Cornerstone to model a chemical process. Develop a way for integration of scheduling and processing activities by using process models as a part of a master recipe
8.	Quality and environmental management systems - The new and old ISO 9000 quality management systems - Environmental management system 14001	Discussion of common pitfalls on the path to registration, false expectations and focal points in external audits and of the role of statistical techniques in quality management systems
9	Introduction to process control - Control configurations - Control algorithms - Control Hierarchy - Formulation of a control problem	Identify goals and the design space for a control problem described for a batch process producing small polymer particles. Translate the goals to objectives and constraints, and develop the tests necessary for mapping the control variables onto objectives and constraints.
10.	Introduction of batch plant modelling standards - ISA-S88 (process cell level) - ISA-S95 (integration of MES level with ERP level)	Demonstration of the software package proCX for recipe management, scheduling, execution of recipes, material flow control, tracking & tracing
11	Optimisation of the plant operation - (Non) Linear programming - Dynamic programming - Integer programming - Heuristic methods	Create a master recipe for the production of an alkyd resin, taking into account the S88 standard terminology for recipe elements. Make a schematic representation of the plant and distinguish process cells, process classes and transfer classes
12	Final discussion on Integrated Plant Management - MES (Manufacturing Execution Systems) as a integration framework	

4. Final remarks

The ICT-supported course is given in an interactive way. In addition to the conventional class, which is given once a week, the students make every week an individual assignment with the business-modelling tool BPWin (BPWin, 2001) and with Cornerstone (Cornerstone, 2001), an exploratory tool for analysing manufacturing and engineering data. They submit them via Blackboard© (Blackboard, 2002), the general course supporting-tool, which gives an excellent environment for communication between teacher and students, for working in groups and for discussion on the topics from class.

By the evaluation of assignments and during the oral exams it becomes clear that the students have assimilated the integrated view on plant management in the right manner. From the other side, it is also clear that the application of the mentioned ideas in a real plant is still difficult to operationalise. However, we are sure that the students understand very well that the industrial systems should be managed in an integrative way and that there is no way to escape from this responsibility. To do so, we still need to improve the scientific basis of the process operation discipline. Hopefully this course has made a small contribution towards this goal.

References

Blackboard, 2002, http://www.blackboard.com/, Blackboard Inc.

BPWin, 2001, http://www.cai.com/products/alm/bpwin.htm, Computer Associates International, Inc.

Cornerstone, 2001, http://www.brooks.com/products/POBU/cstone/index.htm, Brooks Automation Inc.

Foster S.T., 2001, Managing Quality. An Integrative Approach, Prentice Hall, New Jersey.

Herder, P.M., Z. Verwater-Lukszo, M.P.C. Weijnen, 2001, A Novel Perspective on Chemical Engineering Education - Experiences with a Broad, ICT-Supported Course on the Integrated Design of Industrial Systems, 6th World Congress On Chemical Engineering, Melbourne.

ISO 9001:2000, 2000, Quality Management Systems - Requirements, ISO, Geneva.

ISO 14001:1996, 1996, Environmental Management Systems, specification with guidance for use, ISO, Geneva.

Rao, A.,et al., 1996, Total Quality Management: A Cross Functional Perspective, John Wiley & Sons, New York.

Shobry, D.E., D.C. White, 2000, Planning, scheduling and control systems: why can they not work together, Computers and Chem. Eng, 24

Standard Integration Definition for Function Modelling (IDEF0), 1993, Publication 183, FIPS PUBS.

Verwater-Lukszo, Z., P.M.Herder, M.P.C. Weijnen, 1999, Cleaner Production and Process Design. Experiences with a distance–learning module on Internet, Proceedings of ENTREE'99, Tampere.

Verwater-Lukszo, Z., P. Heijnen, 2001, Integrated Plant Management for better economic and ecological business performance, Proceedings of ENTREE'2001, Florence.

European Symposium on Computer Aided Process Engineering – 12
J. Grievink and J. van Schijndel (Editors)
© 2002 Elsevier Science B.V. All rights reserved.

Designing a Multi-user Web-based Distributed Simulator for Process Control eLearning

S. H. Yang and J. L. Alty

Computer Science Department, Loughborough University, Loughborough, Leicestershire, LE11 3TU, United Kingdom

Abstract

Although web-based elearning has been used for various disciplines, web-based experiments are still unusual. A web-based distributed simulator can be a powerful tool for elearning and a good alternative for web-based experiments. This paper discusses the design issues of a web-based distributed simulator for elearning in process control, including architecture selection, communication protocol, interface design, and process modelling. An industrial catalytic reactor is used as a case study to illustrate the methods described here.

1. Introduction

Engineering and Science courses such as control theory and process control courses, are strongly founded on mathematics on one hand, but also need to develop the student's intuition for bridging the gap between theory and practice on the other hand. Obviously, providing opportunities for students to experience the theory which they have learned should form an essential part for the web-based courses (Copinga et. al., 2000). Unfortunately, web-based experiments (in contrast with the use of the web as a simple source of information) are still uncommon in engineering and science courses (Cartwright, 1998).

Web-based dynamic simulators designed to support elearning are currently becoming available on the web (Granlund et. al., 2000). The advantages of learning, supported by dynamic simulations and the availability of the web, form a powerful combination. This combination in terms of speed, potential sophistication and wide availability seems certain to make them widespread and in frequent use within a few years. Developing web-based dynamic simulators for elearning presents special problems in software, networking, and interfacing. This paper will address these problems by reference to an industrial case study. The contents of the rest of this paper are arranged as follows: Section 2 describes the potential architectures of web-based dynamic simulators for elearning Then communication protocol is explored in section 3 for the distributed architecture chosen in section 2. User interface design and process modelling for the web-based dynamic simulator have raised some new issues which are briefly studied in sections 4 and 5. The industrial catalytic reactor is used in section 6 to illustrate the discussion given in pre-sections as case study. Conclusions are presented in section 7.

1034

2. Architectures of the Web-based Simulator

One of the goals of the Internet is the global accessibility, which is a strong reason for designing web-based simulators for elearning. The problems that may occur are slow and unpredictable network performance, and uncertainty as to who the users are. Therefore some mechanism is required to solve time delay and control conflict problems between multi-users. Figure 1 shows the perspective view of the web-based simulator. It consists of a web server running the server application, a web-based instructor interface and several identical web-based student interfaces.

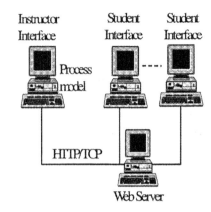

Figure 1. Overview of the web-based distributed simulator.

The instructor interface is designed for the tutor to monitor and assess the operation of the students. The process model runs on the same machine as the instructor interface in order to reduce the working load of the server. The instructor interface acts as a user-interface front end to the process model. The web server acts as a post office and transfer data and control commands between the student interfaces and the process model through a HTTP/TCP connection. At the remote site (student site and instructor site), the HTML browsers display information about the process, including current states, alarm, dynamic trends, et al., and allow the students and the tutor to control the process. These functions are implemented in Java applet. When a student with a Java-capable browser loads the student interface (a web page) that includes applets, the browser downloads the applets from the web server and runs them on the student's own system. The instructor interface and the process model must be loaded and run before the student interface can really work. No specific software is required to be installed at the remote site. At the local site, the server establishes and controls communication between students/tutor and the process model by using the socket techniques. A hash table is built on the server to remain the update registration of the student interfaces. This configuration enables the students as many as possible to operate the same simulator simultaneously to develop their teamwork skill, but requires the students to stay on-line during the session. When no teamwork is required, the simulator can be simplified into a stand-alone application in the Java applet form. In that case once they load the interface front end from the web-server the students can work off-line. Only the distributed architecture will be discussed in the rest of this paper.

3. Communication Protocol

As described above the server acts as a messenger for the instructor interface and the student interfaces, passing message from one to the other according to a special protocol. Two protocols were considered for the student interfaces and the instructor interface connections to the server. The first of these was UDP (Universal Datagram Protocol). UDP is a connectionless protocol and communicates fast. It is however unreliable, and cannot guarantee the delivery of the data in the right order. TCP however, whilst being a slightly slower protocol, guarantees that data delivered will be correct and in the right order. Therefore TCP is chosen as the protocol for all communication in this work.

The server, the student interfaces, and the instructor interface can be a sender and/or a receiver. It enables information pass between them. The protocol defines a strict set of instructions which allow the students to give instructions to the process model to manipulate its operation, allow the student interfaces to regularly update the data they are displaying, and allow the instructor to monitor the operations carried out on the student interfaces as well. The student interface and the instructor interface need to identify themselves to the server program by sending the command "*CONNECT INT*" and "*CONNECT SIM*" respectively. From then on, the server program recognizes the client as a student interface or an instructor interface and treats it as such. The *UPDATE* commands from the instructor interface and the server are not of fixed length however, as the command may have multiple *<device> <value>* pairs, i.e.

UPDATE <device> <value> [<device> <value>]

Once the instructor interface disconnects to the server, the command "*UPDATE sim 0*" is sent from the server to every student interfaces linked with the server and instructs them that the process model is no longer present to control and that they should prevent the student from attempting to make any changes to the process. Likewise, when the instructor interface connects to the server, a similar UPDATE command "*UPDATE sim 1*" is sent to every student interfaces to inform them that they may allow their operators to control the process again. In order for the server to manage the potentially large number of connections, each student interface must be given a unique ID so that when messages need to be sent to individual student interfaces, it must know where to send it. These unique ID's are stored in a hash table shown in the server, along with the actual connection, and whenever a student interface disconnects from the server, the hash table will be shuffled so that there are no gaps in the list, and then the next student interface to connect can be given the next number. The student interface ID is explicitly embedded in the command sent from the server to the instructor interface:

CHANGE <ID> <device> <setting>

4. Web-based User Interface Design

The central design objective for a user interface in the web-based distributed simulator is to enable the students (operators) to appreciate more rapidly what is happening in the process model and to provide a more stimulating problem-solving environment over the Internet. It should be born in mind that media available in the Internet environment will be very limited comparing with those in the central simulation room. Nevertheless, in the Internet environment no specific software and no special type of computer can be provided. Technologies from the areas of "multimedia" and "Virtual Reality" show considerable potential for improving yet further the human-computer interface used in process control technology (Alty, 1999). Using the audio information for warnings and alarms and using animation for displaying information are essential in the user interface design for the web-based simulators. Minimizing the amount of irrelevant information in the interface is another key issue because the irrelevant information may obscure important information by attracting the attention of the user.

5. Requirements of Process Modelling for Web-based Simulators

One of the cores of the web-based simulator is the real-time dynamic model of the actual process. There are several different requirements that must be satisfied in order for the web-based simulator to be used for on-line experiments. A prerequisite is that the process model must be able to reproduce the process dynamic behaviour over a very wide range of operational modes including not just normal operation, but fault and emergency operation, start-up and shutdown as well. In addition, it should be able to handle logical procedures related to actions such as switching on or off valves, purges and so on. Also, main equipment failures should be considered in the process model so that the equipment failure can be invoked for special experiment task. Finally, the computation of the process model must not be time-consuming and complex in order to run the model on-line, and any iterative calculation should be avoided as much as possible. These requirements make web-based simulators distinctive from process simulators, which are concerned with improving optimisation and control. In this paper only a subset of a chemical process plant is simulated and it incorporates a mathematic model that integrates normal operation, start-up operation, and emergency handling operation together using several transition switches. For more details see our recent publication (Yang et al., 2001).

6. Case Study – a Web-based Distributed Simulator for an Industrial Catalytic Reactor

The purpose of the web-based distributed simulator for an industrial catalytic reactor which we implemented here is twofold: (1) to provide students a virtual experiment environment to practice the basics of plant operation, and to build up the teamwork skill, and (2) to provide researchers a test bed for the design of Internet-based process control. The second aspect of the purpose will be addressed in our further publication. The simulator allows students situated in geographically diverse locations to experiment with some aspects of operating an industrial catalytic reactor. The main scenarios include that desired values of the process variables need to be updated and/or malfunctions occur during normal operation of the reactor. The task of the students is to act in response to these requirements and failures and to bring the reactor to a safe and steady state.

Figure 2 shows the student interface running in a remote site. The process consists of a heat exchanger E201, a catalytic reactor R201, and four hand valves for Nitrogen inlet, liquid outlet, gas outlet, and emergency liquid outlet. The inlet temperature of the reactor is controlled by the hot stream flowrate of the heat exchanger E201. The buttons shown at the top of Figure 2 start various web instruments (pop windows) for displaying dynamic trends, alarm panel, controller panel and evaluation panel. These web instruments can be activated as well by clicking a mouse left button over the corresponding components in the process flowchart.

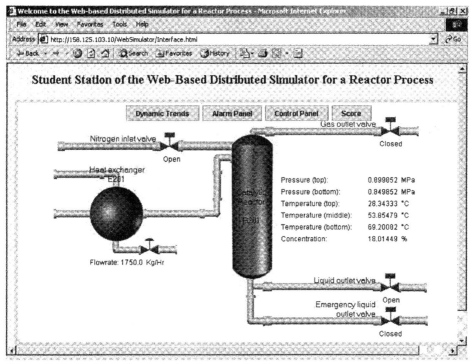

Figure 2. Student interface running in a remote site.

Before developing the web-based simulator several standard web instruments have been implemented for the Internet-based process control system. These web instruments include PID (Proportional Integral, and Differential) controller, manipulator, bar graph alarm panel, dynamic trend panel, and are pure java applet pop windows. Figure 3 shows the web PID controller for the heat exchanger E201 that reappears the appearance of the front panel of a physical digital PID controller. The group radio buttons are used to choose automatic or manual mode for the controller. The bar charts at the top show the current input, output, and setpoint values. The scroll bars at the bottom allow a user to adjust

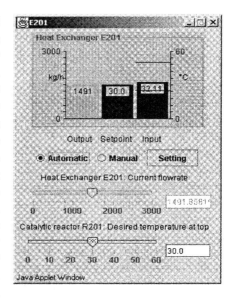

Figure 3. Web PID controller

output value while in the manual mode and setpoint while in the automatic mode. Pressing the setting button allows an authorized user to tune P, I and D parameters and change the output range for the controller. All the web instruments here include security measures in the form of password and user name to identify the authorized users. Another important part of the distributed simulator is the instructor interface. Its main purpose is to provide a tutor a platform to set various malfunctions and operating circumstances for the remote site students, and monitor their operations. The process model is running on the same machine with the instructor interface.

7. Conclusions

This paper has described the design issues of a web-based distributed simulator, such as distributed architecture, communication protocol, interface design and process modelling. A case study is implemented to show the design issues. One of the most important advantages for developing web-based distributed simulators is the availability of the system on the web. The initial motivation of developing this distributed simulator was to provide a test bed for an Internet-based process control system. In this paper we demonstrate that it is straightforward for this system to be used as an online experiment environment for elearning students in their process control practice, especially when teamwork skill is required. We believe that web-based distributed simulators will play an important role in future elearning.

Acknowledgement

The contribution is part of the work of the EPSRC funded project "design of Internet-based process control". The authors would like to thank the support given by the EPSRC, Grant No. GR/R13371/01. Appreciation should be also to our students, R Baxter, C Childs, C George, S Nichols, and P Kay for their valuable programming.

References

Alty, J.L., 1999, Multimedia and Process Control Interfaces: Signals or Noise? Transaction of Inst MC, 21(4/5), 181-190.

Cartwright, H. M., 1998, Remote control: How science students can learn using Internet-based experiments. New Network-based Media in Education; Proceedings of the International CoLoS Conference, Maribor, Slovenia, September, pp. 51-59.

Copinga, G. J. C., Verhaegen, W. H. G., and van de Ven, M. J. J. M., 2000, Toward a web-based study support environment for teaching automatic control. IEEE Control Systems Magazine, August, 8-19.

Granlund, R., Berglund, E., and Eriksson, H., 2000, Designing web-based simulation for learning. Future Generation Computer Systems, 17, 171-185.

Yang, S. H., Yang, L., and He, C. H., 2001, Improve safety of industrial processes using dynamic operator training simulators. Transactions of IchemE, Process Safety and Environmental Protection, 79(B6), 329-338.

Author Index

Printed and bound by CPI Group (UK) Ltd, Croydon, CR0 4YY

03/10/2024

01040333-0007